University Chemistry

List of the Elements with Their Symbols and Atomic Masses*

Element	Symbol	Atomic Number	Atomic Mass†	Element	Symbol	Atomic Number	Atomic Mass†
Actinium	Ac	89	(227)	Mendelevium	Md	101	(256)
Aluminum	Al	13	26.98	Mercury	Hg	80	200.6
Americium	Am	95	(243)	Molybdenum	Mo	42	95.94
Antimony	Sb	51	121.8	Neodymium	Nd	60	144.2
Argon	Ar	18	39.95	Neon	Ne	10	20.18
Arsenic	As	33	74.92	Neptunium	Np	93	(237)
Astatine	At	85	(210)	Nickel	Ni	28	58.69
Barium	Ba	56	137.3	Niobium	Nb	41	92.91
Berkelium	Bk	97	(247)	Nitrogen	N	7	14.01
Beryllium	Be	4	9.012	Nobelium	No	102	(253)
Bismuth	Bi	83	209.0	Osmium	Os	76	190.2
Bohrium	Bh	107	(262)	Oxygen	O	8	16.00
Boron	B	5	10.81	Palladium	Pd	46	106.4
Bromine	Br	35	79.90	Phosphorus	P	15	30.97
Cadmium	Cd	48	112.4	Platinum	Pt	78	195.1
Calcium	Ca	20	40.08	Plutonium	Pu	94	(242)
Californium	Cf	98	(249)	Polonium	Po	84	(210)
Carbon	C	6	12.01	Potassium	K	19	39.10
Cerium	Ce	58	140.1	Praseodymium	Pr	59	140.9
Cesium	Cs	55	132.9	Promethium	Pm	61	(147)
Chlorine	Cl	17	35.45	Protactinium	Pa	91	(231)
Chromium	Cr	24	52.00	Radium	Ra	88	(226)
Cobalt	Co	27	58.93	Radon	Rn	86	(222)
Copper	Cu	29	63.55	Rhenium	Re	75	186.2
Curium	Cm	96	(247)	Rhodium	Rh	45	102.9
Darmstadtium	Ds	110	(269)	Roentgenium	Rg	111	(272)
Dubnium	Db	105	(260)	Rubidium	Rb	37	85.47
Dysprosium	Dy	66	162.5	Ruthenium	Ru	44	101.1
Einsteinium	Es	99	(254)	Rutherfordium	Rf	104	(257)
Erbium	Er	68	167.3	Samarium	Sm	62	150.4
Europium	Eu	63	152.0	Scandium	Sc	21	44.96
Fermium	Fm	100	(253)	Seaborgium	Sg	106	(263)
Fluorine	F	9	19.00	Selenium	Se	34	78.96
Francium	Fr	87	(223)	Silicon	Si	14	28.09
Gadolinium	Gd	64	157.3	Silver	Ag	47	107.9
Gallium	Ga	31	69.72	Sodium	Na	11	22.99
Germanium	Ge	32	72.59	Strontium	Sr	38	87.62
Gold	Au	79	197.0	Sulfur	S	16	32.07
Hafnium	Hf	72	178.5	Tantalum	Ta	73	180.9
Hassium	Hs	108	(265)	Technetium	Tc	43	(99)
Helium	He	2	4.003	Tellurium	Te	52	127.6
Holmium	Ho	67	164.9	Terbium	Tb	65	158.9
Hydrogen	H	1	1.0079	Thallium	Tl	81	204.4
Indium	In	49	114.8	Thorium	Th	90	232.0
Iodine	I	53	126.9	Thulium	Tm	69	168.9
Iridium	Ir	77	192.2	Tin	Sn	50	118.7
Iron	Fe	26	55.85	Titanium	Ti	22	47.88
Krypton	Kr	36	83.80	Tungsten	W	74	183.9
Lanthanum	La	57	138.9	Uranium	U	92	238.0
Lawrencium	Lr	103	(257)	Vanadium	V	23	50.94
Lead	Pb	82	207.2	Xenon	Xe	54	131.3
Lithium	Li	3	6.941	Ytterbium	Yb	70	173.0
Lutetium	Lu	71	175.0	Yttrium	Y	39	88.91
Magnesium	Mg	12	24.31	Zinc	Zn	30	65.39
Manganese	Mn	25	54.94	Zirconium	Zr	40	91.22
Meitnerium	Mt	109	(266)				

*All atomic masses have four significant figures. These values are recommended by the Committee on Teaching of Chemistry, International Union of Pure and Applied Chemistry.

†Approximate values of atomic masses for radioactive elements are given in parentheses.

University **Chemistry**

Brian B. Laird
University of Kansas

With significant contributions by
Raymond Chang
Williams College

McGraw-Hill
Higher Education

Boston   Burr Ridge, IL   Dubuque, IA   New York   San Francisco   St. Louis
Bangkok   Bogotá   Caracas   Kuala Lumpur   Lisbon   London   Madrid   Mexico City
Milan   Montreal   New Delhi   Santiago   Seoul   Singapore   Sydney   Taipei   Toronto

McGraw-Hill
Higher Education

UNIVERSITY CHEMISTRY

Published by McGraw-Hill, a business unit of The McGraw-Hill Companies, Inc., 1221 Avenue of the Americas, New York, NY 10020. Copyright © 2009 by The McGraw-Hill Companies, Inc. All rights reserved. No part of this publication may be reproduced or distributed in any form or by any means, or stored in a database or retrieval system, without the prior written consent of The McGraw-Hill Companies, Inc., including, but not limited to, in any network or other electronic storage or transmission, or broadcast for distance learning.

Some ancillaries, including electronic and print components, may not be available to customers outside the United States.

This book is printed on acid-free paper.

1 2 3 4 5 6 7 8 9 0 DOW/DOW 0 9 8

ISBN 978–0–07–296904–7
MHID 0–07–296904–0

Publisher: *Thomas Timp*
Senior Sponsoring Editor: *Tamara L. Hodge*
Vice-President New Product Launches: *Michael Lange*
Senior Developmental Editor: *Shirley R. Oberbroeckling*
Marketing Manager: *Todd L. Turner*
Senior Project Manager: *Gloria G. Schiesl*
Senior Production Supervisor: *Kara Kudronowicz*
Lead Media Project Manager: *Judi David*
Senior Designer: *David W. Hash*
Cover/Interior Designer: *Elise Lansdon*
(USE) Cover Image: *Water Droplets, ©Masato Tokiwa/Amana Images/Getty Images*
Senior Photo Research Coordinator: *John C. Leland*
Photo Research: *David Tietz/Editorial Image, LLC*
Supplement Producer: *Mary Jane Lampe*
Compositor: *Aptara, Inc.*
Typeface: *10/12 Times Roman*
Printer: *R. R. Donnelley Willard, OH*

The credits section for this book begins on page C-1 and is considered an extension of the copyright page.

Library of Congress Cataloging-in-Publication Data

Laird, Brian B., 1960-
  University chemistry / Brian B. Laird.
     p. cm.
  Includes index.
  ISBN 978–0–07–296904–7 — ISBN 0–07–296904–0 (hard copy : alk. paper) 1. Chemistry—Study and teaching (Higher) 2. Chemistry—Textbooks. I. Title.
  QD40.L275 2009
  540—dc22

                              2007052540

www.mhhe.com

About the **Author**

Brian B. Laird, a native of Port Arthur, Texas, is currently a Professor of Chemistry at the University of Kansas in Lawrence, Kansas. He received Bachelor of Science degrees in Chemistry and Mathematics from the University of Texas, Austin, in 1982, and a Ph.D. in Theoretical Chemistry from the University of California, Berkeley, in 1987. Prior to his current position, he held postdoctoral and lecturer appointments at Columbia University, Forschungszentrum Jülich, Germany (NATO Fellowship), University of Utah, University of Sydney, and the University of Wisconsin. His research interests involve the application of statistical mechanics and computer simulation to the determination of properties of liquid and solids. In addition to honors general chemistry, he regularly teaches undergraduate physical chemistry and graduate courses in quantum and statistical mechanics. In his spare time, he enjoys golfing, bicycling, playing the piano, and traveling.

I dedicate this work to my wife, Uschi, and to the memory of my parents, Don and Nanci Laird.

Brief Contents

Expanded Contents

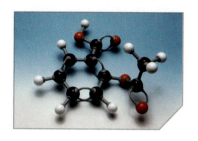

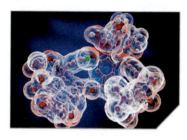

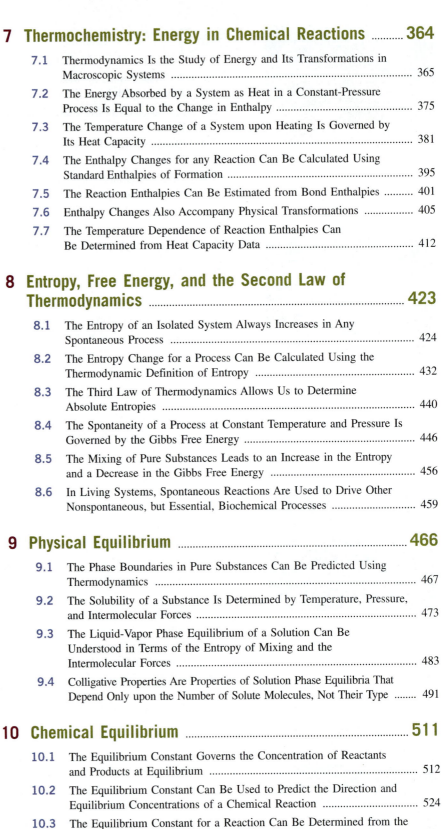

List of **Applications**

Preface

The concept of *University Chemistry* grew from my experiences in teaching Honors General Chemistry at the University of Kansas for a number of semesters. It is my attempt to inform and challenge the well-prepared student to discover and learn the diverse, but related, topics within general chemistry. This text includes the core topics that are necessary for a solid foundation of chemistry.

The Basic Features

▸ *Organization*. In this text, I adopt a "Molecular to Macroscopic" approach, in which the quantum theory of atomic and molecular structure and interaction is outlined in Chapters 1–4. Building on this molecular foundation, the presentation moves to the macroscopic concepts, such as states of matter, thermodynamics, physical and chemical equilibrium, and chemical kinetics. This organization is based on "natural prerequisites"; that is, each topic is positioned relative to what other topics are required to understand it. For example, knowledge of thermodynamics or equilibrium chemistry is not needed to understand the structure and interaction of atoms and molecules; whereas, to understand deeply the application of thermodynamics to chemical systems or the material properties of liquids and solids, knowing how energy is stored in chemical bonds and how molecular structure and bonding affect intermolecular forces is desirable.

▸ *Mathmematical Level*. The presentation in this text assumes that the student has a good working knowledge of algebra, trigonometry and coordinate geometry at the high school level. Knowledge of calculus, while advantageous, is not strictly required for full understanding. Integral and differential calculus is used, where appropriate, in intermediate steps of concept development in quantum theory, thermodynamics, and kinetics. However, the final primary concepts and most major equations (denoted by a blue box) do not depend on an understanding of calculus, nor do the overwhelming majority of end-of-chapter problems. For the interested and advanced student, I have included a small number of calculus-based end-of-chapter problems in the relevant chapters.

The level of calculus used in *University Chemistry* is similar to that used in other general chemistry texts at this level; however, it is not relegated to secondary boxed text, as is often done, but integrated into the primary discussion, so as not to disrupt the linear flow of the presentation. For the interested student, I have included in Appendix 1 a brief review/tutorial of the basic concepts in integral and differential calculus.

▸ *Problem-Solving Model*. Worked Examples are included in every chapter for students to use as a base for applying their problem-solving skills to the concept discussed. The examples present the problem, a strategy, a solution, a check, and a practice problem. Every problem is designed to challenge the student to think logically through the problem. This problem-solving approach is used throughout the text.

Organization and Presentation

▸ **Review.** Students with a strong background in high school chemistry have already been exposed to the concepts of the structure and classification of matter, chemical nomenclature, and stoichiometry. Because of this assumed background, I have condensed the standard introductory chapters of a typical general chemistry text into a single chapter (Chapter 0). Chapter 0 is intended to serve as a refresher of the subject matter students covered in their high school chemistry courses.

▸ **Early Coverage of Quantum Theory.** To provide a molecular-level foundation for the later chapters on states of matter, thermodynamics, and equilibrium, the quantum theory of atoms and molecules is presented early in the text. In Chapters 1 and 2, elementary quantum theory is used to discuss the electronic structure of atoms and the construction of the periodic table. Chapters 3 and 4 cover molecular bonding, structure, and interaction, including molecular-orbital theory. Contrary to the organization of most general chemistry texts, intermolecular forces are discussed at the end of Chapter 4 on molecular structure instead of in a later chapter on liquids and solids. This is a more natural position, which allows for a molecular-level discussion of the forces that influence real gas behavior in Chapter 5.

▶ **States of Matter.** Phase diagrams, equations of state, and states of matter (gases, liquids and solids) are treated in a unified manner in Chapters 5 and 6, with an emphasis on the role of molecular interaction in the determination of material properties.

▶ **Thermochemistry, Entropy, and Free Energy.** The basic principles of thermodynamics are treated together in Chapters 7 "Thermochemistry: Energy in Chemical Reactions," and 8, "Entropy, Free Energy, and the Second Law of Thermodynamics." This

allows for a more sophisticated discussion of physical and chemical equilibrium from a thermodynamic perspective. In particular, the central role of the entropy and free energy of mixing in colligative properties and chemical equilibrium is explored in detail.

▶ **Physical and Chemical Equilibrium.** The principles of physical equilibrium (phase boundary prediction and solubility) are discussed in Chapter 9, followed by a discussion of chemical equilibrium in Chapter 10. Chapters 11, 12, and 13 present applications of chemical equilibrium to acid-base chemistry, aqueous equilibria, and electrochemistry, respectively.

▶ **Chemical Kinetics.** Unlike many general chemistry texts, discussion of chemical kinetics (Chapter 14) follows the presentation of chemical equilibrium, allowing for full discussion of transition-state theory and detailed balance.

▶ **Final chapters.** Chapter 15, "The Chemistry of Transition Metals," Chapter 16, "Organic and Polymer Chemistry," and Chapter 17 "Nuclear Chemistry" are each an entity in itself. Every instructor and student can choose to assign and study the chapters according to time and preference.

Pedagogy

Problem Solving

The development of problem-solving skills is a major objective of this text. Each problem is broken down into learning steps to help students increase their logical critical thinking skills.

Example 1.2

Chlorophyll-a is green because it absorbs blue light at about 435 nm and red light at about 680 nm, so that mostly green light is transmitted. Calculate the energy per mole of photons at these wavelengths.

Strategy Planck's equation (Equation 1.3) gives the relationship between energy and *frequency* (ν). Because we are given wavelength (λ), we must use Equation 1.2, in which u is replaced with c (the speed of light), to convert wavelength to frequency. Finally, the problem asks for the energy per mole, so we must multiply the result we get from Equation 1.3 by Avogadro's number.

Solution The energy of one photon with a wavelength of 435 nm is

$$E = h\nu = h\left(\frac{c}{\lambda}\right) = (6.626 \times 10^{-34}\ \text{J s})\frac{3.00 \times 10^8\ \text{m s}^{-1}}{435\ \text{nm}\ (1 \times 10^{-9}\ \text{m nm}^{-1})}$$
$$= 4.57 \times 10^{-19}\ \text{J}$$

For one mole of photons, we have

$$E = (4.57 \times 10^{-19}\ \text{J})(6.022 \times 10^{23}\ \text{mol}^{-1})$$
$$= 2.75 \times 10^5\ \text{J mol}^{-1}$$
$$= 275\ \text{kJ mol}^{-1}$$

Using an identical approach for the photons at 680 nm, we get $E = 176\ \text{kJ mol}^{-1}$.

Practice Exercise X-rays are convenient to study the structure of crystals because their wavelengths are comparable to the distances between near neighbor atoms (on the order of a few Ångstroms, where $1\text{Å} = 1 \times 10^{-10}$ m). Calculate the energy of a photon of X-ray radiation with a wavelength of 2.00 Å.

There are numerous end-of-chapter problems to continue skill building and then practice solving problems. Many of these same problems appear in the electronic homework program ARIS, providing a seamless homework solution for the student and the instructor.

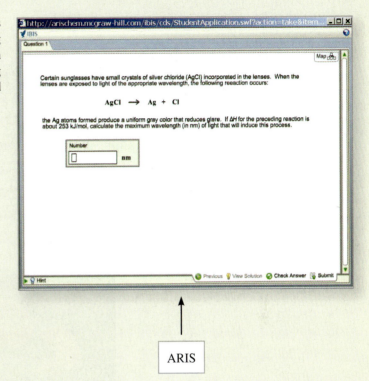

1.58 Certain sunglasses have small crystals of silver chloride (AgCl) incorporated in the lenses. When the lenses are exposed to light of the appropriate wavelength, the following reaction occurs:

$$AgCl \longrightarrow Ag + Cl$$

The Ag atoms formed produce a uniform gray color that reduces the glare. If the energy required for the preceding reaction is 248 kJ mol^{-1}, calculate the maximum wavelength of light that can induce this process.

Text

ARIS

End-of-Chapter Material

At the end of every chapter, you will find a summary of all the material that was presented in the chapter to use as a study tool. The summary highlights each section within the chapter. Key words are also listed and include the page number where the term was introduced.

Summary of Facts and Concepts

Section 1.1

▶ At the end of the nineteenth century, scientists began to realize that the laws of classical physics were incompatible with a number of new experiments that probed the nature of atoms and molecules and their interaction with light. Through the work of a number of scientists over the first three decades of the twentieth century, a new theory—quantum mechanics—was developed that was able to explain the behavior of objects on the atomic and molecular scale.

▶ The quantum theory developed by Planck successfully explains the emission of radiation by heated solids. The quantum theory states that radiant energy is emitted by atoms and molecules in small discrete amounts (quanta), rather than over a continuous range. This behavior is

a moving particle of mass m and velocity u is given by the de Broglie equation $\lambda = h/mu$ (Equation 1.20).

▶ The realization that matter at the atomic and subatomic scale possesses wavelike properties lead to the development of the Heisenberg uncertainty principle, which states that it is impossible to know simultaneously both the position (x) and the momentum (p) of a particle with certainty (see Equation 1.22).

▶ The Schrödinger equation (Equation 1.24) describes the motions and energies of submicroscopic particles. This equation, in which the state of a quantum particle is described by its wavefunction, launched modern quantum mechanics and a new era in physics. The wavefunction contains information about the probability of finding a particle in a given region of space.

Applications

Throughout the text, applications are included to reinforce students' grasp of concepts and principles and to provide grounding to real-world experiences. The applications focus on key industrial chemicals, drugs, and technological advances in chemistry.

Important Experimental Technique: Electron Microscopy

The electron microscope is an extremely valuable application of the wavelike properties of electrons because it produces images of objects that cannot be seen with the naked eye or with light microscopes. According to the laws of optics, it is impossible to form an image of an object that is smaller than half the wavelength of the light used for the observation. Because the range of visible light wavelengths starts at around 400 nm, or 4×10^{-7} m, we cannot see anything smaller than 2×10^{-7} m. In principle, we can see objects on the atomic and molecular scale by using X-rays, whose wavelengths range from about 0.01 nm to 10 nm. X-rays cannot be focused easily, however, so they do not produce crisp images. Electrons, on the other hand, are charged particles, which can be focused in the same way the image on a TV screen is focused (that is, by applying an electric field or a magnetic field). According to Equation 1.20, the wavelength of an electron is inversely proportional to its velocity. By accelerating electrons to very high velocities, we can obtain wavelengths as short as 0.004 nm.

A different type of electron microscope, called the *scanning tunneling microscope (STM)*, uses quantum mechanical tunneling to produce an image of the atoms on the surface of a sample. Because of its extremely small mass, an electron is able to move or "tunnel" through an energy barrier (instead of going over it). The STM consists of a metal needle with a very fine point (the source of the tunneling electrons). A voltage is maintained between the needle and the surface of the sample to induce electrons to tunnel through space to the sample. As the needle moves over the sample at a distance of a few atomic diameters from the surface, the tunneling current is measured. This current decreases with increasing distance from the sample. By using a feedback loop, the vertical position of the tip can be adjusted to a constant distance from the surface. The extent of these adjustments, which profile the sample, is recorded and displayed as a three-dimensional false-colored image. Both the electron microscope and the STM are among the most powerful tools in chemical and biological research.

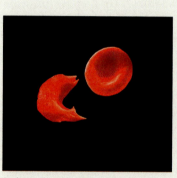

An electron micrograph showing a normal red blood cell and a sickled red blood cell from the same person.

STM image of iron atoms arranged to display the Chinese characters for atom on a copper surface.

360° Development Process

A key factor in developing any chemistry text is the ability to adapt to teaching specifications in a universal way. The only way to do so is by contacting those universal voices—and learning from their suggestions.

We are confident that our book has the most current content the industry has to offer, thus pushing our desire for accuracy and up-to-date information to the highest standard possible. To accomplish this, we have moved along an arduous road to production. Extensive and open-minded advice is critical in the production of a superior text.

Following is a brief overview of the initiatives included in the 360° Development Process of this first edition of *University Chemistry,* by Brian B. Laird.

Symposia Every year McGraw-Hill conducts a general chemistry symposium, which is attended by instructors from across the country. These events provide an opportunity for the McGraw-Hill editors to gather information about the needs and challenges of instructors teaching these courses. The information gleaned from these events helped to create the book plan for *University Chemistry*. In addition, these symposia offer a forum for the attendees to exchange ideas and experiences with colleagues whom they might not have otherwise met.

Manuscript Review Panels Over 50 teachers and academics from across the country and internationally reviewed the various drafts of the manuscript to give feedback on content, pedagogy, and organization. This feedback was summarized by the book team and used to guide the direction of the text.

Developmental Editing In addition to being influenced by a distinguished chemistry author, the development of this manuscript was impacted by three freelance developmental editors. The first edit in early draft stage was completed by an editor who holds a PhD in chemistry, John Murdzek. Katie Aiken and Lucy Mullins went through the manuscript line-by-line offering suggestions on writing style and pedagogy.

Accuracy Check and Class Test Cindy Berrie at the University of Kansas worked closely with the author, checking his work and providing detailed feedback as she and her students did a two-semester class test of the manuscript. The students also provided the author with comments on how to improve the manuscript so that the presentation of content was compatible with their variety of learning styles.

Shawn Phillips at Vanderbilt University reviewed the entire manuscript after the final developmental edit was completed, checked all the content for accuracy, and provided suggestions for further improvement to the author.

A select group reviewed text and art manuscript in draft and final form, reviewed page proofs in first and revised rounds, and oversaw the writing and accuracy check of the instructor's solutions manuals, test bank, and other ancillary materials.

Enhanced Support for the Instructor

McGraw-Hill offers instructors various tools and technology products in support of *University Chemistry*.

ARIS

Assessment, Review, and Instruction System, also known as ARIS, is an electronic homework and course management system designed for greater flexibility, power, and ease of use than any other system. Whether you are looking for a preplanned course or one you can customize to fit your course needs, ARIS is your solution.

In addition to having access to all student digital learning objects, ARIS allows instructors to do the following.

Build Assignments

- Choose from prebuilt assignments or create your own custom content by importing your own content or editing an existing assignment from the prebuilt assignment.
- Assignments can include quiz questions, animations, and videos—anything found on the website.
- Create announcements and utilize full course or individual student communication tools.
- Assign questions that were developed using the same problem-solving strategy as in the textual material, thus allowing students to continue the learning process from the text into their homework assignments.
- Assign algorithmic questions that give students multiple chances to practice and gain skill at problem-solving the same concept.

Track Student Progress

- Assignments are automatically graded.
- Gradebook functionality allows full-course management including:
  —Dropping the lowest grades
  —Weighting grades and manually adjusting grades
  —Exporting your grade book to Excel®, WebCT® or BlackBoard®
  —Manipulating data so that you can track student progress through multiple reports

Offer More Flexibility

- **Sharing Course Materials with Colleagues.** Instructors can create and share course materials and assignments with colleagues with a few clicks of the mouse, allowing for multiple section courses with many instructors and teaching assistants to continually be in synch, if desired.

▶ **Integration with BlackBoard or WebCT.** Once a student is registered in the course, all student activity within McGraw-Hill's ARIS is automatically recorded and available to the instructor through a fully integrated grade book that can be downloaded to Excel, WebCT, or Blackboard.

Presentation Center

The Presentation Center is a complete set of electronic book images and assets for instructors. You can build instructional materials wherever, whenever, and however you want! Accessed from your textbook's ARIS website, the Presentation Center is an online digital library containing selected photos, artwork, animations, and other media types that can be used to create customized lectures, visually enhanced tests and quizzes, compelling course websites, or attractive printed support materials. All assets are copyrighted by McGraw-Hill Higher Education, but can be used by instructors for classroom purposes. The visual resources in this collection include:

▶ **Art** Full-color digital files of all illustrations in the book can be readily incorporated into lecture presentations, exams, or custom-made classroom materials. In addition, all files are preinserted into PowerPoint® slides for ease of lecture preparation.

▶ **Animations** Numerous full-color animations illustrating important processes are also provided. Harness the visual impact of concepts in motion by importing these files into classroom presentations or online course materials.

▶ **PowerPoint Slides** For instructors who prefer to create their lectures from scratch, all illustrations, selected photos, and tables are preinserted by chapter into blank PowerPoint slides.

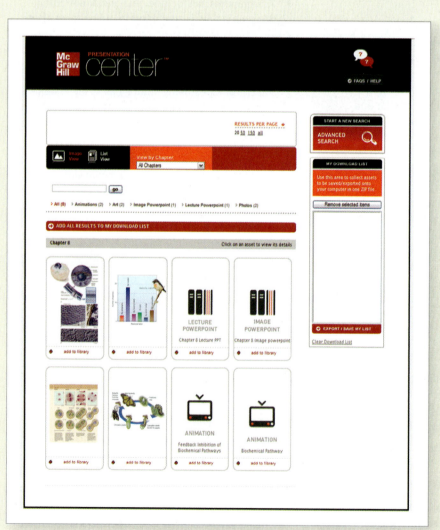

Animations and Media Player Content

Topics in chemistry are available in Media Player format and can be viewed on the text ARIS site. For the instructor, all McGraw-Hill chemistry animations are also available on the Presentation Center for use in lecture.

Access to your book, access to all books! The Presentation Center library includes thousands of assets from many McGraw-Hill titles. This ever-growing resource gives instructors the power to utilize assets specific to an adopted textbook as well as content from all other books in the library.

Nothing could be easier! Accessed from the instructor side of your textbook's ARIS website, Presentation Center's dynamic search engine allows you to explore by discipline, course, textbook chapter, asset type, or keyword. Simply browse, select, and download the files you need to build engaging course materials. All assets are copyright by McGraw-Hill Higher Education but can be used by instructors for classroom purposes. Instructors: To access ARIS, request registration information from your McGraw-Hill sales representative.

Computerized Test Bank Online

A comprehensive bank of test questions, created by Thomas Seery from University of Connecticut, is provided within a computerized test bank powered by McGraw-Hill's flexible electronic testing program EZ Test Online (www.eztestonline.com). EZ Test Online allows you to create paper and online tests or quizzes in this easy to use program!

Imagine being able to create and access your test or quiz anywhere, at any time without installing the testing software. Now, with EZ Test Online, instructors can select questions from multiple McGraw-Hill test banks or author their own, and then either print the test for paper distribution or give it online.

Test Creation

- ▶ Author or edit questions online using the 14 different question-type templates.
- ▶ Create printed tests or deliver online to get instant scoring and feedback.
- ▶ Create question pools to offer multiple versions online—great for practice.
- ▶ Export your tests for use in WebCT, Blackboard, PageOut and Apple's iQuiz.

- ▶ Create tests that are compatible with EZ Test Desktop tests you've already created.
- ▶ Share tests easily with colleagues, adjuncts, and TAs.

Online Test Management

- ▶ Set availability dates and time limits for your quiz or test.
- ▶ Control how your test will be presented.
- ▶ Assign points by question or question type with the drop-down menu.
- ▶ Provide immediate feedback to students or delay until all finish the test.
- ▶ Create practice tests online to enable student mastery.
- ▶ Upload your roster to enable student self-registration.

Online Scoring and Reporting

- ▶ Provides automated scoring for most of EZ Test's numerous question types.
- ▶ Allows manual scoring for essay and other open response questions.
- ▶ Allows manual rescoring and feedback.
- ▶ Is designed so that EZ Test's grade book will easily export to your grade book.
- ▶ Displays basic statistical reports.

Support and Help

The following provide support and help:

- ▶ User's Guide and built-in page-specific help
- ▶ Flash tutorials for getting started on the support site
- ▶ Support Website at www.mhhe.com/eztest

Student Response System

Wireless technology brings interactivity into the classroom or lecture hall. Instructors and students receive immediate feedback through wireless response pads that are easy to use and engage students. This system can be used by instructors to:

- ▶ Take attendance.
- ▶ Administer quizzes and tests.
- ▶ Create a lecture with intermittent questions.
- ▶ Manage lectures and student comprehension through the use of the grade book.
- ▶ Integrate interactivity into their PowerPoint presentations.

Content Delivery Flexibility

University Chemistry by Brian B. Laird is available in many formats in addition to the traditional textbook so that instructors and students have more choices when deciding on the format of their chemistry text. Choices include the following.

Color Custom by Chapter

For greater flexibility, we offer *University Chemistry* in a full-color custom version that allows instructors to pick the chapters they want included. Students pay for only what the instructor chooses.

Cooperative Chemistry Laboratory Manual

This innovative guide by Melanie Cooper (Clemson University) features open-ended problems designed to simulate experience in a research lab. Working in groups, students investigate one problem over a period of several weeks. Thus, they might complete three or four projects during the semester, rather than one preprogrammed experiment per class. The emphasis here is on experimental design, analysis problem solving, and communication.

Enhanced Support for Students

Designed to help students maximize their learning experience in chemistry, we offer the following options to students.

ARIS

Assessment, Review, and Instruction System, known as ARIS, is an electronic study system that offers students a digital portal of knowledge.

Students can readily access a variety of **digital learning objects** which include:

▸ Chapter level quizzing.
▸ Animations.
▸ Interactives.
▸ MP3 and MP4 downloads of selected content.

Student Solutions Manual

In this manual by Jay Shore (South Dakota State University), the student will find detailed solutions and explanations for the even-numbered problems in *University Chemistry*.

Acknowledgments

I would like to thank the following instructors, symposium participants, and students, whose comments were very helpful to me in preparing my first edition text.

Colin D. Abernethy   *Western Kentucky University*

Joseph J. BelBruno   *Dartmouth College*

Philip C. Bevilacqua   *The Pennsylvania State University*

Toby F. Block   *Georgia Institute of Technology*

Robert Bohn   *University of Connecticut*

B. Edward Cain   *Rochester Institute of Technology*

Michelle Chatellier   *University of Delaware*

Charles R. Cornett   *University of Wisconsin–Platteville*

Charles T. Cox   *Georgia Institute of Technology*

Darwin B. Dahl   *Western Kentucky University*

Stephen Drucker   *University of Wisconsin–Eau Claire*

Darcy J. Gentleman   *University of Toronto*

David O. Harris   *University of California–Santa Barbara*

J. Joseph Jesudason   *Acadia University*

Kirk T. Kawagoe   *Fresno City College*

Paul Kiprof   *University of Minnesota–Duluth*

Craig Martens   *University of California–Irvine*

Stephen Mezyk   *California State University at Long Beach*

Matthew L. Miller   *South Dakota State University*

Michael Mombourquette   *Queens University*

Shawn T. Phillips   *Vanderbilt University*

Rozana Abdul Razak   *MARA University of Technology*

Thomas Schleich   *University of California–Santa Cruz*

Thomas A. P. Seery   *University of Connecticut*

Jay S. Shore   *South Dakota University*

Michael S. Sommer   *University of Wyoming*

Larry Spreer   *University of the Pacific*

Marcus L. Steele   *Delta State University*

Mark Sulkes   *Tulane University*

Paul S. Szalay   *Muskingum College*

Michael Topp   *University of Pennsylvania*

Robert B. Towery   *Houston Baptist University*

Thomas R. Webb   *Auburn University*

Stephen H. Wentland   *Houston Baptist University*

John S. Winn   *Dartmouth College*

Paulos Yohannes   *Georgia Perimeter College*

Timothy Zauche   *University of Wisconsin–Platteville*

Lois Anne Zook-Gerdau   *Muskingum College*

Student Class Test, University of Kansas

I would like to thank the following students for using my manuscript for their course throughout the year and providing me insight on student use.

Mirza Nayyar Ahmad	Lizzy Mahoney
Elizabeth Beech	Amber Markey
Amelia Bray	Allison Martin
Thomas Chantz	Ian Mayhugh
Jason Christian	Nicole E. McClure
Connor Dennis	Sara McElhaney
Andrew Dick	Justin Moyes
Ryan Edward Dowell	Chelsea Montgomery
Hollie Farrahi	Ryan Murphy
Lindsey Fisher	Thomas Northup
Stephen Folmsbee	Matthew Oliva
Megan Fowler	Jace Parkhurst
Megan Fracol	Sweta Patel
Jessie Garrett	Megan L. Razak
Casey Gee	Kate Remley
Andy Haverkamp	Thomas Reynolds
Erica Henderson	Richard Robinson
Armand Heyns	Lauren N. Schimming
Allison Ho	Alan Schurle
Kalin Holthaus	Amy Soules
Michael Holtz	Jen Strande
Jake S. Hopkins	Sharayah Stitt
Josh Istas	Jessica Stogsdill
Ladini Jayaratne	Joanna Marie Wakeman
Libby Johnson	Andre W. Wendorff
Sophia Kaska	Thomas K. Whitson
Jim LaRocca	Daniel Zehr
Jenn Logue	Simon Zhang

I wish to acknowledge my colleagues at the University of Kansas for their help and support in preparing this text, with a special thanks to Professor Cindy Berrie, who has been using drafts of the text in her Honors General Chemistry course. The feedback from her and her students was invaluable. Thanks also to Craig Lunte and Robert Dunn who helped me keep my sanity during this project by enticing me to the golf course on many a sunny day.

This text would have not become a reality without the extremely dedicated and competent team at McGraw-Hill Higher Education. For their generous support, I wish to acknowledge Thomas Timp (Publisher), Tami Hodge (Senior Sponsoring Editor), Gloria Schiesl (Senior Project Manager), John Leland (Senior Photo Research Coordinator) and especially my patiently persistent taskmaster Shirley Oberbroeckling (Senior Developmental Editor) for her continual advice, pep talks, and general support at all stages of the project. I would also like to thank former publisher Kent Peterson (VP—Director of Marketing) for talking me into pursuing this project on a balmy (if somewhat blurry) evening in Key West.

Finally, I would like to acknowledge the significant contributions and sage advice of Professor Raymond Chang of Williams College, without which this book would not have been possible.

Brian B. Laird
Lawrence, Kansas
February 2008

0
Chapter

The Language of Chemistry

Since ancient times humans have pondered the nature of matter. Our modern ideas of the structure of matter began to take shape in the early nineteenth century with Dalton's atomic theory. We now know that all ordinary matter is made up of atoms, molecules, and ions. All of chemistry is concerned with the nature of these species and their transformations. Over the past two centuries, chemists have developed a lexicon of terms and concepts that allows them to accurately and efficiently discuss chemical ideas among themselves and communicate these ideas to others. This language of chemistry provides us a way to visualize and quantify chemical transformations at the molecular level while simultaneously understanding the consequences of these transformations in the macroscopic world in which we live.

The Chinese characters for chemistry mean "The study of change."

0.1 | Chemistry Is the Study of Matter and Change

***Chemistry** is the study of matter and the changes it undergoes.* Because a basic knowledge of chemistry is essential for students of biology, physics, astronomy, geology, nutrition, ecology, and many other subjects, chemistry is often referred to as the central science. Indeed, the products of chemistry are central to our way of life; without them, we would be living shorter lives in what we would consider primitive conditions, without automobiles, electricity, computers, DVDs, and numerous other everyday conveniences. Although chemistry is an ancient science, its modern foundation was laid in the nineteenth and twentieth centuries when intellectual and technological advances enabled scientists to break down substances into ever smaller components and consequently to explain many of their physical and chemical characteristics. The rapid development of increasingly sophisticated technology throughout the twentieth century has given us even greater means to study things that cannot be seen with the naked eye. Using computers and special microscopes, for example, chemists can analyze the structure of atoms and molecules—the fundamental units on which chemistry is based—and rationally design new substances with specific properties, such as pharmaceuticals and environmentally friendly consumer products. In the twenty-first century, chemistry will remain an important element in science and technology. This is especially true in emerging fields such as nanotechnology and molecular biology, where the quantum-mechanical behavior of matter cannot be ignored at the molecular level, and in environmental science, where a fundamental understanding of complex chemical-reaction kinetics is crucial to understanding and solving pollution problems. Whatever your reasons for taking college chemistry, knowledge of the subject will better enable you to appreciate its impact on society and the individual.

Chemistry is commonly perceived to be more difficult than other subjects, at least at the introductory level. There is some justification for this perception; for one thing, chemistry has a specialized vocabulary, which to a beginning student may seem quite abstract. Chemistry, however, is so deeply embedded in everyday experience that we all are familiar with the effects of chemical processes, even if we lack precise chemical language to describe them. For example, if you cook, then you are a practicing chemist! From experience gained in the kitchen, you know that oil and water do not mix and that boiling water left on the stove will evaporate. You apply chemical and physical principles when you use baking soda to leaven bread, choose a pressure cooker to shorten cooking time, add meat tenderizer to a pot roast, squeeze lemon juice over sliced pears to prevent them from turning brown or over fish to minimize its odor, and add vinegar to the water in which you are going to poach eggs. Every day we observe such changes without thinking about their chemical nature. The purpose of this course is help you learn to think like a chemist, to look at the *macroscopic world*—the things we can see, touch, and measure directly—and visualize the particles and events of the *molecular world* that we cannot perceive without modern technology and our imaginations. At first, some students find it confusing when their chemistry instructor and textbook seem to continually shift back and forth between the macroscopic and molecular worlds. Just keep in mind that the data for chemical investigations most often come from observations of large-scale phenomena, but the explanations frequently lie in the unseen submicroscopic world of atoms and molecules. In other words, chemists often *see* one thing (in the macroscopic world) and *think* another (in the submicroscopic world). Looking at the rusted nails in Figure 0.1, for example, a chemist might think about the properties of individual atoms of iron and how these units interact with other atoms and molecules to produce the observed change.

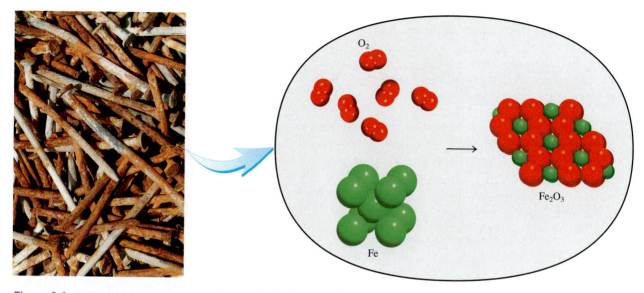

Figure 0.1 A simplified molecular view of rust (Fe$_2$O$_3$) formation from iron atoms (Fe) and oxygen molecules (O$_2$). In reality, the process requires the presence of water and rust also contains water molecules.

Classifications of Matter

We defined chemistry as the study of matter and the changes it undergoes. ***Matter*** is *anything that occupies space and has mass* and includes things we can see and touch (such as water, earth, and trees), as well as things we cannot (such as air). Thus, everything in the universe has a "chemical" connection. Chemists distinguish among several subcategories of matter based on composition and properties. The classifications of matter include substances, mixtures, elements, and compounds, as well as atoms and molecules, which we will consider in Section 0.2.

Substances and Mixtures

A ***substance*** is *a form of matter that has a definite (constant) composition and distinct properties.* Examples include water, ammonia, table salt, gold, and oxygen. Substances differ from one another in composition and can be identified by their appearance, smell, taste, and other properties. A ***mixture*** is *a combination of two or more substances in which the substances retain their distinct identities.* Some familiar examples are air, soft drinks, milk, and cement. Mixtures do not have constant composition. Samples of air collected in different cities differ in composition because of differences in altitude, pollution, weather conditions, and so on. Mixtures are either homogeneous or heterogeneous. When a spoonful of sugar dissolves in water we obtain a ***homogeneous mixture*** in which *the composition of the mixture is the same throughout the sample.* A homogeneous mixture is also called a ***solution.*** If one substance in a solution is present in significantly larger amounts than the other components of the mixture, we refer to the dominant substance as the ***solvent.*** The other components of the solution, present in smaller amounts, are referred to as ***solutes.*** A solution may be gaseous (such as air), solid (such as an alloy), or liquid (seawater, for example). If sand is mixed with iron filings, however, the sand grains and the iron filings remain separate [Figure 0.2(a)]. This type of mixture is called a ***heterogeneous mixture*** because *the composition is not uniform.*

Any mixture, whether homogeneous or heterogeneous, can be created and then separated by physical means into pure components without changing the identities of the components. Thus, sugar can be recovered from a water solution by heating the

Figure 0.2 (a) The mixture contains iron filings and sand. (b) A magnet separates the iron filings from the mixture. The same technique is used on a larger scale to separate iron and steel from nonmagnetic substances such as aluminum, glass, and plastics.

(a)   (b)

solution and evaporating it to dryness. Condensing the vapor will give us back the water component. To separate the iron-sand mixture, we can use a magnet to remove the iron filings from the sand, because sand is not attracted to the magnet [see Figure 0.2(b)]. After separation, the components of the mixture will have the same composition and properties with which they started.

Elements and Compounds

Substances can be either elements or compounds. An ***element*** is *a substance that cannot be separated into simpler substances by chemical means.* To date, 117 elements have been positively identified. Most of them occur naturally on Earth, but scientists have created others artificially via nuclear processes (see Chapter 17). For convenience, chemists use symbols of one or two letters to represent the elements. The first letter of a symbol is *always* capitalized, but any following letters are not. For example, Co is the symbol for the element cobalt, whereas CO is the formula for the carbon monoxide molecule.

Table 0.1 lists the names and symbols of some common elements; a complete list of the elements and their symbols appears inside the front cover of this book. Although

Table 0.1		Some Common Elements and Their Symbols			
Name	**Symbol**	**Name**	**Symbol**	**Name**	**Symbol**
Aluminum	Al	Fluorine	F	Oxygen	O
Arsenic	As	Gold	Au	Phosphorus	P
Barium	Ba	Hydrogen	H	Platinum	Pt
Bismuth	Bi	Iodine	I	Potassium	K
Bromine	Br	Iron	Fe	Silicon	Si
Calcium	Ca	Lead	Pb	Silver	Ag
Carbon	C	Magnesium	Mg	Sodium	Na
Chlorine	Cl	Manganese	Mn	Sulfur	S
Chromium	Cr	Mercury	Hg	Tin	Sn
Cobalt	Co	Nickel	Ni	Tungsten	W
Copper	Cu	Nitrogen	N	Zinc	Zn

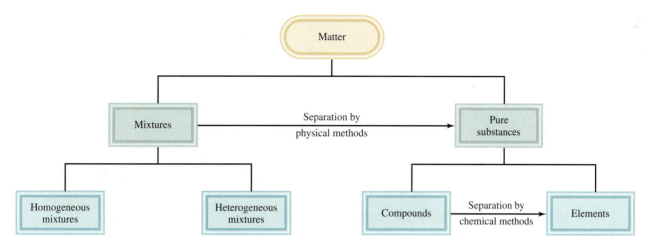

Figure 0.3 Classification of matter by composition.

most symbols for elements are consistent with their English names, some elements have symbols derived from Latin, for example, Au from *aurum* (gold), Fe from *ferrum* (iron), and Na from *natrium* (sodium). In another exception, the symbol W for tungsten is derived from its German name *Wolfram*. Appendix 3 gives the origin of the names and lists the discoverers of most of the elements.

Elements may combine with one another to form compounds. Hydrogen gas, for example, burns in oxygen gas to form water, which has properties that are distinctly different from the elements hydrogen and oxygen. Water is made up of two parts hydrogen and one part oxygen. This composition does not change, regardless of whether the water comes from a faucet in the United States, a lake in Outer Mongolia, or the ice caps on Mars. Thus, water is a ***compound,*** *a substance composed of atoms of two or more elements chemically united in fixed proportions.* Unlike mixtures, compounds can be separated only by chemical means into their elemental components. The relationships among elements, compounds, and other categories of matter are summarized in Figure 0.3.

The Three States of Matter

In addition to composition, matter can also be classified according to its physical state. All matter can, in principle, exist in three physical states: solid, liquid, and gas (or vapor). A ***solid*** is a material that resists changes in both volume *and* shape—the force required to deform a block of solid steel is quite substantial. Solids can either be *crystalline,* possessing a highly ordered periodic array of closely spaced molecules (for example, table sugar), or *amorphous,* with a dense, but disordered, packing of molecules (for example, window glass). A ***liquid*** also resists changes in volume, but not of shape. When a liquid is poured from one container to another, the volume of the liquid does not change, but the shape of the liquid adapts to match the new container. In both solids and liquids, the space between molecules is similar to the sizes of the molecules themselves. In a ***gas,*** though, the distances between atoms or molecules are large compared to molecular size. As a result, a gas resists neither changes in volume nor changes in shape. Both gases and liquids are collectively known as *fluids.* The structural differences between crystalline solids, liquids, and gases are illustrated in Figure 0.4.

The three states of matter can interconvert without changing the composition of the substance. Upon heating, a solid, such as ice, melts to form a liquid. (The

Figure 0.4 Microscopic views of a solid, a liquid, and a gas.

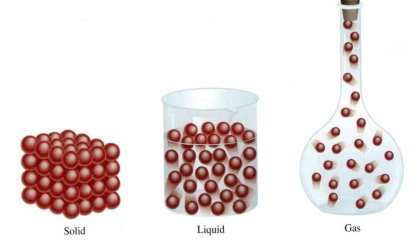

Solid   Liquid   Gas

temperature at which this transition occurs is called the *melting point.*) Further heating converts the liquid into a gas. (The conversion of a liquid to a gas takes place at the *boiling point* of the liquid.) On the other hand, cooling a gas below the boiling point of the substance causes it to condense into a liquid. When the liquid is cooled further, below the melting point of the substance, it freezes to form the solid. Under proper conditions, some solids can convert directly into a gas; this process is called *sublimation.* For example, solid carbon dioxide, commonly known as dry ice, readily sublimes to carbon dioxide gas unless maintained below a temperature of $-78.5°C$.

Physical and Chemical Properties of Matter

Substances are identified by their properties as well as by their composition. Color, melting point, and boiling point are physical properties. A ***physical property*** *can be measured and observed without changing the composition or identity of a substance.* For example, we can measure the melting point of ice by heating a block of ice and recording the temperature at which the ice is converted to water. Water differs from ice only in appearance, not in composition, so this is a physical change; we can freeze the water to recover the ice. Therefore, the melting point of a substance is a physical property. Similarly, when we say that helium gas is less dense than air, we are referring to a physical property.

On the other hand, the statement "Hydrogen gas burns in oxygen gas to form water" describes a ***chemical property*** of hydrogen, because *to observe this property we must carry out a chemical change* (in this case, burning). After the change, the original chemical substance, the hydrogen gas, will have vanished, and all that will be left is a different chemical substance—water. We *cannot* recover the hydrogen from the water by means of a physical change, such as boiling or freezing. Every time we hard-boil an egg, we bring about a chemical change. When subjected to a temperature of about 100°C, the yolk and the egg white undergo changes that alter not only their physical appearance but their chemical makeup as well. When eaten, substances in our bodies called *enzymes* facilitate additional chemical transformations of the egg. This digestive action is another example of a chemical change. What happens during digestion depends on the chemical properties of both the enzymes and the food.

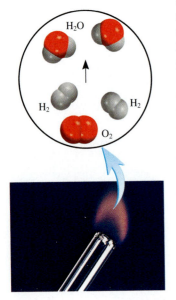

Hydrogen burning in air to form water

All measurable properties of matter may be additionally categorized as *extensive* or *intensive*. The measured value of an **extensive property** *depends on how much matter is being considered.* **Mass,** which is *the quantity of matter in a given sample of a substance,* is an extensive property. More matter means more mass. Values of the same extensive property can be added together. For example, two copper pennies have a combined mass that is the sum of the mass of each penny, and the length of two tennis courts is the sum of the length of each tennis court. **Volume,** defined as *length cubed,* is another extensive property. The value of an extensive property depends on the amount of matter.

The measured value of an **intensive property** *does not depend on how much matter is being considered.* **Density,** defined as *the mass of an object divided by its volume,* is an intensive property. So is temperature. Suppose that we have two beakers of water at the same temperature. If we combine them to make a single quantity of water in a larger beaker, the temperature of the larger quantity of water will be the same as it was in the two separate beakers. Unlike mass, length, volume, and energy, temperature and other intensive properties are nonadditive.

Force and Energy

We have discussed the properties of matter, but chemistry also studies the changes that matter undergoes. These changes are brought about by forces and resulting energy changes, which we examine now.

Force

Anything that happens in the universe is the result of the action of a force. **Force** is the quantity that causes an object to change its course of motion (either in direction or speed). Isaac Newton[1] first quantified the relationship between force and the motion of material objects in his second law of motion:

$$\text{force} = \text{mass} \times \text{acceleration} \tag{0.1}$$

where the *acceleration* of an object is the rate of change (derivative) of the velocity of the object with respect to time. Given the forces on a particle and its initial position and velocity, Equation 0.1 can be used to determine the future motion of the system. The SI[2] units for mass and acceleration are kg and m s^{-2}, respectively, so the SI unit of force is kg m s^{-2}, or the *newton* (N): 1 N = 1 kg m s^{-2}.

Physicists have identified the following four fundamental forces in the universe:

▸ **The electromagnetic force** is the force between electrically charged objects or magnetic materials.

▸ **The gravitational force** is the attractive force between objects caused by their masses.

---

1. Sir Isaac Newton (1643–1727). English physicist and mathematician. One of the most brilliant scientists in history, he founded the fields of classical mechanics and the differential and integral calculus, as well as made major contributions to the field of optics.

2. The **International System of Units** (abbreviated **SI,** from the French *Système Internationale d'Unites*), based on the metric system, was adopted in 1960 by the General Conference of Weights and Measures, the international authority on units. With a few exceptions, we will use SI units throughout this book. A discussion of SI units can be found in Appendix 1.

▸ *The weak force* is the force responsible for some forms of radioactive decay.

▸ *The strong force* is the force binding the protons and neutrons in a nucleus together, overcoming the powerful electromagnetic repulsion between positively charged protons.

In chemistry, by far the most important of these forces is the electromagnetic force. With the exception of nuclear chemistry (discussed in Chapter 17), which involves the strong and weak forces, all of chemistry is a direct result of the action of electromagnetic forces between electrons and protons, and between matter and light (electromagnetic radiation). The gravitational force is far too weak to have an impact at the atomic and molecular scale.

Energy

All chemical or physical transformations in nature are driven by the release, absorption, or redistribution of energy. A detailed description of any such transformation is not possible without an understanding of the role of energy. From a mechanical perspective, *energy* is *the capacity to do work*. **Work** is done whenever an object is moved from one place to another in the presence of some external force. For example, the process of lifting a ball off of the floor and placing it on a table requires work against the gravitational force between Earth and the ball. For a mechanical system, work is the product of the distance d that the object is moved times the applied force F (measured along the object's direction of motion[3]):

$$\text{work} = \text{force} \times \text{distance} = F \times d \qquad (0.2)$$

The SI unit of energy is the *joule* (J), which is the work done in moving an object a distance of 1 m against a force of 1 N. Thus, $1 \text{ J} \equiv 1 \text{ N m} = 1 \text{ kg m}^2 \text{ s}^{-2}$.

The energy possessed by a material object is composed of two basic forms: kinetic and potential. The energy that an object has as a result of its motion is its **kinetic energy.** For an object with mass m and velocity u, the kinetic energy is given by

$$\text{kinetic energy} = \frac{1}{2}mu^2 \qquad (0.3)$$

Example 0.1

Calculate the kinetic energy of a 150-g baseball traveling at a velocity of 50 m s^{-1}.

Solution   Substitute the mass and velocity of the baseball into Equation 0.3 but be careful to convert grams to kilograms, so that the final value is in joules.

$$\text{kinetic energy} = \frac{1}{2}mu^2 = \frac{1}{2}\left[(150 \text{ g})\frac{1 \text{ kg}}{1000 \text{ g}}\right](50 \text{ m s}^{-1})^2$$

$$= 1.9 \times 10^2 \text{ kg m}^2 \text{ s}^{-2}$$

$$= 1.9 \times 10^2 \text{ J}$$

Practice Exercise   Calculate the kinetic energy (in joules) of a car with mass 1000 kg traveling at a velocity of 130 km per hour.

---

3. Both force and distance are vectors, so a more general definition would be *work* = $\boldsymbol{F} \cdot \boldsymbol{d}$, where "·" denotes the usual vector dot product.

Objects can also possess energy as a result of their position in space. This type of energy is called **potential energy (V).** The equation for potential energy depends upon the type of material and its specific environment. Examples include the following:

1. The gravitational potential energy of an object with mass m at a height h above the surface of Earth:

$$\text{potential energy} = V = mgh \qquad\qquad \textbf{(0.4)}$$

where $g = 9.80665$ m s^{-1} is the standard terrestrial gravitational acceleration constant.

2. The electrostatic potential energy of interaction between two particles with electric charges q_1 and q_2 separated by a distance r, which is given by *Coulomb's law:*

$$V = \frac{q_1 q_2}{4\pi\varepsilon_0 r} \qquad\qquad \textbf{(0.5)}$$

where $\varepsilon_0 = 8.8541878 \times 10^{-12}$ C^2 J^{-1} m^{-1} is a fundamental physical constant called the *permittivity of the vacuum,* and q_1 and q_2 are measured in the SI unit of charge, the *coulomb* (C). (The presence of the factor $4\pi\varepsilon_0$ is due to the use of the SI system of units.)

3. The potential energy of a spring obeying Hooke's law

$$V = \frac{1}{2}k(x - x_0)^2 \qquad\qquad \textbf{(0.6)}$$

where x is the length of the spring, x_0 is the equilibrium value of the length, and k is called the spring constant and is a measure of the stiffness of the spring. This potential energy function also represents a good approximation to the potential energy of molecular bond vibrations.

The potential energy is important in determining the future motion of an object because it is directly related to the force on that object. For a one-dimensional potential energy function $V(x)$, the force is given by the negative of the derivative (see Appendix 1) of the potential energy:

$$F(x) = -\frac{dV(x)}{dx} \qquad\qquad \textbf{(0.7)}$$

From this equation, the force on the object is in the direction of decreasing potential energy. Thus, an object placed at the top of a hill will roll down the hill to decrease its gravitational potential energy (Equation 0.4). Equation 0.7 can be used to show that the decrease in potential energy in a process is equal to the work done in that process:

$$\text{work done in process} = -\left[V_{\text{final}} - V_{\text{initial}}\right] = -\Delta V \qquad\qquad \textbf{(0.8)}$$

(The symbol Δ is used to indicate the change in a quantity.)

The total energy of an object is the sum of its kinetic and potential energies. As an object is accelerated (or deaccelerated) under the influence of external forces, its kinetic energy changes with time. Likewise, as the position of the object changes along its trajectory, the potential energy also varies. The total energy, however, remains constant. As the kinetic energy of an object increases, the potential energy must decrease by exactly the same amount to maintain constant total energy. The **law of conservation of energy** summarizes this principle: *the total quantity of energy in the universe is assumed constant.*

As the water falls from the dam, its potential energy is converted into kinetic energy. Use of this energy to generate electricity is called hydroelectric power.

Example 0.2

A 0.295-kg ball, initially at rest, is released from a height of 29.3 m from the ground. How fast will the ball be traveling when it hits the ground? (Ignore any air resistance.)

Strategy The speed of the ball is related to the kinetic energy through Equation 0.3. From the principle of the conservation of energy, the gain in kinetic energy must exactly equal the loss of potential energy, which can be calculated from the change in height using Equation 0.4.

Solution Use Equation 0.4 to calculate the change in potential energy:

$$\Delta(\text{potential energy}) = mgh_{\text{final}} - mgh_{\text{initial}}$$
$$= mg(h_{\text{final}} - h_{\text{initial}})$$
$$= (0.295 \text{ kg})(9.807 \text{ m s}^{-2})(0 \text{ m} - 29.3 \text{ m})$$
$$= -84.8 \text{ kg m}^2 \text{ s}^{-2} = -84.8 \text{ J}$$

From the conservation of energy, $\Delta(\text{kinetic energy}) = -\Delta(\text{potential energy})$, so

$$\frac{1}{2}mu_{\text{final}}^2 - \frac{1}{2}mu_{\text{initial}}^2 = -\Delta(\text{potential energy})$$

$$\frac{1}{2}(0.295 \text{ kg})u_{\text{final}}^2 - \frac{1}{2}(0.295 \text{ kg})(0)^2 = -(-84.8 \text{ kg m}^2 \text{ s}^{-2})$$

$$u_{\text{final}}^2 = \frac{2(84.8 \text{ kg m}^2 \text{ s}^{-2})}{0.295 \text{ kg}} = 575 \text{ m}^2 \text{ s}^{-2}$$

$$u_{\text{final}} = 24.0 \text{ m s}^{-1}$$

Practice Exercise How high above the surface of Earth would you have to drop a ball of mass 10.0 kg for it to reach a speed of 20 m s^{-1} before it hit the ground?

Although all energy can be ultimately identified as kinetic or potential energy or a combination of the two, it is convenient in chemistry to define additional classifications of energy that depend upon the system. Important examples include the following:

▶ *Radiant energy* is the energy contained in electromagnetic radiation, such as X-rays, radio waves, and visible light. On Earth, the primary energy source is radiant energy coming from the sun, which is called *solar energy*. Solar energy heats the atmosphere and surface of Earth, stimulates the growth of vegetation through the process known as photosynthesis, and influences global climate patterns. Radiant energy is a combination of the potential and kinetic energy of the electromagnetic fields that make up light.

▶ *Thermal energy* is *the energy associated with the random motion of atoms and molecules.* Thermal energy can be generally calculated from temperature measurements. The more vigorous the motion of the atoms and molecules in a sample of matter, the hotter the sample is and the greater its thermal energy. Keep in mind, however, thermal energy and temperature are different. A cup of coffee at 70°C has a higher temperature than a bathtub filled with water at 40°C, but the bathtub stores much more thermal energy because it has a much larger volume and greater mass than the coffee. The bathtub water has more water molecules and, therefore, more molecular motion. Put another way, temperature is an intensive property (does not depend upon the amount of matter), whereas thermal energy is an extensive property (does depend proportionally on the amount of matter).

▸ ***Chemical energy*** is a form of potential energy *stored within the structural units of chemical substances;* its quantity is determined by the type and arrangement of the constituent atoms. When substances participate in chemical reactions, chemical energy is released, stored, or converted to other forms of energy.

All forms of energy can be converted (at least in principle) from one form to another. We feel warm when we stand in sunlight because radiant energy is converted to thermal energy on our skin. When we exercise, chemical energy stored in the molecules within our bodies is used to produce kinetic energy. When a ball starts to roll downhill, its potential energy is converted to kinetic energy.

0.2 | Matter Consists of Atoms and Molecules

Atomic Theory

In the fifth century B.C. the Greek philosopher Democritus proposed that all matter consisted of very small, indivisible particles, which he named *atomos* (meaning indivisible). Although Democritus' idea was not accepted by many of his contemporaries (notably Plato and Aristotle), it nevertheless endured. Experimental evidence from early scientific investigations provided support for the notion of "atomism" and gradually gave rise to the modern definitions of elements and compounds. In 1808 an English scientist and schoolteacher, John Dalton,[4] formulated a more precise definition of the indivisible building blocks of matter that we call atoms. Dalton's work marked the beginning of the modern era of chemistry. The hypotheses about the nature of matter on which Dalton's atomic theory is based can be summarized as follows:

1. Elements are composed of extremely small particles called atoms. All atoms of a given element are identical, having the same size, mass, and chemical properties. The atoms of one element are different from the atoms of all other elements.

2. Compounds are composed of atoms of more than one element. In any compound, the ratio of the numbers of atoms of any two of the elements present is either an integer or a simple fraction.

3. A chemical reaction involves only the separation, combination, or rearrangement of atoms; it does not result in the creation or destruction of atoms.

Figure 0.5 shows a schematic representation of the first two hypotheses.

Dalton recognized that atoms of one element were different from atoms of all other elements (the first hypothesis), but he made no attempt to describe the structure or composition of atoms because he had no idea what an atom was really like. He did realize, however, that the different properties shown by elements such as hydrogen and oxygen could be explained by assuming that hydrogen atoms were not the same as oxygen atoms.

The second hypothesis suggests that, to form a certain compound, we need not only atoms of the right kinds of elements, but specific numbers of these atoms as

---

4. John Dalton (1766–1844). English chemist, mathematician, and philosopher. In addition to the atomic theory, he also formulated several gas laws and gave the first detailed description of color blindness, from which he suffered. Dalton was described as an indifferent experimenter, and singularly wanting in the language and power of illustration. His only recreation was lawn bowling on Thursday afternoons. Perhaps it was the sight of those wooden balls that provided him with the idea of atomic theory.

Figure 0.5 (a) According to Dalton's atomic theory, atoms of the same element are identical, but atoms of one element are different from atoms of other elements. (b) Compounds formed from atoms of elements X and Y. In this case, the ratio of the atoms of element X to the atoms of element Y is 2:1. Note that a chemical reaction results only in the redistribution of atoms, not in their destruction or creation.

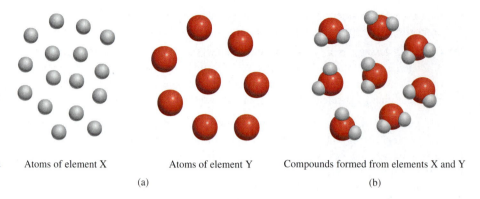

Atoms of element X   Atoms of element Y   Compounds formed from elements X and Y

(a)   (b)

well. This idea is an extension of the ***law of definite proportions,*** first published in 1799 by Joseph Proust.[5] According to the law of definite proportions, *different samples of the same compound always contain its constituent elements in the same proportion by mass.* Thus, if we were to analyze samples of carbon dioxide gas obtained from different sources, we would find in each sample the same ratio by mass of carbon to oxygen. It stands to reason, then, that if the ratio of the masses of different elements in a given compound is fixed, then the ratio of the atoms of these elements in the compound must also be constant.

Dalton's second hypothesis also supports the ***law of multiple proportions.*** According to this law, *if two elements can combine to form more than one compound, the masses of one element that combine with a fixed mass of the other element are in ratios of small whole numbers.* Dalton's theory explains the law of multiple proportions quite simply: Different compounds made up of the same elements differ in the number of atoms of each kind that combine. Carbon, for example, forms two stable compounds with oxygen, namely, carbon monoxide and carbon dioxide. Modern measurement techniques have shown that one atom of carbon combines with one atom of oxygen in carbon monoxide and with two atoms of oxygen in carbon dioxide. Thus, the ratio of oxygen in carbon monoxide to oxygen in carbon dioxide is 1:2. This result is consistent with the law of multiple proportions.

Dalton's third hypothesis is another way of stating the ***law of conservation of mass***[6]—that is, *matter can be neither created nor destroyed.* For chemical reactions this principle had been demonstrated experimentally by Antoine Lavoisier[7] who heated mercury in air to form mercury(II) oxide and showed that the increase in mass of the oxide over the pure mercury was exactly equal to the decrease in the mass of the gas. Because matter is made of atoms that are unchanged in a chemical reaction, it follows that mass must be conserved as well. Dalton's brilliant insight into the nature of matter was the main stimulus for the rapid progress of chemistry during the nineteenth century.

---

5. Joseph Louis Proust (1754–1826). French chemist. Proust was the first person to isolate sugar from grapes.

6. According to Albert Einstein, mass and energy are alternate aspects of a single entity called *mass–energy.* Chemical reactions usually involve a gain or loss of heat and other forms of energy. Thus, when energy is lost in a reaction, mass is lost, too. This is only meaningful, however, for nuclear reactions (see Chapter 17). For chemical reactions the changes of mass are too small to detect. For all practical purposes, therefore, mass is conserved.

7. Antoine Laurent Lavoisier (1743–1794). French chemist, sometimes referred to as the "father of modern chemistry." In addition to his role in verifying the principle of the conservation of mass, he made important contributions to chemical nomenclature and to the understanding of the role of oxygen in combustion, respiration, and acidity. For his role as a tax collector for the king, Lavoisier was guillotined during the French Revolution.

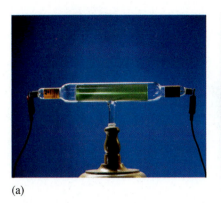
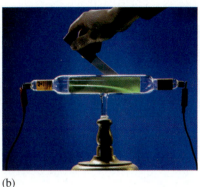
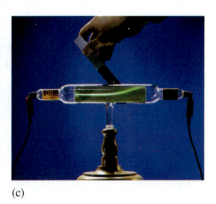

(a)   (b)   (c)

Figure 0.6 (a) A cathode ray produced in a discharge tube travels from left to right. The ray itself is invisible, but the fluorescence of a zinc sulfide coating on the glass causes it to appear green. (b) The cathode ray is bent downward when the south pole of the bar magnet is brought toward it. (c) When the polarity of the magnet is reversed, the ray bends in the opposite direction.

The Building Blocks of the Atom

Based on Dalton's atomic theory, an ***atom*** is *the basic unit of an element that can enter into chemical combination.* Dalton imagined an atom that was both extremely small and indivisible. However, a series of investigations that began in the 1850s and extended into the twentieth century clearly demonstrated that atoms actually possess internal structure; that is, they are made up of even smaller *subatomic particles.* Particle physicists have discovered a complex hierarchy (or "zoo") of such particles, but only electrons, protons, and neutrons are of primary importance in chemical reactions.

The ***electron,*** discovered by J. J. Thomson[8] in 1897, is a tiny, negatively charged particle with a charge of $-1.602177 \times 10^{-19}$ C, where C stands for *coulomb,* the SI unit of electric charge, and a mass $m_e = 9.109383 \times 10^{-31}$ kg. (The magnitude of the electron charge, 1.602177×10^{-19} C, is a fundamental physical constant and is given the symbol e while its charge is denoted as $-e$.) The discovery was made using a device known as a cathode ray tube (Figure 0.6), which consists of two metal plates inside an evacuated glass tube. When the two metal plates are connected to a high-voltage source, the negatively charged plate, called the *cathode,* emits an invisible ray. This *cathode ray* is drawn to the positively charged plate, called the *anode,* where it passes through a hole and continues traveling to the other end of the tube. When the ray strikes the specially coated surface, it produces a strong fluorescence, or bright green light. Thomson showed that the cathode ray was actually a beam of negatively charged particles (electrons). By observing the degree to which the beam was deflected when placed in a magnetic field (see Figure 0.6), Thompson was able to determine the magnitude of the electron charge-to-mass ratio, e/m_e. A decade later, R. A. Millikan[9] succeeded in directly measuring e with great precision by examining the motion of individual tiny drops of oil that picked up static charge from the air. Using the charge-to-mass ratio determined earlier by Thomson, Millikan was also able to calculate the mass of the electron, m_e.

Because atoms are electrically neutral, they must also contain positively charged components in addition to electrons. Thomson proposed that an atom consisted of a diffuse sphere of uniform positively charged matter in which electrons were imbedded like raisins in a plum pudding (a traditional English dessert).

---

8. Joseph John Thomson (1856–1940). British physicist who received the Nobel Prize in Physics in 1906 for the discovery of the electron.

9. Robert Andrews Millikan (1868–1953). American physicist who was awarded the Nobel Prize in Physics in 1923 for determining the charge of the electron.

Figure 0.7 (a) Rutherford's experimental design for measuring the scattering of particles by a piece of gold foil. The vast majority of the particles passed through the gold foil with little or no deflection; however, a few were deflected at wide angles and, occasionally, a particle was turned back. (b) Magnified view of particles passing through and being deflected by nuclei.

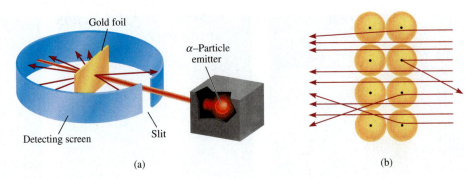

(a)

(b)

Thompsons's "plum pudding" model of atomic structure was disproven by a series of experiments performed in 1910 by the physicist Ernest Rutherford.[10] Together with his associate Hans Geiger[11] and undergraduate student Ernest Marsden,[12] Rutherford bombarded very thin gold foil with *alpha particles* (α), a recently discovered form of radiation that consists of positively charged helium nuclei (Figure 0.7). They observed that the overwhelming majority of α particles passed through the foil with little or no deflection. However, a small number of α particles were strongly scattered through large angles—some as large as 180°. This was inconsistent with Thomson's model of the atom because the positive charge of the atom should have been so diffuse that the α particles should have passed through the foil with very little deflection. Rutherford's initial reaction, when told of this discovery, was to say, "It was as incredible as if you had fired a 15-inch shell at a piece of tissue paper and it came back and hit you." Rutherford was later able to explain these experimental results in terms of a new model for the atom. According to Rutherford, most of the atom must be empty space, with the positive charges concentrated within a very small, but dense, central core called the **nucleus.** Whenever an α particle neared the nucleus in the scattering experiment, it experienced a large repulsive force and therefore a large deflection. Moreover, an α particle traveling directly toward a nucleus would be completely repelled and its direction would be reversed.

The positively charged particles in the nucleus are called **protons.** In separate experiments, it has been found that each proton carries a charge of $+e$ and has a mass $m_p = 1.672622 \times 10^{-27}$ kg, or about 1840 times the mass of the oppositely charged electron. At this stage of the investigation, scientists perceived the atom as follows: The mass of a nucleus constitutes most of the mass of the entire atom, but the nucleus occupies only about $1/10^{13}$ of the volume of the atom. A typical atomic radius is about 100 pm, whereas the radius of an atomic nucleus is only about 5×10^{-3} pm. You can appreciate the relative size of an atomic nucleus by imagining that if an atom were the size of a sports stadium, the volume of its nucleus would be comparable to a small marble. Although protons are confined to the nucleus of the atom, electrons are thought of as being spread out about the nucleus at some distance from it.

A *picometer* (*pm*) $= 10^{-12}$ m and is a convenient unit for measuring atomic and molecular distances. Another commonly used unit for measuring molecular scale distances is the *angstrom* (Å), which is equal to 100 pm or 10^{-10} m.

If the size of an atom were expanded to that of this sports stadium, the size of the nucleus would be that of a marble.

10. Ernest Rutherford (1871–1937). New Zealand physicist and former graduate student of J. J. Thomson. Rutherford did most of his work in England (Manchester and Cambridge Universities). He received the Nobel Prize in Chemistry in 1908 for his investigations into the structure of the atomic nucleus. His often-quoted comment to his students was that "All science is either physics or stamp-collecting."

11. Johannes Hans Wilhelm Geiger (1882–1945). German physicist. Geiger's work focused on the structure of the atomic nucleus and on radioactivity. He invented a device for measuring radiation that is now commonly called the Geiger counter.

12. Ernest Marsden (1889–1970). English physicist. As an undergraduate he performed many of the experiments that led to Rutherford's Nobel Prize. Marsden went on to contribute significantly to the development of science in New Zealand.

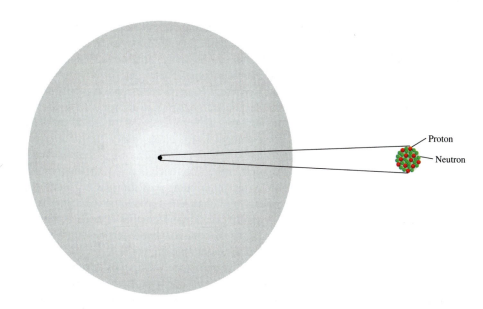

Figure 0.8 The protons and neutrons of an atom are packed in an extremely small nucleus. Electrons are shown as "clouds" around the nucleus.

Rutherford's model of atomic structure left one major problem unsolved. It was known that hydrogen, the simplest atom, contained only one proton and that the helium atom contained two protons. Therefore, the ratio of the mass of a helium atom to that of a hydrogen atom should be 2:1. (Because electrons are much lighter than protons, their contribution to atomic mass can be ignored.) In reality, however, the ratio is 4:1. Rutherford and others postulated that there must exist yet another type of subatomic particle in the atomic nucleus, a particle that had a mass similar to that of a proton, but was electrically neutral. In 1932, James Chadwick[13] provided experimental evidence for the existence of this particle. When Chadwick bombarded a thin sheet of beryllium with α particles, the metal emitted very high-energy radiation. Later experiments showed that the "radiation" consisted of a third type of subatomic particles, which Chadwick named **neutrons,** because they proved to be *electrically neutral particles slightly more massive than protons.* The mystery of the mass ratio could now be explained. In the helium nucleus there are two protons and two neutrons, but in the hydrogen nucleus there is only one proton (no neutrons); therefore, the ratio is 4:1. Figure 0.8 shows the location of the protons, neutrons, and electrons in an atom. Table 0.2 lists the masses and charges of these three elementary particles.

The number of protons in the nucleus of an atom is called the **atomic number (Z),** which, in a neutral atom, is also equal to the number of electrons. The atomic number determines the chemical properties of an atom and the identity of the element;

---

13. James Chadwick (1891–1972). English physicist. In 1935 he received the Nobel Prize in Physics for proving the existence of the neutron.

Table 0.2	Masses and Charges of Subatomic Particles	
Particle	**Mass**	**Charge**
electron	$m_e = 9.109383 \times 10^{-31}$ kg	$-1.602177 \times 10^{-19}$ C $= -e$
proton	$m_p = 1.672622 \times 10^{-27}$ kg	$+1.602177 \times 10^{-19}$ C $= +e$
neutron	$m_n = 1.674927 \times 10^{-27}$ kg	0

that is, *all atoms of a given element have the same atomic number.* For example, the atomic number of nitrogen is 7; therefore, each nitrogen atom has seven protons (and seven electrons because it is neutral). The converse is also true—every atom in the universe that contains seven protons is correctly named "nitrogen."

With the exception of the most common form of hydrogen, which has one proton and no neutrons, all atomic nuclei contain both protons and neutrons. Although each element is specified by the number of protons (atomic number), the mass of an atom can only be determined if we also know the number of neutrons. The **mass number** *(A)* is *the total number of neutrons and protons present in the nucleus of an atom.* Thus, the number of neutrons in an atom is equal to the difference between the mass number and the atomic number. For example, the mass number of fluorine is 19 and the atomic number is 9 (indicating 9 protons in the nucleus). The number of neutrons in an atom of fluorine is $19 - 9 = 10$. Note that the atomic number and mass number must be positive integers. Except in the case of common hydrogen, where they are equal, the mass number is always larger than the atomic number.

Atoms of a given element do not all have the same mass. Most elements have two or more **isotopes,** *atoms that have the same atomic number but different mass numbers.* There are, for example, three isotopes of hydrogen. One, known simply as hydrogen (or, less commonly, as *protium*), has one proton and no neutrons. The *deuterium* isotope contains one proton and one neutron, and *tritium* has one proton and two neutrons. The accepted way to denote the atomic number and mass number of an atom of an element (X) is as follows:

$$_Z^A X$$

Thus, for the isotopes of hydrogen, we write

$$_1^1 H \qquad _1^2 H \qquad _1^3 H$$

Hydrogen   Deuterium   Tritium
(or protium)

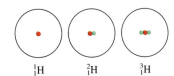

$_1^1 H$ $_1^2 H$ $_1^3 H$

To a very good approximation, different isotopes of the same element have identical chemical properties[14]; however, they can differ in some physical properties, such as density, which depends on mass, and radioactivity, which depends on the ratio of neutrons to protons in the nucleus, as will be discussed in Chapter 17. Two common isotopes of uranium (Z = 92), for example, have mass numbers 235 and 238. The lighter isotope ($_{92}^{235} U$) is used in nuclear reactors and atomic bombs, whereas ($_{92}^{238} U$) lacks the nuclear properties necessary for these applications. With the exception of hydrogen, which has different names for each of its isotopes, isotopes of elements are identified by their mass numbers. Thus, the two common isotopes of uranium are called uranium-235 and uranium-238. A list of the stable and radioactive isotopes for the first 10 elements is given in Appendix 4.

The Periodic Table

More than half of the elements known today were discovered in the nineteenth century. During this period, chemists noted that many elements showed strong similarities to one another. Recognition of periodic regularities in physical and chemical behavior

---

14. The existence of isotopes (atoms of the same element that differ in atomic mass) implies that Proust's law of definite proportions, which states that different sample of the same compound always contain elements in the same proportion by mass, is in reality only an approximation. Two samples of carbon dioxide (CO_2) would not have exactly the same mass ratio of carbon to oxygen if one of the samples were richer in carbon-13.

and the need to organize the volume of available information about the structure and properties of elemental substances led to the development of the ***periodic table,*** *a chart in which elements having similar chemical and physical properties are grouped together.* Figure 0.9 shows the modern periodic table, in which the elements are arranged by atomic number (shown above the element symbol) in *horizontal rows* called ***periods*** and in *vertical columns* known as ***groups*** or *families,* according to similarities in their chemical properties. Elements 112–116 and 118 have recently been synthesized, although they have not yet been named.

The elements can be categorized as metals, nonmetals, or metalloids. A ***metal*** is *a good conductor of heat and electricity,* whereas a ***nonmetal*** is usually *a poor conductor of heat and electricity.* A ***metalloid*** *has properties that are intermediate between those of metals and nonmetals.* Figure 0.9 shows that the majority of the known elements are metals; only 17 elements are nonmetals, and 8 are metalloids. From left to right across any period, the physical and chemical properties of the elements change gradually from metallic to nonmetallic. Elements are often classified by their periodic table group number—Group 1A, Group 2A, and so on. For convenience, some element groups have special names. *The Group 1A elements (Li, Na, K, Rb, Cs, and Fr)* are called ***alkali metals,*** and *the Group 2A elements (Be, Mg, Ca, Sr, Ba, and Ra)* are

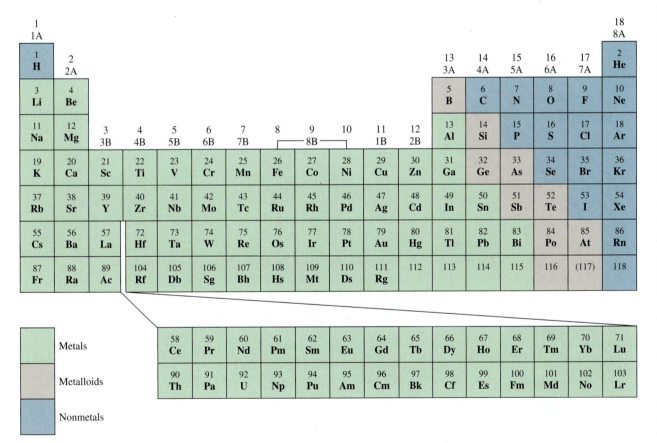

Figure 0.9 The modern periodic table. The elements are arranged according to the atomic numbers above their symbols. With the exception of hydrogen (H), nonmetals appear at the far right of the table. The two rows of metals beneath the main body of the table are conventionally set apart to keep the table from being too wide. Actually, cerium (Ce) should follow lanthanum (La), and thorium (Th) should come right after actinium (Ac). The 1–18 group designation has been recommended by the International Union of Pure and Applied Chemistry (IUPAC). In this text we use the standard U.S. notation for group numbers (1A–8A and 1B–8B) still in wide use. Element 117 has not yet been synthesized.

Distribution of Elements on Earth and in Living Systems

The majority of elements are naturally occurring. How are these elements distributed on Earth, and which are essential to living systems?

Earth's crust extends from the surface to a depth of about 40 km. Because of technical difficulties, scientists have not been able to study the inner portions of Earth as easily as the crust. Nevertheless, it is believed that there is a solid core consisting mostly of iron at the center of Earth. Surrounding the core is a layer called the *mantle,* which consists of hot fluid containing iron, carbon, silicon, and sulfur.

Of the 83 elements that are found in nature, 12 make up 99.7 percent of Earth's crust by mass. They are, in decreasing order of natural abundance, oxygen (O), silicon (Si), aluminum (Al), iron (Fe), calcium (Ca), magnesium (Mg), sodium (Na), potassium (K), titanium (Ti), hydrogen (H), phosphorus (P), and manganese (Mn)—see the following figure. In discussing the natural abundance of the elements, we should keep in mind that (1) the elements are not evenly distributed throughout Earth's crust, and (2) most elements occur in combined forms. These facts provide the basis for most methods of obtaining pure elements from their compounds, as we will see in later chapters.

The accompanying table lists the essential elements in the human body. Of special interest are the *trace elements,* such as iron (Fe), copper (Cu), zinc (Zn), iodine (I), and cobalt (Co), which together make up about 0.1 percent of the mass of the body. These elements are necessary for biological functions such as growth, transport of oxygen for metabolism, and defense against disease. There is a delicate balance in the amounts of these elements in our bodies. Too much or too little over an extended period can lead to serious illness, retardation, or even death.

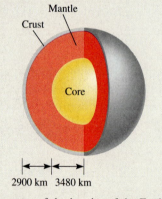

2900 km 3480 km

Structure of the interior of the Earth

Essential Elements in the Human Body

Element	Percent by Mass*	Element	Percent by Mass*
Oxygen	65	Sodium	0.1
Carbon	18	Magnesium	0.05
Hydrogen	10	Iron	<0.05
Nitrogen	3	Cobalt	<0.05
Calcium	1.6	Copper	<0.05
Phosphorus	1.2	Zinc	<0.05
Potassium	0.2	Iodine	<0.05
Sulfur	0.2	Selenium	<0.01
Chlorine	0.2	Fluorine	<0.01

*Percent by mass *gives the mass of the element in grams present in a 100-g sample.*

(a) Natural abundance of the elements in percent by mass. For example, the abundance of oxygen is 45.5 percent. This means that in a 100-g sample of Earth's crust there are, on average, 45.5 g of the element oxygen. (b) Abundance of elements in the human body in percent by mass.

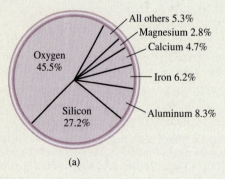

(a)

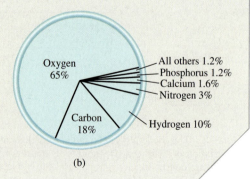

(b)

called **alkaline earth metals.** *Elements in Group 7A (F, Cl, Br, I, and At) are known* as **halogens,** *and elements in Group 8A (He, Ne, Ar, Kr, Xe, and Rn) are called* **noble gases,** *or rare gases.*

The periodic table is a handy tool that correlates the properties of the elements in a systematic way and helps us to predict chemical behavior. We will take a closer look at this keystone of chemistry in Chapter 2.

Molecules and Ions

Of all the elements, only the six noble gases in Group 8A of the periodic table (He, Ne, Ar, Kr, Xe, and Rn) exist in nature as single atoms. For this reason, they are called *monatomic* (meaning a single atom) gases. Most matter is composed of molecules or ions formed by atoms.

A **molecule** is an *aggregate of two or more atoms in a definite arrangement held together by chemical forces* (also called *chemical bonds*). A molecule may contain atoms of the same element or atoms of two or more elements joined in a fixed ratio, in accordance with the law of definite proportions. Thus, a molecule is not necessarily a compound, which, by definition, is made up of two or more elements (see Section 0.1). Hydrogen gas, for example, is a pure element, but it consists of molecules made up of two H atoms each. Water, on the other hand, is a molecular compound that contains hydrogen and oxygen in a ratio of two H atoms to one O atom. Like atoms, molecules are electrically neutral. The hydrogen molecule, symbolized as H_2, is a **diatomic molecule** because it *contains only two atoms.* Other elements that normally exist as diatomic molecules are nitrogen (N_2), oxygen (O_2), and the Group 7A elements, except astatine (At); that is, the *halogens:* fluorine (F_2), chlorine (Cl_2), bromine (Br_2), and iodine (I_2). A diatomic molecule can contain atoms of different elements, too, such as hydrogen chloride (HCl) and carbon monoxide (CO). The vast majority of molecules contain more than two atoms. They can be atoms of the same element, as in ozone (O_3), which consists of three atoms of oxygen, or they can be combinations of two or more different elements. *Molecules containing more than two atoms* are **polyatomic molecules.** Like ozone, water (H_2O) and ammonia (NH_3) are polyatomic molecules.

An **ion** is *an atom or a group of atoms that has a net positive or negative charge.* The number of protons in the nucleus of an atom remains the same during ordinary chemical changes (called chemical reactions), but electrons may be lost or gained. The loss of one or more electrons from a neutral atom or molecule results in a **cation,** *an ion with a net positive charge.* For example, a sodium atom (Na) can readily lose an electron to become a sodium cation with a net charge of $+e$. This ion is represented by the symbol Na^+. On the other hand, an **anion** is *an ion whose net charge is negative* because of an increase in the number of electrons. A chlorine atom (Cl) can gain an electron to become the chloride ion, Cl^-, which is an anion with net charge $-e$. For simplicity, it is customary to state the charge of ions in units of the fundamental electron charge e; so we say that the sodium cation has a charge of $+1$ and the chloride anion has a charge of -1.

An atom can lose or gain more than one electron to form an ion. Examples include Mg^{2+}, Fe^{3+}, S^{2-}, and N^{3-}. These ions, as well as Na^+ and Cl^-, are **monatomic ions** because they *contain only one atom.* Figure 0.10 shows the charges of a number of monatomic ions. With very few exceptions, metals tend to form cations and nonmetals tend to form anions.

In addition, two or more atoms can form an ion that has a net positive or net negative charge. **Polyatomic ions** such as OH^- (hydroxide ion), CN^- (cyanide ion), and NH_4^+ (ammonium ion) are *ions containing more than one atom.*

We will discuss the nature of chemical bonds in Chapters 3 and 4.

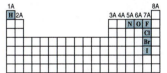

Elements that exist as diatomic molecules

In Chapter 2 we will see why atoms of different elements gain (or lose) a specific number of electrons.

1 1A	2 2A	3 3B	4 4B	5 5B	6 6B	7 7B	8	9 8B	10	11 1B	12 2B	13 3A	14 4A	15 5A	16 6A	17 7A	18 8A
Li^+													C^{4-}	N^{3-}	O^{2-}	F^-	
Na^+	Mg^{2+}											Al^{3+}		P^{3-}	S^{2-}	Cl^-	
K^+	Ca^{2+}				Cr^{2+} Cr^{3+}	Mn^{2+} Mn^{3+}	Fe^{2+} Fe^{3+}	Co^{2+} Co^{3+}	Ni^{2+} Ni^{3+}	Cu^+ Cu^{2+}	Zn^{2+}				Se^{2-}	Br^-	
Rb^+	Sr^{2+}									Ag^+	Cd^{2+}		Sn^{2+} Sn^{4+}		Te^{2-}	I^-	
Cs^+	Ba^{2+}									Au^+ Au^{3+}	Hg_2^{2+} Hg^{2+}		Pb^{2+} Pb^{4+}				

Figure 0.10 Common monatomic ions arranged according to their positions in the periodic table. Note that the Hg_2^{2+} ion is diatomic.

0.3 | Compounds Are Represented by Chemical Formulas

Chemical Formulas

Chemists use **chemical formulas** to *express the composition of molecules and ionic compounds in terms of chemical symbols.* By composition we mean not only the elements present, but also the ratios in which the atoms are combined. At present, we are concerned with three types of formulas: molecular formulas, structural formulas, and empirical formulas.

A **molecular formula** *shows the exact number of atoms of each element in the smallest unit of a substance.* In our discussion of molecules, each example was given with its molecular formula in parentheses. Thus, H_2 is the molecular formula for hydrogen, O_2 for oxygen, O_3 for ozone, and H_2O for water. The subscript numeral indicates the number of atoms of an element present. There is no subscript for O in H_2O because there is only one atom of oxygen in a molecule of water, and so the number "1" is omitted from the formula. Some elements can exist in multiple molecular forms called allotropes. An **allotrope** is *one of two or more distinct forms of an element.* For example, diatomic oxygen (O_2) and ozone (O_3) are the two allotropic forms of oxygen. Two allotropes of the same element can have, despite their similar composition, dramatically different physical and chemical properties. Diamond and graphite, for example, are allotropic forms of the element carbon. Diamond is one of the hardest materials known, is transparent, does not conduct electricity, and is relatively expensive. Graphite, in contrast, is a soft, grey-black material that is electrically conductive and cheap.

The molecular formula shows the number of atoms of each element in a molecule but does not indicate how the atoms are connected to each other in space. A **structural formula** shows *how atoms in a molecule are bonded to one another.* For example, each of the two H atoms is bonded to an O atom in the water molecule, so the structural formula of water is H—O—H. The lines connecting H and O (or any atomic symbols) represent chemical bonds. Some examples of molecular and

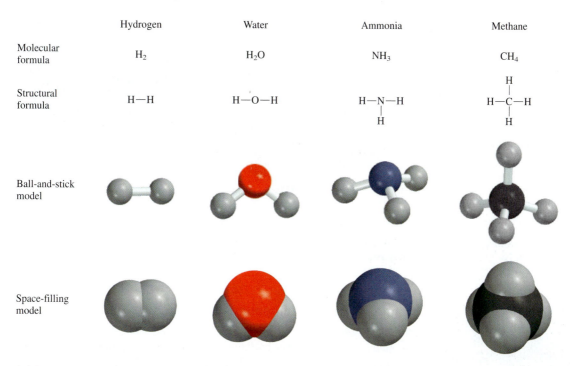

	Hydrogen	Water	Ammonia	Methane
Molecular formula	H_2	H_2O	NH_3	CH_4
Structural formula	H—H	H—O—H	H—N—H with H below	H—C—H with H above and below
Ball-and-stick model				
Space-filling model				

Figure 0.11 Molecular and structural formulas and molecular models of four common molecules.

structural formulas are shown in Figure 0.11. Also shown in Figure 0.11 are two types of three-dimensional **molecular models.** In the *ball-and-stick models,* molecules are represented by balls, which represent atoms, connected by sticks, which represent chemical bonds. The various elements are represented by balls of different colors and sizes. In the *space-filling models,* atoms are represented by overlapping spheres, each of which is proportional in size to the corresponding atoms. Both types of models are used extensively in this text.

Many experiments designed to probe molecular composition can only determine the ratios of the numbers of atoms of each element. For example, chemical analysis of hydrogen peroxide (molecular formula H_2O_2), a substance used as an antiseptic and as a bleaching agent for textiles and hair, only indicates that there is a 1:1 ratio of hydrogen atoms to oxygen atoms. Such an experiment is said to yield only the **empirical formula,** *which tells us which elements are present and the simplest whole-number ratio of their atoms,* and not necessarily the actual number of each kind of atom in the molecule. The empirical formula for hydrogen peroxide is HO. Hydrazine (N_2H_4), which has been used as rocket fuel, has the empirical formula NH_2. Although the ratio of nitrogen to hydrogen is 1:2 in both the molecular formula (N_2H_4) and the empirical formula (NH_2), only the molecular formula accounts for the actual number of N atoms (two) and H atoms (four) present. Empirical formulas are the *simplest* chemical formulas; they are written by reducing the subscripts in molecular formulas to the smallest possible whole numbers. Molecular formulas are the *true* formulas of molecules. For some compounds, such as water (H_2O) and methane (CH_4), the empirical and molecular formulas are identical. Determination of the molecular formula from the empirical formula requires either input from additional experiments or logical reasoning based on knowledge of the rules of chemical bonding (which we discuss in Chapters 3 and 4).

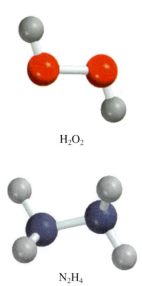

H_2O_2

N_2H_4

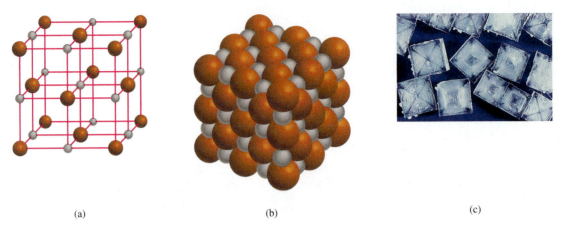

(a)   (b)   (c)

Figure 0.12 (a) Structure of solid NaCl. (b) In reality, the cations are in contact with the anions. In both (a) and (b), the smaller spheres represent Na^+ ions and the larger spheres, Cl^- ions. (c) Crystals of NaCl.

Formulas for Ionic Compounds

Ionic compounds are *compounds that are formed from cations and anions.* Such compounds are held together by the electrostatic attraction between opposite charges. An example of an ionic compound is sodium chloride (NaCl), common table salt. The formulas of ionic compounds are usually the same as their empirical formulas because ionic compounds do not consist of discrete molecular units. For example, a solid sample of sodium chloride (NaCl) consists of equal numbers of Na^+ and Cl^- ions arranged in a three-dimensional network (Figure 0.12).

There is a 1:1 ratio of sodium cations to chlorine anions so that the compound is electrically neutral. According to Figure 0.12, no Na^+ ion in NaCl is associated with just one particular Cl^- ion. In fact, each Na^+ ion is equally held by six surrounding Cl^- ions and vice versa. Thus, NaCl is the empirical formula for sodium chloride. (Note that the charges on the cation and anion are not shown in the formula for an ionic compound.) In other ionic compounds, the actual structure may be different, but the arrangement of cations and anions is such that the compounds are all electrically neutral. For the formula of an ionic compound to be electrically neutral, the sum of the charges on the cation and anion in each formula unit must be zero. If the magnitudes of the charges on the cation and anion are different, use the following rule to make the formula electrically neutral: *The subscript of the cation is numerically equal to the charge on the anion, and the subscript of the anion is numerically equal to the charge on the cation.* If the charges are numerically equal, then no subscripts are necessary. This rule is possible because the formulas of ionic compounds are empirical formulas, so the subscripts must always be reduced to the smallest ratios. Consider the following examples:

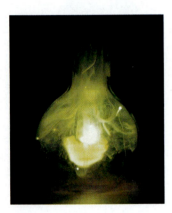

Sodium metal reacting with chlorine gas to form sodium chloride.

▸ **Potassium bromide.** The potassium cation K^+ and the bromine anion Br^- combine to form the ionic compound potassium bromide. The sum of the charges, measured in units of the electron charge is $+1 + (-1) = 0$. The formula is KBr.

▸ **Zinc iodide.** The zinc cation Zn^{2+} and the iodine anion I^- combine to form zinc iodide. To make the charges add up to zero, there have to be twice as many I^- ions as Zn^{2+} ions. Therefore, the formula for zinc iodide is ZnI_2.

▸ **Aluminum oxide.** The cation is Al^{3+} and the oxygen anion is O^{2-}. To make a neutral compound, the ratio of Al^{3+} to O^{2-} must be 2:3, giving the formula Al_2O_3 for aluminum oxide.

Naming Compounds

The number of known compounds is currently well over 20 million. Fortunately, it is unnecessary to memorize their names. Over the years, chemists have devised a system for naming chemical substances. The rules are accepted worldwide, facilitating communication among chemists and providing a useful way of labeling an overwhelming variety of substances. Mastering these rules now will prove beneficial almost immediately as we proceed with our study of chemistry.

To begin our discussion of chemical *nomenclature,* the naming of chemical compounds, we must first distinguish between organic and inorganic compounds. ***Organic compounds*** *contain carbon bonded to hydrogen, sometimes in combination with other elements such as oxygen, nitrogen, or sulfur.* All other compounds are classified as ***inorganic compounds.*** The nomenclature of organic compounds is not discussed in detail until Chapter 16. To organize and simplify our venture into naming compounds, we can categorize inorganic compounds as ionic compounds, molecular compounds, acids and bases, and hydrates.

Naming Ionic Compounds

As discussed earlier, ionic compounds consist of cations (positive ions) and anions (negative ions). With a few exceptions (such as the ammonium ion), all cations of interest to us are derived from metal atoms. Metal cations take their names from their respective elements. For example:

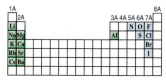

The most reactive metals (green) and the most reactive nonmetals (blue) combine to form ionic compounds.

Element	Name of Cation
Na (sodium)	Na^+ (sodium ion or sodium cation)
K (potassium)	K^+ (potassium ion or potassium cation)
Mg (magnesium)	Mg^{2+} (magnesium ion or magnesium cation)
Al (aluminum)	Al^{3+} (aluminum ion or aluminum cation)

Many ionic compounds, such as NaCl, KBr, ZnI_2, and Al_2O_3, are ***binary compounds,*** that is, compounds *formed from just two elements.* For binary compounds, the first element named is the metal cation, followed by the nonmetallic anion. Thus, NaCl is sodium chloride. We name the anion by taking the first part of the element name (the "chlor-" of chlorine) and adding "-ide." The names of KBr, ZnI_2, and Al_2O_3 are potassium bromide, zinc iodide, and aluminum oxide, respectively. The "-ide" ending is also used in the names of some simple polyatomic anions, such as hydroxide (OH^-) and cyanide (CN^-). Thus, the compounds LiOH and KCN are named lithium hydroxide and potassium cyanide, respectively. These and a number of other such ionic substances are ***ternary compounds,*** meaning *compounds consisting of three elements.* Table 0.3 lists the names of some common cations and anions.

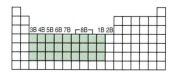

The transition metals are the elements in Groups 1B and 3B–8B.

Certain metals, especially the *transition metals,* can form more than one type of cation. For example, iron can form two cations: Fe^{2+} and Fe^{3+}. An older nomenclature system that is still in limited use assigns the ending "-ous" to the cation with fewer positive charges and the ending "-ic" to the cation with more positive charges:

$$Fe^{2+} \quad \text{ferrous ion}$$
$$Fe^{3+} \quad \text{ferric ion}$$

The names of the compounds that these iron ions form with chlorine would thus be

$$FeCl_2 \quad \text{ferrous chloride}$$
$$FeCl_3 \quad \text{ferric chloride}$$

$FeCl_2$ (left) and $FeCl_3$ (right)

Table 0.3	Names and Formulas of Some Common Inorganic Cations and Anions
Cation	**Anion**
Aluminum (Al^{3+})	Bromide (Br^-)
Ammonium (NH_4^+)	Carbonate (CO_3^{2-})
Barium (Ba^{2+})	Chlorate (ClO_3^-)
Cadmium (Cd^{2+})	Chromate (CrO_4^{2-})
Cesium (Cs^+)	Cyanide (CN^-)
Chromium(III) or chromic (Cr^{3+})	Dichromate ($Cr_2O_7^{2-}$)
Cobalt(II) or cobaltous (Co^{2+})	Dihydrogen phosphate ($H_2PO_4^-$)
Copper(I) or cuprous (Cu^+)	Fluoride (F^-)
Copper(II) or cupric (Cu^{2+})	Hydride (H^-)
Hydrogen (H^+)	Hydrogen carbonate or bicarbonate (HCO_3^-)
Iron(II) or ferrous (Fe^{2+})	Hydrogen phosphate (HPO_4^{2-})
Iron(III) or ferric (Fe^{3+})	Hydrogen sulfate or bisulfate (HSO_4^-)
Lead(II) or plumbous (Pb^{2+})	Hydroxide (OH^-)
Lithium (Li^+)	Iodide (I^-)
Magnesium (Mg^{2+})	Nitrate (NO_3^-)
Manganese(II) or manganous (Mn^{2+})	Nitride (N^{3-})
Mercury(I) or mercurous (Hg_2^{2+})	Nitrite (NO_2^-)
Mercury(II) or mercuric (Hg^{2+})	Oxide (O^{2-})
Potassium (K^+)	Permanganate (MnO_4^-)
Rubidium (Rb^+)	Peroxide (O_2^{2-})
Silver (Ag^+)	Phosphate (PO_4^{3-})
Sodium (Na^+)	Sulfate (SO_4^{2-})
Strontium (Sr^{2+})	Sulfide (S^{2-})
Tin(II) or stannous (Sn^{2+})	Sulfite (SO_3^{2-})
Zinc (Zn^{2+})	Thiocyanate (SCN^-)

This method of naming ions has some distinct limitations. First, the "-ous" and "-ic" suffixes do not provide information regarding the actual charges of the two cations involved. Thus, the ferric ion is Fe^{3+}, but the cation of copper named cupric has the formula Cu^{2+}. In addition, the "-ous" and "-ic" designations provide names for only two different elemental cations. Some metallic elements can form three or more different cations, so it has become increasingly common to designate different cations with Roman numerals. This is called the ***Stock system*** after Alfred Stock,[15] a German inorganic chemist. In this system, a Roman numeral indicates the number of positive charges, for example, II means two positive charges. The three possible cations of manganese (Mn) are:

Mn^{2+}	MnO	manganese(II) oxide
Mn^{3+}	Mn_2O_3	manganese(III) oxide
Mn^{4+}	MnO_2	manganese(IV) oxide

15. Alfred E. Stock (1876–1946). German chemist. Stock did most of his research in the synthesis and characterization of boron, beryllium, and silicon compounds. He was the first scientist to explore the dangers of mercury poisoning.

Using the Stock system, the ferrous ion and the ferric ion are iron(II) and iron(III), respectively, so ferrous chloride becomes iron(II) chloride; and ferric chloride is called iron(III) chloride. In keeping with modern practice, we will use the Stock system of naming compounds in this textbook. Examples 0.3 and 0.4 illustrate how to name and write formulas for ionic compounds based on the information given in Table 0.3.

Example 0.3

Name the following compounds: (a) $Cu(NO_3)_2$, (b) KH_2PO_4, and (c) NH_4ClO_3.

Strategy Our reference for the names of cations and anions is Table 0.3. If a metal can form cations of different charges (see Figure 0.10), we need to use the Stock system.

Solution (a) Each of the two nitrate ions (NO_3^-) bears one negative charge, so the copper ion must have a charge of $+2$. Because copper forms both Cu^+ and Cu^{2+} ions, we need to use the Stock system. The name of the compound is copper(II) nitrate. (b) The cation is K^+ and the anion is $H_2PO_4^-$ (dihydrogen phosphate). Because potassium only forms one type of ion (K^+), there is no need to use the Stock system. The name of the compound is potassium dihydrogen phosphate. (c) The cation is NH_4^+ (ammonium ion) and the anion is ClO_3^- (chlorate). The name of the compound is ammonium chlorate.

Practice Exercise Name the following compounds: (a) PbO, (b) Li_2SO_3, and (c) Fe_2O_3.

Example 0.4

Write chemical formulas for: (a) mercury(I) nitrite, (b) cesium sulfide, and (c) calcium phosphate.

Strategy Refer to Table 0.3 for the formulas of cations and anions. The Roman numerals indicate the charges on the cations.

Solution (a) The Roman numeral indicates that the mercury ion bears a $+1$ charge. According to Table 0.3, however, the mercury(I) ion is diatomic with a $+2$ charge (that is, Hg_2^{2+}) and the nitrite ion is NO_2^-. Therefore, the formula is $Hg_2(NO_2)_2$. (b) Each sulfide ion bears a -2 charge, whereas each cesium ion bears a $+1$ charge. Therefore, the formula is Cs_2S. (c) Each calcium ion (Ca^{2+}) bears a $+2$ charge, and each phosphate ion (PO_4^{3-}) bears a -3 charge. To make the compound electrically neutral (that is, to make the sum of the charges equal zero), we must adjust the numbers of cations and anions $3(+2) + 2(-3) = 0$, so that the formula is $Ca_3(PO_4)_2$.

Practice Exercise Write formulas for the following ionic compounds: (a) rubidium sulfate, (b) chromium(VI) oxide, and (c) barium hydride.

Molecular Compounds

Unlike ionic compounds, molecular compounds contain discrete molecular units. They are usually composed of nonmetallic elements (see Figure 0.9). Many molecular compounds are binary compounds. Naming binary molecular compounds is

Table 0.4	
Greek Prefixes	
Prefix	Meaning
Mono-	1
Di-	2
Tri-	3
Tetra-	4
Penta-	5
Hexa-	6
Hepta-	7
Octa-	8
Nona-	9
Deca-	10

similar to naming binary ionic compounds. We place the name of the first element in the formula first, and the second element is named by adding the suffix "-ide" to the root of the element name:

HCl	hydrogen chloride
HBr	hydrogen bromide
SiC	silicon carbide

It is quite common for one pair of elements to form several different compounds (for example, CO and CO_2). In these cases, confusion in naming the compounds is avoided by the use of Greek prefixes (see Table 0.4) to denote the number of atoms of each element present. Consider the following examples:

CO	carbon monoxide
CO_2	carbon dioxide
SO_2	sulfur dioxide
SO_3	sulfur trioxide
CCl_4	carbon tetrachloride
N_2O_4	dinitrogen tetroxide
P_2O_5	diphosphorus pentoxide

The following guidelines are helpful in naming compounds with prefixes:

▶ The prefix "mono-" can be omitted for the first element because the absence of a prefix is assumed to indicate that only one atom of that element is present. For example, PCl_3 is named phosphorus trichloride, not monophosphorus trichloride. It can also be omitted for the second element if only one compound of that type can be formed. For example, HCl is hydrogen chloride, not hydrogen monochloride, because HCl is the only chloride of hydrogen possible. However, carbon has two oxide forms, CO and CO_2, so CO must be named "carbon monoxide" to distinguish it from CO_2.

▶ For oxides, the ending "a" in the prefix is sometimes omitted. For example, N_2O_4 is named dinitrogen tetroxide rather than dinitrogen tetraoxide.

Molecular compounds containing hydrogen are not named using Greek prefixes. Traditionally, many of these compounds are called either by their common, nonsystematic names or by names that do not specifically indicate the number of H atoms present:

B_2H_6	diborane
CH_4	methane
SiH_4	silane
NH_3	ammonia
PH_3	phosphine
H_2O	water
H_2S	hydrogen sulfide

Even the order of writing the elements in the formulas for hydrogen compounds is irregular. In water and hydrogen sulfide, for example, H is written first, whereas it appears last in the other compounds.

Writing formulas for most molecular compounds is usually straightforward. Thus, the name arsenic trifluoride means that there is one As atom and three F atoms in each molecule, and the molecular formula is AsF_3. Moreover, the order of the elements in the formula is the same as in its name.

Example 0.5

Name the following molecular compounds: (a) $SiCl_4$ and (b) P_4O_{10}.

Strategy Refer to Table 0.4 for prefixes. In (a) there is only one Si atom so we do not use the prefix "mono."

Solution (a) Because there are four chlorine atoms present, the compound is silicon tetrachloride. (b) There are four phosphorus atoms and ten oxygen atoms present, so the compound is tetraphosphorus decoxide. Note that the "a" is omitted from "deca."

Practice Exercise Name the following molecular compounds: (a) NF_3 and (b) Cl_2O_7.

Example 0.6

Write chemical formulas for the following molecular compounds: (a) carbon disulfide and (b) disilicon hexabromide.

Strategy Here we need to convert prefixes to numbers of atoms (see Table 0.4). Because there is no prefix for carbon in (a), it means that there is only one carbon atom present.

Solution (a) Because there are two sulfur atoms (from the prefix *di*) and one carbon atom (implied *mono-*) present, the formula is CS_2. (b) There are two (*di-*) silicon atoms and six (*hexa-*) bromine atoms present, so the formula is Si_2Br_6.

Practice Exercise Write chemical formulas for the following molecular compounds: (a) sulfur tetrafluoride and (b) dinitrogen pentoxide.

Figure 0.13 summarizes the steps for naming ionic and binary molecular compounds.

Naming Acids and Bases

An *acid* can be described as *a substance that yields hydrogen ion (H^+) when dissolved in water* (we will discuss more general definitions in Chapter 10). Formulas for acids contain one or more hydrogen atoms and an anionic group. Anions whose names end in "-ide" form acids with a "hydro-" prefix and an "-ic" ending, as shown in Table 0.5.

In some cases, two different names seem to be assigned to the same chemical formula, for example:

HCl	hydrogen chloride
HCl	hydrochloric acid

The name assigned to the compound depends on its physical state. In the gaseous or pure liquid state, HCl is a molecular compound called hydrogen chloride. When HCl is dissolved in water, the molecules break up into H^+ and Cl^- ions; in this state, the substance is called hydrochloric acid.

Oxoacids are acids that *contain hydrogen, oxygen, and another element (the central element)*. The formulas of oxoacids are usually written with the H first, followed

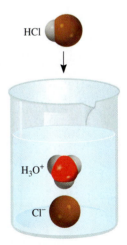

When dissolved in water, the HCl molecule is converted to the H^+ and Cl^- ions. The H^+ ion is associated with one or more water molecules, and is usually represented as H_3O^+.

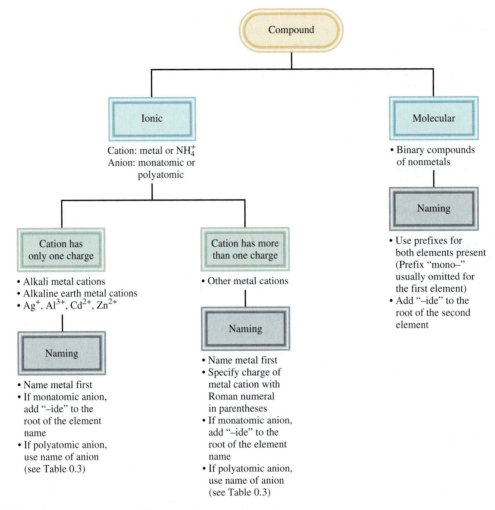

Figure 0.13 Steps for naming ionic and binary molecular compounds.

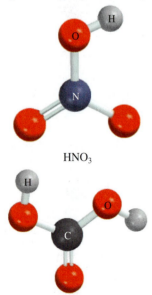

by the central element and then O, as illustrated by the following examples, which all have the suffix "-ic":

HNO_3	nitric acid
H_2CO_3	carbonic acid
H_2SO_4	sulfuric acid
$HClO_3$	chloric acid

Table 0.5	Some Simple Acids
Anion	**Corresponding Acid**
F^- (fluoride)	HF (hydrofluoric acid)
Cl^- (chloride)	HCl (hydrochloric acid)
Br^- (bromide)	HBr (hydrobromic acid)
I^- (iodide)	HI (hydroiodic acid)
CN^- (cyanide)	HCN (hydrocyanic acid)
S^{2-} (sulfide)	H_2S (hydrosulfuric acid)

Often two or more oxoacids have the same central atom but a different number of O atoms. Starting with the oxoacids above, whose names end with "-ic," we use the following rules to name these compounds:

Note that these acids all exist as molecular compounds in the gas phase.

1. Addition of one O atom to the "-ic" acid: The acid is called "per . . . -ic" acid. Thus, adding an O atom to $HClO_3$ changes chloric acid to perchloric acid, $HClO_4$.
2. Removal of one O atom from the "-ic" acid: The acid is called "-ous" acid. Thus, nitric acid (HNO_3) becomes nitrous acid (HNO_2).
3. Removal of two O atoms from the "-ic" acid: The acid is called "hypo . . . -ous" acid. Thus, bromic acid ($HBrO_3$) becomes hypobromous acid ($HBrO$).

The rules for naming **oxoanions,** *anions of oxoacids,* are as follows:

1. When all the H^+ ions are removed from the "-ic" acid, the anion's name ends with "-ate." For example, CO_3^{2-} (the anion derived from H_2CO_3) is called carbonate.
2. When all the H^+ ions are removed from the "-ous" acid, the anion's name ends with "-ite." Thus, ClO_2^-, the anion derived from $HClO_2$, is called chlorite.
3. The names of anions in which one or more but not all the hydrogen ions have been removed must indicate the number of H^+ ions present. Consider, for example, the anions derived from phosphoric acid:

H_3PO_4	phosphoric acid
$H_2PO_4^-$	dihydrogen phosphate
HPO_4^{2-}	hydrogen phosphate
PO_4^{3-}	phosphate

Analogous to the names of molecular compounds, we usually omit the prefix "mono-" when there is only one H in the anion.

A **base** can be described as *a substance that yields hydroxide ions (OH^-) when dissolved in water* (once again a more general definition is discussed in Chapter 11). Commonly, bases are ionic compounds of metal cations with the hydroxide anion and are named according to the usual rules of ionic compounds:

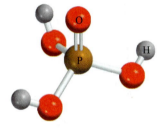

H_3PO_4

NaOH	sodium hydroxide
KOH	potassium hydroxide
$Ba(OH)_2$	barium hydroxide

Ammonia (NH_3), a molecular compound in the gaseous or pure liquid state, is also classified as a common base. Although, at first glance this may seem to be an exception to the definition of a base just given, as long as the substance *yields* hydroxide ions when dissolved in water, it need not contain hydroxide ions in its structure to be considered a base. In fact, when ammonia dissolves in water, some of the NH_3 molecules react with H_2O to yield NH_4^+ and OH^- ions. Thus, it is properly classified as a base.

Naming Hydrates

Many solid ionic compounds have a specific number of water molecules incorporated into their crystal structure. Such compounds are called **hydrates.** For example, in its normal state, each unit of copper(II) sulfate has five water molecules associated with it. In the naming of hydrates, we use Greek prefixes (see Table 0.4) to denote the number of water molecules associated with each ionic compound formula unit,

Figure 0.14 $CuSO_4 \cdot 5H_2O$ (left) is blue; $CuSO_4$ (right) is white.

followed by the suffix "-hydrate." Thus, the systematic name for this compound is copper(II) sulfate pentahydrate, and its formula is written as $CuSO_4 \cdot 5H_2O$. The water molecules can be driven off by heating. When this occurs, the resulting compound is $CuSO_4$, which is sometimes called *anhydrous* copper(II) sulfate; "anhydrous" means that the compound no longer has water molecules associated with it (Figure 0.14). Some other hydrates are

$$BaCl_2 \cdot 2H_2O \text{ barium chloride dihydrate}$$
$$LiCl \cdot H_2O \text{ lithium chloride monohydrate}$$
$$MgSO_4 \cdot 7H_2O \text{ magnesium sulfate heptahydrate}$$
$$Sr(NO_3)_2 \cdot 4H_2O \text{ strontium nitrate tetrahydrate}$$

Familiar Inorganic Compounds
Some compounds are better known by their common names than by their systematic chemical names. Familiar examples are listed in Table 0.6.

Table 0.6	Common and Systematic Names of Some Familiar Compounds	
Formula	**Common Name**	**Systematic Name**
H_2O	Water	Dihydrogen oxide
NH_3	Ammonia	Trihydrogen nitride
CO_2	Dry ice (in solid form)	Carbon dioxide
$NaCl$	Table salt	Sodium chloride
N_2O	Laughing gas	Dinitrogen oxide
$CaCO_3$	Marble, chalk, limestone	Calcium carbonate
CaO	Quicklime	Calcium oxide
$Ca(OH)_2$	Slaked lime	Calcium hydroxide
$NaHCO_3$	Baking soda	Sodium hydrogen carbonate
$Na_2CO_3 \cdot 10H_2O$	Washing soda	Sodium carbonate decahydrate
$MgSO_4 \cdot 7H_2O$	Epsom salt	Magnesium sulfate heptahydrate
$Mg(OH)_2$	Milk of magnesia	Magnesium hydroxide
$CaSO_4 \cdot 2H_2O$	Gypsum	Calcium sulfate dihydrate
$NaOH$	Lye	Sodium hydroxide

0.4 | Reactions Are Represented by Balanced Chemical Equations

Writing Chemical Equations

According to the atomic theory of matter, atoms and molecules react with one another chemically as discrete units and the number of each type of atom does not change in a reaction. Therefore, the number of atoms of a given element must be the same in the products as in the reactants. Consider, for example, the reaction of hydrogen (H_2) and oxygen (O_2) to form water (H_2O). Because a water molecule contains twice as many H atoms as O atoms, the ratio of hydrogen molecules to oxygen molecules participating in the reaction must be 2:1 to preserve atomic balance. Therefore, we could describe this reaction by saying *"two hydrogen molecules react with one oxygen molecule to yield two molecules of water."* However, these kinds of descriptions become quite cumbersome for more complicated reactions and a more compact notation, called a ***chemical equation,*** has been developed to accurately describe chemical reactions. The chemical equation for the described reaction is

$$2H_2 + O_2 \longrightarrow 2H_2O$$

where the "+" sign means "reacts with," the arrow means "to yield," and the reaction is implied to proceed from left to right; that is, the molecules on the left are the ***reactants*** and those on the right are the ***products:***

$$\text{reactants} \longrightarrow \text{products}$$

The numbers in front of each molecular species in a chemical equation are called ***stoichiometric coefficients,*** and the quantitative study of amounts of individual reactants and products in chemical reactions is called *stoichiometry.*

Chemical equations often include the physical state of the reactants and products. For example, we can write

$$2H_2(g) + O_2(g) \longrightarrow 2H_2O(l)$$

to indicate that in this reaction hydrogen and oxygen gas (denoted *g*) react to form liquid (*l*) water. Other abbreviations commonly used are (*s*) for solids and (*aq*) to denote ***aqueous species,*** which are *substances dissolved in a water solution.* (An ***aqueous solution*** is a solution in which water is the solvent.)

Consider the combustion of propane (C_3H_8). Commonly used as fuel in gas grills, propane (a gas) combines with oxygen (O_2, a gas) to form carbon dioxide (CO_2, a gas) and water (H_2O, a liquid). Because each propane molecule contains three carbon atoms and because carbon dioxide (with one carbon atom) is the only product that contains carbon, each propane molecule that reacts must produce three CO_2 molecules. Similarly, the eight hydrogen atoms from each propane molecule should produce four water molecules. These data give ten oxygen atoms in the products (from three CO_2 and four H_2O molecules), so there must be five O_2 molecules in the reactants. Thus, the chemical equation for this reaction is

C_3H_8

$$C_3H_8(g) + 5O_2(g) \longrightarrow 3CO_2(g) + 4H_2O(l)$$

In this equation there are three C atoms, eight H atoms, and ten O atoms on *both* the reactant and product sides of the equation. Such a chemical equation is *balanced;* that is, there are equal numbers of each type of atom on both the reactant and product sides of the equation. Before a chemical equation can be used in any stoichiometric analysis, it must be balanced; therefore, learning how to balance a chemical equation is an essential skill for a chemist.

Balancing Chemical Equations

Suppose we want to write an equation to describe a chemical reaction that we have just carried out in the laboratory. How should we go about doing it? Because we know the identities of the reactants, we can write their chemical formulas. The identities of products are more difficult to establish. For simple reactions, it is often possible to guess the product(s). For more complicated reactions involving three or more products, chemists may need to perform tests to establish the presence of specific compounds. Once we have identified all of the reactants and products and have written the correct formulas for each of them, we can assemble them in the conventional sequence—reactants on the left separated by an arrow from products on the right. The equation written at this point is likely to be *unbalanced;* that is, the number of each type of atom on one side of the arrow differs from the number on the other side.

In general, we can balance a chemical equation by using the following steps:

1. Identify all reactants and products and write their correct formulas on the left side and right side of the equation, respectively.

2. Begin balancing the equation by trying different coefficients to make the number of atoms of each element the same on both sides of the equation. We can change the stoichiometric coefficients (the numbers preceding the formulas), but not the subscripts (the numbers within formulas). Changing the stoichiometric coefficients changes only the amounts of the substances, whereas changing subscripts changes the identity of the substance. For example, $2NO_2$ means "two molecules of nitrogen dioxide," but if we double the subscripts, we have N_2O_4, which is the formula of dinitrogen tetroxide, a completely different compound.

    To do this, first look for elements that appear only once on each side of the equation with the same number of atoms on each side. The formulas containing these elements must have the same coefficient. Therefore, there is no need to adjust the coefficients of these elements at this point. Next, look for elements that appear only once on each side of the equation but in unequal numbers of atoms. Balance these elements. Finally, balance elements that appear in two or more formulas on the same side of the equation.

3. Check your balanced equation to be sure that you have the same total number of each type of atom on both sides of the equation arrow.

    Let's consider a specific example. In the laboratory, small amounts of oxygen gas can be prepared by heating potassium chlorate ($KClO_3$). The products are oxygen gas (O_2) and potassium chloride (KCl). From this information, we write

$$KClO_3 \longrightarrow KCl + O_2$$

(For simplicity, we omit the physical states of the reactants and products at this point.) All three elements (K, Cl, and O) appear only once on each side of the equation, but only for K and Cl do we have equal numbers of atoms on both sides. Thus, $KClO_3$ and KCl must have the same coefficient. The next step is to make the number of O atoms the same on both sides of the equation. Because there are three O atoms on

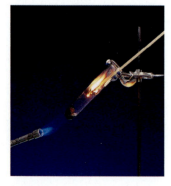

Heating potassium chlorate produces oxygen, which supports the combustion of wood splint.

the left and two O atoms on the right of the equation, we can balance the O atoms by placing a 2 in front of $KClO_3$ and a 3 in front of O_2.

$$2KClO_3 \longrightarrow KCl + 3O_2$$

Finally, we balance the K and Cl atoms by placing a 2 in front of KCl:

$$2KClO_3 \longrightarrow 2KCl + 3O_2$$

As a final check, confirm that the number of atoms of each element present is the same on both sides of the chemical equation. For the heating of $KClO_3$, we have two K atoms, two Cl atoms, and six O atoms on the reactant side and two K atoms, two Cl atoms, and six O atoms on the product side, so the equation is balanced. This equation could also be balanced by multiplying all of the coefficients by a constant. For example, we could multiply all stoichiometric coefficients by 2 to obtain a balanced chemical equation:

$$4KClO_3 \longrightarrow 4KCl + 6O_2$$

It is common practice, however, to use the *smallest* possible set of whole-number coefficients to balance the equation.

Next consider the combustion (that is, the burning) of the natural gas component ethane (C_2H_6) in oxygen or air, which yields carbon dioxide (CO_2) and water. The unbalanced equation is

$$C_2H_6 + O_2 \longrightarrow CO_2 + H_2O$$

The number of atoms is not the same on both sides of the equation for any of the elements (C, H, and O). In addition, C and H appear only once on each side of the equation, whereas O appears in two compounds on the right side (CO_2 and H_2O). To balance the C atoms, place a 2 in front of CO_2:

$$C_2H_6 + O_2 \longrightarrow 2CO_2 + H_2O$$

To balance the H atoms, place a 3 in front of H_2O:

$$C_2H_6 + O_2 \longrightarrow 2CO_2 + 3H_2O$$

At this stage, the C and H atoms are balanced, but the O atoms are not because there are seven O atoms on the right-hand (product) side and only two O atoms on the left-hand (reactant) side of the equation. The inequality of the oxygen atoms is eliminated by placing a stoichiometric coefficient of 7/2 in front of the reactant O_2 molecule:

$$C_2H_6 + \tfrac{7}{2}O_2 \longrightarrow 2CO_2 + 3H_2O$$

Molecules exist as discrete units, however, so it is generally preferable to express the stoichiometric coefficients as whole numbers rather than as fractions; therefore, we multiply the entire equation by 2 to clear the fraction:

$$2C_2H_6 + 7O_2 \longrightarrow 4CO_2 + 6H_2O$$

Quick inspection shows that there are 4 C atoms, 14 O atoms, and 12 H atoms on both sides of the equation, indicating that it is properly balanced. Example 0.7 gives you additional practice balancing chemical equations.

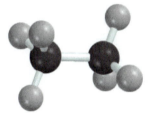

C_2H_6

Example 0.7

When aluminum metal is exposed to air, a protective layer of aluminum oxide (Al_2O_3) forms on its surface. This layer prevents further reaction between aluminum and oxygen, and it is the reason that aluminum beverage cans and aluminum airplanes do not corrode. (In the case of iron exposed to air, the rust, or iron(III) oxide, that forms is too porous to protect the iron metal underneath, so rusting continues.) Write a balanced equation for the formation of Al_2O_3.

Strategy Remember that the formula of an element or compound cannot be changed when balancing a chemical equation. The equation is balanced by placing the appropriate coefficients in front of the formulas.

Solution The unbalanced equation is

$$Al + O_2 \longrightarrow Al_2O_3$$

In a balanced equation, the number and types of atoms on each side of the equation must be the same. There is one Al atom on the reactant side, but two on the product side, so the Al atoms can be balanced by placing a coefficient of 2 in front of reactant Al, yielding

$$2Al + O_2 \longrightarrow Al_2O_3$$

There are two O atoms on the reactant side, and three O atoms on the product side of the equation. We can balance the O atoms by placing a coefficient of 3/2 in front of O_2 on the reactants side. This is a balanced equation. To eliminate the fraction (3/2), multiply both sides of the equation by 2 to obtain whole-number coefficients:

$$4Al + 3O_2 \longrightarrow 2Al_2O_3$$

Check When you are finished balancing an equation, check your answer by making sure that the number of each type of atom is the same on both sides of the equation. For the formation of aluminum oxide, there are four Al and six O atoms on both sides, so the equation is balanced.

Practice Exercise Balance the equation representing the reaction between iron(III) oxide (Fe_2O_3) and carbon monoxide (CO) to yield iron (Fe) and carbon dioxide (CO_2).

0.5 | Quantities of Atoms or Molecules Can Be Described by Mass or Number

Once we have a balanced chemical equation, we can begin to predict the numbers of each type of molecule being reacted or produced in a chemical reaction. For example, in the equation

$$2C_2H_6 + 7O_2 \longrightarrow 4CO_2 + 6H_2O$$

we can infer that if eight ethane (C_2H_6) molecules are reacted, then the number of water molecules produced is $3 \times 8 = 24$ because three water molecules are produced for every ethane molecule that reacts; the multiplier $3 = 6/2$ is simply the ratio of the stoichiometric coefficients. However, such predictive power is of somewhat limited practicality for two reasons:

▸ First, chemists generally measure the mass of chemical samples involved in a reaction and not the number of molecules they contain, so some method for converting mass to number of atoms is necessary in practical calculations.

▶ Second, the number of molecules contained in a laboratory sample is exceedingly large. For example, 1 g of water contains more than 5×10^{22} molecules of water—that is, 5 followed by 22 zeros! Therefore, a simple method for dealing with such large numbers in an efficient way is needed.

These limitations are overcome in modern chemistry through the introduction of two extremely important concepts: the *atomic mass* and the *mole.*

Atomic Mass

The mass of an atom depends on the number of electrons, protons, and neutrons it contains and all atoms of a given isotope are identical in mass. The SI unit of mass (the kilogram) is too large to function as a convenient unit for the mass of an atom, thus a smaller unit is desirable. In 1961, the International Union of Pure and Applied Chemistry (IUPAC) defined the **atomic mass unit** (u)[16] to be *exactly equal to one-twelfth the mass of one carbon-12 atom.*[17] Carbon-12 (^{12}C) is the carbon isotope that has six protons, six neutrons, and six electrons. Using this definition, we have that $1 \text{ u} = 1.660539 \times 10^{-27}$ kg. The **atomic mass** (sometimes called *atomic weight*) of an atom is then defined, relative to this standard, as *the mass of the atom in atomic mass units (u).* For example, the two naturally occurring isotopes of helium, ^{3}He and ^{4}He, have atomic masses of 3.01602931 u and 4.00260324 u, respectively. This means that a helium-4 (^{4}He) atom is $4.00260324/12 = 0.33355027$ times as massive as a carbon-12 atom.[18]

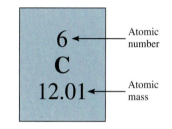

When you look up the atomic mass of carbon in a table, such as the one on the inside front cover of this book, you will find that its value is not 12.00 u but 12.01 u. The reason for the difference is that most elements found in nature (including carbon) have more than one naturally occurring isotope. When measuring the mass of a collection of carbon atoms, we obtain not the individual masses of each atom, but an average of the masses of the different isotopes, weighted by their natural abundances. This weighted average is known as the **average atomic mass** and is the mass most frequently used by chemists. For example, the natural abundances of carbon-12 and carbon-13 are 98.90 percent and 1.10 percent, respectively. The atomic mass of ^{13}C is 13.00335 u and that of ^{12}C is defined to be exactly 12. We can calculate the average atomic mass of carbon by multiplying the atomic mass of each isotope by its relative abundance (expressed as a decimal fraction) and adding them together, as follows:

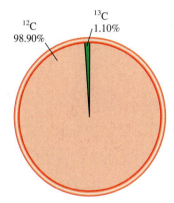

Natural abundances of C-12 and C-13 isotopes

$$\text{Average atomic mass of C} = (0.9890)(12.00000 \text{ u}) + (0.0110)(13.00335 \text{ u})$$
$$= 12.01 \text{ u}$$

---

16. An earlier abbreviation for atomic mass unit (*amu*) is still in common usage. In biochemistry and molecular biology, the atomic mass unit is often called a *dalton* (Da) especially in reference to the atomic mass of proteins.

17. Before 1961, two definitions of the atomic mass unit were used. In physics, the atomic mass unit was defined as one-sixteenth the mass of one ^{16}O atom. In chemistry, the atomic mass unit was defined as one-sixteenth the average atomic mass of oxygen. These units were slightly smaller than the current carbon-12 based unit.

18. Helium-4 (^{4}He) contains two protons, two neutrons, and two electrons, and ^{12}C contains six protons, six neutrons, and six electrons. As a result, you would expect that the mass of ^{4}He would be exactly 1/3 of the mass of ^{12}C, however, three ^{4}He atoms have a mass of $3 \times 4.00260324 \text{ u} = 12.00780972 \text{ u}$—slightly more than a single ^{12}C atom. As we will discuss in Chapter 17, this extra mass is converted into the energy necessary to bind the larger ^{12}C nucleus together. (According to Einstein's theory of relativity, mass can be converted to energy and vice versa.) This *nuclear binding energy* is the source of energy released in nuclear power and nuclear bombs.

Because there are many more carbon-12 atoms than carbon-13 atoms in naturally occurring carbon, the average atomic mass is much closer to 12 u than to 13 u. For simplicity, we will generally use *atomic mass* to mean *average atomic mass*. It is important to understand that when we say that the atomic mass of carbon is 12.01 u, we are referring to the *average* value. If carbon atoms could be examined individually, we would find either an atom of atomic mass 12.00000 u or one of 13.00335 u, but never one of 12.01 u. Examples 0.8 and 0.9 illustrate some calculations involving average atomic mass.

Example 0.8

Copper, a metal known since ancient times, is used in electrical cables and pennies, among other things. The atomic masses of its two stable isotopes, $^{63}_{29}Cu$ (69.09%) and $^{65}_{29}Cu$ (30.91%), are 62.93 u and 64.9278 u, respectively. Calculate the average atomic masses of copper. The relative abundances are given in parentheses.

Strategy Each isotope contributes to the average atomic mass based on its relative abundance. Multiplying the mass of an isotope by its fractional abundance (not percentage) will give the contribution to the average atomic mass of that particular isotope.

Solution First, convert the percentages to decimal fractions: $\dfrac{69.09\%}{100\%} = 0.6909$ and $\dfrac{30.91\%}{100\%} = 0.3091$. Next find the contribution to the average atomic mass for each isotope and then add the contributions together to give the average atomic mass.

$$(0.6909)(62.93\ u) + (0.3091)(64.9278\ u) = 63.55\ u$$

Check The average atomic mass should be between two atomic masses and closer to the mass of the more abundant isotope, copper-63. Therefore, the answer is reasonable.

Practice Exercise The atomic masses of the two stable isotopes of boron, $^{10}_{5}B$ (19.78%) and $^{11}_{5}B$ (80.22%), are 10.0129 u and 11.0093 u, respectively. Calculate the average atomic mass of boron.

Example 0.9

Chlorine has two naturally occurring isotopes, ^{35}Cl and ^{37}Cl, with atomic masses of 34.968852721 u and 36.96590262 u, respectively. The average atomic mass of chlorine is measured to be 35.453 u. Calculate the natural abundances of the two isotopes of Cl.

Strategy There are two unknown quantities in this problem: the relative abundances of ^{35}Cl and ^{37}Cl. A general rule of algebra is that you need as many equations as there are unknown variables to solve such a problem. For two unknowns, therefore, we need two equations that relate these variables. It is also useful to assign the unknowns simple algebraic symbols (such as x or y) to simplify the algebraic manipulations.

Solution Let the fractional abundances of ^{35}Cl and ^{37}Cl be x and y, respectively. The average atomic mass (35.453 u) and the atomic masses of the two isotopes are given in the problem statement; therefore, one equation is given by the definition of the average atomic mass discussed previously.

$$35.453\ u = x(34.968852721\ u) + y(36.96590262\ u) \tag{1}$$

—Continued

Continued—

We need one more equation to solve this problem. This can be obtained by noting that we are assuming that all chlorine atoms are either chlorine-35 or chlorine-37, so their fractional abundances must add to 1:

$$x + y = 1 \qquad (2)$$

Solving Equation (2) for y gives $y = 1 - x$. Substituting this value of y into Equation (1) gives

$$35.453 \text{ u} = x \,(34.968852721 \text{ u}) + (1 - x)(36.96590262 \text{ u})$$

Collecting like terms and rearranging gives

$$(36.96590262 \text{ u} - 34.968852721 \text{ u})\, x = 36.96590262 \text{ u} - 35.453 \text{ u}$$
$$1.997049899 \, x = 1.513 \text{ u}$$

yielding $x = 0.7576$. (In the last two equations, we have used the standard rules of significant figures, as described in Appendix 1, to determine the number of significant figures in the final answer.) Substituting this back into Equation (2) gives

$$y = 1 - x = 1 - 0.7576 = 0.2424$$

Converting the fractions into percentages, the relative abundances of ^{35}Cl and ^{37}Cl are 75.76% and 24.24%, respectively.

Practice Exercise Gallium has two naturally occurring isotopes (^{69}Ga and ^{71}Ga) with atomic masses of 68.925581 u and 70.924701 u, respectively. Given that the average atomic mass of gallium is 69.72 u, determine the relative abundances of ^{69}Ga and ^{71}Ga.

The Mole Concept: Avagadro's Number and the Molar Mass of an Element

Atomic mass units provide a relative scale for the masses of the elements. But because atoms have such small masses, no usable scale can be devised to weigh them in calibrated units of atomic mass units. In any real situation, we deal with macroscopic samples containing enormous numbers of atoms. It is convenient, therefore, to have a special unit to describe a very large number of atoms. The idea of a unit to denote a particular number of objects is not new. For example, the pair (2 items), the dozen (12 items), and the gross (144 items) are all commonly used. Chemists measure atoms and molecules in moles. In the SI system the **mole** (mol) is *the amount of a substance that contains as many elementary entities (atoms, molecules, or other particles) as there are atoms in exactly 12 g (or 0.012 kg) of the carbon-12 isotope.* The actual number of atoms in 12 g of carbon-12 is determined experimentally. This number is called **Avogadro's number** (N_A), in honor of the Italian scientist Amedeo Avogadro.[19] The currently accepted value is

$$N_A = 6.0221415 \times 10^{23}$$

The adjective formed from the noun "mole" is "molar."

---

19. Lorenzo Romano Amedeo Carlo Avagadro di Quaregua e di Cerreto (1776–1856). Italian mathematical physicist. He practiced law for many years before he became interested in science. His most famous work, now known as Avagadro's law (see Chapter 5), was largely ignored during his lifetime, although it became the basis for determining atomic masses in the late nineteenth century.

Figure 0.15 One mole each of several common elements: Carbon (black charcoal powder), sulfur (yellow powder), iron (as nails), copper wires, and mercury (shiny liquid metal).

For most purposes, we can round Avogadro's number to 6.022×10^{23}. Thus, just as one dozen oranges contains 12 oranges, 1 mole of hydrogen atoms contains 6.022×10^{23} H atoms. Figure 0.15 shows samples containing 1 mole each of several common elements. By definition, 1 mole of carbon-12 atoms has a mass of exactly 12 g and contains 6.022×10^{23} atoms. This mass of carbon-12 is its **molar mass** ($\mathcal{M}$), defined as *the mass (usually in grams or kilograms) of* 1 *mole of units* (such as atoms or molecules) of a substance. The molar mass of carbon-12 (in grams) is equal to its atomic mass in u. Likewise, the atomic mass of sodium (Na) is 22.99 u and its molar mass is 22.99 g, the atomic mass of phosphorus is 30.97 u and its molar mass is 30.97 g, and so on. If we know the atomic mass of an element, we also know its molar mass.

Because N_A carbon-12 atoms have a mass of exactly 12 g, the mass of one carbon-12 atom can be calculated by dividing 12 g by the numerical value of N_A:

$$\text{mass of one } {}^{12}\text{C atom} = 12 \text{ g}/(6.0221367 \times 10^{23}) = 1.9926467 \times 10^{-23} \text{ g}$$

By definition, the mass of one ^{12}C atom has a mass of exactly 12 u, so that one u corresponds to $1.9926482 \times 10^{-27}$ kg/12 $= 1.6605369 \times 10^{-27}$ kg.

The notions of Avogadro's number and molar mass enable us to carry out conversions between mass and moles of atoms and between the number of atoms and mass and to calculate the mass of a single atom, as illustrated in Examples 0.10–0.12.

Example 0.10

Helium (He) is a valuable gas used in industry, low-temperature research, deep-sea diving tanks, and balloons. How many moles of He atoms are in 6.46 g of He?

Strategy We are given grams of helium and asked to solve for moles of helium. To do this we need to know the number of grams in one mole of helium; that is we need to know the molar mass of He, which is given in the table at the front of the book.

—Continued

A scientific research helium balloon

Continued—

(Remember that both atomic mass in u and molar mass in grams have the same numerical value.)

Solution The molar mass of He is 4.003 g. Because 1 mole of He and 4.003 g of He represent the same quantity, their ratio is 1; that is

$$\frac{1 \text{ mol He}}{4.003 \text{ g He}} = 1 \quad \text{and} \quad \frac{4.003 \text{ g He}}{1 \text{ mol He}} = 1$$

To convert from grams to moles, we multiply the given number of grams by $\frac{1 \text{ mol He}}{4.003 \text{ g He}}$, so that the grams cancel and we are left only with moles.

$$\text{mol He} = 6.46 \text{ g He} \times \frac{1 \text{ mol He}}{4.003 \text{ g He}} = 1.61 \text{ mol He}$$

Thus, there are 1.61 mol of He atoms in 6.46 g of He. In this problem the ratio $\frac{1 \text{ mol He}}{4.003 \text{ g He}}$ is a *conversion factor* to covert grams of He to moles of He.

Check Because the given mass (6.46 g) is larger than the molar mass of He, we expect to have (and do have) more than 1 mole of He.

Practice Exercise How many moles of magnesium (Mg) are there in 87.3 g of Mg?

Example 0.11

Zinc (Zn) is a silvery metal that is used in making brass (with copper) and in plating iron to prevent corrosion. How many grams of Zn are in 0.356 mole of Zn?

Strategy We are given the number of moles of zinc and need to find the number of grams. Multiplying the number of moles by the number of grams in a mole (molar mass) will give the mass of the sample in grams. To do this, we need to look up the molar mass of zinc.

Solution According to the periodic table on the inside front cover, we see the molar mass of Zn is 65.39 g. Multiplying the number of moles in the sample by the following conversion factor

$$\frac{65.39 \text{ g Zn}}{1 \text{ mol Zn}}$$

yields the mass in grams:

$$0.356 \text{ mol Zn} \times \frac{65.39 \text{ g Zn}}{1 \text{ mol Zn}} = 23.3 \text{ g Zn}$$

(Notice that we write the conversion factor in such a way that the mol unit will cancel, leaving only units of grams in the answer.) Thus, there are 23.3 g of Zn in 0.356 mole of Zn.

Check The mass of one mole of Zn is 65.39 g. The given amount (0.356 mol) is approximately one-third of a mole. 23.3 g is close to one-third of 65.39, so the answer is reasonable.

Practice Exercise Calculate the number of grams of lead (Pb) in 12.4 moles of lead.

Zinc

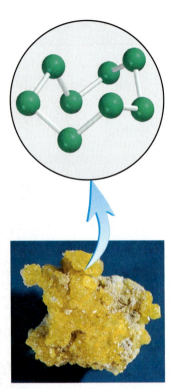

Elemental sulfur (S_8) consists of eight S atoms joined in a ring.

Example 0.12

Sulfur (S) is a nonmetallic element that is present in coal. When coal is burned, sulfur is converted to sulfur dioxide and eventually to sulfuric acid that gives rise to acid rain. How many atoms are in 16.3 g of S?

Strategy We can calculate the number of atoms of sulfur if we know the number of moles—we just multiply by Avogadro's number. We are given the number of grams, however, so we must first convert 16.3 g of S to moles of S and then convert the moles to atoms using Avogadro's number:

$$\text{grams S} \longrightarrow \text{moles S} \longrightarrow \text{atoms of S}$$

Solution The molar mass of S is 32.07 g. To convert from grams of S to moles of S, we multiply by the conversion factor

$$\frac{1 \text{ mol S}}{32.07 \text{ g S}}$$

This will give us the number of moles. Multiplying the number of moles by Avogadro's number (which gives the conversion factor between atoms and moles),

$$N_A = \frac{6.022 \times 10^{23} \text{ atoms of S}}{1 \text{ mol S}}$$

will give the desired answer in number of atoms of S. We can combine these conversions in one step as follows:

$$\text{atoms of S} = 16.3 \text{ g S} \times \frac{1 \text{ mol S}}{32.07 \text{ g S}} \times \frac{6.022 \times 10^{23} \text{ atoms of S}}{1 \text{ mol S}} = 3.06 \times 10^{23} \text{ atoms of S}$$

Both the units "grams" and "mol" cancel, leaving only the desired unit "atoms" in the final answer. Thus, there are 3.06×10^{23} atoms in 16.3 g of S.

Check The mass of sulfur given (16.3 g) is about one-half of the mass of one mole of sulfur (32.07 g), which would contain Avagadro's number of atoms, so we should expect the number of atoms in our sample to be equal to about one-half of Avagadro's number, which is in agreement with our calculation. Our answer is reasonable.

Practice Exercise Calculate the number of atoms in 0.551 g of potassium (K).

Molecular Mass

If we know the atomic masses of the component atoms, we can calculate the mass of a molecule. The ***molecular mass*** (sometimes called *molecular weight*) is *the sum of the atomic masses (in u) in the molecule.* For example, the molecular mass of H_2O is

$$2(\text{atomic mass of H}) + \text{the atomic mass of O}$$

or

$$2(1.008 \text{ u}) + 16.00 \text{ u} = 18.02 \text{ u}$$

In general, we need to multiply the atomic mass of each element by the number of atoms of that element present in the molecule and sum over all the elements. Example 0.13 illustrates this approach.

Example 0.13

Calculate the molecular masses (in u) of the following compounds: (a) sulfur dioxide (SO_2) and (b) caffeine ($C_8H_{10}N_4O_2$).

Strategy Add the atomic masses of the elements that make up each compound, multiplying each by the number of atoms of that element in the compound (as indicated by the subscripts).

Solution We find atomic masses in the periodic table (inside front cover).
(a) There are two O atoms (atomic mass 16.00 u) and one S atom (atomic mass 32.07 u) in SO_2, so that

$$\text{molecular mass of } SO_2 = 32.07 \text{ u} + 2(16.00 \text{ u}) = 64.07 \text{ u}$$

(b) There are eight C atoms, ten H atoms, four N atoms, and two O atoms in caffeine, so the molecular mass of $C_8H_{10}N_4O_2$ is given by

$$8(12.01 \text{ u}) + 10(1.008 \text{ u}) + 4(14.01 \text{ u}) + 2(16.00 \text{ u}) = 194.20 \text{ u}$$

Practice Exercise Calculate the molecular mass of (a) methanol (CH_3OH) and (b) acetylsalicylic acid ($C_9H_8O_4$), commonly known as aspirin.

SO_2

From the molecular mass, we can determine the molar mass of a molecule or compound. The molar mass of a compound (in grams) is numerically equal to its molecular mass (in u). For example, the molecular mass of water is 18.02 u, so its molar mass is 18.02 g. Note that 1 mole of water has a mass of 18.02 g and contains 6.022×10^{23} H_2O *molecules,* just as 1 mole (12.01 g) of elemental carbon contains 6.022×10^{23} carbon *atoms.* As Examples 0.14 shows, knowing the molar mass enables us to calculate the numbers of moles and individual atoms in a given quantity of a compound.

Example 0.14

Methane (CH_4) is the principal component of natural gas. (a) How many moles of CH_4 are present in 6.07 g of CH_4? (b) To how many molecules of methane does this correspond?

Strategy To calculate moles of CH_4 from grams of CH_4 we must determine the conversion factor between grams and moles. To do that, we first need to calculate the molecular mass (and thus the molar mass) of methane by adding the appropriate atomic masses.

Solution (a) The molar mass of CH_4 is determined by the procedure used in Example 0.11. In this case, we add the molar mass of carbon to four times the molar mass of hydrogen:

$$\text{molar mass of } CH_4 = 12.01 \text{ g} + 4(1.008 \text{ g}) = 16.04 \text{ g}$$

CH_4

—Continued

Methane gas burning on a cooking range

Continued—
Thus, 1 mol CH_4 = 16.04 g CH_4. Because we are converting grams to moles, the conversion factor we need should have grams in the denominator so that grams will cancel, leaving the moles in the numerator:

$$\text{mol } CH_4 = 6.07 \text{ g } CH_4 \times \frac{1 \text{ mol } CH_4}{16.04 \text{ g } CH_4} = 0.378 \text{ mol}$$

Thus, there are 0.378 moles of CH_4 in 6.07 g of CH_4. (b) Multiply the number of moles calculated in part (a) by Avogadro's number to get the number of molecules:

$$\text{number of } CH_4 \text{ molecules} = 0.378 \text{ mol} \times \frac{6.022 \times 10^{23} \text{ molecules } CH_4}{1 \text{ mol } CH_4}$$

$$= 2.28 \times 10^{23} \text{ molecules}$$

Practice Exercise (a) Calculate the number of moles of chloroform ($CHCl_3$) in 198 g of chloroform. (b) To how many molecules of $CHCl_3$ does this correspond?

Finally, note that for ionic compounds like NaCl and MgO that do not contain discrete molecular units, we use the term *formula mass* instead. The formula mass of NaCl is the mass of one formula unit in u:

$$\text{formula mass of NaCl} = 22.99 \text{ u} + 35.45 \text{ u} = 58.44 \text{ u}$$

and its molar mass is 58.44 g.

Percent Composition

Knowing the chemical formula and the molecular mass of a compound enables us to calculate the ***percent composition by mass***—*the percent by mass of each element in a compound.* It is useful to know the percent composition by mass if, for example, we needed to verify the purity of a compound for use in a laboratory experiment. From the formula we could calculate what percent of the total mass of the compound is contributed by each element. Then, by comparing the result to the percent composition obtained experimentally for our sample, we could determine the purity of the sample. Mathematically, the percent composition is obtained by dividing the mass of each element in 1 mole of the compound by the molar mass of the compound and multiplying by 100 percent:

$$\text{percent composition of an element} = \frac{n \times \text{molar mass of element}}{\text{molar mass of compound}} \times 100\% \quad \textbf{(0.9)}$$

where n is the number of moles of the element in 1 mole of the compound. For example, in 1 mole of hydrogen peroxide (H_2O_2) there are 2 moles of H atoms and 2 moles of O atoms. The molar masses of H_2O_2, H, and O are 34.02 g, 1.008 g, and 16.00 g, respectively. Therefore, the percent composition of H_2O_2 is calculated as follows:

$$\%H = \frac{2 \times 1.008 \text{ g H}}{34.02 \text{ g } H_2O_2} \times 100\% = 5.926\%$$

$$\%O = \frac{2 \times 16.00 \text{g H}}{34.02 \text{ g } H_2O_2} \times 100\% = 94.06\%$$

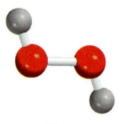

H_2O_2

The sum of the percentages is 5.926% + 94.06% = 99.99%. The small discrepancy from 100 percent is due to round-off error. Note, if we had used the empirical formula for hydrogen peroxide (HO) instead, we would get the same answer, because both the molecular formula and the empirical formula reveal the composition of a compound.

Example 0.15

Phosphoric acid (H_3PO_4) is a colorless, syrupy liquid used in detergents, fertilizers, toothpastes, and in carbonated beverages for a "tangy" flavor. Calculate the percent composition by mass of H, P, and O in H_3PO_4.

Strategy Use Equation 0.9 to calculate a percentage and assume that we have 1 mole of H_3PO_4. The percent by mass of each element (H, P, and O) is given by the combined molar mass of the atoms of the element in 1 mole of H_3PO_4 divided by the molar mass of H_3PO_4, then multiplied by 100 percent.

H_3PO_4

Solution The molar mass of H_3PO_4 is 97.99 g. The percent by mass of each of the elements in H_3PO_4 is calculated as follows:

$$\%H = \frac{3 \times 1.008 \text{ g H}}{97.99 \text{ g } H_3PO_4} \times 100\% = 3.086\%$$

$$\%P = \frac{30.97 \text{ g P}}{97.99 \text{ g } H_3PO_4} \times 100\% = 31.61\%$$

$$\%O = \frac{4 \times 16.00 \text{ g O}}{97.99 \text{ g } H_3PO_4} \times 100\% = 65.31\%$$

Check The percentages are reasonable because they sum to 100 percent: (3.086% + 31.61% + 65.31%) = 100.01%. The small discrepancy from 100 percent is due to the way we rounded off in the calculation.

Practice Exercise Calculate the percent composition by mass of each of the elements in sulfuric acid (H_2SO_4).

The procedure used in Example 0.15 can be reversed, if necessary, to calculate the empirical formula of a compound from the percent composition by mass of the compound (Figure 0.16). Because we are dealing with percentages and the sum of all the percentages is 100 percent, it is convenient to assume that we started with 100 g of a compound, as shown in Example 0.16.

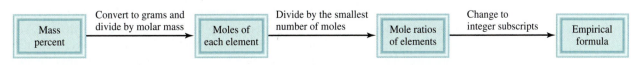

Figure 0.16 Schematic diagram for calculating the empirical formula of a compound from its percent compositions.

The molecular formula of ascorbic acid is $C_6H_8O_6$.

Example 0.16

Ascorbic acid (vitamin C) cures scurvy. It is composed of 40.92 percent carbon (C), 4.58 percent hydrogen (H), and 54.50 percent oxygen (O) by mass. Determine its empirical formula.

Strategy In a chemical formula, the subscripts represent the ratio of the number of moles of each element that combine to form one mole of the compound. If we assume exactly 100-g for the compound sample, we can use the percentage composition to determine the number of moles of each element in that sample. Examining the ratios of these molar quantities gives us the empirical formula.

Solution If we have 100 g of ascorbic acid, then each percentage can be converted directly to grams. In this sample, there will be 40.92 g of C, 4.58 g of H, and 54.50 g of O. Because the subscripts in the formula represent mole ratios, we need to convert the grams of each element to moles. The conversion factor needed is the molar mass of each element. Let n represent the number of moles of each element so that

$$n_C = 40.92 \text{ g} \times \frac{1 \text{ mol C}}{12.01 \text{ g C}} = 3.407 \text{ mol C}$$

$$n_H = 4.58 \text{ g} \times \frac{1 \text{ mol H}}{1.0079 \text{ g H}} = 4.54 \text{ mol H}$$

$$n_O = 54.50 \text{ g} \times \frac{1 \text{ mol O}}{16.00 \text{ g O}} = 3.406 \text{ mol O}$$

Thus, we arrive at the formula $C_{3.407}H_{4.54}O_{3.406}$, which gives the identity and the mole ratios of the atoms present. However, chemical formulas are written with whole numbers. Try to convert to whole numbers by dividing all the subscripts by the smallest subscript (3.406):

$$\text{C: } \frac{3.407}{3.406} \approx 1 \quad \text{H: } \frac{4.54}{3.406} = 1.33 \quad \text{O: } \frac{3.406}{3.406} = 1$$

where the $\approx$ sign means "approximately equal to." This gives $CH_{1.33}O$ as the formula for ascorbic acid. Next, convert 1.33, the subscript for H, into an integer. Note that 1.33 is close to 4/3 so multiplying by 3 gives us $C_3H_4O_3$ as the empirical formula for ascorbic acid.

Check The subscripts in $C_3H_4O_3$ are reasonable because they have been reduced to the smallest whole numbers.

Practice Exercise Determine the empirical formula of a compound having the following percent composition by mass: K: 24.75%; Mn: 34.77%; and O: 40.51%.

Because the percent composition by mass can be obtained experimentally by a number of methods, the procedure outlined in Example 0.16 enables us to determine the empirical formula of an unknown compound. In fact, the word "empirical" literally means "based only on observation and measurement."

Because any two compounds with the same empirical formula will give the same percent composition, it is impossible to uniquely determine the molecular formula from the percent compositions by mass alone. For example, both acetylene (C_2H_2) and benzene (C_6H_6) have the same empirical formula (CH), so any experiment that only gives the percent composition would be unable to distinguish between the two compounds. Therefore, we need to independently determine both the percentage composition by mass (and hence the empirical formula) *and* the molar mass. Because the molar mass of a compound is an integral multiple of the molar mass of its empirical formula, we can use the molar mass, together with the empirical formula, to find the molecular formula, as shown in Example 0.17.

Example 0.17

A sample of a compound contains 1.52 g of nitrogen (N) and 3.47 g of oxygen (O). The molar mass of this compound is between 90 g and 95 g. Determine the molecular formula and the accurate molar mass of the compound.

Strategy To determine the molecular formula, we first need to determine the empirical formula. Comparing the empirical molar mass to the experimentally determined molar mass will reveal the relationship between the empirical formula and molecular formula.

Solution Use molar mass to convert grams of N and O to moles of each element. Let n represent the number of moles of each element.

$$n_N = 1.52 \text{ g N} \times \frac{1 \text{ mol N}}{14.01 \text{ g N}} = 0.108 \text{ mol N}$$

$$n_O = 3.47 \text{ g} \times \frac{1 \text{ mol O}}{16.00 \text{ g O}} = 0.217 \text{ mol O}$$

The mole ratio of O to N is $0.217/0.108 = 2.01$ or 2, so the empirical formula is NO_2. The molar mass based on the empirical formula is $14.01 \text{ g} + 2(16.00 \text{ g}) = 46.01 \text{ g}$. This is smaller than the actual molecular mass (between 90 and 95 g) by about a factor of 2, so the molecular formula is twice the empirical formula or N_2O_4.

Practice Exercise A sample of a compound containing boron (B) and hydrogen (H) is found to contain 6.444 g of B and 1.803 g of H. Based on another experiment, the molar mass of the compound is determined to be about 30 g. What is the molecular formula of the unknown compound?

Concentration Units

A solution consists of one or more solutes dissolved in a solvent (see Section 0.1). In order to apply the quantitative mass-mole relationships to a solution, we must specify its **concentration,** defined as *the amount of particular solute present in a given amount of solution*. Chemists use several different concentration units, each of which has advantages as well as limitations. Let us examine the four most common units of concentration: percent by mass, mole fraction, molarity, and molality.

Types of Concentration Units

▸ **Percent by Mass** The *percent by mass* (also called *percent by weight* or *weight percent*) is *the ratio of the mass of a solute to the mass of the solution, multiplied by 100 percent:*

$$\text{percent by mass of solute} = \frac{\text{mass of solute}}{\text{mass of solute } + \text{ mass of solvent}} \times 100\%$$

$$\text{percent by mass of solute} = \frac{\text{mass of solute}}{\text{mass of solution}} \times 100\% \qquad \textbf{(0.10)}$$

Important Experimental Technique: The Mass Spectrometer

The most direct and most accurate method for determining atomic and molecular masses is mass spectrometry. In a **mass spectrometer,** such as the one depicted schematically in the figure below, a stream of high-energy electrons bombards a gaseous sample. Collisions between the electrons and the gaseous atoms (or molecules) produce positive ions (cations) by dislodging an electron from each neutral atom or molecule. These cations (of mass m and charge $+e$) are accelerated as they pass between two oppositely charged plates. These plates generate a voltage V that is on the order of several thousand volts. The potential energy gained by each cation is given by the product of the charge of the ion (e) and the voltage (V), that is, eV. Because these ions are being accelerated, their potential energies are converted to kinetic energies. Thus, we write

$$eV = \frac{1}{2} mu^2 \qquad (1)$$

where u is the velocity of the cation. Rearranging Equation 1 gives

$$u = \sqrt{\frac{2eV}{m}} \qquad (2)$$

which says that the velocity of the ion is proportional to the square root of the applied voltage. A moving charged particle generates a magnetic field; that is, it behaves somewhat like a magnet. Therefore, by placing an external magnet as shown in the schematic diagram, we can force the cations to move along a circular path of radius r. The magnetic force acting on each ion is given by $H\,e\,u$, where H is the strength of the magnetic field. According to Newton's second law of motion (Equation 0.1),

$$\text{force} = \text{mass} \times \text{acceleration}$$

The acceleration for a curving particle of velocity u moving on a circle of radius r is u^2/r so that

$$Heu = \frac{mu^2}{r} \qquad (3)$$

Substituting Equation 2 into Equation 3 and squaring both sides, we obtain

$$\frac{e}{m} = \frac{2V}{H^2}\left(\frac{1}{r^2}\right) \qquad (4)$$

Thus, Equation 4 predicts that if we fix the voltage V and magnetic field H, ions of smaller charge-to-mass ratio (that is, e/m) should follow a curve of larger radius than those having a large value of e/m. Because the radius of the analyzer

Schematic diagram of one type of mass spectrometer

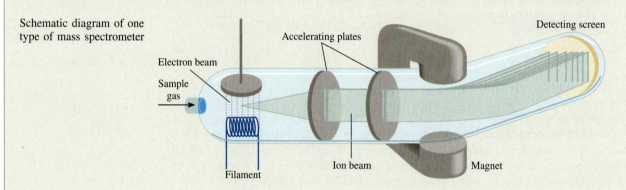

Detecting screen

Accelerating plates

Electron beam

Sample gas

Filament

Ion beam

Magnet

The percent by mass has no units because it is a ratio of two quantities with the same units.

▸ **Mole Fraction** (x)   The **mole fraction** of a component of a solution, say, component A, is written x_A and is defined as *the ratio of the number of moles of A to the total number moles of all components of the solution;* that is,

$$\text{mole fraction of component A} = x_A = \frac{\text{moles of A}}{\text{sum of moles for all components}} \qquad (0.11)$$

tube is normally fixed, different ions can be focused into the collector according to their mass by varying either *V* or *H*. The current measured at the collector is directly proportional to the number of ions, enabling us to determine the relative abundance of particles of different mass in the ion stream. If we begin with a sample of atoms of a single element, the mass spectrometer can measure the relative abundance of each isotope present.

The first mass spectrometer, developed in the 1920s by the English physicist F. W. Aston, was crude by present standards. Nevertheless, it provided indisputable evidence of the existence of the isotopes neon-20 (atomic mass = 19.9924 u and natural abundance = 90.92 percent) and neon-22 (atomic mass = 21.9914 u and natural abundance = 8.82 percent). When more sophisticated and sensitive mass spectrometers became available, scientists were surprised to discover that

neon has a third stable isotope with an atomic mass of 20.9940 u and natural abundance of 0.257 percent (see the following figure). This example illustrates how very important experimental accuracy is to a quantitative science like chemistry. Early experiments failed to detect neon-21 because its natural abundance was just 0.257 percent. In other words, only 26 in 10,000 Ne atoms are neon-21. The masses of molecules can be determined in a similar manner by the mass spectrometer.

Today, *mass spectrometry* is one of the most important analytical tools in chemistry and biochemistry. Advanced mass spectrometric techniques have applications ranging from the detection of trace chemicals in modern forensics laboratories to the determination of protein structure and function in the rapidly growing field of *proteomics*.

The mass spectrum of neon showing the three naturally occurring isotopes.

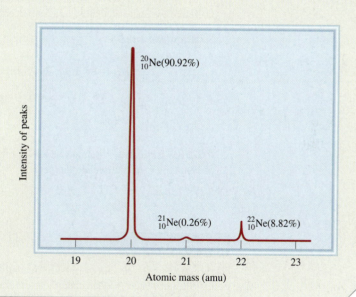

The mole fraction has no units, because it is a ratio of two quantities with identical units (mol).

▶ **Molarity** (*M*) *Molarity* is defined as *the number of moles of solute in 1 L of solution*; that is,

$$\text{molarity} = M = \frac{\text{moles of solute}}{\text{liters of solution}} \tag{0.12}$$

Thus, the units of molarity are mol L^{-1}.

▶ **Molality** (*m*)   *Molality* is *the number of moles of solute dissolved in 1 kg (1000 g) of solvent;* that is,

$$\text{molality} = m = \frac{\text{moles of solute}}{\text{mass of solvent (kg)}} \qquad (0.13)$$

For example, to prepare a 1 molal, or 1 *m*, sodium sulfate (Na_2SO_4) aqueous solution, we need to dissolve 1 mol (142.0 g) of the substance in 1000 g (1 kg) of water. Depending on the nature of the solute-solvent interaction, the final volume of the solution will be either greater or less than 1000 mL. It is also possible, though very unlikely, that the final volume could be equal to 1000 mL.

Examples 0.18–0.20 illustrate typical calculations of percent composition, mole molality, mole fraction and molarity.

Example 0.18

A sample of 0.892 g of potassium chloride (KCl) is dissolved in 54.6 g of water. What is the percent by mass of KCl in the solution?

Strategy We are given the mass of a solute dissolved in a certain amount of solvent. Therefore, we can calculate the mass percent of KCl using Equation (0.10).

Solution We write

$$\begin{aligned}
\text{percent by mass of solute} &= \frac{\text{mass of solute}}{\text{mass of solution}} \times 100\% \\
&= \frac{0.892 \text{ g}}{0.892 \text{ g} + 54.6 \text{ g}} \times 100\% \\
&= 1.61\%
\end{aligned}$$

Practice Exercise A sample of 6.44 g of naphthalene ($C_{10}H_8$) is dissolved in 80.1 g of benzene (C_6H_6). Calculate the percent by mass of naphthalene in this solution.

Example 0.19

(a) Calculate the molality of a sulfuric acid solution containing 24.4 g of sulfuric acid in 198 g of water. (b) Calculate the mole fraction of sulfuric acid for the solution in (a). The molar mass of sulfuric acid is 98.08 g.

Strategy (a) To calculate the molality of a solution, we need to know the number of moles of solute (sulfuric acid) and the mass of the solvent (water) in kilograms. (b) To calculate the mole fraction we need to know the number of moles of both the solute and solute.

Solution (a) The definition of molality (*m*) is

$$\text{molality} = m = \frac{\text{moles of solute}}{\text{mass of solvent (kg)}}$$

—Continued

Continued—

We can find the number of moles of sulfuric acid in 24.4 g of the acid, using its molar mass as the conversion factor.

$$n_{H_2SO_4} = \frac{24.2 \text{ g H}_2SO_4}{98.08 \text{ g H}_2SO_4 \text{ mol}^{-1}} = 0.247 \text{ mol}$$

The mass of water is 198 g, or 0.198 kg. Therefore,

$$m = \frac{0.247 \text{ mol H}_2SO_4}{0.198 \text{ kg H}_2O} = 1.26 \text{ mol kg}^{-1} \text{ (or 1.26 } m\text{)}$$

(b) To find the mole fraction of H_2SO_4 in the solution using Equation 0.11, we need the number of moles of both the solute (H_2SO_4) and solute (H_2O). The number of moles of H_2SO_4 was determined in part (a). The molar mass of water is 18.02 g mol^{-1}, so the number of moles of water in 198 g of water is

$$n_{H_2O} = \frac{198 \text{ g}}{18.02 \text{ g mol}^{-1}} = 11.0 \text{ mol}$$

Using Equation 0.11, the mole fraction of H_2SO_4 is given by

$$\text{mole fraction of H}_2SO_4 = x_{H_2SO_4} = \frac{n_{H_2SO_4}}{n_{H_2SO_4} + n_{H_2O_2}} = \frac{0.247 \text{ mol}}{11.0 \text{ mol} + 0.247 \text{ mol}} = 0.0220$$

Practice Exercise (a) What is the molality of a solution containing 7.78 g of urea [$(NH_2)_2CO$] in 203 g of water? (b) What is the mole fraction of urea in the solution in (a)?

Example 0.20

How many grams of potassium dichromate ($K_2Cr_2O_7$) are required to prepare a 250-mL solution whose concentration is 2.16 M?

Strategy Use the molarity to determine how many moles of $K_2Cr_2O_7$ are needed to make 250 mL of a 2.16 M solution. From this number of moles the mass of $K_2Cr_2O_7$ needed can be determined using the molar mass of $K_2Cr_2O_7$ as a conversion factor.

Solution The first step is to determine the number of moles of $K_2Cr_2O_7$ in 250 mL or 0.250 L of a 2.16 M Solution

$$\text{moles of K}_2Cr_2O_7 = 0.250 \text{ L soln} \times \frac{2.16 \text{ mol K}_2Cr_2O_7}{1 \text{ L soln}} = 0.540 \text{ mol K}_2Cr_2O_7$$

A $K_2Cr_2O_7$ solution

The molar mass of $K_2Cr_2O_7$ is 294.2 g, so we write

$$\text{g of K}_2Cr_2O_7 \text{ needed} = 0.540 \text{ mol K}_2Cr_2O_7 \times \frac{294.2 \text{ g K}_2Cr_2O_7}{1 \text{ mol K}_2Cr_2O_7} = 159 \text{ g K}_2Cr_2O_7$$

Check As a rough estimate, the mass should be given by [molarity (mol L^{-1}) = volume (L) × molar mass (g mol^{-1})] or [2 mol L^{-1} × 0.25 L × 300 g mol^{-1}] = 150 g. So the answer is reasonable.

Practice Exercise What is the molarity of an 85.0-mL ethanol (C_2H_5OH) solution containing 1.77 g of ethanol?

Comparison of Concentration Units

The choice of a concentration unit is based on the purpose of the experiment. For instance, mole fraction would not be used to express the concentrations of solutions for the purposes of determining the amounts of reactants consumed and products produced in a reaction, but it is appropriate for calculating partial pressures of gases (see Section 5.3) and for dealing with vapor pressures of solutions (Section 9.2).

The advantage of molarity is that it is generally easier to measure the volume of a solution, using precisely calibrated volumetric flasks, than to weigh the solvent. For this reason, molarity is often preferred over molality. On the other hand, molality is independent of temperature because the concentration is expressed in number of moles of solute and mass of solvent. The volume of a solution typically increases with increasing temperature, so that a solution that is 1.0 M at 25°C may become 0.97 M at 45°C because of the increase in volume. This concentration dependence on temperature can significantly affect the accuracy of an experiment. In such cases, it is preferable to use molality instead of molarity.

Percent by mass is similar to molality in that it is independent of temperature. Furthermore, because it is defined in terms of ratio of mass of solute to mass of solution, we do not need to know the molar mass of the solute in order to calculate the percent by mass.

Sometimes it is desirable to convert one concentration unit of a solution to another; for example, the same solution may be employed for different experiments that require different concentration units for calculations. Suppose we want to express the concentration of a 0.396 m glucose ($C_6H_{12}O_6$) solution in molarity. We know there is 0.396 mol of glucose in 1000 g of the solvent and we need to determine the volume of this solution to calculate molarity. First, we calculate the mass of the solution from the molar mass of glucose:

$$\left(0.396 \text{ mol } C_6H_{12}O_6 \times \frac{180.2 \text{ g}}{1 \text{ mol } C_6H_{12}O_6}\right) + 1000 \text{ g } H_2O = 1071 \text{ g}$$

The next step is to experimentally determine the density of the solution, which is found to be 1.16 g mL^{-1}. We can now calculate the volume of the solution in liters by writing

$$\text{volume} = \frac{\text{mass}}{\text{density}} = \frac{1071 \text{ g}}{1.16 \text{ g mL}^{-1}} \times \frac{1 \text{ L}}{1000 \text{ mL}} = 0.923 \text{ L}$$

Finally, the molarity of the solution is given by

$$\text{molarity} = \frac{\text{moles solute}}{\text{liters of solution}} = \frac{0.396 \text{ mol}}{0.923 \text{ L}} = 0.429 \text{ mol L}^{-1} = 0.429 \ M$$

As you can see, the density of the solution serves as a conversion factor between molality and molarity.

Examples 0.21 and 0.22 illustrate concentration unit conversions.

Example 0.21

The density of a 2.45 M aqueous solution of methanol (CH_3OH) is 0.976 g mL^{-1}. What is the molality of the solution? The molar mass of methanol is 32.04 g.

—Continued

CH_3OH

Continued—

Strategy To calculate the molality, we need to know the number of moles of methanol and the mass of solvent in kilograms. We assume 1 L of solution, so the number of moles of methanol is 2.45 mol. To calculate the mass of the solvent we use the density to calculate the mass of the solution, from which we subtract the mass of the solute to give the mass of water in the solution.

Solution Our first step is to calculate the mass of water in 1 L of the solution, using density as a conversion factor. The total mass of 1 L of a 2.45 *M* solution of methanol is

$$\text{1 L solution} \times \frac{\text{1000 mL solution}}{\text{1 L solution}} \times \frac{\text{0.976 g}}{\text{1 mL solution}} = \text{976 g}$$

Because this solution contains 2.45 mol of methanol, the mass of water (solvent) in the solution is

$$\text{mass of H}_2\text{O} = \text{mass of solution} - \text{mass of solute}$$
$$= 976 \text{ g} - \left(2.45 \text{ mol CH}_3\text{OH} \times \frac{32.04 \text{ g CH}_3\text{OH}}{1 \text{ mol CH}_3\text{OH}} \right) = 898 \text{ g}$$

The molality of the solution can be calculated by converting 898 g to 0.898 kg:

$$\text{molality} = \frac{2.45 \text{ mol CH}_3\text{OH}}{0.898 \text{ kg H}_2\text{O}} = 2.73 \text{ } m$$

Practice Exercise Calculate the molality of a 5.86 *M* ethanol (C_2H_5OH) solution whose density is 0.927 g mL^{-1}.

Example 0.22

Calculate the molality of a 35.4 percent (by mass) aqueous solution of phosphoric acid (H_3PO_4). The molar mass of phosphoric acid is 97.99 g.

Strategy In solving this type of problem, it is convenient to assume that we start with a 100.0 g of the solution. If the mass of phosphoric acid is 35.4 percent, or 35.4 g, the percent by mass and mass of water must be 100.0% − 35.4% = 64.6% and 64.6 g, respectively.

Solution From the known molar mass of phosphoric acid (97.99 g mol^{-1}) and the mass of the water in kg (64.6 g = 0.0646 kg), we can calculate the molality as shown in Example 0.19:

$$m = \frac{\left(35.4 \text{ g H}_3\text{PO}_4 \times \frac{1 \text{ mol H}_3\text{PO}_4}{97.99 \text{ g H}_3\text{PO}_4} \right)}{0.0646 \text{ kg H}_2\text{O}} = 5.59 \text{ } m$$

Practice Exercise Calculate the molality of a 44.6 percent (by mass) aqueous solution of sodium chloride.

Carbon monoxide burns in air to form carbon dioxide.

0.6 | Stoichiometry Is the Quantitative Study of Mass and Mole Relationships in Chemical Reactions

Stoichiometry

Stoichiometry is *the quantitative study of reactants and products in a chemical reaction.* Although amounts in chemical reactions can be measured in many different units, such as grams, liters, or moles, a chemical equation can only tell us the relative numbers of atoms (or moles) that are produced and consumed in a reaction. Therefore, no matter in what units the initial amounts are given, they must be converted to moles before the chemical equation can be used to determine the quantitative relationships between reactants and products. For example, the combustion of carbon monoxide gas in air produces carbon dioxide according to the chemical equation

$$2CO\ (g) + O_2(g) \longrightarrow 2CO_2(g)$$

The stoichiometric coefficients indicate that two molecules of CO react with one molecule of O_2 to form two molecules of CO_2. Because the number of moles is proportional to the number of atoms, we can just as well say that two *moles* of CO react with one *mole* of O_2 to form two *moles* of CO_2. Thus, 1 mol of O_2 is *stoichiometrically equivalent* to 2 mol of CO_2. As a short hand notation, we use the symbol "$\simeq$" to represent this stoichiometiric equivalence. For the current example, we have

$$1\ mol\ O_2 \simeq 2\ mol\ CO_2$$

However, this same equivalence does *not* hold for masses. Because CO, O_2, and CO_2 have different molar masses, it is *not* true that 2 g of CO will react with 1 g of O_2 to form 2 g of CO_2. If we are given masses of reactants or products in a calculation, we must first convert to moles (number of atoms) *before* using the equation.

To illustrate, let us consider two simple examples. First, suppose that we react completely 4.8 mol of CO with O_2 to form CO_2, and we want to determine how many moles of CO_2 will be produced. We are given moles so we can use the chemical equation for this calculation directly. The chemical equation states that 2 mol of CO_2 are formed for every 2 mol of CO reacted (that is, 2 mol $CO \simeq 2$ mol CO_2), giving a *stoichiometric ratio* of $2/2 = 1$. The number of moles of CO_2 produced is determined by multiplying the number of moles of CO reacted by the stoichiometric ratio as follows

$$\begin{aligned} mol\ CO_2 &= mol\ CO \times \frac{\text{number of moles of } CO_2}{\text{number of moles of } CO} \\ &= 4.8\ mol\ CO \times \frac{2\ mol\ CO_2}{2\ mol\ CO} \\ &= 4.8\ mol\ CO_2 \end{aligned}$$

Now suppose 10.7 g of CO react completely with an excess of O_2 to form CO_2, and we want to know how many grams of CO_2 will be formed. To calculate grams of CO_2, we need to use the mole ratio between CO and CO_2 from the balanced equation. Thus, we need to first convert grams of CO to moles of CO, then to moles of CO_2, and finally to grams of CO_2. To convert grams of CO to moles of CO, divide grams of CO by the molar mass of CO:

$$\text{moles of } CO = 10.7\ g\ CO \times \frac{1\ mol\ CO}{28.01\ g\ CO} = 0.382\ mol\ CO$$

To determine the number of moles of CO_2 produced from the number of moles of CO consumed, use the stoichiometric ratio determined from the balanced chemical equation:

$$\text{moles of } CO_2 = 0.382 \text{ mol CO} \times \frac{2 \text{ mol } CO_2}{2 \text{ mol CO}} = 0.382 \text{ mol } CO_2$$

Finally, multiply the number of moles of CO_2 by the molar mass of CO_2 to determine the number of grams produced:

$$\text{grams of } CO_2 = 0.382 \text{ mol } CO_2 \times \frac{44.01 \text{ g } CO_2}{1 \text{ mol } CO_2} = 16.8 \text{ g } CO_2$$

Thus, 10.7 g of CO reacts in excess oxygen to produce 16.8 g of CO_2.

The three separate calculations in this procedure can be combined in a single step as follows:

$$\text{grams of } CO_2 = 10.7 \text{ g CO} \times \frac{1 \text{ mol CO}}{28.01 \text{ g CO}} \times \frac{2 \text{ mol } CO_2}{2 \text{ mol CO}} \times \frac{44.01 \text{ g } CO_2}{1 \text{ mol } CO_2} = 16.8 \text{ g } CO_2$$

Similarly, we can calculate the mass of O_2 in grams consumed in this reaction. By using the relationship 2 mol CO $\simeq$ 1 mol O_2, we write

$$\text{grams of } O_2 = 10.7 \text{ g CO} \times \frac{1 \text{ mol CO}}{28.01 \text{ g CO}} \times \frac{1 \text{ mol } O_2}{2 \text{ mol CO}} \times \frac{18.02 \text{ g } O_2}{1 \text{ mol } O_2} = 3.44 \text{ g } O_2$$

The general approach for solving stoichiometry problems can be summarized in 4 steps:

1. Write a balanced equation for the reaction.
2. Convert the given amount of the reactant (in grams or other units) to number of moles.
3. Use the mole ratio from the balanced equation to calculate the number of moles of product formed.
4. Convert the moles of product to grams (or other units) of product.

Figure 0.17 shows these steps. Sometimes we may be asked to calculate the amount of a reactant needed to form a specific amount of product. In those cases, we can reverse the steps shown in Figure 0.17.

Example 0.23 illustrates the application of this approach.

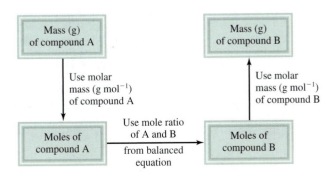

Figure 0.17 The procedure for calculating the amounts of reactants or products in a reaction.

Lithium reacting with water to produce hydrogen gas

Example 0.23

All alkali metals react with water to produce hydrogen gas and the corresponding alkali metal hydroxide. A typical reaction is that between lithium and water:

$$2Li(s) + 2H_2O(l) \longrightarrow 2LiOH(aq) + H_2(g)$$

(a) How many moles of H_2 will be formed by the complete reaction of 6.23 mol of Li with water? (b) How many grams of H_2 will be formed by the complete reaction of 80.57 g of Li with water?

(a) Strategy We can compare the amounts of Li and H_2 based on their *mole ratio* from the balanced equation.

Solution Because the balanced equation is given in the problem, the mole ratio between Li and H_2 is known: 2 mol Li $\simeq$ 1 mol H_2. We calculate moles of H_2 produced as follows:

$$mol\ H_2 = 6.23\ mol\ Li \times \frac{1\ mol\ H_2}{2\ mol\ Li} = 3.12\ mol\ H_2$$

The stoichiometric conversion factor was written with the moles of Li in the denominator so that the unit "moles of Li" cancels, leaving only "moles of H_2."

Check The moles of H_2 produced are half the moles of Li reacted because 2 mol Li $\simeq$ 1 mol H_2.

(b) Strategy Compare Li and H based on the *mole ratio* in the balanced equation. Before we can determine the moles of H_2 produced, however, we need to convert grams of Li to moles of Li using the molar mass of Li. Once we determine moles of H_2, we can convert it to grams of H_2 by using the molar mass of H_2.

Solution The molar mass of Li is 6.941 g. Thus, we can convert from grams of Li to moles of Li by dividing 80.57 g of Li by 6.941 g mol^{-1}. As in part (a), the balanced equation is given, so the mole ratio between Li and H_2 is known; it is 2 mol Li $\simeq$ 1 mol H_2. Finally, the molar mass of H_2 will convert moles of H_2 to grams of H_2. This sequence can be summarized as follows:

$$mass\ of\ H_2 = 80.57\ g\ Li \times \frac{1\ mol\ Li}{6.941\ g\ Li} \times \frac{1\ mol\ H_2}{2\ mol\ Li} \times \frac{2.016\ g\ H_2}{1\ mol\ H_2} \times = 11.7\ g\ H_2$$

So, 80.57 g of Li will react according to this chemical equation to give 11.7 g of H_2.

Check The mass of H_2 produced seems reasonable. It should be less than the mass of Li reacted because the molar mass of H_2 is less than one-third that of Li and because the mole ratio of H_2 to Li is 2:1 in this reaction.

Practice Exercise The reaction between nitrogen monoxide (NO) and oxygen to form nitrogen dioxide (NO_2) is a key step in photochemical smog formation:

$$2NO(g) + O_2(g) \longrightarrow 2NO_2(g)$$

(a) How many moles of NO_2 are formed by the complete reaction of 0.254 mol of O_2? (b) How many grams of NO_2 are formed by the complete reaction of 1.44 g of NO?

Stoichiometric calculations can become quite complex—especially if more than one independent reaction is taking place in a given sample. However, if we understand the basic principle that *stoichiometric coefficients in a reaction determine mole ratios*, these problems can be solved without too much difficulty. Example 0.24 shows how to perform a more complex stoichiometry calculation.

Example 0.24

When heated, both barium carbonate ($BaCO_3$) and calcium carbonate ($CaCO_3$) release carbon dioxide (CO_2), leaving barium oxide (BaO) and calcium oxide (CaO), respectively. In an experiment, a mixture of $BaCO_3$ and $CaCO_3$ with a combined mass of 63.67 g produces 19.67 g of CO_2. Calculate the mole and mass fractions of barium carbonate and calcium carbonate in the original mixture, assuming complete decomposition of the compounds.

Strategy There are two *independent* reactions proceeding here. Although it may be tempting to write a single chemical equation to represent this problem, namely,

$$BaCO_3(s) + CaCO_3(s) \longrightarrow 2CO_2(g) + BaO(s) + CaO(s)$$

this approach is incorrect because it assumes that there are equal amounts of $BaCO_3$ and $CaCO_3$ in the mixture. The correct way is to write two separate equations:

$$BaCO_3(s) \longrightarrow CO_2(g) + BaO(s)$$
$$CaCO_3(s) \longrightarrow CO_2(g) + CaO(s)$$

There are two unknowns in this problem, the amount of $BaCO_3$ and the amount of $CaCO_3$ in the original mixture. To solve for two unknowns requires two equations that relate the unknown variables.

Solution We are given two quantities, the number of grams of CO_2 produced and the total mass of the sample ($BaCO_3 + CaCO_3$). We can calculate the number of moles of CO_2 from the number of grams of CO_2, and the total mass of the sample must equal the sum of the number of moles of $BaCO_3$ and $CaCO_3$, each multiplied by their respective molar masses. Let the number of moles of $BaCO_3$ and $CaCO_3$ be n_{BaCO_3} and n_{CaCO_3}, respectively. One equation can be generated by noting that the number of moles of CO_2 produced must be equal, according to the stoichiometry, to the sum of the number of moles of $BaCO_3$ and $CaCO_3$:

$$n_{CaCO_3} + n_{BaCO_3} = n_{CO_2} = 19.67 \text{ g } CO_2 \times \frac{1 \text{ mol } CO_2}{44.01 \text{ g } CO_2} = 0.447 \text{ mol} \qquad \textbf{(1)}$$

The total mass of the sample gives us another equation:

$$\frac{100.09 \text{ g } CaCO_3}{1 \text{ mol } CaCO_3} \times n_{CaCO_3} + \frac{197.31 \text{ g } BaCO_3}{1 \text{ mol } BaCO_3} \times n_{BaCO_3} = \text{mass of sample} = 63.67 \text{ g} \quad \textbf{(2)}$$

From Equation (1) we have $n_{CaCO_3} = 0.447 \text{ mol} - n_{BaCO_3}$. Substituting this into Equation (2) gives

$$(0.447 \text{ mol} - n_{BaCO_3}) \times 100.09 \frac{\text{g}}{\text{mol}} + 197.31 \frac{\text{g}}{\text{mol}} \times n_{BaCO_3} = 63.67 \text{ g}$$

Collecting like terms gives

$$97.22 \frac{\text{g}}{\text{mol}} \times n_{BaCO_3} = 18.93 \text{ g}$$

—Continued

Continued—

and we obtain $n_{BaCO_3} = 0.195$ mol. Because the total number of moles of $BaCO_3$ and $CaCO_3$ in the original sample was 0.447 mol, the number of moles of calcium carbonate is then 0.447 mol − 0.195 mol = 0.252 mol. The mole fractions are then

$$\text{mole fraction BaCO}_3 = \frac{0.195 \text{ mol BaCO}_3}{0.447 \text{ mol total}} = 0.436$$

$$\text{mole fraction CaCO}_3 = \frac{0.252 \text{ mol CaCO}_3}{0.447 \text{ mol total}} = 0.564$$

The mass fractions can then be determined as follows:

$$\text{mass fraction BaCO}_3 = \frac{0.195 \text{ mol BaCO}_3 \times 197.31 \text{ g mol}^{-1}}{63.67 \text{ g total}} = 0.604$$

$$\text{mass fraction CaCO}_3 = \frac{0.252 \text{ mol CaCO}_3 \times 100.09 \text{ g mol}^{-1}}{63.67 \text{ g total}} = 0.396$$

Check The mass fractions and the mole fractions are reasonable because in both cases they sum to 1.

Practice Exercise A liquid mixture of benzene (C_6H_6) and hexane (C_6H_{14}), with a mass of 80.57 g, is burned in an excess of oxygen to form CO_2 and water (H_2O). If 82.1 g of water are formed in this reaction, calculate the mole fractions and mass fractions of benzene and hexane in the original mixture.

Before reaction has started

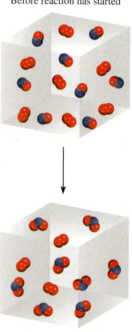

After reaction is complete

Figure 0.18 At the start of the reaction, there were eight NO molecules and seven O_2 molecules. At the end all the NO molecules are gone and three O_2 molecules are left. Therefore, NO is the limiting reagent and O_2 is the excess reagent. Each molecule can also be treated as 1 mol of the substance in this reaction.

Limiting Reagents

When a reaction is carried out, the reactants are usually not present in exact *stoichiometric amounts* (that is, in the proportions indicated by the balanced equation). Because the goal of a reaction is to produce the maximum quantity of a useful compound from the starting materials, a large excess of an inexpensive reactant may be supplied to ensure that a more expensive reactant is completely converted to the desired product. Consequently, some reactant will be left over at the end of the reaction. *The reactant that would be used up first if the reaction were to go to completion is called the* **limiting reagent** *because the maximum amount of product formed depends on how much of this reactant was originally present. When this reactant is used up, no additional product can be formed.* **Excess reagents** *are the reactants present in quantities greater than necessary to react with the available limiting reagent.* The concept of the limiting reagent is analogous to the relationship between men and women in a dance contest at a club. If there are 14 men and only 9 women, then only 9 male/female pairs can compete. A minimum of five men will be left without partners. The number of women thus *limits* the number of men that can dance in the contest, and there is an *excess* of men.

Consider the formation of nitrogen dioxide (NO_2) from nitric oxide (NO) and oxygen:

$$2NO(g) + O_2(g) \longrightarrow 2NO_2(g)$$

Suppose initially we have 8 mol of NO and 7 mol of O_2 (Figure 0.18). One way to determine which of the two reactants is the limiting reagent is to calculate the number of moles of NO_2 obtained based on the initial quantities of NO and O_2. The limiting reagent will yield the smaller amount of the product. Starting with 8 mol of NO, the number of moles of NO_2 produced is

$$8 \text{ mol NO} \times \frac{2 \text{ mol NO}_2}{2 \text{ mol NO}} = 8 \text{ mol NO}_2$$

and starting with 7 mol of O_2, the number of moles of NO_2 formed is

$$7 \text{ mol NO}_2 \times \frac{2 \text{ mol NO}_2}{1 \text{ mol O}_2} = 14 \text{ mol NO}_2$$

Because the amount of NO results in a smaller amount of NO_2, NO must be the limiting reagent, and O_2 must be the excess reagent.

In stoichiometric calculations involving limiting reagents, the first step is to decide which reactant is the limiting reagent. After the limiting reagent has been identified, the rest of the problem can be solved using mole ratios calculated from the stoichiometric coefficients, as in Example 0.23. Example 0.25 illustrates this approach.

Example 0.25

Urea $[(NH_2)_2CO]$ is prepared by reacting ammonia (NH_3) with carbon dioxide:

$$2NH_3(g) + CO_2(g) \longrightarrow (NH_2)_2CO(aq) + H_2O(l)$$

In one process, 637.2 g of NH_3 are treated with 1142 g of CO_2.

(a) Which of the two reactants is the limiting reagent?

(b) Calculate the mass of $(NH_2)_2CO$ formed.

(c) How much excess reagent (in grams) is left at the end of the reaction?

(a) Strategy The reactant that produces fewer moles of product is the limiting reagent because it limits the amount of product that can be formed. We must first convert the amounts of each reactant to moles and then use the stoichiometric ratios to find the number of moles of product formed. Perform this calculation for each reactant, then compare the moles of $(NH_2)_2CO$ formed by the given amounts of NH_3 and CO_2 to determine which reactant is the limiting reagent.

Solution We carry out two separate calculations. First, starting with 637.2 g of NH_3, we calculate the number of moles of $(NH_2)_2CO$ that could be produced if all the NH_3 were to react:

$$\text{moles of } (NH_2)_2CO = 637.2 \text{ g NH}_3 \times \frac{1 \text{ mol NH}_3}{17.03 \text{ g NH}_3} \times \frac{1 \text{ mol } (NH_2)_2CO}{2 \text{ mol NH}_3}$$
$$= 18.71 \text{ mol } (NH_2)_2CO$$

Repeating this calculation for 1142 g of CO_2 gives

$$\text{moles of } (NH_2)_2CO = 1142 \text{ g CO}_2 \times \frac{1 \text{ mol CO}_2}{44.01 \text{ g CO}_2} \times \frac{1 \text{ mol } (NH_2)_2CO}{1 \text{ mol CO}_2}$$
$$= 25.95 \text{ mol } (NH_2)_2CO$$

NH_3 must be the limiting reagent because it produces a smaller amount of $(NH_2)_2CO$, so CO_2 must be the excess reagent.

—Continued

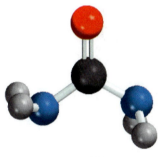

$(NH_2)_2CO$

Continued—

(b) Strategy The number of moles of $(NH_2)_2CO$ produced was determined in part (a) using NH_3 as the limiting reagent, so all we need to do to convert that amount to grams is to multiply by the molar mass of $(NH_2)_2CO$.

Solution The molar mass of $(NH_2)_2CO$ is 60.06 g. Use this value as a conversion factor to convert from moles of $(NH_2)_2CO$ to grams of $(NH_2)_2CO$:

$$\text{mass of } (NH_2)_2CO = 18.71 \text{ mol } (NH_2)_2CO \times \frac{60.06 \text{ g } (NH_2)_2CO}{1 \text{ mol } (NH_2)_2 \text{ CO}}$$

$$= 1124 \text{ g } (NH_2)_2CO$$

Check The molar mass of $(NH_2)_2CO$ is 60.06 g mol^{-1} or about 60 g mol^{-1}. Our result of 1124 g of $(NH_2)_2CO$ formed is reasonable because in part (a) we find that 18.71 mol of $(NH_2)_2CO$ are formed. This amount is slightly less than 20 mol, so we expect that slightly less than 60×20 g of $(NH_2)_2CO$ will be formed—in agreement with our result.

(c) Strategy Working backward, we can determine the amount of CO_2 that reacted to produce 18.71 mol of $(NH_2)_2CO$. The amount of CO_2 left over is the difference between the initial amount and the amount reacted.

Solution Starting with 18.71 mol of $(NH_2)_2CO$, we can determine the mass of CO_2 that reacted using the mole ratio from the balanced equation and the molar mass of CO_2:

$$\text{mass of } CO_2 \text{ reacted} = 18.71 \text{ mol } (NH_2)_2CO \times \frac{1 \text{ mol } CO_2}{1 \text{ mol } (NH_2)_2CO} \times \frac{44.01 \text{ g } CO_2}{1 \text{ mol } CO_2}$$

$$= 823.4 \text{ g } CO_2$$

The amount of CO_2 remaining (in excess) is the difference between the initial amount (1142 g) and the amount reacted (823.4 g):

$$\text{mass of } CO_2 \text{ remaining} = 1142 \text{ g} - 823.4 \text{ g} = 319 \text{ g}$$

Practice Exercise The reaction between aluminum and iron(III) oxide can generate temperatures approaching 3000°C and is used in welding metals:

$$2Al + Fe_2O_3 \longrightarrow Al_2O_3 + 2Fe$$

This reaction is known as the *thermite* reaction. In one process, 124 g of Al are reacted with 601 g of Fe_2O_3. (a) Calculate the mass (in grams) of Al_2O_3 formed. (b) How much of the excess reagent is left at the end of the reaction?

In practice, chemists usually choose the more expensive chemical as the limiting reagent so that all or most of it will be consumed in the reaction. In the synthesis of urea (Example 0.25), NH_3 is invariably the limiting reagent because it is much more expensive than CO_2.

Reaction Yield

The amount of limiting reagent present at the start of a reaction determines the ***theoretical yield*** of the reaction; that is, *the amount of product that would form if all of the limiting reagent reacted.* The theoretical yield, then, is the *maximum* obtainable yield predicted by the balanced equation. In practice, the ***actual yield,*** or *the amount* *of product actually obtained from a reaction,* is almost always less than the theoretical yield. There are many reasons for the difference between actual and theoretical yields. For instance, many reactions are reversible, and so they do not proceed 100 percent from left to right. Even when a reaction is 100 percent complete, it may be difficult to recover all of the product from the reaction medium (say, from an aqueous solution). Some reactions are complex in the sense that the products formed may react further among themselves or with the reactants to form still other products. These additional reactions will reduce the yield of the first reaction. To determine how efficient a given reaction is, chemists often calculate the **percent yield,** which describes *the proportion of the actual yield to the theoretical yield.* It is calculated as follows:

$$\% \text{ yield} = \frac{\text{actual yield}}{\text{theoretical yield}} \times 100\%$$

(0.14)

Percent yields may range from a fraction of 1 percent to 100 percent. Chemists strive to maximize the percent yield in a reaction. Factors that can affect the percent yield include temperature and pressure. We will study these effects later. Example 0.26 shows how to calculate the yield of an industrial process.

Example 0.26

Titanium is a strong, lightweight, corrosion-resistant metal that is used in rockets, aircraft, jet engines, and bicycle frames. It is prepared by the reaction of titanium(IV) chloride with molten magnesium between 950°C and 1150°C:

$$TiCl_4(g) + 2Mg(l) \longrightarrow WTi(s) + 2MgCl_2(l)$$

In a certain industrial operation 3.54×10^7 g of $TiCl_4$ are reacted with 1.13×10^7 g of Mg. (a) Calculate the theoretical yield of Ti in grams. (b) Calculate the percent yield if 7.91×10^6 g of Ti are actually obtained.

(a) Strategy Because there are two reactants, this is likely to be a limiting reagent problem. The reactant that produces fewer moles of product is the limiting reagent. Thus, convert from amount of reactant to amount of product (Ti) for each reactant and compare the moles of product (Ti) formed.

Solution Carry out two separate calculations to see which of the two reactants is the limiting reagent. First, starting with 3.54×10^7 g of $TiCl_4$, calculate the number of moles of Ti that could be produced if all the $TiCl_4$ reacted:

$$\text{moles of Ti} = 3.54 \times 10^7 \text{ g TiCl}_4 \times \frac{1 \text{ mol TiCl}_4}{189.7 \text{ g TiCl}_4} \times \frac{1 \text{ mol Ti}}{1 \text{ mol TiCl}_4} = 1.87 \times 10^5 \text{ mol Ti}$$

Next, calculate the number of moles of Ti formed from 1.13×10^7 g of Mg:

$$\text{moles of Ti} = 1.13 \times 10^7 \text{ g Mg} \times \frac{1 \text{ mol Mg}}{24.31 \text{ g Mg}} \times \frac{1 \text{ mol Ti}}{2 \text{ mol Mg}} = 2.32 \times 10^5 \text{ mol Ti}$$

$TiCl_4$ is the limiting reagent because it produces a smaller amount of Ti than does Mg. The mass of Ti formed is then

$$1.87 \times 10^5 \text{ mol Ti} \times \frac{47.88 \text{ g Ti}}{1 \text{ mol Ti}} = 8.95 \times 10^6 \text{ g Ti}$$

—Continued

This bicycle frame is made of titanium.

Continued—

(b) Strategy The mass of Ti determined in part (a) is the theoretical yield. The amount given in part (b) is the actual yield of the reaction. Calculate the percent yield using Equation 0.14.

Solution The percent yield is given by

$$\% \text{ yield} = \frac{7.91 \times 10^6 \text{ g Ti (actual)}}{8.95 \times 10^6 \text{ g Ti (theoretical)}} \times 100\% = 88.4\%$$

Check The percent yield is reasonable because it is less than 100 percent.

Practice Exercise Industrially, vanadium metal, which is used in steel alloys, can be obtained by reacting vanadium(V) oxide with calcium at high temperatures:

$$5Ca + V_2O_5 \longrightarrow 5CaO + 2V$$

In one process 1.54×10^3 g of V_2O_5 react with 1.96×10^3 g of Ca. (a) Calculate the theoretical yield of V. (b) Calculate the percent yield if 803 g of V are obtained.

Industrial processes usually involve huge quantities (thousands to millions of tons) of products. Thus, even a slight improvement in the yield can significantly reduce the cost of production.

Summary of Facts and Concepts

Section 0.1

▶ Chemists study matter and the changes it undergoes. The substances that make up matter have unique physical properties that can be observed without changing the identity of the substances and unique chemical properties that, when they are demonstrated, do change the identity of the substances.

▶ Mixtures, whether homogeneous or heterogeneous, can be separated into pure components by physical means. A homogeneous mixture is also called a solution.

▶ The simplest substances in chemistry are elements. Compounds are formed by the chemical combination of atoms of different elements in fixed proportions.

▶ All substances, in principle, can exist in three states: solid, liquid, and gas. The interconversion between these states can be effected by changing the external conditions, such as temperature.

▶ Force is the quantity that causes an object to change its course of motion. Work is done whenever an object is moved from one place to another in the presence of a force. The amount of work done is given by the force times the distance moved (Equation 0.2).

▶ Energy is the capacity to do work. There are many forms of energy, and they are interconvertible. The law

of conservation of energy states that the total amount of energy in the universe is constant. The SI unit for energy is the joule.

Section 0.2

▶ Modern chemistry began with Dalton's atomic theory, which states that all matter is composed of tiny, indivisible particles called atoms; that all atoms of the same element are identical; that compounds contain atoms of different elements combined in whole-number ratios; and that atoms are neither created nor destroyed in chemical reactions (the law of conservation of mass).

▶ Atoms of constituent elements in a particular compound are always combined in the same proportions by mass (law of definite proportions). When two elements can combine to form more than one type of compound, the masses of one element that combine with a fixed mass of the other element are in a ratio of small whole numbers (law of multiple proportions).

▶ An atom consists of a very dense central nucleus containing protons and neutrons, with electrons moving about the nucleus at a relatively large distance from it.

▶ Protons are positively charged, neutrons have no charge, and electrons are negatively charged. Protons and neutrons have roughly the same mass, which is about 1840 times greater than the mass of an electron.

▶ The atomic number (Z) of an element is the number of protons in the nucleus of an atom of the element; it determines the identity of an element. The mass number (A) is the sum of the number of protons and the number of neutrons in the nucleus.

▶ Isotopes are atoms of the same element with the same number of protons but different numbers of neutrons.

▶ The periodic table is a chart of the elements in order of increasing atomic number in which elements having similar chemical and physical properties are grouped together.

▶ A molecule is an aggregate of two or more atoms in a definite arrangement held together by chemical forces, which may contain atoms of the same element or atoms of two or more elements joined in a fixed ratio. An ion is an atom or a group of atoms that has a net positive or negative charge.

Section 0.3

▶ Chemical formulas combine the symbols for the constituent elements with whole number subscripts to show the type and number of atoms contained in the smallest unit of a compound.

▶ The molecular formula conveys the specific number and type of atoms combined in each molecule of a compound. The empirical formula shows the simplest ratios of the atoms combined in a molecule.

▶ Chemical compounds are either molecular compounds (in which the smallest units are discrete, individual molecules) or ionic compounds (in which positive and negative ions are held together by mutual attraction). Ionic compounds are made up of cations and anions, formed when atoms lose and gain electrons, respectively.

▶ The names of many inorganic compounds can be deduced from a set of simple rules. The formulas can be written from the names of the compounds.

Section 0.4

▶ Chemical changes, called chemical reactions, are represented by chemical equations. Substances that undergo change—the reactants—are written on the left and the substances formed—the products—appear to the right of the arrow.

▶ Chemical equations must be balanced, in accordance with the law of conservation of mass. The number of atoms of each element in the reactants must equal the number in the products.

Section 0.5

▶ Atomic masses are measured in atomic mass units (u), a relative unit based on a value of exactly 12 for the C-12 isotope. The atomic mass given for the atoms of a particular element is the average of the naturally occurring isotope distribution of that element. The molecular mass of a molecule is the sum of the atomic masses of the atoms in the molecule. Both atomic mass and molecular mass can be accurately determined with a mass spectrometer.

▶ A mole is Avogadro's number (6.022×10^{23}) of atoms, molecules, or other particles.

▶ The molar mass (in grams) of an element or a compound is numerically equal to its mass in atomic mass units (u) and contains Avogadro's number of atoms (in the case of elements), molecules (in the case of molecular substances), or simplest formula units (in the case of ionic compounds).

▶ The percent composition by mass of a compound is the percent by mass of each element present. If we know the percent composition by mass of a compound, we can deduce the empirical formula of the compound and also the molecular formula of the compound if the approximate molar mass is known.

▶ The concentration of a solution can be expressed as percent by mass, mole fraction, molarity, and molality. The choice of units depends on the circumstances.

Section 0.6

▶ Stoichiometry is the quantitative study of products and reactants in chemical reactions. Stoichiometric calculations are best done by expressing both the known and unknown quantities in terms of moles and then converting to other units if necessary.

▶ A limiting reagent is the reactant that is present in the smallest stoichiometric amount. It limits the amount of product that can be formed. The amount of product obtained in a reaction (the actual yield) may be less than the maximum possible amount (the theoretical yield). The ratio of the two multiplied by 100 percent is expressed as the percent yield.

Key Words

acid, p. 27	allotrope, p. 20	atom, p. 13	average atomic mass, p. 35
actual yield, p. 58	anion, p. 19	atomic mass, p. 35	Avogadro's number, p. 37
alkali metal, p. 17	aqueous species, p. 31	atomic mass unit, p. 35	base, p. 29
alkaline earth metal, p. 19	aqueous solution, p. 31	atomic number, p. 15	binary compound, p. 23

Problems

Chemistry Is the Study of Matter and Change

0.1 Do the following statements describe chemical properties or physical properties? (a) Oxygen gas supports combustion. (b) Fertilizers help to increase agricultural production. (c) Water boils below 100°C on top of a mountain. (d) Lead is denser than aluminum. (e) Uranium is a radioactive element.

0.2 Does each of the following describe a physical change or a chemical change? (a) The helium gas inside a balloon tends to leak out after a few hours. (b) A flashlight beam slowly gets dimmer and finally goes out. (c) Frozen orange juice is reconstituted by adding water to it. (d) The growth of plants depends on the energy of the sun in a process called photosynthesis. (e) When heated, a protein molecule "unfolds" (adopts a less compact three-dimensional structure).

0.3 The ancients believed that all substances were made up of the "elements" *earth, air, fire,* and *water*. How would each of these be described according to modern chemical classification schemes?

0.4 Classify each of the following as an element, a compound, a homogeneous mixture, or a heterogeneous mixture: (a) seawater, (b) helium gas, (c) sodium chloride (table salt), (d) a bottle of soft drink, (e) a milkshake, (f) air in a bottle, (g) concrete.

0.5 Describe the interconversions of forms of energy occurring in the following processes: (a) You throw a softball up into the air and catch it. (b) You switch on a flashlight. (c) You ride the ski lift to the top of a hill and then ski down. (d) You strike a match and let it burn down.

0.6 Calculate the work done against the gravitational field of Earth when a 75-kg person climbs a ladder from the ground to a height of 11 m.

0.7 Calculate the kinetic energy of an electron traveling at 5.00 percent of the speed of light. How fast would a proton have to travel (as a percentage of the speed of light) to have the same kinetic energy?

0.8 If a 1000-lb safe is pushed out of a tenth story window (100 ft above the ground), what is the speed of the safe when it hits the ground?

Matter Consists of Atoms and Molecules

0.9 (a) Describe Rutherford's experiment and how it led to the structure of the atom. How was he able to estimate the number of protons in a nucleus from the scattering of the particles? (b) Consider the ^{23}Na atom. Given that the radius and mass of the nucleus are 3.04×10^{-15} m and 3.82×10^{-23} g, respectively, calculate the density of the nucleus in g cm^{-3}. The radius of a ^{23}Na atom is 186 pm. Calculate the density of the space occupied by the electrons in the sodium atom. Do your results support Rutherford's model of an atom? (The volume of a sphere is $\frac{4}{3}\pi r^3$, where r is the radius.)

0.10 For the noble gases (the Group 8A elements), $^{4}_{2}$He, $^{20}_{10}$Ne, $^{40}_{18}$Ar, $^{84}_{36}$Kr, $^{132}_{54}$Xe, and $^{222}_{86}$Rn, (a) determine the number of protons and neutrons in the nucleus of each atom, and (b) determine the ratio of neutrons to protons in the nucleus of each atom. Describe any general trend you discover in the way this ratio changes with increasing atomic number. (c) Does this trend hold for the other elements as well?

0.11 Fluorine reacts with hydrogen (H) and deuterium (D) to form hydrogen fluoride (HF) and deuterium fluoride (DF), where deuterium ($^{2}_{1}$H) is an isotope of hydrogen. Would a given amount of fluorine react with different masses of the two hydrogen isotopes? Does this violate the law of definite proportions? Explain.

0.12 Roughly speaking, the radius of an atom is about 10,000 times greater than that of its nucleus. If an atom were magnified so that the radius of its nucleus became 2.0 cm, about the size of a marble, what would be the radius of the atom in km?

0.13 What is the mass number of an iron atom that has 28 neutrons?

0.14 Calculate the number of neutrons in ^{239}Pu.

0.15 For each of the following species, determine the number of protons and the number of neutrons in the nucleus:

$^{3}_{2}$He, $^{4}_{2}$He, $^{24}_{12}$Mg, $^{25}_{12}$Mg, $^{48}_{22}$Ti, $^{79}_{35}$Br, $^{195}_{78}$Pt

0.16 Indicate the number of protons, neutrons, and electrons in each of the following species:

$^{15}_{7}$N, $^{33}_{16}$S, $^{63}_{29}$Cu, $^{84}_{38}$Sr, $^{130}_{56}$Ba, $^{186}_{74}$W, $^{202}_{80}$Hg

0.17 Write the appropriate symbol for each of the following isotopes: (a) $Z = 11, A = 23$; (b) $Z = 28, A = 64$.

0.18 Write the appropriate symbol for each of the following isotopes: (a) $Z = 74, A = 186$; (b) $Z = 80, A = 201$.

0.19 One isotope of a metallic element has mass number 65 and 35 neutrons in the nucleus. The cation derived from the isotope has 28 electrons. Write the symbol for this cation.

0.20 Give two examples of each of the following: (a) a diatomic molecule containing atoms of the same element, (b) a diatomic molecule containing atoms of different elements, (c) a polyatomic molecule containing atoms of the same element, and (d) a polyatomic molecule containing atoms of different elements.

0.21 Give the number of protons and electrons in each of the following common ions: Na$^+$, Ca^{2+}, Al^{3+}, Fe^{2+}, I$^-$, F$^-$, S^{2-}, O^{2-}, and N^{3-}.

0.22 Give the number of protons and electrons in each of the following common ions: K$^+$, Mg^{2+}, Fe^{3+}, Br$^-$, Mn^{2+}, C^{4-}, Cu^{2+}.

Compounds Are Represented by Chemical Formulas

0.23 Write the molecular formula of glycine, an amino acid present in proteins. The color codes are: black (carbon), blue (nitrogen), red (oxygen), and gray (hydrogen).

0.24 Write the molecular formula for ethanol. The color codes are: black (carbon), red (oxygen), and gray (hydrogen).

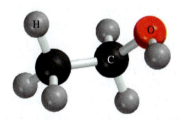

0.25 What are the empirical formulas of the following compounds: (a) C_2N_2, (b) P_4O_{10}, (c) $Na_2S_2O_4$, (d) N_2O_5, (e) B_2H_6, (f) C_4H_8, and (g) $C_6H_{12}O_8$?

0.26 Name the following compounds: (a) KH_2PO_4, (b) K_2HPO_4, (c) HBr, (d) Li_2CO_3, (e) $Na_2Cr_2O_7$, (f) $(NH_4)_2SO_4$, (g) HIO_3, (h) PF_5, (i) P_4O_6, (j) CdI_2, (k) $SrSO_4$, and (l) $Al(OH)_3$.

0.27 Name the following compounds: (a) KClO, (b) Ag_2CO_3, (c) HNO_2, (d) $KMnO_4$, (e) $CsClO_3$, (f) FeO, (g) Fe_2O_3, (h) $TiCl_4$, (i) NaH, (j) Li_3N, (k) Na_2O, and (l) Na_2O_2.

0.28 Name the following compounds: (a) $Co(NO_3)_2 \cdot 6H_2O$, (b) $NiBr_2 \cdot 3H_2O$, (c) $ZnSO_4 \cdot 7H_2O$, and (d) $Na_2SO_4 \cdot H_2O$.

0.29 Write the formulas for the following compounds: (a) rubidium nitrite, (b) potassium sulfide, (c) sodium hydrogen sulfide, (d) magnesium phosphate, (e) calcium hydrogen phosphate, (f) potassium dihydrogen phosphate, (g) iodine heptafluoride, (h) ammonium sulfate, (i) silver perchlorate, (j) boron trichloride, and (k) copper(I) carbonate.

0.30 Write the formulas for the following compounds: (a) sodium sulfate decahydrate, (b) calcium sulfate monohydrate, (c) magnesium chloride hexahydrate, and (d) zinc(II) nitrate hexahydrate.

0.31 Write the formulas for the following compounds: (a) copper(I) cyanide, (b) strontium chlorite, (c) perbromic acid, (d) hydroiodic acid, (e) disodium ammonium phosphate, (f) lead(II) carbonate, (g) tin(II) fluoride, (h) tetraphosphorus decasulfide, (i) mercury(II) oxide, (j) mercury(I) iodide, and (k) selenium hexafluoride.

Reactions Are Represented by Balanced Chemical Equations

0.32 Balance the following chemical equations:

(a) $C + O_2 \longrightarrow CO$

(b) $CO + O_2 \longrightarrow CO_2$

(c) $H_2 + Br_2 \longrightarrow HBr$

(d) $H_2O_2 \longrightarrow H_2O + O_2$

(e) $C_4H_{10} + O_2 \longrightarrow CO_2 + H_2O$

(f) $KOH + H_3PO_4 \longrightarrow K_3PO_4 + H_2O$

(g) $N_2 + H_2 \longrightarrow NH_3$

(h) $(NH_4)NO_3 \longrightarrow N_2O + H_2O$

(i) $C_6H_6 + O_2 \longrightarrow CO_2 + H_2O$

(j) $Na_2O + Al \longrightarrow Al_2O_3 + Na$

0.33 Balance the following chemical equations:

(a) $N_2O_5 \longrightarrow N_2O_4 + O_2$

(b) $P_4O_{10} + H_2O \longrightarrow H_3PO_4$

(c) $NaHCO_3 \longrightarrow Na_2CO_3 + H_2O + CO_2$

(d) $HCl + CaCO_3 \longrightarrow CaCl_2 + H_2O + CO_2$

(e) $Al + H_2SO_4 \longrightarrow Al_2(SO_4)_3 + H_2$

(f) $Be_2C + H_2O \longrightarrow Be(OH)_2 + CH_4$

(g) $C_2H_5OH + O_2 \longrightarrow CO_2 + H_2O$

(h) $S + HNO_3 \longrightarrow H_2SO_4 + NO_2 + H_2O$

(i) $C_3H_5N_3O_9 \longrightarrow N_2 + CO_2 + H_2O + O_2$

(j) $Ca_3(PO_4)_2 + SiO_2 + C \longrightarrow CaSiO_3 + P_4 + CO$

Quantities of Atoms or Molecules Can Be Described by Mass or Number

0.34 The atomic masses of $^{35}_{17}Cl$ (75.53 percent) and $^{37}_{17}Cl$ (24.47 percent) are 34.968 u and 36.956 u,

respectively. Calculate the average atomic mass of chlorine, assuming only these isotopes are present. The percentages in parentheses denote relative abundances.

0.35 The average atomic mass of Li is 6.941 u. Calculate the relative abundances of ^6_3Li (6.0151 u) and ^7_3Li (7.0160 u), assuming only these two isotopes are present. The atomic masses of the isotopes are given in the parentheses.

0.36 Calculate the molecular mass or formula mass (in u) of each of the following substances: (a) CH_4, (b) NO_2, (c) SO_3, (d) C_6H_6, (e) NaI, (f) K_2SO_4, and (g) $Ca_3(PO_4)_2$.

0.37 How many atoms are there in 5.10 mol of sulfur (S)?

0.38 How many moles of cobalt (Co) atoms are there in 6.00×10^9 (6 billion) Co atoms?

0.39 What is the average mass in grams of a single atom of each of the following elements: (a) Hg, (b) Ne, (c) As, and (d) Ni?

0.40 What is the mass in grams of 1.00×10^{12} Pb atoms?

0.41 How many atoms are present in 3.14 g of copper (Cu)?

0.42 Calculate the molar mass of the following substances: (a) Li_2CO_3, (b) CS_2, (c) $CHCl_3$ (chloroform), (d) $C_6H_8O_6$ (ascorbic acid, or vitamin C), (e) KNO_3, (f) UF_6, and (g) $NaHCO_3$ (baking soda).

0.43 How many molecules of ethane (C_2H_6) are present in 0.334 g of C_2H_6?

0.44 Calculate the number of C, H, and O atoms in 1.50 g of glucose ($C_6H_{12}O_6$), a sugar.

0.45 Urea $[(NH_2)_2CO]$ is used for fertilizer (among many other things). Calculate the number of N, C, O, and H atoms in 1.68×10^4 g of urea.

0.46 The density of water is 1.00 g mL^{-1} at 4°C. How many water molecules are present in 2.56 mL of water at this temperature?

0.47 Tin (Sn) exists in Earth's crust as SnO_2. Calculate the percent composition by mass of Sn and O in SnO_2.

0.48 What are the empirical formulas of the compounds with the following compositions: (a) 40.1 percent C, 6.6 percent H, and 53.3 percent O; (b) 18.4 percent C, 21.5 percent N, and 60.1 percent K?

0.49 Monosodium glutamate (MSG), a food-flavor enhancer, has been blamed for "Chinese restaurant syndrome," the symptoms of which are headaches and chest pains. MSG has the following composition by mass: 35.51 percent C, 4.77 percent H, 37.85 percent O, 8.29 percent N, and 13.60 percent Na. What is its molecular formula if its molar mass is about 169 g?

0.50 Allicin is the compound responsible for the characteristic smell of garlic. An analysis of the compound gives the following percent composition by mass: C, 44.4 percent; H, 6.21 percent; S, 39.5 percent; and O, 9.86 percent. Calculate its empirical formula. What is its molecular formula given that its molar mass is about 162 g?

0.51   Cinnamic alcohol is used in perfumery, particularly in soaps and cosmetics. Its molecular formula is $C_9H_{10}O$. (a) Calculate the percent composition by mass of C, H, and O in cinnamic alcohol. (b) How many molecules of cinnamic alcohol are contained in a sample of mass 0.469 g?

0.52   Hydrogen has two stable isotopes, $_1^1H$ and $_1^2H$, and sulfur has four stable isotopes, $_{16}^{32}S$, $_{16}^{33}S$, $_{16}^{34}S$, and $_{16}^{36}S$. How many peaks would you observe in the mass spectrum of the positive ion of hydrogen sulfide (H_2S)? Assume no decomposition of the ion into smaller fragments.

Stoichiometry Is the Quantitative Study of Mass and Mole Relationships in Chemical Reactions

0.53   How many grams of sulfur (S) are needed to react completely with 246 g of mercury (Hg) to form HgS?

0.54   Calculate the mass in grams of iodine (I_2) that will react completely with 20.4 g of aluminum (Al) to form aluminum iodide (AlI_3).

0.55   Each copper(II) sulfate unit is associated with five water molecules in crystalline copper(II) sulfate pentahydrate ($CuSO_4 \cdot 5H_2O$). When this compound is heated in air above 100°C, it loses the water molecules and also its blue color:

$$CuSO_4 \cdot 5H_2O(s) \longrightarrow CuSO_4(s) + 5H_2O(l)$$

If 9.60 g of $CuSO_4$ are left after heating 15.01 g of the blue compound, calculate the number of moles of H_2O originally present in the compound.

0.56   Dinitrogen oxide (N_2O) is also called "laughing gas." It can be prepared by the thermal decomposition of ammonium nitrate (NH_4NO_3). The other product is H_2O. (a) Write a balanced equation for this reaction. (b) How many grams of N_2O are formed if 0.46 mol of NH_4NO_3 is used in the reaction?

0.57   The fertilizer ammonium sulfate [$(NH_4)_2SO_4$] is prepared by the reaction between ammonia (NH_3) and sulfuric acid:

$$2NH_3(g) + H_2SO_4(aq) \longrightarrow (NH_4)_2SO_4(aq)$$

How many kilograms of NH_3 are needed to produce 1.00×10^5 kg of $(NH_4)_2SO_4$?

0.58   The depletion of ozone (O_3) in the stratosphere has been a matter of great concern among scientists in recent years. Although chlorofluorocarbons drew the most attention to this problem, ozone also reacts with nitric oxide (NO) that is discharged from high altitude supersonic jet planes. The reaction is

$$O_3 + NO \longrightarrow O_2 + NO_2$$

If 0.740 g of O_3 reacts with 0.670 g of NO, how many grams of NO_2 will be produced? Which compound is the limiting reagent? Calculate the number of moles of the excess reagent remaining at the end of the reaction.

0.59   Consider the reaction

$$MnO_2 + 4HCl \longrightarrow MnCl_2 + Cl_2 + 2H_2O$$

If 0.86 mol of MnO_2 and 48.2 g of HCl react, which reagent will be used up first? How many grams of Cl_2 will be produced?

0.60   Hydrogen fluoride is used in the manufacture of Freons (refrigerants that are known to destroy ozone in the stratosphere) and in the production of aluminum metal from aluminum ore. It is prepared by the reaction

$$CaF_2 + H_2SO_4 \longrightarrow CaSO_4 + 2HF$$

In one process 6.00 kg of CaF_2 are treated with an excess of H_2SO_4 and yield 2.86 kg of HF. Calculate the percent yield of HF.

0.61   Nitroglycerine ($C_3H_5N_3O_9$) is a powerful explosive. Its decomposition may be represented by

$$4C_3H_5N_3O_9 \longrightarrow 6N_2 + 12CO_2 + 10H_2O + O_2$$

This reaction generates a large amount of energy in the form of heat and many gaseous products. It is the sudden formation of these gases, together with their rapid expansion, that produces the explosion. (a) What is the maximum amount of O_2 in grams that can be obtained from 2.00×10^2 g of nitroglycerine? (b) Calculate the percent yield in this reaction if the amount of O_2 generated is found to be 6.55 g.

0.62   Sulfur, when burned in the presence of oxygen, produces a mixture of SO_2 and SO_3 gases. The average molar mass of the product gas mixture in a particular experiment was found to be 68.07 g. What fraction (in percent) of the product in this experiment is SO_2?

0.63   A common laboratory preparation of oxygen gas is the thermal decomposition of potassium chlorate ($KClO_3$). Assuming complete decomposition, calculate the number of grams of O_2 that can be obtained from 46.0 g of $KClO_3$. (The products are KCl and O_2).

Additional Problems

0.64   According to Newton, the potential energy of gravitational attraction between two bodies of masses m_1 and m_2 separated by a distance r is given by

$$V(r) = -\frac{Gm_1 m_2}{r}$$

where $G = 6.67300 \times 10^{-11}$ m^3 kg^{-1} s^{-2} is the universal gravitational constant. (a) Calculate the ratio of the gravitational potential energy to the electrostatic potential energy (Coulomb's law) between a proton and an electron separated by a distance of 100 pm. What does this ratio tell you about the relative importance of electrostatic and gravitational interactions in chemistry? (b) Calculate the work done in pulling an electron and proton, initially separated by 50 pm, to infinite separation.

0.65 A sample of a compound composed of Cl and O reacts with an excess of H$_2$ to give 0.233 g of HCl and 0.403 g of H$_2$O. Determine the empirical formula of the compound.

0.66 The following diagram represents the products (CO$_2$ and H$_2$O) formed after the combustion of a hydrocarbon (a compound containing only C and H atoms). Write an equation for the reaction. (*Hint:* The molar mass of the hydrocarbon is about 30 g.)

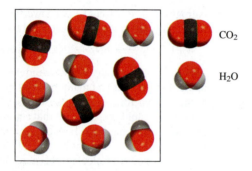

CO$_2$

H$_2$O

0.67 Consider the reaction of hydrogen gas with oxygen gas:

$$2H_2(g) + O_2(g) \longrightarrow 2H_2O(g)$$

H$_2$

O$_2$

H$_2$O

Assuming complete reaction, which of the diagrams shown next represents the amounts of reactants and products left after the reaction?

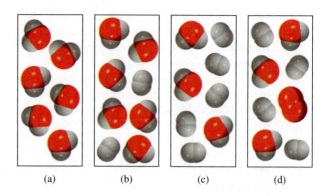

(a)   (b)   (c)   (d)

0.68 Industrially, nitric acid is produced by the Ostwald process, represented by the following equations:

$$4NH_3(g) + 5O_2(g) \longrightarrow 4NO(g) + 6H_2O(l)$$
$$2NO(g) + O_2(g) \longrightarrow 2NO_2(g)$$
$$2NO_2(g) + H_2O(l) \longrightarrow HNO_3(aq) + HNO_2(aq)$$

What mass of NH$_3$ (in g) must be used to produce 1.00 ton of HNO$_3$ by the above procedure, assuming an 80 percent yield in each step? (1 ton = 2000 lb; 1 lb = 453.6 g)

0.69 What is wrong with the name (in parentheses) for each of the following compounds: (a) BaCl$_2$ (barium dichloride), (b) Fe$_2$O$_3$ [iron(II) oxide], (c) CsNO$_2$ (cesium nitrate), and (d) Mg(HCO$_3$)$_2$ [magnesium(II) bicarbonate]?

0.70 What is wrong with the chemical formula for each of the following compounds: (a) (NH$_3$)$_2$CO$_3$ (ammonium carbonate), (b) CaOH (calcium hydroxide), (c) CdSO$_3$ (cadmium sulfide), and (d) ZnCrO$_4$ (zinc dichromate)?

0.71 A common mineral of barium is barytes, or barium sulfate (BaSO$_4$). Because elements in the same periodic group have similar chemical properties, we might expect to find some radium sulfate (RaSO$_4$) mixed with barytes because radium is the last member of Group 2A. However, the only sources of radium compounds in nature are uranium minerals. Why?

0.72 The atomic mass of element X is 33.42 u. A 27.22-g sample of X combines with 84.10 g of another element Y to form a compound XY. Calculate the atomic mass of Y.

0.73 The aluminum sulfate hydrate [Al$_2$(SO$_4$)$_3$ · xH$_2$O] contains 8.20 percent Al by mass. Calculate x, that is, the number of water molecules associated with each Al$_2$(SO$_4$)$_3$ unit.

0.74 Mustard gas (C$_4$H$_8$Cl$_2$S) is a poisonous gas that was used in World War I and banned afterward. It causes general destruction of body tissues, resulting in the formation of large water blisters. There is no effective antidote. Calculate the percent composition by mass of the elements in mustard gas.

0.75 The carat is the unit of mass used by jewelers. One carat is exactly 200 mg. How many carbon atoms are present in a 24-carat diamond?

0.76 An iron bar weighed 664 g. After the bar had been standing in moist air for a month, exactly one-eighth of the iron turned to rust (Fe_2O3). Calculate the final mass of the iron bar and rust.

0.77 In the thermite reaction, aluminum powder is reacted with iron(III) oxide to form elemental iron and aluminum oxide. The reaction produces an extremely large amount of energy in the form of heat, making it useful both as a dramatic chemistry lecture demonstration and in the welding of railroad tracks. If 20.0 g of Al powder are reacted with 50.0 g of iron(III) oxide, what mass (in grams) of Fe will be produced, assuming that at least one of the reactants is completely used up.

0.78 An impure sample of zinc (Zn) is treated with an excess of sulfuric acid to form zinc sulfate and molecular hydrogen. (a) Write a balanced equation for the reaction. (b) If 0.0764 g of H_2 is obtained from 3.86 g of the sample, calculate the percent purity of the sample. (c) What assumptions must you make in (b)?

0.79 A certain metal oxide has the formula MO where M denotes the metal. A 39.46-g sample of the compound is strongly heated in an atmosphere of hydrogen to remove oxygen as water molecules. At the end, 31.70 g of the metal are left over. Calculate the atomic mass of M and identify the element.

0.80 One of the reactions that occurs in a blast furnace, where iron ore is converted to cast iron, is

$$Fe_2O_3 + 3CO \longrightarrow 2Fe + 3CO_2$$

Suppose that 1.64×10^3 kg of Fe are obtained from a 2.62×10^3-kg sample of Fe_2O_3. Assuming that the reaction goes to completion, what is the percent purity of Fe_2O_3 in the original sample?

0.81 Carbon dioxide (CO_2) is the gas that is mainly responsible for global warming. The burning of fossil fuels is a major cause of the increased concentration of CO_2 in the atmosphere. Carbon dioxide is also the end product of the metabolism of food. Using glucose ($C_6H_{12}O_6$) as an example of food, calculate the annual human production of CO_2 in grams, assuming that each person consumes 5.0×10^2 g of glucose per day. The world's population is 6.5 billion. In the metabolism of glucose, glucose reacts with oxygen to produce carbon dioxide and water.

0.82 Hemoglobin ($C_{2952}H_{4664}N_{812}O_{832}S_8Fe_4$) is the oxygen carrier in blood. (a) Calculate its molar mass. (b) An average adult has about 5.0 L of blood. Every milliliter of blood has approximately 5.0×10^9 erythrocytes, or red blood cells, and every red blood cell has about 2.8×10^8 hemoglobin molecules. Calculate the mass of hemoglobin molecules in grams in an average adult.

0.83 Myoglobin stores oxygen for metabolic processes in muscle. Chemical analysis shows that it contains 0.34 percent Fe by mass. What is the molar mass of myoglobin? (There is one Fe atom per molecule.)

0.84 A sample containing NaCl, Na_2SO_4, and $NaNO_3$ gives the following elemental analysis: Na: 32.08 percent, O: 36.01 percent, and Cl: 19.51 percent. Calculate the mass percent of each compound in the sample.

0.85 Calculate the percent composition by mass of all the elements in calcium phosphate [$Ca_3(PO_4)_2$], a major component of bone.

0.86 For molecules having small molecular masses, mass spectrometry can be used to identify their formulas. To illustrate this point, identify the molecule that most likely accounts for the observation of a peak in a mass spectrum at 16 u, 17 u, 18 u, and 64 u. (b) Two molecules that could give rise to a peak at 44 u are C_3H_8 and CO_2. In such cases, a chemist might try to look for other peaks generated when some of the molecules break apart in the spectrometer. For example, if a chemist sees a peak at 44 u and also one at 15 u, which molecule is producing the 44-u peak? Why? (c) Using the following precise atomic masses: ^1H (1.00797 u), ^{12}C (12.00000 u), and ^{16}O (15.99491 u), how precisely must the masses of C_3H_8 and CO_2 be measured to distinguish between them?

0.87 Lysine, an essential amino acid in the human body, contains C, H, O, and N. In one experiment, the complete combustion of 2.175 g of lysine gave 3.94 g CO_2 and 1.89 g H_2O. In a separate experiment, 1.873 g of lysine gave 0.436 g NH_3. (a) Calculate the empirical formula of lysine. (b) The approximate molar mass of lysine is 150 g. What is the molecular formula of the compound?

0.88 The natural abundances of the two stable isotopes of hydrogen (hydrogen and deuterium) are ^1_1H: 99.985 percent and ^2_1H: 0.015 percent. Assume that water exists as either H_2O or D_2O. Calculate the number of D_2O molecules in exactly 400 mL of water. (Density = 1.00 g mL^{-1}.)

0.89 A compound containing only C, H, and Cl was examined in a mass spectrometer. The highest mass peak seen corresponds to an ion mass of 52 u. The most abundant mass peak seen corresponds to an ion mass of 50 u and is about three times as intense as the peak at 52 u. Deduce a reasonable molecular formula for the compound and explain the positions and intensities of the mass peaks mentioned. (*Hint:* Chlorine is the only element that has isotopes in comparable abundances: $^{35}_{17}Cl$: 75.5 percent; $^{37}_{17}Cl$: 24.5 percent. For H, use ^1_1H; for C, use $^{12}_6C$.)

0.90 In the formation of carbon monoxide, CO, it is found that 2.445 g of carbon combine with 3.257 g of oxygen. What is the atomic mass of oxygen if the atomic mass of carbon is 12.01 u?

0.91 What mole ratio of molecular chlorine (Cl_2) to molecular oxygen (O_2) would result from the breakup of the compound Cl_2O_7 into its constituent elements?

0.92 Which of the following substances contains the greatest mass of chlorine: (a) 5.0 g Cl_2, (b) 60.0 g $NaClO_3$, (c) 0.10 mol KCl, (d) 30.0 g $MgCl_2$, or (e) 0.50 mol Cl_2?

0.93 Platinum forms two different compounds with chlorine. One contains 26.7 percent Cl by mass, and the other contains 42.1 percent Cl by mass. Determine the empirical formulas of the two compounds.

0.94 Heating 2.40 g of the oxide of metal X (molar mass of X = 55.9 g mol^{-1}) in carbon monoxide (CO) yields the pure metal and carbon dioxide. The mass of the metal product is 1.68 g. From the data given, show that the simplest formula of the oxide is X_2O_3, and write a balanced equation for the reaction.

0.95 When 0.273 g of Mg is heated strongly in a nitrogen (N_2) atmosphere, a chemical reaction occurs. The product of the reaction weighs 0.378 g. Determine the empirical formula of the compound containing Mg and N. Name the compound.

0.96 A mixture of methane (CH_4) and ethane (C_2H_6) of mass 13.43 g is completely burned in oxygen. If the total mass of CO_2 and H_2O produced is 64.84 g, calculate the fraction of methane in the mixture.

0.97 Potash is any potassium mineral that is used for its potassium content. Most of the potash produced in the United States goes into fertilizer. The major sources of potash are potassium chloride (KCl) and potassium sulfate (K_2SO_4). Potash production is often reported as the potassium oxide (K_2O) equivalent or the amount of K_2O that could be made from a given mineral. (a) If KCl costs \$0.55 per kg, for what price (dollar per kg) must K_2SO_4 be sold to supply the same amount of potassium on a per dollar basis? (b) What mass (in kg) of K_2O contains the same number of moles of K atoms as 1.00 kg of KCl?

0.98 A compound X contains 63.3 percent manganese (Mn) and 36.7 percent O by mass. When X is heated, oxygen gas is evolved and a new compound Y containing 72.0 percent Mn and 28.0 percent O is formed. (a) Determine the empirical formulas of X and Y. (b) Write a balanced equation for the conversion of X to Y.

0.99 The formula of a hydrate of barium chloride is $BaCl_2 \cdot xH_2O$. If 1.936 g of the compound give 1.864 g of anhydrous $BaSO_4$ upon treatment with sulfuric acid, calculate the value of x.

0.100 It is estimated that the day Mt. St. Helens erupted (May 18, 1980), about 4.0×10^5 tons of SO_2 were released into the atmosphere. If all the SO_2 were eventually converted to sulfuric acid, how many tons of H_2SO_4 were produced?

0.101 A mixture of $CuSO_4 \cdot 5H_2O$ and $MgSO_4 \cdot 7H_2O$ is heated until all the water is lost. If 5.020 g of the mixture give 2.988 g of the anhydrous salts, what is the percent by mass of $CuSO_4 \cdot 5H_2O$ in the mixture?

0.102 Leaded gasoline (available in the United States until the mid 1980s) contains an additive to prevent engine "knocking." On analysis, the additive compound is found to contain carbon, hydrogen, and lead (Pb) (hence, "leaded gasoline"). When 51.36 g of this compound are burned, 55.90 g of CO_2 and 28.61 g of H_2O are produced. Determine the empirical formula of the gasoline additive.

0.103 Because of its detrimental effect on the environment, the lead compound described in Problem 0.102 has been replaced in recent years by methyl *tert*-butyl ether (a compound of C, H, and O) to enhance the performance of gasoline. (As of 1999, this compound is also being phased out because of its contamination of drinking water.) When 12.1 g of the compound are burned, 30.2 g of CO_2 and 14.8 g of H_2O are formed. What is the empirical formula of the compound?

0.104 A sample of coal contains 1.6 percent sulfur by mass. When the coal is burned, the sulfur is converted to sulfur dioxide. To prevent air pollution, this sulfur dioxide is treated with calcium oxide (CaO) to form calcium sulfite ($CaSO_3$). Calculate the daily mass (in kilograms) of CaO needed by a power plant that uses 6.60×10^6 kg of coal per day.

0.105 Air is a mixture of many gases. However, in calculating its "molar mass" we need consider only the three major components: nitrogen, oxygen, and argon. Given that 1 mol of air at sea level is made up of 78.08 percent nitrogen, 20.95 percent oxygen, and 0.97 percent argon, what is the molar mass of air?

0.106 A die has an edge length of 1.5 cm. (a) What is the volume of 1 mol of such dice? (b) Assuming that the mole of dice could be packed in such a way that they were in contact with one another, forming stacking layers covering the entire surface of Earth, calculate the height in meters the layers would extend outward. [The radius (r) of Earth is 6371 km and the surface area of a sphere is $4\pi r^3/3$.]

0.107 The following is a crude but effective method for estimating the *order of magnitude* of Avogadro's number using stearic acid ($C_{18}H_{36}O_2$). When stearic acid is added to water, its molecules collect at the surface and form a monolayer; that is, the layer is only one molecule thick. The cross-sectional area of each stearic acid molecule has been measured to be 0.21 nm^2. In one experiment, it is found that 1.4×10^4 g of stearic acid are needed to form a monolayer over water in a dish of diameter 20 cm. Based on these measurements, what is Avogadro's number? (The area of a circle of radius r is πr^2.)

0.108 Suppose you are given a cube made of magnesium (Mg) metal of edge length 1.0 cm. (a) Calculate the number of Mg atoms in the cube. (b) Atoms are spherical; therefore, the Mg atoms in the cube cannot fill all of the available space. If only 74 percent of the space inside the cube is taken up by Mg atoms, calculate the radius in picometers of a Mg atom. (The density of Mg is 1.74 g cm^{-3} and the volume of a sphere of radius r is $4\pi r^3/3$.)

0.109 Octane (C_8H_{18}) is a component of gasoline. Complete combustion of octane yields H_2O and CO_2. Incomplete combustion produces H_2O and CO, which not only reduces the efficiency of the engine using the fuel, but is also toxic. In a certain test run, 1.000 gal of octane is burned in an engine. The total mass of CO, CO_2, and H_2O produced is 11.53 kg. Calculate the efficiency of the process; that is, calculate the fraction of octane converted to CO_2. The density of octane is 0.7001 kg L^{-1} and 1 gallon = 3.7854 L.

0.110 Industrially, hydrogen gas can be prepared by reacting propane gas (C_3H_8) with steam at about 400°C. The products are carbon monoxide (CO) and hydrogen gas (H_2). (a) Write a balanced equation for the reaction. (b) How many kilograms of H_2 can be obtained from 2.84×10^3 kg of propane?

0.111 A reaction having a 90 percent yield may be considered a successful experiment. However, in the synthesis of complex molecules such as chlorophyll and many anticancer drugs, a chemist often has to carry out multiple-step synthesis. What is the overall percent yield for such a synthesis, assuming it is a 30-step reaction with a 90 percent yield at each step?

0.112 You are given a liquid. Briefly describe steps you would take to show whether it is a pure substance or a homogeneous mixture.

0.113 Ethane and acetylene are two gaseous hydrocarbons. Chemical analyses show that in one sample of ethane, 2.65 g of carbon are combined with 0.334 g of hydrogen and in one sample of acetylene, 4.56 g of carbon are combined with 0.191 g of hydrogen. (a) Are these results consistent with the law of multiple proportions? (b) Write reasonable molecular formulas for these compounds.

0.114 On p. 12 it was pointed out that mass and energy are alternate aspects of a single entity called *mass-energy*. The relationship between these two physical quantities is Einstein's famous equation, $E = mc^2$, where E is energy, m is mass, and c is the speed of light. In a combustion experiment, it was found that 12.096 g of hydrogen molecules combined with 96.000 g of oxygen molecules to form water and released 1.715×10^3 kJ of energy as heat. Calculate the corresponding mass change in this process and comment on whether the law of conservation of mass holds for ordinary chemical processes. (*Hint*: The Einstein equation can be used to calculate the change in mass as a result of the change in energy. 1 J = 1 kg m^2 s^{-2} and c = 2.9979×10^8 m s^{-1}.)

0.115 A monatomic ion has a charge of +2. The nucleus of the parent atom has a mass number of 55. If the number of neutrons in the nucleus is 1.2 times that of the number of protons, what are the name and symbol of the element?

0.116 Name the following acids:

0.117 A mixture of NaBr and Na_2SO_4 contains 29.96 percent Na by mass. Calculate the percent by mass of each compound in the mixture.

0.118 Aspirin, or acetyl salicylic acid, is synthesized by reacting salicylic acid with acetic anhydride:

$$C_7H_6O_3 + C_4H_6O_3 \longrightarrow C_9H_8O_4 + C_2H_4O_2$$

salicylic   acetic   aspirin   acetic
acid   anhydride   acid

(a) How much salicylic acid is required to produce 0.400 g of aspirin (about the content in a tablet), assuming acetic anhydride is present in excess? (b) Calculate the amount of salicylic acid needed if only 74.9 percent of salicylic acid is converted to aspirin. (c) In one experiment, 9.26 g of salicylic acid are reacted with 8.54 g of acetic anhydride. Calculate the theoretical yield of aspirin and the percent yield if only 10.9 g of aspirin are produced.

0.119 A compound made up of C, H, and Cl contains 55.0 percent Cl by mass. If 9.00 g of the compound contain 4.19×10^{23} H atoms, what is the empirical formula of the compound?

0.120 A 21.496-g sample of magnesium is burned in air to form magnesium oxide and magnesium nitride. When the products are treated with water, 2.813 g of gaseous ammonia are generated. Calculate the amounts of magnesium nitride and magnesium oxide formed.

0.121 A certain metal M forms a bromide containing 53.79 percent Br by mass. What is the chemical formula of the compound?

0.122 A sample of iron weighing 15.0 g was heated with potassium chlorate ($KClO_3$) in an evacuated container. The oxygen generated from the decomposition of $KClO_3$ converted some of the Fe to Fe_2O_3. If the combined mass of Fe and Fe_2O_3 was 17.9 g, calculate the mass of Fe_2O_3 formed and the mass of $KClO_3$ decomposed.

0.123 A 15.0-mL sample of an oxalic acid ($H_2C_2O_4$) solution requires 25.2 mL of 0.149 M NaOH for neutralization according to the reaction

$$H_2C_2O_4 + 2NaOH \longrightarrow 2H_2O + Na_2C_2O_4$$

Calculate the volume of a 0.122 M $KMnO_4$ solution needed to react with a second 15.0-mL sample of this oxalic acid solution. The balanced equation for this reaction is

$$2MnO_4^- + 16H^+ + 5C_2O_4^{2-} \longrightarrow$$
$$2Mn^{2+} + 10CO_2 + 8H_2O$$

Answers to Practice Exercises

0.1 6.5×10^5 J **0.2** 20.4 m **0.3** (a) lead oxide, (b) lithium sulfite, (c) iron(III) oxide **0.4** (a) Rb_2SO_4, (b) CrO_3, (c) BaH_2 **0.5** (a) nitrogen trifluoride, (b) dichlorine heptoxide **0.6** (a) SF_4, (b) N_2O_5 **0.7** $Fe_2O_3 + 3CO \longrightarrow 2Fe + 3CO_2$ **0.8** 10.81 u **0.9** The relative abundances of ^{69}Ga and ^{71}Ga are 60.26% and 39.74%, respectively. **0.10** 3.59 mol **0.11** 2.57×10^3 g **0.12** 8.49×10^{21} atoms **0.13** (a) 32.04 u, (b) 180.15 u **0.14** (a) 1.66 mol, (b) 2.16×10^{24} molecules **0.15** H: 2.055%; S: 32.69%; O: 65.25% **0.16** $KMnO_4$ (potassium permanganate) **0.17** B_2H_6 **0.18** 7.44% **0.19** 0.638 m, 0.0113 **0.20** 0.452 M **0.21** 8.92 m **0.22** 13.8 m **0.23** (a) 0.508 mol, (b) 2.21 g **0.24** The mole fractions for C_6H_6 and C_6H_{14} are 0.599 and 0.401, respectively. The corresponding mass fractions are 0.576 and 0.424, respectively. **0.25** (a) 234 g, (b) 234 g **0.26** (a) 863 g, (b) 93.0%

1 Chapter

The Quantum Theory of the Submicroscopic World

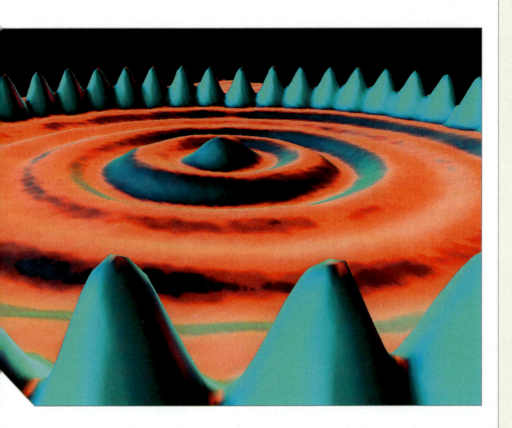

Early attempts by nineteenth-century physicists to understand atoms and molecules met with only limited success. By assuming that molecules behave like rebounding balls, physicists were able to predict and explain some familiar phenomena, such as the pressure exerted by a gas. However, this model could not account for a number of newly discovered phenomena, such as the photoelectric effect or the emission spectra of atoms. The early twentieth century brought the development of quantum mechanics and the realization that the behavior of atoms and molecules cannot be described by the physical laws that work so well for everyday objects.

1.1 | Classical Physics Does Not Adequately Describe the Interaction of Light with Matter

The motion of objects in the everyday world can be well described by the laws of *classical physics*—a description of the physical world that began with the development of the laws of motion by Isaac Newton. These laws were enormously successful as a unifying principle in physics until the end of the nineteenth century.

Science at the End of the Nineteenth Century: The Classical Model

The principal assumptions of classical physics can be summarized as follows:

1. The physical state of any system can be described by a set of quantities called *dynamical variables* that take on well-defined values at any instant of time.

2. The future state of any system is completely determined if the initial state of the system is known.

3. The energy of a system can be varied in a continuous manner over the allowed range.

If these assumptions seem quite reasonable, even obvious, to us, it is because they are consistent with our observations of the world in which we live. Up until 1900, the classical model was successful in describing accurately the motion of known objects up to the planetary scale, so it was widely assumed that this success would extend to the newly discovered submicroscopic world of atoms and molecules. However, as we will see, the classical model was unable to account for a number of experimental results that probed the nature of atoms and molecules and their interaction with light. This led to the discovery of *quantum mechanics,*[1] which represented a fundamental change in our understanding of the way nature works at the atomic scale.

Classical physics just before 1900 consisted of two major theoretical frameworks. The first was *classical mechanics,* which seeks to explain the motion of matter, and the second was the *wave theory of light,* a description of *electromagnetic radiation.*

Classical Mechanics

Classical mechanics as a unifying principle of physics began with Newton's laws of motion. As discussed in Section 0.1, Newton's second law

$$F = ma \tag{1.1}$$

relates the force (F) on an object to the product of its mass, m, and its acceleration[2] (a). Using Equation 1.1, the motion of an object can be determined if the initial velocity, $u_0 = u(t = 0)$, and initial position, $x_0 = x(t = 0)$, are known.

Consider, for example, the motion of a cannonball of mass m being fired through the air and subject to the gravitational pull of Earth. The variables that describe the

---

1. Quantum mechanics was one of two major discoveries that revolutionized the world of physics at the beginning of the twentieth century. The other was the *theory of relativity,* developed by Albert Einstein, which changed the way scientists viewed the behavior of objects that were extremely fast or extremely massive. Unlike quantum mechanics, relativity was consistent with classical physics. Trying to reconcile the theory of relativity with that of quantum mechanics remains a major scientific challenge today.

2. In this and subsequent chapters, boldface type is used to indicate vector quantities, such as force (F) and acceleration (a).

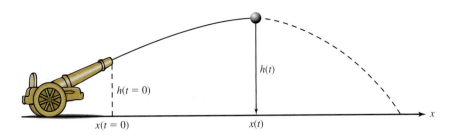

motion of the cannonball are the position and velocity (or momentum) of its center of mass. Using Newton's second law of motion (Equation 1.1), the precise trajectory of the ball (neglecting air friction) can be calculated if you know the initial position and initial velocity (momentum) of the ball. The position of the cannonball is defined by specifying the height above the ground *h* and the forward displacement *x* (see Figure 1.1), both of which are functions of the time *t*. The velocity of the ball is given by the corresponding vertical and horizontal velocity components, u_h and u_x. The gravitational force on the ball is given by *mg*, where *g* is the gravitational acceleration, which at the surface of Earth has the value $g = 9.80665$ m s^{-2}. For this system, Equation 1.1 can be solved as follows to give the trajectory of a cannonball, which is a parabola:

$$h(t) = h_0 + u_{h,0}t - \frac{1}{2}gt^2$$

$$x(t) = x_0 + u_{x,0}t$$

The velocities corresponding to $h(t)$ and $x(t)$ can also be determined:

$$u_h(t) = u_{h,0} - gt$$
$$u_x(t) = u_{x,0}$$

Thus, at any given time the position and velocity of the cannonball can be specified in terms of the initial values, consistent with the first two assumptions of classical physics. The energy of the cannonball is conserved in the motion and can be specified as the sum of the initial kinetic energy and the initial potential energy:

$$E = \left[\frac{1}{2}mu_{x,0}^2 + \frac{1}{2}mu_{h,0}^2\right] + mgh_0$$

(The term in brackets is the total initial kinetic energy, which is the sum of the kinetic energies in the two directions of motion, *x* and *h*.) Because we can independently choose the initial position (by moving the cannon) and initial momentum (by controlling the amount of gunpowder used), the energy can be varied continuously over a wide range—a characteristic of classical theory.

Although a number of generalizations of Newton's laws of motion were developed in the eighteenth and nineteenth centuries, the basic framework of classical mechanics that they represent remained largely unchanged and unchallenged until 1900.

Wave Theory of Light

Parallel to the development of classical mechanics, the seventeenth through nineteenth centuries also saw tremendous progress in the understanding of light. In the seventeenth century, Isaac Newton performed the first quantitative study of the properties of light. He demonstrated that sunlight can be separated by a glass prism into a spectrum of distinct colors and that these colors can be recombined by directing this spectrum

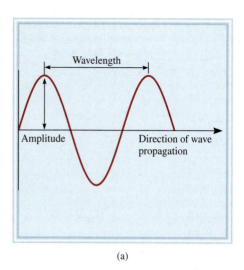

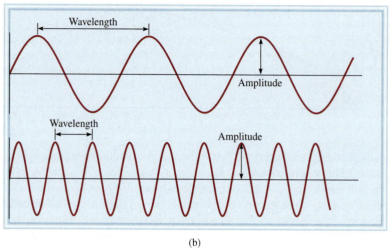

(a)   (b)

Figure 1.2 (a) Wavelength and amplitude. (b) Two waves having different wavelengths and frequencies. The wavelength of the top wave is three times that of the lower wave, but if both waves have the same speed, the frequency of the top wave is only one-third that of the lower wave. Both waves have the same amplitude.

through a second prism, turned opposite to the first, to reproduce white light. Although Newton explained much of this behavior by assuming that light was composed of discrete particles moving in straight lines, further work in the eighteenth and nineteenth centuries established light as a wave phenomenon.

Waves are characterized by their *wavelength, amplitude,* and *frequency* (Figure 1.2). The **wavelength** (λ) is *the distance between identical points on successive waves.* The **amplitude** of a wave is *the vertical distance from the midline of a wave to the peak or trough.* The **frequency** (ν) is the *number of waves that pass through a particular point in one second.* The speed of the wave (u) is the product of its wavelength and its frequency:

$$u = \lambda\nu \tag{1.2}$$

Wavelength is usually expressed in units of meters, centimeters, or nanometers, and frequency is measured in *hertz* (Hz) (after the German physicist Heinrich Rudolf Hertz[3]), where 1 Hz = 1 s^{-1}.

When two (or more) waves interact, interference occurs. To understand this phenomenon, consider the interaction of two waves of equal wavelength, as shown in Figure 1.3. If the waves are *in phase,* that is, the positions the maximum and minimum of wave 1 match those of wave 2, as in Figure 1.3(a), then the two waves will add to give a wave that has twice the amplitude of the original two waves. This is called *constructive interference.* On the other hand, in Figure 1.3(e) the two waves are exactly out of phase, that is, the position of the minimum for wave 1 corresponds to the maximum of wave 2. When these two waves are added together, they exactly cancel one another to give zero. This is called *destructive interference.* If the waves are only partially out of phase, as in Figures 1.3(b)–(d), the waves

3. Heinrich Rudolf Hertz (1857–1894). German physicist. He performed a number of experiments confirming Maxwell's theory of electromagnetic radiation. His discovery of radio waves led to the development of the wireless telegraph and the radio.

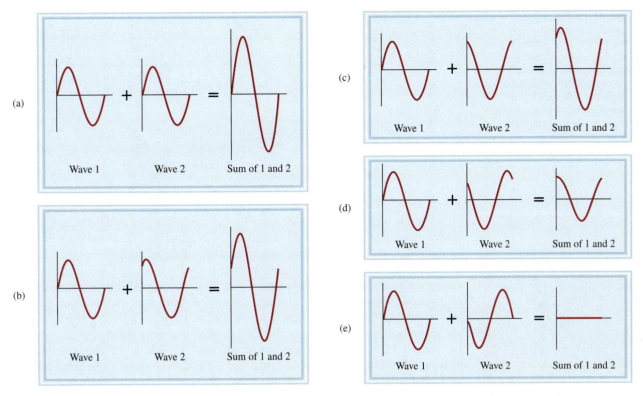

Figure 1.3 Constructive and destructive interference between two waves of equal wavelength and amplitude: (a) two waves completely in phase; (b)–(d) two waves partially out of phase; and (e) two waves exactly out of phase.

will add to give a wave with an amplitude that is intermediate between the two extremes in Figures 1.3(a) and (e). Experimentally, this phenomenon can be observed in a two-slit experiment, such as that shown in Figure 1.4. A light source emitting a single wavelength of light (called *monochromatic* light) is directed at a partition with two openings (slits A and B). These slits are small relative to the distance between them and act as two separate light sources. The two light waves emerging from slits A and B will interfere both constructively and destructively. The resulting interference patterns are observed on a screen as alternating bright and dark regions, respectively.

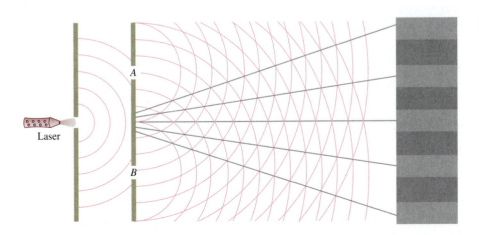

Figure 1.4 Two-slit experiment demonstrating the interference phenomenon. The pattern on the screen consists of alternating bright and dark bands.

Figure 1.5 Electric field component and magnetic field component of an electromagnetic wave. These two components have the same wavelength, frequency, and amplitude, but oscillate in two mutually perpendicular planes. The wave here is traveling along the *x* direction.

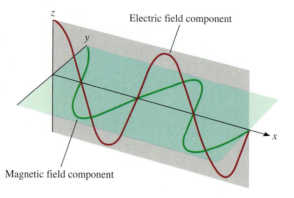

Electric field component

Magnetic field component

James Clerk Maxwell

A major breakthrough in understanding the wave nature of light came in the mid-nineteenth century when James Clerk Maxwell[4] developed the unified theory of the electromagnetic field. This theory, embodied by the Maxwell equations, predicted the existence of electromagnetic waves consisting of an electric-field component and a magnetic-field component oscillating in mutually perpendicular planes, both perpendicular to the direction of travel (see Figure 1.5). The calculated velocity of these waves matched precisely the known speed of light (*c*) in a vacuum (about 3.00×10^8 m s^{-1}), leading Maxwell to be the first to predict that light is just one form of electromagnetic radiation.

Maxwell's theory of electromagnetic radiation fits within the classical doctrine because the electric and magnetic fields (and their rates of change with time) take on well-defined values at all times and the future values of the fields can be predicted with arbitrary precision from their initial state using the Maxwell equations. In Maxwell's theory the energy of the electromagnetic wave depends upon the amplitude of the electromagnetic wave, but not on its frequency, and can be varied continuously.

Visible light was not the only type of radiation described as electromagnetic waves by Maxwell's theory. Figure 1.6 shows various types of electromagnetic radiation, which differ from one another in wavelength and frequency. Long radio waves are those emitted by large antennas, such as those used by broadcasting stations. The motions of electrons within atoms and molecules can produce shorter, visible light waves. The shortest waves, which also have the highest frequency, are associated with γ (or *gamma*)-rays, which result from changes within the nucleus of the atom. As we will discuss shortly, the higher the frequency, the more energetic the radiation, contrary to Maxwell's classical theory. Thus, ultraviolet radiation, X-rays, and γ-rays are high-energy radiation.

Example 1.1

The wavelength of the green light from a traffic signal is centered at 522 nm. What is the frequency of this radiation?

Strategy We are given the wavelength of an electromagnetic wave and asked to calculate its frequency (ν). Rearranging Equation 1.2 and replacing u with c (the speed of light) gives

$$\nu = \frac{c}{\lambda}$$

—Continued

---

4. James Clerk Maxwell (1831–1879). Scottish physicist. Maxwell was one of the great theoretical physicists of the nineteenth century; his work covered many areas of physics, including the kinetic molecular theory of gases, thermodynamics, and electricity and magnetism.

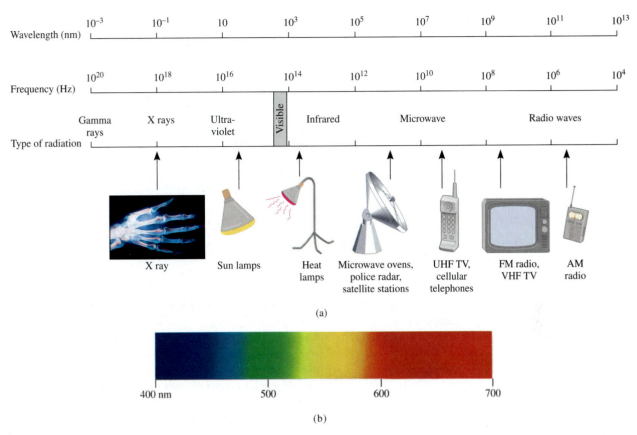

(a)

(b)

Figure 1.6 (a) Types of electromagnetic radiation. Gamma-rays have the shortest wavelength and highest frequency; radio waves have the longest wavelength and the lowest frequency. Each type of radiation is spread over a specific range of wavelengths (and frequencies). (b) Visible light ranges from a wavelength of 400 nm (violet) to 700 nm (red).

Continued—

Solution Because the speed of light is given in meters per second, it is convenient to first convert wavelength to meters. Recall that 1 nm = 1 × 10^{-9} m. We write

$$\nu = \frac{c}{\lambda} = \frac{3.00 \times 10^8 \text{ m s}^{-1}}{522 \text{ nm}} \times \frac{1 \text{ nm}}{1 \times 10^{-9} \text{ m}} = 5.75 \times 10^{14} \text{ s}^{-1} = 5.75 \times 10^{14} \text{ Hz}$$

Practice Exercise The broadcast frequency of a certain radio station is 91.5 MHz. What is the wavelength of these radio waves?

Blackbody Radiation: The Failure of Classical Theory

Any object will radiate energy in the form of electromagnetic radiation purely as a consequence of its temperature. The red glow of an electric heater and the bright white light of the tungsten filament in an incandescent light bulb are familiar examples. This radiation is referred to as **blackbody radiation.**[5] The physical properties

The red glow of an electric heater is an example of blackbody radiation.

5. A *blackbody* is an idealized object that absorbs 100 percent of the radiation that is incident upon it. No real object is a perfect blackbody.

Figure 1.7 The intensity of blackbody radiation as a function of wavelength at various temperatures.

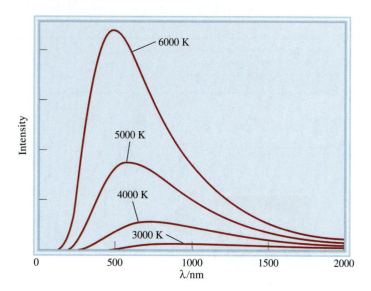

of blackbody radiation depend only on the temperature of the object, not on its composition.

If we measured the intensity of blackbody radiation versus the wavelength emitted at different temperatures, we would obtain a series of curves similar to the ones shown in Figure 1.7. Experiments at the end of the nineteenth century by Josef Stefan[6] and Wilhelm Wien[7] led to two important empirical laws of blackbody radiation, now named the Stefan-Boltzmann law and Wien's law. (An *empirical law* is one that is formulated purely on the basis of experimental data.)

▶ *Stefan-Boltzmann law:* The total intensity of blackbody radiation emitted by an object (obtained by integrating the curves in Figure 1.7 over all wavelengths) is proportional to the fourth power of the absolute temperature (that is, the temperature in kelvins, see Appendix 1):

$$\frac{\text{emitted power}}{\text{surface area of object}} = \sigma T^4$$

where $\sigma = 5.670 \times 10^{-8}$ W m^{-2} K^{-4} is the Stefan-Boltzmann constant and T is the absolute temperature. The law was used by Stefan to estimate the surface temperature of the sun. (Ludwig Boltzmann[8] is associated with this law because he was able to derive it using thermodynamic arguments five years after Stefan's experiments.)

▶ *Wien's law:* The wavelength of maximum intensity (λ_{max}) is inversely proportional to the absolute temperature:

$$T\lambda_{max} = \text{constant} = 1.44 \times 10^{-2} \text{ K m}$$

6. Josef Stefan (1835–1893). Austrian physicist. In addition to his quantitative experiments on blackbody radiation, he made important contributions to the kinetic theory of heat and to the theory of heat conduction in fluids.

7. Wilhelm Wien (1864–1928). German physicist. He received the Nobel Prize in Physics in 1911 for his work on blackbody radiation and also made important contributions to hydrodynamics and radiation theory.

8. Ludwig Boltzmann (1844–1906). Austrian physicist. Although Boltzmann was one of the greatest theoretical physicists of all time, his work was not recognized by other scientists in his lifetime. Suffering from poor health and great depression, he committed suicide in 1906.

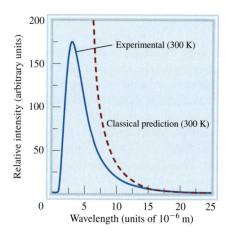

These experimental results caused a sensation among scientists at the time because they could not be explained using classical theory. The classical electromagnetic theory of light developed by Maxwell predicted that the energy of a light wave is a function only of the amplitude of the wave and does not depend on its wavelength (or frequency); therefore, the energy emitted by the object should be distributed equally among all possible electromagnetic waves without regard to frequency. However, for electromagnetic waves in three dimensions, there are many more oscillation modes of electromagnetic radiation possible at high frequency than exist at low frequency;[9] therefore, classical theory predicts that the intensity of blackbody radiation should increase with increasing frequency (decreasing wavelength). As a result, the classical model predicts that even objects at room temperature will emit high intensity ultraviolet, X-ray, and even γ-ray radiation—in effect, all objects in the universe should be infinitely bright! (This erroneous prediction is referred to as the *ultraviolet catastrophe*.) Comparing (Figure 1.8) the classical intensity with that observed experimentally for an object shows that the classical model works well at long wavelengths (low frequencies), but it drastically overestimates the intensity of high frequency electromagnetic waves in the blackbody radiation spectrum. Something fundamental was missing from the laws of classical physics!

Planck's Quantum Hypothesis

In 1900, Max Planck[10] solved the problem of blackbody radiation with an assumption that departed drastically from accepted concepts. Classical physics assumed that atoms and molecules could emit (or absorb) any arbitrary amount of radiant energy. Planck hypothesized that atoms and molecules could emit (or absorb) energy only in discrete quantities, like small packages or bundles. Planck gave the name *quantum* to *the smallest quantity of energy that can be emitted (or absorbed) in the form of electromagnetic radiation*. The energy E of a single quantum of electromagnetic energy is given by

$$E = h\nu \tag{1.3}$$

where h is called *Planck's constant* and ν is the frequency of the radiation. The value of Planck's constant has been determined experimentally to be 6.62608×10^{-34} J s.

Max Planck

---

9. For much the same reason that there are many more ways to write a large integer as the sum of two positive integers than there are ways to represent a smaller integer in the same manner.

10. Max Karl Ernest Ludwig Planck (1858–1947). German physicist. Planck received the Nobel Prize in Physics in 1918 for his quantum theory. He also made significant contributions in thermodynamics and other areas of physics.

According to quantum theory, the energy of electromagnetic radiation of frequency ν is always emitted in multiples of $h\nu$; for example, $h\nu, 2h\nu, 3h\nu, \ldots$ but never, for example, $1.67h\nu$ or $4.98h\nu$. At the time Planck presented his theory, he could not explain why energies should be fixed or quantized in this manner. Starting with this hypothesis, however, he had no trouble correlating the experimental data for the emission by solids over the *entire* range of wavelengths; they all supported the quantum theory.

The idea that energy should be quantized or "bundled" may seem strange, but the concept of quantization has many analogies. For example, an electric charge is quantized; there can be only whole-number multiples of e, the charge of one electron. Matter itself is quantized, because the numbers of electrons, protons, and neutrons and the numbers of atoms in a sample of matter must also be integers.

How does the Planck model explain the decrease in the intensity of blackbody radiation at high frequency (short wavelength)? At any given temperature, there is only a fixed average amount of thermal energy that is available to excite a given electromagnetic oscillation (light wave). In the classical model, where you can put an arbitrary amount of energy into any oscillation, the energy can be distributed evenly among the oscillations, regardless of frequency. In the Planck model, however, there is a minimum amount of energy that can be transferred into an electromagnetic oscillation from the object and this minimum energy (the quantum) increases with increasing frequency. For low-frequency electromagnetic waves, the quantum of energy is much smaller than the average amount of thermal energy available for the excitation of that electromagnetic wave; therefore, this energy can be evenly distributed among these oscillation modes, as in the classical model. For high frequencies, however, the quantum of energy is greater than the average available thermal energy and excitation into high frequency modes is inhibited.

Chlorophyll-a

Example 1.2

Chlorophyll-a is green because it absorbs blue light at about 435 nm and red light at about 680 nm, so that mostly green light is transmitted. Calculate the energy per mole of photons at these wavelengths.

Strategy Planck's equation (Equation 1.3) gives the relationship between energy and *frequency* (ν). Because we are given wavelength (λ), we must use Equation 1.2, in which u is replaced with c (the speed of light), to convert wavelength to frequency. Finally, the problem asks for the energy per mole, so we must multiply the result we get from Equation 1.3 by Avogadro's number.

Solution The energy of one photon with a wavelength of 435 nm is

$$E = h\nu = h\left(\frac{c}{\lambda}\right) = (6.626 \times 10^{-34} \text{ J s})\frac{3.00 \times 10^8 \text{ m s}^{-1}}{435 \text{ nm } (1 \times 10^{-9} \text{ m nm}^{-1})}$$

$$= 4.57 \times 10^{-19} \text{ J}$$

For one mole of photons, we have

$$E = (4.57 \times 10^{-19} \text{ J})(6.022 \times 10^{23} \text{ mol}^{-1})$$

$$= 2.75 \times 10^5 \text{ J mol}^{-1}$$

$$= 275 \text{ kJ mol}^{-1}$$

Using an identical approach for the photons at 680 nm, we get $E = 176$ kJ mol^{-1}.

Practice Exercise X-rays are convenient to study the structure of crystals because their wavelengths are comparable to the distances between near neighbor atoms (on the order of a few Ångstroms, where $1\text{Å} = 1 \times 10^{-10}$ m). Calculate the energy of a photon of X-ray radiation with a wavelength of 2.00 Å.

The Photoelectric Effect

In 1905, only five years after Planck presented his quantum theory, Albert Einstein[11] used the theory to explain the ***photoelectric effect***—a phenomenon in which *electrons are ejected from the surface of certain metals exposed to electromagnetic radiation* (Figure 1.9).

Experimentally, the photoelectric effect is characterized by three primary observations:

1. The number of electrons ejected is proportional to the intensity of the light.

2. No electrons can be ejected if the frequency of the light is lower than a certain *threshold frequency,* which depends upon the identity of the metal.

3. The kinetic energy of the ejected electrons is proportional to the difference between the frequency of the incident light and the threshold frequency.

The photoelectric effect could not be explained by the wave theory of light. In the wave theory the energy of a light wave is proportional to the square of the amplitude (intensity) of the light wave, not its frequency. This contradicts the second observation of the photoelectric effect. Building on Planck's hypothesis, Einstein was able to explain the photoelectric effect by assuming that light consisted of particles (light quanta) of energy $h\nu$, where ν is the frequency of the light. These particles of light are called ***photons.*** Electrons are held in a metal by attractive forces, and so removing them from the metal requires light of a sufficiently high frequency (which corresponds to a sufficiently high energy) to break them free. We can think of electromagnetic radiation (light) striking the metal as a collision between photons and electrons. According to the law of conservation of energy, we have energy input equal to energy output. If ν exceeds the threshold frequency, Einstein's theory predicts

$$h\nu = \Phi + \frac{1}{2} m_e u^2 \tag{1.4}$$

where Φ (called the *work function*) is the energy needed to extract the electron from the metal surface and $\frac{1}{2} m_e u^2$ is the kinetic energy of the ejected electron. The work function measures how strongly the electrons are held in the metal. The threshold frequency is the smallest frequency for which Equation 1.4 has a solution. This occurs when the kinetic energy of the electron is zero, in which case Equation 1.4 gives

$$\nu_{\text{threshold}} = \frac{\Phi}{h}$$

Substituting this expression for $\nu_{\text{threshold}}$ into Equation 1.4 gives, after rearrangement,

$$\text{kinetic energy} = \frac{1}{2} m_e u^2 = h(\nu - \nu_{\text{threshold}})$$

Thus, Einstein's theory predicts that the kinetic energy of the ejected electron is proportional to the difference between the incident and threshold frequencies, as required.

Albert Einstein

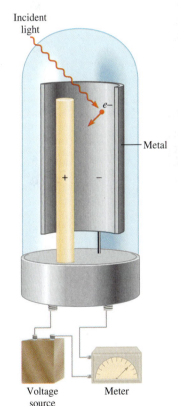

Figure 1.9 An apparatus for studying the photoelectric effect. Light of a certain frequency falls onto a clean metal surface and the ejected electrons are attracted toward the positive electrode. A detecting meter registers the flow of electrons.

---

11. Albert Einstein (1879–1955). German-born American physicist. Regarded by many as one of the two greatest physicists the world has known (the other is Isaac Newton). The three papers (on special relativity, Brownian motion, and the photoelectric effect) that he published in 1905 while employed as a technical assistant in the Swiss patent office in Berne have profoundly influenced the development of physics. He received the Nobel Prize in Physics in 1921 for his explanation of the photoelectric effect.

Figure 1.10 A plot of the kinetic energy of ejected electrons versus the frequency of incident radiation.

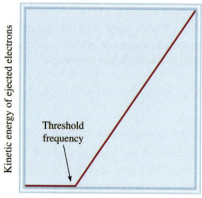

Figure 1.10 A plot of the kinetic energy of ejected electrons versus the frequency of incident radiation.

Figure 1.10 shows a plot of the kinetic energy of ejected electrons versus the frequency of applied electromagnetic radiation.

Example 1.3

When 430-nm wavelength light is shined on a clean surface of sodium metal, electrons are ejected with a maximum kinetic energy of 0.83×10^{-19} J. Calculate the work function for sodium metal and the maximum wavelength of light that can be used to eject electrons from sodium.

Strategy We can use Equation 1.4 to find the work function if we know the excess kinetic energy and the frequency of the incident light. We are given the wavelength so we must use the speed of light (c) in place of u in Equation 1.2 to convert wavelength to frequency (remembering also to convert nanometers to meters). The maximum wavelength (minimum frequency) light that can be used to eject electrons is that which gives a zero electron kinetic energy—that is, all of the energy goes into ejecting the electron.

Solution From Equation 1.4

$$\Phi = h\nu - \text{(kinetic energy)}$$

$$= \frac{hc}{\lambda} - \text{(kinetic energy)}$$

$$= \frac{(6.626 \times 10^{-34} \text{ J s})(3.00 \times 10^{8} \text{ m s}^{-1})}{(430 \text{ nm})(10^{-9} \text{ m nm}^{-1})} - 8.3 \times 10^{-20} \text{ J}$$

$$= 3.8 \times 10^{-19} \text{ J}$$

To find the maximum wavelength, we combine Equation 1.2 with Equation 1.4 except we set the kinetic energy term on the right hand side of Equation 1.4 to zero, giving

$$h\nu_{\min} = h\frac{c}{\lambda_{\max}} = \Phi$$

$$\lambda_{\max} = \frac{hc}{\Phi}$$

$$= \frac{(6.626 \times 10^{-34} \text{ J s})(3.00 \times 10^{8} \text{ m s}^{-1})}{3.8 \times 10^{-19} \text{ J}}$$

$$\lambda_{\max} = 5.20 \times 10^{-7} \text{ m} = 520 \text{ nm}$$

—Continued

Continued—

Practice Exercise The work function for Al is 6.54×10^{-19} J. Calculate the wavelength of light (in nm) that will eject electrons with a maximum kinetic energy of 1.00×10^{-19} J and the maximum wavelength of light (in nm) that can be used to eject electrons from Al.

Einstein's theory posed a dilemma for scientists. On the one hand, it explained the photoelectric effect satisfactorily. On the other hand, the particle theory of light was inconsistent with the known wave behavior of light. The only way to resolve the dilemma was to accept the idea that light possesses *both* particlelike and wavelike properties. Depending on the experiment, light behaves either as a wave or as a stream of particles. This concept was totally alien to the way physicists had thought about radiation and its interaction with matter, and it took a long time for them to accept it. We will see in Section 1.3 that a dual nature (particles and waves) is not unique to light but is characteristic of all matter, including electrons.

1.2 | The Bohr Model Was an Early Attempt to Formulate a Quantum Theory of Matter

The work of Planck and Einstein showed that the energy of electromagnetic radiation at a given frequency (v) is quantized in units of hv. The extension of this quantum hypothesis to matter paved the way for the solution of yet another nineteenth-century mystery in physics: the emission spectra of atoms.

Emission Spectra of Atoms: Evidence of the Energy Quantization of Matter

Ever since the seventeenth century, chemists and physicists have studied the characteristics of **emission spectra,** which are *either continuous or line spectra of the radiation emitted by substances.* The emission spectrum of a substance can be seen by energizing a sample of material either with thermal energy (heating) or with some other form of energy (such as a high-voltage electrical discharge). A "red-hot" or "white-hot" iron bar freshly removed from a high-temperature source produces a characteristic glow. This visible glow is the portion of its emission spectrum that is sensed by eye. The warmth of the same iron bar represents the infrared region of its emission spectrum.

The emission spectra of the sun and of a heated solid are both continuous; that is, all wavelengths of visible light are represented in the spectra (see the visible region in Figure 1.6). The emission spectra of atoms in the gas phase, on the other hand, do not show a continuous spread of wavelengths from red to violet; rather, the atoms produce bright lines in different parts of the visible spectrum. These **line spectra** are *light emissions at specific wavelengths.* Figure 1.11 is a schematic diagram of a discharge tube that is used to study emission spectra, and Figure 1.12 shows the wavelengths of visible light emitted by a hydrogen atom. Every element has a unique emission spectrum.

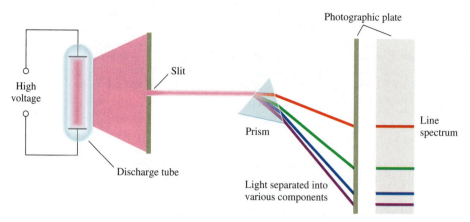

Figure 1.11 An experimental arrangement for studying the emission spectra of atoms and molecules. The gas under study is in a discharge tube containing two electrodes. As electrons flow from the negative electrode to the positive electrode, they collide with the gas. This collision process eventually leads to the emission of light by the atoms (or molecules). The emitted light is separated into its components by a prism. Each component color is focused at a definite position, according to its wavelength, and forms a colored image of the slit on the photographic plate. The colored images are called spectral lines.

Figure 1.12 The line emission spectrum of hydrogen atoms in the visible region.

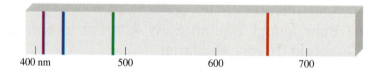

The characteristic lines in atomic spectra can be used in chemical analysis to identify unknown atoms, much as fingerprints are used to identify people. When the lines of the emission spectrum of a known element exactly match the lines of the emission spectrum of an unknown sample, the identity of the sample is established. Although the utility of this procedure had long been recognized in chemical analysis, the origin of these lines was unknown until early in the twentieth century. Figure 1.13 shows the emission spectra of several elements.

The Emission Spectrum of Hydrogen

At the end of the nineteenth century, physicists began exploring the emission spectra of atoms in quantitative detail. Of particular interest, because of the simplicity and importance of the first element, was the emission spectrum of hydrogen (Figure 1.13). The Swedish physicist Johannes Rydberg[12] analyzed the existing experimental data and formulated the following equation for the frequencies of the lines in the hydrogen emission spectrum:

$$\nu = R_H\left(\frac{1}{n_1^2} - \frac{1}{n_2^2}\right) \tag{1.5}$$

where n_1 and n_2 are positive integers with $n_2 > n_1$, and R_H is the **Rydberg constant,** which has the experimental value 3.290×10^{15} s^{-1}.

Color emitted by hydrogen atoms in a discharge tube. The color observed results from the combination of the colors emitted in the visible spectrum.

12. Johannes Robert Rydberg (1854–1919). Swedish mathematician and physicist. Rydberg's major contribution to physics was his study of the line spectra of many elements.

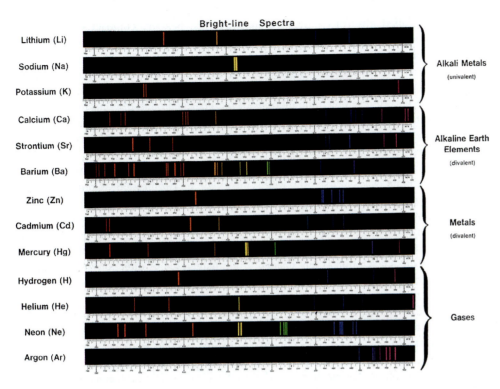

Figure 1.13 The emission spectra of various elements.

Equation 1.5 accurately predicts all of the known lines in the emission spectrum of hydrogen. The emission lines are often classified in terms of the value of n_1. Table 1.1 lists the first five series of emission lines of hydrogen, which are named for their discoverers.

The work of Planck and Einstein showed that the energy of a photon was proportional to its frequency, so the discrete nature of the emission spectrum of atoms suggested that atoms may only transfer energy in the form of electromagnetic radiation at certain well-defined values, which depend only upon the identity of the element. Also, the regularity and simplicity of the Rydberg formula (Equation 1.5) suggested that a mathematical theory of the emission spectrum should be possible—at least for hydrogen.

The Bohr Model of the Hydrogen Atom

The basic building blocks of atoms were fairly well understood at the beginning of the twentieth century. Scientists knew that atoms consisted of negatively charged

Table 1.1	The First Five Spectral Series in the Emission Spectrum of Hydrogen		
Series	n_1	n_2	**Spectral Region**
Lyman	1	2, 3, 4, . . .	UV
Balmer	2	3, 4, 5, . . .	visible, UV
Paschen	3	4, 5, 6, . . .	IR
Brackett	4	5, 6, 7, . . .	IR
Pfund	5	6, 7, 8, . . .	IR

particles called electrons moving around a much heavier, compact nucleus. The nucleus consisted of positively charged particles called protons and some number of neutral particles called neutrons (see Section 0.2). In an atom, the number of electrons and protons was equal. Scientists thought of an atom as an entity in which electrons whirled around the nucleus in circular orbits at high velocities. This was an appealing model because it resembled the motions of the planets around the sun. In this model, the electrostatic attraction between the positively charged proton and the negatively charged electron pulls the electron inward. This force of attraction is balanced exactly by the acceleration due to the orbital motion of the electron.

This classical, planetary view of the atom was at odds with basic physics in two very important ways. First, Maxwell's theory of electromagnetism predicts that a charged particle will radiate energy when undergoing acceleration (this, in fact, is how radio waves are generated by a transmitter). Because of the electrostatic force of attraction between the electron and the protons of the nucleus, an electron revolving around a central nucleus is accelerated as it continually changes direction in its orbit. (Remember, acceleration is any change in the magnitude or direction of the velocity over time.) Thus, the electron would emit electromagnetic radiation with a frequency equal to the frequency of its orbital motion and would lose energy to the electromagnetic field. This continual loss of energy would cause the orbit to decay and the electron to quickly spiral into the nucleus. Thus, the classical atom would be inherently unstable and short lived—contrary to reality. Second, classical mechanics puts no restrictions on the orbital energy of the electron. Decreasing the orbital energy slightly only moves the electron to a slightly lower orbit. Thus, the emission spectrum of the classical atom should be continuous, and not discrete, as is observed.

Extending Planck's quantum hypothesis to the energies of atoms, the Danish physicist Niels Bohr[13] presented a new model of the atom that was able to account for the emission spectrum of hydrogen. Bohr used as his starting point the planetary model of the atom but modified it with restrictions that went beyond classical physics. For a single electron orbiting a nucleus containing Z protons, the **Bohr model** consists of the following assumptions:

Niels Bohr

1. The electron moves in a circular orbit about the nucleus.

2. The energy of the electron can take on only certain well-defined values; that is, it is quantized.

3. The only allowed orbits are those in which the magnitude of the ***angular momentum*** of the electron is equal to an integer multiple of $\hbar$, where $\hbar$ (called *h-bar*) is given by $\dfrac{h}{2\pi}$. The angular momentum of a particle (L) is a vector given by **L = r × p,** where r is the position vector of the particle, measured from the origin, $p = mu$ is the momentum of the particle (with mass m and velocity u), and "×" is the vector cross-product.

4. The electron can only absorb or emit electromagnetic radiation when it moves from one allowed orbit to another. The emitted radiation has an energy $h\nu$ equal to the difference in energy between the two orbits.

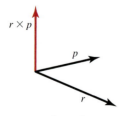

The cross product of two vectors is perpendicular to both.

The first condition of the Bohr model is consistent with the classical model of the atom, but the rest were wholly original. By considering an electron orbiting a nucleus

13. Niels Henrik David Bohr (1885–1962). Danish physicist. One of the founders of modern physics, he received the Nobel Prize in Physics in 1922 for his theory explaining the spectrum of the hydrogen atom.

with Z protons, the Bohr model is not limited to the hydrogen atom ($Z = 1$), but can also describe any one-electron ion, called a **hydrogenlike ion,** such as He^+ ($Z = 2$), Li^{2+} ($Z = 3$), etc.

From these conditions, the allowed energies of the electron can be calculated. Coulomb's law (Equation 0.5) gives the potential energy, V, due to the interaction of two charged particles with charges q_1 and q_2, separated by a distance r:

$$V(r) = \frac{q_1 q_2}{4\pi\varepsilon_0 r} \tag{1.6}$$

where ε_0 is called the *permittivity of the vacuum* ($\varepsilon_0 = 8.854188 \times 10^{-12}\ C^2\ J^{-1}\ m^{-1}$). The attractive interaction between a nucleus of charge $+Ze$ and an electron of charge $-e$ at a distance r away is then

$$V(r) = -\frac{Ze^2}{4\pi\varepsilon_0 r} \tag{1.7}$$

Given a potential energy function, the force on the object is calculated as the negative of the derivative of the potential (see Appendix 1 for the definition of a derivative). For the electron in the atom, this force is

$$F(r) = -\frac{dV(r)}{dr} = -\frac{Ze^2}{4\pi\varepsilon_0 r^2}$$

According to Newton's second law (Equation 1.1), this force is equal to the mass of the electron (m_e) times its acceleration. For an electron in a circular orbit, the acceleration can be calculated to be $-u^2/r$, where u is the magnitude of the electron velocity, and the negative sign indicates that the acceleration is directed toward the nucleus. Newton's second law then gives

$$F = ma$$
$$-\frac{Ze^2}{4\pi\varepsilon_0 r^2} = -\frac{m_e u^2}{r}$$
$$\frac{Ze^2}{4\pi\varepsilon_0 r} = m_e u^2 \tag{1.8}$$

The total energy for the electron is the sum of the kinetic (Equation 0.3) and potential energies (Equation 1.7).

$$E = \frac{1}{2} m_e u^2 - \frac{Ze^2}{4\pi\varepsilon_0 r} \tag{1.9}$$

Substituting the orbit condition (Equation 1.8) into Equation 1.9 gives

$$E = \frac{1}{2} m_e u^2 - m_e u^2 = -\frac{1}{2} m_e u^2 \tag{1.10}$$

At this point, Bohr takes a bold step and postulates the quantum restriction that the angular momentum of the electron can only take on positive integer multiple values of $\hbar$ or $h/2\pi$. For an electron in a circular orbit, the angular momentum is $m_e r u$, so

$$m_e r u = n\hbar, \quad n = 1, 2, 3, \ldots \tag{1.11}$$

where n is referred to as a **quantum number.** Combining Equation 1.11 and Equation 1.8 gives

$$u = \frac{Ze^2}{2h\varepsilon_0 n} \tag{1.12}$$

If we substitute Equation 1.12 for u into the energy equation (Equation 1.10), we get the Bohr expression for the quantized energies, or **energy levels,** of the hydrogen atom.

$$E_n = -\frac{Z^2 e^4 m_e}{8h^2\varepsilon_0^2}\frac{1}{n^2} \quad n = 1, 2, 3, \ldots \tag{1.13}$$

Because the zero of energy was (arbitrarily) defined in Equation 1.7 as the value at infinite separation of the electron from the nucleus (that is, where $n = \infty$), the negative sign in Equation 1.13 implies that all of these states are lower in energy than an infinitely separated proton and electron pair. We call such states the *bound states* of the atom. The most stable state is given by the lowest energy level and is called the **ground state** (or *ground level*), which for the Bohr model corresponds to $n = 1$. The higher energy levels are referred to as **excited states** (or *excited levels*). For the hydrogen atom ($Z = 1$), the ground-state energy is

$$E_1 = -\frac{m_e e^4}{8h^2\varepsilon_0^2} = -2.185 \times 10^{-18}\,\text{J}$$

The negative of this energy, $+2.185 \times 10^{-18}$ J (or 1316 kJ mol^{-1}), is the amount of energy that is needed to completely remove the electron from a ground-state hydrogen atom. This process is called **ionization,** and the energy required for ionization is called the **ionization energy.**

Equations 1.11 and 1.12 can also be used to derive the following expression for the radius of the Bohr orbits:

$$r_n = \frac{\varepsilon_0 h^2 n^2}{Z\pi m_e e^2} \quad n = 1, 2, 3, \ldots \tag{1.14}$$

Thus, like the energy, the orbit radius is quantized and can only take on certain values, which increase proportional to n^2. The value of the radius for the ground state ($n = 1$), calculated in Example 1.4, is called the **Bohr radius** (a_0).

Example 1.4

Calculate the Bohr radius of the ground state ($n = 1$) of the hydrogen atom.

Strategy Use Equation 1.14 with $Z = 1$ (for hydrogen) and $n = 1$ (for the ground state). The other constants in the equation can be found inside the back cover of the book.

Solution Start with Equation 1.14 with $n = Z = 1$:

$$a_0 = \frac{\varepsilon_0 h^2}{\pi m_e e^2}$$

—Continued

Continued—

Now substitute in the values of ε_0, h, m_e, and e:

$$a_0 = \frac{(8.8542 \times 10^{-12}\,\text{C}^2\,\text{J}^{-1}\,\text{m}^{-1})(6.626 \times 10^{-34}\,\text{J s})^2}{\pi(9.109 \times 10^{-31}\,\text{kg})(1.602 \times 10^{-19}\,\text{C})^2}$$

$$= 5.29 \times 10^{-11}\,\text{m}$$

$$= 52.9\,\text{pm}$$

where we have used the conversion factors $1\,\text{J} = 1\,\text{kg m}^2\,\text{s}^{-2}$ and $1\,\text{pm} = 10^{-12}\,\text{m}$. More commonly the Bohr radius is expressed as 0.529 Å. The Bohr radius is generally given the symbol a_0.

Practice Exercise Calculate the radius for the first excited state ($n = 2$) of the He$^+$ ion.

How does Equation 1.13 account for the line spectrum of hydrogen? Radiant energy absorbed by the atom causes the electron to move from a lower-energy quantum state (characterized by a smaller n value) to a higher-energy quantum state (characterized by a larger n value). Conversely, radiant energy (in the form of a photon) is emitted when the electron moves from a higher-energy quantum state to a lower-energy quantum state (Figure 1.14). The conservation of energy requires that the energy of the photon emitted or absorbed be equal to the change in the energy of the electron—that is, equal to the difference in energy (ΔE) between the initial and final energy levels. Consider the case of emission and let n_2 represent the value of the quantum number n in the initial state and n_1 be that of the final state, with $n_2 > n_1$. The energy of the emitted photon is given by

$$E_{\text{photon}} = h\nu = \Delta E = E_{n_2} - E_{n_1}$$

Using Equation 1.13, this gives

$$E_{\text{photon}} = h\nu = \frac{Z^2 e^4 m_e}{8 h^2 \varepsilon_0^2}\left[\frac{1}{n_1^2} - \frac{1}{n_2^2}\right] \tag{1.15}$$

or

$$\nu = \frac{Z^2 e^4 m_e}{8 h^3 \varepsilon_0^2}\left[\frac{1}{n_1^2} - \frac{1}{n_2^2}\right] = Z^2 R_\text{H}\left[\frac{1}{n_1^2} - \frac{1}{n_2^2}\right] \tag{1.16}$$

Setting $Z = 1$, we see that Equation 1.16 is identical to Rydberg's empirical equation for the hydrogen emission spectrum (Equation 1.5) and gives an expression for the Rydberg constant in terms of fundamental physical constants:

$$R_\text{H} = \frac{e^4 m_e}{8 h^3 \varepsilon_0^2} = 3.289832496 \times 10^{15}\,\text{s}^{-1} \tag{1.17}$$

The value of R_H calculated from Equation 1.17 is nearly[14] identical to the experimentally determined value! Equation 1.16 was derived for the case of emission of a photon. For absorption, we have $n_1 > n_2$, so the sign of the frequency would be

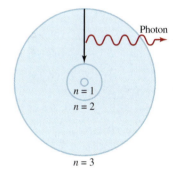

Figure 1.14 The emission process in an excited hydrogen atom, according to Bohr's theory. An electron originally in a higher-energy orbit ($n_2 = 3$) falls back to a lower-energy orbit ($n_1 = 2$). As a result, a photon with energy $h\nu$ is given off. The value of $h\nu$ is equal to the difference in energy between the initial and final electron orbits. For simplicity, only three orbits are shown.

14. See Problem 1.60.

Figure 1.15 The energy levels in the hydrogen atom and the various emission series. Each energy level corresponds to the energy associated with an allowed quantum state for an orbit, as postulated by Bohr and as shown in Figure 1.14. The emission lines are labeled according to the scheme in Table 1.1.

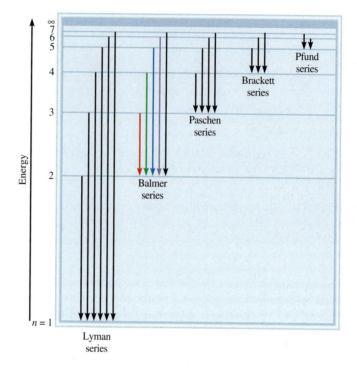

negative (see Equation 1.15), which is not physically meaningful. To ensure that the frequency of transition (whether emission or absorption) is positive, we can take the *absolute value* of $[(1/n_1^2) - (1/n_2^2)]$ in Equation 1.15.

Figure 1.15 shows the various energy levels of the hydrogen atom and the transitions that correspond to the spectral series shown in Table 1.1.

Example 1.5

What is the wavelength of a photon (in nanometers) emitted during a transition from the $n_2 = 5$ state to the $n_1 = 2$ state in the hydrogen atom? To what region of the electromagnetic spectrum does this wavelength correspond?

Strategy We are given the initial and final states in the emission process. We can calculate the frequency of the emitted photon using Equation 1.16. From Equation 1.2 (using the speed of light c as the speed of the wave), we can then calculate the wavelength from the frequency. The region of the electromagnetic spectrum to which the calculated wavelength belongs can be found by consulting Figure 1.6.

Solution From Equation 1.16 we write

$$\nu = R_H Z^2 \left| \frac{1}{n_1^2} - \frac{1}{n_2^2} \right|$$

where the vertical lines indicate that we have used the absolute value to ensure that our frequencies are positive. The Rydberg constant is 3.28983×10^{15} s^{-1} and for hydrogen $Z = 1$, so

$$\nu = 3.28983 \times 10^{15}\ \text{s}^{-1}\ (1)^2 \left| \frac{1}{2^2} - \frac{1}{5^2} \right|$$

$$\nu = 6.90864 \times 10^{14}\ \text{s}^{-1}$$

—*Continued*

Continued—
From Equation 1.2, we have

$$\lambda = \frac{c}{\nu} = \frac{2.99879 \times 10^8 \text{ m s}^{-1}}{6.90864 \times 10^{14} \text{ s}^{-1}}$$

$$= 4.34064 \times 10^{-7} \text{ m} \times \left(\frac{1 \times 10^9 \text{ nm}}{1 \text{ m}}\right)$$

$$= 434.064 \text{ nm}$$

According to Figure 1.6, this wavelength lies in the *visible* region of the electromagnetic spectrum, specifically in the violet/indigo range of the visible spectrum.

Practice Exercise For He^+ ($Z = 2$), calculate the wavelength of light (in nm) absorbed when an electron in the $n = 3$ state is excited to the $n = 6$ level. To what region of the electromagnetic spectrum does this wavelength correspond?

Bohr's idea that the energy states of matter, like light, are quantized was subsequently supported by a series of experiments performed by James Franck[15] and Gustav Hertz[16] in the decade following Bohr's hypothesis. Franck and Hertz collided fast moving electrons with atoms. (Figure 1.16(a) is an illustration of their apparatus.) At low

15. James Franck (1882–1964). German physicist and physical chemist. In addition to his work with Gustav Hertz confirming the existence of energy levels in atoms, he made numerous contributions to the field of photochemistry. He shared the 1925 Nobel Prize in Physics with Gustav Hertz.

16. Gustav Hertz (1887–1975). German physicist. He made major advances in the field of gas-phase spectroscopy and in the separation of isotopes. He shared the 1925 Nobel Prize in Physics with Franck for his work supporting the Bohr hypothesis of energy quantization in matter.

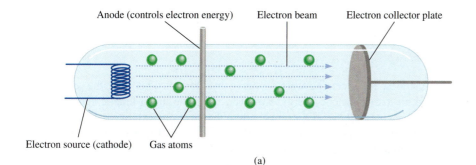

Anode (controls electron energy)   Electron beam   Electron collector plate

Electron source (cathode)   Gas atoms

(a)

Figure 1.16 The Frank-Hertz experiment: (a) The apparatus. (b) Plot of the current through the tube as a function of accelerating voltage.

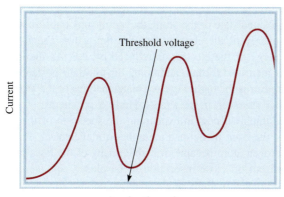

Threshold voltage

Current

Accelerating voltage

(b)

Laser—The Splendid Light

Laser is an acronym for **l**ight **a**mplification by **s**timulated **e**mission of **r**adiation. It is a special type of emission that involves either atoms or molecules. Since the discovery of laser in 1960, it has been used in numerous systems designed to operate in the gas, liquid, and solid states. These systems emit radiation with wavelengths ranging from infrared through visible and ultraviolet. The advent of laser has truly revolutionized science, medicine, and technology.

Ruby laser was the first known laser. Ruby is a deep-red mineral containing corundum, Al_2O_3, in which some of the Al^{3+} ions have been replaced by Cr^{3+} ions. A flashlamp is used to excite the chromium atoms to a higher energy level. The excited atoms are unstable, so at a given instant some of them return to the ground state by emitting a photon in the red region of the spectrum. The photon bounces back and forth many times between mirrors at opposite ends of the laser tube.

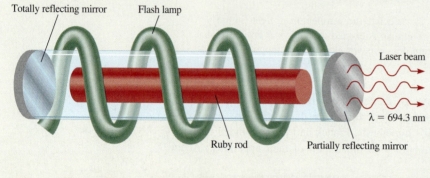

Totally reflecting mirror   Flash lamp

Laser beam

$\lambda = 694.3$ nm

Ruby rod   Partially reflecting mirror

The emission of a laser light from a ruby laser.

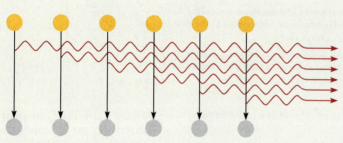

The stimulated emission of one photon in a cascade event that leads to the emission of laser light. The synchronization of the light waves produces an intensely penetrating laser beam.

collision energies, the electrons were scattered by the collisions with the atoms, but experienced no loss of kinetic energy. However, when the initial kinetic energy of the electrons was increased beyond a critical value, the electrons exhibited a large loss of kinetic energy in the collision with an atom [Figure 1.16(b)]. These experiments showed that a certain minimum energy was needed to transfer energy from the electron to the atom. Within the Bohr hypothesis these results can be explained as follows: If the kinetic energy of the electron is smaller than the difference in energy, ΔE, between the ground state and the first excited state of the atom, the atom cannot absorb the energy. However, once the kinetic energy of the electron exceeds this value, energy transfer is possible and the kinetic energy of the electron is reduced by ΔE. As the initial kinetic energy of the electrons is increased further, additional dips in their postcollision kinetic energy are expected to appear as the higher excited state energies of the atom are reached—and this is exactly what was seen in the Franck-Hertz experiments.

In spite of its initial success in explaining the spectrum of the hydrogen atom and hydrogenlike ions, the Bohr model had a number of deficiencies. First, as more accurate spectra of the hydrogen atom became available, many of the lines previously seen to be single lines turned out on close inspection to be closely spaced pairs of lines (called *doublets*). The Bohr model, with its single quantum number n, was neither able to account for this *fine structure* of the atomic spectrum nor able to explain the changes induced in

This photon can stimulate the emission of photons of exactly the same wavelength from other excited chromium atoms; these photons in turn can stimulate the emission of more photons, and so on. Because the light waves are *in phase*—that is, their maxima and minima coincide—the photons enhance one another, increasing their power with each passage between the mirrors. One of the mirrors is only partially reflecting, so that when the light reaches a certain intensity it emerges from the mirror as a laser beam. Depending on the mode of operation, the laser light may be emitted in pulses (as in the ruby laser case) or in continuous waves.

Laser light is characterized by three properties: It is intense, it has precisely known wavelength and hence energy, and it is coherent. By *coherent* we mean that the light waves are all in phase. The applications of lasers are numerous. Their high intensity and ease of focus make them suitable for doing eye surgery, for welding and drilling holes in metals, and for carrying out nuclear fusion. Because they are highly directional and have precisely known wavelengths, they are very useful for telecommunications. Lasers are also used in isotope separation, in holography (three-dimensional photography), in compact disc players, and in supermarket scanners. Lasers have played an important role in the spectroscopic investigation of molecular properties and of many chemical and biological processes. Laser lights are increasingly being used to probe the details of chemical reactions (see Chapter 13).

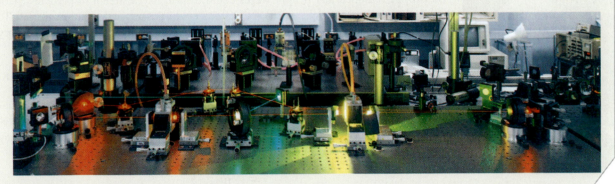

State-of-the-art lasers used in the research laboratory of Dr. A. H. Zewail at the California Institute of Technology.

the spectrum when a magnetic field was applied. In 1916 the German physicist Arnold Sommerfeld[17] rectified these problems by extending the Bohr model by taking into account Einstein's special theory of relativity and included elliptical orbits through the introduction of additional quantization conditions and quantum numbers. This more complete theory is generally referred to as the Bohr-Sommerfeld model of the atom or simply as "Old Quantum Theory." Second, and more serious, was the complete failure of the Bohr-Sommerfeld theory to accurately predict spectral lines in atoms with more than one electron, even those as simple as helium. It became clear that a more general theory was needed to account for atoms other than hydrogen. As we shall see, the problem with the Bohr-Sommerfeld theory was not that it was "wrong." Instead, the real failings of the theory occured because it did not go far enough in moving away from the precepts of classical theory. In the Bohr-Sommerfeld model, the electrons were still viewed as classical point particles orbiting the nucleus, with a well-defined position and a well-defined momentum at all times. As we will see in Section 1.3, the electrons in an

---

17. Arnold Johannes Wilhelm Sommerfeld (1868–1951). German physicist. In addition to his pioneering work on the relationship between atomic structure and spectral lines, he made important contributions to understanding the electronic properties of metals.

atom are more accurately described as wavelike objects for which position and momentum cannot be simultaneously defined with precision.

Despite its shortcomings, however, the Bohr-Sommerfeld model is of enormous significance in the development of the quantum theory and, in a broader sense, as an illustration of the evolution of scientific ideas.

1.3 | Matter Has Wavelike Properties

Physicists were both mystified and intrigued by Bohr's theory and sought to understand *why* the energy levels would be quantized. That is, why is the electron in a Bohr atom restricted to orbiting the nucleus at certain fixed distances? For a decade no one, not even Bohr himself, had a logical explanation. In 1924, though, Louis de Broglie[18] provided a solution to this puzzle.

The Wave-Particle Duality of Matter—The de Broglie Hypothesis

De Broglie reasoned that if light waves could behave like a stream of particles (photons), then perhaps particles such as electrons could possess wavelike properties. To quantify this connection, de Broglie began with the expression (from Einstein's theory of special relativity) for the momentum (p) of the photon:

$$p = E/c \qquad \textbf{(1.18)}$$

where E is the energy of the photon and c is the speed of light. Combining this equation with the Einstein–Planck expression for the photon energy in terms of its frequency (Equation 1.3) and the relationship (from Equation 1.2) between frequency and wavelength, ($\nu = c/\lambda$), we obtain, after rearrangement,

$$p = \frac{h\nu}{c} = \frac{h}{\lambda}$$

or

$$\lambda = \frac{h}{p} \qquad \textbf{(1.19)}$$

Louis de Broglie

Equation 1.19 was derived using equations applicable to the photon, which is massless and has a fixed velocity c. De Broglie postulated that the equation should also apply to particles of matter with mass m and velocity u. Substituting the expression for the momentum of a particle ($p = mu$) into Equation 1.19 gives the de Broglie relation for the wavelength of a particle:

$$\lambda = \frac{h}{mu} \qquad \textbf{(1.20)}$$

The wavelength defined in Equation 1.20 is called the **de Broglie wavelength** of a particle. Equation 1.20 implies that a particle in motion can be treated as a wave and that a wave can exhibit the properties of a particle (that is, its momentum). Thus, the left side of Equation 1.20 addresses the wavelike properties of matter (wavelength), whereas the right side addresses its particle-like properties (mass).

---

18. Louis Victor Pierre Raymond duc de Broglie (1892–1987). French physicist. Member of an old and noble family in France, he held the title of a prince. In his doctoral dissertation, he proposed that matter and radiation have the properties of both waves and particles. For this work, de Broglie was awarded the Nobel Prize in Physics in 1929.

Example 1.6

Calculate the wavelength of the "particle" in the following two cases: (a) A 6.0×10^{-2} kg tennis ball served at 140 miles per hour (63 m s^{-1}). (b) An electron ($m_e = 9.1094 \times 10^{-31}$ kg) moving at 63 m s^{-1}.

Strategy We are given the mass (m) and the speed (u) of the particle in (a) and (b) and asked to calculate the wavelength (λ). We can use Equation 1.20 to do this. Note, however, that Planck's constant has units of J s, so m and u must be in units of kg and m s^{-1}, respectively (1 J = 1 kg m^2 s^{-2}).

(a) Solution Using Equation 1.20 for the tennis ball we write

$$\lambda = \frac{h}{mu}$$
$$= \frac{6.626 \times 10^{-34} \text{ J s}}{(6.0 \times 10^{-2} \text{ kg})(63 \text{ m s}^{-1})}$$
$$= 1.8 \times 10^{-34} \text{ m}$$

Comment This is an exceedingly small wavelength considering that the size of an atom itself is on the order of 1×10^{-10} m. For this reason, the wave properties of a tennis ball cannot be detected by any existing measuring device.

(b) Solution For the electron,

$$\lambda = \frac{h}{mu}$$
$$= \frac{6.626 \times 10^{-34} \text{ J s}}{(9.1094 \times 10^{-31} \text{ kg})(63 \text{ m s}^{-1})}$$
$$= 1.2 \times 10^{-5} \text{ m}$$

Comment This wavelength (1.2×10^{-5} m or 1.2×10^4 nm) is in the infrared region and is much larger than the size of an atom. This calculation shows that only submicroscopic particles (such as electrons) have measurable wavelengths.

Practice Exercise Calculate the wavelength (in nanometers) of a hydrogen atom (mass = 1.674×10^{-27} kg) moving at 7.00×10^2 cm s^{-1}.

Example 1.6 shows that although de Broglie's equation can be applied to diverse systems, the wave properties become observable only for submicroscopic objects. This distinction occurs because Planck's constant, h, which appears in the numerator in Equation 1.20, is so small.

According to de Broglie, an electron bound to the nucleus behaves like a ***standing wave.*** Standing waves can be generated by plucking a guitar string (Figure 1.17). The waves are described as standing, or stationary, because they do not travel along the string. Some points on the string, called ***nodes,*** do not move at all; that is, *the amplitude of the wave at these points is zero.* There is a node at each end, and there may be nodes between the ends. The greater the frequency of vibration, the shorter the

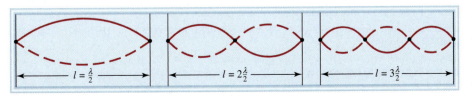

Figure 1.17 The standing waves generated by plucking a guitar string. Each dot represents a node. The length of the string, l, must be equal to a whole number times one-half the wavelength ($\lambda/2$).

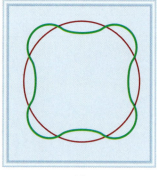

(a)

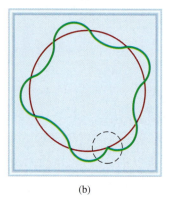

(b)

Figure 1.18 (a) The circumference of the orbit is equal to an integral number of wavelengths. This is an allowed orbit. (b) The circumference of the orbit is not equal to an integral number of wavelengths. As a result, the electron wave does not close in on itself. This is a nonallowed orbit.

wavelength of the standing wave and the greater the number of nodes. As Figure 1.17 shows, there can be only certain wavelengths in any of the allowed motions of the string. *The existence of discrete energy states is the natural consequence of confining a wavelike object to a finite region of space.*

De Broglie argued that if an electron does behave like a standing wave in the hydrogen atom, the length of the wave must fit the circumference of the orbit exactly (Figure 1.18)—otherwise the wave would partially cancel itself on each successive orbit. Eventually the amplitude of the wave would be reduced to zero, and the wave would cease to exist. The relation between the circumference of an allowed orbit ($2\pi r$) and the de Broglie wavelength (λ) of the electron is given by

$$2\pi r = n\lambda \quad n = 1, 2, 3, \ldots \tag{1.21}$$

where r is the radius of the orbit and λ is the wavelength of the electron wave. Using Equation 1.21 together with the expression for λ in Equation 1.20, we obtain

$$2\pi r = n\frac{h}{m_e u}$$

Upon rearrangement,

$$m_e u r = \frac{nh}{2\pi} = n\hbar \quad n = 1, 2, 3, \ldots$$

which is identical to the Bohr angular momentum condition expressed in Equation 1.11. Thus, de Broglie's postulate leads to quantized angular momentum and to the quantized energy levels of the hydrogen atom.

Shortly after de Broglie introduced his equation, Clinton Davisson[19] and Lester Germer[20] in the United States and G. P. Thomson[21] in England demonstrated that electrons do indeed possess wavelike properties. By directing a beam of electrons at a thin piece of gold foil, Thomson obtained a set of concentric rings on a detector screen, similar to the pattern observed when X-rays (which are waves) were used. Figure 1.19 shows the same kind of pattern for aluminum. The wavelike nature of electron beams has application in a number of experimental techniques, such as *electron microscopy* (discussed in the inset on page 109) and *low energy electron diffraction (LEED)*, which is used to study the surfaces of crystalline solids. Another technique, *neutron diffraction*, uses the wavelike properties of neutrons to study the structure and dynamics of dense materials, such as liquids.

The Heisenberg Uncertainty Principle

One of the assumptions of classical physics is that the dynamical variables (positions and momenta) of a particle in motion have well-defined, precise values. However, the concept of a precise position becomes ill-defined when we try to describe a particle as a wavelike object. A wave is an object that is extended over some region of space. To describe the problem of trying to locate a subatomic particle that behaves like a wave, Werner

19. Clinton Joseph Davisson (1881–1958). American physicist. He and G. P. Thomson shared the Nobel Prize in Physics in 1937 for demonstrating the wave properties of electrons.

20. Lester Halbert Germer (1896–1972). American physicist. Discoverer (with Davisson) of the wave properties of electrons.

21. George Paget Thomson (1892–1975). English physicist. Son of J. J. Thomson, he received the Nobel Prize in Physics in 1937, along with Clinton Davisson, for demonstrating the wave properties of electrons.

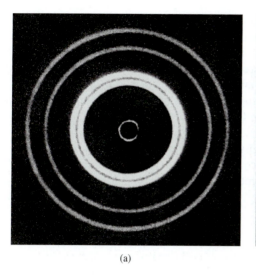
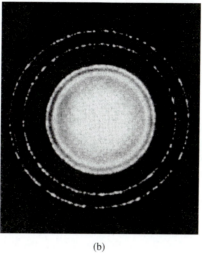

(a)   (b)

Figure 1.19 Left: X-ray diffraction pattern of aluminum foil. Right: Electron diffraction pattern of aluminum foil. The similarity of these two patterns shows that electrons can behave like X-rays and display wave properties.

Heisenberg[22] formulated what is now known as the ***Heisenberg uncertainty principle:*** *It is impossible to know simultaneously both the momentum p* (defined as mass times velocity) *and the position of a particle with certainty.* Stated mathematically we have,

$$\Delta x \, \Delta p \geq \frac{\hbar}{2} \tag{1.22}$$

where Δx and Δp are the uncertainties in measuring the position and momentum, respectively. Thus, if we measure the momentum of a particle more precisely (that is, if we make Δp a small quantity), our knowledge of the position will become correspondingly less precise (that is, Δx will become larger). Similarly, if the position of the particle is known more precisely, then its momentum must be known less precisely. This inverse relationship arises because the position of a wavelike particle is determined by the region of space occupied by the wave, but the momentum, through the de Broglie relationship, is related to the wavelength of the wave.

Figure 1.20 depicts two extreme cases. In Figure 1.20(a) a wave is extended over a large region of space; however, the wavelength of the wave is well defined. Such a particle-wave would have a small uncertainty in the momentum (wavelength), but a large uncertainty in the position. In Figure 1.20(b), on the other hand, the wave is

Werner Heisenberg

22. Werner Karl Heisenberg (1901–1976). German physicist. One of the founders of modern quantum theory. Heisenberg received the Nobel Prize in Physics in 1932.

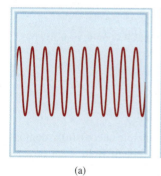

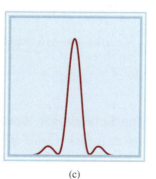

(a)   (b)   (c)

Figure 1.20 Three illustrations of the Heisenberg uncertainty principle. (a) A particle-wave with a large uncertainty in position, but with a well-defined wavelength (momentum). (b) A particle-wave with a well-defined position, but a large uncertainty in wavelength (momentum). (c) A particle-wave with intermediate uncertainty in position and momentum.

highly localized in space (giving a small value of Δx). but the wavelength is difficult to define; that is, it has a large uncertainty in the wavelength. Figure 1.20(c) shows a wave that is intermediate between the two extremes in Figure 1.20(a) and 1.20(b).

When the Heisenberg uncertainty principle is applied to the hydrogen atom, it is found that the electron cannot orbit the nucleus in a well-defined path, as Bohr thought. If it did, we could determine precisely both the position of the electron (from the radius of the orbit) and its momentum (from its kinetic energy) at the same time, a violation of the uncertainty principle.

The uncertainty principle is negligible in the world of macroscopic objects (because of the small size of Planck's constant), but is very important for objects with small masses, such as electrons and protons.

Example 1.7

Recall from Example 1.4 that the Bohr radius of the hydrogen atom is 52.9 pm (or 0.529Å). Assuming that we know the position of an electron in this orbit to an accuracy of 1 percent of the radius, calculate the uncertainty in the velocity of the electron.

Strategy The uncertainty, Δx, in the position of the electron is given. From the Heisenberg uncertainty principle (Equation 1.22), we can calculate the minimum uncertainty in the momentum, Δp, from which the uncertainty in the velocity can be determined.

Solution The uncertainty, Δx, in the position of the electron is

$$\Delta x = \frac{1\%}{100\%} \times 52.9 \text{ pm} \times \frac{1 \times 10^{-12} \text{ m}}{1 \text{ pm}} = 5.29 \times 10^{-13} \text{ m}$$

From the Heisenberg uncertainty principle (Equation 1.22), we have

$$\Delta p \geq \frac{\hbar}{2\Delta x} = \frac{1.054 \times 10^{-34} \text{ J s}}{2(5.29 \times 10^{-13} \text{ m})}$$

$$\geq 9.96 \times 10^{-23} \text{ kg m s}^{-1}$$

Because $\Delta p = m \Delta u$, the uncertainty in the velocity is given by

$$\Delta u = \frac{\Delta p}{m} \geq \frac{9.96 \times 10^{-23} \text{ kg m s}^{-1}}{9.1095 \times 10^{-31} \text{ kg}}$$

$$\geq 1.1 \times 10^{8} \text{ m s}^{-1}$$

The uncertainty in the velocity of the electron is of the same magnitude as the speed of light (3×10^{8} m s^{-1}). At this level of uncertainty, we have virtually no idea what the velocity of the electron is.

Practice Exercise Repeat the calculation in Example 1.7 using a proton instead of an electron.

The Schrödinger Wave Equation

The de Broglie relation and the Heisenberg uncertainty principle successfully demonstrated the major flaw in the Bohr-Sommerfeld model. Although Bohr went beyond classical physics in postulating the quantization of energy levels, his theory still relied on the Newtonian notion of particle trajectories. In the Bohr-Sommerfeld orbits both the momentum and position of the electron had specific well-defined values at all times, in violation of the Heisenberg uncertainty principle. De Broglie postulated that a particle has wavelike properties, but his theory was incomplete, as it did not provide a quantitative determination of the properties of such a system. A general equation was needed for quantum systems, one comparable in predictive power to Newton's second law of motion for classical objects. In 1926, the Austrian physicist Erwin Schrödinger[23] furnished the necessary equation.[24]

In classical mechanics, the state of a particle is defined uniquely by its position and momentum. If you know both of these quantities, then you can predict the future motion of the particle based on the forces acting upon it. According to Heisenberg's uncertainty principle, though, this sort of knowledge is unavailable for a quantum particle, as the position and momentum cannot be simultaneously specified. *Schrödinger postulated that the complete information about the state of a quantum particle was contained in a function $\psi(x)$*, called the **wavefunction,** which is a function of the position of the particle (given by x for a one-dimensional system). One of the most important properties of wavelike objects is the ability to exhibit constructive and destructive interference. For this to be possible, the wave function must be able to take on positive *or* negative values.

We know from everyday experience what is meant by the classical state of a particle—that is, by position and momentum. But what is meant by ψ? The currently accepted physical interpretation of ψ, given in 1926 by German physicist Max Born,[25] is that the wavefunction is related to the *probability* of finding the particle in a specific region of space. Because ψ can take on negative values, and probability is, by definition, a positive quantity, Born postulated that the probability of finding the particle in a particular small region of space was proportional to the *square* of the wavefunction. Specifically, for a one-dimensional system, the probability of finding the particle between positions x and $x + dx$ is given by $\psi^2(x)\,dx$.[26] As shown in Figure 1.21, the probability, P, of finding the particle in a specific region $a < x < b$ is given by the area under the curve, $\psi^2(x)$, between $x = a$ and $x = b$ (which can be calculated from the integral of $\psi^2(x)$ on this interval):

Erwin Schrödinger

$$P = \text{area under } \psi^2(x) \text{ from } a \text{ to } b$$
$$= \int_a^b \psi^2(x)dx \tag{1.23}$$

(Because the probability of finding the particle in a specific region of space is not given directly by $\psi^2(x)$, but instead by an integral of $\psi^2(x)$ over the region, we refer to $\psi^2(x)$ not as a probability, but as a *probability density*).

23. Erwin Schrödinger (1887–1961). Austrian physicist. Schrödinger formulated wave mechanics, which laid the foundation for modern quantum mechanics. He received the Nobel Prize in Physics in 1933.

24. An alternative formulation of quantum mechanics, based on matrices, was developed independently by Heisenberg at about the same time. It was later shown that this "matrix mechanics" is equivalent to Schrödinger's theory.

25. Max Born (1882–1970). German physicist. In addition to being one of the pioneers in modern quantum mechanics, he made major contributions to electrodynamics and the theory of crystals. He received the Nobel Prize in Physics in 1953.

26. Mathematically, the wavefunction can be a complex number, such as $A + iB$ where $i = \sqrt{-1}$. To be physically meaningful, this probability should be given as $\psi^*(x)\psi(x)$, where $\psi^*(x)$ is the *complex conjugate* of $\psi(x)$. If $\psi(x)$ is written as $A + iB$, where A and B are real, then $\psi^*(x)$ is defined as $A - iB$ and $\psi^*(x)\psi(x) = (A + iB)(A - iB) = A^2 + B^2$. This is necessary to ensure that the probability density $\psi^*(x)\psi(x)$ is always positive. If $\psi(x)$ is a real number, then $\psi^*(x) = \psi(x)$ and $\psi^*(x)\psi(x) = \psi^2(x)$, the usual square.

Figure 1.21 (a) An example of a wavefunction [$\psi(x)$] for a particle in a one-dimensional system as a function of position (x). (b) The probability density [$\psi^2(x)$] defined by the wavefunction in (a). The probability of the particle being in the interval ($a < x < b$) is given by the area under the curve over this interval (shaded).

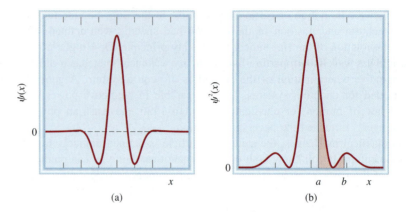

(a)

(b)

In quantum mechanics, we *cannot* specify the exact position of the particle, only the probability that it will be found in some region of interest. This has nothing to do with the inadequacy of our measuring devices—it is a fundamental property of matter!

By analogy with the laws of optics, Schrödinger proposed that the wavefunction for a particle of mass m in one dimension is the solution to the equation

$$-\frac{\hbar^2}{2m}\frac{d^2\psi(x)}{dx^2} + V(x)\psi(x) = E\psi(x) \tag{1.24}$$

where $V(x)$ is the potential energy function and E is the total energy. Like Newton's second law of classical mechanics, the Schrödinger equation (Equation 1.24) is a *postulate* and cannot be derived. The Schrödinger equation marked the beginning of a new era in physics—that of wave mechanics or **quantum mechanics.**

Equation 1.24 does not contain time as a variable and is referred to as the **time-independent Schrödinger equation.** The wavefunctions that are the solutions to Equation 1.24 do not change with time and are called *stationary-state* wavefunctions.[27] For a specific system, the Schrödinger equation can be solved only for certain values of E; that is, the energy of the system is quantized.

To describe a physical system, the wavefunction (ψ) must also be "well-behaved"; that is, it must satisfy the following conditions:

1. ψ must be single-valued at all points.
2. The total area under $\psi^2(x)$ must be equal to unity; that is, $\int_{-\infty}^{\infty}\psi^2(x)\,dx = 1$.
3. ψ must be "smooth"; that is, ψ and its first derivative (slope), $d\psi/dx$, must be continuous at all points.

The first condition ensures that the probability of finding the particle in a given region of space has a unique value. The second condition is a statement that the probability of finding the particle somewhere is equal to one; that is, the particle exists. The third condition is necessary so that the second derivative of ψ, which appears in Equation 1.24, is physically meaningful.

---

27. A more general equation containing time as a variable was also formulated by Schrödinger and is useful in the description of spectroscopic techniques. However, many problems of chemical interest can be adequately described using only the stationary-state wavefunctions.

For an arbitrary system, the solution to the Schrödinger equation can be quite complex. However, it is possible by examining the form of the Schrödinger equation to deduce some qualitative aspects of the wavefunction. The left-hand side of Equation 1.24 contains two terms. The first term,

$$-\frac{\hbar^2}{2m}\frac{d^2\psi(x)}{dx^2}$$

represents the *kinetic energy* part of the equation. This term is proportional to the second derivative of the wave function ($d^2\psi/dx^2$), which describes the *curvature* of the wavefunction (see Appendix 1). Thus, wavefunctions with high curvature have high kinetic energy and those that are relatively flat (low curvature) have low kinetic energy. The second term on the left-hand side of Equation 1.24,

$$V(x)\psi(x)$$

describes the potential energy of the system. Consider the wavefunction for the ground state (lowest energy state) of a given system. (Figure 1.22) The wavefunction of minimum potential energy would be one in which the probability density (and, thus, the wavefunction) for the particle is narrowly peaked in the vicinity of the potential energy minimum [Figure 1.22(b)]. Such a wavefunction, however, would have a high curvature and thus a high kinetic energy. Lowering the kinetic energy involves decreasing the curvature of the wavefunction (that is, spreading it out more over space), which allows the particle to exist in regions that have high potential energy. [Figure 1.22(c)] The true ground-state wavefunction of a particle represents a *compromise* between these two extremes [Figure 1.22(d)].

The Particle in a One-Dimensional Box: A Simple Model

For most problems in nature, the Schrödinger wave equation cannot be solved exactly, and we must use sophisticated computer algorithms to obtain even approximate

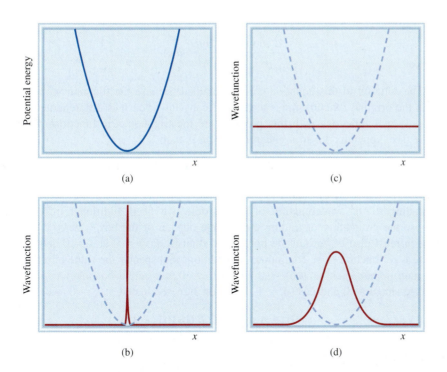

(a)

(b)

(c)

(d)

Figure 1.22 (a) A model potential energy function $V(x)$. (b) An example of a ground-state wavefunction that minimizes the potential energy but has a large kinetic energy (high curvature). (c) A ground-state wavefunction that minimizes kinetic energy (low curvature) but has high potential energy (is nonzero in regions where the potential energy is large). (d) The exact ground-state wavefunction.

Figure 1.23 The potential energy for a one-dimensional particle in a box with infinite energy barriers at $x = 0$ and $x = L$.

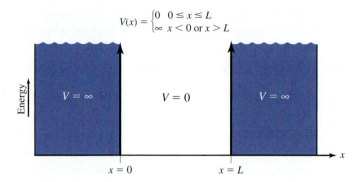

solutions. However, there exist a small number of model systems for which an exact solution is possible. One of these is the particle in a one-dimensional box. Although this system represents a highly idealized situation, the ***particle-in-a-box model*** possesses many of the features of realistic quantum-mechanical systems, and its solution can be applied to real problems of chemical and biological interest.

Consider a particle of mass m confined to a one-dimensional region (or line) of length L. The potential energy inside the box is zero ($V = 0$) and infinite outside ($V = \infty$). If we denote the position of the particle on the line by x, this potential energy can be written as

$$V(x) = \begin{cases} 0 & 0 \le x \le L \\ \infty & x < 0 \text{ or } x > L \end{cases}$$

This potential energy function is shown in Figure 1.23.

Because the potential energy is zero inside the box, the energy of the particle is entirely kinetic. For the region $0 \le x \le L$, the Schrödinger equation (Equation 1.24) becomes

$$-\frac{\hbar^2}{2m}\frac{d^2\psi(x)}{dx^2} = E\psi(x)$$

or

$$\frac{d^2\psi(x)}{dx^2} = -\frac{2mE}{\hbar^2}\psi(x) \qquad (1.25)$$

The solution to this equation can be obtained because the function we are after yields the original function times a negative constant when it is differentiated twice. The only real functions with this property are the trigonometric functions sine and cosine (see Appendix 1), so a *general* solution is then

$$\psi(x) = A\sin(kx) + B\cos(kx) \qquad (1.26)$$

where A, B, and k are constants to be determined. To find the particular solution corresponding to the wavefunction of the particle, we need additional information. Because the probability of finding the particle outside of the box is zero, the wavefunction must also be zero at the boundaries (that is, $\psi(x) = 0$ at $x = 0$ and $x = L$) for the wavefunction to be continuous. (Conditions that specify the value of a function at certain points, such as $\psi(0) = 0$ and $\psi(L) = 0$, are called *boundary conditions*.) Because $\sin(0) = 0$ and $\cos(0) = 1$, Equation 1.26, evaluated at $x = 0$, yields

$$\psi(0) = A\sin(0) + B\cos(0) = 0$$
$$A(0) + B(1) = 0$$

Therefore, $B = 0$, and we have

$$\psi(x) = A \sin(kx)$$

Substituting this expression into Equation 1.25 gives

$$\frac{d^2}{dx^2}[A \sin(kx)] = -\frac{2mE}{\hbar^2} A \sin(kx)$$

$$-Ak^2 \sin(kx) = -\frac{2mE}{\hbar^2} A \sin(kx)$$

Solving for k gives

$$k = \left(\frac{2mE}{\hbar^2}\right)^{1/2}$$

Finally, we have

$$\psi(x) = A \sin\left[\left(\frac{2mE}{\hbar^2}\right)^{1/2} x\right] \tag{1.27}$$

We have yet to use the boundary condition at $x = L$. Requiring $\psi(L) = 0$ in Equation 1.27 gives

$$\psi(L) = A \sin\left[\left(\frac{2mE}{\hbar^2}\right)^{1/2} L\right] = 0 \tag{1.28}$$

One solution to Equation 1.28 is obtained when A is zero; however, this yields a wavefunction that is zero everywhere (that is, it describes a system in which no particle is present), which is trivial, unphysical, and uninteresting. Real, physical solutions are obtained by noting that $\sin(x)$ is zero only when x is a multiple of π (that is, when $x = n\pi$, where n is an integer). Thus, the second boundary condition will be satisfied if the energy E is restricted to values that satisfy[28]

$$\left(\frac{2mE}{\hbar^2}\right)^{1/2} L = n\pi, \quad n = 1, 2, 3, \ldots \tag{1.29}$$

If we define E_n to be the value of E that satisfies Equation 1.29 for a given allowed value of n, we have

$$E_n = \frac{n^2\pi^2\hbar^2}{2mL^2}$$

Using $\hbar = \dfrac{h}{2\pi}$, we have

$$E_n = \frac{n^2h^2}{8mL^2}, \quad n = 1, 2, 3, \ldots \tag{1.30}$$

Because the allowed values of the energy are discrete (not continuous), we say that the energy of this system is *quantized*. This quantization is a direct result of

---

28. The value $n = 0$ was excluded because it gives the trivial and unphysical solution that ψ is zero everywhere.

imposing the boundary conditions. In the absence of the confining walls (which generate the boundary conditions), the system describes a *free particle*. The energy of a free particle is not quantized, but can take on any desired positive value. Thus, we see again that quantization of energy arises when a particle (or wave) is confined to a finite region of space.

The difference between successive energy levels can also be calculated from Equation 1.30. For example, the difference between the energy levels n and $(n + 1)$ is

$$\Delta E = E_{n+1} - E_n = \frac{h^2}{8mL^2}\left[(n + 1)^2 - n^2\right]$$

$$= \frac{h^2}{8mL^2}\left[n^2 + 2n + 1 - n^2\right]$$

which, after simplification, gives

$$\Delta E_{n \to n+1} = \frac{h^2}{8mL^2}(2n + 1) \tag{1.31}$$

The spacing between successive energy levels is inversely proportional to both the particle mass (m) and L^2, the square of the size of the confining region. The energy-level spacing for this problem also increases with increasing n.

The wavefunctions corresponding to the energy level E_n are

$$\psi_n(x) = A\sin\left(\frac{n\pi x}{L}\right) \tag{1.32}$$

Because $\psi_n^2(x)dx$ represents the probability of finding the particle in the interval defined by x and $x + dx$, the integral of $\psi_n^2(x)$ over all possible values of x must equal 1. Thus, we require that

$$\int_0^L \psi_n^2(x)dx = 1 \tag{1.33}$$

Requiring that the wavefunction given in Equation 1.32 satisfies Equation 1.33 makes it possible to specify the value of A. Substituting Equation 1.32 into Equation 1.33 gives

$$\int_0^L A^2 \sin^2\left(\frac{n\pi x}{L}x\right)dx = 1 \tag{1.34}$$

The integral in Equation 1.34 can be obtained from any decent table of integrals. It equals $L/2$, so $A^2(L/2) = 1$ or $A = (2/L)^{1/2}$. A wavefunction that satisfies Equation 1.33 is said to be *normalized*, and the preceding process used to find A (the *normalization constant*) is called *normalization*. The properly normalized wavefunctions for the particle in a box are then

$$\psi_n(x) = \left(\frac{2}{L}\right)^{1/2}\sin\left(\frac{n\pi x}{L}\right) \quad n = 1, 2, 3, \ldots \tag{1.35}$$

The wavefunctions, probability densities [given by $\psi^2(x)$], and energies for the first four energy levels for the particle in a one-dimensional box are plotted in Figure 1.24.

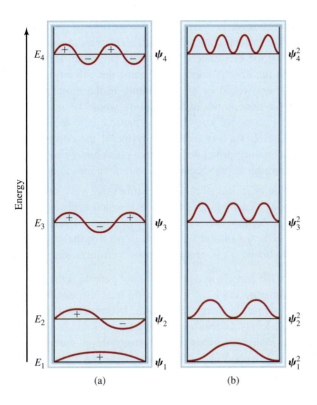

(a)   (b)

Figure 1.24 Plots of (a) ψ and (b) ψ^2 for the first four energy levels (together with their relative energies) of the particle in a one-dimensional box. The "+" and "−" symbols indicate positive and negative regions of the wavefunction, respectively.

Note that the wavefunctions for the one-dimensional particle in a box look just like the standing waves set up in a vibrating string (Figure 1.17). This similarity is not a coincidence—the mathematics that describes the wave behavior of these two seemingly different physical systems is very similar.

The particle in a one-dimensional box illustrates the following points that are true in general about quantum systems:

▶ The quantization of the energy levels of a system is a direct result of the localization of the particle in a finite region of space by the potential energy. In the particle in a one-dimensional box, the infinite potential energy barriers enforce this localization. In an atom, the negatively charged electrons are confined to a small region around the positively charged nucleus by the strong Coulombic attraction between oppositely charged particles. As a result, their energies are also quantized. For the particle in a one-dimensional box, there is one quantum number n that indexes the allowed quantum states. *In general, the number of quantum numbers necessary to describe the quantum state of a particle is equal to the number of dimensions.* For example, the quantum states of a particle in a two-dimensional box require two quantum numbers. We will see in Section 1.4 that the electron in a hydrogen atom, which is an object moving in three dimensions, requires three quantum numbers to specify its wavefunction.

▶ The spacing between successive energy levels is inversely proportional to both the particle mass (m) and L^2. For macroscopic objects, both m and L are large and the resulting spacing between energy levels is vanishingly small, so that the energy spectrum appears continuous, in agreement with the observed classical mechanical behavior of macroscopic objects.

▶ The lowest energy level is not zero. In the particle in a one-dimensional box, the energy of the ground state is $h^2/8mL^2$. This **zero-point energy** can be accounted for by the Heisenberg uncertainty principle. If the lowest energy were zero, then the kinetic energy (and thus, the velocity) of the particle would also be zero. There would be no uncertainty in the momentum of the particle; consequently, the uncertainty in the position would be infinite (according to Heisenberg). However, we know that the particle is in the box, so the maximum uncertainty in x is L. As a result, a zero value of the energy would violate the Heisenberg uncertainty principle. The zero-point energy means, too, that the particle can never be at rest because its lowest energy is not zero.

▶ For a given value of n, Equation 1.35 describes the wave behavior of the particle, but the probability density is given by $\psi_n^2(x)$, which is always positive. For $n = 1$, the maximum probability density is at $x = L/2$ (see Figure 1.24); for $n = 2$, the maxima occur at $x = L/4$ and $x = 3L/4$. Generally, the number of nodes (points at which ψ, and hence ψ^2, is zero) increases with increasing energy.

One property of the particle in a box that is not general is the increase in the spacing between successive energy levels with increasing quantum number n. For example, we will see in Section 1.4 that the energy levels in the hydrogen atom become *closer* together for the higher energy levels. In another important model, the *harmonic oscillator,* which is used to describe molecular vibration, the energy-level spacing between successive levels is constant. The dependence of the energy spacing on the quantum number in a one-dimensional problem is a function of the *shape* of the potential energy function.

Example 1.8

Consider an electron confined within a one-dimensional box of length 0.10 nm, which is close to the size of an atom. (a) Calculate the difference in energy between the $n = 2$ and $n = 1$ states of the electron. (b) Repeat the calculation in (a) for a N_2 molecule in a one-dimensional box of length 10.0 cm. (c) Calculate the probability of finding the electron in (a) between $x = 0$ and $x = 0.05$ nm for the $n = 1$ state.

Strategy In (a) and (b) we are interested in the energy difference between two successive energy levels, a quantity described by Equation 1.31. In (c), we are interested in the probability of finding the particle in a given region. To do this we need to find $\psi^2(x)$ and apply Equation 1.23.

Solution (a) Use Equation 1.31 with $n = 1$:

$$\Delta E_{n \to n+1} = \frac{h^2}{8mL^2}(2n + 1)$$

$$= \frac{(6.626 \times 10^{-34} \text{ J/s})^2[2(1) + 1]}{8(9.109 \times 10^{-31} \text{ kg})[(0.10 \text{ nm})(1 \times 10^{-9} \text{ m/nm})]^2}$$

$$= 1.8 \times 10^{-17} \text{ J}$$

This energy difference is similar in magnitude to the difference between the $n = 1$ and $n = 2$ states of the hydrogen atom (see Equation 1.16).

(b) The mass of a single N_2 molecule (m in Equation 1.31) is calculated by dividing the molar mass of nitrogen by Avogadro's number (see Section 0.5), then converting to kg:

$$m(N_2) = \frac{28.02 \text{ g mol}^{-1}}{6.0221 \times 10^{23} \text{ mol}^{-1}} \times \frac{1 \text{ kg}}{1000 \text{ g}} = 4.65 \times 10^{-26} \text{ kg}$$

—Continued

Continued—

Substituting this value into Equation 1.31, and using $L = 10.0$ cm, gives

$$\Delta E_{n \rightarrow n+1} = \frac{h^2}{8mL^2}(2n + 1)$$

$$= \frac{(6.626 \times 10^{-34} \text{ J/s})^2 [2(1) + 1]}{8(4.65 \times 10^{-26} \text{ kg})[(10.0 \text{ cm})(1 \times 10^{-2} \text{ m cm}^{-1})]^2}$$

$$= 3.5 \times 10^{-40} \text{ J}$$

This result is 23 orders of magnitude smaller than the energy difference calculated in (a). Compared to the electron, the energy levels of the nitrogen molecule in the box are so closely spaced that they appear almost continuous and would be well approximated by classical mechanics. This is an example of the general rule that quantum mechanical effects become smaller as the size of the confining region increases.

(c) The probability (P) that the electron will be found in the region $0 \leq x \leq L/2$ is given by Equation 1.23 (with $a = 0$ and $b = L/2$):

$$P = \int_0^{L/2} \psi^2(x)dx$$

Using the normalized wavefunction in Equation 1.35 and setting $n = 1$,

$$P = \frac{2}{L} \int_0^{L/2} \sin^2\left(\frac{\pi x}{L}\right)dx$$

$$= \frac{2}{L}\left[\frac{x}{2} - \frac{\sin(2\pi x/L)}{4\pi/L}\right]_0^{L/2}$$

$$= \frac{1}{2}$$

which is not an unexpected result, classically or quantum mechanically. We could have also done this problem without having to solve the integral by noting that the $n = 1$ probability density is symmetric about $x = L/2$.

Practice Exercise The highest-energy electrons in the molecule butadiene ($H_2C =$ CH—CH$=$CH$_2$) can be approximated by a particle in a one-dimensional box with a value of L of approximately 580 pm. Calculate the wavelength of light corresponding to an $n = 2$ to $n = 3$ transition in this molecule using the particle-in-a-box model. How does this compare to the experimental value of 217 nm?

Quantum-Mechanical Tunneling

What would happen if the potential walls surrounding the particle in a one-dimensional box were not infinitely high (Figure 1.25)? The particle could escape the box if the kinetic energy of the particle became greater than the potential energy of the walls. What is more surprising, however, is that we might find the particle outside the box even if its kinetic energy is insufficient to reach the top of the barrier! This phenomenon, called *quantum-mechanical tunneling,* has no analog in classical physics. It arises as a consequence of the wave nature of particles. Quantum-mechanical tunneling has many profound consequences in chemistry, physics, and biology.

Figure 1.25 The potential energy for a particle in a one-dimensional box with finite potential walls of height V_0.

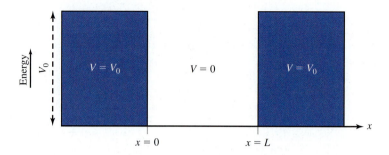

The phenomenon of quantum-mechanical tunneling was introduced in 1928 by the Russian-American physicist George Gamow,[29] and others, to explain α decay, a process in which a nucleus spontaneously decays by emitting an α particle (a helium nucleus, He^{2+}); for example,

$$^{238}_{92}U \longrightarrow\, ^{234}_{90}Th + \alpha$$

Physicists faced a dilemma: For U-238 decay, the measured kinetic energy of the emitted α particle is about 6×10^{-13} J, whereas the potential barrier to escape from the nucleus is on the order of 4×10^{-11} J, or about 70 times greater than the kinetic energy. How does the α particle overcome the barrier and leave the nucleus? Gamow suggested that the α particle, being a quantum-mechanical object, had wavelike properties that allowed it to penetrate a potential barrier. This explanation turned out to be correct. In general, for finite potential barriers, there is some probability of finding the particle outside the box.

Figure 1.26 illustrates this phenomenon for a particle in a one-dimensional box with finite potential walls. In fact, the probability densities for the ground states for three systems with differing particle masses and barrier heights are shown. In general, the effect of tunneling increases with decreasing particle mass and with decreasing height of the potential barrier.

One important practical application of quantum mechanical tunneling is the scanning tunneling microscope described in the inset on page 109.

---

29. Georgy ("George") Antonovich Gamow (1904–1968). Russian-American physicist. In addition to his work on the theoretical nuclear physics, Gamow made important contributions in cosmology and biochemistry.

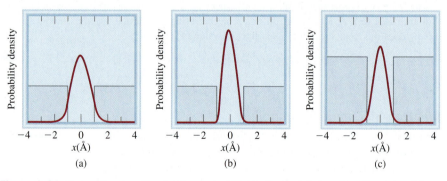

Figure 1.26 Probability densities for the ground state of a particle in a one-dimensional box of length 2 Å with finite potential barriers (as pictured in Figure 1.25). (a) The particle has the mass of an electron and the potential barrier V is equal to the ionization energy of the hydrogen atom. (b) The value of V is the same as in (a), but the mass is that of a hypothetical particle with 10 times the mass of the electron. (c) The particle is an electron, but the potential barrier is twice the hydrogen ionization energy (equal to the ionization energy of He^+). Note that, the total probability outside of the box is significantly less in (b) and (c) than in (a), illustrating that the effect of tunneling is reduced with increasing particle mass and barrier height.

Important Experimental Technique: Electron Microscopy

The electron microscope is an extremely valuable application of the wavelike properties of electrons because it produces images of objects that cannot be seen with the naked eye or with light microscopes. According to the laws of optics, it is impossible to form an image of an object that is smaller than half the wavelength of the light used for the observation. Because the range of visible light wavelengths starts at around 400 nm, or 4×10^{-7} m, we cannot see anything smaller than 2×10^{-7} m. In principle, we can see objects on the atomic and molecular scale by using X-rays, whose wavelengths range from about 0.01 nm to 10 nm. X-rays cannot be focused easily, however, so they do not produce crisp images. Electrons, on the other hand, are charged particles, which can be focused in the same way the image on a TV screen is focused (that is, by applying an electric field or a magnetic field). According to Equation 1.20, the wavelength of an electron is inversely proportional to its velocity. By accelerating electrons to very high velocities, we can obtain wavelengths as short as 0.004 nm.

A different type of electron microscope, called the *scanning tunneling microscope (STM)*, uses quantum mechanical tunneling to produce an image of the atoms on the surface of a sample. Because of its extremely small mass, an electron is able to move or "tunnel" through an energy barrier (instead of going over it). The STM consists of a metal needle with a very fine point (the source of the tunneling electrons). A voltage is maintained between the needle and the surface of the sample to induce electrons to tunnel through space to the sample. As the needle moves over the sample at a distance of a few atomic diameters from the surface, the tunneling current is measured. This current decreases with increasing distance from the sample. By using a feedback loop, the vertical position of the tip can be adjusted to a constant distance from the surface. The extent of these adjustments, which profile the sample, is recorded and displayed as a three-dimensional false-colored image. Both the electron microscope and the STM are among the most powerful tools in chemical and biological research.

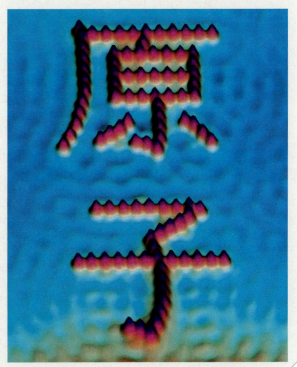

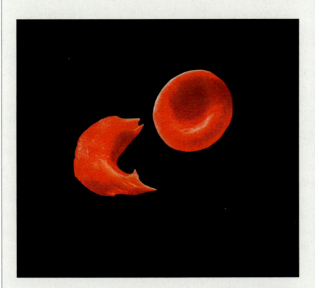

An electron micrograph showing a normal red blood cell and a sickled red blood cell from the same person.

STM image of iron atoms arranged to display the Chinese characters for atom on a copper surface.

1.4 | The Hydrogen Atom Is an Exactly Solvable Quantum-Mechanical System

The simplest atomic system is the hydrogen atom, with its single electron interacting with a positively charged nucleus containing a single proton. Unlike that for the heavier elements, the Schrödinger equation for the electron wavefunctions of a hydrogen atom is exactly solvable. Because the wavefunctions of many-electron atoms

Figure 1.27 The relation between Cartesian coordinates and spherical polar coordinates. For the hydrogen atom, the nucleus is at the origin ($r = 0$), and the electron is at the surface of a sphere of radius r.

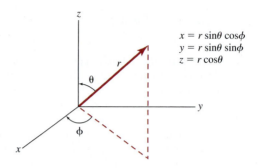

$$x = r \sin\theta \cos\phi$$
$$y = r \sin\theta \sin\phi$$
$$z = r \cos\theta$$

(atoms with more than one electron) share many qualitative properties with those of hydrogen, it is useful to study the hydrogen atom first in some detail.

The Schrödinger Equation for the Hydrogen Atom

By analogy with the one-dimensional Schrödinger equation given in Equation 1.24, the Schrödinger equation for the wavefunction $\psi(x, y, z)$ of a single electron interacting in *three* dimensions with a nucleus of charge $+Ze$ is[30]

$$-\frac{\hbar^2}{2m_e} \left[\frac{\partial^2 \psi}{\partial x^2} + \frac{\partial^2 \psi}{\partial y^2} + \frac{\partial^2 \psi}{\partial z^2} \right] + V(x, y, z)\psi = E\psi \qquad (1.36)$$

where $V(x, y, z)$ was defined previously in Equation 1.7,

$$V(r) = -\frac{Ze^2}{4\pi\varepsilon_0 r}$$

and where $r = \sqrt{x^2 + y^2 + z^2}$ is the distance between the electron and the nucleus. (The symbol ∂ in Equation 1.36 denotes the *partial derivative* and is defined in Appendix 1.)

Because the potential energy depends only on the distance between the nucleus and the electron (that is, it has *spherical symmetry*), Equation 1.36 is most conveniently solved in *spherical polar coordinates*. The relation between Cartesian coordinates (x, y, z) and spherical polar coordinates (r, θ, ϕ) is shown in Figure 1.27. In this coordinate system, the wavefunction is *separable;* that is, it can be written as a product of separate one-dimensional functions of r, θ, and ϕ: $\psi(r, \theta, \phi) = R(r)\Theta(\theta)\Phi(\phi)$. The exact solution to Equation 1.36 in spherical polar coordinates gives rise to three quantum numbers that index the allowed quantum states. (Remember that the number of quantum numbers is generally equal to the dimensionality of the system. In the particle in a one-dimensional box, for instance, there was only one quantum number.) These quantum numbers and their allowed ranges are

▸ The ***principal quantum number:*** $n = 1, 2, 3, \ldots$

▸ The ***angular momentum quantum number:*** $l = 0, 1, \ldots, n - 1$

▸ The ***magnetic quantum number:*** $m_l = -l, \ldots, -1, 0, 1, \ldots, l$

For each value of n, there are n possible values of l, and for each value of l, there are $2l + 1$ possible values of m_l.

---

30. Actually, because the nucleus is not of infinite mass and is not stationary, the electron mass m_e in Equation 1.36 should be replaced with the *reduced mass* $\mu = m_e m_N/(m_e + m_N)$, where m_e and m_N are the electron and nuclear masses, respectively. However, because $m_e \ll m_N$, the difference between the reduced mass and the electron mass is very small (see Problem 1.60).

Each set of these three quantum numbers (n, l, m_l) represents a valid wavefunction for the electron in a hydrogen atom. The wavefunction for a single electron in an atom is called an **atomic orbital.** In quantum mechanics, the position of an electron is described not in terms of orbits, as defined in the Bohr-Sommerfeld model, but in terms of its orbital.

As a result of the spherical symmetry of the potential energy, the energy of the atomic orbitals for hydrogen and hydrogenlike ions depends only upon the value of the principal quantum number (n), and is given by

$$E_n = -\frac{Z^2 e^4 m_e}{8 h^2 \varepsilon_0^2} \frac{1}{n^2} \quad n = 1, 2, 3, \ldots \tag{1.37}$$

which is identical to Equation 1.13, derived using the Bohr model. Note, however, that although the Schrödinger equation for the hydrogen atom is somewhat more complicated than that of a particle in a one-dimensional box, the basic physical origin of the quantization is the same: *Quantization of energy arises when the particle is confined* ("localized") *to a finite region of space.* In the hydrogen atom, the localization of the electron is due to the attractive interaction between the negatively charged electron and the positively charged nucleus.

For $n > 1$, there are multiple orbitals for each value of n, corresponding to different values of l and m_l. Although these combinations of l and m_l represent distinct quantum states of the electron, they have the same energy and are referred to as **degenerate orbitals.** (Equation 1.37 shows that the energy depends only on the principal quantum number n.) Collectively, the set of degenerate orbitals at a particular energy is called an **energy shell.** For example, the orbitals corresponding to ($n = 2$, $l = 0$, $m_l = 0$) and ($n = 2$, $l = 1$, $m_l = -1$) are degenerate and are both members of the second ($n = 2$) energy shell of the hydrogen atom. Within a given energy shell, a set of distinct orbitals that all possess the same value of l form a **subshell.** The subshells are generally designated by the letters s, p, d, . . . , as follows:

l	0	1	2	3	4	5
Name of Subshell	s	p	d	f	g	h

Thus, the set of orbitals with $n = 2$ and $l = 1$ is referred to as a $2p$ subshell and its three orbitals (corresponding to $m_1 = -1$, 0, and $+1$, respectively) are called the $2p$ orbitals. The unusual sequence of letters (s, p, and d) has an historical origin that predates quantum mechanics. Physicists who studied atomic emission spectra tried to correlate the observed spectral lines with the particular quantum states involved in the transitions. They noted that some of the lines were *s*harp; some were rather spread out, or *d*iffuse; and some were very strong and hence were referred to as *p*rincipal lines. Subsequently, the initial letters of each adjective were assigned to the quantum states. However, after the letter d and starting with the letter f (for *f*undamental), the orbital designations follow alphabetical order.

Hydrogenlike Atomic Orbitals

The electron wavefunctions (or atomic orbitals) for the hydrogen atom (and hydrogenlike ions, such as He^+, Li^{2+}, and so on) are given by

$$\psi_{nlm}(r, \theta, \phi) = R_{nl}(r)\, Y_{lm_l}(\theta, \phi) \tag{1.38}$$

where $R_{nl}(r)$ is the radial (r dependent) part of the wavefunction and the angular functions $Y_{lm_l}(\theta, \phi)$ are called *spherical harmonics*.[31] These radial and angular functions for the first three energy shells are shown in Tables 1.2 and 1.3, respectively. A specific

Table 1.2		Radial Part of Hydrogenlike Atomic Orbitals[†]	
n	**l**	**Orbital Name**	$R_{nl}(\bar{r})$; $\bar{r} \equiv Zr/a_0$
1	0	$1s$	$2\left(\dfrac{Z}{a_0}\right)^{\frac{3}{2}} e^{-\bar{r}}$
2	0	$2s$	$\dfrac{\sqrt{2}}{4}\left(\dfrac{Z}{a_0}\right)^{\frac{3}{2}}(2 - \bar{r})e^{-\bar{r}/2}$
	1	$2p$	$\dfrac{\sqrt{6}}{12}\left(\dfrac{Z}{a_0}\right)^{\frac{3}{2}}\bar{r}e^{-\bar{r}/2}$
3	0	$3s$	$\dfrac{2\sqrt{3}}{243}\left(\dfrac{Z}{a_0}\right)^{\frac{3}{2}}(27 - 18\bar{r} + 2\bar{r}^2)e^{-\bar{r}/3}$
	1	$3p$	$\dfrac{2\sqrt{6}}{243}\left(\dfrac{Z}{a_0}\right)^{\frac{3}{2}}(6\bar{r} - \bar{r}^2)e^{-\bar{r}/3}$
	2	$3d$	$\dfrac{4\sqrt{30}}{2430}\left(\dfrac{Z}{a_0}\right)^{\frac{3}{2}}\bar{r}^2 e^{-\bar{r}/3}$

[†]a_0 is the Bohr radius (see section 1.2).

Table 1.3		Angular Part of Hydrogenlike Atomic Orbitals	
l	**m_l**	**Suborbital Name**	$Y_{lm_l}(\theta, \phi)$*
0	0	s	$\left(\dfrac{1}{4\pi}\right)^{\frac{1}{2}}$
1	0	p_z	$\left(\dfrac{3}{4\pi}\right)^{\frac{1}{2}}\cos\theta$
	±1	p_x	$\left(\dfrac{3}{4\pi}\right)^{\frac{1}{2}}\sin\theta\cos\phi$
	±1	p_y	$\left(\dfrac{3}{4\pi}\right)^{\frac{1}{2}}\sin\theta\sin\phi$
2	0	d_{z^2}	$\left(\dfrac{15}{16\pi}\right)^{\frac{1}{2}}[3\cos^2\theta - 1]$
	±1	d_{xz}	$\left(\dfrac{15}{16\pi}\right)^{\frac{1}{2}}\sin\theta\cos\theta\cos\phi$
	±1	d_{yz}	$\left(\dfrac{15}{16\pi}\right)^{\frac{1}{2}}\sin\theta\cos\theta\sin\phi$
	±2	d_{xy}	$\left(\dfrac{15}{16\pi}\right)^{\frac{1}{2}}\sin^2\theta\sin 2\phi$
	±2	$d_{x^2-y^2}$	$\left(\dfrac{15}{16\pi}\right)^{\frac{1}{2}}\sin^2\theta\cos 2\phi$

*For $m_l > 0$, the orbitals shown here are the real orbitals constructed by combining the complex orbitals corresponding to $+m_l$ and $-m_l$.

31. The spherical harmonics can be written as a product of a function of θ and a function of ϕ: $Y_{lm_l}(\theta,\phi) = S_{lm_l}(\theta)\Phi_m(\phi)$, so the wavefunction in Equation 1.38 is separable.

orbital (n, l, m_l) is constructed by multiplying the radial factor $R_{nl}(r)$ in Table 1.2 by the angular function $Y_{lm_l}(\theta, \phi)$ in Table 1.3.

The actual orbitals with $m_l \geq 1$ are complex, making visualization inconvenient. However, because they are equal in energy, the orbitals with $+m_l$ and $-m_l$ within a given subshell can be combined to form two alternate orbitals that are entirely real. It is the real orbitals that are given in Table 1.3. For $n = 2$, for example, the two complex orbitals with $l = 1$ and $m_l = \pm 1$, are

$$\psi_{2,1,1}(r, \theta, \phi) = \left(\frac{3}{8\pi}\right)^{1/2} R_{21}(r) \cos(\theta) e^{+i\phi}$$

$$\psi_{2,1,-1}(r, \theta, \phi) = \left(\frac{3}{8\pi}\right)^{1/2} R_{21}(r) \cos(\theta) e^{-i\phi}$$

where $i = \sqrt{-1}$ and the complex exponentials are given by $e^{i\phi} = \cos(\phi) + i\sin(\phi)$ and $e^{-i\phi} = \cos(\phi) - i\sin(\phi)$. Adding and subtracting these two complex orbitals gives the real $2p_x$ and $2p_y$ orbitals:

$$\psi(2p_x) = \frac{1}{\sqrt{2}}[\psi_{2,1,1} + \psi_{2,1,-1}] = \left(\frac{1}{\sqrt{2}}\right)\left(\frac{3}{8\pi}\right)^{1/2} R_{21}(r) \cos(\theta)[e^{i\phi} + e^{-i\phi}]$$

$$= \left(\frac{3}{4\pi}\right)^{1/2} R_{21}(r) \cos(\theta) \cos(\phi) \tag{1.39}$$

$$\psi(2p_y) = \frac{1}{\sqrt{2}}[\psi_{2,1,1} - \psi_{2,1,-1}] = \left(\frac{1}{\sqrt{2}}\right)\left(\frac{3}{8\pi}\right)^{1/2} R_{21}(r) \cos(\theta)[e^{i\phi} - e^{-i\phi}]$$

$$= \left(\frac{3}{4\pi}\right)^{1/2} R_{21}(r) \cos(\theta) \sin(\phi) \tag{1.40}$$

(The factor $1/\sqrt{2}$ is included to ensure proper normalization.) The $2p_z$ orbital corresponds directly to the orbital with $(n = 2, l = 1, m_l = 0)$, which is real.

Table 1.4 summarizes the relationship between quantum numbers and hydrogen-like atomic orbitals. When $l = 0$, $(2l + 1) = 1$ and there is only one value of m_l, so we have an s orbital. When $l = 1$, $(2l + 1) = 3$, so there are three values of m_l, giving rise to three p orbitals, labeled p_x, p_y, and p_z. When $l = 2$, $(2l + 1) = 5$, so there are five values of m_l, and the corresponding five d orbitals are labeled with more elaborate subscripts. In the following sections we discuss the s, p, and d orbitals separately.

Table 1.4		Relationship between Quantum Numbers and Atomic Orbitals		
n	*l*	m_l	Number of Orbitals	Atomic Orbital Designations
1	0	0	1	$1s$
2	0	0	1	$2s$
	1	$-1, 0, 1$	3	$2p_x, 2p_y, 2p_z$
3	0	0	1	$3s$
	1	$-1, 0, 1$	3	$3p_x, 3p_y, 3p_z$
	2	$-2, -1, 0, 1, 2$	5	$3d_{xy}, 3d_{yz}, 3d_{xz}, 3d_{x^2-y^2}, 3d_{z^2}$

Figure 1.28 (a) A useful measure of the electron density in a 1s orbital is obtained by first dividing the orbital into successive thin spherical shells of thickness dr. (b) A plot of the probability of finding a 1s electron in each shell, called the radial probability function, as a function of distance from the nucleus shows a maximum at 52.9 pm from the nucleus. Interestingly, this is equal to the Bohr radius.

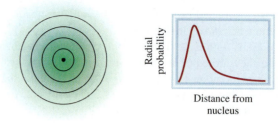

A representation of the electron density distribution surrounding the nucleus in the hydrogen atom. It shows a high probability of finding the electron closer to the nucleus.

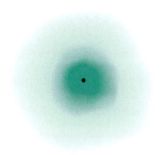

s Orbitals

What are the shapes of the orbitals? Strictly speaking, an orbital does not have a well-defined shape because the wavefunction characterizing the orbital extends from the nucleus to infinity. In that sense, it is difficult to say what an orbital looks like. On the other hand, it is convenient to think of orbitals as having specific shapes, particularly when discussing the formation of chemical bonds between atoms, as we will do in Chapters 3 and 4.

According to Table 1.3, an *s* orbital, defined as any atomic orbital with $l = 0$, has no dependence on θ or ϕ. It depends only on *r*. Such a function is said to be *spherically symmetric* and is defined by the radial function $R_{n0}(r)$. To get a sense of where the electrons are likely to be in these orbitals, we plot the probability of finding the electron in a spherical shell of thickness dr at a distance of r from the nucleus (see Figure 1.27). This probability, denoted by $P(r)$, is given by the expression

$$P(r) = r^2 [R(r)]^2 \tag{1.41}$$

which is called the ***radial probability function***. (The factor r^2 in Equation 1.41 comes from the fact that the surface area of the sphere[32] of radius r increases as r^2.) Figure 1.28 illustrates the radial probability function for the 1s orbital.

The wavefunctions and radial distribution functions for the 1s, 2s, and 3s hydrogenlike atomic orbitals are shown in Figure 1.29 as functions of *r*. Note that the wavefunctions for the 1s, 2s, and 3s orbitals have 0, 1, and 2 *nodes* (points where the wavefunction is zero), respectively. For a general *s* orbital with principal quantum number *n*, the number of nodes is $n - 1$. This increase in the number of nodes as the energy of the particle increases was also seen in the particle in a one-dimensional box and is a general feature of quantum systems.

Also shown in Figure 1.29(c) for the 1s, 2s, and 3s orbitals are the corresponding ***boundary surface diagrams,*** defined as a surface containing 90 percent of the total electron density (defined as ψ^2) in the orbital. The boundary surface diagram serves as a useful representation of the shape of the orbital. All *s* orbitals are spherical but differ in size, which increases as the principal quantum number increases.

p Orbitals

The *p* orbitals come into existence starting with the principal quantum number $n = 2$. If $n = 1$, then the angular momentum quantum number can only assume the value of zero; therefore, there are no 1*p* orbitals, just a 1*s* orbital. As we saw earlier, the magnetic quantum number m_l can have values of -1, 0, and 1 when $l = 1$. Starting with

---

32. The surface area of a sphere of radius r is $4\pi r^2$.

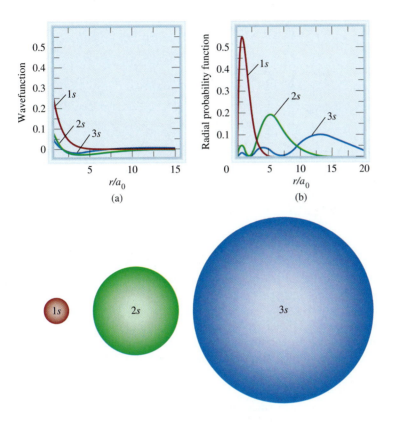

Figure 1.29 (a) The wavefunctions for the 1s, 2s, and 3s orbitals plotted as functions of r in units of the Bohr radius, $a_0 = 52.9$ pm (see Example 1.4). (b) The radial probability functions for the same orbitals. (c) The 90% boundary surfaces for the orbitals in (a). The radii of the boundary surfaces are 2.65 a_0, 9.20 a_0, and 19.5 a_0, for the 1s, 2s and 3s orbitals, respectively. Roughly speaking the size of an orbital is proportional to n^2, where n is the principal quantum number.

$n = 2$ and $l = 1$, the three values of m_l correspond to the three 2p orbitals: $2p_x$, $2p_y$, and $2p_z$. The shapes of orbitals are determined by the angular parts of the wavefunction given in Table 1.3 and Equations 1.39 and 1.40. For p orbitals, the angular part depends on θ and ϕ, so the p orbitals are not spherically symmetric. The letter subscripts x, y, and z in the orbital name indicate the axes along which the orbitals are oriented. The "dumbbell" shape of the three p orbitals is represented in Figure 1.30, where the "+" and "−" signs on the orbital lobes indicate the sign of the wavefunction.

For the wavefunction to change sign, it must go through zero. Each p orbital has a *nodal plane,* defined as a *plane in which the wavefunction is zero,* in the plane perpendicular to the orientation axis. For example, the y-z plane (where $x = 0$) is the nodal plane for the p_x orbital. The three p orbitals are identical in size, shape, and energy; they differ from one another only in orientation. The radial parts of the

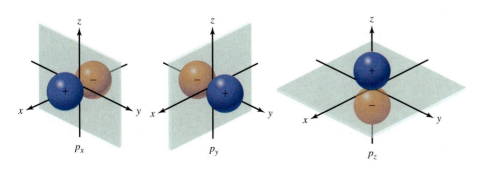

Figure 1.30 The shapes of the three p orbitals showing the sign of the wavefunction and the nodal planes.

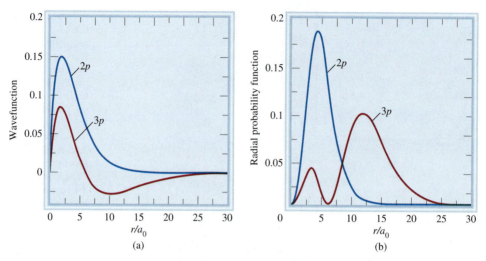

Figure 1.31 (a) The radial part of the wavefunction for the $2p$ and $3p$ hydrogen atomic orbitals. (b) The radial probability function for the orbitals shown in (a).

wavefunctions for the $2p$ and $3p$ orbitals are shown in Figure 1.31, together with the corresponding radial probability functions.[33] Like s orbitals, p orbitals increase in size with increasing principal quantum number n.

d Orbitals and Beyond

When $l = 2$, there are five values of m_l (-2, -1, 0, 1, and 2), which correspond to five d orbitals. The lowest value of n for a d orbital is 3. When $n = 3$ and $l = 2$, we have five $3d$ orbitals, which in their real representations are denoted $3d_{xy}$, $3d_{xz}$, $3d_{yz}$, $3d_{x^2-y^2}$, and $3d_{z^2}$ (the subscripts contain information about their shape and orientation). Representations of the various $3d$ orbitals indicating the shape, sign, and nodal planes are shown in Figure 1.32. Each d orbital has two nodal planes. All the $3d$ orbitals in a hydrogen atom are identical in energy. The d orbitals corresponding to n greater than 3 ($4d$, $5d$, . . .) have similar shapes but are more extended in space (that is, they are larger).

Orbitals with l greater than 2 are labeled f, g, and so on. The f orbitals ($l = 3$) are important in accounting for the behavior of elements with atomic numbers greater than 57, but their shapes are difficult to represent. In general chemistry we are not concerned with orbitals having l values greater than 3 (the g orbitals and beyond).

Example 1.9

List the values of n, l, and ml for orbitals in the $4d$ subshell.

Strategy Use the relationships between n, l, and m_l to solve this problem. What, for example, do the "4" and "d" represent in $4d$?

—Continued

33. From Figure 1.31 you can see that the radial part of the $3p$ orbital has one node, whereas that for $2p$ has no nodes. Nodes in the radial part of the wavefunction are called *radial nodes*. For the hydrogen atom, the number of radial nodes is equal to ($n - l - 1$). The nodal planes, for which a p orbital has one and an s orbital has none, are called *angular nodes*. The number of angular nodes is equal to l. The total number of nodes in a hydrogen atom orbital (angular + radial) is then given by ($n - l - 1$) + $l = n - 1$.

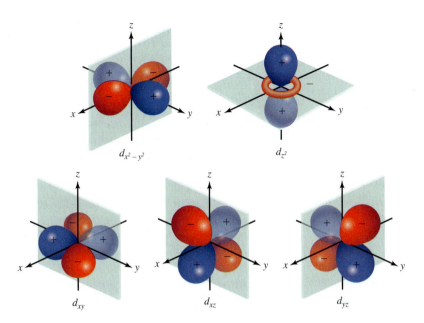

Continued—

Solution The number given in the designation of the subshell is the principal quantum number, so in this case $n = 4$. The letter designates the type of orbital. Because we are dealing with d orbitals, $l = 2$. The values of m_l can vary from $-l$ to $+l$ (that is, from -2 to 2). Therefore, m_l can be -2, -1, 0, 1, or 2.

Check The values of n and l are fixed for $4d$, but m_l can have any one of the five values that correspond to the five d orbitals.

Practice Exercise Give the values of the quantum numbers associated with the orbitals in the $5f$ subshell.

Electron Spin and the Electron Spin Quantum Number (m_s)

Experiments on the emission spectra of hydrogen and sodium atoms indicated that lines in the emission spectra could be split by the application of an external magnetic field. In 1925 Samuel Goudsmit[34] and George Uhlenbeck[35] postulated that these magnetic properties could be explained if electrons possessed an intrinsic angular momentum—as if they were spinning on their own axes, as Earth does. According to electromagnetic theory, a spinning charge generates a magnetic field, and it is this motion that causes an electron to behave like a magnet. These electron magnets are

---

34. Samuel Abraham Goudsmit (1902–1978). Dutch-American physicist. While a student of Paul Ehrenfest at the University of Leiden in 1925, he and fellow student George Uhlenbeck postulated the existence of intrinsic electron spin. Goudsmit was the scientific leader of Operation Alsos at the end of World War II, whose mission was to determine the progress of German efforts toward an atomic bomb.

35. George Eugène Uhlenbeck (1900–1988). Dutch-American physicist. Born in Indonesia (then a Dutch colony), Uhlenbeck studied at the University of Leiden with Paul Ehrenfest, where he, with fellow student Samuel Goudsmit, postulated the existence of intrinsic electron spin. In addition to his work on quantum mechanics, Uhlenbeck made fundamental advances in statistical mechanics and the theory of random processes.

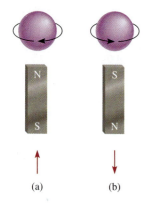

Figure 1.33 The (a) clockwise and (b) counterclockwise spins of an electron. The magnetic fields generated by these two spinning motions are analogous to those from the two magnets. The upward and downward arrows are used to denote the direction of "spin."

highly quantum mechanical and are quantized so that only two possible spinning motions of an electron are possible, clockwise and counterclockwise (Figure 1.33). To take this **electron spin** into account, it is necessary to introduce a fourth quantum number, the **electron spin quantum number** (m_s), which has a value of $+\frac{1}{2}$ or $-\frac{1}{2}$. It is customary to refer to a value $m_s = +\frac{1}{2}$ as "spin up" and to a value $m_s = -\frac{1}{2}$ as "spin down."

One major piece of evidence that led Goudsmit and Uhlenbeck to postulate the existence of electron spin came from an experiment performed by Otto Stern[36] and Walther Gerlach[37] in 1922. Figure 1.34 shows their basic experimental arrangement. In such an experiment, a beam of gaseous hydrogen atoms generated in a hot furnace passes through a nonhomogeneous magnetic field. The interaction between this electron in each atom and the magnetic field causes the atom to be deflected from its straight-line path. Because the spin of each electron can be either "up" or "down," with equal probability, one half of the atoms are deflected in one way; and the other half of the atoms are deflected in the other direction. Thus, two spots of equal intensity are observed on the detecting screen.

It must be noted that the picture of the electron as a tiny spinning magnet, although useful and convenient as a visualization tool, should not be taken literally. On a fundamental level, the electron "spin" can only be understood by going beyond the Schrödinger equation to include the effects of Einstein's theory of special relativity.

To summarize, the quantum state of an electron in hydrogen (or a hydrogenlike ion) is *completely* specified by the four quantum numbers n, l, m_l, and m_s. The energy of this electron is determined only by the value of the *principal quantum number n* (according to Equation 1.37). The *angular momentum quantum number l* determines the basic shape of the orbital, and the *magnetic quantum number m_l* determines its orientation in space. The *electron spin quantum number m_s* determines the intrinsic "spin" of the electron. We will see in Chapter 2 that although an exact solution to the

---

36. Otto Stern (1888–1969). German physicist. He made important contributions to the study of the magnetic properties of atoms and to the kinetic theory of gases. Stern was awarded the Nobel Prize in Physics in 1943.

37. Walther Gerlach (1889–1979). German physicist. Gerlach's main area of research was in quantum theory.

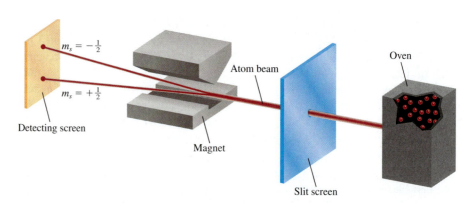

Figure 1.34 Experimental arrangement for demonstrating the existence of electron spin. A beam of atoms is directed through a magnetic field. When a hydrogen atom with a single electron passes through the field, it is deflected in one direction or the other, depending on the direction of the spin of its electron. In a stream consisting of many atoms, there will be equal distribution of the two kinds of spins, so that two spots of equal intensity are detected on the screen.

Schrödinger equation for many-electron atoms is not possible, the four hydrogen quantum numbers, and the basic orbital shapes they represent, retain their usefulness in describing the quantum state of the electrons in those atoms. *Most importantly, the mathematical properties of these four quantum numbers form the basis for the buildup of the elements in the periodic table.*

Summary of Facts and Concepts

Section 1.1

▶ At the end of the nineteenth century, scientists began to realize that the laws of classical physics were incompatible with a number of new experiments that probed the nature of atoms and molecules and their interaction with light. Through the work of a number of scientists over the first three decades of the twentieth century, a new theory—quantum mechanics—was developed that was able to explain the behavior of objects on the atomic and molecular scale.

▶ The quantum theory developed by Planck successfully explains the emission of radiation by heated solids. The quantum theory states that radiant energy is emitted by atoms and molecules in small discrete amounts (quanta), rather than over a continuous range. This behavior is governed by the relationship $E = h\nu$, where E is the energy of the radiation, h is Planck's constant, and ν is the frequency of the radiation. Energy is always emitted in whole-number multiples of $h\nu$ ($1\ h\nu$, $2\ h\nu$, $3\ h\nu$, . . .).

▶ Using quantum theory, Einstein solved another mystery of physics—the photoelectric effect. Einstein proposed that light can behave like a stream of particles (photons).

Section 1.2

▶ The line spectrum of hydrogen, yet another mystery to nineteenth-century physicists, was also explained by applying an early version of quantum theory. Bohr developed a model of the hydrogen atom in which the energy of its single electron is quantized, limited to certain energy values determined by an integer called the quantum number.

▶ The lowest energy state of an electron is its ground state, and states with energies higher than the ground-state energy are excited states. In the Bohr model, an electron emits a photon when it drops from a higher-energy state to a lower-energy state. The release of specific amounts of energy in the form of photons accounts for the lines in the hydrogen emission spectrum.

▶ In spite of its success with the hydrogen atom, the Bohr model was deficient because it was unable to account for the emission spectra of heavier atoms, such as helium.

Section 1.3

▶ De Broglie extended Einstein's wave-particle description of light to all matter in motion. The wavelength of a moving particle of mass m and velocity u is given by the de Broglie equation $\lambda = h/mu$ (Equation 1.20).

▶ The realization that matter at the atomic and subatomic scale possesses wavelike properties lead to the development of the Heisenberg uncertainty principle, which states that it is impossible to know simultaneously both the position (x) and the momentum (p) of a particle with certainty (see Equation 1.22).

▶ The Schrödinger equation (Equation 1.24) describes the motions and energies of submicroscopic particles. This equation, in which the state of a quantum particle is described by its wavefunction, launched modern quantum mechanics and a new era in physics. The wavefunction contains information about the probability of finding a particle in a given region of space.

▶ The Schrödinger equation can be exactly solved for a "particle in a one-dimentional box," an idealized model, which, despite its simplicity, can be applied to understand the behavior of a number of real systems of chemical and biological interest.

▶ Because of the wave nature of matter, a quantum particle can sometimes overcome energy barriers that the particle, if behaving classically, would have insufficient energy to cross. This phenomenon, called quantum tunneling, is the basis for a number of scientific applications, such as the scanning tunneling microscope (STM).

Section 1.4

▶ The Schrödinger equation tells us the possible energy states of the electron in a hydrogen atom and the probability of its location in a particular region surrounding the nucleus.

▶ The quantum state of an electron in a hydrogen atom is given by its wavefunction (or atomic orbital) [$\psi(r, \theta, \phi)$] and the distribution of electron density in space is given by $\psi^2(r, \theta, \phi)$. The sizes and shapes of atomic orbitals can be represented by electron density diagrams or boundary surface diagrams.

▶ Four quantum numbers characterize the electron wavefunction (atomic orbital) in a hydrogen atom: The principal quantum number n identifies the main energy level, or shell, of the orbital; the angular momentum quantum number l indicates the shape of the orbital; the magnetic quantum number m_l specifies the orientation of the orbital in space; and the electron spin quantum number m_s indicates the direction of the electron's spin on its own axis.

▸ The single s orbital for each energy level is spherical and centered on the nucleus. There are three p orbitals present at $n = 2$ and higher; each has two lobes, and the

pairs of lobes are arranged at right angles to one another. Starting with $n = 3$, there are five d orbitals, with more complex shapes and orientations.

▸ The energy of the electron in a hydrogen atom is determined solely by its principal quantum number.

Key Words

amplitude, p. 74	electron spin quantum number, p. 118	line spectra, p. 83	quantum number, p. 88
angular momentum, p. 86	emission spectra, p. 83	magnetic quantum number, p. 110	radial probability function, p. 114
angular momentum quantum number, p. 110	energy levels, p. 88	node, p. 95	Rydberg constant, p. 84
atomic orbital, p. 111	energy shell, p. 111	particle-in-a-box model, p. 102	scanning tunneling microscope, p. 109
blackbody radiation, p. 77	excited state, p. 88	photoelectric effect, p. 81	standing wave, p. 95
Bohr model, p. 86	frequency, p. 74	photon, p. 81	subshell, p. 111
Bohr radius, p. 88	ground state, p. 88	principal quantum number, p. 110	time-independent Schrödinger equation, p. 100
boundary surface diagrams, p. 114	Heisenberg uncertainty principle, p. 97	quantum, p. 79	wavefunction, p. 99
de Broglie wavelength, p. 94	hydrogenlike ion, p. 87	quantum-mechanical tunneling, p. 107	wavelength, p. 74
degenerate orbitals, p. 111	ionization, p. 88	quantum mechanics, p. 100	zero-point energy, p. 106
electron spin, p. 118	ionization energy, p. 88		

Problems

Classical Physics Does Not Adequately Describe the Interaction of Light with Matter

1.1 (a) What is the wavelength (in nanometers) of light having a frequency of 8.6×10^{13} Hz? (b) What is the frequency (in Hz) of light having a wavelength of 566 nm?

1.2 (a) What is the frequency of light having a wavelength of 456 nm? (b) What is the wavelength (in nanometers) of radiation having a frequency of 2.45×10^9 Hz? (This is the type of radiation used in microwave ovens.)

1.3 The SI unit of time is the second, which is defined as 9,192,631,770 cycles of radiation associated with a certain emission process in the cesium atom. Calculate the wavelength of this radiation (to three significant figures). In which region of the electromagnetic spectrum is this wavelength found?

1.4 The SI unit of length is the meter, which is defined as the length equal to 1,650,763.73 wavelengths of the light emitted by a particular energy transition in krypton atoms. Calculate the frequency of the light to three significant figures.

1.5 A photon has a wavelength of 624 nm. Calculate the energy of the photon in joules.

1.6 The blue of the sky results from the scattering of sunlight by air molecules. The blue light has a frequency of about 7.5×10^{14} Hz. (a) Calculate the

wavelength, in nm, associated with this radiation, and (b) calculate the energy, in joules, of a single photon associated with this frequency.

1.7 A photon has a frequency of 6.0×10^4 Hz. (a) Convert this frequency into wavelength (nm). Into what region of the electromagnetic spectrum does this frequency fall? (b) Calculate the energy (in joules) of this photon. (c) Calculate the energy (in joules) of 1 mol of photons all with this frequency.

1.8 What is the wavelength, in nm, of radiation that has an energy content of 1.0×10^3 kJ mol^{-1}? In which region of the electromagnetic spectrum is this radiation found?

1.9 When copper is bombarded with high-energy electrons, X-rays are emitted. Calculate the energy (in joules) associated with the photons if the wavelength of the X-rays is 0.154 nm.

1.10 A particular form of electromagnetic radiation has a frequency of 8.11×10^{14} Hz. (a) What is its wavelength in nanometers? In meters? (b) To what region of the electromagnetic spectrum would you assign it? (c) What is the energy (in joules) of one quantum of this radiation?

1.11 The retina of the human eye can detect light when the radiant energy incident on it exceeds a minimum value of 4.0×10^{-17} J. How many photons does this energy correspond to if the light has a wavelength of 600 nm?

1.12 A microwave oven operating at a wavelength of 1.22×10^8 nm is used to heat 150 mL of water (roughly the volume of a tea cup) from 20°C to 100°C. Given that it takes 4.186 J of energy to heat 1 mL of water by 1°C, calculate the number of photons needed if 92.0 percent of the microwave energy is converted into the thermal energy of water.

1.13 Careful spectral analysis shows that the familiar yellow light of sodium vapor lamps (used in some streetlights) is made up of photons of two wavelengths, 589.0 nm and 589.6 nm. What is the difference in energy (in joules) between photons with these wavelengths?

1.14 Explain how scientists are able to measure the temperature of the surface of stars. (*Hint:* Treat stellar radiation like radiation from a blackbody.)

1.15 Using Wien's law, calculate the temperatures for which the values of λ_{max} for blackbody radiation are 100 nm, 300 nm, 500 nm, and 800 nm. To what region of the electromagnetic spectrum does each of these wavelengths correspond? Using your results, speculate as to how night vision goggles might work.

1.16 A photoelectric experiment was performed by separately shining a laser at 450 nm (blue light) and a laser at 560 nm (yellow light) on a clean metal surface and measuring the number and kinetic energy of the ejected electrons. Which light would generate more electrons? Which light would eject electrons with the greatest kinetic energy? Assume that the same number of photons is delivered to the metal surface by each laser and that the frequencies of the laser lights exceed the threshold frequency.

1.17 In a photoelectric experiment, a student uses a light source whose frequency is greater than that needed to eject electrons from a certain metal. However, after continuously shining the light on the same area of the metal for a long period, the student notices that the maximum kinetic energy of ejected electrons begins to decrease, even though the frequency of the light is held constant. How would you account for this behavior?

1.18 The maximum wavelength of light that can be used to eject electrons from a clean sodium metal surface is 520 nm. (a) Calculate the photoelectric work function (in joules) for sodium. (b) If 450-nm light is used, what is the kinetic energy of the ejected electrons?

1.19 The threshold frequency for dislodging an electron from a clean zinc metal surface is 8.54×10^{14} Hz. Calculate the minimum amount of energy (in joules) required to remove an electron from the zinc surface (that is, calculate the work function for zinc).

The Bohr Model Was an Early Attempt to Formulate a Quantum Theory of Matter

1.20 Some copper compounds emit green light when they are heated in a flame. How would you determine whether the light is of one wavelength or a mixture of two or more wavelengths?

1.21 How is it possible for a fluorescent material to emit radiation in the ultraviolet region after absorbing visible light. Explain your answer.

1.22 Explain how astronomers are able to tell which elements are present in distant stars by analyzing the electromagnetic radiation emitted by the stars.

1.23 Consider the following energy levels of a hypothetical atom:

E_4: -1.0×10^{-19} J
E_3: -5.0×10^{-19} J
E_2: -10×10^{-19} J
E_1: -15×10^{-19} J

(a) What is the wavelength of the photon needed to excite an electron from E_1 to E_4? (b) What is the energy (in joules) a photon must have to excite an electron from E_2 to E_3? (c) When an electron drops from the E_3 level to the E_1 level, the atom is said to undergo emission. Calculate the wavelength of the photon emitted in this process.

1.24 The first line of the Balmer series occurs at a wavelength of 656.3 nm. What is the energy difference between the two energy levels involved in the emission that results in this spectral line?

1.25 Calculate the wavelength (in nm) of a photon that must be absorbed by the electron in a hydrogen atom to excite it from the $n = 3$ to the $n = 5$ energy level.

1.26 Calculate the frequency (in Hz) and wavelength (in nm) of the emitted photon when an electron drops from the $n = 4$ to the $n = 2$ energy level in a hydrogen atom.

1.27 The He$^+$ ion contains only one electron and is therefore a hydrogenlike ion that can be described by the Bohr model. Calculate the wavelengths, in increasing order, of the first four transitions in the Balmer series of the He$^+$ ion. Compare these wavelengths with the same transitions in a hydrogen atom. Comment on the differences.

1.28 Calculate the radii for the Bohr orbits of a hydrogen atom with $n = 2$ and $n = 3$.

1.29 Scientists have found interstellar hydrogen with quantum number n in the hundreds. Calculate the wavelength of light emitted when a hydrogen atom undergoes a transition from $n = 236$ to $n = 235$. In what region of the electromagnetic spectrum does this wavelength fall?

1.30 Calculate the frequency of light necessary to eject an electron from the ground state of a hydrogen atom.

Matter Has Wavelike Properties

1.31 What are the wavelengths associated with (a) an electron moving at 1.50×10^8 cm s^{-1} and (b) a 60.0-g tennis ball moving at 1500 cm s^{-1}?

1.32 Thermal neutrons are neutrons that move at speeds comparable to those of air molecules at room temperature. These neutrons are most effective in initiating a nuclear chain reaction among ^{235}U isotopes. Calculate the wavelength (in nm) associated with a beam of neutrons moving at 7.00×10^2 m s^{-1}. (The mass of a neutron is 1.675×10^{-27} kg.)

1.33 Protons can be accelerated to speeds near that of light in particle accelerators. Estimate the wavelength (in nm) of such a proton moving at 2.90×10^8 m s^{-1}. (The mass of a proton is 1.673×10^{-27} kg.)

1.34 What is the de Broglie wavelength, in cm, of a 12.4-g hummingbird flying at 1.20×10^2 mph? (1 mile = 1.61 km.)

1.35 What is the de Broglie wavelength (in nm) associated with a 2.5-g Ping-Pong ball traveling 35 mph?

1.36 Suppose that the uncertainty in determining the position of an electron circling an atom in an orbit is 0.4 Å. What is the uncertainty in the velocity?

1.37 Sketch the probability densities for the first three energy levels of the particle in a one-dimensional box. Without doing any calculations, determine the average value of the position of the particle (x) corresponding to each distribution.

1.38 Calculate the frequency of light required to promote a particle from the $n = 2$ to the $n = 3$ levels of the particle in a one-dimensional box, assuming that $L = 5.00$ Å and that the mass is equal to that of an electron.

1.39 Suppose we used the particle in a one-dimensional box as a crude model for the electron in a hydrogen atom. What value of L would you have to use so that the energy of the $n = 1$ to $n = 2$ transitions in both models would be the same? Is this a reasonable value for the size of a hydrogen atom?

1.40 Consider an electron in a one-dimensional box subject to an electric field E. The potential energy of this system is now

$$V(x) = \begin{cases} \infty & x \le 0; x \ge L \\ -eEx & 0 < x < L \end{cases}$$

Using what you know about the Schrödinger equation, sketch the wavefunctions for the first three energy levels of this system. What can you

predict about the average value of the position of the particle x?

The Hydrogen Atom Is an Exactly Solvable Quantum-Mechanical System

1.41 Give the values of the quantum numbers associated with the following orbitals: (a) 3s, (b) 4p, and (c) 3d.

1.42 Give the values of the quantum numbers associated with the following orbitals: (a) 2p, (b) 6s, and (c) 5d.

1.43 List all the possible subshells and orbitals associated with the principal quantum number $n = 5$.

1.44 List all the possible subshells and orbitals associated with the principal quantum number $n = 6$.

1.45 Calculate the total number of electrons that can occupy (a) one s orbital, (b) three p orbitals, (c) five d orbitals, (d) seven f orbitals.

1.46 Discuss the similarities and differences between a 1s and a 2s hydrogenlike orbital.

1.47 What are the similarities and differences between a $2p_x$ and a $3p_y$ hydrogenlike orbital?

1.48 The ionization energy is the energy required to completely remove an electron from the ground state of an atom or ion. Calculate the ionization energies (in kJ mol^{-1}) of the He$^+$ and Li^{2+} ions.

1.49 An electron in the hydrogen atom makes a transition from an energy state of principal quantum number n_1 to the $n = 2$ state. If the photon emitted has a wavelength of 434 nm, what is the value of n_1?

1.50 An electron in a hydrogen atom is excited from the ground state to the $n = 4$ state. Which of the following statements are true and which are false?
  (a) The $n = 4$ state is the first excited state.
  (b) It takes more energy to ionize (remove) the electron from $n = 4$ than from the ground state.
  (c) The electron is farther from the nucleus (on average) in $n = 4$ than in the ground state.
  (d) The wavelength of light emitted when the electron drops from $n = 4$ to $n = 1$ is longer than that from $n = 4$ to $n = 2$.
  (e) The wavelength of light the atom absorbs in going from $n = 1$ to $n = 4$ is the same as that emitted as it goes from $n = 4$ to $n = 1$.

Additional Problems

1.51 The radioactive Co-60 isotope is used in nuclear medicine to treat certain types of cancer. Calculate the wavelength and frequency of emitted gamma radiation photons having energy equal to 1.29×10^{11} J mol^{-1}.

1.52 When a compound containing cesium ion is heated in a Bunsen burner flame, photons of energy of 4.30×10^{-19} J are emitted. What color is a cesium flame?

1.53 A ruby laser produces radiation of wavelength 633 nm in pulses whose duration is 1.00×10^{-9} s. (a) If the laser produces 0.376 J of energy per pulse, how many photons are produced in each pulse? (b) Calculate the power (in watts) delivered by the laser per pulse. ($1 \text{ W} = 1 \text{ J s}^{-1}$.)

1.54 A 368-g sample of water absorbs infrared radiation at 1.06×10^4 nm from a carbon dioxide laser. Suppose all the absorbed radiation is converted to energy in the form of heat. Calculate the number of photons at this wavelength required to raise the temperature of the water by 5.00°C. (It takes 4.184 J of heat energy to raise the temperature of 1.00 g of water by 1.00°C.)

1.55 Photodissociation of water

$$H_2O(l) + h\nu \longrightarrow H_2(g) + \frac{1}{2} O_2(g)$$

has been suggested as a source of hydrogen. The energy for the reaction is 285.8 kJ per mole of water decomposed. Calculate the maximum wavelength (in nm) that would provide the necessary energy. In principle, is it feasible to use sunlight as a source of energy for this process?

1.56 Spectral lines of the Lyman and Balmer series do not overlap. Verify this statement by calculating the longest wavelength associated with the Lyman series and the shortest wavelength associated with the Balmer series (in nm).

1.57 Only a fraction of the electrical energy supplied to a tungsten lightbulb is converted to visible light. The rest of the energy shows up as infrared radiation (that is, heat). A 75-W light bulb converts 15.0 percent of the energy supplied to it into visible light (assume the wavelength to be 550 nm). How many photons are emitted by the light bulb per second? ($1 \text{ W} = 1 \text{ J s}^{-1}$.)

1.58 Certain sunglasses have small crystals of silver chloride (AgCl) incorporated in the lenses. When the lenses are exposed to light of the appropriate wavelength, the following reaction occurs:

$$AgCl \longrightarrow Ag + Cl$$

The Ag atoms formed produce a uniform gray color that reduces the glare. If the energy required for the preceding reaction is 248 kJ mol^{-1}, calculate the maximum wavelength of light that can induce this process.

1.59 A student carried out a photoelectric experiment by shining visible light on a clean piece of cesium metal. She determined the kinetic energy of the ejected electrons by applying a retarding voltage such that the current due to the electrons read exactly zero. The condition was reached when $eV = (1/2)m_e u^2$, where e is the electron charge, V is the retarding voltage, and u is the velocity of the electron. Her results were as follows:

λ (nm)   405.0 435.5 480.0 520.0 577.7 650.0
V (volt)   1.475 1.268 1.027 0.886 0.667 0.381

Rearrange Equation 1.4 to read

$$\nu = \frac{\Phi}{h} + \frac{e}{h} V$$

then determine the values of h and Φ graphically.

1.60 Equations 1.13 and 1.16 were derived assuming that the electron orbits a stationary nucleus; however, this would only be true if the nucleus were of infinite mass. Although the nucleus is much more massive than the electron, it is not infinitely so, and both the nucleus and the electron orbit a common point centered very close to the nuclear position. To correct for this you must replace the mass of the electron in Equations 1.13 and 1.16 with the quantity

$$\mu = \frac{m_e m_N}{m_e + m_N},$$ called the *reduced mass*, where m_N is

the mass of the nucleus. Calculate the percent change in the Rydberg constants (R_H) for H and He$^+$ using this correction.

1.61 From the exact solution to the Schrödinger equation, the square of the orbital angular momentum of the electron in a hydrogen atom is equal to $l(l + 1)\hbar^2$. (a) What is the orbital angular momentum for an electron in an s orbital? (b) How does this value compare to the Bohr model prediction?

1.62 The sun is surrounded by a white circle of gaseous material called the corona, which becomes visible during a total eclipse of the sun. The temperature of the corona is in the millions of degrees Celsius, which is high enough to break up molecules and remove some or all of the electrons from atoms. One method by which astronomers can estimate the temperature of the corona is by studying the emission lines of ions of certain elements. For example, the emission spectrum of Fe^{14+} ions has been recorded and analyzed. Knowing that it takes 3.5×10^4 kJ mol^{-1} to convert Fe^{13+} to Fe^{14+}, estimate the temperature of the sun's corona. (*Hint:* The average kinetic energy of 1 mol of a gas is $3/2 RT$, where the gas constant $R = 8.314$ J mol^{-1} K^{-1} and T is the absolute temperature in kelvin.

1.63 An electron in an excited state in a hydrogen atom can return to the ground state in two different ways: (a) via a direct transition in which a photon of wavelength λ_1 is emitted and (b) via an intermediate excited state reached by the emission of a photon of wavelength λ_2. This intermediate excited state then decays to the ground state by emitting another photon of wavelength λ_3. Derive an equation that relates λ_1, λ_2, and λ_3.

1.64 Alveoli are the tiny sacs of air in the lungs whose average diameter is 5.0×10^{-5} m. Consider an oxygen molecule (mass = 5.3×10^{-26} kg) trapped within a sac. Calculate the uncertainty in the velocity of the oxygen molecule. (*Hint:* The maximum uncertainty in the position of the molecule is given by the diameter of the sac.)

1.65 How many photons at 660 nm must be absorbed to melt 5.0×10^2 g of ice? On average, how many H_2O molecules does one photon convert from ice to water? (*Hint:* It takes 334 J to melt 1 g of ice at 0°C.)

1.66 The UV light that is responsible for tanning human skin falls in the 320- to 400-nm region. Calculate the total energy (in joules) absorbed by a person exposed to this radiation for 2.0 h, given that there are 2.0×10^{16} photons hitting the surface of Earth per square centimeter per second over an 80-nm (320 nm to 400 nm) range and that the exposed body area is 0.45 m^2. Assume that only half of the radiation is absorbed and that the remainder is reflected by the body. (*Hint:* Use an average wavelength of 360 nm when calculating the energy of a photon.)

1.67 In 1996 physicists created an antiatom of hydrogen. In such an atom, which is the antimatter equivalent of an ordinary atom, the electrical charges of all the component particles were reversed. Thus, the nucleus of an antiatom is made of an antiproton, which has the same mass as a proton but bears a negative charge, and the electron is replaced by an antielectron (also called positron), which has the same mass as an electron, but bears a positive charge. Would you expect the energy levels, emission spectra, and atomic orbitals of an anti–hydrogen atom to be different from those of a hydrogen atom? What would happen if an antiatom of hydrogen collided with a hydrogen atom?

1.68 In Chapter 5 we will see that the mean velocity of a gas molecule is given by $\sqrt{3RT/\mathcal{M}}$ where T is the absolute temperature, $\mathcal{M}$ is the molar mass, and R is the gas constant ($R = 8.314$ J mol^{-1} K^{-1}). Using this information, calculate the de Broglie wavelength of a nitrogen (N_2) molecule at 300 K.

1.69 When an electron makes a transition between energy levels of a hydrogen atom by absorbing or emitting a photon, there are no restrictions on the initial and final values of the principal quantum number n. However, there is a quantum mechanical rule that restricts the initial and final values of the orbital angular momentum quantum number l. This is the *selection rule,* which states that $\Delta l = \pm 1$, that is, in a transition, the value of l can only increase or decrease by one. According to this rule, which of the following transitions are allowed: (a) $1s \longrightarrow 2s$, (b) $2p \longrightarrow 1s$, (c) $1s \longrightarrow 3d$, (d) $3d \longrightarrow 4f$, (e) $4d \longrightarrow 3s$?

1.70 In an electron microscope, electrons are accelerated by passing them through a voltage difference. The kinetic energy thus acquired by the electrons is equal to the voltage times the charge on the electron. Thus, a voltage difference of 1 volt (V) imparts a kinetic energy of 1.602×10^{-19} volt-coulomb or 1.602×10^{-19} J. Calculate the wavelength associated with electrons accelerated by 5.00×10^3 V.

1.71 (a) An electron in the ground state of the hydrogen atom moves at an average speed of 5×10^6 m s^{-1}. If the speed of the electron is known to an uncertainty of 1 percent, what is the uncertainty in knowing its position? Given that the radius of the hydrogen atom in the ground state is 5.29×10^{-11} m, comment on your result. (b) A 0.15-kg baseball thrown at 100 mph has a momentum of 6.7 kg m s^{-1}. If the uncertainty in measuring the momentum is 1.0×10^{-7} times the momentum, calculate the uncertainty in the baseball's position.

1.72 When two atoms collide, some of their kinetic energy may be converted into electronic energy in one or both atoms. If the average kinetic energy is about equal to the energy for some allowed electronic transition, an appreciable number of atoms can absorb enough energy through an inelastic collision to be raised to an excited electronic state. (a) Calculate the average kinetic energy per atom in a gas sample at 298 K. (b) Calculate the energy difference between the $n = 1$ and $n = 2$ levels in hydrogen. (c) At what temperature is it possible to excite a hydrogen atom from the $n = 1$ level to the $n = 2$ level by collision? (The average kinetic energy of 1 mol of a gas at low pressures is $3/2$ RT, where the gas constant $R = 8.314$ J mol^{-1} K^{-1} and T is the absolute temperature in kelvin.)

1.73 Calculate the energies needed to remove an electron from the $n = 1$ state and the $n = 5$ state in the Li^{2+} ion. What is the wavelength (in nm) of the emitted photon in a transition from $n = 5$ to $n = 1$?

1.74 In the beginning of the twentieth century, some scientists thought that the nucleus contained both electrons and protons. Use the Heisenberg uncertainty principle to show that an electron cannot be confined within a nucleus. Repeat your calculation for a proton. Comment on your results. Assume the radius of a nucleus to be 1.0×10^{-15} m. The masses of an electron and proton are 9.109×10^{-31} kg and 1.673×10^{-27} kg, respectively. (*Hint:* Treat the radius of the nucleus as the uncertainty in position.)

1.75 Given a one-dimensional probability density $p(x)$, the average value, $\bar{f}$, of any function of x, $f(x)$, is given by

$$\bar{f} = \int_a^b f(x)\, p(x)\, dx$$

For the particle in a one-dimensional box, a and b are 0 and L, respectively, and $p(x) = |\psi(x)|^2$. For the first three energy levels of the particle in a one-dimensional box, calculate the average values of x and x^2. (*Hint:* You will need to consult a standard table of integrals for this problem.)

1.76 Using the radial probability function, calculate the average values of r for the $1s$ and $2s$ orbitals of the hydrogen atom. How do these compare with the most probable values? (*Hint:* You will need to consult a standard table of integrals for this problem.)

1.77 An important property of the wavefunctions of the particle in a one-dimensional box is that they are *orthogonal;* that is,

$$\int_0^L \psi_n(x) \, \psi_m(x) \, dx = 0 \text{ for } n \neq m$$

Show that the wavefunctions defined in Equation 1.35 have this property.

1.78 In the particle in a two-dimensional box, the particle is confined to a rectangular box with side dimensions L_x and L_y. The energy levels for this system are given by

$$E(n_x, n_y, n_z) = \frac{h^2}{8m}\left[\frac{n_x^2}{L_x^2} + \frac{n_y^2}{L_y^2}\right]$$

where the three quantum numbers (n_x and n_y) can be any set of two strictly positive integers. The wavefunction corresponding to the state (n_x, n_y) is

$$\psi(x, y, z) = \left(\frac{4}{L_xL_y}\right)^{\frac{1}{2}}\sin\left(\frac{n_x\pi x}{L_x}\right)\sin\left(\frac{n_y\pi y}{L_y}\right)$$

If the box is a square ($L_x = L_y$), determine the set of two quantum numbers for the five lowest energy levels. (Some energy levels may consist of several degenerate quantum states.) Try to represent the corresponding wavefunctions graphically.

Answers to Practice Exercises

1.1 3.28 m **1.2** 9.94×10^{-16} J **1.3** 264 nm, 304 nm **1.4** 105.9 pm **1.5** 273 nm, UV **1.6** 56.54 nm **1.7** 5.95×10^4 m s^{-1} **1.8** 222 nm **1.9** $n = 5$, $l = 3$, $m_l = -3, -2, -1, 0, +1, +2, +3$

Many-Electron Atoms and the Periodic Table

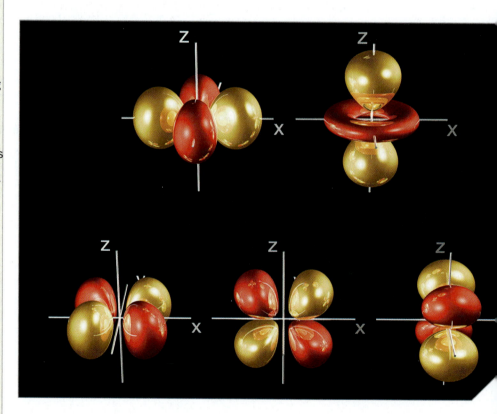

Chemists in the nineteenth century recognized periodic trends in the physical and chemical properties of elements, long before quantum theory came onto the scene. Although these chemists were not aware of the existence of electrons and protons, their efforts to systematize the chemistry of the elements were remarkably successful. Their main sources of information were the atomic masses of the elements and other known physical and chemical properties. Modern quantum theory allows us to understand these periodic trends in terms of the ways in which the electrons are distributed among the atomic orbitals of an atom.

2.1 | The Wavefunctions of Many-Electron Atoms Can Be Described to a Good Approximation Using Atomic Orbitals

The Schrödinger equation works nicely for the simple hydrogen atom (Equation 1.36), but it cannot be solved exactly for atoms containing more than one electron! Fortunately, the atomic orbital concepts from the hydrogen atom can be used to construct approximate (but reliable) solutions to the Schrödinger equation for **many-electron atoms** (that is, *atoms containing two or more electrons*).

Wavefunctions for Many-Electron Atoms

For the hydrogen atom with its single electron, the electronic wavefunction is a function of three variables, namely the x, y, and z coordinates of the electron. For a many-electron atom, the wavefunction is a function of the positions of all electrons. The wavefunction for helium (He), for example, is a function of six position variables: x_1, y_1, and z_1 for electron 1, and x_2, y_2, and z_2 for electron 2:

$$\psi(\text{He}) = \psi(x_1, y_1, z_1, x_2, y_2, z_2)$$

In general, for an atom with N electrons, we have

$$\psi(N\text{-electron atom}) = \psi(x_1, y_1, z_1, x_2, y_2, z_2, \ldots, x_N, y_N, z_N) \qquad \textbf{(2.1)}$$

(where x_i, y_i, and z_i are the Cartesian coordinates of the ith electron relative to the position of the atom nucleus). Thus, the complexity of the wavefunction increases dramatically with the number of electrons. For example, the electronic wavefunction for the neon (Ne) atom with its 10 electrons is a function of $10 \times 3 = 30$ variables and that for uranium (U) with 92 electrons would be a function of $92 \times 3 = 276$ variables! In spite of this complexity, the physical meaning of the wavefunction of a many-electron atom is the same as for that for the hydrogen atom, namely, the square of the wavefunction, ψ^2, gives the probability density of finding electron 1 at (x_1, y_1, z_1), electron 2 at (x_2, y_2, z_2), and so on. As the number of electrons (and number of variables) increases, the Schrödinger equation that determines the wavefunction also becomes increasingly more complicated. For example, the Schrödinger equation for the helium (He) atom with its two electrons is

$$\underbrace{-\frac{\hbar^2}{2m_e}\left[\frac{\partial^2\psi}{\partial x_1^2} + \frac{\partial^2\psi}{\partial y_1^2} + \frac{\partial^2\psi}{\partial z_1^2} + \frac{\partial^2\psi}{\partial x_2^2} + \frac{\partial^2\psi}{\partial y_2^2} + \frac{\partial^2\psi}{\partial z_2^2}\right]}_{\text{kinetic energy term}} + \underbrace{V(x_1, y_1, z_1, x_2, y_2, z_2)}_{\text{potential energy term}}\psi = E\psi \qquad \textbf{(2.2)}$$

The potential energy, V, in Equation 2.2 is given by

$$V(x_1, y_1, z_1, x_2, y_2, z_2) = \underbrace{-\frac{2e^2}{4\pi\varepsilon_0 r_1} - \frac{2e^2}{4\pi\varepsilon_0 r_2}}_{\text{electron–nuclear attraction}} + \underbrace{\frac{e^2}{4\pi\varepsilon_0 r_{12}}}_{\substack{\text{electron–electron} \\ \text{repulsion}}} \qquad \textbf{(2.3)}$$

where r_1 and r_2 are the distances between the nucleus and electrons 1 and 2, respectively, and r_{12} is the distance between the two electrons (see Figure 2.1).

Like that for the hydrogen atom, the Schrödinger equation for the helium atom (Equation 2.2) contains a kinetic energy term as well as potential energy terms arising

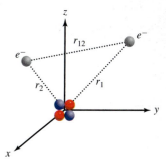

Figure 2.1 The geometry of the helium atom.

from electron-nuclear attraction. However, the potential energy term in Equation 2.3 representing the electron-electron (*e-e*) repulsion is a feature not present in the hydrogen atom. The existence of *e-e* repulsion makes it impossible to obtain an exact solution for the Schrödinger equation for helium (or any other many-electron atom), as we can for a one-electron system such as hydrogen; therefore, approximations are necessary.

To start, let's consider the Schrödinger equation for helium without the *e-e* repulsion term. This equation represents a hypothetical system of two electrons interacting with a nucleus of charge $+2e$ but not with each other. Thus, each electron is identical to the electron in a hydrogenlike atom with $Z = 2$. This is a problem we have already solved. For each separate electron, the ground-state solution to the Schrödinger equation is a hydrogenlike $1s$ orbital, so we can construct the He two-electron wavefunction as a product of two hydrogenlike $1s$ atomic orbitals (with $Z = 2$), one for each electron:

$$\psi = \phi_{1s}(r_1)\,\phi_{1s}(r_2) \tag{2.4}$$

(We use the symbol ϕ to denote an electron orbital for an individual electron to distinguish it from ψ, which represents the overall many-electron wavefunction.)

How well does this represent the He atom? Because the orbitals in Equation 2.4 are hydrogenlike $1s$ orbitals, we can use Equation 1.40 to calculate their energy. Using $Z = 2$ (for He) and $n = 1$ (for the $1s$ orbital), the ground-state energy is -8.740×10^{-18} J or about -5260 kJ mol^{-1}. Therefore, the ionization energy (that is, the amount of energy required to remove that electron) is 5260 kJ mol^{-1}, whereas the experimental value is 2370 kJ mol^{-1}. The large discrepancy is due to neglecting *e-e* repulsion. Because the *e-e* repulsion energy is positive, ignoring it gives an orbital energy that is too low (more negative) and an ionization energy that is too large—here by more than a factor of two! Ignoring the *e-e* repulsion does not lead to a satisfactory model. The question now is, how do we take into account the *e-e* repulsion without destroying the conceptually useful atomic orbital picture?

A practical solution to this problem was developed by the English physicist Douglas Hartree[1] in the mid 1930s. Using physical reasoning, Hartree argued that the many-electron wavefunction could be approximated as the product of single-electron orbitals, one for each electron in the atom. For helium, with two electrons, this approximation is similar to Equation 2.4, except that the orbitals involved are not hydrogenlike:

$$\psi \approx \phi_1(r_1)\,\phi_2(r_2) \tag{2.5}$$

The individual atomic orbitals for the two electrons, $\phi_1(r_1)$ and $\phi_2(r_2)$, are chosen to give the best possible approximation to the actual many-electron wavefunction. The approach that Hartree developed to determine the optimal atomic orbitals is known as the ***self-consistent field (SCF) method.*** The idea is as follows: Start with an initial guess, $\phi_2(r_2)$, for the orbital for electron 2. The square of this function will give the probability that electron 2 is found at a particular position r_2. We can

1. Douglas Rayner Hartree (1897–1958). English physicist. In addition to his work on approximations to the Schrödinger equation for many-electron atoms and molecules, he was an early pioneer in the use of analog and digital computers in numerical analysis. He has been honored with a unit of energy: 1 Hartree = 4.35975×10^{-18} J. The ground-state electronic energy of the hydrogen atom is 0.5 Hartrees.

then solve the Schrödinger equation for electron 1, which is moving in a potential field generated by the nucleus and electron 2 in orbital $\phi_2(r_2)$. The solution to the Schrödinger equation for electron 1 gives a new approximation for the atomic orbital for electron 1, which we call $\widetilde{\phi}_1(r_1)$, where the "~" symbol denotes that the orbital now differs from the original ϕ_1. Next, we do a similar calculation for electron 2, which is moving in the electric field of electron 1 in orbital $\widetilde{\phi}_1(r_1)$. This procedure yields a new approximation, $\widetilde{\phi}_2(r_2)$, for the atomic orbital for electron 2. This is followed by using $\widetilde{\phi}_2(r_2)$ to obtain a new orbital for electron 1. The process is repeated as many times as needed until the approximate orbitals for electron 1 and electron 2 no longer change—that is, until they become *self-consistent*. This type of iteration process is well suited for modern high-speed computers and the SCF approximate wavefunctions for many-electron atoms can be calculated with high precision.

Figure 2.2 shows the optimized ground-state atomic orbital for He that results from the SCF procedure, as compared to a $1s$ hydrogenlike orbital with $Z = 2$. The major difference between the two is that the SCF orbital is larger than the hydrogenlike $1s$ orbital (that is, the value of the SCF orbital exceeds that of the $1s$ hydrogenlike orbital at large distances). This is expected because the electron-electron repulsion (ignored in the hydrogenlike He orbital) acts to push the two electrons apart, thereby expanding the optimal orbital.

The orbital energy for the SCF He orbital shown in Figure 2.2 is $-2410 \text{ kJ mol}^{-1}$, corresponding to an ionization energy for He of 2410 kJ mol^{-1} (which is very close to the experimental value of 2370 kJ mol^{-1}). The SCF orbital for He was constructed using the assumption that there were two electrons in the atom. When one electron is removed to form He^+, the atomic orbital for the remaining electron is altered; it becomes identical to the $Z = 2$ hydrogenlike $1s$ orbital shown in Figure 2.2. The energy to remove this electron (called the *second ionization energy* of He) is identical to that of the $Z = 2$ hydrogenlike orbital (calculated previously): 5260 kJ mol^{-1}. It is harder to remove the second electron because there is no $e\text{-}e$ repulsion term to counteract the electron-nuclear attraction.

For atoms larger than He, the Hartree/SCF procedure is largely the same as for He. For an atom with N electrons, the Hartree wavefunction is

$$\psi \approx \phi_1(r_1)\,\phi_2(r_2)\,\phi_3(r_3)\ldots\phi_N(r_N) \tag{2.6}$$

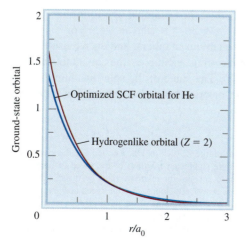

Figure 2.2 The optimized SCF ground-state ($1s$) orbital for helium compared to the $Z = 2$ hydrogenlike orbital.

Figure 2.3 A schematic representation of the SCF method for obtaining the wavefunction of a many-electron atom.

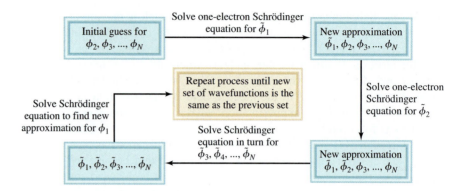

where $\phi_1(r_1), \phi_2(r_2), \ldots \phi_N(r_N)$ are the individual atomic orbitals for electrons $1, 2, \ldots, N$, respectively. The SCF procedure for determining the optimal atomic orbitals[2] in Equation 2.6 is illustrated in Figure 2.3.

The procedure outlined in Figure 2.3 is computationally intensive and requires sophisticated computer programs to accomplish. Fortunately for chemists, however, it is possible to understand a great deal about many-electron atom wavefunctions without having to explicitly solve for the SCF orbitals. Most importantly, the SCF orbitals can be described using the same set of quantum numbers (namely, n, l, and m_l) that index the atomic orbitals of the hydrogen atom. The restrictions on these three quantum numbers are the same as those for hydrogen (namely $n = 1, 2, 3, \ldots$; $l = 0, 1, \ldots, n - 1$; and $m_l = -l, \ldots, l$). In addition, these quantum numbers have the same meaning with regard to orbital shape. For example, the SCF optimized orbital for He shown in Figure 2.2 is spherically symmetric (depends only on r) and is therefore classified as an s-type orbital, with $l = 0$ and $m_l = 0$. In addition, it has no nodes, indicating that $n = 1$. (Remember, in the hydrogen atom, the number of nodes in an s orbital is $n - 1$). Thus, we refer to the ground-state orbital of He as a $1s$ orbital with $n = 1$, $l = 0$, and $m_l = 0$.

Electron Configurations

Within Hartree's SCF approximation, the four quantum numbers n, l, m_l, and m_s enable us to label completely an electron in any orbital in any atom. In a sense, we can regard the set of four quantum numbers as the "address" of an electron in an atom, somewhat in the same way that a street address, city, state, and zip code specify the address of an individual. For example, the four quantum numbers for a $2s$ orbital electron are $n = 2$, $l = 0$, $m_l = 0$, and $m_s = +\frac{1}{2}$ or $-\frac{1}{2}$. It is inconvenient to write out all the individual quantum numbers, and so we use the simplified notation (n, l, m_l, m_s). For the preceding example, the quantum numbers are either $(2, 0, 0, +\frac{1}{2})$ or $(2, 0, 0, -\frac{1}{2})$. The value of the spin quantum number (m_s) has no effect on the energy, size, shape, or orientation of an orbital, but it determines how electrons are arranged in an orbital.

---

2. For very heavy atoms with large Z, such as Hg, Au, and Pb, the electrons in the inner atomic orbitals have average velocities that are significant fractions of the speed of light. For these electrons, corrections to the Schrödinger equation to account for relativity must be made. For example, because the mass of a relativistic object increases as its velocity approaches the speed of light, the inner electrons become heavier and move closer to the nucleus. This shrinking of the inner orbitals is called a *relativistic contraction*. Such relativistic effects are largely responsible for a number of unusual properties of heavy atom materials, such as the yellow color of gold (Au) or the fact that mercury (Hg) is a liquid at room temperature.

Example 2.1 shows how quantum numbers of an electron in an orbital are assigned.

Example 2.1

Write the possible values for the four quantum numbers for an electron in a $3p$ orbital.

Strategy The classification of atomic orbitals for many-electron atoms follows the same rules as for hydrogenlike wavefunctions, which are described in Section 1.4. The designation "$3p$" fixes the values of n and l for this orbital, but does not uniquely specify the values of m_l and m_s. Follow the procedures outlined in Section 1.4 to determine the possible values of m_l and m_s consistent with the values of n and l for this orbital.

Solution To start with, we know that the principal quantum number n is 3 and the angular momentum quantum number l must be 1 (because we are dealing with a p orbital). For $l = 1$, there are three values of m_l, given by -1, 0, and 1. Because the electron spin quantum number m_s can be either $+\frac{1}{2}$ or $-\frac{1}{2}$, there are six possible ways to designate the electron using the (n, l, m_l, m_s) notation:

$$\left(3, 1, -1, +\frac{1}{2}\right) \quad \left(3, 1, -1, -\frac{1}{2}\right)$$

$$\left(3, 1, 0, +\frac{1}{2}\right) \quad \left(3, 1, 0, -\frac{1}{2}\right)$$

$$\left(3, 1, +1, +\frac{1}{2}\right) \quad \left(3, 1, +1, -\frac{1}{2}\right)$$

Check In these six designations the values of n and l are constant, but the values of m_l and m_s can vary.

Practice Exercise Write the four quantum numbers for an electron in a $5p$ orbital.

The hydrogen atom is a particularly simple system because it contains only one electron. The electron may reside in the $1s$ orbital (the ground state), or it may be found in some higher-energy orbital (an excited state). For many-electron atoms, however, we must know the **electron configuration** of the atom (that is, *how the electrons are distributed among the various atomic orbitals*) to understand electronic behavior. In the remainder of this section and in Section 2.2, we use the first 10 elements (hydrogen to neon) to illustrate the rules for writing electron configurations for atoms in the ground state. For this discussion, recall that the number of electrons in an atom is equal to its atomic number Z.

The electron configuration of an atom can be represented using either symbols or diagrams. For example, the ground state of a hydrogen atom consists of a single electron in a $1s$ orbital; the electron configuration of this atom is represented by the symbol $1s^1$:

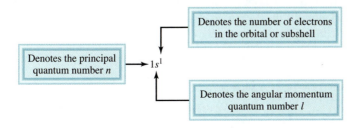

We can also represent the electron configuration of the hydrogen atom using an *orbital diagram,* which is a pictorial representation of the electron configuration that shows the spin of the electron (see Figure 1.30):

$$\text{H} \quad \boxed{\uparrow}$$
$$1s^1$$

The upward arrow denotes one of the two possible orientations of the electron spin ("spin-up"). (Alternatively, we could have represented the electron with a downward arrow to represent a "spin-down" electron.) The box represents an atomic orbital.

Pauli Exclusion Principle

The ground-state electron configuration of a many-electron atom is constructed by placing the electrons into the SCF atomic orbitals in such a way as to minimize the total energy of the system. Quantum mechanics requires that the procedure used to place electrons into atomic orbitals obey the **Pauli[3] exclusion principle.** According to this principle, *no two electrons in an atom can have the same set of four quantum numbers* (n, l, m_l, m_s). If two electrons in the same atom have the same values of n, l, and m_l (that is, the two electrons are in the *same* atomic orbital), then they must have different values of m_s. Because there are only two possible values of m_s ($+\frac{1}{2}$ and $-\frac{1}{2}$), only two electrons may occupy the same atomic orbital, and these two electrons must have opposite spins.

Consider the helium atom, which has two electrons. The three possible ways of placing two electrons into a He $1s$ orbital are

$$\text{He} \quad \boxed{\uparrow\uparrow} \quad \boxed{\downarrow\downarrow} \quad \boxed{\uparrow\downarrow}$$
$$\quad\quad 1s^2 \quad\; 1s^2 \quad\; 1s^2$$
$$\quad\quad (a) \quad\;\; (b) \quad\;\; (c)$$

Diagrams (a) and (b) are ruled out by the Pauli exclusion principle. In (a), both electrons have the same upward spin and would have the quantum numbers $(1, 0, 0, +\frac{1}{2})$; in (b), both electrons have downward spins and would have the quantum numbers $(1, 0, 0, -\frac{1}{2})$. Only configuration (c) is physically acceptable because one electron has the quantum numbers $(1, 0, 0, +\frac{1}{2})$ and the other has $(1, 0, 0, -\frac{1}{2})$. Thus, the helium atom has the following configuration:

$$\text{He} \quad \boxed{\uparrow\downarrow}$$
$$1s^2$$

Note that $1s^2$ is read "one s two," not "one s squared." In He, the $1s$ orbital is full because there are only two possible spin states. If an additional electron is added, it cannot be placed into the $1s$ orbital, but must go into a higher energy orbital, such as $2s$.

The Pauli exclusion principle has a profound effect on the wavefunction of a many-electron atom. If there were no restriction and an orbital could hold an arbitrary number of electrons, the ground state of an atom would simply be constructed by

---

3. Wolfgang Pauli (1900–1958). Austrian physicist. One of the founders of quantum mechanics, Pauli was awarded the Nobel Prize in Physics in 1945.

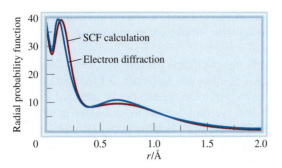

Figure 2.4 The radial probability function for argon as a function of r from both SCF calculations (red line) and experiment (blue line).

placing all of the electrons into the lowest energy orbital (that is, the $1s$). Instead, the electrons must go into progressively higher orbitals as each successive orbital is filled. Argon, for example, has 18 electrons. The $1s$ orbital can only hold two electrons and the four $n = 2$ orbitals ($2s$ and the three $2p$ orbitals) can hold a maximum of 8. The remaining electrons must go into orbitals with $n = 3$. The effect of this filling procedure on the electron density of argon can be seen in Figure 2.4, which shows the electron radial probability function (see Section 1.4) for argon as calculated using the SCF procedure.[4] The three peaks in this function correspond to the regions of space occupied by the $n = 1, 2,$ and 3 electrons, respectively. Thus, the electron density of argon can be described as consisting of two sharply defined inner shells of $1s$ and $2s$ electrons surrounded by a diffuse outer $n = 3$ shell.

Diamagnetism and Paramagnetism

The Pauli exclusion principle is one of the fundamental principles of quantum mechanics. It can be tested by a simple observation. If the two electrons in the $1s$ orbital of a helium atom had the same, or parallel, spins ($\uparrow\uparrow$ or $\downarrow\downarrow$), their net magnetic fields would reinforce each other [see Figure 2.5(a)]. Such an arrangement would make the helium gas *paramagnetic*. **Paramagnetic** substances are those that *contain net unpaired spins and are attracted by a magnet*. On the other hand, if the electron spins are

---

4. Incorporating the Pauli exclusion principle into the SCF procedure requires a slight modification to Hartree's original method. This modified scheme, called the Hartree-Fock method, is the basis of much of modern computational chemistry.

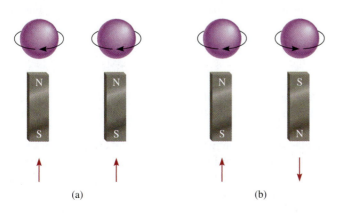

(a)   (b)

Figure 2.5 The (a) parallel and (b) antiparallel spins of two electrons. In (a) the two magnetic fields reinforce each other. In (b) the two magnetic fields cancel each other.

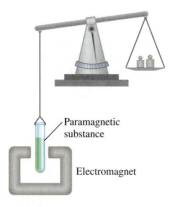

Figure 2.6 Initially, the paramagnetic substance was weighed on a balance. When the electromagnet is turned on, the balance is offset because the sample tube is drawn into the magnetic field. Knowing the concentration and the additional mass needed to reestablish balance, it is possible to calculate the number of unpaired electrons in the substance.

paired, or antiparallel to each other ($\uparrow\downarrow$ or $\downarrow\uparrow$), the magnetic effects cancel out [see Figure 2.5(b)]. ***Diamagnetic*** substances *do not contain net unpaired spins and are slightly repelled by a magnet.*

Measurements of magnetic properties provide the most direct evidence for specific electron configurations of elements. Advances in instrument design during the last 30 years or so enable us to determine the number of unpaired electrons in an atom (Figure 2.6). By experiment, we find that the helium atom in its ground state has no net magnetic field. Therefore, the two electrons in the $1s$ orbital must be paired in accord with the Pauli exclusion principle and helium gas is diamagnetic. A useful rule to keep in mind is that any atom with an *odd* number of electrons will always contain one or more unpaired spins because an even number of electrons is needed for complete pairing. On the other hand, atoms containing an even number of electrons may or may not contain unpaired spins. We will see the reason for this behavior shortly.

As another example, consider the lithium atom ($Z = 3$), which has three electrons. The third electron cannot go into the $1s$ orbital because it would have the same four quantum numbers as one of the first two electrons. Instead, this electron "enters" the next (energetically) higher orbital and is unpaired. Thus, the lithium atom contains one unpaired electron and lithium metal is paramagnetic.

Note that the paramagnetic or diamagnetic behavior of atoms is a consequence of the quantization of electron "spin" and the Pauli exclusion principle, both of which are purely quantum-mechanical phenomena. Thus, the experimental observation of the magnetic behavior of atoms (paramagnetic or diamagnetic) represents further experimental confirmation of the quantum-mechanical nature of atoms and molecules.

2.2 | Electron Configurations of Many-Electron Atoms Are Constructed Using the *Aufbau* (or "Building-Up") Principle

To construct the electron configuration for an atom with atomic number Z, we place the Z electrons into atomic orbitals according to the Pauli exclusion principle, so that the total energy of the configuration is a minimum. To do this requires an understanding of the effect of electron-electron repulsion on the orbital energies.

Orbital Energies

The ***orbital energy*** is *the negative of the amount of energy required to remove an electron from a given orbital.* The lower (more negative) the orbital energy, the more energy is required to remove that electron from the atom. According to Equation 1.37, the energy of an electron in a hydrogen atom is determined solely by its principal quantum number. Thus, the energies of hydrogen orbitals increase as follows (Figure 2.7):

$$1s < 2s = 2p < 3s = 3p = 3d < 4s = 4p = 4d = 4f$$

Although the electron density distributions are different in the $2s$ and $2p$ orbitals, the electron in the hydrogen atom has the same energy whether it is in the $2s$ orbital or a $2p$ orbital.

Because of electron-electron repulsion, the orbital energy picture is more complex for many-electron atoms than for hydrogen. For lithium (three electrons), the first two electrons go into the low-energy $1s$ orbital and fill the $n = 1$ shell. The next electron must go into the next highest energy level. In the hydrogen atom, this would be either $2s$ or $2p$ as those orbitals are equal in energy, but in lithium, the two $n = 2$ orbitals

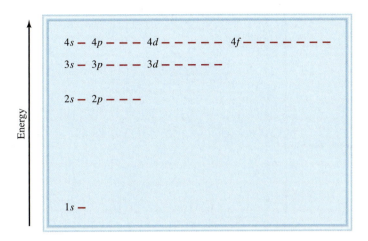

Figure 2.7 Orbital energy
levels in the hydrogen atom.
Each short horizontal line
represents one orbital. Orbitals
with the same principal quantum
number n all have the same
energy.

differ in energy because of electron-electron repulsion with the two $1s$ electrons. Experimentally,[5] the $2s$ orbital is lower in energy than the $2p$ orbital in many-electron atoms. Thus, the electron configuration for Li is $1s^2\,2s^1$, and its orbital diagram is

Li   ⊞ ⊞
   $1s^2$ $2s^1$

The difference in energy between $2s$ and $2p$ orbitals can be explained by examining the electron density of each orbital. Because both the $2s$ and $2p$ orbitals are larger than the $1s$ orbital, an electron in either of these orbitals spends more time away from the nucleus than an electron in the $1s$ orbital. Thus, a $2s$ or $2p$ electron is partly "shielded" from the attractive force of the nucleus by the $1s$ electrons. That is, the shielding effect *reduces* the electrostatic attraction between the protons in the nucleus and the electron in a $2s$ or $2p$ orbital.

The manner in which the electron density varies from the nucleus outward depends on the type of orbital. This can be best seen by looking at the radial probability plots for the $1s$, $2s$, and $2p$ orbitals in Figure 2.8. Although a $2s$ electron spends most of its time (on average) slightly farther from the nucleus than a $2p$ electron, the electron density near the nucleus is actually greater for the $2s$ electron (see the small maximum for the $2s$ orbital in Figure 2.8). For this reason, the $2s$ orbital is said to be more "penetrating" than the $2p$ orbital. Therefore, a $2s$ electron is less shielded by the $1s$ electrons and is more strongly held by the nucleus. In fact, for the same principal quantum number n, the penetrating power decreases as the angular momentum quantum number l increases, or

$$s > p > d > f > \ldots$$

Because the stability of an electron is determined by the strength of its attraction to the nucleus, it follows that a $2s$ electron will be lower in energy than a $2p$ electron. To put it another way, less energy is required to remove a $2p$ electron than a $2s$ electron because a $2p$ electron is not held as strongly by the nucleus. In general, the

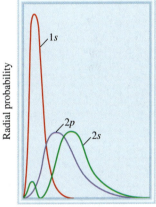

Distance from nucleus

Figure 2.8 Radial probability
plots for the $1s$, $2s$, and $2p$
orbitals. The $1s$ electrons effectively shield both the $2s$ and $2p$
electrons from the nuclear
charge. The $2s$ orbital is more
penetrating than the $2p$ orbital.

---

5. The orbital energies of atoms can be directly measured by a technique called ***photoelectron spectroscopy.*** In this technique, a gas of atoms is subjected to high-energy electromagnetic radiation (UV to X-ray), causing the ejection of electrons from the atoms. By measuring the difference between the energy of the absorbed photons and the kinetic energy of the ejected electrons, the energy to remove the electron from the atom can be determined. This energy corresponds to the orbital energy of the ejected electron.

Figure 2.9 Orbital energy levels in a many-electron atom. The energy level depends upon both n and l values.

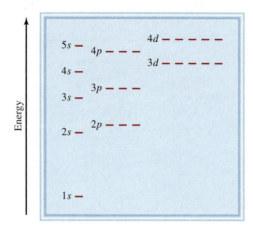

energy of orbitals with a given value of n increases as l increases because of the shielding effect. The hydrogen atom has only one electron and, therefore, is without such a shielding effect. Figure 2.9 shows the relative energies for many-electron orbitals in the first three energy shells ($n = 1, 2,$ and 3). For fixed n and l, the orbital energy has no dependence on the value of m_l. Thus, the $2p_x$, $2p_y$, and $2p_z$ orbitals are degenerate (that is, equal in energy). These three orbitals have exactly the same shape and differ only in orientation. Because the orientation of the orbitals does not affect the relative distances between the electrons or between the electrons and the nucleus, the energies of these orbitals must be identical.

The trend in orbital energy among orbitals with identical values of n is relatively straightforward (energy increases with increasing l) and can be understood in terms of shielding. Energy comparisons for orbitals with different values of n are somewhat more complicated, especially for $n \geq 3$. For example, the energies for the $3d$ and $4s$ orbitals are very close together and which one is lower depends upon the number of electrons in the system. For Z in the range of 6 to 20, the $4s$ orbital energy is lower than that of $3d$, but for $Z > 20$, SCF calculations and spectroscopic evidence show that the situation is reversed. The general order in which atomic orbitals are filled in a many-electron atom can be determined using a procedure known as *Madelung's rule,* which is illustrated graphically in Figure 2.10.

We can now use this knowledge to predict the electron configurations of the next two elements, beryllium (Be; $Z = 4$) and boron (B; $Z = 5$). The ground-state configuration of beryllium is $1s^2 2s^2$, or

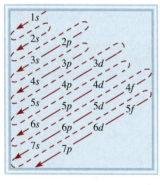

Figure 2.10 Graphical representation of Madelung's rule, showing the order in which atomic subshells are filled in a many-electron atom. Start with the $1s$ orbital and move downward following the direction of the arrows. Thus, the order goes $1s < 2s < 2p < 3s < 3p < 4s < 3d < \ldots .$

Be ⬆⬇ ⬆⬇
$1s^2$ $2s^2$

Beryllium is diamagnetic, because all four electrons are paired.

The $2s$ suborbital is now full, so when we add another electron for boron ($Z = 5$), it must go into a $2p$ orbital. The electron configuration of B is $1s^2 2s^2 2p^1$, or

B ⬆⬇ ⬆⬇ ⬆
$1s^2$ $2s^2$ $2p^1$

The unpaired electron can be in the $2p_x$, $2p_y$, or $2p_z$ orbital. The choice is completely arbitrary because the three p orbitals are equivalent in energy. Also, we show the electron spin of the unpaired electron in B as "up," this is also arbitrary as the energy would be the same if we assigned it a "down" electron spin. As the diagram shows, boron is paramagnetic.

Hund's Rule

The electron configuration of carbon ($Z = 6$) is $1s^2 2s^2 2p^2$. The following are three different ways of distributing two electrons among three $2p$ orbitals:

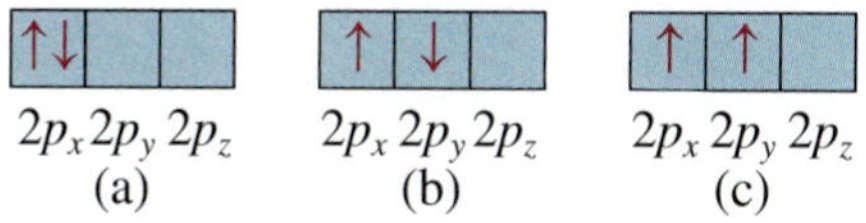

$$2p_x\,2p_y\,2p_z \qquad 2p_x\,2p_y\,2p_z \qquad 2p_x\,2p_y\,2p_z$$
$$\text{(a)} \qquad\qquad \text{(b)} \qquad\qquad \text{(c)}$$

None of the three arrangements violates the Pauli exclusion principle, so we must determine which one will give the greatest stability (that is, the lowest energy). The answer is provided by **Hund's rule,**[6] which states that *the most stable arrangement of electrons in subshells is the one with the greatest number of parallel spins.* The arrangement shown in (c) satisfies this condition. In both (a) and (b) the two spins cancel each other. Thus, the orbital diagram for carbon is

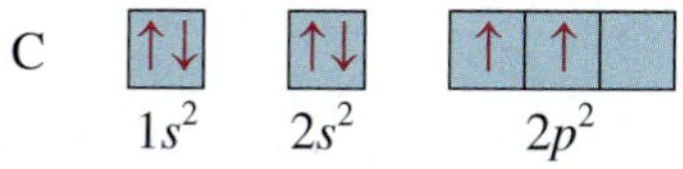

Qualitatively, (c) is more stable than (a) because the two electrons in (a) are in the same $2p_x$ orbital, and their proximity results in greater mutual repulsion than when they occupy two separate orbitals (for example, $2p_x$ and $2p_y$). The preference of (c) over (b) is more subtle, but can be justified on theoretical grounds. Carbon atoms in their ground states contain two unpaired electrons, in accord with Hund's rule. (Note that Hund's rule applies only to the ground-state—lowest energy—electronic configurations. Configurations that violate Hund's rule but obey the Pauli exclusion principle, such as (a) and (b) above for carbon, generally correspond to excited-state electron configurations.)

The electron configuration of nitrogen ($Z = 7$) is $1s^2 2s^2 2p^3$:

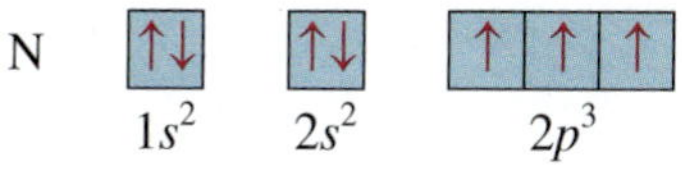

Again, Hund's rule dictates that all three $2p$ electrons have spins parallel to one another; the nitrogen atom contains three unpaired electrons.

The electron configuration of oxygen ($Z = 8$) is $1s^2 2s^2 2p^4$. An oxygen atom has two unpaired electrons:

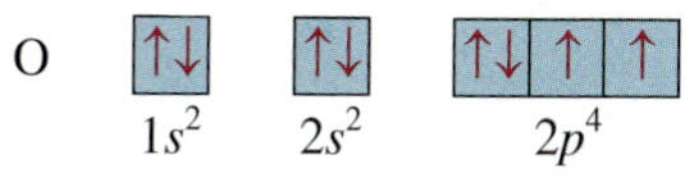

The electron configuration of fluorine ($Z = 9$) is $1s^2 2s^2 2p^5$. The nine electrons are arranged as follows:

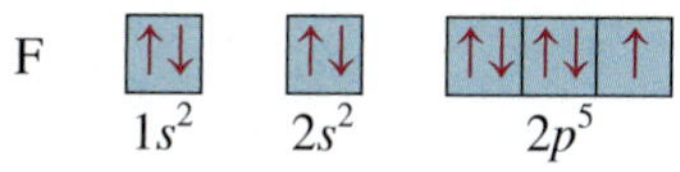

The fluorine atom has one unpaired electron.

---

6. Frederick Hund (1896–1997). German physicist. Hund's work was mainly in quantum mechanics. He also helped to develop the molecular orbital theory of chemical bonding.

In neon ($Z = 10$), the $2p$ subshell is completely filled. The electron configuration of neon is $1s^2 2s^2 2p^6$, and all the electrons are paired, as follows:

Ne   ↑↓   ↑↓   ↑↓ ↑↓ ↑↓
$\quad\;\; 1s^2 \quad 2s^2 \qquad 2p^6$

Neon gas should be diamagnetic, and experimental observation bears out this prediction.

Example 2.2

An oxygen atom has a total of eight electrons. Write the four quantum numbers for each of the eight electrons in the ground state.

Strategy We start with $n = 1$ and proceed to fill orbitals in the order shown in Figure 2.10. For each value of n, we determine the possible values of l. For each value of l, we assign the possible values of m. We can place electrons in the orbitals according to the Pauli exclusion principle and Hund's rule.

Solution We start with $n = 1$, so $l = 0$, a subshell corresponding to the $1s$ orbital. This orbital can accommodate a total of two electrons. Next, $n = 2$, and l may be either 0 or 1. The $l = 0$ subshell contains one $2s$ orbital, which can accommodate two electrons. The remaining four electrons are placed in the $l = 1$ subshell, which contains three $2p$ orbitals. The orbital diagram is given on page 132. The results are summarized in the table. Placement of the eighth electron in the orbital labeled $m_l = 1$ is completely arbitrary. It would be equally correct to assign it to $m_l = 0$ or $m_l = -1$.

Electron	n	l	m_l	m_s	Orbital
1	1	0	0	$+\frac{1}{2}$	$1s$
2	1	0	0	$-\frac{1}{2}$	
3	2	0	0	$+\frac{1}{2}$	$2s$
4	2	0	0	$-\frac{1}{2}$	
5	2	1	-1	$+\frac{1}{2}$	$2p_x$
6	2	1	0	$+\frac{1}{2}$	$2p_y$
7	2	1	1	$+\frac{1}{2}$	$2p_z$
8	2	1	1	$-\frac{1}{2}$	

Practice Exercise Write a complete set of quantum numbers for each of the electrons in boron.

The Aufbau or "Building-Up" Principle

We now extend the rules used in writing electron configurations for the first 10 elements to the rest of the elements. This process is based on the *Aufbau* principle. (*Aufbau* is a German word meaning "building-up.") The ***Aufbau principle*** dictates that *as protons are added one by one to the nucleus to build up the elements, electrons are similarly added to the atomic orbitals in accordance with the Pauli exclusion principle and Hund's rule.* Through this process, we gain a detailed knowledge of the ground-state electron configurations of the elements. As we will see later, knowledge of electron configurations helps us to understand and predict the properties of the elements; it also explains why the periodic table is so useful.

Table 2.1 gives the ground-state electron configurations of elements from H ($Z = 1$) through Rg ($Z = 111$). The electron configurations of all elements except hydrogen and helium are represented by a ***noble gas core,*** which *shows in brackets the noble gas element that most closely precedes the element being considered,* followed by the symbol for the highest filled subshells in the outermost shells. The electron configurations of the highest filled subshells in the outermost shells for the elements sodium ($Z = 11$) through argon ($Z = 18$) follow a pattern similar to those of lithium ($Z = 3$) through neon ($Z = 10$). As mentioned earlier in this section, the $4s$ subshell is filled before the $3d$ subshell in a many-electron atom (see Figure 2.10). Thus, the electron configuration of potassium ($Z = 19$) is $1s^22s^22p^63s^23p^64s^1$. Because $1s^22s^22p^63s^23p^6$ is the electron configuration of argon, we can simplify the representation of the electron configuration of potassium by writing [Ar]$4s^1$, where [Ar] denotes the "argon core." Similarly, we can write the electron configuration of calcium ($Z = 20$) as [Ar]$4s^2$. The placement of the outermost electron in the $4s$ orbital (rather than in the $3d$ orbital) of potassium is strongly supported by experimental evidence. For example, the chemistry of potassium is very similar to that of lithium and sodium, the first two alkali metals. The outermost electron of both lithium and sodium is in an s orbital (there is no ambiguity in assigning their electron configurations); therefore, we expect the last electron in potassium to occupy the $4s$ orbital rather than a $3d$ orbital.

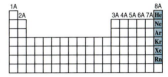

The noble gases

The elements from scandium (Sc, $Z = 21$) to copper (Cu, $Z = 29$) are transition elements. ***Transition elements*** (or ***transition metals***) either *have incompletely filled d subshells or readily give rise to cations that have incompletely filled d subshells.*[7] Although the $3d$ orbital energy dips below that of the $4s$ for $Z > 20$, the electron configuration for scandium ($Z = 21$) is [Ar]$4s^23d^1$. This seems counterintuitive—if the $3d$ orbital energy is *lower* than $4s$, shouldn't the lowest energy configuration be [Ar]$3d^3$? Because of electron-electron repulsion, however, the energies of many-electron atom orbitals also depend upon which orbitals are occupied. Placing three electrons into the $3d$ orbitals raises the energies of both the $4s$ and $3d$ orbitals. So, even though the $4s$ orbital is higher in energy than the $3d$, the total energy of $4s^23d^1$ is lower than both $4s^13d^2$ and $3d^3$. The higher orbital energy of the $4s$ can be seen when an electron is removed from Sc to form Sc$^+$. This electron is removed from the $4s$ orbital (because it has a higher energy than the $3d$), so the configuration of Sc$^+$ is [Ar]$4s^13d^1$.

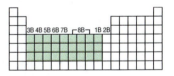

The transition metals

In the rest of the first transition metal series, additional electrons are placed into the $3d$ orbitals according to Hund's rule. However, there are two irregularities. The electron configuration of chromium ($Z = 24$) is [Ar]$4s^13d^5$ and not [Ar]$4s^23d^4$, as

---

7. Many chemistry textbooks include Group 2B (12) elements Zn, Cd, and Hg as transition elements, basing the definition on position within the periodic table rather than on electron configuration. The narrower definition used in this text is the one adopted by the International Union of Pure and Applied Chemists (IUPAC).

| Table 2.1 | | The Ground-State Electron Configurations of the Elements* |

Atomic Number	Symbol	Electron Configuration	Atomic Number	Symbol	Electron Configuration	Atomic Number	Symbol	Electron Configuration
1	H	$1s^1$	38	Sr	$[Kr]5s^2$	75	Re	$[Xe]6s^24f^{14}5d^5$
2	He	$1s^2$	39	Y	$[Kr]5s^24d^1$	76	Os	$[Xe]6s^24f^{14}5d^6$
3	Li	$[He]2s^1$	40	Zr	$[Kr]5s^24d^2$	77	Ir	$[Xe]6s^24f^{14}5d^7$
4	Be	$[He]2s^2$	41	Nb	$[Kr]5s^14d^4$	78	Pt	$[Xe]6s^14f^{14}5d^9$
5	B	$[He]2s^22p^1$	42	Mo	$[Kr]5s^14d^5$	79	Au	$[Xe]6s^14f^{14}5d^{10}$
6	C	$[He]2s^22p^2$	43	Tc	$[Kr]5s^24d^5$	80	Hg	$[Xe]6s^24f^{14}5d^{10}$
7	N	$[He]2s^22p^3$	44	Ru	$[Kr]5s^14d^7$	81	Tl	$[Xe]6s^24f^{14}5d^{10}6p^1$
8	O	$[He]2s^22p^4$	45	Rh	$[Kr]5s^14d^8$	82	Pb	$[Xe]6s^24f^{14}5d^{10}6p^2$
9	F	$[He]2s^22p^5$	46	Pd	$[Kr]4d^{10}$	83	Bi	$[Xe]6s^24f^{14}5d^{10}6p^3$
10	Ne	$[He]2s^22p^6$	47	Ag	$[Kr]5s^14d^{10}$	84	Po	$[Xe]6s^24f^{14}5d^{10}6p^4$
11	Na	$[Ne]3s^1$	48	Cd	$[Kr]5s^24d^{10}$	85	At	$[Xe]6s^24f^{14}5d^{10}6p^5$
12	Mg	$[Ne]3s^2$	49	In	$[Kr]5s^24d^{10}5p^1$	86	Rn	$[Xe]6s^24f^{14}5d^{10}6p^6$
13	Al	$[Ne]3s^23p^1$	50	Sn	$[Kr]5s^24d^{10}5p^2$	87	Fr	$[Rn]7s^1$
14	Si	$[Ne]3s^23p^2$	51	Sb	$[Kr]5s^24d^{10}5p^3$	88	Ra	$[Rn]7s^2$
15	P	$[Ne]3s^23p^3$	52	Te	$[Kr]5s^24d^{10}5p^4$	89	Ac	$[Rn]7s^26d^1$
16	S	$[Ne]3s^23p^4$	53	I	$[Kr]5s^24d^{10}5p^5$	90	Th	$[Rn]7s^26d^2$
17	Cl	$[Ne]3s^23p^5$	54	Xe	$[Kr]5s^24d^{10}5p^6$	91	Pa	$[Rn]7s^25f^26d^1$
18	Ar	$[Ne]3s^23p^6$	55	Cs	$[Xe]6s^1$	92	U	$[Rn]7s^25f^36d^1$
19	K	$[Ar]4s^1$	56	Ba	$[Xe]6s^2$	93	Np	$[Rn]7s^25f^46d^1$
20	Ca	$[Ar]4s^2$	57	La	$[Xe]6s^25d^1$	94	Pu	$[Rn]7s^25f^6$
21	Sc	$[Ar]4s^23d^1$	58	Ce	$[Xe]6s^24f^15d^1$	95	Am	$[Rn]7s^25f^7$
22	Ti	$[Ar]4s^23d^2$	59	Pr	$[Xe]6s^24f^3$	96	Cm	$[Rn]7s^25f^76d^1$
23	V	$[Ar]4s^23d^3$	60	Nd	$[Xe]6s^24f^4$	97	Bk	$[Rn]7s^25f^9$
24	Cr	$[Ar]4s^13d^5$	61	Pm	$[Xe]6s^24f^5$	98	Cf	$[Rn]7s^25f^{10}$
25	Mn	$[Ar]4s^23d^5$	62	Sm	$[Xe]6s^24f^6$	99	Es	$[Rn]7s^25f^{11}$
26	Fe	$[Ar]4s^23d^6$	63	Eu	$[Xe]6s^24f^7$	100	Fm	$[Rn]7s^25f^{12}$
27	Co	$[Ar]4s^23d^7$	64	Gd	$[Xe]6s^24f^75d^1$	101	Md	$[Rn]7s^25f^{13}$
28	Ni	$[Ar]4s^23d^8$	65	Tb	$[Xe]6s^24f^9$	102	No	$[Rn]7s^25f^{14}$
29	Cu	$[Ar]4s^13d^{10}$	66	Dy	$[Xe]6s^24f^{10}$	103	Lr	$[Rn]7s^25f^{14}6d^1$
30	Zn	$[Ar]4s^23d^{10}$	67	Ho	$[Xe]6s^24f^{11}$	104	Rf	$[Rn]7s^25f^{14}6d^2$
31	Ga	$[Ar]4s^23d^{10}4p^1$	68	Er	$[Xe]6s^24f^{12}$	105	Db	$[Rn]7s^25f^{14}6d^3$
32	Ge	$[Ar]4s^23d^{10}4p^2$	69	Tm	$[Xe]6s^24f^{13}$	106	Sg	$[Rn]7s^25f^{14}6d^4$
33	As	$[Ar]4s^23d^{10}4p^3$	70	Yb	$[Xe]6s^24f^{14}$	107	Bh	$[Rn]7s^25f^{14}6d^5$
34	Se	$[Ar]4s^23d^{10}4p^4$	71	Lu	$[Xe]6s^24f^{14}5d^1$	108	Hs	$[Rn]7s^25f^{14}6d^6$
35	Br	$[Ar]4s^23d^{10}4p^5$	72	Hf	$[Xe]6s^24f^{14}5d^2$	109	Mt	$[Rn]7s^25f^{14}6d^7$
36	Kr	$[Ar]4s^23d^{10}4p^6$	73	Ta	$[Xe]6s^24f^{14}5d^3$	110	Ds	$[Rn]7s^25f^{14}6d^8$
37	Rb	$[Kr]5s^1$	74	W	$[Xe]6s^24f^{14}5d^4$	111	Rg	$[Rn]7s^25f^{14}6d^9$

*The symbol [He] is called the helium core and represents $1s^2$. [Ne] is called the neon core and represents $1s^22s^22p^6$. [Ar] is called the argon core and represents [Ne]$3s^23p^6$. [Kr] is called the krypton core and represents [Ar]$4s^23d^{10}4p^6$. [Xe] is called the xenon core and represents [Kr]$5s^24d^{10}5p^6$. [Rn] is called the radon core and represents [Xe]$6s^24f^{14}5d^{10}6p^6$.

we might expect. A similar break in the pattern is observed for copper, whose electron configuration is $[Ar]4s^13d^{10}$ rather than $[Ar]4s^23d^9$. The reason for these irregularities is that a slightly greater stability is associated with half-filled $(3d^5)$ and completely filled $(3d^{10})$ subshells. Electrons in the same subshell (in this case, the d orbitals) have equal energy but different spatial distributions. Consequently, their shielding of one another is relatively small, and the electrons are more strongly attracted by the nucleus when they have the $3d^5$ configuration. According to Hund's rule, the orbital diagram for Cr is

$$\text{Cr} \quad [\text{Ar}] \quad \boxed{\uparrow} \quad \boxed{\uparrow \ \uparrow \ \uparrow \ \uparrow \ \uparrow}$$
$$\qquad\qquad\quad\ 4s^1 \qquad\quad 3d^5$$

Thus, Cr has a total of six unpaired electrons. The orbital diagram for copper is

$$\text{Cu} \quad [\text{Ar}] \quad \boxed{\uparrow} \quad \boxed{\uparrow\downarrow \ \uparrow\downarrow \ \uparrow\downarrow \ \uparrow\downarrow \ \uparrow\downarrow}$$
$$\qquad\qquad\quad\ 4s^1 \qquad\quad 3d^{10}$$

Again, extra stability is gained in this case by having the $3d$ subshell completely filled. From elements Zn ($Z = 30$) through Kr ($Z = 36$), the $4p$ subshells fill in a straightforward manner.

With rubidium (Rb, $Z = 37$, $[Kr]5s^1$) and strontium (Sr, $Z = 38$, $[Kr]5s^2$), electrons begin to enter the $n = 5$ energy level. The electron configurations in the second transition metal series [yttrium (Y, $Z = 39$) to silver (Ag, $Z = 47$)] are also irregular, but a detailed analysis is beyond the scope of this text. From cadmium (Cd, $Z = 48$) through xenon (Xe, $Z = 54$), the filling of the $5p$ subshells follows Hund's rule.

The sixth period of the periodic table begins with cesium (Cs, $Z = 55$) and barium (Ba, $Z = 56$), whose electron configurations are $[Xe]6s^1$ and $[Xe]6s^2$, respectively. Next we come to lanthanum (La, $Z = 57$). From Figure 2.10, we would expect that after filling the $6s$ orbital we would place the additional electrons in $4f$ orbitals. In reality, the energies of the $5d$ and $4f$ orbitals are very close; in fact, for lanthanum $4f$ is slightly higher in energy than $5d$. Thus, lanthanum's electron configuration is $[Xe]6s^25d^1$ and not $[Xe]6s^24f^1$.

Following lanthanum are the 14 elements known as the **lanthanide series,** or **rare earth elements** [cerium (Ce, $Z = 58$) to lutetium (Lu, $Z = 71$)]. The rare earth elements *have incompletely filled 4f subshells or readily give rise to cations that have incompletely filled 4f subshells.* In this series, the added electrons are placed in $4f$ orbitals. After the $4f$ subshell is completely filled, the next electron enters the $5d$ subshell of lutetium. The electron configuration of gadolinium (Gd, $Z = 64$) is $[Xe]$ $6s^24f^75d^1$ rather than $[Xe]6s^24f^8$. Like chromium, gadolinium gains extra stability by having a half-filled subshell $(4f^7)$.

The third transition metal series, including lanthanum and hafnium (Hf, $Z = 72$) and extending through gold (Au, $Z = 79$), is characterized by the filling of the $5d$ subshell. The $6s$ and $6p$ subshells are filled next, which takes us to radon (Rn, $Z = 86$). The *last row of elements* is the **actinide series,** which starts at thorium (Th, $Z = 90$). *Most of the actinides are not found in nature but have been synthesized.* With few exceptions, you should be able to write the electron configuration of any element, using Figure 2.10 as a guide. Elements that require particular care are the transition metals, the lanthanides, and the actinides. As we noted earlier, at larger values of the principal quantum number n, the order of subshell filling may reverse from one element to the next. Figure 2.11 groups the elements according to the type of subshell in which the outermost electrons are placed.

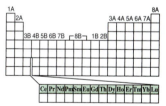

The lanthanides

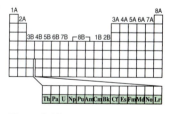

The actinides

Figure 2.11 Classification of groups of elements in the periodic table according to the type of subshell being filled with electrons.

$1s$			
$2s$			$2p$
$3s$			$3p$
$4s$	$3d$		$4p$
$5s$	$4d$		$5p$
$6s$	$5d$		$6p$
$7s$	$6d$		$7p$

$4f$
$5f$

Example 2.3

Write the ground-state electron configurations for (a) sulfur and (b) palladium, which is diamagnetic.

Strategy Because the atoms are neutral, the number of electrons will equal the atomic number. Start with $n = 1$ and proceed to fill orbitals in the order shown in Figure 2.10. For each value of l, assign the possible values of m. Place electrons in the orbitals according to the Pauli exclusion principle and Hund's rule, and then write the electron configuration. The task can be simplified by using the noble gas core preceding the element for the inner electrons.

Solution (a) Sulfur (S, $Z = 16$) has 16 electrons, so the noble gas core in this case is [Ne]. (Ne is the noble gas in the period preceding sulfur.) [Ne] represents $1s^2 2s^2 2p^6$. This leaves us with 6 electrons to fill the $3s$ subshell and partially fill the $3p$ subshell. Thus, the electron configuration of S is $1s^2 2s^2 2p^6 3s^2 3p^4$ or [Ne]$3s^2 3p^4$. (b) Palladium (Pd, $Z = 46$) has 46 electrons. The noble gas core in this case is [Kr]. (Kr is the noble gas in the period preceding palladium.) [Kr] represents $1s^2 2s^2 2p^6 3s^2 3p^6 4s^2 3d^{10} 4p^6$. The remaining 10 electrons are distributed among the $4d$ and $5s$ orbitals. The three choices are (1) $4d^{10}$, (2) $4d^9 5s^1$, and (3) $4d^8 5s^2$. Because palladium is diamagnetic, all the electrons are paired and its electron configuration must be $1s^2 2s^2 2p^6 3s^2 3p^6 4s^2 3d^{10} 4p^6 4d^{10}$ or simply [Kr]$4d^{10}$. The configurations designated (2) and (3) both represent paramagnetic elements.

Check To confirm the answer, write the orbital diagrams for (1), (2), and (3).

Practice Exercise Write the ground-state electron configurations for (a) phosphorus and (b) cobalt.

Electron Configurations for Cations and Anions

The ground-state electron configurations of monatomic cations and anions can be written just as they are for atoms, using the Pauli exclusion principle and Hund's rule.

For nontransition metal atoms, cations are generally formed by removing one or more electrons from the highest energy orbitals to achieve a noble gas configuration.

The following are the electron configurations of some atoms and their corresponding cations:

$$\text{Na: [Ne]}3s^1 \qquad \text{Na}^+\text{: [Ne]}$$
$$\text{Ca: [Ar]}4s^2 \qquad \text{Ca}^{2+}\text{: [Ar]}$$
$$\text{Al: [Ne]}3s^23p^1 \qquad \text{Al}^{3+}\text{: [Ne]}$$

Note that each ion has a stable noble gas configuration. In some heavier nontransition metal atoms, such as indium (In, $Z = 49$) and lead (Pb, $Z = 82$), two cations are possible because of the stability of a filled outer s subshell. Indium has a configuration of $[\text{Kr}]4d^{10}5s^25p^1$ and can lose all three $n = 5$ electrons ($5s^2$ and $5p^1$) to form In^{3+} with a configuration of [Kr], but it can also form In^+ with a configuration $[\text{Kr}]\,5s^2$. Similarly, lead (Pb), can form both $\text{Pb}^{4+}([\text{Xe}]5d^{10})$ and Pb^{2+} ions $([\text{Xe}]5d^{10}6s^2)$. The stability of the outer pair of s electrons (called an *inert pair*) in In^+ and Pb^{2+} is the result of relativistic effects (see the footnote on page 130) affecting only the heavier atoms in Groups 3A and 4A. The lighter atoms in these groups do not exhibit the inert pair effect; aluminum, for example, only forms the $3+$ ion. This tendency of the heavier metallic elements in Group 3A to form both $1+$ and $3+$ ions and those in Group 4A to form both $2+$ and $4+$ ions is called the ***inert pair effect.***

An anion forms when one or more electrons are added to the highest partially filled n shell of an atom, resulting in a noble-gas configuration. Consider the following examples:

$$\text{H: }1s^1 \qquad \text{H}^-\text{: }1s^2 \text{ or [He]}$$
$$\text{F: }1s^22s^22p^5 \qquad \text{F}^-\text{: }1s^22s^22p^6 \text{ or [Ne]}$$
$$\text{O: }1s^22s^22p^4 \qquad \text{O}^{2-}\text{: }1s^22s^22p^6 \text{ or [Ne]}$$
$$\text{N: }1s^22s^22p^3 \qquad \text{N}^{3-}\text{: }1s^22s^22p^6 \text{ or [Ne]}$$

Each of these anions has a stable noble gas electron configuration.[8] In fact, F^-, Na^+, and Ne (as well as Al^{3+}, O^{2-}, and N^{3-}) have the same electron configuration. They are said to be ***isoelectronic*** because they *have the same number of electrons, and hence the same ground-state electron configuration.* H^- and He are also isoelectronic.

For transition metals, the formation of ions is more complicated. In the first-row transition metals (Sc to Cu), recall that total energy considerations dictate that the $4s$ orbital always fills before the $3d$ orbitals, even though the $3d$ orbital energy is lower. When electrons are removed to form ions, however, the higher energy $4s$ electrons are removed first. Consider manganese, whose electron configuration is $[\text{Ar}]4s^23d^5$. When the Mn^{2+} ion is formed, the two electrons are removed from the $4s$ orbital, and the electron configuration of Mn^{2+} is $[\text{Ar}]3d^5$. When a cation is formed from a transition metal atom, electrons are always removed first from the ns orbital and then from the $(n - 1)\ d$ orbitals. Moreover, most transition metals can form more than one cation and frequently the cations are *not* isoelectronic with the preceding noble gases.

Keep in mind that the order of electron filling does not determine or predict the order of electron removal for transition metals.

2.3 | The Periodic Table Predates Quantum Mechanics

We showed in Section 2.2 how the structure of the periodic table arises from a quantum-mechanical description of the electronic structure of elements in terms of atomic orbitals. However, the origin of the periodic table predates the development of quantum

---

8. As will be discussed in Section 2.5, the addition of two electrons to an isolated O atom to form O^{2-} is energetically unfavorable. The O^{2-} anion is unstable by itself, but becomes stable if surrounded by positive ions in an ionic compound. Thus, in the formation of anions, the noble gas configuration is due more to favorable interactions with other ions than to any inherent stability of the noble gas configuration.

Dmitri Mendeleev

Gallium melts in a person's hand
(body temperature is about 37°C).

mechanics by several decades. In the nineteenth century, when chemists had only a vague idea of atoms and molecules and did not know of the existence of electrons and protons, they devised the periodic table using their knowledge of atomic masses, which at that time could be measured quite accurately. Arranging elements according to their atomic masses in a periodic table seemed logical to those chemists, who felt that chemical behavior should somehow be related to atomic mass. In 1864 the English chemist John Newlands[9] noticed that when the elements were arranged in order of atomic mass, every eighth element had similar properties. Newlands referred to this peculiar relationship as the *law of octaves*. This "law" turned out to be inadequate for elements beyond calcium, however, and the scientific community did not accept Newlands' work.

In 1869 the Russian chemist Dmitri Mendeleev[10] and the German chemist Lothar Meyer[11] independently proposed a much more extensive tabulation of the elements based on the regular, periodic recurrence of properties. Mendeleev's classification system was a great improvement over Newlands' for two reasons. First, it grouped the elements together more accurately, according to their properties. Equally important, it made it possible to predict the properties of several elements that had not yet been discovered. For example, Mendeleev proposed the existence of an unknown element that he called eka-aluminum and predicted a number of its properties. (*Eka* is a Sanskrit word meaning "first"; thus eka-aluminum would be the first element under aluminum in the same group.) When gallium was discovered four years later, its properties matched the predicted properties of eka-aluminum remarkably well:

	Eka-Aluminum (Ea)	**Gallium (Ga)**
Atomic mass	68 u	69.9 u
Melting point	Low	30.15°C
Density	5.9 g cm^{-3}	5.94 g cm^{-3}
Formula of oxide	Ea_2O_3	Ga_2O_3

Mendeleev's periodic table included 66 known elements. By 1900, some 30 more had been added to the list, filling in some of the empty spaces. Figure 2.12 charts the discovery of the elements chronologically.

Although this periodic table was a celebrated success, the early versions had some glaring inconsistencies. For example, the atomic mass of argon (39.95 u; $Z = 18$) is *greater* than that of potassium (39.10 u; $Z = 19$). If elements were arranged solely according to increasing atomic mass, argon and potassium would need to switch positions in our modern periodic table (see the inside front cover of your text). But no chemist would place argon, an inert gas, in the same group as lithium and sodium, two very reactive metals. This and other discrepancies suggested that some fundamental property other than atomic mass must be the basis of periodicity. This property turned out to be associated with atomic number, a concept unknown to Mendeleev and his contemporaries.

Using data from scattering experiments (see Section 0.2), Rutherford estimated the number of positive charges in the nucleus of a few elements, but the significance

---

9. John Alexander Reina Newlands (1838–1898). English chemist. Newlands' work was a step in the right direction in the classification of the elements. Unfortunately, because of its shortcomings, he was subjected to much criticism, and even ridicule. At one meeting he was asked if he had ever examined the elements according to the order of their initial letters! Nevertheless, in 1887 the Royal Society honored Newlands for his contribution.

10. Dmitri Ivanovich Mendeleev (1836–1907). Russian chemist. Many regard his work on the periodic classification of elements as the most significant achievement in chemistry in the nineteenth century.

11. Julius Lothar Meyer (1830–1895). German chemist. In addition to his contribution to the periodic table, Meyer also discovered the chemical affinity of hemoglobin for oxygen.

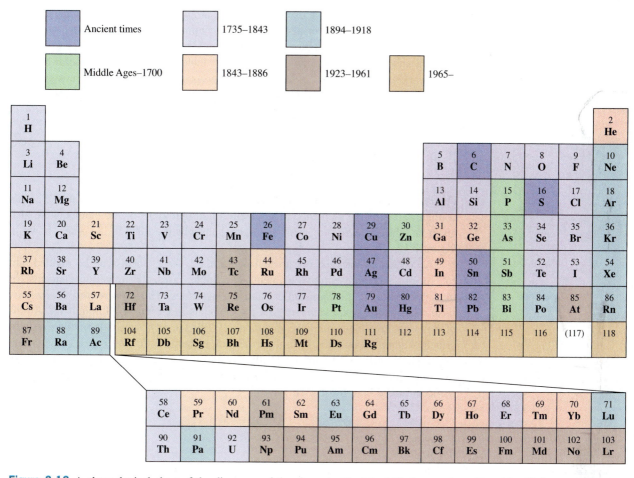

Figure 2.12 A chronological chart of the discovery of the elements. To date, 117 elements have been identified.

of these numbers was overlooked for several years. In 1913 a young English physicist, Henry Moseley,[12] discovered a correlation between what he called *atomic number* and the frequency of X-rays generated by bombarding an element with high-energy electrons. Moseley noticed that the frequencies of X-rays emitted from the elements could be correlated by the equation

$$\sqrt{v} = a(Z - b)$$

where v is the frequency of the emitted X-rays and a and b are constants that are the same for all the elements. Thus, from the square root of the measured frequency of the X-rays emitted, we can determine the atomic number of the element.

With a few exceptions, Moseley found that atomic number increases in the same order as atomic mass. For example, calcium is the twentieth element in order of increasing atomic mass, and it has an atomic number of 20. The discrepancies that had puzzled earlier scientists now made sense. The atomic number of argon is 18 and that of potassium is 19, so potassium should follow argon in the periodic table, not precede it.

12. Henry Gwyn-Jeffreys Moseley (1887–1915). English physicist. Moseley discovered the relationship between X-ray spectra and atomic number. A lieutenant in the Royal Engineers, he was killed in action at the age of 28 during the British campaign in Gallipoli, Turkey.

A modern periodic table usually shows the atomic number along with the element symbol. The atomic number also indicates the number of electrons in the atoms of an element. Electron configurations of elements help to explain the recurrence of physical and chemical properties. The periodic table is important and useful because we can use our understanding of the general properties and trends within a group or a period to predict with considerable accuracy the properties of any element, even though that element may be unfamiliar to us.

The structural organization of the periodic table into groups (columns) and periods (rows) was discussed in Section 0.2.

2.4 | Elements Can Be Classified by Their Position in the Periodic Table

Figure 2.13 shows the periodic table together with the outermost ground-state electron configurations of the elements. (The electron configurations of the elements are also given in Table 2.1.) Starting with hydrogen, subshells are filled in the order shown in Figure 2.10.

According to the type of subshell being filled, the elements can be divided into categories—the representative elements, the noble gases, the transition elements (or transition metals), the lanthanides, and the actinides. The **representative elements** (also called *main-group elements*) are the elements in Groups 1A through 7A. Atoms of representative elements have incompletely filled *s* or *p* subshells of their highest principal quantum number. With the exception of helium, the *noble gases* (the Group 8A elements) all have a completely filled *p* subshell. (The electron configurations are $1s^2$ for helium and ns^2np^6 for the other noble gases, where *n* is the principal quantum number for the outermost shell.)

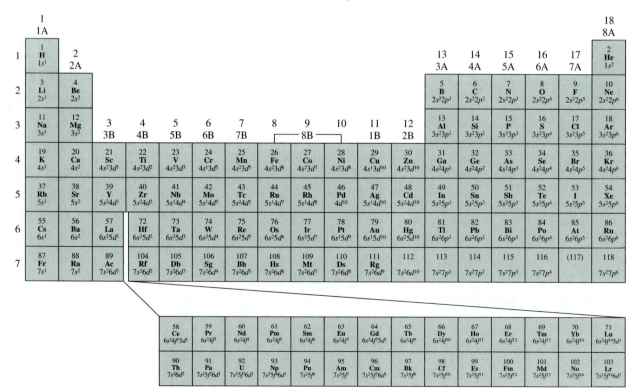

Figure 2.13 The ground-state electron configurations of the elements. For simplicity, only the configurations of the outer electrons are shown.

The transition elements (or transition metals) are the elements in Groups 1B and 3B through 8B (see section 2.2). (These metals are sometimes referred to as the *d*-block transition elements.) The nonsequential numbering of the transition metals in the periodic table (that is, 3B–8B, followed by 1B) acknowledges a correspondence between the outer electron configurations of these elements and those of the representative elements. For example, scandium and gallium both have three outer electrons. However, because the electrons are in different types of atomic orbitals, the elements are placed in different groups (3B and 3A, respectively). The metals iron (Fe), cobalt (Co), and nickel (Ni) do not fit this scheme and are all placed in Group 8B. The Group 2B elements, Zn, Cd, and Hg, are neither representative elements nor transition metals. There is no special name for this group of metals. The designation of A and B groups is not universal. In Europe the practice is to use B for representative elements and A for transition metals, which is just the opposite of the American convention. The International Union of Pure and Applied Chemistry has recommended numbering the columns sequentially with Arabic numerals 1 through 18 (see Figure 2.13). The proposal has sparked much controversy in the international chemistry community, and its merits and drawbacks will be deliberated for some time. In this text, we use the American designation.

The lanthanides and actinides are sometimes called *f*-block transition elements because they have incompletely filled *f* subshells. Figure 2.14 distinguishes these groups of elements discussed here.

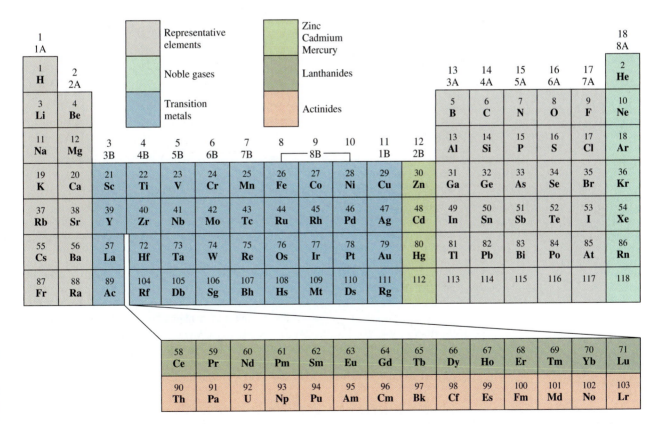

Figure 2.14 Classification of the elements. The Group 2B elements are often classified as transition metals even though they do not exhibit the characteristics of the transition metals.

A clear pattern emerges when we examine the electron configurations of the elements in a particular group (see Table 2.1 and Figure 2.13). All members of the Group 1A alkali metals have similar outer electron configurations; each has a noble gas core and an ns^1 outer electron. Similarly, the Group 2A alkaline earth metals have a noble gas core and an outer electron configuration of ns^2 *The outer electrons of an atom, which are the ones involved in chemical bonding,* are called **valence electrons.** (The inner electrons of an atom are called **core electrons.**) The similarity of the outer electron configurations (that is, the same number and type of valence electrons) is what makes the elements in the same group resemble one another in chemical behavior. This observation holds true for the other representative elements. Thus, the halogens (the Group 7A elements) all have outer electron configurations of ns^2np^5, and they have very similar properties. Be careful, though, when predicting properties for Groups 3A through 6A. The elements in Group 4A, for example, all have the same outer electron configuration (ns^2np^2), but chemical and physical properties vary among these elements: Carbon is a nonmetal, silicon and germanium are metalloids, and tin and lead are metals.

As a group, the noble gases behave very similarly. Although a few compounds of krypton, xenon and, most recently,[13] argon, have been synthesized, the noble gases are chemically inert under most conditions. They are so unreactive because they all have completely filled outer ns and np subshells, a condition that represents great stability. Although the outer electron configuration of the transition metals differs within a group, and there is no pattern in the change of the electron configuration from one metal to the next in the same period, all transition metals share characteristics that set them apart from other elements. These similarities arise because transition metals all have an incompletely filled d subshell. Likewise, the lanthanide (and actinide) elements resemble one another because they have incompletely filled f subshells.

Example 2.4

An atom of a certain element has 15 electrons. Without consulting a periodic table, answer the following questions: (a) What is the ground-state electron configuration of the element? (b) How should the element be classified? (c) Is the element diamagnetic or paramagnetic?

Strategy (a) Use the building-up principle discussed in Section 2.2 to write the electron configuration with principal quantum number $n = 1$ and continue upward until all the electrons are accounted for. (b) To determine whether the atom should be classified as a representative element, a transition metal, or a noble gas, consider its electron configuration characteristics. (c) Examine the pairing scheme of the electrons in the outermost shell. The element will be diamagnetic if all electrons are paired and paramagnetic if some are unpaired.

Solution (a) For $n = 1$, we have a $1s$ orbital (2 electrons); for $n = 2$, we have a $2s$ orbital (2 electrons) and three $2p$ orbitals (6 electrons); for $n = 3$, we have a $3s$ orbital (2 electrons). The number of electrons left is $15 - 12 = 3$, and these three electrons are placed in the $3p$ orbitals. The electron configuration is $1s^2 2s^2 2p^6 3s^2 3p^3$. (b) Because the $3p$ subshell is not completely filled, this is a representative element. Based on the information given, we cannot say whether it is a metal, a nonmetal, or a metalloid. (c) According to Hund's rule, the three electrons in the $3p$ orbitals have parallel spins (three unpaired electrons). The element, therefore, is paramagnetic.

—Continued

---

13. In 2000, researchers at the University of Helsinki, Finland reported production of the compound argon fluorohydride (HArF). This compound was found to be stable up to 40 K.

Continued—

Check For (b), note that a transition metal possesses an incompletely filled *d* subshell and a noble gas has a completely filled outer shell. For (c), recall that if the atoms of an element contain an odd number of electrons, then the element must be paramagnetic.

Practice Exercise An atom of a certain element has 20 electrons. (a) Write the ground-state electron configuration of the element, (b) classify the element, and (c) determine whether the element is diamagnetic or paramagnetic.

2.5 | The Properties of the Elements Vary Periodically Across the Periodic Table

As we have seen, the electron configurations of the elements show a periodic variation with increasing atomic number. Consequently, there are also periodic variations in physical and chemical behavior. In Section 2.5, we examine some physical properties of elements that are in the same group or period, and in Section 2.6 we discuss the chemical behavior of the elements. First, let's look at the concept of effective nuclear charge, which has a direct bearing on atomic size and on the stability of electrons.

Effective Nuclear Charge

In Section 2.2, we discussed the shielding effect electrons close to the nucleus have on outer-shell electrons in many-electron atoms. Shielding electrons reduce the electrostatic attraction between the positively charged protons in the nucleus and the outer electrons. Moreover, the repulsive forces between electrons in many-electron atoms further offset the attractive force exerted by the nucleus. The concept of effective nuclear charge enables us to account for the effects of shielding on periodic properties. Consider the helium atom, for example, which has the ground-state electron configuration $1s^2$. The two protons in helium give a nuclear charge of 2, but the full attractive force of this charge on the two $1s$ electrons is partially offset by electron-electron repulsion. Consequently, the $1s$ electrons shield each other from the nucleus. The *effective nuclear charge* (Z_{eff}), which is the charge felt by an electron, is given by

$$Z_{\text{eff}} = Z - \sigma$$

where Z is the actual nuclear charge (that is, the atomic number of the element) and σ is called the *shielding constant* (also called the *screening constant*). The shielding constant is greater than zero but smaller than Z.

One way to illustrate electron shielding is to consider the amounts of energy required to remove the two electrons from a helium atom. Measurements show that it takes 2373 kJ to remove the first electron from 1 mol of He atoms and 5251 kJ to remove the remaining electron from 1 mole of He^+ ions. It takes much more energy to remove the second electron because there is no shielding with only one electron present, so the electron feels the full effect of the nuclear charge.

For atoms with three or more electrons, the electrons in a given shell are shielded by electrons in inner shells (that is, shells closer to the nucleus) but not by electrons in outer shells. Thus, in a lithium atom, whose electron configuration is $1s^2 2s^1$, the $2s$ electron is shielded by the two $1s$ electrons, but the $2s$ electron does not have a shielding effect on the $1s$ electrons. In addition, filled inner shells shield outer electrons more effectively than electrons in the same subshell shield each other.

The increase in effective nuclear charge from left to right across a period and from bottom to top in a group for representative elements

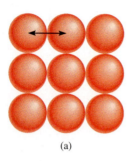

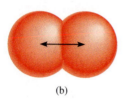

(a)

(b)

Figure 2.15 (a) In metals such as beryllium, the atomic radius is defined as one-half the distance between the nuclei of two adjacent atoms. (b) For elements that exist as diatomic molecules, such as iodine, the radius of the atom is defined as one-half the distance between the nuclei.

Atomic Radius

Several physical properties, including density, melting point, and boiling point, are related to the sizes of atoms, but atomic size is difficult to define. As we saw in Section 2.2, the electron density in an atom extends far beyond the nucleus, but we normally think of atomic size as the volume containing about 90 percent of the total electron density around the nucleus. When we must be more specific, we define the size of an atom in terms of its *atomic radius.* Definitions for atomic radius depend upon the identity and environment of the element in question:

▶ For metals and atoms linked together to form an extensive three-dimensional network, the atomic radius is simply one-half the distance between the nuclei in two neighboring atoms [Figure 2.15(a)].

▶ For elements that exist as simple diatomic molecules, the atomic radius is one-half the distance between the nuclei of the two atoms in a particular molecule [Figure 2.15(b)].

Figure 2.16 shows the atomic radii of many elements according to their positions in the periodic table, and Figure 2.17 shows a plot of the atomic radii of these elements against their atomic numbers.

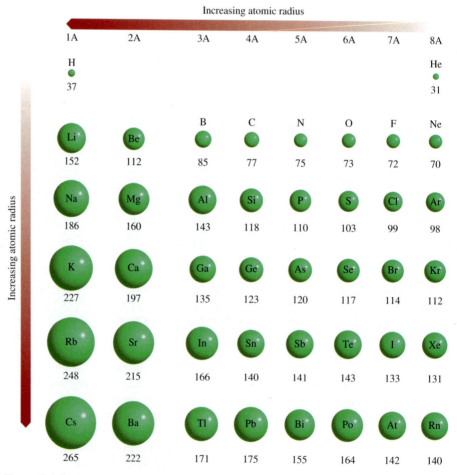

Figure 2.16 Atomic radii (in picometers) of representative elements and noble gases arranged according to their positions in the periodic table.

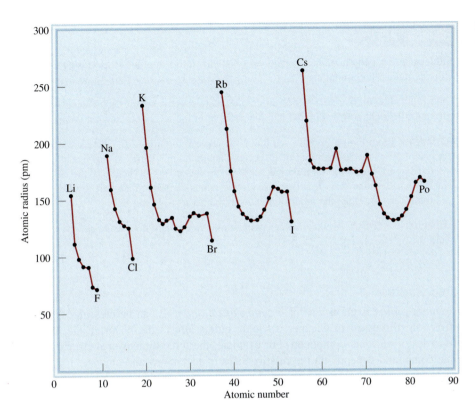

Figure 2.17 A plot of atomic radii (in picometers) of elements as a function of atomic number.

From Figures 2.16 and 2.17, two periodic trends for atomic radius emerge: (1) Atomic radius tends to *decrease* across a row, and (2) atomic radius tends to *increase* down a group. To understand these trends, keep in mind that atomic radius is determined to a large extent by the strength of the attraction between the nucleus and the outer-shell electrons. The larger the effective nuclear charge, the stronger the hold of the nucleus on these electrons, and the smaller the atomic radius. For example, moving from left (Li) to right (F) in the second row, the number of electrons in the inner shell ($1s^2$) remains constant while the nuclear charge increases. The electrons that are added to counterbalance the increasing nuclear charge are all within the $n = 2$ shell, so they ineffectively shield one another from the nuclear charge. Consequently, the effective nuclear charge increases steadily while the principal quantum number remains constant ($n = 2$). For example, the outer $2s$ electron in lithium is shielded from the nucleus (which has three protons) by two $1s$ electrons. As an approximation, we assume that the shielding effect of two $1s$ electrons cancels two positive charges in the nucleus. Thus, the $2s$ electron only "feels" the attraction of one proton in the nucleus; the effective nuclear charge is 1. In beryllium ($1s^2 2s^2$), each of the $2s$ electrons is shielded by the inner two $1s$ electrons, which cancel two of the four positive charges in the nucleus. Because the $2s$ electrons do not shield each other effectively, the effective nuclear charge for each $2s$ electron is greater than 1. Thus, as the effective nuclear charge increases, the atomic radius decreases steadily from lithium to fluorine.

To understand the increase in atomic radius as we move down a group, consider the alkali metals in Group 1A. For elements of this group, the outermost electron resides in the ns orbital. Because orbital size increases with the increasing principal quantum number n, the size of the metal atoms increases from Li to Cs. We can apply the same reasoning to the elements in other groups.

Example 2.5

Referring to a periodic table, arrange the following atoms in order of increasing atomic radius: P, Si, and N.

Strategy Use the periodic trends in atomic radii—namely, atomic radius decreases from left to right across a period and increases from top to bottom down a group.

Solution From Figure 2.13 we see that N is above P in Group 5A. Therefore, the radius of N is smaller than that of P (because atomic radius increases down a group). Both Si and P are in the third period, and Si is to the left of P. Therefore, the radius of P is smaller than that of Si (because atomic radius decreases from left to right across a period). Thus, the order of increasing radius is N, P, and Si.

Practice Exercise Arrange the following atoms in order of decreasing radius: C, Li, and Be.

Ionic Radius

The *ionic radius* is *the radius of a cation or an anion*. It can be measured by X-ray diffraction (discussed in Chapter 6). Ionic radius affects the physical and chemical properties of an ionic compound. For example, the three-dimensional structure of an ionic compound depends on the relative sizes of its cations and anions.

When a neutral atom is converted to an ion, its size changes. If the atom forms an anion, its size (or radius) increases because the nuclear charge remains the same but the repulsion resulting from the additional electron(s) enlarges the domain of the electron cloud. If the atom forms a cation, on the other hand, its size decreases because removing one or more electrons from an atom reduces electron-electron repulsion while the nuclear charge remains the same, thus shrinking the electron cloud. Figure 2.18 shows the changes in size that result when alkali metals are

Figure 2.18 Comparison of atomic radii with ionic radii. (a) Alkali metals and alkali metal cations. (b) Halogens and halide ions.

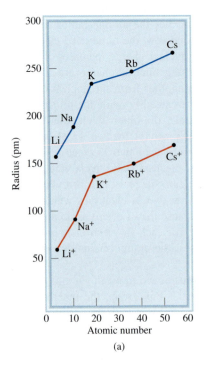

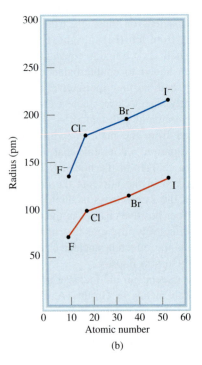

(a)   (b)

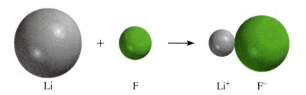

Figure 2.19 Changes in the sizes of Li and F when they react to form LiF.

converted to cations and halogens are converted to anions; Figure 2.19 shows the changes in size that occur when a lithium atom reacts with a fluorine atom to form a unit of LiF.

Figure 2.20 shows the radii of ions derived from some familiar elements, arranged according to the positions of the elements in the periodic table. The trends in ionic radii parallel the trends in atomic radii. For example, from top to bottom both the atomic radius and the ionic radius increase within a group. For ions derived from elements in different groups, a size comparison is meaningful only if the ions are isoelectronic. If we examine isoelectronic ions, we find that cations are smaller than anions. For example, Na^+ is smaller than F^-. Both ions have the same number of electrons, but Na^+ ($Z = 11$) has more protons than F^- ($Z = 9$). The larger effective nuclear charge of Na^+ results in a smaller radius.

Focusing on isoelectronic cations, the radii of *tripositive ions* (ions that bear three positive charges) are smaller than those of *dipositive ions* (ions that bear two positive charges), which in turn are smaller than *unipositive ions* (ions that bear one positive charge). This trend is nicely illustrated by the sizes of three isoelectronic ions in the third period: Al^{3+}, Mg^{2+}, and Na^+ (see Figure 2.20). The Al^{3+} ion has the same number of electrons as Mg^{2+}, but it has one more proton. Thus, the electron

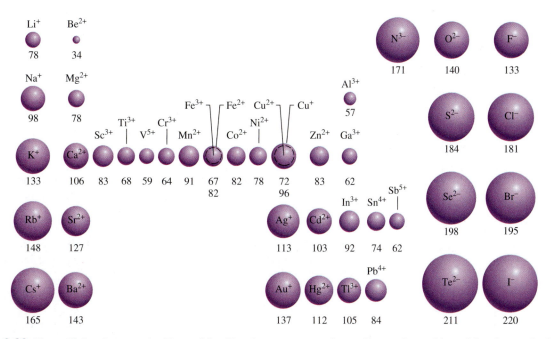

Figure 2.20 The radii (in picometers) of ions of familiar elements arranged according to the position of the element in the periodic table.

cloud in Al^{3+} is pulled inward more than that in Mg^{2+}. The radius of Mg^{2+} is smaller than that of Na^+ for the same reason. Turning to isoelectronic anions, the radius *increases* as we go from ions with uninegative charge $(1-)$ to those with dinegative charge $(2-)$, and so on. Thus, the oxide ion (O^{2-}) is larger than the fluoride ion because oxygen has one fewer proton than fluorine; the electron cloud is spread out more in O^{2-}.

Example 2.6

For each of the following pairs, indicate which of the two species is larger: (a) N^{3-} or F^-; (b) Mg^{2+} or Ca^{2+}; (c) Fe^{2+} or Fe^{3+}.

Strategy When comparing ionic radii, it is useful to classify the ions into the following three categories: (1) isoelectronic ions, (2) ions that carry the same charges and are generated from atoms of the same group, and (3) ions that carry different charges but are generated from the same atom. In case (1), ions carrying a greater negative charge are always larger; in case (2), ions from atoms having a greater atomic number are always larger; in case (3), ions having a smaller positive charge are always larger.

Solution (a) N^{3-} and F^- are isoelectronic anions, both containing 10 electrons. Because N^{3-} has only seven protons and F^- has nine, the weaker attraction exerted by the nucleus of N^{3-} makes it the larger ion. (b) Both Mg and Ca belong to Group 2A (the alkaline earth metals). Thus, Ca^{2+} ion is larger than Mg^{2+} because the outer electrons in Ca^{2+} are in a larger shell ($n = 3$) than are the outer electrons in Mg^{2+} ($n = 2$). (c) Both ions have the same nuclear charge, but Fe^{2+} has one more electron (24 electrons compared to 23 electrons for Fe^{3+}) and hence greater electron–electron repulsion. The radius of Fe^{2+} is larger.

Practice Exercise Select the smaller ion in each of the following pairs: (a) K^+ or Li^+; (b) Au^+ or Au^{3+}; (c) P^{3-} or N^{3-}.

Variation of Physical Properties Across a Period and Within a Group

From left to right across a period there is a transition from metals to metalloids to nonmetals (see Figure 0.9). Consider the third-period elements from sodium to argon (Figure 2.21).

Sodium, the first element in the third period, is a very reactive metal, whereas chlorine, the second-to-last element of that period, is a very reactive nonmetal. In between, the elements show a gradual transition from metallic properties to nonmetallic properties. Sodium, magnesium, and aluminum all have extensive three-dimensional atomic networks, which are held together by forces characteristic of the metallic state. Silicon is a metalloid; it has a giant three-dimensional structure in which the Si atoms are held together very strongly. Starting with phosphorus, the elements exist in simple, discrete molecular units (P_4, S_8, Cl_2, and Ar) that have low melting points and boiling points.

Within a periodic group the physical properties vary more predictably, especially if the elements are in the same physical state. For example, the melting points of argon and xenon are $-189.2°C$ and $-111.9°C$, respectively. We can estimate the

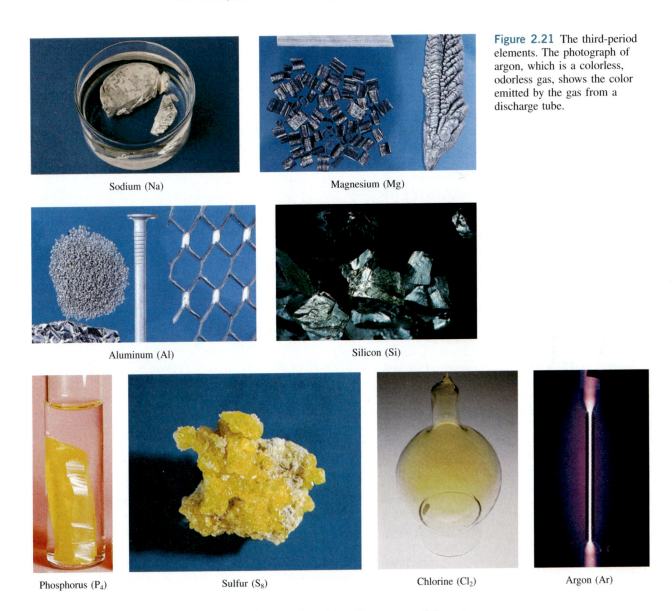

Figure 2.21 The third-period elements. The photograph of argon, which is a colorless, odorless gas, shows the color emitted by the gas from a discharge tube.

Sodium (Na)

Magnesium (Mg)

Aluminum (Al)

Silicon (Si)

Phosphorus (P_4)

Sulfur (S_8)

Chlorine (Cl_2)

Argon (Ar)

melting point of the intermediate element krypton by taking the average of these two values as follows:

$$\text{melting point of Kr} \approx \frac{[(-189.2°C) + (-111.9°C)]}{2} = -150.6°C$$

The actual melting point of krypton is $-156.6°C$.

The inset on page 156 illustrates one interesting application of periodic group properties.

Ionization Energy

The chemical properties of any atom are determined by the configuration of the valence electrons of the atom. The stability of these outermost electrons is reflected directly in the atom's ionization energies. The *ionization energy* (p. 88) is the minimum energy (in kJ mol^{-1}) required to remove an electron from a gaseous atom in its ground state. In other words, the ionization energy is the amount of energy in kilojoules needed to

The Third Liquid Element?

Of the 117 elements currently known, 11 are gases under atmospheric conditions. Six of these are the Group 8A elements (the noble gases He, Ne, Ar, Kr, Xe, and Rn), and the other five are hydrogen (H_2), nitrogen (N_2), oxygen (O_2), fluorine (F_2), and chlorine (Cl_2). Curiously, only two elements are liquids at 25°C: mercury (Hg) and bromine (Br_2).

We do not know the properties of all the known elements because some of them have never been prepared in quantities large enough for investigation. In these cases we must rely on periodic trends to predict their properties. What are the chances, then, of discovering a third liquid element?

Could francium (Fr), the last member of Group 1A, be a liquid at 25°C? All of the isotopes of francium are radioactive. The most stable isotope is francium-223, which has a half-life of 21 min. (*Half-life* is the time it takes for one-half of the nuclei in any given amount of a radioactive substance to disintegrate.) This short half-life means that only very small traces of francium could possibly exist on Earth. And although it is feasible to prepare francium in the laboratory, no weighable quantity of the element has been prepared or isolated. Thus, we know very little about the physical and chemical properties of francium. Nevertheless we can use the group periodic trends to predict some of those properties. Take the melting point of francium as an example. The plot shows how the melting points of the alkali metals vary with atomic number. From lithium to sodium, the melting point drops 81.4°C; from

sodium to potassium, 34.6°C; from potassium to rubidium, 24°C; and from rubidium to cesium, 11°C. On the basis of this trend, the change from cesium to francium should be about 5°C. If so, the melting point of francium would be about 25°C, which would make it a liquid under atmospheric conditions.

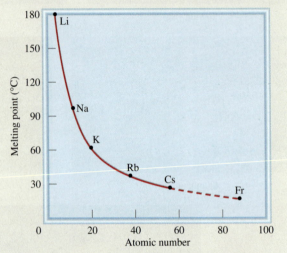

A plot of the melting points of the alkali metals versus their atomic numbers. By extrapolation, the melting point of francium should be about 25°C.

strip 1 mol of electrons from 1 mol of gaseous atoms. Gaseous atoms are specified in this definition because an atom in the gas phase is virtually uninfluenced by its neighbors, and so there are no intermolecular forces (that is, forces between molecules) to take into account.

The magnitude of ionization energy is a measure of how "tightly" the electron is held in the atom. The higher the ionization energy, the more difficult it is to remove the electron. For a many-electron atom, the amount of energy required to remove the first electron from the atom in its ground state,

$$\text{energy} + X(g) \longrightarrow X^+(g) + e^- \tag{2.7}$$

is called the *first ionization energy* (I_1). In Equation 2.7, X represents an atom of any element and e^- is an electron. The second ionization energy (I_2) and the third ionization energy (I_3) are represented by the following equations:

$$\text{energy} + X^+(g) \longrightarrow X^{2+}(g) + e^- \quad \text{second ionization}$$
$$\text{energy} + X^{2+}(g) \longrightarrow X^{3+}(g) + e^- \quad \text{third ionization}$$

The pattern continues for the removal of subsequent electrons.

When an electron is removed from an atom, the repulsions among the remaining electrons decreases. Because the nuclear charge remains constant, more energy is needed to remove another electron from the positively charged ion. Thus, ionization energies always increase in the following order:

$$I_1 < I_2 < I_3 < \ldots$$

Table 2.2	The Ionization Energies (kJ mol^{-1}) of the First Twenty Elements					
Z	Element	First	Second	Third	Fourth	Fifth	Sixth
1	H	1,312					
2	He	2,373	5,251				
3	Li	520	7,300	11,815			
4	Be	899	1,757	14,850	21,005		
5	B	801	2,430	3,660	25,000	32,820	
6	C	1,086	2,350	4,620	6,220	38,000	47,261
7	N	1,400	2,860	4,580	7,500	9,400	53,000
8	O	1,314	3,390	5,300	7,470	11,000	13,000
9	F	1,680	3,370	6,050	8,400	11,000	15,200
10	Ne	2,080	3,950	6,120	9,370	12,200	15,000
11	Na	495.9	4,560	6,900	9,540	13,400	16,600
12	Mg	738.1	1,450	7,730	10,500	13,600	18,000
13	Al	577.9	1,820	2,750	11,600	14,800	18,400
14	Si	786.3	1,580	3,230	4,360	16,000	20,000
15	P	1,012	1,904	2,910	4,960	6,240	21,000
16	S	999.5	2,250	3,360	4,660	6,990	8,500
17	Cl	1,251	2,297	3,820	5,160	6,540	9,300
18	Ar	1,521	2,666	3,900	5,770	7,240	8,800
19	K	418.7	3,052	4,410	5,900	8,000	9,600
20	Ca	589.5	1,145	4,900	6,500	8,100	11,000

Table 2.2 lists the ionization energies of the first 20 elements. Because the removal of an electron from an atom or a stable ion always requires energy, ionization energies are always positive quantities.

Figure 2.22 shows how the first ionization energy varies with atomic number. Apart from small irregularities, the first ionization energies of elements in a period increase with increasing atomic number. This trend is due to the increase in effective nuclear charge from left to right across the period (as in the case of atomic radii). A larger effective nuclear charge means a more tightly held outer electron, and hence a higher first ionization energy. A notable feature of Figure 2.22 is the sequence of peaks that correspond to the noble gases. The high ionization energies of the noble gases, stemming from their stable ground-state electron configurations, account for the fact that most of them are chemically unreactive. In fact, helium ($1s^2$) has the highest first ionization energy of all the elements ($I_1 = 2373$ kJ mol^{-1}).

At the bottom of the graph in Figure 2.22 are the Group 1A elements (the alkali metals), which have the lowest first ionization energies. Each of these metals has one valence electron (the outermost electron configuration is ns^1), which is effectively shielded by the completely filled inner shells. Consequently, it is energetically easy to remove an electron from the atoms of the alkali metals to form unipositive ions (Li^+, Na^+, K^+, . . .) that are isoelectronic with the noble gases that precede them in the periodic table.

The Group 2A elements (the alkaline earth metals) have higher first ionization energies than the alkali metals (Group 1A). The alkaline earth metals have two valence electrons (the outermost electron configuration is ns^2). Because these two s electrons

The increase in first ionization energy from left to right across a period and from bottom to top in a group for representative elements

Figure 2.22 Variation of the first ionization energy with atomic number. The noble gases have high ionization energies, whereas the alkali and alkaline earth metals have low ionization energies.

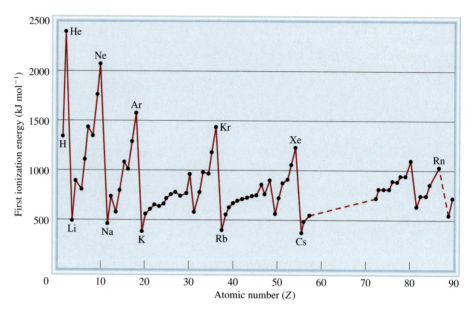

do not shield each other well, the effective nuclear charge for an alkaline earth metal atom is larger than that for the preceding alkali metal. Most alkaline earth compounds contain dipositive ions (Mg^{2+}, Ca^{2+}, Sr^{++}, and Ba^{2+}). The Be^{2+} ion is isoelectronic with Li^+ and with He, Mg^{2+} is isoelectronic with Na^+ and with Ne, and so on.

As Figure 2.22 shows, metals have relatively low ionization energies compared to nonmetals. The ionization energies of the metalloids generally fall between those of metals and nonmetals. The difference in ionization energies suggests why metals always form cations and nonmetals generally form anions in ionic compounds. (An important exception is the ammonium ion, NH_4^+.)

Within a given group, ionization energy decreases with increasing atomic number (that is, from top to bottom down the group). Elements in the same group have similar outer electron configurations. However, as the principal quantum number n increases, so does the average distance of a valence electron from the nucleus. A greater separation between the electron and the nucleus means a weaker attraction, so that it becomes increasingly easier to remove the first electron from element to element down a group. As a result, the metallic character of the elements within a group increases from top to bottom. This trend is particularly noticeable for elements in Groups 3A to 7A. In Group 4A, for example, carbon is a nonmetal, silicon and germanium are metalloids, and tin and lead are metals.

Although the general trend in the periodic table is for first ionization energies to increase from left to right, some irregularities exist. The first exception occurs between Group 2A and 3A elements in the same period (for example, between Be and B and between Mg and Al). The Group 3A elements have lower first ionization energies than 2A elements because they all have a single electron in the outermost p subshell (ns^2np^1), which is well shielded by the inner electrons and the ns^2 electrons. Therefore, less energy is needed to remove a single p electron than to remove a paired s electron from the same principal energy level. The second irregularity occurs between Groups 5A and 6A (for example, between N and O and between P and S). In the Group 5A elements (ns^2np^3) the p electrons are in three separate orbitals according to Hund's rule. In Group 6A (ns^2np^4) the additional electron must be paired with one of the three p electrons. The proximity of two electrons in the same orbital results in greater electrostatic repulsion, which makes it easier to ionize an atom of the Group 6A element, even though the nuclear charge has increased by one unit. Thus, the first ionization energies for Group 6A elements are lower than those for Group 5A elements in the same period.

Example 2.7 compares the ionization energies of some elements.

Example 2.7

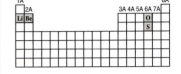

(a) Which atom should have a smaller first ionization energy: oxygen or sulfur?
(b) Which atom should have a higher second ionization energy: lithium or beryllium?

Strategy (a) Oxygen and sulfur are both in Group 6A, and the first ionization energy decreases down a group because the outermost electron for the heavier element is farther away from the nucleus and feels less attraction. (b) Lithium (Group 1A) precedes beryllium (Group 2A) in the second period. Removal of the outermost electron requires less energy if it is shielded by a filled inner shell.

Solution (a) Oxygen and sulfur have the same valence electron configuration (ns^2np^4), but the $3p$ electron in sulfur is farther from the nucleus and experiences less nuclear attraction than the $2p$ electron in oxygen. Thus, sulfur should have a smaller first ionization energy than oxygen. (b) The electron configurations of Li and Be are $1s^22s^1$ and $1s^22s^2$, respectively. The second ionization energy is the minimum energy required to remove an electron from a gaseous unipositive ion in its ground state. For the second ionization process, we write

$$\underset{1s^2}{Li^+(g)} \longrightarrow \underset{1s^1}{Li^{2+}(g)} + e^-$$

and

$$\underset{1s^22s^1}{Be^+(g)} \longrightarrow \underset{1s^2}{Be^{2+}(g)} + e^-$$

Because $1s$ electrons shield $2s$ electrons much more effectively than they shield each other, it should be easier to remove a $2s$ electron from Be than to remove a $1s$ electron from Li. Thus, Li should have a higher second ionization energy than Be.

Check Compare these results with the data in Table 2.2. In (a), the prediction is consistent with the fact that the metallic character of the elements increases down a periodic group. In (b), the prediction is consistent with the fact that alkali metals form 1+ ions and alkaline earth metals form 2+ ions.

Practice Exercise (a) Which of the following atoms should have a larger first ionization energy: N or P? (b) Which of the following atoms should have a smaller second ionization energy: Na or Mg?

Electron Affinity

Another physical property that greatly influences the chemical behavior of atoms is their ability to accept one or more electrons. This property, called *electron affinity,* is *the negative of the energy change that occurs when an electron is accepted by an atom in the gaseous state to form an anion:*

$$X(g) + e^- \longrightarrow X^-(g) \tag{2.8}$$

If an atom has an affinity for an electron, then energy will be released when the atom is converted to an anion. For example, when 1 mol of electrons are added to 1 mol of fluorine atoms, 328 kJ of energy in the form of heat is released. In *thermochemistry* (discussed in Chapter 7) a process that releases energy in the form of heat is said to be *exothermic* and the energy change has a negative sign; that is, −328 kJ. However, we define the electron affinity (E_{ea}) of fluorine as the negative of the energy change, so E_{ea} becomes a positive quantity:

Electron affinity is positive if the reaction is exothermic and negative if the reaction is endothermic.

$$F(g) + e^- \longrightarrow F^-(g) \qquad E_{ea} = 328 \text{ kJ mol}^{-1}$$

The more positive the electron affinity of an element, the greater is the ability of an atom of the element to accept an electron. Another way of viewing electron affinity is to think of it as the energy that must be supplied to remove an electron from the anion. Thus, a large positive electron affinity means that the negative ion is very stable (that is, the atom has a great tendency to accept an electron), just as a high ionization energy for an atom means that the electron in the atom is very stable.

Experimentally, electron affinity is determined by removing an additional electron from an anion. In contrast to ionization energies, however, electron affinities are difficult to measure because the anions of many elements are unstable. Table 2.3 shows the electron affinities of some representative elements and the noble gases, and Figure 2.23 plots the electron affinities of the first 56 elements versus atomic number. The overall trend is an increase in the tendency to accept electrons (electron affinity values become more positive) from left to right across a period. The electron affinities of metals are generally lower than those of nonmetals. The values vary little within a given group. The halogens (Group 7A) have the highest electron affinity values, because accepting an electron means that each halogen atom becomes isoelectronic with the noble gas immediately to its right. For example, the electron configuration of F^- is $1s^2 2s^2 2p^6$ or [Ne]; for Cl^- it is [Ne] $3s^2 3p^6$ or [Ar]; and so on. Calculations show that the noble gases all have electron affinities of less than zero. Thus, the anions of these gases, if formed, would be inherently unstable.

Table 2.3	**Electron Affinities (kJ mol^{-1}) of Some Representative Elements and the Noble Gases***						
1A	**2A**	**3A**	**4A**	**5A**	**6A**	**7A**	**8A**
H							He
73							<0
Li	Be	B	C	N	O	F	Ne
60	≤ 0	27	122	0	141	328	<0
Na	Mg	Al	Si	P	S	Cl	Ar
53	≤ 0	44	134	72	200	349	<0
K	Ca	Ga	Ge	As	Se	Br	Kr
48	2.4	29	118	77	195	325	<0
Rb	Sr	In	Sa	Sb	Te	I	Xe
47	4.7	29	121	101	190	295	<0
Cs	Ba	Tl	Pb	Bi	Po	At	Rn
45	14	30	110	110	?	?	<0

*The electron affinities of the noble gases, Be, and Mg have not been determined experimentally, but they are believed to be close to zero or negative.

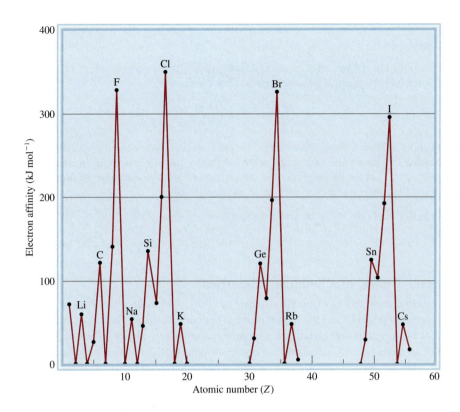

The electron affinity of oxygen has a positive value (141 kJ mol^{-1}), which means that the process

$$O(g) + e^- \longrightarrow O^-(g) \qquad E_{ea} = 141 \text{ kJ mol}^{-1}$$

is energetically favorable. On the other hand, the electron affinity of the O^- ion is highly negative (-780 kJ mol^{-1}), which means the process

$$O^-(g) + e^- \longrightarrow O^{2-}(g) \qquad E_{ea} = -780 \text{ kJ mol}^{-1}$$

is energetically unfavorable, even though the O^{2-} ion is isoelectronic with the highly stable noble gas Ne. This process is unfavorable in the gas phase because the resulting increase in electron–electron repulsion outweighs the stability gained by achieving a noble gas configuration. In solids, however, the O^{2-} ions so common in ionic compounds (for example, Li_2O and MgO) are stabilized by strong electrostatic attractions to the neighboring cations. The stability of ionic compounds is discussed in Chapter 3.

Example 2.8 shows why the alkaline earth metals do not have a great tendency to accept electrons.

Example 2.8

Why are the electron affinities of the alkaline earth metals, shown in Table 2.3, either negative or small positive values?

Strategy Consider the electron configurations of the alkaline earth metals, and whether an electron added to such atoms would be held strongly by the nucleus.

—Continued

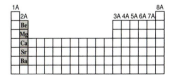

Continued—

Solution The valence electron configuration of the alkaline earth metals is ns^2, where n is the highest principal quantum number. For the process

$$\underset{ns^2}{M\ (g)} + e^- \longrightarrow \underset{ns^2np^1}{M^-(g)}$$

where M denotes a Group 2A element, the extra electron must enter the np subshell, which is effectively shielded by the two ns electrons (the ns electrons are more penetrating than the np electrons) and the inner electrons. Consequently, alkaline earth metals have little tendency to pick up an extra electron.

Practice Exercise Is it likely that Ar will form the anion Ar^-?

General Trends in Chemical Properties

Ionization energy and electron affinity help chemists understand the types of reactions that elements undergo and the nature of the compounds formed. On a conceptual level, these two properties are related in a simple way: Ionization energy indicates the attraction of an atom for its own electrons, whereas electron affinity expresses the attraction of an atom for an additional electron from some other source. Together they provide insight into the general attraction of an atom for electrons. With these concepts we can survey the chemical behavior of the elements systematically, paying particular attention to the relationship between chemical properties and electron configuration.

We have seen that the metallic character of the elements *decreases* from left to right across a period and *increases* from top to bottom within a group. On the basis of these trends and the knowledge that metals usually have low ionization energies and nonmetals usually have high electron affinities, we can frequently predict the outcome of a reaction involving some of these elements.

We have said that elements in the same group resemble one another in chemical behavior because they have similar valence electron configurations. This statement, although correct in the general sense, must be applied with caution. Chemists have long known that the first member of each group (the element in the second period from lithium to fluorine) differs from the rest of the members of the same group. Lithium, for example, exhibits many, but not all, of the properties characteristic of the alkali metals. Similarly, beryllium is a somewhat atypical member of Group 2A, and so on. The difference can be attributed to the unusually small size of the first element in each group (see Figure 2.16).

Another trend in the chemical behavior of the representative elements is the diagonal relationship. *Diagonal relationships* are *similarities between pairs of elements in different groups and periods of the periodic table.* Specifically, the first three members of the second period (Li, Be, and B) exhibit many similarities to Mg, Al, and Si, respectively, the elements located diagonally below them in the periodic table (Figure 2.24). The reason for this phenomenon is the closeness of the charge densities of their cations. (*Charge density* is the charge of an ion divided by its volume.) Cations with comparable charge densities react similarly with anions and therefore form the same type of compounds. Thus, the chemistry of lithium resembles that of magnesium in some ways; the same holds for beryllium and aluminum and for boron and silicon. Each of these pairs is said to exhibit a diagonal relationship.

Comparing the properties of elements in the same group is most valid if the elements are all of the same type (that is, all metals or all nonmetals). This guideline

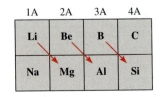

Figure 2.24 Diagonal relationships in the periodic table.

Discovery of the Noble Gases

In the late 1800s, John William Strutt, Third Baron of Rayleigh, who was a professor of physics at the Cavendish Laboratory in Cambridge, England, accurately determined the atomic masses of a number of elements, but he obtained a puzzling result with nitrogen. One of his methods of preparing nitrogen was by the thermal decomposition of ammonia:

$$2NH_3(g) \longrightarrow N_2(g) + 3H_2(g)$$

Another method was to start with air and remove from it oxygen, carbon dioxide, and water vapor. Invariably, the nitrogen from air was a little denser (by about 0.5 percent) than the nitrogen from ammonia.

Lord Rayleigh's work caught the attention of Sir William Ramsay, a professor of chemistry at the University College, London. In 1898 Ramsay passed nitrogen, which he had obtained from air by Rayleigh's procedure, over red-hot magnesium to convert it to magnesium nitride:

$$3Mg(s) + N_2(g) \longrightarrow Mg_3N_2(s)$$

After all of the nitrogen had reacted with magnesium, Ramsay was left with an unknown gas that would not combine with anything.

With the help of Sir William Crookes, the inventor of the discharge tube, Ramsay and Lord Rayleigh found that the emission spectrum of the gas did not match any of the known elements. The gas was a new element! They determined its atomic mass to be 39.95 u and called it argon, which means "the lazy one" in Greek.

Once argon had been discovered, other noble gases were quickly identified. Ramsay, for example, isolated helium from uranium ores (also in 1898). From the atomic masses of helium and argon, their lack of chemical reactivity, and what was then known about the periodic table, Ramsay was convinced that there were other unreactive gases and that they were all members of one periodic group. He and his student Morris Travers set out to find the unknown gases. They used a refrigeration machine to first produce liquid air. Applying a technique called *fractional distillation*, they then allowed the liquid air to warm up gradually and collected components that boiled off at different temperatures. In this manner, they analyzed and identified three new elements—neon, krypton, and xenon—in only three months. Three new elements in three months is a record that may never be broken!

The discovery of the noble gases helped to complete the periodic table. Their atomic masses suggested that these elements should be placed to the right of the halogens. The apparent discrepancy with the position of argon was resolved by Moseley, as discussed in Section 2.3.

Finally, the last member of the noble gases, radon, was discovered by the German chemist Frederick Dorn in 1900. A radioactive element and the heaviest elemental gas known, radon's discovery not only completed the Group 8A elements, but also advanced our understanding about the nature of radioactive decay and the transmutation of elements.

Lord Rayleigh and Ramsay both won Nobel Prizes in 1904 for the discovery of argon. Lord Rayleigh received the prize in physics, and Ramsay's award was in chemistry.

Sir William Ramsay (1852–1916)

applies to the elements in Groups 1A and 2A, which are all metals, and to the elements in Groups 7A and 8A, which are all nonmetals. In Groups 3A through 6A, where the elements change either from nonmetals to metals or from nonmetals to metalloids, there is greater variation in chemical properties, even though the members of the same group have similar outer electron configurations.

Summary of Facts and Concepts

Section 2.1

▸ The Schrödinger equation for a many-electron atom cannot be solved exactly. However, by making the assumption that the wavefunction can be written as a product of single-electron orbitals, a good approximation to the energy levels and electronic wavefunctions for a many-electron atom can be found using the self-consistent field procedure. The resulting atomic orbitals can be described in terms of the same set of four quantum numbers used for the hydrogen atom (n, l, m_l, m_s), but their absolute form and energies will differ from the single-electron hydrogen atom orbitals.

▸ No two electrons in the same atom can have the same four quantum numbers (the Pauli exclusion principle).

▸ The number of unpaired spins determines the behavior of an atom in a magnetic field. Atoms with no unpaired spins are diamagnetic and those with unpaired spins are paramagnetic.

Section 2.2

▸ The Aufbau principle provides the guideline for building up the electronic configurations of the elements.

▸ The most stable arrangement of electrons in a subshell is the one that has the greatest number of parallel spins (Hund's rule).

▸ The ground-state electron configurations of monatomic cations and anions can be written just as they are for atoms, using the Pauli exclusion principle and Hund's rule.

Section 2.3

▸ Nineteenth-century chemists developed the periodic table by arranging elements in order of increasing atomic mass. Discrepancies in early versions of the periodic table were resolved by arranging the elements in order of their atomic numbers.

Section 2.4

▸ Electron configuration determines the properties of an element. The modern periodic table classifies the elements according to their atomic numbers, and thus also by their electron configurations. The configuration of the valence electrons directly affects the properties of the atoms of the representative elements.

Sections 2.5

▸ Periodic variations in the physical properties of the elements reflect differences in atomic structure. The metallic character of elements decreases across a period from metals through the metalloids to nonmetals and increases from top to bottom within a particular group of representative elements.

▸ Atomic radius varies periodically with the arrangement of the elements in the periodic table. It decreases from left to right and increases from top to bottom.

▸ Ionization energy is a measure of the tendency of an atom to resist the loss of an electron. The higher the ionization energy, the stronger the attraction between the nucleus and an electron. Electron affinity is a measure of the tendency of an atom to gain an electron. The more positive the electron affinity, the greater the tendency for the atom to gain an electron. Metals usually have low ionization energies, and nonmetals usually have high electron affinities.

▸ Noble gases are very stable because their outer ns and np subshells are completely filled. The metals among the representative elements (in Groups 1A, 2A, and 3A) tend to lose electrons until their cations become isoelectronic with the noble gases that precede them in the periodic table. The nonmetals in Groups 5A, 6A, and 7A tend to accept electrons until their anions become isoelectronic with the noble gases that follow them in the periodic table.

Key Words

actinide series, p. 141	electron affinity, p. 159	many-electron atom, p. 127	representative elements, p. 146
Aufbau principle, p. 139	electron configuration, p. 131	noble gas core, p. 139	self-consistent field
atomic radius, p. 150	Hund's rule, p. 137	orbital diagram, p. 132	method, p. 128
core electrons, p. 148	inert pair effect, p. 143	orbital energy, p. 134	transition metals, p. 139
diagonal relationship, p. 162	ionic radius, p. 152	paramagnetic, p. 133	valence electrons, p. 148
diamagnetic, p. 134	isoelectronic, p. 143	Pauli exclusion principle, p. 132	
effective nuclear charge, p. 149	lanthanide series, p. 141	rare earth elements, p. 141	

Problems

The Wavefunctions of Many-Electron Atoms Can Be Described to a Good Approximation Using Atomic Orbitals

2.1 Indicate which of the following sets of quantum numbers in an atom are unacceptable and explain why: (a) $(1, 0, 1/2, 1/2)$, (b) $(3, 0, 0, +1/2)$, (c) $(2, 2, 1, +1/2)$, (d) $(4, 3, -2, -1/2)$, and (e) $(3, 2, 1, 1)$.

2.2 How many variables are needed to describe the many-electron wavefunction for sodium? What about for Na^+ ion?

2.3 What do you think would be the major differences between the Hartree SCF $2s$ orbital of Li and that of Be? How does the SCF $2s$ orbital for Li differ from the hydrogenlike $2s$ orbital with $Z = 3$?

2.4 The formula for calculating the energies of an electron in a hydrogenlike ion is given in Equation 1.37. This equation cannot be applied to many-electron atoms. One way to modify it to approximate more complex atoms is to replace the atomic number Z with the effective nuclear charge $Z_{eff} = (Z - \sigma)$, where σ is the shielding constant discussed in Section 2.5. In the helium atom, for example, the physical significance of σ is that it represents the extent of shielding that the two $1s$ electrons exert on each other. Calculate the value of σ given that the first ionization energy of helium is 3.94×10^{-18} J atom^{-1}.

2.5 The atomic number of an element is 73. Is this element diamagnetic or paramagnetic?

2.6 Why do the $3s$, $3p$, and $3d$ orbitals have the same energy in a hydrogen atom but different energies in a many-electron atom?

Electron Configurations of Many-Electron Atoms Are Constructed Using the Aufbau Principle

2.7 Indicate the total number of (a) p electrons in N ($Z = 7$); (b) s electrons in Si ($Z = 14$); and $3d$ electrons in S ($Z = 16$).

2.8 The ground-state electron configurations listed here are incorrect. Explain what mistakes have been made in each and write the correct electron configurations.

Al: $1s^2 2s^2 2p^4 3s^2 3p^3$
B: $1s^2 2s^2 2p^5$
F: $1s^2 2s^2 2p^6$

2.9 Use the Aufbau principle to obtain the ground-state electron configuration of selenium.

2.10 Use the Aufbau principle to obtain the ground-state electron configuration of technetium.

2.11 Write the ground-state electron configurations for the following elements: B, V, Ni, As, I, and Au.

2.12 Write the ground-state electron configurations for the following elements: Ge, Fe, Zn, Ni, W, and Tl.

2.13 The electron configuration of a neutral atom is $1s^2 2s^2 2p^6 3s^2$. Write a complete set of quantum numbers for each of the electrons. Name the element.

2.14 Which of the following species has the most unpaired electrons: S^+, S, or S^-? Explain your answer.

2.15 Considering only the ground-state electron configurations, are there more diamagnetic or paramagnetic elements? Explain.

2.16 A neutral atom of a certain element has 17 electrons. Without consulting a periodic table, (a) write the ground-state electron configuration of the element, and (b) determine whether this element is diamagnetic or paramagnetic.

2.17 Without referring to a periodic table, write the electron configurations of elements with the following atomic numbers: (a) 9, (b) 20, (c) 26, and (d) 33.

2.18 An M^{2+} ion derived from a metal in the first transition element series has four electrons in the $3d$ subshell. What element might M be?

2.19 A metal with a net $3+$ charge in the first transition metal series has five electrons in the $3d$ subshell. Identify the metal.

2.20 Write the ground-state electron configurations for the following ions that play important roles in biochemical processes in our bodies: Na^+, Mg^{2+}, Cl^-, K^+, Ca^{2+}, Cu^{2+}, and Zn^{2+}.

2.21 Write the ground-state electron configurations for the following transition metal ions: Sc^{3+}, Ti^{4+}, V^{5+}, Cr^{3+}, Mn^{2+}, Cu^+, Ag^+, Au^+, Au^{3+}, and Pt^{2+}.

2.22 Which of the following species are isoelectronic with one another: C, Cl^-, Mn^{2+}, B^-, Ar, Zn, Fe^{3+}, and Ge^{2+}?

2.23 Group the species that are isoelectronic: Be^{2+}, F^-, Fe^{2+}, N^{3-}, He, S^{2-}, Co^{3+}, and Ar.

2.24 Indicate the number of unpaired electrons present in each of the following atoms: B, Ne, P, Sc, Mn, Se, Kr, Fe, Cd, I, and Pb.

2.25 A Ti atom and a V^+ ion have the same number of electrons, but have different electron configurations. (a) Write the electron configurations for Ti and V^+ and (b) give a physical explanation as to why these two species are not isoelectronic.

2.26 How many unpaired electrons do the following ions possess: Fe^{2+}, Fe^{3+}, Cr^{+3}, Cr^{4+}, Cu^+, and Cu^{2+}?

2.27 Why are the elements Zn, Cd, and Hg not considered to be transition metals?

2.28 The electron configuration of Pd is $[Kr]4d^{10}$. Why is Pd considered a transition metal even though it has a completely filled $4d$ shell?

Elements Can Be Classified by Their Position in the Periodic Table

2.29 Classify the following elements as metal, nonmetal, or metalloid: As, Xe, Fe, Li, B, Cl, Ba, P, I, and Si.

2.30 Indicate whether the following elements exist as atomic species, molecular species, or extensive three-dimensional structures in their most stable states at 25°C and 1 bar pressure, and write the molecular or empirical formula for each one: phosphorus, iodine, magnesium, neon, carbon, sulfur, cesium, and oxygen.

2.31 You are given a dark shiny solid and asked to determine whether it is iodine or a metallic element. Suggest a nondestructive test that would enable you to arrive at the correct answer.

2.32 Explain why zinc does not fit the definition of a transition element.

The Properties of the Elements Vary Periodically Across the Periodic Table

2.33 In general, atomic radius and ionization energy have opposite trends. Why?

2.34 Explain why the electron affinity of nitrogen is approximately zero, although the elements on either side, carbon and oxygen, have substantial positive electron affinities.

2.35 Although boron lies to the right of beryllium in the periodic table, its ionization energy is lower than that of Be, contrary to the general trend. How can this be explained on the basis of their respective electron configurations?

2.36 On the basis of their positions in the periodic table, select the atom with the larger atomic radius in each of the following pairs: (a) Na or Cs; (b) Be or Ba; (c) N or Sb; (d) F or Br; (e) Ne or Xe.

2.37 Arrange the following atoms in order of decreasing atomic radius: Na, Al, P, Cl, and Mg.

2.38 List the following ions in order of increasing ionic radius: N^{3-}, Na^+, F^-, Mg^{2+}, and O^{2-}.

2.39 Choose which of the following cations is larger, and explain why: Cu^+ or Cu^{2+}.

2.40 Choose which of the following anions is larger, and explain why: Se^{2-} or Te^{2-}.

2.41 The boiling points of neon and krypton are $-245.9°C$ and $-152.9°C$, respectively. Using these data, estimate the boiling point of argon.

2.42 Arrange the following elements in order of increasing first ionization energy: Na, Cl, Al, S, and Cs.

2.43 Arrange the following elements in order of increasing first ionization energy: F, K, P, Ca, and Ne.

2.44 Use the third period of the periodic table as an example to illustrate the change in first ionization energies of the elements from left to right across the row. Explain the trend.

2.45 In general, ionization energy increases from left to right across a given period. Aluminum, however, has a lower ionization energy than magnesium. Explain.

2.46 The first and second ionization energies of K are 419 kJ mol^{-1} and 3052 kJ mol^{-1}, respectively, and those of Ca are 590 kJ mol^{-1} and 1145 kJ mol^{-1}, respectively. Compare their values and comment on the differences.

2.47 Two atoms have the electron configurations $1s^2 2s^2 2p^6$ and $1s^2 2s^2 2p^6 3s^1$. The first ionization energy of one is 2080 kJ mol^{-1}, and that of the other is 496 kJ mol^{-1}. Match each ionization energy with one of the given electron configurations. Justify your choice.

2.48 A hydrogenlike ion is an ion containing only one electron. Using the formula for the energy of the electron in such an ion, given in Section 1.4, calculate (a) the second ionization energy (in kJ mol^{-1}) of He, and (b) the third ionization energy (in kJ mol^{-1}) of lithium.

2.49 Plasma is a state of matter consisting of positive gaseous ions and electrons. In the plasma state, a mercury atom could be stripped of its 80 electrons and would therefore exist as Hg^{80+}. Use the equation for the energy of electrons in hydrogenlike ions to calculate the energy required for the last ionization step—that is,

$$Hg^{79+}(g) \longrightarrow Hg^{80+}(g) + e^-$$

2.50 Arrange the elements in each of the following groups in increasing order of most positive electron affinity: (a) Li, Na, and K; (b) F, Cl, Br, and I.

2.51 Specify which of the following elements you would expect to have the greatest electron affinity: He, K, Co, S, or Cl.

2.52 Considering their electron affinities, do you think it is possible for the alkali metals to form an anion like M^-, where M represents an alkali metal?

2.53 Explain why alkali metals have a greater affinity for electrons than do alkaline earth metals.

2.54 As a group, the noble gases are very stable chemically. (Ar, Kr, Xe and Rn are known to form compounds, but they are few in number.) Why?

2.55 Why are Group 1B elements more stable chemically than Group 1A element even though they seem to have the same outer electron configuration, ns^1, where n is the principle quantum number of the outermost shell?

Additional Problems

2.56 The electron configurations described in Chapter 2 all refer to gaseous atoms in their ground states. An atom may absorb a quantum of energy and promote one of its electrons to a higher-energy orbital. When this happens, we say that the atom is in an excited state. The electron configurations of some excited atoms are given. Identify these atoms and write their ground-state configurations:

(a) $1s^1 2s^1$
(b) $1s^2 2s^2 2p^2 3d^1$
(c) $1s^2 2s^2 2p^6 4s^1$
(d) $[Ar]4s^1 3d^{10} 4p^4$
(e) $[Ne]3s^2 3p^4 3d^1$

2.57 Give the atomic symbol and draw orbital diagrams for atoms with the following electron configurations:

(a) $1s^2 2s^2 2p^5$
(b) $1s^2 2s^2 2p^6 3s^2 3p^3$
(c) $1s^2 2s^2 2p^6 3s^2 3p^6 4s^2 3d^7$

2.58 Calculate the maximum wavelength of light (in nanometers) able to ionize a single sodium atom.

2.59 Write equations representing the following processes:

(a) The electron affinity of S
(b) The third ionization energy of titanium
(c) The electron affinity of Mg^{2+}
(d) The ionization energy of O^{2-}

2.60 Arrange the following isoelectronic species in order of (a) increasing ionic radius and (b) increasing ionization energy: O^{2-}, F^-, Na^+, Mg^{2+}.

2.61 Element M is a shiny and highly reactive metal (melting point 63°C), and element X is a highly reactive nonmetal (melting point −7.2°C). They react to form a compound with the empirical formula MX, a colorless, brittle white solid that melts at 734°C. When dissolved in water or when in the molten state, the substance conducts electricity. When chlorine gas is bubbled through an aqueous solution containing MX, a reddish-brown liquid appears and Cl ions are formed. From these observations, identify M and X. (You may need to consult a handbook of chemistry for the melting-point values.)

2.62 You are given four substances: a fuming red liquid, a dark metallic-looking solid, a pale-yellow gas, and a yellow-green gas that attacks glass. You are told that these substances are the first four members of Group 7A, the halogens. Name each one.

2.63 For each pair of elements, give three properties that show their chemical similarity: (a) sodium and potassium and (b) chlorine and bromine.

2.64 Name the element that forms compounds, under appropriate conditions, with every other element in the periodic table except He and Ne.

2.65 Explain why the first electron affinity of sulfur is 200 kJ mol^{-1}, but the second electron affinity is −649 kJ mol^{-1}.

2.66 Why do noble gases have negative electron affinity values?

2.67 The atomic radius of K is 216 pm and that of K^+ is 133 pm. Calculate the percent decrease in volume that occurs when K(g) is converted to K^+(g). (The volume of a sphere is $\frac{4}{3}\pi r^3$, where r is the radius of the sphere.)

2.68 The atomic radius of F is 72 pm and that of F^- is 136 pm. Calculate the percent increase in volume that occurs when F(g) is converted to F^-(g). (See Problem 2.67 for the volume of a sphere.)

2.69 The H^- ion and He are isoelectronic, with two $1s$ electrons each. Which one is larger? Explain.

2.70 Arrange the following species in isoelectronic pairs: O^+, Ar, S^{2-}, Ne, Zn, Cs^+, N^{3-}, As^{3+}, N, and Xe.

2.71 Experimentally, the electron affinity of an element can be determined by using a laser to eject an electron from the anion of the element in the gas phase:

$$X^-(g) + h\nu \longrightarrow X(g) + e^-$$

Referring to Table 2.3, calculate the photon wavelength (in nanometers) corresponding to the electron affinity for chlorine. In what region of the electromagnetic spectrum does this wavelength fall?

2.72 Write the formulas and names of the hydrides of the following second-period elements: Li, C, N, O, and F. Predict their reaction with water.

2.73 Although it is possible to determine the second, third, and higher ionization energies of an element, the same cannot usually be done with the electron affinities of an element. Explain.

2.74 The only confirmed compound of radon is radon fluoride, RnF. It is difficult to study the chemistry of radon because all isotopes of radon are radioactive and it is dangerous to handle. Can you suggest another reason why there are so few known radon compounds? (*Hint:* Radioactive decays are exothermic processes; that is, they produce heat.)

2.75 Little is known of the chemistry of astatine, the last member of Group 7A. Describe the physical characteristics you would expect this halogen to have.

2.76 As discussed in Section 2.3, the atomic mass of argon is greater than that of potassium. This observation created a problem in the early development of the periodic table because it meant that argon should be placed after potassium. (a) How was this difficulty resolved? (b) From the following data, calculate the average atomic masses of argon and potassium: Ar-36 (35.9675 u; 0.337 percent), Ar-38 (37.9627 u; 0.063 percent), Ar-40 (39.9624 u; 99.60 percent); K-39 (38.9637 u; 93.258 percent), K-40 (39.9640 u; 0.0117 percent), K-41 (40.9618 u, 6.730 percent).

2.77   Predict the atomic number and ground-state electron configuration of the next member of the alkali metals after francium.

2.78   Why do elements that have high ionization energies also have more positive electron affinities? Which group of elements would be an exception to this generalization?

2.79   The first four ionization energies of an element are approximately 738 kJ mol^{-1}, 1450 kJ mol^{-1}, 7.7 × 10^3 kJ mol^{-1}, and 1.1 × 10^4 kJ mol^{-1}. To which periodic group does this element belong? Why?

2.80   Some chemists think that helium should properly be called "helon." Why? What does the "-ium" ending in helium suggest?

2.81   (a) The formula of the simplest hydrocarbon is CH_4 (methane). Predict the formulas of the simplest compounds formed between hydrogen and the following elements: silicon, germanium, tin, and lead. (b) Sodium hydride (NaH) is an ionic compound. Would you expect rubidium hydride (RbH) to be more or less ionic than NaH? (c) Predict the reaction between radium (Ra) and water. (d) When exposed to air, aluminum forms a tenacious oxide coating (consisting of Al_2O_3) that protects the metal from corrosion. Which metal in Group 2A would you expect to exhibit similar properties? Why?

2.82   On the same graph, plot the effective nuclear charge (shown in parentheses in units of e) and atomic radius (see Figure 2.16) versus atomic number for the second-period elements: Li(1.30), Be(1.95), B(2.60), C(3.25), N(3.90), O(4.55), F(5.20), and Ne(5.85). Comment on the trends.

2.83   A technique called photoelectron spectroscopy is used to measure the ionization energy (E_1) of atoms. A sample is irradiated with UV light, and electrons are ejected from the valence shell. The kinetic energies of the ejected electrons are measured. Because the energy of the UV photon and the kinetic energy of the ejected electron are known, we can write

$$h\nu = E_1 + \frac{1}{2}mu^2$$

where ν is the frequency of the UV light, and m and u are the mass and velocity of the electron,

respectively. In one experiment the kinetic energy of the ejected electron from potassium is found to be 5.34 × 10^{-19} J using a UV source of wavelength 162 nm. Calculate the ionization energy of potassium. How can you be sure that this ionization energy corresponds to the electron in the valence shell (that is, the most loosely held electron)?

2.84   The energy needed for the following process is 1.96 × 10^4 kJ mol^{-1}:

$$Li(g) \longrightarrow Li^{3+}(g) + 3e^-$$

If the first ionization energy of lithium is 520 kJ mol^{-1}, calculate the second ionization energy of lithium, that is, the energy required for the process

$$Li^+(g) \longrightarrow Li^{2+}(g) + e^-$$

2.85   What is the electron affinity of the Na^+ ion?

2.86   The ionization energies of sodium (in kJ mol^{-1}), starting with the first and ending with the eleventh, are 495.9, 4560, 6900, 9540, 13,400, 16,600, 20,120, 25,490, 28,930, 141,360, and 170,000. Plot the log of ionization energy (y axis) versus the number of ionization (x axis); for example, log 495.9 is plotted versus 1 (labeled I_1, the first ionization energy), log 4560 is plotted versus 2 (labeled I_2, the second ionization energy), and so on. (a) Label I_1 through I_{11} with the electrons in orbitals such as $1s$, $2s$, $2p$, and $3s$. (b) What can you deduce about electron shells from the breaks in the curve?

2.87   Explain, in terms of their electron configurations, why Fe^{2+} is more easily oxidized to Fe^{3+} than Mn^{2+} is to Mn^{3+}.

2.88   The ionization energy of a certain element is 412 kJ mol^{-1}. When the atoms of this element are in the first excited state, however, the ionization energy is only 126 kJ mol^{-1}. Based on this information, calculate the wavelength of light emitted in a transition from the first excited state to the ground state.

2.89   One way to estimate the effective charge (Z_{eff}) of a many-electron atom is to use the empirical equation $I_1 = (1312 \text{ kJ mol}^{-1})(Z_{eff}^2/n^2)$, where I_1 is the first ionization energy and n is the principal quantum number of the shell in which the electron resides. Use this equation to calculate the effective charges of Li, Na, and K. Also calculate Z_{eff}/n for each metal. Comment on your results.

2.90   Referring to Table 2.2, explain why the first ionization energy of helium is less than twice the ionization energy of hydrogen, but the second ionization energy of helium is greater than twice the ionization energy of hydrogen. (*Hint:* According to Coulomb's law, the energy between two charges q_1 and q_2 separated by a distance r is proportional to q_1q_2/r.)

2.91 A boron atom has an electron configuration of $1s^2 2s^2 2p^1$, with a single unpaired electron in a $2p$ orbital, which is dumbbell shaped. However, the electron probability about a boron atom in free space is found to be spherically symmetric. How do you explain this?

2.92 Assume that an electron in the $1s$ orbital of Hg could be approximately described by the Bohr model with $Z = 80$ and $n = 1$.

    (a) Following the discussion of the Bohr model given in Chapter 1, calculate the velocity of this electron. What fraction of the speed of light does this velocity represent?

    (b) According to Einstein's theory of special relativity, the mass of an object increases as

its velocity approaches the speed of light according to the formula

$$ m = \frac{m_0}{\sqrt{1 - \left(\dfrac{u}{c}\right)^2}} $$

where m_0 is the mass of the object when at rest, u is its velocity, and c is the speed of light in a vacuum. Using the velocity calculated in (a), determine the mass of the electron in the $1s$ orbital in Hg. What implications does this change in mass have for the nature of the atomic orbitals of Hg?

Answers to Practice Exercises

2.1 $n = 5, l = 1, m_l = -1, 0, +1, m_s = -1/2, +1/2$

2.2

Electron	n	l	m_l	m_s	Orbital
1	1	0	0	$+\frac{1}{2}$	$1s$
2	1	0	0	$-\frac{1}{2}$	
3	2	0	0	$+\frac{1}{2}$	$2s$
4	2	0	0	$-\frac{1}{2}$	
5*	2	1	-1	$+\frac{1}{2}$	$2p_x$

*The fifth electron can be in any of the six possible sets of quantum numbers consistent with a $2p$ orbital. The one listed is only an example.

2.3 (a) $1s^2 2s^2 2p^6 3s^2 3p^3$ or [Ne]$3s^2 3p^3$, (b) $1s^2 2s^2 2p^6 3s^2 3p^6 4s^2 3d^7$ or [Ar]$4s^2 3d^7$ **2.4** (a) $1s^2 2s^2 2p^6 3s^2 3p^6 4s^2$ or [Ar]$4s^2$ (b) Ca, an alkaline earth metal (c) diamagnetic **2.5** Li > Be > C **2.6** (a) Li$^+$, (b) Au^{3+}, (c) N^{3-} **2.7** (a) N, (b) Mg **2.8** No

Chapter

3

The Chemical Bond

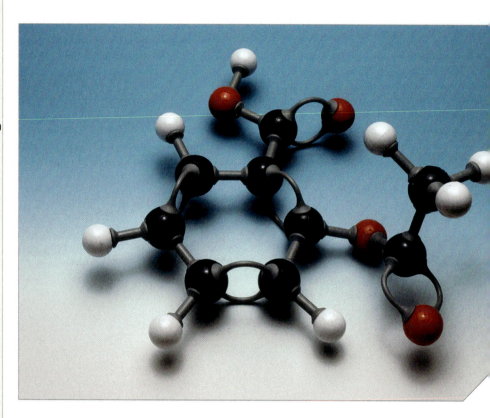

Why do atoms of different elements react? What are the forces that hold atoms together in molecules and ions in ionic compounds? What shapes do these entities assume? These are some of the questions addressed in Chapters 3 and 4. We begin by defining the chemical bond in terms of energy and then examining the two common types of bonds—ionic and covalent—and the forces that stabilize them. Although we can predict the bonding in many molecules using a simple qualitative tool called Lewis dot diagrams, a proper understanding of bonding and the stability of molecules comes from applying the quantum mechanical ideas discussed in Chapters 1 and 2.

3.1 | Atoms in a Molecule Are Held Together by Chemical Bonds

Chapters 1 and 2 dealt primarily with the electronic structure of isolated atoms of the various elements. With the exception of the noble gases, however, it is the molecule, not the individual atom, that is the basic building block of materials in nature. Even elements that occur naturally in their pure state are generally found in molecular form. For example, the oxygen and nitrogen in the atmosphere are made up of the diatomic molecules O_2 and N_2, not discrete O and N atoms. In Chapter 3, we begin our study of molecules and the chemical bonds that hold them together.

To understand the nature of a chemical bond let's examine bonding in the simplest molecule, H_2. If we could place a large number of individual H atoms in a container, they would quickly pair up to form H_2 molecules. This process would release a large amount of energy because the energy of two hydrogen atoms is lower when they are bonded together as a molecule than when they exist as separate atoms.

Figure 3.1 illustrates what happens to the energy of two hydrogen atoms as the nuclei are brought together. When far apart, the two hydrogen atoms do not interact and the electron of each atom is in its ground state $1s$ orbital.

The initial potential energy of this system (that is, the two H atoms) is zero. As the atoms approach each other, each electron is attracted by the nucleus of the other atom; at the same time, the electrons repel each other, as do the nuclei. Responding to these forces, the electron distribution on each atom changes as the internuclear distance decreases. While the atoms are still separated, attraction is stronger than repulsion, so the potential energy of the system *decreases* (that is, becomes negative) as the atoms approach each other. This trend continues until the potential energy reaches a minimum value. At this point, when the system has the lowest potential energy, it is most stable and, in the language of quantum mechanics, the electron distributions merge to form an electron cloud enveloping both nuclei. This condition corresponds to the formation of a stable H_2 molecule. If the distance between nuclei decreases, the potential energy rises steeply and eventually becomes positive because of increased

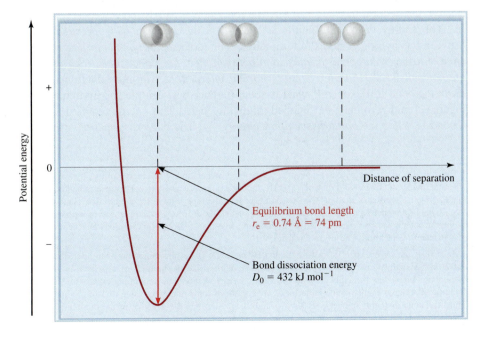

Figure 3.1 Change in the potential energy of two H atoms with their distance of separation. At the point of minimum potential energy (432 kJ mol^{-1}), the H_2 molecule is in its most stable state and the bond length is 74 pm. The spheres represent the $1s$ orbitals.

Potential energy

Distance of separation

Equilibrium bond length
$r_e = 0.74$ Å $= 74$ pm

Bond dissociation energy
$D_0 = 432$ kJ mol^{-1}

Figure 3.2 Top to bottom: As two H atoms approach each other, their $1s$ orbitals begin to interact and each electron begins to feel the attraction of the other proton. Gradually, the electron density builds up in the region between the two nuclei (red color). Eventually, a stable H_2 molecule is formed when the internuclear distance is 74 pm (0.74 Å).

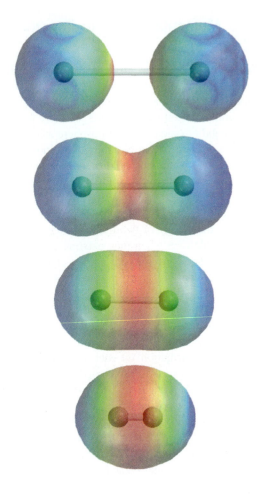

electron-electron and nuclear-nuclear repulsions. Figure 3.2 shows the changes in the electron distribution that occur during the formation of an H_2 molecule.

The energy curve[1] in Figure 3.1 describes the creation of a ***chemical bond*** between two hydrogen atoms to form a hydrogen molecule. *A bond is formed between atoms whenever the forces of attraction between them are sufficiently strong that they are not pulled apart in the course of normal interactions with their environment.* To break the bond in H_2, energy equal to the difference between the energy minimum in Figure 3.1 and zero must be supplied to the molecule.[2] This energy required to break a bond is called the ***bond dissociation energy,*** D_0. The bond dissociation energy for

---

1. Construction of the energy curve in Figure 3.1 requires calculation of the energy of two hydrogen atoms when the nuclei are frozen at a distance R apart, ignoring any nuclear motion. This determination assumes that the relative motion of the nuclei is much slower than the time it takes for the electron cloud to adjust to changes in the nuclear position. This assumption, called the *Born-Oppenheimer approximation,* is valid to high accuracy because the nucleus is far more massive than the electrons (protons and neutrons have masses nearly 2000 times that of an electron).

2. In reality, the lowest energy possible for the H_2 molecule is not the minimum of the potential energy curve, but is slightly higher because of molecular vibration. Like the motions of a particle-in-a-box, the Heisenberg uncertainty principle prevents the kinetic energy of vibration from going exactly to zero, therefore, the vibrational energy of the molecule is always above some minimum value (the zero-point energy). Thus, the bond dissociation energy is always slightly smaller than the negative of the minimum potential energy.

the hydrogen molecule is 432 kJ mol^{-1}, which means that 432 kJ of energy is required to break the bonds in 1 mol of hydrogen molecules. *Typically, two atoms are said to be chemically bonded if the bond energy is greater than about 40 kJ mol^{-1}*. In contrast, the attractive interaction between two helium atoms is only 0.09 kJ mol^{-1}. This interaction is not strong enough for a He$_2$ molecule to survive intact, except in isolated environments at extremely low temperatures, and we do not consider this attractive interaction to represent a chemical bond.

Another important quantitative measure of a chemical bond is the ***bond length,*** r_e, which is *the equilibrium separation between the nuclei in a bond, defined by the distance at which the potential energy is a minimum* (see Figure 3.1). For H$_2$ the bond length is 74 pm (0.74 Å).

Bonds between atoms can arise in a number of different ways. For example, the bonding in H$_2$ results from the sharing of electrons by the two hydrogen nuclei. That is, the electrons forming the bond are not associated with particular individual atoms, as is the case for the separated H atoms, but instead interact equally with both H nuclei. Such a bond is referred to as a *covalent bond*. In contrast, the bond between the Na$^+$ and Cl$^-$ ions in common table salt (NaCl) does not arise from electron sharing but is instead due to the strong electrostatic attraction between the oppositely charged ions, and is referred to as an *ionic bond*. We shall spend the rest of Chapter 3 exploring in more detail the nature of chemical bonding and the formation of molecules.

3.2 | A Covalent Bond Involves the Sharing of Electrons Between Atoms in a Molecule

The development of the periodic table and concept of electron configuration gave chemists a rationale for molecule and compound formation. This explanation, formulated by Gilbert Lewis,[3] is that atoms combine in order to achieve a more stable electron configuration. Maximum stability results when an atom is isoelectronic with a noble gas.

Lewis Dot Symbols

When atoms interact to form a chemical bond, only their outer regions are in contact. For this reason, when we study chemical bonding, we are concerned primarily with the valence electrons of the atoms. To keep track of valence electrons in a chemical reaction, chemists use a system of dots devised by Lewis called the Lewis dot symbols. A ***Lewis dot symbol*** *consists of the symbol of an element and one dot for each valence electron in an atom of the element.* Figure 3.3 shows the Lewis dot symbols for the representative elements and the noble gases.

Except for helium, the number of valence electrons in an atom is the same as the group number of the element. For example, Li is a Group 1A element and has one dot for one valence electron; Be, a Group 2A element, has two valence electrons (two dots); and so on. Elements in the same group have similar outer electron configurations and hence similar Lewis dot symbols. The transition metals, lanthanides, and actinides all have incompletely filled inner shells, so we generally cannot write simple Lewis dot symbols for them.

---

3. Gilbert Newton Lewis (1875–1946). American chemist. Lewis made many significant contributions in the areas of chemical bonding, thermodynamics, acids and bases, and spectroscopy. Despite the significance of Lewis's work, he was never awarded a Nobel Prize.

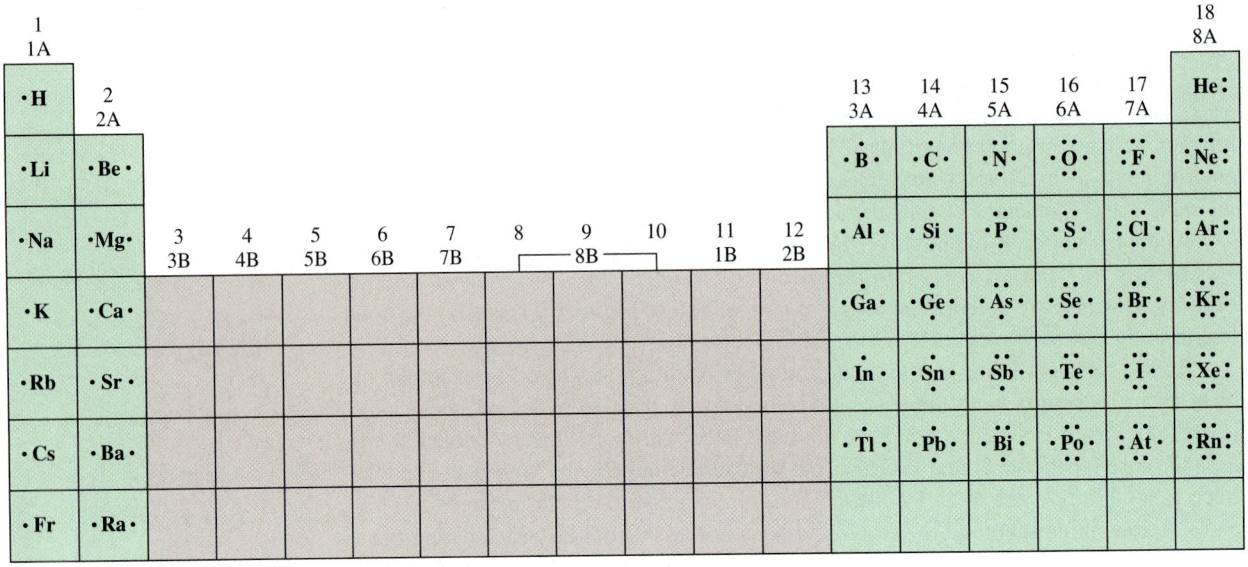

Figure 3.3 Lewis dot symbols for the representative elements and the noble gases.

Covalent Bonds

Although the concept of molecules goes back to the seventeenth century, it was not until early in the twentieth century that chemists began to understand how and why molecules form. The first major breakthrough was a suggestion by Gilbert Lewis that a chemical bond involves electron sharing by atoms. He depicted the formation of a chemical bond in H_2 as

$$H \cdot + \cdot H \longrightarrow H : H$$

This type of electron pairing is an example of a **covalent bond,** *a bond in which a pair of electrons are shared by two atoms.* **Covalent compounds** *are compounds that contain only covalent bonds.* For the sake of simplicity, the shared pair of electrons is often represented by a single line. Thus, the covalent bond in the hydrogen molecule can be written as H—H. In any covalent bond, each electron in a shared pair is attracted to the nuclei of both atoms creating a force that holds the two bonded atoms together.

Covalent bonding between many-electron atoms involves only the valence electrons. Consider the fluorine molecule, F_2. The electron configuration of F is $1s^2 2s^2 2p^5$. The $1s$ electrons are low in energy and stay near the nucleus most of the time. For this reason, they do not participate in bond formation. Thus, each F atom has seven valence electrons (the $2s$ and $2p$ electrons), and there is only one unpaired electron on F. Using Lewis dot symbols, the formation of the F_2 molecule can be represented as follows:

$$: \ddot{F} : \ddot{F} :$$

Only two valence electrons participate in the formation of the covalent bond in F_2. The other, nonbonding electrons, are called **lone pairs**—*pairs of valence electrons that are not involved in covalent bond formation.* Thus, each F in F_2 has three lone pairs of electrons (shown in red):

$$: \ddot{F} : \ddot{F} :$$

A Simplified View of Covalent Bonding: Lewis Structures

We use *Lewis structures* to represent covalent compounds such as H_2 and F_2. A ***Lewis structure*** is *a representation of covalent bonding in which shared electron pairs are represented either as lines or as pairs of dots between two atoms, and lone pairs are represented as pairs of dots on individual atoms.* Only valence electrons are shown in a Lewis structure.

Consider the Lewis structure of the water molecule. Figure 3.3 shows the Lewis dot symbol for oxygen with two unpaired dots (two unpaired electrons), so we expect that O might form two covalent bonds. Because hydrogen has only one electron, it can form only one covalent bond. Thus, the Lewis structure for water is

$$\overset{..}{H:O:H} \quad \text{or} \quad H-\overset{..}{\underset{}{O}}-H$$

In this case, the O atom has two lone pairs. The hydrogen atom has no lone pairs because its only electron is used to form a covalent bond.

In the F_2 and H_2O molecules, the F and O atoms achieve the noble gas configuration by sharing electrons:

$$:\overset{..}{F}\overset{..}{(:)}\overset{..}{F}: \qquad H\overset{..}{(:}\overset{..}{O}\overset{..}{:)}H$$

The formation of these molecules illustrates the ***octet rule,*** formulated by G. N. Lewis: *An atom other than hydrogen tends to form bonds until it is surrounded by eight valence electrons.* In other words, a covalent bond forms when there are not enough electrons for each individual atom to have a complete octet. By sharing electrons in a covalent bond, the individual atoms can complete their octets. The requirement for hydrogen is that it attain the electron configuration of helium, or a total of two electrons (a ***duet***).

The octet rule works mainly for elements in the second period of the periodic table. These elements have only $2s$ and $2p$ outer subshells, which can hold a total of eight electrons. When an atom of one of these elements forms a covalent compound, it can attain the noble gas electron configuration [Ne] by sharing electrons with other atoms in the same compound. Important exceptions to the octet rule, discussed in Section 3.4, provide further insight into the nature of chemical bonding.

Atoms can form different types of covalent bonds. In a ***single bond,*** *two atoms are held together by one shared electron pair.* Many compounds are also held together by ***multiple bonds***—that is, bonds formed when *two atoms share two or more pairs of electrons.* If *two atoms share two pairs of electrons,* the covalent bond is called a ***double bond.*** Carbon-oxygen double bonds are found in molecules of carbon dioxide (CO_2), and carbon-carbon double bonds are found in ethylene (C_2H_4):

$$:\overset{..}{O}::C::\overset{..}{O}: \quad \text{or} \quad :\overset{..}{O}=C=\overset{..}{O}:$$

$$\underset{H}{\overset{H}{\diagdown}}C=C\underset{H}{\overset{H}{\diagup}}$$

ethylene

Note that in each of these structures the octet rule is satisfied. A ***triple bond*** arises when *two atoms share three pairs of electrons,* as in the nitrogen molecule (N_2):

$$:N\overset{..}{(:)}N: \quad \text{or} \quad :N{\equiv}N:$$

$$\underset{8e^- \quad 8e^-}{}$$

The acetylene molecule (C_2H_2) contains a carbon-carbon triple bond:

$$H : C :: C : H \qquad \text{or} \qquad H—C\equiv C—H$$

All the valence electrons in ethylene and acetylene are used in bonding; there are no lone pairs on the carbon atoms. In fact, with the exception of carbon monoxide, stable molecules containing carbon do not have lone pairs on the carbon atoms.

*The number of bonds between two atoms in a molecule is referred to as the **bond order**.* A single bond has a bond order of one; double and triple bonds have bond orders of two and three, respectively.

Valence Bond Theory

Lewis's theory was a significant advance in our understanding of chemical bond formation. Such a simple approach, though, does have its drawbacks. For example, the Lewis theory of chemical bonding does not clearly explain why chemical bonds exist; although relating the formation of a covalent bond to the pairing of electrons was a step in the right direction. Also, the Lewis theory describes the single bond between the H atoms in H_2 and that between the F atoms in F_2 in essentially the same way— as the pairing of two electrons. These two molecules, however, have quite different bond dissociation energies and bond lengths (436.4 kJ mol^{-1} and 74 pm for H_2 and 150.6 kJ mol^{-1} and 142 pm for F_2). For a more complete explanation of chemical bond formation, we must look to quantum mechanics.

At present, two quantum mechanical theories—valence bond theory and molecular orbital theory—are used to describe covalent bond formation and the electronic structure of molecules. *Valence bond theory* assumes that the electrons in a molecule occupy the atomic orbitals of the individual atoms. It enables us to retain a picture of individual atoms taking part in bond formation. *Molecular orbital theory,* on the other hand, assumes the formation of molecular orbitals from atomic orbitals. Neither theory perfectly explains all aspects of bonding, but each has contributed something to our understanding of observed molecular properties.

The ***valence bond (VB) theory*** is an extension of Lewis's concept of electron sharing that originated with the work of Walter Heitler[4] and Fritz London[5] in 1927. To begin our discussion of VB theory, let's consider the formation of an H_2 molecule from two H atoms, as discussed on p. 174. The Lewis theory describes the H—H bond in terms of the pairing of the two electrons on the H atoms. In VB theory, the covalent H—H bond is formed by the *overlap* of the two half-filled $1s$ orbitals in the H atoms. By overlap, we mean that the two orbitals share a common region in space (Figure 3.4).

---

4. Walter Heitler (1904–1981). German physicist. Heitler left Germany in 1933 to escape the Nazi regime and spent the war years at the University of Bristol. He joined the faculty at the University of Zürich in 1949. In addition to his work on valence bond theory, he made important contributions to quantum electrodynamics and quantum field theory.

5. Fritz London (1900–1954). German-American physicist. London was a theoretical physicist whose major work was on the phenomenological theory of superconductivity.

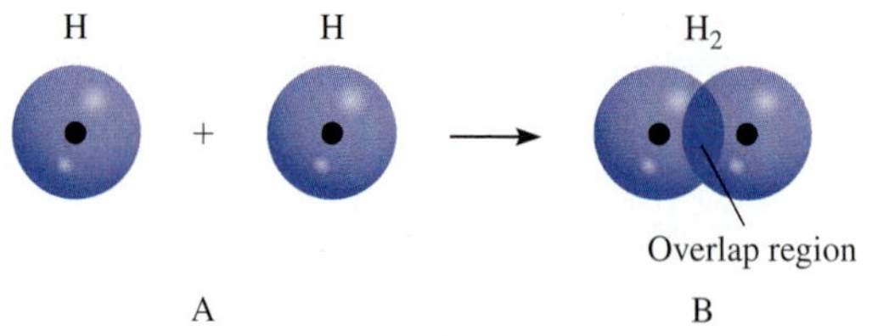

Figure 3.4 Valence bond picture of the formation of a covalent bond in H_2, which results from the overlap of two hydrogen $1s$ orbitals.

H   H H_2

+   →

Overlap region

A   B

The VB approximation to the ground-state electron wavefunction of H_2 is given by the product of the two $1s$ orbitals on the separate atoms (A and B):

$$\psi_{VB} = N[\phi_{A,1s}(1)\ \phi_{B,1s}(2) + \phi_{B,1s}(1)\ \phi_{A,1s}(2)] \qquad (3.1)$$

where the labels "1" and "2" denote "electron 1" and "electron 2," respectively, $\phi_{A,1s}$ and $\phi_{B,1s}$ are hydrogen $1s$ orbitals centered on atoms A and B, respectively, and N is a normalization constant. The VB wavefunction in Equation 3.1 describes a system in which the two electrons (of opposite spin) occupy atomic orbitals centered on each atom. The second term on the right-hand side of Equation 3.1 is necessary because electrons are indistinguishable[6] in quantum mechanics and it is just as likely for electron 1 to be on atom B and electron 2 on atom A as it is for electron 1 to be on atom A and electron 2 on atom B. When the atoms are close together, the overlap of the atomic orbitals places significant electron density between the two nuclei, leading to a lowering of the energy of the molecule and the formation of a bond, as described qualitatively in Section 3.1. Although the electron spin part of the wavefunction is not explicitly shown in Equation 3.1, the construction of the VB ground-state wavefunction and the subsequent formation of the H—H bond depends on the two electrons in Equation 3.1 being of opposite spin, as a consequence of the Pauli exclusion principle. If the two atoms have parallel spins, a stable H_2 atom cannot be formed, and the atoms will repel one another at all distances.

For the covalent bond in H_2, the electron distribution (given by the square of the wavefunction in Equation 3.1) is symmetric with respect to rotation about the H—H bond axis (see Figure 3.4). We refer to objects with such symmetry as *cylindrically symmetric,* because they have the same symmetry as a cylinder—rotating a cylinder about its main axis has no effect on its appearance. Bonds with this property are called ***sigma (σ) bonds.*** By analogy to atomic orbitals, spherically symmetric s orbitals have zero orbital angular momentum ($l = 0$), so the cylindrically symmetric sigma VB orbitals have zero angular momentum about the bond axis.

The VB concept of overlapping atomic orbitals applies equally well to diatomic molecules other than H_2. Thus, a stable F_2 molecule forms when the $2p$ orbitals (containing the unpaired electrons) in the two F atoms overlap to form a covalent bond [see Figure 3.5(a)]. Similarly, the formation of the HF molecule can be explained by the head-on overlap of the $1s$ orbital in H with the $2p$ orbital in F

---

6. Because the orbitals of electrons in atoms and molecules overlap, it is impossible to know at any given time which electron is in which orbital, that is, the electrons are *indistinguishable*. Because of this, the probability density (that is, the wavefunction squared) must remain the same when the identities of the electrons are switched.

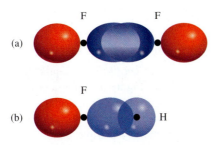

Figure 3.5 (a) Valence bond picture of the formation of the F—F sigma bond in the F_2 molecule in terms of the overlap between two $2p_x$ orbitals centered on each atom. (b) Valence bond picture of the formation of the H—F sigma bond in the HF molecule in terms of the end-on overlap between a $1s$ electron on H and a $2p_x$ electron centered on the F atom. The colors reflect the sign of the wavefunction in each region (blue $= +$, red $= -$).

Figure 3.6 (a) Formation of a sigma bond in O_2 from the head-on overlap of two $2p_x$ orbitals. (b) Formation of a pi bond in O_2 from the side-on overlap of two $2p$ orbitals oriented perpendicular to the bond axis.

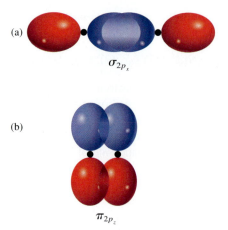

[Figure 3.5(b)]. In each case, VB theory accounts for the changes in potential energy as the distance between the reacting atoms changes. Both the F—F [see Figure 3.5(a)] and H—F [see Figure 3.5(b)] bonds are cylindrically symmetric about the bond axis, and are therefore both examples of sigma bonds. Because the orbitals involved are not the same kind in all cases, the bond dissociation energies and bond lengths in H_2, F_2, and HF might be different. As we stated earlier, Lewis theory treats *all* covalent bonds the same way and offers no explanation for the differences among covalent bonds.

The VB approach can also be applied to multiple bonds in molecules such as oxygen, O_2. The Lewis structure for O_2 contains a double bond:

$$: \overset{..}{O} :: \overset{..}{O} :$$

An oxygen atom has two unpaired electrons in the valence shell available for bonding, each in a separate $2p$ orbital. If we bring the two atoms together along the x axis, the $2p_x$ orbitals can overlap head-on to form a sigma bond, as illustrated in Figure 3.6(a). The other bond in the double bond is formed by the side-on overlap of the other $2p$ orbital, which is perpendicular to the bond axis [see Figure 3.6(b)].

Because the $2p$ orbitals have a node at the atomic nucleus, the electron distribution of the VB wavefunction formed by their side-on overlap does not have cylindrical symmetry, but instead changes sign when rotated by 180° about the bond axis. Bonds that have this property are called *pi (π) bonds.* Thus, the double bond in the oxygen molecule consists of a sigma bond and a pi bond, which are not equivalent. The existence of two different types of bonds in double bond formation is something that is *not* predicted by simple Lewis theory. A triple bond such as that in N_2 is described within the VB approach as a sigma bond formed from the head-on overlap of $2p_x$ orbitals and two pi bonds formed by the overlap of the $2p_y$ and $2p_z$ orbitals on one N atom with their counterparts on the other atom (Figure 3.7).

Bond Lengths and Bond Dissociation Energies

As defined in Section 3.1, the *bond length* is the equilibrium distance of separation between two atoms forming a bond (Figure 3.8). For a covalent bond, the bond length depends primarily upon the identity of the two atoms and on the number of bonds between them. Triple bonds are generally shorter than double bonds, which are shorter than single bonds. For example, the C≡C triple bond length in acetylene (C_2H_2) is

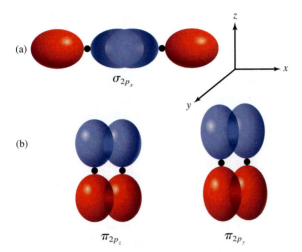

Figure 3.7 Valence bond picture of the triple bond in N_2. (a) A σ bond and (b) two π bonds are formed via overlap of the $2p_z$ and $2p_y$ orbitals on each nitrogen

1.21 Å, whereas the C=C double bond length in ethene (C_2H_4) is 1.34 Å, and the C—C single bond in ethane (C_2H_6) is 1.54 Å.

Table 3.1 lists typical bond lengths for a number of bonds. For bonds that exist only in one molecule, such as the H—H bond in H_2 or the H—Cl bond in HCl, the exact experimental value is given in Table 3.1. For bonds that can occur in a number of different covalent compounds, average values are given, because the actual bond length differs slightly from compound to compound. For example, the average C—H bond length is given in Table 3.1 as 109 pm, but the exact experimental value for the C—H bond length in acetylene (C_2H_2) is 106.1 pm, and in ethane (C_2H_6) it is slightly longer at 110.2 pm.

Note in Table 3.1 that the bond lengths for the heteronuclear diatomic halogen molecules listed (BrCl, ICl, IBr) can be well approximated as the arithmetic mean of the bond distances for the respective homonuclear diatomic molecules. For example, the bond length for IBr is 247 pm. If we average the bond lengths for Br_2 and I_2, we get

$$(228 \text{ pm} + 268 \text{ pm})/2 = 248 \text{ pm}$$

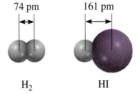

Figure 3.8 Bond length (in pm) in H_2 and HI.

Table 3.1	Average and Exact Bond Lengths for Common Covalent Bonds		
Bond Type	**Bond Length* (pm)**	**Bond Type**	**Bond Length* (pm)**
C—C	154	H_2	74
C=C	133	N_2	110
C≡C	120	O_2	121
C—H	109	F_2	142
C—O	143	Cl_2	199
C=O	121	Br_2	228
C—N	143	I_2	268
C=N	138	BrCl	214
C≡N	116	ICl	232
O—H	96	IBr	247
N—O	136	HCl	127
N=O	122	HF	92

*Average bond lengths (black). Exact Bond lengths (red)

which is very close to the exact value. To apply this useful concept as a general tool for estimating bond lengths in a molecule, we define the **covalent radius.** The length of a bond between atoms A and B in a molecule can be approximated as the sum $r_A + r_B$, where r_A and r_B are the *covalent radii* of atoms A and B, respectively. For example, the average C—C bond distance in covalent compounds is 154 pm, so the covalent radius for a carbon single bond is taken to be $154/2 = 77$ pm. The C—Cl distance in many compounds is approximately 176 pm, so the covalent radius for Cl must be $(176 - 77)$ pm $= 99$ pm. Using this procedure, we can find the covalent radii of many atoms. The covalent radii of several atoms are listed in Table 3.2. The covalent radius is more detailed than the generic atomic radius defined in Section 2.5 because it takes into account the bond order (single, double, or triple) in addition to the identity of the element.

The *bond dissociation energy, D_0,* provides a measure of the stability of a bond in a molecule. As discussed in Section 3.1, the bond dissociation energy is the energy required to break a bond in an isolated molecule. It is generally measured in kJ mol^{-1}, which gives the energy required to break a particular bond in 1 mol of molecules. The experimentally determined bond energy of the diatomic hydrogen molecule, for example, is

$$H_2(g) \longrightarrow H(g) + H(g) \qquad D_0 = 432.0 \text{ kJ mol}^{-1}$$

This equation tells us that breaking the covalent bonds in 1 mole of gaseous H_2 molecules requires 432.0 kJ of energy. For the less stable chlorine molecule,

$$Cl_2(g) \longrightarrow Cl(g) + Cl(g) \qquad D_0 = 240.9 \text{ kJ mol}^{-1}$$

Table 3.2	Covalent Radii for Atoms (in pm)		
	Single-Bond Radius	**Double-Bond Radius**	**Triple-Bond Radius**
H	37		
C	77.2	66.7	60.3
N	70	62	55
O	66	62	
F	64		
Si	117	107	100
P	110	100	93
S	104	94	87
Cl	99	89	
Ge	122	112	
As	121	111	
Se	117	107	
Br	114	104	
Sn	140	130	
Sb	141	131	
Te	137	127	
I	133	123	

Bond energies can also be directly measured for diatomic molecules containing unlike elements, such as HCl,

$$HCl(g) \longrightarrow H(g) + Cl(g) \qquad D_0 = 427.6 \text{ kJ mol}^{-1}$$

as well as for molecules containing double and triple bonds:

$$O_2(g) \longrightarrow O(g) + O(g) \qquad D_0 = 495.8 \text{ kJ mol}^{-1}$$
$$N_2(g) \longrightarrow N(g) + N(g) \qquad D_0 = 942.9 \text{ kJ mol}^{-1}$$

Quantifying the bond energy in polyatomic molecules is more complicated. For example, measurements show that the energy needed to break the first O—H bond in H_2O is different from that needed to break the second O—H bond:

$$H_2O(g) \longrightarrow H(g) + OH(g) \qquad D_0 = 494 \text{ kJ mol}^{-1}$$
$$OH(g) \longrightarrow H(g) + O(g) \qquad D_0 = 424 \text{ kJ mol}^{-1}$$

In each case, an O—H bond is broken, but the first step requires more energy than the second. The difference between the two D_0 values suggests that the second O—H bond itself undergoes a change, because of the changes in chemical environment. Often the bond dissociation energy of polyatomic bonds is reported as an average. For example, the *average bond dissociation energy* for the O—H bond is about 460 kJ mol^{-1}, which is roughly the average of the two successive O—H bond breakings in water. Most commonly, the bond strengths are reported as *bond enthalpies,* which differ from D_0 by only a few kJ mol^{-1}. Bond enthalpies will be discussed in detail in Chapter 7.

Table 3.3 lists the bond dissociation energies of a number of diatomic molecules.

The data in Table 3.3 demonstrate an important general trend in bond dissociation energy. In comparing similar compounds, the bond dissociation energy generally decreases with increasing bond length. For example, the bond dissociation energies for the halogen series HF, HCl, HBr, and HI are 565.7, 427.6, 363.6, and 295.8 kJ mol^{-1}, respectively. Because the covalent radius of a halogen atom increases with its atomic number (I > Br > Cl > F), the bond length of molecules increases as we go down the halogen series, correlating with the decrease in bond dissociation energy. In addition, the bond energy of the compounds listed in Table 3.3 increases with bond order. That is, triple bonds are stronger than double bonds, and double bonds are stronger than single bonds.

Table 3.3	Bond Dissociation Energies for Selected Diatomic Molecules		
Bond	**D_0 kJ mol^{-1}**	**Bond**	**D_0 kJ mol^{-1}**
H—H	432.0	O=O	495.8
H—F	565.7	F—F	154.4
H—Cl	427.6	Cl—Cl	240.2
H—Br	363.6	Br—Br	190.0
H—I	295.8	I—I	148.5
N≡N	942.9		

3.3 | Electronegativity Differences Determine the Polarity of Chemical Bonds

A covalent bond, as we have said, is the sharing of an electron pair by two atoms. In a *homonuclear* diatomic molecule like H_2, in which the atoms are identical, the electrons are equally shared, that is, the electrons spend the same amount of time in the vicinity of each atom. However, in a covalently bonded *heteronuclear* diatomic molecule like HF, the H and F atoms do *not* share the bonding electrons equally because H and F are different atoms

The bond in HF is called a **polar covalent bond,** or simply a *polar bond,* because *the electrons spend more time in the vicinity of one atom than the other.* Experimental evidence indicates that the electrons participating in the H—F bond are more likely to be found near the F atom. We can think of this unequal sharing of electrons as a partial electron transfer or a shift in electron density, as it is more commonly described, from H to F (Figure 3.9). This "unequal sharing" of the bonding electron pair results in a relatively greater electron density near the fluorine atom and a correspondingly lower electron density near hydrogen.

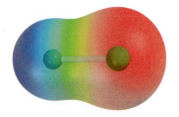

Figure 3.9 Electrostatic potential map of the HF molecule. The electron distribution varies according to the colors of the rainbow. The most electron-rich region is red; the most electron-poor region is blue.

Electronegativity

A property that helps us distinguish a nonpolar covalent bond from a polar covalent bond is **electronegativity,** *the ability of an atom to attract toward itself the electrons in a chemical bond.* Elements with high electronegativity (such as F, O, and N) have a greater tendency to attract electrons than do elements with low electronegativity (such as Li, Be, and B).

Electronegativity is a relative concept, meaning that the electronegativity of an element can be measured only in relation to the electronegativity of other elements. In 1932, Linus Pauling[7] devised a method for calculating *relative* electronegativities of most elements. He observed that the bond energy, $D_0(AB)$, for a bond between two different elements A and B always exceeded the arithmetic mean of the bond energies $[D_0(AA)$ and $D_0(BB)]$ for homonuclear diatomic molecules A—A and B—B, so that the difference, Δ, between these two quantities is always positive:

$$\Delta = D_0(AB) - \frac{1}{2}[D_0(AA) + D_0(BB)] > 0$$

Pauling argued that if the electrons were being shared equally by the two atoms, these two quantities should be equal and the magnitude of Δ should be due to the extra electrostatic attraction of the partially separated charges. Therefore, he postulated that Δ is related to the difference in electronegativity. Pauling defined the difference in electronegativity between two elements A and B as

$$\chi_A - \chi_B = \sqrt{\Delta} \tag{3.2}$$

7. Linus Carl Pauling (1901–1994). American chemist. Regarded by many as the most influential chemist of the twentieth century, Pauling did research in a remarkably broad range of subjects, from chemical physics to molecular biology. Pauling received the Nobel Prize in Chemistry in 1954 for his work on protein structure, and the Nobel Peace Prize in 1962. He is the only person to be the sole recipient of two Nobel Prizes.

Increasing electronegativity →

Increasing electronegativity ↑

1A	2A	3B	4B	5B	6B	7B	8B	8B	8B	1B	2B	3A	4A	5A	6A	7A	8A
H 2.1																	
Li 1.0	**Be** 1.5											**B** 2.0	**C** 2.5	**N** 3.0	**O** 3.5	**F** 4.0	
Na 0.9	**Mg** 1.2											**Al** 1.5	**Si** 1.8	**P** 2.1	**S** 2.5	**Cl** 3.0	
K 0.8	**Ca** 1.0	**Sc** 1.3	**Ti** 1.5	**V** 1.6	**Cr** 1.6	**Mn** 1.5	**Fe** 1.8	**Co** 1.9	**Ni** 1.9	**Cu** 1.9	**Zn** 1.6	**Ga** 1.6	**Ge** 1.8	**As** 2.0	**Se** 2.4	**Br** 2.8	**Kr** 3.0
Rb 0.8	**Sr** 1.0	**Y** 1.2	**Zr** 1.4	**Nb** 1.6	**Mo** 1.8	**Tc** 1.9	**Ru** 2.2	**Rh** 2.2	**Pd** 2.2	**Ag** 1.9	**Cd** 1.7	**In** 1.7	**Sn** 1.8	**Sb** 1.9	**Te** 2.1	**I** 2.5	**Xe** 2.6
Cs 0.7	**Ba** 0.9	**La-Lu** 1.0-1.2	**Hf** 1.3	**Ta** 1.5	**W** 1.7	**Re** 1.9	**Os** 2.2	**Ir** 2.2	**Pt** 2.2	**Au** 2.4	**Hg** 1.9	**Tl** 1.8	**Pb** 1.9	**Bi** 1.9	**Po** 2.0	**At** 2.2	
Fr 0.7	**Ra** 0.9																

Figure 3.10 The Pauling electronegativities of common elements. The color codes are: green (metals), blue (nonmetals), and gray (metalloids). Electronegativities for the noble gases for which no (or very few) known compounds exist (He, Ar, Ne, Rd) are generally not assigned.

where χ_A and χ_B are the electronegativities of element A and element B and the bond energies are given in units of electron volts (1 electron volt = 1 eV = 96.48 kJ mol^{-1}). Equation 3.2 gives only the relative values of electronegativities. To fix a specific value for a given element, the electronegativity of a reference element must be arbitrarily assigned. Assigning hydrogen an electronegativity of 2.10 gives the Pauling electronegativities for the elements, shown in Figure 3.10. These values range from 4.0 for fluorine, the most electronegative element, to 0.7 for francium.

A careful examination of Figure 3.10 reveals trends and relationships among electronegativity values of different elements. In general, electronegativity increases from left to right across a period in the periodic table, as the metallic character of the elements decreases. Within each group, electronegativity decreases as atomic number (and metallic character) increases. The transition metals, though, do not follow these trends. The most electronegative elements (the halogens, oxygen, nitrogen, and sulfur) are found in the upper right-hand corner of the periodic table, and the least electronegative elements (the alkali and alkaline earth metals) are clustered near the lower left-hand corner. (Elements with low electronegativities are often referred to as *electropositive* elements.) These trends are graphed in Figure 3.11.

The electronegativity of an element is related to its electron affinity and ionization energy. Fluorine, for example, which has a high electron affinity (tends to pick up electrons easily) and a high ionization energy (does not lose electrons easily), has a high electronegativity. Sodium, on the other hand, has a low electron affinity (does not gain electrons easily), a low ionization energy (gives up electrons easily), and a low electronegativity.

Two years after Pauling defined electronegativity, the American chemist Robert Mulliken[8] proposed an alternate definition in which the electronegativity for an element

8. Robert Sanderson Mulliken (1896–1986). American chemist. Mulliken was a pioneer in quantum chemistry and made important contributions to the theory of molecular structure and bonding. He was awarded the Nobel Prize in Chemistry in 1966.

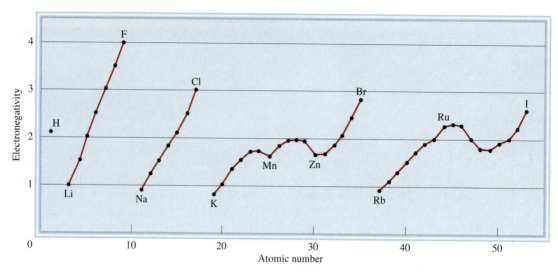

Figure 3.11 Variation of electronegativity with atomic number. The halogens have the highest electronegativities, and the alkali metals the lowest.

A is proportional to the average of the first ionization energy, I_1, and the electron affinity, E_{ea}:

$$\chi_A(\text{Mulliken}) = C\left(\frac{E_{ea} + I_1}{2}\right) \tag{3.3}$$

where C is a constant of proportionality. The Mulliken electronegativity gives largely the same predictions as the Pauling definition regarding the polar nature of covalent bonds, although there are some differences. A number of other definitions have been proposed for electronegativity, but that due to Pauling is the most commonly used.

Electronegativity and electron affinity are related but different concepts. Both indicate the tendency of an atom to attract electrons. However, electron affinity refers to an isolated atom's attraction for an additional electron, whereas electronegativity signifies the ability of an atom in a chemical bond to attract the shared electrons. Furthermore, electron affinity is an experimentally measurable quantity, whereas electronegativity is an estimated number that cannot be measured.

Ionic Bonds

Atoms of elements with widely different electronegativities tend to form bonds in which the electron is completely transferred from the electropositive atom (forming a cation) to the electronegative atom (forming an anion). These interactions are called *ionic bonds* because they do not arise from electron sharing, but instead form as a result of *the Coulombic attraction between oppositely charged ions*. A compound held together by ionic bonds is called an *ionic compound*. As a rule, the elements most likely to form cations in ionic compounds are the alkali and alkaline earth metals and aluminum, and the elements most likely to form anions are the halogens, oxygen, and nitrogen. Consequently, a wide variety of ionic compounds combine a Group 1A or Group 2A metal (or aluminum) with a halogen, oxygen, or nitrogen.

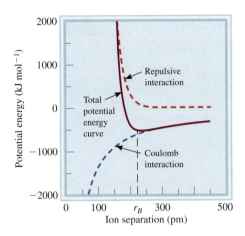

Figure 3.12 The potential energy for an NaCl ion pair versus ion separation in units of kJ mol^{-1}. The total potential energy curve is the sum of the attractive Coulomb interaction between the charges and the repulsive interaction. The equilibrium bond length (r_B = 236 pm) corresponds to the minimum of the total potential energy curve. The potential energy at the minimum is approximately -560 kJ mol^{-1}.

At elevated temperatures, ionic compounds such as NaCl (common table salt) vaporize to form ion pairs in the gas phase. The potential energy of attraction between a sodium cation (Na$^+$) with charge $+e$ and a chlorine anion (Cl$^-$) with charge $-e$ is given by Coulomb's law (Equation 0.5)

$$V_{\text{attraction}} = -\frac{e^2}{4\pi\varepsilon_0 r} \tag{0.5}$$

where r is the separation distance between the ions. However, if the separation distance between the ions becomes small, nuclear-nuclear and electron-electron repulsion become strong. The total potential energy, which is the sum of the attractive Coulomb interaction and the repulsive interaction, is shown in Figure 3.12 in units of kJ mol^{-1}. At the point where the attractive and repulsive forces balance, the total potential energy has a minimum at r_e, which is the equilibrium bond length of this ion pair. For NaCl, the bond length is 236 pm (2.36 Å).

In Figure 3.12, the value of the potential energy at r_B (-560 kJ mol^{-1}) is the energy given off when 1 mol of NaCl ion pairs forms from separated Na$^+$ and Cl$^-$ ions in the gaseous state:

$$\text{Na}^+(g) + \text{Cl}^-(g) \longrightarrow \text{NaCl}(g)$$

In constructing Figure 3.12, we have assumed that sodium and chlorine remain in their ionic state. However, the ground state of the dissociated system consists of separated atoms, not ions. Although we have examined the stability of the ion-bonded NaCl pair relative to the separated *ions,* we have not ruled out the possibility that the ion pair might be unstable relative to dissociation into separated *atoms:*

$$\text{NaCl}(g) \longrightarrow \text{Na}(g) + \text{Cl}(g)$$

We can view this process in two steps:

$$\text{NaCl}(g) \longrightarrow \text{Na}^+(g) + \text{Cl}^-(g) \longrightarrow \text{Na}(g) + \text{Cl}(g) \tag{3.4}$$

The energy of the first step (560 kJ mol^{-1}) is determined in Figure 3.12. The energy of the second step involves the energy required to add an electron to a sodium ion, given by the negative of the ionization energy, I_1, of sodium (495.9 kJ mol^{-1} from Table 2.2), and the energy required to remove an electron from the chloride ion, given

Major Experimental Technique: Microwave Spectroscopy

In Table 3.1, we listed bond lengths for a number of compounds. Given that molecules are extremely small—far smaller than can be seen by conventional optics—how is it possible to determine the lengths of bonds so accurately? Over the past century, chemists and physicists have developed a number of techniques that can quantitatively probe the structure of atoms and molecules by studying how they interact with electromagnetic radiation. For simple gas-phase molecules, one very accurate tool to determine bond distances (and bond angles) is **microwave spectroscopy,** which is *the study of the microwave frequencies that are absorbed by a particular material.* From Figure 1.6, we see that the microwave region of the electromagnetic spectrum corresponds to frequencies in the range 10^8 to 10^{12} Hz. By multiplying these frequencies by Planck's constant, we find that the energy of an individual microwave photon ranges from about 10^{-25} to 10^{-21} J. Such energies, although too low to excite electron transitions, correspond roughly to the energies associated with the rotational motion of molecules.

Consider a heteronuclear diatomic molecule such as HCl or CO. The rotational motion of such a molecule can be described by a simple model called the *rigid rotor*. A rigid rotor consists of two point masses (m_1 and m_2) that are connected by a massless, but rigid, bar of length R. In this model the masses correspond to the two atoms of the diatomic molecule and the bar (R) represents a bond of length R between them:

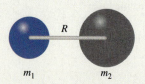

The Schrödinger equation corresponding to the rotations of this rigid rotor can be solved exactly to obtain the following quantized rotational energies:

$$E_{rot} = \frac{J(J+1)h^2}{8\pi^2 I} \qquad J = 0, 1, 2, \ldots \qquad (1)$$

where the integer J is the rotational quantum number. In Equation 1, the quantity I is the *moment of inertia,* which for a diatomic molecule is given by

$$I = \mu R^2 \qquad (2)$$

where the *reduced mass* μ is given by

$$\mu = \frac{m_1 m_2}{m_1 + m_2} \qquad (3)$$

Equation 1 is sometimes written

$$E_J = hBJ(J+1) \qquad J = 0, 1, 2, \ldots \qquad (4)$$

where

$$B = \frac{h}{8\pi^2 I} \qquad (5)$$

is called the *rotational constant.* Because of the nature of the interaction between electromagnetic radiation and a diatomic molecule, absorption of microwave radiation is only possible if $\Delta J = +1$. The frequency of electromagnetic radiation required to excite a molecule from one rotational state (J) to the next higher state ($J + 1$) can be determined using Equation 1

$$\begin{aligned} h\nu_{J \to J-1} = \Delta E &= E_{J+1} - E_J \\ &= hB(J+1)(J+2) - hBJ(J+1) \\ &= 2hB(J+1) \\ \nu_{J \to J+1} &= 2B(J+1) \qquad (6) \end{aligned}$$

So, if the lines of the microwave spectrum are known, then the value of B can be determined, from which we can calculate the moment of inertia and, thus, the bond distance r.

As an example, the $J = 0 \longrightarrow J = 1$ line in the microwave spectrum of $^{12}C^{16}O$ is found experimentally at a frequency of 1.153×10^{11} Hz. From Equation 6, $\nu_{0 \to 1} = 2B(0 + 1) = 2B$, so $B = 5.765 \times 10^{10}$ Hz. The moment of inertia can be then calculated from Equation 5.

$$I = \frac{h}{8\pi^2 B} = \frac{6.62608 \times 10^{-34} \text{ J s}}{8\pi^2(5.765 \times 10^{10} \text{ s}^{-1})} = 1.456 \times 10^{-46} \text{ kg m}^2$$

where we have used the fact that 1 J = 1 kg m^2 s^{-2}. The masses of ^{12}C and ^{16}O are 12.00000 u and 15.99491 u, respectively, giving a reduced mass of

$$\begin{aligned} \mu = \frac{m_1 m_2}{m_1 + m_2} &= \frac{(12.00000 \text{ u})(15.99491 \text{ u})}{(12.00000 \text{ u}) + (15.99491 \text{ u})} = 6.85621 \text{ u} \\ &= 6.85621 \text{ u} \times \frac{1.66054 \times 10^{-27} \text{ kg}}{1 \text{ u}} \\ &= 1.13850 \times 10^{-27} \text{ kg} \end{aligned}$$

so using Equation 2, we have

$$R^2 = \frac{I}{\mu} = \frac{1.456 \times 10^{-46} \text{ kg m}^2}{1.13850 \times 10^{-27} \text{ kg}} = 1.279 \times 10^{-20} \text{ m}^2$$

giving a value for the $^{12}C^{16}O$ bond length of

$$R = 1.131 \times 10^{-10} \text{ m} = 113.1 \text{ pm}$$

It is important to note that because the moment of inertia depends upon the masses of the atoms in the molecule, the microwave spectrum of a diatomic molecule will be sensitive to the isotopic composition of the molecule. The structure of polyatomic molecules can also be studied using microwave spectroscopy, although the procedure is more complicated than that considered here for diatomic molecules.

Electromagnetic radiation can only affect the rotation of a diatomic molecule if the two ends of the molecule are oppositely charged (see the figure). Thus, only diatomic molecules with polar bonds (such as HCl, CO, HF) will absorb microwave radiation. Molecules with nonpolar bonds (N_2, O_2) cannot be studied in this way.

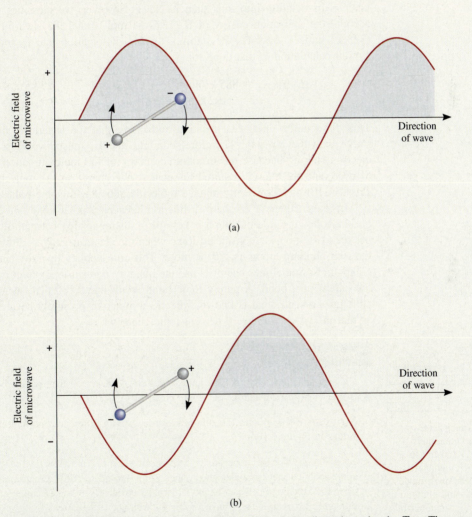

The interaction between the electric field of microwave radiation and a polar diatomic molecule. **Top:** The negative end of the molecule follows the propagation of the wave (the positive region) and rotates in a clockwise direction. **Bottom:** If, after the molecule has rotated to the new position, the radiation has also moved along to its next half cycle, the positive end of the molecule will move into the negative region of the wave, and the negative end will be pushed up. Thus, the molecule will rotate faster. No such interaction can occur in diatomic molecules with nonpolar bonds.

by the electron affinity, E_{ea}, of chlorine (349 kJ mol^{-1} from Table 2.3). Adding these together gives the energy of the process in Equation 3.4:

$$E_{dissociation} = 560 \text{ kJ mol}^{-1} - 495.9 \text{ kJ mol}^{-1} + 349 \text{ kJ mol}^{-1}$$
$$= 413 \text{ kJ mol}^{-1}$$

Therefore, 413 kJ mol^{-1} should be required to break up the NaCl(g) ion pairs into the separated atoms.

The situation is quite different in HBr, a molecule in which the difference in electronegativity is not as great as that in NaCl. The equilibrium bond length in HBr is 1.41 Å (141 pm). Using a reasonable model for the repulsive interaction, the value of the energy of dissociation of a hypothetical HBr ion pair into separated ions is found to be 885 kJ mol^{-1}. Therefore, the HBr ion pair is significantly more stable against dissociation into ions than is NaCl. However, we have not yet taken into account the ionization energy of H (1312 kJ mol^{-1}) and the electron affinity of Br (325 kJ mol^{-1}) to assess the stability against dissociation into H and Br atoms. A similar calculation to that done for NaCl gives

$$E_{dissociation} = 885 \text{ kJ mol}^{-1} - 1312 \text{ kJ mol}^{-1} + 325 \text{ kJ mol}^{-1}$$
$$= -102 \text{ kJ mol}^{-1}$$

This dissociation energy is negative, indicating that the separated atoms are *more* stable than the ion pair. Therefore, bonding in HBr must involve some degree of electron sharing because the Coulombic energy of the ionic attraction is insufficient to overcome the energy required to separate the charge in the molecule. We would classify HBr as a covalently bonded molecule, albeit with some polar character.

In reality, there is no sharp distinction between a polar bond and an ionic bond, but the following rule of thumb is helpful in distinguishing between them: The bond between two atoms is considered ionic when the electronegativity difference between the two bonding atoms is 2.0 or more. This rule applies to most but not all ionic compounds. Sometimes chemists use the quantity *percent ionic character* to describe the nature of a bond. A purely ionic bond would have 100 percent ionic character, although no such bond is known, whereas a nonpolar or purely covalent bond (such as that in H_2) would have 0 percent ionic character.

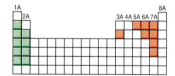

The most electronegative elements are the nonmetals (Groups 5A–7A) and the least electronegative elements are the alkali and alkaline earth metals (Groups 1A–2A) and aluminum. Beryllium, the first member of Group 2A, forms mostly covalent compounds.

Example 3.1

Classify the following bonds as ionic, polar covalent, or covalent: (a) the bond in HCl, (b) the bond in KF, and (c) the C—C bond in H_3CCH_3.

Strategy We follow the 2.0 rule of electronegativity difference and look up the values in Figure 3.10.

Solution (a) The electronegativity difference between H and Cl is 0.9, which is appreciable but not large enough (by the 2.0 rule) to qualify HCl as an ionic compound. Therefore, the bond between H and Cl is polar covalent. (b) The electronegativity difference between K and F is 3.2, which is well above the 2.0 mark; therefore, the bond between K and F is ionic. (c) The two C atoms are identical in every respect—they are bonded to each other and each is bonded to three H atoms. Therefore, the bond between them is purely covalent.

Practice Exercise Which of the following bonds is covalent, which is polar covalent, and which is ionic: (a) the bond in CsCl, (b) the bond in H_2S, (c) the N—N bond in H_2NNH_2?

3.4 | Drawing Correct Lewis Structures Is an Invaluable Skill for a Chemist

Although the octet rule and Lewis structures do not present a complete picture of covalent bonding, they do help to explain the bonding scheme in many compounds and account for the properties and reactions of molecules. For this reason, you should practice drawing Lewis structures of compounds.

Drawing Lewis Structures

The basic steps for drawing Lewis structures are as follows:

1. Draw the skeletal structure of the compound, using chemical symbols and placing bonded atoms adjacent to one another. For simple compounds, this task is fairly easy. For more complex compounds, we must either be given the information or make an intelligent guess. In general, the least electronegative atom occupies the central position. Hydrogen and fluorine usually occupy the terminal (end) positions in the Lewis structure.

2. Count the total number of valence electrons present, referring, if necessary, to Figure 2.14. For polyatomic anions, add the number of negative charges to that total. (For CO_3^{2-}, for example, we would add two electrons because the -2 charge indicates that there are two more electrons than are provided by the atoms.) For polyatomic cations, subtract the number of positive charges from the total number of electrons. (Thus, for NH_4^+ we would subtract one electron because the $+1$ charge indicates a loss of one electron from the group of atoms.)

3. Draw a single covalent bond between the central atom and each of the surrounding atoms. Use the remaining valence electrons to complete the octets of the atoms bonded to the central atom. (Remember that the valence shell of a hydrogen atom is complete with only two electrons.) Electrons belonging to the central or surrounding atoms must be shown as lone pairs if they are not involved in bonding. The total number of electrons to be used is that determined in step 2.

4. If the central atom has fewer than eight electrons after completing steps 1–3, try using lone pairs from the surrounding atoms to form double or triple bonds between the surrounding atoms and the central atom, thus completing the octet of the central atom.

Examples 3.2 and 3.3 illustrate the four-step procedure for drawing the Lewis structures of compounds and polyatomic ions.

Example 3.2

Draw the Lewis structure for nitric acid (HNO_3), in which the three O atoms are bonded to the central N atom and the ionizable H atom is bonded to one of the O atoms.

Strategy We follow the procedure outlined for drawing Lewis structures.

Solution

Step 1 The skeletal structure of HNO_3 is

$$O\ N\ O\ H$$
$$O$$

HNO_3 is a strong electrolyte.

—Continued

Continued—

Step 2 The outer-shell electron configurations of N, O, and H are $2s^2 2p^3$, $2s^2 2p^4$, and $1s^1$, respectively. Thus, there are $5 + (3 \times 6) + 1$, or 24, valence electrons to account for in HNO_3.

Step 3 Draw a single covalent bond between N and each of the three O atoms and between one O atom and the H atom. This uses four pairs of electrons (eight electrons total), leaving 16 valence electrons (eight pairs) to satisfy the octet rule for the O atoms:

$$\ddot{\underset{..}{\text{O}}}\!-\!\text{N}\!-\!\ddot{\underset{..}{\text{O}}}\!-\!\text{H}$$
$$\mid$$
$$:\!\ddot{\underset{..}{\text{O}}}\!:$$

Step 4 The resulting structure satisfies the octet rule for all of the O atoms but not for the N atom. The N atom has only six electrons. Therefore, move a lone pair from one of the end O atoms to form another bond with N. Now the octet rule is also satisfied for the N atom:

$$:\!\ddot{\text{O}}\!=\!\text{N}\!-\!\ddot{\underset{..}{\text{O}}}\!-\!\text{H}$$
$$\mid$$
$$:\!\ddot{\underset{..}{\text{O}}}\!:$$

Check Make sure that all of the atoms (except H) satisfy the octet rule. Count the valence electrons in HNO_3 (in bonds and in lone pairs). The result is 24, the same as the total number of valence electrons on three O atoms ($3 \times 6 = 18$), one N atom (5), and one H atom (1).

Practice Exercise Write the Lewis structure for formic acid (HCOOH).

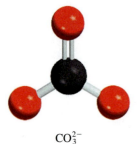

CO_3^{2-}

Example 3.3

Write the Lewis structure for the carbonate ion, CO_3^{2-}.

Strategy Follow the four-step procedure for writing Lewis structures and note that this is an anion with two negative charges.

Solution

Step 1 We can deduce the skeletal structure of the carbonate ion by recognizing that C is less electronegative than O. Therefore, it is most likely to occupy a central position as follows:

$$\begin{array}{c} \text{O} \\ \text{O} \quad \text{C} \quad \text{O} \end{array}$$

Step 2 The outer-shell electron configurations of C and O are $2s^2 2p^2$ and $2s^2 2p^4$, respectively, and the ion itself has two negative charges. Thus, the total number of valence electrons is $4 + (3 \times 6) + 2$, or 24.

—Continued

Continued—

Step 3   Draw a single covalent bond between C and each O and use the remaining valence electrons to satisfy the octet rule for the O atoms:

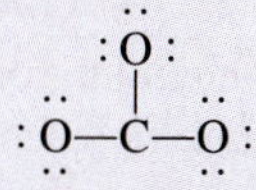

$$: \overset{..}{\underset{}{O}} :$$
$$: \overset{..}{\underset{..}{O}} - C - \overset{..}{\underset{..}{O}} :$$

This structure shows all 24 electrons.

Step 4   Although the octet rule is satisfied for the O atoms, it is not for the C atom. Therefore, move a lone pair from one of the O atoms to form another bond with C. Now the octet rule is also satisfied for the C atom:

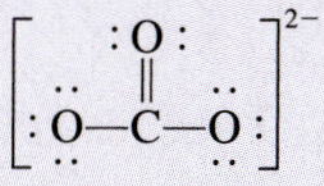

$$\left[: \overset{..}{\underset{..}{O}} - C - \overset{..}{\underset{..}{O}} : \right]^{2-}$$

To indicate that the net charge (-2) belongs to the entire polyatomic ion and not to individual atoms, we enclose the ion in brackets followed by a superscript indicating the net charge.

Check   Make sure that all the atoms satisfy the octet rule. Count the valence electrons in CO_3^{2-} (in chemical bonds and in lone pairs). The result is 24, the same as the total number of valence electrons on three O atoms ($3 \times 6 = 18$), one C atom (4), and two negative charges (2).

Practice Exercise   Draw the Lewis structure for the nitrite ion, NO_2^-.

Formal Charge

In many compounds, the four-step procedure for drawing Lewis structures yields more than one acceptable structure. For example, two possible Lewis structures for CO_2 can be drawn:

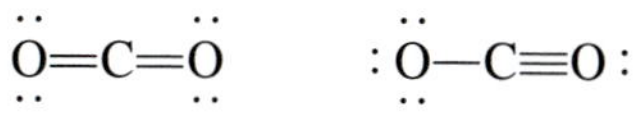

$$\overset{..}{\underset{..}{O}} = C = \overset{..}{\underset{..}{O}} \qquad : \overset{..}{\underset{..}{O}} - C \equiv O :$$
$$\qquad (1) \qquad\qquad (2)$$

Both of these structures have complete octets of electrons surrounding each atom. Which of these two Lewis structures best represents bonding in CO_2? To determine which of these two structures best represents the distribution of electrons in the molecule, we must compare the number of valance electrons in each isolated atom with the number of electrons that are associated with the same atom in a Lewis structure to determine the *formal charge* for each atom, which is *the electrical charge difference between the valence electrons in an isolated atom and the number of electrons assigned to that atom in a Lewis structure*. The number of electrons assigned to a given atom in a Lewis structure is determined by the following procedure:

▸ Assign all nonbonding electrons (lone pairs) entirely to the atom.
▸ Assign one-half of the bonding electrons with the atom, because bonding electrons are shared.

To illustrate the procedure, let us calculate the formal charges for the two Lewis structures for CO_2 given above. From their positions on the periodic table, we find that a carbon atom has four valence electrons and an oxygen atom has six valence electrons. For structure (1)

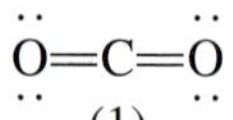

$$\ddot{O}=C=\ddot{O}$$
$$(1)$$

each O atom is surrounded by four nonbonding electrons and four bonding electrons (one double bonds), so the formal charge on each O is

$$\text{valence } e^- - [\text{nonbonded } e^- + \frac{1}{2}(\text{bonded } e^-)] = 6 - (4 + \frac{1}{2} \times 4) = 0$$

For the carbon atom, there are no lone pairs, so the number of nonbonded electrons is zero and there are eight bonding electrons (two double bonds), so the formal charge is

$$\text{valence } e^- - [\text{nonbonded } e^- + \frac{1}{2}(\text{bonded } e^-)] = 4 - (0 + \frac{1}{2} \times 8) = 0$$

So in structure (1), the formal charge is zero on all atoms in the Lewis structure. For structure (2)

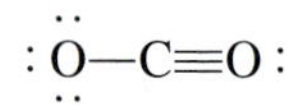

$$:\ddot{O}—C\equiv O:$$
$$(2)$$

the oxygen atom on the left is assigned eight nonbonded electrons and two bonded electrons (one single bond), so the formal charge is $6 - (8 + 2/2) = -1$. The oxygen on the right has two nonbonded electrons and six bonded electrons (one triple bond) and its formal charge is $6 - (2 + 6/2) = +1$. The carbon atom in structure (2) is assigned no nonbonded electrons and eight bonded electrons, so as in structure (1), the carbon atom has a formal charge of zero. We indicate the formal charges in structure (2) by superscripts

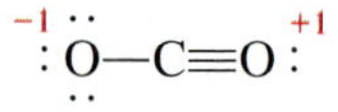

$$\overset{-1}{:}\ddot{O}—C\equiv O\overset{+1}{:}$$

(Formal charges of zero are usually omitted). When writing formal charges, *the sum of the formal charges on a molecule or ion must add up to the total charge on that molecule or ion.* For a molecule, which is neutral, the formal charges should sum to zero. For structure (2) of CO_2, the sum of the formal charges is $(-1) + (0) + (+1) = 0$, as required. For ions, the formal charges should sum to the net charge on the ion.

When more than one acceptable Lewis structure for a given species is possible, the one with the smallest formal charges is generally the best representation of the actual electron distribution in the species. Therefore, we can conclude that the best Lewis structure representation of CO_2 is structure (1)

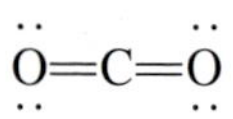

$$\ddot{O}=C=\ddot{O}$$

because the formal charges are all zero, in contrast to structure (2) in which one oxygen atom has a formal charge of -1 and the other oxygen atom has a formal charge of -1. In using formal charges to select the most plausible Lewis structure, the following guidelines are applicable:

▶ For molecules, a Lewis structure in which all formal charges are zero is preferable to one with nonzero formal charges.

▸ Lewis structures with large formal charges (+2, +3, and/or −2, −3, and so on) are less plausible than those with small formal charges.

▸ Among Lewis structures having similar distributions of formal charges, the most plausible structure is the one in which negative formal charges are placed on the more electronegative atoms.

Example 3.4 illustrates the use of formal charges to choose the correct Lewis structure for formaldehyde.

CH_2O

Example 3.4

Formaldehyde (CH_2O), a liquid with a disagreeable odor, has traditionally been used to preserve laboratory specimens. Draw the most likely Lewis structure for formaldehyde.

Strategy A plausible Lewis structure should satisfy the octet rule for all the elements (except H), and have the formal charges (if any) distributed according to electronegativity guidelines.

Solution The two possible skeletal structures are

$$
\begin{array}{ccc}
 & H & \\
H\ \ C\ \ O\ \ H & & C\ \ O \\
 & H & \\
(a) & & (b)
\end{array}
$$

First, draw the Lewis structures for each of these two possibilities:

$$
\overset{-1\ \ +1}{H-C=O-H} \qquad\qquad \overset{H}{\underset{H}{>}}C=\overset{..}{\underset{..}{O}}
$$

(a)   (b)

To determine the formal charges, follow the procedure given in Example 3.4. In (a) the C atom has a total of five electrons (one lone pair plus three electrons from the "breaking" of the single and double bonds). Because C has four valence electrons, the formal charge on the atom is 4 − 5 = −1. The O atom has a total of five electrons (one lone pair and three electrons from the "breaking" of the single and double bonds). Because O has six valence electrons, the formal charge on the atom is 6 − 5 = +1. In (b) the C atom has a total of four electrons from the "breaking" of two single bonds and a double bond, so its formal charge is 4 − 4 = 0. The O atom has a total of six electrons (two lone pairs and two electrons from the "breaking" of the double bond). Therefore, the formal charge on the atom is 6 − 6 = 0. Although both structures satisfy the octet rule, (b) is the more likely structure because it carries no formal charges.

Check In each case, make sure that the total number of valence electrons is 12. Can you suggest two other reasons why (a) is less plausible?

Practice Exercise Draw the most reasonable Lewis structure of a molecule that contains an N atom, a C atom, and an H atom.

It should be noted that formal charges do not represent actual charge separation within the molecule. Assigning these charges to the atoms in the Lewis structure merely helps us keep track of the valence electrons in the molecule.

Example 3.5 illustrates the procedure for determining the formal charges for a polyatomic ion.

Example 3.5

Draw formal charges for the carbonate ion, CO_3^{2-}.

Strategy First draw the Lewis structure and determine the number of electrons assigned to each atom. Then subtract the assigned number from the normal number of valence electrons for each given atom to give the formal charge of that atom in the molecule.

Solution The Lewis structure for the carbonate ion was determined in Example 3.3:

$$\left[\begin{array}{c} : \ddot{O} : \\ \parallel \\ : \ddot{O} - C - \ddot{O} : \end{array} \right]^{2-}$$

The formal charges on the atoms can be calculated using the given procedure.
The C atom: The C atom has four valence electrons. In the Lewis structure there are no nonbonding electrons and eight bonding electrons (two single bonds and one double bond), giving a formal charge of $4 - (0 + 8/2) = 0$.
The O atom in C=O: The O atom has six valence electrons. There are four nonbonding electrons and four bonding electrons (one double bond) on the atom. Here the formal charge is $6 - (4 + 4/2) = 0$.
The O atom in C—O: This atom has six nonbonding electrons and two bonding electrons (one single bond) and the formal charge is $6 - (6 + 2/2) = -1$. Thus, the Lewis structure for CO_3^{2-} with formal charges is

$$\left[\begin{array}{c} : \ddot{O} : \\ \parallel \\ \overset{-1}{:} \ddot{O} - C - \ddot{O} \overset{-1}{:} \end{array} \right]^{2-}$$

Check The sum of the formal charges is -2, the same as the charge on the carbonate ion, so the formal charges are reasonable.

Practice Exercise Write formal charges for the nitrite ion (NO_2^-).

Resonance Structures

In the Lewis structure for ozone (O_3), the octet rule for all atoms can be satisfied by placing one double bond and one single bond between the oxygen atoms:

$$\ddot{O} = \overset{+1}{\ddot{O}} - \overset{-1}{\ddot{O}} :$$

In fact, we can put the double bond at either end of the molecule, as shown by these two equivalent Lewis structures:

$$\ddot{O} = \overset{+1}{\ddot{O}} - \overset{-1}{\ddot{O}} : \qquad \overset{-1}{:} \ddot{O} - \overset{+1}{\ddot{O}} = \ddot{O}$$

However, neither of these two Lewis structures accounts for the known bond lengths in O_3.

Based on these Lewis structures, the O—O bond in O_3 should be longer than the O=O bond because double bonds are shorter than single bonds. Yet experimental evidence shows that both oxygen-oxygen bonds are equal in length (128 pm). We

Eletrostatic potential map of O_3. The electron density is evenly distributed between the two end O atoms.

can resolve this discrepancy by using *both* Lewis structures to represent the ozone molecule:

$$\overset{..}{\underset{..}{O}}=\overset{+1}{\underset{..}{O}}-\overset{..}{\underset{..}{O}}\overset{-1}{:} \longleftrightarrow \overset{-1}{:}\overset{..}{\underset{..}{O}}-\overset{+1}{\underset{..}{O}}=\overset{..}{\underset{..}{O}}$$

Each of these structures is called a resonance structure. A ***resonance structure*** is *one of two or more Lewis structures that can be drawn for a single molecule that cannot be represented accurately by only one Lewis structure.* The double-headed arrow indicates that the structures shown are resonance structures.

The term ***resonance*** itself means *the use of two or more Lewis structures to represent a particular molecule.* Like the medieval European traveler to Africa who described a rhinoceros as a cross between a griffin and a unicorn, two familiar but imaginary animals, we describe ozone, a real molecule, in terms of two familiar but nonexistent structures.

A common misconception about resonance is the notion that a molecule such as ozone somehow shifts quickly back and forth from one resonance structure to the other. Keep in mind that *neither* resonance structure adequately represents the actual molecule, which has its own unique, stable structure. "Resonance" is a human invention, designed to address the limitations in these simple bonding models. To extend the animal analogy, a rhinoceros is a distinct creature, not one that oscillates between a mythical griffin and unicorn!

The carbonate ion can also be represented by resonance structures:

$$\overset{:\overset{..}{O}:}{\underset{..}{\underset{:\overset{..}{\underset{..}{O}}}{\,}}}{}$$

According to experimental evidence, all carbon-oxygen bonds in CO_3^{2-} are equivalent. Therefore, the properties of the carbonate ion are best explained by considering its resonance structures together.

The concept of resonance applies equally well to organic systems, such as the benzene molecule (C_6H_6):

If one of these resonance structures corresponded to the actual structure of benzene, there would be two different bond lengths between adjacent C atoms, one characteristic of the single bond (bond order = 1) and the other of the double bond (bond order = 2). In fact, the distance between all adjacent C atoms in benzene is 140 pm, which is shorter than a C—C bond (154 pm) and longer than a C=C bond (133 pm). To correctly reflect the resonance, we assign a bond order of 3/2 to each of the carbon-carbon bonds in benzene.

A simpler way of drawing the structure of the benzene molecule and other compounds containing the "benzene ring" is to show only the skeleton and not the

Rhinoceros

Griffin

Unicorn

The hexagonal structure of benzene was first proposed by the German chemist August Kekulé (1829–1896).

carbon and hydrogen atoms. By this convention, the resonance structures are represented by

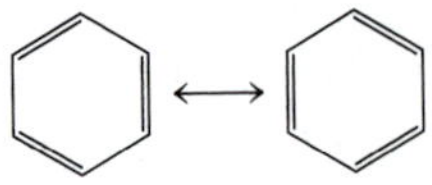

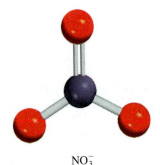

The C atoms at the corners of the hexagon and the H atoms are all omitted, although they are understood to exist. (This is a common convention in organic chemistry.) Only the bonds between the C atoms are shown.

Remember this important rule for drawing resonance structures: The positions of electrons, but not those of atoms, can be rearranged in different resonance structures. In other words, the same atoms must be bonded to one another in all the resonance structures for a given species.

Example 3.6 illustrates the procedure for drawing resonance structures of a molecule or polyatomic ion.

NO_3^-

Example 3.6

Draw all valid resonance structures (including formal charges) for the nitrate ion (NO_3^-), which has the following skeletal arrangement:

$$O$$
$$O \quad N \quad O$$

Strategy Follow the procedure used for drawing Lewis structures and calculating formal charges demonstrated in Examples 3.4 and 3.5.

Solution Just as in the case of the carbonate ion, we can draw three equivalent resonance structures for the nitrate ion:

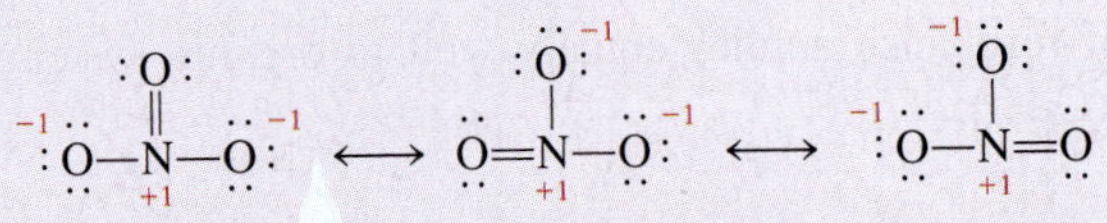

Check The resonance structures are reasonable because the skeletal arrangement of atoms is the same in each of them (only the placement of the electrons is different). Moreover, the formal charges sum to the overall charge on the nitrate ion (-1), and the total number of valence electrons is 24 in each structure (5 from N, 6 from each of the three O atoms, and 1 for the negative charge).

Practice Exercise Draw all the valid resonance structures for the nitrite ion (NO_2^-).

Exceptions to the Octet Rule

As mentioned in Section 3.2, the octet rule applies mainly to the second-period elements. Exceptions to the octet rule fall into three categories, which are characterized by an incomplete octet (that is, fewer than eight valence electrons around the central atom), an odd number of electrons, or more than eight valence electrons around the central atom (an expanded octet).

The Incomplete Octet

In some compounds, the number of electrons surrounding the central atom in a stable molecule is fewer than eight. Beryllium hydride (BeH_2), for example, which exists as discrete molecules in the gas phase, has the following Lewis structure:

$$H—Be—H$$

Beryllium, unlike the other Group 2 elements, forms mostly covalent compounds of which BeH_2 is an example.

Only four electrons surround the Be atom, because beryllium (a Group 2A element) has just two valence electrons (its complete electron configuration is $1s^2 2s^2$) and each of the two hydrogen atoms contributes just one valence electron. There is no way to satisfy the octet rule for beryllium in BeH_2.

Elements in Group 3A, particularly boron and aluminum, also tend to form compounds in which they are surrounded by fewer than eight electrons. Because the electron configuration of boron is $1s^2 2s^2 2p^1$, it has a total of three valence electrons. Boron reacts with the halogens to form a class of compounds having the general formula BX_3, where X is a halogen atom. Thus, in boron trifluoride there are only six electrons around the boron atom:

$$: \ddot{F} :$$
$$|$$
$$: \ddot{F} — B$$
$$|$$
$$: \ddot{F} :$$

The following resonance structures all contain a double bond between B and F and satisfy the octet rule for boron:

$$: \ddot{F} — \overset{-1}{B} — \ddot{F} : \longleftrightarrow \quad \overset{+1}{F} = \overset{-1}{B} — \ddot{F} : \longleftrightarrow \quad : \ddot{F} — \overset{-1}{B} = \overset{+1}{F} \\ \quad \| \qquad\qquad\qquad | \qquad\qquad\qquad\quad | \\ : \ddot{F} : \qquad\qquad\quad : \ddot{F} : \qquad\qquad\quad : \ddot{F} : \\ \;\;{}_{+1}$$

The B—F bond length in BF_3 (130.9 pm) is shorter than a single bond (137.3 pm), indicating that such resonance structures may play some role in the bonding, even though in each case the negative formal charge is placed on the B atom and the positive formal charge on the more electronegative F atom.

Although boron trifluoride is stable, it readily reacts with ammonia. This reaction is better represented by using the Lewis structure in which boron has only six valence electrons around it:

$$: \ddot{F} : \quad H \qquad\qquad\quad : \ddot{F} : \quad H \\ | \qquad\; | \qquad\qquad\qquad | \qquad\; | \\ : \ddot{F} — B \; + \; : N — H \longrightarrow \; : \ddot{F} — \overset{-}{B} — \overset{+}{N} — H \\ | \qquad\; | \qquad\qquad\qquad | \qquad\; | \\ : \ddot{F} : \quad H \qquad\qquad\quad : \ddot{F} : \quad H$$

It seems that the properties of BF_3 are best explained by all four resonance structures.

The B—N bond in $F_3B—NH_3$ is different from the covalent bonds discussed so far because both electrons are contributed by the N atom. This type of bond is called a ***coordinate covalent bond*** (also referred to as a *dative bond*), and is defined as *a covalent bond in which one of the atoms donates both electrons*. Although the properties of a coordinate covalent bond do not differ from those of a covalent bond in which both bonding atoms contribute an electron, the distinction is useful for keeping track of valence electrons and assigning formal charges. In such reactions, the species

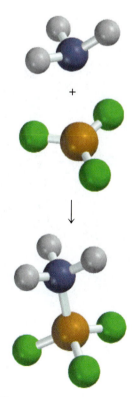

+

$$NH_3 + BF_3 \longrightarrow H_3N - BF_3$$

Just Say NO

Nitric oxide (NO), the simplest nitrogen oxide, is an odd-electron molecule (thus making it paramagnetic). A colorless gas (boiling point: $-152°C$), NO can be prepared in the laboratory by reacting sodium nitrite ($NaNO_2$) with a reducing agent such as Fe^{2+} in an acidic medium:

$$NO_2^- (aq) + Fe^{2+}(aq) + 2H^+(aq) \longrightarrow$$
$$NO(g) + Fe^{3+}(aq) + H_2O(l)$$

Environmental sources of nitric oxide include the burning of fossil fuels containing nitrogen compounds and the reaction between nitrogen and oxygen inside automobile engines at high temperatures:

$$N_2(g) + O_2(g) \longrightarrow 2NO(g)$$

Lightning also contributes to the atmospheric concentration of NO. Exposed to air, nitric oxide quickly forms brown nitrogen dioxide gas:

$$2NO(g) + O_2(g) \longrightarrow 2NO_2(g)$$

Nitrogen dioxide is a major component of smog.

About two decades ago scientists studying muscle relaxation discovered that our bodies produce nitric oxide for use as a neurotransmitter. (A *neurotransmitter* is a small molecule that serves to facilitate cell-to-cell communications.)

Since then, nitric oxide has been detected in at least a dozen cell types in various parts of the body. Cells in the brain, the liver, the pancreas, the gastrointestinal tract, and the blood vessels can synthesize nitric oxide. NO also functions as a cellular toxin to kill harmful bacteria. And that's not all: In 1996 it was reported that NO binds to hemoglobin, the oxygen-carrying protein in the blood, helping to regulate blood pressure.

The discovery of the biological role of nitric oxide has shed light on how nitroglycerin ($C_3H_5N_3O_9$) works as a drug. Nitroglycerin tablets are commonly prescribed for heart patients to relieve the pain (*angina pectoris*) caused by a brief interference in the flow of blood to the heart. It is now believed that nitroglycerin produces nitric oxide, which causes the muscles to relax and allows the arteries to dilate.

That NO evolved as a messenger molecule is entirely appropriate. Nitric oxide is small and so can diffuse quickly from cell to cell. It is a stable molecule but under certain circumstances it is highly reactive, which accounts for its protective function. The enzyme that brings about muscle relaxation contains iron for which nitric oxide has a high affinity. It is the binding of NO to the iron that activates the enzyme. Nevertheless, in the cell, where biological effectors are typically very large molecules, the pervasive effects of one of the smallest known molecules are unprecedented.

Colorless nitric oxide gas is produced by the action of Fe^{2+} on an acidic sodium nitrite solution. The gas is bubbled through water and immediately reacts with oxygen to form the brown NO_2 gas when exposed to air.

that donates the electron pair is referred to as a *Lewis base* and the species that accepts the electron pair to form the coordinate covalent bond is called a *Lewis acid*. In the reaction between NH_3 and BF_3, NH_3 acts as a Lewis base and BF_3 as a Lewis acid. Lewis acids and bases are discussed in more detail in Chapter 10.

Odd-Electron Molecules

Some molecules, such as nitric oxide (NO) and nitrogen dioxide (NO_2), contain an *odd* number of electrons:

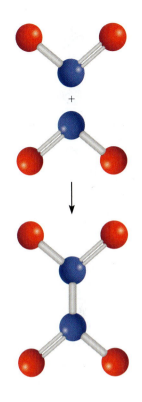

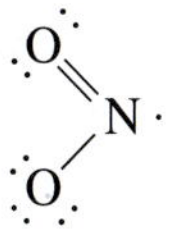

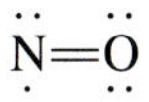

Because we need an even number of electrons for complete pairing (to reach eight), the octet rule cannot be satisfied for all the atoms in any of these molecules.

Odd-electron molecules are sometimes called *radicals*. Many radicals are highly reactive, because there is a tendency for the unpaired electron to form a covalent bond with an unpaired electron on another molecule. For example, when two nitrogen dioxide molecules collide, they form dinitrogen tetroxide, in which the octet rule is satisfied for both the N and O atoms:

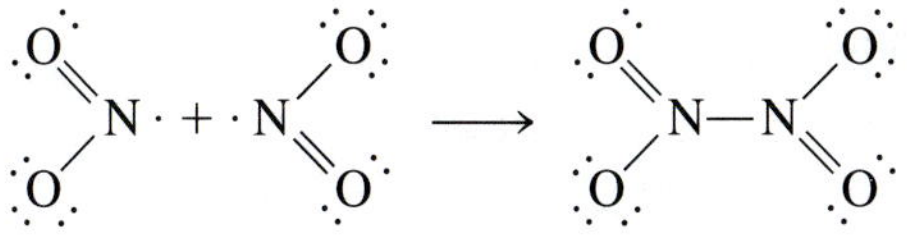

The Expanded Octet

Atoms of the second-period elements cannot have more than eight valence electrons around the central atom, but atoms of elements in and beyond the third period of the periodic table form some compounds in which more than eight electrons surround the central atom. In addition to the 3s and 3p orbitals, elements in the third period also have 3d orbitals that can be used in bonding. These orbitals allow an atom to form an *expanded octet*. One compound in which there is an expanded octet is sulfur hexafluoride, a very stable compound. The electron configuration of sulfur is [Ne] $3s^23p^4$. In SF_6, each of sulfur's six valence electrons forms a covalent bond with a fluorine atom, so there are twelve electrons around the central sulfur atom:

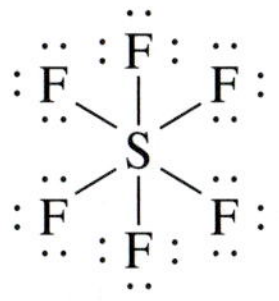

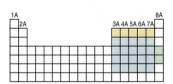

Yellow: second-period elements cannot have an expanded octet. Blue: third-period elements and beyond can have an expanded octet. Green: the noble gases usually only have an expanded octet.

In Chapter 4, we will see that these 12 electrons, or six bonding pairs, are accommodated in six orbitals that originate from the one 3s, the three 3p, and two of the five 3d orbitals. Sulfur also forms many compounds in which it obeys the octet rule. In sulfur dichloride, for instance, S is surrounded by only eight electrons:

Sulfur dichloride is a toxic, foul-smelling cherry-red liquid (boiling point: 59°C).

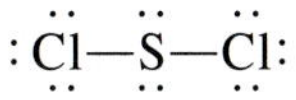

Examples 3.7–3.9 display compounds that do not obey the octet rule.

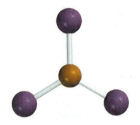

AlI$_3$ has a tendency to dimerize or form two units as Al$_2$I$_6$.

Example 3.7

Draw the Lewis structure for aluminum triiodide (AlI$_3$).

Strategy Follow the procedures used in Examples 3.4 and 3.5 to draw the Lewis structure and calculate formal charges.

Solution The outer-shell electron configurations of Al and I are $3s^2 3p^1$ and $5s^2 5p^5$, respectively. The total number of valence electrons is $3 + 3 \times 7$ or 24. Because Al is less electronegative than I, it occupies a central position and forms three bonds with the I atoms:

$$\ddot{\underset{\cdot\cdot}{:}\mathrm{I}:} \\ | \\ :\ddot{\mathrm{I}}—\mathrm{Al} \\ | \\ :\ddot{\mathrm{I}}:$$

There are no formal charges on the Al and I atoms.

Check Although the octet rule is satisfied for the I atoms, there are only six valence electrons around the Al atom. Thus, the Al atom in AlI$_3$ has an incomplete octet.

Practice Exercise Draw the Lewis structure for BeF$_2$.

Example 3.8

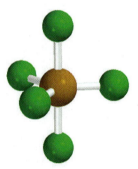

PF$_5$ is a reactive gaseous compound.

Draw the Lewis structure for phosphorus pentafluoride (PF$_5$), in which all five F atoms are bonded to the central P atom.

Strategy P is a third-period element, so it can have an expanded octet, if necessary. Follow the procedures given in Examples 3.4 and 3.5 to draw the Lewis structure and calculate formal charges.

Solution The outer-shell electron configurations for P and F are $3s^2 3p^3$ and $2s^2 2p^5$, respectively, and so the total number of valence electrons is $5 + (5 \times 7)$, or 40. The five F atoms will use 30 of those electrons for lone pairs, leaving 10 electrons for P—F bonds. Phosphorus, therefore, must expand its octet to accommodate the five bonds to F. The Lewis structure of PF$_5$ is

$$\ddot{:}\mathrm{F}: \\ :\ddot{\mathrm{F}}\diagdown \; | \\ \qquad \mathrm{P}—\ddot{\mathrm{F}}: \\ :\ddot{\mathrm{F}}\diagup \; | \\ :\ddot{\mathrm{F}}:$$

There are no formal charges on the P and F atoms.

Check Although the octet rule is satisfied for all five F atoms, there are 10 valence electrons around the P atom, giving it an expanded octet.

Practice Exercise Draw the Lewis structure for chlorine trifluoride (ClF$_3$). (All fluorine atoms are bonded to the central chlorine atom.)

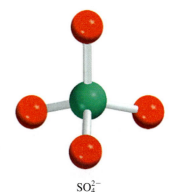

SO_4^{2-}

Example 3.9

Draw a Lewis structure for the sulfate ion (SO_4^{2-}) in which all four O atoms are bonded to the central S atom.

Strategy Follow the procedures given in Examples 3.4 and 3.5 to draw the Lewis structure and calculate formal charges. S is a third-period element, so it may need to expand its octet to accommodate all of the valence electrons in the sulfate ion.

Solution The outer-shell electron configurations of S and O are $3s^2 3p^4$ and $2s^2 2p^4$, respectively.

Step 1 The skeletal structure of SO_4^{2-} is

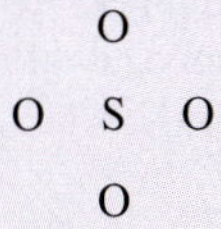

$$O$$
$$O \quad S \quad O$$
$$O$$

Step 2 Both O and S are Group 6A elements and so have six valence electrons each. Including the two negative charges, we must therefore account for a total of $6 + (4 \times 6) + 2$, or 32, valence electrons in SO_4^{2-}.

Step 3 We draw a single covalent bond between all the bonding atoms and add lone pairs on the O atoms:

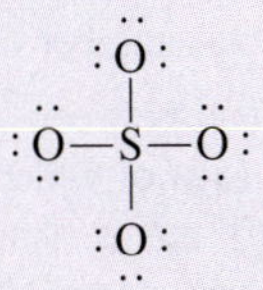

$$
:\!\ddot{O}\!: \\
\,|\, \\
:\!\ddot{O}\!-\!S\!-\!\ddot{O}\!: \\
\,|\, \\
:\!\ddot{O}\!:
$$

Next, show formal charges on the S and O atoms:

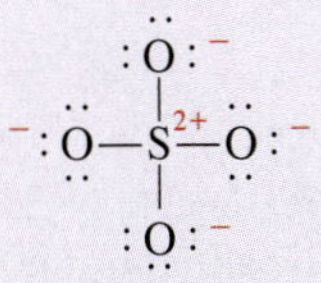

$$
:\!\ddot{O}\!:^{-} \\
\,|\, \\
{}^{-}:\!\ddot{O}\!-\!S^{2+}\!-\!\ddot{O}\!:^{-} \\
\,|\, \\
:\!\ddot{O}\!:^{-}
$$

Check One of six other equivalent structures for SO_4^{2-} is as follows:

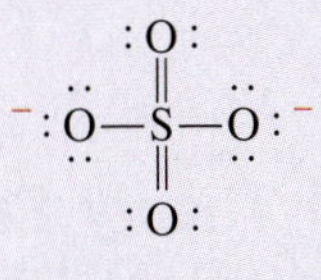

$$
:\!\ddot{O}\!: \\
\,\|\, \\
{}^{-}:\!\ddot{O}\!-\!S\!-\!\ddot{O}\!:^{-} \\
\,\|\, \\
:\!\ddot{O}\!:
$$

This structure involves an expanded octet on S but may be considered more plausible because it bears fewer formal charges. However, detailed theoretical calculations show that the dominant structure is the one that satisfies the octet rule, even though it has greater formal charge separations. The general rule for elements in the third period and beyond is that a resonance structure that obeys the octet rule is preferred over one that involves an expanded octet but bears fewer formal charges.

Practice Exercise Draw the Lewis structure of sulfuric acid (H_2SO_4).

A final note about the expanded octet: When drawing Lewis structures of compounds containing a central atom from the third period and beyond, sometimes we find that the octet rule is satisfied for all the atoms but there are still valence electrons left to place. In these cases, the extra electrons should be placed as lone pairs on the central atom. Example 3.10 shows this approach.

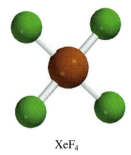

XeF_4

Example 3.10

Draw a Lewis structure of the noble gas compound xenon tetrafluoride (XeF_4), in which all F atoms are bonded to the central Xe atom.

Strategy Follow the procedures in Examples 3.4 and 3.5 for drawing Lewis structures and calculating formal charges. Xe is a fifth-period element, so it may need to expand its octet to accommodate all of the valence electrons.

Solution

Step 1 The skeletal structure of XeF_4 is

$$
\begin{array}{ccc}
\text{F} & & \text{F} \\
& \text{Xe} & \\
\text{F} & & \text{F}
\end{array}
$$

Step 2 The outer-shell electron configurations of Xe and F are $5s^2 5p^6$ and $2s^2 2p^5$, respectively, and so the total number of valence electrons is $8 + (4 \times 7)$ or 36.

Step 3 Draw a single covalent bond between all the bonding atoms. The octet rule is satisfied for the F atoms, each of which has three lone pairs. The sum of the lone pair electrons on the four F atoms (4×6) and the four bonding pairs (4×2) is 32. The remaining four electrons are shown as two lone pairs on the Xe atom:

$$
\begin{array}{ccc}
:\ddot{\text{F}} & & \ddot{\text{F}}: \\
& \text{Xe} & \\
:\ddot{\text{F}} & & \ddot{\text{F}}:
\end{array}
$$

Check The Lewis structure is reasonable because the formal charges on Xe and the F atoms are all zero. Additionally, all 36 of the valence electrons are accounted for because Xe, a fifth-period element, can expand its octet.

Practice Exercise Write the Lewis structure of sulfur tetrafluoride (SF_4).

3.5 | Molecular Orbital Theory Provides a Detailed Description of Chemical Bonding

The valence bond and Lewis structure approaches provide chemists with enormously useful tools for the qualitative prediction of chemical bonding and molecular structure. Both approaches are built upon the intuitively appealing concept of electron pairing to form covalent bonds. As a consequence, they are unable to predict correctly the true nature of bonding in many paramagnetic molecules (molecules with unpaired spins). For example, experiments show that the oxygen molecule is paramagnetic (Figure 3.13), implying that it has at least two unpaired electrons. (Because O_2 has an even number of electrons, the number of unpaired electrons must be even.) For such systems, a more sophisticated, but somewhat less intuitive, approach known as *molecular orbital theory* is more appropriate. In addition, even for diamagnetic systems,

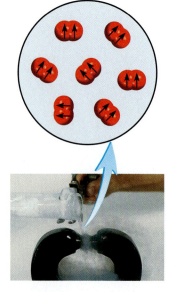

Figure 3.13 Liquid oxygen can be caught between the poles of a magnet, because O_2 molecules are paramagnetic.

the construction of valence bond wavefunctions can be quite cumbersome for large molecules, and the molecular orbital approach is generally more useful.

In *molecular orbital (MO) theory,* the electrons in a molecule are placed, not in localized atomic orbitals, but into *molecular orbitals,* so called because they are associated with the entire molecule. In contrast, an atomic orbital is associated with only one atom.

The Simplest Molecular Species: The H_2^+ Ion

In Chapter 2, the development of atomic orbitals for many-electron atoms was built upon an understanding of the orbitals obtained from the exact solution for the one-electron hydrogen atom. Similarly, it is useful to examine the electronic struc- ture for the simplest molecular species—the hydrogen molecule ion (H_2^+)—to begin our discussion of molecular orbitals. Like the H atom, the ion H_2^+ contains only one electron. The difference is the presence of two nuclei instead of one. The potential energy of the system is a sum of the Coulomb attraction of the electron to each of the two nuclei and Coulomb repulsion between the positively charged nuclei,

$$V(r_A, r_B; R) = \underbrace{-\frac{e^2}{4\pi\varepsilon_0 r_A} - \frac{e^2}{4\pi\varepsilon_0 r_B}}_{\substack{\text{electron-nuclear} \\ \text{attraction}}} + \underbrace{\frac{e^2}{4\pi\varepsilon_0 R}}_{\substack{\text{internuclear} \\ \text{repulison}}} \qquad (3.5)$$

where r_A and r_B are the distances between the electron and each of the two nuclei (A and B) and R is the distance between the nuclei (Figure 3.14).

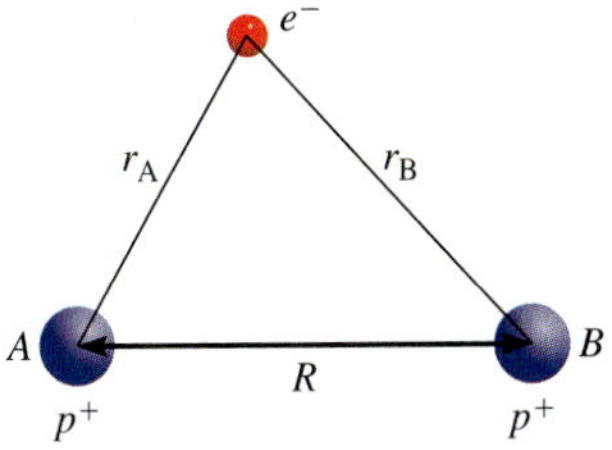

Using the Born-Oppenheimer approximation (Section 3.1) to ignore nuclear motion, the Schrödinger equation for the single electron in H_2^+ can be solved exactly to yield a set of wavefunctions $\psi\ (x, y, z; R)$ describing the quantum state of the electron, where x, y, and z are the Cartesian coordinates of the electron (Figure 3.14). By analogy to the atomic orbitals of the H atom, these wavefunctions are called *molecular orbitals because they are not associated with individual nuclei, but extend over the entire molecule or ion.* For each molecular orbital, there will be an associated orbital energy $E(R)$. As indicated, both the molecular orbitals and orbital energies depend upon the parameter R, the distance between the nuclei.

Even though the Schrödinger equation can be solved exactly for H_2^+, the mathe- matical form of the wavefunctions (molecular orbitals) is quite complicated and dif- ficult to use and interpret qualitatively. Fortunately, it is possible to approximate the molecular orbitals in this system in a way that gives excellent physical insight into how and why bonds form. In this approximation, the molecule orbitals are constructed from *linear combinations of atomic orbitals (LCAO)* and we refer to them as *LCAO molecular orbitals.*[9] To construct the LCAO molecular orbitals, begin with the atomic orbitals of the individual electrons and mix them linearly to form the molecular orbit- als. A general rule is that only atomic orbitals that are similar in energy and have appreciable overlap will mix to form molecular orbitals.

The construction of the ground-state molecular orbital for H_2^+ provides a useful illustration of the LCAO method. The lowest-energy orbital for the H atom is the $1s$ orbital. The $1s$ orbital on nucleus A, $\phi_{A,1s}$, is identical in energy to that on nucleus B, $\phi_{B,1s}$, and when the nuclei are close together, these two atomic orbitals overlap sig- nificantly; therefore, they should effectively mix to form the lowest-energy molecular

Figure 3.14 Geometry of the hydrogen molecule ion, H_2^+, consisting of two protons (p^+) and one electron (e^-).

---

9. A *linear combination* of two functions $f(x)$ and $g(x)$ is any function of the form $a\ f(x) + b\ g(x)$, where a and b are constants. The term "linear" in this context is used to imply that $f(x)$ and $g(x)$ are raised to the power 1 in the combination.

orbital. We then construct the molecular orbital as a linear combination of these two atomic orbitals,

$$\psi_{MO}(A,\ B) = c_A\phi_{A,1s} + c_B\phi_{B,1s} \tag{3.6}$$

where c_A and c_B are constants. Normally the determination of the optimal values of the linear coefficients, c_A and c_B, is nontrivial. In this case, however, we can use symmetry to determine the allowed values. The two nuclei of the hydrogen molecule ion are identical, so the electron distribution should be unchanged if the two nuclei were to exchange places. Because the electron distribution is given by the square of the wavefunction, we have

$$\psi_{MO}^2(A,\ B) = \psi_{MO}^2(B,\ A) \tag{3.7}$$

which implies that ψ_{MO} must satisfy the conditions

$$\psi_{MO}(A,\ B) = \pm\psi_{MO}(B,\ A) \tag{3.8}$$

There are only two possible ways for the LCAO molecular orbital in Equation 3.6 to satisfy the condition specified by Equation 3.8—namely, $c_A = c_B$ or $c_A = -c_B$. Thus, two molecular orbitals can be formed from linear combinations of the two $1s$ atomic orbitals:

$$\begin{aligned}
\psi_+ &= \frac{1}{\sqrt{2}}[\phi_{A,1s} + \phi_{B,1s}] \\
\psi_- &= \frac{1}{\sqrt{2}}[\phi_{A,1s} - \phi_{B,1s}]
\end{aligned} \tag{3.9}$$

(The factor $1/\sqrt{2}$ is included so that the molecular orbitals are properly normalized.) These two orbitals are pictured in Figure 3.15. We started with two atomic orbitals and constructed two molecular orbitals from their combinations. In general, the number of molecular orbitals obtained from this procedure is equal to the number of atomic orbitals being combined. Both of the molecular orbitals in Equation 3.9 are classified as *sigma* (σ) orbitals because they are unchanged by rotation about the bond axis. The ψ_+ orbital is formed by the *constructive interference* (see Figure 1.3) of the two $1s$ orbitals and the ψ_- orbital is formed by the *destructive interference* of the two $1s$ orbitals. The constructive interference of the $1s$ orbitals in ψ_+ results in a significant build-up of electron density in the overlap region between the two nuclei, thereby lowering the energy below that of the individual $1s$ orbitals. Such an orbital is referred to as a **bonding molecular orbital** because it has *lower energy and greater stability than the atomic orbitals from which it was formed*. The ψ_- orbital, on the other hand, has a node in the electron density between the two nuclei. Such an orbital is called an **antibonding molecular orbital** *because it has higher energy and lower stability than the atomic orbitals from which it was formed*. As the names suggest, placing electrons into a bonding orbital yields a stable covalent bond, whereas placing electrons

Figure 3.15 The molecular orbitals from (a) constructive and (b) destructive interference of $1s$ atomic orbitals centered on the two nuclei of the hydrogen molecule ion. The sign of the wavefunction is indicated by color (blue = +, red = −).

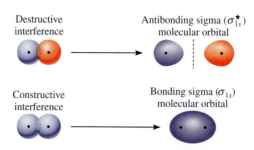

Destructive interference

Antibonding sigma ($\sigma_{1s}^{\star}$) molecular orbital

Constructive interference

Bonding sigma (σ_{1s}) molecular orbital

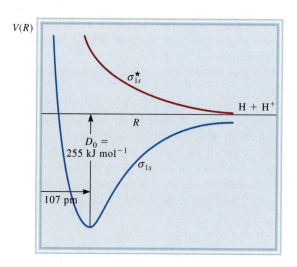

Figure 3.16 The energy of the hydrogen molecule ion as a function of the internuclear distance for an electron in the σ_{1s} bonding molecular orbital (blue curve) and an electron in the $\sigma_{1s}^{\star}$ antibonding molecular orbital (red curve). The zero of energy corresponds to the energy of infinite nuclear separation.

into an antibonding orbital results in a unstable bond. We label our two H_2^+ molecular orbitals formed from $1s$ atomic orbitals in the following way:

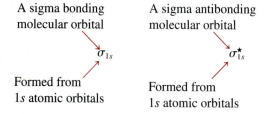

where the star denotes an antibonding orbital. The formation of the σ_{1s} and $\sigma_{1s}^{\star}$ orbitals from the constructive and destructive interference of two $1s$ orbitals, respectively, is illustrated in Figure 3.15.

Figure 3.16 shows the energy of the H_2^+ ion as a function of internuclear separation for the electron in the ground-state σ_{1s} bonding orbital and in the excited state $\sigma_{1s}^{\star}$ antibonding orbital. Figure 3.16 shows that the energy curve for the bonding orbital has a well-defined minimum, indicating a stable bond, whereas no minimum is observed for the antibonding orbital, indicating that bond formation is not possible in this state. If light were used to excite a ground-state hydrogen molecule ion into the antibonding excited state, the ion would rapidly fall apart (dissociate) into a hydrogen atom (H) and a proton (H^+) in a process called *photodissociation*.

A useful illustration of the electron configuration of a molecule within the molecular orbital picture is the ***molecular orbital energy-level diagram (MO diagram)***, which shows the relative energy levels of the orbitals produced in the formation of the molecule and the occupation of those orbitals by the electrons in the ground-state molecule. The MO diagram for H_2^+ shown in Figure 3.17 indicates that the H_2^+ ground

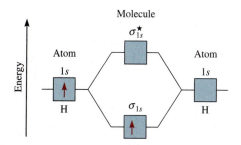

Figure 3.17 Molecular orbital diagram for the H_2^+ ion.

state has one electron in a σ_{1s} bonding orbital. Just as we would write the ground-state electron configuration of the hydrogen atom as $1s^1$, we designate the ground-state electron configuration for the H_2^+ ion as $(\sigma_{1s})^1$.

Note that the MO diagram also shows the atomic orbitals from which the molecular orbitals are derived.

Before going on to more complicated molecules, two comments are in order. First, within MO theory, it is insufficient to say that a bond is formed by electron sharing. An electron in an antibonding orbital is equally shared between the two nuclei; however, the nature of the orbital prevents a bonding interaction. It is the sharing of electrons in bonding orbitals that gives rise to covalent bonding in molecules. It is also unnecessary for electrons to be paired to generate a bonding interaction. In the case of H_2^+, the bond is generated by the sharing of a single electron in a bonding orbital.

Molecular Orbitals for Other Hydrogen and Helium Molecules and Ions

Just as in many-electron atoms, the presence of two or more electrons in a molecule prevents the exact solution of the Schrödinger equation. In such systems, approximation of the wavefunction is no longer just convenient, but becomes absolutely necessary. Fortunately, the LCAO procedure discussed for H_2^+ provides a framework with which to construct the wavefunctions for many-electron diatomic molecules and ions, much in the same way that the orbitals for the H atom provide a framework to construct the wavefunction for many-electron atoms. To start, let's consider H_2, the simplest diatomic molecule. The hydrogen molecule contains two electrons. Just as we constructed the helium ground-state wavefunction by placing each of the two electrons into a $1s$ orbital (Equation 2.4), we construct an approximate hydrogen molecule wavefunction by placing the two electrons (1 and 2) with opposite spins into the lowest energy molecular orbital,

$$\psi_{H_2} \approx \sigma_{1s}(1)\,\sigma_{1s}(2) = \frac{1}{2}[\phi_{A,1s}(1) + \phi_{B,1s}(1)][\phi_{A,1s}(2) + \phi_{B,1s}(2)] \quad \textbf{(3.10)}$$

where 1 and 2 denote the two electrons. (For simplicity, Equation 3.10 does not explicitly include the spin parts of the wavefunction.) Thus, the ground-state molecular orbital configuration of H_2 is $(\sigma_{1s})^2$ and the molecular orbital diagram for the hydrogen molecule (Figure 3.18) is identical to that of H_2^+ (see Figure 3.17) except that the σ_{1s} bonding orbital contains two electrons of opposite spin:

Figure 3.18 Molecular orbital diagram for H_2. Note that the two electrons in the σ_{1s} bonding orbital must have opposite spins in accord with the Pauli exclusion principle.

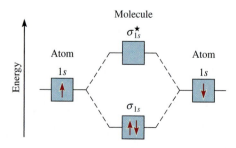

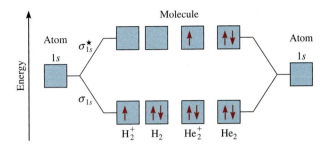

Figure 3.19 Molecular orbital diagrams for H_2^+, H_2, He_2^+, and He_2. In all these species, the molecular orbitals are formed by the mixing of two $1s$ orbitals.

Just as in the case of many-electron atoms, the exact mathematical form of the molecular orbitals in Equation 3.9 can be optimized to include electron-electron repulsion using the Hartree-Fock procedure described in Section 2.1; however, the overall shape and symmetry of the orbitals remain the same, so the qualitative picture is unchanged. With the addition of the second electron in H_2 the ground-state molecular orbital, σ_{1s}, is now full, according to the Pauli exclusion principle. Addition of another electron to form the helium cation He_2^+ requires the additional electron to be placed into the antibonding $\sigma_{1s}^\star$ orbital, giving an electron configuration of $(\sigma_{1s})^2(\sigma_{1s}^\star)^1$. Similarly, we can write the electron configuration for He_2 as $(\sigma_{1s})^2(\sigma_{1s}^\star)^2$. The molecular orbital diagrams for all of these species are summarized in Figure 3.19.

We can evaluate the relative stabilities of these species using their *bond order*. As defined earlier, the bond order tells us the number of pairs of electrons that form the bond between two atoms. In molecular orbital theory, this concept has the following more general definition:

$$\text{bond order} = \frac{1}{2}\left(\begin{array}{c}\text{number of electrons} \\ \text{in bonding MOs}\end{array} - \begin{array}{c}\text{number of electrons} \\ \text{in antibonding MOs}\end{array}\right) \quad (3.11)$$

Qualitatively, the bond order is an indication of the strength of a bond. For example, if there are two electrons in the bonding molecular orbital and none in the antibonding molecular orbital, the bond order is one, which means that there is one covalent bond and the molecule is stable. The bond order can be a fraction, but a bond order of zero (or a negative value) means the bond has no stability and the molecule cannot exist. Thus, H_2 has a bond order of one and both H_2^+ and He_2^+ have bond orders of $1/2$. Diatomic helium has a bond order of zero because it has an equal number of electrons in bonding and antibonding orbitals. For these species the bond order correlates quite well with the bond energies and bond lengths. The bond energies of the two species with bond order $1/2$ are quite similar: 268 kJ mol^{-1} and 241 kJ mol^{-1} for H_2^+ and He_2^+, respectively. Similarly, the bond lengths for H_2^+ and He_2^+ are nearly the same: 106 and 108 pm, respectively. The bond energy for H_2, with a bond order of one, is 457 kJ mol^{-1}, nearly twice that of H_2^+ and He_2^+. The bond length of H_2 is 74 pm, considerably shorter than the bonds in H_2^+ and He_2^+. This simple molecular orbital picture predicts that the helium molecule is not stable because it has a bond order of zero. Although He_2 has been shown to exist under very specialized conditions, it has a bond energy that is less than 1 kJ mol^{-1}, making it extremely unstable, as predicted.[10]

In using approximate models such as this one, it is important to realize that all models have their limits. Based on the preceding discussion, for example, we would

10. At room temperature, the random collisions with other atoms and molecules can supply an average energy of about 2.5 kJ mol^{-1} into a bond, which is much greater than the energy required to break up the He_2 molecule, so He_2 is unstable under these conditions.

predict that the hydrogen molecule anion (H_2^-) would have a bond order of 1/2 and would be isoelectronic with He_2^+; therefore, you might expect the two species to exhibit similar stability. However, the hydrogen molecule anion has never been observed and is believed to be unstable. Accounting for this fact requires a more sophisticated treatment of electron-electron repulsion than is included within the simple MO treatment presented here.

General Rules Governing the Construction of Molecular Orbital Electron Configurations

Given a set of molecular orbitals, the procedure for determining the electron configuration of a molecule is analogous to the Aufbau process used to determine the electron configuration of many-electron atoms (Section 2.2). To write the electron configuration of a molecule, we must first arrange the molecular orbitals in order of increasing energy. Then we can use the following guidelines to fill the molecular orbitals with electrons. The rules also help us understand the stabilities of the molecular orbitals.

1. The number of molecular orbitals formed is always equal to the number of atomic orbitals combined.

2. The filling of molecular orbitals proceeds from low energies to high energies. In a stable molecule, the number of electrons in bonding molecular orbitals is always greater than that in antibonding molecular orbitals because we place electrons first in the lower-energy bonding molecular orbitals.

3. Like an atomic orbital, each molecular orbital can accommodate up to two electrons with opposite spins in accordance with the Pauli exclusion principle.

4. When electrons are added to molecular orbitals of the same energy, the most stable arrangement is predicted by Hund's rule, that is, electrons enter these molecular orbitals with parallel spins.

5. The number of electrons in the molecular orbitals is equal to the sum of all the electrons on the bonding atoms.

Molecular Orbitals for Homonuclear Diatomic Molecules of the Second-Period Elements

In the H_2 and He_2 molecules and ions, we considered only the interaction between $1s$ orbitals, but the molecular orbitals formed can only accommodate four electrons. To examine bonding in more complex diatomic molecules we need to consider additional atomic orbitals as well. For many-electron atoms, the next highest energy atomic orbital after the $1s$ orbital is the $2s$ orbital followed by the three degenerate $2p$ orbitals. The $2s$ orbitals are different enough in energy from the $2p$ orbitals, that we would not expect them to mix when forming the next level of molecular orbitals. So, the next highest energy set of molecular orbitals for homonuclear diatomic molecules is formed from constructive and destructive interference of $2s$ orbitals centered on the two atoms (A and B) to form the σ_{2s} bonding orbital and the $\sigma_{2s}^{\star}$ antibonding orbital. By analogy to the formation of the σ_{1s} and $\sigma_{1s}^{\star}$ from the 1s orbitals we have

$$\sigma_{2s} = \frac{1}{\sqrt{2}}[\phi_{A,2s} + \phi_{B,2s}] \quad \text{constructive interference}$$

$$\sigma_{2s}^{\star} = \frac{1}{\sqrt{2}}[\phi_{A,2s} - \phi_{B,2s}] \quad \text{destructive interference}$$

The formation of the next set of molecular orbitals from the $2p$ atomic orbitals is more complex. The p orbitals are at right angles to one another, so there will only be overlap between p orbitals with similar orientations and each pair of $2p$ orbitals will add constructively and destructively to form a pair of bonding and antibonding molecular orbitals, respectively. If we let the x direction be along the bond axis, head-on overlap of the $2p_x$ orbitals centered on the two nuclei produces a sigma bonding σ_{2p_x} and a sigma antibonding $\sigma^{\star}_{2p_x}$ orbital. The other two $2p$ orbitals can only overlap side-on to produce a pair of pi bonding orbitals (π_{2p_y} and π_{2p_z}) and a pair of pi anti-bonding orbitals ($\pi^{\star}_{2p_y}$ and $\pi^{\star}_{2p_z}$). These six molecular orbitals produced from a total of six $2p$ atomic orbitals on the two nuclei are shown in Figure 3.20.

To use these molecular orbitals to construct electron configurations and MO diagrams for the second-period homonuclear diatomic molecules, we need to know their order of increasing energy. Just as was the case in many-electron atoms where the ordering of the $4s$ and $3d$ atomic orbitals depended upon the number of electrons that were present, the ordering of the molecular orbitals for second-period homonuclear diatomics is not constant across the period. Geometrically, the head-on overlap of the $2p_x$ orbitals to form the σ_{2p} orbital should be greater than the side-on overlap of the

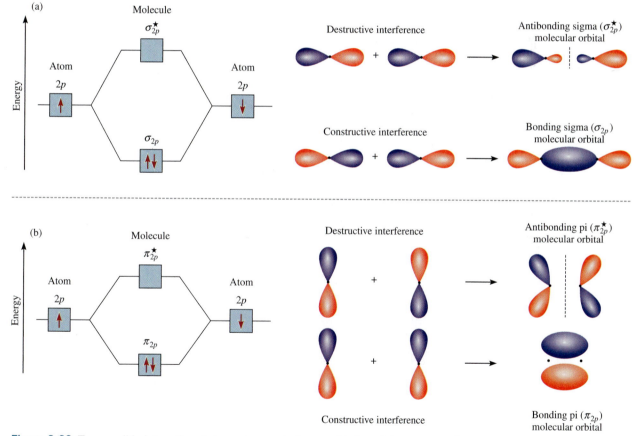

Figure 3.20 Two possible interactions between two equivalent p orbitals and the corresponding molecular orbitals. (a) When the p orbitals overlap head on, a sigma bonding and a sigma antibonding molecular orbital form. (b) When the p orbitals overlap side on, a pi bonding and a pi antibonding molecular orbital form. Normally, a sigma bonding molecular orbital is more stable than a pi bonding molecular orbital, because side-on interaction leads to a smaller overlap of the p orbitals than does head-on interaction. We assume that the $2p_x$ orbitals take part in the sigma molecular orbital formation. The $2p_y$ and $2p_z$ orbitals can interact to form only pi molecular orbitals. The behavior shown in (b) represents the interaction between the $2p_y$ orbitals or the $2p_z$ orbitals. In both cases, the dashed line represents a nodal plane between the nuclei, where the electron density is zero, and the sign of the orbital is represented by color: blue (+) and red (−).

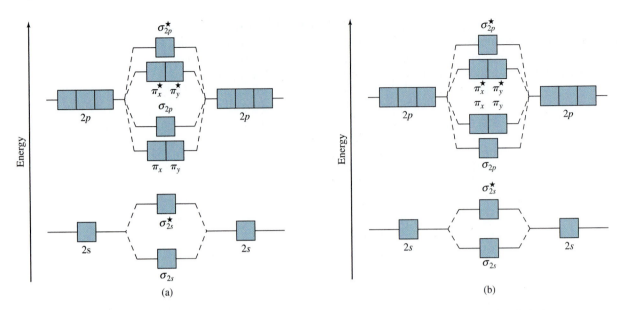

Figure 3.21 General molecular orbital energy level diagrams for homonuclear diatomics of the second-period elements. (a) Diagram for Li_2 through N_2. (b) Diagram for O_2 and F_2. For simplicity we omit the σ_{1s} and $\sigma_{1s}^\star$ orbitals.

other $2p$ orbitals to form the π_{2p} orbital, resulting in a lower energy for the sigma orbital. However, for Li_2 to N_2, the π_{2p} orbitals are actually lower in energy than the σ_{2p} orbital [Figure 3.21(a)]. For these elements, the difference in energy between the $2s$ orbital on one atom and $2p$ atomic orbitals on the other is small enough that some mixing of these orbitals to form molecular orbitals can occur, which has been ignored in the simple picture presented here. It is this mixing that accounts for the lower energy of the π_{2p} orbitals relative to the σ_{2p_x} orbital. For O_2 and F_2, where the difference in energy between the $2s$ and $2p$ orbitals is larger (hence less mixing), this order is reversed and the σ_{2p_x} orbital is lower in energy than the π_{2p} orbitals. The general molecular orbital diagram relevant to O_2 and F_2 is shown in Figure 3.21(b).

Given this description, we can now examine the electron configuration of the second-period homonuclear diatomic molecules:

Li_2

The electron configuration of Li is $1s^2 2s^1$ so the Li_2 molecule contains six electrons. Four of these electrons will go into the σ_{1s} and $\sigma_{1s}^\star$ orbitals, leaving two electrons that will go into σ_{2s}, the next lowest molecular orbital. The electron configuration of Li_2 is given by

$$(\sigma_{1s})^2 (\sigma_{1s}^\star)^2 (\sigma_{2s})^2$$

There are four electrons in bonding orbitals and two in antibonding orbitals, so the bond order is $(4 - 2)/2 = 1$, and we predict that the molecule is stable and diamagnetic.

Be_2

The electron configuration of Be is $1s^2 2s^2$ so Be_2 contains eight electrons (two more than Li_2). In Be_2, these two additional electrons go into the $\sigma_{2s}^\star$ antibonding orbital, giving an electron configuration of

$$(\sigma_{1s})^2 (\sigma_{1s}^\star)^2 (\sigma_{2s})^2 (\sigma_{2s}^\star)^2$$

Having equal numbers of bonding and antibonding orbitals gives a bond order of zero for diatomic beryllium, which means it should be unstable. The Be_2 molecule does exist, but its bond dissociation energy is extremely small (~ 10 kJ mol^{-1}), and it is stable only under very special conditions, consistent with our prediction.

B_2, C_2, and N_2

With Be_2, the MOs formed from $1s$ and $2s$ orbitals are full. Using the molecular orbital scheme pictured in Figure 3.21(a), the next highest molecular orbitals are the two degenerate π_{2p} orbitals. The electron configuration of B is $1s^2 2s^2 2p^1$. In B_2, the next two electrons are placed in the two degenerate π_{2p} orbitals with parallel spins, in accordance with Hund's rule. Thus, the configuration of B_2 is

$$(\sigma_{1s})^2 (\sigma_{1s}^\star)^2 (\sigma_{2s})^2 (\sigma_{2s}^\star)^2 (\pi_{2p_y})^1 (\pi_{2p_z})^1$$

giving a bond order of 1, which indicates that the B_2 molecule should be stable in the gas phase. The presence of the two unpaired electrons suggests that B_2 is paramagnetic, in agreement with experiment.

The electron configuration of C is $1s^2 2s^2 2p^3$. For C_2, the additional two electrons fill the two π_{2p} orbitals, giving an electron configuration of

$$(\sigma_{1s})^2 (\sigma_{1s}^\star)^2 (\sigma_{2s})^2 (\sigma_{2s}^\star)^2 (\pi_{2p_y})^2 (\pi_{2p_z})^2$$

with a bond order of 2 (double bond). The molecule should be diamagnetic. C_2 molecules have been detected in the gaseous state and have a bond dissociation energy (630.1 kJ mol^{-1}) that is slightly greater than twice that of B_2 (291.2 kJ mol^{-1}), as expected from the difference in bond order.

The electron configuration of N is $1s^2 2s^2 2p^3$. For N_2, the next two electrons go into the next higher energy level (σ_{2p}), giving the electronic configuration

$$(\sigma_{1s})^2 (\sigma_{1s}^\star)^2 (\sigma_{2s})^2 (\sigma_{2s}^\star)^2 (\pi_{2p_y})^2 (\pi_{2p_z})^2 (\sigma_{2p_x})^2$$

Thus, N_2 should be stable and diamagnetic, with a bond order of 3, which is consistent with the Lewis prediction of a triple bond.

O_2

As stated earlier in Section 3.5, Lewis structures and valence bond theory do not account for the magnetic properties of the oxygen molecule, which is known to be paramagnetic (it possesses two unpaired electrons). The following Lewis structure can be drawn with two unpaired electrons:

$$: \overset{\cdot\cdot}{\underset{\cdot}{O}} : \overset{\cdot\cdot}{\underset{\cdot}{O}} :$$

However, this structure is unsatisfactory on at least two counts. First, it implies the presence of a single covalent bond, but experimental evidence strongly suggests that there is a double bond in O_2 (bond order $= 2$). Second, it places seven valence electrons around each oxygen atom, a violation of the octet rule.

The ground-state electron configuration of O is $1s^2 2s^2 2p^4$; thus, there are 16 electrons in O_2. Using the order of increasing energies of the molecular orbitals shown in Figure 3.21(b), the ground-state electron configuration of O_2 is

$$(\sigma_{1s})^2 (\sigma_{1s}^\star)^2 (\sigma_{2s})^2 (\sigma_{2s}^\star)^2 (\sigma_{2p_x})^2 (\pi_{2p_y})^2 (\pi_{2p_z})^2 (\pi_{2py}^\star)^1 (\pi_{2p_z}^\star)^1$$

Table 3.4 Properties of Stable Homonuclear Diatomic Molecules of the Second-Period Elements*

	Li_2	B_2	C_2	N_2	O_2	F_2	
$\sigma^{\star}_{2p_x}$	☐	☐	☐	☐	☐	☐	$\sigma^{\star}_{2p_x}$
$\pi^{\star}_{2p_y}, \pi^{\star}_{2p_z}$	☐☐	☐☐	☐☐	☐☐	↑ ↑	↑↓ ↑↓	$\pi^{\star}_{2p_y}, \pi^{\star}_{2p_z}$
σ_{2p_x}	☐	☐	☐	↑↓	↑↓ ↑↓	↑↓ ↑↓	π_{2p_y}, π_{2p_z}
π_{2p_y}, π_{2p_z}	☐☐	↑ ↑	↑↓ ↑↓	↑↓ ↑↓	↑↓	↑↓	σ_{2p_x}
$\sigma^{\star}_{2s}$	☐	↑↓	↑↓	↑↓	↑↓	↑↓	$\sigma^{\star}_{2s}$
σ_{2s}	↑↓	↑↓	↑↓	↑↓	↑↓	↑↓	σ_{2s}
Bond order	1	1	2	3	2	1	
Bond length (pm)	267	159	131	110	121	142	
Bond Dissociation Energy (kJ mol^{-1})	110.0	290	613.8	955.4	502.9	118.8	
Magnetic properties	Diamagnetic	Paramagnetic	Diamagnetic	Diamagnetic	Paramagnetic	Diamagnetic	

*For simplicity, the σ_{1s} bonding and antibonding orbitals are omitted. These two orbitals hold a total of four electrons.

According to Hund's rule, the last two electrons enter the $\pi^{\star}_{2p_y}$ and $\pi^{\star}_{2p_z}$ antibonding orbitals with parallel spins. The bond order of O_2 is equal to 2 and the unpaired spins indicate that it is paramagnetic, a prediction that corresponds to experimental observations. The successful explanation of the paramagnetism of diatomic oxygen was one of the early successes of MO theory.

Table 3.4 summarizes the general properties of the stable diatomic molecules of the second period.

Example 3.11 shows how MO theory can be used to predict the molecular properties of ions.

Example 3.11

The N_2^+ ion can be prepared by bombarding the N_2 molecule with fast-moving electrons. Predict the following properties of N_2^+: (a) electron configuration, (b) bond order, (c) magnetic properties, and (d) bond length relative to the bond length of N_2 (is it longer or shorter?).

Strategy Use the molecular orbital diagrams in Table 3.4 to deduce the properties of ions generated from the homonuclear molecules.

Solution

(a) Because N_2^+ has one fewer electron than N_2, its electron configuration is

$$(\sigma_{1s})^2(\sigma^{\star}_{1s})^2(\sigma_{2s})^2(\sigma^{\star}_{2s})^2(\pi_{2p_y})^2(\pi_{2p_z})^2(\sigma_{2p_x})^1$$

(b) The bond order of N_2^+ is found by using Equation 3.11:

$$\text{bond order} = (9 - 4)/2 = 2.5$$

—Continued

Continued—

(c) N_2^+ has one unpaired electron, so it is paramagnetic.

(d) Because the electrons in the bonding molecular orbitals are responsible for holding the atoms together, N_2^+ should have a weaker and, therefore, longer bond than N_2. (In fact, the bond length of N_2^+ is 112 pm, compared with 110 pm for N_2.)

Check Because an electron is removed from a bonding molecular orbital, the bond order should decrease. The N_2^+ ion has an odd number of electrons (13), so it should be paramagnetic. The bond order in N_2^+ is less than in N_2, so the bond length in N_2^+ should be greater than in N_2.

Practice Exercise Which of the following species has a longer bond length: F_2 or F_2^-?

Heteronuclear Diatomic Molecules of the First- and Second-Period Elements

For heteronuclear diatomic molecules containing elements of the first two periods, the approach is basically the same as that for homonuclear diatomic molecules, except that the molecular orbitals are no longer symmetrically positioned relative to the atomic orbitals because the atoms are different. In particular, the bonding molecular orbitals have a greater contribution from the atomic orbitals of the more electronegative element. The opposite is true for the antibonding orbitals—they have a greater contribution from the more electropositive element.

The Lewis structure for hydrogen fluoride (HF), for example, contains a single bond and three lone pairs on the F atom:

$$H\!-\!\overset{\displaystyle ..}{\underset{\displaystyle ..}{F}}:$$

Figure 3.22 shows the relative energies of atomic and molecular orbitals in HF. Several features are worth noting. First, the $1s$ orbital of H overlaps with the $2p_x$ orbital of F to form a σ molecular orbital that contains a large component of the $2p_x$ orbital on F. This is reasonable because the H—F bond is polar, and the greater electron

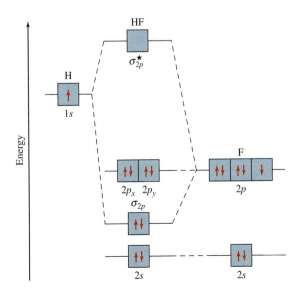

Figure 3.22 Molecular energy-level diagram for the HF molecule. The placements of hydrogen $1s$ and fluorine $2p$ are based on their first ionization energy values (see Table 2.2). For simplicity, the fluorine $1s$ orbital has been omitted.

density is in the vicinity of fluorine. The $\sigma^{\star}$ molecular orbital lies closer to the hydrogen $1s$ orbital and resembles mostly that orbital. The $1s$ orbital has no net interaction with the $2p_y$ and $2p_z$ orbitals of fluorine. Consequently, these two p orbitals are *nonbonding,* that is, two of the three lone pairs reside in them. The energy of the fluorine $2s$ orbital is too low to mix with the hydrogen $1s$ orbital. For this reason, it is also nonbonding and holds the remaining lone pair on the F atom.

Mathematically, the wave functions for the bonding and antibonding sigma molecular orbitals are

$$\psi_\sigma = c_1\phi_{1s,H} + c_2\phi_{2p_x,F}$$

and

$$\psi_\sigma^{\star} = c_3\phi_{1s,H} - c_4\phi_{2p_x,F}$$

respectively, where the coefficients c_1 and c_2 give the relative contributions of the hydrogen $1s$ and fluorine $2p_x$ atomic orbitals, respectively, in the σ molecular orbital, and c_3 and c_4 are the contributions of the same atomic orbitals, respectively, for the $\sigma^{\star}$ antibonding molecular orbital. The large difference between the electronegativities of H and F means that $c_2 \gg c_1$, so the electron density is much greater near the F in the HF molecule. The opposite is true for the $\sigma^{\star}$ antibonding orbital, which is empty, but could be occupied in certain excited electronic states. This orbital has largely hydrogen $1s$ character, that is, $c_3 \gg c_4$.

In carbon monoxide (CO), which is isoelectronic with N_2, the energies of the $2s$ and $2p$ orbitals on both C and O are close enough that they mix in a manner similar to those in N_2. Figure 3.23 shows the molecular orbital diagram for CO. The placement of the atomic orbitals of the C and O atoms is determined by the values of the first ionization energy. As for HF, the orbital coefficients for the more electronegative atom (O) will be larger in the bonding molecular orbitals. The less electronegative

Figure 3.23 Molecular orbital energy level diagram for the CO molecule. The placements of carbon and oxygen $2p$ orbitals are based on their first ionization energy values (see Table 2.2). For simplicity, the $1s$ orbitals have been omitted.

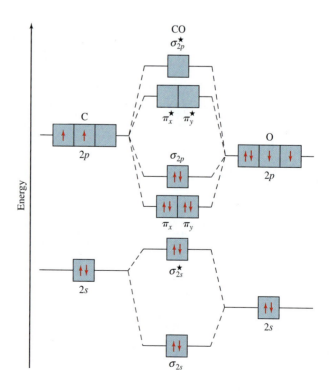

atom (C) will have larger coefficients in the antibonding molecular orbitals. The electron configuration for CO is analogous to that of N_2 (Table 3.4).

Example 3.12

The molecule NO is important in atmospheric and biological chemistry (see the inset on p. 198). (a) Write the electron configuration of NO, (b) calculate its bond order, and (c) predict its magnetic properties.

Strategy The electronegativity of N is close to that of O, so the molecular orbital diagram should be similar to that of N_2 and CO. Use Equation 3.11 to calculate the bond order, and determine whether it has unpaired electrons (paramagnetic) or all its electrons are paired (diamagnetic).

Solution

(a) Nitric oxide has one more electron than N_2, so its electron configuration is

$$(\sigma_{1s})^2(\sigma_{1s}^{\star})^2(\sigma_{2s})^2(\sigma_{2s}^{\star})^2(\pi_{2p_y})^2(\pi_{2p_z})^2(\sigma_{2p_x})^2(\pi_{2p_y}^{\star})^1$$

(b) The bond order is given by

$$\text{bond order} = (10 - 5)/2 = 2.5$$

(c) NO is paramagnetic because it has one unpaired electron.

Check NO is an odd-electron molecule, so it should be paramagnetic.

Practice Exercise Draw a molecular orbital diagram and write the electron configuration for the cyanide ion (CN^-).

With today's high-speed computing and efficient quantum chemistry algorithms for the solution of the Schrödinger equation, chemists can calculate the electronic structure of atoms and molecules with high accuracy. Given these advanced computational methods, you might wonder why we do not simply abandon the very basic Lewis dot, VB, and MO approaches discussed in Chapters 3 and 4. Even though they are highly approximate representations of the actual electronic structure of molecules, simple VB and MO theories have an advantage over more accurate computational approaches in that they provide a clear physical picture of bonding in molecules that is easy to visualize and interpret. Such simple theories provide chemists with surprisingly accurate predictive tools for the qualitative behavior of atoms and molecules in chemical reactions.

Summary of Facts and Concepts

Section 3.1

▶ A chemical bond forms between two atoms whenever the forces of attraction between them are sufficiently strong that they are not pulled apart in the course of normal interactions with their environment.

▶ The bond dissociation energy (bond energy) is the energy required to break a bond. The bond energy measures the strength of a chemical bond. The distance between two bonded atoms at equilibrium is the bond length.

Section 3.2

▶ In a covalent bond, two electrons (one pair) are shared by two atoms. In multiple covalent bonds, two or three pairs of electrons are shared by two atoms. Some covalently bonded atoms also have lone pairs, that is, pairs of valence electrons that are not involved in bonding. The arrangement of bonding electrons and lone pairs around atoms in a molecule is represented by a Lewis structure.

▶ The octet rule predicts that atoms form enough covalent bonds to surround themselves with eight electrons each.

▸ In valence bond theory, a chemical bond is described in terms of overlap between atomic orbitals on the bonded atoms.

▸ Bonds in which the electron wavefunction is cylindrically symmetric about the bond axis are called sigma bonds and bonds in which the electron wavefunction changes sign when the molecule is rotated 180° about the bond axis are called pi bonds.

▸ Triple bonds are stronger than double bonds, and double bonds are stronger than single bonds. The length of a bond decreases as the number of bonds between two atoms increases.

Section 3.3

▸ A polar covalent bond is one in which the bonding electrons are shared unequally between the two bonded atoms.

▸ The electronegativity of an atom is a measure of its ability to pull electron density to itself in a bond. A bond between two atoms of differing electronegativity is polar.

▸ An ionic bond is one in which the electrons are not shared, and the bond is formed by the electrostatic attraction between two oppositely charged ions. An ionic compound is held together with ionic bonds. An ionic bond forms between two atoms when the difference in electronegativity is large (>2).

Section 3.4

▸ When one atom in a covalently bonded pair donates two electrons to the bond, the Lewis structure can include the formal charge on each atom as a means of keeping track of the valence electrons.

▸ For some molecules or polyatomic ions, two or more Lewis structures based on the same skeletal structure satisfy the octet rule and appear chemically reasonable. Taken together, such resonance structures represent the molecule or ion more accurately than any single Lewis structure does.

▸ There are exceptions to the octet rule, particularly for covalent beryllium compounds, elements in Group 3A, and elements in the third period and beyond in the periodic table.

Section 3.5

▸ Molecular orbital theory describes bonding in terms of the combination and rearrangement of atomic orbitals to form orbitals that are associated with the molecule as a whole.

▸ Bonding molecular orbitals increase electron density between the nuclei and are lower in energy than individual atomic orbitals. Antibonding molecular orbitals have a region of zero electron density between the nuclei, and an energy level higher than that of the individual atomic orbitals.

▸ We write electron configurations for molecular orbitals as we do for atomic orbitals, filling in electrons in the order of increasing energy levels. The number of molecular orbitals always equals the number of atomic orbitals that were combined. The Pauli exclusion principle and Hund's rule govern the filling of molecular orbitals.

▸ Molecules are stable if the number of electrons in bonding molecular orbitals is greater than that in antibonding molecular orbitals.

Key Words

antibonding molecular orbital, p. 204	covalent bond, p. 174	linear combination of atomic orbitals, p. 203	octet rule, p. 175
bond dissociation energy, p. 172	covalent compound, p. 174		pi bond, p. 178
bond length, p. 173	covalent radius, p. 180	lone pair, p. 174	polar covalent bond, p. 182
bond order, p. 176	double bond, p. 175	microwave spectroscopy, p. 186	resonance, p. 195
bonding molecular orbital, p. 204	duet, p. 175	molecular orbital, p. 203	resonance structure, p. 195
chemical bond, p. 172	electronegativity, p. 182	molecular orbital energy-level diagram, p. 205	sigma bond, p. 177
coordinate covalent bond, p. 197	formal charge, p. 191	molecular orbital theory, p. 203	single bond, p. 175
	ionic bonds, p. 184		triple bond, p. 175
	Lewis dot symbol, p. 173	multiple bond, p. 175	valence bond theory, p. 176
	Lewis structure, p. 175		

Problems

A Covalent Bond Involves the Sharing of Electrons Between Atoms in a Molecule

3.1 Write Lewis dot symbols for the following elements: Na, Ca, N, B, F, I, Te, and S.

3.2 Write Lewis dot symbols for the following elements: Sr, Si, O, Br, Be, K, Se, and Al.

3.3 Write a Lewis structure for the HCN molecule showing specifically how the octet rule is satisfied.

3.4 Write a Lewis structure for the OCS molecule showing specifically how the octet rule is satisfied.

3.5 Explain the difference between the Lewis structure concept and VB theory.

3.6 Describe the bonding in CN^- using the VB model.

3.7 A reasonable Lewis structure for carbon monoxide (CO) is

$$: C \equiv O :$$

Sketch the orbital overlap in the VB description corresponding to this Lewis structure. In this description, where do the lone pairs reside?

3.8 Discuss how electron spin affects the number of covalent bonds that an atom can form.

Electronegativity Differences Determine the Polarity of Chemical Bonds

3.9 List the following bonds in order of increasing ionic character: the lithium-to-fluorine bond in LiF, the potassium-to-oxygen bond in K_2O, the nitrogen-to-nitrogen bond in N_2, the sulfur-to-oxygen bond in SO_2, the chlorine-to-fluorine bond in ClF_3.

3.10 Arrange the following bonds in order of increasing ionic character: carbon to hydrogen, fluorine to hydrogen, bromine to hydrogen, sodium to chlorine, potassium to fluorine, lithium to chlorine.

3.11 Four atoms are arbitrarily labeled D, E, F, and G. Their electronegativities are as follows: D = 3.8, E = 3.3, F = 2.8, and G = 1.3. If the atoms of these elements form the molecules DE, DG, EG, and DF, how would you arrange these molecules in order of increasing covalent bond character?

3.12 List the following bonds in order of increasing ionic character: cesium to fluorine, chlorine to chlorine, bromine to chlorine, silicon to carbon.

3.13 Classify the following bonds as ionic, polar covalent, or covalent, and give your reasons: (a) the CC bond in H_3CCH_3, (b) the KI bond in KI, (c) the NB bond in H_3NBCl_3, and (d) the CF bond in CF_4.

3.14 Classify the following bonds as ionic, polar covalent, or covalent, and give your reasons: (a) the SiSi bond in $Cl_3SiSiCl_3$, (b) the SiCl bond in $Cl_3SiSiCl_3$, (c) the CaF bond in CaF_2, and (d) the NH bond in NH_3.

3.15 Using the bond energies for the H_2, Cl_2, and HCl molecules from Table 3.3, calculate the electronegativity difference between H and Cl using the Pauling formula. How does it compare to that obtained from Figure 3.10? Explain any differences.

3.16 Using the bond energies for H_2 and Br_2 from Table 3.3, together with the electronegativity difference between H and Br given in Figure 3.10, estimate the bond energy of HBr. Compare your answer to the bond energy for HBr given in Table 3.3.

3.17 Which of the following are ionic compound, and which are covalent compounds: RbCl, PF_5, BrF_3, KO_2, and CI_4?

3.18 An ionic bond is formed between a cation A^+ and an anion B^-. How would the energy of the ionic bond be affected by the following changes: (a) doubling the radius of A^+, (b) tripling the charge on A^+, (c) doubling the charges on A^+ and B^-, and (d) decreasing the radii of A^+ and B^- to half of their original values?

3.19 For each of the following pairs of elements, state whether the binary compound they form is likely to be ionic or covalent. Write the empirical formula and name of the compound: (a) I and Cl, (b) Mg and F, (c) B and F, and (d) K and Br.

3.20 Use ionization energies and electron affinity values to calculate the energy change (in $kJ \ mol^{-1}$) for the following reactions:
(a) $Li(g) + I(g) \longrightarrow Li^+(g) + I^-(g)$
(b) $Na(g) + F(g) \longrightarrow Na^+(g) + F^-(g)$
(c) $K(g) + Cl(g) \longrightarrow K^+(g) + Cl^-(g)$

Drawing Correct Lewis Structures Is an Invaluable Skill for a Chemist

3.21 Write Lewis structures for the following molecules and ions: (a) NCl_3, (b) OCS, (c) H_2O_2, (d) CH_3COO^-, (e) CN^-, and (f) $CH_3CH_2NH_3^+$.

3.22 Write Lewis structures for the following molecules and ions: (a) OF_2, (b) N_2F_2, (c) Si_2H_6, (d) OH^-, (e) CH_2ClCOO^-, and (f) $CH_3NH_3^+$.

3.23 Write Lewis structures for the following molecules: (a) ICl, (b) PH_3, (c) P_4 (each P is bonded to three other P atoms), (d) H_2S, (e) N_2H_4, (f) $HClO_3$, and (g) $COBr_2$ (C is bonded to O and Br atoms).

3.24 Write Lewis structures for the following ions: (a) O_2^{2-}, (b) C_2^{2-}, (c) NO^+, and (d) NH_4^+. Show formal charges.

3.25 Write Lewis structures for the following species, including all resonance forms, and show formal charges: (a) HCO_2, and (b) CH_2NO_2. The relative positions of the atoms are as follows:

O	H	O
H C		C N
O	H	O

3.26 Draw three resonance structures for the chlorate ion, ClO_3^-. Show formal charges.

3.27 Write three resonance structures for hydrazoic acid, HN_3. The atomic arrangement is HNNN. Show formal charges.

3.28 Draw two resonance structures for diazomethane, CH_2N_2. Show formal charges. The skeletal structure of the molecule is

H

C N N

H

3.29 Draw three resonance structures for the OCN$^-$ ion. Show formal charges.

3.30 Draw three resonance structures for the molecule N_2O in which the atoms are arranged in the order NNO. Indicate formal charges.

3.31 Draw resonance structures for the following ions: (a) HSO_4^-, (b) PO_4^{3-}, (c) HSO_3^-, and (d) SO_3^{2-}.

3.32 Draw four reasonable resonance structures for the PO_3F^{2-} ion. The central P atom is bonded to the three O atoms and to the F atom. Show formal charges.

3.33 The triiodide ion (I_3^-) in which the I atoms are arranged along a straight line is stable, but the corresponding F_3^- ion does not exist. Explain.

Molecular Orbital Theory Provides a Detailed Description of Chemical Bonding

3.34 Explain in molecular orbital terms the changes in H—H internuclear distance that occur as the molecular H_2 is ionized first to H_2^+ and then to H_2^{2+}.

3.35 Draw a molecular orbital energy level diagram for each of the following species: He_2, HHe, and He_2^+. Compare their relative stabilities in terms of bond orders. (Treat HHe as a diatomic molecule with three electrons.)

3.36 Arrange the following species in order of increasing stability: Li_2, Li_2^+, and Li_2^-. Justify your choice with a molecular orbital energy-level diagram.

3.37 Which of these species has a longer bond, B_2 or B_2^+? Explain in terms of molecular orbital theory.

3.38 Acetylene (C_2H_2) has a tendency to lose two protons (H) and form the carbide ion C_2^{2-}, which is present in a number of ionic compounds, such as CaC_2 and MgC_2. Describe the bonding scheme in the C_2^{2-} ion in terms of molecular orbital theory. Compare the bond order in C_2^{2-} with that in C_2.

3.39 Compare the Lewis and molecular orbital treatments of the oxygen molecule.

3.40 Explain why the bond order of N_2 is greater than that of N_2^+, but the bond order of O_2 is less than that of O_2^+.

3.41 Compare the relative stability of the following species and indicate their magnetic properties (that is, diamagnetic or paramagnetic): O_2, O_2^+, O_2^- (superoxide ion), and O_2^{2-} (peroxide ion).

3.42 Use MO theory to compare the relative stabilities of F_2 and F_2^+. Which of these species will have the shorter bond length?

3.43 A single bond is almost always a sigma bond, and a double bond is almost always made up of a sigma bond and a pi bond. There are very few exceptions to this rule. Show that the B_2 and C_2 molecules are examples of the exceptions.

Additional Problems

3.44 The equilibrium C—H bond distances in C_2H_6, C_2H_4, and C_2H_2 are 1.102 Å, 1085 Å, and 1.061 Å, respectively. Can you offer an explanation of this trend?

3.45 Write Lewis structures for BrF_3, ClF_5, and IF_7. Identify those in which the octet rule is not obeyed.

3.46 Write three resonance structures for the azide ion, N_3^-, in which the atoms are arranged as NNN. Show formal charges.

3.47 The amide group plays an important role in determining the structure of proteins:

$$\underset{\underset{H}{|}}{-C}\overset{\overset{O}{\|}}{}-N-$$

Draw another resonance structure for this group. Show formal charges.

3.48 Give an example of an ion or molecule containing Al that (a) obeys the octet rule, (b) has an expanded octet, and (c) has an incomplete octet.

3.49 Attempts to prepare CF_2, LiO_2, $CsCl_2$, and PI_5 as stable species under atmospheric conditions have failed. Suggest possible reasons for the failure.

3.50 A rule for drawing plausible Lewis structures is that the central atom is invariably less electronegative than the surrounding atoms. Explain why this is so.

3.51 Which of the following molecules has the shortest nitrogen-to-nitrogen bond: N_2H_4, N_2O, N_2, and N_2O_4. Explain.

3.52 Most organic acids can be represented as RCOOH, where COOH is the carboxyl group and R is the rest of the molecule. (For example, R is CH_3 in acetic acid, CH_3COOH.) (a) Draw a Lewis structure for the carboxyl group. (b) Upon ionization, the carboxyl group is converted to the carboxylate group, COO$^-$. Draw resonance structures for the carboxylate group.

3.53 Which of the following species are isoelectronic: NH_4^+, C_6H_6, CO, CH_4, N_2, and $B_3N_3H_6$?

3.54 The following species have been detected in interstellar space: (a) CH, (b) OH, (c) C_2, (d) HNC, and (e) HCO. Draw Lewis structures for these species and indicate whether they are diamagnetic or paramagnetic.

3.55 Draw Lewis structures for the following organic molecules: (a) tetrafluoroethylene (C_2F_4), (b) propane (C_3H_8), (c) butadiene ($CH_2CHCHCH_2$), (d) propyne (CH_3CCH), and (e) benzoic acid (C_6H_5COOH). (To draw C_6H_5COOH, replace an H atom in benzene with a COOH group.)

3.56 Compare the bond energy of F_2 with the energy change for the following process:

$$F_2(g) \longrightarrow F^+(g) + F^-(g)$$

Which is the preferred dissociation for F_2, energetically speaking?

3.57 Methyl isocyanate (CH_3NCO) is used to make certain pesticides. In December 1984, water leaked into a tank containing this substance at a chemical plant, producing a toxic cloud that killed thousands of people in Bhopal, India. Draw Lewis structures for CH_3NCO, showing formal charges.

3.58 The chlorine nitrate molecule ($ClONO_2$) is believed to be involved in the destruction of ozone in the Antarctic stratosphere. Draw a plausible Lewis structure for this molecule.

3.59 Several resonance structures for the molecule CO_2 are shown. Explain why some of them are likely to be of little importance in describing the bonding in this molecule.

(a) $\overset{..}{O}{=}C{=}\overset{..}{O}$ (b) $:O{\equiv}\overset{+}{C}{-}\overset{.. \; -}{\underset{..}{O}}:$

(c) $:O{\equiv}\overset{+}{C}{-}\overset{\; -}{\underset{..}{O}}:$ (d) $:\overset{.. \; -}{\underset{..}{O}}{-}\overset{2+}{C}{-}\overset{.. \; -}{\underset{..}{O}}:$

3.60 For each of the following organic molecules draw a Lewis structure in which the carbon atoms are bonded to each other by single bonds: (a) C_2H_6, (b) C_4H_{10}, and (c) C_5H_{12}. For (b) and (c), show only structures in which each C atom is bonded to no more than two other C atoms.

3.61 Draw Lewis structures for the following chlorofluorocarbons (CFCs), which are partly responsible for the depletion of ozone in the stratosphere: (a) $CFCl_3$, (b) CF_2Cl_2, (c) CHF_2Cl, and (d) CF_3CHF_2.

3.62 Draw Lewis structures for the following organic molecules: C_2H_3F, C_3H_6, and C_4H_8. In each there is one C=C bond, and the rest of the carbon atoms are joined by C—C bonds.

3.63 Draw Lewis structures for the following organic molecules: (a) methanol (CH_3OH); (b) ethanol (CH_3CH_2OH); (c) tetraethyllead [$Pb(CH_2CH_3)_4$], which is used in "leaded gasoline"; (d) methylamine (CH_3NH_2), which is used in tanning animal skins; (e) mustard gas ($ClCH_2CH_2SCH_2CH_2Cl$), a poisonous gas used in World War I; (f) urea [$(NH_2)_2CO$], a fertilizer; and (g) glycine (NH_2CH_2COOH), an amino acid.

3.64 Write Lewis structures for the following four isoelectronic species: (a) CO, (b) NO^+, (c) CN^-, and (d) N_2. Show formal charges.

3.65 Oxygen forms three types of ionic compounds, in which the anions are oxide (O^{2-}), peroxide (O_2^{2-}), and superoxide (O_2^-). Draw Lewis structures of these ions.

3.66 Comment on the correctness of the statement, "All compounds containing a noble gas atom violate the octet rule."

3.67 Write three resonance structures for (a) the cyanate ion (NCO^-) and (b) the isocyanate ion (CNO^-). In each case, rank the resonance structures in order of increasing importance.

3.68 The resonance concept is sometimes described by analogy to a mule, which is a cross between a horse and a donkey. Compare this analogy with the one used in Chapter 3, that is, the description of a rhinoceros as a cross between a griffin and a unicorn. Which description is more appropriate? Why?

3.69 The N—O bond distance in nitric oxide is 115 pm, which is intermediate between a triple bond (106 pm) and a double bond (120 pm). (a) Draw two resonance structures for NO and comment on their relative importance. (b) Is it possible to draw a resonance structure having a triple bond between the atoms?

3.70 Although nitrogen dioxide (NO_2) is a stable compound, there is a tendency for two such molecules to combine to form dinitrogen tetroxide (N_2O_4). Why? Draw four resonance structures of N_2O_4, showing formal charges.

3.71 Another possible skeletal structure for the CO_3^{2-} (carbonate) ion besides the one presented in Example 3.3 is O C O O. Why would we not use this structure to represent CO_3^{2-}.

3.72 Draw a Lewis structure for dinitrogen pentoxide (N_2O_5) in which each N is bonded to three O atoms.

3.73 In the gas phase, aluminum chloride exists as a dimer (a unit of two) with the formula Al_2Cl_6. Its skeletal structure is given by

$$
\begin{array}{ccccc}
 & Cl & & Cl & \\
Cl & Al & & Al & Cl \\
 & Cl & & Cl &
\end{array}
$$

Complete the Lewis structure and indicate the coordinate covalent bonds in the molecule.

3.74 The hydroxyl radical (OH) plays an important role in atmospheric chemistry. It is highly reactive and has a tendency to combine with an H atom from other compounds, causing them to break up. Thus, OH is sometimes called a "detergent" radical because it helps to clean up the atmosphere. (a) Write the Lewis structure for the radical. (b) Refer to the discussion on p. 181 and explain why the radical has a high affinity for H atoms. (c) The radical is generated when sunlight hits water vapor. Calculate the maximum wavelength (in nanometers) required to break one O—H bond in H_2O to form the OH radical.

3.75 As discussed in Section 3.3, the total potential energy of interaction, V_{total}, between two ions of charge $+e$ and $-e$ is the sum of the attractive Coulomb interaction between the charges

$$V_{attraction} = -\frac{e^2}{4\pi\varepsilon_0 r}$$

and a repulsive interaction, $V_{repulsive}(r)$, where r is the ionic separation distance. Often the repulsive interaction is modeled by a term of the form b/r^n, where b is a constant specific for a given ion pair and n is an integer typically between 8 and 12. For this model we have

$$V_{total} = V_{attraction} + V_{repulsion} = -\frac{e^2}{4\pi\varepsilon_0 r} + \frac{b}{r^n}$$

(a) Sketch this function and identify the equilibrium bond length.

(b) If the equilibrium bond length (r_B) is given, show that the minimum value of the potential energy, which gives the energy released when an ion pair forms from the separated ions, is given by

$$V_B = -\frac{e^2}{4\pi\varepsilon_0 r_B}\left(1 - \frac{1}{n}\right)$$

(c) Using the result in (b) and data from Figure 2.20, estimate the energy required to separate 1 mole of NaBr gas-phase ion pairs into separated sodium and bromine atoms.

3.76 Draw three resonance structures of sulfur dioxide (SO_2). Indicate the most plausible structure(s).

3.77 Vinyl chloride (C_2H_3Cl) differs from ethylene (C_2H_4) in that one of the H atoms is replaced with a Cl atom. Vinyl chloride is used to prepare poly(vinyl chloride), which is an important polymer used in pipes. (a) Draw the Lewis structure of vinyl chloride. (b) The repeating unit in poly(vinyl chloride) is —CH_2—$CHCl$—. Draw a portion of the molecule showing three such repeating units.

3.78 In 1998 scientists using a special type of electron microscope were able to measure the force needed to break a *single* chemical bond. If 2.0×10^{-9} N was needed to break a C—Si bond, estimate the bond dissociation energy in kJ mol^{-1}. Assume that the bond had to be stretched by a distance of 2 Å (2×10^{-10} m) before it was broken.

3.79 Among the common inhaled anesthetics are:

halothane: $CF_3CHClBr$

enflurane: $CHFClCF_2OCHF_2$

isoflurane: $CF_3CHClOCHF_2$

methoxyflurane: $CHCl_2CF_2OCH_3$.

Draw Lewis structures of these molecules.

3.80 A student in your class claims that magnesium oxide actually consists of Mg^+ and O^- ions, not Mg^{2+} and O^{2-} ions. Suggest some experiments one could do to show that your classmate is wrong.

3.81 Use the molecular orbital energy-level diagram for O_2 to show that the following Lewis structure corresponds to an excited state:

$$\ddot{O}\!=\!\ddot{O}$$

3.82 Assume that the third-period element phosphorus forms a diatomic molecule, P_2, in an analogous way as nitrogen does to form N_2. (a) Write the electron configuration for P_2. Use [Ne_2] to represent the electron configuration for the first two periods. (b) Calculate its bond order. (c) What are its magnetic properties (diamagnetic or paramagnetic)?

3.83 Use molecular orbital theory to explain the differences between the bond energies and bond lengths of F_2 and F_2^-.

3.84 Predict which of the following species will have the longest bond: CN^+, CN, or CN^-.

3.85 Describe the bonding in NO^+, NO, and NO^- using MO theory. How will their bond dissociation energies and bond lengths compare?

3.86 Compare the MO theory description for the H_2 molecule given by Equation 3.10 with the VB theory treatment given in Equation 3.1. Under what condition do they become identical?

3.87 Draw MO diagrams and give the bond order for the diatomic molecules BN, CO, OF, and LiO. Which are paramagnetic?

3.88 Draw MO diagrams and give the bond order for the diatomic ions BN^-, CO^-, and OF^+.

3.89 Recalling the particle-in-a-box discussion in Chapter 1, explain why the energy of the VB wavefunction for H_2 shown in Equation 3.1 is lower than the energy that would be obtained if we neglected the second term on the right-hand side of Equation 3.1.

3.90 The $J = 0 \rightarrow J = 1$ line in the microwave spectrum of $^{35}Cl^{19}F$ is found experimentally to have a frequency of 3.06×10^{10}Hz. Following the procedure outlined in the inset on page 186–187, calculate the bond length of ClF from these data.

3.91 Sulfuric acid (H_2SO_4), the most important industrial chemical in the world, is prepared by oxidizing sulfur to sulfur dioxide and then to sulfur trioxide. Although sulfur trioxide reacts with water to form sulfuric acid, it forms a mist of fine droplets of H_2SO_4 with water vapor that is hard to condense. Instead, sulfur trioxide is first dissolved in 98 percent sulfuric acid to form oleum ($H_2S_2O_7$). On treatment with water, concentrated sulfuric acid can be generated. Write equations for all of the steps and draw Lewis structures of oleum based on the discussion in Example 3.9.

3.92 The species H_3^+ is the simplest polyatomic ion. The geometry of the ion is that of an equilateral triangle. Draw three resonance structures to represent the ion.

3.93 The bond energy of the C—N bond in the amide group of proteins (see Problem 3.48) can be treated as an average of C—N and C=N bonds. Calculate the maximum wavelength of light needed to break the bond.

3.94 In 1999 an unusual cation containing only nitrogen (N_5^+) was prepared. Draw three resonance structures of the ion, showing formal charges. (*Hint:* The N atoms are joined in a linear fashion.)

3.95 The Mulliken electronegativity was defined in Equation 3.3. Calculate the electronegativities of O, F, and Cl using the Mulliken definition. Compare the electronegativities of these elements on the Mulliken and Pauling scales. (To convert to the Pauling scale, divide each Mulliken electronegativity value by 230 kJ mol^{-1}.)

3.96 For each pair listed, state which one has a higher first ionization energy and explain your choice: (a) H or H_2, (b) N or N_2, (c) O or O_2, and (d) F or F_2.

3.97 The molecule benzyne (C_6H_4) is a very reactive species. It resembles benzene in that it has a six-membered ring of carbon atoms. Draw a Lewis structure of the molecule and account for the molecule's high reactivity.

3.98 Consider an N_2 molecule in its first excited electronic state, that is, when an electron in the highest occupied molecular orbital is promoted to the lowest unoccupied molecular orbital. (a) Identify the molecular orbitals involved and sketch a diagram for the transition. (b) Compare the bond order and bond length of $N_2^\star$ with N_2, where the asterisk denotes the excited molecule. (c) Is $N_2^\star$ diamagnetic or paramagnetic? (d) When $N_2^\star$ loses its excess energy and converts to the ground state, it emits a photon of wavelength 470 nm, which makes up part of the auroras lights. Calculate the energy difference between these levels.

3.99 Mathematically, the valence bond and molecular orbital descriptions of bonding in H_2 are described in Equations 3.1 and 3.10, respectively. The MO expression contains terms that are not contained in the VB description. What are these terms? Physically, what do they represent?

Answers to Practice Exercises

3.1 (a) ionic, (b) polar covalent, (c) covalent

3.2 H—C(=O)—O—H **3.3** $\left[:\ddot{O}—N=\ddot{O}\right]^-$ **3.4** H—C≡N:

3.5 :Ö—N=Ö **3.6** $\left[\ddot{O}=N—\ddot{O}: \longleftrightarrow :\ddot{O}—N=\ddot{O}\right]^-$

3.7 :F̈—Be—F̈: **3.8** structure of Cl with F atoms

3.9 O—S—O—H (with O and H substituents) **3.10** :F̈—S—F̈: (with F substituents) **3.11** F_2^-

3.12

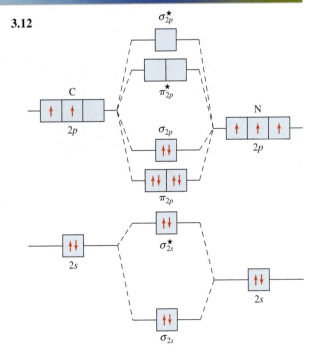

Molecular Structure and Interaction

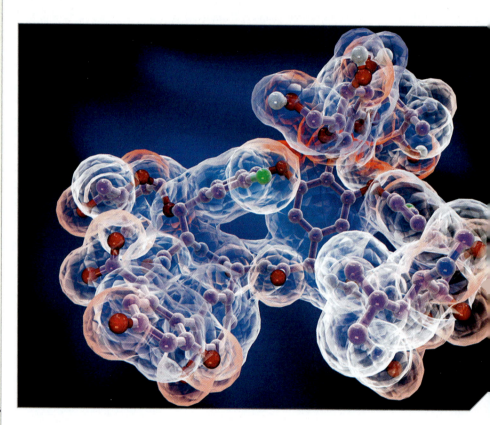

In Chapter 3, we introduced Lewis dot structures as a tool with which to predict the bonding patterns in a molecule. In Chapter 4, we will study the shape, or geometry, of molecules. Geometry has an important influence on the physical and chemical properties of molecules, such as melting point, boiling point, and reactivity. We will see that we can predict the shapes of molecules with considerable accuracy using a simple method based on Lewis structures.

Also, in Chapter 3 we introduced the concept of molecular-orbital theory to explain bonding in diatomic molecules. In Chapter 4, we will extend this useful quantum-mechanical concept to polyatomic molecules. In addition, in the final section of Chapter 4, we will examine how molecular shape and bonding affect the interactions of molecules with one another.

4.1 | The Basic Three-Dimensional Structure of a Molecule Can Be Predicted Using the VSEPR Model

Molecular geometry is the three-dimensional arrangement of atoms in a molecule. The geometry of a molecule has a profound effect on its physical and chemical properties, such as melting point, boiling point, density, and the types of reactions it undergoes. In general, the exact geometry of a compound (bond lengths, bond angles, etc.) must be determined by experiment. However, there is a simple procedure that enables us to predict with considerable success the overall geometry of a molecule or ion if we know the number of electron pairs in the valence shell surrounding a central atom in its Lewis structure. The *valence shell* is *the outermost electron-occupied shell of an atom; it holds the electrons that are usually involved in chemical bonding.* The basis of this approach is the idea that the geometry of a molecule corresponds to the arrangement of its atoms that minimizes the total energy of the molecule. In a polyatomic molecule, there are two or more covalent bonds between the central atom and the surrounding atoms. Because of the strong electrostatic repulsion between the electron pairs in different bonds, minimizing the total energy leads to a geometry in which these electron pairs are kept as far apart as possible so as to minimize the repulsion. This qualitative approach to the study of molecular geometry is called the *valence-shell electron-pair repulsion (VSEPR) model* because *it accounts for the geometric arrangements of electron pairs around a central atom in terms of the electrostatic repulsion between electron pairs.*

Within the VSEPR model, the basic molecular geometry about a given central atom is uniquely determined from two numbers:

▶ The *steric*[1] *number* (N_S), which gives the total number of electron pairs surrounding the central atom. Each lone pair counts as an individual nonbonding electron pair. As far as the VSEPR method is concerned, the multiple electron pairs in double and triple bonds are counted as a single electron pair, so the number of bonding electron pairs is equal to the number of atoms bonded to the central atom. The steric number is then the sum of the number of lone pairs and the number of atoms bonded to the central atom:

$$N_S = (\text{\# of lone pairs}) + (\text{\# of atoms bonded to central atom})$$

▶ The number of lone pairs (nonbonding valence electron pairs) on the central atom.

How, then, do you arrange N_S objects about a central point so as to maximize the distances between the objects? The minimum energy (maximum distance) geometries for values of N_S from two to six are shown in Figure 4.1.

VSEPR for Molecules with No Lone Pairs on the Central Atom

For simplicity we will consider molecules that contain atoms of only two elements, A and B, of which A is the central atom. These molecules have the general formula AB_x, where x is an integer 2, 3, ... (If $x = 1$, we have the diatomic molecule AB, which is linear by definition.) In the vast majority of cases, x is between two and six. If there are no lone pairs, the *molecular geometry* (the arrangement of neighboring

---

1. The word *steric* is an adjective meaning "of or related to the three-dimensional arrangement of atoms in a molecule."

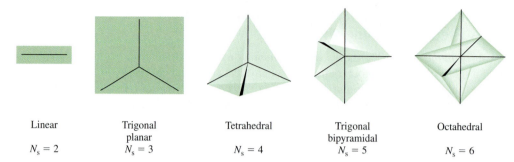

Linear	Trigonal planar	Tetrahedral	Trigonal bipyramidal	Octahedral
$N_s = 2$	$N_s = 3$	$N_s = 4$	$N_s = 5$	$N_s = 6$

Figure 4.1 The five basic three-dimensional arrangements of electron pair sets around a central atom according to the VSEPR model.

atoms around the central atom) is the same as the *electron-pair geometry* (the arrangement of the electron pairs about the central atom) shown in Figure 4.1, and x is equal to the steric number N_s. Using Figure 4.1 as a reference, let us now take a close look at the geometry of molecules with the formulas AB_2, AB_3, AB_4, AB_5, and AB_6:

▶ **AB_2, $N_S = 2$:** *beryllium chloride* ($BeCl_2$). The Lewis structure of the molecule beryllium chloride is

$$:\overset{..}{\underset{..}{Cl}}—Be—\overset{..}{\underset{..}{Cl}}:$$

Maximizing the distance between the electrons in the Be—Cl bonds is accomplished by placing them on opposite sides of the central beryllium atom. Thus, the Cl—Be—Cl bond angle is 180° and the molecule is linear—consistent with Figure 4.1.

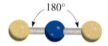

Another example of a linear molecule with no lone pairs on the central atom is carbon dioxide (CO_2).

▶ **AB_3, $N_S = 3$:** *boron trifluoride* (BF_3). The Lewis structure for BF_3 is

$$
\begin{array}{c}
:\overset{..}{F}: \\
| \\
:\overset{..}{\underset{..}{F}}—B \\
| \\
:\overset{..}{\underset{..}{F}}:
\end{array}
$$

The molecule has three B—F single bonds and no lone pairs so $N_S = 3$, which from Figure 4.1 gives a *trigonal planar* geometry as the most stable arrangement. The three B—F bonds point to the corners of an equilateral triangle with boron in the center.

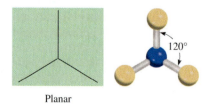

Planar

Each of the three F—B—F angles is 120°, and all four atoms lie in the same plane.

▶ **AB₄, $N_S = 4$:** *methane* (CH_4). The Lewis structure of methane is

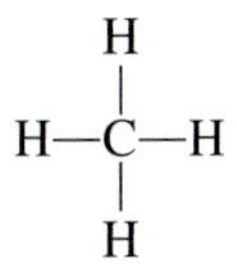

$$
\begin{array}{c}
\text{H} \\
| \\
\text{H—C—H} \\
| \\
\text{H}
\end{array}
$$

Because there are four bonding pairs, the geometry of methane is tetrahedral. A *tetrahedron* has four faces (the prefix *tetra* means "four"), all of which are equilateral triangles. In a tetrahedral bonding geometry, the central atom (carbon in this case) is located at the center of the tetrahedron and the other four atoms are located at the corners (or *vertices*). All H—C—H bond angles are 109.5°.

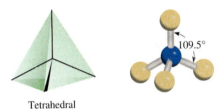

Tetrahedral

▶ **AB₅, $N_S = 5$:** *phosphorus pentachloride* (PCl_5). The Lewis structure of PCl_5 is

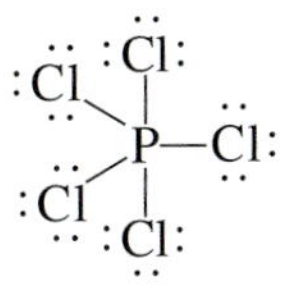

$$
\begin{array}{c}
:\overset{..}{\text{Cl}}: \\
:\overset{..}{\text{Cl}}\diagdown \overset{|}{} \\
\text{P—}\overset{..}{\underset{..}{\text{Cl}}}: \\
:\overset{..}{\text{Cl}}\diagup \overset{|}{} \\
:\overset{..}{\text{Cl}}:
\end{array}
$$

From Figure 4.1, the minimum energy geometry of an atom surrounded by five bonding pairs is *trigonal bipyramidal,* which can be constructed by joining two tetrahedra at a common triangular base:

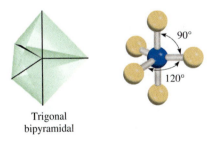

Trigonal
bipyramidal

The central atom (phosphorous in this case) is at the center of the common triangle with the surrounding atoms positioned at the five corners of the trigonal bipyramid. The atoms that are above and below the triangular plane are said to occupy *axial* positions, and those that are in the triangular plane are said to occupy *equatorial* positions. The angle between any two equatorial bonds is 120°, that between an axial bond and an equatorial bond is 90°, and that between the two axial bonds is 180°. In general, axial bonds are longer than equatorial bonds.

▶ **AB_6, $N_S = 6$:** *sulfur hexafluoride* (SF_6). The Lewis structure of sulfur hexafluoride is

$$
\begin{array}{c}
\ddot{\text{F}} \quad \ddot{\text{F}} \\
\vdots\ddot{\text{F}} \diagdown \mid \diagup \ddot{\text{F}}\vdots \\
\text{S} \\
\vdots\ddot{\text{F}} \diagup \mid \diagdown \ddot{\text{F}}\vdots \\
\ddot{\text{F}}
\end{array}
$$

The most stable arrangement of the six S—F bonding pairs is the octahedron shown in Figure 4.1. An octahedron has eight sides (the prefix *octa* means "eight") and can be generated by joining two square pyramids on a common base. The central atom (sulfur in this case) is at the center of the square base and the surrounding atoms are at the six corners. All bond angles are 90° except the one made by the bonds between the central atom and the pairs of atoms that are opposite each other. That angle is 180°. The six bonds in an octahedral molecule are equivalent, so the terms "axial" and "equatorial" do not apply.

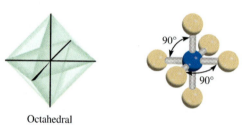

Octahedral

The molecular geometry of molecules with no lone pairs on the central atom is summarized in Table 4.1.

VSEPR for Molecules with Lone Pairs on the Central Atom

Determining the geometry of a molecule is more complicated if the central atom has lone pair electrons. For such molecules the electron-pair geometry (see Figure 4.1), which describes the arrangement of both bonds *and* lone pairs, differs from the molecular geometry, which describes the arrangement of the bonded atoms only. Another complication is that lone pairs occupy a larger region of space than do bonding electron pairs, so they exert greater repulsion on neighboring lone pairs and bonding pairs. To keep track of the total number of bonding pairs and lone pairs, we designate molecules with lone pairs as AB_xE_y, where A is the central atom, B is a surrounding atom, and E is a lone pair on A. Both x and y are integers; $x = 2, 3, \ldots$, and $y = 1, 2, \ldots$. Thus, the values of x and y indicate the number of surrounding atoms and number of lone pairs on the central atom, respectively. The simplest such molecule is a triatomic molecule with one lone pair on the central atom and the formula AB_2E.

For $x = 1$, we have a diatomic molecule, which is by definition linear.

As the following examples illustrate, in most cases, the presence of lone pairs on the central atom makes it difficult to predict the bond angles precisely.

▶ **AB_2E, $N_S = 3$, one lone pair:** *sulfur dioxide* (SO_2). The Lewis structure of sulfur dioxide is

$$
\ddot{\text{O}}{=}\ddot{\text{S}}{-}\ddot{\text{O}}{:} \longleftrightarrow {:}\ddot{\text{O}}{-}\ddot{\text{S}}{=}\ddot{\text{O}}
$$

Because VSEPR treats double bonds as a single electron pair set, the steric number for SO_2 is three (two bonding electron pair sets and one lone pair).

Table 4.1	Electron-Pair and Molecular Geometries of Simple Molecules and Ions in Which the Central Atom Has No Lone Pairs		
Number of Electron Pairs	**Arrangement of Electron Pairs***	**Molecular Geometry***	**Examples**
2	180° :—A—: Linear	B—A—B Linear	$BeCl_2$, $HgCl_2$
3	120° A Trigonal planar	B A B B Trigonal planar	BF_3
4	109.5° A Tetrahedral	B A B B B Tetrahedral	CH_4, NH_4^+
5	90° A 120° Trigonal bipyramidal	B B A B B B Trigonal bipyramidal	PCl_5
6	90° 90° A Octahedral	B B A B B B Octahedral	SF_6

*The colored lines are used only to show the overall shapes; they do not represent bonds.

From Figure 4.1 the electron pair geometry is trigonal planar. Because one of the electron pairs is a lone pair, however, which is not counted in the molecular geometry, the SO_2 molecule has a "bent" geometry around the S atom.

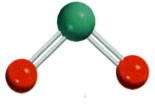

SO_2

Because the repulsion between the lone pair and the bonding pairs is greater than that between the two bonding pairs, the S—O bonds are pushed together slightly and the O—S—O angle is less than 120°. *Although the VSEPR model can predict that the bond angle in SO_2 is somewhat less than the ideal trigonal angle of 120°, it cannot predict the exact value of the bond angle. Experiments or sophisticated quantum mechanical calculations must be used to find the exact bond angle.*

▶ **AB_3E, $N_S = 4$, one lone pair:** *ammonia* (NH_3). The ammonia molecule contains three bonding pairs and one lone pair:

$$H-\overset{\displaystyle ..}{\underset{\displaystyle |}{N}}-H$$
$$H$$

According to Figure 4.1, the overall arrangement of four electron pairs is tetrahedral. One of the electron pairs in NH_3 is a lone pair, however, so the molecular geometry of NH_3 is trigonal pyramidal (so called because it looks like a pyramid, with the central nitrogen atom at the apex).

$$\overset{\displaystyle \ddot{N}}{H\diagup \underset{\displaystyle H}{|} \diagdown H}$$

(Note that, in this drawing different types of bond lines are used to indicate a three-dimensional perspective—dashed lines represent bond axes extending behind the plane of the paper and wedged lines represent bond axes extending in front of the plane of the paper, and the thin solid lines represent bonds in the plane of the paper.) Because the lone pair repels the bonding pairs more strongly, the three N—H bonding pairs are pushed closer together. Thus, the H—N—H angle in ammonia is smaller than the ideal tetrahedral angle of 109.5° (Figure 4.2).

Figure 4.2 (a) The relative sizes of bonding pairs and lone pairs in CH_4, NH_3, and H_2O. (b) The bond angles in CH_4, NH_3, and H_2O.

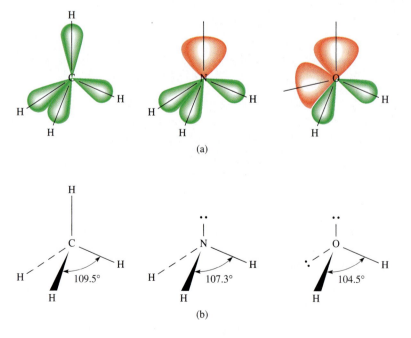

▶ **AB₂E₂, N_S = 4, two lone pairs:** *water* (H₂O). A water molecule contains two bonding pairs and two lone pairs:

The overall arrangement of the four electron pairs in water is tetrahedral, the same as in ammonia. Unlike ammonia, however, water has two lone pairs on the central oxygen atom. These lone pairs tend to be as far from each other as possible. Consequently, the two O—H bonding pairs are pushed toward each other, and we predict an even greater deviation from the tetrahedral angle than in NH₃. As Figure 4.2 shows, the H—O—H angle is 104.5°, and the molecular geometry of H₂O is bent:

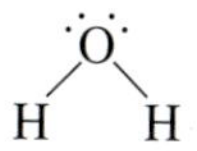

Water, H₂O, H—O—H

▶ **AB₄E, N_S = 5, one lone pair:** *sulfur tetrafluoride* (SF₄). The Lewis structure of SF₄ is

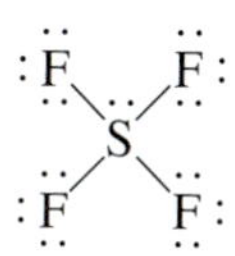

The central sulfur atom has five electron pairs whose arrangement, according to Figure 4.1, is trigonal bipyramidal. In the SF₄ molecule, however, one of the electron pairs is a lone pair, so the molecule must have one of the following geometries:

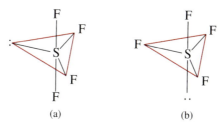

(a)   (b)

In (a) the lone pair occupies an equatorial position, and in (b) it occupies an axial position. The axial position has three neighboring pairs at 90° and one at 180°, whereas the equatorial position has two neighboring pairs at 90° and two at 120°. The electron pair repulsion is smaller if the lone pair is in the equatorial position (a) and indeed (a) is the structure observed experimentally. The molecular geometry is generally referred to as a "see-saw" shape. The angle between the axial fluorine atoms and the sulfur is 173° and that between the equatorial F atoms is 102°.

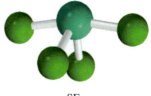

SF₄

Similar arguments can be used to predict other geometries:

▶ **AB₃E₂, N_S = 5, two lone pairs:** chlorine trifluoride (ClF₃).

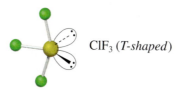

ClF₃ (*T-shaped*)

▶ **AB₂E₃, N_S = 5, three lone pairs:** The *triiodide Ion* (I_3^-).

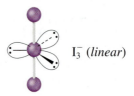

I_3^- (*linear*)

▶ **AB₅E, N_S = 6, one lone pair:** *bromine pentafluoride* (BrF₅).

BrBF₅ (*square pyramidal*)

▶ **AB₄E₂, N_S = 6, two lone pairs:** *xenon tetrafluoride* (XeF₄).

XeF₄ (*square planar*)

The VSEPR geometries of molecules with lone pairs around the central atom are summarized in Table 4.2.

Geometry of Molecules with More Than One Central Atom

So far we have discussed the geometry of molecules having only one central atom. The overall geometry of molecules with more than one central atom is difficult to define in most cases. Often we can only describe the shape around each of the central atoms. Methanol (CH₃OH), for example, has the following Lewis structure:

$$
\begin{array}{c}
\text{H} \\
| \\
\text{H—C—Ö—H} \\
| \\
\text{H}
\end{array}
$$

The two central (nonterminal) atoms in methanol are carbon and oxygen. The three C—H and the C—O bonding pairs are tetrahedrally arranged about the carbon atom. The H—C—H and O—C—H bond angles are approximately 109.5°. The oxygen atom here is like the one in water in that it has two lone pairs and two bonding pairs. Therefore, the H—O—C portion of the molecule is bent, and the H—O—C bond angle is approximately equal to 105° (Figure 4.3).

Guidelines for Applying the VSEPR Model

Having studied the geometries of molecules in two categories (central atoms with and without lone pairs), it is now time to consider the following rules for applying the VSEPR model to all types of molecules:

1. Write the Lewis structure of the molecule, considering only the electron pairs around the central atom (that is, the atom that is bonded to more than one other atom).

Figure 4.3 The geometry of CH₃OH (methanol).

Table 4.2	Geometry of Simple Molecules and Ions in Which the Central Atom Has One or More Lone Pairs					
Class of Molecule	Total Number of Electron Pairs	Number of Bonding Pairs	Number of Lone Pairs	Arrangement of Electron Pairs*	Geometry of Molecule or Ion	Examples
AB_2E	3	2	1	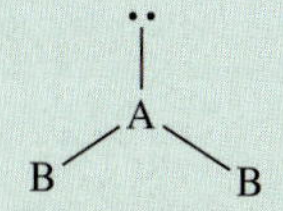 Trigonal planar	Bent	 SO_2
AB_3E	4	3	1	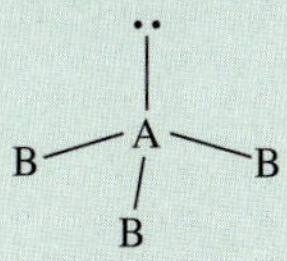 Tetrahedral	Trigonal pyramidal	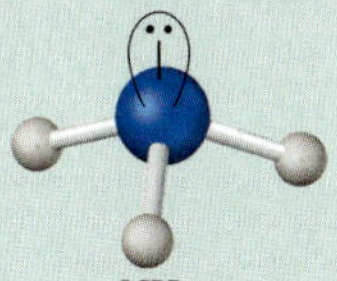 NH_3
AB_2E_2	4	2	2	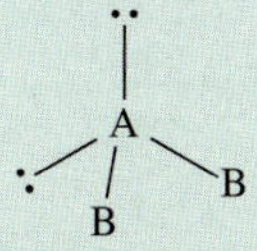 Tetrahedral	Bent	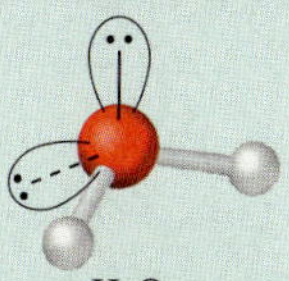 H_2O
AB_4E	5	4	1	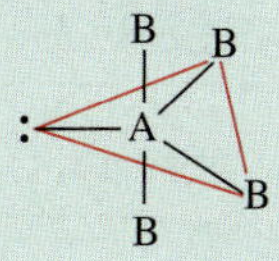 Trigonal bipyramidal	Distorted tetrahedron (or seesaw)	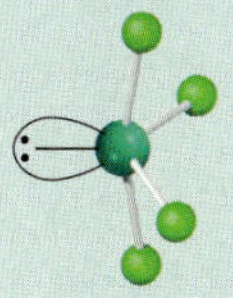 SF_4
AB_3E_2	5	3	2	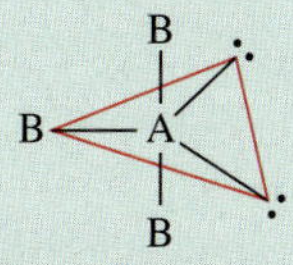 Trigonal bipyramidal	T-shaped	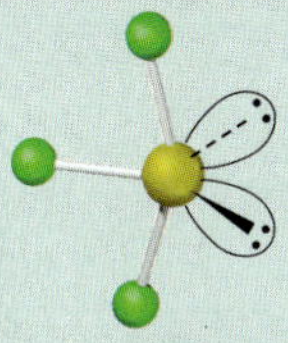 ClF_3
AB_2E_3	5	2	3	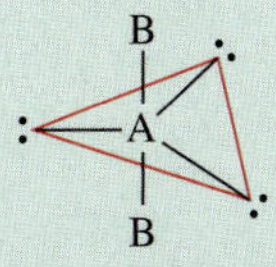 Trigonal bipyramidal	Linear	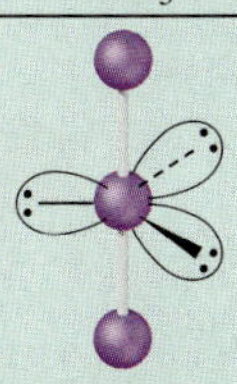 I_3^-
AB_5E	6	5	1	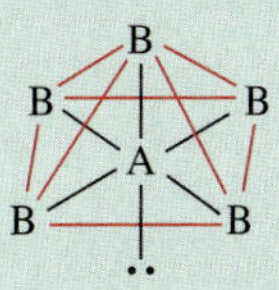 Octahedral	Square pyramidal	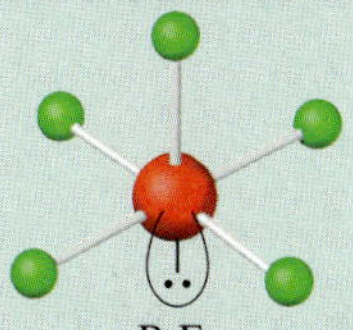 BrF_5
AB_4E_2	6	4	2	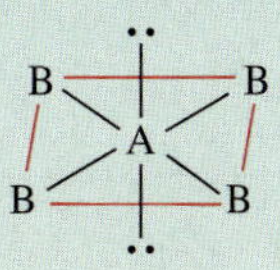 Octahedral	Square planar	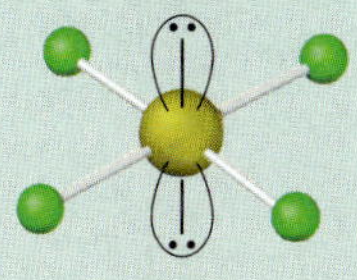 XeF_4

*The colored lines are used to show the overall shape, not bonds.

2. Determine the steric number (N_S) by counting the number of electron pairs around the central atom (bonding pair sets and lone pairs). The electron pairs in double and triple bonds are treated as a single bonding electron pair. Refer to Figure 4.1 to predict the overall arrangement of the electron pairs.

3. Use Tables 4.1 and 4.2 to predict the geometry of the molecule.

4. When predicting bond angles, remember that a lone pair repels another lone pair or a bonding pair more strongly than a bonding pair repels another bonding pair. In general, there is no easy way to predict bond angles accurately when the central atom possesses one or more lone pairs.

5. If a molecule has two or more resonance structures, we can apply the VSEPR model to any one of them to determine the molecular geometry.

The VSEPR model reliably predicts the geometry of many molecules and polyatomic ions. Chemists use the VSEPR approach because of its simplicity. Although there are some theoretical concerns about whether "electron-pair repulsion" actually determines molecular shapes, the assumption that it does leads to useful (and generally reliable) predictions. Example 4.1 illustrates the application of VSEPR.

Example 4.1

Use the VSEPR model to predict the geometry of (a) AsH_3, (b) NO_2, (c) SF_5^-, (d) CCl_3F, and (e) C_2H_4.

Strategy The sequence of steps in determining molecular geometry is as follows:

draw Lewis structure $\longrightarrow$ determine N_S $\longrightarrow$ find arrangement of electron pairs $\longrightarrow$

find arrangement of bonding pairs $\longrightarrow$ determine geometry based on bonding pairs

Solution (a) The Lewis structure of AsH_3 is

$$H-\overset{..}{\underset{|}{As}}-H$$
$$H$$

AsH_3

The steric number of arsenic hydride is four (three bonding pairs and one lone pair), so the electron-pair geometry is tetrahedral (see Figure 4.2). The molecular geometry does not include the lone pair, which gives a trigonal pyramidal geometry for the bonding pairs, as in NH_3 (see Table 4.1). We cannot accurately predict the H—As—H bond angle, but we know that it is less than the ideal tetrahedral angle of 109.5° because the repulsion of the bonding electron pairs in the As—H bonds by the lone pair on As is greater than repulsion between the bonding pairs.

(b) The Lewis structure of NO_2 is

$$\overset{..}{O}=N-\overset{..}{\underset{..}{O}}:$$

There are two equivalent resonance structures for NO_2; for the purposes of VSEPR, we can use either one. NO_2 is different from the examples considered up to now in that it has an odd number of electrons and the Lewis structure has an unpaired lone electron on the nitrogen. For the purposes of determining the basic VSEPR

—Continued

Continued—

geometry, we treat this single lone electron in the same way as lone pair electrons are considered. Thus, $N_S = 3$, and the electron "pair" geometry is trigonal planar. The molecular geometry is bent, similar to SO_2.

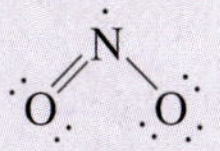

The actual bond angle cannot be predicted from the VSEPR model as outlined above. Experimentally, the O—N—O angle is found to be 134°, which is larger than the ideal trigonal planar angle of 120°. This implies that, unlike the situation with lone pairs, the repulsion between a single lone electron and the bonding pairs is *less* than that between the bonding pairs. Because of the resonance, the two N—O bond distances in NO_2 are equal.

(c) The Lewis structure for the sulfur pentafluoride ion, SF_5^-, is

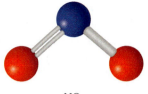

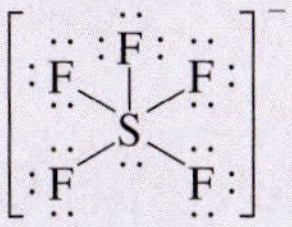

NO_2

There are five S—F bonding pairs and one lone pair around the central sulfur, giving $N_S = 6$. The overall electron-pair geometry from Figure 4.1 is octahedral. Ignoring the lone pair, the molecular geometry from Table 4.2 is square pyramidal (as in BrF_5). Because of lone pair repulsion, the adjacent F—S—F bond angles are slightly less than the ideal angle of 90°.

(d) The Lewis structure for CCl_3F (trichlorofluoromethane) is

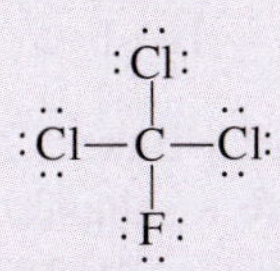

There are four bonding pairs around the central carbon atom and no lone pairs, so the electron-pair geometry and the molecular geometry are both tetrahedral:

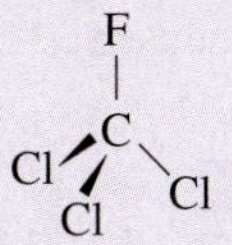

Because the four bonds to carbon are not all the same—there are three C—Cl bonds and one C—F bond—the bond angles are not all exactly 109.5°, the ideal tetrahedral angle. Chlorine is larger than fluorine, so the Cl—C—Cl bond angles are slightly larger than 109.5° and the F—C—Cl bond angles are slightly smaller.

(e) The Lewis structure of C_2H_4 is

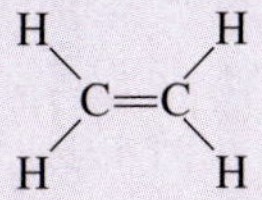

The C=C bond is treated as though it were a single bond in the VSEPR model. Because there are three electron pairs around each carbon atom and there are no

—Continued

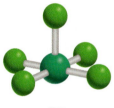

SF_5^-

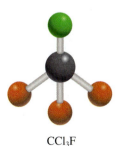

CCl_3F

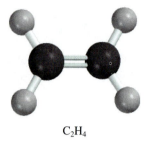

C_2H_4

Continued—

lone pairs present, the arrangement around each carbon atom has a trigonal planar shape like BF_3 (see Table 4.1).

$$H \overset{120°}{\diagdown}C = C \overset{H}{\diagup} 120°$$
$$H \overset{120°}{\diagup} \qquad H$$

Thus, the predicted bond angles in C_2H_4 are close to 120°, but not exactly equal to the ideal trigonal planar angle because the bonds are not all equivalent. Also, all six of the atoms in C_2H_4 lie in the same plane. The planar overall geometry is not predicted by the VSEPR model, but we shall see why the molecule "prefers" to be planar in Section 4.3.

Practice Exercise Use the VSEPR model to predict the geometry of (a) $SiBr_4$, (b) OCS, and (c) NO_3.

4.2 | The Polarity of a Molecule Can Be Described Quantitatively by Its Dipole Moment

Recall from Section 3.4 that the hydrogen fluoride molecule is a covalent compound with a polar bond. There is a shift of electron density from hydrogen to fluorine because the fluorine atom is more electronegative than the hydrogen atom (see Figure 3.9). The consequent charge separation can be represented as

$$\overset{\delta^+ \quad \delta^-}{H—F}$$

where δ (delta) denotes a positive partial charge. This separation of charge can be confirmed by applying an electric field to a sample of HF in the gas phase (Figure 4.4). When the field is turned on, HF molecules orient their negatively charged ends toward the positive plate and their positively charged ends toward the negative plate. This alignment of molecules can be detected experimentally.

A quantitative measure of the polarity of a bond is its ***dipole moment*** (μ). Consider a system of two charges $+q$ and $-q$ separated by a distance r, where q is assumed to

Figure 4.4 Behavior of polar molecules (a) in the absence of an electric field and (b) when the field is turned on. Nonpolar molecules are not affected by an electric field.

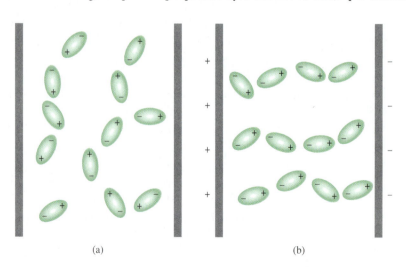

(a)   (b)

be positive.[2] Because the positive and negative charges are equal in magnitude, the system is electrically neutral. The dipole moment of this system is a *vector quantity,* so it has both magnitude and direction. The magnitude of the dipole moment is equal to the *product of the charge q and the distance r between the charges:*

$$\mu = q \times r \qquad \textbf{(4.1)}$$

In a diatomic molecule, q is equal to δ^+ and δ^-.

The direction of the dipole moment vector points from the negative charge to the positive charge.[3]

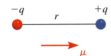

The SI unit of dipole moment is the coulomb-meter (C m), but dipole moments for molecules are often more conveniently expressed in debye units (D), named for Peter Debye.[4] The conversion factor is

$$1\ D = 3.336 \times 10^{-30}\ C\ m$$

Diatomic molecules containing atoms of *different* elements (for example, HF, HCl, CO, and NO) *have nonzero dipole moments* and are called **polar molecules.** Diatomic molecules containing atoms of the *same* element (for example, H_2, O_2, and F_2) are **nonpolar molecules** because they *have zero dipole moments*. For a molecule made up of three or more atoms, both the polarity of the bonds and the molecular geometry determine whether there is a nonzero dipole moment. Even if polar bonds are present, the molecule will not necessarily have a nonzero dipole moment. Carbon dioxide (CO_2), for example, is a triatomic molecule, so its geometry is either linear or bent:

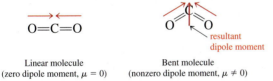

Linear molecule
(zero dipole moment, $\mu = 0$)

Bent molecule
(nonzero dipole moment, $\mu \neq 0$)

The dipole moment of the entire molecule is made up of two *bond moments*—that is, individual dipole moments in the polar C=O bonds (shown as arrows above). These bond moments result from the shift of electron density from the less electronegative carbon atom to the more electronegative oxygen atom. The measured dipole moment is equal to the vector sum of the bond moments. The two bond moments in CO_2 are equal in magnitude, but they point in opposite directions in a linear CO_2 molecule, so the sum or

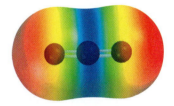

Each carbon-to-oxygen bond is polar, with the electron density shifted toward the more electronegative oxygen atom. However, the linear geometry of the molecule results in the cancellation of the two bond moments.

---

2. In a real molecule, the charges are not point charges because the electron density is spread out over space. In such a system we do not talk about point charges, but instead about a charge density $\rho(\mathbf{r})$, which gives the density of charge at any point in space. The dipole moment is then defined in terms of an integral over this charge density. However, for the qualitative analysis of molecular polarity, the point charge approximation is often sufficient.

3. The convention for the direction of the dipole moment is based on the standard mathematical expression for the dipole moment and is consistent with the definition recommended by IUPAC and that used in most physical chemistry texts. Many introductory and organic chemistry texts use an alternate convention in which the dipole moment points from the positive to the negative end of the bond—opposite to the direction used here. When reading other texts, it is important to know which convention is being used.

4. Peter Joseph William Debye (1884–1966). American chemist and physicist of Dutch origin. Debye made many significant contributions in the study of molecular structure, polymer chemistry, X-ray analysis, and electrolyte solutions. He was awarded the Nobel Prize in Chemistry in 1936.

Figure 4.5 Bond moments and resultant dipole moments in NH_3 and NF_3. The electrostatic potential maps show the electron density distributions in these molecules.

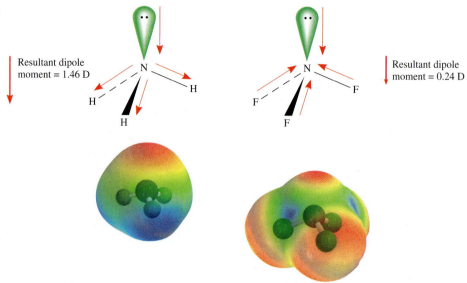

Resultant dipole moment = 1.46 D

Resultant dipole moment = 0.24 D

resultant dipole moment for this molecular geometry would be zero. On the other hand, if the CO_2 molecule were bent, the two bond moments would partially reinforce each other, so that the molecule would have a nonzero dipole moment. Experimentally it is found that the dipole moment of carbon dioxide is zero. Therefore, we conclude that the carbon dioxide molecule is linear, consistent with the VSEPR model. The linear geometry of carbon dioxide has been confirmed through other experimental measurements.

Next, consider the NH_3 and NF_3 molecules shown in Figure 4.5. In both cases, the central nitrogen atom has a lone pair, whose electron density is away from the nitrogen atom, resulting in a dipole moment contribution that points toward nitrogen from the lone pair. From Figure 3.9, we know that fluorine is more electronegative than nitrogen and nitrogen is more electronegative than hydrogen. For this reason, the shift of electron density in NH_3 is toward nitrogen, yielding bond dipole moment contributions that add constructively to the lone-pair contribution, whereas the N—F bond moments are directed toward the nitrogen atom and so together they offset the contribution of the lone pair to the dipole moment. Thus, the resultant dipole moment in NH_3 is larger than that in NF_3, even though the N—F bonds are more polar than the N—H bonds.

Dipole moments can be used to distinguish between molecules that have the same formula but different structures. For example, the following molecules both exist; they have the same molecular formula ($C_2H_2Cl_2$), and the same number and type of bonds, but different molecular structures:

Resultant dipole moment

Cl   Cl
   C═C
H   H

cis-dichloroethylene
$\mu = 1.89$ D

Cl   H
   C═C
H   Cl

trans-dichloroethylene
$\mu = 0$

In *cis*-dichloroethylene, the bond dipole moments do not cancel, and *cis*-dichloroethylene is a polar molecule. However, in *trans*-dichloroethylene, the bond moments do cancel, giving a zero dipole moment for the molecule. Thus, these two molecules can readily be distinguished by a dipole moment measurement. Additionally, we will see in Section 4.6,

Table 4.3	Dipole Moments of Some Polar Molecules	
Molecule	**Geometry**	**Dipole Moment (D)**
HF	Linear	1.92
HCl	Linear	1.08
HBr	Linear	0.78
HI	Linear	0.38
H_2O	Bent	1.87
H_2S	Bent	1.10
NH_3	Trigonal pyramidal	1.46
SO_2	Bent	1.60

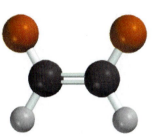

cis-dichloroethylene

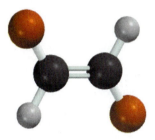

trans-dichloroethylene

the strength of intermolecular forces is partially determined by whether molecules possess a dipole moment. Table 4.3 lists the dipole moments of several polar molecules.

Example 4.2 shows how we can predict whether a molecule possesses a dipole moment if we know its molecular geometry.

Example 4.2

Predict whether each of the following molecules has a nonzero dipole moment: (a) IBr, (b) BF_3, and (c) CH_2Cl_2.

Strategy The dipole moment of a molecule depends on both the difference in electronegativities of the elements present and its geometry. A molecule can have polar bonds (if the bonded atoms have different electronegativities), but its dipole moment may be zero if it has a highly symmetrical geometry.

Solution (a) Because IBr (iodine bromide) is diatomic, it has a linear geometry. Bromine is more electronegative than iodine (see Figure 3.9), so IBr is a polar molecule with bromine at the negative end.

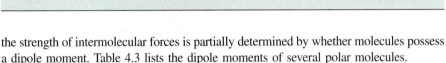

$$\overset{\longleftarrow}{\text{I—Br}}$$

Thus, IBr has a nonzero dipole moment, which points from the Br atom to the I atom.

(b) Because fluorine is more electronegative than boron, each B—F bond in BF_3 (boron trifluoride) is polar, and the three bond moments are equal. According to the VSEPR model, however, BF_3 should be trigonal planar and the symmetry of the trigonal planar shape means that the three bond moments exactly cancel one another:

(c) The Lewis structure of CH_2Cl_2 (methylene chloride) is

$$\text{H—}\overset{\overset{\displaystyle Cl}{|}}{\underset{\underset{\displaystyle Cl}{|}}{C}}\text{—H}$$

—Continued

Major Experimental Technique: Infrared Spectroscopy

All molecules vibrate, even at the lowest temperature of 0 K. Much like the electronic energies of atoms and molecules, the energy of molecular vibration is quantized. To vibrate more energetically, a molecule can absorb a photon of electromagnetic radiation with an energy ($h\nu$) that corresponds to the difference between two vibrational energy levels. The electromagnetic radiation frequencies required to excite molecular vibrations are much lower than those required for electronic excitation and typically lie in the infrared (IR) region (See Figure 1.6). The specific IR frequency needed to excite a particular molecular vibration depends strongly on the types of bond(s) involved and on the molecular environment. For example, the excitation of the C≡O bond vibration in carbon monoxide gas requires a vibrational frequency of 2170 cm^{-1}, but a frequency of 2990 cm^{-1} is required to excite vibrations of the H—Cl bond in hydrogen chloride gas.[5] As a result, *infrared spectroscopy, the study of the infrared frequencies that are absorbed by a particular material,* can give us information as to its molecular structure and identity.

To understand IR spectroscopy, it is necessary to examine the vibrations of molecules from a quantum mechanical perspective. Consider a diatomic molecule (for example, HCl) consisting of two atoms (masses m_1 and m_2) separated by a distance r.

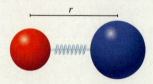

To a good approximation, we can treat the vibrational motion of such a molecule to be that of two masses separated by a spring obeying Hooke's law (see Section 0.1), which states that the force (F) between the two masses is proportional to the displacement ($x = r - r_e$) from the equilibrium bond length (r_e):

$$F = -k(r - r_e) = -kx$$

The constant of proportionality (k) is called the *force constant* and is a measure of the stiffness of the spring. The negative sign in this equation indicates that the force operates in the opposite direction as the displacement, that is, it is a restoring force. When the bond between the two atoms is stretched beyond R_0 ($x > 0$), the force acts to shorten the bond, and when the bond is compressed ($x < 0$), the force acts to lengthen the bond.

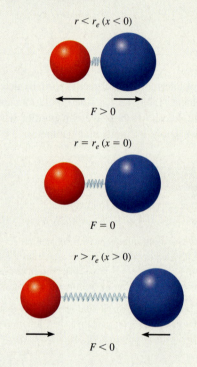

The potential energy function corresponding to this Hooke's law force is

$$V(x) = \frac{1}{2}kx^2$$

This model for molecular vibrations is referred to as the *harmonic oscillator model.* The Schrödinger Equation (Equation 1.24) for this potential can be solved exactly to give a quantized set of vibrational energy levels

$$E_{\text{vib}} = \left(n + \frac{1}{2}\right)h\nu \quad n = 0, 1, 2, \ldots$$

where n is the *vibrational quantum number* and ν is the fundamental vibrational frequency given by

$$\nu = \frac{1}{2\pi}\sqrt{\frac{k}{\mu}}$$

5. For convenience, chemists generally report infrared frequencies as *wavenumbers*. For a given vibrational frequency (ν) the wavenumber ($\tilde{\nu}$) is given by $\tilde{\nu} = \nu/c$, where c is the speed of light in cm s^{-1}. Because frequencies are in units of s^{-1} (or Hz), the wavenumber corresponding to that frequency will have units of cm^{-1} (or *inverse centimeters*). To see the usefulness of this transformation, consider the vibrational frequency of CO, which is 6.503×10^{13} s^{-1} when reported in Hz, but only 2170 cm^{-1} when reported in wavenumbers.

The *reduced mass* (μ) is defined as $\mu = m_1 m_2 / (m_1 + m_2)$. These energy levels are shown in the figure.

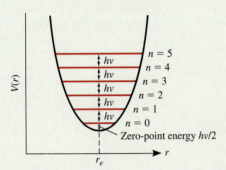

When a molecule in a vibrational state n absorbs a photon of IR radiation with the correct energy $h\nu$, it is excited to a higher vibrational state $n + 1$. The lowest possible energy state of a harmonic oscillator lies above the zero of energy by $h\nu/2$. As with the one-dimensional particle-in-a-box, this *zero-point energy* is a consequence of the Heisenberg uncertainty principle. If the energy (and thus both the potential and kinetic energies) were exactly zero, then the two atoms would be exactly r_e apart and would have zero velocity, allowing us to specify simultaneously the position and momentum of the particles, in violation of the uncertainty principle.

As mentioned above, different types of bonds absorb infrared radiation at different frequencies related to the local molecular environment, allowing the use of IR spectroscopy in the identification of compounds. Because of the complexity of molecular vibrations, no two compounds will have exactly the same IR spectrum. By matching the IR spectrum of an unknown compound to that of a compound in a database, a process known as *fingerprinting*, is a definitive method of identification. The figure shows the IR spectrum of a relatively simple molecule ($CH_2{=}CHC{\equiv}N$ or 2–propenenitrile) and an assignment of the major peaks to different molecular vibrations.

In order for a molecule to interact directly with an electromagnetic field (such as IR radiation) and absorb a photon, the vibrational motion of the molecule must be such that the molecule possesses a nonzero dipole moment sometime during the course of the vibration. Electromagnetic radiation interacts with the charges on a molecule—in an electric field positive charges are pushed in the opposite direction as negative ones. So, the different ends of a dipole will experience forces of opposite directions in an oscillating electromagnetic field. If the frequency of the radiation matches the fundamental frequency of the molecular vibration, the molecule can absorb an IR photon and be promoted to a higher vibrational energy level. Because of their symmetry, molecules such as N_2 and O_2 cannot absorb IR photons and are said to be IR inactive. Molecules with permanent dipoles such as HCl and H_2O are strong IR absorbers. Nonpolar molecules can absorb IR radiation if the molecular vibrations produce instantaneous dipoles. For example, there are three possible types of molecular vibrations for CO_2—a symmetric stretch, an asymmetric stretch and a bend.

O=C=O   O=C=O   O=C=O

O=C=O   O=C=O

Symmetric stretch   Asymmetric stretch   Bend

The symmetric stretch does not produce an instantaneous dipole moment and is IR inactive. The asymmetric stretch and bend motions do produce instantaneous dipole moments and are IR active. It is these IR active vibrations that make CO_2 a strong IR absorber and greenhouse gas (see Chapter 13).

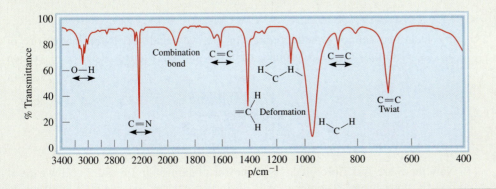

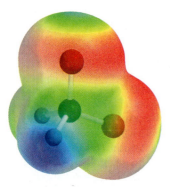

Electrostatic potential map of CH_2Cl_2. The electron density is shifted toward the electronegative Cl atoms.

Continued—

CH_2Cl_2 is similar to CH_4 in that it has an overall tetrahedral shape, as predicted by the VSEPR model. Not all the bonds are identical, however, so there are three different bond angles: H—C—H, H—C—Cl, and Cl—C—Cl. These bond angles are close to, but not equal to 109.5°. Because chlorine is more electronegative than carbon, which is more electronegative than hydrogen, the bond moments do not cancel, and the molecule possesses a nonzero dipole moment:

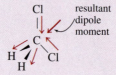

Thus, CH_2Cl_2 is a polar molecule.

Practice Exercise Does the $AlCl_3$ molecule have a nonzero dipole moment?

4.3 | Valence Bond Theory for Polyatomic Molecules Requires the Use of Hybrid Orbitals

The valence bond concepts of atomic orbital overlap discussed in Chapter 3 also apply to polyatomic molecules. A satisfactory bonding scheme, however, must account for molecular geometry. We will discuss three examples of VB treatment of bonding in polyatomic molecules in terms of hybridization.

sp Hybridization

Consider the triatomic molecule beryllium chloride ($BeCl_2$). Experimental evidence shows that the $BeCl_2$ molecule in the gas phase is linear with two equivalent Be—Cl bonds, consistent with the VSEPR model prediction. The construction of a valence bond picture of bonding in linear $BeCl_2$ requires the existence of unpaired electrons in Be atomic orbitals that overlap with the half-filled $3p$ orbital on each Cl to form two equivalent bonds that are 180° apart. However, the ground state electron configuration for Be is $1s^2 2s^2$, which contains no unpaired electrons.

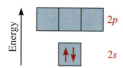

To obtain two unfilled bonding orbitals on Be, we could start by promoting one of the $2s$ electrons to a $2p$ orbital, resulting in the excited state configuration $[He]2s^1 2p^1$.

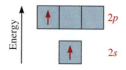

Now there are two Be orbitals available for bonding: the $2s$ and $2p$. However, if two Cl atoms were to combine with Be in this excited state, one Cl atom would

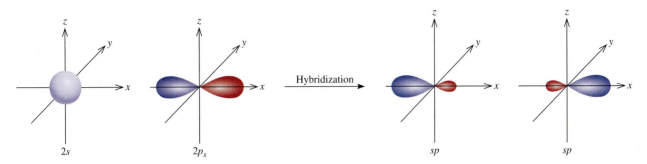

Figure 4.6 Formation of *sp* hybrid orbitals. The colors indicate the sign of the wavefunction: blue (+), red (−).

share a 2*s* electron and the other Cl would share a 2*p* electron, making two nonequivalent Be—Cl bonds and a different geometry. This scheme contradicts experimental evidence. In reality, the two Be—Cl bonds are identical in every respect. To construct equivalent valence-bond orbitals from these nonequivalent atomic orbitals we employ a process called **hybridization,** which is the *mixing of nonequivalent atomic orbitals in an atom (usually a central atom) to generate a set of hypothetical bonding orbitals, called **hybrid orbitals,** which have the proper geometry to form equivalent bonds.[6]

The hybrid orbitals for Be in $BeCl_2$ can be constructed by mixing (or *hybridizing*) the 2*s* and one of the 2*p* orbitals to form two equivalent *sp* hybrid orbitals in the following way:

$$\text{First } sp \text{ orbital} = \frac{1}{\sqrt{2}}[\phi_{2s} + \phi_{2p_x}]$$

$$\text{Second } sp \text{ orbital} = \frac{1}{\sqrt{2}}[\phi_{2s} - \phi_{2p_x}]$$

where the *x* direction is taken to be the direction of bonding. These orbitals are called *sp* orbitals because they are formed from one *s* and one *p* orbital. The shape and orientation of these hybrid orbitals are shown in Figure 4.6. These two hybrid orbitals lie on the same line, the *x* axis, so that the angle between them is 180°.

The orbital diagram for Be now consists of one electron in each of the two *sp* hybrid orbitals with the two remaining 2*p* orbitals remaining unoccupied.

sp orbitals   Empty 2*p* orbitals

Each of the Be—Cl bonds is then formed by the overlap of a Be *sp* hybrid orbital and a Cl 3*p* orbital, and the resulting $BeCl_2$ molecule has a linear geometry (Figure 4.7).

Energy is required for the promotion of an electron from the filled $[He]2s^2$ ground state configuration of Be to form the excited state and the subsequent hybridization of the 2*s* and 2*p* orbitals to form the two hybrid *sp* orbitals; however, this energy input is more than compensated for by the energy release accompanying the formation of the Be—Cl bonds.

Figure 4.7 The linear geometry of $BeCl_2$ can be explained by assuming that Be is *sp* hybridized. In the valence bond model, the two *sp* hybrid orbitals overlap with the two chlorine 3*p* orbitals to form two covalent bonds.

---

6. Both hybridization and molecular orbital theory (discussed in Section 3.5) involve the mixing of atomic orbitals; however, the construction of hybrid orbitals involves mixing of atomic orbitals on the same atom, whereas the construction of molecular orbitals involves atomic orbitals on different atoms.

sp^2 Hybridization

Next, consider the BF₃ (boron trifluoride) molecule, which has trigonal planar geometry based on VSEPR and experimental evidence. The ground state electron configuration of B is $[He]2s^2 2p^1$, so it has only one unpaired valence electron.

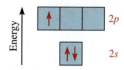

To generate the three unpaired electrons that are required to form bonds with the $2p$ orbitals on the F atoms, we first promote a $2s$ electron to an empty $2p$ orbital to generate the excited state configuration $[He]2s^1 2p^2$.

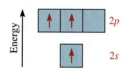

There are now three unpaired electrons—one in the $2s$ orbital and two in separate $2p$ orbitals. Because these are in nonequivalent orbitals, they cannot be used in this form to form the three equivalent B—F bonds; instead, they must be hybridized. Mixing the $2s$ orbital with the two $2p$ orbitals generates three hybrid orbitals. These orbitals are referred to as sp^2 because they are formed from the mixing of one s orbital and two p orbitals.

sp^2 is pronounced "s-p two."

Mathematically, the hybrid orbitals are

$$\text{First } sp^2 \text{ orbital} = \frac{1}{\sqrt{3}}[\phi_{2s} + \sqrt{2}\phi_{2p_x}]$$

$$\text{Second } sp^2 \text{ orbital} = \frac{1}{\sqrt{3}}\left[\phi_{2s} - \frac{1}{\sqrt{2}}\phi_{2p_x} + \sqrt{\frac{3}{2}}\phi_{2p_y}\right]$$

$$\text{Third } sp^2 \text{ orbital} = \frac{1}{\sqrt{3}}\left[\phi_{2s} - \frac{1}{\sqrt{2}}\phi_{2p_x} - \sqrt{\frac{3}{2}}\phi_{2p_y}\right]$$

These three sp^2 orbitals lie in the same plane (the x-y plane), and the angle between any two of them is 120° (Figure 4.8).

Each of the B—F bonds is formed by the overlap of a boron sp^2 hybrid orbital and a fluorine $2p$ orbital (Figure 4.9). This result conforms to experimental findings and also to VSEPR predictions.

sp^3 Hybridization

Methane is the simplest hydrocarbon. Physical and chemical studies show that all four C—H bonds are identical in length and strength and the molecule has a tetrahedral geometry with bond angles of 109.5°—consistent with the VSEPR model. How can we explain the tetravalency of carbon within the VB approach? In its ground state, the electron configuration of carbon is $[He]2s^2 2p^2$. Because the carbon

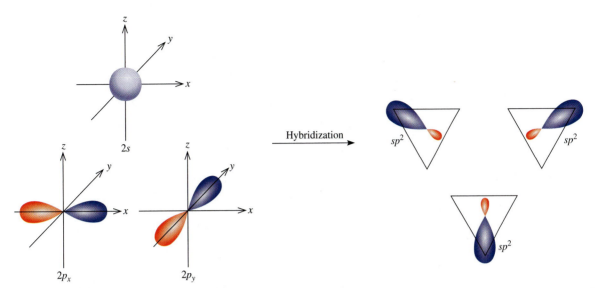

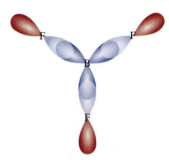

Figure 4.8 Formation of sp^2 hybrid orbitals. The colors indicate the sign of the wavefunction: blue (+), red (−).

atom has two unpaired electrons (one in each of the two $2p$ orbitals), it can form only two bonds with hydrogen in its ground state. Although the species CH_2 is known, it is very unstable. To account for the four C—H bonds in methane, we can try to promote an electron from the $2s$ orbital to the $2p$ orbital to give the excited configuration $[He]2s^12p^3$. This would give C four unpaired electrons that could form four C—H bonds.

However, if this were the case, methane would contain three C—H bonds of one type (involving the three $2p$ orbitals) and one of another (involving the s orbital). This is contrary to the experimental evidence that all C—H bonds are identical, suggesting that the bonding orbitals are all equivalent and that hybrid orbitals are necessary. We can generate four equivalent hybrid orbitals for carbon by mixing the $2s$ orbital and the three $2p$ orbitals to form four equivalent sp^3 hybrid orbitals.

Figure 4.9 The sp^2 hybrid orbitals on boron overlap with the $2p$ orbitals of fluorine to form the covalent bonds in BF_3. The BF_3 molecule is planar, and all F—B—F angles are 120°.

Mathematically, these orbitals are given by

$$\text{First } sp^3 \text{ orbital} = \frac{1}{2}[\phi_{2s} + \phi_{2p_x} + \phi_{2p_y} + \phi_{2p_z}]$$

$$\text{Second } sp^3 \text{ orbital} = \frac{1}{2}[\phi_{2s} + \phi_{2p_x} - \phi_{2p_y} - \phi_{2p_z}]$$

$$\text{Third } sp^3 \text{ orbital} = \frac{1}{2}[\phi_{2s} - \phi_{2p_x} + \phi_{2p_y} - \phi_{2p_z}]$$

$$\text{Fourth } sp^3 \text{ orbital} = \frac{1}{2}[\phi_{2s} - \phi_{2p_x} - \phi_{2p_y} + \phi_{2p_z}]$$

Figure 4.10 shows the shape and orientations of the sp^3 orbitals.

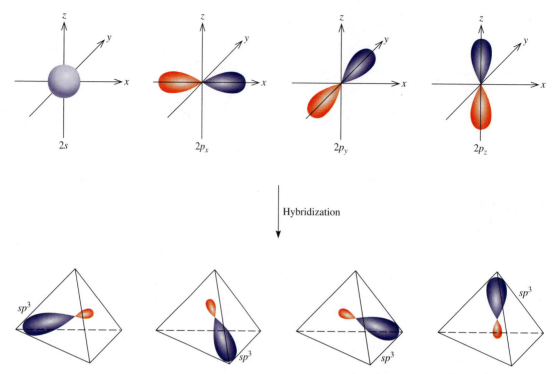

Figure 4.10 Formation of sp^3 hybrid orbitals. The colors indicate the sign of the wavefunction: blue ($+$), red ($-$).

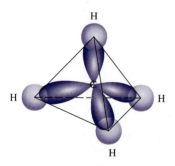

Figure 4.11 Formation of four covalent bonds between the carbon sp^3 hybrid orbitals and the hydrogen $1s$ orbitals in CH_4.

These four hybrid orbitals are directed toward the four corners of a regular tetrahedron. Figure 4.11 shows the formation of four covalent bonds between the carbon sp^3 hybrid orbitals and the hydrogen $1s$ orbitals in CH_4.

Thus, CH_4 has a tetrahedral shape, and all the H—C—H angles are 109.5°.

Another example of sp^3 hybridization is ammonia (NH_3). Table 4.1 shows that the arrangement of four electron pairs is tetrahedral, so that the bonding in NH_3 can be explained by assuming that N, like C in CH_4, is sp^3 hybridized. Nitrogen ($[He]2s^2 2p^3$) has five valence electrons, so three of the four hybrid sp^3 orbitals are half-filled and will form covalent N—H bonds. The fourth hybrid orbital does not participate in bonding and accommodates the lone pair on nitrogen (Figure 4.12). Repulsion between the lone-pair electrons and electrons in the bonding orbitals decreases the H—N—H bond angles from 109.5° to 107.3°.

It is important to understand the relationship between hybridization and the VSEPR model. We use hybridization to describe the bonding scheme only when the arrangement of electron pairs has been predicted using VSEPR. If the VSEPR model predicts a tetrahedral arrangement of electron pairs, then we assume that one s and three p orbitals are hybridized to form four sp^3 hybrid orbitals.

There is a connection between hybridization and the octet rule. Regardless of the type of hybridization, an atom starting with one s and three p orbitals would still possess four orbitals, enough to accommodate a total of eight electrons in a compound. For elements in the second period of the periodic table, eight is the maximum number of electrons that an atom of any of these elements can accommodate in the valence shell. It is for this reason that the octet rule is usually obeyed by the second-period elements.

The situation is different for an atom of a third-period element. If we use only the $3s$ and $3p$ orbitals of the atom to form hybrid orbitals in a molecule, then the octet rule applies. However, in some molecules the same atom may use one or more $3d$ orbitals, in addition to the $3s$ and $3p$ orbitals, to form hybrid orbitals. In these cases the octet rule does not hold. We will see specific examples of the participation of the $3d$ orbital in hybridization shortly.

Although our discussion has focused on the role of hybridization in the bonding of polyatomic molecules, hybridization can also be employed in the discussion of bonding in diatomic molecules. For example, in Chapter 3 we described the bonding in F_2 in terms of a sigma bond formed from the overlap between the $2p_z$ orbitals on the two F atoms. The lone pair electrons then go into the remaining $2s$, $2p_x$ and $2p_x$ orbitals. As an alternative description, we could also consider the sigma bond to be formed by the overlap of two sp^3 orbitals on each of the F atoms. The lone pairs would then occupy the remaining sp^3 orbitals on each atom. The advantage of this description is that the lone pair electrons occupy equivalent orbitals. However, for our purposes, both of these descriptions are acceptable in that they both provide the same basic valence bond description of the bonding in F_2—two F atoms surrounded by lone pair electrons and bonded by a sigma bond. Unlike a polyatomic molecule, the diatomic molecule F_2 has no bond angles so the molecular structure we predict does not change whether we use unhybridized or hybridized orbitals in bond formation.

The discussion of hybridization can be summarized as follows:

▶ The concept of hybridization is primarily employed as a theoretical model to explain covalent bonding in polyatomic molecules.

▶ Hybridization is the mixing of at least two inequivalent atomic orbitals (for example, s and p orbitals) on the same atom. Therefore, a hybrid orbital is not a pure atomic orbital. Hybrid orbitals and pure atomic orbitals have very different shapes.

▶ The number of hybrid orbitals generated is equal to the number of pure atomic orbitals that participate in the hybridization process.

▶ Hybridization requires an input of energy; however, the system more than recovers this energy during bond formation.

▶ Covalent σ bonds in polyatomic molecules and ions are formed by the overlap of hybrid orbitals or of hybrid orbitals with unhybridized ones. Therefore, the hybridization bonding scheme is still within the framework of valence bond theory; electrons in a molecule are assumed to occupy hybrid orbitals of the individual atoms.

Table 4.4 summarizes sp, sp^2, and sp^3 hybridization (as well as other types that we will discuss shortly).

Procedure for Hybridizing Atomic Orbitals

Before going on to discuss the hybridization of d orbitals, let us specify what we need to know in order to apply hybridization to bonding in polyatomic molecules in general. In essence, hybridization simply extends Lewis theory and the VSEPR model. To assign a suitable state of hybridization to the central atom in a molecule, we must first have some idea about the geometry of the molecule. The steps are as follows:

1. Draw the Lewis structure of the molecule.

2. Predict the overall arrangement of the electron pairs (both bonding pairs and lone pairs) using the VSEPR model (see Tables 4.1 and 4.2).

3. Deduce the hybridization of the central atom by matching the arrangement of the electron pairs with those of the hybrid orbitals shown in Table 4.4.

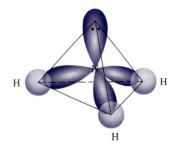

Figure 4.12 The N atom in NH_3 is sp^3 hybridized. Three sp^3 orbitals form bonds with the H atoms. The fourth is occupied by nitrogen's lone pair.

Table 4.4	Important Hybrid Orbitals and Their Shapes			
Pure Atomic Orbitals of the Central Atom	**Hybridization of the Central Atom**	**Number of Hybrid Orbitals**	**Combined Shape of Hybrid Orbitals**	**Examples**
s, p	sp	2	 180° Linear	$BeCl_2$
s, p, p	sp^2	3	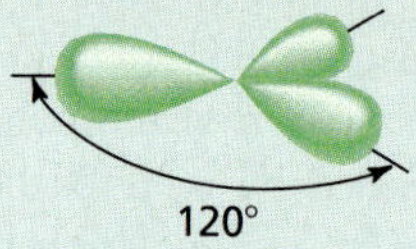 120° Trigonal planar	BF_3
s, p, p, p	sp^3	4	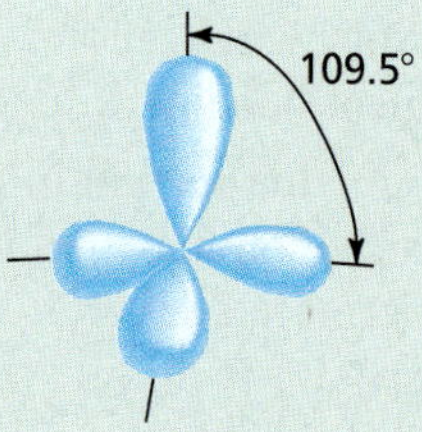 109.5° Tetrahedral	CH_4, NH_4^+
s, p, p, p, d	sp^3d	5	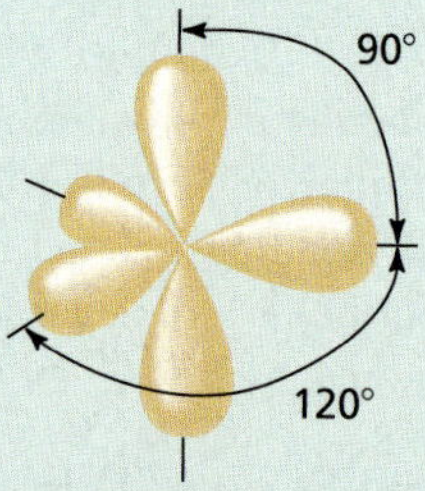 90° 120° Trigonal bipyramidal	PCl_5
s, p, p, p, d, d	sp^3d^2	6	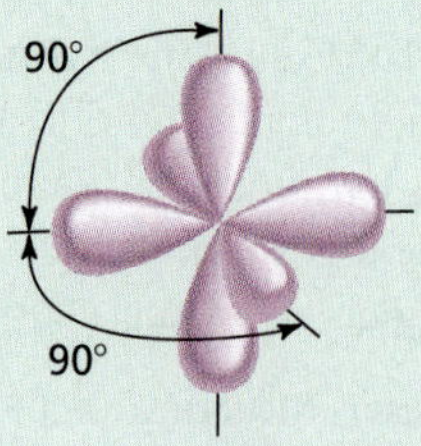 90° 90° Octahedral	SF_6

Example 4.3 illustrates this procedure.

Example 4.3

Determine the hybridization state of the central atom in each of the following molecules: (a) BeH_2, (b) AlI_3, and (c) PF_3. Describe the hybridization process and determine the molecular geometry in each case.

Strategy The steps for determining the hybridization of the central atom in a molecule are given in the preceding paragraph.

Solution (a) The central atom is beryllium (Be). The Be atom has two valence electrons with a ground-state electron configuration of $[He]2s^2$. The Lewis structure of BeH_2 is

H—Be—H

There are two bonding pairs around Be; therefore, the electron pair arrangement is linear. Be must use sp hybrid orbitals when bonding with H, because sp orbitals have a linear arrangement (see Table 4.3). The hybridization process can be imagined as follows. We first promote a $2s$ electron into one of the $2p$ orbitals to get the excited state configuration $[He]2s^1 2p^1$. The $2s$ and $2p$ orbitals then mix to form two sp hybrid orbitals. The two Be—H bonds are formed by the overlap of the Be sp orbitals with the $1s$ orbitals of the H atoms. Thus, BeH_2 is a linear molecule.

BeH$_2$

(b) The ground-state electron configuration of Al is $[Ne]3s^2 3p^1$, so the Al atom has three valence electrons. The Lewis structure of AlI_3 is

:Ï—Al—Ï:
|
:Ï:

There are three pairs of electrons around Al; therefore, the electron-pair arrangement is trigonal planar. Al must use sp^2 hybrid orbitals when bonding with I because sp^2 orbitals have a trigonal planar arrangement (see Table 4.4). The $3s$ and two $3p$ orbitals mix to form three sp^2 hybrid orbitals. The sp^2 hybrid orbitals overlap with the $5p$ orbitals of I to form three covalent Al—I bonds. The AlI_3 molecule is trigonal planar and all the I—Al—I angles should be 120°.

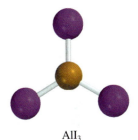

AlI$_3$

(c) The ground-state electron configuration of P is $[Ne]3s^2 3p^3$, so P has five valence electrons. The Lewis structure of PF_3 is

:Ë—P—Ë:
|
:Ë:

There are four pairs of electrons around P; therefore, the electron pair arrangement is tetrahedral. P must use sp^3 hybrid orbitals when bonding to F, because sp^3 orbitals have a tetrahedral arrangement (see Table 4.4). By mixing the $3s$ and $3p$ orbitals, we obtain four sp^3 hybrid orbitals. As in the case of NH_3, one of the sp^3 hybrid orbitals is used to accommodate the lone pair on P. The other three sp^3 hybrid orbitals form covalent P—F bonds with the $2p$ orbitals of F. The geometry of the molecule should be trigonal pyramidal, and the F—P—F angle should be somewhat less than 109.5° because of the greater repulsion between the lone pair and bonding pairs.

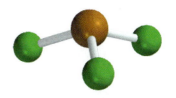

PF$_3$

Practice Exercise Determine the hybridization state of the central atoms in (a) $SiBr_4$ and (b) BCl_3.

Hybridization Involving *d* Orbitals

Hybridization neatly explains bonding that involves *s* and *p* orbitals. For elements in the third period and beyond, however, we cannot always account for molecular geometry by assuming that only *s* and *p* orbitals hybridize. To understand the formation of molecules with trigonal bipyramidal and octahedral geometries, for instance, we must include *d* orbitals in the hybridization scheme.

Recall from Section 4.1, for example, that the SF_6 molecule has octahedral geometry, which is also the arrangement of the six electron pairs. Table 4.4 shows that the S atom in SF_6 is sp^3d^2 hybridized. The ground-state electron configuration of S is $[Ne]3s^23p^4$.

Because the $3d$ level is quite close in energy to the $3s$ and $3p$ levels, we can promote $3s$ and $3p$ electrons to two of the $3d$ orbitals to give the excited state $[Ne]3s^13p^33d^2$.

Mixing the $3s$, three $3p$, and two $3d$ orbitals generates six sp^3d^2 hybrid orbitals that are oriented at 90° to one another toward the vertices of an octahedron (see Table 4.4).

The six S—F bonds are formed by the overlap of the hybrid orbitals of the S atom with the $2p$ orbitals of the F atoms. Because there are 12 electrons around the S atom, the octet rule is violated. The use of *d* orbitals in addition to *s* and *p* orbitals to form an expanded octet (see Section 3.4) is an example of *valence-shell expansion*. Second-period elements, unlike third-period elements, do not have $2d$ energy levels, so they can never expand their valence shells. (Recall, from Section 1.4, that when $n = 2$, $l = 0$ and 1. Thus, we can only have $2s$ and $2p$ orbitals.) Hence, atoms of second-period elements can never be surrounded by more than eight electrons in any of their compounds.

Example 4.4 shows how to use valence-shell expansion to describe the hybridization in a third-period element.

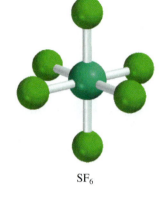

SF_6

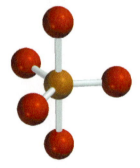

PBr_5

Example 4.4

Describe the hybridization state of phosphorus in phosphorus pentabromide (PBr_5).

Strategy Follow the same procedure used in Example 4.3.

Solution The ground-state electron configuration of P is $[Ne]3s^23p^3$, so the P atom has five valence electrons. The Lewis structure of PBr_5 is

$$
\begin{array}{c}
: \overset{\displaystyle ..}{Br} : \\
: \overset{..}{Br} \diagdown \;\; | \;\;\;\; .. \\
\qquad \diagup P{-}\overset{..}{Br} : \\
: \overset{..}{Br} \diagup \;\; | \\
\overset{..}{} : \overset{..}{Br} : \\
..
\end{array}
$$

—Continued

Continued—
There are five pairs of electrons around P; therefore, the electron pair arrangement is trigonal bipyramidal. P must use sp^3d hybrid orbitals when bonding to Br, because sp^3d hybrid orbitals have a trigonal bipyramidal arrangement (see Table 4.3). The hybridization process can be imagined as follows.

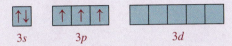

Promoting a $3s$ electron into a $3d$ orbital gives the excited state $[Ne]3s^13p^33d^1$, which has five unpaired electrons in nonequivalent orbitals.

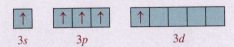

Mixing the one $3s$, three $3p$, and one $3d$ orbitals generates five sp^3d hybrid orbitals that point to the vertices of a trigonal bipyramid, as required.

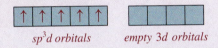

These hybrid orbitals overlap the $4p$ orbitals of Br to form five covalent P—Br bonds. Because there are no lone pairs on the P atom, the geometry of PBr_5 is trigonal bipyramidal.

Practice Exercise Describe the hybridization state of Xe in XeF_4.

Hybridization in Molecules Containing Double and Triple Bonds

The concept of hybridization is useful for molecules with double and triple bonds, too. Recall from Example 4.1, for example, that the ethylene molecule (C_2H_4) contains a carbon–carbon double bond and has planar geometry. Both the geometry and the bonding can be understood if we assume that each carbon atom is sp^2 hybridized. We assume that only the $2p_x$ and $2p_y$ orbitals combine with the $2s$ orbital, and that the $2p_z$ orbital remains unchanged (or *unhybridized*). Figure 4.13 shows that the $2p_z$ orbital is perpendicular to the plane of the hybrid orbitals.

How, then, do we account for the bonding of the C atoms? As Figure 4.14(a) shows, each carbon atom uses the three sp^2 hybrid orbitals to form two *sigma* (σ) bonds with the two hydrogen $1s$ orbitals and one σ bond with the sp^2 hybrid orbital of the adjacent C atom. In addition, the two unhybridized $2p_z$ orbitals of the C atoms form a pi (π) bond by overlapping sideways [Figure 4.14(b)].

It is this π-bond formation that gives ethylene its planar geometry because rotation of a CH_2 group out of the plane would necessitate the breaking of the C—C π bond. Figure 4.14(c) shows the orientation of the σ and π bonds. Figure 4.15 is yet another way of looking at the planar C_2H_4 molecule and the formation of the π bond. Although we normally represent the carbon-carbon double bond as C=C (as in a Lewis structure), the two bonds are different types: one is a σ bond and the other is a π bond. In fact, the bond energies of the carbon-carbon π and σ bonds are about 270 kJ mol^{-1} and 350 kJ mol^{-1}, respectively.

The acetylene molecule (C_2H_2) contains a carbon-carbon triple bond. Because the molecule is linear, we can explain its geometry and bonding by assuming that each

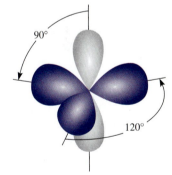

Figure 4.13 Each carbon atom in the C_2H_4 molecule has three sp^2 hybrid orbitals (blue) and one unhybridized $2p_z$ orbital (gray), which is oriented perpendicular to the plane of the hybrid orbitals.

Figure 4.14 Bonding in ethylene, C_2H_4. (a) Top view of the sigma bonds between carbon atoms and between carbon and hydrogen atoms. All of the atoms lie in the same plane, making C_2H_4 a linear molecule. (b) Side view showing how the two $2p_z$ orbitals on the two carbon atoms overlap, leading to the formation of a pi bond. Blue and red colors denote + and − regions of the orbitals. (c) The interactions in (a) and (b) lead to the formation of the sigma bonds and pi bond in ethylene. Note that the pi bond lies above and below the plane of the molecule.

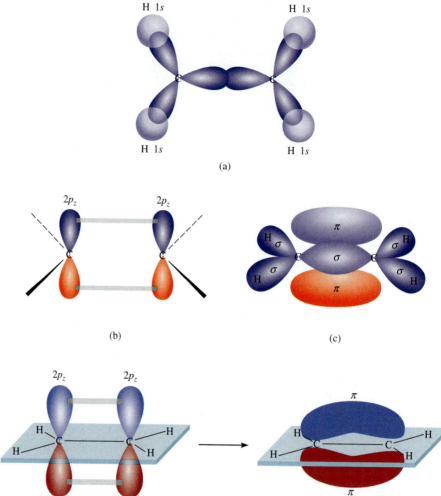

Figure 4.15 (a) Another view of pi bond formation in the C_2H_4 molecule. Note that all six atoms are in the same plane. It is the overlap of the $2p_z$ orbitals that causes the molecule to assume a planar structure. (b) Electrostatic potential map of C_2H_4.

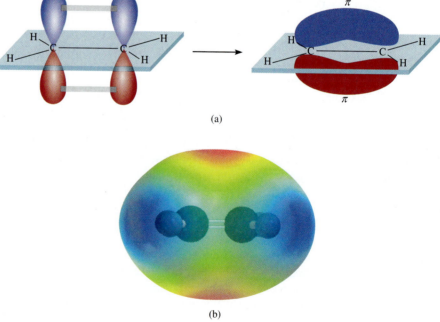

C atom is *sp* hybridized by mixing the 2*s* with the $2p_x$ orbital (Figure 4.16). As Figure 4.16 shows, the two *sp* hybrid orbitals of each C atom form one σ bond with a hydrogen 1*s* orbital and another σ bond with the other C atom. In addition, two π bonds are formed by the sideways overlap of the unhybridized $2p_y$ and $2p_z$ orbitals. Thus, the C≡C bond consists of one σ bond and two π bonds.

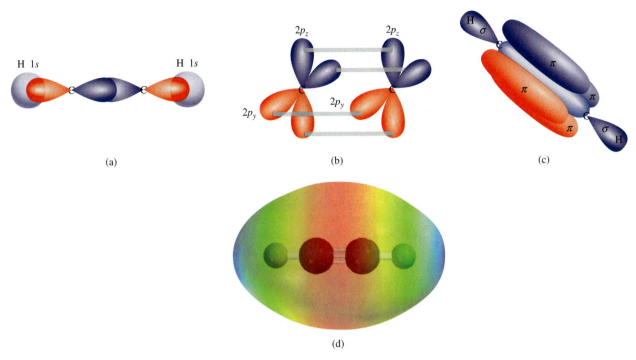

(a)   (b)   (c)

(d)

Figure 4.16 Bonding in acetylene, C_2H_2. (a) Top view showing the overlap of the *sp* orbitals between the C atoms and the overlap between the *sp* orbital on the C atom and the 1*s* orbital on H. All of the atoms lie along a straight line; therefore, acetylene is a linear molecule. (b) Side view showing the overlap of the two $2p_y$ orbitals and of the two $2p_z$ orbitals of the carbon atoms, which leads to the formation of the two pi bonds. Blue and red colors denote + and − regions of the orbitals, respectively. (c) Formation of the sigma and pi bonds as a result of the interaction in (a) and (b). (d) Electrostatic potential map of C_2H_2.

Example 4.5

Describe the bonding in the formaldehyde molecule, whose Lewis structure is

$$\overset{\text{H}}{\underset{\text{H}}{\diagdown}}\text{C}=\ddot{\text{O}}\!:$$

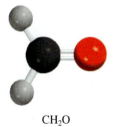

CH₂O

Assume that the O atom is sp^2 hybridized.

Strategy Follow the procedure shown in Example 4.3.

Solution There are three pairs of electrons around the C atom, so the electron-pair arrangement is trigonal planar. (Recall that a double bond is treated as a single group of electrons in the VSEPR model.) C must use sp^2 hybrid orbitals when bonding, because sp^2 hybrid orbitals have a trigonal planar arrangement (see Table 4.4). We can imagine the hybridization processes for C and O as follows

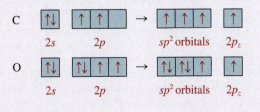

—Continued

Figure 4.17 Bonding in the formaldehyde molecule. A sigma bond is formed by the overlap of the sp^2 hybrid orbital of carbon and the sp^2 hybrid orbital of oxygen; a pi bond is formed by the overlap of the unhybridized $2p_z$ orbitals of the carbon and oxygen atoms. The two lone pairs are placed in the other two sp^2 orbitals of oxygen. The blue and red colors denote $+$ and $-$ regions of the pi orbital, respectively.

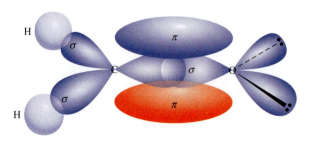

Continued—

Carbon has one electron in each of the three sp^2 orbitals, which are used to form σ bonds with the H atoms and the O atom. There is also an electron in the $2p_z$ orbital, which forms a π bond with one of the sp^2 orbitals on oxygen. Oxygen has two electrons in two of its sp^2 hybrid orbitals. These are the lone pairs on oxygen. Its third sp^2 hybrid orbital with one electron is used to form a σ bond with carbon. The $2p_z$ orbital (with one electron) overlaps with the $2p_z$ orbital of C to form a π bond (Figure 4.17).

Practice Exercise Describe the bonding in the hydrogen cyanide molecule, HCN. Assume that the N is sp hybridized.

4.4 | Isomers Are Compounds That Have the Same Molecular Formula but Different Atomic Arrangements

Merely knowing the molecular formula of a compound is often insufficient to uniquely specify its identity and three-dimensional structure. It is possible for two or more compounds to share the same molecular formula. Such compounds are called *isomers.* Isomers can be distinguished from one another either by their *bond connectivity*—the specific way the atoms of the molecule are bonded to one another—or by its three-dimensional structure.

Structural Isomers

Structural isomers are molecules with the same molecular formula but different bond connectivity. For example, the hydrocarbon C_4H_{10} has two structural isomers:

n-butane

2-methylpropane (or isobutane)

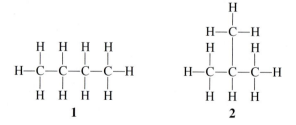

Both of these molecules have 4 carbon atoms and 10 hydrogen atoms, but they differ in the way in which these atoms are bonded to one another. Compound **1** is called *n*-butane (or simply butane) and is a *straight-chain hydrocarbon* (all of the carbon atoms are joined along one line). Compound **2** is 2-methylpropane (or isobutane). It is a *branched hydrocarbon* because at least one carbon atom is bonded to three other carbon atoms. In hydrocarbons, the number of possible structural isomers increases rapidly as the number of carbon atoms increases. For example, butane (C_4H_{10}) has two isomers, decane ($C_{10}H_{22}$) has 75 isomers, and the hydrocarbon $C_{30}H_{62}$ has over 400 million (or 4×10^8) possible isomers!

Example 4.6

Draw all possible structural isomers for C_2H_6O.

Strategy Carbon and oxygen form four and two covalent bonds with neighboring atoms, respectively, but hydrogen can only form one bond. We must use these bonding rules to logically construct all possible structures.

Solution Because hydrogen can only form one covalent bond with neighboring atoms, it will always be found in terminal positions. Thus, the central atoms of the molecule must be the two carbons and the oxygen. If we arrange them in a chain, we can either place the oxygen in the middle or at the end, which, after filling out the remaining bond positions with the six hydrogens, gives two possible structures:

$$
\begin{array}{cc}
\underset{1}{\text{H}-\overset{\displaystyle H}{\underset{\displaystyle H}{\text{C}}}-\overset{\displaystyle H}{\underset{\displaystyle H}{\text{C}}}-\text{O}-\text{H}} & \underset{2}{\text{H}-\overset{\displaystyle H}{\underset{\displaystyle H}{\text{C}}}-\text{O}-\overset{\displaystyle H}{\underset{\displaystyle H}{\text{C}}}-\text{H}}
\end{array}
$$

Structure **1** is ethanol and structure **2** is dimethylether. The properties of the two compounds are quite different. For example, ethanol is a liquid at room temperature with a boiling temperature of 352 K, whereas dimethylether is a gas at room temperature and its boiling point is 249 K.

Practice Exercise Draw all the possible structural isomers for pentane, C_5H_{12}.

ethanol

dimethylether

Stereoisomers

For some compounds, knowing the molecular formula and the bond connectivity is still not enough to uniquely specify the identity of an isomer. ***Stereoisomers*** *are compounds that have identical molecular formulas and identical bond connectivity, but have different three-dimensional structures.* There are two types of stereoisomers: geometric isomers and optical isomers.

Geometric isomers *are stereoisomers that differ in the spatial arrangement of the atoms relative to one another.* Such isomers cannot be interconverted without breaking a chemical bond. Geometric isomers come in pairs. We use the terms *cis* and *trans* to distinguish one geometric isomer of a compound from the other: *cis* means that two particular atoms (or groups of atoms) are on the same side in the structural formula across the double bond, and *trans* means that the atoms (or groups of atoms) are on opposite sides across the double bond. For example, dichloroethylene (ClHC=CHCl) can exist as one of the two geometric isomers called *cis*-dichloroethylene and *trans*-dichloroethylene:

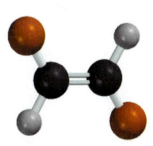

cis-dichloroethylene

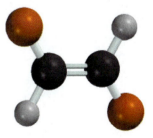

trans-dichloroethylene

cis-dichloroethylene   *trans*-dichloroethylene

cis-trans Isomerization in the Vision Process

The conversion of a *cis* isomer to the *trans* isomer (or vice versa) is a process called *cis-trans isomerization*. Such conversions are crucial in the vision process. The molecules in the retina that respond to light are rhodopsin, which has two components called 11-*cis*-retinal and opsin. Retinal is the light-sensitive component, and opsin is a protein molecule. Upon receiving a photon in the visible region, 11-*cis*-retinal isomerizes to all-trans retinal by breaking a carbon-carbon π bond. With the π bond broken, the atoms connected by the carbon–

carbon σ bond are free to rotate and the molecule transforms into all-*trans* retinal when the double bond reforms. At this point, an electrical impulse is generated and transmitted to the brain, which forms a visual image. The all-*trans* retinal does not fit into the binding site on opsin and eventually separates from the protein. In time, the *trans* isomer is converted back to 11-*cis*-retinal by an enzyme (in the absence of light), and rhodopsin is regenerated by the binding of the *cis* isomer to opsin and the visual cycle can begin again.

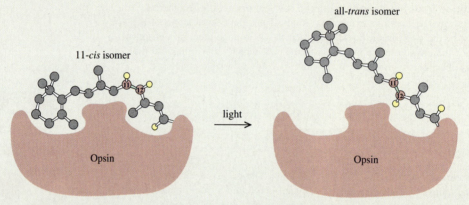

The primary event in the vision process is the conversion of 11-*cis*-retinal to the all-*trans* isomer of rhodopsin. The double bond at which the isomerization occurs is between carbon number 11 and carbon number 12. For simplicity, most of the H atoms are omitted. In the absence of light, this transformation takes place about once in a 1000 years!

Figure 4.18 A left hand and its mirror image, which looks the same as the right hand.

These two structures could only be interconverted through rotation of one of the —CHCl groups by 180° about the C=C bond axis. This would necessitate the breaking of the π bond between the two carbon atoms. The properties of the two molecules are quite different. As discussed in Section 4.2, the *cis* isomer has a relatively large dipole moment (1.89 D), whereas the more symmetric *trans* isomer has a dipole moment of zero. This difference is reflected in the boiling points, which are 333.5 K and 320.7 K for the *cis* and *trans* isomers, respectively.

Another class of stereoisomers are **optical isomers,** or **enantiomers,** which are *non-superimposable mirror images*. ("Superimposable" means that if one structure is laid over the other, the positions of all the atoms will match.) Like geometric isomers, optical isomers come in pairs. However, the optical isomers of a compound have *identical* physical and chemical properties, such as melting point, boiling point, dipole moment, and chemical reactivity toward molecules that are not enantiomers themselves. Enantiomers differ from each other in their interactions with plane-polarized light, as we will see.

The structural relationship between two enantiomers is analogous to the relationship between your left and right hands. If you place your left hand in front of a mirror, the image you see will look like your right hand (Figure 4.18). We say that your left hand and right hand are mirror images of each other. They are nonsuperimposable, however, because when you place your left hand over your right hand (with both palms facing down), they do not match.

The most common occurrences of optical isomerism occur in molecules that contain carbon atoms bonded to four different elements or molecular fragments. In the compound CHFClBr, for example, the geometry about the central carbon atom is tetrahedral ($N_S = 4$, no lone pairs). Two enantiomers are possible:

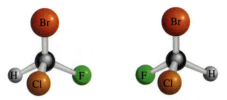

These two structures are mirror images, and it is impossible to rotate the structure on the left so that it is identical to the structure on the right, that is, they are not super-imposable. Contrast this to the molecule CH_2ClBr for which the mirror images are superimposible (Figure 4.19). Optical isomerism can also be found in many transition metal compounds and will be discussed in Chapter 14.

Optical isomers are described as ***chiral*** (from the Greek word for "hand") because, like your left and right hands, chiral molecules are nonsuperimposable. Isomers that are superimposable with their mirror images are said to be *achiral*. The central atom in a achiral molecule (for example, the carbon atom in CHFClBr) is called a *chiral center*. Chiral molecules play a vital role in enzyme reactions in biological systems. Many drug molecules are chiral, though only one of the pair of chiral isomers is biologically effective.

Chiral molecules are said to be optically active because they rotate the plane of polarization of polarized light as it passes through them. Unlike ordinary light, which vibrates in all directions, *plane-polarized light* vibrates only in a single plane. We use a ***polarimeter*** to *measure the rotation of polarized light by optical isomers* (Figure 4.20).

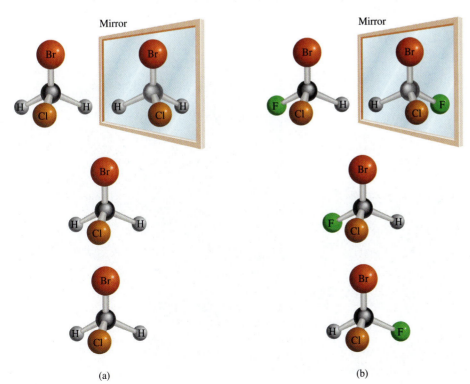

(a)   (b)

Figure 4.19 (a) The CH_2ClBr molecule and its mirror image. Because the molecule and its mirror image are superimposable, the molecule is said to be achiral. (b) The CHFClBr molecule and its mirror image. Because the molecule and its mirror image are not superimposable, no matter how we rotate one with respect to the other, the molecule is said to be chiral.

Figure 4.20 Operation of a polarimeter. Initially, the tube is filled with a solution of an achiral compound. The analyzer is rotated so that its plane of polarization is perpendicular to that of the polarizer. Under this condition, no light reaches the observer. Next, a chiral compound is placed in the tube as shown. The plane of polarization of the polarized light is rotated as it travels through the tube so that some light reaches the observer. Rotating the analyzer (either to the left or to the right) until no light reaches the observer again allows the angle of optical rotation to be measured.

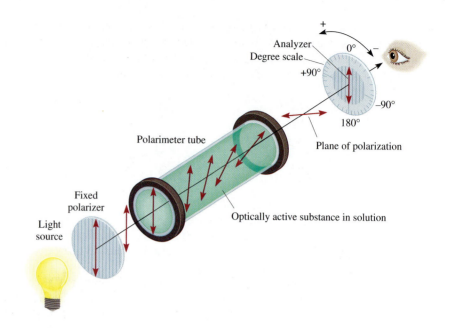

A beam of unpolarized light first passes through a sheet of film, called the polarizer, which polarizes the light, and then through a sample tube containing a solution of an optically active, chiral compound. As the polarized light passes through the sample tube, its plane of polarization is rotated either to the counterclockwise (relative to the observer in Figure 4.20) or clockwise. This rotation can be measured directly by turning the analyzer in the appropriate direction until minimal light transmission is achieved (Figure 4.21). If the plane of polarization is rotated to the right, the isomer is said to be *dextrorotatory (d)*; it is said to be *levorotatory (l)* if the rotation is to the

Figure 4.21 With one polarizing sheet over a picture, light passes through. With a second sheet of polarizing material placed over the first so that the axes of polarization of the two sheets are perpendicular, little or no light passes through. If the axes of polarization of the two sheets were parallel, light would pass through.

left. *The d and l isomers of a chiral substance,* called *enantiomers,* always rotate the light by the same amount, but in opposite directions. Thus, in *an equimolar mixture of two enantiomers,* called a **racemic mixture,** the net rotation is zero.

4.5 | Bonding in Polyatomic Molecules Can Be Explained Using Molecular Orbitals

Thus far we have discussed the chemical bonding in polyatomic molecules in terms of the VB model, or more crudely in terms of Lewis structures. These two treatments are related in that they focus on chemical bonding in terms of the sharing of electron pairs by adjacent atoms. In many cases, however, a more sophisticated approach based on *molecular orbital* concepts is needed to accurately picture the electronic structure of polyatomic molecules—even on a qualitative level.

The Ozone Molecule, O_3

A case in point is the ozone molecule, O_3. The two O—O bond lengths in ozone are found experimentally to be equivalent (equal in length and bond energy). However, the Lewis structure for O_3 is

$$\ddot{O}{=}\ddot{O}{-}\ddot{\underset{..}{O}}:$$

which implies that one O—O bond should be shorter than the other, contrary to experiment. In Section 3.4 we overcame this dilemma by introducing the concept of resonance. The two resonance structures for O_3 are

$$\ddot{O}{=}\ddot{O}{-}\ddot{\underset{..}{O}}: \longleftrightarrow :\ddot{\underset{..}{O}}{-}\ddot{O}{=}\ddot{O}$$

In Section 4.5, we will tackle the problem in another way—by applying the molecular orbital approach.

The two σ bonds in ozone can each be well described within the VB picture as the overlap between an sp^2 hybrid orbital on the central oxygen atom and an sp^2 hybrid orbital on one of the terminal oxygen atoms. The remaining five sp^2 orbitals (one on the central atom and two on each terminal oxygen) contain lone pairs. This leaves four electrons [18 valence electrons $-$ 4 σ-bonding electrons $-(5 \times 2)sp^2$ lone pairs] for π bonding. To describe the π bonding without the use of resonance requires the construction of molecular orbitals. To do this, we will use the LCAO (linear combination of atomic orbitals) approach introduced in Section 3.5. Designating the z direction as the direction perpendicular to the plane of the molecule, the orbitals involved in π bonding are the $2p_z$ orbitals on each of the three oxygen atoms: $\phi_{A,2p_z}$, $\phi_{B,2p_z}$, and $\phi_{C,2p_z}$, where the index "B" denotes the central oxygen and "A" and "C" denote the two terminal oxygen atoms. Using these atomic orbitals, we can construct the molecular orbitals as linear combinations of atomic orbitals:

$$\psi_{MO} = c_A\phi_{A,2p_z} + c_B\phi_{B,2p_z} + c_C\phi_{C,2p_z}$$

Because we start with three atomic orbitals, we can construct three molecular orbitals. The general procedure for determining the values of the coefficients c_A, c_B, and c_C for these three molecular orbitals is mathematically involved; in the present case, however, we can use what we know about the symmetry of the problem and the general behavior of electron wavefunctions to obtain a very good qualitative picture of the molecular orbitals. First, the ozone molecule is symmetric, with a mirror plane bisecting the O—O—O bond angle. Reflection of the molecular orbital through this mirror plane cannot change the electron density (given by ψ^2). For this to be true, c_A and c_C must be equal in magnitude, but may have the same or opposite signs. The symmetry puts no restriction on c_B. Also, in analogy to the particle in a one-dimensional box, the number of nodes should increase as the energy of the molecular orbital increases. (Because all atomic orbitals being considered here are p orbitals, there will be, at the very minimum, a nodal plane in the plane of the molecule for any combination of coefficients.) The lowest energy state is the linear combination that fits the symmetry with the fewest number of nodes. This can be achieved by assuming that all coefficients are positive and $c_A = c_C$, thus yielding

$$\psi_1 = c_A \phi_{A,2p_z} + c_B \phi_{B,2p_z} + c_A \phi_{C,2p_z}$$

This combination has no nodes, except for the one in the plane of the molecule. The next highest energy molecular orbital should have one additional node. By symmetry, this node must pass through the central oxygen, thus implying that c_B must be zero and $c_A = -c_C$:

$$\psi_2 = c_A(\phi_{A,2p_z} - \phi_{C,2p_z})$$

Finally, the next highest energy molecular orbital should have two additional nodes, and by symmetry these must bisect the O—O bonds, which implies that c_B is of opposite sign to c_A and c_C, and which are equal:

$$\psi_3 = c_A \phi_{A,2p_z} - c_B \phi_{B,2p_z} + c_A \phi_{C,2p_z}$$

Figure 4.22 shows three-dimensional renderings of each of these molecular orbitals. These orbitals are *delocalized molecular orbitals* because they extend over the entire molecule. The lowest energy orbital, ψ_1, is a bonding molecular orbital (π_z) in that it puts significant electron density between the oxygen atoms. The highest energy orbital, ψ_3, is antibonding (π_z^*) because there are nodal planes that bisect each of the O—O bonds. The intermediate energy state (ψ_2) has no bonding character—it neither has a node between the oxygen atoms in the O—O bonds, nor does it place significant electron density in the bonding region. Such an orbital is referred to as a *nonbonding* orbital ($\pi_z^{nonbonding}$).

Placing the four π electrons into these molecular orbitals gives the following MO diagram:

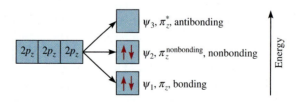

The nonbonding orbital $\pi_z^{nonbonding}$ is the **highest occupied molecular orbital** (or HOMO) and the antibonding orbital π_z^* is the **lowest unoccupied molecular orbital** (or LUMO). There are two electrons in bonding orbitals and none in antibonding orbitals, so the total bond order is 1 (giving a bond order per bond of ½). This is consistent with the idea of resonance in that a single π-bonding pair is spread over two bonds. In fact, the ground state bonding MO represents what we really mean when we refer to "resonance" between the two Lewis (or VB) structures, only in the MO picture it arises naturally, without having to specifically invoke the concept of resonance. Within the VB picture, the nonbonding molecular orbital would represent a "resonance" between a lone pair of electrons localized in the $2p_z$ orbital on oxygen

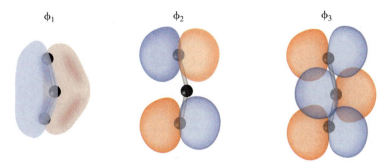

ϕ_1　　　　ϕ_2　　　　ϕ_3

Figure 4.22 Surface boundary plots of the three π molecular orbitals in ozone discussed in text. Left to right: ψ_1, ψ_2, and ψ_3. ψ_1 is an occupied bonding orbital, ψ_2 is an occupied nonbonding orbital, and the ψ_3 antibonding orbital is unoccupied. Blue and red denote $+$ and $-$ regions of the orbitals, respectively.

1 and a lone pair localized in the $2p_z$ orbital on oxygen 3. (Recall that the other lone pairs are located in sp^2 orbitals of the oxygen atoms.)

This simple MO picture for ozone can also be employed to describe the bonding in other bent triatomic molecules of second-row elements, such as NO_2. Example 4.7 applies molecular orbital theory to CO_2, a *linear* second-row triatomic molecule.

Example 4.7

Use molecular orbital theory to describe the π bonding in CO_2.

Strategy Identify the atomic orbitals involved in π bonding in carbon dioxide and use them to construct molecular orbitals that fit the symmetry of the molecule.

Solution Because CO_2 is a linear molecule, the carbon atom is sp hybridized and the σ bonding can be well described by the overlap of the sp hybrid orbitals on C with the $2p_x$ orbitals on the oxygen atoms. The $2s$ orbitals on each oxygen are filled with a nonbonding ("lone") electron pair.[7] The total number of valence electrons in CO_2 is $4 + 2(6) = 16$. So far we have accounted for eight electrons (four in the σ bonds and four non-bonded lone pair electrons on oxygen), leaving eight electrons to place in π molecular orbitals.

The six atomic orbitals available for π bonding are the $2p_z$ and $2p_y$ orbitals on the carbon atom and two oxygen atoms. Because we are starting with six atomic orbitals, we obtain six molecular orbitals. The CO_2 molecule is symmetric about the central carbon atom, so any molecular orbital must be either symmetric or antisymmetric with respect to switching the two oxygen atoms. Also, because the $2p_z$ and $2p_y$ orbitals are at right angles to one another, there will be negligible mixing of these atomic orbitals in forming the molecular orbitals. The qualitative construction of the lowest energy molecular orbitals can be accomplished by finding the combinations of the atomic orbitals that give the fewest nodes:

$$\pi_z = c_1\phi_{O_1,2p_z} + c_2\phi_{C,2p_z} + c_1\phi_{O_2,2p_z}$$
$$\pi_y = c_1\phi_{O_1,2p_y} + c_2\phi_{C,2p_y} + c_1\phi_{O_2,2p_y}$$

These two molecular orbitals are equal in energy (*degenerate*) and are bonding orbitals.

—Continued

7. Alternatively, consider the valence electrons on the oxygen atoms to be sp hybridized. Each sigma bond would then be formed from the overlap between an sp-hybridized orbital on C and one sp-hybridized orbital on an oxygen. The lone pair on each O atom would occupy the remaining sp orbitals. For the purposes of this problem, either description is acceptable because the molecular orbital description of the pi bonding is unaffected by the choice.

Continued—

The next lowest energy MOs are obtained by constructing the combination of available atomic orbitals with one more node than the lowest energy π MOs. By symmetry this node must be at the carbon atom, giving two degenerate MOs:

$$\pi_z^{\text{nonbonding}} = c_1\phi_{O_1,2p_z} - c_1\phi_{O_2,2p_z}$$
$$\pi_y^{\text{nonbonding}} = c_1\phi_{O_1,2p_y} - c_1\phi_{O_2,2p_y}$$

These two orbitals are nonbonding because the electron density is localized on the oxygen atoms.

The two remaining highest energy orbitals have nodes bisecting the C—O bond and are therefore antibonding:

$$\pi_z^{*} = c_1\phi_{O_1,2p_z} - c_2\phi_{C,2p_z} + c_1\phi_{O_1,2p_z}$$
$$\pi_y^{*} = c_1\phi_{O_1,2p_y} - c_2\phi_{C,2p_y} + c_1\phi_{O_1,2p_y}$$

Placing the eight remaining valence electrons into these orbitals gives the following molecular orbital diagram:

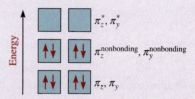

Practice Exercise Use molecular orbital theory to describe the π bonding in NO_2.

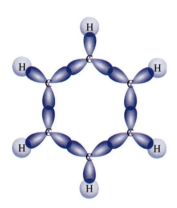

Figure 4.23 The sigma bond framework in the benzene molecule. Each carbon is sp^2 hybridized and forms a sigma bond with two adjacent carbon atoms and another sigma bond with the $1s$ orbital on hydrogen.

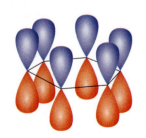

Figure 4.24 The six $2p_z$ orbitals on the carbon atoms in benzene. Blue and red colors denote regions where the orbitals are positive or negative in sign, respectively.

The Benzene Molecule

Benzene (C_6H_6) is a planar hexagonal molecule with carbon atoms situated at the six corners. All carbon-carbon bonds are equal in length and strength, as are all carbon-hydrogen bonds, and the C—C—C and H—C—C bond angles are all 120°. Each carbon atom, therefore, is sp^2 hybridized, and each forms three σ bonds with two adjacent carbon atoms and a hydrogen atom (Figure 4.23).

This arrangement leaves an unhybridized $2p_z$ orbital on each carbon atom perpendicular to the plane of the benzene molecule, or *benzene ring,* as it is often called. So far the description resembles the configuration of ethylene (C_2H_4), discussed in Section 4.3, except that in this case there are six unhybridized $2p_z$ orbitals in a cyclic arrangement (Figure 4.24). Because of their similar shape and orientation, each $2p_z$ orbital overlaps two others, one on each adjacent carbon atom.

We can explain the π bonding in benzene by constructing molecular orbitals from linear combinations of these unhybridized $2p_z$ atomic orbitals. Because we start with six atomic orbitals, we will form six molecular orbitals, which are shown in Figure 4.25.

The lowest energy state (π_1) has equal contributions from each of the atomic orbitals and has no nodes around the ring (not counting the nodal plane in the plane of the benzene ring that exists because we are dealing with p orbitals). By symmetry, the next two molecular orbitals (π_2 and π_3) are degenerate (as are π_4 and π_5) and have a nodal plane perpendicular to the ring plane. These first three molecular orbitals are bonding orbitals. The next three orbitals (π_4, π_5, and π_6)

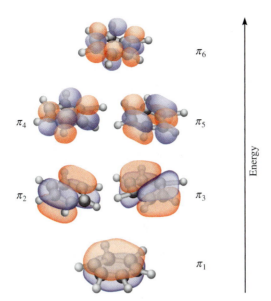

Figure 4.25 Molecular orbitals describing pi bonding in benzene. The three lowest energy molecular orbitals are bonding molecular orbitals and are occupied in the ground state. The three higher unoccupied orbitals in the ground state are antibonding.

π_6

π_4 π_5

π_2 π_3

π_1

Energy

have multiple nodal planes perpendicular to the plane of the benzene ring. Because these nodal planes also bisect the C—C bonds, they are classified as antibonding orbitals. Orbital π_6 has the greatest number of nodal planes, so it is the highest energy orbital.

Placing the six π electrons into these molecular orbitals gives the following MO diagram:

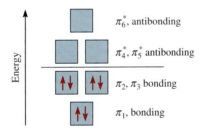

π_6^*, antibonding

π_4^*, π_5^* antibonding

π_2, π_3 bonding

π_1, bonding

Energy

Electrostatic potential map of benzene shows the electron density (red color) above and below the plane of the molecule. For simplicity, only the framework of the molecule is shown.

In the ground state, all six pi electrons go into bonding orbitals, leaving the antibonding orbitals unoccupied. This gives a total bond order for the pi electrons of 3, which when shared over six bonds gives an average bond order per pi bond of 1/2. (Each sigma bond has a bond order of 1, so the total bond order per bond is $1 + 1/2 = 3/2$.) Delocalized molecular orbitals give a much more detailed and accurate picture of bonding in molecules like benzene than does the simple (but nevertheless useful) concept of resonant Lewis structures, which is based on valence bond theory.

Molecules with delocalized molecular orbitals are generally more stable than those containing molecular orbitals localized on only two atoms. The benzene molecule, for example, which contains delocalized molecular orbitals, is chemically less reactive (and hence more stable) than molecules containing localized C=C bonds, such as ethylene. Benzene is so stable because the energy of the pi electrons is lower when the electrons are delocalized over the entire molecule than when they are localized in individual bonds, much as the energy of the particle in a one-dimensional box is lowered when the length of the box is increased (see Section 1.3).

Resonance structures of benzene

Buckyball, Anyone?

In 1985 chemists at Rice University in Texas used a high-powered laser to vaporize graphite in an effort to create unusual molecules believed to exist in interstellar space. Mass spectrometry revealed that one of the products was an unknown species with the formula C_{60}. Because of its size and the fact that it is pure carbon, this molecule has an exotic shape, which the researchers worked out using paper, scissors, and tape. Subsequent spectroscopic and X-ray measurements confirmed that C_{60} is shaped like a hollow sphere with a carbon atom at each of 60 vertices. The molecule was named buckminsterfullerene (or "buckyball" for short) after the British architect and futurist Buckminster Fuller. Geometrically, buckminsterfullerene is the most symmetrical molecule known. In spite of its unique features, however, its bonding scheme is straightforward. Each carbon is sp^2 hybridized, and there are extensive delocalized molecular orbitals over the entire structure. The discovery of buckyball generated tremendous interest within the scientific community. Here was a new allotrope of carbon with an intriguing geometry and unknown properties to investigate. Since 1985 chemists have created a whole class of *fullerenes,* with 70, 76, and even larger numbers of carbon atoms. Moreover, buckyball has been found to be a natural component of soot, and the fullerenes C_{60} and C_{70} have turned up in a rock sample from northwestern Russia.

Fullerenes represent a whole new concept in molecular architecture with far-reaching implications. Studies have shown that the fullerenes and their compounds can act as high-temperature superconductors and lubricants, as well as catalysts. One fascinating discovery, made in 1991 by Japanese scientists, was the identification of structural relatives of buckyball. These molecules are hundreds of nanometers long with a tubular shape and an internal cavity about 15 nm in diameter. Dubbed "buckytubes" or "nanotubes" (because of their size), these molecules have two distinctly different structures. One is a single sheet of graphite that is capped at both ends with a kind of truncated buckyball. The

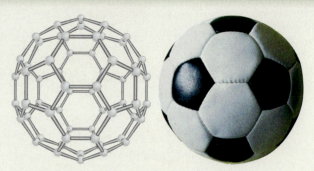

The geometry of a buckyball (C_{60}) (left) resembles a soccer ball (right). Scientists arrived at this structure by fitting together paper cutouts of enough hexagons and pentagons to accommodate 60 carbon atoms at the points where they intersect.

other is a scroll-like tube having anywhere from 2 to 30 graphite-like layers. Buckytubes are several times stronger than steel wires of similar dimensions, and one day they might be used to make ultralight bicycles, rocket motor casings, and tennis racquets. They could also serve as casting molds for very thin metal wires to be used in microscopic integrated circuits or as "bottles" for storing molecules. Recently, the discovery that fluids could be made to flow through nanotube "pipes" has spawned the field of "nanofluidics."

In the first biological application of buckyball, chemists at the University of California at San Francisco and Santa Barbara made a discovery in 1993 that could help in designing drugs to treat AIDS. The human immunodeficiency virus (HIV) that causes AIDS reproduces by synthesizing a long protein chain, which is cut into smaller segments by an enzyme called HIV-protease.

One way to stop AIDS, then, might be to inactivate the enzyme. When the chemists reacted a water-soluble derivative of buckyball with HIV-protease, they found that the buckyball derivative binds to the portion of the enzyme that

4.6 | The Interactions Between Molecules Greatly Affect the Bulk Properties of Materials

Intermolecular interactions are *forces between molecules,* whereas *intramolecular interactions* are *forces that hold atoms together in a molecule* (that is, they are responsible for chemical bonding). Intramolecular interactions stabilize individual molecules, whereas intermolecular interactions are primarily responsible for the bulk properties of matter (for example, melting point and boiling point). Intermolecular interactions are generally much weaker than intramolecular interactions. It usually requires much less energy to evaporate a liquid than to break the bonds in the molecules of the

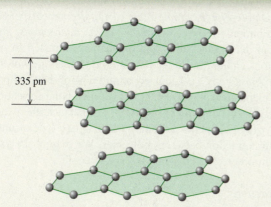

335 pm

Graphite is made up of layers of six-membered rings of carbon.

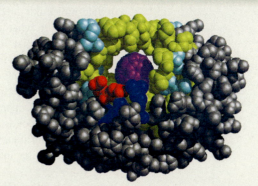

Computer-generated model of the binding of a buckyball derivative to the site of an HIV-protease that normally attaches to a protein needed for the reproduction of HIV. The bucykball structure (purple) fits tightly into the active site, thus preventing the enzyme from carrying out its function.

would ordinarily cleave the reproductive protein, thereby preventing HIV from reproducing. Consequently, the virus could no longer infect the human cells grown in the laboratory. The buckyball compound itself is an unsuitable drug for use against AIDS because of potential side effects and delivery difficulties, but it does provide a model for the development of such drugs.

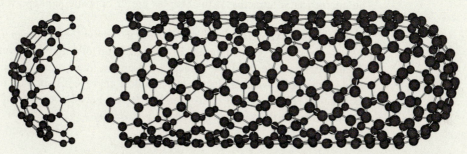

The structure of a buckytube that consists of a single layer of carbon atoms. Note that the truncated buckyball "cap," which has been separated from the rest of the buckytube in this view, has a different structure than the graphitelike cylindrical portion of the tube. Chemists have devised ways to open the cap to place other molecules inside the tube.

liquid. For example, it takes about 41 kJ of energy to vaporize 1 mol of water at its boiling point, but about 930 kJ of energy to break the two O—H bonds in 1 mol of water molecules.

At the boiling point, enough energy must be supplied to overcome the attractive forces among molecules before they can enter the vapor phase. If it takes more energy to separate molecules of substance A than of substance B because the A molecules are held together by stronger intermolecular forces, then the boiling point of A is higher than that of B. The same principle applies to the melting points of the substances. In general, the melting points of substances increase as the strength of their intermolecular interactions increases.

Types of Intermolecular Interactions

To discuss the properties of condensed matter, we must understand the different types of intermolecular interactions. *Dipole-dipole, dipole-induced dipole,* and *dispersion forces* make up what chemists commonly refer to as *van der Waals interactions,* after the Dutch physicist Johannes van der Waals.[8] Ions and dipoles are attracted to one another by electrostatic interactions called *ion-dipole interactions,* which are *not* van der Waals interactions. *Hydrogen bonding* is a particularly strong type of dipole-dipole interaction. Because only a few elements can participate in hydrogen bond formation, it is treated as a separate category. Depending on the phase of a substance, the nature of its chemical bonds, and the types of elements present, more than one type of interaction may contribute to the total attraction between molecules. We present the various intermolecular interactions in order of decreasing average strength.

Ion-Ion and Ion-Dipole Interactions

The strongest intermolecular interactions are those between ions and between ions and dipoles. We have encountered this **ion-ion interaction** before in the context of chemical bonding in ionic solids, but it is also a major intermolecular interaction in solutions of ionic compounds, such as aqueous sodium chloride. As discussed in Section 0.1, the interaction between two ions of charge q_A and q_B separated by a distance r is given by the Coulomb potential,

$$V = \frac{q_A q_B}{4\pi\varepsilon_0 r} \tag{4.2}$$

The interaction is attractive if the charges are of opposite sign and repulsive for like charges.

    Ion-dipole interactions are *the dominant interactions between ions and polar molecules* and are very important in the solvation of ionic compounds in a polar solvent. For example, it is the attractive ion-dipole interactions of the ions Na^+ and Cl^- with the polar water molecule that stabilize an aqueous solution of NaCl against precipitation. The potential energy of interaction between an ion of charge q at a distance r from a dipole of magnitude μ is given by

$$V = -\frac{q\mu\cos\theta}{4\pi\varepsilon_0 r^2} \tag{4.3}$$

where θ is the angle between the direction of the dipole moment and the axis between the dipole and the charge (Figure 4.26).

---

8. Johannes Diderck van der Waals (1837–1923). Dutch physicist. Van der Waals received the Nobel Prize in Physics in 1910 for his work on the properties of gases and liquids.

Figure 4.26 Schematic diagram of the interaction between an ion and a polar molecule (dipole) showing the angle θ between the dipole orientation vector and the ion-dipole separation axis.

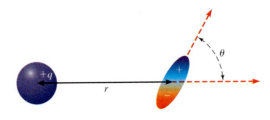

For a positively charged ion (cation), the interaction is attractive if the dipole is pointing away from the ion, (that is, if the negative end of the polar molecule is closer to the cation). If the dipole is pointing toward the cation, the potential energy is repulsive. If the dipole is oriented perpendicular to the ion-dipole axis, the interaction is zero.

$$+q \quad \mu \qquad\qquad +q \quad \mu \qquad\qquad\qquad +q$$
$$\bullet \;\longrightarrow \qquad\qquad \bullet \;\longleftarrow \qquad\qquad \mu \uparrow \quad \bullet$$

| attractive | repulsive | zero interaction |

The strength of this interaction depends on the charge and size of the ion and on the magnitude of the dipole moment and size of the molecule. The charges on cations are generally more concentrated because cations are usually smaller than anions. Therefore, a cation interacts more strongly with dipoles than does an anion having a charge of the same magnitude. Also, smaller ions interact more strongly than larger ones with neighboring polar molecules because of the inverse dependence of the interaction strength with separation. For example, the interaction between a Na^+ cation and water molecules in an aqueous solution of a sodium salt is considerably weaker than that between a Mg^{2+} ion and water molecules in an aqueous magnesium salt solution. This is due both to the smaller size of the magnesium ion and to its greater charge (Figure 4.27). Similar differences exist for anions of different charges and sizes.

Example 4.8

A sodium ion (Na^+) is situated in air at a distance of 4.0 Å (400 pm) from an H_2O molecule with a dipole moment of 1.87 D. Calculate the potential energy of interaction in kJ mol^{-1} for the dipole orientation with the lowest potential energy.

Strategy Use Equation 4.3. The lowest energy occurs when the dipole is pointing away from the cation ($\theta = 0$, so $\cos\theta = 1$). The conversion factor is $1\,D = 3.336 \times 10^{-30}$ C m.

Solution From Equation 4.3 we have

$$V = -\frac{q\mu\cos\theta}{4\pi\varepsilon_0 r^2}$$

$$= -\frac{(1.602 \times 10^{-19}\,C)\left(\dfrac{3.336 \times 10^{-30}\,C\,m}{1\,D}\right)(1.87\,D)}{4\pi(8.854 \times 10^{-12}\,C^2 N^{-1} m^{-2})(4.0 \times 10^{-10}\,m)^2}$$

$$= -5.6 \times 10^{-20}\,J$$

Converting to kJ mol^{-1} gives

$$V = (-5.6 \times 10^{-20}\,J)(6.022 \times 10^{23}\,mol^{-1})\frac{1\,kJ}{1000\,J}$$

$$= -33.8\,kJ\,mol^{-1}$$

Practice Exercise A potassium ion (K^+) is situated in air at a distance of 4.8 Å from an HCl molecule ($\mu = 1.08$ D). Calculate the potential energy of interaction in kJ mol^{-1} for the dipole orientation with the highest potential energy.

Figure 4.27 Ion-dipole interactions between Mg^{2+} and Na^+ cations and a water molecule.

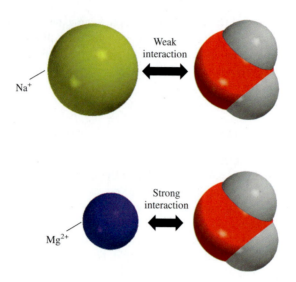

Dipole-Dipole Interactions

Dipole-dipole interactions occur between polar molecules (that is, molecules that possess permanent dipole moments). Consider the electrostatic interaction between two molecules with dipoles μ_A and μ_B separated by a distance r. In extreme cases, these two dipoles can be aligned as shown in Figure 4.28.

For the top example, the potential energy of interaction is given by

$$V = -\frac{2\mu_A\mu_B}{4\pi\varepsilon_0 r^3} \tag{4.4}$$

and for the bottom pair,

$$V = -\frac{\mu_A\mu_B}{4\pi\varepsilon_0 r^3} \tag{4.5}$$

where the negative sign indicates that this is an attractive interaction, that is, energy is released when the two molecules interact. If we reverse the direction of one of the dipoles then V becomes a positive quantity, indicating that the interaction between the two molecules is repulsive. In general, the potential energy between two fixed dipoles is proportional to the product of the dipole moments and inversely proportional to r^3,

Figure 4.28 Schematic drawing of two extreme attractive orientations of two permanent dipolar molecules. (a) Dipoles aligned end-to-end. (b) Dipoles antialigned side-to-side.

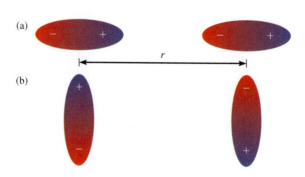

that is, $V \propto \mu_A\mu_B/r^3$. The constant of proportionality (which can be positive or negative) is determined by the positions and relative orientation of the dipoles.

Example 4.9

Two HCl molecules ($\mu = 1.08$ D) are separated by 4.5 Å (450 pm) in air. Calculate the dipole-dipole interaction energy in kJ mol^{-1} if they are oriented end-to-end, that is, H—Cl H—Cl.

Strategy Use Equation 4.4 and the conversion factor 1 D $= 3.336 \times 10^{-30}$ C m.

Solution Using Equation 4.4

$$V = -\frac{2\mu_A\mu_B}{4\pi\varepsilon_0 r^3}$$

$$V = -\frac{2\left[(1.08\text{ D})\left(\dfrac{3.336 \times 10^{-30}\text{ C m}}{1\text{ D}}\right)\right]^2}{4\pi(8.854 \times 10^{-12}\text{ C}^2\text{N}^{-1}\text{m}^{-2})(4.5 \times 10^{-10}\text{ m})^3} = -2.6 \times 10^{-21}\text{ N m}$$

$$= -2.6 \times 10^{-21}\text{J}$$

To express the potential energy on a per mole basis, we write

$$V = (-2.6 \times 10^{-21}\text{J})(6.022 \times 10^{23}\text{ mol}^{-1})\frac{1\text{ kJ}}{1000\text{ J}}$$

$$= -1.6\text{ kJ mol}^{-1}$$

Comment The value calculated for the dipole-dipole interaction energy illustrates the difference in magnitude between intermolecular and intramolecular interactions. The energy required to break the H—Cl intermolecular bond is 439.9 kJ mol^{-1} (from Table 3.3), which is significantly larger than the dipole-dipole interaction considered here.

Practice Exercise Two H_2O molecules ($\mu = 1.87$ D) are separated by 4.0 Å (400 pm) in air. Calculate the dipole-dipole interaction energy in kJ mol^{-1} if the dipole moments are antialigned as in Figure 4.28(b).

Equations 4.4 and 4.5 assume that the dipoles are fixed in orientation, which is generally appropriate for molecules in the solid phase. In liquids and gases, however, the molecules are free to rotate, so you might expect the average value of V to be zero because there would be as many repulsive interactions as attractive interactions. Because of their lower potential energy, however, the orientations giving rise to attractive interactions are favored over the repulsive orientations, which results in a higher potential energy. A calculation using *statistical mechanics* (discussed in Chapter 8) shows that the average or net energy of interaction of freely rotating permanent dipoles is given by

$$V = -\frac{2}{3\,k_B T}\frac{\mu_A^2\mu_B^2}{(4\pi\varepsilon_0)^2 r^6} \tag{4.6}$$

where T is the absolute temperature (in Kelvin) and k_B is Boltzmann's constant. The potential energy V is inversely proportional to the sixth power of r, so the energy of interaction falls off rapidly with distance. Attractive interactions that fall off as r^{-6} are collectively referred to as **van der Waals interactions.** Also, V is inversely proportional to the temperature, T, because at higher temperatures the kinetic energy of the molecules is greater, a condition unfavorable to aligning dipoles for attractive interaction. In other words, the dipole-dipole interaction will gradually average out to zero with increasing temperature.

Ion-Induced Dipole and Dipole-Induced Dipole Interactions

Thus far in this section, all of the species we have considered have been charged or polar. What types of intermolecular forces are present when one or both interacting species is a neutral atom or nonpolar molecule? As an example, consider the interaction of an ion or an polar molecule with an atom (such as helium) or nonpolar molecule. If we place an ion or polar molecule near the atom (or nonpolar molecule), the electron distribution of the atom will be distorted by the electrostatic force exerted by the charge of the ion or the dipole moment of the polar molecule, causing the formation of a dipole on the atom (or nonpolar molecule) (Figure 4.29).

The dipole generated in the atom or nonpolar molecule is said to be an **induced dipole** because *the separation of positive and negative charges in the atom (or nonpolar molecule) is due to the proximity of an ion of a polar molecule.* The magnitude of the induced dipole moment, μ_{ind}, is proportional to the strength of the electric field, E, generated by the ion or polar molecule

$$\mu_{ind} \propto E$$
$$= \alpha E \qquad (4.7)$$

where the proportionality constant, α, is called the **polarizability** and is a measure of the ease with which the electron distribution in the atom, molecule, or ion can be distorted by an external electric field. Because μ_{ind} has units of coulomb-meters (C m) and an electric field has units of $J\ m^{-1}\ C^{-1}$, the polarizability as defined by Equation 4.7 will have SI units of $C^2\ m^2\ J^{-1}$. Generally, the larger the number of electrons and the more diffuse the electron charge cloud in an atom or molecule, the greater its polarizability. By *diffuse cloud,* we mean an electron cloud that is spread over an appreciable volume, so that the electrons are not held tightly by the nucleus. For example, Xe is more polarizable than Kr, which is more polarizable than He. Also molecules with highly delocalized molecular orbitals, such as benzene or ozone, will also be very polarizable.

The attractive interaction between an ion and an induced dipole in a neutral, nonpolar species with polarizability (α) is called an **ion-induced dipole interaction** and is given mathematically by

$$V = -\frac{1}{2}\frac{\alpha q^2}{(4\pi\varepsilon_0)^2\, r^4} \qquad (4.8)$$

Similarly, the interaction between the dipole moment of a polar molecule and the induced dipole of a neutral, nonpolar species is called a **dipole-induced dipole interaction** (see Figure 4.29). If the dipole moment of the polar molecule is of magnitude μ, then the potential energy for the dipole-induced dipole interaction is given by

$$V = -\frac{\alpha\mu^2}{(4\pi\varepsilon_0)^2\, r^6} \qquad (4.9)$$

where α is the polarizability of the nonpolar molecule. The dipole-induced dipole interaction is another example of the van der Waals interaction because it falls off as

For simplicity, we use the term "intermolecular forces" for both atoms and molecules.

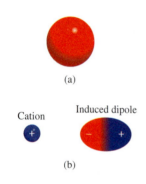

(a)

(b)

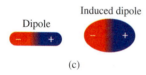

(c)

Figure 4.29 (a) An isolated helium atom has spherically symmetric electron density. (b) Induced dipole moment in helium caused by electrostatic interaction with a cation. (c) Induced dipole moment in helium caused by electrostatic interaction with the permanent dipole of a polar molecule. For the dipoles blue indicates electron deficient regions (positively charged) and red denotes electron rich regions (negatively charged).

Table 4.5	Polarizabilities of Some Atoms and Molecules		
Atom	$\dfrac{\alpha}{4\pi\varepsilon_0}/10^{-30}\text{m}^3$	Molecule	$\dfrac{\alpha}{4\pi\varepsilon_0}/10^{-30}\text{m}^3$
He	0.22	H_2	0.89
Ne	0.44	N_2	1.93
Ar	1.85	CO_2	2.93
Kr	2.82	NH_3	2.51
Xe	5.11	CH_4	2.90
I	5.51	C_6H_6	11.6
Cs	46.6	CCl_4	11.7

the inverse sixth power of distance. Both the ion-induced dipole and dipole-induced dipole interactions (Equation 4.8 and 4.9) are independent of temperature because induced dipole moment is formed instantaneously and the value of potential energy (V) is unaffected by the thermal motion of the molecules. Table 4.5 lists the polarizability of some atoms and simple molecules.

In general, both ion-induced dipole and dipole-induced dipole interactions are weaker than ion-dipole or dipole-dipole interactions. For this reason, ionic compounds such as NaCl and polar molecules like SO_2 and HCl are not soluble in nonpolar solvents such as benzene, hexane, or carbon tetrachloride—the interactions between the solvent and solute in these cases are not strong enough to overcome the ionic or dipole-dipole interactions between the solutes.

Dispersion (or London) Interactions

The cases considered thus far consist of at least one charged ion or one permanent dipole (polar molecule) among the interacting species. We must also consider the case where both interaction species are nonpolar molecules. If there were no attractive interactions between such species, then nonpolar substances such as helium, nitrogen (N_2), carbon tetrachloride, or benzene would only exist as gases, which is certainly not the case. We expect such an interaction to be weak; indeed, the low boiling point (4 K) of helium suggests that only very weak interactions hold the atoms together in the liquid state. For molecules with larger polarizabilities, however, this interaction can be significant and can be stronger than some dipole-dipole and dipole-induced dipole interactions. For example, the nonpolar carbon tetrachloride molecule (CCl_4) is liquid at room temperature (b.p. 76.5°C) because has a large polarizability (see Table 4.5) and thus strongly attractive dispersion forces. The boiling point of CCl_4 is significantly higher than the polar molecule methyl fluoride (CH_4F, $\mu = 1.86$ D) (b.p. -141.8°C), indicating that the dispersion forces in carbon tetrachloride are stronger than the dipole-dipole forces in methyl fluoride.

In 1930 Fritz London showed using quantum mechanics that the potential energy arising from the interactions of two identical atoms or nonpolar molecules is given by

$$V = -\frac{3}{4}\frac{\alpha^2 I_1}{(4\pi\varepsilon_0)^2 r^6} \tag{4.10}$$

where α is the polarizability of the atom or molecule (Equation 4.7) and I_1 is its first ionization. For nonidentical atoms or molecules A and B, Equation 4.10 becomes

$$V = -\frac{3}{2}\frac{I_{1,A}I_{1,B}}{I_{1,A} + I_{1,B}}\frac{\alpha_A\alpha_B}{(4\pi\varepsilon_0)^2 r^6} \tag{4.11}$$

CCl_4

CH_3F

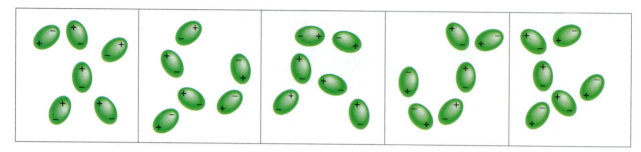

Figure 4.30 Induced dipoles interacting with each other. Such patterns exist only momentarily; new arrangements are formed in the next instant. This type of interaction is responsible for the condensation of nonpolar gases.

The forces that arise from this type of interaction are called **_dispersion_** (or **_London_**) **_forces._** Because the forces decrease with increasing distance as r^{-6}, they are also classified as van der Waals forces.

The London interaction is a quantum mechanical phenomenon; however, its physical origin can be understood as follows: Consider two atoms separated by only a few Ångstroms. The electron clouds on each atom are not static objects but are constantly fluctuating. At any time, it is possible that there is slightly more electron density on one side of an atom than the other, which generates a small dipole moment, called an *instantaneous dipole moment*. Averaged over time (that is, the time it takes to make a dipole moment measurement), however, the dipole moment of the atom is zero because the instantaneous dipoles all cancel one another. When an instantaneous dipole forms in one atom, the electron cloud of the other atom is affected and forms an induced dipole, giving an instantaneous attractive interaction between the atoms (Figure 4.30). When the instantaneous dipole on the first atom changes, so does the induced dipole maintaining the attractive interaction.

Dispersion forces usually increase with molar mass. Molecules with larger molar mass tend to have more electrons, and dispersion forces increase in strength with the number of electrons because of increases in atomic or molecular polarizability. Furthermore, larger molar mass often means a bigger atom whose electron distribution is more easily disturbed because the outer electrons are less tightly held by the nuclei. Table 4.6 compares the melting points of similar substances that consist of nonpolar molecules. As expected, the melting point increases as the number of electrons in the molecule increases. Because these are all nonpolar molecules, the only attractive intermolecular forces present are the dispersion forces.

Example 4.10 shows that if we know the kind of species present, we can readily determine the types of intermolecular interactions that exist between the species.

Table 4.6

Melting Points of Similar Nonpolar Compounds

Compound	Melting Point (°C)
CH_4	−182.5
CF_4	−150.0
CCl_4	−23.0
CBr_4	90.0
CI_4	171.0

Example 4.10

What type(s) of intermolecular interactions exist between the following pairs: (a) HBr and H_2S, (b) Cl_2 and CBr_4, (c) I_2 and NO_3^-, and (d) NH_3 and C_6H_6?

Strategy Classify the species into three categories: ionic, polar (possessing a permanent dipole moment), and nonpolar. Keep in mind that dispersion forces exist between *all* species.

—Continued

Continued—

Solution (a) Both HBr and H_2S are polar molecules; therefore, the intermolecular interactions present are dipole-dipole forces, as well as dispersion forces.

(b) Both Cl_2 and CBr_4 are both nonpolar, so there are only dispersion forces between these molecules.

(c) I_2 is a homonuclear diatomic molecule and therefore nonpolar, so the forces between it and the ion NO_3^- are ion-induced dipole forces and dispersion forces.

(d) NH_3 is polar, and C_6H_6 is nonpolar. The forces are dipole-induced dipole forces and dispersion forces.

Practice Exercise Name the type(s) of intermolecular forces that exists between molecules (or basic units) in each of the following species: (a) LiF, (b) CH_4, and (c) SO_2.

The Hydrogen Bond

Normally, the boiling points of a series of similar compounds containing elements in the same periodic group increase with increasing molar mass. This increase in boiling point is due to the increase in dispersion forces for molecules with more electrons. Hydrogen compounds of Group 4A follow this trend, as Figure 4.31 shows.

The lightest compound, CH_4, has the lowest boiling point, and the heaviest compound, SnH_4, has the highest boiling point. However, hydrogen compounds of the elements in Groups 5A, 6A, and 7A do not follow this trend. In each of these series, the lightest compound (NH_3, H_2O, and HF) has the highest boiling point, contrary to our expectations based on molar mass. This observation must mean that there are stronger intermolecular attractions in NH_3, H_2O, and HF, compared to other molecules in the same groups. In fact, this particularly strong type of intermolecular attraction is called a **hydrogen bond**, which is *a special type of dipole-dipole interaction between*

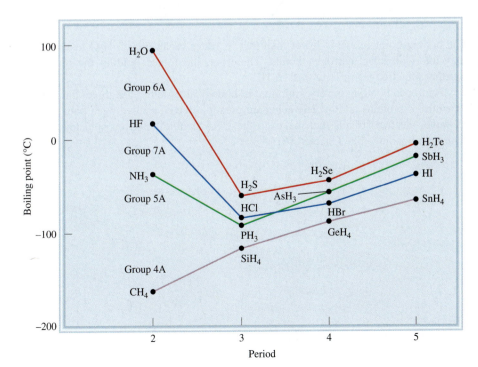

Figure 4.31 Boiling points of the hydrogen compounds of Groups 4A, 5A, 6A, and 7A elements. Although normally we expect the boiling point to increase as we move down a group, we see that three compounds (NH_3, H_2O, and HF) behave differently. The anomaly can be explained in terms of intermolecular hydrogen bonding.

the hydrogen atom in a polar bond, such as N—H, O—H, or F—H, and an N, O, or F atom. A hydrogen bond is much stronger than a typical dipole-dipole interaction because a hydrogen atom is small and, when bonded to very electronegative N, O, or F atoms, highly electron deficient (that is, positively charged). Thus, the positively charged hydrogen atom is able to get very close to the negatively charged lone pairs on small electronegative N, O, or F atoms, resulting in an especially strong electrostatic interaction.

The hydrogen-bond interaction is written

$$A—H\cdots B \text{ or } A—H\cdots A$$

where A and B represent O, N, or F; A—H is one molecule or part of a molecule; B is a part of another molecule; and the dotted line represents the hydrogen bond. The three atoms usually lie in a straight line, but the angle A—H—B (or A—H—A) can deviate as much as 30° from linearity. Note that the O, N, and F atoms all possess at least one lone pair that can interact with the hydrogen atom in hydrogen bonding.

The average energy of a hydrogen bond is quite large for a dipole-dipole interaction (up to 40 kJ mol^{-1}). Thus, hydrogen bonds have a powerful effect on the structures and properties of many compounds. Figure 4.32 shows several examples of hydrogen bonding.

The strength of a hydrogen bond is determined by the Coulombic interaction between the lone-pair electrons of the electronegative atom and the hydrogen nucleus (proton). For example, fluorine is more electronegative than oxygen, and so we would expect a stronger hydrogen bond to exist in liquid HF than in H_2O. In the liquid phase, the HF molecules form zigzag chains:

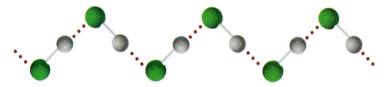

The boiling point of HF is lower than that of water because each H_2O takes part in *four* intermolecular hydrogen bonds. Therefore, the forces holding the molecules together are stronger in H_2O than in HF.

Hydrogen bonding is very important in biological systems. The three-dimensional structure of proteins and nucleic acids are greatly affected by hydrogen bonding. For example, the two strands that make up the DNA double helix are held together by hydrogen bonding (See Chapter 16)

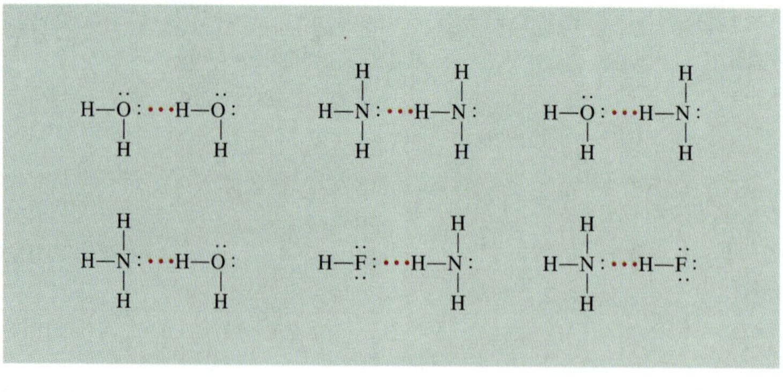

Figure 4.32 Hydrogen bonding in water, ammonia, and hydrogen fluoride. Solid lines represent covalent bonds, dotted lines represent hydrogen bonds.

Example 4.11 shows the type of species that can form hydrogen bonds with water.

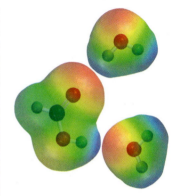

Example 4.11

Which of the following can form hydrogen bonds with water: CH_3OCH_3, CH_4, F^-, HCOOH, or Na^+?

Strategy A species can form hydrogen bonds with water if it contains one of the three electronegative elements (F, O, or N) or it has an H atom bonded to one of these three elements.

Solution There are no electronegative elements (F, O, or N) in either CH_4 or Na^+. Therefore, only CH_3OCH_3, F^-, and HCOOH can form hydrogen bonds with water.

Check Note that HCOOH (formic acid) can form hydrogen bonds with water in two different ways.

Practice Exercise Which of the following species are capable of hydrogen bonding among themselves: (a) H_2S, (b) C_6H_6, or (c) CH_3OH?

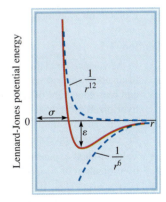

HCOOH forms hydrogen bonds with two H_2O molecules.

Including Repulsive Interactions

Up to now, we have only considered attractive intermolecular forces; however, atoms and molecules must also repel one another at short distances because otherwise, they would eventually collapse onto one another. Such a collapse is prevented by the strong repulsive electrostatic forces between electron clouds and between nuclei and by the Pauli exclusion principle, which prevents electrons from occupying the same physical space. The repulsive force between atoms typically dominates the interaction of atoms and molecules at short range. The British physicist Lennard-Jones[9] proposed the following expression to model both the attractive and repulsive interactions in nonionic systems:

$$V = 4\varepsilon\left[\left(\frac{\sigma}{r}\right)^{12} - \left(\frac{\sigma}{r}\right)^{6}\right] \tag{4.12}$$

Equation 4.12 is called the ***Lennard-Jones potential.*** The first term in brackets in Equation 4.12, which is very short ranged, represents repulsion between the particles. The attractive second term in brackets is of the van der Waals form and applicable to dipole-dipole, dipole-induced dipole, and dispersion interactions.

Figure 4.33 shows a plot of the Lennard-Jones potential between two atoms or molecules. The quantity ε in Equation 4.12 is equal to the depth of the potential well and is a measure of the strength of the intermolecular force. The quantity σ is a measure of the size of the molecules and is defined as the separation at which $V = 0$. Table 4.7 lists the Lennard-Jones parameters, ε and σ, for a few atoms and molecules. The Lennard-Jones potential is a good representation of generic intermolecular interactions in nonionic systems and is used extensively in the modeling and simulation of chemical systems.

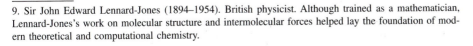

Figure 4.33 The Lennard-Jones potential energy curve (red) between two nonpolar molecules separated by a distance r. The upper and lower curves (blue) show the repulsive $1/r^{12}$ and attractive $1/r^6$ contributions, respectively.

---

9. Sir John Edward Lennard-Jones (1894–1954). British physicist. Although trained as a mathematician, Lennard-Jones's work on molecular structure and intermolecular forces helped lay the foundation of modern theoretical and computational chemistry.

Table 4.7	Lennard-Jones Parameters for Atoms and Molecules	
Atom or Molecule	ε (kJ mol^{-1})	σ (Å)
Ar	0.0997	3.40
Xe	1.77	4.10
H_2	0.307	2.93
N_2	0.765	3.92
O_2	0.943	3.65
CO_2	1.65	4.33
CH_4	1.23	3.82
C_6H_6	2.02	8.60

Summary of Facts and Concepts

Section 4.1

▶ The VSEPR model for predicting molecular geometry is based on the assumption that valence-shell electron pairs repel one another and tend to stay as far apart as possible.

▶ According to the VSEPR model, molecular geometry can be predicted from the number of bonding electron pairs and lone pairs. Lone pairs repel other pairs more forcefully than bonding pairs do and thus distort bond angles from the ideal geometry.

Section 4.2

▶ Dipole moment is a measure of the charge separation in molecules containing atoms of different electronegativities. The dipole moment of a molecule is the resultant of whatever bond moments are present. Information about molecular geometry can be obtained from dipole moment measurements.

Section 4.3

▶ In the valence bond theory of polyatomic molecules, hybridized atomic orbitals are formed by the combination and rearrangement of orbitals from the same atom. The hybridized orbitals are all of equal energy and electron density, and the number of hybridized orbitals is equal to the number of pure atomic orbitals that combine.

▶ Valence-shell expansion can be explained by assuming hybridization of s, p, and d orbitals.

▶ In sp hybridization, the two hybrid orbitals lie in a straight line; in sp^2 hybridization, the three hybrid orbitals are directed toward the corners of a triangle; in sp^3 hybridization, the four hybrid orbitals are directed toward the corners of a tetrahedron; in sp^3d hybridization, the five hybrid orbitals are directed toward the corners of a trigonal bipyramid; and in sp^3d^2 hybridization, the six hybrid orbitals are directed toward the corners of an octahedron.

▶ In an sp^2-hybridized atom (for example, carbon), the one unhybridized p orbital can form a pi bond with a p orbital on another atom. A carbon-carbon double bond consists of a sigma bond and a pi bond. In an sp-hybridized carbon atom, the two unhybridized p orbitals can form two pi bonds with two p orbitals on another atom (or atoms). A carbon-carbon triple bond consists of one sigma bond and two pi bonds.

Section 4.4

▶ Chemically distinct compounds that have the same molecular formula are called isomers. Structural isomers are isomers that have different bond connectivity.

▶ Stereoisomers are isomers that have identical bond connectivity but have different three-dimensional structures. There are two types of stereoisomers: geometric isomers and optical isomers.

Section 4.5

▶ Bonding in many polyatomic molecules, such as ozone or benzene, can often be best described in terms of delocalized molecular orbitals.

Section 4.6

▶ Intermolecular forces act between molecules or between molecules and ions. Generally, these attractive forces are much weaker than bonding forces.

▶ Dipole-dipole forces and ion-dipole forces attract molecules with dipole moments to other polar molecules or ions.

▶ Dispersion forces are the result of temporary dipole moments induced in ordinarily nonpolar molecules. The extent to which a dipole moment can be induced in a molecule is called its polarizability. The term "van der Waals forces" refers to dipole-dipole, dipole-induced dipole, and dispersion forces.

▶ Hydrogen bonding is a relatively strong dipole-dipole interaction between a polar bond containing a hydrogen atom and an electronegative O, N, or F atom. Hydrogen bonds between water molecules are particularly strong.

Key Words

Problems

The Basic Three-Dimensional Structure of a Molecule Can Be Predicted Using the VSEPR Model

4.1 Predict the geometries of the following species using the VSEPR method: (a) PCl_3, (b) $CHCl_3$, (c) SiH_4, and (d) $TeCl_4$.

4.2 Predict the geometries of the following species: (a) $AlCl_3$, (b) $ZnCl_2$, and (c) $ZnCl_4^{2-}$.

4.3 Predict the geometry of the following molecules and ions using the VSEPR model: (a) CBr_4, (b) BCl_3, (c) NF_3, (d) H_2Se, and (e) NO_2^-.

4.4 Predict the geometry of the following molecules and ion using the VSEPR model: (a) CH_3I, (b) ClF_3, (c) H_2S, (d) SO_3, and (e) SO_3^{2-}.

4.5 Predict the geometry of the following molecules using the VSEPR method: (a) $HgBr_2$, (b) N_2O (arrangement of atoms is NNO), and (c) SCN (the arrangement of atoms is SCN).

4.6 Predict the geometries of the following ions: (a) NH_4^+, (b) NH_2^-, (c) CO_3^{2-}, (d) ICl_2^-, (e) ICl_4^-, (f) AlH_4^-, (g) $SnCl_5^-$, (h) H_3O^+, and (i) BeF_4^{2-}.

4.7 Predict the geometries of XeO_2 and XeO_4.

4.8 Describe the geometry about each of the three central atoms in the CH_3COOH (acetic acid) molecule.

4.9 Which of the following species are tetrahedral? $SiCl_4$, SeF_4, XeF_4, CI_4, or $CdCl_4^{2-}$.

4.10 Predict the bond angles for the following molecules: (a) $BeCl_2$, (b) BCl_3, (c) CCl_4, (d) Hg_2Cl_2 (arrangement of atoms: ClHgHgCl), (e) $SnCl_2$, (f) H_2O_2, and (g) SnH_4.

4.11 Which of the following molecules is linear ICl_2^-, IF_2^+, OF_2, SnI_2, or $CdBr_2$.

4.12 Which of the following species is not likely to have a tetrahedral shape? (a) $SiBr_4$, (b) NF_4^+, (c) SF_4, (d) $BeCl_4^{2-}$, (e) BF_4^-, or (f) $AlCl_4^-$.

4.13 Predict the geometry of sulfur dichloride (SCl_2) and the hybridization of the sulfur atom.

The Polarity of a Molecule Can Be Described Quantitatively by Its Dipole Moment

4.14 Arrange the following molecules in order of increasing dipole moment: H_2O, H_2S, H_2Te, and H_2Se.

4.15 The dipole moments of the hydrogen halides decrease from HF to HI. Explain this trend.

4.16 Which of the following molecules have nonzero dipole moments? OCS, $BeCl_2$, OCl_2, BCl_3, PCl_3, SF_6, or SF_4. Explain your answers using three-dimensional molecular structures.

4.17 List the following molecules in order of increasing dipole moment: H_2O, CBr_4, H_2S, HF, NH_3, and CO_2.

4.18 Does the molecule OCS have a higher or lower dipole moment than CS_2?

4.19 Sketch the bond moments and resultant dipole moments for the following molecules: H_2O, PCl_3, XeF_4, PCl_5, and SF_6.

4.20 Consider the molecule CF_2Cl_2. Does it have a nonzero dipole moment? If so, draw the molecule and indicate the direction of the dipole moment vector.

4.21 The dipole moment of carbon monoxide, CO, is very small (0.12 D) even though the electronegativity difference between oxygen and carbon is rather large. Can you give an explanation of this from the Lewis structure? (Hint: think about possible resonance structures.)

4.22 The dipole moment of HF is 1.92 D. Given that the H—F bond length is 91.7 pm, calculate the separated charge q, assuming the point charge model.

4.23 Which of the following molecules has a higher dipole moment?

(a)   (b)

4.24 Arrange the following compounds in order of increasing dipole moment.

(a)   (b)   (c)   (d)

Valence Bond Theory for Polyatomic Molecules Requires the Use of Hybrid Orbitals

4.25 Describe the bonding scheme of the AsH_3 molecule in terms of hybridization.

4.26 What is the hybridization state of Si in SiH_4 and in $H_3Si{-}SiH_3$?

4.27 Describe the change in hybridization (if any) of the Al atom in the following reaction:

$$AlCl_3 + Cl^- \longrightarrow AlCl_4^-$$

4.28 Consider the reaction

$$BF_3 + NH_3 \longrightarrow F_3B{-}NH_3$$

Describe the changes in hybridization (if any) of the B and N atoms as a result of this reaction.

4.29 Describe the bonding in XeF_4 in terms of hybridization.

4.30 What hybrid orbitals are used by the nitrogen atoms in the following species? (a) NH_3, (b) $H_2N{-}NH_2$, and (c) NO_3^-.

4.31 What are the hybrid orbitals of the carbon atoms in the following molecules?

(a) $H_3C{-}CH_3$

(b) $H_3C{-}CH{=}CH_2$

(c) $H_3C{-}C{\equiv}C{-}CH_2OH$

(d) $CH_3CH{=}O$

(e) CH_3COOH

4.32 Specify which hybrid orbitals are used by carbon atoms in the following species: (a) CO, (b) CO_2, and (c) CN^-.

4.33 What is the hybridization state of the central N atom in the azide ion N_3^-? (Arrangement of atoms: NNN.)

4.34 The allene molecule $H_2C{=}C{=}CH_2$ is linear (the three carbons lie on a straight line). What are the hyridization states of the carbon atoms? Draw diagrams to show the formation of σ and π bonds.

4.35 Describe the hybridization of phosphorous in PF_5.

4.36 How many sigma and pi bonds are there in each of the following:

(a)   (b)   (c)

4.37 How many pi bonds and sigma bonds are there in the tetracyanoethylene molecule?

Isomers Are Compounds That Have the Same Molecular Formula but Different Atomic Arrangements

4.38 Draw all the possible structural isomers of hexane (C_6H_{14}).

4.39 Draw all the possible structural isomers of pentanol $(C_5H_{11}OH)$.

4.40 Cyclopropane (C_3H_6) has the shape of a triangle in which a C atom is bonded to two H atoms and two other C atoms at each corner. Substituting Cl atoms for two of the hydrogen atoms gives dichlorocyclopropane. Draw all the possible isomers of dichlorocyclopropane and label them as geometric, structural, or optical isomers.

4.41 Draw three possible structural isomers for C_3H_8O.

Bonding in Polyatomic Molecules Can Be Explained Using Molecular Orbitals

4.42 Describe the lowest energy molecular orbital for π bonding in the nitrate ion, NO_3^-.

4.43 Using ozone as a model, describe the π bonding in NO_2 using a molecular orbital scheme.

4.44 Describe π bonding in NO_2^+ and NO_2^- using molecular orbitals.

4.45 Determine which of these molecules has a more delocalized molecular orbital and justify your choice.

biphenyl   *naphthalene*

(*Hint*: Both molecules contain two benzene rings. In naphthalene, the two rings are fused together. In biphenyl, the two rings are joined by a single bond, around which the two rings can rotate.)

4.46 Describe the bonding in the nitrate ion NO_3^- in terms of delocalized molecular orbitals.

4.47 A single bond is usually a σ bond and a double bond is usually made up of a σ and a π bond. Can you identify the exceptions in the homonuclear diatomic molecules of the second period?

Interactions Between Molecules Greatly Affect the Bulk Properties of Materials

4.48 The compounds Br_2 and ICl have the same number of electrons, yet Br_2 melts at $-7.2°C$, whereas ICl melts at $27.2°C$. Explain.

4.49 List the types of intermolecular forces that exist between molecules (or basic units) of the following species: (a) benzene (C_6H_6), (b) CH_3Cl, (c) PF_3, (d) $NaCl$, and (e) CS_2.

4.50 Arrange the following in order of increasing boiling point: RbF, CO_2, CH_3OH, and CH_3Br. Explain your reasoning.

4.51 Which member of the following pairs would you expect to have the highest boiling point? (a) O_2 and Cl_2, (b) SO_2 and CO_2, and (c) HF and HI.

4.52 Which substance in each of the following pairs would you expect to have the lower boiling point? (a) Ne or Xe, (b) CO_2 and CS_2, (c) CH_4 or Cl_2, (d) F_2 or LiF, and (e) NH_3 or PH_3.

4.53 Name the kinds of intermolecular interactions that must be overcome in order to (a) boil liquid ammonia, (b) melt solid phosphorus (P_4), and (c) dissolve CsI in liquid HF.

4.54 The fluorides of the second-period elements and their melting points are (a) LiF, $845°C$; BeF_2, $800°C$; BF_3, $-127.7°C$; CF_4, $-184°C$; NF_3, $-206.6°C$; OF_2, $-223.8°C$; and F_2, $-219.6°C$. Classify the type(s) of intermolecular interactions present in each compound.

4.55 Using data from Tables 4.3 and 4.5, calculate the dipole-induced dipole interaction energy between an H_2O molecule and an Ar atom that are separated by 6.0 Å in the gas phase.

4.56 Using the data from Table 4.3, calculate the energy of interaction between two NH_3 molecules separated by 4.0 Å in the gas phase, assuming that the NH_3 molecular dipoles are antialigned as in Figure 4.27.

4.57 Use Equation 4.10 and data from Tables 2.2 and 4.5 to calculate the dispersion forces between two argon atoms separated by 5.00Å (500 pm).

4.58 Using the Lennard-Jones model (Equation 4.12), plot on the same graph the potential energy of interaction between (a) two Ar atoms and (b) two Xe atoms. If you knew nothing about these two gases, what could you conclude about the relative physical properties of Ar and Xe based on this plot?

Additional Problems

4.59 Draw the Lewis structure of mercury(II) bromide. Is this molecule linear or bent? How would you establish its geometry?

4.60 Draw Lewis structures and give the other information requested for the following molecules or ions: (a) BF_3. Shape: planar or nonplanar? (b) ClO_3^-. Shape: planar or nonplanar? (c) H_2O. Show the direction of the resultant dipole moment. (d) OF_2. Polar or nonpolar molecule? (e) NO_2. Estimate the $O—N—O$ bond angle.

4.61 Predict the bond angles for the following molecules: (a) $BeCl_2$, (b) BCl_3, (c) CCl_4, (d) CH_3Cl, (e) Hg_2Cl_2 (arrangement of atoms: $ClHgHgCl$), (f) $SnCl_2$, (g) H_2O_2, and (h) SnH_4.

4.62 Describe the hybridization state of arsenic in arsenic pentafluoride (AsF_5).

4.63 The compound 1,2-dichloroethane ($C_2H_4Cl_2$) is nonpolar, while *cis*-dichloroethylene ($C_2H_2Cl_2$) has a dipole moment:

1,2-dichloroethane   *cis*-dichloroethylene

The reason for the difference is that groups connected by a single bond can rotate with respect to each other, but no rotation occurs when a double bond connects the groups. On the basis of bonding considerations, explain why rotation occurs in 1,2-dichloroethane but not in *cis*-dichloroethylene.

4.64 Does the following molecule have a dipole moment?

4.65 So-called greenhouse gases, which contribute to global warming, have a dipole moment or can be bent or distorted into shapes that have a dipole moment. Which of the following gases are greenhouse gases? N_2, O_2, O_3, CO, CO_2, NO_2, N_2O, CH_4, or $CFCl_3$.

4.66 Draw Lewis structures and give the other information requested for the following: (a) SO_3. Polar or nonpolar molecule? (b) PF_3. Polar or nonpolar molecule? (c) F_3SiH. Show the direction of the resultant dipole moment. (d) SiH_3^-. Planar or pyramidal shape? (e) Br_2CH_2. Polar or nonpolar molecule?

4.67 Draw the Lewis structure for the $BeCl_4^{2-}$ ion. Predict its geometry and describe the hybridization state of the Be atom.

4.68 The chlorine nitrate molecule $ClONO_2$ has been implicated in the destruction of ozone in the Antarctic stratosphere. Draw a plausible Lewis structure for the molecule and predict its three dimensional shape.

4.69 The dipole moment of *cis*-dichloroethylene ($C_2H_2Cl_2$) is 1.81 D at 25°C. On heating, its dipole moment begins to decrease. Give a reasonable explanation for this observation.

4.70 Antimony pentafluoride, SbF_5, reacts with XeF_4 and XeF_6 to form ionic compounds $XeF_3^+ SbF_6^-$ and $XeF_5^+ SbF_6^-$, respectively. Describe the geometries of the cations and anions in these two compounds.

4.71 The bond angle of SO_2 is very close to 120°, even though there is a lone pair on sulfur. Explain.

4.72 Which of the following molecules or ions are linear? ICl_2^-, IF_2^+, OF_2, SnI_2, or $CdBr_2$.

4.73 Briefly compare the VSEPR and hybridization approaches to the study of molecular geometry.

4.74 Use molecular orbital theory to explain the bonding in the azide ion (N_3^-). The arrangement of atoms is NNN.

4.75 The ionic character of the bond in a diatomic molecule can be estimated by the formula

$$\frac{\mu}{ed} \times 100\%$$

where μ is the experimentally measured dipole moment (in C m), e the electronic charge, and d the bond length in meters. (The quantity ed is the hypothetical dipole moment for the case in which the transfer of an electron from the less electronegative to the more electronegative atom is complete.) Given that the dipole moment and bond length of HF are 1.92 D and 91.7 pm, respectively, calculate the percent ionic character of the molecule.

4.76 Draw three Lewis structures for compounds with the formula $C_2H_2F_2$. Indicate which of the compounds are polar.

4.77 Aluminum trichloride ($AlCl_3$) is an electron deficient molecule. It has a tendency to form a dimer (a molecule made of two $AlCl_3$ units).

$$AlCl_3 + AlCl_3 \longrightarrow Al_2Cl_6$$

(a) Draw a Lewis structure for the dimer.
(b) Describe the hybridization state of Al in $AlCl_3$ and Al_2Cl_6.
(c) Sketch the geometry of the dimer.
(d) Do either of these molecules posses a dipole moment?

4.78 The stable allotropic form of phosphorous is P_4, in which each P atom is bonded to three other P atoms. Draw a Lewis structure of this molecule and describe its geometry. At high temperatures, P_4 dissociates to form P_2 molecules containing a P=P bond. Explain why P_4 is more stable than P_2.

4.79 Although both carbon and silicon are in Group 4A, very few Si=Si bonds are known. Account for the instability of silicon-to-silicon double bonds in general. (*Hint:* Compare the covalent radii of C and Si.)

4.80 Using Equations 4.10 and 4.11 and the data in Tables 2.2 and 4.5, calculate the magnitude of the dispersion interaction between (a) two Ar atoms separated by 5.00 Å and (b) an Ar atom and a Ne atom separated by the same distance as in (a).

4.81 3-azido-3-deoxythymidine, commonly known as AZT, is one of the drugs used to treat acquired immune deficiency syndrome (AIDS). What are the hybridization states of the C and N atoms in this molecule?

4.82 What are the hybridization states of the C and N atoms in this molecule?

4.83 The compounds carbon tetrachloride (CCl_4) and silicon tetrachloride ($SiCl_4$) are similar in geometry and hybridization. However, CCl_4 does not react with water but $SiCl_4$ does. Explain the difference in their chemical reactivities. (*Hint:* The first step of the reaction is believed to be the addition of a water molecule to the Si atom in $SiCl_4$.)

4.84 Which of the two isomers of C_4H_{10} discussed in Section 4.4 would you expect to have the highest boiling point? Explain.

4.85 The geometries discussed in this chapter all lend themselves to fairly straightforward elucidation of bond angles. The exception is the tetrahedron, because its bond angles are hard to visualize. Consider the CCl_4 molecule, which has a tetrahedral geometry and is nonpolar. By equating the bond moment of a particular C—Cl bond to the resultant bond moments of the other three C—Cl bonds in opposite directions, show that the bond angles are all equal to $109.5°$.

4.86 Carbon and silicon belong to Group 4A of the periodic table and have the same valence electron configuration ($ns^2 np^2$). Why does silicon dioxide (SiO_2) have a much higher melting point than carbon dioxide (CO_2)?

4.87 Nitryl fluoride (FNO_2) is very reactive chemically. The fluorine and oxygen atoms are bonded to the nitrogen atom. (a) Write a Lewis structure for FNO_2. (b) Indicate the hybridization of the nitrogen atom. (c) Describe the bonding in terms of molecular orbital theory. Where would you expect delocalized molecular orbitals to form?

4.88 The N_2F_2 molecule can exist in either of the following two forms:

(a) What is the hybridization of N in the molecule?

(b) Which structure has a dipole moment?

4.89 Both ethylene (C_2H_4) and benzene (C_6H_6) contain the C=C bond. The reactivity of ethylene is greater than that of benzene. For example, ethylene readily reacts with molecular bromine, whereas benzene is normally quite inert toward molecular bromine and many other compounds. Explain this difference in reactivity.

4.90 In Chapter 3, we described the bonding in O_2 within the valence bond approach in terms of overlap between unhybridized $2s$ and $2p$ orbitals. Give a valence bond description of bonding in O_2 that employs hybrid orbitals. How do these descriptions differ?

4.91 TCDD, or 2,3,7,8-tetrachlorodibenzo-p-dioxin, is a highly toxic compound.

It gained considerable notoriety in 2004 when it was implicated in the murder plot of a Ukranian politician. (a) Describe its geometry and state whether the molecule has a nonzero dipole moment. (b) How many pi bonds and sigma bonds are there in the molecule?

4.92 How close together would an Na^+ and Cl^- ion have to be for their ion-ion interaction energy to equal to the dipole-dipole interaction between two H_2O molecules that are 3.0Å apart and that are antialigned as in Figure 4.27.

4.93 Describe the Lewis structure of ketene (C_2H_2O) and describe the hybridization states of the C atoms. The molecules does not contain O—H bonds. On separate diagrams, sketch the formation of sigma and pi bonds.

4.94 Progesterone is a hormone responsible for female sex characteristics. In the usual shorthand structure, each point where lines meet represent a C atom, and most H atoms are not shown. Draw the complete structure of the molecule, showing all C and H atoms. Indicate which C atoms are sp^2- and sp^3-hybridized.

4.95 Carbon monoxide (CO) is a poisonous compound due to its ability to bind to strongly to Fe^{2+} in the hemoglobin molecule. The molecular orbitals of CO have the same order as those of the N_2 molecule. (a) Draw a Lewis structure of CO and assign formal charges. Explain why CO has a rather small dipole moment of 0.12 D. (b) Compare the bond order of CO predicted from the Lewis structure with that from molecular orbital theory. (c) Which of the atoms (C or O) is more likely to form bonds with the Fe^{2+} ion in hemoglobin.

Answers to Practice Exercises

4.1 (a) tetrahedral, (b) linear, (c) trigonal planar **4.2** No.
4.3 (a) sp^3, (b) sp^2 **4.4** sp^3d^2 **4.5** The C and N atoms are sp-hybridized. The sp orbital on C forms a sigma bond with the H atom and another sigma bond with one sp orbital on the N atom. The two unhybridized p orbitals on the C atom are used to form two pi bonds with the N atom. The lone pair on the N atom is placed in the remaining sp orbital on N.

4.6

4.7 Because it is a bent molecule, the pi-bonding molecular orbital diagram for NO_2 will be the same as for O_3 (see p. 258); however, NO_2 has one fewer electron than O_3 and so will have only one electron in the nonbonding orbital. **4.8** 13.6 kJ mol^{-1} **4.9** -3.3 kJ mol^{-1} **4.10** (a) ion-ion, dispersion, (b) dispersion, (c) dipole-dipole, dispersion **4.11** only (c)

The States of Matter I: Phase Diagrams and Gases

Under certain conditions of pressure and temperature, most substances can exist in any one of the three states of matter: solid, liquid, or gas. Water, for example, can be solid ice, liquid water, or water vapor (steam). The physical properties and chemical reactivity of a substance often depend on its state. In the first part of Chapter 5, we will introduce the concept of the phase diagram, which provides us with an understanding of the conditions under which a given substance is gas, liquid, or solid, and allows us to determine the changes in pressure and/or temperature needed to transform the substance from one state into another. The remainder of Chapter 5 is devoted to a discussion of gases, whereas the properties of liquids and solids are explored in Chapter 6.

Gases are simpler than liquids or solids in many ways. Molecular motion in gases is totally random, and the forces of attraction between gas molecules are so small that each molecule moves freely and essentially independently of other molecules. It is easier to predict the behavior of gases, when subjected to changes in pressure or temperature, than that of liquids and solids. The laws that govern the behavior of gases have played an important role in the development of the atomic theory of matter and the kinetic molecular theory of gases.

5.1 | Pressure and Temperature Are Two Important Macroscopic Properties of Chemical Systems

In Chapters 1 through 4, we focused on a description of matter at the molecular and atomic levels. In such a description, the state of the system is described quantum mechanically in terms of the wave function, which is a function of the positions of all the particles. However, without highly specialized equipment, the observable world is far removed from the molecular realm both in terms of the number of atoms or molecules ($\sim 10^{23}$ instead of just a few) and length scale (centimeters and meters instead of Ångstroms and nanometers). Objects that are very large compared to the molecular scale are referred to as **macroscopic**.[1] It is both inconvenient and impossible to describe a macroscopic system in terms of the detailed atomic-scale variables of the constituent molecules—there are simply too many. Instead we characterize the state of macroscopic systems using a relatively small set of quantities, called *macroscopic properties* (or *thermodynamic properties*). Two important examples of such properties are pressure and temperature.

Pressure

Pressure is one of the most readily measurable properties of a material. In order to understand how we measure pressure, it is helpful to know how the units of measurement are defined. **Pressure** (P) is defined as

$$P = \frac{\text{force}}{\text{area}} \tag{5.1}$$

One newton is roughly equivalent to the force exerted by Earth's gravity on an apple.

The SI unit of pressure is the **pascal (Pa),** named after the French mathematician and physicist Blaise Pascal,[2] and defined as *one newton of force per square meter*:

$$1 \text{ Pa} = 1 \text{ N m}^{-2}$$

(Because $1 \text{ J} = 1 \text{ N m}$, the pascal can also be expressed as $1 \text{ Pa} = 1 \text{ J m}^{-3}$.) Another unit, based on the pascal, is the **bar,** defined as

$$1 \text{ bar} = 10^5 \text{ Pa} = 0.1 \text{ MPa}$$

The most familiar example of pressure in everyday life is *atmospheric pressure,* which results from the force experienced by any area exposed to Earth's atmosphere due to the weight of the air above it. At the surface of Earth, the atmospheric pressure averages about 1 bar, but its actual value will vary depending

---

1. For the purposes of the present discussion, we consider a system to be macroscopic if it is larger than roughly $1\mu m$ on a side. Such a system would contain roughly 10^{11} atoms.

2. Blaise Pascal (1623–1662). French mathematician and physicist. Pascal's work ranged widely in mathematics and physics, but his specialty was in the area of hydrodynamics (the study of the motion of fluids). He also invented a calculating machine.

upon location, temperature, and weather conditions. Atmospheric pressure is the basis for two common units of pressure: the ***standard atmospheric pressure (atm)***, defined as

$$1 \text{ atm} = 1.01325 \text{ bar, exactly}$$

and the ***torr***, named for Evangelista Torricelli,[3]

$$1 \text{ torr} = 1/760 \text{ atm, exactly}$$

One instrument that measures atmospheric pressure is the ***barometer***, which was invented by Evangelista Torricelli. A simple barometer can be constructed by filling a long glass tube, closed at one end, with mercury and then carefully inverting the tube in a dish of mercury, making sure that no air enters the tube. Some mercury will flow down into the dish, creating a vacuum at the top of the tube (Figure 5.1).

The weight of the mercury column remaining in the tube is supported by atmospheric pressure acting on the surface of the mercury in the dish. Thus, the atmospheric pressure can then be measured by the height of the mercury column and is expressed in millimeters of mercury (torr): 1 mmHg is the pressure exerted by a column of mercury 1 mm high when its density is 13.5951 g cm^{-3}. The relation between mmHg and torr is

$$1 \text{ mmHg} = 1.000000142 \text{ torr}$$

Except in the most precise work, we shall treat these two units as the same; that is

$$1 \text{ mmHg} \approx 1 \text{ torr} = 1/760 \text{ atm}$$

A ***manometer*** is *a device used to measure the pressure of gases other than the atmosphere.* The principle of operation of a manometer is similar to that of a barometer. There are two types of manometers, shown in Figure 5.2. The *closed-tube manometer* is normally used to measure pressures below atmospheric pressure [Figure 5.2(a)], whereas the *open-tube manometer* is better suited for measuring pressures equal to or greater than atmospheric pressure [Figure 5.2(b)].

Nearly all barometers and most manometers use mercury as the working fluid, despite the fact that it is a toxic substance with a harmful vapor. The reason is that mercury has a very high density (13.6 g mL^{-1}) compared with most other liquids. Because the height of the liquid in a column is inversely proportional to the density of the liquid, this property enables the construction of manageably small barometers and manometers.

If a solid or liquid surface is in contact with a gas, the pressure of solid or liquid will be the same as that of the surrounding gas. If this were not the case, the forces on either side of the surface would not be balanced, and the solid or liquid would either expand (if the gas pressure were less than that of the solid or liquid) or contract (if the gas pressure were greater than that of the solid or liquid). Thus, we can measure the pressure of a solid or liquid by measuring the pressure of the surrounding gas.

At the molecular level, the pressure of a gas results from collisions of the gas molecules and any surface with which they come in contact (as will be discussed in Section 5.4). The magnitude of the pressure depends upon the frequency and strength of the collisions.

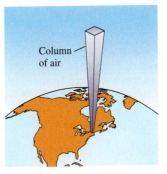

A column of air extending from sea level to the upper atmosphere

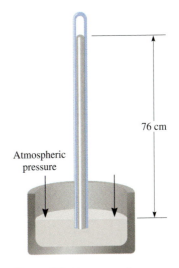

Figure 5.1 Barometer for measuring atmospheric pressure. Above the mercury in the tube is a vacuum. The column of mercury is supported by the atmospheric pressure.

---

3. Evangelista Torricelli (1608–1674). Italian mathematician. Torricelli was supposedly the first person to recognize the existence of atmospheric pressure.

Figure 5.2 Two types of manometers used to measure gas pressures: (a) Closed tube: gas pressure is less than atmospheric pressure. (b) Open tube: gas pressure is greater than atmospheric pressure.

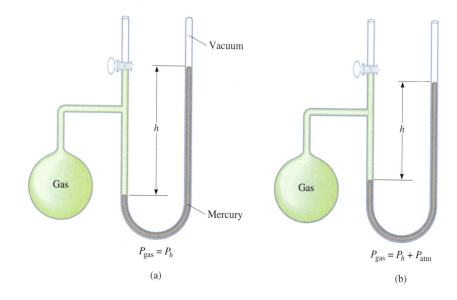

Vacuum

h

Gas

Mercury

Gas

h

$P_{gas} = P_h$

$P_{gas} = P_h + P_{atm}$

(a)

(b)

Temperature

Temperature is a very important and familiar quantity in science and everyday life, but it can be difficult to define. In daily experience, we use temperature as a measure of "hotness" or "coldness," but for our purposes we need a more precise operational definition.

Consider two systems A and B that are placed in contact so that energy can be exchanged between the two. The containers enclosing systems A and B are *rigid*, meaning no volume changes can occur, and *impermeable,* that is, no matter exchange between systems A and B is possible. Under these conditions, **temperature** (T) is *the quantity that controls the direction of energy flow between the two systems*. If the temperature of system A (T_A) is less than that of system B (T_B), then energy will flow from system B into system A, as shown in Figure 5.3(a). Conversely, if $T_A > T_B$, then energy will flow from system A into system B [Figure 5.3(b)]. If the temperatures are equal, $T_B = T_A$, then no net energy flow will occur between the two systems [Figure 5.3(c)], and we say that systems A and B are in **thermal equilibrium.** An important property of thermal equilibrium (and temperature) is that, *if two systems are in thermal equilibrium with a third system, then they are also in thermal equilibrium with one another.* This statement is generally known as the **zeroth law of thermodynamics.**

Figure 5.3 Two systems A and B are enclosed in containers that are rigid and impermeable, but that allow energy (heat) exchange when placed in contact. (a) If $T_B > T_A$, there is a net flow of energy from system B into system A. (b) If $T_B < T_A$, there is a net flow of energy from system A into system B. (c) If $T_B = T_A$, there is no net energy flow between the systems and the two systems are in thermal equilibrium.

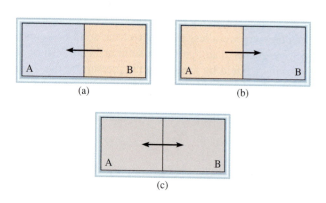

A   B

(a)

A   B

(b)

A   B

(c)

Changes in temperature can affect the properties of materials significantly. These changes can be used to define temperature scales. For example, the Celsius temperature scale, named for Anders Celsius,[4] defines 0°C to be the temperature at which water freezes (at 1 atm pressure) and 100°C to be the temperature at which water boils (at the same pressure). The earlier Fahrenheit scale, developed by Daniel Fahrenheit[5] and still in everday (but not scientific) usage in the United States, sets the freezing point of water (at 1 atm) at 32°F and the boiling point at 212°F. In the early nineteenth century, Lord Kelvin[6] postulated the existence of a fundamental limiting temperature, called the **absolute zero of temperature,** below which any material cannot be cooled. Based on studies of gases (Section 5.3), Kelvin determined absolute zero to be −273.15°C. This temperature provides a natural reference temperature that is independent of material properties. The Kelvin scale is the standard unit of temperature in the SI system and is defined in terms of the Celsius scale by

$$T \text{ (kelvins)} = 273.15 + T \text{ (Celsius)} \qquad \textbf{(5.2)}$$

so that absolute zero corresponds to 0 K and the freezing point of water is 273.15 K. (Note that the degree symbol ° is not used for kelvins). See Figure 5.4 for a comparison between the three temperature scales discussed above.

A **thermometer** is a *device for measuring temperature.* Thermometers operate by measuring the temperature-dependent changes in material properties and calibrating them to a predefined temperature scale. For example, liquid mercury expands as

Under special experimental conditions, scientists have succeeded in approaching absolute zero to within a nanokelvin (one billionth of a kelvin).

---

4. Anders Celsius (1701–1744). Swedish astronomer and meterologist. Celsius's original temperature scale had 100° for the freezing point of water and 0° for the boiling point, which were reversed after his death.

5. Daniel Gabriel Fahrenheit (1686–1736). German instrument maker. Fahrenheit invented the first practical mercury thermometer. In developing his temperature scale, Fahrenheit chose 0° to be the temperature of the coldest sample he could make in his laboratory, namely, a mixture of snow, water and ammonium chloride. The high end of his scale was chosen to make body temperature equal to 96° (more precise modern measurements give body temperature as 98.6°F).

6. William Thompson, Lord Kelvin (1824–1907). Scottish mathematician and physicist. Kelvin did important work in many branches of physics.

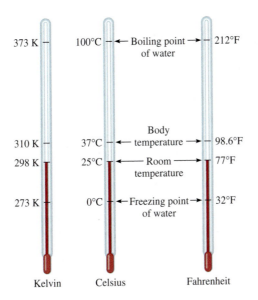

Figure 5.4 Comparison of the three temperature scales: Celsius, Fahrenheit, and the absolute temperature (Kelvin) scale. The Celsius scale was formerly called the centigrade scale.

temperature increases, so a mercury thermometer can be constructed by placing mercury in a sealed glass tube. The height of the column of mercury when the thermometer is immersed in an ice-water mixture is marked as 0°C and when immersed in boiling water, as 100°C. The scale between the two reference temperatures is obtained by dividing the distance between the reference marks into 100 equal parts, assuming that the expansion of mercury is a linear function of temperature over the temperature range of the thermometer.

5.2 | Substances and Mixtures Can Exist as Solid, Liquid, or Gas, Depending upon the External Conditions

As discussed in Section 0.1, most substances and mixtures can exist in three states: solid, liquid, and gas. The physical properties of these states of matter are very different. In gases, the distances between molecules are so large compared to molecular size that the effect of molecular interaction is negligibly small at ordinary temperatures and pressures. Because there is a great deal of empty space in a gas—that is, space that is not occupied by molecules—gases can be readily compressed. The lack of strong forces between molecules also allows a gas to expand to fill the volume of its container. Furthermore, the large amount of empty space results in very low densities under normal conditions.

Liquids and solids are quite a different story. The principal difference between the condensed states (liquids and solids) and the gaseous state is the distance between molecules. In a liquid, the molecules are so close together that there is very little empty space. Thus, liquids are much more difficult to compress than gases, and they are also much denser under normal conditions. Molecules in a liquid are held together by one or more types of attractive forces, which were discussed in Section 4.6. A liquid also has a definite volume because molecules in a liquid cannot overcome the attractive forces that bind them to one another. The molecules can, however, move past one another freely, and so a liquid can flow, can be poured, and will assume the shape of its container.

In a solid, molecules are held rigidly in position with little freedom of motion. Many solids are characterized by long-range order, that is, the molecules are arranged in regular configurations in three dimensions. Like liquids, solids are much denser than gases and are nearly incompressible, but, unlike liquids, solids possess a definite shape and volume. With very few exceptions (water being the most important), the density of the solid form is higher than that of the liquid form for a given substance.

Phases and Phase Transitions

The specific state of matter (solid, liquid or gas) that is observed for a particular material depends upon the pressure, temperature, and, in the case of mixtures, the composition. For example, at 1 bar pressure and 25°C, water is a liquid, but if the temperature is lowered below 0°C at 1 bar, the liquid water becomes ice, a solid. Under certain conditions it is even possible for two different states of the same material to coexist. An ice cube (solid) floating in a glass of water (liquid) at 0°C is a familiar example. We refer to the different states of a substance that are present in a system as phases. A *phase* is *a homogeneous part of the system in contact with other parts of the system but separated from them by well-defined boundaries.* Thus, our glass of ice water contains both the solid phase and the liquid phase of water. The conversion of one phase into another due to changes in conditions is called a ***phase transition*** (or *phase transformation*). Examples of phase transitions are

▶ ***Evaporation*** (or ***vaporization***): *The transformation of a liquid into a gas.* The reverse process in which a gas changes into a liquid is called ***condensation.***

▶ ***Freezing:*** *The transformation of a liquid into a solid.* The reverse transformation of a solid into a liquid is called ***melting.***

▶ ***Sublimation:*** *The direct transformation of a solid into a gas (vapor).* The reverse process is called ***deposition.***

Liquid water and ice in coexistence

A system in which the amounts of matter in each phase present do not change with time is said to be in ***physical equilibrium.*** As an example, consider the vaporization of water in a closed container at a given temperature. Such a system will quickly reach a state of physical equilibrium in which the number of moles of water in the gas and liquid phases no longer change with time. Although, there is no net change in the amounts of water in each phase, this equilibrium state is not static, and water molecules are constantly being exchanged between the liquid and vapor phases. The amount of material in each phase is constant over time because the number of H_2O molecules leaving and the number returning to the liquid phase in any given time period are equal. The coexistence between liquid and vapor is an example of ***dynamic equilibrium,*** in which *the rate of a forward process is exactly balanced by the rate of the reverse process.* A dynamic equilibrium can be likened to the movement of skiers at a busy ski resort, where the number of skiers carried up the mountain on the chair lift is equal to the number coming down the slopes. Although there is a constant transfer of skiers, the number of people at the top of the slope and the number at the bottom do not change.

Phase Diagrams

In practical applications of chemistry, it is important to know the phases of reactants and products at the conditions of interest. Such information can be obtained using a ***phase diagram,*** *which is a plot summarizing the conditions at which the various phases of a material can exist.* The regions in which different phases exist are separated by ***phase boundaries,*** which represent *the conditions under which the two (or more) phases separated by the boundary can coexist in equilibrium.*

Liquid water in equilibrium with its vapor in a closed system at room temperature

For a pure substance, the phase diagram is simply a graph of temperature versus pressure. For mixtures, the phase diagram also includes variables that describe the composition of the substance. To illustrate the information contained in a phase diagram, we will examine the phase diagrams of two pure substances: water and carbon dioxide.

Water

Figure 5.5(a) shows the low-pressure phase diagram of water. The graph is divided into regions, each of which represents a pure phase. The line separating any two regions indicates the conditions under which these two phases coexist. For example, the phase boundary curve between the liquid and vapor (gas) phases, called the *liquid-vapor coexistence line,* shows how the pressure of the coexisting vapor above a liquid, called the ***equilibrium vapor pressure,*** depends upon the temperature. At the equilibrium vapor pressure, the rate of evaporation is equal to the rate of condensation.

Every liquid has a temperature at which it begins to boil. The ***boiling point*** is *the temperature at which the vapor pressure of a liquid is equal to the external pressure.*

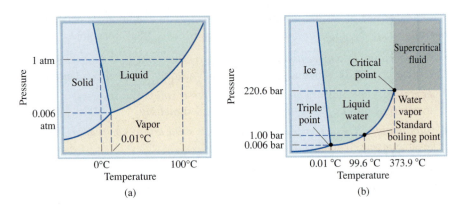

The **normal boiling point** is defined as the boiling point when the external pressure is equal to 1 atm. The normal boiling point for water is 100°C. From Figure 5.5(a), we can determine that the normal boiling point of water is 100°C, but if the pressure is reduced to 0.5 atm, water boils at only 82°C because the vapor pressure of water is 0.5 atm at this temperature. The **standard boiling point** [shown in Figure 5.5(b)] is defined as the boiling point when the external pressure is equal to the standard pressure of 1 bar (0.987 atm). The standard boiling point of water is 99.6°C.

The other two curves in Figure 5.5(a) similarly indicate conditions for coexistence between ice and liquid water and between ice and water vapor. The **melting point** of a solid (or the **freezing point** of a liquid) is *the temperature at which solid and liquid phases coexist at a given pressure.* The **normal melting** (or *freezing*) **point** refers to the temperature at which a substance melts (freezes) at 1 atm pressure. When the pressure is equal to 1 bar, the melting (or freezing) point is referred to as the **standard melting** (or *freezing*) **point.** (Because the melting point changes very little with pressure, the standard and normal melting points for most materials are identical.) Note that the solid-liquid coexistence curve has a negative slope. Water is unusual in that the melting point decreases with increasing pressure, so it is possible to melt ice by increasing the pressure. The point at which all three curves meet is called the **triple point,** which is *the only condition under which all three phases can be in equilibrium with one another.* For water, this point is at 0.01°C and 0.006 bar.

Figure 5.5(b) shows the phase diagram of water at intermediate pressures. From this graph we see that the liquid-vapor coexistence line does not continue indefinitely, but stops abruptly at a pressure of 220.6 bar and 647.1 K. This point is called the **critical point,** and represents the highest pressure (called the **critical pressure, P_c**) and temperature (called the **critical temperature, T_c**) at which water can exist as a liquid. Above this point, there is no fundamental distinction between a liquid and a gas—we simply have a fluid, called a **supercritical fluid (SCF).**

A number of practical applications have been found for supercritical fluids. Because of its industrial importance, supercritical carbon dioxide is the most studied SCF. Depending upon the conditions of pressure and temperature, supercritical carbon dioxide (*sc*-CO_2) is used as a solvent in a number of industrial processes, such as the decaffeination of coffee and the extraction of fragrant oils from natural materials for the perfume industry, as a replacement for volatile organic solvents in dry cleaning, and in the removal of chlorinated hydrocarbons from contaminated soil. Recent efforts, primarily in the pharmaceutical industry, are aimed at using *sc*-CO_2 in the production of nanoparticle materials. At the present time, the widespread use of *sc*-CO_2 in large-scale industrial processes is limited somewhat, however, by the safety issues involving the operation of industrial chemical reactors at the relatively critical pressure of CO_2 (~70 bar).

The relation between the slope of the phase boundary curve and properties of the substance is discussed in Section 9.2

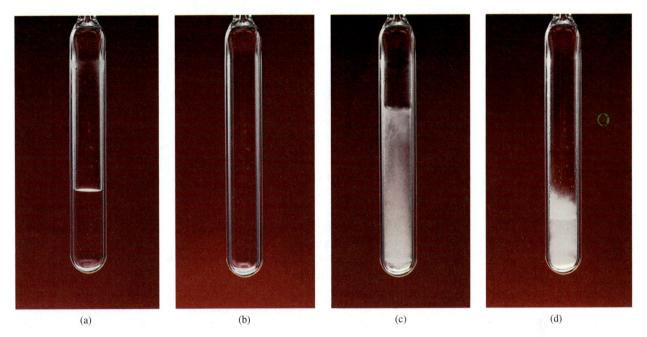

(a) (b) (c) (d)

Figure 5.6 The critical behavior of sulfur hexafluoride. (a) Below the critical temperature the clear liquid phase is visible. (b) Above the critical temperature the liquid phase has disappeared. (c) The substance is cooled to just below the critical temperature. The fog is a phenomenon known as critical opalescence and is due to scattering of light by large density fluctuations in the critical fluid. (d) Finally, the liquid phase reappears.

Figure 5.6 shows what happens when sulfur hexafluoride is heated above its critical temperature (45.5°C) and then cooled down to 45°C. The existence of a critical point is a common feature of the liquid-vapor coexistence for any substance, but the specific values of P_c and T_c vary greatly from substance to substance. Table 5.1 shows the critical temperatures and critical pressures for a variety of substances. The critical

Table 5.1	Critical Temperatures and Critical Pressures of Selected Substances	
Substance	T_c **(°C)**	P_c **(bar)**
Ammonia (NH_3)	132.4	112.9
Argon (Ar)	−186	6.4
Benzene (C_6H_6)	288.9	48.5
Carbon dioxide (CO_2)	31.0	74.0
Ethanol (C_2H_5OH)	243	63.8
Diethyl ether ($C_2H_5OC_2H_5$)	192.6	36.1
Mercury (Hg)	1462	1050
Methane (CH_4)	−83.0	46.2
Molecular hydrogen (H_2)	−239.9	13.0
Molecular nitrogen (N_2)	−147.1	33.9
Molecular oxygen (O_2)	−118.8	50.4
Sulfur hexafluoride (SF_6)	45.5	38.1
Water (H_2O)	374.4	222.4

Figure 5.7 High-pressure phase diagram of water showing various polymorphs of ice. Additional polymorphs of ice (Ic, IV, VII–XIII) exist at higher pressures (or lower temperatures than shown here). Note: Only the liquid boundary with ice-Ih has negative slope. The liquid boundaries with ice-III, ice-V, ice-VI, and ice-VII all have positive slope.

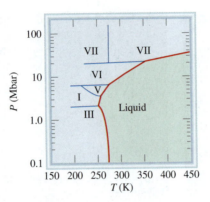

temperature reflects the strength of the intermolecular forces. Benzene, ethanol, mercury, and water, which have strong intermolecular forces, also have high critical temperatures when compared with other substances listed in the table. The critical point will be discussed in more detail in Section 5.5.

Because gases have no well-defined boundaries other than those of a container, for any given material there can only be one gas phase present. For liquids and solids, it is possible to have more than one phase present. For example, when oil and water are placed together in the same container, they separate into two distinct liquid phases—one that is oil-rich and one that is primarily water. In addition, a given solid substance, depending on the pressure and temperature, can exist in a number of different forms each with a distinct crystal structure. This phenomenon is known as *polymorphism— the existence of a solid in more than one crystal form.* A good example of this can be seen in the phase diagram of water at very high pressures (Figure 5.7). There are currently 14 known different forms of ice. The familiar form, known as ice Ih, is the stable form up to 2 kbar (2000 bar). Another example of polymorphism is carbon, which can exist in two solid forms—diamond and graphite.

Carbon Dioxide

The phase diagram of carbon dioxide (Figure 5.8) is generally similar to that of water, with one important exception—the slope of the curve between solid and liquid is positive. In fact, this holds true for almost all other substances. Water behaves differently because ice is less dense than liquid water. The triple point of carbon dioxide is at 5.16 bar and $-56.6°C$. The critical pressure of carbon dioxide is 73.6 bar and the critical temperature is $31.0°C$—just above standard room temperature.

Figure 5.8 Phase diagram of carbon dioxide. Note that the solid-liquid boundary line has a positive slope, which is true of most substances. The liquid phase is not stable below the triple point which is at 5.2 bar and 216.6 K, so only the solid and vapor phases can exist under normal atmospheric conditions. The pressure axis is shown on a logarithmic scale so that the triple point and critical point can be easily seen on the same graph.

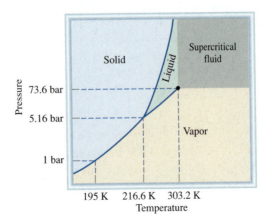

An interesting observation can be made about the phase diagram in Figure 5.8. As you can see, the entire liquid phase lies well above 1 bar pressure; therefore, it is impossible for solid carbon dioxide to melt at 1 bar. Instead, when solid CO_2 is heated to $-78°C$ at 1 bar, it sublimes. In fact, solid carbon dioxide is called dry ice because it looks like ice and *does not melt* (Figure 5.9). Because of this property, dry ice is useful as a refrigerant.

The phase diagrams for pure substances, like those for water and carbon dioxide shown in Figure 5.5 and 5.8, respectively, can be expressed on a single graph because the state of such a system only depends upon pressure and temperature. However, for mixtures, it is necessary to also show the dependence of the phase on composition. For example, pure water at 1 bar and 0°C can be either in the liquid or solid phase, but if propylene glycol (a compound used in antifreeze) is dissolved in the water, the only phase possible is the liquid phase, as the melting point is decreased by the addition of the antifreeze. In general, to specify the phase, we would need to know how the concentration of propylene glycol affects the phase diagram. To specify the phase(s) present in a two-component system, we need to know the pressure, the temperature, and the concentration of one of the components. A full phase diagram would require a three-dimensional plot. To simplify such phase diagrams, the phase diagram is usually represented as a series of two-dimensional plots in which one of the variables (P, T, or concentration) is held constant. We will see examples of such multicomponent phase diagrams when we examine the phase behavior of solutions in Chapter 9.

In spite of the complexity of phase diagrams for multicomponent systems, it is possible to get some information as to the nature of phase coexistence in such systems without knowing the specifics of the phase diagram. To see this, consider the phase diagram of water presented in Figure 5.5(a). The regions in which only one phase (solid, liquid, or gas) exists are two-dimensional (that is, they are areas in the P-T plane). In contrast, two-phase coexistence is only possible along one of the two-phase coexistence lines (liquid-solid, liquid-vapor, *or* solid-vapor), which are one-dimensional geometrical objects. The conditions under which three-phase coexistence (solid-liquid-vapor) is possible is confined to a single point (the triple point), which has zero dimensions. This pattern can be generalized to multicomponent equilibrium. For a system with c components and r phases in equilibrium, the dimensionality (or number of degrees of freedom) of the phase coexistence region, f, is given by

$$f = c - r + 2 \tag{5.3}$$

Equation 5.3 was first recognized by the American physicist Josiah Willard Gibbs[7] and is known as the **Gibbs phase rule.** For example, consider two-phase coexistence ($r = 2$) in a one-component system ($c = 1$). The Gibbs phase rule predicts the dimensionality of this coexistence region to be

$$f = 1 - 2 + 2 = 1$$

That is, the region of two-phase coexistence in this system is a line—consistent with our observations from the water phase diagram (Figure 5.5). Similarly, the region of three-phase coexistence has a dimensionality of $f = 1 - 3 + 2 = 0$, which describes

Figure 5.9 Under atmospheric conditions, solid carbon dioxide does not melt; it only sublimes. The cold carbon dioxide gas causes nearby water vapor to condense and form a fog.

---

7. Josiah Willard Gibbs (1839–1903). American physicist. One of the founders of thermodynamics and statistical mechanics, Gibbs was a modest and private individual who spent most of his professional life at Yale University. Because he published most of his works in obscure journals, Gibbs never gained the eminence that his contemporary and admirer James Clerk Maxwell enjoyed. Even today, very few people outside of chemistry and physics have ever heard of Gibbs.

a single point. For multicomponent systems, application of the Gibbs phase rule (Equation 5.3) is straightforward. For example, regions of three-phase coexistence in two-component ($c = 2$) systems have a dimensionality of $f = 2 - 3 + 2 = 1$, which describes a line. Thus, three-phase coexistence in two-component systems occurs along *triple lines*. The Gibbs phase rule can also tell us the maximum number of phases that can coexist in a particular system. This maximum will occur when $f = 0$. Using Equation 5.3 gives $f = c - r_{max} + 2 = 0$, yielding

$$r_{max} = c + 2 \qquad\qquad (5.4)$$

The value of r cannot exceed r_{max} for a given system because then f would be negative, which is unphysical. For example, for a one-component system, the maximum number of phases that can coexist is $r_{max} = 1 + 2 = 3$. Therefore, four-phase coexistence in a system with only one component is impossible, although it can occur in a two-component system, where r_{max} is equal to 4.

5.3 | The Ideal-Gas Equation Describes the Behavior of All Gases in the Limit of Low Pressure

Given the pressure, temperature, and composition of a substance, we can use the phase diagram to tell us whether it is a solid, liquid, or gas. However, the phase diagram does not represent a complete description of the physical state of the substance. Consider, for example, two samples of carbon dioxide at 300 K, one at a pressure of 1 bar and one at a pressure of 5 bar. The sample at 1 bar is measured to have a density of 1.76 g L^{-1} whereas that at 5 bar has a density that is significantly larger (8.82 g L^{-1}). From the phase diagram of carbon dioxide (see Figure 5.8), we can infer only that both samples are gases, but the phase diagram does not provide information, for example, about how the two gas samples differ in density. For this information, the phase diagram must be supplemented by an **equation of state**, which is *an equation that relates the pressure of a substance to its temperature, volume, and mole numbers.*

Equations of State

For a single-component system with temperature T, volume V, and number of moles n, the *equation of state* has the form

$$\text{Pressure} = P(T, V, n) \qquad\qquad (5.5)$$

Once the equation of state is known for a system, one can calculate any one of the four quantities P, V, T, and n from knowledge of the other three. For example, if one knows the temperature, the number of moles, and the pressure, the equation of state can be used to determine the volume occupied by the system.

In general, the equation of state for a substance depends specifically upon its composition. For example, the equation of state for liquid water is different than that for liquid ethanol. However, as we shall see, in the limit of low densities the properties of gases can be described by an equation of state that is independent of the nature and composition of the gas.

The Ideal Gas Equation of State

For gases at low density, the average distance between the atoms and molecules is large compared to their size. Because intermolecular interactions decrease rapidly with an increased separation distance, the strength of those interactions in a low-density gas is expected to be very small, perhaps even negligible. As a consequence, the behavior of all gases in the limit of very low density (or low pressure) approaches that of a hypothetical model gas consisting of noninteracting point particles, referred to as an ***ideal gas.*** The equation of state for an ideal gas is independent of the composition of the particles and has the form

$$PV = nRT \tag{5.6}$$

where R is a constant called the ***universal gas constant*** and T is the absolute temperature. The gas constant can be determined experimentally and takes on a variety of values depending upon the units used:

$$R = 0.0831447\frac{\text{L bar}}{\text{mol K}} = 0.0820574\frac{\text{L atm}}{\text{mol K}} = 62.3636\frac{\text{L torr}}{\text{mol K}} = 8.31447\frac{\text{J}}{\text{mol K}}$$

Equation 5.6 is called the ***ideal gas equation of state.***

The ideal gas equation is an example of a ***limiting law,*** that is, a *scientific law that becomes exact only in some well-defined limit,* which, in the case of Equation 5.6, is the limit when $P \rightarrow 0$. Any real gas will show deviations from the ideal gas equation, but these deviations become progressively smaller as the pressure approaches zero. One way to see this behavior is to examine the ***compression factor***[8] (Z), defined as

$$Z = \frac{PV}{nRT} \tag{5.7}$$

From Equation 5.6, the compression factor is equal to 1 for an ideal gas, independent of pressure. Figure 5.10 shows the compression factor as a function of pressure for several gases at 273 K. Note that the deviation from ideal behavior is small for pressures near 1 bar or below. *For gases at pressures of up to a few bar, the ideal gas law is generally of sufficient accuracy for all but the most precise calculations.*

A useful quantity to calculate for a gas is the molar volume $\overline{V} = V/n$, which is the volume occupied by one mole of gas molecules. For an ideal gas, the molar volume is given by

$$\overline{V} = \frac{V}{n} = \frac{RT}{P} \tag{5.8}$$

---

8. Sometimes Z is called the *compressibility factor.*

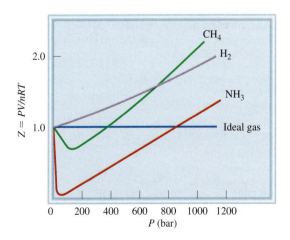

Figure 5.10 Plot of the compression factor Z versus pressure P of various gases at 0°C. For an ideal gas, $Z = 1$, no matter what the pressure is. For real gases, we observe various deviations from ideality at high pressures. At very low pressures, all gases exhibit ideal behavior, that is Z approaches 1 as P approaches zero.

Figure 5.11 Comparison of the molar volume at STP (which is about 22.4 L) with a basketball.

At a temperature of 25°C (273.15 K) and 1 bar pressure, the molar volume of an ideal gas can be calculated from Equation 5.7

$$\bar{V} = \frac{RT}{P} = \frac{(0.0831447 \text{ L bar mol}^{-1} \text{ K}^{-1})(298.15 \text{ K})}{1 \text{ bar}}$$
$$= 24.790 \text{ L}$$

The conditions of 25°C and 1 bar pressure are called **standard ambient temperature and pressure (SATP).** Thus, the molar volume of an ideal gas at SATP is 24.790 L. The SATP conditions are used as standard states in the construction of thermodynamic data. An older standard, still in common usage, is called the **standard temperature and pressure (STP),** which represents a temperature of 0°C and a pressure of 1 atm. At STP, the molar volume of an ideal gas is 22.414 L. Figure 5.11 shows a comparison of the molar volume of a gas at STP with a basketball.

1 atm = 1.01325 bar, exactly.

Example 5.1 shows that if we know the quantity, volume, and temperature of a gas, we can calculate its pressure using the ideal gas equation.

Example 5.1

Sulfur hexafluoride (SF_6) is a colorless, odorless, very unreactive gas. Calculate the pressure (in bar) exerted by 0.374 moles of the gas in a steel vessel of volume 5.43 L at 69.5°C, assuming ideal behavior.

Strategy The problem gives *n*, *V*, and *T* for this sample of gas. From this information the ideal gas equation (Equation 5.5) can be used to calculate the pressure. Temperature here is given in degrees Celsius, so it is important to convert to absolute temperature in kelvins before using the ideal gas law.

Solution Rearranging Equation 5.6, we write

$$P = \frac{nRT}{V}$$
$$= \frac{(0.374 \text{ mol})(0.08314 \text{ L bar mol}^{-1} \text{ K}^{-1})(69.5 + 273.15 \text{ K})}{5.43 \text{ L}}$$
$$= 1.96 \text{ bar}$$

Comment We are given the mole amount and volume data to only three significant figures (see Appendix 1), so it is unnecessary to use all the significant figures of *R*

—Continued

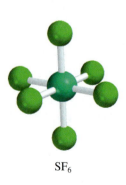

SF_6

Continued—

that are available. Thus, we used $R = 0.08314$ L bar mol^{-1} K^{-1}, instead of the more precise value $R = 0.0831447$ L bar mol^{-1} K^{-1}.

Practice Exercise Calculate the volume (in liters) occupied by 2.12 moles of nitric oxide (NO) at 6.63 bar and 76°C.

The density, d, of a gas can be obtained from the ideal gas equation by recognizing that the number of moles, n, of a gas is given by

$$n = \frac{m}{\mathcal{M}}$$

where m is the mass of the gas in grams and $\mathcal{M}$ is the molar mass. Substituting this relationship into the ideal gas equation (Equation 5.6) gives, after some rearrangement,

$$d = \frac{m}{V} = \frac{P\mathcal{M}}{RT} \tag{5.9}$$

where the density is defined as the mass per unit volume. Unlike molecules in condensed matter (that is, in liquids and solids), gaseous molecules are separated by distances that are large compared with their size. Consequently, the density of gases is very low under atmospheric conditions. For this reason, gas densities are usually expressed in grams per liter (g L^{-1}) rather than grams per milliliter (g mL^{-1}), as Example 5.2 shows.

Example 5.2

Calculate the density of carbon dioxide (CO_2) in grams per liter (g L^{-1}) at 0.990 bar and 55.0°C. Assume ideal behavior.

Strategy We need Equation 5.9 to calculate the gas density. Be careful to use the correct units for temperature.

CO_2

Solution To use Equation 5.9 we convert temperature to kelvins ($T = 273.15 + 55 = 328.15$ K) and use 44.01 g mol^{-1} for the molar mass of CO_2

$$d = \frac{\mathcal{M}P}{RT}$$

$$= \frac{(44.01 \text{ g mol}^{-1})(0.990 \text{ bar})}{(0.08314 \text{ L bar mol}^{-1}\text{ K}^{-1})(55.0 + 273.15 \text{ K})}$$

$$= 1.60 \text{ g L}^{-1}$$

Alternatively we could use the ideal gas equation (Equation 5.6) directly by assuming that we have one mole (or 44.01 g) of CO_2. The volume of the gas can be obtained from Equation 5.6

$$V = \frac{nRT}{P}$$

$$= \frac{(1 \text{ mol})(0.08314 \text{ L bar mol}^{-1}\text{ K}^{-1})(328.15 \text{ K})}{0.990 \text{ bar}}$$

$$= 27.5 \text{ L}$$

—Continued

Continued—
Therefore, the density of CO_2 is given by

$$d = \frac{44.01 \text{ g}}{27.5 \text{ L}} = 1.60 \text{ g L}^{-1}$$

Comment In units of grams per milliliter, the gas density is 1.60×10^{-3} g mL^{-1}, which is a very small number. In comparison, the density of water is 1.0 g mL^{-1} and that of gold is 19.3 g mL^{-1}.

Practice Exercise What is the mass density (in g L^{-1}) of uranium hexafluoride (UF_6) at 779 torr and 62°C?

Equation 5.9 can also be used to determine the molar mass of an unknown gas sample. All that is needed is an experimentally determined density value (or mass and volume data) for the gas at a known temperature and pressure. By rearranging Equation 5.9, we get

$$\mathcal{M} = \frac{dRT}{P} \tag{5.10}$$

In a typical experiment, a bulb of known volume is filled with the gaseous substance under study. The temperature and pressure of the gas sample are recorded, and the total mass of the bulb plus gas sample is determined (Figure 5.12). The bulb is then evacuated (emptied) and weighed again. The difference in mass is the mass of the gas. The density of the gas is equal to its mass divided by the volume of the bulb. Once we know the density of a gas, we can calculate the molar mass of the substance using Equation 5.10. Of course, a mass spectrometer (see the inset on p. 46) would be the ideal instrument to determine the molar mass, but not every chemist can afford one.

Example 5.3 shows the density method for molar mass determination.

Figure 5.12 Apparatus for measuring the density of a gas. A bulb of known volume is filled with the gas under study at a certain temperature and pressure. First the bulb is weighed, and then it is emptied (evacuated) and weighed again. The difference in masses gives the mass of the gas. Knowing the volume of the bulb, we can calculate the density of the gas. Under ambient conditions, 100 mL of air weighs about 0.12 g, an easily measured quantity.

Example 5.3

A chemist has synthesized a greenish-yellow gaseous compound of chlorine and oxygen and finds that its density is 7.71 g L^{-1} at 36°C and 2.92 bar. Calculate the molar mass of the compound and determine its molecular formula.

Strategy From Equation 5.10, we can calculate the molar mass of a gas if we know its density, temperature, and pressure. The molecular formula of the compound must be consistent with its molar mass. Be careful to use the appropriate units for temperature.

Solution From Equation 5.10

$$\mathcal{M} = \frac{dRT}{P}$$

$$= \frac{(7.71 \text{ g L}^{-1})(0.08314 \text{ L bar mol}^{-1} \text{ K}^{-1})(36 + 273.15 \text{ K})}{2.92 \text{ bar}}$$

$$= 67.8 \text{ g mol}^{-1}$$

—Continued

Continued—

Alternatively, we can solve for the molar mass directly from the ideal gas law. We assume that there is 1.00 L of the gas, so the mass of the gas is 7.71 g. The number of moles of the gas can be obtained from the ideal gas equation (Equation 5.6)

$$n = \frac{PV}{RT}$$

$$= \frac{(2.92\ \text{bar})(1.00\ \text{L})}{(0.08314\ \text{L atm mol}^{-1}\ \text{K}^{-1})(36 + 273.15\ \text{K})}$$

$$= 0.114\ \text{mol}$$

Therefore, the molar mass is given by

$$\mathcal{M} = \frac{7.71\ \text{g}}{0.114\ \text{mol}} = 67.6\ \text{g mol}^{-1}$$

We can determine the molecular formula of the compound by trial and error, using only the knowledge of the molar masses of chlorine (35.45 g) and oxygen (16.00 g). We know that a compound containing one Cl atom and one O atom would have a molar mass of 51.45 g, which is too low, while the molar mass of a compound made up of two Cl atoms and one O atom is 86.90 g, which is too high. Thus, the compound must contain one Cl atom and two O atoms and have the formula ClO_2, which has a molar mass of 67.45 g.

Practice Exercise The density of a gaseous organic compound is 3.38 g L^{-1} at 40°C and 2.00 bar. What is its molar mass?

ClO_2

Because Equation 5.10 is derived from the ideal gas equation, we can also calculate the molar mass of a gaseous substance using the ideal gas equation, as shown in Example 5.4.

Example 5.4

Chemical analysis of a gaseous compound showed that it contained 33.0 percent silicon (Si) and 67.0 percent fluorine (F) by mass. At 35°C, 0.210 L of the compound exerted a pressure of 1.72 bar. If the mass of 0.210 L of the compound was 2.38 g, calculate the molecular formula of the compound.

Strategy This problem can be divided into two parts. First, it asks for the empirical formula of the compound from the percent by mass of Si and F. Second, the information provided enables us to calculate the molar mass of the compound and hence determine its molecular formula.

Solution We follow the procedure in Example 0.16 to calculate the empirical formula by assuming that we have 100 g of the compound, so the percentages are converted to grams. The number of moles of Si and F are given by

$$n_{Si} = 33.0\ \text{g Si} \times \frac{1\ \text{mol Si}}{28.09\ \text{g Si}} = 1.17\ \text{mol Si}$$

$$n_{F} = 67.0\ \text{g Si} \times \frac{1\ \text{mol F}}{19.0\ \text{g F}} = 3.53\ \text{mol F}$$

—Continued

Continued—

Therefore, the empirical formula is $Si_{1.17}F_{3.53}$, or SiF_3. To calculate the molar mass of the compound, we need first to calculate the number of moles contained in 2.38 g of the compound. From the ideal gas equation (Equation 5.6),

$$n = \frac{PV}{RT} = \frac{(1.72 \text{ bar})(0.210 \text{ L})}{(0.08314 \text{ L bar mol}^{-1}\text{ K}^{-1})(308 \text{ K})} = 0.0141 \text{ mol}$$

Because there are 2.38 g in 0.0141 mole of the compound, the mass in 1 mol, or the molar mass, is given by

$$\mathcal{M} = \frac{2.38 \text{ g}}{0.0141 \text{ mol}} = 169 \text{ g mol}^{-1}$$

The molar mass of the empirical formula SiF_3 is 85.09 g. Recall that the ratio (molar mass/empirical molar mass) is always an integer ($169/85.09 \approx 2$). Therefore, the molecular formula of the compound must be $(SiF_3)_2$ or Si_2F_6.

Practice Exercise   A gaseous compound is 78.14 percent boron and 21.86 percent hydrogen. At 27°C, 74.3 mL of the gas exerted a pressure of 1.13 bar. If the mass of gas was 0.934 g, calculate the molecular formula of the compound.

Si_2F_6

Historical Development of the Ideal Gas Equation

The ideal gas equation was not the scientific development of a single person, but represents the combination of a number of laws concerning the behavior of gases inferred from experimental observation over a 200-year period from 1650 to 1850. These laws, while completely contained within the ideal gas equation, are of tremendous importance in the history of science and also provide practical insight into the application of the ideal gas equation to real problems. For these reasons, they are summarized here.

Boyle's Law

If the temperature and number of moles of a gas are held constant, then the product nRT on the right-hand side of the ideal gas equation (Equation 5.6) is a constant. Therefore,

$$PV = \text{constant} \tag{5.11}$$

for any changes in a gas in which the temperature and number of moles are fixed. Equation 5.11, developed by Robert Boyle[9] in the seventeenth century through a systematic and quantitative study of the behavior of gases, is known as *Boyle's law,* which states that *the pressure of a fixed amount of gas at a constant temperature is inversely proportional to the volume of the gas.* Figure 5.13 shows two conventional ways of expressing Boyle's law (Equation 5.11) graphically. Plotting P versus V at a fixed temperature yields the hyperbolic curve shown in Figure 5.13(a); however, a linear relationship is obtained when P is plotted versus $1/V$ [Figure 5.13(b)].

---

9. Robert Boyle (1627–1691). British chemist and natural philosopher. Although Boyle is commonly associated with the gas law that bears his name, he made many other significant contributions in chemistry and physics. Despite the fact that Boyle was often at odds with contemporary scientists, his book *The Skeptical Chymist* (1661) influenced generations of chemists.

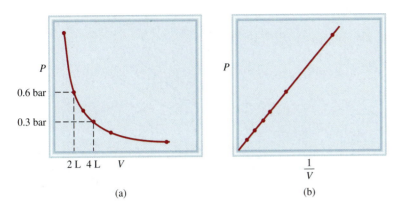

Figure 5.13 Graphs showing variation of the volume of a gas sample with the pressure exerted on the gas, at constant temperature. (a) P versus V. Note that the volume of the gas doubles as the pressure is halved. (b) P versus $1/V$.

Boyle's law is useful for calculating the changes in pressure or volume of an ideal gas that are the result of an isothermal process. If an ideal gas initially at pressure P_1 and volume V_1 undergoes a change at constant temperature to a final state with pressure P_2 and volume V_2, Equation 5.11 gives

$$P_1 V_1 = P_2 V_2 \qquad (5.12)$$

Thus, if any three of the variables P_1, V_1, P_2, and V_2 are known, then the remaining variable can be determined from Equation 5.12.

Charles's and Gay-Lussac's Law

Boyle's law depends on the temperature of the system remaining constant. But suppose the temperature changes while the pressure is held fixed? Rearranging the ideal gas equation (Equation 5.6) gives

$$\frac{V}{T} = \frac{nR}{P}$$

For fixed n and P, we obtain

$$\frac{V}{T} = \text{constant} \qquad (5.13)$$

This relationship was the result of early investigations by the French scientists Jacques Charles[10] and Joseph Gay-Lussac.[11] Equation 5.13 is known as ***Charles's and Gay-Lussac's law*** (or often simply *Charles's law*), and states that *the volume of a fixed amount of gas maintained at constant pressure is directly proportional to the absolute temperature of the gas.*

The work of Charles and Gay-Lussac played a central role in the development of the absolute temperature scale. At any given pressure that is sufficiently low for a gas to be treated as ideal, a plot of volume versus temperature in degrees Celsius yields a straight line. By extrapolating the line to zero volume, we find the intercept on the temperature axis to be $-273.15°C$ (Figure 5.14). The slope of the line changes

---

10. Jaques Alexandre Cesar Charles (1746–1823). French physicist. He was a gifted lecturer and inventor of scientific apparatus, and the first person to use hydrogen to inflate balloons.

11. Joseph Louis Gay-Lussac (1778–1850). French chemist and physicist. Like Charles, Gay-Lussac was a balloon enthusiast. Once he ascended to an altitude of 6000 meters to collect air samples for analysis.

Figure 5.14 Variation of the volume of a gas sample with temperature, at constant pressure. Each line represents the variation at a certain pressure. The pressures increase from P_1 to P_4. All gases ultimately condense (become liquids) if they are cooled to sufficiently low temperatures; the solid portions of the lines represent the temperature region above the condensation point. When these lines are extrapolated, or extended (the dashed portions), they all intersect at the point representing zero volume and a temperature of $-273.15°C$.

In practice, we can measure the volume of a gas only over a limited temperature range, because all gases condense at low temperatures to form liquids.

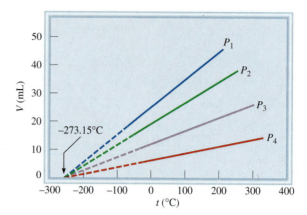

when the pressure is varied, but the zero-volume intercept is unchanged. In 1848, Kelvin realized the significance of this phenomenon. He identified $-273.15°C$ as *absolute zero,* the theoretical lowest value of temperature. He then set up the *absolute temperature scale,* now called the *Kelvin temperature scale,* with absolute zero as the starting point.

Just as we did for pressure-volume relationships at constant temperature, we can compare two sets of volume-temperature conditions for a given sample of gas at constant pressure. From Equation 5.13 we can write

$$\frac{V_1}{T_1} = \frac{V_2}{T_2} \tag{5.14}$$

where V_1 and V_2 are the volumes of the gas at temperatures T_1 and T_2 (both in kelvin), respectively.

The ideal gas law can also be rearranged to show that *at a constant amount of gas and volume, the pressure of a gas is proportional to temperature* or

$$\frac{P}{T} = \text{constant} \tag{5.15}$$

Equation 5.15 is known as **Gay-Lussac's law**. From the ideal gas equation (Equation 5.6), we see that the constant in Equation 5.15, is equal to nR/V. Starting with Equation 5.15, we have

$$\frac{P_1}{T_1} = \frac{P_2}{T_2} \tag{5.16}$$

where P_1 and P_2 are the pressures of the gas at temperatures T_1 and T_2, respectively.

Avogadro's Law

The work of the Italian scientist Amedeo Avogadro complemented the studies of Boyle, Charles, and Gay-Lussac. In 1811 he published a hypothesis stating that at the same temperature and pressure, equal volumes of different gases contain the same number of molecules. It follows that the volume of any given gas must be proportional to the number of moles of molecules present; that is,

$$V = \text{constant} \times n \tag{5.17}$$

where n represents the number of moles. Equation 5.17 is the mathematical expression of *Avogadro's law,* which states that at *constant pressure and temperature, the volume of a gas is directly proportional to the number of moles of the gas present.* From the ideal gas equation (Equation 5.6), we can identify the constant in Equation 5.17 as RT/P. Like the ideal gas equation itself, Equation 5.17 only holds strictly in the limit of low pressures.

According to Avogadro's law, we see that when two gases react with each other, their reacting volumes have a simple ratio to each other. If the product is a gas, its volume is related to the volumes of the reactants by a simple ratio (a fact demonstrated earlier by Gay-Lussac). For example, consider the synthesis of ammonia from molecular hydrogen and molecular nitrogen:

$$3H_2(g) + N_2(g) \longrightarrow 2NH_3(g)$$

Because at the same temperature and pressure, the volumes of gases are directly proportional to the number of moles of the gases present, we can now write the volume ratio of molecular hydrogen to molecular nitrogen as 3:1, and that of ammonia (the product) to molecular hydrogen and molecular nitrogen (the reactants) combined is 2:4, or 1:2.

Boyle's law (Equation 5.11), Charles's law (Equation 5.13), and Avogadro's law (Equation 5.17) combine to give the ideal gas equation (Equation 5.6), where R is defined as the overall constant of proportionality. Just as Boyle's law and Charles's law can be written in a form that is useful for predicting changes in P, T, V, or n (Equations 5.12, 5.14, and 5.16), the ideal gas law can be modified to take into account the initial and final conditions. Consider a gas whose initial state has pressure P_1, volume V_1, absolute temperature T_1, and number of moles n_1. If the gas undergoes a change to a new state with P_1, V_2, T_2, and n_2, Equation 5.6 gives

$$\frac{P_1 V_1}{n_1 T_1} = R = \frac{P_2 V_2}{n_2 T_2} \tag{5.18}$$

If $n_1 = n_2$, as is usually the case because the amount of gas normally does not change, the equation then becomes

$$\frac{P_1 V_1}{T_1} = \frac{P_2 V_2}{T_2} \tag{5.19}$$

Applications of Equation 5.19 are the subject of Examples 5.5 and 5.6.

Example 5.5

An inflated helium balloon with a volume of 0.55 L at sea level (1.0 atm) is allowed to rise to a height of 6.5 km, where the pressure is about 0.40 atm. Assuming that the temperature remains constant, what is the final volume of the balloon?

Strategy The amount of gas inside the balloon and its temperature remain constant, but both the pressure and the volume change, so we need to use Equation 5.19 with $T_1 = T_2$.

Solution If $T_1 = T_2$, Equation 5.19 becomes

$$P_1 V_1 = P_2 V_2$$

—Continued

A scientific research helium balloon

Continued—

which is Boyle's law. From the given data, $P_1 = 1.0$ atm, $V_1 = 0.55$ L, $P_2 = 0.40$ atm, and V_2 is unknown. Rearranging Boyle's law to isolate V_2 gives

$$V_2 = \frac{P_1 V_1}{P_2}$$

$$= \frac{(1.0 \text{ atm})(0.55 \text{ L})}{0.40 \text{ atm}}$$

$$= 1.4 \text{ L}$$

Check When pressure applied on the balloon is reduced (at constant temperature), the helium expands and the volume of the balloon increases. The final volume is greater than the initial volume, so the answer is reasonable.

Practice Exercise A sample of oxygen gas initially at 0.98 bar is cooled from 21°C to −68°C at constant volume. What is the final pressure of the gas?

Example 5.6

A small bubble rises from the bottom of a lake, where the temperature and pressure are 8°C and 6.4 bar, to the surface of the water, where the temperature is 25°C and the pressure is 1.0 bar. Calculate the final volume (in mL) of the bubble if its initial volume was 2.1 mL.

Strategy We are given the initial conditions P_1, V_1, T_1 as well as the final temperature T_2 and the final pressure, P_2. To find the final volume, rearrange Equation 5.19 to isolate V_2. Be sure to use the correct units for temperature.

Solution Assuming that the amount of the air in the bubble remains constant, we can use Equation 5.19:

$$\frac{P_1 V_1}{T_1} = \frac{P_2 V_2}{T_2}$$

Rearranging to isolate the unknown quantity V_2 gives

$$V_2 = \frac{P_1}{P_2} \times \frac{T_2}{T_1} \times V_1$$

Substituting in the given values

$$V_2 = \frac{6.4 \text{ bar}}{1.0 \text{ bar}} \times \frac{25 + 273.15 \text{ K}}{8 + 273.15 \text{ K}} \times 2.1 \text{ mL}$$

$$= 14 \text{ mL}$$

Check We see that the final volume involves multiplying the initial volume by a ratio of pressures (P_1/P_2) and a ratio of temperatures (T_2/T_1). Recall that volume is inversely proportional to pressure and volume is directly proportional to temperature. Because the pressure decreases and the temperature increases as the bubble rises, we expect the volume of the bubble to increase. In fact, here the change in pressure plays a greater role than temperature in the volume change.

—Continued

Continued—

Practice Exercise A gas initially at 4.0 L, 1.2 bar, and 66°C undergoes a change so that its final volume and temperature are 1.7 L and 42°C. What is the final pressure? Assume that the number of moles stays the same.

Gas Stoichiometry

In Chapter 0, we used relationships between amounts (in moles) and masses (in grams) of reactants and products to solve stoichiometry problems. When the reactants and/or products are gases, we can also use the ideal gas equation to relate the number of moles n to the volume V or pressure P to solve such problems. Examples 5.7 and 5.8 show how the ideal gas equation is used in such calculations.

An air bag can protect the driver in an automobile collision.

The symbol $\simeq$, meaning "stoichiometrically equivalent to" was defined in Section 0.6.

Example 5.7

Sodium azide (NaN_3) is used in some automobile air bags. The impact of a collision triggers the decomposition of NaN_3 as follows:

$$2NaN_3(s) \longrightarrow 2Na(s) + 3N_2(g)$$

The nitrogen gas produced quickly inflates the bag between the driver and the windshield and dashboard. Calculate the volume of N_2 generated at 80°C and 823 torr by the decomposition of 60.0 g of NaN_3.

Strategy From the balanced equation we see that 2 mol $NaN_3 \simeq$ 3 mol N_2. Because the mass of NaN_3 is given, we can calculate the number of moles of NaN_3 and hence the number of moles of N_2 produced. Finally, we can calculate the volume of N_2 using the ideal gas equation. Because pressure is given in mmHg (torr), make sure to use the correct value for R.

Solution First we calculate the number of moles of N_2 produced by 60.0 g NaN_3

$$\text{moles of } N_2 = 60.0 \text{ g NaN}_3 \times \frac{1 \text{ mol NaN}_3}{65.02 \text{ g NaN}_3} \times \frac{3 \text{ mol N}_2}{2 \text{ mol NaN}_3}$$

$$= 1.38 \text{ mol N}_2$$

The volume of 1.38 moles of N_2 can be obtained by using the ideal gas equation (Equation 5.6):

$$V = \frac{nRT}{P} = \frac{(1.38 \text{ mol})(62.36 \text{ L torr mol}^{-1} \text{ K}^{-1})(353 \text{ K})}{(832 \text{ torr})}$$

$$= 36.5 \text{ L}$$

where we have used the value of R corresponding to the pressure units of torr (mmHg). One could alternatively use the values of R for pressure units of bar or atm, but one must first convert the pressure to those units.

Practice Exercise The equation for the metabolic breakdown of glucose ($C_6H_{12}O_6$) is the same as the equation for the combustion of glucose in air:

$$C_6H_{12}O_6(s) + 6O_2(g) \longrightarrow 6CO_2(g) + 6H_2O(l)$$

Calculate the volume of CO_2 produced at 37°C and 1.00 bar when 5.60 g of glucose is used up in the reaction.

The air in submerged submarines and space vehicles needs to be purified continuously.

Example 5.8

An aqueous solution of lithium hydroxide is used to purify air in spacecrafts and submarines because it absorbs carbon dioxide, which is an end product of metabolism, according to the equation

$$2\text{LiOH}(aq) + \text{CO}_2(g) \longrightarrow \text{Li}_2\text{CO}_3(aq) + \text{H}_2\text{O}(l)$$

The pressure of carbon dioxide inside the cabin of a submarine having a volume of 2.4×10^5 L is 8.0×10^{-3} bar at 312 K. A solution of lithium hydroxide (LiOH) of negligible volume is introduced into the cabin. Eventually the pressure of CO_2 falls to 1.2×10^{-4} bar. How many grams of lithium carbonate are formed by this process?

Strategy Using the initial and final pressures, calculate the number of moles of CO_2 reacted from the drop in CO_2 pressure using the ideal gas law. Use the reaction equation given to determine the number of moles of lithium carbonate produced per mole of carbon dioxide consumed.

Solution First we calculate the number of moles of CO_2 consumed in the reaction. The drop in CO_2 pressure, which is $(8.0 \times 10^{-3}$ bar$) - (1.2 \times 10^{-4}$ bar$)$, or 7.9×10^{-3} bar, corresponds to the consumption of CO_2. Using the ideal gas equation, we write

$$n = \frac{PV}{RT} = \frac{(7.9 \times 10^{-3}\,\text{bar})(2.4 \times 10^5\,\text{L})}{(0.08314\,\text{L bar mol}^{-1}\,\text{K}^{-1})(312\,\text{K})} = 73\,\text{mol}$$

From the chemical equation, we see that 1 mol $\text{CO}_2 \triangleq$ 1 mol Li_2CO_3, so the amount of Li_2CO_3 formed is also 73 moles. Then, with the molar mass of Li_2CO_3 (73.89 g), we calculate its mass:

$$\text{mass of Li}_2\text{CO}_3 \text{ formed} = 73\,\text{mol Li}_2\text{CO}_3 \times \frac{73.89\,\text{g Li}_2\text{CO}_3}{1\,\text{mol Li}_2\text{CO}_3}$$

$$= 5.4 \times 10^3\,\text{g Li}_2\text{CO}_3$$

Practice Exercise A 5.0-g sample of benzene (C_6H_6) of negligible volume is combusted in a closed 10.0-L vessel, initially containing a 50:50 mixture of O_2 and N_2 gas at 1.00 bar pressure and 298 K. The reaction is

$$\text{C}_6\text{H}_6(l) + 15\text{O}_2(g) \longrightarrow 6\text{H}_2\text{O}(l) + 12\text{CO}_2(g)$$

Is all the benzene combusted in this process? If not, how many grams of benzene remain after all the oxygen has been consumed?

Ideal Gas Mixtures: Dalton's Law of Partial Pressures

Thus far we have concentrated on the behavior of pure gaseous substances, but experimental studies very often involve mixtures of gases. For example, for a study of air pollution, we may be interested in the equation of state for a sample of air, which contains several gases. Consider a mixture of gases in a volume V at temperature T. The total number of moles of gas n_T in the sample is equal to the sum of the number of moles of each of the constituent gases:

$$n_\text{T} = n_1 + n_2 + n_3 + \cdots \tag{5.20}$$

In the ideal gas approximation, *all* molecular sizes and interactions are considered to be negligible, so the contribution of each molecule to the equation of state is equal, independent of molecule type. Thus, the total pressure P_T of an ideal gas mixture at fixed V and T will depend only upon n_T and not on the specific composition, that is

$$P_T = \frac{n_T RT}{V} \tag{5.21}$$

Substituting Equation 5.20 into Equation 5.21 gives

$$P_T = \frac{n_1 RT}{V} + \frac{n_2 RT}{V} + \frac{n_3 RT}{V} + \cdots \tag{5.22}$$

The individual terms in the right-hand side of Equation 5.22 represent the **partial pressures** of each component gas, or *the pressures that each gas would exert if it were present alone in the container.* The partial pressure of component i in the mixture is

$$P_i = \frac{n_i RT}{V} \tag{5.23}$$

thus, the total pressure of an ideal gas mixture is given by

$$P_T = P_1 + P_2 + P_3 + \cdots \tag{5.24}$$

Equation 5.24, formulated in 1801 by Dalton, is known as **Dalton's law of partial pressures,** which states that *the total pressure of a mixture of gases is just the sum of the partial pressures of the individual gas components* (Figure 5.15). Like the ideal gas law, Dalton's law is only strictly valid in the limit of zero pressure.

The partial pressures (Equation 5.23) can also be expressed in terms of the total pressure by multiplying the numerator and denominator of Equation 5.23 by n_T:

$$P_i = \frac{n_T}{n_T}\frac{n_i RT}{V} = \frac{n_i}{n_T}\frac{n_T RT}{V}$$

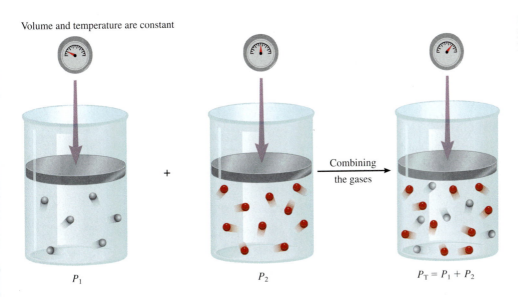

Volume and temperature are constant

$+$

Combining the gases

P_1

P_2

$P_T = P_1 + P_2$

Figure 5.15 Schematic illustration of Dalton's law of partial pressures.

Using Equation 5.21, we obtain

$$P_i = x_i P \tag{5.25}$$

where x_i is the *mole fraction* of component *i*, defined earlier in Equation 0.11,

$$x_i = \frac{n_i}{n_T} \tag{5.26}$$

The mole fraction of any component is unitless and always smaller than 1, and the sum of the mole fractions for all components is equal to 1. For example, if three components are present, then $x_1 + x_2 + x_3 = 1$. Thus, the partial pressures of a gaseous mixture can be determined by measuring the total pressure and the mole fractions of each component. One practical way of obtaining mole fractions of a gas is through the use of a mass spectrometer (see inset on page 46). The relative intensities of the peaks in a mass spectrum are directly proportional to the amounts, and hence the mole fractions, of the gases present.

From mole fractions and total pressure, we can calculate the partial pressure of individual components, as illustrated in Example 5.9.

Example 5.9

A mixture of gases contains 4.46 moles of neon (Ne), 0.74 mole of argon (Ar), and 2.15 moles of xenon (Xe). Calculate the partial pressures of the gases if the total pressure is 2.00 bar at a certain temperature.

Strategy Determine the mole fractions of each gas using Equation 5.26 and then use Equation 5.25 to determine the partial pressures.

Solution According to Equation 5.25 the partial pressure of Ne (P_{Ne}) is equal to the product of its mole fraction (x_{Ne}) and the total pressure (P_T). Using Equation 5.26, we calculate the mole fraction of each of the components as follows:

$$x_{Ne} = \frac{n_{Ne}}{n_T} = \frac{n_{Ne}}{n_{Ne} + n_{Ar} + n_{Xe}} = \frac{4.46 \text{ mol}}{4.46 \text{ mol} + 0.74 \text{ mol} + 2.15 \text{ mol}}$$
$$= 0.607$$

$$x_{Ar} = \frac{n_{Ar}}{n_T} = \frac{n_{Ar}}{n_{Ne} + n_{Ar} + n_{Xe}} = \frac{0.74 \text{ mol}}{4.46 \text{ mol} + 0.74 \text{ mol} + 2.15 \text{ mol}}$$
$$= 0.10$$

$$x_{Xe} = \frac{n_{Xe}}{n_T} = \frac{n_{Xe}}{n_{Ne} + n_{Ar} + n_{Xe}} = \frac{2.15 \text{ mol}}{4.46 \text{ mol} + 0.74 \text{ mol} + 2.15 \text{ mol}}$$
$$= 0.293$$

Using Equation 5.25 then gives

$$P_{Ne} = x_{Ne} P_T = (0.607)(2.00 \text{ bar}) = 1.21 \text{ bar}$$
$$P_{Ar} = x_{Ar} P_T = (0.10)(2.00 \text{ bar}) = 0.20 \text{ bar}$$
$$P_{Xe} = x_{Xe} P_T = (0.293)(2.00 \text{ bar}) = 0.586 \text{ bar}$$

Check Make sure that the sum of the partial pressures is equal to the given total pressure; that is, $(1.21 + 0.20 + 0.586)$ bar $= 2.00$ bar.

—*Continued*

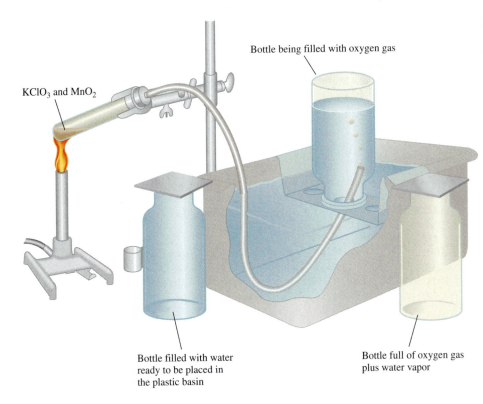

Bottle being filled with oxygen gas

KClO₃ and MnO₂

Bottle filled with water
ready to be placed in
the plastic basin

Bottle full of oxygen gas
plus water vapor

Figure 5.16 Apparatus for collecting gas over water. The oxygen generated by heating potassium chlorate ($KClO_3$) in the presence of a small amount of manganese dioxide (MnO_2), which speeds up the reaction, is bubbled through water and collected in a bottle as shown. Water originally present in the bottle is pushed into the trough by the oxygen gas.

Continued—

Practice Exercise A sample of natural gas contains 8.24 moles of methane (CH_4), 0.421 mole of ethane (C_2H_6), and 0.116 mole of propane (C_3H_8). If the total pressure of the gases is 1.37 bar, what are the partial pressures of the gases?

Dalton's law of partial pressures is useful for calculating volumes of gases collected over water. For example, when potassium chlorate ($KClO_3$) is heated, it decomposes to KCl and O_2:

$$2KClO_3(s) \longrightarrow 2KCl(s) + 3O_2(g)$$

The oxygen gas can be collected over water, as shown in Figure 5.16. Initially, the inverted bottle is completely filled with water. As oxygen gas is generated, the gas bubbles rise to the top and displace water from the bottle. This method of collecting a gas is based on the assumptions that the gas does not react with water and that it is not appreciably soluble in it. These assumptions are valid for oxygen gas, but not for gases such as NH_3, which dissolves readily in water. The oxygen gas collected in this way is not pure, however, because water vapor is also present in the bottle. The total gas pressure is equal to the sum of the pressures exerted by the oxygen gas and the water vapor:

$$P_T = P_{O_2} + P_{H_2O}$$

Consequently, we must allow for the pressure caused by the presence of water vapor when we calculate the amount of O_2 generated. Table 5.2 shows the pressure of water (in mm Hg) vapor at various temperatures.

Table 5.2	
Pressure of Water Vapor at Various Temperatures	
Temperature (°C)	**Water Vapor Pressure (mmHg)**
0	4.58
5	6.54
10	9.21
15	12.79
20	17.54
25	23.76
30	31.82
35	42.18
40	55.32
45	71.88
50	92.51
55	118.04
60	149.38
65	187.54
70	233.7
75	289.1
80	355.1
85	433.6
90	525.76
95	633.90
100	760.00

Example 5.10 shows a typical application of Dalton's law in calculating the amount of gas collected over water.

Example 5.10

Oxygen gas generated by the decomposition of potassium chlorate, according to the reaction

$$2KClO_3(s) \longrightarrow 2KCl(s) + 3O_2(g)$$

is collected over water as shown in Figure 5.16. The volume of the mixture of oxygen and water vapor collected at 24°C and an ambient pressure of 762 torr is 128 mL. Calculate the mass (in grams) of oxygen gas obtained. The equilibrium vapor pressure of water at 24°C is 22.4 torr.

Strategy The gas will be a mixture of O_2 and water vapor. To solve for the mass of O_2 generated, we must first calculate the partial pressure of O_2 in the mixture, taking into account the presence of the water vapor. Use the value of R corresponding to the pressure units given (torr).

Solution From Dalton's law of partial pressures (Equation 5.24), we know that

$$P_T = P_{O_2} + P_{H_2O}$$

Therefore,

$$\begin{aligned} P_{O_2} &= P_T - P_{H_2O} \\ &= 762 \text{ torr} - 22.4 \text{ torr} \\ &= 740 \text{ torr} \end{aligned}$$

From the ideal gas equation (Equation 5.6), we write

$$PV = nRT = \frac{m}{\mathcal{M}}RT$$

where m and $\mathcal{M}$ are the mass of O_2 collected and the molar mass of O_2, respectively. Rearranging the equation we obtain

$$\begin{aligned} m = \frac{PV\mathcal{M}}{RT} &= \frac{(740 \text{ torr})(0.128 \text{ L})(32.00 \text{ g mol}^{-1})}{(62.36 \text{ L torr mol}^{-1} \text{ K}^{-1})(273.15 \text{ K} + 24 \text{ K})} \\ &= 0.164 \text{ g} \end{aligned}$$

Practice Exercise Hydrogen gas generated when calcium metal reacts with water is collected as shown in Figure 5.16. The volume of gas collected at 30°C and a pressure of 988 torr is 641 mL. What is the mass (in grams) of the hydrogen gas obtained? The pressure of water vapor at 30°C is 31.82 torr.

5.4 | The Kinetic Theory of Gases Provides a Molecular Explanation for the Behavior of Gases

The gas laws help us predict the behavior of gases, but they do not explain what happens at the molecular level to cause the changes we observe in the macroscopic world. For example, the gas laws do not tell us how the pressure of a gas arise from the motion and collisions of individual molecules, nor can they explain why a gas expands upon heating. In the nineteenth century, a number of physicists, notably Ludwig Boltzmann and James Clerk Maxwell, found that the physical properties of gases can be explained in terms of the motion of individual molecules. These findings resulted in a number of generalizations about gas behavior that have since been known as the **kinetic molecular theory of gases,** or simply the *kinetic theory of gases.* Central to the kinetic theory are the following assumptions:

1. A gas is composed of a very large number of molecules (or atoms) that are separated from each other by distances that are large compared to their own size. The molecules (or atoms) can be considered to be "points," that is, they possess mass, but have negligible volume.

2. Gases are in constant motion in random directions and frequently collide. Collisions among molecules and between molecules and the walls of the container are assumed to be perfectly *elastic,* that is, kinetic energy may be transferred from one molecule to another, but the total kinetic energy remains constant.

3. Gas molecules are sufficiently far apart that they exert neither attractive nor repulsive forces upon one another.

Assumptions 1 and 3 were introduced earlier in our discussion of ideal gases. In the kinetic theory of gases, these three assumptions are used in an explicit manner to derive expressions for macroscopic properties, such as pressure and temperature, in terms of the average motion of individual molecules.

The Pressure of a Gas

Using the kinetic theory of gases, it is possible to derive an expression for the pressure of a gas in terms of its underlying random molecular motion. Consider an ideal gas consisting of N particles of mass m occupying a cubic box of length L (Figure 5.17). The pressure on the walls is generated through the collisions of the individual molecules on the wall. Because each of the six walls of the cube will experience the same pressure, it suffices to calculate the pressure on just one of them. Consider the wall perpendicular to the x direction on the positive x side of the cube (shaded in Figure 5.17). Let us assume for simplicity that the x component of the velocity, u_x, is the same for each colliding molecule.

Figure 5.18 shows the effect of the elastic collision of a molecule with this wall. The velocity in the x direction changes from $+u_x$ to $-u_x$, whereas the y and z components are unchanged.

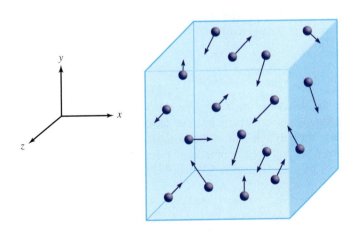

Figure 5.17 Molecular view of an ideal gas occupying a cubic box of side length L. The pressure on the shaded wall is calculated explicitly in this section.

Figure 5.18 Effect of an elastic collision of a molecule with a wall perpendicular to the x axis. The y and z components of the velocity remain unchanged, but the x component is reversed: $u_x \longrightarrow -u_x$.

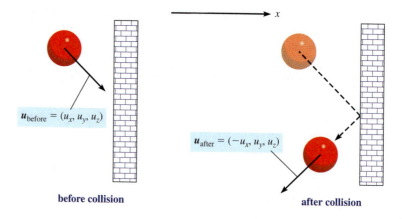

before collision   after collision

The change in the x component of the velocity of the colliding molecule is then

$$\Delta u_x = u_{x,\text{after}} - u_{x,\text{before}} = (-u_x) - u_x = -2u_x \tag{5.27}$$

From Newton's second law of motion (Equation 1.1), the force on an object is equal to the rate of change of the velocity (acceleration) multiplied by the mass. For these the x component of the force gives

$$F_x = m\frac{\Delta u_x}{\Delta t} \tag{5.28}$$

Combining Equations 5.27 and 5.28 gives the force on the molecule, over time Δt, due to collision with the wall: $-2m\, u_x/\Delta t$. The forces that the molecule and the wall experience during the collision are equal in magnitude, but opposite in direction, so the force on the wall during the collision with a single molecule is

$$F_{\text{wall}} = +\frac{2mu_x}{\Delta t} \tag{5.29}$$

Pressure is force per unit area, so to calculate the pressure on the wall we need to multiply the force in Equation 5.29 by the number of such particles that collide with the wall in time Δt per unit area. The distance in the x direction that a molecule with velocity u_x can travel in time Δt is $u_x\Delta t$, so only those molecules within a distance of $u_x\Delta t$ of the wall will collide with the wall in the time interval Δt. The number of colliding molecules is then given by

$$N_{\text{collision}} = \frac{1}{2}\left(\frac{N}{L^3}\right)(u_x\Delta t L^2) = \frac{Nu_x\Delta t}{2L} \tag{5.30}$$

where N/L^3 is the number of molecules per unit volume and $u_x\Delta t\, L^2$ is the volume of the region containing colliding particles. The factor of $1/2$ in Equation 5.30 takes into account the fact that only one-half of the particles are traveling *toward* the wall at any given time—the other half are traveling away from the wall and will not collide with it. The definition of pressure (Equation 5.1) together with Equations 5.29 and 5.30 gives

$$P = \frac{F_{\text{total}}}{area} = \frac{N_{\text{collision}}F_{\text{collision}}}{L^2} = \frac{Nmu_x^2}{V} \tag{5.31}$$

where the small, but arbitrary, time interval Δt cancels from the equations and we have used the fact that $V = L^3$. Of course, the large number of molecules in a gas do not all have the same value of u_x^2, so it is appropriate to replace u_x^2 in Equation 5.31 with its average (or mean) value $\langle u_x^2 \rangle$, where the angle brackets $\langle \ \rangle$ denote averaging. The velocity is a vector

$$\mathbf{u} = u_x\mathbf{i} + u_y\mathbf{j} + u_z\mathbf{k}$$

where $\mathbf{i}$, $\mathbf{j}$, and $\mathbf{k}$ are unit vectors in the x, y, and z directions, respectively, and its magnitude, u, called the **molecular speed,** is given by

$$u = |\mathbf{u}| = \sqrt{u_x^2 + u_y^2 + u_z^2}$$

Because there is no difference between the x, y, and z directions, it is correct to assume that the average square of the velocity components in each the three directions will be equal. This means that

$$\langle u_x^2 \rangle = \langle u_y^2 \rangle = \langle u_z^2 \rangle = \frac{\langle u^2 \rangle}{3}$$

The quantity $\langle u^2 \rangle$ is called the *mean square velocity*. Using this definition, Equation 5.29 can be rewritten as

$$P = \frac{N m \langle u^2 \rangle}{3V} \tag{5.32}$$

Equation 5.33 shows that pressure exerted by N gas molecules on a wall is directly proportional to the mean square velocity of the individual particles. The physical meaning of this dependence is that the larger the average velocity, the more frequent the collisions and the larger the force of collision. These two independent terms give us the quantity $\langle u^2 \rangle$ in the kinetic theory expression for the pressure.

Temperature and Kinetic Energy

Using the definition of the kinetic energy (Equation 0.3), we can rewrite Equation 5.32 in terms of the average translational kinetic energy of a molecule, $\langle E_{\text{trans}} \rangle = m\langle u^2 \rangle/2$, to give

$$P = \frac{2N}{3V}\left(\frac{1}{2}m \langle u^2 \rangle\right) = \frac{2N}{3V} \langle E_{\text{trans}} \rangle \tag{5.33}$$

From the kinetic theory expression for the pressure (Equation 5.33), we have

$$PV = \frac{2N}{3} \langle E_{\text{trans}} \rangle$$

Compare this equation with the ideal gas equation (Equation 5.6):

$$PV = nRT$$

The left-hand sides of these two equations are identical so we have

$$\frac{2N}{3} \langle E_{\text{trans}} \rangle = nRT$$

Recognizing that the ratio of the number of molecules to the number of moles, N/n, is equal to Avogadro's number, N_A, we have

$$\langle E_{\text{trans}} \rangle = \frac{3}{2}\frac{R}{N_A}T = \frac{3}{2}k_B T \tag{5.34}$$

where $R = k_B N_A$ and k_B is *Boltzmann's constant*, equal to 1.380650×10^{-23} J K^{-1}. We see that the mean kinetic energy of a molecule is proportional to the temperature. As a consequence, two ideal gas samples at the same temperature T will have identical average kinetic energies, independent of composition.

Equation 5.34 gives us a way to determine $\langle u^2 \rangle$, a difficult quantity to measure directly, simply from knowledge of the temperature and molecular mass:

$$\langle u^2 \rangle = \frac{3RT}{mN_A} = \frac{3k_B T}{m}$$

Because the molar mass $\mathcal{M}$ is equal to $m N_A$, where m is the mass of one molecule, we can get one estimate of the ***root-mean-square (rms) speed***[12] ($\langle u \rangle_{\text{rms}}$) of a molecule

$$\langle u \rangle_{\text{rms}} = \sqrt{\langle u^2 \rangle} = \sqrt{\frac{3k_B T}{m}} = \sqrt{\frac{3RT}{\mathcal{M}}} \tag{5.35}$$

The quantity $\langle u \rangle_{\text{rms}}$ is directly proportional to the square root of temperature, so that molecules in a gas at high temperature are faster on average than in a lower temperature gas. In addition, Equation 5.35 shows that $\langle u \rangle_{\text{rms}}$ is inversely proportional to the square root of molar mass; therefore, the heavier molecules move more slowly on average than lighter molecules at a given temperature.

Although we derived Equations 5.34 and 5.35 from the kinetic theory of gases, these equations apply to solids and liquids as well as long as the temperature is high enough that quantum mechanical effects are negligible (above ~20 K for most materials). The kinetic theory presented here is based on classical mechanics (Newton's equations). At very low temperatures, close to absolute zero, quantum effects render Equation 5.34 invalid, especially for light molecules, such as hydrogen and helium. A classic example is helium, which remains a liquid at 1 bar pressure even in the limit of 0 K. For such a system, the molecules remain in motion even arbitrarily close to absolute zero, so temperature is no longer a direct measurement of average kinetic energy. This phenomenon can be understood by recognizing that the motion of molecules can never completely cease without violating the Heisenberg uncertainty principle—if the kinetic energy were to go to zero then we would exactly know the momentum of the particle, which means that its position would have to be completely unknown.

The Maxwell-Boltzmann Distribution of Molecular Speeds

The rms speed is a measure of the average speed of the molecules in a gas and does not represent the speed of any particular molecule, which can be significantly different (faster or slower) than the average speed. Because the number of molecules in a macroscopic gas sample is extremely large, it is impossible to know the exact speed

---

12. Because velocity is a vector quantity, the average molecular velocity, $\langle \mathbf{u} \rangle$ must be zero; there are just as many molecules moving in the positive direction as there are in the negative direction. On the other hand, $\langle u_{\text{rms}} \rangle$ is a scalar quantity, that is, it has magnitude but no direction.

of any particular molecule at a given time. We can, however, predict, at a given instant, the fraction of molecules in a gas moving at a particular speed—that is, we can predict the distribution of molecular speeds.

The distribution of molecular speeds in a gas was first studied by Maxwell in 1860 and later refined by Boltzmann. For a gas sample containing N molecules of mass m at a temperature T, they showed that the number of molecules (N_u) in the sample with molecular speeds in the range u to $u + du$ is given by

$$N_u = N f(u) du$$

The function $f(u_x)$ is called the **Maxwell-Boltzmann distribution of molecular speeds**

$$f(u) = 4\pi u^2 \left(\frac{m}{2\pi k_B T} \right)^{3/2} e^{-mu^2/2k_B T} \qquad \textbf{(5.36)}$$

The distribution of speeds given in Equation 5.36 holds for all matter (solid, liquid, or gas) in thermal equilibrium, except for light molecules and atoms at very low temperatures, where quantum effects are large and the precise specification of the speed of a particle is limited by the Heisenberg uncertainty principle.

Figure 5.19(a) shows the distribution of molecular speeds $f(u)$ (Equation 5.36) for nitrogen (N_2) at three different temperatures. From this distribution we see that the number of particles with very high or very low speeds is small. Figure 5.19(b) shows the speed distributions for three different gases at the *same* temperature. The difference in the curves can be explained by noting that lighter molecules move faster, on average, than heavier ones.

The distribution of molecular speeds can be demonstrated with the apparatus shown in Figure 5.20. A beam of atoms (or molecules) exits from an oven at a known temperature and passes through a pinhole (to collimate the beam). Two circular plates

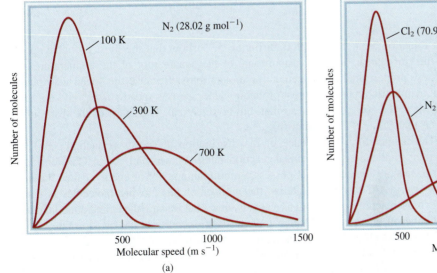

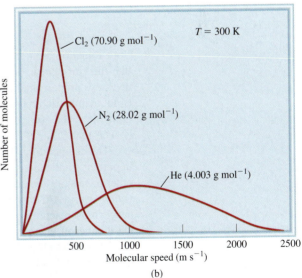

(a)

(b)

Figure 5.19 (a) The distribution of speeds $f(u)$ for nitrogen gas at three different temperatures. At the higher temperatures, more molecules are moving at higher speeds. (b) The distribution of speeds for three different gases at 300 K. At a given temperature, the lighter molecules are moving faster, on average.

Figure 5.20 (a) Apparatus for studying molecular speed distribution at a certain temperature. The vacuum pump causes the molecules to travel from left to right as shown. (b) The spread of the deposit on the detector gives the range of molecular speeds, and the density of the deposit is proportional to the fraction of molecules.

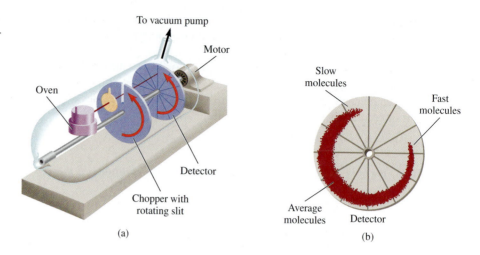

(a)   (b)

mounted on the same shaft are rotated by a motor. The first plate is called the "chopper," and the second is the detector. The purpose of the chopper is to allow small bursts of atoms (or molecules) to pass through it whenever the slit is aligned with the beam. Within each burst, the faster-moving molecules will reach the detector earlier than the slower-moving ones. Eventually, a layer of deposit will accumulate on the detector. Because the two plates are rotating at the same speed, molecules in the next burst will hit the detector plate at approximately the same place as molecules from the previous burst having the same speed. In time, the molecular deposition will become visible. The density of the deposition indicates the distribution of molecular speeds at that particular temperature.

The speed at which $f(u)$ is maximum is called the ***most probable molecular speed*** $\langle u \rangle_{mp}$ and can be determined by taking the derivative of $f(u)$ with respect to u and setting this derivative to zero (see Problem 5.53) to get

$$\langle u \rangle_{mp} = \sqrt{\frac{2k_BT}{m}} = \sqrt{\frac{2RT}{\mathcal{M}}} \qquad (5.37)$$

Both Equation 5.37 and the plots in Figure 5.19 show that the most probable speed, like the rms speed, $\langle u \rangle_{mp}$, is proportional to the square root of the temperature and inversely proportional to molecular (or molar) mass. Because the speed distribution is not symmetric around the most probable value, the rms speed ($\langle u \rangle_{rms}$) and the most probable speed ($\langle u \rangle_{mp}$) will be similar, but not identical. The shape of the distribution is such that $\langle u \rangle_{rms}$ is larger than $\langle u \rangle_{mp}$.

The temperature dependence of the distribution of speeds (Equation 5.36) has important implications in chemical reaction rates. As we shall see in Chapter 14, in order to react, a molecule must possess a minimum amount of energy, called the *activation energy*. At low temperatures, the number of fast-moving molecules is small; hence, most reactions proceed at a relatively slow rate. Raising the temperature increases the number of energetic molecules and causes an increase in the reaction rate.

The Maxwell-Boltzmann distribution can be used to determine the fraction of molecules with speeds in a given interval. The fraction of molecules with speeds between a and b is given by

$$\text{fraction } (a < x < b) = \int_a^b f(u) \, du = 4\pi \left(\frac{m}{2\pi k_BT} \right)^{3/2} \int_a^b u^2 e^{-mu^2/2k_BT} du \qquad (5.38)$$

With modern graphing calculators, this integral can be solved numerically for specific cases with little difficulty. If the interval is small enough that $f(u)$ is approximately constant over the interval, the integral can be approximated as

$$\text{fraction } (a < x < b) = \int_a^b f(u)\, du \approx f(a) \times (b - a) \qquad \textbf{(5.39)}$$

Example 5.11

Calculate the values of $\langle u \rangle_{rms}$ and $\langle u \rangle_{mp}$ for N_2 at 298.15 K.

Strategy The equations for the two measures of average speed are Equations 5.35 and 5.38. In general, the molar mass is easier to obtain for a compound than the molecular mass, so the versions of Equations 5.35 and 5.38 employing R and $\mathcal{M}$ are most often used for this type of calculation.

Solution The constants and parameters needed are $R = 8.314$ J mol^{-1} K^{-1}, $T = 298.15$ K and $\mathcal{M} = 28.02$ g mol$^{-1} = 0.02802$ kg mol^{-1}. The most probable speed is given by (Equation 5.38)

$$\langle u \rangle_{mp} = \sqrt{\frac{2RT}{\mathcal{M}}} = \sqrt{\frac{2 \times 8.314 \text{ J K}^{-1}\text{ mol}^{-1} \times 298.15 \text{ K}}{0.02802 \text{ kg mol}^{-1}}}$$
$$= 420.6 \text{ m s}^{-1}$$

For the rms speed, we have (Equation 5.35)

$$\langle u \rangle_{rms} = \sqrt{\frac{3RT}{\mathcal{M}}} = \sqrt{\frac{3 \times 8.314 \text{ J K}^{-1}\text{ mol}^{-1} \times 298.15 \text{ K}}{0.02802 \text{ kg mol}^{-1}}}$$
$$= 515 \text{ m s}^{-1}$$

Check As required, $\langle u \rangle_{mp}$ is smaller than $\langle u \rangle_{rms}$.

Practice Exercise Calculate the values of $\langle u \rangle_{mp}$ and $\langle u \rangle_{rms}$ for H_2 at 500 K.

The path traveled by a single gas molecule. Each change in direction represents a collision with another molecule.

It is important to note that, even though the rms speed of N_2 is 515 m s^{-1} at 298.15 K (Example 5.11), the time it would take for an individual N_2 molecule to cross a room is much longer than this speed would indicate. The reason for this is that, although molecular speeds in a gas are quite large, molecules do not move in straight-line paths, but are continually changing direction due to collisions with other molecules.

Calculations of molecular speeds like those in Example 5.11 have an interesting relationship to the composition of Earth's atmosphere. Unlike Jupiter, Earth does not have appreciable amounts of hydrogen or helium in its atmosphere. Why is this the case? A smaller planet than Jupiter, Earth has a weaker gravitational attraction for these lighter molecules. A fairly straightforward calculation shows that to escape Earth's gravitational field, a molecule must possess an escape velocity equal to or greater than 1.1×10^4 m s^{-1}. Because the average speed of helium is considerably greater than that of molecular nitrogen or molecular oxygen, more helium atoms escape from Earth's atmosphere into outer space. Consequently, only a trace amount of helium is present in our atmosphere. On the other hand, Jupiter, with a mass about 320 times greater than that of Earth, retains both heavy and light gases in its atmosphere.

Jupiter, the interior of which consists mainly of hydrogen

Super Cold Atoms

What happens to a gas when cooled to nearly absolute zero? More than 70 years ago, Albert Einstein, extending work by the Indian physicist Satyendra Nath Bose, predicted that at extremely low temperatures gaseous atoms of certain elements would "merge" or "condense" to form a single entity and a new form of matter. Unlike ordinary gases, liquids, and solids, this supercooled substance, which was named a *Bose–Einstein condensate* (*BEC*), would contain no individual atoms because the wave functions of the original atoms would overlap one another, leaving no space in between.

Einstein's hypothesis inspired an international effort to produce the BEC. But, as sometimes happens in science, the necessary technology was not available until fairly recently, and so early investigations were fruitless. Lasers, which use a process based on another of Einstein's ideas, were not designed specifically for BEC research, but they became a critical tool for this work.

Finally, in 1995, physicists found the evidence they had sought for so long. A team at the University of Colorado was the first to report success. They created a BEC by cooling a sample of gaseous rubidium (Rb) atoms to about 1.7×10^{-7} K using a technique called "laser cooling," a process in which laser light is directed at a beam of atoms, hitting them head on and dramatically slowing them down. The Rb atoms were further cooled in an "optical molasses" produced by the intersection of six lasers. The slowest, coolest atoms were trapped in a magnetic field while the faster-moving, "hotter" atoms escaped, thereby removing more energy from the gas. Under these conditions, the kinetic energy of the trapped atoms was virtually zero, which accounts for the extremely low temperature of the gas. At this point, the Rb atoms formed the condensate, just as Einstein had predicted. Although this BEC was invisible to the naked eye (it measured only 5×10^{-3} cm across), the scientists were able to capture its image on a computer screen by focusing another laser beam on it. The laser caused the BEC to break up after about 15 seconds, but that was long enough to record its existence.

The figure shows the Maxwell velocity distribution of the Rb atoms at this temperature. The colors indicate the number of atoms having velocity specified by the two horizontal axes. The blue and white portions represent atoms that have merged to form the BEC.

Within weeks of the Colorado team's discovery, a group of scientists at Rice University, using similar techniques, succeeded in producing a BEC with lithium atoms, and in 1998 at the Massachusetts Institute of Technology were able to produce a BEC using hydrogen atoms. Since then, many advances have been made in understanding the properties of the BEC in general, and experiments are being extended to molecular systems. It is expected that studies of the BEC will shed light on atomic properties that are still not fully understood and on the mechanism of superconductivity. An additional benefit might be the development of better lasers. Other applications will depend on further study of the BEC itself. Nevertheless, the discovery of a new form of matter has to be one of the foremost scientific discoveries of the twentieth century.

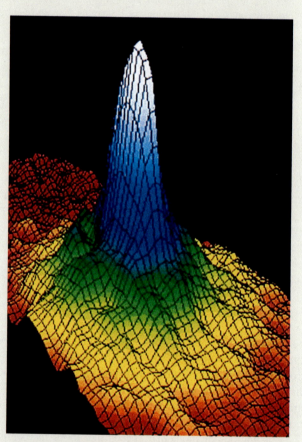

Maxwell velocity distribution of Rb atoms at about 1.7×10^{-7} K. The velocity increases from the center (zero) outward along the two axes. The red color represents the lowest number of Rb atoms and the white color the highest. The average speed in the white region is about 0.5 mm s^{-1}.

The kinetic theory of gases can be used to understand a number of phenomena relevant to chemistry. In Chapter 14, we will use the kinetic theory to determine the collision frequency of gas-phase reacting molecules, which will be used to calculate the rate of gas-phase reactions. Problems 5.57 and 5.58 illustrate the use of kinetic theory in gas **diffusion** (*the gradual mixing of molecules of one gas with molecules of another by virtue of their kinetic motion*) and **effusion** (*escape of a gas through a small hole in a container*).

5.5 | Real Gases Exhibit Deviations from Ideal Behavior at High Pressures

In Section 5.3, we saw that the ideal gas equation is a limiting law that is strictly valid only in the limit of zero pressure. At higher pressures, real gases will exhibit deviations from ideal behavior indicated by a value of the compression factor Z (Equation 5.7) that deviates from unity (see Figure 5.10). In contrast to the universality of the ideal gas equation of state, the deviations from ideality at high pressure are strongly dependent on the identity of the gas being studied. We can understand the nonideal behavior of real gases illustrated in Figure 5.10 in terms of intermolecular forces.

As discussed in Section 4.6, attractive intermolecular forces are the dominant interactions at intermediate intermolecular separations, whereas repulsive forces dominate when molecules are very close. When molecules are very far apart, as is the case at very low pressure, the attractive force is negligible, and the ideal gas approximation works well. As the pressure is increased, the average intermolecular separation decreases and attractive forces become significant, leading to a lower gas pressure than predicted by the ideal gas law—that is $Z < 1$, as seen for CH_4 and NH_3 in Figure 5.10. The degree to which the attractive forces affect the pressure is proportional to the strength of the intermolecular attractions. For example, the induced-dipole-induced dipole forces in nonpolar CH_4 are weaker than the dipole-dipole forces between polar NH_3 molecules, which is consistent with Figure 5.10, which shows smaller deviations from ideality for CH_4 than for NH_3 at moderate pressure. At very high pressures, the average intermolecular separation becomes small enough that repulsive interactions begin to dominate and the pressure will be higher than the ideal gas prediction, that is $Z > 1$.

Another way to observe the nonideal behavior of gases is to lower the temperature. Cooling a gas decreases the average kinetic energy of the molecules, which in a sense deprives the molecules of the drive they need to break away from their mutual attraction.

Because of the intermolecular interactions, the ideal gas equation of state does not adequately describe the behavior of gases at high pressure, and alternate equations of state must be developed for real gases. In this section, we will discuss two important such equations of state for real gases: the van der Waals equation and the virial equation of state.

The van der Waals Equation of State

In 1873, the Dutch physicist Johannes van der Waals[13] developed an equation of state for real gases that explicitly took into account the effect of molecular size

---

13. Johannes Diderck van der Waals (1837–1923). Dutch physicist. Van der Waals received the Nobel Prize in Physics in 1910 for his work on the properties of gases and liquids.

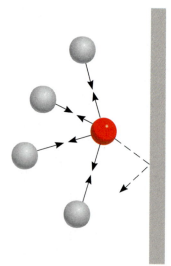

Figure 5.21 Effect of intermolecular forces on the pressure exerted by a gas. The speed of a molecule that is moving toward the container wall (red sphere) is reduced by the attractive forces exerted by its neighbors (gray spheres). Consequently, the impact this molecule makes with the wall is not as great as it would be if no intermolecular forces were present. In general, the measured gas pressure is lower than the pressure the gas would exert if it behaved ideally.

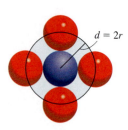

Figure 5.22 The volume excluded by a spherical particle of diameter d is a larger sphere of radius d. The volume of the excluded region is $4\pi d^3/3$.

and intermolecular attraction. Besides being mathematically simple, van der Waals's treatment provides us with an interpretation of real gas behavior at the molecular level.

Consider the approach of a particular molecule toward the wall of a container (Figure 5.21). The pressure exerted by the individual molecules on the walls of the container depends on both the frequency of molecular collisions to the walls and the force of the collision. Both contributions are diminished by the attractive intermolecular forces. The overall effect is a lower gas pressure than we would expect for an ideal gas. Van der Waals suggested that the pressure exerted by an ideal gas, $P_{ideal} = nRT/V$, is related to the experimentally measured pressure, P_{real}, by the equation

$$P_{real} = P_{ideal} - a\left(\frac{n}{V}\right)^2 \tag{5.40}$$

where a is a constant and n and V are the number of moles and volume of the gas, respectively. The correction term for pressure, $-a(n/V)^2$, can be understood as follows: The intermolecular interaction that gives rise to nonideal behavior depends on how frequently any two molecules approach each other closely. The number of such "encounters" increases with the square of the number of molecules per unit volume, $(n/V)^2$, because the presence of each of the two molecules in a particular region is proportional to n/V. The quantity P_{ideal} is the pressure we would measure if there were no intermolecular attractions, and so a is a proportionality constant, which is a measure of the strength of intermolecular attractions.

Another correction concerns the volume occupied by the gas molecules. In the ideal gas equation, V represents the volume of the container. However, if the size of the molecules is nonnegligible, the actual volume available to any molecule is reduced by the volume occupied by the other molecules in the system. Figure 5.22 shows that the volume excluded by a spherical particle is

$$V_{excluded} = \frac{4\pi d^3}{3}$$

where d is the diameter of the particle. Thus, the total volume excluded by N such particles is

$$V_{excluded} = \left(\frac{1}{2}\right)\frac{4\pi d^3}{3}N = BN \tag{5.41}$$

where $B = 2\pi d^3/3$ is the excluded volume per particle, and the factor of $(1/2)$ in Equation 5.41 is included to take into account the fact that the excluded volume shown in Figure 5.22 is shared by two particles. On a per mole basis, we have

$$V_{excluded} = bn$$

where $b = BN_A$ is the excluded volume per mole. Thus, we can correct Equation 5.40 to take into account molecular size by replacing V in the ideal gas contribution with $V - nb$

$$P = \frac{nRT}{V - nb} - a\left(\frac{n}{V}\right)^2 \tag{5.42}$$

Equation 5.42, *relating P, V, T, and n for a nonideal gas,* is known as the ***van der Waals equation of state*** and is valid over a wider range of pressure and temperature than is the ideal gas equation. The van der Waals equation is also commonly written in terms of the molar volume $\overline{V} = V/n$:

$$P = \frac{RT}{\overline{V} - b} - a\overline{V}^{-2} \tag{5.43}$$

The van der Waals constants a and b are selected to give the best possible agreement between Equation 5.42 and the observed behavior of a particular gas. Table 5.3 lists the values of a and b for a number of gases. The value of a indicates how strongly molecules of a given type of gas attract one another. For example, of the substances listed in Table 5.3, helium is expected to have the weakest interatomic interactions as it is nonpolar and has a low polarizability (see Section 4.6). Consequently, it also has the smallest a value.

There is also a rough correlation between molecular (or atomic) size and b. Generally, the larger the molecule (or atom), the greater b is, but the relationship between b and molecular (or atomic) size is not a simple one. An exception to this trend is He and Ne. The atomic radius of Ne is larger than that of He, but the value of b for Ne $(0.0174\,\text{L mol}^{-1})$ is smaller than that for He $(0.0237\,\text{L mol}^{-1})$. The reason for this exception is that He is very light and highly quantum mechanical in nature, so the large uncertainty in position because to the Heisenberg uncertainty principle makes it appear larger than it actually is. The values of a and b can be determined by a variety of methods, but the most common is to use the critical properties of the material $(P_c, T_c, \text{ and } V_c)$ to estimate a and b, as will be discussed in Section 6.1.

Example 5.12 compares the pressure of a gas calculated using the ideal gas equation and the van der Waals equation.

Table 5.3	Van der Waals Constants of Some Common Gases	
Gas	**a (bar L^2 mol^{-2})**	**b (L mol^{-1})**
He	0.034	0.0237
Ne	0.0213	0.0171
Ar	1.35	0.0322
Kr	2.34	0.0398
Xe	4.23	0.0266
H_2	0.246	0.0266
N_2	1.40	0.0391
O_2	1.37	0.0318
Cl_2	6.55	0.0562
CO_2	3.63	0.0427
CH_4	2.27	0.0428
CCl_4	20.6	0.138
NH_3	4.21	0.0371
H_2O	5.32	0.0305

NH$_3$

Example 5.12

Given that 3.50 moles of NH$_3$ occupy 5.20 L at 47°C, calculate the pressure of the gas (in bar) using (a) the ideal gas equation and (b) the van der Waals equation.

Strategy To calculate the pressure of NH$_3$ using the ideal gas equation, we proceed as in Example 5.1. What corrections are made to the pressure and volume terms in the van der Waals equation?

Solution (a) We have the following data:

$$V = 5.20\ \text{L}$$
$$T = (47 + 273)\ \text{K} = 320\ \text{K}$$
$$n = 3.50\ \text{mol}$$
$$R = 0.08314\ \text{L bar K}^{-1}\,\text{mol}^{-1}$$

Substituting these values in the ideal gas equation (Equation 5.6), we write

$$P = \frac{nRT}{V} = \frac{(3.50\ \text{mol})(0.8314\ \text{L bar mol}^{-1}\,\text{K}^{-1})(320\ \text{K})}{5.20\ \text{L}}$$
$$= 17.9\ \text{bar}$$

(b) We need Equation 5.42. From Table 5.3, we have

$$a = 4.21\ \text{bar L}^2\,\text{mol}^{-2}$$
$$b = 0.0371\ \text{L mol}^{-1}$$

Substituting these values into the van der Waals equation gives

$$P = \frac{nRT}{V - nb} - a\left(\frac{n}{V}\right)^2$$
$$= \frac{(3.50\ \text{mol})(0.08314\ \text{L bar mol}^{-1}\,\text{K}^{-1})(320\ \text{K})}{5.20\ \text{L} - (3.50\ \text{mol})(0.0371\ \text{L mol}^{-1})} - (4.21\ \text{bar L}^2\,\text{mol}^{-2})\left(\frac{3.50\ \text{mol}}{5.20\ \text{L}}\right)^2$$
$$= 16.4\ \text{bar}$$

Check Based on your understanding of nonideal gas behavior, is it reasonable that the pressure calculated using the van der Waals equation should be smaller than that using the ideal gas equation? Why?

Practice Exercise Using the data shown in Table 5.3, calculate the pressure (in bar) exerted by 4.37 moles of molecular chlorine confined in a volume of 2.45 L at 38°C. Compare the pressure with that calculated using the ideal gas equation.

The Virial Equation of State

An alternate way of representing the nonideality of real gases to expand the compression factor (Equation 5.7) as a power series expansion in inverse powers of the molar volume $\bar{V}$:

$$Z = \frac{PV}{nRT} = 1 + \frac{B}{\bar{V}} + \frac{C}{\bar{V}^2} + \frac{D}{\bar{V}^3} + \cdots \tag{5.44}$$

This expression is called the *virial equation of state,* and the coefficients B, C, D, . . ., are the second, third, fourth, . . ., virial coefficients, respectively. Except for the first virial coefficient, which is 1, the virial coefficients depend upon temperature. The

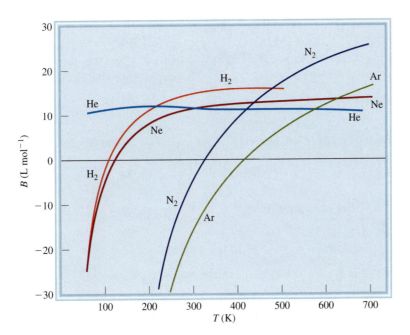

virial coefficients can be determined theoretically, if accurate intermolecular potentials are known for a given gas, but most often they are calculated so as to give optimal agreement with experimental data. For an ideal gas, all virial coefficients other than the first are zero, and Equation 5.47 reduces to the ideal-gas equation (Equation 5.6).

Figure 5.23 shows the second virial coefficient B for several gases as a function of temperature. The sign of B depends upon what type of intermolecular forces dominate at the temperature of interest—B is negative when attractive forces are dominant and positive when repulsive forces dominate. This can be seen by comparison to the van der Waals equation. Because the van der Waals equation and the virial equation are both descriptions of the equation of state for a real gas, their coefficients must be related. If you arrange the van der Waals equation into the same form as the virial equation (Problem 5.118), you can show that the second virial coefficient is related to the a and b of the van der Waals equation by $B = b - a/RT$. At low temperatures, a/RT is large and attractive forces dominate the second virial coefficient, but at high temperatures a/RT is small and B is dominated by molecular size (that is, repulsive interactions). For a given gas, the temperature at which the second virial coefficient B becomes zero is of special importance. At this temperature, called the **Boyle temperature**, the gas behaves nearly ideally over a wider density range than at higher or lower temperatures.

Example 5.13

Calculate the molar volume of carbon dioxide (CO_2) at 300 K and 20 bar, and compare your result with that obtained using the ideal gas equation. The second virial coefficient for CO_2 at this temperature (B) is -0.123 L mol^{-1}.

Strategy Because we are only given the second virial coefficient, we must truncate the virial equation (Equation 5.44) after the second term and assume that the effect of $C, D, \ldots$, are negligible.

—Continued

Continued—

Solution Neglecting terms containing virial coefficients higher than *B*, the virial equation (Equation 5.44) becomes

$$Z = \frac{P\overline{V}}{RT} = 1 + \frac{B}{\overline{V}}$$

Multiplying by $\overline{V}RT/P$ and rearranging gives a quadratic equation for $\overline{V}$:

$$\overline{V}^2 - \frac{RT}{P}\overline{V} - \frac{BRT}{P} = 0$$

Substituting in the given values ($T = 300$ K, P = 100 bar, $R = 0.08314$ L bar mol^{-1} K^{-1}) gives

$$\overline{V}^2 - 1.247\overline{V} + 0.1534 = 0$$

where $\overline{V}$ is in L mol^{-1}. Using the quadratic equation (see Appendix 1) gives

$$\overline{V} = \frac{1.247 \pm \sqrt{1.247^2 - 4(1)(0.1534)}}{2}$$
$$\overline{V} = 1.11 \text{ L mol}^{-1} \quad \text{or} \quad 0.138 \text{ L mol}^{-1}$$

Because we have two solutions, we must choose the more physical one. The ideal gas equation ($\overline{V} = RT/P$) gives a value of 1.247 L mol^{-1}, which is obtained from the coefficient of $\overline{V}$ in the quadratic. For the truncated virial equation to be valid, the molar volume must not be too small, so we can reject the smaller value above and conclude that $\overline{V} = 1.11$ L mol^{-1}.

Check The second virial coefficient for CO_2 at 300 K is negative, indicating that attractive forces are dominant and that we would expect the molar volume at a given pressure would be smaller than the ideal gas prediction.

Practice Exercise Calculate the molar volume of N_2 at 50.0 bar and 600 K given that the second virial coefficient at that temperature is 0.0217 L mol^{-1}.

Summary of Facts and Concepts

Section 5.1

▶ Pressure and temperature are macroscopic quantities that characterize the state of a material.

▶ Pressure is the external force exerted on the material per unit area.

▶ Temperature is the quantity that controls the flow of energy between two systems. If two objects with different temperatures are placed in contact, energy will flow from the object with the higher temperature into the lower temperature object.

Section 5.2

▶ A phase is a homogeneous part of the system in contact with other parts of the system but separated from them by well-defined boundaries. If the external conditions (temperature, pressure, etc.) change, a system may undergo a phase transition. Examples of phase transitions are evaporation/condensation, freezing/melting, and sublimation/deposition.

▶ A phase diagram is a plot summarizing the conditions at which the various phases of a material can exist.

Section 5.3

▶ An equation of state is an equation that relates the pressure of a substance to its temperature, volume, and mole numbers.

▶ The ideal gas equation ($PV = nRT$) is an accurate equation of state for gases at low pressure and high temperatures.

▶ Absolute zero (273.15°C) is the lowest theoretically attainable temperature. The Kelvin temperature scale takes 0 K as absolute zero. In all gas law calculations, temperature must be expressed in kelvins.

▶ A number of earlier laws were combined to obtain the ideal gas law:

— Boyle's law [volume is inversely proportional to pressure (at constant T and n)]

— Charles's and Gay-Lussac's law [volume is directly proportional to temperature (at constant P and n)]

— Avogadro's law [Equal volumes of gases contain equal numbers of molecules (at the same T and P)]

▶ Dalton's law of partial pressures states that each gas in a mixture of gases exerts the same pressure that it would if it were alone and occupied the same volume.

Section 5.4

▶ The kinetic molecular theory, a mathematical way of describing the behavior of gas molecules, is based on the following assumptions: Gas molecules are separated by distances far greater than their own dimensions, they possess mass (but negligible volume), they are in constant motion, and they frequently collide with one another. The molecules neither attract nor repel one another.

▶ A Maxwell speed distribution curve shows how many gas molecules are moving at various speeds at a given temperature. As temperature increases, more molecules move at greater speeds.

Section 5.5

▶ The van der Waals equation is a modification of the ideal gas equation that takes into account the nonideal behavior of real gases. It corrects for the fact that real gas molecules do exert forces on each other and that they do have volume. The van der Waals constants are determined experimentally for each gas.

▶ Another useful equation of state for real gases is the virial equation of state.

Key Words

absolute zero of temperature, p. 285	effusion, p. 317	melting point, p. 288	standard temperature and pressure (STP), p. 294
Avogadro's law, p. 301	equation of state, p. 292	molecular speed, p. 311	sublimation, p. 287
bar, p. 282	equlibrium vapor pressure, p. 287	most probable molecular speed, p. 314	supercritical fluid (SCF), p. 288
barometer, p. 283	evaporation, p. 287	normal boiling point, p. 288	temperature, p. 284
boiling point, p. 287	freezing point, p. 288	normal melting point, p. 288	thermal equilibrium, p. 284
Boyle's law, p. 298	Gay-Lussac's law, p. 300	partial pressure, p. 305	thermometer, p. 285
Boyle temperature, p. 321	Gibbs phase rule, p. 291	pascal (Pa), p. 282	triple point, p. 288
Charles's and Gay-Lussac's law, p. 299	ideal gas, p. 293	phase, p. 286	torr, p. 283
compression factor, p. 293	ideal gas equation of state, p. 293	phase diagram, p. 287	universal gas constant, p. 293
condensation, p. 287	kinetic molecular theory of gases, p. 309	phase transition, p. 287	van der Waals equation of state, p. 319
critical point, p. 288	limiting law, p. 293	physical equilibrium, p. 287	vaporization p. 287
critical pressure (P_c), p. 288	macroscopic, p. 282	pressure, p. 282	virial equation of state, p. 320
critical temperature (T_c), p. 288	manometer, p. 283	root-mean-square speed, p. 312	zeroth law of thermodynamics, p. 284
Dalton's law of partial pressures, p. 305	Maxwell-Boltzmann distribution of molecular speeds, p. 313	standard ambient temperature and pressure (SATP), p. 294	
deposition, p. 287	melting, p. 287	standard atmospheric pressure, p. 283	
diffusion, p. 317		standard boiling point, p. 288	
dynamic equilibrium, p. 287		standard melting point, p. 288	

Problems

Pressure and Temperature Are Two Important Macroscopic Properties of Chemical Systems

5.1 The atmospheric pressure at the summit of Mt. McKinley is 606 torr on a certain day. What is the pressure in atm, bar, and kPa?

5.2 Calculate the pressure underneath a car tire assuming that the car has a mass of 1000 kg and that the area of contact of the tire with the ground is 2.0×10^{-2} m^3.

Substances and Mixtures Can Exist as Solid, Liquid, or Gas, Depending upon the External Conditions

5.3 The following compounds, listed with their boiling points, are liquid at −10°C: butane, −0.5°C; ethanol, 78.3°C; toluene, 110.6°C. At −10°C, which of these liquids would you expect to have the highest vapor pressure? Which the lowest? Explain.

5.4 Freeze-dried coffee is prepared by freezing brewed coffee and then removing the ice component with a vacuum pump. Describe the phase changes taking place during these processes.

5.5 A student hangs wet clothes outdoors on a winter day when the temperature is $-15°C$. After a few hours, the clothes are found to be fairly dry. Describe the phase changes in this drying process.

5.6 A length of wire is placed on top of a block of ice. The ends of the wire extend over the edges of the ice, and a heavy weight is attached to each end. It is found that the ice under the wire gradually melts, so the wire slowly moves through the ice block. At the same time, the water above the wire refreezes. Explain the phase changes that accompany this phenomenon.

5.7 The boiling point and freezing point of sulfur dioxide at 1 bar are $-10°C$ and $-72.7°C$, respectively. The triple point is at $-75.5°C$ and 1.67×10^3 bar, and its critical point is at $157°C$ and 79 bar. On the basis of this information, draw a rough sketch of the phase diagram of SO_2.

5.8 A phase diagram of water is shown. Label the regions. Predict what would happen as a result of the following changes: (a) Starting at A, we raise the temperature at constant pressure. (b) Starting at C, we lower the temperature at constant pressure. (c) Starting at B, we lower the pressure at constant temperature.

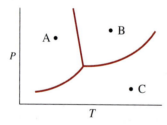

5.9 What is the maximum number of phases that can coexist in a system with five components present?

5.10 What is the dimensionality of the three-phase coexistence region in a mixture of Al, Ni, and Cu? What type of geometrical region does this define?

5.11 Liquid-liquid two-phase coexistence in pure (that is, one-component) systems is extremely rare. Using library and Internet resources, find two examples of liquid-liquid coexistence in a pure material.

The Ideal-Gas Equation Describes the Behavior of All Gases in the Limit of Low Pressure

5.12 The volume of a gas is 5.80 L, measured at 1.00 bar. What is the pressure of the gas in torr if the volume is changed to 9.65 L at constant temperature?

5.13 A sample of air occupies 3.8 L when the pressure is 1.2 atm. (a) What volume does it occupy at 6.6 atm?

(b) What pressure is required in order to compress it to 0.075 L? (The temperature is kept constant.)

5.14 (a) A 36.4-L volume of methane gas is heated from 25°C to 88°C at constant pressure. What is the final volume of the gas? (b) Under constant-pressure conditions, a sample of hydrogen gas initially at 88°C and 9.6 L is cooled until its final volume is 3.4 L. What is its final temperature?

5.15 A sample of nitrogen gas kept in a container of volume 2.3 L and at a temperature of 32°C exerts a pressure of 4.7 bar. Calculate the number of moles of gas present.

5.16 Given that 6.9 moles of carbon monoxide gas are present in a container of volume 30.4 L, what is the pressure of the gas (in bar) if the temperature is 62°C?

5.17 What volume will 5.6 moles of sulfur hexafluoride (SF_6) gas occupy if the temperature and pressure of the gas are 128°C and 9.4 atm?

5.18 A certain amount of gas at 25°C and at a pressure of 0.800 bar is contained in a glass vessel. Suppose that the vessel can withstand a pressure of 2.00 bar. How high can you raise the temperature of the gas without bursting the vessel?

5.19 Consider the following gaseous sample in a cylinder fitted with a movable piston. Initially, there are n moles of the gas at temperature T, pressure P, and volume V.

Choose the cylinder that correctly represents the gas after each of the following changes. (1) The pressure on the piston is tripled at constant n and T. (2) The temperature is doubled at constant n and P. (3) n moles of another gas are added at constant T and P. (4) T is halved and pressure on the piston is reduced to a quarter of its original value.

(a)   (b)   (c)

5.20 A gas occupying a volume of 725 mL at a pressure of 0.970 bar is allowed to expand at constant temperature until its pressure reaches 0.541 bar. What is its final volume?

5.21 At 46°C a sample of ammonia gas exerts a pressure of 5.3 bar. What is the molar volume?

5.22 A gas-filled balloon having a volume of 2.50 L at 1.2 bar and 25°C is allowed to rise to the stratosphere (about 30 km above the surface of Earth), where the temperature and pressure are -23°C and 3.00×10^{-3} bar, respectively. Calculate the final volume of the balloon.

5.23 The temperature of 2.5 L of a gas initially at SATP is raised to 250°C at constant volume. Calculate the final pressure of the gas in bar.

5.24 The pressure of 6.0 L of an ideal gas in a flexible container is decreased to one-third of its original pressure, and its absolute temperature is decreased by one-half. What is the final volume of the gas?

5.25 A gas evolved during the fermentation of glucose (wine making) has a volume of 0.78 L at 20.1°C and 1.00 atm. What was the volume of this gas at the fermentation temperature of 36.5°C and 1.00 atm pressure?

5.26 An ideal gas originally at 0.85 bar and 66°C was allowed to expand until its final volume, pressure, and temperature were 94 mL, 0.60 bar, and 45°C, respectively. What was its initial volume?

5.27 Calculate the volume (in liters) of 88.4 g of CO_2 at SATP.

5.28 Dry ice is solid carbon dioxide. A 0.050-g sample of dry ice is placed in an evacuated 4.6-L vessel at 30°C. Calculate the pressure inside the vessel after all the dry ice has been converted to CO_2 gas.

5.29 At STP, 0.280 L of a gas weighs 0.400 g. Calculate the molar mass of the gas.

5.30 At 741 torr and 44°C, 7.10 g of a gas occupy a volume of 5.40 L. What is the molar mass of the gas?

5.31 Ozone molecules in the stratosphere absorb much of the harmful radiation from the sun. Typically, the temperature and pressure of ozone in the stratosphere are 250 K and 1.0×10^{-3} bar, respectively. How many ozone molecules are present in 1.0 L of air under these conditions?

5.32 Assuming that air contains 78 percent N_2, 21 percent O_2, and 1.0 percent Ar, all by volume, how many molecules of each type of gas are present in 1.0 L of air at SATP?

5.33 A 2.10-L vessel contains 4.65 g of a gas at 1.00 bar and 27.0°C. (a) Calculate the density of the gas in g L^{-1}. (b) What is the molar mass of the gas?

5.34 Calculate the density of hydrogen bromide (HBr) gas (in g L^{-1}) at 733 torr and 46°C.

5.35 A certain anesthetic contains 64.9 percent C, 13.5 percent H, and 21.6 percent O by mass. At 120°C and 750 torr, 1.00 L of the gaseous compound has a mass of 2.30 g. What is the molecular formula of the compound?

5.36 A compound has the empirical formula SF_4. At 20°C, 0.100 g of the gaseous compound occupies a volume of 22.1 mL and exerts a pressure of 1.02 bar. What is the molecular formula of the gas?

5.37 Consider the formation of nitrogen dioxide from nitric oxide and oxygen:

$$2NO(g) + O_2(g) \longrightarrow 2NO_2(g)$$

If 9.0 L of NO are reacted with excess O_2 at SATP, what is the volume in liters of the NO_2 produced?

5.38 Methane, the principal component of natural gas, is used for heating and cooking. The combustion process is

$$CH_4(g) + 2O_2(g) \longrightarrow CO_2(g) + 2H_2O(l)$$

If 15.0 moles of CH_4 are reacted, what is the volume of CO_2 (in liters) produced at 23.0°C and 0.985 bar?

5.39 When coal is burned, the sulfur present in coal is converted to sulfur dioxide (SO_2), which is responsible for the acid rain phenomenon.

$$S(s) + O_2(g) \longrightarrow SO_2(g)$$

If 2.54 kg of S are reacted with oxygen, calculate the volume of SO_2 gas (in mL) formed at 30.5°C and 1.12 bar.

5.40 In alcohol fermentation, yeast converts glucose to ethanol and carbon dioxide:

$$C_6H_{12}O_6(s) \longrightarrow 2C_2H_5OH(l) + 2CO_2(g)$$

If 5.97 g of glucose are reacted and 1.44 L of CO_2 gas are collected at 293 K and 0.984 bar, what is the percent yield of the reaction?

5.41 A compound of P and F was analyzed as follows: Heating 0.2324 g of the compound in a 378-cm^3 container turned it to all gas, which had a pressure of 97.3 torr at 77°C. Then the gas was mixed with a calcium chloride solution, which turned all of the F to 0.2631 g of CaF_2. Determine the molecular formula of the compound.

5.42 A quantity of 0.225 g of a metal M (molar mass $=$ 27.0 g mol^{-1}) liberated 0.303 L of molecular hydrogen (measured at 17°C and 741 torr) from an excess of hydrochloric acid. Deduce from these data the corresponding equation, and write formulas for the oxide and sulfate of M.

5.43 What is the mass of the solid NH_4Cl formed when 73.0 g of NH_3 are mixed with an equal mass of HCl? What is the volume of the gas remaining, measured at 14.0°C and 752 torr? What gas is it?

5.44 Dissolving 3.00 g of an impure sample of calcium carbonate in hydrochloric acid produced 0.656 L of carbon dioxide (measured at 20.0°C and 792 torr). Calculate the percent by mass of calcium carbonate in the sample. State any assumptions.

5.45 Calculate the mass in grams of hydrogen chloride produced when 5.6 L of molecular hydrogen measured at SATP react with an excess of molecular chlorine gas.

5.46 Ethanol (C_2H_5OH) burns in air:

$$C_2H_5OH(l) + O_2(g) \longrightarrow CO_2(g) + H_2O(l)$$

Balance the equation and determine the volume of air in liters at 35.0°C and 790 torr required to burn 227 g of ethanol. Assume that air is 21.0 percent O_2 by volume.

5.47 A mixture of gases contains 0.31 mol CH_4, 0.25 mol C_2H_6, and 0.29 mol C_3H_8. The total pressure is 1.50 bar. Calculate the partial pressures of the gases.

The Kinetic Theory of Gases Provides a Molecular Explanation for the Behavior of Gases

5.48 Is temperature a microscopic or macroscopic concept? Explain.

5.49 Compare the rms speeds of O_2 and UF_6 at 65°C.

5.50 The temperature in the stratosphere is −23°C. Calculate the rms speeds of N_2, O_2, and O_3 molecules in this region.

5.51 If 2.0×10^{23} argon (Ar) atoms strike 4.0 cm^2 of a wall per second at a 90° angle when moving with a speed of 45,000 cm s^{-1}, what pressure (in bar) do they exert on the wall?

5.52 A square box contains He at 25°C. If the atoms are colliding with the walls perpendicularly (at 90°) at the rate of 4.0×10^{22} times per second, calculate the force and the pressure exerted on the wall given that the area of the wall is 100 cm^2 and the speed of the atoms is 600 m s^{-1}.

5.53 Derive Equation 5.37 for the most probable speed $\langle u \rangle_{mp}$ from the Maxwell-Boltzmann distribution of molecular speeds (Equation 5.36). [*Hint:* The derivative of a function is zero at its maximum (or minimum).]

5.54 To what temperature must He atoms be cooled so that they have the same rms speed as O_2 at 25°C? (*Hint:* You do not have to calculate the value of the rms speeds to solve this problem.)

5.55 The rms speed of CH_4 is 846 m s^{-1}. What is the temperature of the gas?

5.56 Calculate the average translational kinetic energy for an N_2 molecule and for 1 mole of N_2 molecules at 20°C.

5.57 The relative rates of diffusion of different molecules in a gas can be expressed in terms of Graham's law of diffusion

$$\frac{r_1}{r_2} = \sqrt{\frac{M_2}{M_1}}$$

where r_1 and r_2 are the rates of diffusion for molecules with molar mass M_1 and M_2, respectively. Using the kinetic theory of gases, derive Graham's law.

5.58 Gas effusion is the process by which a gas under pressure escapes from one compartment of a container to another through a small opening, as illustrated here.

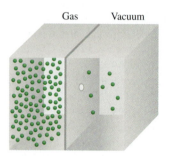

The relative rates of effusion for gases obey Graham's law (see Problem 5.57). The ^{235}U isotope undergoes fission when bombarded with neutrons. However, its natural abundance is only 0.72 percent. To separate it from the more abundant ^{238}U isotope, uranium is first converted to UF_6, which is easily vaporized above room temperature. The mixture of the $^{235}UF_6$ and $^{238}UF_6$ gases is then subjected to many stages of effusion. Calculate the separation factor, that is, the enrichment of ^{235}U relative to ^{238}U after one stage of effusion.

Real Gases Exhibit Deviations from Ideal Behavior at High Pressures

5.59 Which of the following combinations of conditions most influences a gas to behave ideally: (a) low pressure and low temperature, (b) low pressure and high temperature, (c) high pressure and high temperature, or (d) high pressure and low temperature?

5.60 Use the data from Table 5.3 to calculate the pressure exerted by 2.50 moles of CO_2 confined in a volume of 5.00 L at 450 K. Compare the pressure (in bar) with that predicted by the ideal gas equation.

5.61 At 27°C, 10.0 moles of a gas in a 1.50-L container exert a pressure of 130 bar. Is this an ideal gas?

5.62 Under the same conditions of temperature and pressure, which of the following gases would you expect to behave less ideally: CH_4 or SO_2? Explain.

5.63 Without referring to a table, select from the following list the gas that has the largest value of *b* in the van der Waals equation: CH_4, O_2, H_2O, CCl_4, or Ne.

5.64 Calculate the molar volume of carbon dioxide at 400 K and 30 bar, given that the second virial coefficient (*B*) of CO_2 is −0.0605 L mol^{-1}. Compare your result with that obtained using the ideal gas equation.

5.65 Calculate the molar volume of carbon dioxide at 400 K and 30 bar, assuming that CO_2 is well described as a van der Waals gas under these conditions. Compare your result with that obtained using the ideal gas equation.

Additional Problems

5.66 Under the same conditions of temperature and pressure which of the following gases would behave most ideally: Ne, N_2, or CH_4? Explain.

5.67 Nitroglycerin, an explosive compound, decomposes according to the equation

$$4C_3H_5(NO_3)_3(s) \longrightarrow$$
$$12CO_2(g) + 10H_2O(g) + 6N_2(g) + O_2(g)$$

Calculate the total volume of gases when collected at 1.2 bar and 25°C from 2.6×10^2 g of nitroglycerin. What are the partial pressures of the gases under these conditions?

5.68 The empirical formula of a compound is CH. At 200°C, 0.145 g of this compound occupies 97.2 mL at a pressure of 0.74 atm. What is the molecular formula of the compound?

5.69 When ammonium nitrite (NH_4NO_2) is heated, it decomposes to give nitrogen gas. This property is used to inflate some tennis balls. (a) Write a balanced equation for the reaction. (b) Calculate the quantity (in grams) of NH_4NO_2 needed to inflate a tennis ball to a volume of 86.2 mL at 1.20 bar and 22°C.

5.70 A 2.5-L flask at 15°C contains a mixture of N_2, He, and Ne at partial pressures of 0.32 bar for N_2, 0.15 bar for He, and 0.42 bar for Ne. (a) Calculate the total pressure of the mixture. (b) Calculate the volume in liters at SATP occupied by He and Ne if the N_2 is removed selectively.

5.71 Dry air near sea level has the following composition by volume: N_2, 78.08 percent; O_2, 20.94 percent; Ar, 0.93 percent; and CO_2, 0.05 percent. The atmospheric pressure is 1.00 atm. Calculate (a) the partial pressure of each gas in atm and (b) the concentration of each gas in mol L^{-1} at 0°C.

5.72 A mixture of helium and neon gases is collected over water at 28.0°C and 745 torr. If the partial pressure of helium is 368 torr, what is the partial pressure of neon? (Vapor pressure of water at 28°C is 28.3 torr.)

5.73 A piece of sodium metal reacts completely with water as follows:

$$2Na(s) + 2H_2O(l) \longrightarrow 2NaOH(aq) + H_2(g)$$

The hydrogen gas generated is collected over water at 25.0°C. The volume of the gas is 246 mL measured at 1.00 bar. Calculate the number of grams of sodium used in the reaction. (Vapor pressure of water at 25°C is 0.0310 bar.)

5.74 Helium is mixed with oxygen gas for deep-sea divers. Calculate the percent by volume of oxygen gas in the mixture if the diver has to submerge to a depth where the total pressure is 4.2 bar. The partial pressure of oxygen is maintained at 0.20 bar at this depth.

5.75 A sample of zinc metal reacts completely with an excess of hydrochloric acid:

$$Zn(s) + 2HCl(aq) \longrightarrow ZnCl_2(aq) + H_2(g)$$

The hydrogen gas produced is collected over water at 25.0°C. The volume of the gas is 7.80 L, and the pressure is 0.980 bar. Calculate the amount of zinc metal (in grams) consumed in the reaction. (The vapor pressure of water at 25°C is 23.8 torr.)

5.76 The percent by mass of bicarbonate (HCO_3^-) in a certain Alka-Seltzer product is 32.5 percent. Calculate the volume of CO_2 generated (in mL) at 37°C and 1.00 bar when a person ingests a 3.29-g tablet. (*Hint:* The reaction is between HCO_3^- and HCl acid in the stomach.)

5.77 In the metallurgical process of refining nickel, the metal is first combined with carbon monoxide to form tetracarbonylnickel, which is a gas at 43°C:

$$Ni(s) + 4CO(g) \longrightarrow Ni(CO)_4(g)$$

This reaction separates nickel from other solid impurities. (a) Starting with 86.4 g of Ni, calculate the pressure of $Ni(CO)_4$ in a container of volume 4.00 L. (Assume the reaction goes to completion.) (b) At temperatures above 43°C, the pressure of the gas is observed to increase much more rapidly than predicted by the ideal gas equation. Explain.

5.78 The partial pressure of carbon dioxide varies with seasons. Would you expect the partial pressure in the northern hemisphere to be higher in the summer or winter? Explain.

5.79 The measured pressures of NO_2 or NF_2 gases are significantly smaller than the pressure predicted by the ideal gas equation for these systems at *all* pressures. Is the ideal gas equation invalid for these systems, or is there another explanation?

5.80 A healthy adult exhales about 5.0×10^2 mL of a gaseous mixture with each breath. Calculate the number of molecules present in this volume at 37°C and 1.1 bar. List the major components of this gaseous mixture.

5.81 Sodium bicarbonate ($NaHCO_3$) is called baking soda because, when heated, it releases carbon dioxide gas, which is responsible for the rising of cookies, doughnuts, and bread. (a) Calculate the volume (in liters) of CO_2 produced by heating 5.0 g of $NaHCO_3$ at 180°C and 1.3 bar. (b) Ammonium bicarbonate (NH_4HCO_3) has also been used for the same purpose. Suggest one advantage and one disadvantage of using NH_4HCO_3 instead of $NaHCO_3$ for baking.

5.82   A mixture of methane (CH_4) and ethane (C_2H_6) is stored in a container at 294 mmHg. The gases are burned in air to form CO_2 and H_2O. If the pressure of CO_2 is 356 mmHg measured at the same temperature and volume as the original mixture, calculate the mole fractions of the gases.

5.83   Use the kinetic theory of gases to explain why hot air rises.

5.84   A 5.00-mole sample of NH_3 gas is kept in a 1.92-L container at 300 K. If the van der Waals equation of state is assumed to give the correct answer for the pressure of the gas, calculate the percent error made in using the ideal gas equation to calculate the pressure.

5.85   The root-mean-square speed of a certain gaseous oxide is 493 m s^{-1} at 20°C. What is the molecular formula of the compound?

5.86   A barometer having a cross-sectional area of 1.00 cm^2 at sea level measures a pressure of 76.0 cm of mercury. The pressure exerted by this column of mercury is equal to the pressure exerted by all the air on 1 cm^2 of Earth's surface. Given that the density of mercury is 13.6 g mL^{-1} and the average radius of Earth is 6371 km, calculate the total mass of Earth's atmosphere in kilograms. (*Hint:* The surface area of a sphere is $4\pi r^2$, where r is the radius of the sphere.)

5.87   Some commercial drain cleaners contain a mixture of sodium hydroxide and aluminum powder. When the mixture is poured down a clogged drain, the following reaction occurs:

$$2NaOH(aq) + 2Al(s) + 6H_2O(l) \longrightarrow$$
$$2NaAl(OH)_4(aq) + 3H_2(g)$$

The heat generated in this reaction helps melt away obstructions such as grease, and the hydrogen gas released stirs up the solids clogging the drain. Calculate the volume of H_2 formed at 23°C and 1.00 bar if 3.12 g of Al are treated with an excess of NaOH.

5.88   The volume of a sample of pure HCl gas was 189 mL at 25°C and 108 mmHg. It was completely dissolved in about 60 mL of water and titrated with a NaOH solution; 15.7 mL of the NaOH solution were required to neutralize the HCl. Calculate the molarity of the NaOH solution.

5.89   Propane (C_3H_8) burns in oxygen to produce carbon dioxide gas and water vapor. (a) Write a balanced equation for this reaction. (b) Calculate the number of liters of carbon dioxide measured at SATP that could be produced from 7.45 g of propane.

5.90   Consider the following apparatus.

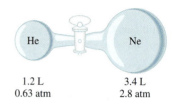

1.2 L
0.63 atm

3.4 L
2.8 atm

Calculate the partial pressures of helium and neon after the stopcock is open. The temperature remains constant at 16°C.

5.91   Nitric oxide (NO) reacts with molecular oxygen as follows:

$$2NO(g) + O_2(g) \longrightarrow 2NO_2(g)$$

Initially NO and O_2 are separated as shown here.

4.00 L at
0.500 bar

2.00 L at
1.00 bar

When the valve is opened, the reaction quickly goes to completion. Determine what gases remain at the end and calculate their partial pressures. Assume that the temperature remains constant at 25°C.

5.92   Consider the apparatus shown here. When a small amount of water is introduced into the flask by squeezing the bulb of the medicine dropper, water is squirted upward out of the long glass tubing. Explain this observation. (*Hint:* Hydrogen chloride gas is soluble in water.)

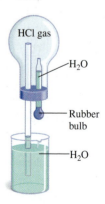

5.93   Describe how you would measure, by either chemical or physical means, the partial pressures of a mixture of gases of the following composition: (a) CO_2 and H_2, and (b) He and N_2.

5.94   A certain hydrate has the formula $MgSO_4 \cdot xH_2O$. A quantity of 54.2 g of the compound is heated in an oven to drive off the water. If the steam generated exerts a pressure of 25.1 bar in a 2.00-L container at 120°C, calculate x.

5.95 A mixture of Na_2CO_3 and $MgCO_3$ of mass 7.63 g is reacted with an excess of hydrochloric acid. The CO_2 gas generated occupies a volume of 1.67 L at 1.26 bar and 26°C. From these data, calculate the percent composition by mass of Na_2CO_3 in the mixture.

5.96 The following apparatus can be used to measure atomic and molecular speed. Suppose that a beam of metal atoms is directed at a rotating cylinder in a vacuum. A small opening in the cylinder allows the atoms to strike a target area. Because the cylinder is rotating, atoms traveling at different speeds will strike the target at different positions. In time, a layer of the metal will deposit on the target area, and the variation in its thickness is found to correspond to Maxwell's speed distribution. In one experiment it is found that at 850°C some bismuth (Bi) atoms struck the target at a point 2.80 cm from the spot directly opposite the slit. The diameter of the cylinder is 15.0 cm and it is rotating at 130 revolutions per second. (a) Calculate the speed (m/s) at which the target is moving. (*Hint:* The circumference of a circle is given by $2\pi r$, where r is the radius.) (b) Calculate the time (in seconds) it takes for the target to travel 2.80 cm. (c) Determine the speed of the Bi atoms. Compare your result in (c) with the $\langle u \rangle_{rms}$ of Bi at 850°C. Comment on the difference.

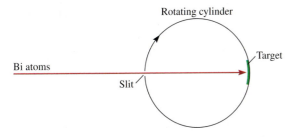

5.97 Commercially, compressed oxygen is sold in metal cylinders. If a 120-L cylinder is filled with oxygen to a pressure of 134 bar at 22°C, what is the mass of O_2 present? How many liters of O_2 gas at 1.00 bar and 22°C could the cylinder produce? (Assume ideal behavior.)

5.98 Ethylene gas (C_2H_4) is emitted by fruits and is known to be responsible for their ripening. Based on this information, explain why a bunch of bananas ripens faster in a closed paper bag than in a bowl.

5.99 About 8.0×10^6 tons of urea $[(NH_2)_2CO]$ are used annually as a fertilizer. The urea is prepared at 200°C and under high-pressure conditions from carbon dioxide and ammonia (the products are urea and steam). Calculate the volume of ammonia (in liters) measured at 150 bar needed to prepare 1.0 ton of urea.

5.100 Some ballpoint pens have a small hole in the main body of the pen. What is the purpose of this hole?

5.101 Nitrous oxide (N_2O) can be obtained by the thermal decomposition of ammonium nitrate (NH_4NO_3). (a) Write a balanced equation for the reaction. (b) In a certain experiment, a student obtains 0.340 L of the gas at 718 mmHg and 24°C. If the gas weighs 0.580 g, calculate the value of the gas constant.

5.102 Two vessels are labeled A and B. Vessel A contains NH_3 gas at 70°C, and vessel B contains Ne gas at the same temperature. If the average kinetic energy of NH_3 is 7.1×10^{-21} J molecule^{-1}, calculate the mean square speed of Ne atoms in $m^2 s^{-2}$.

5.103 Which of the following molecules has the largest a value: CH_4, F_2, C_6H_6, or Ne? Explain your answer in terms of intermolecular forces.

5.104 The following procedure is a simple though somewhat crude way to measure the molar mass of a gas. A liquid of mass 0.0184 g is introduced into a syringe like the one shown here by injection through the rubber tip using a hypodermic needle. The syringe is then transferred to a temperature bath heated to 45°C, and the liquid vaporizes. The final volume of the vapor (measured by the outward movement of the plunger) is 5.58 mL and the atmospheric pressure is 760 mmHg. Given that the compound's empirical formula is CH_2, determine the molar mass of the compound.

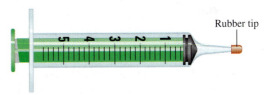

5.105 In 1995 a man suffocated as he walked by an abandoned mine in England. At that moment there was a sharp drop in atmospheric pressure due to a change in the weather. Suggest what might have caused the man's death.

5.106 Acidic oxides such as carbon dioxide react with basic oxides like calcium oxide (CaO) and barium oxide (BaO) to form salts (metal carbonates). (a) Write equations representing these two reactions. (b) A student placed a mixture of BaO and CaO of combined mass 4.88 g in a 1.46-L flask containing carbon dioxide gas at 35°C and 746 mmHg. After the reactions were complete, she found that the CO_2 pressure had dropped to 252 mmHg. Calculate the percent composition by mass of the mixture. Assume volumes of the solids are negligible.

5.107   (a) What volume of air at 1.0 bar and 22°C is needed to fill a 0.98-L bicycle tire to a pressure of 5.0 bar at the same temperature? (Note that the 5.0 bar is the gauge pressure, which is the difference between the pressure in the tire and atmospheric pressure. Before filling, the pressure in the tire was 1.0 bar.) (b) What is the total pressure in the tire when the gauge pressure reads 5.0 bar? (c) The tire is pumped by filling the cylinder of a hand pump with air at 1.0 bar and then, by compressing the gas in the cylinder, adding all the air in the pump to the air in the tire. If the volume of the pump is 33 percent of the volume of the tire, what is the gauge pressure in the tire after three full strokes of the pump? Assume constant temperature.

5.108   The running engine of an automobile produces carbon monoxide (CO), a toxic gas, at the rate of about 188 g CO per hour. A car is left idling in a poorly ventilated garage that is 6.0 m long, 4.0 m wide, and 2.2 m high at 20°C. (a) Calculate the rate of CO production in moles per minute. (b) How long would it take to build up a lethal concentration of CO of 1000 ppmv (parts per million by volume)?

5.109   Interstellar space contains mostly hydrogen atoms at a concentration of about 1 atom/cm^3. (a) Calculate the pressure of the H atoms. (b) Calculate the volume (in liters) that contains 1.0 g of H atoms. The temperature is 3 K.

5.110   Atop Mt. Everest, the atmospheric pressure is 210 mmHg and the air density is 0.426 kg m^{-3}. (a) Calculate the air temperature, given that the molar mass of air is 29.0 g mol^{-1}. (b) Assuming no change in air composition, calculate the percent decrease in oxygen gas from sea level to the top of Mt. Everest.

5.111   Under the same conditions of temperature and pressure, why does one liter of moist air weigh less than one liter of dry air? In weather forecasts, an oncoming low-pressure front usually means imminent rainfall. Explain.

5.112   Air entering the lungs ends up in tiny sacs called alveoli. It is from the alveoli that oxygen diffuses into the blood. The average radius of the alveoli is 0.0050 cm, and the air inside contains 14 percent oxygen. Assuming that the pressure in the alveoli is 1.0 atm and the temperature is 37°C, calculate the number of oxygen molecules in one of the alveoli. (*Hint:* The volume of a sphere of radius r is $4/3\,\pi r^3$.)

5.113   A student breaks a thermometer and spills most of the mercury (Hg) onto the floor of a laboratory that measures 15.2 m long, 6.6 m wide, and 2.4 m high. (a) Calculate the mass of mercury vapor (in grams) in the room at 20°C. The vapor pressure of mercury at 20°C is 1.7×10^{-6} bar. (b) Does the concentration of mercury vapor exceed the air quality regulation of 0.050 mg Hg m^{-3} of air? (c) One way to treat

small quantities of spilled mercury is to spray sulfur powder over the metal. Suggest a physical and a chemical reason for this action.

5.114   Nitrogen forms several gaseous oxides. One of them has a density of 1.33 g L^{-1} measured at 764 mmHg and 150°C. Write the formula of the compound.

5.115   Nitrogen dioxide (NO_2) cannot be obtained in a pure form in the gas phase because it exists as a mixture of NO_2 and N_2O_4. At 25°C and 0.99 bar, the density of this gas mixture is 2.7 g L^{-1}. What is the partial pressure of each gas?

5.116   The insert on p. 316 describes the cooling of rubidium vapor to 1.7×10^{-7} K. Calculate the root-mean-square speed and average kinetic energy of a Rb atom at this temperature.

5.117   A container of volume 2.00 L is filled with an unknown gas to a pressure of 5.00 bar at 298 K. When the system is heated to 400 K, the pressure increases. When enough gas is let out of the container so that the pressure returns to its original value, the container with gas is found to have decreased in mass by 4.53 g. Calculate the molecular weight of the gas, assuming ideal behavior.

5.118   (a) Express the van der Waals equation in the form of Equation 5.44 (the virial equation). Derive relationships between the van der Waals constants (a and b) and the virial coefficients (B, C, and D), given that

$$\frac{1}{1-x} = 1 + x + x^2 + x^3 + \cdots \quad \text{for} \quad |x| < 1$$

(b) Using your results, determine the Boyle temperature for a van der Waals gas in terms of a and b. (c) Using your result from part (b) and the data in Table 5.3, calculate the Boyle temperature for ammonia (NH_3).

5.119   At 250°C, the mean molecular speed of a certain gaseous compound is 500 m s^{-1}. At what temperature (in °C) would the mean molecular speed for this substance equal 800 m s^{-1}?

5.120   A chemistry instructor performed the following mystery demonstration. Just before the students arrived in class, she heated some water to boiling in an Erlenmeyer flask. She then removed the flask from the flame and closed the flask with a rubber stopper. After the class commenced, she held the flask in front of the students and announced that she could make the water boil simply by rubbing an ice cube on the outside walls of the flask. To the amazement of everyone, it worked. Give an explanation for this phenomenon.

5.121   In 2.00 min, 29.7 mL of He effuse through a small hole. Under the same conditions of pressure and temperature, 10.0 mL of a mixture of CO and CO_2 effuse through the hole in the same amount of time. Calculate the percent composition by volume of the mixture. See Problems 5.57 and 5.58 for information on gas effusion.

5.122 Lithium hydride reacts with water as follows:

$$LiH(s) + H_2O(l) \longrightarrow LiOH(aq) + H_2(g)$$

During World War II, U.S. pilots carried LiH tablets. In the event of a crash landing at sea, the LiH would react with the seawater and fill their life belts and lifeboats with hydrogen gas. How many grams of LiH are needed to fill a 4.1-L life belt at 0.97 bar and 12°C?

5.123 The atmosphere on Mars is composed mainly of carbon dioxide. The surface temperature is 220 K and the atmospheric pressure is about 6.0 mmHg. Taking these values as Martian "SATP," calculate the molar volume in liters of an ideal gas on Mars.

5.124 The atmosphere of Venus is composed of 96.5 percent CO_2, 3.5 percent N_2, and 0.015 percent SO_2 by volume. Its average atmospheric pressure is 9.0×10^6 Pa. Calculate the partial pressures of the gases in Pa.

5.125 A student tries to determine the volume of a bulb like the one shown on p. 296 These are her results: Mass of the bulb filled with dry air at 23°C and 744 mmHg 91.6843 g; mass of evacuated bulb, 91.4715 g. Assume the composition of air is 78 percent N_2, 21 percent O_2, and 1 percent argon. What is the volume (in mL) of the bulb? (*Hint:* First calculate the average molar mass of air.)

5.126 Apply your knowledge of the kinetic theory of gases to the following situations. (a) Two flasks of volumes V_1 and V_2 ($V_2 > V_1$) contain the same number of helium atoms at the same temperature. (i) Compare the root-mean-square (rms) speeds and average kinetic energies of the helium (He) atoms in the flasks. (ii) Compare the frequency and the force with which the He atoms collide with the walls of their containers. (b) Equal numbers of He atoms are placed in two flasks of the same volume at temperatures T_1 and T_2 ($T_2 > T_1$). (i) Compare the rms speeds of the atoms in the two flasks. (ii) Compare the frequency and the force with which the He atoms collide with the walls of their containers. (c) Equal numbers of He and neon (Ne) atoms are placed in two flasks of the same volume, and the temperature of both gases is 74°C. Comment on the validity of the following statements: (i) The rms speed of He is equal to that of Ne. (ii) The average kinetic energies of the two gases are equal. (iii) The rms speed of each He atom is 1.47×10^3 m s^{-1}.

5.127 It has been said that every breath we take, on average, contains molecules that were once exhaled by Wolfgang Amadeus Mozart (1756–1791). The following calculations demonstrate the validity of this statement. (a) Calculate the total number of molecules in the atmosphere. (*Hint:* Use the result in Problem 5.86 and 29.0 g mol^{-1} as the molar mass of air.) (b) Assuming the volume of every breath (inhale or exhale) is 500 mL, calculate the number of molecules exhaled in each breath at 37°C, which is the body temperature. (c) If Mozart's lifespan was exactly 35 years, what is the number of molecules he exhaled in that period? (Given that an average person breathes 12 times per minute.) (d) Calculate the fraction of molecules in the atmosphere that was exhaled by Mozart. How many of Mozart's molecules do we breathe in with every breath of air? Round off your answer to one significant figure. (e) List three important assumptions in these calculations.

5.128 Referring to Figure 5.10, explain the following: (a) Why do the curves dip below the horizontal line labeled ideal gas at low pressures and then why do they rise above the horizontal line at high pressures? (b) Why do the curves all converge to 1 at very low pressures? (c) Each curve intercepts the horizontal line labeled ideal gas. Does it mean that at that point the gas behaves ideally?

5.129 A relation known as the barometric formula is useful for estimating the change in atmospheric pressure with altitude. The formula is given by

$$P = P_0 e^{-Mgh/RT}$$

where P and P_0 are the pressures at height h and sea level ($h = 0$), respectively, g is the acceleration due to gravity (9.8 m s^{-2}), M is the average molar mass of air (29.0 g mol^{-1}), and R is the gas constant. Calculate the atmospheric pressure in bar at a height of 5.0 km, assuming the temperature is constant at 5°C and $P_0 = 1.0$ bar.

5.130 A 5.72-g sample of graphite was heated with 68.4 g of O_2 in an 8.00-L flask. The reaction that took place was

$$C(graphite) + O_2(g) \longrightarrow CO_2(g)$$

After the reaction was complete, the temperature in the flask was 182°C. What was the total pressure inside the flask?

5.131 A 6.11-g sample of a Cu-Zn alloy reacts with HCl acid to produce hydrogen gas. If the hydrogen gas has a volume of 1.26 L, temperature of 22°C, and pressure of 728 mmHg, what is the percent of Zn in the alloy? (*Hint:* Cu does not react with HCl.)

5.132 Given the phase diagram of carbon shown here, answer the following questions: (a) How many triple points are there and what are the phases that can coexist at each triple point? (b) Which has a higher density, graphite or diamond? (c) Synthetic diamond can be made from graphite. Based on the phase diagram, how would you do this?

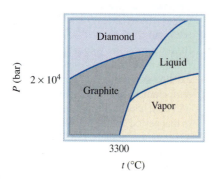

5.133 A closed vessel of volume 9.6 L contains 2.0 g of water. Calculate the temperature (in °C) at which only half of the water remains in the liquid phase. (See Table 5.2 for vapor pressures of water at different temperatures.)

5.134 The rms speed of a certain gaseous oxide is 493 m s^{-1} at 0°C. What is the molecular formula of the compound?

5.135 Another measure of the average speed of a molecule is the mean molecular speed defined as

$$\langle u \rangle = \int_{-\infty}^{+\infty} u\, f(u)\, du$$

where $f(u)$ is the Maxwell-Boltzmann distribution of molecular speeds (Equation 5.36). Evaluate the integral to show that

$$\langle u \rangle = \sqrt{\frac{8RT}{\pi \mathcal{M}}}$$

(You will need a standard table of integrals to do this problem.) Is the mean molecular speed larger or smaller than the rms molecular speed?

5.136 The following diagram shows the Maxwell-Boltzmann speed distributions for gas at two different temperatures T_1 and T_2. Calculate the value of T_2.

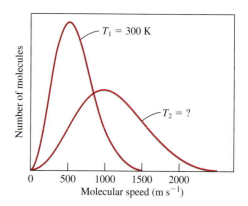

5.137 A gaseous reaction takes place at constant volume and constant pressure in a cylinder as shown. Which of the following equations best describes the reaction? The initial temperature (T_1) is twice that of the final temperature (T_2).

(a) A + B $\longrightarrow$ C

(b) AB $\longrightarrow$ C + D

(c) A + B $\longrightarrow$ C + D

(d) A + B $\longrightarrow$ 2C + D

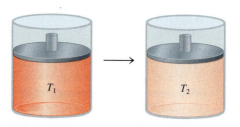

Answers to Practice Exercises

5.1 9.38 L **5.2** 13.1 g L^{-1} **5.3** 44.1 g mol^{-1} **5.4** B$_2$H$_6$
5.5 0.69 bar **5.6** 2.6 bar **5.7** 4.81 L **5.8** no, 0.8 g will be
left **5.9** CH$_4$: 1.29 bar; C$_2$H$_6$: 0.0657 bar; C$_3$H$_8$: 0.0181 bar

5.10 0.0653 g **5.11** 2.031 × 10^3 m s^{-1}, 2.49 × 10^3 m s^{-1}
5.12 30.4 bar, 46.1 bar using the ideal gas equation.
5.13 1.02 L mol^{-1}

6

C h a p t e r

The States of Matter II:
Liquids and Solids

Although we live immersed in a mixture of gases that make up Earth's atmosphere, we are more familiar with the behavior of liquids and solids because they are more visible. Every day we use water and other liquids for drinking, bathing, cleaning, and cooking, and we handle, sit upon, and wear solids.

Molecular motion is more restricted in liquids than in gases; and in solids, the atoms and molecules are packed even more tightly together. In fact, in a solid they are held in well-defined positions and are capable of little free motion relative to one another. In Chapter 6, we examine the structure of liquids and solids and discuss some of the fundamental properties of these two states of matter.

6.1 | The Structure and Properties of Liquids Are Governed by Intermolecular Interactions

A liquid is formed when the temperature of a gas is lowered sufficiently so that condensation occurs. Both the nature of the condensation and the properties of the resulting liquid are governed by intermolecular forces. In this section, we will examine the condensation process and look at the factors that determine the structure and properties of liquids.

The Condensation of Liquids from the Gas Phase

Most gases, when compressed or cooled sufficiently, will condense into a liquid. This process can by illustrated by plotting the pressure of gas versus molar volume at constant temperature. Such constant-temperature curves are called *isotherms*. Figure 6.1 shows a series of typical isotherms for a gaseous material.

At high temperatures (T_6 and higher), the system is well approximated by an ideal gas and the isotherms follow Boyle's Law (that is, PV = constant). As the temperature is lowered, the system begins to show significant deviation from ideal gas behavior. At temperatures below T_4, the deviations become extreme. If we move along the T_3 isotherm from high to low volumes, the pressure increases at first with decreasing volume, as is expected for a gas, but when the T_3 isotherm intersects the shaded region at point A in Figure 6.1, there is a region over which the pressure remains constant as the volume is decreased further. The pressure ceases to change with compression at this point because the gas begins to condense into a liquid. Decreasing the volume in the region of constant pressure does not compress the gas or liquid present, but instead causes more of the gas to condense to the liquid phase. The system described by the part of the T_3 isotherm in the shaded region is a coexisting mixture of liquid and vapor both at a pressure equal to the equilibrium vapor pressure (Section 5.2) of the liquid at temperature T_3. Figure 6.2 illustrates this condensation process. At point B of the T_3 isotherm, the pressure begins to increase again with decreasing volume. At this point, all the gas has condensed into liquid and the curve reflects the compression of a liquid. Note that the slope of the P-V curve in the liquid region is quite steep. This is because liquids are much less compressible than gases, so small changes in volume correspond to large changes in pressure.

Figure 6.1 Pressure versus molar volume isotherms of a gas at various temperatures (increasing from T_1 to T_6). On this plot, T_4 corresponds to the critical temperature T_c defined in Section 5.2. Above this temperature, the gas cannot be liquefied by isothermal compression.

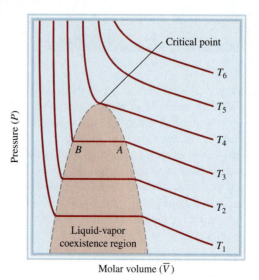

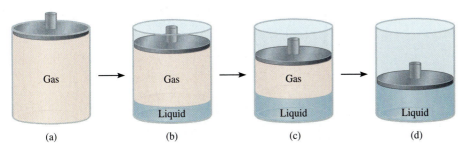

For temperatures above T_4, there is no longer a liquid-vapor coexistence region with constant pressure. Thus, we see from the *P-V* plot that T_4 corresponds to the critical temperature T_c, which we defined in Section 5.1 as the temperature above which there was no distinction between liquid and vapor. The pressure and volume at which the T_4 isotherm intersects the shaded region correspond then to the critical pressure P_c and the critical volume V_c. The isotherm T_4 is called the *critical isotherm.*

In Section 5.5, we discussed the van der Waals equation (Equation 5.42). In addition to its role as an equation of state for real gases, this equation was also the first equation of state that could describe both condensation and critical phenomena. For example it can be shown (Problem 6.59) that the critical conditions of a gas obeying the van der Waals equation can be determined from the van der Waals parameters a and b as follows:

$$\overline{V}_c = 3b \tag{6.1}$$

$$P_c = \frac{a}{27\,b^2} \tag{6.2}$$

$$T_c = \frac{8a}{27\,Rb} \tag{6.3}$$

Therefore, if we know the van der Waals constants, we can estimate the critical parameters for a given system. Conversely, if we are given any pair of the three critical constants P_c, T_c and $\overline{V}_c$, then we can estimate the van der Waals parameters of the system. If the van der Waals equation were an exact equation of state for a real system, it would not matter which pair of critical constants (P_c and T_c, or P_c and $\overline{V}_c$, or T_c and $\overline{V}_c$) one choses to determine a and b—they would all yield the same results. However, the van der Waals equation of state is only approximate, so the specific values of a and b depend significantly on which pair of critical constants one uses. In common practice, P_c and T_c are chosen because $\overline{V}_c$ is difficult to measure accurately.

The Properties of Liquids

Intermolecular forces give rise to a number of structural features and properties of liquids. Here we explore the structure of liquids in general and examine two important properties of the liquid state: surface tension and viscosity. Then we will discuss the structure and properties of the most important liquid: water.

The Molecular Structure of Liquids

Although using the word *structure* may seem strange when discussing liquids because they lack long-range order, all liquids possess some degree of short-range order. A

Figure 6.3 Radial distribution function, $g(r)$, where r is interatomic separation, for liquid argon at 85 K.

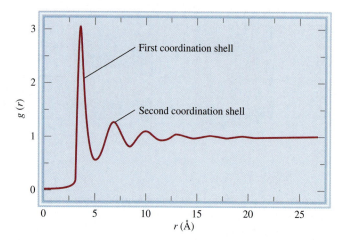

convenient measure of the structure of liquids is the ***radial distribution function,***[1] $g(r)$. This function is defined such that, for a liquid with density $\rho = N/V$, the average number of particles that are found in a spherical shell of radius r (and thickness dr) centered around any particle in the system is given by $4\pi\rho r^2 g(r)\,dr$. If there were no short-range order in the liquid, then the molecules would be completely randomly distributed and the number of molecules expected would be equal to the molecule density multiplied by the volume of the shell, which is equal to $4\pi\rho r^2 dr$, giving a $g(r)$ that is equal to 1 everywhere. Thus, $g(r)$ measures the deviation from a completely random distribution of molecules. Experimentally, $g(r)$ can be determined from X-ray or neutron diffraction patterns of the liquid.

Figure 6.3 shows the radial distribution function, $g(r)$, for liquid argon. To understand the details of this plot of $g(r)$, examine the environment around a given Ar atom in the liquid, as illustrated in Figure 6.4.

Because of atomic size, the number of particles found a distance r around the central Ar atom is not random. For example, the number of particles closer than the molecular diameter is zero, as the repulsive forces prevent the molecules from overlapping—this leads to a value of $g(r)$ that is zero for distances smaller than the Ar diameter (about 3.3Å). Just beyond this distance of closest separation, the high density of the liquid leads

---

1. This radial distribution function is similar to the one applied to the hydrogen atom in Section 1.4 except that the function in Section 1.4 deals with the distribution of electrons around the nucleus, and the function defined here concerns the distribution of molecules around a chosen molecule in a liquid.

Figure 6.4 Two-dimensional representation of the environment around an argon atom in liquid argon. The coordination shells correspond to the peaks in the radial distribution shown in Figure 6.3.

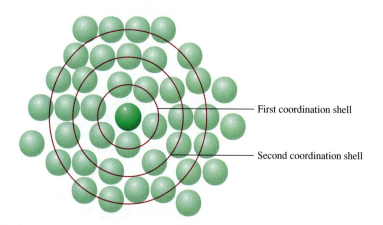

to a large number of particles packed just around the central Ar, leading to an enhancement in $g(r)$ at distances between about 3.3 and 3.8 Å. This group of atoms is called the *first coordination shell*. Just beyond this shell, $g(r)$ drops substantially below 1, implying a density *deficit* in this region. This deficit is due to the nonzero size of the atoms in the first coordination shell, which excludes atoms from getting too close. Beyond this dip in $g(r)$ is another peak that indicates a *second coordination shell* packed around the first. At large r, $g(r)$ approaches 1 indicating that there is no longer any correlation between the position of the central particle and those far away from it.

Surface Tension

Molecules within a liquid are pulled in all directions by intermolecular forces; there is no tendency for them to be pulled in any one particular way. However, molecules at the surface are pulled downward and sideways by other molecules, but not upward away from the surface (Figure 6.5). These intermolecular attractions thus tend to pull the molecules into the liquid and cause the surface to tighten like an elastic film. Because there is little or no attraction between polar water molecules and, say, the nonpolar wax molecules on a freshly waxed car, a drop of water assumes the shape of a small round bead, because a sphere minimizes the surface area of a liquid. The waxy surface of a wet apple also produces this effect (Figure 6.6).

A measure of the elastic force in the surface of a liquid is surface tension. The **surface tension** is *the amount of energy required to stretch or increase the surface of a liquid by a unit area* (for example, by 1 cm^2). Liquids that have strong intermolecular forces also have high surface tensions. Thus, because of hydrogen bonding, water has a considerably greater surface tension than most other liquids.

Another example of surface tension is *capillary action*. Figure 6.7(a) shows water rising spontaneously in a capillary tube. A thin film of water adheres to the wall of the glass tube. The surface tension of water causes this film to contract, and as it does, it pulls the water up the tube. Two types of forces bring about capillary action. One is **cohesion,** which is *the intermolecular attraction between like molecules* (in this case, the water molecules). The second force, called **adhesion,** is *an attraction between*

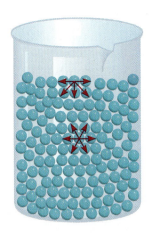

Figure 6.5 Intermolecular forces acting on a molecule in the surface layer of a liquid and in the interior region of the liquid.

Surface tension enables the water strider to "walk" on water.

Figure 6.6 Water beads on an apple that has a waxy surface.

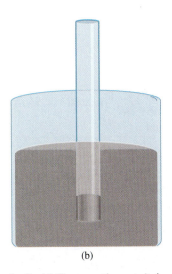

(a)   (b)

Figure 6.7 (a) When adhesion is greater than cohesion, the liquid (for example, water) rises in the capillary tube. (b) When cohesion is greater than adhesion, as it is for mercury, a depression of the liquid in the capillary tube results. Note that the meniscus in the tube of water is concave, or rounded downward, whereas that in the tube of mercury is convex, or rounded upward.

unlike molecules, such as those in water and in the sides of a glass tube. If adhesion is stronger than cohesion, as it is in Figure 6.7(a), the contents of the tube will be pulled upward. This process continues until the adhesive force is balanced by the weight of the water in the tube. This action is by no means universal among liquids. As Figure 6.7(b) shows, in mercury cohesion is greater than the adhesion between mercury and glass, so that when a capillary tube is dipped in mercury, the result is a depression or lowering of the mercury level—that is, the height of the liquid in the capillary tube is below the surface of the mercury.

Viscosity

The expression "slow as molasses in January" owes its truth to another physical property of liquids called viscosity. *Viscosity* (η) is *a measure of a fluid's resistance to flow.* The greater the viscosity, the more slowly the liquid flows. The viscosity of a liquid usually decreases as temperature increases; thus hot molasses flows much faster than cold molasses. The SI unit of viscosity is the *pascal-second* (Pa s):

$$1\,\text{Pa s} = 1\,\text{kg s}^{-1}\,\text{m}^{-1}$$

Another common unit is the *poise* (P), equal to $1\,\text{g cm}^{-1}\text{s}^{-1}$. The units are related by

$$1\,\text{Pa s} = 10\,\text{P}$$

A physical interpretation of these units can be obtained as follows: If a liquid with a viscosity of 1 Pa s is placed between two flat plates separated by a distance d, and one of the plates is subjected to a sideways force equal to 1 N for every m^2 of plate area, the plate will move at a velocity corresponding to the distance d per second.

Liquids that have strong intermolecular forces have higher viscosities than those that have weak intermolecular forces (Table 6.1). Water has a higher viscosity than many other liquids because of its ability to form hydrogen bonds. Interestingly, the viscosity of glycerol is significantly higher than that of all the other liquids listed in Table 6.1. Glycerol has the structure

$$\begin{array}{l} CH_2\!-\!OH \\ | \\ CH\!-\!OH \\ | \\ CH_2\!-\!OH \end{array}$$

Table 6.1	Viscosity of Some Common Liquids at 20°C
Liquid	**Viscosity (Pa s)**
Acetone (C_3H_6O)	3.16×10^{-4}
Benzene (C_6H_6)	6.25×10^{-4}
Blood	4×10^{-3}
Carbon tetrachloride (CCl_4)	9.69×10^{-4}
Ethanol (C_2H_5OH)	1.20×10^{-3}
Diethyl ether ($C_2H_5OC_2H_5$)	2.33×10^{-4}
Glycerol ($C_3H_8O_3$)	1.49
Mercury (Hg)	1.55×10^{-3}
Water (H_2O)	1.01×10^{-3}

Figure 6.8 Left: Ice cubes float on water. Right: Solid benzene sinks to the bottom of liquid benzene.

Like water, glycerol can form hydrogen bonds. Each glycerol molecule has three —OH groups, called *hydroxyl* groups, that can participate in hydrogen bonding with other glycerol molecules. Furthermore, because of their shape, the molecules have a great tendency to become entangled rather than to slip past one another as the molecules of less viscous liquids do. These interactions contribute to its high viscosity.

Glycerol is a clear, odorless, syrupy liquid used to make explosives, ink, and lubricants.

The Structure and Properties of Water

Water is so common a substance on Earth that we often overlook its unique nature. All life processes involve water. Water is an excellent solvent for many ionic compounds, as well as for other substances capable of forming hydrogen bonds with water.

If water did not have the ability to form hydrogen bonds, it would be a gas at room temperature.

The most striking property of water is that its solid form is less dense than its liquid form: ice floats at the surface of liquid water (Figure 6.8). With a few exceptions, (for example, silicon, gallium, germanium, bismuth, and pure acetic acid), the density of almost all other substances is greater in the solid state than in the liquid state. To understand why water is different, we have to examine the electronic structure of the H_2O molecule. As we saw in Chapters 2 and 3, there are two pairs of nonbonding electrons, or two lone pairs, on the oxygen atom:

Although many compounds can form intermolecular hydrogen bonds, the difference between H_2O and other polar molecules, such as NH_3 and HF, is that each oxygen atom can form *two* hydrogen bonds, the same as the number of lone electron pairs on the oxygen atom. Thus, water molecules are joined together in an extensive three-dimensional network in which each oxygen atom is approximately tetrahedrally bonded to four hydrogen atoms, two by covalent bonds and two by hydrogen bonds. This equality in the number of hydrogen atoms and lone pairs is not characteristic of NH_3 or HF or, for that matter, of any other molecule capable of forming hydrogen bonds. Consequently, these other molecules can form rings or chains, but not three-dimensional structures.

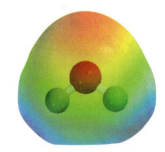

Electrostatic potential map of water

The highly ordered three-dimensional structure of ice (Figure 6.9) prevents the molecules from getting too close to one another. But consider what happens when ice melts. At the melting point, a number of water molecules have enough kinetic energy to pull free of the intermolecular hydrogen bonds. These molecules become trapped in the cavities of the three-dimensional structure, which is broken down into smaller

Why Do Lakes Freeze from the Top Down?

The fact that ice is less dense than water has a profound ecological significance. Consider, for example, the temperature changes in the fresh water of a lake in a cold climate. As the temperature of the water near the surface drops, the density of this water increases. The colder water then sinks toward the bottom, while warmer water, which is less dense, rises to the top. This normal convection motion continues until the temperature throughout the water reaches 4°C. Below this temperature, the density of water begins to decrease with decreasing temperature (see Figure 6.10), so it no longer sinks. On further cooling, the water begins to freeze at the surface. The ice layer formed does not sink because it is less dense than the liquid; it even acts as a thermal insulator for the water below it. Were ice heavier, it would sink to the bottom of the lake and eventually the water would freeze upward. Most living organisms in the body of water could not survive being frozen in ice. Fortunately, lake water does not freeze upward from the bottom. This unusual property of water makes the sport of ice fishing possible.

Ice fishing. The ice layer that forms on the surface of a lake insulates the water beneath and maintains a high enough temperature to sustain aquatic life.

Figure 6.9 The three-dimensional structure of ice. Each O atom is bonded to four H atoms. The covalent bonds are shown by short solid lines and the weaker hydrogen bonds by long dotted lines between O and H. The empty space in the structure accounts for the low density of ice.

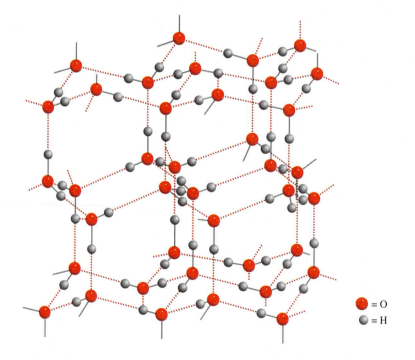

● = O
● = H

clusters. As a result, there are more molecules per unit volume in liquid water than in ice. Thus, because density = mass/volume, the density of water is greater than that of ice. With further heating, more water molecules are released from intermolecular hydrogen bonding, so the density of water tends to increase with rising temperature just above the melting point. Of course, at the same time, water expands as it is being heated so that its density is decreased. These two processes—the trapping of free water molecules in cavities and thermal expansion—act in opposite directions. From 0°C to 4°C, the trapping prevails and water becomes progressively denser. Beyond 4°C, however, thermal expansion predominates and the density of water decreases with increasing temperature (Figure 6.10).

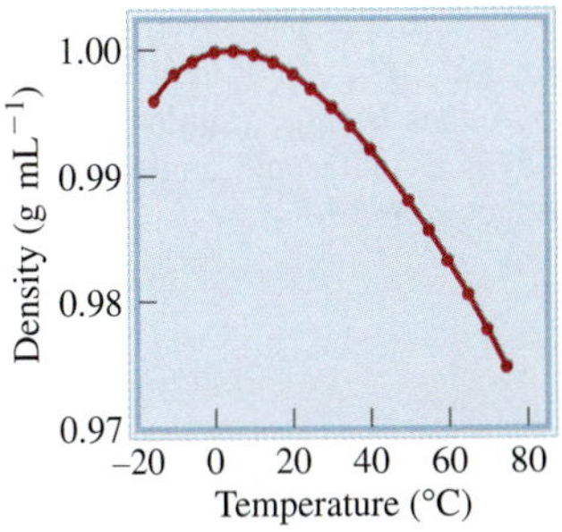

Figure 6.10 Plot of the density as a function of temperature for liquid water at 1 bar pressure. The maximum density of water is reached at 4°C. The density of ice at 0°C is about 0.92 g cm.

6.2 | Crystalline Solids Can Be Classified in Terms of Their Structure and Intermolecular Interactions

Solids can be divided into two categories: crystalline and amorphous. A *crystalline solid* is one that *possesses long-range order, that is, its atoms, molecules or ions occupy specific and regular positions.* The arrangement of such particles in a crystalline solid is such that the potential energy of the system is at a minimum. This arrangement gives the solid its rigidity, because any deformation of the crystal moves the system away from the potential energy minimum and is thus energetically unfavorable. The interactions responsible for the stability of a crystal can be ionic interactions, covalent bonds, van der Waals interactions, hydrogen bonds, or a combination of these. In contrast, *amorphous solids* such as glass lack a long-range molecular order and have atomic arrangements resembling those of liquids.

Crystal Structure

A *unit cell* is *the basic repeating structural unit of a crystalline solid.* Figure 6.11 shows a unit cell and its extension in three dimensions. Each vertex in Figure 6.11 represents a *lattice point.* In a crystal lattice, every lattice point has an identical environment. In simple crystals, such as metals, the lattice point is occupied by an atom. For more complex crystals, however, there may be several atoms, molecules, or ions arranged around each lattice point. There are only seven basic unit cell shapes (*crystal systems*) that can be used to form a crystalline solid (Figure 6.12).

Figure 6.11 (a) A unit cell and (b) its extension in three dimensions. The black spheres represent either atoms or molecules.

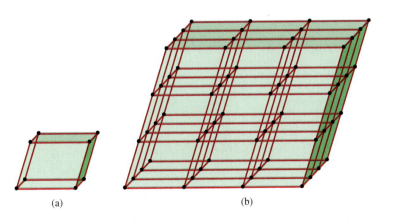

(a)   (b)

Figure 6.12 The seven types of unit cells. Angle α is defined by edges b and c, angle β by edges a and c, and angle γ by edges a and b.

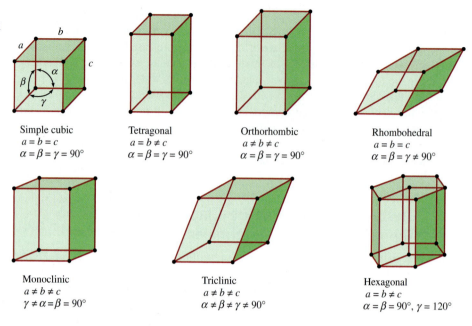

Simple cubic
$a = b = c$
$\alpha = \beta = \gamma = 90°$

Tetragonal
$a = b \neq c$
$\alpha = \beta = \gamma = 90°$

Orthorhombic
$a \neq b \neq c$
$\alpha = \beta = \gamma = 90°$

Rhombohedral
$a = b = c$
$\alpha = \beta = \gamma \neq 90°$

Monoclinic
$a \neq b \neq c$
$\gamma \neq \alpha = \beta = 90°$

Triclinic
$a \neq b \neq c$
$\alpha \neq \beta \neq \gamma \neq 90°$

Hexagonal
$a = b \neq c$
$\alpha = \beta = 90°, \gamma = 120°$

These crystal systems[2] are characterized by the three edge lengths, or **lattice constants,** a, b, and c and by the angles between these edges, α, β, and γ. (For the special case of the cubic unit cell, the geometry is particularly simple because all sides are equal and all angles are 90°.) A unit cell that contains only one lattice point is called a *primitive unit cell*. Often, however, the overall symmetry of the crystal is better represented by *nonprimitive* unit cells containing more than one lattice point. Such nonprimitive cells (Figure 6.13) can be *body centered*, with one additional lattice point at the center of the unit cell; *face centered,* with additional lattice points at the center of each unit cell face, or *side centered*, with additional lattice points at the center of two of the unit cell faces.

Taking into account such nonprimitive unit cells, the structure of any crystalline solid can be represented in terms of one of 14 possible basic types called **Bravais lattices** (Figure 6.14).

Packing of Spheres

We can understand the general geometric structures of simple crystals, such as metals, by considering the different ways of packing identical spheres to form an ordered three-dimensional structure. The way the spheres are stacked in layers determines the type of Bravais lattice.

---

2. The requirement that the unit cells can be stacked to fill space in a regular pattern puts severe restrictions on the type of polyhedra that can represent a unit cell. For example, it is not possible to stack polyhedra that have pentagonal sides to generate a crystal lattice with long-ranged order.

Figure 6.13 Three classes of non-primitive cells (a) body centered, (b) face centered, and (c) side centered.

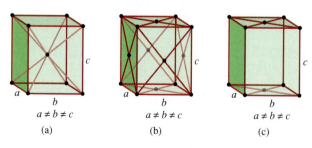

$a \neq b \neq c$

(a)

$a \neq b \neq c$

(b)

$a \neq b \neq c$

(c)

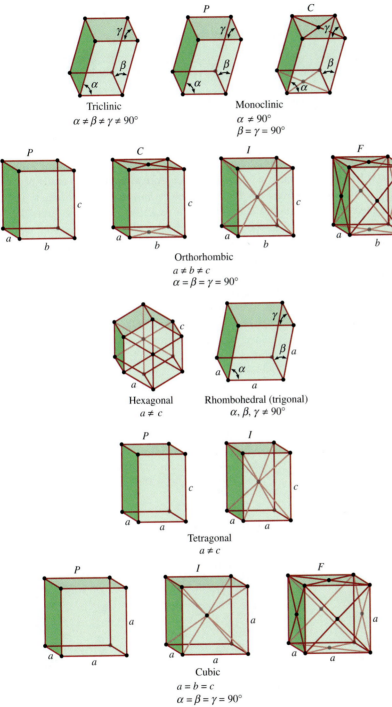

Figure 6.14 The 14 Bravais lattice types. The symbols are defined as follows: primitive (*P*), face centered (*F*), body centered (*I*), side centered (*C*).

Triclinic
$\alpha \neq \beta \neq \gamma \neq 90°$

Monoclinic
$\alpha \neq 90°$
$\beta = \gamma = 90°$

Orthorhombic
$a \neq b \neq c$
$\alpha = \beta = \gamma = 90°$

Hexagonal
$a \neq c$

Rhombohedral (trigonal)
$\alpha, \beta, \gamma \neq 90°$

Tetragonal
$a \neq c$

Cubic
$a = b = c$
$\alpha = \beta = \gamma = 90°$

In the simplest case, a layer of spheres can be arranged as shown in Figure 6.15(a). The three-dimensional structure can be generated by placing a layer above and below this layer in such a way that spheres in one layer are directly over the spheres in the layer below it. This procedure can be extended to generate many, many layers, as in the case of a crystal. Focusing on the sphere labeled with an "*x*," we see that it is in contact with four spheres in its own layer, one sphere in the layer above, and one sphere in the

Figure 6.15 Arrangement of identical spheres in a simple cubic cell. (a) Top view of one layer of spheres. (b) Definition of a simple cubic cell. (c) Because each sphere is shared by eight unit cells and there are eight corners in a cube, there is the equivalent of one complete sphere inside a simple cubic unit cell.

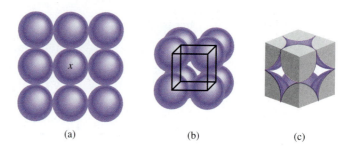

(a)   (b)   (c)

layer below. Each sphere in this arrangement is said to have a *coordination number* of 6 because it has six immediate neighbors. The **coordination number** is defined as *the number of atoms (or ions) surrounding an atom (or ion) in a crystal lattice*. Its value gives us a measure of how tightly the spheres are packed together—the larger the coordination number, the closer the spheres are to each other. The basic, repeating unit in the array of spheres described here is called a **simple cubic (sc)** cell [Figure 6.15(b)].

The other types of cubic cells are the **body-centered cubic (bcc)** cell and the **face-centered cubic (fcc)** cell (Figure 6.16). A body-centered cubic arrangement differs from a simple cube in that the second layer of spheres fits into the depressions of the first layer and the third layer into the depressions of the second layer (Figure 6.17). The coordination number of each sphere in this structure is 8 (each sphere is in contact with four spheres in the layer above and four spheres in the layer below). In the face-centered cubic cell, there are spheres at the center of each of the six faces of the cube, in addition to the eight corner spheres.

Because every unit cell in a crystalline solid is adjacent to other unit cells, most atoms in a cell are shared by neighboring cells. For example, in all types of cubic

Figure 6.16 Three types of cubic cells. In reality, the spheres representing atoms, molecules, or ions are in contact with one another in these cubic cells.

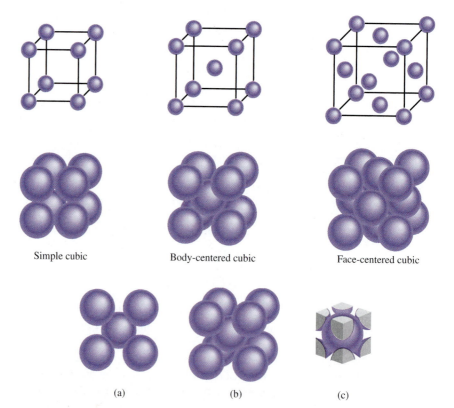

Simple cubic   Body-centered cubic   Face-centered cubic

Figure 6.17 Arrangement of identical spheres in a body-centered cube. (a) Top view. (b) Definition of a body-centered cubic unit cell. (c) There is the equivalent of two complete spheres inside a body-centered cubic unit cell.

(a)   (b)   (c)

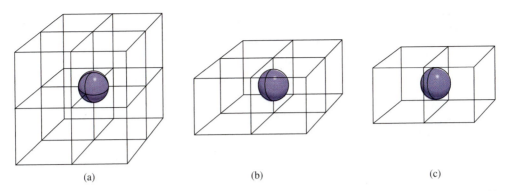

(a)   (b)   (c)

Figure 6.18 (a) A corner atom in any cell is shared by eight unit cells. (b) An edge atom is shared by 4 unit cells. (c) A face-centered atom in a cubic cell is shared by two unit cells.

cells, each corner atom belongs to eight unit cells [Figure 6.18(a)]; an edge atom is shared by four unit cells [Figure 6.18(b)], and a face-centered atom is shared by two unit cells [Figure 6.18(c)]. Because each corner sphere is shared by eight unit cells and there are eight corners in a cube, there will be the equivalent of only one complete sphere inside a simple cubic unit cell [see Figure 6.16]. A body-centered cubic cell contains the equivalent of two complete spheres, one in the center and eight shared corner spheres. A face-centered cubic cell contains four complete spheres—three from the six face-centered atoms and one from the eight shared corner spheres.

Closest Packing

Clearly there is more empty space in the simple cubic and body-centered cubic cells than in the face-centered cubic cell. **Closest packing,** *the most efficient arrangement of spheres,* starts with the structure shown in Figure 6.19(a), which we call layer A. Focusing on the only enclosed sphere, we see that it has six immediate neighbors in that layer. In the second layer (which we call layer B), spheres are packed into the depressions between the spheres in the first layer, so all the spheres are as close together as possible [Figure 6.19(b)].

There are two ways that a third-layer sphere may cover the second layer to achieve closest packing. The spheres may fit into the depressions so that each third layer sphere is directly over a first-layer sphere [Figure 6.19(c)]. Because there is no difference between the arrangement of the first and third layers, the third layer is also called layer A. Alternatively, the third-layer spheres may fit into the depressions that lie directly over the depressions in the first layer [Figure 6.19(d)]. In this case, we call the third layer C. Figure 6.20 shows the "exploded views" and the structures resulting from these two arrangements. The ABA arrangement is known as the *hexagonal close-packed (hcp) structure,* and the ABC arrangement is the *cubic close-packed (ccp) structure,* which corresponds to the face-centered cube already described. Note that in the hcp structure, the spheres in every other layer occupy the same vertical position (ABABAB . . .), while in the ccp structure, the spheres in every fourth layer occupy the same vertical position (ABCABCA . . .). In both structures, each sphere has a coordination number of 12 (each sphere is in contact with six spheres in its own layer, three spheres in the layer above, and three spheres in the layer below). Both the hcp and ccp structures represent the most efficient way of packing identical spheres in a unit cell, and there is no way to increase the coordination number to beyond 12.

Many metals and noble gases, which are monatomic, form crystals with hcp or ccp structures. For example, magnesium, titanium, and zinc crystallize with their

These oranges are in a closest packed arrangement, as shown in Figure 6.19(a).

Figure 6.19 (a) In a close-packed layer, each sphere is in contact with six others. (b) Spheres in the second layer fit into the depressions between the first-layer spheres. (c) In the hexagonal close-packed structure, each third-layer sphere is directly over a first-layer sphere. (d) In the cubic close-packed structure, each third-layer sphere fits into a depression that is directly over a depression in the first layer.

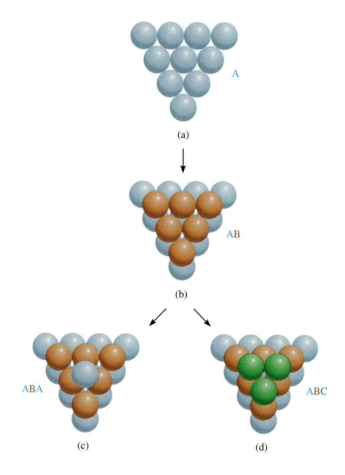

atoms in an hcp array, while aluminum, nickel, and silver crystallize in the ccp arrangement. All solid noble gases have the ccp structure except helium, which crystallizes in the hcp structure. It is natural to ask why a series of related substances, such as the transition metals or the noble gases, would form different crystal structures. The answer lies in the relative stability of a particular crystal structure, which is governed by intermolecular forces. Thus, magnesium metal has the hcp structure because this arrangement of Mg atoms results in the greatest stability of the solid.

Figure 6.21 summarizes the relationship between the atomic radius r and the edge length a of a simple cubic cell, a body-centered cubic cell, and a face-centered cubic cell. This relationship can be used to determine the atomic radius of a sphere if the density of the crystal is known, as Example 6.1 shows.

Example 6.1

Gold (Au) crystallizes in a cubic close-packed structure (the face-centered cubic unit cell) and has a density of 19.3 g cm^{-3}. Calculate the atomic radius of gold in picometers.

Strategy We want to calculate the radius of a gold atom. For a face-centered cubic unit cell, the relationship between radius (r) and edge length (a), according to Figure 6.21, is $a = \sqrt{8}r$. Therefore, to determine r of a gold atom, we need to find a. The volume of a

—Continued

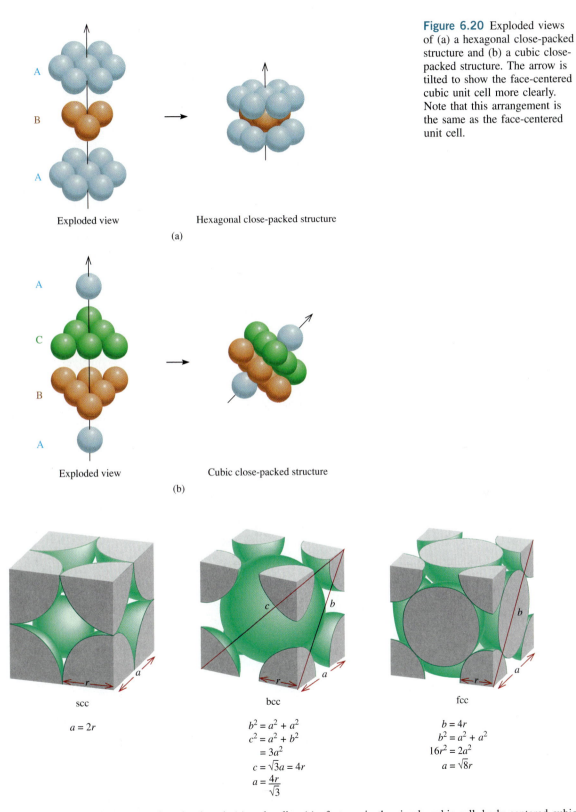

Figure 6.20 Exploded views of (a) a hexagonal close-packed structure and (b) a cubic close-packed structure. The arrow is tilted to show the face-centered cubic unit cell more clearly. Note that this arrangement is the same as the face-centered unit cell.

Exploded view   Hexagonal close-packed structure
(a)

Exploded view   Cubic close-packed structure
(b)

scc

$$a = 2r$$

bcc

$$b^2 = a^2 + a^2$$
$$c^2 = a^2 + b^2$$
$$= 3a^2$$
$$c = \sqrt{3}a = 4r$$
$$a = \frac{4r}{\sqrt{3}}$$

fcc

$$b = 4r$$
$$b^2 = a^2 + a^2$$
$$16r^2 = 2a^2$$
$$a = \sqrt{8}r$$

Figure 6.21 The relationship between the edge length (a) and radius (r) of atoms in the simple cubic cell, body-centered cubic cell, and face-centered cubic cell.

Continued—

cube is $V = a^3$ or $a = \sqrt[3]{V}$. Thus, if we can determine the volume of the unit cell, we can calculate a. We are given the density in the problem.

$$\text{density} = \frac{\text{mass}}{\text{volume}}$$

The sequence of steps is summarized as follows:

$$\begin{array}{ccccccc} \text{density of} & \longrightarrow & \text{volume of} & \longrightarrow & \text{edge length} & \longrightarrow & \text{radius of} \\ \text{unit cell} & & \text{unit cell} & & \text{of unit cell} & & \text{Au atom} \end{array}$$

Solution

Step 1: We know the density, so to determine the volume, we find the mass of the unit cell. Each unit cell has eight corners and six faces. The total number of atoms within such a cell, according to Figures 6.16 and 6.18, is

$$\left(8 \times \frac{1}{8}\right) + \left(6 \times \frac{1}{2}\right) = 4$$

The mass of a unit cell in grams is

$$m = \frac{4 \text{ atoms}}{1 \text{ unit cell}} \times \frac{1 \text{ mol}}{6.022 \times 10^{23} \text{ atoms}} \times \frac{197.0 \text{ g Au}}{1 \text{ mol Au}}$$
$$= 1.31 \times 10^{-21} \text{ g per unit cell}$$

Remember that density is an intensive quantity, so that it is the same for one unit cell and 1 cm³ of the substance.

From the definition of density ($d = m/V$), we calculate the volume of the unit cell as follows:

$$V = \frac{m}{d} = \frac{1.31 \times 10^{-21} \text{ g}}{19.3 \text{ g cm}^{-3}} = 6.79 \times 10^{-23} \text{ cm}^3$$

Step 2: Because volume is length cubed, we take the cubic root of the volume of the unit cell to obtain the edge length (a) of the cell

$$a = \sqrt[3]{V} = \sqrt[3]{6.79 \times 10^{-23} \text{ cm}^3}$$
$$= 4.08 \times 10^{-8} \text{ cm}$$

Step 3: From Figure 6.21 we see that the radius of an Au sphere (r) is related to the edge length by

$$a = \sqrt{8}r$$

Therefore,

$$r = \frac{a}{\sqrt{8}} = \frac{4.08 \times 10^{-8} \text{ cm}}{\sqrt{8}}$$
$$= 1.44 \times 10^{-8} \text{ cm} = 144 \text{ pm}$$

Practice Exercise When silver crystallizes, it forms face-centered cubic cells. The unit cell edge length is 408.7 pm. Calculate the density of silver.

In this section, we have concentrated primarily on crystals with simple spherical units (such as metals) for which packing considerations dominate the crystal structure. For molecular materials, such as water, which has a hexagonal structure (see Figure 6.9), the complex electrostatic interactions between the molecules lead to more complex crystal structures than those predicted by simple packing.

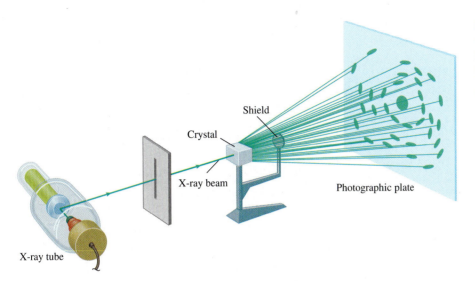

Figure 6.22 Arrangement for obtaining the X-ray diffraction pattern of a crystal. The shield prevents the strong undiffracted X-rays from damaging the photographic plate.

Shield

Crystal

X-ray beam

Photographic plate

X-ray tube

X-Ray Diffraction by Crystals

Virtually all we know about crystal structure has been learned from X-ray diffraction studies. *X-ray diffraction* refers to *the scattering of X-rays by the units of a crystalline solid.* The scattering, or diffraction, patterns produced are used to deduce the arrangement of particles in the solid lattice.

In Section 1.1, we discussed the interference phenomenon associated with waves (see Figure 1.4). Because X-rays are one form of electromagnetic radiation, and therefore waves, we would expect them to exhibit such behavior under suitable conditions. In 1912 the German physicist Max von Laue[3] suggested that, because the wavelength of X-rays is comparable in magnitude to the distances between lattice points in a crystal, the lattice should be able to *diffract* X-rays. An X-ray diffraction pattern is the result of interference in the waves associated with X-rays.

Figure 6.22 shows a typical X-ray diffraction setup. A beam of X-rays is directed at a mounted crystal. Atoms in the crystal absorb some of the incoming radiation and then reemit it; the process is called the *scattering of X-rays.*

To understand how a diffraction pattern may be generated, consider the scattering of X-rays by atoms in two parallel planes (Figure 6.23). Initially, the two incident rays are *in phase* with each other (their maxima and minima occur at the same positions). The upper wave is scattered, or reflected, by an atom in the first layer, while the lower wave is scattered by an atom in the second layer. In order for these two scattered waves to be in phase again, the extra distance traveled by the lower wave must be an integral multiple of the wavelength (λ) of the X-ray, that is,

$$BC + CD = 2d\sin\theta = n\lambda \qquad (6.6)$$

where θ is the angle between the X-rays and the plane of the crystal and d is the distance between adjacent planes. Equation 6.6 is known as the Bragg equation after William H. Bragg[4] and Sir William L. Bragg.[5] The reinforced waves produce a dark

---

3. Max Theodor Felix von Laue (1879–1960). German physicist. Von Laue received the Nobel Prize in Physics in 1914 for his discovery of X-ray diffraction.

4. William Henry Bragg (1862–1942). English physicist. Bragg's work was mainly in X-ray crystallography. He shared the Nobel Prize in Physics with his son Sir William Bragg in 1915.

5. Sir William Lawrence Bragg (1890–1972). English physicist. Bragg formulated the fundamental equation for X-ray diffraction and shared the Nobel Prize in Physics with his father in 1915.

Figure 6.23 Reflection of X-rays from two layers of atoms. The lower wave travels a distance $2d \sin \theta$ longer than the upper wave. For the two waves to be in phase again after reflection, it must be true that $2d \sin \theta = n\lambda$, where λ is the wavelength of the X-ray and $n = 1, 2, 3, \ldots$ The sharply defined spots in Figure 6.22 are observed only if the crystal is large enough to consist of hundreds of parallel layers.

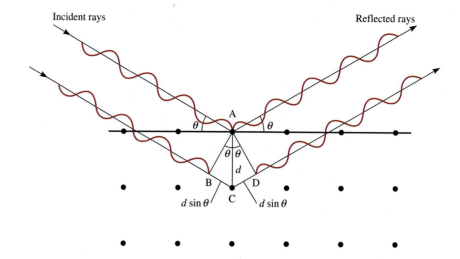

spot on a photographic film for each value of θ that satisfies the Bragg equation. Example 6.2 illustrates the use of Equation 6.6.

Example 6.2

X-rays of wavelength 0.154 nm strike an aluminum crystal; the X-rays are reflected at an angle of 19.3°. Assuming that $n = 1$, calculate the spacing between the planes of aluminum atoms (in pm) that is responsible for this angle of reflection. The conversion factor is obtained from 1 nm = 1000 pm.

Strategy This is an application of Equation 6.6.

Solution Converting the wavelength to picometers and using the angle of reflection (19.3°), we write

$$d = \frac{n\lambda}{2\sin\theta} = \frac{\lambda}{2\sin\theta}$$

$$= \frac{0.154 \text{ nm} \times \dfrac{1000 \text{ pm}}{1 \text{ nm}}}{2\sin 19.3°}$$

$$= 223 \text{ pm}$$

Practice Exercise X-rays of wavelength 0.154 nm are diffracted from a crystal at an angle of 14.17°. Assuming that $n = 1$, calculate the distance (in pm) between layers in the crystal.

The X-ray diffraction technique offers the most accurate method for determining bond lengths and bond angles in molecules in the solid state. Because X-rays are scattered by electrons, chemists can construct an electron-density contour map from the diffraction patterns by using a complex mathematical procedure. Basically, an *electron-density contour map* tells us the relative electron densities at various locations in a molecule. The densities reach a maximum near the center of each atom. In this manner, we can determine the positions of the nuclei and hence the geometric parameters of the molecule.

6.3 | The Properties of Crystalline Solids Are Determined Largely by the Intermolecular Interactions

The structures and properties of crystals, such as melting point, density, and hardness, are determined by the kinds of forces that hold the particles together. We can classify any crystal as one of four types: ionic, covalent, molecular, or metallic.

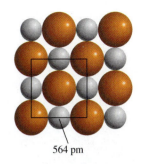

Ionic Crystals

Ionic crystals have two important characteristics: (1) They are composed of charged species, and (2) anions and cations are generally quite different in size. Knowing the radii of the ions is helpful in understanding the structure and stability of these compounds.

There is no way to measure the radius of an individual ion, but sometimes it is possible to come up with a reasonable estimate. For example, if we know the radius of the I^- in KI is about 216 pm, we can determine the radius of the K^+ ion in KI, and from that, the radius of the Cl^- ion in KCl, and so on. The ionic radii in Figure 2.20 are average values derived from many different compounds. Let us consider the NaCl crystal, which has a face-centered cubic lattice (see Figure 0.12). Figure 6.24 shows that the edge length of the unit cell of NaCl is twice the sum of the ionic radii of Na^+ and Cl^-. Using the values given in Figure 2.20, we calculate the edge length to be 2(95 + 181) pm, or 552 pm. But the edge length shown in Figure 6.24 was determined by X-ray diffraction to be 564 pm. The discrepancy between these two values tells us that the radius of an ion actually varies slightly from one compound to another.

Figure 6.25 shows the crystal structures of three ionic compounds: CsCl, ZnS, and CaF_2. Because Cs^+ is considerably larger than Na^+, CsCl has the simple cubic lattice. ZnS has the *zincblende* structure, which is based on the face-centered cubic lattice. If the S^{2-} ions occupy the lattice points, the Zn^{2+} ions are located one-fourth of the distance along each body diagonal. Other ionic compounds that have the zincblende structure include CuCl, BeS, CdS, and HgS. CaF_2 has the *fluorite* structure. The Ca^{2+} ions occupy the lattice points, and each F^- ion is surrounded by four Ca^{2+} ions in a tetrahedral arrangement. The compounds SrF_2, BaF_2, $BaCl_2$, and PbF_2 also have the fluorite structure.

Examples 6.3 and 6.4 show how to calculate the number of ions contained in and the density of a unit cell.

Figure 6.24 Relation between the radii of Na^+ and Cl^- ions and the unit cell dimensions. Here the cell edge length is equal to twice the sum of the two ionic radii.

These giant potassium dihydrogen phosphate crystals were grown in the laboratory. The largest one weighs 701 lb!

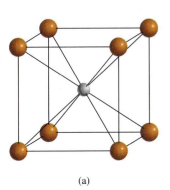

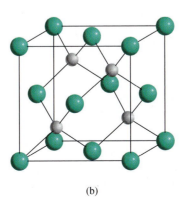

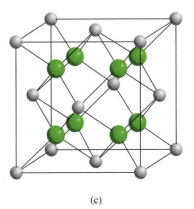

(a)   (b)   (c)

Figure 6.25 Crystal structures of (a) CsCl, (b) ZnS, and (c) CaF_2. In each case, the cation is the smaller sphere.

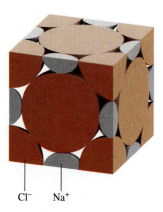

Cl⁻   Na⁺

Figure 6.26 Portions of Na^+ and Cl^- ions within a face-centered cubic unit cell.

Example 6.3

How many Na^+ and Cl^- ions are in each NaCl unit cell?

Solution NaCl has a structure based on a face-centered cubic lattice. As Figure 0.12 shows, one whole Na^+ ion is at the center of the unit cell, and there are twelve Na^+ ions at the edges. Because each edge Na^+ ion is shared by four unit cells, the total number of Na^+ ions is $1 + (12 \times 1/4) = 4$. Similarly, there are six Cl^- ions at the face centers and eight Cl^- ions at the corners. Each face-centered ion is shared by two unit cells, and each corner ion is shared by eight unit cells (see Figure 6.16), so the total number of Cl^- ions is $(6 \times 1/2) + (8 \times 1/8) = 4$. Thus, there are four Na^+ ions and four Cl^- ions in each NaCl unit cell. Figure 6.26 shows the portions of the Na^+ and Cl^- ions *within* a unit cell.

Practice Exercise How many atoms are in a body-centered cube, assuming that all atoms occupy lattice points?

Example 6.4

The edge length of the NaCl unit cell is 564 pm. What is the density of NaCl in $g\ cm^{-3}$?

Strategy To calculate the density, we need to know the mass of the unit cell. The volume can be calculated from the given edge length because $V = a^3$. How many Na^+ and Cl^- ions are in a unit cell? What is the total mass in atomic mass units (u)? What are the conversion factors between u and g and between pm and cm?

Solution From Example 6.3 we see that there are four Na^+ ions and four Cl^- ions in each unit cell. So the total mass (in u) of a unit cell is

$$\text{mass} = 4(22.99\ u + 35.45\ u) = 233.8\ u$$

Converting u to grams, we write

$$233.8\ u \times \frac{1\ g}{6.022 \times 10^{23}\ u} = 3.882 \times 10^{-22}\ g$$

The volume of the unit cell is

$$V = a^3 = (564\ pm)^3 = (564 \times 10^{-10}\ cm)^3 = 1.794 \times 10^{-22}\ cm^3$$

Finally, from the definition of density

$$\text{density} = \frac{\text{mass}}{\text{volume}} = \frac{3.882 \times 10^{-22}\ g}{1.794 \times 10^{-22}\ cm^3}$$
$$= 2.16\ g\ cm^{-3}$$

Practice Exercise Copper crystallizes in a face-centered cubic lattice (the Cu atoms are at the lattice points only). If the density of the metal is $8.96\ g\ cm^{-3}$, what is the unit cell edge length in pm?

Most ionic crystals have high melting points, an indication of the strong cohesive forces holding the ions together. These solids do not conduct electricity because the ions are fixed in position. However, in the molten state (that is, when melted) or dissolved in water, the ions are free to move and the resulting liquid is electrically conducting.

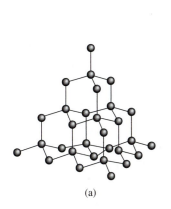

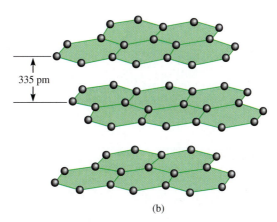

(a)

(b)

Figure 6.27 (a) The structure of diamond. Each carbon is tetrahedrally bonded to four other carbon atoms. (b) The structure of graphite. The distance between successive layers is 335 pm.

Covalent Crystals

In *covalent* (or *network*) *crystals,* atoms are held together in an extensive three-dimensional network entirely by covalent bonds. Well-known examples are the two allotropes of carbon: diamond and graphite. In diamond, each carbon atom is sp^3-hybridized; it is bonded to four other atoms [Figure 6.27(a)]. The strong covalent bonds in three dimensions contribute to diamond's unusual hardness (it is the hardest material known) and very high melting point (3550°C). In graphite [Figure 6.27(b)], carbon atoms are arranged in six-membered rings. The atoms are all sp^2-hybridized; each atom is covalently bonded to three other atoms. The remaining unhybridized $2p$ orbital is used in pi bonding. In fact, each layer of graphite has the kind of delocalized molecular orbital that is present in benzene (see Section 4.5). Because electrons are free to move around in this extensively delocalized molecular orbital, graphite is a good conductor of electricity in directions along the planes of carbon atoms. The layers are held together by weak van der Waals forces. The covalent bonds in graphite account for its hardness; however, because the layers can slide over one another, graphite is slippery to the touch and is effective as a lubricant. It is also used in pencils and in ribbons made for computer printers and typewriters.

Another covalent crystal is quartz (SiO_2). The arrangement of silicon atoms in quartz is similar to that of carbon in diamond, but in quartz there is an oxygen atom between each pair of Si atoms. Because Si and O have different electronegativities, the Si—O bond is polar. Nevertheless, SiO_2 is similar to diamond in many respects, such as hardness and high melting point (1610°C).

Quartz

Molecular Crystals

In a *molecular crystal*, the lattice points are occupied by molecule. The attractive forces that help stabilize the lattice are typically van der Waals forces and/or hydrogen bonding. An example of a molecular crystal is solid sulfur dioxide (SO_2), in which the predominant attractive force is a dipole-dipole interaction. Intermolecular hydrogen bonding is mainly responsible for maintaining the three-dimensional lattice of ice (see Figure 6.9). Other examples of molecular crystals are I_2, P_4, and S_8.

In most molecular crystals, the molecules are in a fixed orientation relative to their neighbors and rotation is restricted; however, in some crystals, called *plastic crystals,* the molecules can rotate freely and are orientationally disordered. Examples of plastic crystals are the *beta* phase of solid molecular nitrogen, which is stable between 35 and 64 K, and the high-temperature solid phase of succinonitrile (N≡C—C_2H_4—C≡N), which is stable between 238 K and its melting point at 335 K. Typically, plastic crystals

Sulfur

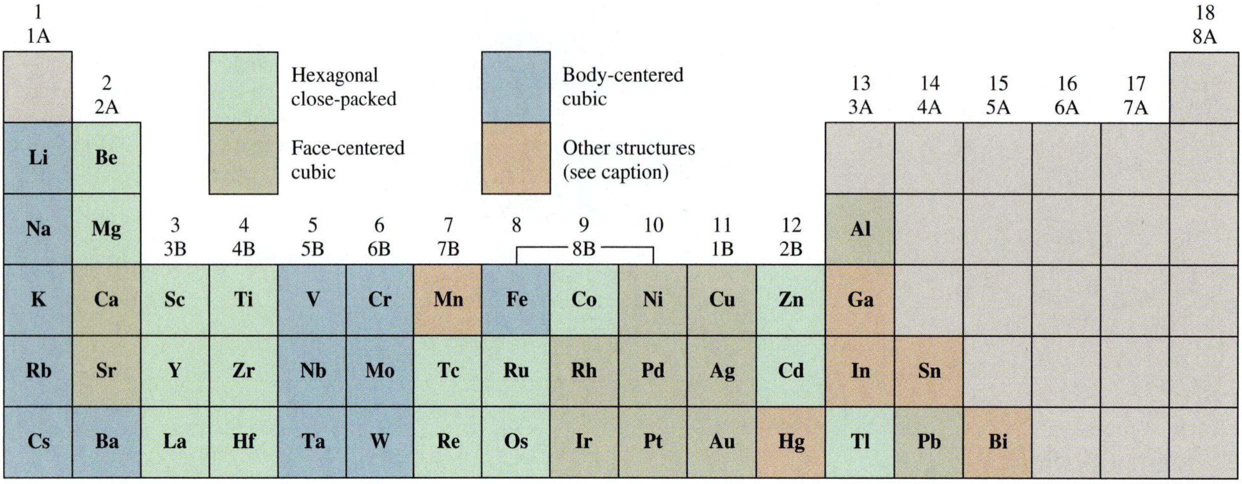

1 1A																	18 8A
	2 2A		Hexagonal close-packed			Body-centered cubic					13 3A	14 4A	15 5A	16 6A	17 7A	
Li	Be		Face-centered cubic			Other structures (see caption)											
Na	Mg	3 3B	4 4B	5 5B	6 6B	7 7B	8	9 8B	10	11 1B	12 2B	Al					
K	Ca	Sc	Ti	V	Cr	Mn	Fe	Co	Ni	Cu	Zn	Ga					
Rb	Sr	Y	Zr	Nb	Mo	Tc	Ru	Rh	Pd	Ag	Cd	In	Sn				
Cs	Ba	La	Hf	Ta	W	Re	Os	Ir	Pt	Au	Hg	Tl	Pb	Bi			

Figure 6.28 Crystal structures of metals. The metals are shown in their positions in the periodic table. Mn has a cubic structure, Ga an orthorhombic structure, In and Sn a tetragonal structure, and Hg and Bi a rhombohedral structure (see Figure 6.12).

will undergo a solid-solid phase transition to an orientationally ordered crystal structure at a transition temperature significantly below the melting point.

In general, except in ice, molecules in molecular crystals are packed together as closely as their size and shape allow. Because van der Waals forces and hydrogen bonding are generally quite weak compared with covalent and ionic bonds, molecular crystals are more easily broken apart than ionic and covalent crystals. Indeed, most molecular crystals melt at temperatures below 100°C.

Metallic Crystals

Figure 6.29 A cross section of a metallic crystal. Each circled positive charge represents the nucleus and inner electrons of a metal atom. The gray area surrounding the positive metal ions indicates the mobile sea of electrons.

In a sense, the structure of *metallic crystals* is the simplest because every lattice point in a crystal is occupied by an atom of the same metal. Metallic crystals are generally body-centered cubic, face-centered cubic, or hexagonal close-packed (Figure 6.28). Consequently, metallic elements are usually very dense.

The bonding in metals is quite different from that in other types of crystals. In a metal, the bonding electrons are delocalized over the entire crystal. In fact, metal atoms in a crystal can be imagined as an array of positive ions immersed in a sea of delocalized valence electrons (Figure 6.29). The great cohesive force resulting from delocalization is responsible for a metal's strength. The mobility of the delocalized electrons makes metals good conductors of heat and electricity. The role of electron delocalization in the properties of metals (and semiconductors) will be discussed in more detail in Section 6.4.

Table 6.2 summarizes the properties of the four different types of crystals discussed.

Amorphous Solids

Solids are most stable in crystalline form. However, if a solid is formed rapidly (for example, when a liquid is cooled quickly), its atoms or molecules do not have time to align themselves and may become locked in positions other than those of a regular crystal. The resulting solid is said to be *amorphous*. **Amorphous solids** are *solids that lack a regular three-dimensional arrangement of atoms*. X-ray diffraction studies show that amorphous solids lack long-range periodic order. In this respect, the structures of amorphous

Table 6.2	Types of Crystals and General Properties		
Type of Crystal	**Force(s) Holding the Units Together**	**General Properties**	**Examples**
Ionic	Electrostatic attraction	Hard, brittle, high melting point, poor conductor of heat and electricity	NaCl, LiF, MgO, $CaCO_3$
Covalent	Covalent bond	Hard, high melting point, poor conductor of heat and electricity	C(diamond),[†] SiO_2 (quartz)
Molecular*	Dispersion forces, dipole-dipole forces, hydrogen bonds	Soft, low melting point, poor conductor of heat and electricity	Ar, CO_2, I_2, H_2O, $C_{12}H_{22}O_{11}$ (sucrose)
Metallic	Metallic bond	Soft to hard, low to high melting point, good conductor of heat and electricity	All metallic elements; for example, Na, Mg, Fe, Cu

*Included in this category are crystals made up of individual atoms.

[†]Diamond is a good thermal conductor.

solids are more similar to liquids than to solids. The term **glass** is reserved for *amorphous solids produced by the rapid cooling of a liquid,* although commonly the term glass is used to denote glasses whose major component is silicon dioxide, called *silica glass* or window glass. In this section, we will discuss briefly the properties of silica glass.

Silica glass is one of civilization's most valuable and versatile materials. It is also one of the oldest—glass articles date back as far as 1000 B.C. Silica glass is formed by mixing molten silicon dioxide (SiO_2), its chief component, with compounds such as sodium oxide (Na_2O), boron oxide (B_2O_3), and certain transition metal oxides for color and other properties. There are about 800 different types of glass in common use today. Figure 6.30 shows two-dimensional schematic representations of crystalline SiO_2 (quartz) and amorphous silica glass.

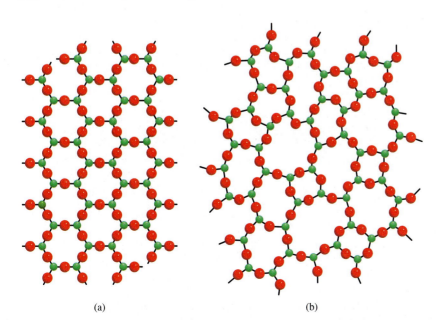

(a)   (b)

Figure 6.30 Two-dimensional representations of (a) crystalline quartz and (b) noncrystalline silica glass. The small spheres represent silicon. In reality, the structure of quartz is three-dimensional. Each Si atom is tetrahedrally bonded to four O atoms.

The color of glass is due largely to the presence of metal ions (as oxides). For example, green glass contains iron(III) oxide, Fe_2O_3, or copper(II) oxide, CuO; yellow glass contains uranium(IV) oxide, UO_2; blue glass contains cobalt(II) and copper(II) oxides, CoO and CuO, respectively; and red glass contains small particles of gold and copper. Note that most of the ions mentioned here are derived from the transition metals.

6.4 | Band Theory Accurately Explains the Conductivity of Metals, Semiconductors, and Insulators

In Section 6.3, we saw that the ability of metals to conduct heat and electricity can be explained by considering the metal to be an array of positive ions immersed in a sea of highly mobile delocalized valence electrons. In this section, we will use our knowledge of quantum mechanics and molecular orbital theory to develop a more detailed model for the conductivity of metals. The model we will use to study metallic bonding is the **band theory of conductivity,** so called because it states that *delocalized electrons move freely through "bands" formed by overlapping molecular orbitals.* We will also apply band theory to understand the properties of semiconductors and insulators.

Conductors

Metals are characterized by high electrical conductivity. Consider magnesium, for example. The electron configuration of a single magnesium atom is $1s^2 2s^2 2p^6 3s^2$, so each atom has two valence electrons in the $3s$ orbital. In a metallic crystal, the atoms are packed closely together, so the energy levels of each magnesium atom are affected by the immediate neighbors of the atom as a result of orbital overlaps. In Section 3.5, we saw that, in terms of molecular orbital theory, the interaction between two atomic orbitals leads to the formation of a bonding and an antibonding molecular orbital. Because the number of atoms in even a small piece of lithium is enormously large (on the order of 10^{20} atoms), the number of molecular orbitals they form is also very large. These molecular orbitals are so closely spaced on the energy scale that they are more appropriately described as a "band" (Figure 6.31). The closely spaced *filled* energy levels make up the **valence band.** The upper half of the energy levels corresponds to the empty, delocalized molecular orbitals formed by the overlap of the $3p$ orbitals. This set of closely spaced *empty* levels is called the **conduction band.**

Figure 6.31 Formation of conduction bands in magnesium. The electrons in the $1s$, $2s$, and $2p$ orbitals are localized on each Mg atom. However, the $3s$ and $3p$ orbitals overlap to form delocalized molecular orbitals. Electrons in these orbitals can travel throughout the metal, and this accounts for the electrical conductivity of the metal.

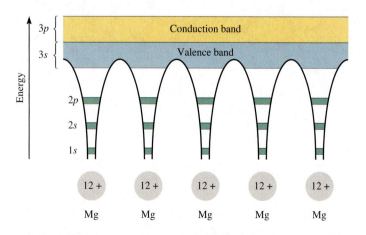

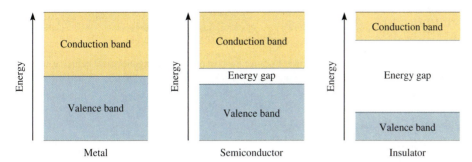

Figure 6.32 Comparison of the energy gaps between the valence band and conduction band in a metal, a semiconductor, and an insulator. In a metal the energy gap is virtually nonexistent; in a semiconductor the energy gap is small; and in an insulator the energy gap is very large, thus making the promotion of an electron from the valence band to the conduction band difficult.

We can imagine a metallic crystal as an array of positive ions immersed in a sea of delocalized valence electrons (see Figure 6.29). The great cohesive force resulting from the delocalization is partly responsible for the strength noted in most metals.

Because the valence band and the conduction band are adjacent to each other, the amount of energy needed to promote a valence electron to the conduction band is negligible. There the electron can travel freely through the metal because the conduction band is void of electrons. This freedom of movement accounts for the fact that metals are good ***conductors***, that is, they are *capable of conducting electric current.*

Why don't substances like glass conduct electricity as metals do? Figure 6.32 provides an answer to this question. Basically, the electrical conductivity of a solid depends on the spacing and the state of occupancy of the energy bands. In magnesium and other metals, the valence bands are adjacent to the conduction bands, and, therefore, these metals readily act as conductors. In glass, on the other hand, the gap between the valence band and the conduction band is considerably greater than that in a metal. Consequently, much more energy is needed to excite an electron into the conduction band. Lacking this energy, electrons cannot move freely. Therefore, glass is an ***insulator***, that is*, an ineffective conductor of electricity.*

Semiconductors

A number of elements and compounds are ***semiconductors***, that is, they *normally are not conductors, but will conduct electricity at elevated temperatures or when combined with a small amount of certain other elements.* The Group 4A elements silicon and germanium are especially suited for this purpose. The use of semiconductors in transistors and solar cells, to name two applications, has revolutionized the electronic industry in recent decades, leading to increased miniaturization of electronic equipment.

The energy gap between the filled and empty bands of semiconductors is much smaller than that for insulators (see Figure 6.32). If the temperature is high enough, thermal energy can provide sufficient energy to promote electrons from the valence band into the conduction band, and the solid can conduct electricity. Increasing the temperature further will increase the number of electrons promoted into the conduction band, so the conductivity of a semiconductor *increases* with increasing temperature. This behavior is opposite to that of metals. The ability of a metal to conduct electricity *decreases* with increasing temperature because the enhanced vibration of atoms at higher temperatures tends to disrupt the flow of electrons.

The ability of a semiconductor to conduct electricity can also be enhanced by adding small amounts of certain impurities to the element, a process called *doping.* Let us consider what happens when a trace amount of boron or phosphorus is added to solid silicon. (Only about five out of every million Si atoms are replaced by B or P atoms.) The structure of solid silicon is similar to that of diamond; each Si atom is

High-Temperature Superconductors

Metals such as copper and aluminum are good conductors of electricity, but they do possess some electrical resistance. In fact, up to about 20 percent of electrical energy may be lost in the form of heat when cables made of these metals are used to transmit electricity. Wouldn't it be marvelous if we could produce cables that possessed no electrical resistance?

Actually, it has been known for over 90 years that certain metals and metal alloys, when cooled to very low temperatures (around the boiling point of helium, or 4 K), lose their resistance totally. However, it is not practical to use these substances, called *superconductors,* for transmission of electric power because the cost of maintaining electrical cables at such low temperatures is prohibitive and would far exceed the savings from more efficient electricity transmission.

In 1986, two physicists in Switzerland discovered a new class of materials that are superconducting at around 30 K. Although 30 K is still a very low temperature, the improvement over the 4 K range was so dramatic that their work generated immense interest and triggered a flurry of research activity. Within months, scientists synthesized compounds that are superconducting around 95 K, which is well above the boiling point of liquid nitrogen (77 K). The first accompanying figure shows the crystal structure of one of these compounds, a mixed oxide of yttrium, barium, and copper with the formula $YBa_2Cu_3O_x$ (where x = 6 or 7). The next accompanying figure

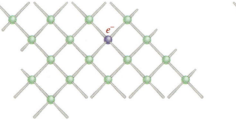

Cu
O
Y
Ba

Crystal structure of $YBa_2Cu_3O_x$ (x = 6 or 7). Because some of the O atom sites are vacant, the formula is not constant.

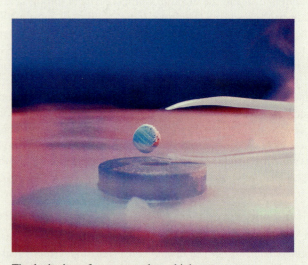

The levitation of a magnet above high-temperature superconductor immersed in liquid nitrogen

covalently bonded to four other Si atoms. Phosphorus ($[Ne]3s^2 3p^3$) has one more valence electron than silicon ($[Ne]3s^2 3p^2$), so there is a valence electron left over after four of them are used to form covalent bonds with silicon (Figure 6.33).

Figure 6.33 (a) Silicon crystal doped with phosphorus. (b) Silicon crystal doped with boron. Note the formation of a negative center in (a) and of a positive center in (b).

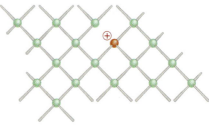

(a)

(b)

shows a magnet being levitated above such a superconductor, which is immersed in liquid nitrogen.

Despite the initial excitement, this class of high-temperature superconductors has not fully lived up to its promise. After more than 20 years of intense research and development, scientists still puzzle over how and why these compounds superconduct. It has also proved difficult to make wires of these compounds, and other technical problems have limited their large-scale commercial applications so far.

In another encouraging development, in 2001 scientists in Japan discovered that magnesium diboride (MgB_2) becomes superconducting at about 40 K.

Although liquid neon (boiling point = 27 K) must be used as a coolant instead of liquid nitrogen, it is still much cheaper than using liquid helium. Magnesium diboride has several advantages as a high-temperature superconductor. First, it is an inexpensive compound (about \$2 per gram) so large quantities are available for testing. Second, the mechanism of superconductivity in MgB_2 is similar to the well-understood metal alloy superconductors at 4 K. Third, it is much easier to fabricate this compound, that is, to make it into wires or thin films. With further research effort, it is hoped that someday soon different types of high-temperature superconductors will be used to build supercomputers, whose speeds are limited by how fast electric current flows; more powerful particle accelerators; efficient devices for nuclear fusion; and more accurate magnetic resonance imaging (MRI) machines for medical use. The progress in high-temperature superconductivity is just warming up!

An experimental levitation train that operates on superconducting material at the temperature of liquid helium.

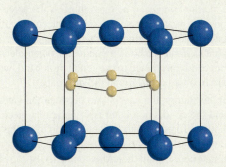

Crystal structure of MgB_2. The Mg atoms (blue) form a hexagonal layer, while the B atoms (gold) form a graphitelike honeycomb layer.

This extra electron can be removed from the phosphorus atom by applying a voltage across the solid. The free electron can move through the structure and function as a conduction electron. Impurities of this type are known as ***donor impurities*** because they *provide conduction electrons. Solids containing donor impurities are called **n-type semiconductors***, where *n* stands for negative (the charge of the "extra" electron). The opposite effect occurs if boron is added to silicon. A boron atom has three valence electrons ($1s^2 2s^2 2p^1$). Thus, for every boron atom in the silicon crystal there is a single *vacancy* in a bonding orbital. It is possible to excite a valence electron from a nearby Si into this vacant orbital. A vacancy created at that Si atom can then be filled by an electron from a neighboring Si atom, and so on. In this manner, electrons can move through the crystal in one direction while the vacancies, or "positive holes," move in the opposite direction, and the solid becomes an electrical conductor.

Impurities that are electron deficient are called ***acceptor impurities.*** Semiconductors that *contain acceptor impurities* are called ***p-type semiconductors***, where *p* stands for positive. In both the *p*-type and *n*-type semiconductors, the energy gap between the valence band and the conduction band is effectively reduced, so only a small amount of energy is needed to excite the electrons. Typically, the conductivity of a semiconductor is increased by a factor of 100,000 or so by the presence of impurity atoms.

The growth of the semiconductor industry since the early 1960s has been truly remarkable. Today semiconductors are essential components of nearly all electronic equipment, ranging from radios and television sets to pocket calculators and computers. One of the main advantages of solid-state devices over vacuum-tube electronics is that the former can be made on a single "chip" of silicon no larger than the cross section of a pencil eraser. Consequently, much more equipment can be packed into a small volume—a point of particular importance in space travel, as well as in handheld calculators and microprocessors (computers-on-a-chip).

Summary of Facts and Concepts

Section 6.1

▸ Intermolecular forces govern the condensation process as well as the structure and properties of liquids.

▸ The radial distribution function describes the local structure in a liquid.

▸ Liquids tend to assume a geometry that minimizes surface area. Surface tension is the energy needed to expand a liquid surface area; strong intermolecular forces lead to greater surface tension.

▸ Viscosity is a measure of the resistance of a liquid to flow; it decreases with increasing temperature.

Section 6.2

▸ All solids are either crystalline (with a regular structure of atoms, ions, or molecules) or amorphous (without a regular structure). Glass is an example of an amorphous solid.

▸ The basic structural unit of a crystalline solid is the unit cell, which is repeated to form a three-dimensional crystal lattice. X-ray diffraction has provided much of our knowledge about crystal structure.

Section 6.3

▸ The four types of crystals and the forces that hold their particles together are ionic crystals, held together by ionic bonding; covalent crystals, covalent bonding; molecular crystals, van der Waals forces and/or hydrogen bonding; and metallic crystals, metallic bonding.

Section 6.4

▸ Metallic bonds can be thought of as the force between positive ions immersed in a sea of electrons. In terms of band theory, the atomic orbitals merge to form energy bands. A substance is a conductor when electrons can be readily promoted to the conduction band, where they are free to move through the substance.

▸ In insulators, the energy gap between the valence band and the conduction band is so large that electrons cannot be promoted into the conduction band. In semiconductors, electrons can cross the energy gap at higher temperatures, and therefore conductivity increases with increasing temperature as more electrons are able to reach the conduction band.

▸ *n*-Type semiconductors contain donor impurities and extra electrons. *p*-Type semiconductors contain acceptor impurities and "positive holes."

Key Words

acceptor impurities, p. 359
adhesion, p. 337
amorphous solid, p. 354
band theory of conductivity, p. 356
body-centered cubic (bcc), p. 344
Bravais lattice, p. 342
closest packing, p. 345

cohesion, p. 337
conduction band, p. 356
conductor, p. 357
coordination number, p. 344
covalent crystal, p. 353
crystalline solid, p. 341
donor impurities, p. 359
face-centered cubic (fcc), p. 344

glass, p. 355
insulator, p. 357
ionic crystal, p. 351
lattice constants, p. 342
metallic crystal, p. 354
molecular crystal, p. 353
n-type semiconductor, p. 359
p-type semiconductor, p. 359

radial distribution function, p. 336
semiconductor, p. 357
simple cubic (sc) cell, p. 344
surface tension, p. 337
unit cell, p. 341
valence band, p. 356
viscosity, p. 338
X-ray diffraction, p. 349

Problems

The Structure and Properties of Liquids Are Governed by Intermolecular Interactions

6.1 Ethanol (C_2H_5OH) and dimethyl ether (CH_3OCH_3) have the same empirical formula but different molecular structures. Predict which will have the higher surface tension.

6.2 What is the significance of the critical point in the liquefaction of gases?

6.3 Predict the viscosity of ethylene glycol relative to that of ethanol and glycerol (see Table 6.1).

$$CH_2—OH$$
$$|$$
$$CH_2—OH$$ *ethylene glycol*

6.4 Water has an unusually large surface tension. Explain.

6.5 Give a molecular interpretation of the decrease in surface tension of a liquid with increasing temperature.

6.6 How do you expect the viscosity of a liquid to change when the temperature is increased? Explain.

6.7 The compound dichlorodifluoromethane (CCl_2F_2) has a normal boiling point of −30°C, a critical temperature of 112°C, and a corresponding critical pressure of 40.5 bar. If the gas is compressed to 18 bar at 20°C, will the gas condense? Your answer should be based on a graphical interpretation.

6.8 The radial distribution function, $g(r)$, for liquid argon was shown in Figure 6.4. How would you expect the $g(r)$ for liquid krypton to differ from that for argon?

6.9 Explain the difference in the melting points of the following compounds: (*Hint:* Only one of them can form intermolecular hydrogen bonds.)

m.p. 45°C   m.p. 115°C

Crystalline Solids Can Be Classified in Terms of Their Structure and Intermolecular Interactions

6.10 What is the coordination number of each sphere in (a) a simple cubic cell, (b) a body-centered cubic cell, and (c) a face-centered cubic cell? Assume the spheres are all the same.

6.11 Calculate the number of spheres that would be found within simple cubic, body-centered cubic, and face-centered cubic cells. Assume that the spheres are the same.

6.12 Metallic iron crystallizes in a cubic lattice. The unit cell edge length is 287 pm. The density of iron is 7.87 g cm^{-3}. How many iron atoms are within a unit cell?

6.13 Barium metal crystallizes in a body-centered cubic lattice (the Ba atoms are at the lattice points only). The unit cell edge length is 502 pm, and the density of the metal is 3.50 g cm^{-3}. Using this information, calculate Avogadro's number. (*Hint:* First calculate the volume (in cm^3) occupied by 1 mole of Ba atoms in the unit cells. Next calculate the volume (in cm^3) occupied by one Ba atom in the unit cell. Assume that 68 percent of the unit cell is occupied by Ba atoms.)

6.14 Vanadium crystallizes in a body-centered cubic lattice (the V atoms occupy only the lattice points). How many V atoms are present in a unit cell?

6.15 Europium crystallizes in a body-centered cubic lattice (the Eu atoms occupy only the lattice points). The density of Eu is 5.26 g cm^{-3}. Calculate the unit cell edge length (in pm).

6.16 Crystalline silicon has a cubic structure. The unit cell edge length is 543 pm. The density of the solid is 2.33 g cm^{-3}. Calculate the number of Si atoms in one unit cell.

6.17 Metallic iron crystallizes into a body-centered lattice. The unit cell edge is 287 pm. The density of iron is 7.87 g cm^{-3}. How many iron atoms are there within a unit cell?

6.18 A face-centered cubic cell contains 8 X atoms at the corners of the cell and 6 Y atoms at the faces. What is the empirical formula of the solid?

6.19 When X-rays of wavelength 0.85 Å are diffracted by a metallic crystal, the angle of first-order diffraction ($n = 1$) is measured to be 14.8°. What is the distance (in pm) between the layers of atoms responsible for the diffraction?

6.20 Iron crystallizes in a body-centered cubic lattice. The cell length as determined by X-ray diffraction is 286.8 pm. Given that the density of iron is 7.874 g cm^{-3}, calculate Avogadro's number.

The Properties of Crystalline Solids Are Determined Largely by Intermolecular Interactions

6.21 A solid is hard, brittle, and electrically nonconducting. Its melt (the liquid form of the substance) and an aqueous solution containing the substance conduct electricity. Classify the solid.

6.22 A solid is soft and has a low melting point (below 100°C). The solid, its melt, and an aqueous solution containing the substance are all nonconductors of electricity. Classify the solid.

6.23 A solid is very hard and has a high melting point. Neither the solid nor its melt conducts electricity. Classify the solid.

6.24 Which of the following are molecular solids and which are covalent solids: Se_8, HBr, Si, CO_2, C, P_4O_6, or SiH_4?

6.25 Classify the solid state of the following substances as ionic crystals, covalent crystals, molecular crystals, or metallic crystals: (a) CO_2, (b) B_{12}, (c) S_8, (d) KBr, (e) Mg, (f) SiO_2, (g) LiCl, and (h) Cr.

6.26 Classify the solids states in terms of crystal types of the elements in the third period of the periodic table. Predict the trends in their melting points and boiling points.

6.27 The melting points of the oxides of the third-period elements are given in parentheses: $Na_2O(1275°C)$, $MgO(2800°C)$, $Al_2O_3(2045°C)$, $SiO_2(1610°C)$, $P_4O_{10}(580°C)$, $SO_3(16.8°C)$, and $Cl_2O_7(-91.5°C)$. Classify these solids in terms of crystal type.

6.28 Explain why diamond is harder than graphite. Why is graphite an electrical conductor but diamond is not?

6.29 Without referring to a chemistry handbook, decide which of the following is denser: diamond or graphite.

Band Theory Accurately Explains the Conductivity of Metals, Semiconductors, and Insulators

6.30 The conductivity of metals generally decreases with increasing temperature. Give an explanation of this in terms of the band theory of conductivity.

6.31 The conductivity of a semiconductor will increase with increasing temperature. Give an explanation of this in terms of the band theory of conductivity.

6.32 State whether silicon would form *n*-type or *p*-type semiconductors with the following elements: Ga, Sb, Al, or As.

6.33 The band gap in silicon is 1.14 eV. Calculate the frequency of electromagnetic radiation that would be required to promote an electron from the valence to the conduction band.

6.34 Why are metals shiny?

Additional Problems

6.35 Which of the following properties indicates very strong intermolecular forces in a liquid: (a) very low surface tension, (b) very low critical temperture, (c) very low boiling point, or (d) very low vapor pressure?

6.36 Based on the following properties of elemental boron, classify it as one of the crystalline solids discussed in Section 6.3: high melting point (2300°C), poor conductor of heat and electricity, insoluble in water, and very hard substance.

6.37 Which has a greater density, crystalline SiO_2 or amorphous SiO_2? Why?

6.38 A small drop of oil in water assumes a spherical shape. Explain?

6.39 Which liquid would you expect to have a greater viscosity: water or diethyl ether (CH_3—CH_2—O—CH_2CH_3)?

6.40 Silicon used in computer chips must have an impurity level below 10^{-9} (that is, fewer than one impurity atom for every 10^9 Si atoms). Silicon is prepared by the reduction of quartz (SiO_2) with coke (a form of carbon made by the destructive distillation of coal) at about 2000°C:

$$SiO_2(s) + 2C(s) \longrightarrow Si(l) + 2CO(g)$$

Next, solid silicon is separated from other solid impurities by treatment with hydrogen chloride at 350°C to form gaseous trichlorosilane ($SiCl_3H$):

$$Si(s) + 3HCl(g) \longrightarrow SiCl_3H(g) + H_2(g)$$

Finally, ultrapure Si can be obtained by reversing the above reaction at 1000°C:

$$SiCl_3H(g) + H_2(g) \longrightarrow Si(s) + 3HCl(g)$$

(a) What types of crystals do Si and SiO2 form? (b) Silicon has a diamond crystal structure (see Figure 6.27). Each cubic unit cell (edge length $a = 543$ pm) contains eight Si atoms. If there are 1.0×10^{13} boron atoms per cubic centimeter in a sample of pure silicon, how many Si atoms are there for every B atom in the sample? Does this sample satisfy the 10^{-9} purity requirement for the electronic grade silicon?

6.41 Carbon and silicon belong to Group 4A of the periodic table and have the same valence electron configuration (ns^2np^2). Why does silicon dioxide (SiO_2) have a much higher melting point than carbon dioxide (CO_2)?

6.42 An early view of metallic bonding assumed that bonding in metals consisted of localized, shared electron-pair bonds between metal atoms. What evidence would help you to argue against this viewpoint?

6.43 Silver crystallizes into a face-centered cubic arrangement; the edge length is 4.08 Å, and the density of the metal is 10.5 g cm^{-3}. From these data, calculate Avogadro's number.

6.44 When X-rays of wavelength 0.090 nm are diffracted by a metallic crystal, the angle of first-order diffraction ($n = 1$) is measured to be 15.2°. What is the distance (in pm) between the layers of atoms responsible for the diffraction?

6.45 The distance between layers in a NaCl crystal is 282 pm. X-rays are diffracted from these layers at an angle of 23.0°. Assuming that $n = 1$, calculate the wavelength of the X-rays (in nm).

6.46 Europium crystallizes in a body-centered cubic lattice (the Eu atoms occupy only the lattice points). The density of Eu is 5.26 g cm^{-3}. Calculate the unit cell edge in pm.

6.47 A quantitative measure of how efficiently spheres pack into unit cells is called *packing efficiency,* which is the percentage of the cell space occupied by the spheres. Calculate the packing efficiencies of a simple cubic cell, a body-centered cubic cell, and a face-centered cubic cell. (*Hint:* Refer to Figure 6.21 and use the relationship that the volume of a sphere is $4\pi r^3/3$, where r is the radius of the sphere.)

6.48 Argon crystallizes in the face-centered cubic arrangement at 40K. Given that the atomic radius of argon is 191 pm, calculate the density of solid argon.

6.49 When choosing motor oil, we are confronted with viscosity ratings like SAE 30 or 10W-40. What do these designations mean in terms of the viscosity of the motor oil? How do the differences between single-grade and multigrade motor oils on the molecular level give rise to their differences in performance? (You will need to do some outside research to answer this problem.)

6.50 Estimate the distance (in Å) between molecules of water vapor at 100°C and 1.0 bar. Assume ideal-gas behavior. Repeat the calculation for liquid water at 100°C, given that the density of water at 100°C and 1 bar is 0.96 g cm^{-3}. Comment on your results. (The diameter of an H_2O molecule is approximately 3 Å, 1 Å = 10^{-8} cm.)

6.51 Given that the density of solid CsCl is 3.97 g cm^{-3}, calculate the distance between adjacent Cs^+ and Cl^- ions.

6.52 The distance between Li^+ and Cl^- is 257 pm in solid LiCl and 203 pm in the LiCl unit in the gas phase. Explain the difference in bond lengths.

6.53 Metallic iron can exist in the β form (bcc, cell dimension = 2.90 Å) and the γ form (fcc, cell dimension = 3.68 Å). The β form can be converted to the γ form by applying high pressures. Calculate the ratio of the densities of the β form to the γ form.

6.54 An alkali metal in the form of a cube of edge length 0.171 cm is vaporized in a 0.843-L container at 1235 K. The vapor pressure is 19.2 mmHg. Identify the metal by calculating the atomic radius in pm and the density. (You may need to consult a handbook of chemical and physical properties.)

6.55 The electrical conductance of copper metal decreases with temperature, but that of a $CuSO_4$ solution increases with temperature. Explain.

6.56 Polonium forms a simple cubic lattice. If the density of polonium is 9.196 g cm^{-3}, calculate the lattice spacing in pm.

6.57 An alkali metal in the form of a cube of edge length 0.215 cm is vaporized in a 0.843-L container at 1235 K. The vapor pressure is 19.2 mmHg. Identify the metal by calculating the atomic radius in picometers.

6.58 Why do critical points exist for liquid-gas transitions, but not for solid-liquid transitions?

6.59 At the critical point, both the slope $\left[\left(\frac{\partial P}{\partial V}\right)_T\right]$ and the curvature $\left[\left(\frac{\partial^2 P}{\partial V^2}\right)_T\right]$ along the critical isotherm are zero. Use these conditions to derive Equations 6.1, 6.2 and 6.3 for the critical volume, pressure, and temperature of a system obeying the van der Waals equation of state (Equation 5.42).

Answers to Practice Exercises

6.1 10.50 g cm^{-3} **6.2** 315 pm **6.3** two **6.4** 361 pm

7

Thermochemistry: Energy in Chemical Reactions

Every chemical reaction obeys two fundamental laws: the law of conservation of mass and the law of conservation of energy. We discussed the mass relationships between reactants and products in Chapter 0; here we will look at the energy changes that accompany physical and chemical reactions.

7.1 | Thermodynamics Is the Study of Energy and Its Transformations in Macroscopic Systems

Any process that occurs within a system in nature is accompanied by a redistribution of energy, either between the system and its surroundings or within the system itself. When a ball rolls down a hill, for example, gravitational potential energy is converted to kinetic energy of motion as the ball accelerates and, in the process, some energy is lost to the surroundings (the hill) due to friction. Although energy is redistributed, the total amount remains fixed. Energy is conserved in chemical reactions, too, where an input of energy is needed to break the bonds in the reactant molecules, but energy is released when the bonds in the product molecules are formed.

The redistribution of energy in chemical reactions is exemplified by the reaction of hydrogen and chlorine gas to form hydrogen chloride:

$$H_2(g) + Cl_2(g) \longrightarrow 2HCl(g)$$

In this reaction, the energy released during the formation of the two H—Cl bonds in the products is larger than the energy required to break the H—H and Cl—Cl bonds in the reactants. We can estimate this energy using the bond dissociation energy data from Table 3.3. The energy released when two moles of H—Cl product molecules are formed is given by

$$
\begin{aligned}
\text{energy released} &= 2\, D_0(\text{H—Cl}) \\
&= 2\,(427.6 \text{ kJ mol}^{-1}) \\
&= 855.2 \text{ kJ mol}^{-1}
\end{aligned}
$$

The energy needed to break the reactant bonds is

$$
\begin{aligned}
\text{energy input} &= D_0(\text{H—H}) + D_0(\text{Cl—Cl}) \\
&= 432.0 \text{ kJ mol}^{-1} + 240.2 \text{ J mol}^{-1} \\
&= 672.2 \text{ kJ mol}^{-1}
\end{aligned}
$$

The total energy released is then

$$\text{energy released} = 855.2 \text{ kJ mol}^{-1} - 672.2 \text{ kJ mol}^{-1} = 183.0 \text{ kJ mol}^{-1}$$

This energy is either absorbed by the surroundings or will act to increase the temperature (average kinetic energy) of the reaction system.

The redistribution of energy in chemical reactions is governed by ***thermodynamics***, which is *the study of energy and its transformations from one form to another in macroscopic systems.* Thermodynamics not only describes the energy changes that occur in macroscopic processes, but it also provides information about whether a process will be possible under a given set of conditions. In this chapter, we will primarily study a subfield of thermodynamics called ***thermochemistry***, which is *the study of the energy released or absorbed during chemical reactions.*

Energy, Work, and Heat

When studying thermodynamics, we are not interested in the energy that an object has due to its overall motion—a cup of coffee at 60°C and 1 bar pressure in your car is exactly the same, in a thermodynamic sense, whether the car is traveling at 100 km h^{-1} or stopped at an intersection. *The energy of a system that is left when the kinetic energy of overall motion is subtracted out is called the **internal energy**, U:*

$$U = E_{\text{total}} - E_{\text{motion}}$$

On a molecular level, the internal energy of a system is the sum of the kinetic and potential energies. The kinetic energy contribution consists of various types of molecular motion and the movement of electrons within the molecules. The potential energy contribution results from the attractive interactions between electrons and nuclei and from the repulsive interactions between electrons and between nuclei in individual molecules, as well as by interactions between molecules.

Recall from Section 0.1 that we can change the energy of an object by doing work on it and that the amount of work done on the object is the product of the applied force F and the distance d by which the object is displaced (Equation 0.2):

$$w = F \times d$$

We could imagine using this equation to determine the change in the internal energy of a macroscopic system in terms of the work done during the process on each individual atom or molecule. This is an impossible task, however, because the ***molecular state*** (or ***microstate***) of a macroscopic system is given in terms of the positions and velocities of *all* the atoms and molecules within the system. Determining the work done would require knowing the individual forces on all the 10^{23} or so atoms and molecules over the course of the process, as well as the corresponding displacements. Not only would such a description be impossible to formulate because of the large number of atoms and molecules that make up macroscopic systems, it would also be largely irrelevant because we do not describe macroscopic objects in terms of their molecular state.

From a thermodynamic perspective, knowing the pressure P, the volume V, and the temperature T of a beaker of water is enough to completely specify its macroscopic properties. Two samples of water with exactly the same pressure, volume, and temperature are thermodynamically identical, and we say that they are in the same ***thermodynamic state*** (or ***macrostate***). (The quantities P, V, and T are called *thermodynamic variables*.) As noted before, we would require on the order of 10^{23} variables to precisely describe the molecular state of a beaker of water, but only a handful of variables (in this case, three) to determine its thermodynamic state. This reduction of variables (Figure 7.1) is a tremendous simplification; however, it does have consequences in how we keep track of energy changes in the system. For thermodynamic systems, we must distinguish between energy transferred to the system in the form of *work* and energy transferred in the form of *heat*.

Work, in the thermodynamic sense, is *energy that is transferred to the system through the change in a macroscopic mechanical variable, such as volume, against an opposing force, such as pressure.* It has a broad meaning that includes mechanical work (for example, compressing a gas), electrical work (a battery supplying electrons to light the bulb of a flashlight), and surface work (blowing up a soap bubble). In this section, we will concentrate on mechanical work; in Chapter 13, we will discuss electrical work.

Figure 7.2 illustrates the mechanical work involved in the expansion (or compression) of a gas. Suppose that a gas is in a cylinder fitted with a weightless, frictionless movable piston, at a certain temperature, pressure, and volume. As it expands, the gas pushes the piston upward a distance Δx against a constant opposing external atmospheric pressure P_{ext}, which is defined as the external force F_{ext} per unit area A (Equation 5.1). Using Equation 0.2, the work done on the gas by the piston is

$$w = -F_{ext} \times \Delta x = -(F_{ext}/A) \times (A \times \Delta x)$$

giving

$$w = -P_{ext}\Delta V \qquad (7.1)$$

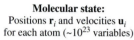

Molecular state:
Positions $\mathbf{r}_i$ and velocities $\mathbf{u}_i$ for each atom (~10^{23} variables)

Thermodynamic state:
P, V, n, T, etc.
Only a few variables

Figure 7.1 Molecular state and thermodynamic (macroscopic) descriptions of a beaker of water.

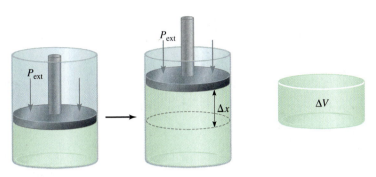

Figure 7.2 The expansion of a gas against a constant external pressure (such as atmospheric pressure). The gas in a cylinder fitted with a weightless, movable piston. The work done is given by $-P_{ext}\Delta V$.

where ΔV, the change in volume, is given by $A\,\Delta x$. The minus sign in Equation 7.1 is necessary because we have defined w as the work done *on* the system. For gas expansion, $\Delta V > 0$, so $-P_{ext}\Delta V$ is a negative quantity. For gas compression, $\Delta V < 0$, so $-P_{ext}\Delta V$ is a positive quantity. If P is zero (that is, if the gas is expanding against a vacuum), the work done must also be zero. In the limit of infinitesimally small volume changes ($\Delta V \longrightarrow dV$), we have

$$dw = -P_{ext}\,dV \qquad (7.2)$$

Thus, if P_{ext} is not constant, the work done can be calculated by integrating Equation 7.2 (see Appendix 1) to give

$$w = -\int_{V_1}^{V_2} P_{ext}\,dV \qquad (7.3)$$

Although we have used gases as an example, Equations 7.1 and 7.3 also apply to the compression or expansion of liquids and solids by an external pressure.

According to Equation 7.1, the units for work done on a gas are liter-bars (L bar). To express the work done in the more familiar unit of joules, use the fact that 1 bar = 10^5 pascals (Pa) and that 1 Pa = 1 N m^{-2}. A liter, moreover, is a cubic decimeter (1 L = 1.0 dm^3 = 10^{-3} m^3), so

$$1\text{ L bar} = (10^{-3}\text{ m}^3)(10^5\text{ N m}^{-2}) = 10^2\text{ N m} = 100\text{ J}$$

where 1 J = 1 N m.

Example 7.1

A sample of gas expands in volume from 2.0 L to 6.0 L at constant temperature. Calculate the work (in joules) done by the gas if it expands (a) against a vacuum and (b) against a constant external pressure of 1.2 bar.

Strategy First, organize the data from the problem statement as follows

$\quad$ *Initial volume* = 2.0 L $\qquad \Delta V = (6.0 - 2.0)\text{ L} = 4.0\text{ L}$
$\quad$ *Final volume* = 6.0 L

$\quad$ (a) $P_{ext} = 0$ $\qquad$ (b) $P_{ext} = 1.2$ bar

Next, use these data in Equation 7.1 because the work done when a gas expands is equal to the product of the external, opposing pressure and the change in volume. What is the conversion factor between L bar and J?

—Continued

Continued—

Solution

(a) Because the external pressure of a vacuum is zero, no work is done in the expansion.

$$w = -P\Delta V = -(0 \text{ bar})(4.0 \text{ L}) = 0$$

(b) The external, opposing pressure is 1.2 bar, so

$$w = -P\Delta V = -(1.2 \text{ bar})(4.0 \text{ L}) = -4.8 \text{ L bar}$$

To convert the answer to joules, we write

$$w = -4.8 \text{ L bar} \times \frac{100 \text{ J}}{1 \text{ L bar}} = -4.8 \times 10^2 \text{ J}$$

Check Because this problem describes a gas expansion (work is done by the system on the surroundings), the work done on the system in part (b) has a negative sign. Although the initial and final volumes of the gas are the same in both parts, the work done in the two processes is different. In other words, the work depends on the path taken.

Practice Exercise A gas expands from 264 mL to 971 mL at constant temperature. Calculate the work done (in joules) by the gas if it expands (a) against a vacuum and (b) against a constant pressure of 4.00 bar.

Thermodynamically, the term *work* applies only to energy changes due to the displacement of macroscopic mechanical variables against an opposing force. However, this does not account for all the energy that is transferred to a system in a process. Some of the energy is transferred into the many nonmacroscopic molecular variables that we have ignored in the macroscopic description of the system. Such energy transfer is referred to as ***heat,*** which can be defined as *the transfer of thermal energy between two bodies solely due to a difference in temperature.* The quantity of energy transferred as heat is generally given the symbol q.

Suppose, for example, that we have a sample of gas at a temperature T_1 in a rigid, but thermally conducting container (Figure 7.3). If this container is put in thermal contact with an object with a temperature T_2, where $T_2 > T_1$, energy will flow from the hotter object into the gas container, increasing the temperature (average kinetic energy) of the gas. Because the container is rigid, ΔV is zero, and no work is done on the gas. The energy goes directly into increasing the random thermal motion of the gas molecules, and we say that energy has been added to the gas in the form of heat. Although we often speak of the "heat flow" from a hot object to a cold one, it is important to keep in mind that "heat" only refers to the process by which energy is transferred, not to the actual energy itself. However, it is customary to talk of "heat absorbed" or "heat released" when describing the energy changes that occur during a process.

If a process absorbs heat from the surroundings ($q > 0$), it is ***endothermic.*** If it releases heat to the surroundings ($q < 0$), it is ***exothermic.***

Figure 7.3 When a sample of gas at a temperature T_1 in a rigid, but thermally conducting container is placed in contact with another system at a higher temperature $T_2 > T_1$, energy in the form of heat flows from the higher temperature object into the gas sample, but no work is done.

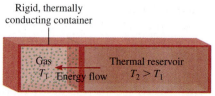

Rigid, thermally conducting container

Gas T_1 Energy flow   Thermal reservoir $T_2 > T_1$

The First Law of Thermodynamics

According to the law of conservation of energy, *energy can be converted from one form to another, but cannot be created or destroyed.*[1] Thus, the change in the internal energy of an object during a process must always be exactly equal to the energy that is added to it from the outside. For a macroscopic object, we can add energy to a system in the form of work (w) or heat (q), or both. Thus, the total change in internal energy for a process is given by

$$\Delta U = w + q \qquad\qquad (7.4)$$

Equation 7.4 is known as the ***first law of thermodynamics***. The first law states not only that energy is conserved but also that energy changes in macroscopic systems can be due to work being done on the system or by the addition of energy in the form of heat.[2]

According to Equation 7.4, the change in the internal energy ΔU of a system is the sum of the heat exchange q between the system and the surroundings and the work w done on the system. The sign conventions for q and w are as follows:

▶ q is positive for an endothermic process and negative for an exothermic process.

▶ w is positive for work done on the system by the surroundings and negative for work done by the system on the surroundings.

We can think of the first law of thermodynamics as an energy balance sheet, much like a balance sheet kept in a bank for currency exchange. We can withdraw or deposit money in either of two different currencies (analogous to energy change due to heat exchange and work done). However, the value of our bank account depends only on the net amount of money left in it after these transactions, not on which currency we used.

Equation 7.4 may seem abstract, but it is actually quite logical. If a system loses heat to its surroundings or does work on the surroundings, its internal energy decreases because both are energy-depleting processes. As a result, both q and w are negative. Conversely, if heat is added to the system or if work is done on the system (both energy-replenishing processes), then the internal energy of the system increases. In this case, both q and w are positive.

State Functions

The internal energy of an object is entirely determined by its thermodynamic state— two glasses of water with the same temperature, pressure, and volume will have identical internal energies because they are in identical thermodynamic states. Consequently, the change in the internal energy of an object in a process depends only upon the energies of the initial and final thermodynamic states, not on any details of the process. Properties such as internal energy that depend only upon the thermodynamic state of a system, and not upon how that condition was created, are called ***state functions***. Other state functions are pressure, volume, and temperature. Potential energy is a state function, too, because the net increase in gravitational potential energy when we go from the same starting point at the base of a mountain to the summit is always the same, regardless of the path we take to get there (Figure 7.4).

---

1. See pages 8–12 of Chapter 0 for a discussion of the mass and energy relationship in chemical reactions.

2. Equation 7.4 can also be viewed as providing a mechanical definition of heat—*heat is the difference between the change in internal energy of a system and the work done on the system:* $q = \Delta U - w$.

Figure 7.4 The gain in gravitational potential energy that occurs when a person climbs from the base to the top of a mountain is independent of the path taken.

Unlike internal energy, heat and work are properties only of the process, not of the state, so they are not state functions. In Example 7.1, for instance, the initial and final states are the same in parts (a) and (b), but the amount of work done is different because the external, opposing pressures are different. We *cannot* write $\Delta w = w_f - w_i$ for a change. Work done depends not only on the initial state and final state, but also on how the process is carried out, that is, the amount of work done depends on the path.

The amount of heat transferred depends on the path, too. Suppose that we raise the temperature of 100.0 g of water at 1 bar from 20.0°C to 30.0°C. We cannot know the amount of heat transferred to the water without knowing how the heat was transferred. If we heated the water using a Bunsen burner or an immersion heater, for example, the amount of heat transferred would be 4184 J.[3]

Alternatively, we can stir the water with a magnetic stirring bar until the desired temperature is reached as a result of friction. The heat transferred in this case is zero. On the other hand, if we first raised the temperature of the water from 20°C to 25°C by direct heating, and then used the friction of the stirring bar to bring it up to 30°C, then q would be somewhere between zero and 4184 J. This simple example shows that the heat associated with a given process, like work, depends on how the process is carried out.

Although neither heat nor work is a state function, their sum ($q + w$) is equal to ΔU and the internal energy U is a state function. Thus, if changing the path from the initial state to the final state increases the value of q, it will decrease the value of w by the same amount, and vice versa, because ΔU remains the same for both paths. In summary, heat and work are not state functions because they are not properties of a system. They manifest themselves only during a process (during a change). As a result, their values depend on the path of the process and vary accordingly.

Systems and Processes

To analyze energy changes associated with chemical reactions, we must first distinguish between the system and the surroundings. The **system** is the *specific part of the universe that is of interest to us.* For chemists, the system usually includes substances involved in chemical and physical changes. In an acid-base neutralization experiment, for example, the system may be a beaker containing 50 mL of 0.1 *M* HCl to which 50 mL of 0.1 *M* NaOH is added. The **surroundings** are the *rest of the universe outside the system.* In general, in thermodynamics we study the properties of a system in an *equilibrium state* or the processes that connect equilibrium states. A system is

---

3. The heat transferred to water is $q = m\ C_s\ \Delta T$, where m is the mass of water in grams, C_s the specific heat capacity of water (4.184 J/g °C), and ΔT the temperature change. Thus, $q = (100.0\ \text{g})(4.184\ \text{J g}^{-1}\ {}^{\circ}\text{C}^{-1})(10°\text{C}) = 4184$ J. Specific heat capacities will be discussed in Section 7.3.

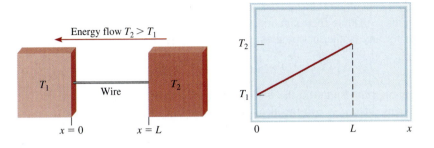

Figure 7.5 A system in steady state, but not in equilibrium. A metal wire of length L is connected at one end to a large metal block with temperature T_1 and on the other to a similar large metal block with fixed temperature $T_2 > T_1$. The steady state temperature as a function of distance (x) along the wire is shown on the right. The system is not in equilibrium because the temperature profile is maintained by the continual transfer of heat through the wire from the hot block to the colder block.

considered to be in an **equilibrium state** *if all macroscopic properties of the system are unchanging in time and remain so even if the system is disconnected from its surroundings.* The latter condition is necessary to distinguish an equilibrium system from one in *steady state*. A system in **steady state** has properties that are unchanging with time, but it requires a continual inflow and outflow of energy and/or matter to remain unchanging. If a system in steady state is disconnected from the surroundings, the steady state is disrupted. A system that is in steady state, but not in equilibrium, is shown in Figure 7.5.

Systems can be classified by the manner in which they exchange energy and matter with the surroundings. An **open system** *can exchange mass and energy, usually in the form of heat, with its surroundings.* For example, the open container of water shown in Figure 7.6(a) is an open system because it can exchange mass (in the form of water vapor) and energy with its surroundings. If we close the flask, as in Figure 7.6(b), so that no water vapor can escape from or condense into the container, we obtain a **closed system,** which *allows the transfer of energy (work or heat) but not mass.* By placing the water in a totally insulated container, we can construct an **isolated system,** which *does not allow the transfer of either mass or energy,* as shown in Figure 7.6(c). For any process in an isolated system, the work w, heat q, and change in internal energy ΔU must all be zero.

Just as we can classify systems, we can also classify processes in terms of the conditions under which they take place. Processes that take place under conditions of constant temperature, such as the ones described in Example 7.1, are called **isothermal processes.** As discussed in Chapter 5, the temperature of a gas can be viewed as a measure of the average kinetic energy of the molecules making up the gas. For gases that are well described by the ideal gas model, the internal energy U depends only

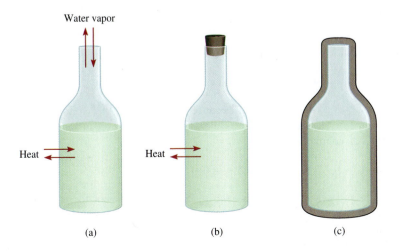

(a)   (b)   (c)

Figure 7.6 Three systems represented by water in a flask: (a) an open system, which allows the exchange of both energy and mass with surroundings; (b) a closed system, which allows the exchange of energy but not mass; and (c) an isolated system, which allows neither energy nor mass to be exchanged (here the flask is enclosed by a vacuum jacket).

upon the temperature (kinetic energy) because ideal gas molecules do not interact with one another (making their potential energy zero). Thus, the change in internal energy is zero for any isothermal process involving only ideal gases:

$$\Delta U = 0 \quad \text{(ideal gas, isothermal process)} \qquad \text{(7.5)}$$

Combining this result with the first law of thermodynamics ($\Delta U = w + q$, Equation 7.4) means that $w = -q$ for ideal gases undergoing an isothermal process.

Constant-pressure (or *isobaric*) **processes** occur under conditions where $\Delta P = 0$. They are common in chemistry because reactions are often run in vessels that are open to the atmosphere, so the reaction proceeds at a constant pressure equal to that of the ambient atmospheric pressure. **Constant-volume** (or *isochoric*) **processes,** on the other hand, occur under conditions where $\Delta V = 0$. If only expansion work is possible, the work done in a constant-volume process is zero (because $w = -P_{ext}\Delta V = 0$). Combining this result with the first law of thermodynamics yields

$$q_V = \Delta U \qquad \text{(constant volume, expansion work only)} \qquad \text{(7.6)}$$

where the subscript "V" indicates a constant-volume process. Thus, the heat evolved by a reaction under constant-volume conditions equals the change in internal energy for that reaction.

If the system is thermally isolated to prevent energy flow in the form of heat (but possibly allowing for work to be done on the system), then any process occurring within this system will have $q = 0$. Any process with $q = 0$ is called an **adiabatic process.**

Reversible and Irreversible Processes

Processes can also be *reversible* or *irreversible*. To illustrate the difference between them, consider a sample of gas placed in a cylinder fitted with a weightless and frictionless piston. Placed on top of the piston are five weights that, together with the atmospheric pressure, exert a total initial external pressure $P_{ext,5}$ on the gas [Figure 7.7(a)]. The subscript "5" indicates the number of weights present. When the system is in equilibrium, the pressure (P) of the gas will equal $P_{ext,5}$. If one of the weights is removed [Figure 7.7(b)], the external pressure is decreased to a new value, $P_{ext,4}$. Because the pressure of the gas is greater than the external pressure the gas will expand until $P = P_{ext,4}$. The external pressure is constant during this expansion, so the work done on the gas can be calculated using Equation 7.1:

$$w(5 \rightarrow 4) = -P_{ext,4}\,\Delta V_{expansion}$$

If we then return the removed weight, the gas will contract to its original volume under a constant external pressure $P_{ext,5}$. The work done in this process is

$$w(4 \rightarrow 5) = -P_{ext,4}\Delta V_{compression} = P_{ext,5}\,\Delta V_{expansion}$$

because $\Delta V_{compression} = -\Delta V_{expansion}$. Because $P_{ext,5} > P_{ext,4}$, the work done *by* the gas in the expansion, $-w(5 \rightarrow 4)$, is less than the work required to recompress the gas to the original volume, $w(4 \rightarrow 5)$:

$$-w(5 \rightarrow 4) < w(4 \rightarrow 5)$$

This process is **irreversible** because the reverse process is not the exact opposite of the forward process. This is because the gas is out of equilibrium with the surroundings (that is, $P_{ext} \neq P$) during the expansion (compression).

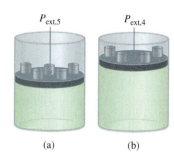

$P_{ext,5}$ $P_{ext,4}$

(a)   (b)

Figure 7.7 (a) A sample of gas placed in a cylinder fitted with a frictionless, massless piston. The gas is compressed under an external pressure ($P_{ext,5}$) exerted by the combination of atmospheric pressure and five weights. (b) Same as in (a) with one of the weights removed. The gas expands because the external pressure is lowered to $P_{ext,4}$.

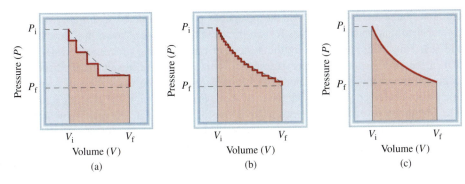

Figure 7.8 (a) The *P-V* curve for the expansion of the gas in Figure 7.7(a) from an initial volume V_i to a final volume V_f by the successive removal of the five weights. (b) The *P-V* curve for the expansion of the gas in Figure 7.7(a), except that the five weights are replaced with twenty smaller weights of equal total mass. (c) The *P-V* curve for the reversible expansion of the gas in Figure 7.7(a) to a final volume V_f. For comparison, the curve in (c) is reproduced as a dashed line in (a) and (b).

Next, consider removing each of the five weights in succession. The total work done by the system in the expansion can be calculated as the area under the pressure-volume (*P-V*) curve shown in Figure 7.8(a). Now suppose we replace the 5 original weights with 20 weights such that the total mass of the weights is unchanged. Removing each of these weights in succession produces the *P-V* curve shown in Figure 7.8(b). Because the two processes shown in Figure 7.8(a) and (b) are performed at constant temperature, and the initial and final pressures are the same for both, the initial and final volumes will be the same. However, comparing the areas under the *P-V* curves shows that the work done by the gas is greater for slower process (removing 20 weights) than it is for the faster process (removing 5 weights). If we then let the number of weights approach infinity (while keeping the total mass fixed), the volume change upon the removal of each weight will be infinitesimal (that is $\Delta V \longrightarrow dV$). Also, at each step the pressure of the gas will be only infinitesimally greater than the external pressure ($P - P_{ext} = dP$), so the work can be calculated using the following integral:

$$w = -\int_{V_i}^{V_f} P_{ext}\, dV = -\int_{V_i}^{V_f}(P - dP)dV$$

where V_i and V_f are the initial and final volumes of the gas, respectively. Because dP and dV are both infinitesimal quantities, their product $dP\, dV$ is negligible, and we can write

$$w = -\int_{V_i}^{V_f} P\, dV$$

Thus, the work done by the gas in the expansion will be the area under the smooth curve shown in Figure 7.8(c). Because the same arguments apply for the reverse process (compression from V_f to V_i), the work done by the gas in the expansion will be the same as the work required to compress the gas back to its original volume. Under these conditions, the expansion of the gas is **reversible** because *it proceeds so slowly and smoothly that the system always remains infinitesimally close to equilibrium with itself and its surroundings.*

In reality, a reversible process would take an infinite amount of time to complete, so no process is ever truly reversible. Any *real* process will be irreversible to some degree. The concept of a reversible process is useful, however, because simple formulas for calculating changes in thermodynamic quantities often do not exist for "real" processes, but do exist for reversible processes. We can perform the calculation assuming a reversible process as long as the initial and final states are the same as for the actual process. As long as the quantity of interest is a state function, we are guaranteed to get the same answer for the reversible path as for the actual process. Additionally, the area under the *P-V* curve for the reversible process depicted in Figure 7.8(c) is greater than that under the curves for the irreversible processes shown in Figure 7.8(a) and (b). Thus, the work done by the system in the reversible process is greater than that of the two irreversible processes shown. This is a general result that is discussed in greater detail in Chapter 8. For now, keep in mind that the work done *by* a system in a process between given initial and final states is a *maximum* if the process is reversible. Therefore, reversible processes make it possible to calculate the maximum amount of work that can be extracted from a process. This quantity is important in estimating the efficiency of chemical and biological processes, as we shall see in Chapter 8.

As an example of a calculation involving a reversible process, consider the isothermal, reversible expansion of *n* moles of an ideal gas from a volume V_i to a volume V_f. Because the temperature *T* is fixed, the pressure will vary, and we need to use Equation 7.3 to do the calculation:

$$w = -\int_{V_i}^{V_f} P_{ext}\, dV$$

For an arbitrary process, we would need to know the exact P_{ext} that had been applied; however, if the process is known to be reversible, then $P_{ext} = P$, where *P* is the pressure of the system (the gas). For an ideal gas, we can use the ideal gas equation of state (Equation 5.6) to express *P* as a function of *n*, *T*, and *V*:

$$P = \frac{nRT}{V}$$

Using this equation in Equation 7.3 gives

$$w = -\int_{V_i}^{V_f} P_{ext}\, dV = -\int_{V_i}^{V_f} P\, dV \quad \text{(reversible process)}$$

$$= -\int_{V_i}^{V_f} \frac{nRT}{V} dV = -nRT \int_{V_i}^{V_f} \frac{dV}{V}$$

$$= -nRT \ln V \big|_{V_i}^{V_f} = -nRT\big[\ln V_f - \ln V_i\big]$$

and

$$w = -nRT \ln \frac{V_f}{V_i} \quad \text{(ideal gas, reversible-isothermal process)} \qquad \textbf{(7.7)}$$

Example 7.2 illustrates the use of Equation 7.7 for calculating the work done in the isothermal, reversible expansion of an ideal gas. Because $q = -w$ for an ideal gas, Equation 7.7 can also be written as follows:

$$q = +nRT \ln \frac{V_f}{V_i} \quad \text{(ideal gas, reversible-isothermal process)} \qquad \textbf{(7.8)}$$

This procedure can be used to calculate the work done in a reversible process for real gases, as well, by using the appropriate equation of state—such as the van der Waals equation (Equation 5.45) or the virial equation (Equation 5.47)—to determine the pressure–volume relationship.

Example 7.2

Calculate the work, heat, and change in internal energy involved in the expansion of 3.50 moles of CO_2 gas from 20.0 L to 40.0 L in an isothermal process at $T = 400$ K. Assume ideal gas behavior.

Strategy Because this is an isothermal process on an ideal gas, we know from Equation 7.5 that the change in internal energy is zero ($\Delta U = 0$). We can use Equation 7.7 to calculate the work, and the relationship $q = -w$ to calculate the heat. What value of R should you use in Equation 7.7?

Solution Equation 7.7 with $V_1 = 20.0$ L and $V_2 = 40.0$ L gives

$$w = -nRT \ln \frac{V_2}{V_1}$$

$$= -(3.50 \text{ mol})(8.314 \text{ J mol}^{-1}\text{K}^{-1})(400 \text{ K}) \ln \frac{40.0 \text{ L}}{20.0 \text{ L}}$$

$$= -8.07 \times 10^3 \text{ J}$$

$$= -8.07 \text{ kJ}$$

The heat absorbed by the system is then

$$q = -w = +8.07 \text{ kJ}$$

Because q is positive, heat is absorbed from the surroundings and this process is endothermic. As mentioned earlier, this is an ideal gas undergoing an isothermal process, so $\Delta U = 0$.

Check The sign of w is negative, indicating that net work is being done by the system, which is consistent with an expansion.

Comment Suppose the expansion were done irreversibly against a constant pressure of 1 bar (which is always smaller than the internal pressure of this sample of gas during the expansion). In this case, the work would be $w = -P \Delta V = -(1 \text{ bar})$ (40 L − 20 L) = −20 L bar = −2 kJ. Thus, the amount of work done by the system in the reversible process (+8.07 kJ) is greater than that done in the irreversible process (+2 kJ), as predicted.

Practice Exercise Calculate the work, heat, and internal energy change (in kJ) associated with the reversible and isothermal compression of 10.0 g of argon gas at 298 K from an initial volume of 12.0 L to a final volume of 2.0 L. Assume ideal gas behavior.

7.2 | The Energy Absorbed by a System as Heat in a Constant-Pressure Process Is Equal to the Change in Enthalpy

Constant-volume conditions are often inconvenient and sometimes impossible to achieve. Most reactions occur under constant pressure (isobaric) conditions (usually at atmospheric pressure). If such a reaction results in a net increase in the number of moles of gas, then the system does work on the surroundings (expansion). Work is done (expansion occurs) because the gas formed must push back the surrounding air in order to enter the atmosphere. Conversely, if more gas molecules are consumed than are produced, work is done on the system by the surroundings (compression).

Finally, no work is done if there is no net change in the number of moles of gases from reactants to products.[4]

Enthalpy

In general, for a constant-pressure process we write

$$\Delta U = q + w$$
$$= q_P - P\Delta V$$

or

$$q_P = \Delta U + P\Delta V \qquad (7.9)$$

where the subscript "P" denotes constant-pressure conditions.

We now introduce a new thermodynamic function called **_enthalpy_** (H) which is defined by the equation

$$H = U + PV \qquad (7.10)$$

where U is the internal energy of the system and P and V are the pressure and volume of the system, respectively. Because U and PV have units of energy, enthalpy also has units of energy. Furthermore, U, P, and V are all state functions, so the changes in $(U + PV)$ depend only on their initial and final states. Therefore, enthalpy is also a state function and the change in H (that is, ΔH) depends only on the initial and final states.

For any process, the change in enthalpy according to Equation 7.10 is given by

$$\Delta H = \Delta U + \Delta(PV) \qquad (7.11)$$

If the pressure is held constant, then

$$\Delta H = \Delta U + P\Delta V \qquad (7.12)$$

Comparing Equation 7.11 with Equation 7.9, we see that for a constant-pressure process,

$$q_P = \Delta H \qquad \text{(constant pressure, expansion work only)} \qquad (7.13)$$

Although q is not a state function, the heat change at constant pressure is equal to H because the "path" is defined and therefore it can have only a specific value.

We now have two quantities—ΔU and ΔH—that can be associated with a reaction. If the reaction occurs under constant-volume conditions, then the heat change, q_V, is equal to ΔU. When the reaction is carried out at constant pressure, though, the heat change, q_P, is equal to ΔH.

For a fixed sample of ideal gases, both U and PV depend only on temperature. As a result, the enthalpy of the gas (calculated using Equation 7.10) will also depend only upon temperature. Like changes in internal energy (Equation 7.5), the change in enthalpy for an ideal gas undergoing an isothermal process is identically zero:

$$\Delta H = 0 \qquad \text{(ideal gas, isothermal process)} \qquad (7.14)$$

---

4. This statement is only strictly true for gases behaving ideally because the molar volume of an ideal gas at a given pressure and temperature is the same for reactant and product gases. For real gases (and liquids and solids), the molar volume of reactant and product species will vary due to interparticle interactions and the volume of the system may change even if the number of reactant and product molecules in the reaction are equal. However, volume changes (and the subsequent work done) in such situations are generally quite small and can often be neglected.

Enthalpy of Reaction

Because most reactions occur under constant-pressure conditions, we can equate the heat change in these cases with the change in enthalpy. For any reaction of the type

$$\text{reactants} \longrightarrow \text{products}$$

the change in enthalpy, called the **enthalpy of reaction** (ΔH), is *the difference between the enthalpies of the products and the enthalpies of the reactants:*

$$\Delta H = H(\text{products}) - H(\text{reactants}) \qquad \textbf{(7.15)}$$

The enthalpy of reaction can be positive or negative, depending on the process. For an *endothermic* reaction (where heat is absorbed by the system from the surroundings), ΔH is positive (that is, $\Delta H > 0$). For an *exothermic* reaction (where heat is released by the system to the surroundings), ΔH is negative (that is, $\Delta H < 0$).

The change in enthalpy for a reaction is analogous to the change in the balance of your bank account as you make deposits and withdrawals. After a transaction (deposit or withdrawal), the change in your bank balance, ΔX, is given by

$$\Delta X = X_f - X_i$$

where X_i and X_f represent the initial and final bank balances, respectively. If your initial balance is \$100 and you deposit \$80 into your account, then $\Delta X = \$180 - \$100 = \$80$. A deposit corresponds, therefore, to an endothermic reaction. (The balance increases and so does the enthalpy of the system.) On the other hand, a withdrawal of \$60 means $\Delta X = \$40 - \$100 = -\$60$. The negative sign for ΔX means your account balance has decreased. Similarly, a negative value of ΔH reflects a decrease in enthalpy of the system as a result of an exothermic process. The difference between this analogy and Equation 7.15 is that, while you always know your exact bank balance, there is no way to know the enthalpies of individual products and reactants. In practice, we can only measure the *difference* in their values. Now let us apply the idea of enthalpy changes to two common processes, the first involving a physical change, the second a chemical change.

Thermochemical Equations

At 0°C and a pressure of 1 atm, ice melts to form liquid water. Measurements show that for every mole of ice converted to liquid water under these conditions, 6.01 kJ of heat energy are absorbed by the system (the ice). Because the pressure is constant, the heat change is equal to the enthalpy change, ΔH. Furthermore, this is an endothermic process ($\Delta H > 0$) because the ice must absorb energy to melt [Figure 7.9(a)]. The equation for this physical change is

$$H_2O(s) \longrightarrow H_2O(l) \qquad \Delta H = 6.01 \text{ kJ mol}^{-1}$$

The value of ΔH is *per mole of the reaction (or process) as it is written*—that is, when 1 mole of ice is converted to 1 mole of liquid water.

The combustion of methane (CH_4), the principal component of natural gas, is an exothermic process ($\Delta H < 0$) because it releases heat to the surroundings:

$$CH_4(g) + 2O_2(g) \longrightarrow CO_2(g) + 2H_2O(l) \qquad \Delta H = -890.4 \text{ kJ mol}^{-1}$$

Under constant-pressure conditions this heat change is equal to the enthalpy change [Figure 7.9(b)]. The value of ΔH given in the balanced chemical equation means that

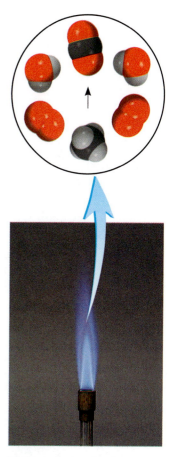

Methane gas burning from a Bunsen burner.

Figure 7.9 (a) Melting 1 mole of ice at 0°C (an endothermic process) results in an enthalpy increase in the system of 6.01 kJ. (b) Burning 1 mole of methane in oxygen gas (an exothermic process) results in an enthalpy decrease in the system of 890.4 kJ.

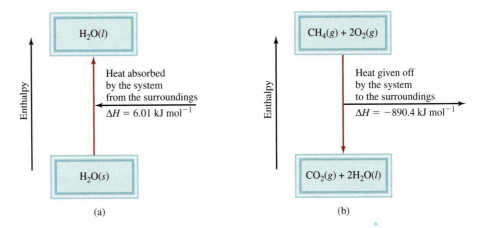

890.4 kJ of heat are released when 1 mole of CH_4 reacts with 2 moles of O_2 to yield 1 mole of CO_2 and 2 moles of liquid H_2O.

The equations for the melting of ice and the combustion of methane are called **thermochemical equations** because they *show the enthalpy changes as well as the molar relationships*. A correct thermochemical equation is always balanced, and the enthalpy change is always specified on a per-mole basis. The following guidelines are helpful in writing and interpreting thermochemical equations:

1. Always specify the physical states of all reactants and products in a thermochemical equation because they affect the enthalpy changes. In the combustion of methane, for example, if we specify water vapor rather than liquid water as a product,

$$CH_4(g) + 2O_2(g) \longrightarrow CO_2(g) + 2H_2O(g) \qquad \Delta H = -802.4 \text{ kJ mol}^{-1}$$

the enthalpy change is -802.4 kJ mol^{-1} rather than -890.4 kJ mol^{-1}, because 88.0 kJ are needed to convert 2 moles of liquid water to water vapor:

$$2H_2O(l) \longrightarrow 2H_2O(g) \qquad \Delta H = 88.0 \text{ kJ mol}^{-1}$$

2. If we multiply or divide both sides of a thermochemical equation by a factor n, then ΔH must also change by the same factor. Thus, if $n = 2$ for the melting of ice, the resulting thermochemical equation is

$$2H_2O(s) \longrightarrow 2H_2O(l) \qquad \Delta H = 2(6.01 \text{ kJ mol}^{-1}) = 12.0 \text{ kJ mol}^{-1}$$

The factor of 2 difference is necessary because enthalpy is an extensive quantity.

3. When we reverse a chemical equation, we change reactants to products and products to reactants. The magnitude of ΔH for the equation remains the same, but its sign changes. Thus, an endothermic reaction becomes exothermic and an exothermic reaction becomes endothermic, as shown by reversing the thermochemical equations for the melting of ice and the combustion of methane:

$$H_2O(l) \longrightarrow H_2O(s) \qquad \Delta H = -6.01 \text{ kJ mol}^{-1}$$
$$CO_2(g) + 2H_2O(l) \longrightarrow CH_4(g) + 2O_2(g) \qquad \Delta H = +890.4 \text{ kJ mol}^{-1}$$

Example 7.3

During the manufacture of sulfuric acid, SO_2 is oxidized to SO_3 according to the following thermochemical equation:

$$SO_2(g) + \frac{1}{2}O_2(g) \longrightarrow SO_3(g) \qquad \Delta H = -99.1 \text{ kJ mol}^{-1}$$

Calculate the heat released when 74.6 g of SO_2 (molar mass = 64.07 g mol^{-1}) is converted to SO_3 under constant-pressure (1 bar) conditions.

Strategy Under constant-pressure conditions, the heat absorbed (q_P) per mole of a reaction is given by ΔH. According to the thermochemical equation, 99.1 kJ of heat is given off (note the negative sign) for every mole of SO_2 burned. To calculate the total amount of heat produced we first need to use the mass of SO_2 given in the problem statement and the molar mass of SO_2 to calculate the number of moles of SO_2 reacted.

Solution Beginning with the balanced thermochemical equation and 74.6 g of SO_2, the amount of heat evolved can be calculated from the following sequence of conversions:

$$\text{grams of } SO_2 \longrightarrow \text{moles of } SO_2 \longrightarrow \text{kJ of heat generated}$$

Therefore, the heat *absorbed* by the reaction is given by

$$q_P = 74.6 \text{ g } SO_2 \times \frac{1 \text{ mol } SO_2}{64.07 \text{ g } SO_2} \times \frac{-99.1 \text{ kJ}}{1 \text{ mol } SO_2} = -115 \text{ kJ}$$

The heat *released* by this exothermic reaction is then $+115$ kJ.

Check Because 74.6 g is greater than the molar mass of SO_2, we expect the heat released to be larger than 99.1 kJ.

Practice Exercise Calculate the heat released when 266 g of white phosphorus (P_4) burns in air (at a constant pressure of 1 bar) according to the equation

$$P_4(s) + 5O_2(g) \longrightarrow P_4O_{10}(s) \qquad \Delta H = -3013 \text{ kJ mol}^{-1}$$

A Comparison of ΔH and ΔU

What is the relationship between ΔH and ΔU for a process? To find out, consider the reaction between sodium metal and water:

$$2Na(s) + 2H_2O(l) \longrightarrow 2NaOH(aq) + H_2(g) \qquad \Delta H = -367.5 \text{ kJ mol}^{-1}$$

According to this thermochemical equation, 2 moles of sodium react with 2 moles of water under constant-pressure conditions to produce 2 moles of aqueous NaOH, 1 mole of hydrogen gas, and 367.5 kJ of heat. Some of the energy produced by the reaction is used to do the work of pushing back a volume of air (V) against atmospheric pressure (P) (Figure 7.10), so the hydrogen gas can enter the atmosphere.

To calculate the change in internal energy, we rearrange Equation 7.11 as follows:

$$\Delta U = \Delta H - \Delta(PV)$$

If we assume the temperature is 25°C and ignore the small change in the volume of the solution, we can use the ideal gas equation to show that the volume of 1 mole of

Sodium reacting with water to form hydrogen gas.

Figure 7.10 (a) A beaker of water inside a cylinder fitted with a movable piston. The pressure inside is equal to the atmospheric pressure. (b) As the sodium metal reacts with water, hydrogen gas pushes the piston upward (doing work on the surroundings) until the pressure inside is again equal to that outside.

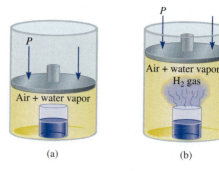

(a)   (b)

H_2 gas at 1.0 bar and 298 K is 24.8 L, so then $\Delta(PV) = 24.8$ L bar or 2.48 kJ. Finally,

$$\Delta U = -367.5 \text{ kJ mol}^{-1} - 2.48 \text{ kJ mol}^{-1} = -370.0 \text{ kJ mol}^{-1}$$

Based on this calculation, ΔU and ΔH are approximately the same for this reaction. ΔH is slightly larger (less negative) than ΔU because some of the internal energy released is used to do gas expansion work, so less heat is evolved. For reactions that do not involve gases, ΔV is usually very small and so ΔU is practically the same as ΔH.

Another way to calculate the internal energy change of a gaseous reaction is to assume ideal gas behavior and constant temperature. In this case,

$$\Delta U = \Delta H - \Delta(PV)$$
$$= \Delta H - \Delta(nRT)$$

$$\Delta U = \Delta H - RT\Delta n \tag{7.16}$$

where Δn is defined as = number of moles of product gases − number of moles of reactant gases for one mole of reaction. Note that, if the number of moles of product gases and the number of moles of reactant gases are equal, as in the reaction

$$H_2(g) + F_2(g) \longrightarrow 2HF(g)$$

the difference between ΔU and ΔH will be very small because $\Delta n = 0$.

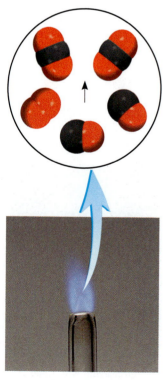

Carbon monoxide burns in air to form carbon dioxide.

Example 7.4

Calculate the change in internal energy when 2 moles of CO are converted to 2 moles of CO_2 at 1 bar and 25°C:

$$2CO(g) + O_2(g) \longrightarrow 2CO_2(g) \qquad \Delta H = -566.0 \text{ kJ mol}^{-1}$$

(Carbon monoxide in auto exhaust is the result of incomplete combusting of gasoline in the engine. The removal of this major pollutant from auto emissions by conversion to CO_2 via this reaction is one of the primary functions of catalytic converters in automobiles.)

Strategy We are given the molar enthalpy change, ΔH, for a reaction involving gases and are asked to calculate the change in molar internal energy, ΔU. Therefore, we need Equation 7.16. What is the change in the number of moles of gases? ΔH is given in kJ mol^{-1}, so what units should we use for R?

—Continued

Continued—

Solution According to the balanced chemical equation, 3 moles of gases are converted to 2 moles of gases, so

Δn = number of moles of product gas − number of moles of reactant gases
    = 2 − 3 = −1

Using R = 8.314 J K^{-1} mol^{-1} and T = 298 K in Equation 7.16, we write

$$\Delta U = \Delta H - RT\Delta n$$

$$= -566.0 \text{ kJ mol}^{-1} - (8.314 \text{ J mol}^{-1} \text{ K}^{-1})\left(\frac{1 \text{ kJ}}{1000 \text{ J}}\right)(298 \text{ K})(-1 \text{ mol})$$

$$= -563.0 \text{ kJ mol}^{-1}$$

Check The reacting gaseous system undergoes a compression (3 moles to 2 moles), so $\Delta V < 0$. Because $\Delta U = \Delta H - P\,\Delta V$ and ΔV is negative, we expect $\Delta U > \Delta H$, in agreement with our result.

Practice Exercise What is ΔU for the formation of 1 mole of CO from graphite at 1 bar and 25°C?

$$C(\text{graphite}) + \frac{1}{2}O_2(g) \longrightarrow CO(g) \qquad \Delta H = -110.5 \text{ kJ mol}^{-1}$$

7.3 | The Temperature Change of a System Upon Heating Is Governed by Its Heat Capacity

When energy in the form of heat is added to any material, the temperature of the material rises.[5] However, the amount of this temperature rise per unit of input heat energy depends both upon the material and the nature of the process.

Heat Capacities

For any heating process, adding an amount of energy in the form of heat (q) results in a temperature rise (ΔT) that is given by

$$q = C\,\Delta T \qquad (7.17)$$

or

$$C = \frac{q}{\Delta T} \qquad (7.18)$$

where the proportionality constant, C, is called the **heat capacity,** and is usually given in units of J K^{-1}. The heat capacity depends on the physical properties of the material, the amount of material present, and the precise nature of the process. Heat capacity is an extensive quantity, so doubling the size of a system also doubles the heat capacity. The **molar heat capacity** ($\overline{C}$) is the heat capacity per 1 mole of substance

$$\overline{C} = \frac{C}{n} = \frac{q}{n\Delta T} \qquad (7.19)$$

---

5. This is true assuming that no phase transitions are occurring in the system. As we will see in Section 7.6, when heat is added to a system at a phase transition temperature (boiling point, freezing point, etc.), the temperature remains fixed as all the heat is used to convert one phase of the system into another.

Table 7.1	
The Specific Heat Capacities of Some Common Substances at 25°C.	
Substance	Specific Heat ($J\ g^{-1}\ °C^{-1}$)
Al	0.900
Au	0.129
C (graphite)	0.720
C (diamond)	0.502
Cu	0.385
Fe	0.444
Hg	0.139
H_2O	4.184
C_2H_5OH (ethanol)	2.46

where n is the number of moles of substance. The **specific heat capacity** (C_s) is the heat capacity per 1 gram of substance:

$$C_s = \frac{C}{m} = \frac{q}{m\Delta T} \tag{7.20}$$

where m is the mass (in grams) of the substance. The specific heat capacities for several substances are listed in Table 7.1. The molar and specific heat capacities are both intensive properties and have units of $J\ mol^{-1}\ K^{-1}$ and $J\ g^{-1}\ K^{-1}$, respectively.

Unlike energy or enthalpy, the heat capacity of a system can be directly measured. From Equation 7.19, we can readily calculate the value of $\overline{C}$ by adding a known amount of energy in the form of heat to a given amount of substance and measuring the temperature rise. However, the precise value of $\overline{C}$ that we determine also depends upon how the heating process is performed. Although many different conditions can be realized in practice, we shall consider only the two most commonly encountered cases here: constant volume and constant pressure. If the substance is heated at constant volume, then the heat capacity determined using Equation 7.19 is called the **constant-volume heat capacity** (C_V). Similarly, C_P denotes the **constant-pressure heat capacity,** which describes heating under constant-pressure conditions.

Recall from Section 7.1 that the heat energy absorbed by a system in a constant-volume process is equal to the change in internal energy (that is, $\Delta U = q_V$). Hence, the constant-volume heat capacity is given by

$$C_V = \frac{q_V}{\Delta T} = \frac{\Delta U}{\Delta T} \tag{7.21}$$

or

$$\Delta U = C_V \Delta T = n\overline{C}_V \Delta T \quad \text{(constant } V \text{ process)} \tag{7.22}$$

Similarly, the heat energy absorbed by a system in a constant-pressure process is equal to the change in enthalpy—that is, $\Delta H = q_P$ (Equation 7.13). Hence, the constant-pressure heat capacity is given by

$$C_P = \frac{q_P}{\Delta T} = \frac{\Delta H}{\Delta T} \tag{7.23}$$

or

$$\Delta H = C_P \Delta T = n\overline{C}_P \Delta T \quad \text{(constant } P \text{ process)} \tag{7.24}$$

Equations 7.22 and 7.24 are valid as long as the heat capacities can be assumed to be independent of temperature, which is often a good approximation if the temperature change is small, say 50 K or less. For calculations in which the temperature dependence of the heat capacity cannot be ignored, we must use the differential forms of Equations 7.22 and 7.24:

$$dU = C_V\,dT \tag{7.25}$$

and

$$dH = C_P\,dT \tag{7.26}$$

which when integrated give

$$\Delta U = \int_{T_1}^{T_2} C_V \, dT \tag{7.27}$$

and

$$\Delta H = \int_{T_1}^{T_2} C_P \, dT \tag{7.28}$$

respectively. Studies of the temperature dependence of heat capacity for many substances show that it can be well approximated by an equation $C_P = a + bT$, where a and b are constants for a given substance over a particular temperature range. A similar equation also holds for C_V. For the special case of ideal gases, Equations 7.22, 7.24, 7.27, and 7.28 can be used for any process (constant P, constant V, etc.) because both the energy and enthalpy of ideal gases depend only on temperature (Equations 7.5 and 7.14).

In general, C_V and C_P are not equal to each other. In a constant-volume process the expansion work is zero (because $\Delta V = 0$) and all energy goes into changing the temperature of the system. However, heating at constant pressure leads, in most systems, to an expansion of the system ($\Delta V > 0$), so some of the energy input must be used to do work on the surroundings. Therefore, it takes more energy to increase the temperature of a system at constant pressure than at constant volume and we have $C_P > C_V$. For gases, the difference between C_P and C_V is significant because the change in volume on heating can be significant. For solids and liquids, however, the volume change is quite small, so C_P and C_V can usually be assumed to be identical.

For ideal gases, the difference between C_P and C_V can be determined. From the definition of enthalpy (Equation 7.10) we have

$$H = U + PV = U + nRT$$

where we have used the ideal gas equation ($PV = nRT$). For a given change in temperature, ΔT, the change in enthalpy for a fixed amount of ideal gas (n moles) is then

$$\Delta H = \Delta U + \Delta(nRT)$$
$$= \Delta U + nR\Delta T$$

Substituting $\Delta U = C_V \, \Delta T$ (Equation 7.22) and $\Delta H = C_P \Delta T$ into the preceding equation gives

$$C_P \Delta T = C_V \, \Delta T + nR\Delta T$$

Dividing by ΔT and rearranging gives

$$C_P - C_V = nR$$

or

$$\overline{C}_P - \overline{C}_V = R \qquad \text{(for an ideal gas)} \tag{7.29}$$

Therefore, for an ideal gas, the constant-pressure molar heat capacity always exceeds the constant-volume molar heat capacity by exactly R. The values of $\overline{C}_P$ for many substances are given in Appendix 2.

Example 7.5

Calculate ΔU and ΔH for heating of 35.36 g of krypton ($\overline{C}_V = 12.47\ \text{J mol}^{-1}\text{K}^{-1}$) from 300 K to 400 K. Assume ideal gas behavior and that the heat capacities at constant volume and constant pressure are independent of temperature.

Strategy Because we assume that the heat capacities are independent of temperature, we can use Equations 7.22 and 7.24 to calculate ΔH and ΔU, but to do so we need both $\overline{C}_V$ (which is given) and $\overline{C}_P$. It is an ideal gas, so we only need $\overline{C}_V$ because $\overline{C}_P$ can be obtained using Equation 7.29. Note that, because both the energy and enthalpy for ideal gases depend only upon temperature, it is not necessary to know the exact nature of the heating process (constant P, constant V, etc.) to apply Equations 7.22 and 7.24.

Solution The quantity 35.36 g of Kr corresponds to 35.36 g/83.80 g mol^{-1} = 0.4219 mol. Thus, from Equation 7.21 we have

$$\Delta U = n\overline{C}_V\Delta T$$
$$= (0.4219\ \text{mol})(12.47\ \text{J mol}^{-1}\text{K}^{-1})(400\ \text{K} - 300\ \text{K})$$
$$= 526\ \text{J}$$

For ΔH we need $\overline{C}_P$, which from Equation 7.29 is

$$\overline{C}_P = \overline{C}_V + R = 12.47\ \text{J mol}^{-1}\text{K}^{-1} + 8.314\ \text{J mol}^{-1}\text{K}^{-1} = 20.79\ \text{J mol}^{-1}\text{K}^{-1}$$

so using Equation 7.24 gives

$$\Delta H = n\overline{C}_P\Delta T$$
$$= (0.4219\ \text{mol})(20.79\ \text{J mol}^{-1}\text{K}^{-1})(400\ \text{K} - 300\ \text{K})$$
$$= 877\ \text{J}$$

Check The answer is consistent with the fact that the enthalpy change on heating should be larger than energy change because $\overline{C}_P > \overline{C}_V$.

Practice Exercise: A quantity of 10.0 L of carbon dioxide (CO_2) gas, initially at a pressure of 1.00 bar, is heated from 250 K to 300 K. Calculate the values of ΔH and ΔH for this process. Assume ideal-gas behavior and that the heat capacities at constant volume and constant pressure are independent of temperature. For carbon dioxide, $\overline{C}_V = 28.80\ \text{J mol}^{-1}\text{K}^{-1}$.

Calorimetry

In the laboratory, heat changes in physical and chemical processes are measured with a *calorimeter,* a closed container designed specifically for this purpose. Such experimental studies are called **calorimetry,** which *the measurement of heat changes for processes.*

Constant-Volume Calorimetry

The heat of a reaction, such as combustion, is usually measured by placing a known mass of a compound in a steel container called a *constant-volume bomb calorimeter.* The calorimeter is then filled with oxygen at about 30 bar of pressure. The closed bomb is immersed in a known amount of water, as shown in Figure 7.11. The sample is ignited electrically, and the heat produced by the combustion reaction can be calculated accurately by recording the rise in temperature of the water. The heat given off by the sample is absorbed by the water and the bomb. The special design of the

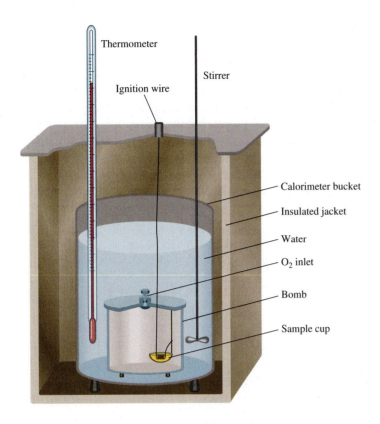

Thermometer

Stirrer

Ignition wire

Calorimeter bucket

Insulated jacket

Water

O_2 inlet

Bomb

Sample cup

Figure 7.11 A constant-volume bomb calorimeter. The bomb is filled with oxygen gas before it is placed in the bucket. The sample is ignited electrically, and the heat produced by the reaction can be accurately determined by measuring the temperature increase in the known amount of surrounding water.

calorimeter enables us to assume that no heat (or mass) is lost to the surroundings during the time it takes to make measurements. Therefore, we can call the bomb and the water in which it is submerged an isolated system. Because no heat enters or leaves the system throughout the process, the heat change of the system (q_{system}) must be zero and we can write

$$q_{\text{system}} = q_{\text{cal}} + q_{\text{rxn}} = 0 \qquad (7.30)$$

where q_{cal} and q_{rxn} are the heat changes for the calorimeter and the reaction, respectively. Thus,

$$q_{\text{rxn}} = -q_{\text{cal}} \qquad (7.31)$$

To calculate q_{cal}, we need to know the heat capacity of the calorimeter (C_{cal}) and the temperature rise, that is,

$$q_{\text{cal}} = C_{\text{cal}} \, \Delta T \qquad (7.32)$$

The quantity C_{cal} is calibrated by burning a substance with an accurately known heat of combustion. For example, it is known that the combustion of 1 g of benzoic acid (C_6H_5COOH) releases 26.42 kJ of heat. If the temperature rise is 4.673°C, then the heat capacity of the calorimeter is given by

$$C_{\text{cal}} = \frac{q_{\text{cal}}}{\Delta T}$$

$$= \frac{26.42 \text{ J}}{4.673\,°C} = 5.654 \text{ kJ } °C^{-1}$$

Once C_{cal} has been determined, the calorimeter can be used to measure the heat of combustion of other substances.

Note that because reactions in a bomb calorimeter occur under constant-volume rather than constant-pressure conditions, the heat changes *do not* correspond to the enthalpy change ΔH (see Section 7.2). It is possible to correct the measured heat changes so that they correspond to H values, but the corrections usually are quite small so we will not concern ourselves with the details here. Finally, it is interesting to note that the energy contents of food and fuel (usually expressed in calories where 1 cal = 4.184 J) are measured with constant-volume calorimeters. (See the inset on page 390.)

Example 7.6

A quantity of 1.435 g of naphthalene ($C_{10}H_8$), a pungent-smelling substance used in moth repellents, was burned in a constant-volume bomb calorimeter. Consequently, the temperature of the water rose from 20.28°C to 25.95°C. If the heat capacity of the bomb plus water was 10.17 kJ °C^{-1}, calculate the heat of combustion of naphthalene on a molar basis, that is, find the molar heat of combustion.

Strategy Knowing the heat capacity and the temperature rise, we can use Equation 7.32 to calculate the heat absorbed by the calorimeter. This gives us the heat generated by the combustion of 1.435 g of naphthalene, which must be converted to molar heat of combustion using the molar mass of naphthalene.

Solution The heat absorbed by the bomb and water is equal to the product of the heat capacity and the temperature change. From Equation 7.32, assuming no heat is lost to the surroundings, we write

$$q_{cal} = C_{cal}\,\Delta T$$
$$= (10.17 \text{ kJ °C}^{-1})(25.95°C - 20.28°C)$$
$$= 57.66 \text{ kJ}$$

Because $q_{sys} = q_{cal} - q_{rxn} = 0$, $q_{cal} = -q_{rxn}$. The heat change of the reaction is -57.66 kJ. This is the heat released by the combustion of 1.435 g of $C_{10}H_8$; therefore, we can write the conversion factor as

$$\frac{-57.66 \text{ kJ}}{1.435 \text{ g } C_{10}H_8}$$

The molar mass of naphthalene is 128.2 g, so the heat of combustion of 1 mole of naphthalene is

$$\text{molar heat of combustion} = \frac{-57.66 \text{ kJ}}{1.435 \text{ g } C_{10}H_8} \times \frac{128.2 \text{ g } C_{10}H_8}{1 \text{ mol } C_{10}H_8}$$
$$= -5.151 \times 10^3 \text{ kJ mol}^{-1}$$

Check Knowing that the combustion reaction is exothermic and that the molar mass of naphthalene is much greater than 1.4 g, is the answer reasonable? Under the reaction conditions, can the heat change (-57.66 kJ) be equated to the enthalpy change of the reaction?

Practice Exercise A quantity of 1.922 g of methanol (CH_3OH) was burned in a constant-volume bomb calorimeter. Consequently, the temperature of the water rose by 4.20°C. If the heat capacity of the bomb plus water was 10.4 kJ °C^{-1}, calculate the molar heat of combustion of methanol.

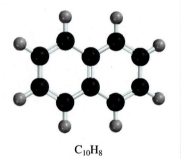

$C_{10}H_8$

Constant-Pressure Calorimetry

A simpler device than the constant-volume calorimeter is the constant-pressure calorimeter, which is used to determine the heat changes for noncombustion reactions. A crude constant-pressure calorimeter can be constructed from two Styrofoam coffee cups, as shown in Figure 7.12. This device measures the heat effects of a variety of reactions, such as acid-base neutralization, as well as the heat of solution and heat of dilution.

Because the pressure is constant, the heat change for the process (q_{rxn}) is equal to the enthalpy change (ΔH). As in the case of a constant-volume calorimeter, we treat the calorimeter as an isolated system. Furthermore, we neglect the small heat capacity of the coffee cups in our calculations. Table 7.2 lists some reactions that have been studied with the constant-pressure calorimeter.

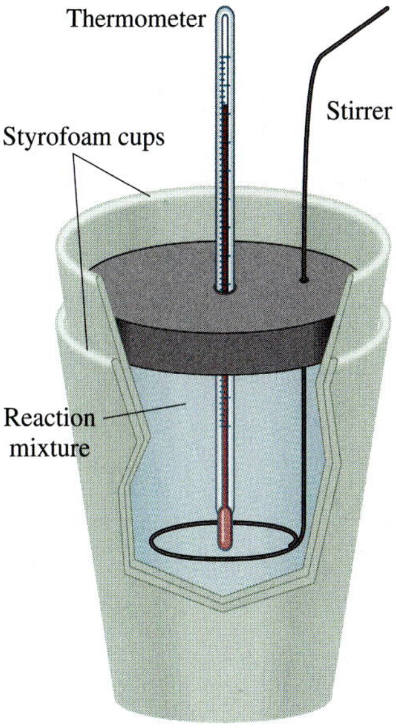

Figure 7.12 A constant-pressure calorimeter made of two Styrofoam coffee cups. The outer cup helps to insulate the reacting mixture from the surroundings. Two solutions of known volume containing the reactants at the same temperature are carefully mixed in the calorimeter. The heat produced or absorbed by the reaction can be determined by measuring the temperature change.

Example 7.7

A lead pellet having a mass of 26.47 g at 89.98°C was placed in a constant-pressure calorimeter of negligible heat capacity containing 100.0 mL of water. The water temperature rose from 22.50°C to 23.17°C. What is the specific heat capacity of the lead pellet?

Strategy We know the masses of water and the lead pellet as well as the initial and final temperatures. Assuming no heat is lost to the surroundings, we can equate the heat lost by the lead pellet to the heat gained by the water. Knowing the specific heat of water, we can then calculate the specific heat of lead.

Solution Treating the calorimeter as an isolated system (no heat lost to the surroundings), we write

$$q_{Pb} + q_{H_2O} = 0$$

or

$$q_{Pb} = -q_{H_2O}$$

The heat gained by the water is given by

$$q_{H_2O} = mC_s \Delta t$$

—Continued

Table 7.2	Heats of Some Typical Reactions Measured at 25°C and Constant Pressure	
Type of Reaction	**Example**	**ΔH (kJ mol^{-1})**
Heat of neutralization	$HCl(aq) + NaOH(aq) \longrightarrow NaCl(aq) + H_2O(l)$	−56.2
Heat of ionization	$H_2O(l) \longrightarrow H^+(aq) + OH^-(aq)$	56.2
Heat of fusion	$H_2O(s) \longrightarrow H_2O(l)$	6.01
Heat of vaporization	$H_2O(l) \longrightarrow H_2O(g)$	44.0*
Heat of reaction	$MgCl_2(s) + 2Na(l) \longrightarrow 2NaCl(s) + Mg(s)$	−180.2

*Measured at 25°C. At 100°C, the value is 40.79 kJ.

Continued—

where m and C_s are the mass and specific heat capacity, respectively, and $\Delta t = t_f - t_i$. Therefore,

$$q_{H_2O} = (100.0 \text{ g})(4.184 \text{ J g}^{-1} \text{ °C}^{-1})(23.17\text{°C} - 22.50\text{°C})$$
$$= 280.3 \text{ J}$$

Because the heat lost by the lead pellet is equal to the heat gained by the water, so $q_{Pb} = -280.3$ J. To determine the specific heat of Pb, we write

$$q_{Pb} = mC_s \Delta t$$

which gives, after rearrangement

$$C_s = \frac{q_{Pb}}{m \, \Delta t}$$
$$= \frac{-280.3 \text{ J}}{(26.47 \text{ g})(23.17\text{°C} - 89.98\text{°C})}$$
$$= 0.158 \text{ J g}^{-1} \text{ °C}^{-1}$$

Practice Exercise A 30.14-g stainless steel ball bearing at 117.82°C is placed in a constant-pressure calorimeter containing 120.0 mL of water at 18.44°C. If the specific heat of the ball bearing is 0.474 J g^{-1} °C^{-1}, calculate the final temperature of the water. Assume the calorimeter to have negligible heat capacity.

Example 7.8

A quantity of 1.00×10^2 mL of 0.500 *M* HCl was mixed with 1.00×10^2 mL of 0.500 *M* NaOH in a constant-pressure calorimeter of negligible heat capacity. The initial temperature of the HCl and NaOH solutions was the same, 22.50°C, and the final temperature of the mixed solution was 25.86°C. Calculate the heat change for the neutralization reaction on a molar basis

$$\text{NaOH}(aq) + \text{HCl}(aq) \longrightarrow \text{NaCl}(aq) + \text{H}_2\text{O}(l)$$

Assume that the densities and specific heats of the solutions are the same as for water (1.00 g mL^{-1} and 4.184 J g^{-1} °C^{-1}, respectively).

Strategy Because the temperature rose, the neutralization reaction is exothermic. Find the total mass of the solution and use Equation 7.20 to determine the heat absorbed given the specific heat. Determine the relationship between the heat absorbed by the solution and the heat of reaction.

Solution Assuming no heat is lost to the surroundings, $q_{sys} = q_{soln} + q_{rxn} = 0$, so $q_{rxn} = -q_{soln}$, where q_{soln} is the heat absorbed by the combined solution. Because the density of the solution is 1.00 g mL^{-1}, the mass of a 100-mL solution is 100 g. To find the heat absorbed by the solution given the specific heat, we use Equation 7.20:

$$q_{soln} = mC_s \Delta T$$
$$= (1.00 \times 10^2 \text{ g} + 1.00 \times 10^2 \text{ g})(4.184 \text{ J g}^{-1} \text{ °C}^{-1})(25.86\text{°C} - 22.50\text{°C})$$
$$= 2.81 \times 10^3 \text{ J} = 2.81 \text{ kJ}$$

Because $q_{rxn} = -q_{soln}$, $q_{rxn} = -2.81$ kJ.

—Continued

Continued—

From the molarities given, the number of moles of both HCl and NaOH in 1.00×10^2 mL solution is

$$\frac{0.500 \text{ mol}}{1 \text{ L}} \times 0.100 \text{ L} = 0.0500 \text{ mol}$$

Therefore, the heat of neutralization when 1.00 mole of HCl reacts with 1.00 mole of NaOH is

$$\text{heat of neutralization} = \frac{-2.81 \text{ kJ}}{0.0500 \text{ mol}} = -56.2 \text{ kJ mol}^{-1}$$

Check Is the sign consistent with the nature of the reaction? Under the reaction condition, can the heat change be equated to the enthalpy change?

Practice Exercise A quantity of 4.00×10^2 mL of 0.600 *M* HNO_3 is mixed with 4.00×10^2 mL of 0.300 *M* $Ba(OH)_2$ in a constant-pressure calorimeter of negligible heat capacity. The initial temperature of both solutions is the same at 18.46°C. What is the final temperature of the solution? (Use the density and specific heat data given in Example 7.8.)

The Molecular Interpretation of Heat Capacities

We defined heat as that part of the energy that is added to the multitudes of molecular variables that we do not explicitly observe in a system (the positions and velocities of individual molecules, for example). The way in which this energy is distributed can be understood in terms of a concept based on classical mechanics and directly applicable to gas behavior.

This concept, called the *equipartition of energy theorem,* states *that the energy in a molecule is, on average, distributed evenly among all types of molecular motion (degrees of freedom).* The number of degrees of freedom in a molecule is the number of variables that are required to describe its motion. For monatomic gases, we require three coordinates (*x, y,* and *z*) to completely define the position of each atom—we say that each atom possesses three translational degrees of freedom. For molecules, on the other hand, additional degrees of freedom are required to describe rotational and vibrational motion. The equipartition of energy theorem is a fundamental result from *statistical mechanics,* the *study of the molecular origin of thermodynamics.*

To see how the equipartition theorem works, first consider the simple case of a monatomic ideal gas (which would be a good model for argon or neon at low pressure). We saw in Section 5.4 that each of the N atoms in this system contributes $3/2\, k_B T$ to the total translational kinetic energy (Equation 5.34), where k_B is Boltzmann's constant. Because the system is ideal and there are no interatomic interactions, the potential energy is zero and the total energy of the gas is

$$U(\text{monatomic}) = \frac{3}{2}Nk_B T = \frac{3}{2}nRT \tag{7.33}$$

where we have used the fact that $R = N_A k_B$, where N_A is Avagadro's number. Thus, if we change the temperature by ΔT, the change in energy is given by

$$\Delta U = \frac{3}{2}Nk_B \Delta T = \frac{3}{2}nR\Delta T \tag{7.34}$$

Fuel Values of Foods and Other Substances

The food we eat is broken down (metabolized) in stages by a group of complex biological molecules called *enzymes*. Most of the energy released at each stage is captured for function and growth. One interesting aspect of metabolism is that the overall change in energy is the same as it is in combustion. For example, the total enthalpy change for the conversion of glucose ($C_6H_{12}O_6$) to carbon dioxide and water is the same whether we burn the substance in air or digest it in our bodies:

$$C_6H_{12}O_6(s) + 6O_2(g) \longrightarrow 6CO_2(g) + 6H_2O(l)$$
$$\Delta H = -2801 \text{ kJ mol}^{-1}$$

The important difference between metabolism and combustion is that the latter is usually a one-step, high-temperature process. Consequently, much of the energy released by combustion is lost to the surroundings.

Various foods have different compositions and hence different energy contents. The energy content of food is generally measured in calories. The *calorie (cal)* is a non-SI unit of energy that is defined as the amount of energy required to heat 1 g of liquid water by 1°C and is equivalent to 4.184 J. In the context of nutrition, however, the calorie we speak of (sometimes called a *nutritional calorie* or "*big calorie*") is actually equal to a *kilocalorie*, that is,

$$1 \text{ Cal} = 1000 \text{ cal} = 4184 \text{ J}$$

Note the use of a capital "C" to represent the "big calorie." The bomb calorimeter described in Section 7.3 is ideally suited for measuring the energy content, or "fuel value," of foods. Fuel values are just the enthalpies of combustion (see the table). In order to be analyzed in a bomb calorimeter, food must be dried first because most foods contain a considerable amount of water. Because the composition of particular foods is often not known, fuel values are expressed in terms of kJ g^{-1} rather than kJ mol^{-1}.

Fuel Values of Foods and Some Common Fuels

Substance	$\Delta H_{combustion}$ (kJ g^{-1})
Apple	−2
Beef	−8
Beer	−1.5
Bread	−11
Butter	−34
Cheese	−18
Eggs	−6
Milk	−3
Potatoes	−3
Charcoal	−35
Coal	−30
Gasoline	−34
Kerosene	−37
Natural Gas	−50
Wood	−20

Nutrition Facts

Serving Size 6 cookies (28g)
Servings Per Container about 11

Amount Per Serving

Calories 120   Calories from Fat 30

	% Daily Value*
Total Fat 4g	**6%**
Saturated Fat 0.5g	**4%**
Polyunsaturated Fat 0g	
Monounsaturated Fat 1g	
Cholesterol 5mg	**2%**
Sodium 105mg	**4%**
Total Carbohydrate 20g	**7%**
Dietary Fiber Less than 1gram	**2%**
Sugars 7g	
Protein 2g	

The labels on food packages reveal the calorie content of the food inside.

Comparing Equations 7.34 and 7.22 yields

$$\overline{C}_V(\text{monatomic}) = \frac{3}{2}Nk_B = \frac{3}{2}nR \tag{7.35}$$

giving a molar constant-volume heat capacity

$$\overline{C}_V(\text{monatomic}) = \frac{3}{2}R = 12.471 \text{ J mol}^{-1} \text{ K}^{-1} \tag{7.36}$$

using $R = 8.314 \text{ J mol}^{-1} \text{ K}^{-1}$. Each atom in this system has three translational degrees of freedom, so each translational degree of freedom contributes $1/2\ k_BT$ to the total energy and $1/2\ k$ to the constant volume heat capacity. On a molecular level, the heat capacity can be understood as *a measure of the number of degrees of freedom of a system available for the storage of energy*. The equipartition theorem also gives a useful alternative definition of temperature as *a measure of the amount of energy available, on average, in each molecular degree of freedom.*

The equipartition theorem prediction for the constant-pressure heat capacity $\overline{C_P}$ of a monatomic ideal gas can be determined from $\overline{C_V}$ (Equation 7.36) using the relationship between $\overline{C_P}$ and $\overline{C_V}$ given in Equation 7.29.

$$\overline{C_P}(\text{monatomic}) = \overline{C_V} + R = \frac{3}{2}R + R = \frac{5}{2}R = 20.785 \text{ J mol}^{-1} \text{ K}^{-1} \quad \textbf{(7.37)}$$

For molecules containing two or more atoms, other types of motion, such as rotation and vibration, are also present. For a molecule containing M atoms, we require $3M$ coordinates to describe the motion of every atom. Three coordinates are needed to describe the translational motion of the center of mass. The remaining $3M - 3$ coordinates are needed to describe rotation and vibration. For linear molecule, such as CO_2 or HCl, two angles are required to specify the orientation of the molecule relative to two coordinate axes through the center of mass and perpendicular to the bond axes—rotation about the bond axis leaves the molecule unchanged, so it does not constitute a degree of freedom. This leaves $(3M - 5)$ vibrational degrees of freedom for a linear molecule. For example, Figure 7.13 shows the translational, rotational, and vibrational motions of a diatomic molecule.

In contrast, three angles about three mutually perpendicular axes through the center of mass are needed to describe the rotation of a nonlinear molecule, such as H_2O or CH_4, leaving $(3M - 6)$ degrees of freedom for vibration.

The equipartition theorem states that the energy will be distributed equally among the degrees of freedom of the system; therefore, each rotational degree of freedom will contain the same amount of energy $(1/2\ k_BT)$ as each translational degree of freedom. Because the energy of molecular vibration consists of contributions from both kinetic energy (due to vibrational motion) and potential energy (due to bond stretching and compression), the equipartition theorem predicts that each vibrational degree of freedom will contribute k_BT to the energy $(1/2\ k_BT$ each from kinetic and

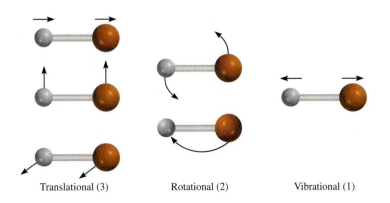

Translational (3)   Rotational (2)   Vibrational (1)

Figure 7.13 The three translational, two rotational, and one vibrational motions of a diatomic molecule, such as HCl.

Table 7.3	Equipartition Theorem Contributions to C_V for Ideal Gases		
Type of Molecule	**Translation**	**Rotation**	**Vibration**
Atom	$\frac{3}{2}R$	—	—
Linear Molecule	$\frac{3}{2}R$	R	$(3M - 5)R$
Nonlinear Molecule	$\frac{3}{2}R$	$\frac{3}{2}R$	$(3M - 6)R$

potential energy). Thus, for an ideal gas containing n moles of nonlinear molecules, the equipartition theorem gives an energy of

$$U(\text{nonlinear}) = \left[\underbrace{\frac{3}{2}}_{\text{translational}} + \underbrace{\frac{3}{2}}_{\text{rotational}} + \underbrace{(3M - 6)}_{\text{vibrational}} \right] nRT \qquad (7.38)$$

corresponding to a molar constant-volume heat capacity of

$$\bar{C}_v(\text{nonlinear}) = \left[\underbrace{\frac{3}{2}}_{\text{translational}} + \underbrace{\frac{3}{2}}_{\text{rotational}} + \underbrace{(3M - 6)}_{\text{vibrational}} \right] R \qquad (7.39)$$

For a linear molecule, we have

$$C_V(\text{linear}) = \left[\underbrace{\frac{3}{2}}_{\text{translational}} + \underbrace{\frac{2}{2}}_{\text{rotational}} + \underbrace{(3M - 5)}_{\text{vibrational}} \right] R \qquad (7.40)$$

The contributions to $\bar{C}_V$ by the translational, rotational, and vibrational degrees of freedom for ideal gases of atoms, linear molecules, and nonlinear molecules are summarized in Table 7.3.

The values of $\bar{C}_P$ for nonlinear and linear molecules can be obtained by adding R to the $\bar{C}_V$ values in Equations 7.39 and 7.40, respectively.

In Table 7.4, we compare, for a number of gases, the values $\bar{C}_P$ predicted from the equipartition theorem with experimentally measured heat capacities. For monatomic gases the agreement between experiment and theory is excellent; however, considerable deviations are observed for molecular gases. To understand the origin of these discrepancies, consider diatomic hydrogen (H_2). For a diatomic molecule, $M = 2$ and Equation 7.40, together with Equation 7.29 ($\bar{C}_P = \bar{C}_V + R$), yields

$$\bar{C}_P(\text{diatomic}) = \left[\underbrace{\frac{3}{2}}_{\text{translational}} + \underbrace{1}_{\text{rotational}} + \underbrace{1}_{\text{vibrational}} \right] R + R = \frac{9}{2}R = 37.41 \text{ J mol}^{-1} \text{ K}^{-1}$$

However, from Table 7.4 we see that the calculated value is larger than the experimental value of 28.85 J mol^{-1} K^{-1}, which is very close to $7/2R$. This would be the

Table 7.4	Calculated and Experimentally Measured Values of $\overline{C}_P$ for Several Gases at 298 K and 1 Bar	
Gas	Calculated $\overline{C}_P$ (J mol^{-1} K^{-1})	Measured $\overline{C}_P$ (J mol^{-1} K^{-1})
He	20.79	20.79
Ne	20.79	20.79
Ar	20.79	20.79
H_2	37.41	28.81
N_2	37.41	29.12
O_2	37.41	29.36
F_2	37.41	31.30
Cl_2	37.41	33.91
Br_2	37.41	36.02
I_2	37.41	36.90
CO_2	62.37	37.13
H_2O	58.18	33.54
SO_2	58.18	39.82

value if one ignored the vibrational contribution. The absence of the vibrational contribution can be understood using quantum mechanics (Chapter 1), which predicts that the translational, rotational, and vibrational energies of a molecule are quantized. The spacing between vibrational energy levels is typically on the order of 10^{-20} J, which is larger than that between typical rotational energy levels (about 10^{-23} J), and very much larger than the typical translational energy-level spacing (about 10^{-34} J). At room temperature (298 K), $k_BT/2$ corresponds to 2.0×10^{-21} J, which is larger than the typical rotational and translational energy-level spacings, but smaller than vibrational energy-level spacings. Thus, at room temperature, energy can easily be distributed into the translational and rotational degrees of freedom, but the average amount of energy available for a single degree of freedom ($k_BT/2$) is below the minimum amount of energy required to excite most vibrational and nearly all electronic degrees of freedom. Thus, these degrees of freedom are not readily available for energy storage and so will not contribute significantly to the heat capacity. At higher temperatures, the available thermal energy, $k_BT/2$, becomes larger, and excitation of vibrational (and even electronic motion) becomes possible, leading to an increase in $\overline{C}_V$ with increasing temperature. For example, at 3000 K, $\overline{C}_P$ for hydrogen increases to 37.00 J mol^{-1} K^{-1}, which is very close to the equipartition prediction of 37.41 J mol^{-1} K^{-1} with vibrational motion fully activated.

The preceding example of H_2 was somewhat extreme because the vibrational frequency of H_2 is very high, so at room temperature there was no contribution from vibrational modes to the heat capacity. In most other molecules, some vibrational frequencies will be low enough so that vibrational motion will contribute at least a small amount to the heat capacity at room temperature; however, this contribution will still represent only a fraction of the total vibrational contribution given in Equations 7.39 and 7.40. This is especially true in polyatomic molecules with low-frequency vibrational modes corresponding to bending motions. In some cases, it is possible to make conclusions as to the relative vibrational frequencies of two molecules simply by looking at the heat capacity. For example, the constant-pressure heat capacities of F_2 and I_2 at 298 K are 31.30 and 36.90 J mol^{-1} K^{-1}, respectively (see Table 7.4). Both of these are linear diatomics, so translation and rotation alone account for $7/2\ R = 29.10$ J mol^{-1} K^{-1}. Thus, the vibrational contributions to $\overline{C}_P$ for F_2 and I_2 are 2.2 and 7.8 J mol^{-1} K^{-1}, respectively. Becase I_2 has a larger vibrational contribution to $\overline{C}_P$ than F_2, we can conclude that the vibrational frequency of I_2 is lower than that of F_2, allowing for easier excitation of vibrational degrees of freedom given the available thermal energy. This conclusion can be verified by IR spectroscopy, which gives the vibrational frequencies for F_2 and I_2 as 919 and 214 cm^{-1}, respectively. This conclusion is interesting because we were able to make a conclusion about a molecular property of a material (vibrational frequency) solely from an examination of a macroscopic property (heat capacity).

In summary, the heat capacity is a measure of the degrees of freedom *available* for energy storage. Because quantum mechanics requires that a minimum amount of energy is needed to excite a given motion, not all degrees of freedom will be available to store energy because the energy spacing corresponding to some degrees of freedom will be larger than the available thermal energy. For gases, we should keep in mind that at room temperature, the dominant contributions to the heat capacity come from translational and rotational motion. At elevated temperatures, vibrational motion must also be taken into account. Only at very high temperatures does electronic motion play a role in determining the $\overline{C}_V$ values.

Example 7.9

The heat capacity for CO_2 is 37.1 J mol^{-1} K^{-1} at 298 K. What percentage of this value is due to vibrational motion? What percentage of the total possible vibrational contribution does this represent?

Strategy Use Table 7.3 to determine the $C_P = C_V + R$ for CO_2 in the absence and presence of the vibrational contribution.

Solution: CO_2 is a linear triatomic ($M = 3$) molecule. The number of possible vibrational degrees of freedom is $3M - 5 = 4$. Using Table 7.3 (or Equation 7.40), we have for translation and rotation:

CO_2

$$\overline{C}_P = \overline{C}_V + R = \left[\frac{3}{2} + 1\right]R + R = \frac{7}{2}R$$
$$= 3.5(8.314 \text{ J mol}^{-1}\text{ K}^{-1}) = 29.1 \text{ J mol}^{-1}\text{ K}^{-1}$$

For translation, rotation, and vibration:

$$\overline{C}_P = \overline{C}_V + R = \left[\frac{3}{2} + 1 + 4\right]R + R = 7.5R$$
$$= 15(8.314 \text{ J mol}^{-1}\text{ K}^{-1}) = 62.4 \text{ J mol}^{-1}\text{ K}^{-1}$$

Ignoring electronic contributions to $\overline{C}_P$, which should be negligible, the vibrational contribution to the heat capacity is the difference between the experimental value and that obtained from the equipartition theorem including only translation and rotation:

$$\text{vibrational contribution to } \overline{C}_P = 37.1 \text{ J mol}^{-1}\text{ K}^{-1} - 29.1 \text{ J mol}^{-1}\text{ K}^{-1}$$
$$= 8.0 \text{ J mol}^{-1}\text{ K}^{-1}$$

$$\text{fraction due to vibration} = \frac{8.0 \text{ J mol}^{-1}\text{ K}^{-1}}{37.1 \text{ J mol}^{-1}\text{ K}^{-1}} = 0.215 = 21.5\%$$

—Continued

Continued—

total possible vibrational contribution $= 4R = 4(8.314 \text{ J mol}^{-1}\text{K}^{-1}) = 33.3 \text{ J mol}^{-1}\text{K}^{-1}$

fraction of total possible vibrational contribution $= \dfrac{8.0 \text{ J mol}^{-1}\text{K}^{-1}}{33.3 \text{ J mol}^{-1}\text{K}^{-1}} = 0.240 = 24.0\%$

So only 24.0% of the total possible vibrational contribution to the heat capacity is available at $T = 298$ K.

Practice Exercise Use the equipartition theorem to predict the heat capacity for ClF using only rotational and translational contributions. Compare this with the experimental value of 32.1 J mol^{-1} K^{-1}. What fraction of this value is due to vibrational motion?

7.4 | The Enthalpy Change for Any Reaction Can Be Calculated Using Standard Enthalpies of Formation

From Equation 7.15, we see that ΔH can be calculated if we know the actual enthalpies of all reactants and products. However, as mentioned earlier, there is no way to measure the *absolute* value of the enthalpy of a substance. Only values *relative* to an arbitrary reference can be determined. This problem is similar to the one geographers face in expressing the elevations of specific mountains or valleys. Rather than trying to devise some type of "absolute" elevation scale (perhaps based on distance from the center of Earth?), by common agreement all geographic heights and depths are expressed relative to sea level, an arbitrary reference with a defined elevation of "zero" meters or feet. Similarly, chemists have agreed on an arbitrary reference point for enthalpy.

Standard Enthalpy of Formation

The reference point for all enthalpy expressions is called the ***standard molar enthalpy of formation*** (ΔH_f°) which is defined as *the heat change that results when 1 mole of a compound in its standard state is formed from its elements in their standard states. The **standard state** of a liquid or solid substance is its most thermodynamically stable pure form at 1 bar pressure.*[6] The standard state for gases is similar, except that standard state gases are assumed to obey the ideal gas law exactly. The standard state for solutes dissolved in solution will be discussed in Chapter 10. In the notation ΔH_f°, the superscript "$\circ$" represents standard-state conditions (1 bar), and the subscript "f" stands for formation. Although the standard state does not specify a temperature, we will assume, unless otherwise stated, ΔH_f° values are measured at 25°C.

Table 7.5 lists the standard enthalpies of formation for a number of elements and compounds. (For a more complete list of ΔH_f° values, see Appendix 2.) By definition, *the standard enthalpy of formation of any element in its most stable allotropic form is zero.* Take the element oxygen as an example. Molecular oxygen (O_2) is more stable than the other allotropic form of oxygen, ozone (O_3), at 1 bar and 25°C. Thus, we can write ΔH_f° (O_2) = 0, but ΔH_f° (O_3) = 142.7 kJ mol^{-1} $\neq 0$. Similarly,

---

6. Many texts still employ the older standard pressure of 1 atm. The change to the current standard pressure of 1 bar was recommended by the International Union of Pure and Applied Chemists (IUPAC) in 1982. Because 1 bar only differs from 1 atm by slightly more than one percent, the values of thermodynamic quantities using the two standards will generally differ by a very small amount that is often negligible.

Table 7.5	Standard Enthalpies of Formation of Some Inorganic Substances at 25°C		
Substance	**ΔH_f° kJ mol^{-1}**	**Substance**	**ΔH_f° kJ mol^{-1}**
$Ag(s)$	0	$H_2O_2(l)$	-187.8
$AgCl(s)$	-127.0	$Hg(l)$	0
$Al(s)$	0	$I_2(s)$	0
$Al_2O_3(s)$	-1675.7	$HI(g)$	25.9
$Br_2(l)$	0	$Mg(s)$	0
$HBr(g)$	-36.2	$MgO(s)$	-601.8
$C(graphite)$	0	$MgCO_3(s)$	-1112.9
$C(diamond)$	1.90	$N_2(g)$	0
$CO(g)$	-110.5	$NH_3(g)$	-45.94
$CO_2(g)$	-393.5	$NO(g)$	90.4
$Ca(s)$	0	$NO_2(g)$	33.85
$CaO(s)$	-634.9	$N_2O_4(g)$	9.66
$CaCO_3(s)$	-1206.9	$N_2O(g)$	81.56
$Cl_2(g)$	0	$O(g)$	249.2
$HCl(g)$	-92.3	$O_2(g)$	0
$Cu(s)$	0	$O_3(g)$	142.7
$CuO(s)$	-155.2	$S(rhombic)$	0
$F_2(g)$	0	$S(monoclinic)$	0.30
$HF(g)$	-273.3	$SO_2(g)$	-296.84
$H(g)$	218.2	$SO_3(g)$	-395.7
$H_2(g)$	0	$H_2S(g)$	-20.6
$H_2O(g)$	-241.83	$ZnO(s)$	-350.5
$H_2O(l)$	-285.83		

Graphite (top) and diamond (bottom).

graphite is a more stable allotropic form of carbon than diamond at 1 bar and 25°C, so we have ΔH_f° (C, graphite) $= 0$ and $\Delta \overline{H}_f^\circ$ (C, diamond) $= 1.9$ kJ mol^{-1} $\neq 0$.

Standard Enthalpy of Reaction

The importance of the standard enthalpies of formation is that once we know their values, we can readily calculate the ***standard enthalpy of reaction*** (ΔH_{rxn}°), defined as *the enthalpy of a reaction carried out such that all reactants and products are in their standard states.* For example, consider the hypothetical reaction

$$a\,A + b\,B \longrightarrow c\,C + d\,D$$

where a, b, c, and d are stoichiometric coefficients. For this reaction, ΔH_{rxn}° is given by

$$\Delta H_{rxn}^\circ = \left[c\,\Delta H_f^\circ + d\,\Delta H_f^\circ \right] - \left[a\,\Delta H_f^\circ + b\,\Delta H_f^\circ \right]$$

We can generalize this equation as:

$$\Delta H_{rxn}^\circ = \sum_{products} v_P\,\Delta H_f^\circ(\text{products}) - \sum_{reactants} v_R\,\Delta H_f^\circ(\text{reactants}) \qquad (7.41)$$

where v_P and v_R denote the stoichiometric coefficients for the reactants and products, respectively, and Σ (sigma) means "the sum of."

The advantage of Equation 7.41 is that, given a table of values for ΔH_f° for a set of compounds, we can calculate ΔH_{rxn}° for *any* possible reaction involving compounds from this set. The ΔH_f° value for a compound can be determined experimentally using either a direct method or an indirect method, depending upon the identity of the compound of interest.

The Direct Method

This method of measuring ΔH_f° works for compounds that can be readily synthesized from their elements. Suppose we want to know the enthalpy of formation of carbon dioxide. We must measure the enthalpy of the reaction when carbon (graphite) and molecular oxygen in their standard states are converted to carbon dioxide in its standard state:

$$C(\text{graphite}) + O_2(g) \longrightarrow CO_2(g) \qquad \Delta H_{rxn}^\circ = -393.5 \text{ kJ mol}^{-1}$$

We know from experience that this combustion easily goes to completion. Thus, from Equation 7.41 we can write

$$\Delta H_{rxn}^\circ = \Delta H_f^\circ(CO_2, g) - \left[\Delta H_f^\circ(C, \text{graphite}) - \Delta H_f^\circ(O_2, g)\right]$$
$$= -393.5 \text{ kJ mol}^{-1}$$

Because both graphite and O_2 are stable allotropic forms of the elements, it follows that both $\Delta H_f^\circ(C, \text{graphite})$ and $\Delta H_f^\circ(O_2, g)$ are zero. Therefore

$$\Delta H_{rxn}^\circ = \Delta \overline{H}_f^\circ(CO_2, g) = -393.5 \text{ kJ mol}^{-1}$$

or $\qquad \Delta \overline{H}_f^\circ(CO_2, g) = -393.5 \text{ kJ mol}^{-1}$

Note that arbitrarily assigning zero $\Delta \overline{H}_f^\circ$ for each element in its most stable form at the standard state does not affect our calculations in any way. Remember, in thermochemistry we are interested only in enthalpy *changes* because they can be determined experimentally whereas the absolute enthalpy values cannot. The choice of a zero "reference level" for enthalpy makes calculations easier to handle. Again referring to the terrestrial altitude analogy, we find that Mt. Everest is 2650 m higher than Mt. McKinley. This difference in altitude is unaffected by the decision to set sea level at 0 m or at 1000 m.

Some other compounds that can be studied by the direct method are SF_6, P_4O_{10}, and CS_2. The equations representing their syntheses are

$$S(\text{rhombic}) + 3F_2(g) \longrightarrow SF_6(g)$$
$$P_4(\text{white}) + 5O_2(g) \longrightarrow P_4O_{10}(s)$$
$$C(\text{graphite}) + 2S(\text{rhombic}) \longrightarrow CS_2(l)$$

Note that S(rhombic) and P(white) are the most stable allotropes of sulfur and phosphorus, respectively, at 1 bar and 25°C, so their ΔH_f° values are zero.

The Indirect Method

Many compounds cannot be directly synthesized from their elements. In some cases, the reaction proceeds too slowly, or side reactions produce substances other than the desired compound. In these cases ΔH_f° can be determined by an indirect approach, which is based on *Hess's law of summation* for extensive thermodynamic quantities,

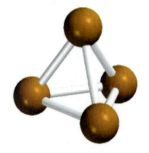

P_4

White phosphorus burns in air to form P_4O_{10}.

or simply **Hess's law.**[7] This law can be stated as follows: *When reactants are converted to products, the change in any extensive thermodynamic state function, such as enthalpy, is the same whether the reaction takes place in one step or in a series of steps.* In other words, if we can break down the reaction of interest into a series of reactions for which ΔH°_{rxn} can be measured, we can calculate ΔH°_{rxn} for the overall reaction. Hess's law is based on the fact that because H is a state function, ΔH depends only on the initial and final state (that is, only on the nature of reactants and products). The enthalpy change would be the same whether the overall reaction takes place in one step or many steps.

An analogy for Hess's law is as follows. Suppose you go from the first floor to the sixth floor of a building by elevator. The gain in your gravitational potential energy (which corresponds to the enthalpy change for the overall process) is the same whether you go directly there or stop at each floor on your way up (breaking the trip into a series of steps).

Let's say we are interested in the standard enthalpy of formation of carbon monoxide (CO). We might represent the reaction as

$$C(graphite) + \frac{1}{2}O_2(g) \longrightarrow CO(g)$$

However, burning graphite also produces some carbon dioxide (CO_2), so we cannot measure the enthalpy change for CO directly as shown. Instead, we must employ an indirect route, based on Hess's law. It is possible to carry out the following two-step reactions, which do go to completion:

(a) $C(graphite) + O_2(g) \longrightarrow CO_2(g)$ $\Delta H^{\circ}_{rxn} = -393.5 \text{ kJ mol}^{-1}$

(b) $CO(g) + \frac{1}{2}O_2(g) \longrightarrow CO_2(g)$ $\Delta H^{\circ}_{rxn} = -283.0 \text{ kJ mol}^{-1}$

First, we reverse reaction (b) to get

(c) $CO_2(g) \longrightarrow CO(g) + \frac{1}{2}O_2(g)$ $\Delta H^{\circ}_{rxn} = +283.0 \text{ kJ mol}^{-1}$

Because chemical reactions can be added and subtracted just like algebraic equations, we carry out the operation (a) + (c) and obtain

(a) $C(graphite) + O_2(g) \longrightarrow CO_2(g)$ $\Delta H^{\circ}_{rxn} = -393.5 \text{ kJ mol}^{-1}$

(c) $CO_2(g) \longrightarrow CO(g) + \frac{1}{2}O_2(g)$ $\Delta H^{\circ}_{rxn} = +283.0 \text{ kJ mol}^{-1}$

(d) $C(graphite) + \frac{1}{2}O_2(g) \longrightarrow CO(g)$ $\Delta H^{\circ}_{rxn} = -110.5 \text{ kJ mol}^{-1}$

Because reaction (d) represents the synthesis of 1 mole of CO from its elements, we have $\Delta H^{\circ}_{f}(CO) = -110.5 \text{ kJ mol}^{-1}$.

The general rule in applying Hess's law is to arrange a series of chemical equations (corresponding to a series of steps) in such a way that, when added together, all species will cancel except for the reactants and products that appear in the overall reaction. This means that we want the elements on the left and the compound of

---

7. Germain Henri Hess (1802–1850). Swiss chemist. Hess was born in Switzerland but spent most of his life in Russia. For formulating Hess's law, he is called the *father of thermochemistry.*

interest on the right of the arrow. Further, we often need to multiply some or all of the equations representing the individual steps by the appropriate coefficients.

C_2H_2

Example 7.10

Calculate the standard enthalpy of formation of acetylene (C_2H_2) from its elements:

$$2C(graphite) + H_2(g) \longrightarrow C_2H_2(g)$$

given that the standard enthalpies of formation of $CO_2(g)$ and $H_2O(g)$ are -393.5 and -285.8 kJ mol^{-1}, respectively, and the standard enthalpy change for the reaction

$$2C_2H_2(g) + 5O_2(g) \longrightarrow 4CO_2(g) + 2H_2O(l)$$

is -2598.8 kJ mol^{-1}.

Strategy The information given corresponds to the following set of reactions:

(a) $\quad C(graphite) + O_2(g) \longrightarrow CO_2(g)$ $\qquad \Delta H^\circ_{rxn} = -393.5$ kJ mol^{-1}

(b) $\quad H_2(g) + \dfrac{1}{2}O_2(g) \longrightarrow H_2O(l)$ $\qquad \Delta H^\circ_{rxn} = -285.8$ kJ mol^{-1}

(c) $\quad 2C_2H_2(g) + 5O_2(g) \longrightarrow 4CO_2(g) + 2H_2O(l) \quad \Delta H^\circ_{rxn} = -2598.8$ kJ mol^{-1}

Our goal here is to calculate the enthalpy change for the formation of C_2H_2 from its elements C and H_2. The reaction does not occur directly, however, so we must use an indirect route using the reactions (a), (b), and (c).

Solution Looking at the synthesis of C_2H_2, we need 2 moles of graphite as reactant. So we multiply Equation (a) by 2 to get

(d) $\quad 2C(graphite) + 2O_2(g) \longrightarrow 2CO_2(g)$ $\qquad \begin{aligned} \Delta H^\circ_{rxn} &= 2(-393.5 \text{ kJ mol}^{-1}) \\ &= -787.0 \text{ kJ mol}^{-1} \end{aligned}$

Next, we need 1 mole of H_2 as a reactant and this is provided by Equation (b). Last, we need 1 mole of C_2H_2 as a product. Equation (c) has 2 moles of C_2H_2 as a reactant so we need to reverse the equation and divide it by 2:

(e) $\quad 2CO_2(g) + H_2O(l) \longrightarrow C_2H_2(g) + \dfrac{5}{2}O_2(g)$ $\qquad \begin{aligned} \Delta H^\circ_{rxn} &= \dfrac{1}{2}(-2598.8 \text{ kJ mol}^{-1}) \\ &= -1299.4 \text{ kJ mol}^{-1} \end{aligned}$

Adding Equations (b), (d), and (e) together, we get

(b) $\quad H_2(g) + \dfrac{1}{2}O_2(g) \longrightarrow H_2O(l)$ $\qquad \Delta H^\circ_{rxn} = -285.8$ kJ mol^{-1}

(d) $\quad C(graphite) + O_2(g) \longrightarrow CO_2(g)$ $\qquad \Delta H^\circ_{rxn} = -787.0$ kJ mol^{-1}

(e) $\quad 2CO_2(g) + H_2O(l) \longrightarrow C_2H_2(g) + \dfrac{5}{2}O_2(g)$ $\qquad \Delta H^\circ_{rxn} = -1299.4$ kJ mol^{-1}

$\overline{}$

$\quad 2C(graphite) + H_2(g) \longrightarrow C_2H_2(g)$ $\qquad \Delta H^\circ_{rxn} = +226.6$ kJ mol^{-1}

Therefore, $\Delta H^\circ_f = \Delta H^\circ_{rxn} = 226.6$ kJ mol^{-1}. The ΔH°_f value means that when 1 mole of C_2H_2 is synthesized from 2 moles of C(graphite) and 1 mole of H_2, $+226.6$ kJ of energy in the form of heat is absorbed by the reacting system from the surroundings. Thus, this is an endothermic process.

—Continued

An oxyacetylene torch has a high flame temperature (3000°C) and is used to weld metals.

Continued—

Practice Exercise Calculate the standard enthalpy of formation of carbon disulfide (CS_2) from its elements, given that the standard enthalpies of formation of $CO_2(g)$ and $SO_2(g)$ are -393.5 and -296.4 kJ mol^{-1}, respectively and that the standard enthalpy of the reaction $CS_2(l) + 3O_2(g) \longrightarrow CO_2(g) + 2SO_2(g)$ is -1073.6 kJ mol^{-1}.

We can calculate the enthalpy of reactions from the values of ΔH_f° as shown in Example 7.11.

B_5H_9

Example 7.11

Pentaborane-9 (B_5H_9) is a colorless, highly reactive liquid that will burst into flame or even explode when exposed to oxygen. The reaction is

$$B_5H_9(l) + 6O_2(g) \longrightarrow \frac{5}{2}B_2O_3(s) + \frac{5}{2}H_2O(l)$$

In the 1950s, pentaborane-9 was considered as a potential rocket fuel because it produces a large amount of heat per gram. However, the solid B_2O_3 formed by the combustion of B_5H_9 is an abrasive that would quickly destroy the nozzle of the rocket, and so the idea was abandoned. Calculate the kilojoules of heat released per gram of the compound reacted with oxygen. The standard enthalpy of formation of B_5H_9 is 73.2 kJ mol^{-1}.

Strategy The enthalpy of a reaction is the difference between the sum of the enthalpies of the products and the sum of the enthalpies of the reactants each multiplied by their respective stoichiometric coefficients (Equation 7.41).

Solution Using the ΔH_f° values in Appendix 2 and Equation 7.41, we write

$$\Delta H_{rxn}^\circ = \left\{ \frac{5}{2}\left[\Delta H_f^\circ(B_2O_3) + \frac{9}{2} \right] \Delta H_f^\circ[H_2O(l)] \right\} - [\Delta H_f^\circ(B_5H_9) + 6\Delta H_f^\circ(O_2)]$$

$$= \left[\frac{5}{2}(-1273.5\,\text{kJ mol}^{-1}) + \frac{9}{2}(-285.8\,\text{kJ mol}^{-1}) \right] - [(73.2\,\text{kJ mol}^{-1}) + 6(0\,\text{kJ mol}^{-1})]$$

$$= -4543.1\,\text{kJ mol}^{-1}$$

The molar mass of B_5H_9 is 63.12 g so the energy released in the form of heat per gram is

$$\text{Heat released per gram} = \frac{-4543.1\,\text{kJ}}{1\,\text{mol } B_5H_9} \times \frac{1\,\text{mol } B_5H_9}{63.12\,\text{g } B_5H_9}$$

$$= -71.97\,\text{kJ g}^{-1}$$

Check Is the negative sign consistent with the fact that B_5H_9 was considered to be used as a rocket fuel?

Practice Exercise Benzene (C_6H_6) burns in air to produce carbon dioxide and liquid water. Calculate the heat released (in kJ) per gram of the compound reacted with oxygen. The standard enthalpy of formation of benzene is 49.04 kJ mol^{-1}.

Pentaborane-9 (B_5H_9) (see Example 7.11) is called a high-energy compound because of the large amount of energy it releases upon combustion. In general, compounds that have positive ΔH_f° values tend to release more heat as a result of combustion and to be less stable than those with negative ΔH_f° values.

7.5 | The Reaction Enthalpies Can Be Estimated from Bond Enthalpies

Recall from Section 3.2 that the bond dissociation energy, D_0, of a particular bond in a molecule is the energy required to break that bond. For example, the bond dissociation energy of the H—H bond in H_2 is the energy required to cleave an H_2 molecule into two well-separated H atoms. In the language of the current chapter, bond dissociation energy for a diatomic molecule such as H_2 is equal to the change in energy, ΔU, for the reaction

$$H_2(g) \longrightarrow H(g) + H(g) \qquad\qquad \Delta U = D_0 = 433.9 \text{ kJ mol}^{-1}$$

If the bond-breaking process occurs under constant-pressure conditions, however, then the energy required for bond breaking is better described by the *bond enthalpy,* rather than the bond energy. The **bond enthalpy** (ΔH_B) is *the enthalpy change, per mole of gaseous molecules, required to break a particular bond in a molecule.*

For diatomic molecules, bond enthalpies differ slightly from bond dissociation energies, and this difference can be determined using Equation 7.16 from Section 7.2:

$$\Delta H = \Delta U + RT\Delta n$$

Consider two atoms X and Y that are connected by a single bonding set of electrons to form the molecule XY. The bond energy (or enthalpy) is the energy (or enthalpy) of the reaction

$$X—Y(g) \longrightarrow X(g) + Y(g)$$

For this reaction $\Delta n = 1$, so

$$\Delta H_B = D_0 + RT$$

which for $T = 298$ K gives

$$\Delta H_B (298 \text{ K}) = D_0 + (8.314 \text{ J mol}^{-1} \text{ K}^{-1})(298 \text{ K})(1 \text{ kJ}/1000 \text{ J})$$
$$= D_0 + 2.48 \text{ kJ mol}^{-1}$$

Thus, the bond enthalpy of a diatomic molecule can be determined from the bond dissociation energy (such as those listed in Table 3.3) at 298 K by adding 2.48 kJ mol^{-1}.

For example, the bond dissociation energy for H_2 is 432.9 kJ mol^{-1}, so the bond enthalpy for H_2 is $(432.9 + 2.48)$ kJ mol^{-1} = 436.4 kJ mol^{-1}. This conversion is useful because experiments on gases typically measure the bond enthalpy, but theoretical calculations often determine the bond dissociation energy directly. For polyatomic molecules, the enthalpy required to break a particular bond type (say C—H or C=O) will vary among the many compounds that contain these bond types. For this reason, bond enthalpies for polyatomic molecules are reported as *average bond* enthalpies. For example, we can measure the energy of the O—H bond in 10 different polyatomic molecules and obtain the average O—H bond energy by dividing the sum of the bond energies by 10. Table 7.6 lists the exact or average bond enthalpies of a number of diatomic and polyatomic molecules. Although bond energies and bond enthalpies are generally quite similar in magnitude (at least for diatomic molecules), the difference can be important in very accurate work.

Nearly all chemical reactions involve the breaking and/or formation of bonds. Therefore, knowing the bond enthalpies and hence the stability of molecules reveals something about the thermochemical nature of the reactions that molecules undergo. In many cases, it is possible to predict the approximate enthalpy of reaction by using

Table 7.6	Some Bond Enthalpies of Diatomic Molecules and Average Bond Energies for Bonds in Polyatomic Molecules*		
Bond	**Bond Enthalpy (kJ mol^{-1})**	**Bond**	**Bond Enthalpy (kJ mol^{-1})**
H—H	436.4	N—N	193
H—N	393	N=N	418
H—O	460	N≡N	941.4
H—S	368	N—O	176
H—P	326	N—P	209
H—F	568.2	O—O	142
H—Cl	431.9	O=O	498.7
H—Br	366.1	O—P	502
H—I	298.3	O=S	469
C—H	414	P—P	197
C—C	347	P=P	489
C=C	620	S—S	268
C≡C	812	S=S	352
C—N	276	F—F	156.9
C=N	615	Cl—Cl	242.7
C≡N	891	Br—Br	192.5
C—O	351	I—I	151.0
C=O†	745	C—F	484
C—P	263	C—Cl	338
C—S	255	C—Br	276
C=S	477	C—I	238

*Bond enthalpies for diatomic molecules (in red) have more significant figures than for bonds in polyatomic molecules because the bond energies of diatomic molecules are directly measurable quantities and not averaged over many compounds.
†The C=O bond enthalpy in CO_2 is 799 kJ mol^{-1}.

bond enthalpy data, such as that listed in Table 7.6. Because energy is always required to break chemical bonds and chemical bond formation is always accompanied by the release of energy, we can estimate the enthalpy of a reaction by counting the total number and kind of bonds broken and formed in the reaction and recording the corresponding energy changes. The enthalpy of reaction in the *gas phase* is given by

$$\Delta H^{\circ}_{rxn} = \sum_{\substack{\text{bonds} \\ \text{broken}}} \Delta H_B - \sum_{\substack{\text{bonds} \\ \text{formed}}} \Delta H_B$$

$$= \text{total energy input} - \text{total energy released} \qquad \textbf{(7.42)}$$

Equation 7.42 is consistent with the usual sign convention for ΔH°_{rxn}. Thus, if the total energy input exceeds the total energy released, ΔH°_{rxn} is positive and the reaction is endothermic [Figure 7.14(a)]. On the other hand, if more energy is released than is absorbed, ΔH°_{rxn} is negative and the reaction is exothermic [Figure 7.14(b)]. If the reactants and products are all diatomic molecules, then Equation 7.42 will yield accurate results because the bond enthalpies of diatomic molecules are accurately known. However, if some or all of the reactants and products are polyatomic

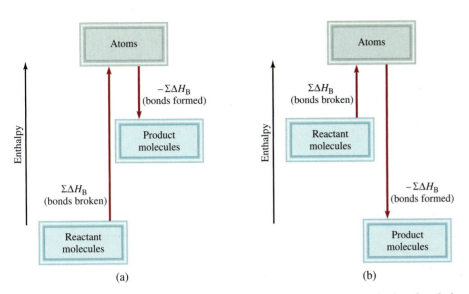

Figure 7.14 Bond enthalpy changes in (a) an endothermic reaction and (b) an exothermic reaction.

molecules, Equation 7.42 will yield only approximate results because the bond enthalpies used will be average values.

For diatomic molecules, Equation 7.42 is equivalent to Equation 7.41, so the results obtained from these two equations should correspond, as Example 7.12 illustrates.

Example 7.12

(a) Use Equation 7.42 to calculate the enthalpy of reaction for the process

$$H_2(g) + Cl_2(g) \longrightarrow 2HCl(g)$$

(b) Compare your result with that obtained using standard enthalpies of formation (Equation 7.41).

Strategy Bond breaking is an energy-absorbing (endothermic) process, and bond making is an energy-releasing (exothermic) process. Therefore, the overall energy change is the difference between these two opposing processes, as described by Equation 7.42.

Solution (a) We start by counting the number of bonds broken and the number of bonds formed and the corresponding energy changes. This is best done by creating a table:

Type of Bonds Broken	Number of Bonds Broken	Bond Enthalpy (kJ mol^{-1})	Enthalpy Change (kJ mol^{-1})
H—H	1	436.4	436.4
Cl—Cl	1	242.7	242.7

Type of Bonds Formed	Number of Bonds Formed	Bond Enthalpy (kJ mol^{-1})	Enthalpy Change (kJ mol^{-1})
H—Cl	2	431.9	836.8

Next, we obtain the total enthalpy input and total enthalpy released:

$$\text{total enthalpy input} = \sum_{\substack{\text{bonds} \\ \text{broken}}} \Delta H_B = 436.4 \text{ kJ mol}^{-1} + 242.7 \text{ kJ mol}^{-1} = 679.1 \text{ kJ mol}^{-1}$$

$$\text{total enthalpy released} = \sum_{\substack{\text{bonds} \\ \text{formed}}} \Delta H_B = 2(431.9 \text{ kJ mol}^{-1}) = 863.8 \text{ kJ mol}^{-1}$$

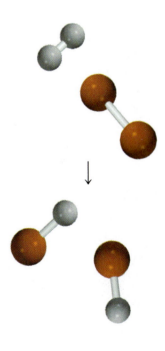

—Continued

Continued—
Using Equation 7.42, we write

$$\Delta H^{\circ}_{rxn} = \underset{\substack{bonds \\ broken}}{\sum \Delta H_{B}} - \underset{\substack{bonds \\ formed}}{\sum \Delta H_{B}} = 679.1 \text{ kJ mol}^{-1} - 863.8 \text{ kJ mol}^{-1} = -184.7 \text{ kJ mol}^{-1}$$

(b) Alternatively, we can use Equation 7.41 and the data in Appendix 2 to calculate the enthalpy of reaction:

$$\begin{aligned}
\Delta H^{\circ}_{rxn} &= 2\Delta \overline{H}^{\circ}_{f} (\text{HCl}) - [\Delta \overline{H}^{\circ}_{f} (\text{H}_2) + \Delta \overline{H}^{\circ}_{f} (\text{Cl}_2)] \\
&= (2)(-92.3 \text{ kJ mol}^{-1}) - [0 + 0] \\
&= -184.6 \text{ kJ mol}^{-1}
\end{aligned}$$

Check Because the reactants and products are all diatomic molecules, we expect the results of Equations 7.42 and 7.41 to be the same. The small discrepancy here is due to different ways of rounding off.

Practice Exercise Calculate the enthalpy of the reaction

$$\text{H}_2(g) + \text{F}_2(g) \longrightarrow 2\text{HF}(g)$$

using (a) Equation 7.42 and (b) Equation 7.41.

Example 7.13 uses Equation 7.42 to estimate the enthalpy of a reaction involving a polyatomic molecule.

Example 7.13

Estimate the enthalpy change for the combustion of hydrogen gas:

$$2\text{H}_2(g) + \text{O}_2(g) \longrightarrow 2\text{H}_2\text{O}(g)$$

Strategy We follow the same procedure as was used in Example 7.10. Note, however, that H_2O is a polyatomic molecule, so we need to use the average bond energy value for the O—H bond.

Solution Construct the following table:

Type of Bonds Broken	Number of Bonds Broken	Bond Enthalpy (kJ mol^{-1})	Enthalpy Change (kJ mol^{-1})
H—H	2	436.4	872.8
O=O	1	498.7	498.7

Type of Bonds Formed	Number of Bonds Formed	Bond Enthalpy (kJ mol^{-1})	Enthalpy Change (kJ mol^{-1})
O—H	4	460	1840

Next, obtain the total enthalpy input and total enthalpy released:

$$\text{total enthalpy input} = \underset{\substack{bonds \\ broken}}{\sum \Delta H_{B}} = 872.8 \text{ kJ mol}^{-1} + 498.7 \text{ kJ mol}^{-1} = 1371.5 \text{ kJ mol}^{-1}$$

$$\text{total enthalpy released} = \underset{\substack{bonds \\ formed}}{\sum \Delta H_{B}} = 1840 \text{ kJ mol}^{-1}$$

—Continued

Continued—
Using Equation 7.42, we write

$$\Delta H^{\circ}_{\text{rxn}} = \sum_{\substack{\text{bonds} \\ \text{broken}}} \Delta H_{\text{B}} - \sum_{\substack{\text{bonds} \\ \text{formed}}} \Delta H_{\text{B}} = 1371.5 \text{ kJ mol}^{-1} - 1840 \text{ kJ mol}^{-1} = -469 \text{ kJ mol}^{-1}$$

This result is only an estimate because the bond enthalpy of an O—H bond is an average quantity. Alternatively, we can use Equation 7.40 and the data in Appendix 2 to calculate the enthalpy of reaction:

$$\begin{aligned}
\Delta H^{\circ}_{\text{rxn}} &= 2\Delta \overline{H}^{\circ}_{\text{f}}\,[\text{H}_2\text{O}(g)] - \{2\Delta H^{\circ}_{\text{f}}\,[\text{H}_2(g)] + \Delta H^{\circ}_{\text{f}}\,[\text{O}_2(g)]\} \\
&= 2(-241.8 \text{ kJ mol}^{-1}) - [0 + 0] \\
&= -483.6 \text{ kJ mol}^{-1}
\end{aligned}$$

Check The estimated value based on average bond enthalpies is close, but not exactly equal, to the value calculated using $\Delta H^{\circ}_{\text{f}}$ data. In general, Equation 7.42 works best for reactions that are either quite endothermic or quite exothermic, that is, reactions for which $\Delta H^{\circ}_{\text{rxn}} > 100 \text{ kJ mol}^{-1}$ or for which $\Delta H^{\circ}_{\text{rxn}} < -100 \text{ kJ mol}^{-1}$.

Practice Exercise For the reaction

$$\text{H}_2(g) + \text{C}_2\text{H}_4(g) \longrightarrow \text{C}_2\text{H}_6(g)$$

(a) Use the bond energy values in Table 7.6 to estimate the enthalpy of reaction.

(b) Use standard enthalpies of formation to calculate the enthalpy of reaction. ($\Delta H^{\circ}_{\text{f}}$ values for H_2, C_2H_4, and C_2H_6 are 0, 52.3, and -84.7 kJ mol^{-1}, respectively.)

7.6 | Enthalpy Changes Also Accompany Physical Transformations

Although we have focused so far primarily on the thermal energy effects resulting from chemical reactions, many physical processes also involve the absorption or release of heat. Examples include phase transformations, such as the melting of ice or the condensation of water vapor. Enthalpy changes occur as well when a solute dissolves in a solvent or when a solution is diluted.

Enthalpy Changes in Phase Transitions

Phase transitions generally involve changes in the density and/or structural rearrangement of the molecules in a system. Because such changes alter the spatial arrangement of the molecules relative to one another, and thus the intermolecular interactions, phase transitions are generally accompanied by the evolution or absorption of energy in the form of heat. Because phase transitions take place at constant pressure, this heat energy is equal to the change in enthalpy for the transition process of interest. In Chapter 5, we discussed three common types of phase transformations: vaporization, fusion, and sublimation. Here we discuss the magnitude of enthalpy changes for these transformations, paying particular attention to the role played by intermolecular forces.

Enthalpy of Vaporization

Vaporization is the process by which a liquid transforms into a vapor (gas). The **molar enthalpy** (or *heat*) **of vaporization** (ΔH_{vap}) is *the energy required to vaporize 1 mole*

Table 7.7	Molar Enthalpies of Vaporization for Selected Substances at Their Normal Boiling Points	
Substance	**Boiling Point* (°C)**	ΔH_{vap} **(kJ mol^{-1})**
Argon (Ar)	−186	6.3
Benzene (C_6H_6)	80.1	31.0
Ethanol (C_2H_5OH)	78.3	39.3
Diethyl ether ($C_2H_5OC_2H_5$)	34.6	26.0
Mercury (Hg)	357	59.0
Methane (CH_4)	−164	9.2
Water (H_2O)	100	40.79

*Measured at 1 atm.

of liquid.[8] The molar heat of vaporization is directly related to the strength of the intermolecular forces that exist in a liquid. If the intermolecular attractions are strong, the energy required to free the molecules from the liquid phase is large. Such a liquid has a high molar enthalpy of vaporization and a low vapor pressure (see Section 6.1).

A practical way to demonstrate the molar heat of vaporization is to rub an alcohol such as ethanol (C_2H_5OH) or isopropanol (C_3H_7OH) on your hands. These alcohols have a lower $\Delta H°_{vap}$ than water, so the heat from your hands is enough to increase the kinetic energy of the alcohol molecules and evaporate them. As a result of the loss of heat, your hands feel cool. This process is similar to perspiration, which is one of the means by which the human body maintains a constant temperature. Because of the strong intermolecular hydrogen bonding that exists in water, a considerable amount of energy is needed to vaporize the water in perspiration from the surface of the skin. This energy is supplied by the heat generated in various metabolic processes.

Table 7.7 lists ΔH_{vap} values for a number of liquids at their normal boiling points (at 1 atm). These data show that substances with high boiling points tend to have large enthalpies of vaporization. To understand this trend, you must recognize that both the boiling point and the heat of vaporization are measures of intermolecular attraction in the liquid. For example, argon (Ar) and methane (CH_4), which have weak dispersion forces, have low boiling points and small molar enthalpies of vaporization. Diethyl ether ($C_2H_5OC_2H_5$) has a dipole moment, and the dipole-dipole forces account for its moderately high boiling point and ΔH_{vap}. Both ethanol (C_2H_5OH) and water have strong hydrogen bonding, which accounts for their high boiling points and large ΔH_{vap} values. Strong metallic bonding causes mercury to have the highest boiling point and ΔH_{vap} of this group of liquids. Interestingly, the boiling point of benzene, which is nonpolar, is comparable to that of ethanol. Benzene has a high polarizability due to the distribution of its electrons in the delocalized π molecular orbitals, and the dispersion forces among benzene molecules can be as strong as or even stronger than dipole-dipole forces and/or hydrogen bonds.

---

8. The enthalpy change in any phase transition is sometimes referred to as the ***latent heat*** of that transition. For example, ΔH_{vap} is often called the *latent heat of vaporization*.

Table 7.8	Molar Enthalpies of Fusion for Select Substances at Their Normal Melting Points	
Substance	**Melting Point* (°C)**	**ΔH_{fus} (kJ mol^{-1})**
Argon (Ar)	−190	1.3
Benzene (C_6H_6)	5.5	10.9
Ethanol (C_2H_5OH)	−117.3	7.61
Diethyl ether ($C_2H_5OC_2H_5$)	−116.2	6.90
Mercury (Hg)	−39	23.4
Methane (CH_4)	−183	0.84
Water (H_2O)	0	6.01

*Measured at 1 atm.

Enthalpy of Fusion

The **molar enthalpy** (or *heat*) **of fusion** (ΔH_{fus}) is *the energy required to melt 1 mole of solid.* Table 7.8 lists the molar enthalpies of fusion at the melting points (ΔH_{fus}) for the substances in Table 7.7. Notice that ΔH_{fus} is smaller than ΔH_{vap} for each substance. This occurs because considerably more energy is needed to overcome the attractive forces between liquid molecules to completely separate them into vapor molecules than is needed to rearrange solid molecules into liquid molecules.

Based on Table 7.8, there is also a rough correlation between melting temperature and the enthalpy of fusion. Weakly interacting substances, such as argon and methane, generally have lower melting points and enthalpies of fusion than substances with stronger intermolecular attractions, such as benzene and water. However, the correlation between the melting temperature and ΔH_{fus} is not as strong as that between the boiling temperature and ΔH_{fus}. For example, mercury melts at −39°C, but has a much larger enthalpy of fusion than benzene (23.4 kJ mol^{-1} vs. 10.9 kJ mol^{-1}) despite the fact that benzene has the higher melting point (5.5°C). During fusion, the molecules are simply rearranged and not completely pulled apart, so the enthalpy change will depend upon factors other than the strength of intermolecular forces, such as crystal structure and liquid density.

Enthalpy of Sublimation

Sublimation is the direct conversion of a solid to a vapor, so the **molar enthalpy of sublimation** (ΔH_{sub}) is *the energy required to sublime 1 mole of a solid.* In general, the molecules in a solid are more tightly held together by intermolecular forces than those in a liquid, so ΔH_{sub} is larger than ΔH_{vap}. If ΔH_{sub}, ΔH_{vap}, and ΔH_{fus} could all be measured at the same temperature (such as is possible at the triple point of water), we would have

$$\Delta H_{sub} = \Delta H_{fus} + \Delta H_{vap} \tag{7.43}$$

Equation 7.43 is an application of Hess's law (Section 7.5), because the enthalpy change for the overall process is the same whether the substance changes directly from the solid to the vapor (sublimation) or from the solid to the liquid and then to the vapor (fusion + vaporization). If the enthalpies for the three processes (sublimation, fusion, and vaporization) are measured at different temperatures, which they often are, then Equation 7.43 can only be used as an approximation.

Heating Curves

When a liquid or solid is far from a phase transition temperature, any addition of energy in the form of heat at constant pressure (q_p) leads to an increase in the temperature (ΔT) of the system, as shown by rearranging Equation 7.18:

$$\Delta T = \frac{q_p}{C_p} \qquad (7.44)$$

where C_p is the constant-pressure heat capacity of the solid or liquid. When the substance reaches a transition temperature, however, energy added as heat to the system does not raise the temperature. Instead, it supplies the energy necessary to change the phase of the material. The amount of heat energy required to convert n moles of substance from one phase to another is given by

$$q_p = n\,\Delta H_{trans} \qquad (7.45)$$

where ΔH_{trans} is the molar enthalpy of the transition. Consider, for example, the heating of a block of ice, originally at $-30°C$, to a final temperature of $130°C$, where the heating is done at a constant pressure of 1 atm. Figure 7.15 shows the *heating curve* for this process, which is a plot of the temperature as a function of the amount of heat added.

Initially, the addition of heat raises the temperature of the ice. The slope of the heating curve is the reciprocal of the heat capacity of ice (Equation 7.44). When the temperature reaches the melting point of ice ($0°C$), further addition of heat goes entirely into converting the ice into water according to Equation 7.45, using ΔH°_{fus} for the enthalpy of transition. During this conversion, the temperature remains constant. After the ice is completely transformed to liquid water, the temperature will begin to rise again upon the further addition of heat according to Equation 7.44, except now the slope is determined by the reciprocal of the heat capacity of liquid water. Because the heat capacity of liquid water ($\overline{C}_p = 75.3\ \text{J mol}^{-1}\ \text{K}^{-1}$) is greater than that of ice ($\overline{C}_p = 38.0\ \text{J mol}^{-1}\ \text{K}^{-1}$), the slope of the heating curve is less in the region between $0°C$ and $100°C$ than in the region from $-30°C$ to zero. The heating

Figure 7.15 The heating curve (temperature versus energy added as heat) for the heating of water from $-30°C$ to a final temperature of $130°C$. Because $\Delta \overline{H}^\circ_{fus}$ is less than $\Delta \overline{H}^\circ_{vap}$, less heat is required to melt the ice at $0°C$ than to boil the liquid water at $100°C$. This explains why AB is shorter than CD. The steepness of the solid, liquid, and vapor heating lines is determined by the heat capacity of water in each state.

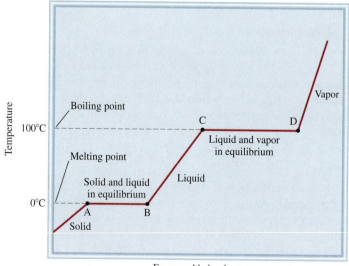

of the liquid continues until the temperature reaches the boiling point of water (100°C) at which point the temperature remains constant as the liquid is converted to vapor. Following complete vaporization, the temperature rises again with a slope that is greater than that for heating ice or water because the constant pressure molar heat capacity of water vapor is 33.6 J mol^{-1} K^{-1}, which is smaller than that for both liquid water and ice.

Example 7.14 shows how to calculate the total energy required to heat a system through a phase change at constant pressure.

Example 7.14

Calculate the amount of energy (in kJ) needed to heat 346 g of liquid water from 0°C to 182°C. Assume that the specific heat capacity of water is 4.184 J g^{-1} °C^{-1} over the entire liquid range and that the specific heat of steam is 1.99 J g^{-1} °C^{-1}.

Strategy The process consists of three steps, heating water from 0°C to 100°C, evaporating water at 100°C, and heating steam from 100°C to 182°C. The heat change (q) at each stage is given by $q = m\,C_s\,\Delta T$ (Equation 7.19), where m is the mass of water, C_s is the specific heat, and ΔT is the temperature change. At the phase change (vaporization), q is given by $n\,\Delta H_{vap}$, where n is the number of moles of water.

Solution The calculation can be broken down into three steps:

Step 1: Heating water from 0°C to 100°C
Using Equation 7.20, we write

$$q_1 = mC_s\Delta T$$
$$= (346\ \text{g})(4.184\ \text{J g}^{-1}\ \text{°C}^{-1})(100\text{°C} - 0\text{°C}) = 1.45 \times 10^5\ \text{J} = 145\ \text{kJ}$$

Step 2: Evaporating 346 g of water at 100°C (a phase change)
According to Table 7.7, $\Delta H_{vap} = 40.79$ kJ mol^{-1} for water, so

$$q_2 = n\Delta H_{vap}^{\circ}$$
$$= \left(\frac{346\ \text{g}}{18.02\ \text{g mol}^{-1}}\right)(40.79\ \text{kJ mol}^{-1}) = 783\ \text{kJ}$$

Step 3: Heating steam from 100°C to 182°C

$$q_3 = mC_s\Delta T$$
$$= (346\ \text{g})(1.99\ \text{J g}^{-1}\ \text{°C}^{-1})(182\text{°C} - 100\text{°C}) = 5.65 \times 10^4\ \text{J} = 56.5\ \text{kJ}$$

The overall energy required is given by

$$q_{overall} = q_1 + q_2 + q_3$$
$$= 145\ \text{kJ} + 783\ \text{kJ} + 56.5\ \text{kJ} = 985\ \text{kJ}$$

Note that the dominant contribution to this heating process comes from the vaporization step.

Check All three qs have a positive sign, because heat is absorbed at all three stages to raise the temperature from 0°C to 182°C.

Practice Exercise Calculate the heat released when 68.0 g of steam at 124°C is condensed to water at 45°C.

Figure 7.16 Hydration of Na^+ and Cl^- ions. Note that the arrangement of the water molecules surrounding the ion are considerably different for these two ions. Around the Na^+ ions, the surrounding water molecules orient so that their negatively charged ends (the oxygen atoms) are adjacent to the cation. Around Cl^-, the positively charged end of the water molecules (the hydrogen atom) is adjacent to the anion.

Enthalpy of Solution and Dilution

In addition to the enthalpies of phase transitions just discussed, enthalpy changes also occur when two substances are mixed, such as when a solute is dissolved in a solvent (the enthalpy of solution) or when a solution is diluted (the enthalpy of dilution).

Enthalpy of Solution

In the vast majority of cases, dissolving a solute in a solvent produces measurable heat change. At constant pressure, the heat change is equal to the enthalpy change. The **enthalpy** (or *heat*) **of solution** (ΔH_{soln}) is *the heat absorbed when a certain amount of solute dissolves in a certain amount of solvent*. When this quantity is measured at standard pressure (1 bar), it is called the *standard enthalpy of solution* (ΔH°_{soln}).

The quantity ΔH_{soln} represents the difference between the enthalpy of the final solution and the enthalpies of its original components (that is, the solute and solvent) before they are mixed. Thus,

$$\Delta H_{soln} = H_{soln} - H_{components} \tag{7.46}$$

Neither H_{soln} nor $H_{components}$ can be measured, but their difference (ΔH_{soln}) can be readily determined using calorimetry. Like other enthalpy changes, ΔH_{soln} is positive for endothermic (heat-absorbing) processes and negative for exothermic (heat-generating) processes.

To better understand the heat of solution, consider what happens when solid NaCl (an ionic compound) dissolves in water. In solid NaCl, the Na^+ and Cl^- ions are held together by strong positive-negative (electrostatic) forces, but when a small crystal of NaCl dissolves in water, the three-dimensional network of ions breaks into its individual units. (The structure of solid NaCl is shown in Figure 0.12.) The separated Na^+ and Cl^- ions are stabilized in solution by their interaction with water molecules, which completely surround the ions (Figure 7.16).

These ions are said to be *hydrated*. Water molecules shield the Na^+ and Cl^- ions from each other and effectively reduce the electrostatic attraction that held them together in the solid state. Thus, the water molecules act like a good electrical insulator. The heat of solution is defined by the following process:

$$NaCl(s) \xrightarrow{H_2O} Na^+(aq) + Cl^-(aq) \qquad \Delta H_{soln} = ?$$

Dissolving an ionic compound such as NaCl in water involves complex interactions among the solute and solvent species. For the sake of analysis, however, we can imagine that the solution process takes place in the two separate steps illustrated in Figure 7.17.

First, the Na^+ and Cl^- ions in the solid crystal are separated from each other and converted to the gaseous state:

$$NaCl(s) \longrightarrow Na^+(g) + Cl^-(g)$$

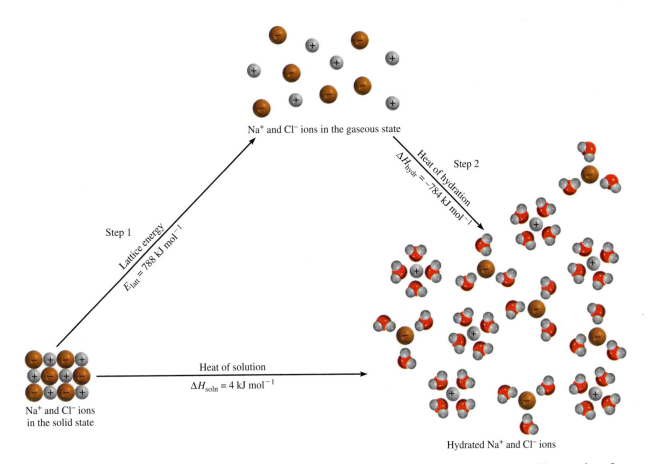

Figure 7.17 The solution process for NaCl. The process can be considered to occur in two separate steps: (1) separation of ions from the crystal state to the gaseous state and (2) hydration of the gaseous ions. The heat of solution is equal to the energy changes for these two steps, $\Delta H_{sol} = E_{latt} + \Delta H_{hydr}$.

The energy required to completely separate 1 mole of a solid ionic compound into gaseous ions is called the **lattice energy** (E_{latt}). The lattice energy of NaCl is 788 kJ mol^{-1}. In other words, we need 788 kJ of energy to break 1 mole of solid NaCl into 1 mole of Na^+ ions and 1 mole of Cl^- ions. Next, the "gaseous" Na^+ and Cl^- ions enter the water and become hydrated:

$$Na^+(g) + Cl^-(g) \xrightarrow{H_2O} Na^+(aq) + Cl^-(aq) + \text{energy}$$

The enthalpy change associated with hydration is called the **enthalpy** (or heat) **of hydration** (ΔH_{hydr}). The heat of hydration is a negative quantity for cations and anions. Applying Hess's law, it is possible to treat ΔH_{soln} as the sum of two related quantities, the lattice energy (E_{latt}) and the heat of hydration ΔH_{hydr}:

$$NaCl(s) \longrightarrow Na^+(g) + Cl^-(g) \qquad\qquad E_{latt} = 788 \text{ kJ mol}^{-1}$$
$$Na^+(g) + Cl^-(g) \xrightarrow{H_2O} Na^+(aq) + Cl^-(aq) \qquad \Delta H_{hydr} = -784 \text{ kJ mol}^{-1}$$

$$\overline{NaCl(s) \xrightarrow{H_2O} Na^+(aq) + Cl^-(aq) \qquad\qquad \Delta H_{soln} = 4 \text{ kJ mol}^{-1}}$$

Therefore, when 1 mole of NaCl dissolves in water, 4 kJ of heat is absorbed from the surroundings (the beaker containing the solution becomes slightly colder). The ΔH_{soln}

of 4 kJ mol^{-1} calculated for NaCl is valid if the solution formed is dilute. When NaCl is added to a concentrated solution the enthalpy change will differ due to interactions of the Na$^+$ and Cl$^-$ ions already in the solution. For this reason, the ΔH_{soln} just calculated is often referred to as the *infinite dilution* (or *limiting*) *enthalpy of solution*. Table 7.9 lists the infinite dilution enthalpies of solution of several ionic compounds in water. Depending on the nature of the cation and anion involved, ΔH_{soln} for an ionic compound may be either negative (exothermic) or positive (endothermic).

The enthalpy of solution for ammonium nitrate (NH$_4$NO$_3$) is very positive (highly endothermic) and much larger than the other values in Table 7.9. As a result, the dissolution of NH$_4$NO$_3$ is used in many commercial disposable cold packs. These cold packs contain two separate chambers, one containing solid NH$_4$NO$_3$ and one containing water. When the seal between the two chambers is broken, the ammonium nitrate dissolves in the water. This endothermic process absorbs a great deal of heat energy, leading to a significant drop in the temperature of the pack. (For an example of the use of exothermic reactions in the construction of hot packs, see Problem 7.69.)

Heat of Dilution

When a solution is *diluted*—that is, when more solvent is added to lower the overall concentration of the solute—additional heat is usually given off or absorbed. The **enthalpy** (or *heat*) **of dilution** (ΔH_{dil}) is *the heat change associated with the dilution process*. If a certain solution process is endothermic and the solution is subsequently diluted, *more* heat will be absorbed by the same solution from the surroundings. In an exothermic solution process, more heat will be liberated if additional solvent is added to dilute the solution. Therefore, always be cautious when diluting a solution in the laboratory.

Because of its highly exothermic heat of dilution, concentrated sulfuric acid (H$_2$SO$_4$) can be particularly hazardous if it must be diluted with water. Concentrated H$_2$SO$_4$ consists of 98 percent acid and 2 percent H$_2$O by mass. Diluting it with water releases so much heat to the surroundings (the process is so exothermic) that you must *never* attempt to dilute the concentrated acid by adding water to it. The heat generated could cause the acid solution to boil and splatter. The recommended procedure is to add the concentrated acid slowly to the water (while constantly stirring).

7.7 | The Temperature Dependence of Reaction Enthalpies Can Be Determined from Heat Capacity Data

So far all our calculations for ΔH_{rxn}° have been for room temperature (298 K). Suppose, however, that our reaction of interest takes place at 400 K. How would we determine the heat energy absorbed or released by the reaction at this temperature? If high precision is not required, it is often sufficient to assume that ΔH_{rxn}° is approximately independent of temperature and to use the value of ΔH_{rxn}° (298 K) for ΔH_{rxn}° (350 K). For high-precision calculations, however, the change in ΔH_{rxn}° as we go from $T_1 = 298$ K to our temperature of interest (T_2) might be significant. One way to find the value at T_2 would be to repeat the measurement of ΔH_{rxn}° directly at T_2 or to find a table of ΔH_f° values measured at T_2, but doing experiments is time consuming and finding an appropriate table of data at exactly the temperature we want is rarely possible. Fortunately, using thermodynamics, it is possible to obtain ΔH_{rxn}° values at T_2 from data tabulated at another temperature, T_1, without having to do another experiment.

For any reaction, the standard enthalpy change at a particular temperature is

$$\Delta H_{rxn}^\circ = \sum_{products} H^\circ - \sum_{reactants} H^\circ \tag{7.47}$$

To see how the standard enthalpy of reaction (ΔH°_{rxn}) changes with temperature, recall that $dH = C_P\, dT$ (Equation 7.26) at constant pressure. Thus,

$$H^\circ(T_2) = H^\circ(T_1) + \int_{T_1}^{T_2} C_P\, dT$$

$$\approx H^\circ(T_1) + C_P(T_2 - T_1) = H^\circ(T_1) + C_P\,\Delta T \qquad \textbf{(7.48)}$$

if C_P can be treated as constant over the temperature range of interest. Combining Equations 7.47 and 7.48 gives

$$\Delta H^\circ_{rxn}(T_2) = \sum_{\text{products}}\left[H^\circ(T_1) + C_P\Delta T\right] - \sum_{\text{reactants}}\left[H^\circ(T_1) + C_P\Delta T\right]$$

$$= \left[\sum_{\text{products}} H^\circ(T_1) - \sum_{\text{reactants}} H^\circ(T_1)\right] - \left[\sum_{\text{products}} C_P^\circ - \sum_{\text{reactants}} C_P^\circ\right]\Delta T$$

$$= \Delta H^\circ_{rxn}(T_1) + \left(\sum_{\text{products}} C_P^\circ - \sum_{\text{reactants}} C_P^\circ\right)\Delta T$$

or

$$\Delta H^\circ_{rxn}(T_2) = \Delta H^\circ_{rxn}(T_1) + \Delta C^\circ_{P,rxn}\Delta T \qquad \textbf{(7.49)}$$

where we have defined the standard reaction heat capacity as

$$\Delta C^\circ_{P,rxn} = \sum_{\text{products}} v_P \overline{C_P^\circ} - \sum_{\text{reactants}} v_R \overline{C_P^\circ} \qquad \textbf{(7.50)}$$

and where v_P and v_R are the stoichiometric coefficients for the products and reactants, respectively, and $\overline{C_P^\circ}$ is the standard molar heat capacity.

Equation 7.49 is known as **Kirchhoff's law,** after the German physicist Gustav-Robert Kirchhoff.[9] According to Kirchhoff's law, the difference between the enthalpies of a reaction at two different temperatures is just the difference in the enthalpies of heating the products and reactants from T_1 to T_2. Note that in deriving Equation 7.48 we have assumed that the constant-pressure heat capacities are independent of temperature. Otherwise, they must be expressed as functions of T, and the integral in Equation 7.48 must be done explicitly.

Example 7.15 shows how to use Kirchhoff's law to calculate ΔH°_{rxn} for a reaction occuring at a temperature other than 298 K.

Example 7.15

The standard enthalpy change for the reaction

$$2H_2(g) + O_2(g) \longrightarrow 2H_2O(l)$$

is -571.6 kJ mol^{-1} at 25°C. Calculate the value of ΔH°_{rxn} at 100°C, assuming that all $\overline{C_P^\circ}$ values are independent of temperature.

—Continued

---

9. Gustav-Robert Kirchhoff (1824–1887). German physicist. Kirchhoff made important contributions to electrical circuit theory, spectroscopy, and thermodynamics. Together with colleague Robert Bunsen (inventor of the Bunsen burner), he codiscovered the elements rubidium and cesium.

Continued—

Strategy We can solve this problem using Kirchhoff's law (Equation 7.49). To do this, though, we need the molar heat capacities, $\overline{C}_P^\circ$, of the reactants and products, so we can use Equation 7.50 to calculate $\Delta C_{P,\text{rxn}}^\circ$. The molar heat capacity values are listed in Appendix 2.

Solution In Appendix 2, $\overline{C}_P^\circ$ is 29.4, 38.2 J and 75.3 mol^{-1} K^{-1} for O$_2(g)$, H$_2(g)$, and H$_2$O(l), respectively. Using Equation 7.50, we obtain

$$\Delta C_{P,\text{rxn}}^\circ = \sum_{\text{products}} v_P \overline{C}_P^\circ - \sum_{\text{reactants}} v_R \overline{C}_P^\circ$$

$$= 2\overline{C}_P^\circ[\text{H}_2\text{O}(l)] - 2\overline{C}_P^\circ[\text{H}_2(g)] - \overline{C}_P^\circ[\text{O}_2(g)]$$

$$= 2(75.3 \text{ J mol}^{-1}\,\text{K}^{-1}) - 2(28.8 \text{ J mol}^{-1}\,\text{K}^{-1}) - 29.4 \text{ J mol}^{-1}\,\text{K}^{-1}$$

$$= 63.8 \text{ J mol}^{-1}\,\text{K}^{-1}$$

Using Kirchhoff's law (Equation 7.49) gives

$$\Delta H_{\text{rxn}}^\circ(T_2) = \Delta H_{\text{rxn}}^\circ(T_1) + \Delta C_{P,\text{rxn}}^\circ \Delta T$$

$$= -571.6 \text{ kJ mol}^{-1} + (63.8 \text{ J mol}^{-1}\,\text{K}^{-1})(373.15 \text{ K} - 298.15 \text{ K})\left(\frac{1 \text{ kJ}}{1000 \text{ J}}\right)$$

$$= -566.8 \text{ kJ mol}^{-1}$$

(Heat capacities are generally given in units of J mol^{-1} K^{-1}, whereas enthalpies are usually in units of kJ mol^{-1}, so we must always be careful to include the conversion from joules to kilojoules in these types of calculations.)

Practice Exercise Using data from Appendix 2, calculate $\Delta H_{\text{rxn}}^\circ$ for the following reaction at 450 K:

$$2\text{NO}_2(g) \longrightarrow \text{N}_2\text{O}_4(g)$$

Summary of Facts and Concepts

Section 7.1

▶ Thermodynamics is the study of energy and its transformations from one form to another in macroscopic systems. Thermochemistry is the study of the energy released or absorbed during chemical reactions.

▶ The first law of thermodynamics states that energy is conserved and the change in internal energy of a macroscopic system is the sum of the work done on the system and the energy transferred to the system as heat. In chemistry we are concerned mainly with thermal energy, electrical energy, and mechanical energy, which is usually associated with pressure-volume work.

▶ The state of a macroscopic system is defined by a small number of properties such as composition, volume, temperature, and pressure. These properties are called state functions.

▶ The change in a state function for a system depends only on the initial and final states of the system, and not on the path by which the change is accomplished. Energy is a state function; work and heat are not.

Section 7.2

▶ Enthalpy is a state function. A change in enthalpy ΔH is equal to $\Delta U + \Delta(PV)$, which for a constant-pressure process is equal to $\Delta U + P\Delta V$.

▶ The change in enthalpy (ΔH, usually given in kJ) is a measure of the heat absorbed in a process at constant pressure.

Section 7.3

▶ The heat capacity of a system tells us how much heat energy must be transferred to a system to achieve a given temperature change.

▶ Constant-volume and constant-pressure calorimeters are used to measure heat changes that occur in physical and chemical processes.

▶ The equipartition of energy theorem states that the energy in a molecule is, on average, distributed evenly among all molecular degrees of freedom. This theorem can be used to predict the heat capacities of gases.

Section 7.4

▶ Hess's law states that the overall enthalpy change in a reaction is equal to the sum of enthalpy changes for individual steps in the overall reaction. Hess's law holds for any extensive thermodynamic quantity.

▶ The standard enthalpy of a reaction can be calculated from the standard enthalpies of formation of reactants and products.

Section 7.5

▶ The bond enthalpy is the enthalpy change, per mole of gaseous molecules, required to break a particular bond in a molecule. Exact and average bond enthalpies can be used to estimate the standard enthalpy change for reactions.

Section 7.6

▶ Enthalpy changes also accompany physical transformations such as vaporization, freezing, and sublimation.

▶ The heat of solution of an ionic compound in water is the sum of the lattice energy of the compound and the heat of hydration. The relative magnitudes of these two quantities determine whether the solution process is endothermic or exothermic. The heat of dilution is the heat absorbed or evolved when a solution is diluted.

Section 7.7

▶ The temperature dependence of reaction enthalpies can be determined from heat capacity data using Kirchhoff's law.

Key Words

adiabatic process, p. 372
bond enthalpy, p. 401
calorimetry, p. 384
closed system, p. 371
constant-pressure heat capacity, p. 382
constant-pressure process, p. 372
constant-volume heat capacity, p. 382
constant-volume process, p. 372
endothermic, p. 368
enthalpy, p. 376
enthalpy of dilution, p. 412
enthalpy of hydration, p. 411
enthalpy of reaction, p. 377

enthalpy of solution, p. 410
equilibrium state, p. 371
equipartition of energy theorem, p. 389
exothermic, p. 368
first law of thermodynamics, p. 369
heat, p. 368
heat capacity, p. 381
Hess's law, p. 398
internal energy, p. 365
irreversible process, p. 372
isolated system, p. 371
isothermal process, p. 371
Kirchhoff's law, p. 413
latent heat, p. 406

lattice energy, p. 411
macrostate, p. 366
microstate, p. 366
molar enthalpy of fusion, p. 407
molar enthalpy of sublimation, p. 407
molar enthalpy of vaporization, p. 405
molar heat capacity, p. 381
molecular state, p. 366
open system, p. 371
reversible process, p. 373
specific heat capacity, p. 382
standard enthalpy of reaction, p. 396

standard molar enthalpy of formation, p. 395
standard state, p. 395
state function, p. 369
statistical mechanics, p. 389
steady state, p. 371
surroundings, p. 370
system, p. 370
thermochemical equation, p. 378
thermochemistry, p. 365
thermodynamic state, p. 366
thermodynamics, p. 365
work, p. 366

Problems

Thermodynamics Is the Study of Energy and Its Transformations in Macroscopic Systems

7.1 A sample of nitrogen gas expands in volume from 1.6 L to 5.4 L at constant temperature. Calculate the work done (in joules) if the gas expands (a) against a vacuum, (b) against a constant pressure of 0.80 bar, and (c) against a constant pressure of 3.7 bar.

7.2 A gas expands in volume from 26.7 mL to 89.3 mL at constant temperature. Calculate the work done (in joules) if the gas expands (a) against a vacuum, (b) against a constant pressure of 1.5 bar, and (c) against a constant pressure of 2.8 atm.

7.3 A gas expands and does P-V work on the surroundings equal to 325 J. At the same time, it absorbs 127 J

of heat from the surroundings. Calculate the change in energy of the gas.

7.4 Calculate the work done when 50.0 g of tin dissolves in excess acid at 1.00 bar and 25°C

$$Sn(s) + 2H^+(aq) \longrightarrow Sn^{2+}(aq) + H_2(g)$$

Assume ideal gas behavior.

7.5 Calculate the work done (in joules) when 1.0 mole of water vaporizes at 1.0 atm and 100°C. Assume ideal gas behavior and that the volume of liquid water is negligible compared with that of steam at 100°C.

7.6 Show that the work required for the reversible and isothermal expansion of n moles of a van der Waals gas (Equation 5.42) from an initial volume V_1 to a final volume V_2 is given by

$$w = nRT \ln\frac{V_1 - nb}{V_2 - nb} + an^2\left(\frac{1}{V_1} - \frac{1}{V_2}\right)$$

7.7 Using the result from Problem 7.6, calculate the work required to compress 10.0 moles of CO_2 from a volume of 2500 L to 100 L at a fixed temperature of 298 K. The van der Waals parameters a and b for carbon dioxide can be found in Table 5.3. Compare this to the value obtained assuming that CO_2 behaves ideally. What does the difference between these two predictions tell you about the relative importance of attractive and repulsive molecular interactions in CO_2 under these conditions?

7.8 If energy is conserved, how can there be an energy crisis?

Energy Absorbed by a System as Heat in a Constant-Pressure Process Is Equal to Change in Enthalpy

7.9 The first step in the industrial recovery of zinc from the zinc sulfide ore is roasting—that is, the conversion of ZnS to ZnO by heating:

$$2ZnS(s) + 3O_2(g) \longrightarrow 2ZnO(s) + 2SO_2(g)$$
$$\Delta H = -879 \text{ kJ mol}^{-1}$$

Calculate the heat evolved (in kJ) per gram of ZnS roasted.

7.10 Determine the amount of heat (in kJ) given off when 1.26×10^4 g of NO_2 are produced according to the equation

$$2NO(g) + O_2(g) \longrightarrow 2NO_2(g)$$
$$\Delta H = -114.6 \text{ kJ mol}^{-1}$$

7.11 Consider the reaction

$$2H_2O(g) \longrightarrow 2H_2(g) + O_2(g)$$
$$\Delta H = 483.6 \text{ kJ mol}^{-1}$$

If 2.0 moles of $H_2O(g)$ are converted to $H_2(g)$ and $O_2(g)$ against a pressure of 1.0 bar at 125°C, what is ΔU for this reaction?

7.12 Consider the reaction

$$H_2(g) + Cl_2(g) \longrightarrow 2HCl(g)$$
$$\Delta H = -184.6 \text{ kJ mol}^{-1}$$

If 3.0 moles of H_2 react with 3.0 moles of Cl_2 to form HCl, calculate the work done (in joules) against a pressure of 1.0 bar at 25°C. What is ΔU for this reaction? Assume the reaction goes to completion.

The Temperature Change of a System Upon Heating Is Governed by Its Heat Capacity

7.13 A 6.22-kg slab of copper metal is heated from 20.5°C to 324.3°C. Calculate the heat absorbed (in kJ) by the metal.

7.14 Calculate the amount of heat liberated (in kJ) from 366 g of mercury when it cools from 77.0°C to 12.0°C.

7.15 A sheet of gold weighing 10.0 g and at a temperature of 18.0°C is placed flat on a sheet of iron weighing 20.0 g and at a temperature of 55.6°C. What is the final temperature of the combined metals? Assume that no heat is lost to the surroundings. (*Hint:* The heat gained by the gold must be equal to the heat lost by the iron. The specific heat capacities of the metals are given in Table 7.1.)

7.16 A 40.0-g sample of copper at 200°C is dropped into a well-insulated vessel containing 100.0 g of H_2O initially at 25°C. Calculate the final temperature of the system given that the specific heat capacity of water is 4.184 J g^{-1} °C^{-1} and the specific heat capacity of copper is 0.385 J g^{-1} °C^{-1}.

7.17 A 0.1375-g sample of solid magnesium is burned in a constant-volume bomb calorimeter that has a heat capacity of 3024 J °C^{-1}. The temperature increases by 1.126°C. Calculate the heat given off by the burning Mg, both in kJ g^{-1} and in kJ mol^{-1}.

7.18 A quantity of 2.00×10^2 mL of 0.862 M HCl is mixed with 2.00×10^2 mL of 0.431 M Ba(OH)$_2$ in a constant-pressure calorimeter of negligible heat capacity. The initial temperatures of the HCl and Ba(OH)$_2$ solutions are both 20.48°C. The heat of neutralization is 56.2 kJ mol^{-1} for

$$H^+(aq) + OH^-(aq) \longrightarrow H_2O(l)$$

What is the final temperature of the mixed solution?

7.19 Determine the equipartition predictions for the heat capacities of CO, N_2O, O_3, and H_2CO both including and excluding the vibrational contribution. Compare these to the experimental values at 25°C. How important are the vibrational contributions in each case?

7.20 The average temperature in deserts is high during the day but quite cool at night, whereas that in regions along the coastline is more moderate. Explain.

The Enthalpy Change for Any Reaction Can Be Calculated Using Standard Enthalpies of Formation

7.21 Which is the more negative quantity at 25°C: ΔH_f° for $H_2O(l)$ or ΔH_f° for $H_2O(g)$?

7.22 Predict the value of ΔH_f° (greater than, less than, or equal to zero) for the following elements at 25°C: (a) Br$_2(g)$, Br$_2(l)$; and (b) I$_2(g)$, I$_2(s)$.

7.23 In general, compounds with negative ΔH_f° values are more stable than those with positive ΔH_f° values. $H_2O_2(l)$ has a negative ΔH_f° (see Table 7.5). Why, then, does $H_2O_2(l)$ have a tendency to decompose to $H_2O(l)$ and $O_2(g)$?

7.24 Suggest ways (with appropriate equations) that would allow you to measure the ΔH_f° values of Ag$_2O(s)$ and CaCl$_2(s)$ from their elements. No calculations are necessary.

7.25 Calculate the heat of decomposition for the following process at constant pressure and 25°C:

$$CaCO_3(s) \longrightarrow CaO(s) + CO_2(g)$$

(*Hint:* You can find the standard enthalpy of formation of the reactant and products in Table 7.5.)

7.26 The standard enthalpies of formation of ions in aqueous solutions are obtained by arbitrarily assigning a value of zero to H^+ ions, that is, ΔH_f° $[H^+(aq)] = 0$.

(a) For the following reaction

$$HCl(g) \xrightarrow{\text{H}_2\text{O}} H^+(aq) + Cl^-(aq)$$
$$\Delta H_f^\circ = -74.9 \text{ kJ mol}$$

calculate ΔH_f° for the Cl^- ions.

(b) Given that ΔH_f° for OH^- ions is -229.6 kJ mol^{-1}, calculate the enthalpy of neutralization when 1 mole of a strong monoprotic acid (such as HCl) is neutralized by 1 mole of a strong base (such as KOH) at 25°C. Assume complete neutralization.

7.27 Calculate the heats of combustion for the following reactions from the standard enthalpies of formation listed in Appendix 2:

(a) $2H_2(g) + O_2(g) \longrightarrow 2H_2O(l)$

(b) $2C_2H_2(g) + 5O_2(g) \longrightarrow 4CO_2(g) + 2H_2O(l)$

(c) $2C_4H_{10}(g) + 13O_2(g) \longrightarrow 8CO_2(g) + 10H_2O(l)$

7.28 Calculate the heats of combustion for the following reactions from the standard enthalpies of formation listed in Appendix 2:

(a) $C_2H_4(g) + 3O_2(g) \longrightarrow 2CO_2(g) + 2H_2O(l)$

(b) $2H_2S(g) + 3O_2(g) \longrightarrow 2H_2O(l) + 2SO_2(g)$

(c) $C_3H_8(g) + 5O_2(g) \longrightarrow 3CO_2(l) + 4H_2O(g)$

7.29 Methanol, ethanol, and *n*-propanol are three common alcohols. When 1.00 g of each of these alcohols is burned in air, heat is liberated as shown by the following data: (a) methanol (CH_3OH), -22.6 kJ; (b) ethanol (C_2H_5OH), -29.7 kJ; and (c) *n*-propanol (C_3H_7OH), -33.4 kJ. Calculate the heats of combustion of these alcohols in kJ mol^{-1}.

7.30 The standard enthalpy change for the following reaction is 436.4 kJ mol^{-1}:

$$H_2(g) \longrightarrow H(g) + H(g)$$

Calculate the standard enthalpy of formation of atomic hydrogen (H).

7.31 At 850°C, $CaCO_3$ undergoes substantial decomposition to yield CaO and CO_2. Assuming that the ΔH_f° values of the reactant and products are the same at 850°C as they are at 25°C, calculate the enthalpy change (in kJ) if 66.8 g of CO_2 is produced in one reaction.

7.32 From these data,

$$S(\text{rhombic}) + O_2(g) \longrightarrow SO_2(g)$$
$$\Delta H_{rxn}^\circ = -296.84 \text{ kJ mol}^{-1}$$

$$S(\text{monoclinic}) + O_2(g) \longrightarrow SO_2(g)$$
$$\Delta H_{rxn}^\circ = -297.16 \text{ kJ mol}^{-1}$$

calculate the enthalpy change for the transformation

$$S(\text{rhombic}) \longrightarrow S(\text{monoclinic})$$

(Monoclinic and rhombic are different allotropic forms of elemental sulfur.)

7.33 From the following data,

$$C(\text{graphite}) + O_2(g) \longrightarrow CO_2(g)$$
$$\Delta H_{rxn}^\circ = -393.5 \text{ kJ mol}^{-1}$$

$$H_2(g) + \frac{1}{2}O_2(g) \longrightarrow H_2O(l)$$
$$\Delta H_{rxn}^\circ = -285.8 \text{ kJ mol}^{-1}$$

$$2C_2H_6(g) + 7O_2(g) \longrightarrow 4CO_2(g) + 6H_2O(l)$$
$$\Delta H_{rxn}^\circ = -3119.6 \text{ kJ mol}^{-1}$$

calculate the enthalpy change for the reaction $2C(\text{graphite}) + 3H_2(g) \longrightarrow C_2H_6(g)$.

7.34 From the following heats of combustion,

$$CH_3OH(l) + \frac{3}{2}O_2(g) \longrightarrow CO_2(g) + 2H_2O(l)$$
$$\Delta H_{rxn}^\circ = -726.4 \text{ kJ mol}^{-1}$$

$$C(\text{graphite}) + O_2(g) \longrightarrow CO_2(g)$$
$$\Delta H_{rxn}^\circ = -393.5 \text{ kJ mol}^{-1}$$

$$H_2(g) + \frac{1}{2}O_2(g) \longrightarrow H_2O(l)$$
$$\Delta H_{rxn}^\circ = -285.8 \text{ kJ mol}^{-1}$$

calculate the standard enthalpy of formation of methanol (CH_3OH) from its elements.

7.35 The convention of arbitrarily assigning a zero enthalpy value for the most stable form of each element in the standard state at 25°C is a convenient way of dealing with enthalpies of reactions. Explain why this convention cannot be applied to nuclear reactions.

The Reaction Enthalpies Can Be Estimated from Bond Enthalpies

7.36 Explain why the bond enthalpy of a molecule is always defined in terms of a gas-phase reaction.

7.37 The bond enthalpy of F_2 is 150.6 kJ mol^{-1}. Calculate the value of ΔH_f° for $F(g)$.

7.38 Use the bond enthalpy data in Table 7.6 to calculate the enthalpies of formation of the following gases: (a) $H_2O(g)$, (b) NH_3, (c) H_2O_2, (d) H_2NCl.

7.39 Use the bond enthalpy values in Table 7.6 to calculate the enthalpy of combustion for ethane (C_2H_6):

$$C_2H_6(g) + \frac{7}{2}O_2(g) \longrightarrow 2CO_2(g) + 3H_2O(g)$$

7.40 Alkanes are hydrocarbons with the general formula C_nH_{2n+2}, where *n* is a positive integer. Using the data from Table 7.6, derive general formulas for the enthalpies of combustion (in kJ g^{-1}) for gas-phase alkanes as functions of *n*.

Enthalpy Changes Also Accompany Physical Transformations

7.41 Calculate the amount of heat (in kJ) required to convert 74.6 g of water to steam at 100°C.

7.42 How much heat (in kJ) is needed to convert 866 g of ice at -10°C to steam at 126°C? (The specific heat capacities of ice and steam are 2.03 and 1.99 J g^{-1} °C^{-1}, respectively.)

7.43 The molar enthalpies of fusion and sublimation of molecular iodine are 15.27 and 62.30 kJ mol^{-1}, respectively. Estimate the molar heat of vaporization of liquid iodine.

7.44 The molar enthalpies of fusion and vaporization of water at 298 K are 6.01 and 44.01 kJ mol^{-1}, respectively. Use these values to estimate the molar enthalpy of sublimation of ice.

7.45 The enthalpy of vaporization is found to decrease with increasing temperature for a certain substance. What information does this give you about the relative properties of the liquid and vapor phases?

The Temperature Dependence of Reaction Enthalpies Can Be Determined from Heat Capacity Data

7.46 The hydrogenation of ethylene is given by the following equation:

$$C_2H_4(g) + H_2(g) \longrightarrow C_2H_6(g)$$

Using data from Appendix 2, calculate the change in the enthalpy of hydrogenation from 298 K to 398 K. Assume the heat capacities are temperature independent.

7.47 Use the data in Appendix 2 to calculate the value of ΔH°_{rxn} for the following reaction at 298 K:

$$N_2O_4(g) \longrightarrow 2NO_2(g)$$

What is its value at 350 K? State any assumptions in your calculation.

7.48 The heat of fusion of ice at the standard melting point (0°C) is 6.01 kJ mol^{-1}. Determine the heat of fusion at -10°C using data from Appendix 2.

Additional Problems

7.49 From the enthalpy of vaporization of water at its standard boiling point and the bond enthalpies of H_2 and O_2, calculate the average O—H bond energy in liquid water, given that

$$H_2(g) + \frac{1}{2}O_2(g) \longrightarrow H_2O(l)$$

$$\Delta H^\circ_{rxn} = -285.8 \text{ kJ mol}^{-1}$$

7.50 Calculate the fraction of the enthalpy of vaporization of water used for the expansion of steam at its standard boiling point.

7.51 The combustion of what volume of ethane (C_2H_6), measured at 23.0°C and 752 mm Hg, would be

required to heat 855 g of water from 25.0°C to 98.0°C?

7.52 The standard enthalpy change (ΔH°_{rxn}) for the thermal decomposition of silver nitrate according to the following equation is 78.67 kJ mol^{-1}:

$$AgNO_3(s) \longrightarrow AgNO_2(s) + \frac{1}{2}O_2(g)$$

The standard enthalpy of formation of $AgNO_3(s)$ is -123.02 kJ mol^{-1}. Calculate the standard enthalpy of formation of $AgNO_2(s)$.

7.53 Hydrazine (N_2H_4) decomposes according to the following equation:

$$3N_2H_4(l) \longrightarrow 4NH_3(g) + N_2(g)$$

(a) Given that the standard enthalpy of formation of hydrazine is 50.63 kJ mol^{-1}, calculate ΔH°_{rxn} for its decomposition. (b) Both hydrazine and ammonia burn in oxygen to produce $H_2O(l)$ and $N_2(g)$. Write balanced equations for each of these processes and calculate ΔH°_{rxn} for each of them. On a mass basis (per kg), would hydrazine or ammonia be the better fuel?

7.54 Consider the equation

$$N_2(g) + 3H_2(g) \longrightarrow 2NH_3(g)$$
$$\Delta H^\circ_{rxn} = -91.8 \text{ kJ mol}^{-1}$$

If 2.0 moles of N_2 react with 6.0 moles of H_2 to form NH_3, calculate the work done (in joules) against a pressure of 1.0 bar at 25°C. What is ΔU° for this reaction? Assume the reaction goes to completion.

7.55 Calculate the heat released when 2.00 L of $Cl_2(g)$ (density = 1.88 g L^{-1}) reacts with an excess of sodium metal as 25°C and 1 bar to form sodium chloride.

7.56 Photosynthesis produces glucose ($C_6H_{12}O_6$) and oxygen from carbon dioxide and water:

$$6CO_2 + 6H_2O \longrightarrow C_6H_{12}O_6 + 6O_2$$

(a) How would you determine experimentally the ΔH°_{rxn} value for this reaction? (b) Solar radiation produces about 7.0×10^{14} kg of glucose a year on Earth. What is the corresponding ΔH° change?

7.57 A 2.10-mole sample of crystalline acetic acid, initially at 17.0°C, is allowed to melt at 17.0°C and is then heated to 118.1°C (its normal boiling point) at 1.00 atm. The sample is allowed to vaporize at 118.1°C and is then rapidly quenched to 17.0°C, so that it recrystallizes. Calculate ΔH° for the total process as described.

7.58 Calculate the work done in joules by the following equation:

$$2Na(s) + 2H_2O(l) \longrightarrow 2NaOH(aq) + H_2(g)$$

when 0.34 g of Na reacts with water to form hydrogen gas at 0°C and 1.0 bar.

7.59 You are given the following data:

$$H_2(g) \longrightarrow 2H(g) \qquad \Delta H^\circ_{rxn} = 436.4 \text{ kJ mol}^{-1}$$
$$Br_2(g) \longrightarrow 2Br(g) \qquad \Delta H^\circ_{rxn} = 192.5 \text{ kJ mol}^{-1}$$
$$H_2(g) + Br_2(g) \longrightarrow 2HBr(g)$$
$$\Delta H^\circ_{rxn} = -72.4 \text{ kJ mol}^{-1}$$

Calculate ΔH°_{rxn} for the reaction

$$H(g) + Br(g) \longrightarrow HBr(g)$$

7.60 A 44.0-g sample of an unknown metal at 99.0°C was placed in a constant-pressure calorimeter containing 80.0 g of water at 24.0°C. The final temperature of the system was found to be 28.4°C. Calculate the specific heat capacity of the metal. (The heat capacity of the calorimeter is 12.4 J °C^{-1}.)

7.61 A 1.00-mole sample of ammonia at 14.0 bar and 25°C in a cylinder fitted with a movable piston expands against a constant external pressure of 1.00 bar. At equilibrium, the pressure and volume of the gas are 1.00 bar and 23.5 L, respectively. (a) Calculate the final temperature of the sample. (b) Calculate q, w, and ΔU for the process. The specific heat capacity of ammonia is 0.0258 J g °C^{-1}.

7.62 Producer gas (carbon monoxide) is prepared by passing air over red-hot coke:

$$C(s) + \frac{1}{2}O_2(g) \longrightarrow CO(g)$$

Water gas (a mixture of carbon monoxide and hydrogen) is prepared by passing steam over red-hot coke:

$$C(s) + H_2O(g) \longrightarrow CO(g) + H_2(g)$$

For many years, both producer gas and water gas were used as fuels in industry and for domestic cooking. The large-scale preparation of these gases was carried out alternately, that is, first producer gas, then water gas, and so on. Using thermochemical reasoning, explain why this procedure was chosen.

7.63 Compare the heat produced by the complete combustion of 1.00 mole of methane (CH$_4$) with 1.00 mol of water gas (0.50 mol H$_2$ and 0.50 mol CO) under the same conditions. On the basis of your answer, would you prefer methane over water gas as a fuel? Can you suggest two other reasons why methane is preferable to water gas as a fuel?

7.64 The so-called hydrogen economy is based on hydrogen produced from water using solar energy. The gas is then burned as a fuel:

$$2H_2(g) + O_2(g) \longrightarrow 2H_2O(l)$$

A primary advantage of hydrogen as a fuel is that it is nonpolluting. A major disadvantage is that it is a gas and therefore is harder to store than liquids or solids. Calculate the volume of hydrogen gas at 25°C and 1.00 bar required to produce an amount of energy equivalent to that produced by the combustion of a liter of octane (C$_8$H$_{18}$). The density of octane is 2.66 kg gal^{-1}, and its standard enthalpy of formation is -249.9 kJ mol^{-1}.

7.65 Ethanol (C$_2$H$_5$OH) and gasoline [assumed to be all octane (C$_8$H$_{18}$)] are both used as automobile fuel. If gasoline is selling for $3.00 per gal, what would the price of ethanol have to be in order to provide the same amount of heat per dollar? The density and ΔH°_f of octane are 0.7025 g mL^{-1} and -249.9 kJ mol^{-1}, respectively, and of ethanol are 0.7894 g mL^{-1} and -277.0 kJ mol^{-1}, respectively. (1 gal = 3.785 L.)

7.66 The molar enthalpy of vaporization of a liquid (ΔH_{vap}) is the energy required to vaporize 1.00 mol of the liquid. In one experiment, 60.0 g of liquid nitrogen (boiling point -196°C) is poured into a Styrofoam cup containing 2.00×10^2 g of water at 55.3°C. Calculate the molar enthalpy of vaporization of liquid nitrogen if the final temperature of the water is 41.0°C.

7.67 A quantity of 0.020 mole of a gas initially at 0.050 L and 20°C undergoes a constant-temperature expansion until its volume is 0.50 L. Calculate the work done (in joules) by the gas if it expands (a) against a vacuum and (b) against a constant pressure of 0.20 bar. (c) If the gas in (b) is allowed to expand unchecked until its pressure is equal to the external pressure, what would its final volume be before it stopped expanding, and what would be the work done?

7.68 (a) For most efficient use, refrigerator freezer compartments should be fully packed with food. What is the thermochemical basis for this recommendation? (b) Starting at the same temperature, tea and coffee remain hot longer in a thermal flask than chicken noodle soup. Explain.

7.69 Portable hot packs are available for skiers and people engaged in other outdoor activities in a cold climate. The air-permeable paper packet contains a mixture of powdered iron, sodium chloride, and other components, all moistened by a little water. The exothermic reaction that produces the heat (the rusting of iron) is

$$4Fe(s) + 3O_2(g) \longrightarrow 2Fe_2O_3(s)$$

When the outside plastic envelope is removed, O$_2$ molecules penetrate the paper, causing the reaction to begin. A typical packet contains 250 g of iron to warm your hands or feet for up to 4 hours. How much heat (in kJ) is produced by this reaction? (*Hint:* See Appendix 2 for ΔH_f values.)

7.70 A man ate 0.50 pound of cheese (an energy intake of 4000 kJ). Suppose that none of the energy was stored in his body. What mass (in grams) of water would he need to perspire in order to maintain his original temperature? (It takes 44.0 kJ of heat to vaporize 1 mole of water.)

7.71 The total volume of the Pacific Ocean is estimated to be 7.2×10^8 km^3. A medium-sized atomic bomb produces 1.0×10^{15} J of energy upon explosion. Calculate the number of atomic bombs needed to release enough energy to raise the temperature of the water in the Pacific Ocean by 1°C.

7.72 A 19.2-g quantity of dry ice (solid carbon dioxide) is allowed to sublime in an apparatus like the one shown in Figure 7.2. Calculate the expansion work done against a constant external pressure of 0.995 bar and at a constant temperature of 22°C. Assume that the initial volume of dry ice is negligible and that CO_2 behaves like an ideal gas.

7.73 The enthalpy of combustion of benzoic acid (C_6H_5COOH) is commonly used as the standard for calibrating constant-volume bomb calorimeters; its value has been accurately determined to be -3226.7 kJ mol^{-1}. When 1.9862 g of benzoic acid are burned in a calorimeter, the temperature rises from 21.84°C to 25.67°C. What is the heat capacity of the bomb? (Assume that the quantity of water surrounding the bomb is exactly 2000 g.)

7.74 *Lime* is a term that includes calcium oxide (CaO, also called quicklime) and calcium hydroxide [Ca(OH)$_2$, also called slaked lime]. It is used in the steel industry to remove acidic impurities, in air-pollution control to remove acidic oxides such as SO_2, and in water treatment. Quicklime is made industrially by heating limestone ($CaCO_3$) above 2000°C:

$$CaCO_3(s) \longrightarrow CaO(s) + CO_2(g)$$
$$\Delta H^\circ_{rxn} = 177.8 \text{ kJ mol}^{-1}$$

Slaked lime is produced by treating quicklime with water:

$$CaO(s) + H_2O(l) \longrightarrow Ca(OH)_2(s)$$
$$\Delta H^\circ_{rxn} = -65.2 \text{ kJ mol}^{-1}$$

The exothermic reaction of quicklime with water and the rather small specific heats of both quicklime (0.946 J g^{-1} °C^{-1}) and slaked lime (1.20 J g^{-1} °C^{-1}) make it hazardous to store and transport lime in vessels made of wood. Wooden sailing ships carrying lime would occasionally catch fire when water leaked into the hold. (a) If a 500-g sample of water reacts with an equimolar amount of CaO (both at an initial temperature of 25°C), what is the final temperature of the product, Ca(OH)$_2$? Assume that the product absorbs all the heat released in the reaction. (b) Given that the standard enthalpies of formation of CaO and H$_2$O are -635.6 and -285.8 kJ mol^{-1}, respectively,

calculate the standard enthalpy of formation of Ca(OH)$_2$.

7.75 Calcium oxide (CaO) is used to remove sulfur dioxide generated by coal-burning power stations: $2CaO(s) + 2SO_2(g) + O_2(g) \longrightarrow 2CaSO_4(s)$. Calculate the enthalpy change for this process if 6.6×10^5 g of SO_2 are removed by this process every day.

7.76 Glauber's salt, sodium sulfate decahydrate (Na$_2$SO$_4$ · 10H$_2$O), undergoes a phase transition (that is, melting or freezing) at a convenient temperature of about 32°C

$$Na_2SO_4 \cdot 10H_2O(s) \longrightarrow Na_2SO_4 + 10H_2O(l)$$
$$\Delta H^\circ_{rxn} = 74.4 \text{ kJ mol}^{-1}$$

As a result, this compound is used to regulate the temperature in homes. It is placed in plastic bags in the ceiling of a room. During the day, the endothermic melting process absorbs heat from the surroundings, cooling the room. At night, it gives off heat as it freezes. Calculate the mass of Glauber's salt in kilograms needed to lower the temperature of air in a room by 8.2°C at 1.0 bar. The dimensions of the room are 2.80 m × 10.6 m × 17.2 m, the specific heat of air is 1.2 J g^{-1} °C^{-1}, and the molar mass of air may be taken as 29.0 g mol^{-1}.

7.77 A balloon 16 m in diameter is inflated with helium at 18°C. (a) Calculate the mass of He in the balloon, assuming ideal behavior. (b) Calculate the work done (in joules) during the inflation process if the atmospheric pressure is 98.7 kPa.

7.78 An excess of zinc metal is added to 50.0 mL of a 0.100 M AgNO$_3$ solution in a constant-pressure calorimeter like the one depicted in Figure 7.12. As a result of the reaction

$$Zn(s) + 2Ag^+(aq) \longrightarrow Zn^{2+}(aq) + 2Ag(s)$$

the temperature rises from 19.25°C to 22.17°C. If the heat capacity of the calorimeter is 98.6 J °C^{-1}, calculate the enthalpy change for this reaction on a molar basis. Assume that the density and specific heat capacity of the solution are the same as those for water, and ignore the specific heat capacities of the metals.

7.79 (a) A person drinks four glasses of cold water (3.0°C) every day. The volume of each glass is 2.5×10^2 mL. How much heat (in kJ) does the body have to supply to raise the temperature of the water to 37°C (the normal body temperature)? (b) How much heat would your body lose if you were to ingest 8.0×10^2 g of snow at 0°C to quench your thirst? (The amount of heat necessary to melt snow is 6.01 kJ mol^{-1}.)

7.80 Equal amounts of heat are added to two containers. One contains 1.0 mol of carbon dioxide and the other contains 1.0 mol of ammonia. If both are originally at the same temperature, which container (if any) will have the higher temperature at the end? Use molecular reasoning to justify your answer.

7.81 A driver's manual states that the stopping distance quadruples as the speed (u) doubles, that is, if it takes 30 ft to stop a car moving at 25 mph, then it would take 120 ft to stop a car moving at 50 mph. Justify this statement by using mechanics and the first law of thermodynamics. (Assume that when a car is stopped, its kinetic energy ($mu^2/2$) is completely converted to heat.)

7.82 At 25°C the standard enthalpy of formation of HF(aq) is 320.1 kJ mol^{-1}; of OH$^-$(aq), it is -229.6 kJ mol^{-1}; of F$^-$(aq), it is -329.1 kJ mol^{-1}; and of H$_2$O(l), it is -285.8 kJ mol^{-1}. (a) Calculate the standard enthalpy of neutralization of HF(aq):

$$\text{HF}(aq) + \text{OH}^-(aq) \longrightarrow \text{F}^-(aq) + \text{H}_2\text{O}(l)$$

(b) Using the value of -56.2 kJ mol^{-1} as the standard enthalpy change for the reaction

$$\text{H}^+(aq) + \text{OH}^-(aq) \longrightarrow \text{H}_2\text{O}(l)$$

calculate the standard enthalpy change for the reaction

$$\text{HF}(aq) \longrightarrow \text{H}^+(aq) + \text{F}^-(aq)$$

7.83 Why are cold, damp air and hot, humid air more uncomfortable than dry air at the same temperatures? (The specific heat capacities of water vapor and air are approximately 1.9 and 1.0 J g^{-1} °C^{-1}, respectively.)

7.84 A 46-kg person drinks 500 g of milk, which has a "caloric" value of approximately 3.0 kJ g^{-1}. If only 17 percent of the energy in milk is converted to mechanical work, how high (in meters) can the person climb based on this energy intake? [*Hint:* The work done in ascending is given by mgh, where m is the mass (in kg), g is the gravitational acceleration (9.8 m s^{-2}), and h is the height (in meters).]

7.85 The height of Niagara Falls on the American side is 51 m. (a) Calculate the potential energy of 1.0 g of water at the top of the falls relative to the ground level. (b) What is the speed of the falling water if all the potential energy is converted to kinetic energy? (c) What would be the increase in temperature of the water if all the kinetic energy were converted to heat? (See the Problem 7.84 for suggestions.)

7.86 In early 2005, the European Space Agency's Huygens Probe landed on Saturn's largest moon Titan. Soon after the 319-kg probe impacted the surface at a speed of 5 m s^{-1}, a sudden increase in methane gas concentration was detected. The methane is believed to have been produced by the conversion of liquid methane in the soil to vapor as a result of heating from the impact. Determine the maximum amount of liquid methane that could have been evaporated in this process. State any assumptions that you make.

7.87 In the nineteenth century two scientists named Dulong and Petit noticed that for a solid element, the product of its molar mass and its specific heat capacity is approximately 25 J °C^{-1}. This observation, now called Dulong and Petit's law, was used to estimate the specific heat capacity of metals. Verify the law for the metals listed in Table 7.1. The law does not apply to one of the metals. Which one is it? Why?

7.88 Determine the standard enthalpy of formation of ethanol (C$_2$H$_5$OH) from its standard enthalpy of combustion (-1367.4 kJ mol^{-1}).

7.89 Acetylene (C$_2$H$_2$) and benzene (C$_6$H$_6$) have the same empirical formula. In fact, benzene can be made from acetylene as follows:

$$3\text{C}_2\text{H}_2(g) \longrightarrow \text{C}_6\text{H}_6(l)$$

The enthalpies of combustion for C$_2$H$_2$ and C$_6$H$_6$ are -1299.4 kJ mol^{-1} and -3267.4 kJ mol^{-1}, respectively. Calculate the standard enthalpies of formation of C$_2$H$_2$ and C$_6$H$_6$ and hence the enthalpy change for the formation of C$_6$H$_6$ from C$_2$H$_2$.

7.90 Ice at 0°C is placed in a Styrofoam cup containing 361 g of a soft drink at 23°C. The specific heat of the drink is about the same as that of water. Some ice remains after the ice and soft drink reach an equilibrium temperature of 0°C. Determine the mass of ice that has melted. Ignore the heat capacity of the cup. (*Hint:* It takes 334 J to melt 1 g of ice at 0°C.)

7.91 A gas company in Massachusetts charges $1.30 for 15 ft^3 of natural gas (CH$_4$) measured at 20°C and 1.0 bar. Calculate the cost of heating 200 mL of water (enough to make a cup of coffee or tea) from 20°C to 100°C. Assume that only 50 percent of the heat generated by the combustion is used to heat the water; the rest of the heat is lost to the surroundings.

7.92 Suppose 100 g of Al metal at 250°C is added to 100 g of ice at -15°C. Assuming no heat is lost to the surroundings, determine (a) the final temperature and (b) the amounts of each phase present.

7.93 Calculate the internal energy of a Goodyear blimp filled with helium gas at 1.2×10^5 Pa. The volume of the blimp is 5.5×10^3 m^3. If all the energy were used to heat 10.0 tons of copper at 21°C, calculate the final temperature of the metal (1 ton = 9.072×10^5 g).

7.94 Decomposition reactions are usually endothermic, whereas combination reactions are usually exothermic. Give a qualitative explanation for these trends.

7.95 Acetylene (C$_2$H$_2$) can be made by reacting calcium carbide (CaC$_2$) with water. (This reaction was once employed in lamps used in cave exploration.) (a) Write an equation for the reaction. (b) What is the maximum amount of heat (in joules) that can be obtained from the combustion of acetylene, starting with 74.6 g of CaC$_2$?

7.96 When 1.034 g of naphthalene ($C_{10}H_8$) is burned in a constant-volume bomb calorimeter at 298 K, 41.56 kJ of heat is evolved. Calculate ΔU and ΔH for the reaction on a molar basis.

7.97 From a thermochemical point of view, explain why a carbon dioxide fire extinguisher or water should not be used on a magnesium fire.

7.98 Consider the reaction

$$CO(g) + \frac{1}{2}O_2(g) \longrightarrow CO_2(g)$$

Without doing any calculation and basing your answer on molecular considerations alone, will the enthalpy change for this reaction increase or decrease with increasing temperature?

7.99 A 4.117-g impure sample of glucose ($C_6H_{12}O_6$) was burned in a constant-volume calorimeter having a heat capacity of 19.65 kJ °C^{-1}. If the rise in temperature is 3.134°C, calculate the percent by mass of the glucose in the sample. Assume that the impurities are unaffected by the combustion process. See Appendix 2 for thermodynamic data.

7.100 Construct a table with the headings q, w, ΔU, and ΔH. For each of the following processes, deduce whether each of the quantities listed is positive (+), negative (−), or zero. (a) Freezing of benzene. (b) Compression of an ideal gas at constant temperature. (c) Reaction of sodium with water. (d) Boiling liquid ammonia. (e) Heating a gas at constant volume. (f) Melting of ice.

7.101 The combustion of 0.4196 g of a hydrocarbon releases 17.55 kJ of heat. The masses of the products are $CO_2 = 1.419$ g and $H_2O = 0.290$ g. (a) What is the empirical formula of the compound? (b) If the approximate molar mass of the compound is 76 g, calculate its standard enthalpy of formation.

7.102 Metabolic activity in the human body releases approximately 1.0×10^4 kJ of heat per day. Assuming the body contains 50 kg of water, how much would the body temperature rise if it were an isolated system? How much water must the body eliminate as perspiration to maintain the normal body temperature (37.0 °C)? Comment on your results. The heat of vaporization of water is 2.41 kJ g^{-1}.

7.103 For a diatomic molecule, the vibrational contribution to the heat capacity can be calculated using statistical mechanics, assuming that the vibrations are well modeled by a harmonic oscillator:

$$\bar{C}_V(\text{vibration}) = \frac{\Theta_v^2}{T^2}\left(\frac{e^{-\Theta_v/2T}}{1 - e^{-\Theta_v/T}}\right)^2 R$$

The quantity $\Theta_v = hc\bar{v}/k$ is called the *vibrational temperature,* where h is Planck's constant, c is the speed of light, k is Boltzmann's constant, and $\bar{v}$ is the vibrational frequency in wave numbers (cm^{-1}). The vibrational frequencies for F_2 and I_2 are 919 and 214 cm^{-1}, respectively. Use the formula above, together with the equipartition values for the translational and rotational contributions, to calculate the constant-pressure heat capacity for these two diatomic molecules. Compare your results with the experimental values given in Table 7.4.

7.104 Starting at A, an ideal gas undergoes a cyclic process involving expansion and compression at constant temperature as shown. Calculate the total work done. Does your result support the notion that work is not a state function?

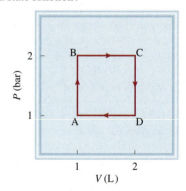

7.105 Calculate the ΔU for the following reaction at 298 K

$$2H_2(g) + O_2(g) \longrightarrow 2H_2O(l)$$

7.106 From the following data, calculate the heat of solution for KI:

	NaCl	NaI	KCl	KI
Lattice energy (kJ mol^{-1})	788	686	699	632
Heat of solution (kJ mol^{-1})	4.0	−5.1	17.2	?

Answers to Practice Exercises

7.1 (a) 0; (b) 283 J **7.2** $w = 1.11$ kJ; $q = -1.11$ kJ; $\Delta U = 0$
7.3 -6.47×10^3 kJ **7.4** -111.7 kJ mol^{-1} **7.5** $\Delta U = 693$ J;
$\Delta H = 893$ J **7.6** -728 kJ mol^{-1} **7.7** 21.19°C **7.8** 22.49°C

7.9 3.0 J mol^{-1} K^{-1}; 9.3% **7.10** 87.3 kJ mol^{-1} **7.11** 40.1 kJ g^{-1}
7.12 (a) -543.1 kJ mol^{-1}; (b) -543.2 kJ mol^{-1} **7.13** (a)
-119 kJ mol^{-1}; (b) -137 kJ mol^{-1} **7.14** -57.60 kJ mol^{-1}

8

Entropy, Free Energy, and the Second Law of Thermodynamics

According to the first law of thermodynamics (Chapter 7), energy can be neither created nor destroyed, but flows from one part of the universe to another or is converted from one form to another. The total amount of energy in the universe remains constant. Despite its immense value in the study of energy changes in chemical reactions, the first law is still limited because it cannot predict the direction of change. It helps us to do the bookkeeping of energy balance, such as the energy input, energy released, and work done, but it reveals nothing about whether a particular process can indeed occur. For this kind of information, we must turn to the second law of thermodynamics and the concept of entropy.

8.1 | The Entropy of an Isolated System Always Increases in Any Spontaneous Process

One of the main objectives in studying thermodynamics, as far as chemists are concerned, is to be able to predict whether or not a process or reaction will occur under a specific set of conditions (such as temperature, pressure, and concentration). This knowledge is useful whether we are synthesizing compounds in a research laboratory, manufacturing chemicals on an industrial scale, or trying to understand the intricate biological processes in a cell.

Spontaneous Processes

A **spontaneous process** *occurs without external intervention under the given set of conditions.* A process is *nonspontaneous* if it does *not* occur without external intervention under the specified set of conditions.

We observe spontaneous physical and chemical processes every day. For example, a lump of sugar when placed in hot coffee will dissolve on its own. Because no external intervention is required to make the sugar dissolve, this is a spontaneous process. The reverse process—precipitation of the sugar from the hot coffee—is nonspontaneous. It will never occur—no matter how long you wait. *In general, processes that occur spontaneously in one direction cannot, under the same conditions, also take place spontaneously in the reverse direction.* Other examples of spontaneous processes include the following:

▶ Heat flows from a hotter object to a colder one, but the reverse never happens spontaneously.

▶ The expansion of a gas into an evacuated bulb is spontaneous [Figure 8.1(a)], but the reverse process (the concentration of all the molecules into one bulb) is nonspontaneous [Figure 8.1(b)].

▶ A piece of sodium metal reacts violently with water to form sodium hydroxide and hydrogen gas, but hydrogen gas does not react with sodium hydroxide to form water and sodium metal.

▶ Iron exposed to water and oxygen forms rust, but rust does not spontaneously change back to iron.

Figure 8.1 (a) A spontaneous process. (b) A nonspontaneous process—the reverse of (a).

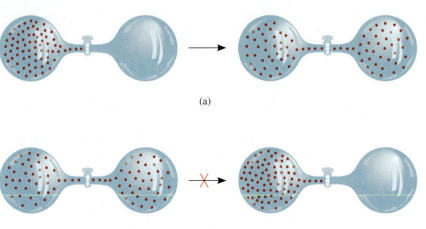

(a)

(b)

The term *spontaneous* says nothing about the speed with which a process takes place—only that it will take place *if given enough time*. For example, it may take days or weeks for wet iron to rust, but the process is still spontaneous.

A spontaneous process does not necessarily mean an instantaneous one.

If we assume that spontaneous processes occur so as to decrease the energy of a system, we can explain why a ball rolls downhill and why springs in a clock unwind. Similarly, many exothermic reactions are spontaneous, such as the combustion of methane and acid-base neutralization reactions:

$$CH_4(g) + 2O_2(g) \longrightarrow CO_2(g) + 2H_2O(l) \qquad \Delta H° = -890.4 \text{ kJ mol}^{-1}$$
$$H^+(aq) + OH^-(aq) \longrightarrow H_2O(l) \qquad \Delta H° = -56.2 \text{ kJ mol}^{-1}$$

However, consider a solid-to-liquid phase transition such as the melting of ice,

$$H_2O(s) \longrightarrow H_2O(l) \qquad \Delta H° = +6.01 \text{ kJ mol}^{-1}$$

It is spontaneous above 0°C even though the process is endothermic. Another process that is spontaneous, as well as endothermic, is dissolution of ammonium nitrate in water,

$$NH_4NO_3(s) \xrightarrow{H_2O} NH_4^+(aq) + NO_3^-(aq) \qquad \Delta H° = 25 \text{ kJ mol}^{-1}$$

The decomposition of mercury(II) oxide is endothermic and nonspontaneous at room temperature, but it becomes spontaneous when the temperature is raised:

$$2HgO(s) \longrightarrow 2Hg(l) + O_2(g) \qquad \Delta H° = 90.7 \text{ kJ mol}^{-1}$$

A spontaneous and nonspontaneous process

Based on these examples and many others, the most we can conclude is that exothermicity *favors* the spontaneity of a reaction but does *not* guarantee it. Just as it is possible for an endothermic reaction to be spontaneous, an exothermic reaction can be nonspontaneous. In other words, we cannot determine whether a chemical reaction will occur spontaneously solely on the basis of energy changes in the system. To determine the spontaneity of a reaction, we need another thermodynamic quantity called *entropy.*

The Statistical Definition of Entropy

Whether a given macroscopic process is spontaneous or not depends upon what is happening at the molecular level. Consider, for example, the expansion of a gas into an evacuated bulb as depicted in Figure 8.1(a). When the stopcock is opened, gas molecules (all initially confined to bulb A) will flow spontaneously into the evacuated bulb B until the pressure of the gas is the same in both bulbs. Why does this occur? The motions of the gas molecules in bulb A are random; at any given time roughly the same number of molecules are moving in every direction. When the stopcock is opened, the molecules near the stopcock that are moving to the right will enter bulb B. As long as there are fewer molecules in bulb B, there will be more molecules moving from A to B than from B to A. This spontaneous process will continue until the pressures equalize.

When heated, HgO decomposes to give Hg and O_2.

Figure 8.1(b) depicts the reverse process—that is, the spontaneous movement of all the gas molecules in bulb B back into bulb A. For this to happen, all the molecules in B would have to move to the left *at the same time*. Many of the molecules are indeed moving to the left at any given time; however, because molecular motion is random, just as many molecules will be moving to the right on average. The probability that all the molecules in bulb B will move *simultaneously* to the left is vanishingly small. This process is nonspontaneous not because it is strictly impossible, but because it is so *improbable* that it would in practice never be observed. In trying to understand spontaneous processes, we must focus on the *statistical* behavior of a very large number of molecules, not on the motion of just a few of them.

Recall from Chapter 7 that the thermodynamic state (or *macrostate*) of a system is described by relatively few variables (P, T, V, n, etc.), whereas the molecular state (or *microstate*) represents *all* the variables associated with the individual atoms or molecules that make up the system. When we completely specify the macrostate (by specifying, for example, the pressure, temperature, and volume of a sample of argon gas), there are still a very large number of distinct microstates possible. The number of microstates possible for a given macrostate can be represented by the variable W. The value of W, which determines the relative probability that a macrostate will be observed, varies widely from macrostate to macrostate. Macrostates for which W is large are more probable than those for which W is small and there are relatively few possible microstates.

The connection between the number of microstates and probability can be illustrated by rolling dice. In this analogy, we define the microstate as the numbers on each individual die and the macrostate as the sum of the numbers on all the dice thrown. If we have only one die, there are six possible outcomes given by 1, 2, 3, 4, 5, and 6. In this case, the microstates are the same as macrostates because there is only one way to achieve each total number by rolling die. If we have two dice, however, there are 36 (6 × 6) possible microstates, but only 11 possible macrostates (the sum of both dice) represented by the integers from 2 to 12. Therefore, at least some of the macrostates must correspond to multiple microstates. The distribution of the 36 microstates among the 11 macrostates is shown in Figure 8.2. The macrostate with the largest number of microstates, and thus the most probable corresponds to the number 7. This macrostate (7) can be achieved by six different dice combinations (microstates); that is, $W = 6$. The probability of getting 7 is equal to 1/6, which is obtained by dividing the number of microstates that sum to 7 by the total number of possible microstates: 6/36 = 1/6. The least probable microstates correspond to 2 and 12. For each of these macrostates there is only one microstate [(1,1) for 2 and (6,6) for 12]. For these macrostates, $W = 1$, and they each have a probability of 1/36. The probability of rolling a 7 is much greater than rolling a 2 or 12 not because the dice are loaded—each of the 36 possible rolls are equally probable—but because there are more ways to achieve 7 than either 2 or 12.

If we increase the number of dice, the distribution in probability among the macrostates becomes much more dramatic. For example, with six dice there are 46,656 (or 6^6) possible microstates ranging from 6 to 36, for a total of 31 macrostates. For the macrostates 6 and 36 there is only one possible microstate [(1,1,1,1,1,1) and (6,6,6,6,6,6), respectively], so they both have $W = 1$ and a probability of 1/46,656. There are, however, 4332 possible ways of rolling a 21 (the most probable macrostate) giving a probability of 4332/46,656. Thus, we are over 4000 times more likely to roll a 21 than a 6 or 36. As we increase the number of dice, the probability of the most probable macrostate increases rapidly relative to less probable macrostates

Figure 8.2 The microstates that arise from the combination of two dice and the corresponding macrostates.

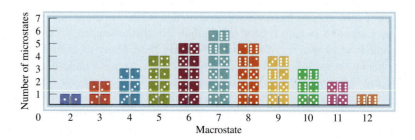

(Figure 8.3). As the number of dice approaches values corresponding to the number of molecules in a macroscopic system (Avogadro's number), the number of microstates corresponding to the most probable macrostate is so large compared to the rest that the probability of rolling anything that deviates more than very slightly from the most probable value is vanishingly small. This example shows how something that is completely random at the microstate level can generate quite predictible behavior when viewed macroscopicly—just as the random motion of molecules in a gas at the molecular level gives rise to the very nonrandom ideal gas law at the macroscopic level.

For a more chemically relevant example, consider a simple system of four molecules moving rapidly between two equal compartments, as shown in Figure 8.4. There is only one way for all four molecules to be in the left compartment, four ways to have three molecules in the left compartment and one in the right compartment, and six ways to have two molecules in each of the two compartments. Thus, there are 11 possible microstates that give rise to 3 possible macrostates.[1] Of the three, macrostate III is the most probable because there are six microstates or six ways to achieve it ($W = 6$). On the other hand, macrostate I is the least probable because it has only one microstate ($W = 1$), so there is only one way to achieve it. As the number of molecules approaches macroscopic scale (Avogadro's number), the molecules will be evenly distributed between the two compartments because this macrostate has many, many more microstates than all the other macrostates.

Microstates, Entropy, and the Second Law of Thermodynamics

The probability of a particular macrostate occurring depends on the number of ways (microstates) in which the given macrostate can be achieved. If a macroscopic system

---

1. Actually there are still other possible ways to distribute the four molecules between the two compartments. We can have all four molecules in the right compartment (one way) or three molecules in the right compartment and one molecule in the left compartment (four ways). However, the macrostates shown in Figure 8.4 are sufficient for our discussion.

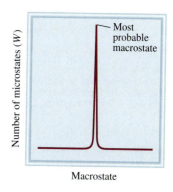

Figure 8.3 For a large number of dice, the most probable macrostate has an overwhelmingly large number of microstates compared to other macrostates.

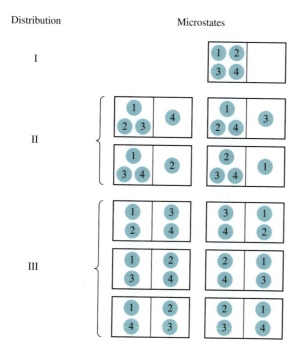

Figure 8.4 Some possible ways of distributing four molecules between two equal compartments. Distribution I can be achieved in only one way (all four molecules in the left compartment) and has one microstate. Distribution II can be achieved in four ways and has four microstates. Distribution III can be achieved in six ways and has six microstates.

is prepared initially in a macrostate with a low probability (low W), then the system will naturally evolve to the macrostate with the highest possible probability. This process is spontaneous. The reverse process, however, in which a macrostate with high probability (large W) evolves toward one with low probability (low W) will not be observed because this process is nonspontaneous.

In thermodynamics, the spontaneity of a macroscopic process can be quantified by a thermodynamic state function called **entropy**, which is a *measure of the number of possible microstates available to a system in a given macrostate*. In 1877, Ludwig Boltzmann showed that the entropy of a system in a given macrostate is directly related to the natural log of the number of microstates (W):

$$S = k_B \ln W \tag{8.1}$$

where k_B is the Boltzmann constant (1.380658×10^{-23} J K^{-1}). Thus, the entropy of the system increases as the value of W increases. Like enthalpy, entropy is an extensive property.

The entropy change (ΔS) for a process in a system is

$$\Delta S = S_f - S_i \tag{8.2}$$

where S_i and S_f are the entropies of the system in the initial and final states, respectively. By combining Equations 8.1 and 8.2, we obtain

$$\Delta S = k_B \ln W_f - k_B \ln W_i$$

$$\Delta S = k_B \ln \frac{W_f}{W_i} \tag{8.3}$$

where W_i and W_f are the corresponding numbers of microstates in the initial and final macrostates, respectively. Thus, if $W_f > W_i$, then $\Delta S > 0$ and the entropy of the system increases.

The connection between entropy and the spontaneity of a process is expressed by the **second law of thermodynamics:** *The entropy of any isolated system increases in any spontaneous process and remains unchanged in an equilibrium (reversible) process.* We can consider the universe to be an isolated system; therefore, the entropy of the universe can never decrease in any spontaneous process: $\Delta S_{univ} \geq 0$, where the equality holds for a reversible process. Because the universe is made up of the system and the surroundings, the entropy change in the universe (ΔS_{univ}) for any process is the *sum* of the entropy changes in the system (ΔS_{sys}) and in the surroundings (ΔS_{surr}). Mathematically, we can express the second law of thermodynamics as follows:

for a spontaneous process: $\Delta S_{univ} = \Delta S_{sys} + \Delta S_{surr} > 0$ (8.4)
for a reversible (or equilibrium) process: $\Delta S_{univ} = \Delta S_{sys} + \Delta S_{surr} = 0$ (8.5)

For a spontaneous process, ΔS_{univ} must be greater than zero, but there are no restrictions on either ΔS_{sys} or ΔS_{surr}. Thus, either ΔS_{sys} or ΔS_{surr} can be negative, as long as the sum of these two quantities is greater than zero. For an equilibrium process, ΔS_{univ} is zero. In this case, ΔS_{sys} and ΔS_{surr} must be equal in magnitude, but opposite in sign. If ΔS_{univ} is negative, then the process is nonspontaneous in the direction described. Instead, it is spontaneous in the *opposite* direction.

The word *entropy* was derived from the Greek work *trope* meaning "transformation." The prefix *en-* was added to make it similar to *energy*.

Engraved on Ludwig Boltzmann's tombstone in Vienna is his famous equation: The "log" stands for log$_e$, which is the natural logarithm or ln.

Example 8.1

Calculate the entropy change for the process (in $J\ K^{-1}$) depicted in Figure 8.1(a) where initially there is 1.000 mole of gas in the left bulb. Assume that the gas behaves ideally and that the process is isothermal.

Strategy Use Equation 8.3 to calculate the entropy change. To do so, we need to determine the ratio W_f/W_i for the process in Figure 8.1(a).

Solution For a general process, determining ΔS using Equation 8.3 is complicated; however, the properties of ideal gases greatly simplify the matter. First, because the process is isothermal, the temperature of the gas does not change. From the kinetic theory of gases (Section 5.4), constant temperature implies that there is no change in the velocity distribution of the gas molecules. Thus, only the properties of the microstates involving position can change as we go from the initial state to the final state.

Suppose, for example, that there is only one gas particle present in the system. Because the two bulbs have equal volumes, the particle has twice as many position microstates available in the final state as in the initial state: $W_f/W_i = 2$. If two particles are present to start with, then $W_f/W_i = (2 \times 2)/1 = 4/1 = 4$ because each of the two particles in the final state can be independently either in the left bulb or the right bulb ($2 \times 2 = 4$ total possibilities), but in the initial state both particles are constrained to be in the left bulb (1 possibility). If we have N particles in the system, the ratio of microstates becomes

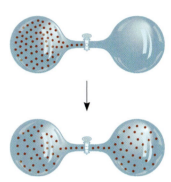

Figure 8.1 (a)

$$\frac{W_f}{W_i} = 2 \times 2 \times 2 \ldots = 2^N$$

We can now apply Equation 8.3:

$$\Delta S = k_B \ln \frac{W_f}{W_i}$$
$$= k_B \ln 2^N$$
$$= N k_B \ln 2$$

where we have used the fact that $\ln a^b = b \ln a$. Recall from Section 5.4 that the Boltzmann constant is related to the gas constant by $k_B = R/N_A$, so the entropy change becomes

$$\Delta S = \frac{N}{N_A} R \ln 2$$
$$= nR \ln 2$$

where $n = N/N_A$ is the number of moles of gas present. So for 1.000 mol of gas we have

$$\Delta S = (1.000\ \text{mol})(8.314\ \text{J mol}^{-1}\ \text{K}^{-1}) \ln 2$$
$$= 5.763\ \text{J K}^{-1}$$

Check The process depicted in Figure 8.1(a) is spontaneous, so the entropy change should be positive (according to the second law of thermodynamics), which our calculation confirms.

Comment Example 8.1 shows why entropy is a more convenient way to describe the process than W. For 1 mole of particles, the ratio of the number of microstates in the

—Continued

Continued—
final and initial states is $2^{N_A} = 2^{6.02 \times 10^{23}} \approx 10^{1.8 \times 10^{23}}$, which is a very large number indeed!

Practice Exercise Repeat the calculation in Example 8.1 except assume the volume of the right-hand bulb in Figure 8.1(a) has twice the volume of the left-hand bulb.

The Physical Interpretation of Entropy

Entropy is often described as a measure of disorder or randomness. These terms should be used with a great deal of caution, however, because they are subjective and can lead to erroneous conclusions.[2] It is preferable, instead, to view the change in entropy of a system in terms of the change in the number of microstates of the system.

Consider the situations shown in Figure 8.5. In the solid in Figure 8.5(a), the atoms or molecules are held relatively rigidly in the crystal lattice. In the liquid, though, these atoms or molecules can occupy many more positions as they move away from the lattice points. Also, for most materials (water is an exception), the liquid phase is less dense than the solid phase, so molecules in a liquid have a greater available volume in which to move. Consequently, the number of microstates increases when a solid transitions into a liquid because there are now many more ways to arrange the particles. Therefore, the solid-to-liquid phase transition should result in

---

2. See, for example, B. Laird, "Entropy, Disorder and the Freezing Transition," *Journal of Chemical Education,* **76**, *255* (1998).

Figure 8.5 Processes that lead to an increase in entropy of the system: (a) melting: $S_{liquid} > S_{solid}$, (b) vaporization: $S_{vapor} > S_{liquid}$, (c) dissolving, and (d) heating: $S_{T_2} > S_{T_1}$ (if $T_2 > T_1$).

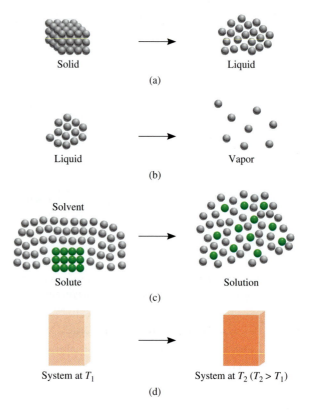

Solid   Liquid
(a)

Liquid   Vapor
(b)

Solvent   Solution
Solute
(c)

System at T_1 System at T_2 ($T_2 > T_1$)
(d)

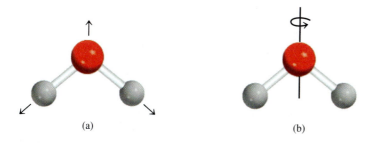

(a)   (b)

Figure 8.6 (a) A vibrational motion in a water molecule. The atoms are displaced, as shown by the arrows, and then reverse their directions to complete a cycle of vibration. (The vibration shown is one of three possible vibrational modes of water.) (b) A rotational motion of a water molecule about an axis through the oxygen atom. The molecule can also rotate with respect to the other two mutually perpendicular axes.

an increase in entropy because the number of microstates has increased. Similarly, the vaporization process [Figure 8.5(b)] should also lead to an increase in the entropy of the system. The increase will be considerably greater than that for melting, however, largely because the change in available volume in going from liquid to vapor is much greater than that corresponding to a solid-liquid transition.

The solution process [Figure 8.5(c)] usually leads to an increase in entropy, too. When a sugar crystal (a nonelectrolyte) dissolves in water, the highly ordered structure of the solid and part of the ordered structure of water break down. Thus, the solution has a greater number of microstates than the pure solute and pure solvent combined. When an ionic solid (an electrolyte) such as NaCl dissolves in water, two factors contribute to an increase in entropy: the solution process (the mixing of the solute with solvent) and the dissociation of the compound into ions:

$$NaCl(s) \xrightarrow{\text{H}_2\text{O}} Na^+(aq) + Cl^-(aq)$$

More particles lead to a greater number of microstates. Hydration, however, causes water molecules to become more structured around the ions. This process decreases entropy because it reduces the number of microstates available to the solvent molecules. For small, highly charged ions such as Al^{3+} and Fe^{3+}, the decrease in entropy due to hydration can exceed the increase in entropy due to mixing and dissociation, so the entropy change for the overall process can actually be negative.

Heating also increases the entropy of a system [Figure 8.5(d)]. In addition to translational motion (that is, the motion through space of the whole molecule), molecules can also rotate and vibrate (Figure 8.6). As the temperature increases, the energies associated with all types of molecular motion increase. This increase in energy is distributed or dispersed among the quantized energy levels. Consequently, more microstates become available at higher temperatures, which means the entropy of the system increases with increasing temperature.

Example 8.2

Predict whether the entropy change is greater or less than zero for each of the following processes: (a) freezing ethanol, (b) evaporating a beaker of liquid bromine at room temperature, (c) dissolving glucose in water, and (d) cooling nitrogen gas from 80°C to 20°C.

Strategy To determine the direction of the entropy change in each case, we must decide whether the number of microstates of the system increases or decreases. The sign of ΔS will be positive if the number of microstates increases and negative if the number decreases.

—Continued

Bromine is a fuming liquid at room temperature.

Continued—

Solution (a) Upon freezing, the ethanol molecules are held rigid in position. This phase transition reduces the number of microstates and therefore the entropy decreases, that is, $\Delta S < 0$.

(b) Evaporating bromine increases the number of microstates because the Br_2 molecules can occupy many more positions in nearly empty space. Therefore, $\Delta S > 0$.

(c) Glucose is a nonelectrolyte (that is, it does not disassociate into ions in aqueous solution). The solution process disperses the glucose molecules among the water molecules, so we expect $\Delta S > 0$.

(d) The cooling process decreases various molecular motions. This leads to a decrease in microstates, and so $\Delta S < 0$.

Practice Exercise How does the entropy of a system change for each of the following processes: (a) condensing water vapor, (b) forming sucrose crystals from a supersaturated solution, (c) heating hydrogen gas from 60°C to 80°C, and (d) subliming dry ice?

8.2 | The Entropy Change for a Process Can Be Calculated Using the Thermodynamic Definition of Entropy

Equation 8.1 is a statistical definition of entropy. Defining entropy in terms of probability provides a molecular interpretation of entropy changes as well as allowing for the calculation of entropy changes in simple systems, such as that of an ideal gas. In general, though, Equation 8.1 is not used to determine entropy changes in complex systems, such as those in which chemical reactions occur, because the calculation of W is too difficult. Fortunately, in addition to the statistical definition, there is a thermodynamic definition of entropy that allows us to determine entropy changes from changes in experimentally measurable thermodynamic quantities, such as heat, energy, and enthalpy.

The Thermodynamic Definition of Entropy

In 1865, building on the work of Kelvin, Sadi Carnot,[3] and others who studied the efficiency of heat engines, Rudolf Clausius[4] postulated that the entropy change of a system undergoing an infinitesimal reversible process was the heat absorbed (dq_{rev}) divided by the absolute temperature T (in kelvins) of the system:

$$dS = \frac{dq_{rev}}{T} \tag{8.6}$$

Thus, the entropy change for any process between two states A and B can be calculated by integrating Equation 8.6 to give

$$\Delta S = \int_{\text{state A}}^{\text{state B}} \frac{dq_{rev}}{T} \tag{8.7}$$

---

3. Nicolas Léonard Sadi Carnot (1796–1832). French physicist. Son of a famous French general, Carnot was a pioneer in the study of the relationship between heat and mechanical work.

4. Rudolf Julius Emanuel Clausius (1822–1888). German physicist. The work of Clausius was mainly in electricity, the kinetic theory of gases, and thermodynamics.

If the process takes place at constant temperature, Equation 8.7 can be integrated to give

$$\Delta S = \frac{q_{rev}}{T} \qquad \text{constant temperature} \qquad (8.8)$$

Clausius's analysis showed that, although q_{rev} is not a state function and is path dependent, the entropy changes calculated using Equation 8.7 are independent of the path taken, so entropy is a state function.

We saw in Section 7.1 that the work done by a system $(-w)$ is maximum if the process is reversible, that is, $-w \le -w_{rev}$. Therefore, because $\Delta U = q + w$ is independent of the nature of the process (U is a state function), the heat absorbed during a process is always less than or equal to that absorbed reversibly: $q \le q_{rev}$. Combining this with Equation 8.6 gives

$$dS \ge \frac{dq}{T} \qquad (8.9)$$

where the equality holds if the process is reversible. Equation 8.9 is known as the **Clausius inequality.** If the system is isolated then $dq = 0$ and Equation 8.9 yields $dS \ge 0$ in accordance with our earlier statement of the second law of thermodynamics.

In general, the thermodynamic definition of entropy (Equations 8.6 to 8.8) yields the same value for the entropy change of a process as Boltzmann's statistical definition (Equation 8.3) for the same process. Consider, for example, the entropy change in the reversible and isothermal (constant temperature) expansion of n moles of an ideal gas from an initial volume V_1 to a final volume V_2. We saw in Section 7.1 that the heat absorbed by the gas in such a reversible process is given by (Equation 7.8)

Although we began our discussion of entropy with the statistical view, the thermodynamic view was developed first.

$$q_{rev} = nRT \ln\frac{V_2}{V_1}$$

Because this process is isothermal, we can use Equation 8.8 to calculate the thermodynamic entropy change

$$\Delta S = \frac{q_{rev}}{T} = \frac{nRT \ln(V_2/V_1)}{T}$$

$$\Delta S = nR \ln\frac{V_2}{V_1} \qquad \text{for an ideal gas} \qquad (8.10)$$

If $V_2 = 2V_1$, as in the process depicted in Figure 8.1, then we have

$$\Delta S = nR \ln\frac{2V_1}{V_1} = nR \ln 2$$

This is exactly the same value that we obtained for this process in Example 8.1 using the statistical definition of entropy.

Equation 8.9 can also be written in terms of pressure changes. For an ideal gas at a fixed temperature, Boyle's law (Equation 5.11) states that $V_2/V_1 = P_1/P_2$, so Equation 8.10 becomes

$$\Delta S = nR \ln\frac{P_1}{P_2} \qquad \text{for an ideal gas} \qquad (8.11)$$

Example 8.3

At a constant temperature of 300 K, 1.00 mole of argon gas is expanded from an initial volume of 20.0 L to a final volume of 50.0 L. Assuming ideal behavior, calculate the entropy change of the gas for this process.

Strategy Because entropy is a state function, changes in entropy are independent of the path taken; therefore, we can assume that the expansion occurs reversibly for the purposes of calculating ΔS, allowing us to use Equation 8.9.

Solution Plugging the values of n, R, V_1 ($V_{initial}$), and V_2 (V_{final}) into Equation 8.9, we obtain the following value for ΔS:

$$\Delta S = (1.00 \text{ mol})(8.314 \text{ J mol}^{-1} \text{ K}^{-1}) \ln\frac{50.0 \text{ L}}{20.0 \text{ L}}$$

$$= 7.62 \text{ J K}^{-1}$$

Check The expansion of a gas is a spontaneous process, so we expect that the entropy change should be positive. Our result here is consistent with that fact.

Practice Exercise Calculate the entropy change for 50.0 g of N_2 gas that is compressed at a constant temperature of 50°C from an initial pressure of 1.00 bar to a final pressure of 3.00 bar. Assume ideal behavior.

Entropy Change Due to Heating

We learned at the end of Section 8.1 [recall Figure 8.5(d)] that the entropy of a system increases when the temperature of the system is raised from T_1 to T_2. Using the thermodynamic definition of entropy, we can calculate the change in entropy for a system upon heating (or cooling). Beginning with Equation 8.6,

$$dS = \frac{dq_{rev}}{T}$$

then integrating this equation from an initial state with temperature T_1 to a final temperature T_2 gives

$$\Delta S = \int_{T_1}^{T_2} \frac{dq_{rev}}{T} \tag{8.12}$$

As discussed in Section 7.3, the amount of heat added in the course of a temperature change can be determined from the heat capacity C using the differential version of Equation 7.16:

$$dq = C \, dT \tag{8.13}$$

Combining Equations 8.12 and 8.13 gives

$$\Delta S = \int_{T_1}^{T_2} \frac{C}{T} \, dT \tag{8.14}$$

The value of C to use in Equation 8.14 depends upon the nature of the specific process under consideration. If the process occurs at constant volume, then the constant-volume heat capacity C_V should be used, giving

$$\Delta S = \int_{T_1}^{T_2} \frac{C_V}{T} \, dT \quad \text{(constant } V) \tag{8.15}$$

Similarly, the constant-pressure heat capacity C_P should be used for a constant pressure process, giving

$$\Delta S = \int_{T_1}^{T_2} \frac{C_P}{T} \, dT \quad \text{(constant } P) \tag{8.16}$$

If the temperature range is small, we can assume that the heat capacity (C_P or C_V) is independent of temperature, giving (after integration)

$$\Delta S = C_V \ln \frac{T_2}{T_1} \quad \text{(constant } V, T_2 - T_1 \text{ small)} \tag{8.17}$$

and

$$\Delta S = C_P \ln \frac{T_2}{T_1} \quad \text{(constant } P, T_2 - T_1 \text{ small)} \tag{8.18}$$

Example 8.4

Calculate the increase in entropy when 250 g of water is heated at constant pressure from 20°C to 50°C. The molar heat capacity of water at constant pressure is 75.3 J K^{-1} mol^{-1}. Assume that C_P is constant over this temperature range.

Strategy Use Equation 8.18 to solve this problem, but remember that the total heat capacity of the water (or any material) is its molar heat capacity multiplied by the number of moles of substance. Also, we must convert all Celsius temperatures to absolute temperatures (kelvins).

Solution We are given the molar heat capacity of water ($\overline{C}_P$), not the heat capacity (C_P), so we must determine the number of moles of water (n) that are being heated and use $n\overline{C}_P$ in Equation 8.18 instead of C_P.

$$n_{\text{water}} = \frac{250 \text{ g}}{18.02 \text{ g mol}^{-1}} = 13.9 \text{ mol}$$

From Equation 8.18, we have

$$\Delta S = C_P \ln \frac{T_2}{T_1} = n\overline{C}_P \ln \frac{T_2}{T_1}$$

$$= (13.9 \text{ mol})(75.3 \text{ J mol}^{-1} \text{K}^{-1}) \ln \frac{323.15 \text{ K}}{293.15 \text{ K}} = 1.02 \times 10^2 \text{ J K}^{-1}$$

Practice Exercise Calculate the entropy change that arises when a 30.0-L sample of argon gas, initially at a pressure of 1.00 bar and a temperature of 300 K, is heated at constant volume to a final temperature of 400 K. Assume ideal gas behavior.

Entropy Change Due to a Phase Transition

When a solid melts at its equilibrium melting temperature, the system absorbs heat reversibly, because the solid and liquid are in equilibrium under these conditions. During the melting process, both temperature and pressure remain fixed. Because this is a constant-pressure process, the heat absorbed by the system can be equated to the enthalpy of fusion, ΔH_{fus} (see Section 6.6), that is, $q_{rev} = \Delta H_{fus}$. Because this process is isothermal, the entropy of fusion ΔS_{fus} can be obtained using Equation 8.8:

$$\Delta S_{fus} = \frac{\Delta H_{fus}}{T_m} \tag{8.19}$$

where T_m is the melting temperature (for example, 273.15 K for ice at 1 atm). Using a similar analysis for the boiling process, the entropy of vaporization ΔS_{vap} is

$$\Delta S_{vap} = \frac{\Delta H_{vap}}{T_b} \tag{8.20}$$

where ΔH_{vap} is the enthalpy of vaporization and T_b is the boiling temperature (for example, 373.15 K for water at 1 atm). A similar formula holds for sublimation.

Example 8.5

Calculate the molar entropies of fusion and vaporization for water at its normal melting point and boiling point. The molar enthalpies of vaporization and fusion of water are given in Tables 7.7 and 7.8, respectively.

Strategy Use Equation 8.19 to calculate the entropy change for fusion and Equation 8.20 to calculate the entropy change for vaporization. From Tables 7.7 and 7.8, we find $\Delta H_{vap} = 40.79 \text{ kJ mol}^{-1}$ and $\Delta H_{fus} = 6.01 \text{ kJ mol}^{-1}$, for water at its normal boiling and melting points, respectively.

Solution Using Equation 8.19, we have for the entropy of fusion

$$\Delta S_{fus} = \frac{\Delta H_{fus}}{T_m} = \frac{6.01 \text{ kJ mol}^{-1}}{273.15 \text{ K}}$$
$$= 0.0220 \text{ kJ K}^{-1} \text{ mol}^{-1} = 22.0 \text{ J K}^{-1} \text{ mol}^{-1}$$

Similarly, using Equation 8.20, we have for the entropy of vaporization

$$\Delta S_{vap} = \frac{\Delta H_{vap}}{T_b} = \frac{40.79 \text{ kJ mol}^{-1}}{373.15 \text{ K}}$$
$$= 0.1094 \text{ kJ K}^{-1} \text{ mol}^{-1} = 109.4 \text{ J K}^{-1} \text{ mol}^{-1}$$

Practice Exercise Calculate the entropy change when 1.00 mol of dry ice (solid CO_2) sublimes at the equilibrium sublimation temperature of $-78.5°C$. The enthalpy of sublimation for CO_2 is 26.2 kJ mol^{-1}.

In the operation of the Carnot heat engine, a given amount of energy is extracted from the high-temperature heat source. Some of this energy is converted into work, and some is lost to the low-temperature heat sink. We can define the efficiency (ε) of a heat engine as the ratio of the work output ($w_{\text{delivered}}$) to the heat input (q_{h}):

$$\varepsilon = \frac{\text{net work done by heat engine}}{\text{heat absorbed by engine}}$$

$$= \frac{w_{\text{delivered}}}{q_{\text{h}}}$$

Using Equation (1) above we have

$$\varepsilon = \frac{q_{\text{h}} - q_{\text{c}}}{q_{\text{h}}}$$

giving

$$\varepsilon = 1 - \frac{q_{\text{c}}}{q_{\text{h}}} \qquad (2)$$

Thus, based on first-law considerations, the efficiency of a heat engine is reduced according to the amount of heat energy transferred to the low temperature heat sink. However, nothing in the first law prevents q_{c} from being zero giving an efficiency of 1 (or 100 percent efficiency). To understand the restrictions on the efficiency of a heat engine, we must turn to the second law.

According to the second law (Equations 8.4 and 8.5), the total entropy change of the system plus surroundings must be greater than or equal to zero:

$$\Delta S_{\text{sys}} + \Delta S_{\text{surr}} \geq 0$$

where the equal sign applies if the engine is operating reversibly. Because steps 1 through 4 of the Carnot heat engine operation represent a cyclic process, $\Delta S_{\text{sys}} = 0$, giving

$$\Delta S_{\text{surr}}(\text{cycle}) \geq 0 \qquad (3)$$

Furthermore, if we assume that the heat source, heat sink, and work sink are much larger than the system, then energy transfer to these external bodies can be assumed to take place reversibly, and the entropy changes can be calculated using Equation 8.8 ($\Delta S = q_{\text{rev}}/T$); Equation (3) then becomes

$$\Delta S_{\text{surr}}(\text{cycle}) = -\frac{q_{\text{h}}}{T_{\text{h}}} + \frac{q_{\text{c}}}{T_{\text{c}}} \geq 0$$

which, after some rearrangement, gives

$$\frac{q_{\text{c}}}{q_{\text{h}}} \geq \frac{T_{\text{c}}}{T_{\text{h}}}$$

Substitution of this inequality into Equation (2) above yields

$$\varepsilon \leq 1 - \frac{T_{\text{c}}}{T_{\text{h}}} \qquad (4)$$

Thus, the thermodynamic efficiency of the Carnot heat engine is at a maximum when the engine is operating reversibly and can never be 1, or 100 percent, because T_{c} can never be zero and T_{h} cannot be infinite. In other words, we can never convert heat totally into work; some of it escapes into the surroundings as waste heat.

Although we have derived the thermodynamic efficiency inequality [Equation (4) above] specifically for the Carnot heat engine, it can be shown that no heat engine operating between a hot and a cold heat source or sink can be more efficient that the Carnot engine when operating reversibly. In addition, all such heat engines when operating reversibly have the same efficiency as the Carnot engine.

If we do work *on* a heat engine instead of extracting work, it is possible to remove heat from a low temperature heat source and deposit it in a high temperature heat sink, against the natural tendency for heat to flow from hot to cold bodies. Two familiar devices that reverse the direction of heat flow are refrigerators and air conditioners. We can model such devices using a Carnot engine operating in reverse (see the figure). The performance of a refrigerator or air conditioner is measured by the *coefficient of performance* (COP), which is the ratio of the heat extracted from the cold heat source to the amount of work applied:

$$\text{COP} = \frac{q_{\text{c}}}{w_{\text{done}}}$$

Using an analysis similar to that used to determine the efficiency of a Carnot heat engine, the maximum COP of a refrigerator or air conditioner is given by

$$\text{COP} \leq \frac{T_{\text{c}}}{T_{\text{h}} - T_{\text{c}}} \qquad (5)$$

Consider, for example, a refrigerator set at 273 K in a room with an ambient temperature of 293 K. Equation (5) above yields a COP of about 14. In reality, the COP values of commercially available refrigèrators are in the range of only two to six because these devices do not operate under optimal reversible conditions.

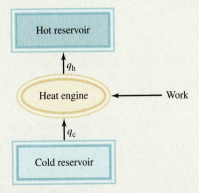

A Carnot engine working in the reverse direction. Refrigerators and air conditioners supply work to remove heat q_{c} from a cold heat source and deposit heat q_{h} into a hot heat sink.

8.3 | The Third Law of Thermodynamics Allows Us to Determine Absolute Entropies

We cannot measure the absolute internal energy U or enthalpy H because the zero of energy is arbitrary. As a result, we are usually only interested in determining changes in these properties (ΔU and ΔH) during a process. However, it is possible to determine the absolute entropy of a substance. This is because of the ***third law of thermodynamics,*** which states that *the entropy of a pure substance in its thermodynamically most stable form is zero at the absolute zero of temperature, independent of pressure.* For the vast majority of substances, the thermodynamically most stable form at 0 K is a perfect crystal. An important exception is helium, which remains liquid, due to its large quantum zero-point motion, at 0 K for pressures below about 10 bar.

The third law is a consequence of the statistical nature of entropy as reflected in Boltzmann's formula (Equation 8.1), which relates the entropy to W, the number of molecular quantum states (microstates) consistent with the macroscopic conditions. At $T = 0$, there is no available thermal energy, and the thermodynamically most stable state is the lowest possible energy state (the ground state). In general, this state is unique, so $W = 1$ for a system at 0 K. From Boltzmann's formula we then have

$$S(T = 0 \text{ K}) = k_B \ln W = k_B \ln (1) = 0$$

We can never reach absolute zero, however (although we can get very, very close), so we can state the third law mathematically as

$$\lim_{T \to 0 \text{ K}} S = 0$$

for a pure substance in its thermodynamically most stable state.

At temperatures above absolute zero, the entropy will be greater than zero due to thermal motion—even for perfect crystals. If the material is impure or if it has defects, then its entropy will be greater than zero, even at 0 K. The significance of the third law is that it enables us to calculate the absolute value of the entropy of a substance.

Third Law or Absolute Entropies

If we heat a system at constant pressure P from an initial temperature T_1 to a final temperature T, the entropy at T can be related to that at T_1 using Equation 8.16, assuming that no phase transitions intervene:

$$S(T) = S(T_1) + \int_{T_1}^{T} \frac{C_P}{T} dT$$

If we take $T_1 = 0$ K, then we have

$$S(T) = \int_{0}^{T} \frac{C_P}{T} dT \tag{8.21}$$

because $S(T = 0 \text{ K})$ is zero. Equation 8.21 defines the absolute entropy S of the substance at the pressure P. If the interval $T = 0$ K to T contains a phase transition at a temperature T_{trans}, then we add $\Delta S = \Delta H_{trans}/T_{trans}$ to the entropy calculated in Equation 8.21 to account for the entropy change due to the phase transition. Because C_P and ΔH_{trans} are measurable quantities, this procedure makes it possible to experimentally determine S for any substance. Figure 8.7 shows a typical graph of the increase in S with temperature from absolute zero to a temperature above the boiling temperature, noting the contributions of the melting and boiling transitions.

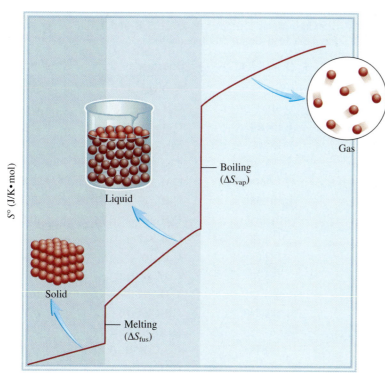

Figure 8.7 The increase in the entropy of a substance at constant pressure from absolute zero to its gaseous state at some temperature. The vertical jumps at the melting and boiling points represent the contributions to S due to the solid-liquid and liquid-vapor phase transitions.

Entropies calculated using Equation 8.21 (with phase transitions) are called ***third-law*** (or) ***entropies*** because these values are not measured relative to some reference state. Third-law entropies per mole of material measured at the standard pressure of 1 bar are referred to as ***standard molar entropies,*** denoted by $S°$. Table 8.2 lists standard molar entropies for a variety of inorganic and organic substances—values for many other substances are given in Appendix 2. The units of $S°$ are J mol^{-1} K^{-1}, in contrast to $\Delta H_f°$ values, which are generally given in kJ mol^{-1}. Entropies of elements and compounds are all positive (that is, $S° > 0$) for all $T > 0$ K. By contrast, the standard enthalpy of formation ($\Delta H_f°$) for elements in their stable form is arbitrarily set equal to zero, and for compounds it may be positive or negative.

According to Table 8.2, the standard entropy of water vapor is greater than that of liquid water. Similarly, bromine vapor has a higher standard entropy than liquid bromine, and iodine vapor has a greater standard entropy than solid iodine. For different substances in the same phase, molecular complexity determines which ones have higher entropies. Both diamond and graphite are solids, but diamond has a more regular structure[6] and hence a smaller number of microstates (see Figure 6.27). Therefore, diamond has a smaller standard entropy than graphite. Ethane (C_2H_6, a component of natural gas) has a more complex structure than methane (CH_4, also a component of natural gas), and hence ethane has more ways to execute molecular motions, which also increases its number of microstates. Ethane, therefore, has a

---

6. Although both are crystalline, diamond and graphite have quite different structures. Diamond is a crystalline solid with a three-dimensional tetrahedral network of bonds. The structure of graphite consists of sheets of carbon atoms arranged in a two-dimensional network of hexagons. The interaction between these graphite sheets is relatively weak, which gives graphite both its lubricating property and a higher entropy than that of the more rigidly packed diamond.

Table 8.2	Standard Molar Entropies at 298.15 K and 1 Bar for Select Inorganic and Organic Substances.		
Substance	$S°$ (J mol^{-1} K^{-1})	Substance	$S°$ (J mol^{-1} K^{-1})
He(g)	126.15	N_2O (g)	219.99
Ne(g)	146.33	O_2 (g)	205.0
Ar(g)	154.84	O_3 (g)	238.92
Kr(g)	164.08	SO_2 (g)	248.22
C(graphite)	5.69	CH_4 (g)	186.25
C(diamond)	2.4	C_2H_6 (g)	229.2
CO(g)	197.66	C_3H_8 (g)	270.3
CO_2(g)	213.6	C_2H_2 (g)	200.94
HF(g)	173.78	C_2H_4 (g)	219.3
HCl(g)	187.0	C_6H_6 (l)	173.2
HBr(g)	198.70	CH_3OH (l)	126.8
HI(g)	206.3	C_2H_5OH (l)	160.7
H_2O(g)	188.84	CH_3CHO (l)	263.8
H_2O(l)	69.95	HCOOH (l)	129.0
NH_3(g)	192.77	CH_3COOH (l)	159.8
NO(g)	210.76	$C_6H_{12}O_6$ (s)	212.1
N_2O_4(g)	304.38	$C_{12}H_{22}O_{11}$ (s)	360.2

larger standard entropy than methane. Both helium and neon are monatomic gases, so neither undergoes rotational or vibrational motions, but neon has a larger standard entropy than helium because its molar mass is greater. As we saw in the case of the one-dimensional particle-in-a box (Section 1.3), heavier atoms have more closely spaced energy levels, so there is a broader distribution of the energy of the atom among the energy levels. Consequently, there are more microstates associated with heavier atoms.

For simple molecules it is possible to calculate absolute entropies theoretically, and for most systems the theoretical value very closely agrees with third-law entropies calculated experimentally. However, in some situations the third-law entropy (determined using Equation 8.27 and accounting for any phase transitions) does not agree with the theoretical value. The origin of this discrepancy is generally due to defects or impurities that are frozen in the system at low temperatures. The third law does not apply for such a system because the system is not in its thermodynamically most stable state. The difference between the experimental entropy and the theoretical entropy for such a system is referred to as the *residual entropy,* which is *the value of the entropy at 0 K for systems for which the third law is not applicable.*

As an example, consider a crystal of carbon monoxide (CO). The dipole moment of CO is quite small (0.12 D) and carbon and oxygen are very similar in size, so the CO molecule is very nearly symmetrical. In a perfect crystal of CO [Figure 8.8(a)], the CO molecules are all aligned in an ordered fashion. However, because the two ends of the molecule are so similar, the molecules in a real crystal may be randomly oriented [Figure 8.8(b)].

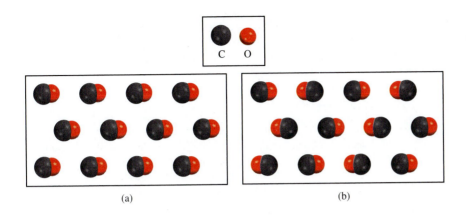

(a)   (b)

If the orientation of the CO molecules is completely random, then each molecule has 2 possible orientations. Two molecules would have $2^2 = 4$ possibilities; three molecules would have $2^3 = 8$ possibilities, and so on. Using Boltzmann's formula (Equation 8.1) we can estimate the residual entropy for one mole ($N = N_A$) of CO molecules as

$$
\begin{aligned}
S_{res} &= k_B \ln W \\
&= k_B \ln 2^{N_A} = k_B N_A \ln 2 = R \ln 2 \\
&= (8.314 \text{ J mol}^{-1} \text{ K}^{-1}) \ln 2 \\
&= 5.8 \text{ J mol}^{-1} \text{ K}^{-1}
\end{aligned}
$$

where we have used $R = k_B N_A$. The experimental value of the residual entropy for CO is about 4.2 J mol^{-1} K^{-1}. This value is slightly lower than the estimate from Boltzmann's formula, so the orientation of the CO molecules in the crystal is not completely random.

Entropy of Reaction

Suppose that the system is represented by the following chemical reaction:

$$a\text{A} + b\text{B} \longrightarrow c\text{C} + d\text{D}$$

Analogous to the standard enthalpy change of a reaction (see Equation 7.41), the **standard entropy change of reaction** is given by *the difference in standard entropies between the products and reactants:*

$$\Delta S^\circ_{rxn} = \left[c\, S^\circ(\text{C}) + d\, S^\circ(\text{D}) \right] - \left[a\, S^\circ(\text{A}) + b\, S^\circ(\text{B}) \right] \qquad \textbf{(8.22)}$$

We can generalize Equation 8.22 as:

$$\Delta S^\circ_{rxn} = \sum_{products} \upsilon_P\, S^\circ(\text{products}) - \sum_{reactants} \upsilon_R\, S^\circ(\text{reactants}) \qquad \textbf{(8.23)}$$

where υ_P and υ_R denote the stoichiometric coefficients for the reactants and products, and Σ (sigma) means "the sum of." Note that in contrast to the standard enthalpy of formation, the standard entropy of an element in its most stable form is not zero, so all reactants and products (including elements) must be included in Equation 8.29.

Example 8.6 illustrates the calculation of ΔS°_{rxn} from standard molar entropy data (Table 8.2 and Appendix 2).

Example 8.6

From the standard entropy values in Appendix 2, calculate the standard entropy changes for the following reactions at 25°C:

(a) $CaCO_3(s) \longrightarrow CaO(s) + CO_2(g)$

(b) $N_2(g) + 3H_2(g) \longrightarrow 2NH_3(g)$

(c) $H_2(g) + Cl_2(g) \longrightarrow 2HCl(g)$

Strategy To calculate the standard entropy of a reaction, we look up the standard entropies of the reactants and products in Appendix 2 and apply Equation 8.23. As in the calculation of enthalpy of reaction [see Equation (7.41)], the stoichiometric coefficients have no units, so ΔS°_{rxn} is expressed in units of J mol^{-1} K^{-1}.

Solution

(a) $\Delta S^\circ_{rxn} = [S^\circ(CaO) + S^\circ(CO_2)] - [S^\circ(CaCO_3)]$

$= [39.8 \text{ J mol}^{-1} \text{ K}^{-1} + 213.79 \text{ J mol}^{-1} \text{ K}^{-1}] - [92.9 \text{ J mol}^{-1} \text{ K}^{-1}]$

$= 160.7 \text{ J mol}^{-1} \text{ K}^{-1}$

Thus, when 1 mole of $CaCO_3$ decomposes to form 1 mole of CaO and 1 mole of gaseous CO_2, there is an increase in entropy equal to 160.7 J mol^{-1} K^{-1}.

(b) $\Delta S^\circ_{rxn} = [2S^\circ(NH_3)] - [S^\circ(N_2) + 3S^\circ(H_2)]$

$= (2)(192.77 \text{ J mol}^{-1} \text{ K}^{-1}) - [(191.61 \text{ J mol}^{-1} \text{ K}^{-1}) + (3)(130.68 \text{ J mol}^{-1} \text{ K}^{-1})]$

$= -198.11 \text{ J mol}^{-1} \text{ K}^{-1}$

This result shows that when 1 mole of gaseous nitrogen reacts with 3 moles of gaseous hydrogen to form 2 moles of gaseous ammonia, there is a decrease in entropy equal to 198.11 J mol^{-1} K^{-1}.

(c) $\Delta S^\circ_{rxn} = [2S^\circ(HCl)] - [S^\circ(H_2) + S^\circ(Cl_2)]$

$= (2)(186.90 \text{ J mol}^{-1} \text{ K}^{-1}) - [(130.68 \text{ J mol}^{-1} \text{ K}^{-1}) + (223.08 \text{ J mol}^{-1} \text{ K}^{-1})]$

$= 20.04 \text{ J mol}^{-1} \text{ K}^{-1}$

Thus, the formation of 2 moles of gaseous HCl from 1 mole of gaseous H_2 and 1 mole of gaseous Cl_2 results in a small increase in entropy equal to 20.04 J mol^{-1} K^{-1}.

Practice Exercise Calculate the standard entropy change for the following reactions at 25°C:

(a) $2CO(g) + O_2(g) \longrightarrow 2CO_2(g)$

(b) $3O_2(g) \longrightarrow 2O_3(g)$

(c) $2NaHCO_3(s) \longrightarrow Na_2CO_3(s) + H_2O(l) + CO_2(g)$

The results of Example 8.6 are consistent with those observed for many other reactions. Taken together, they support the following general rules:

▸ If a reaction produces more gas molecules than it consumes [Example 8.6 (a)], ΔS°_{rxn} is positive.

▸ If the total number of gas molecules diminishes [Example 8.6(b)], ΔS°_{rxn} is negative.

▸ If there is no net change in the total number of gas molecules [Example 8.6(c)], then ΔS°_{rxn} may be positive or negative, but it will be relatively small numerically.

These conclusions make sense, given that gases invariably have greater entropy than liquids and solids. For reactions involving only liquids and solids, predicting the sign of ΔS°_{rxn} is more difficult, but in many such cases an increase in the total number of molecules and/or ions is accompanied by an increase in entropy.

Example 8.7 shows how knowing the nature of reactants and products makes it possible to predict entropy changes.

Example 8.7

Predict whether the entropy change of the system in each of the following reactions is positive or negative:

(a) $2H_2(g) + O_2(g) \longrightarrow 2H_2O(l)$

(b) $NH_4Cl(s) \longrightarrow NH_3(g) + HCl(g)$

(c) $H_2(g) + Br_2(g) \longrightarrow 2HBr(g)$

Strategy We are asked to predict, not calculate, the sign of the entropy change in the reactions. The factors that lead to an increase in entropy are: (1) a transition from a condensed phase to the vapor phase and (2) a reaction that produces more product molecules than reactant molecules in the same phase. It is also important to compare the relative complexity of the product and reactant molecules. In general, the more complex the molecular structure, the greater the entropy of the compound.

Solution (a) Two gaseous reactant molecules combine to form one liquid product molecule. Even though H_2O is more complex structurally than either H_2 or O_2, the net decrease of one molecule and conversion of gases to a liquid decreases the number of microstates; hence, ΔS°_{rxn} is negative.

(b) A solid reactant is converted to two gaseous products. Therefore, ΔS°_{rxn} is positive.

(c) Two diatomic gas molecules combine to form two different diatomic gas molecules. Thus, the number of reactant molecules is the same as the number of product molecules and both the reactants and products are of similar complexity (that is, all are diatomic). As a result, we cannot easily predict the sign of ΔS°_{rxn}, but the magnitude of the change must be quite small.

Practice Exercise Discuss qualitatively the sign of the entropy change expected for each of the following processes:

(a) $I_2(s) \longrightarrow 2I(g)$

(b) $2Zn(s) + O_2(g) \longrightarrow 2ZnO(s)$

(c) $N_2(g) + O_2(g) \longrightarrow 2NO(g)$

Temperature Dependence of Standard Entropy Changes

For most substances, the values of standard entropies are not available for a wide range of temperatures. Most commonly, tables present only data at 25°C. For most purposes, it is sufficient to use the standard entropies at 25°C to calculate reaction entropy changes at other temperatures because ΔS_{rxr} does not generally depend highly on temperature. If accurate work is needed, however, or if the temperature of interest is well removed from the temperature for which data are available, it is necessary to correct the tabulated values for the change in temperature. To do this, we can use a procedure analogous to that outlined in Section 7.7 for the calculation of ΔH_{rxn} at alternate temperatures.

Like enthalpy, the change in entropy due to temperature change is governed by the heat capacity. Because standard entropies are defined as being at fixed pressure (1 bar), we can use Equation 8.18 to determine the value of the standard entropy at a temperature T_2 relative to its known value at temperature T_1:

$$S^\circ_f(T_2) = S^\circ_f(T_1) + \overline{C}_P \ln \frac{T_2}{T_1}$$

assuming that the standard constant pressure molar heat capacity, $\overline{C}_P^\circ$, is temperature independent. Substituting this expression for $S_f^\circ(T_2)$ into Equation 8.23 and using algebra similar to that used to derive Equation 7.49 for reaction enthalpy changes, yields

$$\Delta S_{rxn}^\circ(T_2) = \Delta S_{rxn}^\circ(T_1) + \Delta \overline{C}_{P,\,rxn}^\circ \ln \frac{T_2}{T_1} \qquad (8.24)$$

where $\Delta \overline{C}_{P,\,rxn}^\circ$ is defined as in Equation 7.50:

$$\Delta C_{p,\,rxn}^\circ = \sum_{products} v_P \overline{C}_P^\circ - \sum_{reactants} v_R \overline{C}_P^\circ$$

Equation 8.24 describes the temperature dependence of the reaction entropy change and is analogous to Kirchhoff's law for the temperature dependence of the reaction enthalpy change (Equation 7.49).

Example 8.8 shows how to use Equation 8.24 to calculate the standard entropy change for a reaction that occurs well above 25°C.

Example 8.8

The standard entropy change for the reaction

$$3O_2(g) \longrightarrow 2O_3(g)$$

is $-137.2 \text{ J K}^{-1} \text{ mol}^{-1}$ at 298 K and 1 bar. Calculate the value of ΔS_{rxn}° at 380 K, assuming that all $\overline{C}_P^\circ$ values are independent of temperature.

Strategy   To use Equation 8.30, we need $\Delta \overline{C}_{P,\,rxn}^\circ$ for this reaction. This value was calculated to be $-9.8 \text{ kJ mol}^{-1} \text{ K}^{-1}$ in Example 7.13 using data from Appendix 2 and Equation 7.50.

Solution   We are given ΔS_{rxn}° in the problem statement and $\Delta \overline{C}_{P,\,rxn}^\circ$ in Example 7.13, so using Equation 8.24 gives

$$\Delta S_{rxn}^\circ(T_2) = \Delta S_{rxn}^\circ(T_1) + \Delta \overline{C}_{p,\,rxn}^\circ \ln \frac{T_2}{T_1}$$

$$= -137.2 \text{ J mol}^{-1} \text{ K}^{-1} + (-9.8 \text{ J mol}^{-1} \text{ K}^{-1}) \ln \frac{380 \text{ K}}{298 \text{ K}}$$

$$= -139.6 \text{ J mol}^{-1} \text{ K}^{-1}$$

Practice Exercise   Using data from Appendix 2, calculate ΔS_{rxn}° for the reaction

$$2NO_2(g) \longrightarrow N_2O_4(g)$$

at 450 K.

8.4 | The Spontaneity of a Process at Constant Temperature and Pressure Is Governed by the Gibbs Free Energy

The second law of thermodynamics tells us that a spontaneous reaction increases the entropy of the universe, that is, $\Delta S_{univ} > 0$. In order to determine the sign of ΔS_{univ} for a reaction, however, we would need to calculate both ΔS_{sys} and ΔS_{surr}. In general, we are usually concerned only with what happens in a particular system and not with the entire universe. Therefore, we need another thermodynamic function to help us determine whether a reaction will occur spontaneously if we consider only the system itself.

Spontaneity in Nonisolated Systems: Gibbs Free Energy

When a isothermal process occurs in a system that is not isolated, but is instead coupled to its surroundings, we cannot apply the second law of thermodynamics directly to the system alone. Instead, we must consider the total entropy change in the universe, which from Equations 8.4 and 8.5 is given by

$$\Delta S_{\text{univ}} = \Delta S_{\text{sys}} + \Delta S_{\text{surr}} \geq 0$$

Because the surroundings are so much bigger than the system, we can assume that the surroundings are unperturbed by the transfer of heat, so the transfer can be considered to be reversible. This reversibility combined with the condition that T is constant implies that $\Delta S_{\text{surr}} = q_{\text{surr}}/T$ (Equation 8.8), giving

$$\Delta S_{\text{univ}} = \Delta S_{\text{sys}} + \frac{q_{\text{surr}}}{T} \geq 0$$

Because $q_{\text{sys}} = -q_{\text{surr}}$, we have

$$\Delta S_{\text{univ}} = \Delta S_{\text{sys}} - \frac{q_{\text{sys}}}{T} \geq 0 \qquad\qquad \textbf{(8.25)}$$

If we make a further assumption that the process occurs at constant pressure, then

$$\Delta S_{\text{sys}} - \frac{\Delta H_{\text{sys}}}{T} \geq 0 \qquad\qquad \textbf{(8.26)}$$

because $q_{\text{sys}} = \Delta H_{\text{sys}}$ for a constant pressure process (Equation 7.13). Thus, an exothermic process ($\Delta H_{\text{sys}} < 0$) will increase the entropy of the surroundings ($\Delta S_{\text{surr}} > 0$), and an endothermic process ($\Delta H_{\text{sys}} > 0$) will decrease the entropy of the surroundings ($\Delta S_{\text{surr}} < 0$) (Figure 8.9). Any spontaneous process that occurs in a nonisolated system at constant T and P must satisfy Equation 8.26. For convenience, we can multiply Equation 8.26 above by $-T$ to obtain

$$\Delta H_{\text{sys}} - T\Delta S_{\text{sys}} \leq 0 \qquad\qquad \textbf{(8.27)}$$

Equation 8.27 is useful because it only requires information about the system, eliminating the need to consider the surroundings specifically. In order to express the spontaneity of a process more directly, we introduce a new thermodynamic function called the **Gibbs free energy** (*G*) (after the American physicist Josiah Willard Gibbs[7]) as

$$G = H - TS \qquad\qquad \textbf{(8.28)}$$

---

7. Josiah Willard Gibbs (1839–1903). American physicist. One of the founders of thermodynamics and statistical mechanics, Gibbs was a modest and private individual who spent most of his professional life at Yale University. Because he published most of his works in obscure journals, Gibbs never gained the eminence that his contemporary and admirer James Clerk Maxwell enjoyed. Even today, very few people outside of chemistry and physics have ever heard of Gibbs.

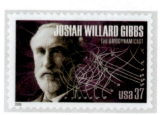

A 2005 commemorative stamp honoring Gibbs

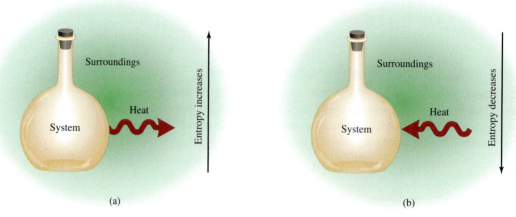

Figure 8.9 (a) An exothermic process transfers heat from the system to the surroundings and results in an increase in the entropy of the surroundings. (b) An endothermic process absorbs heat from the surroundings and thereby decreases the entropy of the surroundings.

Like both H and TS, G has units of energy. Also, because H, T, and S are all state functions, G is also a state function.

The change in the Gibbs free energy (ΔG) for a constant temperature process is

$$\Delta G = \Delta H - T\Delta S \qquad (8.29)$$

From Equations 8.27 and 8.29, we see that the spontaneity criterion for process occurring under constant pressure and temperature conditions is

$$\Delta G \leq 0 \qquad (8.30)$$

($\Delta G < 0$ holds for a spontaneous process and $\Delta G = 0$ for a reversible (or equilibrium) process.) Because G, H, T, and S are state functions, Equations 8.27 and 8.30 hold even if the temperature and pressure are not strictly constant throughout the process as long as the initial and final states are at the same temperature and pressure.

Gibbs Free Energy and Nonexpansion Work

Equation 8.30 provides an extremely useful criterion for determining the direction of spontaneous changes and the nature of physical and chemical equilibrium. In addition, the changes in the Gibbs free energy function enable us to determine the amount of work that can be done in a process at constant temperature and pressure. To see this, start with the expression for ΔG for a process at constant temperature and pressure (Equation 8.29):

$$\Delta G = \Delta H - T\Delta S = \Delta U + P\Delta V - T\Delta S \qquad (8.31)$$

where we have used the fact that $\Delta H = \Delta U + P\Delta V$ at constant pressure (Equation 7.12). The second law of thermodynamics, in the form of the Clausius inequality (Equation 8.9), states that for any isothermal process $q \leq T\Delta S$. Combining this inequality with Equation 8.31 gives

$$\Delta G \geq \Delta U + P\Delta V - q$$

Using the first law of thermodynamics (Equation 7.4), we can replace ΔU with $q + w$ to give

$$\Delta G \geq q + w + P\Delta V - q$$
$$\Delta G \geq w + P\Delta V$$

The quantity $P\,\Delta V$ represents the work delivered by the system due to expansion, so the quantity $w_{delivered} - P\Delta V$ represents that part of the work delivered by the system that is not due to expansion. Thus, the maximum nonexpansion work, $w_{delivered,\,nonexp}$, that can be delivered by a system in a process at constant temperature and pressure is equal to the *decrease* in the Gibbs free energy of the process, that is,

$$w_{delivered,\,nonexp} \leq -\Delta G \qquad (8.32)$$

where the equality holds only if the process is reversible.[8] Equation 8.31 is very important in electrochemistry (Chapter 13) because the electrical work that is produced by the chemical reactions in electrochemical cells, such as batteries, is a type of nonexpansion work.

The word *free* in *free energy* does not mean "without cost," but instead refers to amount of energy "available" to do work.

Standard Gibbs Free-Energy Changes

The **standard Gibbs free energy of reaction** (ΔG_{rxn}°) is *the Gibbs free energy change for a reaction that occurs under standard-state conditions when reactants in their standard states are converted to products in their standard states.* The calculation of ΔG_{rxn}° is very similar to that for ΔH_{rxn}° (Section 7.4). We start with the generic reaction

$$a\text{A} + b\text{B} \longrightarrow c\text{C} + d\text{D}$$

The standard Gibbs free energy change for this reaction is given by

$$\Delta G_{rxn}^{\circ} = \left[c\Delta G_f^{\circ}(\text{C}) + d\Delta G_f^{\circ}(\text{D})\right] - \left[a\Delta G_f^{\circ}(\text{A}) + b\Delta G_f^{\circ}(\text{B})\right] \qquad (8.33)$$

or, in general,

$$\Delta G_{rxn}^{\circ} = \sum_{products} v_P \Delta G_f^{\circ}(\text{products}) - \sum_{reactants} v_R \Delta G_f^{\circ}(\text{reactants}) \qquad (8.34)$$

where v_P and v_R are the stoichiometric coefficients of the products and reactants, respectively. The ΔG_f° term is the **standard Gibbs free energy of formation** of a compound, that is, *the Gibbs free-energy change that occurs when 1 mole of the compound is synthesized from its elements in their standard states.* For the combustion of graphite,

$$\text{C(graphite)} + \text{O}_2(g) \longrightarrow \text{CO}_2(g)$$

---

8. In addition to the Gibbs free energy, another similar thermodynamic state function commonly used in thermodynamics is the *Helmholtz free energy* (A), defined as $A = U - TS$. At constant temperature, the change in the Helmholtz free energy is given by $\Delta A = \Delta U - T\,\Delta S$. In analogy to Equation 8.32, the maximum total work that can be delivered by a system in a constant temperature process is equal to the negative of the change in the Helmholtz energy: $w_{delivered} \leq -\Delta A$. The Helmholtz free energy is important because it gives the criterion for spontaneity for processes at constant temperature and *volume,* that is, for any spontaneous process at constant T and V, the change in the Helmholtz free energy must be negative: $\Delta A < 0$. In chemistry, the Gibbs free energy is used far more often than the Helmholtz free energy because most processes of chemical interest occur under constant-pressure (and not constant-volume) conditions.

the standard free-energy change [from Equation 8.34] is

$$\Delta G^\circ_{rxn} = \Delta G^\circ_f(CO_2) - [\Delta G^\circ_f(C, graphite) + \Delta G^\circ_f(O_2)]$$

As in the case of the standard enthalpy of formation (Section 7.4), we define the standard Gibbs free energy of formation of any element in its stable allotropic form at 1 bar and 25°C as zero. Thus,

$$\Delta G^\circ_f(C, graphite) = 0 \quad \text{and} \quad \Delta G^\circ_f = (O_2) = 0$$

Therefore, the standard Gibbs free-energy change for the combustion of graphite equals the standard free energy of formation of CO_2:

$$\Delta G^\circ_{rxn} = \Delta G^\circ_f(CO_2)$$

Elements not in their most stable form will have nonzero Gibbs free energies of formation. For example, carbon in the form of diamond has a nonzero Gibbs free energy of formation $[\Delta G^\circ_f(diamond) = 2.87 \text{ kJ mol}^{-1}]$ at 25°C, which represents the standard Gibbs free energy change for the conversion of graphite to diamond at that temperature. Appendix 2 lists the values of ΔG°_f for a number of substances.

In the event that tables for standard Gibbs energies of formation are not readily available, we can also obtain the Gibbs free energy of the reaction by first determining the standard enthalpy change of a reaction using the methods discussed in Section 7.4 and the standard entropy change for the reaction as discussed in Section 8.3. Once ΔH°_{rxn} and ΔS°_{rxn} are known, ΔG°_{rxn} can be calculated using Equation 8.29 applied to standard states:

$$\Delta G^\circ_{rxn} = \Delta H^\circ_{rxn} - T\Delta S^\circ_{rxn} \tag{8.35}$$

Example 8.9 shows how to calculate standard Gibbs free energy changes.

Example 8.9

Calculate the standard Gibbs free energy changes for the following reactions at 25°C:

(a) $CH_4(g) + 2O_2(g) \longrightarrow CO_2(g) + 2H_2O(l)$
(b) $2MgO(s) \longrightarrow 2Mg(s) + O_2(g)$

Strategy To calculate the standard Gibbs free energy change of a reaction, we look up the standard free energies of formation of reactants and products in Appendix 2 and use these values in Equation 8.40. Remember that ΔG°_f for elements such as $O_2(g)$ and $Mg(s)$ is zero because they are stable allotropes of their respective elements at 1 bar and 25°C. Check that the chemical equations are balanced so that you use the correct stoichiometric coefficients in Equation 8.40. Finally, all stochiometric coefficients are unitless, so ΔG°_{rxn} is expressed in units of kJ mol^{-1}.

Solution

(a) According to Equation 8.40,

$$\Delta G^\circ_{rxn} = [\Delta G^\circ_f(CO_2) + 2\Delta G^\circ_f(H_2O)] - [\Delta G^\circ_f(CH_4) + 2\Delta G^\circ_f(O_2)]$$

—Continued

Continued—

Inserting the appropriate values from Appendix 2 gives

$$\Delta G^{\circ}_{\text{rxn}} = [(-394.4 \text{ kJ mol}^{-1}) + (2)(-237.1 \text{ kJ mol}^{-1})]$$
$$- [(-50.8 \text{ kJ mol}^{-1}) + (2)(0 \text{ kJ mol}^{-1})]$$
$$= -817.8 \text{ kJ mol}^{-1}$$

(b) The equation is

$$\Delta G^{\circ}_{\text{rxn}} = [2\Delta G^{\circ}_{\text{f}}(\text{Mg}) + \Delta G^{\circ}_{\text{f}}(\text{O}_2)] - [2\Delta G^{\circ}_{\text{f}}(\text{MgO})]$$

Plugging in the data from Appendix 2 yields

$$\Delta G^{\circ}_{\text{rxn}} = [(2)(0 \text{ kJ mol}^{-1}) + (0 \text{ kJ mol}^{-1})] - [(2)(-569.6 \text{ kJ mol}^{-1})]$$
$$= +1139 \text{ kJ mol}^{-1}$$

Practice Exercise Calculate the standard free-energy changes for the following reactions at 25°C:

(a) $H_2(g) + Br_2(l) \longrightarrow 2HBr(g)$

(b) $2C_2H_6(g) + 7O_2(g) \longrightarrow 4CO_2(g) + 6H_2O(l)$

At this point, it is useful to distinguish between ΔG and ΔG°. In any given experiment, it is rare for the reactants and products to all be present at their standard-state pressures or concentrations. Even if the reaction is set up initially under standard-state conditions for all substances involved, as the reaction proceeds to equilibrium, the system may evolve to non–standard-state conditions. Under these conditions, it is the sign of ΔG not ΔG° that predicts the direction of the reaction. As discussed in Chapter 10, however, the sign of ΔG° does indicate whether the reactants or the products are favored when the reacting system reaches equilibrium.

Temperature Dependence of ΔG

In all but the most accurate work, it is usually a good approximation to assume that ΔH and ΔS are constant for a chemical reaction even if the temperature changes. The Gibbs free energy change (ΔG) generally depends strongly upon temperature, however, as was shown by Equation 8.29

$$\Delta G = \Delta H - T\Delta S$$

Thus, the magnitude and sign of ΔG as a function of temperature depends upon the relative magnitudes of ΔH and ΔS. Based on Equation 8.29, the change in ΔG with temperature primarily depends on the sign of ΔS. If $\Delta S > 0$, then ΔG will *decrease* with increasing temperature, and if $\Delta S < 0$, ΔG will *increase* with increasing temperature. Based on the signs of ΔH and ΔS, there are four possibilities for ΔG:

▶ If ΔH and ΔS are both positive, then ΔG will be positive at low temperatures (where enthalpy dominates) and become negative at high temperatures (where entropy dominates). The temperature at which ΔG crosses over from positive to negative (that is, when $\Delta H = T \Delta S$) depends upon the relative magnitudes of ΔH and ΔS. For example: $2HgO(s) \longrightarrow 2Hg(l) + O_2(g)$; $\Delta H^{\circ} = 181.6 \text{ kJ mol}^{-1}$, $\Delta S^{\circ} = 219.3 \text{ J mol}^{-1}$.

▶ If ΔH is positive and ΔS is negative, ΔG will always be positive regardless of temperature. For example: $3O_2(g) \longrightarrow 2O_3(g)$; $\Delta H° = 285.4$ kJ mol^{-1}, $\Delta S° = -137.5$ J mol^{-1}.

▶ If ΔH is negative and ΔS is positive, ΔG will always be negative regardless of temperature. For example: $2H_2O_2(l) \longrightarrow 2H_2O(l) + O_2(g)$; $\Delta H° = -196.5$ kJ mol^{-1}, $\Delta S° = 166.8$ J mol^{-1}.

▶ If both ΔH and ΔS are both negative, then ΔG will be negative at low temperatures and become positive at high temperatures. The temperature at which ΔG crosses over from negative to positive depends upon the relative magnitude of ΔH and ΔS. For example: $NH_3(g) + HCl(g) \longrightarrow NH_4Cl(s)$; $\Delta H° = -177.0$ kJ mol^{-1}, $\Delta S° = -285.11$ J mol^{-1}.

As an example, consider the preparation of calcium oxide (CaO), also called quicklime. Calcium oxide is an extremely valuable inorganic substance used in steel-making, the production of calcium metal, the paper industry, water treatment, and pollution control. It is prepared industrially on a large scale by decomposing limestone (CaCO$_3$) in a kiln at a high temperature:

$$CaCO_3(s) \rightleftharpoons CaO(s) + CO_2(g)$$

The reaction is reversible, and CaO readily combines with CO$_2$ to form CaCO$_3$. The pressure of CO$_2$ in equilibrium with CaCO$_3$ and CaO increases with temperature. In the industrial preparation of quicklime, the system is never maintained at equilibrium; rather, CO$_2$ is constantly removed from the kiln to shift the equilibrium from left to right, promoting the formation of calcium oxide.

The important information for the practical chemist to know is the temperature at which the decomposition of CaCO$_3$ becomes appreciable (that is, the temperature at which the reaction begins to favor products). We can make a reliable estimate of that temperature as follows. First, use the data in Appendix 2 to calculate $\Delta H°$ and $\Delta S°$ for the reaction at 25°C. Using Equation 7.41 to determine $\Delta H°$:

$$\begin{aligned}\Delta H° &= \left[\Delta H_f° (CaO) + \Delta H_f° (CO_2)\right] - \left[\Delta H_f° (CaCO_3)\right] \\ &= \left[(-634.9 \text{ kJ mol}^{-1}) + (-393.5 \text{ kJ mol}^{-1})\right] - (-1207.6 \text{ kJ mol}^{-1}) \\ &= 179.2 \text{ kJ mol}^{-1}\end{aligned}$$

Next, use Equation 8.23 to determine $\Delta S°$

$$\begin{aligned}\Delta S° &= \left[S°(CaO) + S°(CO_2)\right] - S°(CaCO_3) \\ &= \left[38.1 \text{ J K}^{-1} \text{ mol}^{-1} + 213.6 \text{ J mol}^{-1} \text{ K}^{-1}\right] - 91.7 \text{ J K}^{-1} \text{ mol}^{-1} \\ &= 160.0 \text{ J mol}^{-1} \text{ K}^{-1}\end{aligned}$$

From Equation 8.35

$$\Delta G° = \Delta H° - T\Delta S°$$

we obtain

$$\begin{aligned}\Delta G° &= 179.2 \text{ kJ mol}^{-1} - (298 \text{ K})(160.0 \text{ J mol}^{-1} \text{ K}^{-1})(1 \text{ kJ}/1000 \text{ J}) \\ &= 131.5 \text{ kJ mol}^{-1}\end{aligned}$$

Because $\Delta G°$ is a large positive quantity, products are not favored for this reaction at 25°C (298 K). Indeed, the pressure of CO$_2$ is so low at room temperature that it cannot be measured.

In order to make $\Delta G°$ negative, we first have to find the temperature at which $\Delta G°$ is zero, that is,

$$0 = \Delta H° - T\Delta S°$$

or

$$T = \frac{\Delta H°}{\Delta S°}$$
$$= \frac{(179.2 \text{ kJ mol}^{-1})(1000 \text{ J kJ}^{-1})}{160.0 \text{ J mol}^{-1} \text{ K}^{-1}}$$
$$= 1120 \text{ K or } 830°\text{C}$$

At a temperature higher than 847°C, $\Delta G°$ becomes negative, indicating that the reaction now favors the formation of CaO and CO_2. At 860°C (1133 K), for example,

$$\Delta G° = 179.2 \text{ kJ mol}^{-1} - (1133 \text{ K})(160.0 \text{ J mol}^{-1} \text{ K}^{-1})(1 \text{ kJ}/1000 \text{ J})$$
$$= -2.1 \text{ kJ mol}^{-1}$$

Note that we used the $\Delta H°$ and $\Delta S°$ values at 25°C to calculate the changes that occur at a much higher temperature. Because both $\Delta H°$ and $\Delta S°$ change with temperature, this approach will not yield an accurate value of $\Delta G°$, but it is good enough for a rough estimate. Additionally, do not be misled into thinking that nothing happens below 847°C and that at 847°C $CaCO_3$ suddenly begins to decompose. Just because $\Delta G°$ is positive at some temperature below 847°C does not mean that CO_2 is not produced; on the contrary, it means that the pressure of the CO_2 gas formed at that temperature will be below 1 bar (its standard-state value). As Figure 8.10 shows, the pressure of CO_2 at first increases very slowly with temperature and becomes easily measurable above 700°C. The significance of 847°C is that this is the temperature at which the equilibrium pressure of CO_2 reaches 1 bar. Above 847°C, the equilibrium pressure of CO_2 exceeds 1 bar.

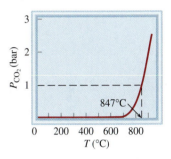

Figure 8.10 The equilibrium pressure of CO_2 from the decomposition of $CaCO_3$, as a function of temperature. This curve is calculated by assuming that $\Delta H°$ and $\Delta S°$ of the reaction do not change with temperature.

Example 8.10

For the reaction $NH_3(g) + HCl(g) \longrightarrow NH_4Cl(s)$, calculate $\Delta G°$ at 25°C and 1000°C. At what temperature (if any) does $\Delta G°$ become zero? Assume that $\Delta H°$ and $\Delta S°$ are independent of temperature.

Strategy First determine $\Delta H°$ and $\Delta S°$ using Equation 7.41 and 8.26, respectively, and the data in Appendix 2. Then $\Delta G°$ can be calculated at the two temperatures using Equation 8.35.

Solution Plugging the appropriate $\Delta H_f°$ values from Appendix 2 into Equation 7.41 yields

$$\Delta H° = \Delta H_f°(NH_4Cl) - [\Delta H_f°(NH_3) + \Delta H_f°(HCl)]$$
$$= -314.4 \text{ kJ mol}^{-1} - [-45.9 \text{ kJ mol}^{-1} + (-92.3 \text{ kJ mol}^{-1})]$$
$$= -176.2 \text{ kJ mol}^{-1}$$

Similarly, plugging the appropriate $S°$ values into Equation 8.23 yields

$$\Delta S° = S°(NH_4Cl) - [S°(NH_3) + S°(HCl)]$$
$$= 94.6 \text{ J mol}^{-1} \text{ K}^{-1} - [192.8 \text{ J mol}^{-1} \text{ K}^{-1} + 187.0 \text{ J mol}^{-1} \text{ K}^{-1}]$$
$$= -285.2 \text{ J K}^{-1} \text{ mol}^{-1}$$

—Continued

Continued—

Equation 8.38 then gives the following for $\Delta G°$ at 25°C and at 1000°C:

$$\Delta G°(25°C) = -176.2 \text{ kJ mol}^{-1} - (298.15 \text{ K})(-285.2 \text{ J mol}^{-1} \text{ K}^{-1})(1 \text{ kJ}/1000 \text{ J})$$
$$= -91.32 \text{ kJ mol}^{-1}$$
$$\Delta G°(1000°C) = -176.2 \text{ kJ mol}^{-1} - (1273.15 \text{ K})(-285.2 \text{ J mol}^{-1} \text{ K}^{-1})(1 \text{ kJ}/1000 \text{ J})$$
$$= +186.9 \text{ kJ mol}^{-1}$$

The temperature at which $\Delta G°$ is zero can be found using the same method used for the decomposition of $CaCO_3$:

$$T = \frac{\Delta H°}{\Delta S°}$$
$$= \frac{(-176.2 \text{ kJ mol}^{-1})(1000 \text{ J kJ}^{-1})}{-285.2 \text{ J mol}^{-1} \text{ K}^{-1}}$$
$$= 618 \text{ K or } 347°C$$

Check Because both $\Delta H°$ and $\Delta S°$ are negative, $\Delta G°$ should be negative at low temperatures and positive at sufficiently high temperatures. This is consistent with the values calculated at 25°C and 1000°C.

Practice Exercise For the reaction $2HgO(s) \longrightarrow 2Hg(l) + O_2(g)$, calculate $\Delta G°$ at 25°C and 800°C. At what temperature (if any) does $\Delta G°$ become zero? Assume that $\Delta H°$ and $\Delta S°$ are independent of temperature.

In some cases, if accurate work is required, it is necessary to take into account the temperature dependence of $\Delta H°$ and $\Delta S°$ in the calculation of $\Delta G°$ at temperatures other than that assumed in the data tables (usually 25°C). The most straightforward method to determine the effect of temperature change on $\Delta G°$ is to use Kirchhoff's law (Equation 7.49) to calculate $\Delta H°$ at the new temperature and to use Equation 8.24 to calculate $\Delta S°$. Then, use these values for $\Delta H°$ and $\Delta S°$ in Equation 8.35 to calculate the new value of $\Delta G°$. Example 8.11 shows how this is done.

Example 8.11

For the reaction $2NO_2(g) \longrightarrow N_2O_4(g)$, use the data in Appendix 2 to determine the temperature at which $\Delta G° = 0$, assuming that $\Delta H°$ and $\Delta S°$ are temperature independent. To estimate the error caused by this assumption, use the heat capacity data in Appendix 2 to better estimate $\Delta G°$ at this temperature.

Strategy The first part of the calculation proceeds in a similar fashion as Example 8.10. That is, determine $\Delta H°$ and $\Delta S°$ using Equations 7.41 and 8.23, respectively, and the data in Appendix 2. Then, use these values of $\Delta H°$ and $\Delta S°$ to calculate the temperature at which $\Delta G°$ should equal zero by setting Equation 8.35 equal to zero and solving for T. Next, use this value for T in Equation 7.49 (Kirchhoff's law) and Equation 8.24 to calculate the corrections in $\Delta H°$ and $\Delta S°$, respectively, due to the nonzero heat capacity. Both Kirchhoff's law (Equation 7.49) for the temperature dependence of enthalpy change and the corresponding equation (Equation 8.24) for entropy change require the calculation of $\Delta \overline{C}_P$ using Equation 7.50. Finally, use these new values of $\Delta H°$ and $\Delta S°$ in Equation 8.35 to calculate a more accurate value of $\Delta G°$ at T.

—Continued

Continued—

Solution Plugging the appropriate ΔH_f° from Appendix 2 into Equation 7.41 yields

$$\begin{aligned}
\Delta H^\circ &= \Delta H_f^\circ(N_2O_4) - 2\,\Delta H_f^\circ(NO_2) \\
&= 9.66 \text{ kJ mol}^{-1} - 2(33.85 \text{ kJ mol}^{-1}) \\
&= -58.04 \text{ kJ mol}^{-1}
\end{aligned}$$

Similarly, plugging the appropriate S° values into Equation 8.23 yields

$$\begin{aligned}
\Delta S^\circ &= S^\circ(N_2O_4) - 2S^\circ(NO_2) \\
&= 304.3 \text{ J mol}^{-1}\text{ K}^{-1} - 2(240.46 \text{ J mol}^{-1}\text{ K}^{-1}) \\
&= -176.6 \text{ J mol}^{-1}\text{ K}^{-1}
\end{aligned}$$

The temperature at which ΔG° is zero can be found, as in Example 8.10

$$\begin{aligned}
T &= \frac{\Delta H^\circ}{\Delta S^\circ} \\
&= \frac{(-58.04 \text{ kJ mol}^{-1})(1000 \text{ J kJ}^{-1})}{-176.6 \text{ J mol}^{-1}\text{ K}^{-1}} \\
&= 328.6 \text{ K or } 55.5^\circ\text{C}
\end{aligned}$$

To get a better estimate of ΔG° at this temperature, we need to calculate the corrections in ΔH° and ΔS° due to the nonzero heat capacity using Equations 7.49 (Kirchhoff's law) and 8.33, respectively. We need $\Delta \overline{C}_P$ for both of these equations, which can be obtained using Equation 7.50 as follows:

$$\begin{aligned}
\Delta \overline{C}_P &= \overline{C}_P(N_2O_4) - 2\overline{C}_P(NO_2) \\
&= 79.2 \text{ J mol}^{-1}\text{ K}^{-1} - 2(37.2 \text{ J mol}^{-1}\text{ K}^{-1}) \\
&= -4.8 \text{ J mol}^{-1}\text{ K}^{-1}
\end{aligned}$$

Using Kirchhoff's law (Equation 7.49), we have

$$\begin{aligned}
\Delta H^\circ(325.5 \text{ K}) &= \Delta H^\circ(298.15 \text{ K}) + \Delta \overline{C}_P(325.5 \text{ K} - 298.15 \text{ K}) \\
&= -58.04 \text{ kJ mol}^{-1} + (+4.8 \text{ J K}^{-1}\text{ mol}^{-1})(27.35 \text{ K})(1 \text{ kJ}/1000 \text{ J}) \\
&= -57.91 \text{ kJ mol}^{-1}
\end{aligned}$$

Equation 8.33 gives the new entropy change:

$$\begin{aligned}
\Delta S^\circ(328.6 \text{ K}) &= \Delta S^\circ(298.15 \text{ K}) + \Delta \overline{C}_P \ln \frac{325.5 \text{ K}}{298.15 \text{ K}} \\
&= -176.6 \text{ J K}^{-1}\text{ mol}^{-1} + (+4.8 \text{ J mol}^{-1}\text{ K}^{-1}) \ln (1.092) \\
&= -176.2 \text{ J K}^{-1}\text{ mol}^{-1}
\end{aligned}$$

With these new values for ΔH° and ΔS° we can calculate ΔG° at 328.6 K using Equation 8.38:

$$\begin{aligned}
\Delta G^\circ &= \Delta H^\circ - T\Delta S^\circ \\
&= -57.91 \text{ kJ mol}^{-1} - 328.6 \text{ K } (-176.2 \text{ J K}^{-1}\text{ mol}^{-1})(1 \text{ kJ}/1000 \text{ J}) \\
&= -0.011 \text{ kJ mol}^{-1}
\end{aligned}$$

Comment If entropy and enthalpy changes are temperature independent, then $\Delta G^\circ = 0$ at 325.5 K for the reaction in this problem. If we account for the temperature dependence of ΔH° and ΔS°, then we find that ΔG° is slightly negative at this temperature. Because both ΔH° and ΔS° are negative, ΔG° will be negative at low temperatures and positive at high temperatures. Because ΔG° is negative at 328.6 K, then that temperature must be slightly too low.

—Continued

The Thermodynamics of a Rubber Band

The common rubber band has some very interesting thermodynamic properties due to its molecular structure. To observe these properties, quickly stretch a rubber band (at least 0.5 cm wide) and then press it against your lips. You should feel a slight warming effect. Next, stretch a rubber band and hold it that way for a few seconds. Then quickly release the tension and press the rubber band against your lips again. This time you should feel a slight cooling effect. A thermodynamic analysis of these two experiments reveals a lot about the molecular structure of rubber.

Rearranging Equation 8.35 ($\Delta G = \Delta H - T\Delta S$) gives

$$T\Delta S = \Delta H - \Delta G$$

The warming effect (an exothermic process) due to stretching means that $\Delta H < 0$, and since stretching is nonspontaneous (that is, $\Delta G > 0$ and $-\Delta G < 0$), $T\Delta S$ must be negative, too. Because T, the absolute temperature, is always positive, the ΔS due to stretching must be negative. As a result, rubber in its natural state is more entangled (higher entropy) than when it is

under tension. When the tension is removed, the stretched rubber band spontaneously snaps back to its original shape, that is, ΔG is negative (so $-\Delta G$ is positive). The cooling effect means that it is an endothermic process ($\Delta H > 0$), so $T\Delta S$ is positive, too. Thus, the entropy of the rubber band increases when it goes from the stretched state to the natural state.

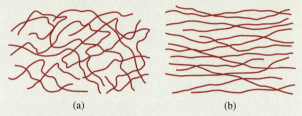

(a)   (b)

(a) Rubber molecules in their normal state. Note the high degree of entanglement and hence a large number of microstates (high entropy). (b) Under tension, the molecules line up in an orderly fashion and the number of microstates decreases (low entropy).

Continued—

Practice Exercise   Use the data in Appendix 2 to determine the Gibbs free energy of formation, ΔG_f°, for CO at 400 K both with and without the assumption that ΔH° and ΔS° are independent of temperature. What is the percentage difference between the two values of ΔG_f° that you obtain?

8.5 | The Mixing of Pure Substances Leads to an Increase in the Entropy and a Decrease in the Gibbs Free Energy

Chemical reactions nearly always involve the mixing of two or more substances, so to understand the thermodynamics of a chemical reaction it is insufficient to know the entropy, enthalpy, and Gibbs free energy of the substances involved in their pure forms. We must also understand the thermodynamic consequences of mixing. We begin by examining the thermodynamics of mixing ideal gases.

Figure 8.11 shows a container in which n_A moles of ideal gas A at temperature T, pressure P, and volume V_A are separated by a partition from n_B moles of ideal gas B at the same temperature and pressure as gas A, but with volume V_B. When the partition is removed, the gases mix spontaneously and the entropy of the system increases. To calculate the entropy of mixing, ΔS_{mix}, we can treat the process as two separate isothermal ideal gas expansions using Equation 8.10. For gas A:

$$\Delta S_A = n_A R \ln \frac{V_A + V_B}{V_A}$$

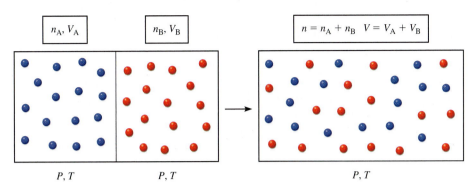

Figure 8.11 When two ideal gases at the same temperature and pressure are mixed, an increase in the entropy, called the entropy of mixing, is observed.

For gas B:

$$\Delta S_{B} = n_{B}R \ln \frac{V_{A} + V_{B}}{V_{B}}$$

(We can treat these expansions separately because the two gases are ideal and do not interact. That is, each gas expands without being affected by the presence of the other gas.) Thus, we have

$$\Delta S_{mix} = \Delta S_{A} + \Delta S_{B} = n_{A}R \ln \frac{V_{A} + V_{B}}{V_{A}} + n_{B}R \ln \frac{V_{A} + V_{B}}{V_{B}}$$

According to Avogadro's law (Equation 5.17), the volume of an ideal gas is directly proportional to the number of moles of the gas at constant T and P, so the previous equation can be rewritten as

$$
\begin{aligned}
\Delta S_{mix} &= n_{A}R \ln \frac{n_{A} + n_{B}}{n_{A}} + n_{B}R \ln \frac{n_{A} + n_{B}}{n_{B}} \\
&= -n_{A}R \ln \frac{n_{A}}{n_{A} + n_{B}} + n_{B}R \ln \frac{n_{B}}{n_{A} + n_{B}} \\
&= -n_{A}R \ln x_{A} + n_{B}R \ln x_{B}
\end{aligned}
$$

$$\Delta S_{mix} = -R(n_{A} \ln x_{A} + n_{B} \ln x_{B}) \tag{8.36}$$

where x_{A} and x_{B} are the mole fractions of A and B, respectively (see Equation 0.11). Because $x < 1$, it follows that $\ln x < 0$ and that the right-hand side of Equation 8.36 is a positive quantity, which is consistent with the spontaneous nature of the process. Because the individual ideal gas molecules do not interact, the ***entropy of mixing*** depends only upon the number of moles of each type of gas involved and not on the molecular identity of the two gases.

To calculate the ***Gibbs free energy of mixing,*** we use Equation 8.29 to give

$$\Delta G_{mix} = \Delta H_{mix} - T\Delta S_{mix}$$

Because there are no interactions between the molecules and the pressure is unchanged by the mixing process, the change in the enthalpy due to mixing, ΔH_{mix}, will be zero. Therefore,

$$\Delta G_{mix} = -T\Delta S_{mix}$$

$$\Delta G_{mix} = RT(n_{A} \ln x_{A} + n_{B} \ln x_{B}) \tag{8.37}$$

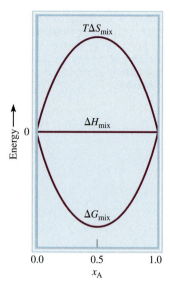

Figure 8.12 Plots of $T\Delta S_{mix}$, ΔH_{mix}, and ΔG_{mix} as functions of composition x_A for an ideal solution.

Because the right-hand side of Equation 8.37 is negative, the mixing of two ideal gases leads to a *decrease* in the Gibbs free energy. This has important consequences for the thermodynamics of chemical equilibrium, as discussed in Chapter 10. Both equations 8.36 and 8.37 can also be used to approximate the entropy and Gibbs free energy changes associated with the mixing of liquids if the interactions among the A and B molecules in the mixture are nearly equal (that is, if ΔH_{mix} is small). An idealized mixture, gas or liquid, for which $\Delta H_{mix} = 0$ and ΔG_{mix} is given by Equation 8.42 is referred to as an **ideal solution.**

Figure 8.12 shows the plots of ΔG_{mix}, ΔH_{mix}, and $T\Delta S_{mix}$ for the ideal gas mixture in Figure 8.11 as a function of composition. Both the maximum in $T\Delta S_{mix}$ and the minimum in ΔG_{mix} occur at $x_A = 0.5$. The shape of the ΔG_{mix} curve has important implications for the purification of mixtures. Suppose one sets out to separate component A from a mixture that is initially a 50:50 mixture of A and B, that is, we wish to increase the mole fraction of A in the solution. The curve for ΔG_{mix} as a function of mole fraction is relatively flat for equimolar mixtures ($x_A \approx 0.5$). Therefore, the initial steps of the purification (say from $x_A = 0.5$ to $x_A = 0.6$) require relatively little work (ΔG is small). However, as x_A approaches 1, the ΔG_{mix} curve becomes quite steep and the amount of work required to further purify the mixture becomes quite substantial (large ΔG). For this reason, obtaining very high purity in separations is often prohibitively difficult.

Example 8.12

Calculate the Gibbs free energy and entropy due to mixing 2.5 moles of argon with 3.5 moles of oxygen, both at 1 bar and 25°C. Assume ideal gas behavior.

Strategy First calculate the mole fractions of argon and oxygen, and then use them in Equation 8.36 to calculate the entropy of mixing. Next, use the calculated value of ΔS_{mix} and the relationship $\Delta G_{mix} = -T\Delta S_{mix}$ to obtain ΔG_{mix}. (Remember $\Delta H_{mix} = 0$ for an ideal solution.)

Solution The mole fractions of argon and oxygen are

$$x_{Ar} = \frac{2.5 \text{ mol}}{2.5 \text{ mol} + 3.5 \text{ mol}} = 0.42 \qquad x_{O_2} = \frac{3.5 \text{ mol}}{2.5 \text{ mol} + 3.5 \text{ mol}} = 0.58$$

From Equation 8.36 (with A = Ar and B = O_2),

$$\begin{aligned}
\Delta S_{mix} &= -R(n_{Ar} \ln x_{Ar} + n_{O_2} \ln x_{O_2}) \\
&= -(8.314 \text{ J mol}^{-1} \text{ K}^{-1})[(2.5 \text{ mol}) \ln 0.42 + (3.5 \text{ mol}) \ln 0.58] \\
&= 40 \text{ J K}^{-1}
\end{aligned}$$

Because $\Delta G_{mix} = -T\Delta S_{mix}$, we have

$$\begin{aligned}
\Delta G_{mix} &= -T\Delta S_{mix} \\
&= -(298 \text{ K})(40 \text{ J K}^{-1}) \\
&= -1.2 \times 10^4 \text{ J} \\
&= -12 \text{ kJ}
\end{aligned}$$

Check Our results correctly predict that entropy of mixing is positive and the Gibbs free energy of mixing is negative. Also, the calculated mole fractions (0.42 and 0.58) sum to 1, as required.

Practice Exercise Calculate the Gibbs free energy and entropy of mixing when 10 g of CO_2 is mixed with 20 g of O_2, both at 1.5 bar pressure and 400°C. Assume ideal behavior.

8.6 | In Living Systems, Spontaneous Reactions Are Used to Drive Other Nonspontaneous, but Essential, Biochemical Processes

Many biochemical reactions have a positive $\Delta G°$ value, yet they still need to occur because they are essential to the maintenance of life. In living systems, these unfavorable reactions are coupled to energetically favorable processes, ones that have negative $\Delta G°$ values. The principle of **coupled reactions** is based on a simple concept: We can use a thermodynamically favorable reaction to drive an unfavorable one. Consider an industrial process. Suppose, for example, that we want to extract zinc from the ore sphalerite (ZnS). The following reaction will not work because it has a large positive $\Delta G°$ value:

$$ZnS(s) \longrightarrow Zn(s) + S(s) \qquad \Delta G° = 198.3 \text{ kJ mol}^{-1}$$

On the other hand, the combustion of sulfur to form sulfur dioxide is favored because of its large negative $\Delta G°$ value:

$$S(s) + O_2(g) \longrightarrow SO_2(g) \qquad \Delta G° = -300.1 \text{ kJ mol}^{-1}$$

By coupling these two processes we can separate zinc from zinc sulfide. In practice, this means heating ZnS in air so that the tendency of S to form SO_2 will promote the decomposition of ZnS:

$$
\begin{aligned}
ZnS(s) &\longrightarrow Zn(s) + S(s) & \Delta G° &= 198.3 \text{ kJ mol}^{-1} \\
S(s) + O_2(g) &\longrightarrow SO_2(g) & \Delta G° &= -300.1 \text{ kJ mol}^{-1} \\
\hline
ZnS(s) + O_2(g) &\longrightarrow Zn(s) + SO_2(g) & \Delta G° &= -101.8 \text{ kJ mol}^{-1}
\end{aligned}
$$

Coupled reactions play a crucial role in our survival. In biological systems, enzymes facilitate a wide variety of nonspontaneous reactions. In the human body, for example, food molecules, represented by glucose ($C_6H_{12}O_6$), are converted to carbon dioxide and water during metabolism with a substantial release of free energy:

$$C_6H_{12}O_6(s) + 6O_2(g) \longrightarrow 6CO_2(g) + 6H_2O(l) \qquad \Delta G° = -2880 \text{ kJ mol}^{-1}$$

In a living cell, this reaction does not take place in a single step (as burning glucose in a flame would); rather, the glucose molecules are broken down with the aid of enzymes in a series of steps. Much of the free energy released along the way is used to synthesize adenosine triphosphate (ATP) from adenosine diphosphate (ADP) and phosphoric acid (Figure 8.13):

$$ADP + H_3PO_4 \longrightarrow ATP + H_2O \qquad \Delta G° = +31 \text{ kJ mol}^{-1}$$

A mechanical analog for coupled reactions. We can make the smaller weight move upward (a nonspontaneous process) by coupling it with the falling of a larger weight.

Figure 8.13 Structures of ATP and ADP in their ionized forms. The adenine group is in blue, the ribose group in black, and the phosphate group in red. ADP has one fewer phosphate group than ATP.

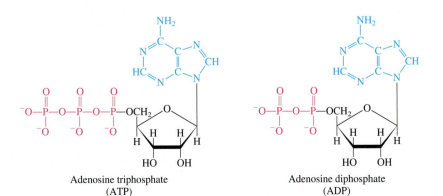

Adenosine triphosphate (ATP)

Adenosine diphosphate (ADP)

Figure 8.14 A schematic representation of ATP synthesis and coupled reactions in living systems. The conversion of glucose to carbon dioxide and water during metabolism releases free energy. The released free energy is used to convert ADP into ATP. The ATP molecules are then used as an energy source to drive unfavorable reactions, such as protein synthesis from amino acids.

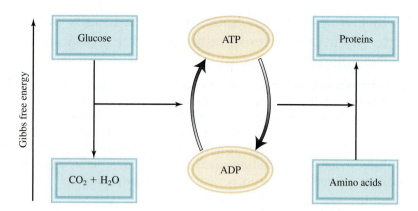

The function of ATP is to store free energy until it is needed by cells. Under appropriate conditions, ATP is hydrolyzed to give ADP and phosphoric acid, with a release of 31 kJ mol^{-1} of free energy, which can be used to drive energetically unfavorable reactions, such as protein synthesis.

Proteins are polymers made of amino acids. The stepwise synthesis of a protein molecule involves the joining of individual amino acids. Consider, for example, the formation of the dipeptide (a two-amino-acid unit) alanylglycine from alanine and glycine. This reaction represents the first step in the synthesis of a protein molecule:

$$\text{alanine} + \text{glycine} \longrightarrow \text{alanylglycine} \qquad \Delta G° = +29 \text{ kJ mol}^{-1}$$

This reaction does not favor the formation of product, so only a little of the dipeptide would be formed at equilibrium. With the aid of an enzyme, however, the reaction is coupled to the hydrolysis of ATP as follows:

$$\text{ATP} + \text{H}_2\text{O} + \text{alanine} + \text{glycine} \longrightarrow \text{ADP} + \text{H}_3\text{PO}_4 + \text{alanylglycine}$$

The overall free-energy change is given by $G° = -31$ kJ mol^{-1} + 29 kJ mol^{-1} = -2 kJ mol^{-1}, which means that the coupled reaction now favors the formation of product, and an appreciable amount of alanylglycine will be formed under this condition. Figure 8.14 shows the ATP-ADP interconversions where energy is stored (from metabolism) and free energy is released (from ATP hydrolysis) to drive essential reactions.

Summary of Facts and Concepts

Section 8.1

▶ Entropy is an extensive state function that measures the number of ways that the macroscopic state of a system can be realized microscopically. Any spontaneous process must lead to a net increase in entropy in the universe (second law of thermodynamics).

Section 8.2

▶ Changes in entropy for a process can be defined thermodynamically in terms of the reversible work absorbed during the process.

▶ The thermodynamic definition of entropy can be used to calculate changes in entropy on expansion or contraction, heating or cooling, or as a result of a phase change.

Section 8.3

▶ The third law of thermodynamics states that the entropy of a system in its thermodynamically most stable form is zero at 0 K. The thermodynamically most stable form for most substances at 0 K is a perfect crystal. The third law enables us to measure the absolute entropies of substances.

▶ The standard entropy of a chemical reaction can be calculated from the absolute entropies of reactants and products.

Section 8.4

▶ Under conditions of constant temperature and pressure, the Gibbs free-energy change ΔG is less than zero for a spontaneous process and greater than zero for a nonspontaneous process. For an equilibrium process, $\Delta G = 0$.

▶ For a chemical or physical process at constant temperature and pressure, $\Delta G = \Delta H - T\Delta S$. This equation can be used to predict the spontaneity of a process in which the initial and final states are at the same temperature and pressure.

▶ The standard Gibbs free-energy change for a reaction, $\Delta G°$, can be calculated from the standard Gibbs free energies of formation of reactants and products.

Section 8.5

▶ The entropy change for the mixing of two ideal gases can be calculated in terms of the mole fractions of each component. The enthalpy of mixing for two ideal gases is identically zero. Any solution that has the same formula for entropy change and enthalpy change on mixing is called an ideal solution.

Section 8.6

▶ Many biological reactions are nonspontaneous. They are driven by the hydrolysis of ATP, for which $\Delta G°$ is negative.

Key Words

Clausius inequality, p. 433
coupled reactions, p. 459
entropy, p. 428
entropy of mixing, p. 457
Gibbs free energy, p. 447

Gibbs free energy of mixing, p. 457
ideal solution, p. 458
residual entropy, p. 442
second law of thermodynamics, p. 428

spontaneous process, p. 424
standard entropy change of reaction, p. 443
standard Gibbs free energy of formation, p. 449

standard Gibbs free energy of reaction, p. 449
standard molar entropy, p. 441
third law of thermodynamics, p. 440

Problems

The Entropy of an Isolated System Always Increases in Any Spontaneous Process

8.1 Determine the probability that all the molecules of a gas will be found in one half of a container when the gas consists of (a) 3 molecules, (b) 6 molecules, (c) 60 molecules, and (d) 2 million molecules.

8.2 Comment on the statement: "Even thinking about entropy increases its value in the universe."

8.3 When rolling four dice, consider the microstates as corresponding to the individual numbers on each die and the macrostates as corresponding to the sum of these numbers. (a) How many microstates and macrostates are they? (b) What are the most probable and least probable macrostates. (c) What is the ratio of the number of microstates corresponding to the most probable macrostate versus the least probable macrostate.

8.4 Consider 12 balls placed randomly in 3 different compartments. Consider the macrostate to correspond to the number of balls in each compartment. Show that the most probable macrostate corresponds to four balls in each compartment. What is the least probable macrostate(s)?

8.5 According to the second law of thermodynamics, the entropy of an irreversible process in an isolated system must always increase. The entropy of living systems, however, remains small. (For example, the synthesis of highly complex protein molecules from individual amino acids is a process that leads to a decrease in entropy.) Is the second law invalid for living systems? Explain.

The Entropy Change for a Process Can Be Calculated Using the Thermodynamic Definition of Entropy

8.6 Suppose that your friend told you of the following extraordinary event. A block of metal with a mass of 500 g was seen rising spontaneously from the table on which it was resting to a height of 1.00 cm above the table. She stated that the metal had absorbed thermal energy from the table that was then used to raise itself against the acceleration of gravity. (a) Does this process violate the first law of thermodynamics? (b) Does it violate the second law? Assume that the room temperature was 298 K and that the table was large enough that its temperature was unaffected by this transfer of energy. (*Hint:* First calculate the decrease in entropy as a result of this process, and then estimate the probability for the occurrence of such a process. The acceleration due to gravity is 9.81 m s^{-2}.)

8.7 A refrigerator is contained within a small well-insulated room and receives power from a wall socket. If the door to the refrigerator is left open, does the temperature in the room decrease, increase, or stay the same? Explain your answer.

8.8 The molar heat of vaporization of ethanol is 39.3 kJ mol^{-1}, and the boiling point of ethanol is 78.3°C.

Calculate the value of ΔS_{vap} for the vaporization of 0.50 mole of ethanol.

8.9   Use the thermodynamic definition of entropy and the second law to explain why heat cannot flow spontaneously from a region of low temperature to one of higher temperature.

8.10   Calculate the value of ΔS in heating 3.5 moles of a monatomic ideal gas from 50°C to 77°C at constant pressure.

8.11   Repeat the calculation in Problem 8.10 for constant-volume heating. Which one results in the greater entropy change? Give a physical explanation for the difference.

8.12   One mole of an ideal gas is first heated at constant pressure from T to $3T$ and then it is cooled back to T at constant volume. (a) Determine an expression for ΔS for the overall process. (b) Show that the overall process is equivalent to an isothermal expansion of the gas at T from V to $3V$, where V is the original volume. (c) Show that the value of ΔS for the process in part (a) is the same as that for part (b).

8.13   A sample of 1.00 g of Ne gas initially at 20°C is expanded from 1.2 to 2.6 L and simultaneously heated to 40°C. Calculate the entropy change for the process.

8.14   One mole of an ideal gas at 298 K expands isothermally from 1.0 to 2.0 L (a) reversibly and (b) against a constant external pressure of 12.2 bar. Calculate ΔS_{sys}, ΔS_{surr}, and ΔS_{univ} in both cases. Are your results consistent with the nature of the process?

8.15   The entropy of vaporization for HF is considerably smaller than the 88 J mol^{-1} K^{-1} predicted by Trouton's rule. Give a physical explanation for this exception to the rule.

8.16   The enthalpy of vaporization of CCl_4 at its normal boiling point of 77.6°C is 32.54 kJ mol^{-1}. Determine the entropy of vaporization at this temperature.

The Third Law of Thermodynamics Allows Us to Determine Absolute Entropies

8.17   Choose the substance with the greater molar entropy in each of the following pairs: (a) $H_2O(l)$, $H_2O(g)$; (b) NaCl(s), $CaCl_2(s)$; (c) N_2 (0.1 bar), N_2 (1 bar); (d) $O_2(g)$, $O_3(g)$; (e) ethanol (C_2H_5OH), dimethyl ether (CH_3OCH_3); (f) $N_2O_4(g)$, $2NO_2(g)$; and (g) Fe(s) at 298 K, Fe(s) at 398 K. (Unless otherwise stated, assume the temperature is 298 K.)

8.18   Calculate the molar residual entropy of a solid in which the molecules can adopt (a) three, (b) four, and (c) five orientations of equal energy at absolute zero.

8.19   Account for the measured residual entropy of 10.1 J mol^{-1} K^{-1} for the CH_3D molecule.

8.20   Explain why the value of $S°$ for graphite is greater than that for diamond at 298 K (see Appendix 2). Would this inequality hold at 0 K?

8.21   Using the data in Appendix 2, calculate the standard entropy changes for the following reactions at 25°C:

(a) $S(s) + O_2(g) \longrightarrow SO_2(g)$
(b) $MgCO_3(s) \longrightarrow MgO(s) + CO_2(g)$

8.22   Using the data in Appendix 2, calculate the standard entropy changes for the following reactions at 25°C:

(a) $H_2(g) + CuO(s) \longrightarrow Cu(s) + H_2O(g)$
(b) $2Al(s) + 3ZnO(s) \longrightarrow Al_2O_3(s) + 3Zn(s)$
(c) $CH_4(g) + 2O_2(g) \longrightarrow CO_2(g) + 2H_2O(l)$

8.23   Without consulting Appendix 2, predict whether the standard entropy change is positive or negative for each of the following reactions. Give reasons for your predictions.

(a) $2KClO_4(s) \longrightarrow 2KClO_3(s) + O_2(g)$
(b) $H_2O(g) \longrightarrow H_2O(l)$
(c) $2Na(s) + 2H_2O(l) \longrightarrow 2NaOH(aq) + H_2(g)$
(d) $N_2(g) \longrightarrow 2N(g)$

8.24   Without consulting Appendix 2, predict whether the standard entropy change is positive or negative for each of the following reactions. Give reasons for your predictions.

(a) $4Fe(s) + 3O_2(g) \longrightarrow 2Fe_2O_3(s)$
(b) $2O(g) \longrightarrow O_2(g)$
(c) $NH_4Cl(s) \longrightarrow NH_3(g) + HCl(g)$
(d) $H_2(g) + Cl_2(g) \longrightarrow 2HCl(g)$

8.25   Use Appendix 2 to calculate $\Delta S°$ for the reactions in Problem 8.24.

8.26   Consider the reaction

$$N_2(g) + O_2(g) \longrightarrow 2NO(g)$$

Calculate $\Delta S°$ for the reaction mixture, the surroundings, and the universe at 298 K. Why is your result reassuring to the inhabitants of Earth?

8.27   Using the data from Appendix 2, determine the standard entropy for $CO_2(g)$ at 350 K.

8.28   Use the following data to determine the normal boiling point of mercury (in kelvins). What assumptions must you make to do this calculation?

Hg(l) $S° = 77.4$ J mol^{-1} K^{-1}
Hg(g) $\Delta H_f° = 60.78$ kJ mol^{-1}, $S° = 174.7$ J mol^{-1} K^{-1}

8.29   Using the data from Appendix 2, calculate the standard entropy change for the reaction

$$H_2(g) + Cl_2(g) \longrightarrow 2HCl(g)$$

at 500°C. State any assumptions that you make.

The Spontaneity of a Process at Constant Temperature and Pressure Is Governed by Gibbs Free Energy

8.30 Using the data from Appendix 2, calculate $\Delta G°$ for the following reactions at 25°C:

(a) $N_2(g) + O_2(g) \longrightarrow 2NO(g)$

(b) $H_2O(l) \longrightarrow H_2O(g)$

(c) $2C_2H_2(g) + 5O_2(g) \longrightarrow 4CO_2(g) + 2H_2O(l)$

8.31 Using the data from Appendix 2, calculate $\Delta G°$ for the following reactions at 25°C:

(a) $2Mg(s) + O_2(g) \longrightarrow 2MgO(s)$

(b) $2SO_2(g) + O_2(g) \longrightarrow 2SO_3(g)$

(c) $2C_2H_6(g) + 7O_2(g) \longrightarrow 4CO_2(g) + 6H_2O(l)$

8.32 From the given values of ΔH and ΔS, predict which of the following reactions would be spontaneous at 25°C:

(a) $\Delta H = 10.5$ kJ mol^{-1}, $\Delta S = 30$ J mol^{-1} K^{-1}

(b) $\Delta H = 1.8$ kJ mol^{-1}, $\Delta S = 113$ J mol^{-1} K^{-1}

If either of the reactions is nonspontaneous at 25°C, at what temperature might it become spontaneous? State any assumptions that you make.

8.33 Find the temperatures at which reactions with the following ΔH and ΔS values would become spontaneous (state any assumptions that you make):

(a) $\Delta H = 126$ kJ mol^{-1}, $\Delta S = 84$ J mol^{-1} K^{-1}

(b) $\Delta H = 11.7$ kJ mol^{-1}, $\Delta S = 105$ J mol^{-1} K^{-1}

8.34 At one time, the domestic gas used for cooking, called "water gas," was prepared as follows:

$$H_2O(g) + C(graphite) \longrightarrow CO(g) + H_2(g)$$

From the thermodynamic quantities listed in Appendix 2, predict whether this reaction will occur to an appreciable extent at 298 K. If not, at what temperature will the reaction begin to occur at a noticeable level? Assume $\Delta H°$ and $\Delta S°$ are temperature independent.

8.35 Calculate $\Delta G°$ for the process

$$C(diamond) \longrightarrow C(graphite)$$

Is the formation of graphite from diamond favored at 25°C? If so, why is it that diamonds do not become graphite on standing?

8.36 Predict the signs of ΔH, ΔS, and ΔG of the system for the following processes at 1 bar: (a) ammonia melts at 60°C, (b) ammonia melts at 77.7°C, (c) ammonia melts at 100°C. (The normal melting point of ammonia is 77.7°C.)

8.37 Ammonium nitrate (NH_4NO_3) dissolves spontaneously and endothermically in water. What can you deduce about the sign of ΔS for the solution process?

8.38 Using data from Appendix 2, calculate the standard Gibbs free energy of formation for $NO_2(g)$ at 400 K by (a) assuming that standard enthalpy and entropy changes are independent of temperature, and (b) by assuming that standard enthalpy and entropy changes depend on temperature but that constant-pressure heat capacities are independent of temperature.

8.39 Calculate $\Delta G°$ for the melting of water: $H_2O(s) \longrightarrow H_2O(l)$ at –10.0°C (a) assuming that the enthalpy and entropy of fusion are constant over the temperature range and (b) assuming that the heat capacities are constant over the temperature range, but that the enthalpy and entropy of fusion are not.

The Mixing of Pure Substances Leads to an Increase in the Entropy and a Decrease in the Gibbs Free Energy

8.40 Calculate the changes in entropy and Gibbs free energy for the following processes: (a) the mixing of 1.00 mole of nitrogen and 1 mole of oxygen, and (b) the mixing of 3.00 moles of argon, 1 mole of helium, and 3.00 moles of hydrogen. Both parts (a) and (b) are carried out under conditions of constant temperature (298 K) and constant pressure. Assume ideal behavior.

8.41 At 25°C and 1 bar pressure, the absolute third-law entropies of methane and ethane are 186.19 and 229.49 J mol^{-1} K^{-1}, respectively, in the gas phase. Calculate the total absolute third-law entropy of a gaseous mixture containing 1 mole of each gas. Assume ideal behavior.

8.42 What is the Gibbs free energy change for a process in which a mixture of 1 mole of argon gas and 1 mole of nitrogen gas is separated into two containers, each with a volume equal to half the original—one containing a mixture of 0.3 mole of argon and 0.7 mole of nitrogen, and the other containing the remainder? Repeat the calculation with 0.001 mol Ar and 0.999 mol nitrogen in the first container. Assume ideal behavior and a constant temperature of 298 K for this problem.

In Living Systems, Spontaneous Reactions Are Used to Drive Other Nonspontaneous, but Essential, Biochemical Processes

8.43 In the metabolism of glucose, the first step is the conversion of glucose to glucose 6-phosphate:

$$\text{glucose} + H_3PO_4 \longrightarrow \text{glucose 6-phosphate} + H_2O$$
$$\Delta G° = 13.4 \text{ kJ mol}^{-1}$$

Because $\Delta G°$ is positive, this reaction does not favor the formation of products. Show how this reaction can be made to proceed by coupling it with the hydrolysis of ATP. Write an equation for the coupled reaction, and determine the overall $\Delta G°$ for the coupled process.

8.44 Certain bacteria in the soil obtain the necessary energy for growth by oxidizing nitrite to nitrate:

$$2NO_2^-(aq) + O_2(g) \longrightarrow 2NO_3^-(aq)$$

Given that the standard Gibbs free energies of formation (298 K) of $NO_2^-(aq)$ and $NO_3^-(aq)$ are -32.2 and -111.3 kJ mol^{-1}, respectively, calculate the amount of Gibbs free energy released when 1 mole of $NO_2^-(aq)$ is oxidized to 1 mole of $NO_3^-(aq)$.

Additional Problems

8.45 Water freezes spontaneously at $-5°C$ and 1 atm, but ice has a lower entropy than liquid water. Explain how a spontaneous process can lead to a decrease in entropy.

8.46 A certain reaction is known to have a $\Delta G°$ value of -122 kJ mol^{-1}. Will the reaction necessarily occur if the reactants are mixed together?

8.47 Entropy has sometimes been described as "time's arrow" because it is the property that determines the forward direction of time. Explain.

8.48 A certain reaction is spontaneous at $72°C$. If the enthalpy change for the reaction is 19 J, what is the *minimum* value of ΔS (in joules per kelvin) for the reaction?

8.49 Predict whether the entropy change is positive or negative for each of these reactions:
  (a) $Zn(s) + 2HCl(aq) \longrightarrow ZnCl_2(aq) + H_2(g)$
  (b) $O(g) + O(g) \longrightarrow O_2(g)$
  (c) $NH_4NO_3(s) \longrightarrow N_2O(g) + 2H_2O(g)$
  (d) $2H_2O_2(l) \longrightarrow 2H_2O(l) + O_2(g)$

8.50 A student looked up the $\Delta G_f°$, $\Delta H_f°$, and $S°$ values for CO_2 in Appendix 2. Plugging these values into Equation 8.35, he found that $\Delta G_f° \neq \Delta H_f° - T S°$ at 298 K. What is wrong with his approach?

8.51 Consider the following reaction at 298 K:

$$2H_2(g) + O_2(g) \longrightarrow 2H_2O(l)$$
$$\Delta H° = -571.6 \text{ kJ mol}^{-1}$$

Calculate ΔS_{sys}, ΔS_{surr}, and ΔS_{univ} for the reaction.

8.52 As an approximation, we can assume that proteins exist either in the native (physiologically functioning) state or the denatured state. The standard molar enthalpy and entropy of the denaturation of a certain protein are 512 kJ mol^{-1} and 1.60 kJ mol^{-1} K^{-1}, respectively. Comment on the signs and magnitudes of these quantities, and calculate the temperature at which the process favors the denatured state.

8.53 Which of the following are not state functions: $S, H, q, w,$ and T?

8.54 Which of the following is not accompanied by an increase in the entropy of the system: (a) mixing of two gases at the same temperature and pressure, (b) mixing of ethanol and water, (c) discharging a battery, and (d) expansion of a gas followed by compression to its original temperature, pressure, and volume?

8.55 Hydrogenation reactions (for example, the process of converting C=C bonds to C—C bonds by the food industry) are facilitated by the use of a transition metal catalyst, such as Ni or Pt. The initial step is the adsorption, or binding, of hydrogen gas onto the metal surface. Predict the signs of $\Delta H, \Delta S,$ and ΔG when hydrogen gas is absorbed onto the surface of Ni metal.

8.56 Calculate the entropy change for the following process: 10.0 g of oxygen (O_2) gas at a pressure of 1.0 bar and $25°C$ is combined in a 30.0-L container with 20.0 g of nitrogen (N_2) gas initially at a pressure of 2.0 bar and $50°C$. The resulting mixture is then heated at constant pressure to $100°C$.

8.57 When an amount of ammonium nitrate is dissolved in water, the solution becomes colder. What conclusion can you draw about $\Delta S°$ for this process? Would you expect the solubility of ammonium nitrate to increase or decrease with increasing temperature?

8.58 A rubber band under tension will contract when heated. Explain.

8.59 Older thermodynamic tables used a standard pressure of 1 atm (1.01325 bar) instead of 1 bar as is currently used. For a temperature of $25°C$, calculate the difference between the standard entropy of CO at the current standard of 1 bar and its value using the older standard. What fraction of the value of $S°$ for CO does this difference represent?

8.60 Comment on the correctness of the analogy sometimes used to relate a student's dormitory room becoming untidy to an increase in entropy.

8.61 A heat engine operates between $210°C$ and $35°C$. Calculate the minimum amount of heat that must be withdrawn from the hot source to obtain 2000 J of work. (See the inset on page 438–439.)

8.62 The internal engine of a 1200-kg car is designed to run on octane (C_8H_{18}), whose enthalpy of combustion is 5510 kJ mol^{-1}. If the car is moving up a slope, calculate the maximum height (in meters) to which the car can be driven on 1.0 gal of fuel. Assume that the engine cylinder temperature is $2200°C$ and that the exit temperature is $760°C$, and neglect all forms of friction. The mass of 1 gallon of fuel is 3.1 kg. [*Hint:* The work done in moving the car over a vertical distance is mgh, where m is the mass of the car in kilograms, g is the acceleration due to gravity (9.8 m s^{-2}), and h is the height in meters.] (See the inset on pp. 438–439.)

Answers to Practice Exercises

465

8.63 Calculate the entropy change for the conversion of a 100.0-g sample of ice at –20.0°C to water at 37°C.

8.64 At very low temperatures (<20 K), the molar heat capacity of crystals can be well approximated by the equation

$$\overline{C}_P = \alpha T^3$$

where α is a constant that depends upon the material and T is the absolute temperature in kelvin. This equation is often used to estimate the heat capacity and standard entropy for materials at temperatures below liquid helium temperatures (4 K).

(a) Use this equation to show that the standard entropy for a crystalline material at very low temperatures is given by

$$S° = \frac{\alpha}{3}T^3$$

(b) In an experiment, the $\overline{C}_P$ for NaCl is measured at 12.0 K to be 0.1016 J mol^{-1} K^{-1}. Use this data and the given equation for $\overline{C}_P$ to determine the value of α for NaCl.

(c) Using your results from parts (a) and (b), determine the standard molar entropy of NaCl at 1.0 K, 5.0 K, and 12.0 K.

8.65 (a) Using the data in Appendix 2, calculate the standard entropy changes ($\Delta S°$) for the following two reactions at 25°C and 80°C:

(i) $2H_2(g) + O_2(g) \longrightarrow 2H_2O(l)$
(ii) $2H_2(g) + O_2(g) \longrightarrow 2H_2O(g)$

(b) Do the $\Delta S°$ values for reactions (i) and (ii) change in the same way when temperature is increased? If not, give a physical explanation for the differences.

8.66 Which of the three laws of thermodynamics cannot be understood without quantum mechanics? Explain.

8.67 The standard enthalpy of formation and the standard entropy of gaseous benzene are 82.93 kJ mol^{-1} and 269.2 J mol^{-1} K^{-1}, respectively. Calculate $\Delta H°$, $\Delta S°$, and $\Delta G°$ for the following process at 25°C. Comment on your answers.

$$C_6H_6(l) \longrightarrow C_6H_6(g)$$

Answers to Practice Exercises

8.1 9.134 J K^{-1} **8.2** (a) $\Delta S < 0$, (b) $\Delta S < 0$, (c) $\Delta S > 0$, (d) $\Delta S > 0$ **8.3** -16.3 J K^{-1} **8.4** 4.31 J K^{-1} **8.5** 135 J K^{-1} **8.6** (a) -172.84 J mol^{-1} K^{-1}, (b) -137.49 J mol^{-1} K^{-1}, (c) 215.54 J mol^{-1} K^{-1} **8.7** (a) $\Delta S > 0$, (b) $\Delta S < 0$, (c) $\Delta S \approx 0$ **8.8** -176.6 J mol^{-1} K^{-1} **8.9** (a) -106.4 kJ mol^{-1}, (b) -2936 kJ mol^{-1} **8.10** at 25°C: $\Delta G° = 117.1$ kJ mol^{-1}; at 800°C: $\Delta G° = -50.2$ kJ mol^{-1}; at $T = 840$ K, $\Delta G° = 0$ **8.11** -146.4 J mol^{-1}, -146.5 J mol^{-1}, 0.07% **8.12** $\Delta S_{mix} = 5.35$ J K^{-1}, $\Delta G_{mix} = 3.60$ kJ

Chapter 9

Physical Equilibrium

The physical state of a substance (whether it is a solid, liquid, or gas) can have a profound effect on its chemical properties. In Chapter 5, we examined the general properties of gases, liquids and solids. In this chapter, we focus on the factors that influence the transformations of a substance from one phase of matter to another. While phase transformations in pure systems are examined in Section 9.1, the remainder of the chapter deals with physical changes in solutions.

9.1 | The Phase Boundaries in Pure Substances Can Be Predicted Using Thermodynamics

In designing and implementing any chemical process, it is important to know the physical phase (solid, liquid, or gas) of each of the reactants and products under the conditions of interest (pressure, temperature, etc.). For example, the design of an industrial chemical reactor is very different for reactions in which all reactants and products are present in the same phase (*homogeneous reactions*) than for those in which at least one reactant or product is in a different phase than the others (*heterogeneous reactions*). To keep track of phase behavior, we introduced the phase diagram in Chapter 5. Recall that a phase diagram is a plot summarizing the temperature, pressure, and composition at which the various phases of a material can exist.

Figure 9.1 shows a typical phase diagram for a pure (one-component) substance. The three phase boundary lines (liquid-solid, liquid-gas, and solid-gas) meet at the triple point (the one temperature and pressure where all three phases coexist). The liquid-gas boundary line terminates at the critical point. These phase boundaries indicate how the coexistence pressure between two phases changes as the temperature changes. In this section, we will examine how we can use thermodynamics to predict the shape of phase coexistence curves for pure substances. For mixtures, more complicated phase diagrams can be constructed that indicate the dependence of the coexistence pressure and temperature upon the composition of the various phases. Phase diagrams for mixtures are discussed in Sections 9.2 and 9.3.

Two (or more) phases can be in physical equilibrium with one another only if they have the same temperature and pressure. In addition, the Gibbs free energies per mole, $\overline{G} = G/n$, of the coexisting phases at the phase boundary must also be equal. Therefore, any point of coexistence between two phases (phase α and phase β)

$$\Delta \overline{G} = \overline{G}(\text{phase } \alpha) - \overline{G}(\text{phase } \beta) = 0$$

Using this restriction and the thermodynamic concepts developed in Chapters 7 and 8, we can predict the shape of the phase boundaries.

The Clapeyron Equation

Consider points 1 and 2 along the boundary line separating phases α and β in Figure 9.2.

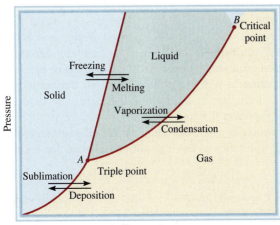

Temperature

Figure 9.1 Phase diagram for a typical pure substance showing the liquid-gas, solid-liquid, and solid-gas phase coexistence boundaries. Point *A* where these three coexistence lines meet is the triple point. Point *B* at which the liquid-gas coexistence line terminates is the critical point.

Figure 9.2 The phase coexistence line between two phases, α and β. At any point along the coexistence line (e.g., point 1 or point 2), the Gibbs free energies of the two phases must be equal. This condition makes it possible to determine the slope, dP/dT, of the coexistence line from the thermodynamic properties of the two phases.

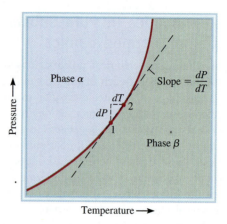

Because the phases α and β are in equilibrium, their Gibbs free energies must be equal at both points 1 and 2, that is, $G_\alpha(1) = G_\beta(1)$ and $G_\alpha(2) = G_\beta(2)$. As a result, the slope, dP/dT, of the P-T coexistence curve must be such that the change in the Gibbs free energy of phase α from point 1 to point 2 equals the corresponding change in the Gibbs free energy for phase β, that is,

$$G_\alpha(2) - G_\alpha(1) = G_\beta(2) - G_\beta(1) \tag{9.1}$$

A thermodynamic analysis of the pressure and temperature dependence of the Gibbs free energy shows that for Equation 9.1 to be satisfied, the slope, dP/dT, must obey Equation 9.2,

$$\left(\frac{dP}{dT}\right)_{\text{coex}} = \frac{\Delta S}{\Delta V} \tag{9.2}$$

where ΔS and ΔV are, respectively, the molar entropy and molar volume changes for the α to β transition, respectively: $\Delta S = S_\beta - S_\alpha$ and $\Delta V = V_\beta - V_\alpha$. Because $\Delta G = \Delta H - T\Delta S = 0$ across the phase coexistence line, we have $\Delta H = T\Delta S$, which means Equation 9.2 can be rewritten as

$$\left(\frac{dP}{dT}\right)_{\text{coex}} = \frac{\Delta H}{T\Delta V} \tag{9.3}$$

This is the **Clapeyron**[1] **equation.** This compact expression makes it possible to predict the change in the coexistence pressure for a given change in coexistence temperature in terms of the easily measurable thermodynamic quantities ΔH and ΔV. Equation 9.3 is a general equation governing phase coexistence and can be applied to fusion, vaporization, and sublimation, as well as to solid-solid phase transitions, such as the conversion of diamond and graphite or the fcc-to-bcc transition in iron.

The form of Equation 9.3 has important implications for melting transitions. The enthalpy of fusion is a positive quantity for all substances (i.e., heat must be added to melt all substances), but the volume change for melting can be either

---

1. Benoit Paul Emile Clapeyron (1799–1864). French engineer. Clapeyron made contributions to the thermodynamics of steam engines.

positive or negative depending upon whether the substance expands ($\Delta V_{\text{fus}} > 0$) or contracts ($\Delta V_{\text{fus}} < 0$) on melting. The vast majority of substances expand on melting ($\Delta V_{\text{fus}} > 0$), so the liquid is less dense than the solid. For these systems, both ΔV_{fus} and ΔH_{fus} are positive, and based on Equation 9.3, the slope of the solid-liquid coexistence curve will therefore be positive, as well. The solid-liquid coexistence curve for carbon dioxide (Figure 5.8) has a positive slope. Because dP/dT is positive, a liquid below the coexistence curve can be frozen by increasing the pressure at constant temperature.

A small number of materials contract when they melt ($\Delta V_{\text{fus}} < 0$), so the liquid is denser than the solid phase. Examples include water, gallium, antimony, bismuth, and many important semiconductors, such as silicon, germanium and gallium arsenide. (Anyone who has accidentally left a closed bottle of water in the freezer has experienced this effect!) For these systems, the slope of the solid-liquid P-T coexistence curve is negative, as seen in the phase diagram of water [Figure 5.5(a)]. In these systems, a solid state lying below the solid-liquid coexistence line melts if the pressure is increased sufficiently.

The volume changes that accompany melting are generally small (usually only a few percent). Because ΔV_{fus} is in the denominator of Equation 9.3, a small value for ΔV_{fus} means that the P-T slope of the phase boundary is large. That is, large changes in pressure are required to cause even small changes in melting temperature. Also, because both ΔV_{fus} and ΔH_{fus} do not change significantly with either temperature or pressure, the slope of the melting curve is nearly constant and the solid-liquid boundary is very close to a straight line, as seen in Figure 5.5(a).

Example 9.1 shows how to use the Clapeyron equation to determine the melting point of water at high pressure.

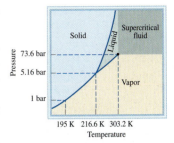

Phase diagram of carbon dioxide (Figure 5.8)

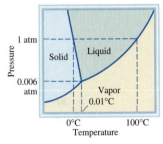

Low pressure phase diagram of water [Fig. 5.5(a)]

Example 9.1

Determine the melting temperature of pure water at 500-bar pressure. The molar volume of ice and liquid water at 273 K and 1 atm pressure are 0.0196 and 0.0180 L mol^{-1}, respectively.

Strategy The normal melting temperature of ice at 1 atm (1.01325 bar) is 273.15 K. To find the melting temperature at 500-bar pressure, we need to find the slope of the P-T curve using the Clapeyron equation (Equation 9.3). To do this, we need both ΔV_{fus} and ΔH_{fus}. We are given the molar volumes of ice and water from which we can determine ΔV_{fus}. The value of ΔH_{fus} for water is listed in Table 7.8 ($\Delta H_{\text{fus}} = 6.01$ kJ mol^{-1}). We can assume that the P-T melting curve is a straight line (constant slope).

Solution The Clapeyron equation, applied to fusion (melting), is

$$\left(\frac{dP}{dT}\right)_{\text{fusion}} = \frac{\Delta H_{\text{fus}}}{T \Delta V_{\text{fus}}}$$

From the given data the change in molar volume on fusion is

$$\Delta V_{\text{fus}} = V_{\text{liquid}} - V_{\text{solid}}$$
$$= 0.0180 \text{ L mol}^{-1} - 0.0196 \text{ L mol}^{-1}$$
$$= -0.0016 \text{ L mol}^{-1}$$

—Continued

Continued—
Thus,

$$\left(\frac{dP}{dT}\right)_{\text{fusion}} \approx \frac{\Delta P}{\Delta T} = \frac{6010 \text{ J mol}^{-1}}{(273.15 \text{ K})(-0.0016 \text{ L mol}^{-1})}$$

$$= -1.38 \times 10^4 \text{ J L}^{-1} \text{ K}^{-1}$$

To convert to pressure units, use the conversion 1 L bar = 100 J:

$$\frac{\Delta P}{\Delta T} = -(1.38 \times 10^4 \text{ J L}^{-1} \text{ K}^{-1}) \times \frac{1 \text{ L bar}}{100 \text{ J}}$$

$$= -138 \text{ bar K}^{-1}$$

Thus, an increase in pressure of 138 bar is required to lower the melting point of water by 1 K.

The change in the melting point in going from 1 atm to 500 bar can then be calculated as follows:

$$\Delta T = \frac{\Delta P}{-138 \text{ bar K}^{-1}}$$

$$= \frac{500 \text{ bar} - 1.01 \text{ bar}}{-138 \text{ bar K}^{-1}}$$

$$= -3.6 \text{ K}$$

Thus, the new melting point is $T_{\text{m}} = 273.15 \text{ K} - 3.6 \text{ K} = 269.5 \text{ K}$ (or $-3.6°C$).

Comment The lowering of the melting point of water at high pressure is sometimes used to explain why ice-skating is possible. When the ice under the skates melts due to the pressure, a film of water is formed that then acts as a lubricant. The pressure exerted by a human on skates, however, is only on the order of 500 bar—the same as in this problem. Therefore, if pressure-induced melting were the sole mechanism lubricating the ice, it would be impossible to ice skate in temperatures below about $-4°C$ (the melting point of ice pressurized by a human skater). It is possible to skate below $-4°C$, however, so other processes must be operating here. More detailed studies indicate that the main causes of lubrication in skating are ice premelting (a natural formation of a thin film of water at the surface of the ice) and melting due to frictional heating as the skates slide across the ice.

Practice Exercise What is the melting point of mercury at 500-bar pressure? The densities of liquid and solid mercury at its normal melting point ($-38.9°C$) are 13.69 and 14.38 g mL^{-1}, respectively. The heat of fusion for Hg is 9.75 kJ mol^{-1}.

The Clausius-Clapeyron Equation

For phase transition, such as vaporization or sublimation, in which one phase is a gas, the Clapeyron equation can be expressed in a simple approximate form. For such transitions, the molar volume of the gas phase (V_g) is so much greater than that of the liquid or solid ($V_{\text{liquid or solid}}$) that we can ignore the molar volume of the dense phase in determining the molar volume change for the transition. For example, for the molar volume change on vaporization, we can write

$$\Delta V_{\text{vap}} = V_g - V_{\text{liquid or solid}} \approx V_g$$

If the pressures are not too large, we can assume ideal gas behavior, giving

$$\Delta V_{vap} \approx V_g = \frac{RT}{P}$$

Substituting this value of ΔV_{vap} into Equation 9.3 (applied to vaporization) gives

$$\frac{dP}{dT} = \frac{\Delta H_{vap}}{T \Delta V_{vap}} \approx \frac{P \Delta H_{vap}}{RT^2}$$

where ΔH_{vap} is the molar enthalpy of vaporization (Section 7.6). After rearrangement, we have

$$\frac{dP}{P} = \frac{\Delta H_{vap}}{RT^2} dT$$

Integrating both sides from the initial pressure and temperature (P_1 and T_1) to the final state (P_2 and T_2) gives

$$\int_{P_1}^{P_2} \frac{dP}{P} = \int_{T_1}^{T_2} \frac{\Delta H_{vap}}{RT^2} dT$$

$$\ln \frac{P_2}{P_1} \approx \frac{\Delta H_{vap}}{R} \left(\frac{1}{T_1} - \frac{1}{T_2} \right) \tag{9.4}$$

where we have made the additional approximation that ΔH_{vap} is independent of T. Equation 9.4, which describes the temperature dependence of vapor pressure, is known as the ***Clausius-Clapeyron equation.*** As written here, Equation 9.4 applies to the vaporization of a liquid. For applications to sublimation, the enthalpy of vaporization must be replaced by ΔH_{sub}, the enthalpy of sublimation. Because the enthalpy of vaporization (or sublimation) is a positive quantity, Equation 9.4 shows that the vapor pressure of a liquid or solid increases with increasing temperature.

Example 9.2 shows how to use Equation 9.4 to determine the temperature dependence of the vapor pressure of a liquid.

Example 9.2

Diethyl ether is a volatile, highly flammable organic liquid that is used mainly as a solvent. The vapor pressure of diethyl ether is 401 mmHg at 18°C. Calculate the vapor pressure at 32°C.

Strategy We are given the vapor pressure ($P_1 = 401$ mmHg) at a given temperature ($T_1 = 18$°C $= 291.15$ K). We are asked to find the vapor pressure (P_2) at a different temperature ($T_2 = 32$°C $= 305.15$ K). To use the Clausius-Clapeyron equation, we need to know the molar enthalpy of vaporization of diethyl ether. This value is listed in Table 7.7 ($\Delta H_{vap} = 26.0$ kJ mol^{-1}). (The values of the molar enthalpies of vaporization given in Table 7.7 are taken at the normal boiling point, which for diethyl ether is 34.6°C. We can use this value in Equation 9.4 because we derived this equation making the assumption that ΔH_{vap} is independent of temperature over the temperature range of interest.)

—Continued

Continued—

Solution Substituting these values into Equation 9.4 gives

$$\ln\frac{P_2}{P_1} \approx \frac{\Delta H_{vap}}{R}\left(\frac{1}{T_1} - \frac{1}{T_2}\right)$$

$$\ln\frac{P_2}{401\ \text{mmHg}} \approx \frac{26{,}000\ \text{J mol}^{-1}}{8.314\ \text{J mol}^{-1}\ \text{K}^{-1}}\left(\frac{1}{291\ \text{K}} - \frac{1}{305\ \text{K}}\right)$$

$$\ln\frac{P_2}{401\ \text{mmHg}} \approx 0.493$$

Taking the exponential of both sides gives

$$P_2 = (401\ \text{mmHg}) \times e^{0.493}$$
$$P_2 = 656\ \text{mmHg}$$

Check The answer is reasonable because the vapor pressure should be greater at the higher temperature.

Practice Exercise The vapor pressure of ethanol is 100 mm Hg at 34.9°C. What is its vapor pressure at 63.5°C?

When P_1 and T_1 are known, the Clausius-Clapeyron equation (Equation 9.4) has the form

$$\ln P = -\frac{\Delta H_{vap}}{R}\left(\frac{1}{T}\right) + \text{constant} \tag{9.5}$$

which has the form of a straight line ($y = mx + b$). Therefore, a plot of $\ln P$ versus $1/T$ (assuming that ΔH_{vap} is independent of T) will give a straight line whose slope is equal to $-\Delta H_{vap}/R$. This method can be used to determine the heat of vaporization for a substance. Figure 9.3 shows plots of $\ln P$ versus $1/T$ for water and diethyl ether. Both are straight lines, but the one for water has a steeper slope because water has a larger ΔH_{vap}.

Example 9.3 shows how to determine the molar enthalpy of vaporization from data for the vapor pressure of a substance as a function of temperature.

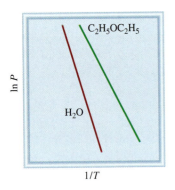

Figure 9.3 Plots of $\ln P$ versus $1/T$ for water and diethyl ether. The slopes for both graphs are equal to $-\Delta H_{vap}/R$.

Example 9.3

An experiment to determine the vapor pressure of water at a variety of temperatures between 15°C and 80°C yields the following data:

T (°C)	15.0	25.0	37.0	60.0	80.0
P (mm Hg)	12.79	23.76	47.07	149.4	355.1

Use these data to determine the molar enthalpy of vaporization of water.

Strategy According to Equation 9.5, a plot of $\ln P$ versus $1/T$ yields a straight line whose slope is equal to $-\Delta H_{vap}/R$. The temperature data are given in degrees Celsius, so they must first be converted to absolute temperature values (in kelvins) before inverting and plotting. Also, the pressure data must be converted to $\ln P$ data.

—Continued

Continued—

Solution To use Equation 9.5, we must first convert the data into a suitable form for plotting:

$1/T$ (K^{-1})	3.470×10^{-3}	3.354×10^{-3}	3.224×10^{-3}	3.002×10^{-3}	2.832×10^{-3}
ln P	2.549	3.168	3.852	5.006	5.872

We can then plot ln P versus $1/T$, which gives a straight line.

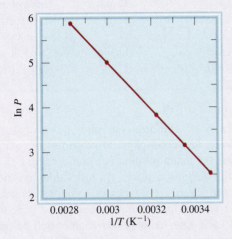

Using a graphing calculator or computer spreadsheet computer program, least-squares linear regression (Appendix 1) gives a slope of -5.21×10^3 K. Thus,

$$-\Delta H_{vap}/R = -5210 \text{ K}$$
$$\Delta H_{vap} = -(8.314 \text{ J mol K}^{-1})(-5.21 \times 10^3 \text{ K})$$
$$= 43.3 \text{ kJ mol}^{-1}$$

Comment This estimate of the molar heat of vaporization of water is somewhat lower than the value measured at the normal boiling point (100°C) listed in Table 7.7 (40.79 kJ mol^{-1}). This apparent discrepancy arises because ΔH_{vap} is temperature dependent, so the value determined graphically in this exercise actually represents an average over the temperature interval 15°C to 80°C.

Practice Exercise The vapor pressure of mercury at various temperatures has been determined as follows:

T (K)	323	353	393.5	413	433
P (mmHg)	0.0127	0.0888	0.7457	1.845	4.189

Calculate the value of $\Delta \overline{H}_{vap}$ for mercury.

9.2 | The Solubility of a Substance Is Determined by Temperature, Pressure, and Intermolecular Forces

The discussion in Section 9.1 focused on phase equilibrium in pure substances. Most chemical reactions take place, though, not between pure solids, liquids, or gases, but among ions and molecules dissolved in water or other solvents. In Sections 9.2 to 9.4,

Table 9.1	Types of Solutions		
Component 1	**Component 2**	**State of Resulting Solution**	**Examples**
Gas	Gas	Gas	Air
Gas	Liquid	Liquid	Soda water (CO_2 in water)
Gas	Solid	Solid	H_2 gas in palladium
Liquid	Liquid	Liquid	Ethanol in water
Liquid	Solid	Solid	Dental Amalgam (Hg/Ag and other metals)
Solid	Liquid	Liquid	NaCl in water
Solid	Solid	Solid	Brass (Cu/Zn), Solder (Sn/Pb)

we examine the properties of solutions, concentrating mainly on the role of intermolecular forces in solubility and other physical properties of solutions.

Recall from Section 0.1 that a solution is a homogeneous mixture of two or more substances. Because this definition places no restrictions on the nature of the substances involved, we can identify six types of solutions, depending on the original states (solid, liquid, or gas) of the solution components. Table 9.1 lists all six types and cites examples of each. Our focus in this chapter will be on solutions involving at least one liquid component, that is, gas-liquid, liquid-liquid, and solid-liquid solutions. And, in most cases, the liquid solvent will be water.

Solutions can be characterized in terms of their capacity to dissolve a solute. A *saturated solution contains the maximum amount of a solute that will dissolve in a given solvent at a specific temperature.* This maximum solute concentration is referred to as the *solubility* of the solute in the given solvent. An *unsaturated solution contains less solute than it has the capacity to dissolve.* A third type, a *supersaturated solution, contains more solute than is present in a saturated solution.* Supersaturated solutions are unstable. In time, some of the solute will come out of a supersaturated solution as crystals.

Crystallization is *the process in which dissolved solute comes out of solution and forms crystals* (Figure 9.4). Both precipitation and crystallization describe the separation of an excess solid substance from a supersaturated solution. However, solids

Figure 9.4 Sodium acetate crystals rapidly form in a supersaturated solution when a small seed crystal is added (left).

formed by the two processes differ in appearance. Precipitates are usually made up of small particles, whereas crystals may be large and well formed.

Chemists qualitatively refer to substances as soluble, slightly soluble, or insoluble. A substance is said to be soluble in a given solvent if an observable amount of it dissolves when added to that solvent. If not, the substance is described as slightly soluble or insoluble in the given solvent. As a general rule of thumb, a substance that has a solubility less than about 0.1 mol L^{-1} can be considered to be insoluble.

A Molecular View of the Solution Process

The intermolecular attractions that hold molecules together in liquids and solids also play a central role in the formation of solutions. When one substance (the solute) dissolves in another (the solvent), particles of the solute disperse throughout the solvent. The solute particles occupy positions that are normally taken by solvent molecules. The ease with which a solute particle replaces a solvent molecule depends on the relative strengths of three types of interactions:

▸ Solvent-solvent interactions

▸ Solute-solute interactions

▸ Solvent-solute interactions

Although the exact mechanism of the solution process can be complex, for the purposes of calculating the enthalpy changes associated with the dissolution process, we can imagine this process as taking place in the three distinct steps shown in Figure 9.5. (We can use any path to calculate enthalpy changes in a system because enthalpy is a state function.) The solvent molecules separate in step 1, and the solute molecules separate in step 2. (Although the solute is shown here as a crystalline solid, it could also be a liquid or gas.) These steps require an input of energy to break the attractive intermolecular forces, so they are endothermic. In step 3, the solvent and solute molecules mix. This process can be exothermic or endothermic. The heat of solution (ΔH_{soln}) is given by

$$\Delta H_{\text{soln}} = \Delta H_1 + \Delta H_2 + \Delta H_3$$

If the solute-solvent attraction (ΔH_3) is stronger than the solvent-solvent attraction (ΔH_1) and the solute-solute attraction (ΔH_2), the solution process is favorable, or exothermic ($\Delta H_{\text{soln}} < 0$).

If the solute-solvent interaction is weaker than the solvent-solvent and solute-solute interactions, then the solution process is endothermic ($\Delta H_{\text{soln}} > 0$).

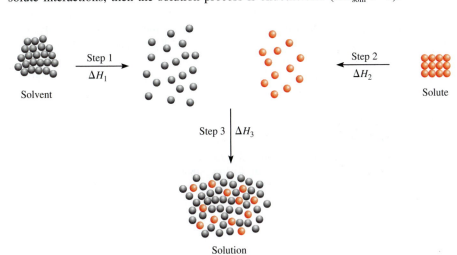

Figure 9.5 A molecular view of the solution process portrayed as taking place in three steps: First the solvent and solute molecules are separated (steps 1 and 2), and then the solvent and solute molecules mix (step 3).

CH₃OH
(methanol)

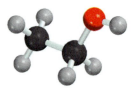

C₂H₅OH
(ethanol)

Why would a solute dissolve in a solvent at all if the attraction for its own molecules is stronger than the solute-solvent attraction? The solution process, like all physical and chemical processes, is governed by an enthalpy factor and an entropy factor. The enthalpy of solution (ΔH_{sol}) determines whether a solution process is exothermic or endothermic. The entropy of solution (ΔS_{soln}), which is related to the entropy of mixing discussed in Section 8.6, generally favors dissolution, so a solute may have significant solubility in a solvent even if the solution process is endothermic.

The concept "like dissolves like" can be helpful in predicting the solubility of a substance in a given solvent. What this expression means is that two substances with intermolecular forces of similar type and magnitude are likely to be soluble in each other. For example, both carbon tetrachloride (CCl_4) and benzene (C_6H_6) are nonpolar liquids. The only intermolecular forces present in these substances are the dispersion forces discussed in Section 4.6. When these two liquids are mixed, they readily dissolve in each other, because the CCl_4-C_6H_6 attractions are comparable in magnitude to the CCl_4-CCl_4 forces and to the C_6H_6-C_6H_6 forces.

When two liquids are *completely soluble in each other in all proportions* they are said to be **miscible.** Although CCl_4 and C_6H_6 are miscible in each other, neither is miscible (or even soluble) in water, which is a strongly polar solvent. Acetone and alcohols such as methanol, ethanol, and 1,2-ethylene glycol are miscible with water because they can form hydrogen bonds with water molecules:

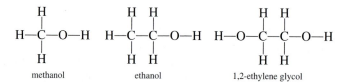

methanol   ethanol   1,2-ethylene glycol

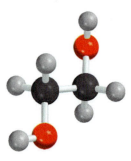

CH₂(OH)CH₂(OH)
(1,2-ethylene glycol)

Polar substances, which dissolve easily in water, are often referred to as **hydrophilic** (*water loving*), whereas nonpolar substances that are insoluble in water are referred to as **hydrophobic** (*water fearing*).

In general, ionic compounds (e.g., NaCl) are much more soluble in polar solvents, such as water, liquid ammonia, and liquid hydrogen fluoride, than in nonpolar solvents, such as benzene and carbon tetrachloride. Because the molecules of nonpolar solvents lack a dipole moment, they cannot effectively solvate cations (e.g., Na^+) and anions (e.g., Cl^-). (**Solvation** is *the process in which an ion or a molecule is surrounded by solvent molecules arranged in a specific manner.* The process is called **hydration** when the solvent is water.) The predominant intermolecular interaction between ions and nonpolar compounds is an ion-induced dipole interaction, which is much weaker than an ion-dipole interaction (see Section 4.6). Consequently, ionic compounds usually have extremely low solubilities in nonpolar solvents. (The qualitative and quantitative prediction of the solubility of specific ionic compounds in water is discussed in Chapter 10.)

Example 9.4 shows how to qualitatively predict solubility based on the intermolecular forces in the solute and the solvent.

CH₂O
(formaldehyde)

Example 9.4

Predict the relative solubilities in the following cases: (a) bromine (Br_2) in benzene (C_6H_6, $\mu = 0$ D) and in water ($\mu = 1.87$ D), (b) KCl in carbon tetrachloride (CCl_4, $\mu = 0$ D) and in liquid ammonia (NH_3, $\mu = 1.46$ D), and (c) formaldehyde (CH_2O) in carbon disulfide (CS_2, $\mu = 0$) and in water.

—Continued

Continued—

Strategy To predict the solubility of a particular compound in a particular solvent, remember that like dissolves like. Thus, a nonpolar solute will dissolve in a nonpolar solvent; ionic compounds will generally dissolve in polar solvents due to favorable ion-dipole interaction; and solutes that can form hydrogen bonds with the solvent will be very soluble in the solvent.

Solution (a) Br_2 is a nonpolar molecule, so it should be more soluble in C_6H_6 (also nonpolar) than in water (very polar). The only intermolecular forces between Br_2 and C_6H_6 are dispersion forces.

(b) KCl is an ionic compound. For it to dissolve, the individual K^+ and Cl^- ions must be stabilized by ion-dipole interactions. Because CCl_4 has no dipole moment, KCl should be more soluble in liquid NH_3, a polar molecule with a large dipole moment.

(c) Because CH_2O is a polar molecule and CS_2 (a linear molecule) is nonpolar, the forces between molecules of CH_2O and CS_2 are weak dipole-induced dipole interactions and dispersion forces. On the other hand, CH_2O can act as a hydrogen-bond acceptor with water, so it should be more soluble in H_2O than in CS_2.

Practice Exercise Is iodine (I_2) more soluble in water or in carbon disulfide (CS_2)?

The concept of like dissolves like helps explain how soap works to remove oily substances. Sodium stearate, a typical soap molecule, has a polar head group and a long nonpolar hydrocarbon tail that is nonpolar (Figure 9.6).

The cleansing action of soap is due to the dual nature of the soap molecule (hydrophobic tail and hydrophilic head). The hydrocarbon tail is readily soluble in oily substances, which are also nonpolar, while the ionic—COO^- group remains outside the oily surface. When enough soap molecules have surrounded an oil droplet, as shown in Figure 9.7, the entire system becomes stabilized in water because the exterior portion is now largely hydrophilic. This is how soap removes greasy substances from surfaces such as hands, dishes, and clothing.

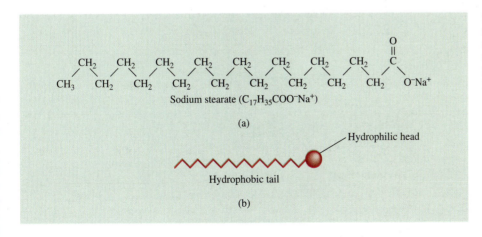

(a)

Hydrophilic head

Hydrophobic tail

(b)

Figure 9.6 (a) A sodium stearate molecule. (b) The simplified representation of the molecule that shows a hydrophilic head and a hydrophobic tail.

Figure 9.7 The cleansing action of soap. (a) Grease (oily substance) is insoluble in water. (b) When soap is added to water, the nonpolar tails of the soap molecules dissolve in the grease, leaving the polar head groups on the surface, exposed to water. (c) The grease is removed in the form of an emulsion in which each oily droplet has an ionic exterior that is hydrophilic.

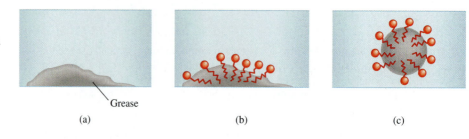

(a)   (b)   (c)

The Effect of Temperature on Solubility

Recall that solubility is defined as the maximum amount of a solute that will dissolve in a given quantity of solvent *at a specific temperature and pressure.* Figure 9.8 shows how the solubility of some ionic compounds in water depends on the temperature. In most but certainly not all cases, the solubility of a solid substance increases with temperature.

For reasons that will be discussed in Section 10.6, the temperature dependence of the solubility can be predicted from the enthalpy change associated with adding solute to a saturated solution ΔH_{sat}. If ΔH_{sat} is positive (i.e., if the addition of solute is an endothermic process), then the solubility of the solute will *increase* with increasing temperature. If ΔH_{sat} is negative (i.e., if the addition of solute is an exothermic process), then the solubility of the solute will *decrease* with increasing temperature.

The value of ΔH_{sat} can differ from the infinite dilution enthalpy of solution ΔH_{soln} defined in Section 7.6 for two reasons:

▸ The enthalpy of solution at infinite dilution ΔH_{soln} is the enthalpy change associated with the addition of a solute to a *pure solvent,* whereas ΔH_{sat} is the enthalpy change associated with the addition of solute to a *saturated solution.*

▸ The stable form of an ionic solid at saturation is often a hydrate. The enthalpy of solution for hydrates is generally positive (endothermic), especially at

Figure 9.8 Temperature dependence of the solubility of some ionic compounds in water.

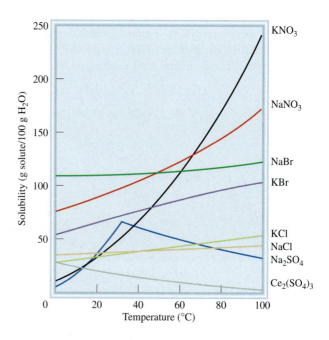

saturation. For this reason, the solubility of an ionic compound may increase with temperature, even though the ΔH_{soln} for the anhydrous form is negative (exothermic).

The infinite dilution enthalpy ΔH_{soln} for anhydrous Na_2SO_4, for example, is -23 kJ mol^{-1}. Based on this value, the solubility of sodium sulfate should *decrease* with increasing temperature. It does at high temperatures, according to Figure 9.8, but at low temperatures the solubility *increases* with increasing temperature. This occurs because the stable form between 0°C and about 30°C is not the anhydrous compound, but the decahydrate $Na_2SO_4 \cdot 10H_2O$, for which the enthalpy of solution is positive (endothermic) at saturation. Above about 30°C, the stable solid form is anhydrous Na_2SO_4, and the solubility decreases with temperature as predicted by the negative value of ΔH_{soln}.

The dependence of the solubility of a solid on temperature varies considerably, as shown in Figure 9.9. For example, the solubility of KNO_3, for example, increases sharply with temperature, while the solubilities of NaBr and NaCl change very little. This wide variation makes it possible to obtain pure substances from mixtures through the process of ***fractional crystallization,*** defined as *the separation of a mixture of substances into pure components on the basis of their differing solubilities.*

Suppose we have 90 g of KNO_3 that is contaminated with 10 g of NaCl. To purify the KNO_3 sample, we dissolve the mixture in 100 mL of water at 60°C and then gradually cool the solution to 0°C. At this temperature the solubilities of KNO_3 and NaCl are 12.1 g/100 g H_2O and 34.2 g/100 g H_2O, respectively. Thus, 78 g of KNO_3 (that is, 90 g -12 g) will crystallize out of the solution but all the NaCl will remain dissolved (Figure 9.9). In this manner, we can obtain about 90 percent of the original amount of KNO_3 in pure form. The KNO_3 crystals can be separated from the solution by filtration. Many of the solid inorganic and organic compounds that are used in the laboratory are purified by fractional crystallization. The method usually works best if

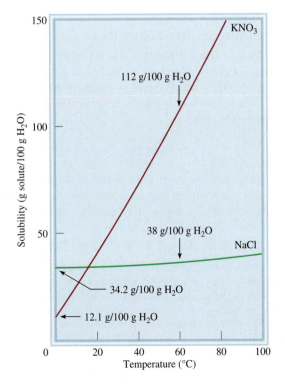

Figure 9.9 Graph of the temperature dependence of the solubilities of KNO_3 and NaCl in water. KNO_3 is considerably more soluble at high temperatures (that is, 60°C) than at low temperatures (that is, 0°C), whereas the solubility of NaCl increases only slightly as the temperature increases. The difference in temperature dependence makes it possible to isolate pure KNO_3 from a solution containing both salts via fractional crystallization.

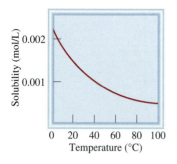

Figure 9.10 Graph of the temperature dependence of the solubility of O_2 gas in water. The solubility decreases as the temperature increases. The pressure of the gas over the solution is 1 bar.

the compound to be purified has a steep solubility curve, that is, if it is considerably more soluble at high temperatures than at low temperatures. Otherwise, much of it will remain dissolved as the solution is cooled. Fractional crystallization also works well if the amount of impurity in the solution is relatively small.

Temperature and Gas Solubility

When a gas is put in contact with a solvent, some of the gas molecules will dissolve in the liquid. This process is usually exothermic, so the solubility of gases in water usually *decreases* with increasing temperature (Figure 9.10). When water is heated in a beaker, bubbles of air form on the side of the glass before the water boils. As the temperature rises, the dissolved air molecules begin to "boil out" of the solution long before the water itself boils.

The reduced solubility of molecular oxygen in hot water contributes to the problem of *thermal pollution,* that is, the heating of the environment (usually waterways) to temperatures that are harmful to its living inhabitants. It is estimated that every year in the United States some 100,000 billion gallons of water are used for industrial cooling, mostly in electric power and nuclear power production. This process heats the water, which is then returned to the rivers and lakes from which it was taken. Fish, like all other cold-blooded animals, have much more difficulty coping with rapid temperature fluctuations in the environment than humans do. An increase in water temperature accelerates their rate of metabolism, which generally doubles with each 10°C rise. The speedup of metabolism increases the fish's need for oxygen at the same time that the supply of oxygen is diminished because of its lower solubility in heated water.

Understanding how gas solubility varies with temperature can improve your chances of catching a fish, too. On a hot summer day, an experienced angler usually picks a deep spot in the river or lake to cast their bait. Because the oxygen content is greater in the deeper, cooler region, most fish will be found there.

The Effect of Pressure on the Solubility of Gases

For all practical purposes, external pressure has no influence on the solubilities of liquids and solids, but it greatly affects the solubility of gases. The quantitative relationship between gas solubility and pressure is given by **Henry's[2] law,** which states that *the equilibrium concentration of a gas dissolved in a liquid is proportional to the pressure of the gas over the solution:*[3]

$$c = k_H P \tag{9.6}$$

Here c is the molar concentration (mol L^{-1}) of the dissolved gas; P is the pressure (in bar) of the gas over the solution; and k_H is the *Henry's law constant,* which for a given gas, depends only on the temperature and the identity of the solvent. The

---

2. William Henry (1775–1836). English chemist. Henry's major contribution to science was his discovery of the law describing the solubility of gases, which now bears his name.

3. In advanced studies, Henry's law is often written as $P_i = x_i K_{H,i}$, where x_i is the mole fraction of the solute gas in the solution and P_i is the partial pressure of the gas in the vapor phase. In this form, Henry's law constant $K_{H,i}$ has units of pressure. Equation 9.6 can be derived from this equation by noting that, for dilute solutions, the mole fraction of a solute is proportional to its concentration. In Section 9.3, we will demonstrate how this form of Henry's law results from the vapor-liquid equilibrium of real binary liquids.

Table 9.2	Henry's Law Constants for Select Gases in Water at 298 K
Gas	**Henry's Law Constant k_H (mol L^{-1} bar^{-1})**
H_2	7.8×10^{-4}
He	3.7×10^{-4}
Ne	4.5×10^{-4}
Ar	1.4×10^{-3}
Kr	2.4×10^{-3}
Xe	4.3×10^{-3}
N_2	6.2×10^{-4}
O_2	1.3×10^{-3}
CO_2	3.4×10^{-2}
CH_4 (methane)	1.4×10^{-3}
C_2H_6 (ethane)	1.9×10^{-3}
C_3H_8 (propane)	1.4×10^{-3}
C_4H_8 (n-butane)	1.1×10^{-3}

constant k_H has the units mol L^{-1} bar^{-1}. When the pressure of the gas is 1 bar, c is *numerically* equal to k_H. If several gases are present, P is the partial pressure. Henry's law constants for several common gases at 298 K are listed in Table 9.2. Because the solubility of gases generally decreases with increasing temperature, the Henry's law constants also generally decrease with temperature.

A practical demonstration of Henry's law is the effervescence of a soft drink when the cap of the bottle is removed. Before the beverage bottle is sealed, it is pressurized with a mixture of air and CO_2 saturated with water vapor. Because of the high partial pressure of CO_2 in the pressurizing gas mixture, the amount dissolved in the soft drink is many times the amount that would dissolve under normal atmospheric conditions. When the cap is removed, the pressurized gases escape. Eventually the pressure in the bottle falls to atmospheric pressure, and the amount of CO_2 remaining in the beverage is determined only by the normal atmospheric partial pressure of CO_2 (0.0003 bar). The excess dissolved CO_2 comes out of solution, causing the effervescence. Example 9.5 applies Henry's law to nitrogen gas.

Example 9.5

The solubility of nitrogen gas in water at 25°C and 1 bar is 6.3×10^{-4} mol L^{-1}. What is the concentration of nitrogen dissolved in water under atmospheric conditions? The partial pressure of nitrogen gas in the atmosphere is 0.79 bar.

Strategy The given solubility enables us to calculate Henry's law constant (k_H), which can then be used to determine the concentration of the solution.

—Continued

Continued—

Solution The first step is to calculate k_H in Equation 9.6:

$$c = k_H P$$
$$6.3 \times 10^{-1}\,\text{mol L}^{-1} = k_H\,(1\,\text{bar})$$
$$k_H = 6.3 \times 10^{-4}\,\text{mol L}^{-1}\,\text{bar}^{-1}$$

Then, plug the calculated value of k_H and the given value of P into Equation 9.6 to determine the solubility of nitrogen gas in water under normal atmospheric conditions:

$$c = (6.3 \times 10^{-4}\,\text{mol L}^{-1}\,\text{bar}^{-1})(0.79\,\text{bar})$$
$$= 5.0 \times 10^{-4}\,\text{mol L}^{-1}$$
$$= 5.0 \times 10^{-4}\,M$$

The solubility of N_2 decreases because the pressure was lowered from 1 bar to 0.79 bar.

Check The ratio of the concentrations $[(5.0 \times 10^{-4}\,M/6.3 \times 10^{-4}\,M) = 0.79]$ should be equal to the ratio of the pressures (0.79 bar/1.0 bar = 0.79).

Practice Exercise Calculate the molar concentration of oxygen in water at 25°C for a partial pressure of 0.22 bar.

For most gases, the relationship between Henry's law and solubility is straightforward, however, in cases where the dissolved gas can undergo a reaction with the solvent, the solubility will be underestimated by Henry's law. For example, the solubility of ammonia in water is much higher than predicted by Henry's law because NH_3 *reacts* with H_2O:

$$NH_3(aq) + H_2O(l) \rightleftharpoons NH_4^+(aq) + OH^-(aq)$$

Henry's law predicts the amount of dissolved $NH_3(aq)$ present in the solution, but the total amount of NH_3 from the gas phase that dissolves in solution is given by the sum of $NH_3(aq)$ and $NH_4^+(aq)$. Similarly, the solubility of carbon dioxide will be underestimated using Henry's law because CO_2 also reacts with water, as follows:

$$CO_2 + H_2O(l) \rightleftharpoons H_2CO_3(aq)$$

For such gases, the solubility requires the combined use of Henry's law and the equations governing the reaction equilibrium in solution, which will be discussed in Chapter 10.

Another important example is the dissolution of molecular oxygen in blood. Normally, oxygen gas is only sparingly soluble in water, the solvent in blood (see the Practice Exercise in Example 9.5). The solubility of O_2 in blood is significantly higher, however, because blood contains a high concentration of hemoglobin (Hb) molecules. Each hemoglobin molecule can bind up to four oxygen molecules, which are eventually delivered to the tissues for use in metabolism:

$$Hb + 4O_2 \rightleftharpoons Hb(O_2)_4$$

Thus, molecular oxygen is highly soluble in blood because it binds to hemoglobin.

The box text applies Henry's law to a natural disaster that occurred at Lake Nyos in Cameroon (west Africa).

The Killer Lake

Disaster struck swiftly and without warning. On August 21, 1986, Lake Nyos in Cameroon, a small nation on the west coast of Africa, suddenly belched a dense cloud of carbon dioxide. Speeding down a river valley, the cloud asphyxiated over 1700 people and many livestock.

How did this tragedy happen? Lake Nyos is stratified into layers that do not mix. A boundary separates the fresh-water at the surface from the deeper, denser solution containing dissolved minerals and gases, including CO_2. The CO_2 gas comes from springs of carbonated groundwater that percolate upward into the bottom of the volcanically formed lake. Given the high water pressure at the bottom of the lake, the concentration of CO_2 gradually accumulated to a dangerously high level, in accordance with Henry's law.

What triggered the release of CO_2 is not known for certain. It is believed that an earthquake, landslide, or even strong winds may have upset the delicate balance within the lake, creating waves that overturned the water layers. When the deep water rose, dissolved CO_2 came out of solution, just as a soft drink fizzes when the bottle is uncapped. Being heavier than air, the CO_2 traveled close to the ground and literally smothered an entire village 15 miles away. Now, more than 18 years after the incident, scientists are concerned that the CO_2 concentration at the bottom of Lake Nyos is again reaching saturation level. To prevent a recurrence of the earlier tragedy, an attempt has been made to pump up the deep water, thus releasing the dissolved CO_2. In addition to being costly, this approach is controversial because it might disturb the waters near the bottom of the lake, leading to an uncontrollable release of CO_2 to the surface. In the meantime, a natural time bomb is ticking away.

Deep waters in Lake Nyos are pumped to the surface to remove dissolved CO_2 gas.

9.3 | The Liquid-Vapor Phase Equilibrium of a Solution Can Be Understood in Terms of the Entropy of Mixing and the Intermolecular Forces

Sections 9.3 and 9.4 examine phase equilibrium in mixtures, focusing primarily on systems in which at least one of the coexisting phases is a liquid solution (i.e., vapor-liquid, liquid-liquid, and solid-liquid solution phase equilibrium).

Liquid-Vapor Equilibrium of Ideal Solutions

A liquid solution that is a mixture of two components, A and B, is called a *binary liquid mixture* (or *binary solution*). Like a single-component liquid, this solution will have a vapor pressure (P) that depends upon the temperature. For a binary solution, though, the vapor pressure will also depend upon the composition of the liquid, defined by the mole fractions x_A^l and x_B^l, where the superscript "l" denotes the liquid phase. The composition of the vapor above a binary liquid mixture, denoted by x_A^v and x_B^v (v for "vapor"), will usually differ from that of the liquid.

If the vapor is considered ideal, then the total vapor pressure, described by Dalton's law, is the sum of the partial pressures of components A and B.

$$P_T = P_A + P_B \tag{9.7}$$

In the late nineteenth century, François-Marie Raoult[4] observed that for many binary liquid mixtures, the partial pressure of component i in the equilibrium vapor phase is given by the mole fraction of the component in the solution (x_i^l) times the vapor pressure of i in its pure liquid form (P_i^*):

$$P_i = x_i^l P_i^* \tag{9.8}$$

This is known as **Raoult's law.** In Section 8.6, we defined an *ideal solution* of two liquids as one in which the enthalpy of mixing is zero and the entropy of mixing is given by Equation 8.48. It can be shown using thermodynamics that Raoult's law is exactly satisfied for ideal solutions. In fact, the converse is also true (i.e., any solution that follows Raoult's law can be shown to be ideal), so an ideal solution can be alternatively defined as any solution that follows Raoult's law.

For an ideal binary solution, the total vapor pressure (P_T) can be determined by combining Equations 9.7 and Raoult's Law (Equation 9.8):

$$P_T = x_A^l P_A^* + x_B^l P_B^* \tag{9.9}$$

For a given value of x_A^l, however, the mole fraction of B is given by $x_B^l = 1 - x_A^l$, so Equation 9.9 can be written as

$$P_T = x_A^l P_A^* + (1 - x_A^l)P_B^*$$

$$P_T = P_B^* + x_A^l(P_A^* - P_B^*) \tag{9.10}$$

which defines a straight line with slope $(P_A^* - P_B^*)$ and intercept P_B^*.

On the molecular level, the interactions between A and B molecules in an ideal binary solution are identical to the interactions between A molecules and the interactions between B molecules. Although no solution is strictly ideal, many come close. For example, benzene (C_6H_6) and toluene (C_7H_8) have very similar structures, and therefore very similar intermolecular interactions, and are known to form a very nearly ideal solution when mixed:

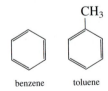

benzene   toluene

The graph in Figure 9.11 shows how the total vapor pressure (P_T) in a benzene–toluene solution at 80.1°C depends on the composition of the solution (as expressed by the mole fraction of benzene).

As predicted by Equation 9.10 for ideal solutions, the total vapor pressure above the benzene-tol uene solution as a function of the mole fraction of benzene is a straight

Benzene (C_6H_6) and toluene (C_7H_8).

---

4. François-Marie Raoult (1830–1901). French chemist. Raoult's work was mainly in solution properties and electrochemistry.

line connecting the vapor pressure of pure toluene (at $x^l_{benzene} = 0$) and the vapor pressure of pure benzene (at $x^l_{benzene} = 1$).

Example 9.6 shows how to use Equation 9.10 to calculate the vapor pressure of an ideal solution.

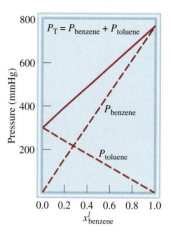

Figure 9.11 Graph of the dependence of the total and partial pressures of a benzene-toluene binary liquid mixture as functions of the mole fractions of each component ($x_{toluene} = 1 - x_{benzene}$) at 80.1°C. This solution can be considered to be nearly ideal because the vapor pressures are well approximated by Raoult's law.

Example 9.6

Benzene (C_6H_6) and toluene (C_7H_8) form a nearly ideal solution. The vapor pressures of pure benzene and pure toluene are 94.6 and 29.1 torr at 298 K, respectively. (a) Calculate the total vapor pressure above a solution formed by mixing 25.0 g of benzene and 25.0 g of toluene. (b) Calculate the mole fractions of benzene and toluene in the vapor phase.

Strategy Assuming that the mixture of benzene and toluene is an ideal solution, the partial pressures of both benzene and toluene can be calculated using Raoult's law (Equation 9.8). To do this we need to convert grams of both benzene and toluene to moles and then determine the mole fractions. (a) The total vapor pressure is then obtained by adding together the partial pressures of benzene and toluene. (b) The mole fractions of benzene and toluene in the vapor phase can be determined from the partial pressures using Dalton's law.

Solution The number of moles of benzene and toluene are calculated as follows:

$$\text{moles of benzene} = \frac{25.0 \text{ g benzene}}{78.1 \text{ g mol}^{-1}} = 0.320 \text{ mol}$$

$$\text{moles of toluene} = \frac{25.0 \text{ g toluene}}{92.1 \text{ g mol}^{-1}} = 0.271 \text{ mol}$$

The mole fraction of benzene ($x^l_{benzene}$) is then given by

$$x^l_{benzene} = \frac{\text{moles of benzene}}{(\text{moles of benzene}) + (\text{moles of toluene})}$$

$$= \frac{0.320 \text{ mol}}{0.320 \text{ mol} + 0.271 \text{ mol}} = 0.541$$

The mole fraction of toluene ($x^l_{toluene}$) could be calculated in a similar fashion, but it is easier to use the following relationship, which works because this is a two-component mixture: $x^l_{benzene} + x^l_{toluene} = 1$. Thus,

$$x^l_{toluene} = 1 - x^l_{benzene} = 1 - 0.541 = 0.459$$

Using Raoult's law and the vapor pressures of pure benzene and toluene, the partial pressure of each component can be calculated as follows:

$$P_{benzene} = x^l_{benzene}P^*_{benzene} = (0.541)(94.6 \text{ torr}) = 51.2 \text{ torr}$$
$$P_{toluene} = x^l_{toluene}P^*_{toluene} = (0.459)(29.1 \text{ torr}) = 13.3 \text{ torr}$$

(a) The total vapor pressure is then $P_{total} = 51.2 \text{ torr} + 13.3 \text{ torr} = 64.5 \text{ torr}$.

(b) According to Dalton's law, the partial pressure of a gas is equal to its mole fraction (x^v) times the total pressure (P_{total}), so

$$x^v_{benzene} = \frac{P_{benzene}}{P_{total}} = \frac{51.2 \text{ torr}}{64.5 \text{ torr}} = 0.794$$

$$x^v_{toluene} = 1 - x^v_{benzene} = 1 - 0.794 = 0.206$$

—Continued

Check Benzene has a higher vapor pressure than toluene, so the mole fraction of benzene in the vapor phase should be larger than that in the liquid phase, which is consistent with our result.

Practice Exercise Hexane (C_6H_{14}) and heptane (C_7H_{16}) form a nearly ideal solution. The vapor pressures of pure hexane and pure heptane are 151 and 46.0 torr at 298 K, respectively. (a) Calculate the total vapor pressure above a solution formed by mixing 50.0 g of hexane and 100.0 g of heptane. (b) Calculate the mole fractions of hexane and heptane in the vapor phase.

Equation 9.10 gives the total vapor pressure as a function of the composition of the liquid phase, expressed as x_A^l. We can also express the total vapor pressure as a function of the vapor composition, x_A^v. According to Dalton's law of partial pressures (Equation 5.24), the mole fraction of A in the vapor phase is

$$x_A^v = \frac{P_A}{P_T}$$

Using Raoult's law and Equation 9.10, we have

$$x_A^v = \frac{x_A^l P_A^*}{P_B^* + x_A^l(P_A^* - P_B^*)} \qquad (9.11)$$

Multiplying Equation 9.11 through by the right-hand-side denominator and solving for x_A^l gives

$$x_A^l = \frac{x_A^v P_B^*}{P_A^* - x_A^v(P_A^* - P_B^*)} \qquad (9.12)$$

Applying Raoult's law ($P_A = x_A^l P_A^*$) and Dalton's law ($P_A = x_A^v P_T$) to Equation 9.12 gives the total pressure as a function of the vapor composition:

$$P_T = \frac{x_A^l P_A^*}{x_A^v} = \frac{P_A^* P_B^*}{P_A^* - x_A^v(P_A^* - P_B^*)} \qquad (9.13)$$

The plots of the total vapor pressure as functions of the mole fraction of A in both the liquid and vapor phases are shown in Figure 9.12(a) and (b), respectively. The combined plot shown in Figure 9.12(c) is a liquid-vapor phase diagram for an ideal binary solution at a fixed temperature *T*—often called a ***pressure-composition diagram***. At any pressure and composition above the upper curve (the liquid line) the mixture is a liquid. Below the lower curve (the vapor line), the mixture is entirely vapor. The region between the two curves is a region of phase coexistence, that is, both liquid and vapor phases are present in the system.

The line *a-e* in Figure 9.12(c) represents what happens if we start with a liquid solution above the upper curve and slowly lower the pressure at fixed temperature and overall composition from point *a* to point *e*. The solution remains entirely liquid until point *b* where we enter the coexistence region and some of the liquid begins to evaporate to form a vapor phase. From point *b* to point *d*, the system consists of both liquid

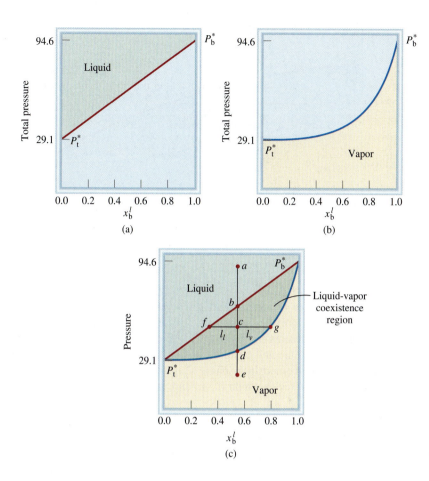

Figure 9.12 Vapor-liquid phase equilibrium in a benzene-toluene solution as a function of pressure at 23°C. (a) The total vapor pressure as a function of the mole fraction of benzene in the liquid. (b) The total vapor pressure as a function of the mole fraction of benzene in the vapor. (c) The pressure-composition phase diagram constructed by combining plots (a) and (b). The line *f-g* is the tie line corresponding to the system at point *c*.

and vapor in coexistence. At any given pressure in the coexistence region, the composition of the liquid and vapor is given by the points where a horizontal line crosses the upper and lower curves, respectively. This line is called a *tie line*. At point *c*, therefore, the system consists of a liquid with a composition given by point *f* and a vapor with a composition given by point *g*. The tie line for this system is the line *f-g* in Figure 9.12(c). The fraction of the system that is liquid or vapor can be found using the **lever rule:**

$$n_l l_l = n_v l_v \qquad\qquad (9.14)$$

where n_l and n_v are the number of moles in the liquid and vapor phases, respectively, and l_l and l_v are the distances between point *c* and the upper (liquid) curve (*f-c*)) and the lower (vapor) curve (*c-g*), respectively [see Figure 9.12(c)].

In constructing Figure 9.12 we have assumed that component A is the more volatile component (i.e., $P_A^* > P_B^*$). At a given total vapor pressure *P*, the mole fraction of A in the vapor (x_A^v) is larger than the mole fraction of A in the coexisting liquid (x_A^l). This is a general result for liquid-vapor coexistence in ideal and nearly ideal solutions—the vapor phase is richer in the more volatile component than in the liquid phase.

Another common way of representing a binary liquid-vapor equilibrium is through a **temperature-composition phase diagram,** in which the pressure is held fixed and phase coexistence is examined as a function of temperature and composition. Figure 9.13 shows the temperature-composition phase diagram for the benzene-toluene system at a pressure of 1 atm. In Figure 9.13, the lower curve (the *boiling-point curve*)

Figure 9.13 Temperature-composition phase diagram for the liquid-vapor equilibrium in benzene-toluene mixtures at 1 atm. The boiling points of toluene and benzene are 110.6°C and 80.1°C, respectively.

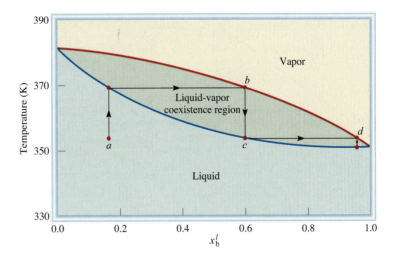

gives the liquid composition as a function of temperature. For temperatures and compositions below this curve, only the liquid phase is possible. Similarly, the upper curve (the *dew-point curve*) represents the vapor composition as a function of temperature. Any system with a temperature and composition above this curve will consist of vapor only. As for the pressure-composition phase diagram, the region between the two curves corresponds to systems in which liquid and vapor phases coexist.

Fractional Distillation

The temperature-composition phase diagram can be used to illustrate *fractional distillation, a procedure for separating liquid components of a solution based on their different boiling points.* Suppose we begin by heating a mixture of benzene and toluene with a temperature and composition corresponding to point *a* of Figure 9.13. When the temperature reaches about 370 K, the liquid will begin to boil. The vapor that results has a composition given by point *b*, where the tie line intersects the dew point curve. If the vapor is separated from the liquid and condensed by cooling to point *c*, the resulting liquid solution is richer in benzene than the original solution. If the condensate at point *c* is heated further, the vapor that results has the composition corresponding to point *d*. At this point the mixture is very nearly pure benzene, so this process separates benzene from the initial solution. For nearly ideal solutions such as benzene-toluene, this procedure can be repeated until the desired purity is reached.

In practice, chemists use an apparatus like that shown in Figure 9.14 to separate volatile liquids. The round-bottomed flask containing the benzene-toluene solution is fitted with a long column packed with small glass beads. When the solution boils, the vapor condenses on the beads in the lower portion of the column, and the liquid falls back into the distilling flask. As time goes on, the beads gradually heat up, allowing the vapor to move upward slowly and not condense back into the flask. In essence, the packing material causes the benzene–toluene mixture to be subjected continuously to numerous vaporization-condensation steps. At each step, the composition of the vapor in the column will be richer in the more volatile, or lower boiling-point, component (in this case, benzene). The vapor that rises to the top of the column is essentially pure benzene, which is then condensed and collected in a receiving flask.

Fractional distillation is as important in industry as it is in the laboratory. The petroleum industry employs fractional distillation on a large scale to separate the components of crude oil.

Typical industrial fractional distillation columns in an oil refinery.

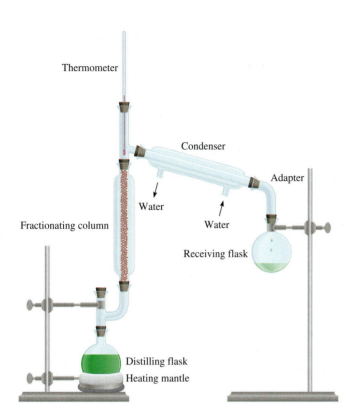

Thermometer

Condenser

Adapter

Water

Water

Fractionating column

Receiving flask

Distilling flask

Heating mantle

Figure 9.14 An apparatus for small-scale fractional distillation. The fractionating column is packed with tiny glass beads. The longer the fractionating column, the more complete the separation of the volatile liquids.

Vapor-Liquid Equilibrium in Nonideal Solutions

Although Raoult's law is an important starting point in understanding vapor-liquid equilibrium in solutions, most real solutions deviate from ideal behavior. The nature and magnitude of these deviations can be understood in terms of the intermolecular interactions between two volatile substances A and B. Consider the following two cases:

Case 1: If the intermolecular forces between A and B molecules are weaker than those between A molecules and those between B molecules, then there is a greater tendency for these molecules to leave the solution than in the case of an ideal solution. Consequently, the vapor pressure of the solution is greater than the sum of the vapor pressures as predicted by Raoult's law for the same concentration. This behavior gives rise to the *positive deviation* shown in Figure 9.15(a). In this case, the enthalpy of solution is positive (that is, mixing is an endothermic process). An ethanol-water solution is an example of a system with a positive deviation from Raoult's law because the hydrogen bonding interactions between ethanol molecules and water molecules are weaker than those between water molecules.

Case 2: If A molecules attract B molecules more strongly than they do their own kind, the vapor pressure of the solution will be less than the sum of the vapor pressures as predicted by Raoult's law. This gives rise to the *negative deviation* shown in Figure 9.15(b). In this case, the enthalpy of solution is negative (i.e., mixing is an exothermic process).

Binary solutions that deviate significantly from ideal behavior (as exemplified in their temperature-composition phase diagrams) have important consequences for fractional distillation processes. Figure 9.16 shows two types of such phase diagrams for

Figure 9.15 Pressure-composition curves for nonideal solutions. (a) Positive deviation occurs when P_T is greater than that predicted by Raoult's law (the solid black line). (b) Negative deviation occurs when P_T is less than that predicted by Raoult's law (the solid black line).

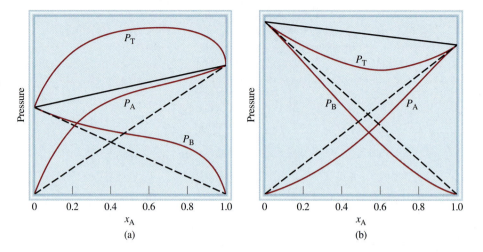

nonideal solutions. The phase diagram in Figure 9.16(a) is for a system that exhibits a strong *positive* deviation from Raoult's law. This system will show a *minimum* boiling point. Examples of these systems are ethanol-water, ethanol-benzene, and *n*-propanol-water. The phase diagram in Figure 9.16(b) is for a system that exhibits a strong *negative* deviation from Raoult's law, which will yield a *maximum* boiling point. These solutions, which include nitric acid-water and acetone-chloroform, are less common that those with minimum boiling points.

Both of these phase diagrams exhibit a point (at the minimum or the maximum) where the liquid and vapor curves coincide. At these points, the solutions form an *azeotrope*, which is a solution for which the liquid phase and the vapor phase have the same composition. A solution with a *low-boiling azeotrope* [Figure 9.16(a)] cannot be purified completely by distillation because the distillation process will always converge to the azeotrope, as illustrated by the fractional distillation pathways *a-d* and *e-h* in

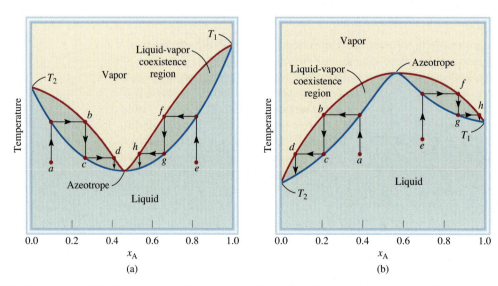

Figure 9.16 Different types of liquid-vapor phase diagrams for a binary liquid mixture of component A and B as functions of x_A, the mole fraction of the component with the higher boiling temperature. (a) The phase diagram for a system with a low-boiling azeotrope (minimum boiling point) and (b) the phase diagram for a system with a high-boiling azeotrope (maximum boiling point). The arrows show how the paths for various distillation processes depend upon the position of the initial composition relative to the azeotrope.

Figure 9.16(a). Solutions with a ***high-boiling azeotrope*** [Figure 9.16(b)] can be purified by distillation, but the component that is separated depends upon which side of the azeotrope the initial concentration lies. For example, following the fractional distillation pathway *a-d* in Figure 9.16(b) will eventually lead to a solution that is nearly pure component B and the pathway *e-h* will lead to nearly pure component A.

9.4 | Colligative Properties Are Properties of Solution Phase Equilibria That Depend Only upon the Number of Solute Molecules, Not Their Type

Four general properties of dilute liquid solutions are vapor-pressure lowering, boiling-point elevation, freezing-point depression, and osmotic pressure. These properties are commonly referred to as ***colligative*** (or *collective*) ***properties*** because they are bound together through their common origin—they depend *only on the number of solute molecules present in the solution, not on the nature (size, molecular mass, etc.) of the solute molecules.* For solutes that are ***nonelectrolytes*** (*i.e., they do not dissociate into ions in solution*), colligative properties typically hold for concentrations less than about 0.1 *M*.

All of the colligative properties can be understood in terms of the free energy of mixing for ideal solutions. A dilute liquid solution usually behaves as an ideal solution (or approximately so), that is, the solution obeys Raoult's law and has a free energy of mixing given by Equation 8.54:

$$\Delta G_{mix} = RT(n_A \ln x_A + n_B \ln x_B)$$

Equation 8.54 can be rewritten as

$$\Delta G_{mix} = nRT(x_A \ln x_A + x_B \ln x_B) \tag{9.15}$$

where n ($= n_A + n_B$) is the total number of moles in the solution. If we let the solvent be represented by A and the solute by B, then $\Delta G_{mix} = 0$ for the pure solvent ($x_A = 1$, $x_B = 0$). If we add a small amount of solute (B) to the pure solvent to form a dilute solution, then $\Delta G_{mix} < 0$ because both the solvent and solute mole fractions (x_A and x_B) will be less than one and greater than zero and their natural logarithms will both be negative. Thus, a solution formed by adding a small amount of solute to a solvent will have a lower free energy (and thus be more stable) than the original pure solvent. Because the enthalpy of mixing, ΔH_{mix}, is zero for an ideal solution, this decrease of the free energy is due to the increase in entropy accompanying the dissolution process.

In discussing colligative properties, we shall initially make two important approximations: (1) the solutions are dilute enough that they can be considered to be ideal and (2) the solutes are nonelectrolytes. Also, for simplicity, we will only consider two-component systems (that is, systems containing a solvent and a single solute).

Vapor-Pressure Lowering

If a solute is ***nonvolatile*** (that is, if it *does not have a measurable vapor pressure*), the vapor pressure of its solution is always less than that of the pure solvent. Thus, the relationship between solution vapor pressure and solvent vapor pressure depends on the concentration of the solute in the solution. This phenomenon is known as ***vapor-pressure lowering.*** Consider a solution that contains a solvent (A) and a nonvolatile solute (B), such as a solution of sucrose (table sugar) in water. Because the solution is ideal, Raoult's law (Equation 9.8) can be used to determine the vapor pressure of the solution:

$$P_A = x_A P_A^* \tag{9.8}$$

where P_A^* is the vapor pressure of the pure solvent. Because $x_A = 1 - x_B$, Equation 9.8 can be rewritten as

$$P_A = (1 - x_B)P_A^* \tag{9.16}$$

Rearranging Equation 9.16 gives

$$P_A^* - P_A = \Delta P = x_B P_A^* \tag{9.17}$$

where ΔP, the decrease in the vapor pressure from that of the pure solvent, is directly proportional to the mole fraction of the solute.

   Why does the vapor pressure of a solvent decrease when a solute is added? It cannot be due to the modification of intermolecular forces because vapor-pressure lowering occurs even in ideal solutions, in which the solute-solvent and solvent-solvent interactions are considered to be identical. The correct explanation is based on an analysis of the Gibbs free energy of mixing. When a pure solvent is in coexistence with its vapor phase, the molar Gibbs free energies of the solvent in the vapor and liquid phases must be equal for the system to be in equilibrium. As discussed previously, adding a small amount of solute to the solvent decreases the Gibbs free energy due to the increase in entropy accompanying the dissolution process. To maintain equilibrium, the Gibbs free energy of the vapor must also decrease. Because the solute is nonvolatile and does not enter the vapor phase, there is no Gibbs free energy of mixing in the vapor. At a fixed temperature, the only way to lower the free energy of a pure vapor is to reduce its pressure (which reduces the density of solvent molecules in the vapor). Thus, the vapor pressure of the solution must decrease to maintain equilibrium as solute is added.

   Example 9.7 shows how to calculate the vapor pressure of a solution of water and a nonvolatile solute.

Example 9.7

Calculate the vapor pressure of a solution made by dissolving 218 g of glucose (molar mass = 180.2 g mol^{-1}) in 460 mL of water at 30°C. What is the vapor-pressure lowering value? The vapor pressure of pure water at 30°C is 31.82 mmHg. Assume the density of the solution is 1.00 g mL^{-1}.

Strategy Use Equation 9.8 (Raoult's law) to determine the vapor pressure of the solution. Glucose is a nonvolatile solute.

Solution The vapor pressure of a solution (P_A) is given by Raoult's law:

$$P_A = x_A P_A^*$$

Because the vapor pressure of pure water (P_A^*) is given, we only need to find x_A, the mole fraction of water in the solution, to calculate P_A. To determine x_A, we must first

—Continued

$C_6H_{12}O_6$ (glucose)

Continued—
calculate the number of moles of glucose and the number of moles of water in the solution:

$$n_A(\text{water}) = 460 \text{ mL} \times \frac{1.00 \text{ g}}{1 \text{ mL}} \times \frac{1 \text{ mol}}{18.02 \text{ g}} = 25.5 \text{ mol}$$

$$n_B(\text{glucose}) = 218 \text{ g} \times \frac{1 \text{ mol}}{180.2 \text{ g}} = 1.21 \text{ mol}$$

The mole fraction of water (x_A) is given by

$$x_A = \frac{n_A}{n_A + n_B}$$

$$= \frac{25.5 \text{ mol}}{25.5 \text{ mol} + 1.21 \text{ mol}} = 0.955$$

The vapor pressure of pure water at 30°C (given in the problem statement) is 31.82 mmHg, so the vapor pressure of the glucose solution is

$$P_A = 0.955 \times 31.82 \text{ mmHg}$$
$$= 30.4 \text{ mmHg}$$

Finally, the vapor-pressure lowering is 31.82 mmHg − 30.4 mmHg = 1.4 mmHg.

Check We can also calculate the vapor-pressure lowering by using Equation 9.17 (i.e., $\Delta P = x_B P_A^*$). Because the mole fraction of glucose (x_B) is $1 - 0.955 = 0.045$, the vapor-pressure lowering is given by $(0.045)(31.82 \text{ mmHg}) = 1.4 \text{ mmHg}$.

Practice Exercise A solution is prepared by dissolving 82.4 g of urea [$(NH_2)_2CO$] in 212 g of water at 35°C. What is the vapor pressure of this solution? What is the vapor-pressure lowering? (Urea is a nonvolatile solute.)

Boiling-Point Elevation

The normal boiling point of a solution is the temperature at which its vapor pressure equals 1 atm (see Section 5.1). When a nonvolatile solute is added to a pure solvent at its normal boiling point, its vapor pressure is lowered below 1 atm. To bring the solution back to the boiling point, we must increase the temperature until the vapor pressure of the solution is 1 atm. As a result, the boiling temperature of the solution will be higher than that of the pure solvent. This phenomenon is known as **boiling-point elevation.**

Like vapor-pressure lowering, boiling-point elevation can be understood in terms of the free energy of mixing. At 1 atm, the boiling temperature is the temperature at which the molar Gibbs free energy of the liquid phase is equal to that of the vapor phase. When a small amount of a nonvolatile solute is added to a solvent, its Gibbs free energy is lowered due to the entropy of mixing. The Gibbs free energy of the vapor phase is unaffected, however, because no solute is present. As Figure 9.17 shows, the liquid and vapor curves intersect at a higher temperature when the Gibbs free energy of the liquid phase is lowered, consistent with an increase in the boiling temperature due to mixing.

Numerically, the *boiling-point elevation* (ΔT_b) is defined as *the boiling point of the solution* (T_b) *minus the boiling point of the pure solvent* (T_b^*):

$$\Delta T_b = T_b - T_b^* \qquad (9.18)$$

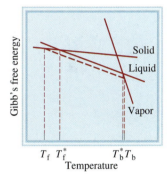

Figure 9.17 The solid red curves show the Gibbs free energies for the solid, liquid, and vapor phases of a pure solvent at 1 atm pressure, all as functions of temperature. The boiling point is given by the point of intersection of the liquid and vapor curves, and the freezing point is the point of intersection of the liquid and solid curves. The * is used to indicate values corresponding to the pure solvent. When a solute is added to the liquid phase, the Gibbs free energy of the resulting liquid solution (dotted curve) is lowered due to the entropy of mixing, causing the liquid and vapor curves to intersect at a higher temperature, that is, the boiling point is raised by the addition of the solute. In contrast, the lowering of the liquid curve causes the liquid and solid lines to intersect at a lower temperature, that is, the freezing point is lowered by the addition of the solute.

Because $T_b > T_b^*$, ΔT_b is a positive quantity. The value of ΔT_b is proportional to the vapor-pressure lowering, so it is also proportional to the mole fraction of the solute, x_B. It can be shown that, as long as the solution is dilute and the temperature change is small, the boiling-point elevation is given by

$$\Delta T_b = \frac{RT_b^2}{\Delta H_{vap}} x_B \tag{9.19}$$

where ΔH_{vap} is the enthalpy change due to vaporization. To convert the mole fraction x_B to a more practical concentration unit, such as molality (m_B), we write

$$x_B = \frac{n_B}{n_A + n_B} \approx \frac{n_B}{n_A} = \frac{n_B}{w_A/M_A}$$

where w_A is the mass of the solvent in kg and M_A is the molar mass of the solvent in kg mol^{-1}. Because n_B/w_A gives the molality of the solution, m_B, it follows that $x_B = M_A m_B$, so

$$\Delta T_b = \frac{RT_b^2 M_A}{\Delta H_{vap}} m_B \tag{9.20}$$

All the quantities in the first term on the right-hand side of Equation 9.20 are constants for a given solvent, so we have

$$\Delta T_b = K_b m_B \tag{9.21}$$

where K_B, the **molal boiling-point elevation constant** (or *ebullioscopic constant*), is given by

$$K_b = \frac{RT_b^2 M_A}{\Delta H_{vap}} \tag{9.22}$$

The units of K_b are K mol^{-1} kg. The concentration portion of K_b is based on molality, not molarity, because the temperature of the system (the solution) is *not* constant. Recall from Section 0.3 that molality is independent of temperature, whereas molarity changes with temperature.

Table 9.3 lists values of K_b for several common solvents. Use the boiling-point elevation constant for water and Equation 9.21 to prove to yourself that the boiling

Table 9.3	Molal Boiling-Point Elevation and Freezing-Point Depression Constants of Several Common Liquids			
Solvent	Normal Freezing Point (°C)*	K_f (K kg mol^{-1})	Normal Boiling Point (°C)*	K_b (K kg mol^{-1})
Water	0	1.86	100	0.52
Benzene	5.5	5.12	80.1	2.53
Ethanol	−117.3	1.99	78.4	1.22
Acetic acid	16.6	3.90	117.9	2.93
Cyclohexane	6.6	20.0	80.7	2.79

*Measured at 1 atm.

9.4 *Colligative Properties Are Properties of Solution Phase Equilibria That Depend Only upon the Number of Solute Molecules*

495

point of water will be 100.52°C if the molality of an aqueous nonelectrolyte solution is 1.00 *m*.

De-icing of airplanes is based on freezing-point depression.

Freezing-Point Depression

Ice on frozen roads and sidewalks melts when sprinkled with salts such as NaCl or $CaCl_2$. This method of thawing succeeds because it depresses the freezing point of water. As in boiling-point elevation, the Gibbs free energy of the solvent is decreased due to mixing by the addition of a small amount of solvent. The Gibbs free energy of the solid is unaffected as long as no solvent is incorporated into the crystal lattice. As a result, the temperature at which the solution Gibbs free energy curve crosses the solid free energy curve (i.e., the melting temperature) will be lower than that for the pure solvent (see Figure 9.17). The lowering of the freezing point of a solvent when a solute is added is called *freezing-point depression*.

Quantitatively, the ***freezing-point depression*** (ΔT_f) is defined as *the freezing point of the pure solvent* (T_b^*) *minus the freezing point of the solution* (T_b):

$$\Delta T_f = T_t^* - T_f \qquad (9.23)$$

Because $T_f^* > T_f$, ΔT_f is a positive quantity. Just as in boiling-point elevation, the freezing point depression for dilute solutions is proportional to the molality of the solute (m_B):

$$\Delta T_f = K_f m_B \qquad (9.24)$$

where K_f, the ***molal freezing-point depression constant*** (or *cryoscopic constant*), is given by

$$K_f = \frac{RT_f^2 M_A}{\Delta H_{fus}} \qquad (9.25)$$

In Equation 9.25, ΔH_{fus} is the molar enthalpy change of fusion for the solvent. Like K_b, K_f has the units K mol^{-1} kg. Values of K_f for some common solvents are listed in Table 9.3.

The phenomena of freezing-point depression and boiling-point elevation can also be understood by studying the phase diagram in Figure 9.18. At 1 atm, the normal

In cold climate regions, antifreeze must be used in car radiators in winter.

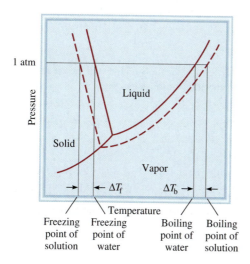

Figure 9.18 Phase diagram illustrating the boiling-point elevation and freezing-point depression of aqueous solutions. The dashed curves pertain to the solution and the solid curves to the pure solvent. The boiling point of the solution is *higher* than that of water, whereas the freezing point of the solution is *lower* than that of water.

freezing point of the solution lies at the intersection of the dashed curve (between the solid and liquid phases) and the horizontal line at 1 atm. Although the solute must be nonvolatile in the case of boiling-point elevation, no such restriction applies to freezing-point depression. Ethanol (C_2H_3OH), for example, a fairly volatile liquid that boils at only 78.5°C, is sometimes used as an antifreeze additive to lower the freezing point of water in automobile radiators.

A practical application of freezing-point depression and boiling-point elevation is described in Example 9.8.

Ethylene glycol
$CH_2(OH)CH_2(OH)$

Example 9.8

Ethylene glycol (EG) [$CH_2(OH)CH_2(OH)$] is a common automobile antifreeze. It is water soluble and fairly nonvolatile (b.p. 197°C). Calculate the freezing point of a solution containing 651 g of EG in 2505 g of water. Should you keep this substance in your car radiator during the summer? The molar mass of EG is 62.01 g.

Strategy We must use Equation 9.24 to calculate the depression in freezing point of the solution. First, though, we must calculate the molality of the solution, which we can determine from the information given in the problem statement. Refer to Table 9.3 for the value of K_f for the solvent (water).

Solution To solve for the molality of the solution, we need to know the number of moles of EG and the mass of the solvent in kilograms.

$$\text{moles of solute} = 651 \text{ g EG} \times \frac{1 \text{ mol EG}}{62.07 \text{ g EG}} = 10.5 \text{ mol EG}$$

The mass of the solvent is given in the problem statement as 2505 g. This can be converted to 2.505 kg, so the molality is calculated as follows:

$$m = \frac{\text{moles of solute}}{\text{mass of solvent (kg)}}$$

$$= \frac{10.5 \text{ mol EG}}{2.505 \text{ kg}} = 4.19 \text{ mol kg}^{-1}$$

$$= 4.19 \, m$$

Using Equation 9.24 and Table 9.3, we can write

$$\Delta T_f = K_f m$$
$$= (1.86°C \, m^{-1})(4.19 \, m)$$
$$= 7.79°C$$

Because pure water freezes at 0°C, the solution will freeze at −7.79°C. We can calculate boiling-point elevation in the same way:

$$\Delta T_b = K_b m$$
$$= (0.52°C \, m^{-1})(4.19 \, m)$$
$$= 2.2°C$$

Because the solution will boil at 100°C + 2.2°C = 102.2°C, it makes sense to leave the antifreeze in your car radiator in the summer to prevent the solution from boiling.

Practice Exercise Calculate the boiling point and freezing point of a solution containing 478 g of ethylene glycol in 3202 g of water.

Osmotic Pressure

Many chemical and biological processes depend on *osmosis, the selective passage of solvent molecules through a porous membrane from a dilute solution to a more concentrated one.* Figure 9.19 illustrates this phenomenon. The left compartment of the apparatus contains pure solvent, whereas the right compartment contains a solution of the solvent and some solute. The two compartments are separated by a *semipermeable membrane,* which *allows the passage of solvent molecules but blocks the passage of solute molecules.* At the start, the water levels in the two tubes are equal [see Figure 9.19(a)]. After some time, the level in the right tube begins to rise and continues to increase until equilibrium is reached (i.e., until no further change can be observed). The *osmotic pressure* (Π) of a solution is *the pressure required to stop osmosis.* As shown in Figure 9.19(b), this pressure can be measured directly from the difference in the final fluid levels.

Like freezing-point depression and boiling-point elevation, osmotic pressure can be understood in terms of the free energy of mixing.

If pure solvent is placed on both sides of the semipermeable membrane depicted in Figure 9.19(a), the pressures and molar Gibbs free energies of the liquid in both sides of the container will be equal at equilibrium because the solvent is free to move across the membrane. If we add solute to the solvent on the right-hand side of the membrane, the Gibbs free energy of this solution will be lowered relative to the pure solvent on the left, and the system will no longer be in equilibrium. To restore equilibrium, solvent will move across the membrane from the pure solvent into the solution (i.e., from the left-hand side to the right-hand side). This transfer of matter lowers the Gibbs free energy of the pure solvent, while raising the Gibbs free energy of the solvent in the solution. The transfer stops when the molar Gibbs free energy of the solvent is the same on both sides of the membrane. Because the volume of the solution (right-hand side) is fixed, addition of matter (solvent) at constant T causes a rise in pressure. Similarly, removal of matter from the pure solvent (left-hand side) causes a decrease in pressure, resulting in a pressure imbalance across the membrane at equilibrium.

The osmotic pressure of a solution is given by

$$\Pi = cRT \tag{9.26}$$

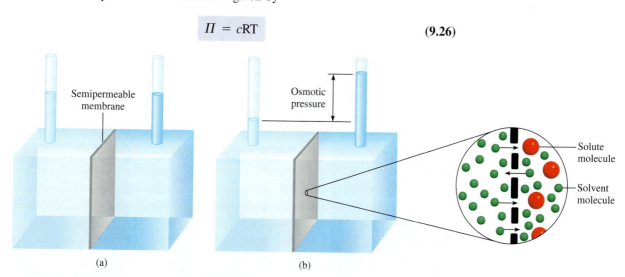

(a)   (b)

Figure 9.19 Osmotic pressure. (a) The levels of the pure solvent (left) and of the solution (right) are equal at the start. (b) During osmosis, the level of the solution on the solution side rises as a result of the net flow of solvent from left to right. The osmotic pressure is equal to the hydrostatic pressure exerted by the column of fluid in the right tube at equilibrium. The same effect occurs when the pure solvent is replaced by a more dilute solution than that on the right.

where c is the concentration of the solution, R is the gas constant (0.08314 L bar $mol^{-1} K^{-1}$), and T is the absolute temperature. The osmotic pressure (Π), is expressed in bar. Because osmotic pressure measurements are carried out at constant temperature, the concentration (c) is expressed in units of molarity rather than molality. Because the molarity of the solution is given by the number of solute molecules (n_{solute}) divided by the volume (V) of solution ($c = n_{solute}/V$), Equation 9.26 can also be written as

$$\Pi V = n_{solute} R T \tag{9.27}$$

which is often easier to remember than Equation 9.26 because this form is similar to the ideal gas law.

If two solutions have the same concentration and, hence, the same osmotic pressure, they are said to be *isotonic*. If two solutions have different osmotic pressures, the more concentrated solution is said to be *hypertonic* and the more dilute solution is said to be *hypotonic* (Figure 9.20).

Although osmosis is a common and well-studied phenomenon, relatively little is known about how the semipermeable membrane stops some molecules yet allows others to pass. In some cases, it is simply a matter of size. A semipermeable membrane may have pores small enough to let only the solvent molecules through. In other cases, a different mechanism may be responsible for the membrane's selectivity, such as the solvent's greater "solubility" in the membrane.

The osmotic pressure phenomenon manifests itself in many interesting applications. To study the contents of red blood cells, which are protected from the external environment by a semipermeable membrane, biochemists use a technique called *hemolysis*. The red blood cells are placed in a hypotonic solution. Because the hypotonic solution is less concentrated than the interior of the cell, water moves into the cells,

Figure 9.20 A cell in (a) an isotonic solution, (b) a hypotonic solution, and (c) a hypertonic solution. The cell remains unchanged in (a), swells in (b) and shrinks in (c). (d) From left to right: a red blood cell in an isotonic solution, a hypotonic solution, and a hypertonic solution.

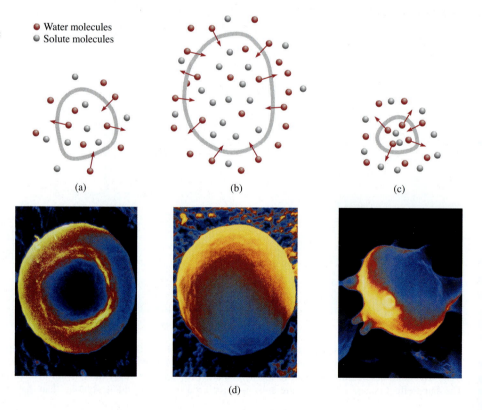

as shown in the middle photo of Figure 9.20(d). The cells swell and eventually burst, releasing hemoglobin and other molecules, which are collected and studied. Home preserving of jam and jelly provides another example of the use of osmotic pressure. A large quantity of sugar is actually essential to the preservation process because the sugar helps to kill bacteria that may cause botulism. When a bacterial cell is in a hypertonic (high-concentration) sugar solution, as shown in Figure 9.20(c), the intracellular water tends to move out of the bacterial cell by osmosis to the more concentrated solution. This process, known as *crenation,* causes the cell to shrink and, eventually, to cease functioning. The natural acidity of fruits also inhibits bacteria growth.

Osmotic pressure is the major mechanism for transporting water upward in plants. Because leaves constantly lose water to the air, in a process called *transpiration,* the solute concentrations in leaf fluids increase. Water is pulled up through the trunk, branches, and stems of trees by osmotic pressure. Up to 10 to 15 bar pressure is necessary to transport water to the leaves at the tops of California's redwoods, which reach about 120 m in height.

California redwoods

Example 9.9 shows how an osmotic pressure measurement can be used to determine the concentration of a solution.

Example 9.9

The average osmotic pressure of seawater, measured in the kind of apparatus shown in Figure 9.19, is about 30.0 bar at 25°C. Calculate the molar concentration of an aqueous solution of sucrose ($C_{12}H_{22}O_{11}$) that is isotonic with seawater.

Strategy The sucrose solution is isotonic with seawater, so it has the same concentration and the same osmotic pressure as the seawater.

Solution The aqueous sucrose solution has an osmotic pressure of 30.0 bar because it is isotonic with seawater that has an osmotic pressure of 30.0 bar. Using Equation 9.26,

$$\Pi = cRT$$

$$c = \frac{\Pi}{RT} = \frac{30.0 \text{ bar}}{(0.08314 \text{ L bar K}^{-1})(298 \text{ K})}$$

$$= 1.23 \text{ mol L}^{-1}$$

$$= 1.23 \ M$$

Practice Exercise What is the osmotic pressure (in bar) of a 0.884 *M* urea solution at 16°C?

Using Colligative Properties to Determine Molar Mass

The colligative properties of nonelectrolyte solutions provide a means of determining the molar mass of a solute. Theoretically, any of the four colligative properties is suitable for this purpose. In practice, however, only freezing-point depression and osmotic pressure are used because they exhibit the most pronounced colligative effects, that is, they exhibit the largest temperature changes for a given molarity and generate measurable effects for even very small concentrations. The procedure is as follows. From the experimentally determined freezing-point depression or osmotic pressure, we can calculate the molality or molarity of the solution. Knowing the mass of the solute, we can readily determine its molar mass, as shown in Examples 9.10 and 9.11.

Example 9.10

A 7.85-g sample of a compound with the empirical formula C_5H_4 is dissolved in 301 g of benzene. The freezing point of the solution is 1.05°C below that of pure benzene. What are the molar mass and molecular formula of this compound?

Strategy Solving this problem requires three steps. First, we calculate the molality of the solution from the freezing-point depression. Next, from the molality we determine the number of moles in 7.85 g of the compound and hence its molar mass. Finally, comparing the experimental molar mass with the empirical molar mass enables us to write the molecular formula.

Solution The sequence of conversions for calculating the molar mass of the compound is as follows:

$$\text{freezing-point} \longrightarrow \text{molality} \longrightarrow \text{number of} \longrightarrow \text{molar mass}$$
$$\text{depression} \qquad\qquad\qquad\qquad \text{moles}$$

To calculate the molality of the solution, use Equation 9.24 and the appropriate value of K_f from Table 9.3:

$$\text{molality} = \frac{\Delta T_f}{K_f} = \frac{1.05°C}{5.12°C\ m^{-1}} = 0.205\ m$$

Because there is 0.205 mole of the solute in 1 kg of solvent, the number of moles of solute in 301 g, or 0.301 kg, of solvent is

$$0.301\ \text{kg} \times \frac{0.205\ \text{mol}}{1\ \text{kg}} = 0.0617\ \text{mol}$$

Thus, the molar mass of the solute is

$$\text{molar mass} = \frac{\text{grams of compound}}{\text{moles of compound}} = \frac{7.85\ \text{g}}{0.0617\ \text{mol}} = 127\ \text{g mol}^{-1}$$

Now we can determine the ratio of the molar mass to the empirical molar mass:

$$\frac{\text{molar mass}}{\text{empirical molar mass}} = \frac{127\ \text{g mol}^{-1}}{64\ \text{g mol}^{-1}} \approx 2$$

Thus, the molecular formula is $(C_5H_4)_2$ or $C_{10}H_8$. This does not give us, however, enough information to identify the compound as there are at least two compounds that have this molecular formula, namely, naphthalene and azulene:

naphthalene   azulene

Practice Exercise A solution of 0.85 g of an organic compound in 100.0 g of benzene has a freezing point of 5.16°C. What are the molality of the solution and the molar mass of the solute?

Example 9.11

A solution is prepared by dissolving 35.0 g of hemoglobin (Hb) in enough water to make up 1 L in volume. If the osmotic pressure of the solution is found to be 10.0 mmHg at 25°C, calculate the molar mass of hemoglobin.

Strategy The steps needed to calculate the molar mass of Hb are similar to those outlined in Example 9.10, except we use osmotic pressure instead of freezing-point depression. First, we must calculate the molarity of the solution from the osmotic pressure of the solution. Then, from the molarity, we can determine the number of moles in 35.0 g of Hb and hence its molar mass. Because the pressure is given in mmHg, it is more convenient to use R in terms of L atm instead of L bar because the conversion factor from mm Hg to atm is simpler.

Solution The sequence of conversions is as follows:

$$\text{osmotic pressure} \longrightarrow \text{molarity} \longrightarrow \text{number of moles} \longrightarrow \text{molar mass}$$

First, calculate the molarity using Equation 9.26:

$$\Pi = cRT$$
$$c = \frac{\Pi}{RT}$$

$$= \frac{10.0 \text{ mmHg} \times \dfrac{1 \text{ atm}}{760 \text{ mmHg}}}{(0.0821 \text{ L atm mol}^{-1} \text{ K}^{-1})(298 \text{ K})} = 5.38 \times 10^{-4} M$$

The volume of the solution is 1 L, so it must contain 5.38×10^{-4} mol of Hb. We use this quantity to calculate the molar mass:

$$\text{molar mass of Hb} = \frac{\text{grams of Hb}}{\text{moles of Hb}} = \frac{35.0 \text{ g}}{5.38 \times 10^{-4} \text{ mol}} = 6.51 \times 10^{4} \text{ g mol}^{-1}$$

Practice Exercise A 202-mL benzene solution containing 2.47 g of an organic polymer has an osmotic pressure of 8.63 mmHg at 21°C. Calculate the molar mass of the polymer.

A pressure of 10.0 mmHg, like the one in Example 9.11, can be measured easily and accurately. As a result, osmotic pressure measurements are very useful for determining the molar masses of large molecules, such as proteins. To see how much more practical the osmotic pressure technique is than freezing-point depression would be, let us estimate the change in freezing point of the hemoglobin solution from Example 9.11. If an aqueous solution is quite dilute, we can assume that molarity is roughly equal to molality. (Molarity would be equal to molality if the density of the aqueous solution were 1 g mL^{-1}.) Hence, from Equation 9.24 we can write

$$\Delta T_f = (1.86°\text{C } m^{-1})(5.38 \times 10^{-4} m) = 1.00 \times 10^{-3}°\text{C}$$

The freezing-point depression of one-thousandth of a degree is too small of a temperature change to measure accurately. For this reason, the freezing-point depression technique is more suitable for determining the molar mass of smaller and more soluble molecules, those having molar masses of 500 g or less, because the freezing-point depressions of their solutions are much greater.

Colligative Properties of Electrolyte Solutions

The study of the colligative properties of *electrolytes* (*substances that dissociate into ions in solution*) requires a slightly different approach than the one used for the colligative properties of nonelectrolytes. One unit of an electrolyte compound yields two or more particles when it dissolves. Because it is the number of solute particles that determines the colligative properties of a solution, it is important to know which compounds dissociate and to what extent. For example, each unit of NaCl dissociates into two ions, Na^+ and Cl^-, so the colligative properties of a 0.1 m NaCl solution should be twice as great as those of a 0.1 m solution containing a nonelectrolyte, such as sucrose. Similarly, a 0.1 m $CaCl_2$ solution depresses the freezing point by three times as much as a 0.1 m sucrose solution because each $CaCl_2$ produces three ions. To account for this effect, there is a quantity called the *van't Hoff*[5] *factor* (i) given by the following:

$$i = \frac{\text{actual number of particles in solution after dissociation}}{\text{number of formula units initially dissolved in solution}} \tag{9.28}$$

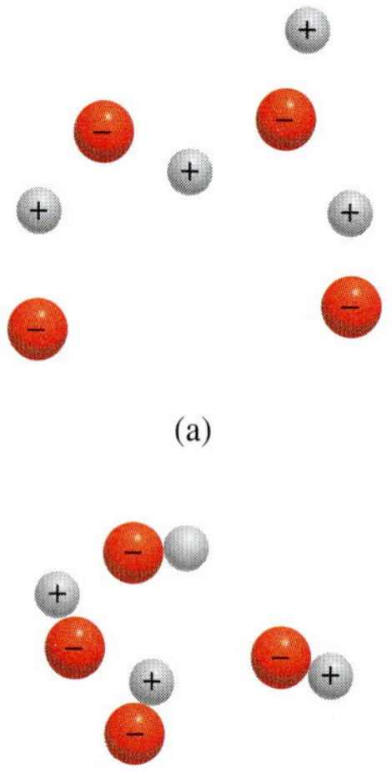

Thus, i should be 1 for all nonelectrolytes. For strong electrolytes such as NaCl and KNO_3, i should be 2, and for strong electrolytes such as Na_2SO_4 and $CaCl_2$, i should be 3. Consequently, the equations for colligative properties must be modified as follows:

$$\Delta T_b = iK_b m_B \tag{9.29}$$

$$\Delta T_f = iK_f m_B \tag{9.30}$$

$$\Pi = i c R T \tag{9.31}$$

(a)

(b)

Figure 9.21 (a) Free ions and (b) ion pairs in solution. An ion pair bears no net charge, so it cannot conduct electricity in solution.

In reality, the colligative properties of electrolyte solutions are usually less than anticipated because at higher concentrations ($> \sim 0.01$ M), electrostatic forces come into play and bring about the formation of ion pairs. An *ion pair* is made up of *one or more cations and one or more anions held together by electrostatic forces*. The presence of an ion pair reduces the effective number of particles in solution, causing a reduction in the colligative properties (Figure 9.21). Electrolytes containing multicharged ions such as Mg^{2+}, Al^{3+}, and Fe^{2+} have a greater tendency to form ion pairs than electrolytes such as NaCl and KNO_3, which are made up of singly charged ions. Noninteger values of the van't Hoff factor are necessary to compensate for the deviation from ideal-solution behavior exhibited by electrolyte solutions because of ion-pair formation.

Table 9.4 lists the experimentally measured values of i and those calculated assuming complete dissociation. The agreement is close but not perfect, indicating that the extent of ion-pair formation in these solutions at 0.0500 M concentration is appreciable.

---

5. Jacobus Hendricus van't Hoff (1852–1911). Dutch chemist. One of the most prominent chemists of his time, van't Hoff did significant work in thermodynamics, molecular structure and optical activity, and solution chemistry. In 1901, he received the first Nobel Prize in Chemistry.

Table 9.4	The van't Hoff Factor of 0.0500 *M* Electrolyte Solutions at 25°C	
Electrolyte	***i* (Measured)**	***i* (Calculated)**
Sucrose*	1.0	1.0
HCl	1.9	2.0
NaCl	1.9	2.0
$MgSO_4$	1.3	2.0
$MgCl_2$	2.7	3.0
$FeCl_3$	3.4	4.0

*Sucrose is a nonelectrolyte. It is listed here for comparison only.

Example 9.12 shows how the van't Hoff factor can be determined from colligative properties measurements.

Example 9.12

The osmotic pressure of a 0.010 *M* potassium iodide (KI) solution at 25°C is 0.471 bar. Calculate the van't Hoff factor for KI at this concentration.

Strategy KI is a strong electrolyte, so it should dissociate completely in solution. If so, its osmotic pressure would be

$$2(0.010\ M)(0.08314\ \text{L bar mol}^{-1}\ \text{K}^{-1})(298\ \text{K}) = 0.496\ \text{bar}$$

The measured osmotic pressure, however, is only 0.471 bar. This value is less than predicted, indicating that ion pairs are forming, thus reducing the number of solute particles (K^+ and I^- ions) in solution.

Solution From Equation 9.31, we have

$$i = \frac{\Pi}{MRT} = \frac{0.471\ \text{bar}}{(0.010\ M)(0.08314\ \text{L bar mol}^{-1}\ \text{K}^{-1})(298\ \text{K})} = 1.90$$

Practice Exercise The freezing-point depression of a 0.100 *m* $MgSO_4$ solution is 0.225°C. Calculate the van't Hoff factor for $MgSO_4$ at this concentration.

Summary of Facts and Concepts

Section 9.1

▶ The shapes of the phase boundaries in a phase diagram can be determined from thermodynamics using the Clapeyron equation. If one of the phases is a vapor, then the Clausius-Clapeyron approximation can be used.

Section 9.2

▶ The ease of dissolving a solute in a solvent is governed by intermolecular forces. The enthalpy and entropy of

mixing are the forces driving the solution process. These effects can be roughly summarized by the concept of *like dissolves like*.

▶ Increasing temperature usually increases the solubility of solid and liquid substances and usually decreases the solubility of gases in water.

▶ According to Henry's law, the solubility of a gas in a liquid is directly proportional to the partial pressure of the gas over the solution.

Section 9.3

▶ Raoult's law states that the partial pressure of a substance A over a solution is equal to the mole fraction (x_A) of A times the vapor pressure (P_A^*) of pure A. An ideal solution obeys Raoult's law over the entire range of concentration. In practice, very few solutions exhibit ideal behavior.

▶ Raoult's law can be used to determine the vapor pressure and composition of the vapor above a solution of two volatile liquids.

▶ Deviations from ideal solution behavior can be understood in terms of the intermolecular forces.

Section 9.4

▶ Vapor-pressure lowering, boiling-point elevation, freezing-point depression, and osmotic pressure are colligative properties of solutions, that is, they depend only on the number of solute particles that are present and not on their nature.

▶ In electrolyte solutions, the interaction between ions leads to the formation of ion pairs. The van't Hoff factor provides a measure of the extent of dissociation of electrolytes in solution.

Key Words

azeotrope, p. 490
boiling-point elevation, p. 493
Clapeyron equation, p. 468
Clausius-Clapeyron
   equation, p. 471
colligative properties, p. 491
crystallization, p. 474
electrolytes, p. 502
fractional crystallization, p. 479
fractional distillation, p. 488
freezing-point depression, p. 495

Henry's law, p. 480
high-boiling azeotrope, p. 491
hydration, p. 476
hydrophilic, p. 476
hydrophobic, p. 476
ion pair, p. 502
lever rule, p. 487
low-boiling azeotrope, p. 490
miscible, p. 476
molal boiling-point elevation
   constant, p. 494

molal freezing-point
   depression constant, p. 495
nonelectrolytes, p. 491
nonvolatile, p. 491
osmotic pressure, p. 497
osmosis, p. 497
pressure-composition
   diagram, p. 486
Raoult's law, p. 484
saturated solution, p. 474

semipermeable
   membrane, p. 497
solubility, p. 474
solvation, p. 476
supersaturated solution, p. 474
temperature-composition
   phase diagram, p. 487
unsaturated solution, p. 474
van't Hoff factor, p. 502
vapor-pressure lowering, p. 491

Problems

The Phase Boundaries in Pure Substances Can Be Predicted Using Thermodynamics

9.1 A length of wire is placed on top of a block of ice. The ends of the wire extend over the edges of the ice, and a heavy weight is attached to each end. It is found that the ice under the wire gradually melts, and that the wire then slowly moves through the ice block. At the same time, the water above the wire refreezes. Give an explanation for this observation.

9.2 Vapor pressure measurements at several different temperatures are shown here for mercury. (a) Determine graphically the molar heat of vaporization for mercury.

T (°C)	200	250	300	320	340
P (mmHg)	17.3	74.4	246.8	376.3	557.9

(b) Use the results from this experiment to predict the normal boiling temperature of mercury. Does this temperature agree with the experimental value given in Table 7.7? If not, give some possible origins for this discrepancy.

9.3 The vapor pressure of benzene (C_6H_6) is 40.1 mmHg at 7.6°C. What is its vapor pressure at 60.6°C? The molar heat of vaporization of benzene is 31.0 kJ mol^{-1}.

9.4 The vapor pressure of liquid X is lower than that of liquid Y at 20°C, but higher at 60°C. What can you deduce about the relative magnitude of the molar heats of vaporization of X and Y?

9.5 Estimate the molar heat of vaporization of a liquid whose vapor pressure doubles when the temperature is raised from 85°C to 95°C.

9.6 In Example 9.1, we used the Clapeyron equation to predict that the melting point of water decreases by 1°C for every 138 bar increase in pressure. Using this, one could predict that at a pressure of about 3.8×10^4 bar ($= 273°C \times 138$ bar $°C^{-1}$) the melting point of water would be $-273.15°C$ or 0 K (absolute zero). Is this prediction reasonable? If not, explain where it went wrong.

The Solubility of a Substance Is Determined by Temperature, Pressure, and Intermolecular Forces

9.7 Why is naphthalene ($C_{10}H_8$) more soluble in benzene than CsF?

9.8 Why is ethanol (C_2H_5OH) insoluble in cyclohexane (C_6H_{12})?

9.9 Arrange the following compounds in order of increasing solubility in water: O_2, LiCl, Br_2, and methanol (CH_3OH).

9.10 Explain the variations in solubility in water of the alcohols listed here:

Compound	Solubility in Water (g/100 g) at 20°C
CH_3OH	∞
CH_3CH_2OH	∞
$CH_3CH_2CH_2OH$	∞
$CH_3CH_2CH_2CH_2OH$	9
$CH_3CH_2CH_2CH_2CH_2OH$	2.7

9.11 A 3.20-g sample of a salt dissolves in 9.10 g of water to give a saturated solution at 25°C. What is the solubility (in g salt/100 g of H_2O) of the salt?

9.12 The solubility of KNO_3 is 155 g/100 g of water at 75°C but only 38.0 g/100 g of water at 25°C. What mass (in grams) of KNO_3 will crystallize out of solution if exactly 100 g of its saturated solution at 75°C is cooled to 25°C?

9.13 A 50-g sample of impure $KClO_3$ (solubility = 7.1 g per 100 g H_2O at 20°C) is contaminated with 10 percent KCl (solubility = 25.5 g per 100 g of H_2O at 20°C). Calculate the minimum quantity of water at 20°C that is needed to dissolve all the KCl from the sample. How much $KClO_3$ will be left after this treatment? (Assume that the solubilities are unaffected by the presence of the other compound.)

9.14 A beaker of water is initially saturated with dissolved air. Explain what happens when He gas at 1 bar is bubbled through the solution for a long time.

9.15 A miner working 260 m below sea level opened a carbonated soft drink during a lunch break. To his surprise, the soft drink tasted rather "flat." Shortly afterward, the miner took an elevator to the surface. During the trip up, he could not stop belching. Why?

9.16 The solubility of CO_2 in water at 25°C and 1 bar is 0.034 mol L^{-1}. What is its solubility under atmospheric conditions? (The partial pressure of CO_2 in air is 0.0003 atm.) Assume that CO_2 obeys Henry's law.

9.17 The solubility of N_2 in blood at 37°C and at a partial pressure of 0.80 bar is 5.6×10^{-4} mol/L. A deep-sea diver breathes compressed air with the partial pressure of N_2 equal to 4.0 bar. Assume that the total

volume of blood in the diver's body is 5.0 L. Calculate the amount of N_2 gas released (in liters at 37°C and 1 bar) when the diver returns to the surface of the water where the partial pressure of N_2 is 0.80 bar.

The Liquid-Vapor Phase Equilibrium of a Solution Can Be Understood in Terms of the Entropy of Mixing and Intermolecular Forces

9.18 The vapor pressures of ethanol (C_2H_5OH) and 1-propanol (C_3H_7OH) at 35°C are 100 and 37.6 mmHg, respectively. Assume ideal behavior and calculate the partial pressures of ethanol and 1-propanol at 35°C over a solution of ethanol in 1-propanol, in which the mole fraction of ethanol is 0.300.

9.19 At 20°C, the vapor pressure of ethanol (C_2H_5OH) is 44 mmHg and the vapor pressure of methanol (CH_3OH) is 94 mmHg. A mixture of 30.0 g of methanol and 45.0 g of ethanol is prepared (and can be assumed to behave as an ideal solution). (a) Calculate the vapor pressure of methanol and ethanol above this solution at 20°C. (b) Calculate the mole fraction of methanol and ethanol in the vapor above this solution at 20°C. (c) Suggest a method for separating the two components of the solution.

9.20 A mixture of liquids A and B exhibits ideal behavior. At 84°C, the total vapor pressure of a solution containing 1.2 moles of A and 2.3 moles of B is 331 mm Hg. Upon the addition of another mole of B to the solution, the vapor pressure increases to 347 mmHg. Calculate the vapor pressures of pure A and pure B at 84°C.

9.21 Explain in terms of molecular interactions why the temperature-composition phase diagram for chloroform and acetone possesses a high-boiling azeotrope.

9.22 If you try to purify ethanol from an ethanol-water mixture by fractional distillation, the maximum purity obtainable is 95 percent. Explain.

Colligative Properties Are Properties of Solution Phase Equilibria That Depend Only upon the Number of Solute Molecules, Not Their Type

9.23 A solution is prepared by dissolving 396 g of sucrose ($C_{12}H_{22}O_{11}$) in 624 g of water. What is the vapor pressure of this solution at 30°C? (The vapor pressure of water is 31.8 mmHg at 30°C.)

9.24 How many grams of sucrose ($C_{12}H_{22}O_{11}$) must be added to 552 g of water to give a solution with a vapor pressure 2.0 mmHg less than that of pure water at 20°C? (The vapor pressure of water at 20°C is 17.5 mmHg.)

9.25 The vapor pressure of benzene is 100.0 mmHg at 26.1°C. Calculate the vapor pressure of a solution containing 24.6 g of camphor ($C_{10}H_{16}O$) dissolved in 98.5 g of benzene. (Camphor is a low-volatility solid.)

9.26 How many grams of urea [$(NH_2)_2CO$] must be added to 450 g of water to give a solution with a vapor pressure 2.50 mmHg less than that of pure water at 30°C? (The vapor pressure of water at 30°C is 31.8 mmHg.)

9.27 What are the boiling point and freezing point of a 2.47 m solution of naphthalene in benzene? (The boiling point and freezing point of benzene are 80.1°C and 5.5°C, respectively.)

9.28 An aqueous solution contains the amino acid glycine (NH_2CH_2COOH). Assuming that the acid does not ionize in water, calculate the molality of the solution if it freezes at −1.1°C.

9.29 Pheromones are compounds secreted by the females of many insect species to attract males. One of these compounds contains 80.78 percent C, 13.56 percent H, and 5.66 percent O. A solution of 1.00 g of this pheromone in 8.50 g of benzene freezes at 3.37°C. What are the molecular formula and molar mass of the compound? (The normal freezing point of pure benzene is 5.50°C.)

9.30 The elemental analysis of an organic solid extracted from gum arabic (a gummy substance used in adhesives, inks, and pharmaceuticals) showed that it contained 40.0 percent C, 6.7 percent H, and 53.3 percent O. A solution of 0.650 g of the solid in 27.8 g of the solvent diphenyl gave a freezing-point depression of 1.56°C. Calculate the molar mass and molecular formula of the solid. (K_f for diphenyl is 8.00°C m^{-1}.)

9.31 How many liters of the antifreeze ethylene glycol [$CH_2(OH)CH_2(OH)$] would you add to a car radiator containing 6.50 L of water if the coldest winter temperature in your area is −20°C? Calculate the boiling point of this water/ethylene glycol mixture. (The density of ethylene glycol is 1.11 g mL^{-1}.)

9.32 A solution is prepared by condensing 4.00 L of a gas, measured at 27°C and 748 mmHg pressure, into 58.0 g of benzene. Calculate the freezing point of this solution.

9.33 The molar mass of benzoic acid (C_6H_5COOH) determined by measuring the freezing-point depression in benzene is twice what we would expect for the molecular formula $C_7H_6O_2$. Explain this apparent anomaly.

9.34 A solution of 2.50 g of a compound having the empirical formula C_6H_5P in 25.0 g of benzene is observed to freeze at 4.3°C. Calculate the molar mass of the solute and determine its molecular formula.

9.35 What is the osmotic pressure (in bar) of a 1.36 M aqueous solution of urea [$(NH_2)_2CO$] at 22.0°C?

9.36 A solution containing 0.8330 g of a polymer of unknown structure in 170.0 mL of an organic solvent was found to have an osmotic pressure of 5.20 mmHg at 25°C. Determine the molar mass of the polymer.

9.37 A quantity of 7.480 g of an organic compound is dissolved in water to make 300.0 mL of solution. The solution has an osmotic pressure of 1.45 bar at 27°C. The analysis of this compound shows that it contains 41.8 percent C, 4.7 percent H, 37.3 percent O, and 16.3 percent N. Determine the molecular formula of the compound.

9.38 A solution of 6.85 g of a carbohydrate in 100.0 g of water has a density of 1.024 g mL^{-1} and an osmotic pressure of 4.670 bar at 20.0°C. Calculate the molar mass of the carbohydrate.

9.39 Which aqueous solution, 0.35 m $CaCl_2$ or 0.90 m urea, has (a) the higher boiling point, (b) the higher freezing point, and (c) the lower vapor pressure? Explain. Assume complete dissociation.

9.40 Aqueous solutions of sucrose ($C_{12}H_{22}O_{11}$) and of nitric acid (HNO_3) both freeze at 1.5°C. What other properties do these solutions have in common?

9.41 Arrange the following solutions in order of decreasing freezing point: 0.10 m Na_3PO_4, 0.35 m NaCl, 0.20 m $MgCl_2$, 0.15 m $C_6H_{12}O_6$, and 0.15 m CH_3COOH.

9.42 Arrange the following aqueous solutions in order of decreasing freezing point: 0.50 m HCl, 0.50 m glucose, and 0.50 m acetic acid. Explain your reasoning.

9.43 What are the normal freezing points and boiling points of (a) 21.2 g NaCl in 135 mL of water and (b) 15.4 g of urea in 66.7 mL of water?

9.44 At 25°C, the vapor pressure of pure water is 23.76 mmHg and that of seawater is 22.98 mmHg. Assuming that seawater contains only NaCl, estimate its molal concentration.

9.45 Both NaCl and $CaCl_2$ are used to melt ice on roads and sidewalks in winter. What advantages do these substances have over sucrose or urea in lowering the freezing point of water?

9.46 A 0.86 percent by mass solution of NaCl is called "physiological saline" because its osmotic pressure is equal to that of the solution in blood cells. Calculate the osmotic pressure of this solution at normal body temperature (37°C). The density of the saline solution is 1.005 g mL^{-1}.

9.47 The osmotic pressure of 0.010 M solutions of $CaCl_2$ and urea at 25°C are 0.613 and 0.247 bar, respectively. Calculate the van't Hoff factor for the $CaCl_2$ solution.

9.48 Calculate the osmotic pressure of a 0.0500 M $MgSO_4$ solution at 25°C. (*Hint:* See Table 9.4.)

Additional Problems

9.49 A beaker of water is placed in a closed container. Predict the effect on the vapor pressure of water when (a) the temperature is lowered, (b) the volume of the container is doubled, and (c) more water is added to the beaker.

9.50 The south pole of Mars is covered with dry ice, which partially sublimes during the summer. The CO_2 vapor recondenses in the winter when the temperature drops to 150 K. Given that the heat of sublimation of CO_2 is 25.9 kJ mol^{-1}, calculate the atmospheric pressure on the surface of Mars. The normal sublimation temperature of CO_2 is $-78°C$.

9.51 A pressure cooker is a sealed container that allows steam to escape when it exceeds a predetermined pressure. How does this device reduce the time needed for cooking?

9.52 Lysozyme is an enzyme that cleaves bacterial cell walls. A sample of lysozyme extracted from egg white has a molar mass of 13,930 g. A quantity of 0.100 g of this enzyme is dissolved in 150 g of water at 25°C. Calculate the vapor-pressure lowering, the depression in freezing point, the elevation in boiling point, and the osmotic pressure of this solution. (The vapor pressure of water at 25°C is 23.76 mmHg.)

9.53 Solutions A and B have osmotic pressures of 2.4 and 4.6 bar, respectively, at a certain temperature. What is the osmotic pressure of a solution prepared by mixing equal volumes of A and B at the same temperature?

9.54 A cucumber placed in concentrated brine (saltwater) shrivels into a pickle. Explain.

9.55 Two liquids A and B have vapor pressures of 76 and 132 mmHg, respectively, at 25°C. What is the total vapor pressure of the ideal solution made up of (a) 1.00 mole of A and 1.00 mole of B, and (b) 2.00 moles of A and 5.00 moles of B?

9.56 Calculate the van't Hoff factor of Na_3PO_4 in a 0.40 m solution whose freezing point is 2.6°C.

9.57 A 262-mL sample of a sugar solution containing 1.22 g of the sugar has an osmotic pressure of 30.3 mmHg at 35°C. What is the molar mass of the sugar?

9.58 Consider the three mercury manometers shown in the diagram. One of them has 1 mL of water on top of the mercury, another has 1 mL of a 1 m urea solution on top of the mercury, and the third one has 1 mL of a 1 m NaCl solution placed on top of the mercury. Which of these solutions is in the tube labeled X, which is in Y, and which is in Z?

9.59 A forensic chemist is given a white powder for analysis. She dissolves 0.50 g of the substance in 8.0 g of benzene. The solution freezes at 3.9°C. Can the chemist conclude that the compound is cocaine ($C_{17}H_{21}NO_4$)? What assumptions are made in the analysis?

9.60 "Time-release" drugs have the advantage of releasing the drug to the body at a constant rate so that the drug concentration at any time is not too high as to have harmful side effects or too low as to be ineffective. A schematic diagram of a pill that works on this basis is shown here. Explain how it works.

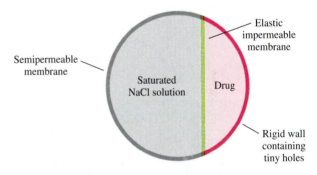

9.61 A solution of 1.00 g of anhydrous aluminum chloride ($AlCl_3$) in 50.0 g of water freezes at $-1.11°C$. Does the molar mass determined from this freezing point agree with that calculated from the formula? Why?

9.62 Explain why reverse osmosis is (theoretically) more desirable as a desalination method than distillation or freezing. What minimum pressure must be applied to seawater at 25°C in order for reverse osmosis to occur? (Treat seawater as a 0.70 M NaCl solution.)

9.63 A protein has been isolated as a salt with the formula $Na_{20}P$ (this notation means that there are 20 Na^+ ions associated with a negatively charged protein P^{20-}). The osmotic pressure of a 10.0-mL solution containing 0.225 g of the protein is 0.260 bar at

25.0°C. (a) Calculate the molar mass of the protein from these data. (b) Calculate the actual molar mass of the protein.

9.64 A nonvolatile organic compound Z was used to make up two solutions. Solution A contains 5.00 g of Z dissolved in 100 g of water, and solution B contains 2.31 g of Z dissolved in 100 g of benzene. Solution A has a vapor pressure of 754.5 mmHg at the normal boiling point of water, and solution B has the same vapor pressure at the normal boiling point of benzene. Calculate the molar mass of Z in solutions A and B and account for the difference.

9.65 State which of the alcohols listed in Problem 9.10 you would expect to be the best solvent for each of the following substances, and explain why: (a) I_2, (b) KBr, and (c) $CH_3CH_2CH_2CH_2CH_3$.

9.66 Before a carbonated beverage bottle is sealed, it is pressurized with a mixture of air and carbon dioxide. (a) Explain the effervescence that occurs when the cap of the bottle is removed. (b) What causes the fog to form near the mouth of the bottle right after the cap is removed?

9.67 Iodine (I_2) is only sparingly soluble in water (see left photo). Yet upon the addition of iodide ions (for example, from KI), iodine is converted to the triiodide ion, which readily dissolves (right photo):

$$I_2(s) + I^-(aq) \rightleftharpoons I_3^-(aq)$$

Describe the change in solubility of I_2 in terms of the change in intermolecular forces.

9.68 Two beakers, one containing a 50-mL aqueous 1.0 M glucose solution and the other a 50-mL aqueous 2.0 M glucose solution, are placed under a tightly sealed bell jar at room temperature. What are the volumes in these two beakers at equilibrium?

9.69 In the apparatus shown here, what will happen if the membrane is (a) permeable to both water and the Na^+ and Cl^- ions, (b) permeable to water and Na^+ ions but not to Cl^- ions, (c) permeable to water but not to Na^+ and Cl^- ions?

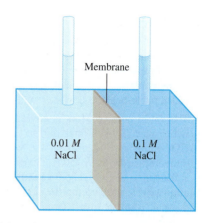

9.70 Explain why it is essential that fluids used in intravenous injections have approximately the same osmotic pressure as blood.

9.71 Explain each of the following statements: (a) The boiling point of seawater is higher than that of pure water. (b) Carbon dioxide escapes from the solution when the cap is removed from a carbonated soft drink bottle. (c) Molal and molar concentrations of dilute aqueous solutions are approximately equal. (d) In discussing the colligative properties of a solution (other than osmotic pressure), it is preferable to express the concentration in units of molality rather than in molarity. (e) Methanol (b.p. 65°C) is useful as an antifreeze, but it should be removed from the car radiator during the summer season.

9.72 A mixture of NaCl and sucrose ($C_{12}H_{22}O_{11}$) of combined mass 10.2 g is dissolved in enough water to make up a 250 mL solution. The osmotic pressure of the solution is 7.42 bar at 23°C. Calculate the mass percent of NaCl in the mixture.

9.73 A 1.32-g sample of a mixture of cyclohexane (C_6H_{12}) and naphthalene ($C_{10}H_8$) is dissolved in 18.9 g of benzene (C_6H_6). The freezing point of the solution is 2.2°C. Calculate the mass percent of the mixture. (See Table 9.3 for constants.)

9.74 How does each of the following affect the solubility of an ionic compound: (a) lattice energy, (b) solvent (polar versus nonpolar), and (c) enthalpies of hydration of cation and anion?

9.75 A solution contains two volatile liquids A and B. Complete the following table, in which the symbol $\longleftrightarrow$ indicates attractive intermolecular forces.

Attractive Forces	Deviation from Raoult's Law	ΔH_{soln}
$A \longleftrightarrow A, B \longleftrightarrow B >$ $A \longleftrightarrow B$		
	Negative	
		Zero

9.76 A mixture of ethanol and 1-propanol behaves ideally at 36°C and is in equilibrium with its vapor. If the mole fraction of ethanol in the solution is 0.62, calculate its mole fraction in the vapor phase at this temperature. (The vapor pressures of pure ethanol and 1-propanol at 36°C are 108 and 40.0 mmHg, respectively.)

9.77 For ideal solutions, the volumes are additive. This means that if 5 mL of A and 5 mL of B form an ideal solution, the volume of the solution is 10 mL. Provide a molecular interpretation for this observation. When 500 mL of ethanol (C_2H_5OH) are mixed with 500 mL of water, the final volume is less than 1000 mL. Why?

9.78 Ammonia (NH_3) is very soluble in water, but nitrogen trichloride (NCl_3) is not. Explain.

9.79 Aluminum sulfate [$Al_2(SO_4)_3$] is sometimes used in municipal water treatment plants to remove undesirable particles. Explain how this process works.

9.80 Acetic acid is a weak acid that ionizes in solution as follows:

$$CH_3COOH(aq) \rightleftharpoons CH_3COO^-(aq) + H^+(aq)$$

If the freezing point of a 0.106 m CH_3COOH solution is 0.203°C, calculate the percent of the acid that has undergone ionization.

9.81 Making mayonnaise involves beating oil into small droplets in water, in the presence of egg yolk. What is the purpose of the egg yolk? (*Hint:* Egg yolk contains lecithins, which are molecules with a polar head and a long nonpolar hydrocarbon tail.)

9.82 Acetic acid is a polar molecule and can form hydrogen bonds with water molecules. Therefore, it has a high solubility in water. Yet acetic acid is also soluble in benzene (C_6H_6), a nonpolar solvent that lacks the ability to form hydrogen bonds. A solution of 3.8 g of CH_3COOH in 80 g C_6H_6 has a freezing point of 3.5°C. Calculate the molar mass of the solute and suggest what its structure might be. (*Hint:* Acetic acid molecules can form hydrogen bonds between themselves.)

9.83 Fish in the Antarctic Ocean swim in water at about −2°C. (a) To prevent their blood from freezing, what must be the concentration (in molality) of the blood? Is this a reasonable physiological concentration? (b) In recent years scientists have discovered a special type of protein in these fishes' blood which, although present in quite low concentrations (≤ 0.001 m), has the ability to prevent the blood from freezing. Suggest a mechanism for its action.

9.84 As we know, if a soft drink can is shaken and then opened, the drink escapes violently. However, if after shaking the can we tap it several times with a metal spoon, no such "explosion" of the drink occurs. Why?

9.85 Why are ice cubes (for example, those you see in the trays in the freezer of a refrigerator) cloudy inside?

9.86 Using Henry's law and the ideal gas equation, prove the statement that the volume of a gas that dissolves in a given amount of solvent is *independent* of the pressure of the gas dissolved in the solvent.

9.87 At 27°C, the vapor pressure of pure water is 23.76 mmHg and that of an urea solution is 22.98 mmHg. Calculate the molality of the solution.

9.88 An example of the positive deviation from Raoult's law is a solution made of acetone (CH_3COCH_3) and carbon disulfide (CS_2). (a) Draw Lewis structures of these molecules. Explain the deviation from ideal behavior in terms of intermolecular forces. (b) A solution composed of 0.60 mole of acetone and 0.40 mole of carbon disulfide has a vapor pressure of 615 mmHg at 35.2°C. What would be the vapor pressure if the solution behaved ideally? The vapor pressure of the pure solvents at the same temperature are: 349 mmHg for acetone and 501 mmHg for carbon disulfide. (c) Predict the sign of ΔH_{soln}.

9.89 Liquids A (molar mass 100 g mol^{-1}) and B (molar mass 110 g mol^{-1}) form an ideal solution. At 55°C, A has a vapor pressure of 95 mmHg and B has a vapor pressure of 42 mmHg. A solution is prepared by mixing equal masses of A and B. (a) Calculate the mole fraction of each component in the solution. (b) Calculate the partial pressures of A and B over the solution at 55°C. (c) Suppose that some of the vapor described in part (b) is condensed to a liquid. Calculate the mole fraction of each component in this liquid and the vapor pressure of each component above this liquid at 55°C.

9.90 A very long pipe is capped at one end with a semipermeable membrane. How deep (in meters) must the pipe be immersed into the sea for freshwater to begin to pass through the membrane? Assume the water to be at 20°C and treat it as a 0.70 M NaCl solution. The density of seawater is 1.03 g cm^{-3}, and the acceleration due to gravity is 9.81 m s^{-2}.

9.91 Benzene and toluene form an ideal solution. Prove that the maximum entropy of mixing in the liquid is obtained when the mole fraction of each component is 0.5.

9.92 The vapor pressures of pure benzene and pure toluene at 298 K are 94.6 and 29.1 torr, respectively. If 5.00 g each of benzene and toluene are placed in an evacuated container with a volume of 5.0 L and allowed to reach equilibrium, what fraction of the mixture will be in the liquid phase? What is the final pressure? Describe any assumptions that you make.

9.93   Two beakers, 1 and 2, containing 50 mL of 0.10 M urea and 50 mL of 0.20 M urea, respectively, are placed under a tightly sealed container at 298 K. Calculate the mole fraction of urea in the solutions at equilibrium. Assume ideal behavior. Urea is a nonvolatile compound.

9.94   Two miscible liquids A and B have vapor pressures of 123 mmHg and 564 mmHg at 25°C, respectively, when pure. If 1.4 moles of A are mixed with an unknown amount of B, the resulting solution has a vapor pressure of 427 mmHg at 25°C. Determine the amount of B that was added to form this solution.

9.95   At 298 K, the osmotic pressure of a glucose solution is 10.50 atm. Calculate the freezing point of the solution. The density of the solution is 1.16 g mL^{-1}.

9.96   A closed vessel of volume 9.6 L contains 2.0 g of water. Calculate the temperature (in °C) at which half of the water is in the vapor phase and half is liquid.

9.97   The normal boiling point of methanol is 65.0°C, and the standard enthalpy of formation of methanol vapor is −202.2 kJ mol^{-1}. Calculate the vapor pressure of methanol (in mmHg) at 25°C. (*Hint:* See Appendix 2 for other thermodynamic data of methanol.)

9.98   Henry's law constants for the noble gases in water are listed in Table 9.2. Do these values exhibit a trend with increasing atomic number? If so, give an explanation for this trend in terms of intermolecular forces.

9.99   Using Equations 9.21 and 9.24 and data from Tables 7.7 and 7.8, calculate the freezing-point depression and boiling-point elevation constants for diethyl ether.

9.100 In Example 9.3, we used data for the vapor pressure of water at several temperatures to estimate the molar enthalpy of vaporization of water. In this example, the pressures were in units of mmHg. The numerical values of P (and thus ln P) will depend upon the choice of units; however, the calculated molar enthalpy of vaporization should be independent of this choice. (a) Convert the pressure values in Example 9.3 to bar and repeat the calculation of ΔH_{vap} for water to demonstrate that the value obtained does not depend upon the pressure units. (b) Demonstrate mathematically that the slope of Equation 9.5 (and thus ΔH_{vap}) is independent of the units used for pressure.

Answers to Practice Exercises

9.1 −37.6°C **9.2** 369 mmHg **9.3** 61.3 kJ mol^{-1} [This is the average over the temperature range 323 K to 433 K, and therefore, it differs slightly from the value of 59.0 given in Table 7.7, which is measured at the boiling point of Hg (630 K)]. **9.4** CS$_2$ **9.5** 2.9 × 10^{-4} mol L^{-1} **9.6** (a) 85 torr; (b) for benzene 0.654, for heptane 0.346 **9.7** 37.8 mmHg; 4.4 mmHg **9.8** T_b: 101.3°C; T_f: −4.48°C **9.9** 21.3 bar **9.10** 0.066 m; 1.3 × 10^2 g mol^{-1} **9.11** 2.60 × 10^4 g **9.12** 1.21

10 Chapter

Chemical Equilibrium

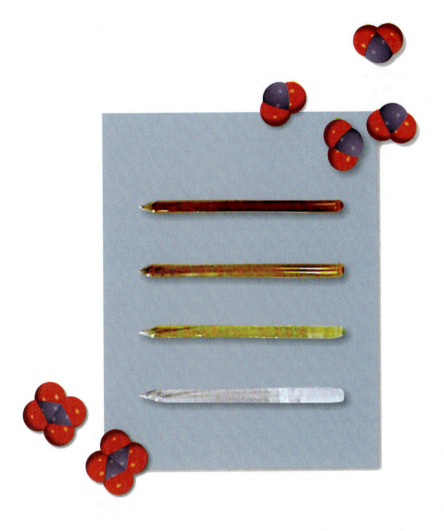

When a chemical reaction has reached the equilibrium state, the concentrations of reactants and products remain constant over time, and there are no visible changes in the system. However, there is much activity at the molecular level because reactant molecules continue to form product molecules while product molecules react to yield reactant molecules. This dynamic process is the subject of this chapter. We will discuss different types of equilibrium reactions, the meaning of the equilibrium constant and its relationship to thermodynamics, and how chemical equilibrium shifts in response to changes in external conditions.

10.1 | The Equilibrium Constant Governs the Concentration of Reactants and Products at Equilibrium

Few chemical reactions proceed in only one direction. Most are, at least to some extent, *reversible reactions,* that is, *they can proceed in either the forward or reverse direction.* At the start of a reversible reaction, the reaction proceeds toward the formation of products. As soon as some product molecules are formed, the reverse reaction begins to take place and reactant molecules are formed from product molecules. Eventually, the rates of product and reactant formation will equalize, and the system will have attained a state of *chemical equilibrium in which the concentrations of the reactants and products no longer change.* Like physical equilibrium (discussed in Chapters 5 and 8), chemical equilibrium is a dynamic process. However, the transformations involved in chemical equilibrium require the breaking and formation of chemical bonds, whereas physical equilibrium involves the transfer of intact molecules between different physical phases of a substance and no chemical bonds are broken in the process.

NO$_2$ and N$_2$O$_4$ gases at equilibrium.

The reversible reaction between dinitrogen tetroxide (N$_2$O$_4$) and nitrogen dioxide (NO$_2$), depicted with space-filling models in Figure 10.1, is an example of a chemical equilibrium. The progress of the reaction

$$N_2O_4(g) \rightleftharpoons 2NO_2(g)$$

can be monitored easily because N$_2$O$_4$ is a colorless gas, whereas NO$_2$, a component of smog, has a dark brown color. (The reaction is written with a double arrow to show that it is a reversible reaction.) Suppose that N$_2$O$_4$ is injected into an evacuated flask. Some brown color appears immediately, indicating the formation of NO$_2$ molecules. The color intensifies as the dissociation of N$_2$O$_4$ continues until eventually an equilibrium is reached. Beyond that point, no further change in color is evident because the partial pressures (and, by Dalton's law, the concentrations) of both N$_2$O$_4$ and NO$_2$ remain constant. Individual molecules of N$_2$O$_4$ continue to dissociate into NO$_2$ molecules, and individual NO$_2$ molecules continue to combine to form N$_2$O$_4$ molecules, but the net concentrations of reactants and products does not change once equilibrium is established.

An equilibrium state can also be established by starting with pure NO$_2$. As NO$_2$ molecules combine to form N$_2$O$_4$, the color fades. The color never completely disappears, though, because some of the newly formed N$_2$O$_4$ molecules dissociate into NO$_2$ molecules again. Yet another way to create an equilibrium state is to start with a mixture of NO$_2$ and N$_2$O$_4$ and monitor the system until the color stops changing. These studies demonstrate that the reaction between N$_2$O$_4$ and NO$_2$ is indeed reversible, because a pure component (N$_2$O$_4$ or NO$_2$) reacts to give the other gas. The important thing to keep in mind is that at equilibrium, the conversions of N$_2$O$_4$ to

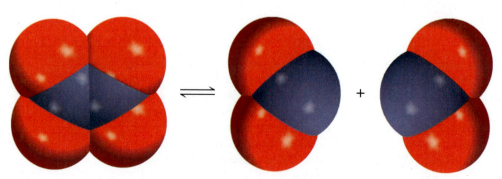

Figure 10.1 Space-filling models depicting the reversible reaction between N$_2$O$_4$ and NO$_2$ molecules.

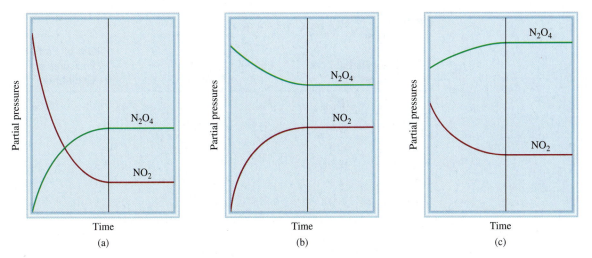

Figure 10.2 Graphs showing the change in the partial pressures of NO_2 and N_2O_4 with time, beginning with three different initial partial pressures of reactants and products. (a) Initially only NO_2 is present. (b) Initially only N_2O_4 is present. (c) Initially a mixture of NO_2 and N_2O_4 is present. Even though equilibrium is reached in all three cases, the equilibrium partial pressures of NO_2 and N_2O_4 are not the same.

NO_2 and of NO_2 to N_2O_4 are still going on. We do not see a color change because the two rates are equal—the removal of NO_2 molecules takes place as fast as the production of NO_2 molecules, and N_2O_4 molecules are formed as quickly as they dissociate. The graphs in Figure 10.2 summarize these three situations.

The Equilibrium Constant

Table 10.1 lists some experimental data for the NO_2-N_2O_4 system at 25°C. The gas concentrations are expressed as partial pressures, which can be calculated at a given temperature (using the ideal gas law) from the number of moles of the gases present initially and at equilibrium and the volume of the flask in liters. The last two columns of Table 10.1 use the equilibrium data to compare the ratios of $P_{NO_2}/P_{N_2O_4}$ and $P_{NO_2}^2/P_{N_2O_4}$, where pressures are measured in bar. Although the ratio $P_{NO_2}/P_{N_2O_4}$ gives scattered values, the ratio $P_{NO_2}^2/P_{N_2O_4}$ gives a nearly constant value that averages 0.115:

$$K = \frac{(P_{NO_2}/P°)^2}{(P_{N_2O_4}/P°)} = 0.115 \qquad \textbf{(10.1)}$$

where K is a constant for the equilibrium reaction $N_2O_4(g) \rightleftharpoons 2NO_2(g)$ at 25°C. (To make K unitless, we have divided each partial pressure in Equation 10.1 by the standard pressure $P° = 1$ bar.) The exponent 2 for P_{NO_2} in this expression is the same as the stoichiometric coefficient for NO_2 in the reversible reaction.

Equation 10.1 is an example of the ***law of mass action,*** which holds that *for a reversible reaction at equilibrium and at a constant temperature, a certain ratio of reactant and product concentrations has a constant value K (called the **equilibrium constant**).* This law was first formulated by two Norwegian chemists, Cato Guldberg[1] and Peter Waage,[2] in 1864.

---

1. Cato Maximilian Guldberg (1836–1902). Norwegian chemist and mathematician. Guldberg's research was mainly in thermodynamics.

2. Peter Waage (1833–1900). Norwegian chemist. Like that of his coworker, Guldberg, Waage's research was primarily in thermodynamics.

Table 10.1	NO₂-N₂O₄ System at 25°C				
Initial Partial Pressures (bar)		Equilibrium Partial Pressures (bar)		Ratio of Partial Pressures at Equilibrium	
P_{NO_2}	$P_{N_2O_4}$	P_{NO_2}	$P_{N_2O_4}$	$P_{NO_2}/P_{N_2O_4}$	$P^2_{NO_2}/P_{N_2O_4}$
0.000	16.60	1.355	15.93	0.0851	0.115
1.238	11.05	1.132	11.10	0.102	0.115
0.7433	12.39	1.177	12.16	0.0968	0.114
0.9910	14.86	1.296	14.72	0.0880	0.114
4.955	0.000	0.5054	2.225	0.227	0.115

The law of mass action can be generalized in its modern form for the following reversible reaction:

$$b\text{B} + c\text{C} \rightleftharpoons d\text{D} + e\text{E}$$

where b, c, d, and e are stoichiometric coefficients. For this reaction at equilibrium at a particular temperature,

$$K = \frac{a_D^d a_E^e}{a_B^b a_C^c} \qquad (10.2)$$

where a_i is called the **activity** of component i. The activity of a species in a reaction is generally a complex function of the pressures and concentrations of all the components present in the reaction mixture. In many very important cases, however, the activity of a species takes on a simple form:

▸ **Ideal gases**: Gases at low to moderate pressures (less than a few bar) behave ideally, at least to a good approximation. The activity of an ideal gas of species i is given by its partial pressure in units of the standard pressure $P° = 1$ bar:

$$a_i = \frac{P_i}{P°} \qquad \text{(ideal gas)} \qquad (10.3)$$

We will often omit the $P°$ from equilibrium constant expressions involving gases and write $a_i = P_i$, with the understanding that P_i is a dimensionless quantity numerically equal to the partial pressure in bar.

▸ **Solutes in dilute solution**: A solute present in a solution in dilute concentrations will exhibit nearly ideal solution behavior. Here "dilute" is typically taken to be less than 0.1 M for nonelectrolyte solutions and less than 0.01 M for electrolyte solutions. The activity of a solute i in an ideal solution is given by its molarity relative to the standard-state concentration $c° = 1$ mol L^{-1}:

$$a_i = \frac{[i]}{c°} \qquad \text{(dilute solute)} \qquad (10.4)$$

We will often omit the $c°$ from equilibrium constant expressions involving solutes and write $a_i = [i]$, with the understanding that $[i]$ is a dimensionless quantity numerically equal to the molarity.

▸ *Pure solids and liquids*: The activity of a pure (or nearly pure) solid or liquid in a reaction can be well approximated by assuming that

$$a_i = 1 \qquad \text{(pure solid or liquid)} \qquad \textbf{(10.5)}$$

The activity of a species is a dimensionless quantity, so the equilibrium constant K is dimensionless, too.

   Both the partial pressure of an ideal gas and the molarity of a solute are measures of the concentration of a material in a reaction mixture, so the activity can be viewed as a measure of the effective amount of a substance available for reaction. At high gas pressures or high solute concentrations, the activities of reacting species may differ somewhat from the expressions in Equations 10.3 and 10.4 due to the influence of strong intermolecular interactions. For the purposes of the most of the reactions considered in this book, however, using Equations 10.3 to 10.5 will be sufficient to describe equilibrium. An example where the deviation from ideal solution behavior is important was discussed in Section 9.4, where it was necessary to use nonintegral values of the van't Hoff factor to compensate for ion pairing in determining the colligative properties of electrolyte solutions. The presence of ion pairing decreases the effective number of solute particles in solution; in other words, the activity of the solute is smaller than the numerical value of the solute molarity.

   The validity of the law of mass action (illustrated by Equation 10.2) has been established by studying many reversible reactions. The equilibrium constant, then, is defined by a *quotient*. The numerator of the quotient is obtained by multiplying together the equilibrium activities of the *products,* each raised to a power equal to its stoichiometric coefficient in the balanced equation. The denominator of the quotient is obtained by multiplying together the equilibrium activities of the *reactants,* each raised to a power equal to its stoichiometric coefficient in the balanced equation. As will be discussed in Section 10.3, the form of the equilibrium constant, as well as the specific value of K for a given reaction, can be determined from thermodynamics.

   The magnitude of the equilibrium constant indicates whether an equilibrium reaction favors the products or the reactants. If K is much greater than 1 (i.e., $K \gg 1$), the equilibrium lies to the right and favors the products. Conversely, if the equilibrium constant is much smaller than 1 (i.e., $K \ll 1$), the equilibrium lies to the left and favors the reactants. In this context, any number greater than 10 is considered to be much greater than 1, and any number less than 0.1 is much less than 1.

The signs $\ll$ and $\gg$ mean "much less than" and "much greater than," respectively.

   Although the law of mass action and the equilibrium constant can be used to predict the amounts of reactants and products that are present in the system when equilibrium is established, these concepts have no bearing on the *kinetics* of the reaction, that is, how long it takes a reaction to reach equilibrium. The kinetics of chemical reactions are discussed in Chapter 13.

Writing Equilibrium Constant Expressions

Equilibrium constants are the key to solving a wide variety of stoichiometry problems involving equilibrium systems. For example, an industrial chemist who wants to maximize the yield of sulfuric acid must have a clear understanding of the equilibrium constants for all the steps in the process, starting from the oxidation of sulfur and ending with the formation of H_2SO_4. A physician specializing in clinical cases of acid-base imbalance needs to know the equilibrium constants of weak acids and bases. And, knowledge of equilibrium constants of pertinent gas-phase reactions will help an atmospheric chemist better understand the process of ozone destruction in the stratosphere.

To use equilibrium constants, we must express them in terms of the reactant and product concentrations. Our only guide is the law of mass action (Equation 10.2), which is the general formula for finding equilibrium activities in terms of the equilibrium constant, and the expressions for the relationship of activities to concentrations (or partial pressures) given by Equations 10.3 to 10.5. Because the concentrations of the reactants and products can often be expressed in different ways, however, and because the reacting species are not always in the same phase, there may be more than one way to express the equilibrium constant for the *same* reaction. To begin with, we will consider reactions in which the reactants and products are in the same phase.

Homogeneous Equilibria

Homogeneous equilibria are *equilibrium reactions in which all reacting species are in the same phase.* The gas-phase equilibrium between NO_2 and N_2O_4 is a homogeneous equilibrium as is the reaction of hydrogen with nitrogen to form ammonia:

$$3H_2(g) + N_2(g) \rightleftharpoons 2NH_3(g)$$

The equilibrium constant, as given in Equation 10.2, is

$$K = \frac{a_{NH_3}^2}{a_{H_2}^3 a_{N_2}} = \frac{P_{NH_3}^2}{P_{H_2}^3 P_{N_2}}$$

where we have assumed ideal gas behavior for the activities (Equation 10.3 with the $P°$ term omitted for simplicity).

The ionization of acetic acid (CH_3COOH) in water is a homogeneous equilibrium:

$$CH_3COOH(aq) + H_2O(l) \rightleftharpoons CH_3COO^-(aq) + H_3O^+(aq)$$

The equilibrium constant expression is

$$K = \frac{a_{CH_3COO^-(aq)} a_{H_3O^+(aq)}}{a_{CH_3COOH(aq)} a_{H_2O(l)}} = \frac{[CH_3COO^-(aq)][H_3O^+(aq)]}{[CH_3COOH(aq)]}$$

where we have assumed that the solutes $CH_3COOH(aq)$, $CH_3COO^-(aq)$, and $H_3O^+(aq)$ are sufficiently dilute that their activities can be expressed as solute molarities. $H_2O(l)$ represents a nearly pure liquid in this system, so its activity is equal to unity and it does not appear in the final equilibrium expression.

Example 10.1 shows how to write equilibrium constant expressions for homogeneous reactions.

Example 10.1

Write appropriate expressions for K for the following reversible homogeneous reactions at equilibrium:

(a) $HF(aq) + H_2O(l) \rightleftharpoons H_3O^+(aq) + F^-(aq)$

(b) $2NO(g) + O_2(g) \rightleftharpoons 2NO_2(g)$

(c) $CH_3COOH(aq) + C_2H_5OH(aq) \rightleftharpoons CH_3COOC_2H_5(aq) + H_2O(l)$

Strategy Use the law of mass action (Equation 10.2) to obtain the form of K, and then use Equations 10.3, 10.4, and/or 10.5 to obtain the appropriate expressions for the

—Continued

Continued—
activities, depending upon whether the reactants and products are gases, dilute solutes, or nearly pure liquids.

Solution (a) Equation 10.2 gives

$$K = \frac{a_{F^-(aq)}a_{H_3O^+(aq)}}{a_{HF(aq)}a_{H_2O(l)}} = \frac{[F^-(aq)][H_3O^+(aq)]}{[HF(aq)]}$$

where we have assumed that the concentrations of HF, F^-, and H_3O^+ are dilute enough that the dilute solute expression for the activity (Equation 10.4) holds. H_2O is the solvent, so it can be treated as a nearly pure liquid in a dilute solution, giving $a_{H_2O} = 1$ (Equation 10.5).

(b) All of the reacting species are in the gas phase, so the activities can be equated to the partial pressures (as long as the pressures are sufficiently low that the ideal gas law holds).

$$K = \frac{a_{NO_2}^2}{a_{NO}^2 a_{O_2}} = \frac{P_{NO_2}^2}{P_{NO}^2 P_{O_2}}$$

(c) All of the reacting species are either solutes or a nearly pure solvent (water). From Equations 10.2 and 10.4, the equilibrium constant K is given by

$$K = \frac{a_{CH_3COOC_2H_5(aq)}a_{H_2O(l)}}{a_{CH_3COOH(aq)}a_{C_2H_5OH(aq)}} = \frac{[CH_3COOC_2H_5(aq)]}{[CH_3COOH(aq)][C_2H_5OH(aq)]}$$

Practice Exercise Write K for

(a) the decomposition of nitrogen pentoxide:

$$2N_2O_5(g) \rightleftharpoons 4NO_2(g) + O_2(g)$$

(b) $NH_3(aq) + H_2O(l) \rightleftharpoons NH_4^+(aq) + OH^-(aq)$

Sometimes it is convenient to express the equilibrium constant for a reaction in terms of alternate units. Consider, for example, the following generalized equilibrium in the gas phase:

$$aA(g) \rightleftharpoons bB(g)$$

where a and b are stoichiometric coefficients. The equilibrium constant K, assuming ideal gas behavior, is given by

$$K = \frac{P_B^b}{P_A^a}$$

Suppose that it is easier to measure the concentration (in mol L^{-1}) than it is the pressure. We can convert the partial pressure of A in this expression to molarity using the ideal gas law:

$$P_A V = n_A RT$$

$$P_A = \frac{n_A}{V}RT = [A]RT$$

where [A] is the molarity of A. Similarly, for B, we have $P_B = [B]RT$. Substituting into the expression for K gives

$$K_P = \frac{P_B^b}{P_A^a} = \frac{([B]RT)^b}{([A]RT)^a} = \frac{[B]^b}{[A]^a} \times (RT)^{b-a}$$

where we have used the subscript "P" in the equilibrium constant to indicate specifically that the gas activities are measured as partial pressures, not concentrations. The **concentration equilibrium constant** for this reaction, K_c, can be defined as

$$K_c = \frac{[B]^b}{[A]^a}$$

The two equilibrium constant expressions are related by

In this equation, the value of R should be that corresponding to units of L bar $mol^{-1} K^{-1}$ and T should be in K. However, so that K_P and K_c remain dimensionless, the units on R and T should be omitted in any calculation.

$$K_P = K_c(RT)^{\Delta n} \tag{10.6}$$

where $\Delta n = b - a =$ moles of gaseous products $-$ moles of gaseous reactants.

In general, $K_P \neq K_c$ except in the special case in which $\Delta n = 0$ as in the equilibrium mixture of molecular hydrogen, molecular bromine, and hydrogen bromide:

$$H_2(g) + Br_2(g) \rightleftharpoons 2HBr(g)$$

In this case, Equation 10.6 can be written as

$$K_P = K_c(RT)^0 = K_c$$

Example 10.2 shows how to calculate K_P and K_c from a specific reaction.

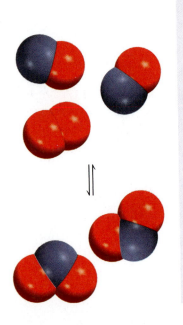

Example 10.2

The following equilibrium process has been studied at 230°C:

$$2NO(g) + O_2(g) \rightleftharpoons 2NO_2(g)$$

In one experiment the concentrations of the reacting species at equilibrium were found to be [NO] = 2.71×10^{-3} M, [O_2] = 0.127 M, and [NO_2] = 0.775 M. Calculate the equilibrium constants K_c and K_P for this reaction at 230°C.

Strategy The concentrations given are equilibrium concentrations. They have units of mol L^{-1}, so we can calculate the concentration equilibrium constant K_c as a ratio of molarities. The equilibrium constant K_P can be calculated from the value of K_c using Equation 10.6 with $\Delta n = 2 - (2 + 1) = -1$.

Solution The concentration equilibrium constant is given by

$$K_c = \frac{[NO_2]^2}{[NO]^2[O_2]}$$

where the concentrations are in M. Substituting the concentrations, we find that

$$K_c = \frac{(0.775)^2}{(2.71 \times 10^{-3})^2(0.127)} = 6.44 \times 10^5$$

—Continued

Continued—

For K_P we use Equation 10.6:

$$K_P = K_c(RT)^{\Delta n} = (6.44 \times 10^5)[(0.08314)(503)]^{-1} = 1.54 \times 10^4$$

We must use the gas constant ($R = 0.08314$ L bar mol^{-1} K^{-1}) and the temperature (503 K) without units so that K_P will remain dimensionless.

Check The large magnitudes of K_c and K_P are consistent with the high product (NO_2) concentration relative to the concentrations of the reactants (NO and O_2).

Practice Exercise Carbonyl chloride ($COCl_2$), also called phosgene, was used in World War I as a poisonous gas. The equilibrium concentrations for the reaction between carbon monoxide and molecular chlorine to form carbonyl chloride,

$$CO(g) + Cl_2(g) \rightleftharpoons COCl_2(g)$$

at 74°C are [CO] = 1.2×10^{-3} M, [Cl_2] = 0.054 M, and [$COCl_2$] = 0.14 M. Calculate the equilibrium constants K_c and K_P.

Heterogeneous Equilibrium

Heterogeneous equilibria are *equilibrium reactions in which the reactants and products are in different phases.* When solid calcium carbonate is heated in a closed vessel, for example, the following equilibrium is attained:

$$CaCO_3(s) \rightleftharpoons CaO(s) + CO_2(g)$$

The two solids and one gas constitute three separate phases. At equilibrium, the equilibrium constant expression can be written as

$$K = \frac{a_{CaO(s)}\, a_{CO_2(g)}}{a_{CaCO_3(s)}}$$

The mineral calcite is made of calcium carbonate, as are chalk and marble.

According to Equation 10.5, the activities of the pure solids (CaO and $CaCO_3$) are equal to unity. According to Equation 10.3, the activity for the gas (CO_2) is given by the numerical value of its partial pressure in bar, thus,

$$K = \frac{1 \times P_{CO_2}}{1} = P_{CO_2}$$

The equilibrium constant in this case is numerically equal to the pressure of CO_2 gas in bar, which is an easily measurable quantity. In other words, the value of the equilibrium constant is independent of the amount of $CaCO_3$ or CaO that is present, as long as some of each is present at equilibrium (Figure 10.3).

Example 10.3 illustrates the calculation of equilibrium constants for a heterogeneous reaction.

Example 10.3

Consider the heterogeneous equilibrium shown in Figure 10.3:

$$CaCO_3(s) \rightleftharpoons CaO(s) + CO_2(g)$$

—Continued

Figure 10.3 The equilibrium pressure of CO_2 is the same in (a) and (b) at the same temperature, despite the presence of different amounts of $CaCO_3$ (shown in brown) and CaO (shown in green).

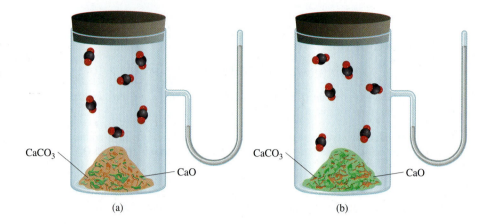

(a)   (b)

Continued—

At 800°C, the pressure of CO_2 is 0.239 bar. Calculate (a) K_P and (b) K_c for the reaction at this temperature.

Strategy Remember that the activity of pure solids can be well approximated as 1, so concentrations of pure solids will not appear in the final equilibrium constant expression. The relationship between K_P and K_c is given by Equation 10.6

Solution (a) Using Equation 10.2 we have

$$K = \frac{a_{CaO(s)} a_{CO_2(g)}}{a_{CaCO_3(s)}}$$

Using Equations 10.3 and 10.5 to approximate the activities gives

$$K_P = \frac{1 \times P_{CO_2}}{1} = P_{CO_2} = 0.239$$

where again we have used the subscript "P" to specifically indicate that partial pressures are being used for the gas phase activities.

(b) From Equation 10.6, we know

$$K_P = K_c(RT)^{\Delta n}$$

In this case, $T = 800 + 273 = 1073$ K and $\Delta n = 1$, so we substitute these values in the equation and obtain

$$0.239 = K_c(0.08314 \times 1073)$$
$$K_c = 2.68 \times 10^{-3}$$

Practice Exercise Consider the following equilibrium at 395 K:

$$NH_4HS(s) \rightleftharpoons NH_3(g) + H_2S(g)$$

The partial pressure of each gas is 0.268 bar. Calculate K_P and K_c for the reaction.

Multiple Equilibria

The reactions we have considered so far are all relatively simple. A more complicated case is one in which the product molecules from one equilibrium system are involved in a second equilibrium process. Consider the following pair of gas-phase equilibria:

$$A + B \rightleftharpoons C + D$$
$$C + D \rightleftharpoons E + F$$

The products formed in the first reaction, C and D, react further to form products E and F. At equilibrium we can write two separate equilibrium constants:

$$K_1 = \frac{P_C P_D}{P_A P_B}$$

and

$$K_2 = \frac{P_E P_F}{P_C P_D}$$

The overall reaction is given by the sum of the two reactions

$$A + B \rightleftharpoons C + D \quad K_1$$
$$C + D \rightleftharpoons E + F \quad K_2$$

Overall reaction: $A + B \rightleftharpoons E + F \quad K_{12}$

and the equilibrium constant K for the overall reaction is

$$K_{12} = \frac{P_E P_F}{P_A P_B}$$

We obtain the same expression if we take the product of the expressions for K_1 and K_2

$$K_1 \times K_2 = \frac{P_C P_D}{P_A P_B} \times \frac{P_E P_F}{P_C P_D} = \frac{P_E P_F}{P_A P_B}$$

Therefore,

$$K_{12} = K_1 K_2 \tag{10.7}$$

We can now make an important statement about multiple equilibria: *If a reaction can be expressed as the sum of two or more reactions, the equilibrium constant for the overall reaction is given by the product of the equilibrium constants for the individual reactions.*

Among the many known examples of multiple equilibria is the ionization of diprotic acids in aqueous solution. The following equilibrium constants have been determined for carbonic acid (H_2CO_3) at 25°C:

$$H_2CO_3(aq) \rightleftharpoons H^+(aq) + HCO_3^-(aq) \quad K_1 = \frac{[H^+][HCO_3^-]}{[H_2CO_3]} = 4.2 \times 10^{-7}$$

$$HCO_3^-(aq) \rightleftharpoons H^+(aq) + CO_3^{2-}(aq) \quad K_2 = \frac{[H^+][CO_3^{2-}]}{[HCO_3^-]} = 4.8 \times 10^{-11}$$

The overall reaction is the sum of these two reactions

$$H_2CO_3(aq) \rightleftharpoons 2H^+(aq) + CO_3^{2-}(aq)$$

and the corresponding equilibrium constant is given by

$$K = \frac{[H^+]^2[CO_3^{2-}]}{[H_2CO_3]}$$

Using Equation 10.7,

$$K = K_1 K_2 = (4.2 \times 10^{-7})(4.8 \times 10^{-11}) = 2.0 \times 10^{-17}$$

The Form of K *and the Equilibrium Expression*

Besides Equation 10.7, there are two more important rules for writing equilibrium constants:

1. When the equation for a reversible reaction is written in the opposite direction, the equilibrium constant becomes the reciprocal of the original equilibrium constant. Thus, if we write the NO_2-N_2O_4 equilibrium as

   $$N_2O_4(g) \rightleftharpoons 2NO_2(g)$$

   then at 25°C

   $$K_1 = \frac{P_{NO_2}^2}{P_{N_2O_4}} = 0.115$$

   However, we can represent the equilibrium equally well as

   $$2NO_2(g) \rightleftharpoons N_2O_4(g)$$

   and the equilibrium constant is now given by

   $$K_{-1} = \frac{P_{N_2O_4}}{P_{NO_2}^2} = 8.70$$

   Thus, $K_{-1} = 1/K_1$ or $K_1 \times K_{-1} = 1$. Both K_1 and K_{-1} are valid equilibrium constants, but it is meaningless to say that the equilibrium constant for the NO_2-N_2O_4 system is 0.115 or 8.70 unless we also specify how the equilibrium equation is written.

2. The value of K also depends upon how the equilibrium equation is balanced. Consider the following ways of describing the same equilibrium:

   $$\frac{1}{2}N_2O_4(g) \rightleftharpoons NO_2(g) \quad K_2 = \frac{P_{NO_2}}{P_{N_2O_4}^{1/2}}$$

   $$N_2O_4(g) \rightleftharpoons 2NO_2(g) \quad K_1 = \frac{P_{NO_2}^2}{P_{N_2O_4}}$$

   The second reaction can be obtained by adding the first reaction to itself, so according to Equation 10.7, $K_1 = K_2 \times K_2 = K_2^2$. The equilibrium constant for the second reaction (K_1) was given at the beginning of this section as 0.115 at 25°C, so the equilibrium constant for the first reaction (K_2) can be obtained from the square root of K_1 or 0.339.

According to the law of mass action, each activity term in the equilibrium constant expression is raised to a power equal to its stoichiometric coefficient. Thus, if you double the stoichiometric coefficients in a chemical equation throughout, the corresponding equilibrium constant will be the square of the original value; if you triple the stoichiometric coefficients in a chemical equation, the equilibrium constant will be the cube of the original value, and so on. The NO_2-N_2O_4 example also demonstrates why you need to write the chemical equation when quoting the numerical value of the equilibrium constant.

Example 10.4 shows how to use equilibrium constant data from two reactions to calculate the equilibrium constant of a third related reaction.

Example 10.4

Consider the following gas-phase reaction equilibria involving N_2O_4:

Reaction 1: $\quad N_2O_4(g) \rightleftharpoons 2NO_2(g) \qquad\qquad K_1 = 0.115$

Reaction 2: $\quad 2N_2O_4(g) \rightleftharpoons 2N_2O(g) + 3O_2(g) \quad K_2 = 1.37 \times 10^{-2}$

Use this information to determine the equilibrium constant for the following reaction:

Reaction 3: $\quad 4NO_2(g) \rightleftharpoons 2N_2O(g) + 3O_2(g) \quad K_3 = ?$

Strategy If we can write reaction 3 in terms of reactions 1 and 2, we can use Equation 10.7 and the other two rules for writing equilibrium constants to express K_3 as a function of K_1 and K_2.

Solution We can write reaction 3 as the sum of reaction 2 and twice the reverse of Reaction 1:

$$
\begin{array}{lll}
\text{Reaction 2:} & 2N_2O_4(g) \rightleftharpoons 2N_2O(g) + 3O_2(g) & K_2 = 1.37 \times 10^{-2} \\
(-2) \times \text{Reaction 1:} & 4NO_2(g) \rightleftharpoons 2N_2O_4(g) & K_{(-2)\times 1} = ? \\
\hline
\text{Reaction 3:} & 4NO_2(g) \rightleftharpoons 2N_2O(g) + 3O_2(g) & K_3 = K_2 \times K_{(-2)\times 1}
\end{array}
$$

[Here we use the notation $K_{(-2)\times 1}$ to denote the equilibrium constant for the reaction obtained by multiplying reaction 1 by (-2).] Using Equation 10.7, the equilibrium constant for Reaction 3 is then given by the product of the equilibrium constant for Reaction 2 (K_2) and the equilibrium constant for the reaction that is twice the reverse of Reaction 1 ($K_{(-2)\times 1}$). The equilibrium constant for the reverse of Reaction 1 is the reciprocal of the equilibrium constant for the forward reaction,

$$(-1) \times \text{Reaction 1:} \quad 2NO_2(g) \rightleftharpoons N_2O_4(g) \quad K_{(-1)\times 1} = 1/K_1 = 8.70$$

and multiplying a reaction through by 2 means that the equilibrium constant must be squared:

$$(-2) \times \text{Reaction 1:} \quad 4NO_2(g) \rightleftharpoons 2N_2O_4(g) \quad K_{(-2)\times 1} = (K_{(-1)\times 1})^2 = 8.70^2 = 75.7$$

Combining these two equilibrium constants yields K_3:

$$K_3 = K_2 \times K_{(-2)\times 1} = (1.37 \times 10^{-2})(75.7) = 1.04$$

—Continued

Continued—

Practice Exercise The following equilibrium constants have been determined for hydrosulfuric acid at 25°C:

Reaction 1: $H_2S(aq) \rightleftharpoons H^+(aq) + HS^-(aq)$ $K_1 = 9.5 \times 10^{-8}$

Reaction 2: $HS^-(aq) \rightleftharpoons H^+(aq) + S^{2-}(aq)$ $K_2 = 1.0 \times 10^{-19}$

Calculate the equilibrium constant for the following reaction at the same temperature:

Reaction 3: $H_2S(aq) \rightleftharpoons 2H^+(aq) + S^{2-}(aq)$ $K_3 = ?$

Summary of Guidelines for Writing Equilibrium Expressions

1. The equilibrium constant is written as a ratio of the activities of the products to the activities of the reactants (Equation 10.2). The activities of each species in this ratio are raised to a power equal to the stochiometric coefficients of the species in the balanced chemical equation.

2. For reactions in which the solute concentrations and partial pressures are low, the gases and solutes are well approximated by ideal gas or ideal solution behavior. For such systems, the activities are given by

    a. *Ideal gases*: $a_i = P_i/P°$, where P_i = partial pressure of gas i and $P° = 1$ bar.

    b. *Dilute solutes*: $a_i = [i]/c°$, where $[i]$ = molarity of solute in solution and $c° = 1$ mol L^{-1}.

    c. *Pure or nearly pure solids and liquids*: $a_i = 1$.

3. The equilibrium constant K is a dimensionless quantity.

4. In quoting a value for the equilibrium constant, we must specify the balanced chemical equation and the temperature.

5. If the equilibrium equation is written for a reaction in the opposite direction, the equilibrium constant becomes the reciprocal of the original equilibrium constant.

6. If the equilibrium equation is multiplied through by a factor n, the equilibrium constant for the new equation becomes the original equilibrium constant raised to the power of n.

7. If a reaction can be expressed as the sum of two or more reactions, the equilibrium constant for the overall reaction is given by the product of the equilibrium constants of the individual reactions.

10.2 | The Equilibrium Constant Can Be Used to Predict the Direction and Equilibrium Concentrations of a Chemical Reaction

We saw in Section 10.1 that the equilibrium constant for a given reaction can be calculated from known equilibrium concentrations. Once we know the value of the equilibrium constant, we can use Equation 10.2 to calculate the unknown equilibrium concentrations (keep in mind, though, that the equilibrium constant has a constant value only if the temperature does not change). In general, the equilibrium constant makes it possible to predict the direction in which a reaction mixture will proceed to achieve equilibrium and to calculate the concentrations of reactants and products once equilibrium has been reached. Both of these applications are discussed in this section.

Predicting the Direction of a Reaction

In the gas phase, hydrogen iodide forms from molecular hydrogen and molecular iodine:

$$H_2(g) + I_2(g) \rightleftharpoons 2HI(g)$$

The equilibrium constant for this reaction is 54.3 at 430°C. Suppose that in a certain experiment the initial partial pressures of H_2, I_2, and HI are 0.710 bar, 0.427 bar, and 5.78 bar, respectively, at 430°C. Will there be a net reaction to form more H_2 and I_2 or to form more HI? Inserting these initial pressures into the equilibrium constant expression, assuming ideal-gas behavior, we obtain

$$H_2 + I_2 \rightleftharpoons 2HI$$

$$\left(\frac{P_{HI}^2}{P_{H_2}P_{I_2}}\right)_0 = \frac{(5.78)^2}{(0.710)(0.427)} = 110$$

where the subscript 0 indicates that we are using initial partial pressures (before equilibrium is reached). Because the quotient $(P_{HI}^2/P_{H_2}P_{I_2})_0$ is greater than K, this system is not at equilibrium. Consequently, some of the HI will react to form more H_2 and I_2 (decreasing the value of the quotient). Thus, the net reaction proceeds from right to left to reach equilibrium.

For reactions that have not reached equilibrium, such as the formation of HI just considered, we obtain, not the equilibrium constant, but instead the **reaction quotient** **(Q),** *which is the quantity obtained when nonequilibrium activities are substituted into the equilibrium constant expression.* In this case, the nonequilibrium activities are the initial partial pressures of the reactants and products. To determine the direction in which the net reaction will proceed to achieve equilibrium, we must compare the values of Q and K. The three possible cases are as follows:

- ▸ $Q < K$ The ratio of initial activities of products to reactants is too small. To reach equilibrium, reactants must be converted to products. The system proceeds from left to right (consuming reactants, forming products) to reach equilibrium.

- ▸ $Q = K$ The initial activities are equilibrium activities. The system is at equilibrium.

- ▸ $Q > K$ The ratio of initial activities of products to reactants is too large. To reach equilibrium, products must be converted to reactants. The system proceeds from right to left (consuming products, forming reactants) to reach equilibrium.

Figure 10.4 shows a comparison of K with Q.

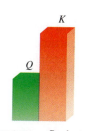

Reactants → Products

Equilibrium : no net change

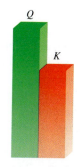

Reactants ← Products

Figure 10.4 The direction of a reversible reaction to reach equilibrium depends on the relative magnitudes of Q and K.

Example 10.5 shows how to use the value of Q to determine the direction of net reaction toward equilibrium.

$N_2 + 3H_2 \rightleftharpoons 2NH_3$

Example 10.5

At the start of a reaction, there are 0.249 mol N_2, 3.21×10^{-2} mol H_2, and 6.42×10^{-4} mol NH_3 in a 3.50-L reaction vessel at 375°C. If the equilibrium constant (K_c) for the reaction

$$N_2(g) + 3H_2(g) \rightleftharpoons 2NH_3(g)$$

is 1.2 at this temperature, decide whether the system is at equilibrium. If it is not, predict which way the net reaction will proceed to achieve equilibrium.

Strategy The value given for the equilibrium constant is K_c not K_P, so we must use concentrations instead of pressures in the equilibrium expression. The initial amounts of the gases (in moles) in a vessel of known volume (in liters) are given, so we can calculate their molar concentrations and hence the reaction quotient (Q_c). The magnitude of Q_c relative to K_c will indicate the direction of the reaction.

Solution The initial concentrations of the reacting species are

$$[N_2]_0 = \frac{0.249 \text{ mol}}{3.50 \text{ L}} = 0.0711 \; M$$

$$[H_2]_0 = \frac{3.21 \times 10^{-2} \text{ mol}}{3.50 \text{ L}} = 9.17 \times 10^{-3} \; M$$

$$[NH_3]_0 = \frac{6.42 \times 10^{-4} \text{ mol}}{3.50 \text{ L}} = 1.83 \times 10^{-4} \; M$$

Next, write the equilibrium constant expression and plug these values into it to obtain Q_c:

$$Q_c = \frac{[NH_3]_0^2}{[N_2]_0[H_2]_0^3} = \frac{(1.83 \times 10^{-4})^2}{(0.0711)(9.17 \times 10^{-3})^3} = 0.611$$

Because $Q_c < K_c$ (1.2), the system is not at equilibrium. The net result will be an increase in the concentration of NH_3 and a decrease in the concentrations of N_2 and H_2. That is, the net reaction will proceed from left to right until equilibrium is reached.

Practice Exercise Nitrosyl chloride, an orange-yellow compound, forms from nitric oxide and molecular chlorine:

$$2NO(g) + Cl_2(g) \rightleftharpoons 2NOCl(g)$$

The equilibrium constant (K_c) for this reaction is 6.5×10^4 at 35°C. In a certain experiment, 2.0×10^{-2} mole of NO, 8.3×10^{-3} mole of Cl_2, and 0.80 moles of NOCl are mixed in a 2.0-L flask. In which direction will the system proceed to reach equilibrium?

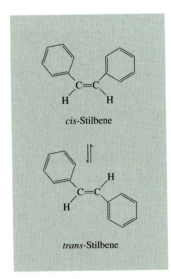

cis-Stilbene

trans-Stilbene

Figure 10.5 The equilibrium between *cis*-stilbene and trans-stilbene. Both molecules are geometric isomers, that is, they have the same molecular formula ($C_{14}H_{12}$) and the same bond connectivity. In *cis*-stilbene, however, the benzene rings are on one side of the C=C double bond and the H atoms are on the other side, whereas in *trans*-stilbene the benzene rings (and the H atoms) are across the C=C double bond from each other. *Cis*- and *trans*-stilbene, like most geometric isomers, have different melting points and dipole moments.

Calculating Equilibrium Concentrations

If we know the equilibrium constant for a particular reaction, we can calculate the concentrations in the equilibrium mixture from the initial concentrations. Commonly, only the initial reactant concentrations are given. Consider, for example, the equilibrium between two organic compounds, *cis*-stilbene and *trans*-stilbene, in a nonpolar hydrocarbon solvent (Figure 10.5):

$$cis\text{-stilbene} \rightleftharpoons trans\text{-stilbene}$$

The equilibrium constant (K) for this system is 24.0 at 200°C. Suppose that initially only *cis*-stilbene is present at a concentration of 0.850 mol L^{-1}. How do we calculate the concentrations of *cis*- and *trans*-stilbene at equilibrium? According to the stoichiometry of the balanced chemical equation, 1 mole of *trans*-stilbene forms for every mole of *cis*-stilbene that reacts. If x is the equilibrium concentration of *trans*-stilbene in mol L^{-1}, then the equilibrium concentration of *cis*-stilbene must be $(0.850 - x)$ mol L^{-1}. The changes in concentration can be summarized as follows:

	cis-stilbene $\rightleftharpoons$	*trans*-stilbene
Initial (M):	0.850	0
Change (M):	$-x$	$+x$
Equilibrium (M):	$0.850 - x$	x

A positive ($+$) change represents an increase and a negative ($-$) change represents a decrease in concentration at equilibrium. Next, we set up the equilibrium constant expression, substitute the corresponding equilibrium concentrations, and solve for x:

$$K = \frac{[trans\text{-stilbene}]}{[cis\text{-stilbene}]}$$

$$24.0 = \frac{x}{0.850 - x}$$

$$x = 0.816 \, M$$

Having solved for x, we can calculate the equilibrium concentrations of *cis*-stilbene and *trans*-stilbene as follows:

$$[cis\text{-stilbene}] = (0.850 - x) \, M = (0.850 - 0.816) \, M = 0.034 \, M$$
$$[trans\text{-stilbene}] = x \, M = 0.816 \, M$$

Prove to yourself that these results are correct by substituting the calculated values for [*cis*-stilbene] and [*trans*-stilbene] back into the equilibrium constant expression and see if you get 24.0, the value given for K.

The approach to solving equilibrium constant problems can be summarized as follows:

1. Express the equilibrium concentrations (or partial pressures for gases) of all species in terms of the initial concentrations and a single unknown x, which represents the change in concentration.

2. Write the equilibrium constant expression in terms of the equilibrium concentrations (or partial pressures). Knowing the value of the equilibrium constant, solve for x.

3. Having solved for x, calculate the equilibrium concentrations of all species.

Example 10.6 shows how to apply this three-step procedure to equilibrium problems.

Example 10.6

A mixture of 0.0623 mol H_2, 0.0414 mol I_2, and 0.224 mol HI was placed in a 10.00-L stainless-steel flask at 430°C. The equilibrium constant K_P for the reaction,

$$H_2(g) + I_2(g) \rightleftharpoons 2HI(g)$$

—Continued

Continued—

is 54.3 at this temperature. Is this system at equilibrium? If not, determine the direction of the reaction and calculate the amounts of H_2, I_2, and HI present at equilibrium.

Strategy From the initial amounts of each species and the volume, we can use the ideal gas law to determine the initial pressures. From these initial pressures, we can calculate the reaction quotient (Q_P) to see if the system is already at equilibrium or, if not, in which direction the net reaction will proceed to reach equilibrium. If the system is not in equilibrium, we must use the equilibrium constant given and the procedure above to determine the equilibrium amounts.

Solution First we need to find the initial partial pressures using the ideal gas law:

$$P_i = n_i \frac{RT}{V}$$

For this temperature and volume, $RT/V = (0.08314 \text{ L bar mol}^{-1} \text{ K}^{-1})(703.15 \text{ K})/ (10.0 \text{ L}) = 5.846 \text{ bar mol}^{-1}$, so

$$(P_{H_2})_0 = (0.0623 \text{ mol})(5.846 \text{ bar mol}^{-1}) = 0.364 \text{ bar}$$
$$(P_{I_2})_0 = (0.0414 \text{ mol})(5.846 \text{ bar mol}^{-1}) = 0.242 \text{ bar}$$
$$(P_{HI})_0 = (0.224 \text{ mol})(5.846 \text{ bar mol}^{-1}) = 1.31 \text{ bar}$$

Substituting the initial partial pressures into the appropriate reaction quotient expression gives

$$Q_P = \frac{(P_{HI})_0^2}{(P_{H_2})_0(P_{I_2})_0} = \frac{(1.31)^2}{(0.364)(0.242)} = 19.5$$

Because Q_P (19.5) is smaller than K_P (54.3), we conclude that the net reaction will proceed from left to right until equilibrium is reached (see Figure 10.4), that is, there will be a depletion of H_2 and I_2 and a gain in HI.

We can now use the three-step procedure to calculate the equilibrium concentrations.

Step 1: The stoichiometry of the reaction is 1 mol H_2 reacting with 1 mol I_2 to yield 2 mol HI. Let x be the depletion in pressure (in bar) of H_2 and I_2 at equilibrium. Based on the stoichiometry of the balanced chemical equation, then, the change in the equilibrium partial pressure of HI must be $2x$. The changes in concentrations can be summarized as follows:

	$H_2(g)$	$+$	$I_2(g)$	$\rightleftharpoons$	$2HI(g)$
Initial (bar):	0.364		0.242		1.31
Change (bar):	$-x$		$-x$		$+2x$
Equilibrium (bar):	$0.364 - x$		$0.242 - x$		$1.31 + 2x$

Step 2: The equilibrium constant is given by

$$K_P = \frac{P_{HI}^2}{P_{H_2} P_{I_2}}$$

Substituting, we get

$$K_P = \frac{(1.31 + 2x)^2}{(0.364 - x)(0.242 - x)} = 54.3$$

To solve this equation, we begin by clearing the denominator to give

$$(1.31 + 2x)^2 = (54.3)(0.364 - x)(0.242 - x)$$
$$1.72 + 5.24x + 4x^2 = 4.78 - 32.9x + 54.3x^2$$

—Continued

Continued—

Grouping like terms, we get

$$50.3x^2 - 38.1x + 3.06 = 0$$

This is a quadratic equation of the form $ax^2 + bx + c = 0$. The solution for a quadratic equation (see Appendix 1) is

$$x = \frac{-b \pm \sqrt{b^2 - 4ac}}{2a}$$

Here we have $a = 50.3$, $b = -38.1$, and $c = 3.06$, so

$$x = 0.0913 \text{ bar} \quad \text{or} \quad x = 0.666 \text{ bar}$$

The second solution is physically impossible because the amounts of H_2 and I_2 reacted would be more than those originally present. The first solution gives the correct answer. When solving quadratic equations of this type, one answer is always physically impossible, so choosing a value for x is relatively straightforward.

Step 3: At equilibrium, the partial pressures are

$$P_{H_2} = (0.364 - 0.0913) \text{ bar} = 0.273 \text{ bar}$$
$$P_{I_2} = (0.242 - 0.0913) \text{ bar} = 0.151 \text{ bar}$$
$$P_{HI} = (1.31 + 2 \times 0.0913) \text{ bar} = 1.49 \text{ bar}$$

The amounts of each gas present at equilibrium can be calculated using the ideal gas law in the form

$$n_i = \frac{P_i V}{RT}$$

$$n_{H_2} = \frac{P_{H_2} V}{RT} = \frac{(0.273 \text{ bar})(10.0 \text{ L})}{(0.08314 \text{ L bar mol}^{-1}\text{ K}^{-1})(703.15 \text{ K})} = 0.0467 \text{ mol}$$

$$n_{I_2} = \frac{P_{I_2} V}{RT} = \frac{(0.151 \text{ bar})(10.0 \text{ L})}{(0.08314 \text{ L bar mol}^{-1}\text{ K}^{-1})(703.15 \text{ K})} = 0.0258 \text{ mol}$$

$$n_{HI} = \frac{P_{HI} V}{RT} = \frac{(1.49 \text{ bar})(10.0 \text{ L})}{(0.08314 \text{ L bar mol}^{-1}\text{ K}^{-1})(703.15 \text{ K})} = 0.255 \text{ mol}$$

Check We can check the answers by calculating K_P using the equilibrium pressures. Substituting the calculated equilibrium partial pressures into the equilibrium expression yields a value of K of 53.8, which is very close to the given value of 54.3. The discrepancy here is due to round-off error.

Comment One can alternatively approach this problem using concentrations instead of pressures by first calculating K_c from the given value of K_c using Equation 10.6. For this particular problem, $\Delta n = 0$, so K_c and K_P would be identical.

Practice Exercise At 1280°C, the equilibrium constant (K_c) is 1.1×10^{-3} for the following reaction:

$$Br_2(g) \rightleftharpoons 2Br(g)$$

If the initial concentrations are $[Br_2]_0 = 6.3 \times 10^{-2} M$ and $[Br]_0 = 1.2 \times 10^{-2} M$, calculate the concentrations of these species at equilibrium.

Example 10.6 shows that we can calculate the concentrations of all the reacting species at equilibrium if we know the equilibrium constant and the initial concentrations. This information is valuable if we need to estimate the yield of a reaction.

Suppose, for example, that we started with 0.500 mol of H_2, 0.500 mol of I_2, and no HI in the reaction between H_2 and I_2 to form HI at 430°C. If the reaction were to go to completion, then 2×0.500 mol or 1.00 mol HI would form. Because of the equilibrium, however, the actual amount of HI formed when the system reaches equilibrium can be calculated to be 0.786 mol—a 78.6 percent yield.

Example 10.7 gives another example of calculating equilibrium concentrations given the equilibrium constant.

Example 10.7

Consider the following heterogeneous equilibrium:

$$AgCl(s) \rightleftharpoons Ag^+(aq) + Cl^-(aq)$$

At 25°C, the equilibrium constant for this reaction is 1.6×10^{-10}. If 5.0 g of AlCl(s) is added to 10.0 L of water initially containing no silver or chloride ions, what will be the equilibrium concentrations of $Ag^+(aq)$ and $Cl^-(aq)$ in the solution at this temperature?

Strategy The equilibrium equation contains both ions and a pure solid (AgCl). To set up the equilibrium expression, use Equation 10.2 together with the appropriate approximation for the activities—Equation 10.4 for the ions and Equation 10.5 for AgCl solid.

Solution The equilibrium constant expression (Equation 10.2 together with Equations 10.4 and 10.5) for this equilibrium is

$$K = \frac{a_{Ag^+} a_{Cl^-}}{a_{AgCl(s)}} = [Ag^+][Cl^-] = 1.6 \times 10^{-10}$$

The equilibrium amounts of Ag^+ and Cl^- will not depend upon the initial amount of AgCl because the amount of solid does not appear in the equilibrium expression (as long as there is a sufficient amount of AgCl so that some is present at equilibrium). Furthermore, from the stoichiometry of the balanced chemical equation, the concentrations of silver ion and chloride ion will be equal because no ions were present in the initial solution and all must come from the dissolution of AgCl. Thus,

$$K = [Ag^+][Cl^-] = (x)(x) = x^2 = 1.6 \times 10^{-10}$$

where we have let $x = [Ag^+] = [Cl^-]$. Therefore,

$$x = \sqrt{K} = \sqrt{1.6 \times 10^{-10}} = 1.26 \times 10^{-5}$$

The concentrations of Ag^+ and Cl^- in the solution are both 1.26×10^{-5} M. This concentration is low enough so only a small fraction of the initial amount of solid AgCl will dissolve, and there will be AgCl present at equilibrium.

Comment The equilibrium constant for the dissolution of an ionic compound in aqueous solution is called the *solubility product* and is discussed in more detail in Chapter 12.

Practice Exercise The equilibrium constant K_P is 0.0721 for the following equilibrium at 395 K:

$$NH_4HS(s) \rightleftharpoons NH_3(g) + H_2S(g)$$

If 20.0 g of ammonium hydrogen sulfide [$NH_4HS(s)$] is placed in a 10.0 L evacuated container at 395 K, calculate the equilibrium partial pressures of ammonia (NH_3) and hydrogen sulfide (H_2S) and determine the amount of $NH_4HS(s)$ left in the container.

10.3 | The Equilibrium Constant for a Reaction Can Be Determined from the Standard Gibbs Energy Change

In Chapter 7, we learned that the spontaneity of any process under constant pressure and temperature conditions is determined by the sign of ΔG, the change in the Gibbs energy for the process. If $\Delta G < 0$, the process will occur spontaneously; if $\Delta G > 0$, on the other hand, the process will be nonspontaneous. If a system is at a minimum of G, all processes away from that state will be nonspontaneous, so the system will remain at the minimum Gibbs energy state in equilibrium. Thus, all systems in chemical equilibrium represent minima in Gibbs energy relative to changes in the amounts of reactants and products. In this section, we examine the Gibbs energy changes in a reacting system and derive the relationship between the equilibrium constant and the standard Gibbs energy change for a reaction. As was the case for the equations governing binary liquid-vapor equilibria and colligative properties, the mathematical form of the equilibrium constant is a consequence of the Gibbs energy of mixing.

To find the relationship between K and the Gibbs free energy, let's begin with a simple gas-phase reaction:

$$A(g) \rightleftharpoons B(g)$$

A sample of pure A gas is placed in the reaction vessel at an initial partial pressure of 1 bar. The reaction will proceed to the right, forming B, until an equilibrium is established. At equilibrium, the system is a mixture of A and B molecules in the gas phase with partial pressures $P_{A,eq}$ and $P_{B,eq}$. To determine the equilibrium partial pressures using thermodynamics, we need to determine the values for the partial pressures of A and B that minimize the Gibbs energy.

The total Gibbs energy for the reacting system is the sum of the Gibbs energy for pure A and pure B (namely, G_A° and G_B°, respectively) and the Gibbs energy of mixing the two (ΔG_{mix}):

$$G = n_A G_A^\circ + n_B G_B^\circ + \Delta G_{mix}$$

Recall from Section 8.5 that the Gibbs energy of mixing for ideal gases was given by Equation 8.37:

$$\Delta G_{mix} = RT(n_A \ln x_A + n_B \ln x_B)$$

Thus, the total Gibbs energy can be rewritten as follows:

$$G = n_A G_A^\circ + n_B G_B^\circ + RT(n_A \ln x_A + n_B \ln x_B) \tag{10.8}$$

Based on the stochiometry of the balanced chemical equation, the total number of moles ($n = n_A + n_B$) is fixed throughout the reaction. Dividing Equation 10.8 by n gives

$$G/n = x_A G_A^\circ + x_B G_B^\circ + RT(x_A \ln x_A + x_B \ln x_B)$$

where we have used the relationships $x_A = n_A/n$ and $x_B = n_B/n$. Because $x_A = 1 - x_B$, we can eliminate x_A to give

$$G/n = \underbrace{(1 - x_B)G_A^\circ + x_B G_B^\circ}_{\text{Gibbs free energy for pure substances}} + \underbrace{RT\big[(1 - x_B)\ln(1 - x_B) + x_B \ln x_B\big]}_{\text{Gibbs free energy of mixing}} \tag{10.9}$$

For the reaction $A(g) \rightleftharpoons B(g)$, x_B is a measure of the *extent of reaction,* that is, to monitor the reaction from start to finish, it is only necessary to know x_B. If B is the component with the lowest standard Gibbs energy (i.e., $G_B^\circ < G_A^\circ$), then a graph of

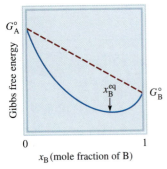

Figure 10.6 Graph of the total Gibbs free energy for the ideal gas reaction A(g) $\rightleftharpoons$ B(g) as a function of x_B, the mole fraction of B, in the reaction mixture (blue line). The red dotted line is the same quantity assuming that there is no Gibbs energy of mixing. The value of x_B corresponding to the minimum in the blue line is the equilibrium composition x_B^{eq}!

Equation 10.9 as a function of x_B will give the blue curve shown in Figure 10.6. The value of x_B at equilibrium (x_B^{eq}) corresponds to the minimum value of G on this curve.

The minimum value of G in Equation 10.9 can be found by finding the value of x_B at which the derivative (slope) of $G(x_B)$ is equal to zero. Taking the derivative of Equation 10.9 gives

$$\frac{d(G/n)}{dx_B} = \frac{d}{dx_B}\{(1 - x_B)G_A^\circ + x_B G_B^\circ + RT[(1 - x_B)\ln(1 - x_B) + x_B \ln x_B]\}$$

$$= G_B^\circ - G_A^\circ + RT\ln\frac{x_B}{1 - x_B}$$

$$\frac{d(G/n)}{dx_B} = \Delta G_{rxn}^\circ + RT\ln\left(\frac{x_B}{x_A}\right) \qquad (10.10)$$

where $\Delta G_{rxn}^\circ = G_B^\circ - G_A^\circ$ is the standard Gibbs free energy of the reaction. For these two gases, the ratio (x_B/x_A) is equal (by Dalton's law) to the ratio of the partial pressures (P_B/P_A), which for this reaction [A(g) $\rightleftharpoons$ B(g)] is equal to the reaction quotient.

$$Q = \frac{P_B}{P_A} = \frac{x_B}{x_A}$$

For a given initial value of x_A, the direction in which the reaction will spontaneously proceed depends upon the slope of the curve (dG/dx_B) in Figure 10.6 because the reaction will always proceed in the direction that lowers G. With this in mind, we define

$$\Delta G_{rxn} \equiv \frac{d(G/n)}{dx_B} = \text{slope of } G \text{ versus extent of reaction curve}$$

Using this definition, Equation 10.10 becomes

$$\Delta G_{rxn} = \Delta G_{rxn}^\circ + RT\ln Q \qquad (10.11)$$

Although we have derived Equation 10.11 for the ideal gas-phase reaction A(g) $\rightleftharpoons$ B(g), this result can be shown by thermodynamics to be generally true for all reactions. In short, ΔG_{rxn} depends on two quantities: ΔG_{rxn}° and $RT\ln Q$. For a given reaction at temperature T, the value of ΔG_{rxn}° is fixed, but that of $RT\ln Q$ is not, because Q varies according to the composition of the reaction mixture. Let us consider two special cases:

Case 1: A large negative value of ΔG_{rxn}° will tend to make ΔG_{rxn} negative, too. Thus, the net reaction will proceed from left to right until a significant amount of product has been formed. At that point, the $RT\ln Q$ term will become positive enough to match the negative ΔG_{rxn}° term.

Case 2: A large positive ΔG_{rxn}° term will tend to make ΔG_{rxn} positive, too. Thus, the net reaction will proceed from right to left until a significant amount of reactant has been formed. At that point, the $RT\ln Q$ term will become negative enough to match the positive ΔG_{rxn}° term.

At equilibrium, $\Delta G_{rxn} = 0$ (the total Gibbs energy is a minimum by definition) and $Q = K$, where K is the equilibrium constant. Thus,

$$0 = \Delta G_{rxn}^\circ + RT\ln K$$

or

$$\Delta G^{\circ}_{\text{rxn}} = -RT \ln K \qquad\qquad (10.12)$$

In Equation 10.12, it is important that the quantities that go into making up K are measured relative to the standard states used to determine $\Delta G^{\circ}_{\text{rxn}}$. For gases that means partial pressures in units of bar, and for solutes it means molarity. In a gas-phase reaction, for example, the K in Equation 10.12 refers to K_P, whereas in a reaction in solution it would be K_c. The larger the value of K, the more negative $\Delta G^{\circ}_{\text{rxn}}$ is. For chemists, Equation 10.12 is one of the most important equations in thermodynamics because it enables us to determine the equilibrium constant of a reaction if we know the change in standard free energy and vice versa.

Equation 10.12 relates the equilibrium constant to the *standard* free-energy change ($\Delta G^{\circ}_{\text{rxn}}$) rather than to the *actual* free-energy change (ΔG_{rxn}). The actual free-energy change of the system varies as the reaction progresses and becomes zero at equilibrium. On the other hand, $\Delta G^{\circ}_{\text{rxn}}$ is a constant for a particular reaction at a given temperature. Figure 10.7 shows plots of the free energy of a reacting system versus the extent of the reaction for two types of reactions. Based on the graphs in Figure 10.7 and Equation 10.11,

▶ If $\Delta G^{\circ}_{\text{rxn}} < 0$, the products are favored over reactants at equilibrium.

▶ If $\Delta G^{\circ}_{\text{rxn}} > 0$, reactants are favored over products at equilibrium.

▶ If $\Delta G^{\circ}_{\text{rxn}} = 0$, the system is at equilibrium if all reactants and products are in their standard states.

It is the sign of ΔG_{rxn}, not the sign of $\Delta G^{\circ}_{\text{rxn}}$, that determines the direction of reaction spontaneity. The sign of $\Delta G^{\circ}_{\text{rxn}}$ only tells us the relative amounts of products and reactants when equilibrium is reached, not the direction of the net reaction.

For reactions having very large or very small equilibrium constants, it is generally very difficult, if not impossible, to measure the K values by monitoring the

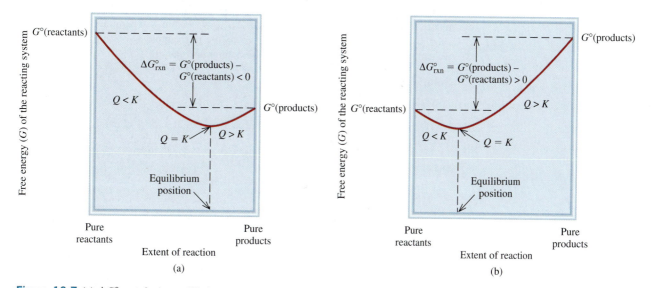

Figure 10.7 (a) $\Delta G^{\circ}_{\text{rxn}} < 0$. At equilibrium, products are favored over reactants. (b) $\Delta G^{\circ}_{\text{rxn}} > 0$. At equilibrium, reactants are favored over products. In both cases, the net reaction toward equilibrium is from left to right (products to reactants) if $Q < K$ and right to left (reactants to products) if $Q > K$. At equilibrium $Q = K$.

concentrations of all the reacting species. Consider, for example, the formation of nitric oxide from molecular nitrogen and molecular oxygen:

$$N_2(g) + O_2(g) \rightleftharpoons 2NO(g)$$

At 25°C, the equilibrium constant K is

$$K = \frac{P_{NO}^2}{P_{N_2}P_{O_2}} = 1.9 \times 10^{-31}$$

The very small value of K means that the concentration of NO at equilibrium will be exceedingly low, making it difficult to measure. In such a case, the equilibrium constant is more conveniently obtained from ΔG_{rxn}°. [As discussed in Section 8.6, ΔG_{rxn}° can be calculated from ΔH_{rxn}° and ΔS_{rxn}° or determined directly from Gibbs free energy of formation (ΔG_f°) data.] On the other hand, the equilibrium constant for the formation of hydrogen iodide from molecular hydrogen and molecular iodine is near unity at room temperature:

$$H_2(g) + I_2(g) \rightleftharpoons 2HI(g)$$

For this reaction, it is easier to measure K and then calculate ΔG_{rxn}° using Equation 10.12 than to measure ΔH_{rxn}° and ΔS_{rxn}° and use Equation 8.35.

Examples 10.8 to 10.10 show how to use Equations 10.11 and 10.12 to calculate K and ΔG_{rxn}°.

Example 10.8

Using data listed in Appendix 2, calculate the equilibrium constant K for the following reaction at 25°C:

$$2H_2O(l) \rightleftharpoons 2H_2(g) + O_2(g)$$

Strategy According to Equation 10.12, the equilibrium constant for the reaction is related to the standard free energy change: $\Delta G_{rxn}^\circ = -RT \ln K$. Therefore, we first need to calculate ΔG_{rxn}° by following the procedure in Example 8.9. Using that value, we can then calculate K. Remember to use appropriate units for T.

Solution According to Equation 8.34,

$$\Delta G_{rxn}^\circ = [2\Delta G_f^\circ(H_2) + \Delta G_f^\circ(O_2)] - [2\Delta G_f^\circ(H_2O)]$$
$$= [2(0 \text{ kJ mol}^{-1}) + (0 \text{ kJ mol}^{-1})] - [2(-237.1 \text{ kJ mol}^{-1})]$$
$$= +474.2 \text{ kJ mol}^{-1}$$

Using Equation 10.12,

$$\Delta G_{rxn}^\circ = -RT \ln K$$
$$\ln K = -\frac{\Delta G_{rxn}^\circ}{RT}$$
$$\ln K = -\frac{474.2 \text{ kJ mol}^{-1}}{(8.314 \text{ J mol}^{-1} \text{ K}^{-1})(298.15 \text{ K})} \times \frac{1000 \text{ J}}{1 \text{ kJ}} = -191.3$$
$$K = e^{-191.3} = 8.3 \times 10^{-84}$$

—Continued

Continued—

Comment This extremely small equilibrium constant is consistent with the fact that water does not decompose into hydrogen and oxygen gases at 25°C. Thus, a large positive ΔG_{rxn}° favors reactants over products at equilibrium.

Practice Exercise Calculate the equilibrium constant K at 25°C for the reaction

$$2O_3(g) \rightleftharpoons 3O_2(g)$$

Example 10.9

If solid AgCl is placed into water, some amount will dissolve according to the equilibrium

$$AgCl(s) \rightleftharpoons Ag^+(aq) + Cl^-(aq)$$

The equilibrium constant for this dissolution reaction (called the *solubility product, K_{sp}*) is 1.6×10^{-10} at 25°C. Calculate ΔG_{rxn}°.

Strategy According to Equation 10.12, the equilibrium constant for the reaction is related to the standard Gibbs free energy change: $\Delta G_{rxn}^{\circ} = -RT \ln K$.

Solution The solubility equilibrium for AgCl is

$$AgCl(s) \rightleftharpoons Ag^+(aq) + Cl^-(aq)$$

$$K = K_{sp} = \frac{a_{Ag^+} a_{Cl^-}}{a_{AgCl(s)}} \approx \frac{[Ag^+][Cl^-]}{1} = [Ag^+][Cl^-] = 1.6 \times 10^{-10}$$

Using Equation 10.12 we obtain

$$\Delta G_{rxn}^{\circ} = -(8.314 \text{ J mol}^{-1} \text{ K}^{-1})(298.15 \text{ K}) \ln(1.6 \times 10^{-10})$$
$$= 5.6 \times 10^4 \text{ J mol}^{-1} = 56 \text{ kJ mol}^{-1}$$

Comment Solubility products for ionic compounds are discussed in more detail in Chapter 11.

Practice Exercise Calculate ΔG_{rxn}° for the following process at 25°C:

$$BaF_2(s) \rightleftharpoons Ba^{2+}(aq) + 2F^-(aq)$$

The K_{sp} of BaF_2 is 1.7×10^{-6}.

Example 10.10

The equilibrium constant K for the reaction

$$N_2O_4(g) \rightleftharpoons 2NO_2(g)$$

is 0.115 at 298 K, which corresponds to a standard free-energy change of 5.36 kJ mol^{-1}. In a certain experiment, the initial pressures are $P_{NO_2} = 0.124$ bar and $P_{N_2O_4} = 0.459$ bar. Calculate ΔG_{rxn} for the reaction at these pressures, and predict the direction of the net reaction.

—Continued

Continued—

Strategy According to the information given in the problem statement, neither the reactant nor the product is at its standard state of 1 bar. To determine the direction of the net reaction, we need to calculate the free-energy change under nonstandard-state conditions (ΔG_{rxn}) using Equation 10.11 and the given ΔG_{rxn}° value. The partial pressures are expressed as dimensionless quantities in the reaction quotient Q.

Solution Equation 10.11 can be written as

$$\Delta G_{rxn} = \Delta G_{rxn}^{\circ} + RT \ln Q$$

$$= \Delta G_{rxn}^{\circ} + RT \ln \frac{P_{NO_2}^2}{P_{N_2O_4}}$$

$$= 5.40 \times 10^3 \, \text{J mol}^{-1} + (8.314 \, \text{J mol}^{-1} \, \text{K}^{-1})(298 \, \text{K}) \ln \frac{(0.123)^2}{0.459}$$

$$= 5.40 \times 10^3 \, \text{J mol}^{-1} - 8.46 \times 10^3 \, \text{J mol}^{-1}$$

$$= -3.06 \times 10^3 \, \text{J mol}^{-1} = -3.06 \, \text{kJ mol}^{-1}$$

Because $\Delta G_{rxn} < 0$, the net reaction proceeds from left to right to reach equilibrium.

Check Although $\Delta G_{rxn}^{\circ} > 0$, the reaction can be made to favor product formation initially by having a small concentration (pressure) of the product compared to that of the reactant. Confirm this prediction by showing that $Q < K$.

Practice Exercise The ΔG_{rxn}° for the reaction

$$H_2(g) + I_2(s) \rightleftharpoons 2HI(g)$$

is 3.40 kJ mol^{-1} at 25°C. In one experiment, the initial pressures are $P_{H_2} = 0.426$ bar, $P_{I_2} = 0.240$ bar, and $P_{HI} = 0.235$ bar. Calculate ΔG_{rxn} for the reaction, and predict the direction of the net reaction.

10.4 | The Response of an Equilibrium System to a Change in Conditions Can Be Determined Using Le Châtelier's Principle

There is a general rule that helps us to predict the direction in which an equilibrium reaction will move when a change in concentration, pressure, volume, or temperature occurs. The rule, known as *Le Châtelier's*[3] *principle,* states that *if an external stress is applied to a system at equilibrium, the system adjusts in such a way that the stress is partially offset as the system reaches a new equilibrium position.* The word *stress* here means a change in concentration, pressure, volume, or temperature that removes the system from the equilibrium state. We will use Le Châtelier's principle to assess the effects of such changes.

---

3. Henri Louis Le Châtelier (1850–1936). French chemist. Le Châtelier contributed to our understanding of metallurgy, cements, glasses, fuels, and explosives. He was also noted for his skills in industrial management.

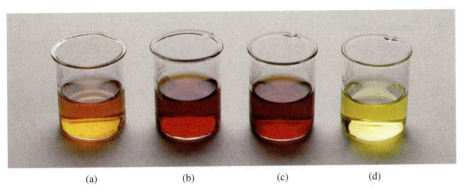

Figure 10.8 Effect of concentration change on the position of equilibrium. (a) An aqueous Fe(SCN)$_3$ solution. The color of the solution is due to both the red FeSCN^{2+} and the yellow Fe^{3+} species. (b) After the addition of some NaSCN to the solution in (a), the equilibrium shifts to the left. (c) After the addition of some Fe(NO$_3$)$_3$ to the solution in (a), the equilibrium shifts to the left as well. (d) After the addition of some H$_2$C$_2$O$_4$ to the solution in (a), the equilibrium shifts to the right. The yellow color is due to the presence of Fe(C$_2$O$_4$)$_3^{3-}$ ions.

(a)   (b)   (c)   (d)

Changes in Concentration

Iron(III) thiocyanate [Fe(SCN)$_3$] dissolves readily in water to give an equilibrium mixture of undissociated FeSCN^{2+} and Fe^{3+} and SCN$^-$ ions:

$$\text{FeSCN}^{2+} \rightleftharpoons \text{Fe}^{3+}(aq) + \text{SCN}^-(aq)$$
$$\quad\text{red}\qquad\qquad\text{pale yellow}\qquad\text{colorless}$$

The solution appears orange [Figure 10.8(a)] because hydrated FeSCN^{2+} ions are red and Fe^{3+}(aq) ions are pale yellow. What happens if we add some sodium thiocyanate (NaSCN) to this solution? In this case, the stress applied to the equilibrium system is an increase in the concentration of SCN$^-$ (from the dissociation of NaSCN). To offset this stress, some Fe^{3+} ions react with the added SCN$^-$ ions, and the equilibrium shifts from right to left:

$$\text{FeSCN}^{2+}(aq) \longleftarrow \text{Fe}^{3+}(aq) + \text{SCN}^-(aq)$$

Consequently, the red color of the solution deepens [Figure 10.8(b)]. Similarly, if we added iron(III) nitrate [Fe(NO$_3$)$_3$] to the original solution, the red color would also deepen [Figure 10.8(c)] because the additional Fe^{3+} ions [from Fe(NO$_3$)$_3$] would shift the equilibrium from right to left.

Now suppose we add some oxalic acid (H$_2$C$_2$O$_4$) to the original solution. Oxalic acid ionizes in water to form the oxalate ion (C$_2$O$_4^{2-}$), which binds strongly to the Fe^{3+} ions:

$$\text{Fe}^{3+} + 3\text{C}_2\text{O}_4^{2-} \longrightarrow \text{Fe}\,(\text{C}_2\text{O}_4)_3^{3-}$$

The formation of the stable yellow ion [Fe(C$_2$O$_4$)$_3^{3-}$] removes free Fe^{3+} ions in solution. Consequently, more FeSCN^{2+} units dissociate and the equilibrium shifts from left to right:

$$\text{FeSCN}^{2+}(aq) \longrightarrow \text{Fe}^{3+}(aq) + \text{SCN}^-(aq)$$

The red solution will turn yellow due to the formation of Fe(C$_2$O$_4$)$_3^{3-}$ ions along with the production of more Fe^{3+} ions [Figure 10.8(d)].

This experiment demonstrates that all reactants and products are present in the reacting system at equilibrium. Additionally, increasing the concentrations of the products (Fe^{3+} or SCN$^-$) shifts the equilibrium to the left, whereas decreasing the concentration of the product Fe^{3+} shifts the equilibrium to the right. These results are just as predicted by Le Châtelier's principle.

The effect of a change in concentration on the equilibrium position is shown in Example 10.11.

Oxalic acid is sometimes used to remove bathtub rings that consist of rust, or Fe$_2$O$_3$.

Le Châtelier's principle simply summarizes the observed behavior of equilibrium systems; therefore, it is incorrect to say that a given equilibrium shift occurs "because of" Le Châtelier's principle.

$$N_2 + 3H_2 \rightleftharpoons 2NH_3$$

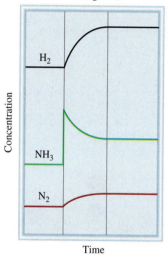

Figure 10.9 Changes in the concentrations of H_2, N_2, and NH_3 after the addition of NH_3 to the equilibrium mixture. The changes are consistent with Le Châtelier's principle.

Example 10.11

At 720°C, the equilibrium constant K_c for the reaction

$$N_2(g) + 3H_2(g) \rightleftharpoons 2NH_3(g)$$

is 2.37×10^{-3}. In a certain experiment, the equilibrium concentrations are $[N_2] = 0.683\ M$, $[H_2] = 8.80\ M$, and $[NH_3] = 1.05\ M$. Suppose some NH_3 is added to the mixture so that its concentration is increased to $3.65\ M$. (a) Use Le Châtelier's principle to predict the shift in direction of the net reaction to reach a new equilibrium. (b) Confirm your prediction by calculating the reaction quotient Q_c and comparing its value with K_c.

Strategy (a) Determine the stress that is applied to the system, and then apply Le Châtelier's principle to determine the direction in which the equilibrium will respond. (b) At the instant when some NH_3 is added, the system is no longer at equilibrium. Calculate Q_c for the reaction at this point, and compare its value to K_c to determine the direction of the net reaction to reach equilibrium.

Solution (a) The stress applied to the system is the addition of NH_3 (the reaction product). To offset this stress, some NH_3 must react to produce N_2 and H_2 (the reactants) until a new equilibrium is established. The net reaction therefore shifts from right to left:

$$N_2(g) + 3H_2(g) \longleftarrow 2NH_3(g)$$

(b) The reaction quotient is given by

$$Q_c = \frac{[NH_3]_0^2}{[N_2]_0[H_2]_0^3} = \frac{(3.65)^2}{(0.683)(8.80)^3} = 2.86 \times 10^{-2}$$

Because this value is greater than 2.37×10^{-3} (the value of K_c), the net reaction shifts from right to left (towards reactants) until Q_c equals K_c.

Figure 10.9 shows qualitatively the changes in concentrations of the reacting species.

Check Both the analysis using Le Châtelier's principle and the analysis comparing the values of Q_c and K_c give the same result: the reaction proceeds from left to right until equilibrium is reestablished.

Practice Exercise At 430°C, the equilibrium constant (K_P) for the reaction

$$2NO(g) + O_2(g) \rightleftharpoons 2NO_2(g)$$

is 1.5×10^5. In one experiment, the initial pressures of NO, O_2, and NO_2 at equilibrium are 1.1×10^{-2}, 1.1×10^{-3}, and 0.14 bar, respectively. Suppose enough O_2 is added to the system to increase the partial pressure of O_2 to 5.3×10^{-3} bar. (a) Use Le Châtelier's principle to predict which way the reaction will shift to restore equilibrium. (b) Confirm your answer in part (a) by calculating Q_P and comparing it to K_P.

Response of Equilibrium to Changes in Pressure

Changes in pressure ordinarily do not affect the concentrations of reacting species in condensed phases (e.g., in an aqueous solution) because liquids and solids are virtually incompressible. Concentrations of gases, on the other hand, are greatly affected by changes in pressure. Suppose, for example, that the equilibrium system

$$N_2O_4(g) \rightleftharpoons 2NO_2(g)$$

is in a cylinder fitted with a movable piston. What happens if we increase the pressure on the gases by pushing down on the piston at constant temperature? The equilibrium constant expression for this system (assuming ideal gas behavior) is

$$K_P = \frac{(P_{NO_2}/P_o)^2}{P_{N_2O_4}/P_o}$$

This expression contains only the partial pressures, not the total pressure. Using Dalton's law of partial pressures ($P_i = x_i P$) gives

$$K_P = \frac{(x_{NO_2}P/P_o)^2}{x_{N_2O_4}P/P_o} = \frac{x_{NO_2}^2}{x_{N_2O_4}}\left(\frac{P}{P_o}\right)$$

$$K_P = K_x \frac{P}{P_o} \tag{10.13}$$

where we have introduced a new kind of equilibrium constant, K_x, which is the equilibrium constant in terms of mole fractions. For a general ideal gas reaction, Equation 10.13 becomes

$$\boxed{K_P = K_x\left(\frac{P}{P_o}\right)^{\Delta n}} \tag{10.14}$$

where Δn is as previously defined in Equation 10.6:

$$\Delta n = \text{moles of gaseous products} - \text{moles of gaseous reactants}$$

For a given temperature, K_P is a constant, independent of total pressure. From Equation 10.14, therefore, the changes in concentration due to a change in the pressure (as measured by K_x) will depend upon the value of Δn for the specific reaction. We can identify the following three cases:

▸ **$\Delta n > 0$:** The reaction as written results in a net *increase* in the total number of molecules in going from reactant to product. For this case, increasing the pressure increases the term $(P/P_o)^{\Delta n}$, so for K_P to remain constant, K_x must decrease, resulting in the reaction shifting toward the reactants. (A smaller K_x means that the mole fractions of the products decrease and those of the reactants increase.) The reaction $N_2O_4(g) \rightleftharpoons 2NO_2(g)$ falls into this category (Figure 10.10).

▸ **$\Delta n < 0$:** The reaction as written results in a net *decrease* in the total number of molecules in going from reactant to product. For this case, increasing the pressure decreases the term $(P/P_o)^{\Delta n}$, so for K_P to remain constant, K_x must increase, resulting in the reaction shifting toward the products. (A larger K_x means that the mole fractions of the reactants decrease and those of the products increase.) An example of this type of reaction is

$$3H_2(g) + N_2(g) \rightleftharpoons 2NH_3(g)$$

▸ **$\Delta n = 0$:** The reaction as written results in no net change in the total number of molecules present as the reaction proceeds from left to right. For this case, $K_P = K_x$ and changing the pressure has no effect on the relative amount of reactants and products. An example of this type of reaction is

$$H_2(g) + Br_2(g) \rightleftharpoons 2HBr(g)$$

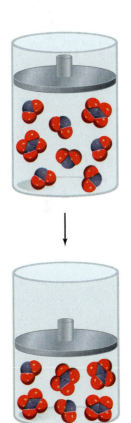

Figure 10.10 Qualitative illustration showing the effect of an increase in pressure on the $N_2O_4(g) \rightleftharpoons 2NO_2(g)$ equilibrium. The reaction shifts toward the reactants in order to reestablish equilibrium.

The response of a gas-phase equilibrium to pressure changes is an example of Le Châtelier's principle, where the stress is the change in pressure. When the pressure of a reaction mixture at equilibrium is increased, the system will respond by shifting the equilibrium in a direction that decreases the pressure, that is, in the direction that reduces the number of gas-phase molecules in the reaction mixture.

It is possible to change the pressure of a system without changing its volume. Suppose the NO_2-N_2O_4 system is contained in a stainless-steel vessel whose volume is constant. We can increase the total pressure in the vessel by adding an inert gas (e.g., helium) to the equilibrium system. Adding helium to the equilibrium mixture at constant volume increases the total gas pressure and decreases the mole fractions of both NO_2 and N_2O_4, but the partial pressure of each gas, given by the product of its mole fraction and total pressure, does not change. Thus, the presence of the inert gas does not affect the equilibrium.

Examples 10.12 and 10.13 show how pressure changes affect the equilibrium position.

Example 10.12

Consider the following equilibrium systems:

(a) $2PbS(s) + 3O_2(g) \rightleftharpoons 2PbO(s) + 2SO_2(g)$
(b) $PCl_5(g) \rightleftharpoons PCl_3(g) + Cl_2(g)$
(c) $H_2(g) + CO_2(g) \rightleftharpoons H_2O(g) + CO(g)$

In each case, predict the direction of the net reaction when the pressure of the system is increased (the volume is decreased) at constant temperature.

Strategy A change in pressure can affect the volume of a gas, but not the volume of a solid or liquid because solids and liquids are far less compressible than gases. The stress applied is an increase in pressure. According to Le Châtelier's principle, the system will adjust to partially offset this stress. In other words, the system will adjust to decrease the pressure. This can be achieved by shifting the equilibrium to the side of the equation that has fewer moles of gas.

Solution (a) Consider only the gaseous molecules. In the balanced equation, there are 3 moles of gaseous reactants and 2 moles of gaseous products. Therefore, $\Delta n = 2 - 3 = -1$ and the net reaction will shift toward the products (to the right) when the pressure is increased.

(b) The number of moles of products is 2 and that of reactants is 1; therefore, $\Delta n = 2 - 1 = +1$ and the net reaction will shift to the left, toward the reactant.

(c) The number of moles of products is equal to the number of moles of reactants ($\Delta n = 0$), so a change in pressure will have very little effect (if any) on the equilibrium.[4]

Check In each case, the prediction is consistent with Le Châtelier's principle.

Practice Exercise Consider the following equilibrium reaction involving nitrosyl chloride, nitric oxide, and molecular chlorine:

$$2NOCl(g) \rightleftharpoons 2NO(g) + Cl_2(g)$$

Predict the direction of the net reaction when the pressure of the system is decreased (the volume is increased) at constant temperature.

---

4. Equation 10.14 assumes that all gases behave ideally. For real gases, some pressure dependence on the gas-phase equilibrium, due to nonideality, is possible if $\Delta n = 0$, but such effects are usually very small.

Example 10.13

Consider a gas-phase reaction vessel in which only NO_2 and N_2O_4 are present. These species undergo the reversible reaction

$$N_2O_4(g) \rightleftharpoons 2NO_2(g)$$

The equilibrium constant K_P for this reaction is 0.115 at 298 K. Calculate the mole fractions of NO_2 and N_2O_4 if the total pressure is (a) 1.00 bar and (b) 5.00 bar. Are your conclusions consistent with Le Châtelier's principle? (For this particular reaction, it is unnecessary to know the initial amounts of each species.)

Strategy The mole fractions can be obtained if we know K_x. The relationship between K_P and K_x is given in Equation 10.14. Here $\Delta n = 1$.

Solution (a) Equation 10.14 with $\Delta n = 1$ gives

$$K_P = K_x \left(\frac{P}{P_o} \right) \quad \text{or} \quad K_x = K_P \left(\frac{P_o}{P} \right)$$

For $P = 1.00$ bar, $K_x = K_P = 0.115$. For this reaction,

$$K_x = \frac{x_{NO_2}^2}{x_{N_2O_4}} = \frac{x_{NO_2}^2}{1 - x_{NO_2}} = 0.115$$

where we have used the relationship $x_{N_2O_4} = 1 - x_{NO_2}$. Multiplying through by the denominator on both sides gives

$$x_{NO_2}^2 = 0.115 \, (1 - x_{NO_2})$$
$$x_{NO_2}^2 + 0.115 x_{NO_2} - 0.115 = 0$$

The solutions to this quadratic equation are $x_{NO_2} = -0.401$ and $+0.286$. Mole fractions must be positive, so the first solution is physically impossible. At $P = 1.00$ bar, therefore,

$$x_{NO_2} = 0.286$$
$$x_{N_2O_4} = 1 - x_{NO_2} = 0.714$$

(b) At $P = 5$ bar, we have

$$K_x = K_P \frac{P_o}{P} = 0.115 \times \frac{1 \text{ bar}}{5.00 \text{ bar}} = 0.0230$$

Proceeding as in (a), the quadratic equation is

$$x_{NO_2}^2 + 0.0230 x_{NO_2} - 0.0230 = 0$$

The solutions to this quadratic equation are $x_{NO_2} = -0.164$ and $+0.140$. Mole fractions must be positive, so the second solution is the plausible one. At $P = 5.00$ bar, therefore,

$$x_{NO_2} = 0.140$$
$$x_{N_2O_4} = 1 - x_{NO_2} = 0.860$$

According to Le Châtelier's principle, increasing the pressure from 1 bar to 5 bar will shift the reaction in the direction that contains fewer molecules (in this case, to the left), so the equilibrium mole fraction of N_2O_4 will increase, as is observed.

Check Make sure the mole fractions for the two species add to 1. If the mole fractions and pressures are substituted into Equation 10.14, the correct value of K_P (0.115) should be obtained.

—Continued

Continued—

Practice Exercise The equilibrium constant for the reaction

$$PCl_5(g) \rightleftharpoons PCl_3(g) + Cl_2(g)$$

is 0.411 at 500 K. If only pure $PCl_5(g)$ is present in the container initially, calculate the mole fractions of PCl_5, PCl_3, and Cl_2 at equilibrium at a pressure of (a) 1 bar and (b) 7 bar. Are your conclusions consistent with Le Châtelier's principle? (*Hint:* Think about the stoichiometric relationships between the mole fractions for this system.)

Changes in Temperature

A change in concentration, pressure, or volume may alter the equilibrium position (that is, the relative amounts of reactants and products), but it does not change the value of the equilibrium constant. Only a change in temperature can alter the equilibrium constant. To see why, consider again the following reaction:

$$N_2O_4(g) \rightleftharpoons 2NO_2(g)$$

The forward reaction is endothermic (absorbs heat, $\Delta H° > 0$)

$$\text{heat} + N_2O_4(g) \rightleftharpoons 2NO_2(g) \quad \Delta H° = 58.0 \text{ kJ mol}^{-1}$$

so the reverse reaction is exothermic (releases heat, $H° < 0$):

$$2NO_2(g) \rightleftharpoons N_2O_4(g) + \text{heat} \quad \Delta H° = -58.0 \text{ kJ mol}^{-1}$$

At equilibrium at a certain temperature, the heat effect is zero because there is no net reaction. If we treat heat as though it were a chemical reagent, then a rise in temperature "adds" heat to the system and a drop in temperature "removes" heat from the system. As with a change in any other parameter (concentration, pressure, or volume), the system shifts to reduce the effect of the change. Therefore, a temperature increase favors the endothermic direction (from left to right in the equilibrium equation) and a temperature decrease favors the exothermic direction (from right to left in the equilibrium equation). Consequently, the equilibrium constant increases when the system is heated and decreases when the system is cooled (Figure 10.11).

As another example, consider the equilibrium between the following ions:

$$CoCl_4^{2-} + 6H_2O \rightleftharpoons Co(H_2O)_6^{2+} + 4Cl^-$$
$$\text{blue} \qquad\qquad\qquad \text{pink}$$

Figure 10.11 (a) Two bulbs containing a mixture of NO_2 and N_2O_4 gases at equilibrium. (b) When one bulb is immersed in ice water (left), its color becomes lighter, indicating the formation of colorless N_2O_4 gas. When the other bulb is immersed in hot water, its color darkens, indicating an increase in NO_2.

(a)   (b)

Figure 10.12 (Left) Heating favors the formation of the blue $CoCl_4^{2-}$ ion. (Right) Cooling favors the formation of the pink $Co(H_2O)_6^{2+}$ ion.

The formation of $CoCl_4^{2-}$ is endothermic. On heating, the equilibrium shifts to the left and the solution turns blue. Cooling favors the exothermic reaction [the formation of $Co(H_2O)_6^{2+}$, and the solution turns pink (Figure 10.12). In summary, *a temperature increase favors an endothermic reaction, and a temperature decrease favors an exothermic reaction.*

We can quantify the change in the equilibrium constant due to changes in temperature using the relationship between ΔG°_{rxn} and K (Equation 10.12):

$$\Delta G^\circ_{rxn} = -RT \ln K$$

or

$$\ln K = -\frac{\Delta G^\circ}{RT}$$

If we subtract the values for $\ln K$ at two different temperature (T_1 and T_2), we get

$$\ln K_2 - \ln K_1 = \ln \frac{K_2}{K_1} = -\left(\frac{\Delta G^\circ_{T_2}}{RT_2} - \frac{\Delta G^\circ_{T_1}}{RT_1} \right)$$

Using the relation $\Delta G^\circ = \Delta H^\circ - T\Delta S^\circ$ (Equation 8.35), we have

$$\ln \frac{K_2}{K_1} = -\left(\frac{\Delta H^\circ_{T_2} - T_2\Delta S^\circ_{T_2}}{RT_2} - \frac{\Delta H^\circ_{T_1} - T_1\Delta S^\circ_{T_1}}{RT_1} \right)$$

Rearranging gives

$$\ln \frac{K_2}{K_1} = \frac{1}{R}\left(\frac{\Delta H^\circ_{T_2}}{T_1} - \frac{\Delta H^\circ_{T_1}}{T_2} + \Delta S^\circ_{T_2} - \Delta S^\circ_{T_1} \right) \qquad \textbf{(10.15)}$$

As discussed in Section 8.4, we can often assume, to a good approximation, that ΔH° and ΔS° are independent of temperature as long as the range of temperatures is not too great. Applying this assumption to Equation 10.15 gives

$$\ln \frac{K_2}{K_1} = \frac{\Delta H^\circ}{R}\left(\frac{1}{T_1} - \frac{1}{T_2} \right) \qquad \textbf{(10.16)}$$

This equation, which relates the equilibrium constant at one temperature to that at another temperature assuming that $\Delta H°$ and $\Delta S°$ are approximately constant over the temperature range, is called the *van't Hoff equation.*

Equation 10.16 is a mathematical statement of Le Châtelier's principle for the change in the equilibrium constant with changes in temperature. Suppose that $T_2 > T_1$. If the reaction is endothermic ($\Delta H° > 0$), the right-hand side of the equation will be positive and $K_2 > K_1$, that is, the reaction shifts toward the products in an endothermic reaction, consistent with Le Châtelier's principle. If the reaction is exothermic ($\Delta H° < 0$), the right-hand side of the equation will be negative and $K_2 < K_1$, that is, the reaction shifts toward the reactants in an exothermic reaction.

Example 10.14 shows how to use Equation 10.16 to determine the change in the equilibrium constant with a change in temperature.

Example 10.14

The equilibrium constant for the reaction

$$3H_2(g) + N_2(g) \rightleftharpoons 2NH_3(g)$$

is 5.58×10^5 at 25°C. Determine the equilibrium constant for this reaction at 200°C.

Strategy We are given K_1, T_1, and T_2, so we need to know $\Delta H°$ for this reaction in order to use Equation 10.16 to determine K_2. We can determine $\Delta H°$ using the data in Appendix 2. Don't forget to convert the temperatures from degrees Celsius to kelvins before using Equation 10.16.

Solution The standard enthalpy change for this reaction ($\Delta H°$) can be found from the standard enthalpies of formation for the reactants and products.

$$\Delta H° = 2\Delta H_f°(NH_3) - 3\Delta H_f°(H_2) - \Delta H_f°(N_2)$$
$$= 2(-45.9 \text{ kJ mol}^{-1}) - 3(0) - (0) = -91.8 \text{ kJ mol}^{-1}$$

For this problem $T_1 = 298.15$ K and $T_2 = 473.15$ K. Substitution into Equation 10.16 gives

$$\ln \frac{K_2}{K_1} = \frac{\Delta H°}{R}\left(\frac{1}{T_1} - \frac{1}{T_2}\right)$$
$$= \frac{-91.8 \times 10^3 \text{ J mol}^{-1}}{8.314 \text{ J mol}^{-1} \text{ K}^{-1}}\left(\frac{1}{298.15 \text{ K}} - \frac{1}{473.15 \text{ K}}\right)$$
$$= -13.7$$

The equilibrium constant K_1 is 5.58×10^5 at 25°C (T_1), so

$$\frac{K_2}{K_1} = e^{-13.7} = 1.12 \times 10^{-6}$$
$$K_2 = (1.12 \times 10^{-6})K_1 = (1.12 \times 10^{-6})(5.58 \times 10^5) = 0.625$$

Thus, the equilibrium constant for this reaction decreases with increasing temperature; this is consistent with Le Châtelier's principle for an exothermic reaction.

Practice Exercise The equilibrium constant for the reaction $N_2O_4(g) \rightleftharpoons 2NO_2(g)$ is 0.115 at 298 K. At what temperature would the equilibrium constant be equal to 1?

Life at High Altitudes and Hemoglobin Production

In the human body, countless chemical equilibria must be maintained to ensure physiological well-being. If environmental conditions change, the body must adapt to keep functioning. The consequences of a sudden change in altitude dramatize this fact. Flying from San Francisco, which is at sea level, to Mexico City, where the elevation is 2.3 km (1.4 mi), or scaling a 3-km mountain in two days can cause headache, nausea, extreme fatigue, and other discomforts. These conditions are all symptoms of hypoxia, a deficiency in the amount of oxygen reaching body tissues. In serious cases, the victim may slip into a coma and die if not treated quickly enough. A person living at a high altitude for weeks or months, however, gradually recovers from altitude sickness and adjusts to the low oxygen content in the atmosphere, so that he or she can function normally.

The combination of oxygen with the hemoglobin (Hb) molecule, which carries oxygen through the blood, is a complex reaction, but for our purposes it can be represented by the following simplified equation:

$$Hb(aq) + O_2(g) \rightleftharpoons HbO_2(aq)$$

where HbO_2 is oxyhemoglobin, the hemoglobin-oxygen complex that actually transports oxygen to tissues. The equilibrium constant for this reaction is

$$K = \frac{[HbO_2]}{[Hb]\, P_{O_2}}$$

At an altitude of 3 km, the partial pressure of oxygen is only about 0.14 bar, compared with 0.20 bar at sea level. According to Le Châtelier's principle, a decrease in the oxygen partial pressure will shift the hemoglobin-oxyhemoglobin equilibrium from right to left. This change depletes the supply of oxyhemoglobin, causing hypoxia. Given enough time, the body copes with this condition by producing more hemoglobin molecules. The equilibrium will then gradually shift back toward the formation of oxyhemoglobin. It takes two to three weeks for the increase in hemoglobin production to meet the body's basic needs adequately. Studies show that long-time residents of high altitude areas have high hemoglobin levels in their blood—sometimes as much as 50% more than individuals living at sea level!

Mountaineers need weeks or even months to become acclimatized to the reduced level of oxygen in the air before scaling summits such as Mount Everest.

Summary of Facts and Concepts

Section 10.1

▶ Chemical equilibrium is a dynamic process in which there is no *net* change in the concentrations of reactants and products.

▶ For the general chemical reaction

$$aA + bB \rightleftharpoons cC + dD$$

the activities of reactants and products at equilibrium are related by the equilibrium constant expression (Equation 10.2).

▶ In many important cases, the activity of a species can be approximated by a simple expression

　—*Gases at low to moderate pressure:* $a_i \approx P_i/P°$ (Equation 10.3), where P_i is the partial pressure of the gas and $P°$ is the standard pressure (1 bar).

　—*Dilute solutes:* $a_i \approx [i]/c°$ (Equation 10.4), where $[i]$ is the concentration of species i and $c°$ is the standard concentration (1 mol L^{-1}).

　—*Pure solids and liquids:* $a_i = 1$.

▶ A chemical equilibrium process in which all reactants and products are in the same phase is homogeneous. If the reactants and products are not all in the same phase, the equilibrium is heterogeneous.

▶ If a reaction can be expressed as the sum of two or more reactions, the equilibrium constant for the overall reaction is given by the product of the equilibrium constants of the individual reactions.

▶ The value of K depends on how the chemical equation is balanced, and the equilibrium constant for the reverse of a particular reaction is the reciprocal of the equilibrium constant of that reaction.

▶ The equilibrium constant is the ratio of the rate constant for the forward reaction to that for the reverse reaction.

Section 10.2

▶ The reaction quotient Q has the same form as the equilibrium constant, but it applies to a reaction that may not be at equilibrium. If $Q > K$, the reaction will proceed from right to left to achieve equilibrium. If $Q < K$, the reaction will proceed from left to right to achieve equilibrium.

▶ The equilibrium constant can be used to determine the equilibrium concentrations (or pressures) if the initial concentrations (or pressures) are known.

Section 10.3

▶ The form of the equilibrium constant is a consequence of the Gibbs free energy of mixing

▶ The equilibrium constant of a reaction and the standard Gibbs free energy change are related by the equation $\Delta G^\circ = -RT \ln K$ (Equation 10.12).

Section 10.4

▶ Le Châtelier's principle states that if an external stress is applied to a system at chemical equilibrium, the system will adjust to partially offset the stress.

▶ Only a change in temperature changes the value of the equilibrium constant for a particular reaction. Changes in concentration, pressure, or volume may change the equilibrium concentrations of reactants and products.

▶ The change in the equilibrium constant with temperature is determined by the standard enthalpy change for a reaction using the van't Hoff equation (Equation 10.16).

Key words

activity, p. 514
chemical equilibrium, p. 512
concentration equilibrium
    constant, p. 518

equilibrium
    constant, p. 513
heterogeneous
    equilibrium, p. 519

homogeneous
    equilibrium, p. 516
law of mass action, p. 513
Le Châlelier's principle, p. 536

reaction quotient, p. 525
reversible reactions, p. 512
van't Hoff equation, p. 544

Problems

The Equilibrium Constant Governs the Concentration of Reactants and Products at Equilibrium

10.1 The equilibrium constant for the reaction A $\rightleftharpoons$ B is $K_c = 10$ at a certain temperature. (a) Starting with only reactant A, which of the following diagrams best represents the system at equilibrium? (b) Which of the diagrams best represents the system at equilibrium if $K_c = 0.10$? Explain why you can calculate K_c in each case without knowing the volume of the container. The gray spheres represent the A molecules and the green spheres represent the B molecules.

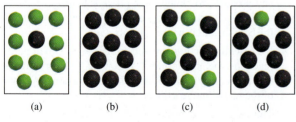

(a)   (b)   (c)   (d)

10.2 The following diagrams represent the equilibrium state for three different reactions of the type A + X $\rightleftharpoons$ AX (X = B, C, or D):

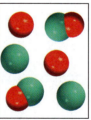
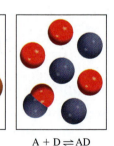

A + B $\rightleftharpoons$ AB   A + C $\rightleftharpoons$ AC   A + D $\rightleftharpoons$ AD

(a) Which reaction has the largest equilibrium constant?

(b) Which reaction has the smallest equilibrium constant?

10.3 The equilibrium constant K_c for the reaction

$$2HCl(g) \rightleftharpoons H_2(g) + Cl_2(g)$$

is 4.17×10^{-34} at 25°C. What is the equilibrium constant for the reaction

$$H_2(g) + Cl_2(g) \rightleftharpoons 2HCl(g)$$

at the same temperature?

10.4 Consider the following equilibrium process at 700°C:

$$2H_2(g) + S_2(g) \rightleftharpoons 2H_2S(g)$$

Analysis shows that there are 2.50 moles of H_2, 1.35×10^{-5} mole of S_2, and 8.70 moles of H_2S present in a 12.0-L flask. Calculate the equilibrium constants K_P and K_c for the reaction.

10.5 What is K_P at 1273°C for the reaction

$$2CO(g) + O_2(g) \rightleftharpoons 2CO_2(g)$$

if K_c is 2.24×10^{-22} at the same temperature?

10.6 The equilibrium constant K_P for the reaction

$$2SO_3(g) \rightleftharpoons 2SO_2(g) + O_2(g)$$

is 1.8×10^{-5} at 350°C. What is K_c for this reaction?

10.7 Consider the following reaction:

$$N_2(g) + O_2(g) \rightleftharpoons 2NO(g)$$

If the equilibrium partial pressures of N_2, O_2, and NO at 2200°C are 0.15, 0.33, and 0.051 bar, respectively, what is K_P?

10.8 A reaction vessel contains NH_3, N_2, and H_2 at equilibrium at a certain temperature. The equilibrium concentrations are $[NH_3] = 0.25\,M$, $[N_2] = 0.11\,M$, and $[H_2] = 1.91\,M$. Calculate the equilibrium constants K_c and K_P for the synthesis of ammonia if the reaction is represented as
(a) $N_2(g) + 3H_2(g) \rightleftharpoons 2NH_3(g)$
(b) $\frac{1}{2}N_2(g) + \frac{3}{2}H_2(g) \rightleftharpoons NH_3(g)$

10.9 The equilibrium constant K_c for the reaction $I_2(g) \rightleftharpoons 2I(g)$ is 3.8×10^{-5} at 727°C. Calculate K_c and K_P for the equilibrium $2I(g) \rightleftharpoons I_2(g)$ at the same temperature.

10.10 At equilibrium, the pressure of the reacting mixture

$$CaCO_3(s) \rightleftharpoons CaO(s) + CO_2(g)$$

is 0.106 bar at 350°C. Calculate K_P and K_c for this reaction.

10.11 The equilibrium constant K_P for the reaction

$$PCl_5(g) \rightleftharpoons PCl_3(g) + Cl_2(g)$$

is 1.05 at 250°C. The reaction starts with a mixture of PCl_5, PCl_3, and Cl_2 at pressures of 0.17, 0.226, and 0.112 bar, respectively, at 250°C. When the mixture comes to equilibrium at that temperature, which pressures will have decreased and which will have increased? Explain why.

10.12 Ammonium carbamate ($NH_4CO_2NH_2$) decomposes as follows:

$$NH_4CO_2NH_2(s) \rightleftharpoons 2NH_3(g) + CO_2(g)$$

Starting with only the solid, it is found that at 40°C the total gas pressure (NH_3 and CO_2) is 0.367 bar. Calculate the equilibrium constant K_P.

10.13 Consider the following reaction at 1600°C:

$$Br_2(g) \rightleftharpoons 2Br(g)$$

When 1.05 moles of Br_2 are put in a 0.980-L flask, 1.20 percent of the Br_2 undergoes dissociation. Calculate the equilibrium constants K_P and K_c for the reaction.

10.14 Pure phosgene gas ($COCl_2$), 3.00×10^{-2} mol, was placed in a 1.50-L container. It was heated to 800 K, and at equilibrium the pressure of CO was found to be 0.503 bar. Calculate the equilibrium constant K_P for the reaction

$$CO(g) + Cl_2(g) \rightleftharpoons COCl_2(g)$$

10.15 Consider the equilibrium

$$2NOBr(g) \rightleftharpoons 2NO(g) + Br_2(g)$$

If nitrosyl bromide (NOBr) is 34 percent dissociated at 25°C and the total pressure is 0.25 bar, calculate K_P and K_c for the dissociation at this temperature.

10.16 A 2.50-mole quantity of NOCl was initially in a 1.50-L reaction chamber at 400°C. After equilibrium was established, it was found that 28.0 percent of the NOCl had dissociated:

$$2NOCl(g) \rightleftharpoons 2NO(g) + Cl_2(g)$$

Calculate the equilibrium constant K_P for the reaction.

10.17 The following equilibrium constants have been determined for hydrosulfuric acid at 25°C:

$$H_2S(aq) \rightleftharpoons H^+(aq) + HS^-(aq)$$
$$K = 9.5 \times 10^{-8}$$
$$HS^-(aq) \rightleftharpoons H^+(aq) + S^{2-}(aq)$$
$$K = 1.0 \times 10^{-19}$$

Calculate the equilibrium constant for the following reaction at the same temperature:

$$H_2S(aq) \rightleftharpoons 2H^+(aq) + S^{2-}(aq)$$

10.18 The following equilibrium constants have been determined for oxalic acid at 25°C:

$$C_2H_2O_4(aq) \rightleftharpoons H^+(aq) + C_2HO_4^-(aq)$$
$$K = 6.5 \times 10^{-2}$$
$$C_2HO_4^-(aq) \rightleftharpoons H^+(aq) + C_2O_4^{2-}(aq)$$
$$K = 6.1 \times 10^{-5}$$

Calculate the equilibrium constant for the following reaction at the same temperature:

$$C_2H_2O_4(aq) \rightleftharpoons 2H^+(aq) + C_2O_4^{2-}(aq)$$

10.19 The following equilibrium constants were determined at 1123 K:

$$C(s) + CO_2(g) \rightleftharpoons 2CO(g)$$
$$K_P = 1.3 \times 10^{14}$$
$$CO(g) + Cl_2(g) \rightleftharpoons COCl_2(g)$$
$$K_P = 6.0 \times 10^{-3}$$

Write the equilibrium constant expression K_P, and calculate the equilibrium constant at 1123 K for

$$C(s) + CO_2(g) + 2Cl_2(g) \rightleftharpoons 2COCl_2(g)$$

10.20 At a certain temperature, the following reactions have the constants shown:

$$S(s) + O_2(g) \rightleftharpoons SO_2(g) \quad K_c = 4.2 \times 10^{52}$$
$$2S(s) + 3O_2(g) \rightleftharpoons 2SO_3(g) \quad K_c = 10.8 \times 10^{128}$$

Calculate the equilibrium constant K_c for the following reaction at that temperature:

$$2SO_2(g) + O_2(g) \rightleftharpoons 2SO_3(g)$$

The Equilibrium Constant Can Be Used to Predict the Direction and Equilibrium Concentrations of a Chemical Reaction

10.21 The equilibrium constant K_P for the reaction

$$2SO_2(g) + O_2(g) \rightleftharpoons 2SO_3(g)$$

is 5.53×10^4 at 350°C. The initial pressures of SO_2 and O_2 in a mixture are 0.354 and 0.772 bar, respectively, at 350°C. When the mixture equilibrates, is the total pressure less than or greater than the sum of the initial pressures (1.127 bar)?

10.22 For the synthesis of ammonia

$$N_2(g) + 3H_2(g) \rightleftharpoons 2NH_3(g)$$

the equilibrium constant K_c at 375°C is 1.2. Starting with $[H_2]_0 = 0.76\ M$, $[N_2]_0 = 0.60\ M$, and $[NH_3]_0 = 0.48\ M$, which gases will have increased in concentration and which will have decreased in concentration when the mixture comes to equilibrium?

10.23 For the reaction $H_2(g) + CO_2(g) \rightleftharpoons H_2O(g) + CO(g)$ at 700°C, $K_c = 0.534$. Calculate the number of moles of H_2 that are present at equilibrium if a mixture of 0.300 mole each of CO and H_2O is heated to 700°C in a 10.0-L container.

10.24 At 1000 K, a sample of pure NO_2 gas decomposes:

$$2NO_2(g) \rightleftharpoons 2NO(g) + O_2(g)$$

The equilibrium constant K_P is 160. Analysis shows that the partial pressure of O_2 is 0.25 bar at equilibrium. Calculate the pressure of NO and NO_2 in the mixture.

10.25 The equilibrium constant K_c for the reaction

$$H_2(g) + Br_2(g) \rightleftharpoons 2HBr(g)$$

is 2.18×10^6 at 730°C. Starting with 3.20 moles of HBr in a 12.0-L reaction vessel, calculate the concentrations of H_2, Br_2, and HBr at equilibrium.

10.26 The dissociation of molecular iodine into iodine atoms is represented as

$$I_2(g) \rightleftharpoons 2I(g)$$

At 1000 K, the equilibrium constant K_c for the reaction is 3.80×10^{-5}. Suppose you start with 0.0456 mole of I_2 in a 2.30-L flask at 1000 K. What are the concentrations of the gases at equilibrium?

10.27 The equilibrium constant K_c for the decomposition of phosgene ($COCl_2$) is 4.69×10^{-3} at 527°C:

$$COCl_2(g) \rightleftharpoons CO(g) + Cl_2(g)$$

Calculate the equilibrium partial pressure of all the components, starting with pure phosgene at 0.785 bar.

10.28 Consider the following equilibrium process at 686°C:

$$CO_2(g) + H_2(g) \rightleftharpoons CO(g) + H_2O(g)$$

The equilibrium concentrations of the reacting species are $[CO] = 0.050\ M$, $[H_2] = 0.045\ M$, $[CO_2] = 0.086\ M$, and $[H_2O] = 0.040\ M$. (a) Calculate K_c for the reaction at 686°C. (b) If we add CO_2 to increase its concentration to 0.50 mol L^{-1}, what will the concentrations of all the gases be when equilibrium is reestablished?

10.29 Consider the heterogeneous equilibrium process:

$$C(s) + CO_2(g) \rightleftharpoons 2CO(g)$$

At 700°C, the total pressure of the system is found to be 4.56 bar. If the equilibrium constant K_P is 1.54, calculate the equilibrium partial pressures of CO_2 and CO.

10.30 The equilibrium constant K_c for the reaction

$$H_2(g) + CO_2(g) \rightleftharpoons H_2O(g) + CO(g)$$

is 4.2 at 1650°C. Initially 0.80 mol H_2 and 0.80 mol CO_2 are injected into a 5.0-L flask. Calculate the concentration of each species at equilibrium.

Thermodynamics and the Equilibrium Constant

10.31 Calculate K_P for the following reaction at 25°C:

$$H_2(g) + I_2(g) \rightleftharpoons 2HI(g) \quad \Delta G^\circ_{rxn} = 2.60\ \text{kJ mol}^{-1}$$

10.32 For the autoionization of water at 25°C,

$$H_2O(l) \rightleftharpoons H^+(aq) + OH^-(aq)$$

K is 1.0×10^{-14}. What is ΔG°_{rxn} for the process?

10.33 For the following reaction at 25°C:

$$Fe(OH)_2(s) \rightleftharpoons Fe^{2+}(aq) + 2OH^-(aq)$$

K is 1.6×10^{-14}. Calculate ΔG°_{rxn} for the reaction.

10.34 Using data from Appendix 2, calculate ΔG°_{rxn} and K for the following equilibrium reaction at 25°C.

$$2H_2O(g) \rightleftharpoons 2H_2(g) + O_2(g)$$

10.35 (a) Using the data from Appendix 2, calculate ΔG°_{rxn} and K for the following equilibrium reaction at 25°C.

$$PCl_5(g) \rightleftharpoons PCl_3(g) + Cl_2(g)$$

(b) Calculate ΔG_{rxn} for this reaction if the partial pressures of the initial mixture are $P_{PCl_5} = 0.0029$ bar, $P_{PCl_3} = 0.27$ bar, and $P_{Cl_2} = 0.40$ bar.

10.36 The equilibrium constant K for the reaction

$$H_2(g) + CO_2(g) \rightleftharpoons H_2O(g) + CO(g)$$

is 4.40 at 2000 K. (a) Calculate ΔG°_{rxn} for the reaction. (b) Calculate ΔG_{rxn} for the reaction when the partial pressures are $P_{H_2} = 0.25$ bar, $P_{CO_2} = 0.78$ bar, $P_{H_2O} = 0.67$ bar, and $P_{CO} = 1.21$ bar.

10.37 Consider the decomposition of calcium carbonate:

$$CaCO_3(s) \rightleftharpoons CaO(s) + CO_2(g)$$

Calculate the pressure in bar of CO_2 in an equilibrium process (a) at 25°C and (b) at 800°C. Assume that $\Delta H^\circ_{rxn} = 177.8$ kJ mol^{-1} and $\Delta S^\circ_{rxn} = 160.5$ J mol^{-1} K^{-1} for the temperature range.

10.38 The equilibrium constant K for the reaction

$$CO(g) + Cl_2(g) \rightleftharpoons COCl_2(g)$$

is 5.62×10^{35} at 25°C. Calculate ΔG°_f for $COCl_2$ at 25°C.

10.39 Using the data from Appendix 2, calculate ΔG°_{rxn} for the process

$$C(\text{diamond}) \rightleftharpoons C(\text{graphite})$$

Is the formation of graphite from diamond favored at 25°C? If so, why is it that diamonds do not become graphite on standing?

10.40 At 25°C, ΔG°_{rxn} for the process

$$H_2O(l) \rightleftharpoons H_2O(g)$$

is 8.6 kJ mol^{-1}. Calculate the vapor pressure of water at this temperature.

The Response of an Equilibrium System to a Change in Conditions Can Be Determined Using Le Châtelier's Principle

10.41 Consider the following equilibrium system involving SO_2, Cl_2, and SO_2Cl_2 (sulfuryl dichloride):

$$SO_2(g) + Cl_2(g) \rightleftharpoons SO_2Cl_2(g)$$

Predict how the equilibrium position would change if (a) Cl_2 gas were added to the system; (b) SO_2Cl_2 were removed from the system; or (c) SO_2 were removed from the system. The temperature remains constant.

10.42 Heating solid sodium bicarbonate in a closed vessel establishes the following equilibrium:

$$2NaHCO_3(s) \rightleftharpoons Na_2CO_3(s) + H_2O(g) + CO_2(g)$$

What would happen to the equilibrium position if (a) some of the CO_2 were removed from the system; (b) some solid Na_2CO_3 were added to the system; or (c) some of the solid $NaHCO_3$ were removed from the system? The temperature remains constant.

10.43 Consider the following equilibrium systems:

(a) $\quad$ A $\rightleftharpoons$ 2B $\quad \Delta H^\circ = 20.0$ kJ mol^{-1}
(b) A + B $\rightleftharpoons$ C $\quad \Delta H^\circ = -5.4$ kJ mol^{-1}
(c) $\quad$ A $\rightleftharpoons$ B $\quad \Delta H^\circ = 0.0$ kJ mol^{-1}

Predict the change in the equilibrium constant K_P that would occur in each case if the temperature of the reacting system were raised.

10.44 What effect does an increase in pressure have on each of the following systems at equilibrium? The temperature is kept constant and, in each case, the reactants are in a cylinder fitted with a movable piston.

(a) $A(s) \rightleftharpoons 2B(s)$
(b) $2A(l) \rightleftharpoons B(l)$
(c) $A(s) \rightleftharpoons B(g)$
(d) $A(g) \rightleftharpoons B(g)$
(e) $A(g) \rightleftharpoons 2B(g)$

10.45 Consider the equilibrium

$$2I(g) \rightleftharpoons I_2(g)$$

What would be the effect on the position of equilibrium of (a) increasing the total pressure on the system by decreasing its volume; (b) adding I_2 to the reaction mixture; or (c) decreasing the temperature?

10.46 Consider the following equilibrium process:

$$PCl_5(g) \rightleftharpoons PCl_3(g) + Cl_2(g)$$
$$\Delta H^\circ = 92.5 \text{ kJ mol}^{-1}$$

Predict the direction of the shift in equilibrium when (a) the temperature is raised; (b) more chlorine gas is added to the reaction mixture; (c) some PCl_3 is removed from the mixture; or (d) the pressure on the gases is increased.

10.47 Consider the reaction

$$2SO_2(g) + O_2(g) \rightleftharpoons 2SO_3(g)$$
$$\Delta H° = -197.8 \text{ kJ mol}^{-1}$$

Comment on the changes in the concentrations of SO_2, O_2, and SO_3 at equilibrium if we were to (a) increase the temperature; (b) increase the pressure; (c) increase SO_2; or (d) add helium at constant volume.

10.48 Consider the gas-phase reaction

$$2CO(g) + O_2(g) \rightleftharpoons 2CO_2(g)$$

Predict the shift in the equilibrium position when helium gas is added to the equilibrium mixture (a) at constant pressure and (b) at constant volume.

10.49 Consider the following equilibrium reaction in a closed container:

$$CaCO_3(s) \rightleftharpoons CaO(s) + CO_2(g)$$

What will happen if (a) the volume is increased; (b) some CaO is added to the mixture; (c) some $CaCO_3$ is removed; (d) some CO_2 is added to the mixture; (e) a few drops of aN NaOH solution are added to the mixture; (f) a few drops of a HCl solution are added to the mixture (ignore the reaction between CO_2 and water); or (g) the temperature is increased?

10.50 At a certain temperature and a total pressure of 1.2 bar, the partial pressures of an equilibrium mixture

$$2A(g) \rightleftharpoons B(g)$$

are $P_A = 0.60$ bar and $P_B = 0.60$ bar. (a) Calculate the K_P for the reaction at this temperature. (b) If the total pressure were increased to 1.5 bar, what would be the partial pressures of A and B at equilibrium?

10.51 When heated calcium carbonate decomposes according to the reaction

$$CaCO_3(s) \rightleftharpoons CaO(s) + CO_2(g)$$

The rate of decomposition is slow until the partial pressure of CO_2 reaches 1 bar. Using the data in Appendix 2, determine the temperature at which the equilibrium partial pressure of CO_2 for this reaction is equal to 1 bar. Assume that $\Delta S°$ and $\Delta H°$ for this reaction are both temperature independent.

10.52 Use the data in Appendix 2 to calculate the equilibrium constant, K_P, for the reaction

$$2SO_2(g) + O_2(g) \rightleftharpoons 2SO_3(g)$$

at (a) 25°C and (b) 75°C. State any assumptions that you make.

10.53 The following table gives the equilibrium constant (K_P) for the reaction

$$2NO(g) + O_2(g) \rightleftharpoons 2NO_2(g)$$

at various temperatures:

K_P	138	5.12	0.436	0.0626	0.0130
T(K)	600	700	800	900	1000

Determine graphically the standard enthalpy change for this reaction ($\Delta H°$). State any assumptions that you make.

10.54 For the reaction

$$CO_2(g) + H_2(g) \rightleftharpoons CO(g) + H_2O(g)$$

the equilibrium constant K_P is 0.534 at 960 K and 1.571 at 1260 K. From these data, determine the standard enthalpy change ($\Delta H°$) for this reaction. State any assumptions that you make.

Additional Problems

10.55 Consider the statement: "The equilibrium constant of a reacting mixture of solid NH_4Cl and gaseous NH_3 and HCl is 0.316." List three important pieces of information that are missing from this statement.

10.56 Pure nitrosyl chloride (NOCl) gas was heated to 240°C in a 1.00-L container. At equilibrium, the total pressure was 1.01 bar and the NOCl pressure was 0.65 bar.

$$2NOCl(g) \rightleftharpoons 2NO(g) + Cl_2(g)$$

(a) Calculate the partial pressures of NO and Cl_2 in the system. (b) Calculate the equilibrium constant K_P.

10.57 The equilibrium constant K_P for the formation of the air pollutant nitric oxide (NO) in an automobile engine at 530°C is 2.9×10^{-11}:

$$N_2(g) + O_2(g) \rightleftharpoons 2NO(g)$$

(a) Calculate the partial pressure of NO under these conditions if the partial pressures of nitrogen and oxygen are 3.0 and 0.012 bar, respectively. (b) Repeat the calculation for atmospheric conditions where the partial pressures of nitrogen and oxygen are 0.79 and 0.21 bar, respectively, and the temperature is 25°C. (The K_P for the reaction is 4.0×10^{-31} at this temperature.) (c) Is the formation of NO endothermic or exothermic? (d) What natural phenomenon promotes the formation of NO? Why?

10.58 Baking soda (sodium bicarbonate) undergoes thermal decomposition as follows:

$$2NaHCO_3(s) \rightleftharpoons Na_2CO_3(s) + CO_2(g) + H_2O(g)$$

Would we obtain more CO_2 and H_2O by adding extra baking soda to the reaction mixture in (a) a closed vessel or (b) an open vessel?

10.59 Consider the following reaction at equilibrium:

$$A(g) \rightleftharpoons 2B(g)$$

From the following data, calculate the equilibrium constant (both K_P and K_c) at each temperature. Is the reaction endothermic or exothermic?

Temperature (°C)	[A] (*M*)	[B] (*M*)
200	0.0125	0.843
300	0.171	0.764
400	0.250	0.724

10.60 The equilibrium constant K_P for the reaction

$$2H_2O(g) \rightleftharpoons 2H_2(g) + O_2(g)$$

is 2×10^{-42} at 25°C. (a) What is K_c for the reaction at the same temperature? (b) The very small value of K_P (and K_c) indicates that the reaction overwhelmingly favors the formation of water molecules. Explain why, despite this fact, a mixture of hydrogen and oxygen gases can be kept at room temperature without any change.

10.61 Consider the following reacting system:

$$2NO(g) + Cl_2(g) \rightleftharpoons 2NOCl(g)$$

What combination of temperature and pressure would maximize the yield of nitrosyl chloride (NOCl)? [*Hint:* $\Delta H_f^\circ(NOCl) = 51.7$ kJ mol^{-1}. You will also need to consult Appendix 2.]

10.62 The decomposition of ammonium hydrogen sulfide

$$NH_4HS(s) \rightleftharpoons NH_3(g) + H_2S(g)$$

is an endothermic process. A 6.1589-g sample of the solid is placed in an evacuated 4.000-L vessel at exactly 24°C. After equilibrium has been established, the total pressure inside is 0.732 bar. Some solid NH₄HS remains in the vessel. (a) What is the K_P for the reaction? (b) What percentage of the solid has decomposed? (c) If the volume of the vessel were doubled at constant temperature, what would happen to the amount of solid in the vessel?

10.63 Consider the reaction

$$2NO(g) + O_2(g) \rightleftharpoons 2NO_2(g)$$

At 430°C, an equilibrium mixture consists of 0.020 mole of O₂, 0.040 mole of NO, and 0.96 mole of NO₂. Calculate K_P for the reaction, given that the total pressure is 0.20 bar.

10.64 When heated, ammonium carbamate decomposes as follows:

$$NH_4CO_2NH_2(s) \rightleftharpoons 2NH_3(g) + CO_2(g)$$

At a certain temperature, the equilibrium pressure of the system is 0.322 bar. Calculate K_P for the reaction.

10.65 A mixture of 0.47 mole of H₂ and 3.59 moles of HCl is heated to 2800°C. Calculate the equilibrium partial pressures of H₂, Cl₂, and HCl if the total pressure is 2.03 bar. For the reaction

$$H_2(g) + Cl_2(g) \rightleftharpoons 2HCl(g)$$

K_P is 193 at 2800°C.

10.66 Consider the reaction between NO₂ and N₂O₄ in a closed container:

$$N_2O_4(g) \rightleftharpoons 2NO_2(g)$$

Initially there is 1 mole of N₂O₄ present. At equilibrium, α mole of N₂O₄ has dissociated to form NO₂. (a) Derive an expression for K_P in terms of α and P, the total pressure. (b) How does the expression in part (a) help you predict the shift in the equilibrium concentrations due to an increase in P? Does your prediction agree with Le Châtelier's principle?

10.67 One mole of N₂ and three moles of H₂ are placed in a flask at 375°C. Calculate the total pressure of the system at equilibrium if the mole fraction of NH₃ is 0.21. The K_P for the reaction is 4.37×10^{-1}.

10.68 At 1130°C, the equilibrium constant K_c for the reaction

$$2H_2S(g) \rightleftharpoons 2H_2(g) + S_2(g)$$

is 2.25×10^{-4}. If [H₂S] = 4.84×10^{-3} *M* and [H₂] = 1.50×10^{-3} *M*, calculate [S₂].

10.69 A quantity of 6.75 g of SO₂Cl₂ was placed in a 2.00-L flask. At 648 K, there is 0.0345 mole of SO₂ present. Calculate K_c for the reaction

$$SO_2Cl_2(g) \rightleftharpoons SO_2(g) + Cl_2(g)$$

10.70 The formation of SO₃ from SO₂ and O₂ is an intermediate step in the manufacture of sulfuric acid, and it is also responsible for the acid rain phenomenon. The equilibrium constant K_P for the reaction

$$2SO_2(g) + O_2(g) \rightleftharpoons 2SO_3(g)$$

is 0.13 at 830°C. In one experiment, 2.00 mol SO₂ and 2.00 mol O₂ were initially present in a flask. What must the total pressure at equilibrium be in order to have an 80.0 percent yield of SO₃?

10.71 Consider the dissociation of iodine:

$$I_2(g) \rightleftharpoons 2I(g)$$

A 1.00-g sample of I₂ is heated to 1200°C in a 500-mL flask. At equilibrium the total pressure is 1.53 bar. Calculate K_P for the reaction. (*Hint:* Use the result in Problem 10.66(a). The degree of dissociation α can be obtained by first calculating the ratio of observed pressure over calculated pressure, assuming no dissociation.)

10.72 Eggshells are composed mostly of calcium carbonate ($CaCO_3$) formed by the reaction

$$Ca^{2+}(aq) + CO_3^{2-}(aq) \rightleftharpoons CaCO_3(s)$$

The carbonate ions are supplied by carbon dioxide produced as a result of metabolism. Explain why eggshells are thinner in the summer when the rate of panting by chickens is greater. Suggest a remedy for this situation.

10.73 The equilibrium constant K_P for the following reaction is 4.26×10^{-4} at 375°C:

$$N_2(g) + 3H_2(g) \rightleftharpoons 2NH_3(g)$$

In a certain experiment, a student starts with 0.884 bar of N_2 and 0.383 bar of H_2 in a constant-volume vessel at 375°C. Calculate the partial pressures of all species when equilibrium is reached.

10.74 A quantity of 0.20 mole of carbon dioxide was heated to a certain temperature with an excess of graphite in a closed container until the following equilibrium was reached:

$$C(s) + CO_2(g) \rightleftharpoons 2CO(g)$$

Under these conditions, the average molar mass of the gases was 35 g mol^{-1}. (a) Calculate the mole fractions of CO and CO_2. (b) What is K_P if the total pressure is 11 bar? (*Hint:* The average molar mass is the sum of the products of the mole fraction of each gas and its molar mass.)

10.75 Heated calcium carbonate decomposes according to the reaction

$$CaCO_3(s) \rightleftharpoons CaO(s) + CO_2(g)$$

Given that the equilibrium partial pressures of CO_2 at 700°C and 950°C are 22.6 and 1829 mmHg, respectively, determine the standard enthalpy change ($\Delta H°$) for this reaction. State any assumptions that you make.

10.76 When dissolved in water, glucose (corn sugar) and fructose (fruit sugar) exist in equilibrium as follows:

$$\text{fructose} \rightleftharpoons \text{glucose}$$

A chemist prepared a 0.244 M fructose solution at 25°C. At equilibrium, it was found that its concentration had decreased to 0.113 M. (a) Calculate the equilibrium constant for the reaction. (b) At equilibrium, what percentage of fructose was converted to glucose?

10.77 At room temperature, solid iodine is in equilibrium with its vapor through sublimation and deposition. Describe how you would use radioactive iodine, in either solid or vapor form, to show that there is a dynamic equilibrium between these two phases.

10.78 At 1024°C, the pressure of oxygen gas from the decomposition of copper(II) oxide (CuO) is 0.50 bar:

$$4CuO(s) \rightleftharpoons 2Cu_2O(s) + O_2(g)$$

(a) What is K_P for the reaction? (b) Calculate the fraction of CuO that will decompose if 0.16 mole of it is placed in a 2.0-L flask at 1024°C. (c) What would the fraction be if a 1.0-mole sample of CuO were used? (d) What is the smallest amount of CuO (in moles) that would establish the equilibrium?

10.79 A mixture containing 3.9 moles of NO and 0.88 mole of CO_2 was allowed to react in a flask at a certain temperature according to the equation

$$NO(g) + CO_2(g) \rightleftharpoons NO_2(g) + CO(g)$$

At equilibrium, 0.11 mole of CO_2 was present. Calculate the equilibrium constant K_c of this reaction.

10.80 The equilibrium constant K_c for the reaction

$$H_2(g) + I_2(g) \rightleftharpoons 2HI(g)$$

is 54.3 at 430°C. At the start of the reaction, there are 0.714 mole of H_2, 0.984 mole of I_2, and 0.886 mole of HI in a 2.40-L reaction chamber. Calculate the concentrations of the gases at equilibrium.

10.81 When heated, a gaseous compound A dissociates as follows:

$$A(g) \rightleftharpoons B(g) + C(g)$$

In an experiment, A was heated at a certain temperature until its equilibrium pressure reached 0.14P, where P is the total pressure in bar. Calculate the equilibrium constant K_P of this reaction.

10.82 The solubility of the salt $MgCO_3$ is governed by the following equilibrium equation:

$$MgCO_3(s) \rightleftharpoons Mg^{2+}(aq) + CO_3^{2-}(aq)$$

The equilibrium constant K for this reaction is 6.8×10^{-6} at 25°C. Use the data in Appendix 2 to determine K for this reaction at 75°C.

10.83 When a gas was heated under atmospheric conditions, its color deepened. Heating above 150°C caused the color to fade, and at 550°C the color was barely detectable. However, at 550°C, the color was partially restored by increasing the pressure of the system. Which of the following best fits the preceding description? Justify your choice. (a) A mixture of hydrogen and bromine, (b) pure bromine, or (c) a mixture of nitrogen dioxide and dinitrogen tetroxide. (*Hint:* Bromine has a reddish color and nitrogen dioxide is a brown gas. The other gases are colorless.)

10.84 The equilibrium constant K_c for the following reaction is 1.2 at 375°C:

$$N_2(g) + 3H_2(g) \rightleftharpoons 2NH_3(g)$$

(a) What is the value of K_P for this reaction?

(b) What is the value of the equilibrium constant K_c for $2NH_3(g) \rightleftharpoons N_2(g) + 3H_2(g)$?

(c) What is K_c for $N_2(g) + H_2(g) \rightleftharpoons NH_3(g)$?

(d) What are the values of K_P for the reactions described in parts (b) and (c)?

10.85 A sealed glass bulb contains a mixture of NO_2 and N_2O_4 gases. Describe what happens to the following properties of the gases when the bulb is heated from 20°C to 40°C: (a) color, (b) pressure, (c) average molar mass, (d) degree of dissociation (from N_2O_4 to NO_2), and (e) density. Assume that volume remains constant. (*Hint:* NO_2 is a brown gas; N_2O_4 is colorless.)

10.86 At 20°C, the vapor pressure of water is 0.0234 bar. Calculate K_P and K_c for the process

$$H_2O(l) \rightleftharpoons H_2O(g)$$

10.87 Industrially, sodium metal is obtained by electrolyzing molten sodium chloride. The reaction at the cathode is $Na^+ + e^- \longrightarrow Na$. We might expect that potassium metal would also be prepared by electrolyzing molten potassium chloride. However, potassium metal is soluble in molten potassium chloride and therefore is hard to recover. Furthermore, potassium vaporizes readily at the operating temperature, creating hazardous conditions. Instead, potassium is prepared by the distillation of molten potassium chloride in the presence of sodium vapor at 892°C:

$$Na(g) + KCl(l) \rightleftharpoons NaCl(l) + K(g)$$

In view of the fact that potassium is a stronger reducing agent than sodium, explain why this approach works. (The boiling points of sodium and potassium are 892°C and 770°C, respectively.)

10.88 In the gas phase, nitrogen dioxide is actually a mixture of nitrogen dioxide (NO_2) and dinitrogen tetroxide (N_2O_4). If the density of such a mixture is 2.3 g L^{-1} at 74°C and 1.3 bar, calculate the partial pressures of the gases and K_P for the dissociation of N_2O_4.

10.89 About 75 percent of hydrogen for industrial use is produced by the *steam-reforming* process. This process is carried out in two stages called primary and secondary reforming. In the primary stage, a mixture of steam and methane at about 30 bar is heated over a nickel catalyst at 800°C to give hydrogen and carbon monoxide:

$$CH_4(g) + H_2O(g) \rightleftharpoons CO(g) + 3H_2(g)$$
$$\Delta H° = 206 \text{ kJ mol}^{-1}$$

The secondary stage is carried out at about 1000°C, in the presence of air, to convert the remaining methane to hydrogen:

$$CH_4(g) + O_2(g) \rightleftharpoons CO(g) + 2H_2(g)$$
$$\Delta H° = 35.7 \text{ kJ mol}^{-1}$$

(a) What conditions of temperature and pressure would favor the formation of products in both the primary and secondary stage? (b) The equilibrium constant K_c for the primary stage is 18 at 800°C. (i) Calculate K_P for the reaction. (ii) If the partial pressures of methane and steam were both 15 bar at the start, what are the pressures of all the gases at equilibrium?

10.90 Photosynthesis can be represented by

$$6CO_2(g) + 6H_2O(l) \rightleftharpoons C_6H_{12}O_6(s) + 6O_2(g)$$
$$\Delta H° = 2801 \text{ kJ mol}^{-1}$$

Explain how the equilibrium would be affected by the following changes: (a) partial pressure of CO_2 is increased, (b) O_2 is removed from the mixture, (c) $C_6H_{12}O_6$ (glucose) is removed from the mixture, (d) more water is added, or (e) temperature is decreased.

10.91 Consider the decomposition of ammonium chloride at a certain temperature:

$$NH_4Cl(s) \rightleftharpoons NH_3(g) + HCl(g)$$

Calculate the equilibrium constant K_P if the total pressure is 2.2 bar at that temperature.

10.92 At 25°C, the equilibrium partial pressures of NO_2 and N_2O_4 are 0.15 and 0.20 bar, respectively. If the volume is doubled at constant temperature, calculate the partial pressures of the gases when a new equilibrium is established.

10.93 In 1899, the German chemist Ludwig Mond developed a process for purifying nickel by converting it to the volatile nickel tetracarbonyl [$Ni(CO)_4$] (b.p. 42.2°C):

$$Ni(s) + 4CO(g) \rightleftharpoons Ni(CO)_4(g)$$

(a) Describe how you can separate nickel and its solid impurities. (b) How would you recover nickel? [$\Delta H_f°$ for $Ni(CO)_4$ is -602.9 kJ mol^{-1}.]

10.94 Consider the equilibrium reaction described in Problem 10.11. A quantity of 2.50 g of PCl_5 is placed in an evacuated 0.500-L flask and heated to 250°C. (a) Calculate the pressure of PCl_5, assuming it does not dissociate. (b) Calculate the partial pressure of PCl_5 at equilibrium. (c) What is the total pressure at equilibrium? (d) What is the degree of dissociation of PCl_5? (The degree of dissociation is given by the fraction of PCl_5 that has undergone dissociation.)

10.95 Consider the equilibrium system $3A \rightleftharpoons B$. Sketch the changes in the concentrations of A and B over time for the following situations: (a) initially only A is present; (b) initially only B is present; and (c) initially both A and B are present (with A in higher concentration). In each case, assume that the concentration of B is higher than that of A at equilibrium.

10.96 The vapor pressure of mercury is 0.0020 mmHg at 26°C. (a) Calculate K_c and K_P for the process $Hg(l) \rightleftharpoons Hg(g)$. (b) A chemist breaks a thermometer measuring 6.1 m long, 5.3 m wide, and 3.1 m high and spills mercury onto the floor of a laboratory. Calculate the mass of mercury (in grams) vaporized at equilibrium and the concentration of mercury vapor in mg m^{-3}. Does this concentration exceed the safety limit of 0.05 mg m^{-3}? (Ignore the volume of furniture and other objects in the laboratory.)

10.97 At 25°C, a mixture of NO_2 and N_2O_4 gases are in equilibrium in a cylinder fitted with a movable piston. The concentrations are $[NO_2] = 0.0475\ M$ and $[N_2O_4] = 0.487\ M$. The volume of the gas mixture is halved by pushing down on the piston at constant temperature. Calculate the concentrations of the gases when equilibrium is reestablished. Will the color become darker or lighter after the change? (*Hint:* K_c for the dissociation of N_2O_4 is 4.63×10^{-3}. $N_2O_4(g)$ is colorless, and $NO_2(g)$ has a brown color.)

10.98 A student placed a few ice cubes in a drinking glass with water. A few minutes later, she noticed that some of the ice cubes were fused together. Explain what happened.

10.99 The K_P for the reaction

$$SO_2Cl_2(g) \rightleftharpoons SO_2 + Cl_2(g)$$

is 2.07 at 648 K. A sample of SO_2Cl_2 is placed in a container and heated to 648 K while the total pressure is kept constant at 9.00 bar. Calculate the partial pressures of the gases at equilibrium.

10.100 The equilibrium constant K_c for the reaction

$$2NH_3(g) \rightleftharpoons N_2(g) + 3H_2(g)$$

is 0.83 at 375°C. A 14.6-g sample of ammonia is placed in a 4.00-L flask and heated to 375°C. Calculate the concentrations of all the gases when equilibrium is reached.

10.101 Consider the following reaction at a certain temperature:

$$A_2 + B_2 \rightleftharpoons 2AB$$

The mixing of 1 mole of A_2 with 3 moles of B_2 gives rise to x moles of AB at equilibrium. The addition of 2 more moles of A_2 produces another x moles of AB. What is the equilibrium constant for the reaction?

10.102 The van't Hoff equation for the temperature dependence of equilibrium constants (Equation 10.16) is very similar in form to the Clausius-Clapeyron equation (Equation 9.4) for the temperature dependence of vapor pressure. Give an explanation for this similarity. (*Hint:* Write the process of vaporization of a liquid as an equilibrium reaction. What is the equilibrium constant?)

10.103 Iodine is sparingly soluble in water but is much more soluble in carbon tetrachloride (CCl_4). The equilibrium constant, also called the partition coefficient, for the distribution of I_2 between these two phases

$$I_2(aq) \rightleftharpoons I_2(CCl_4)$$

is 83 at 20°C. (a) A student adds 0.030 L of CCl_4 to 0.200 L of an aqueous solution containing 0.032 g I_2. The mixture is shaken, and the two phases are then allowed to separate. Calculate the fraction of I_2 remaining in the aqueous phase. (b) The student now repeats the extraction of I_2 with another 0.030 L of CCl_4. Calculate the fraction of the I_2 from the original solution that remains in the aqueous phase. (c) Compare the result in part (b) with a single extraction using 0.060 L of CCl_4. Comment on the difference.

10.104 The following diagram shows the variation of the equilibrium constant with temperature for the reaction

$$I_2(g) \rightleftharpoons 2I(g)$$

Calculate $\Delta G°$, $\Delta H°$, and $\Delta S°$ for the reaction at 872 K.

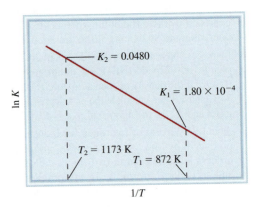

Answers to Practice Exercises

10.1 (a) $K_P = \dfrac{a_{NO_2}^4 a_{O_2}}{a_{N_2O_5}^2} \approx \dfrac{P_{NO_2}^4 P_{O_2}}{P_{N_2O_5}^2}$;

(b) $K = \dfrac{a_{NH_4^+} a_{H_3O^+}}{a_{NH_3} a_{H_2O(l)}} \approx \dfrac{[NH_4^+][OH^-]}{[NH_3]}$

10.2 $K_c = 2.2 \times 10^2$; $K_P = 7.6$ **10.3** $K_P = 0.0718$; $K_c = 6.7 \times 10^{-5}$ **10.4** 9.5×10^{-17} **10.5** $Q > K$, so the reaction proceeds from right to left (towards reactants). **10.6** $[Br_2] = 0.065\ M$, $[Br] = 8.4 \times 10^{-3}\ M$. **10.7** $P_{NH_3} = P_{H_2S} = 0.268$ bar; 15.8 g of $NH_4HS(s)$ remain in the container. **10.8** 1.6×10^{57}

10.9 33 kJ mol^{-1} **10.10** $\Delta G_{rxn} = 1.87$ kJ mol$^{-1} > 0$, so the reaction will proceed from right to left (toward reactants). **10.11** (a) Reaction will shift toward products (left to right). (b) $Q_P = 3.1 \times 10^4 < K_P$ indicating that the reaction will shift toward products, as seen in part (a). **10.12** The reaction shifts from left to right (toward products). **10.13** (a) $x_{PCl_3} = x_{Cl_2} = 0.350$, $x_{PCl_5} = 0.300$, (b) $x_{PCl_3} = x_{Cl_2} = 0.191$, $x_{PCl_5} = 0.612$; at higher pressure, the reaction shifts toward the reactant, which is consistent with Le Châtelier's principle because $\Delta n = 1 > 0$. **10.14** 329 K

11

Acids and Bases

Some of the most important processes in chemical and biological systems are acid-base reactions in aqueous solutions. In this chapter on the properties of acids and bases, we will study the definitions of acids and bases, the pH scale, the ionization of weak acids and weak bases, and the relationship between acid strength and molecular structure. We will also look at oxides that can act as acids and bases.

11.1 | Many Processes in Chemistry Are Acid-Base Reactions

Common acids and bases include aspirin (acetylsalicylic acid) and milk of magnesia (magnesium hydroxide). Besides providing the basis of many medicinal and household products, though, acid-base chemistry is important in industrial processes and essential in sustaining biological systems. Before we can discuss acid-base reactions, we need to know more about acids and bases themselves.

Arrhenius Definition of Acids and Bases

The first definitions of acids and bases were formulated in the late nineteenth century by the Swedish chemist Svante Arrhenius.[1] According to Arrhenius, an ***acid*** is a *substance that yields hydrogen ions (H^+) when dissolved in water.* For example, HCl fits the Arrhenius definition of an acid because, when dissolved in water, HCl dissociates into $H^+(aq)$ and $Cl^-(aq)$:

$$HCl(aq) \longrightarrow H^+(aq) + Cl^-(aq)$$

Arrhenius defined a ***base,*** on the other hand, as *a substance that yields hydroxide ions (OH^-) when dissolved in water.* For example, when sodium hydroxide is dissolved in water, it dissociates into $Na^+(aq)$ and $OH^-(aq)$

$$NaOH(aq) \longrightarrow Na^+(aq) + OH^-(aq)$$

so NaOH fits the Arrhenius definition of a base.

Acids have a number of distinguishing properties:

▶ Acids taste sour; vinegar, for example, owes its sour taste to acetic acid, whereas lemons and other citrus fruits owe theirs to citric acid.

▶ Acids cause color changes in plant dyes, for example, they change the color of litmus from blue to red.

▶ Acids react with certain metals, such as zinc, magnesium, and iron, to produce hydrogen gas. A typical reaction is that between hydrochloric acid and magnesium:

$$2HCl(aq) + Mg(s) \longrightarrow MgCl_2(aq) + H_2(g)$$

▶ Acids react with carbonates and bicarbonates, such as Na_2CO_3, $CaCO_3$, and $NaHCO_3$, to produce carbon dioxide gas. Figure 11.1 shows the reaction between hydrochloric acid and $CaCO_3$ (in the form of chalk):

$$2HCl(aq) + CaCO_3(s) \longrightarrow CaCl_2(aq) + H_2O(l) + CO_2(g)$$

▶ Aqueous acid solutions conduct electricity.

In contrast, bases have the following characteristic properties:

▶ Bases taste bitter.

▶ Bases feel slippery; soaps, for example, which contain bases, are slippery.

▶ Bases cause color changes in plant dyes, for example, they change the color of litmus from red to blue.

▶ Aqueous base solutions conduct electricity.

Figure 11.1 A piece of blackboard chalk, which is mostly $CaCO_3$, reacts with hydrochloric acid.

---

1. Svante August Arrhenius (1859–1927). Swedish chemist. Arrhenius made important contributions in the study of chemical kinetics and electrolyte solutions. He also speculated that life had come to Earth from other planets, a theory now known as *panspermia*. Arrhenius was awarded the Nobel Prize in Chemistry in 1903.

Brønsted Definition of Acids and Bases

Arrhenius's definitions of acids and bases are limited because they apply only to aqueous solutions. Broader definitions were proposed by the Danish chemist Johannes Brønsted[2] in 1932; a ***Brønsted acid*** is *a proton donor,* and a ***Brønsted base*** is *a proton acceptor.* These definitions do not require acids and bases to be in aqueous solution, just that acids donate a proton and bases accept one.

Hydrochloric acid is a Brønsted acid because it donates a proton in water:

$$HCl(aq) \longrightarrow H^+(aq) + Cl^-(aq)$$

The H^+ ion is a hydrogen atom that has lost its electron, that is, it is just a bare proton. The size of a proton is about 10^{-15} m, compared to a diameter of 10^{-10} m for an average atom or ion. Such an exceedingly small charged particle cannot exist as a separate entity in aqueous solution owing to its strong attraction for the negative pole (the O atom) in H_2O. Consequently, the proton exists in the hydrated form as shown in Figure 11.2, and therefore the ionization of hydrochloric acid should be written as

$$HCl(aq) + H_2O(l) \longrightarrow H_3O^+(aq) + Cl^-(aq)$$

The *hydrated proton* (H_3O^+) is called the ***hydronium ion.*** This equation shows a reaction in which a Brønsted acid (HCl) donates a proton to a Brønsted base (H_2O). Experiments show that the hydronium ion is further hydrated, so the proton may have several water molecules associated with it. Because the acidic properties of the proton are unaffected by the degree of hydration, in this text we will often use $H^+(aq)$ to represent the hydrated proton. This notation is for convenience, but H_3O^+ is closer to reality. Keep in mind that both notations represent the same species in aqueous solution.

Acids commonly used in the laboratory include hydrochloric acid (HCl), nitric acid (HNO_3), acetic acid (CH_3COOH), sulfuric acid (H_2SO_4), and phosphoric acid (H_3PO_4). The first three are ***monoprotic acids,*** that is, *each unit of the acid yields one hydrogen ion upon ionization:*

$$HCl(aq) \longrightarrow H^+(aq) + Cl^-(aq)$$
$$HNO_3(aq) \longrightarrow H^+(aq) + NO_3^-(aq)$$
$$CH_3COOH(aq) \rightleftharpoons H^+(aq) + CH_3COO^-(aq)$$

The double arrow in the acetic acid ionization reaction indicates that the dissociation is incomplete, that is, a certain amount of undissociated acetic acid exists in solution with the dissociated acetic acid. Acetic acid is called a ***weak acid*** because it *ionizes only to a limited extent in water.* HCl and HNO_3 are ***strong acids*** because they *are, for all practical purposes, completely ionized in water.* The relative strengths of acids and bases are discussed in detail in Section 11.2.

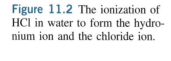

Electrostatic potential map of the H_3O^+ ion. In the rainbow color spectrum representation, the most electron-rich region is red and the most electron-poor region is blue.

In reality, no acids are known to ionize completely in water.

---

2. Johannes Nicolaus Brønsted (1879–1947). Danish chemist. In addition to his theory of acids and bases, Brønsted worked on thermodynamics and the separation of mercury into isotopes.

Figure 11.2 The ionization of HCl in water to form the hydronium ion and the chloride ion.

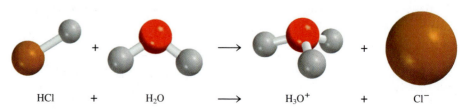

HCl   + H_2O $\longrightarrow$ H_3O^+ + Cl^-

Sulfuric acid (H_2SO_4) is a ***diprotic acid*** because *each unit of the acid can give up two H^+ ions,* in two separate steps:

$$H_2SO_4(aq) \longrightarrow H^+(aq) + HSO_4^-(aq)$$
$$HSO_4^-(aq) \rightleftharpoons H^+(aq) + SO_4^{2-}(aq)$$

H_2SO_4 is a strong acid (the first step of ionization is complete), but HSO_4^- is a weak acid or weak electrolyte, so we use a double arrow to represent its incomplete ionization.

Triprotic acids, which *yield three H^+ ions,* are relatively few in number. The best-known triprotic acid is phosphoric acid, whose ionizations are

$$H_3PO_4(aq) \rightleftharpoons H^+(aq) + H_2PO_4^-(aq)$$
$$H_2PO_4^-(aq) \rightleftharpoons H^+(aq) + HPO_4^{2-}(aq)$$
$$HPO_4^-(aq) \rightleftharpoons H^+(aq) + PO_4^{3-}(aq)$$

All three species (H_3PO_4, $H_2PO_4^-$, and HPO_4^{2-}) are weak acids, so we use the double arrows to represent each ionization step. Anions such as $H_2PO_4^-$ and HPO_4^- are found in aqueous solutions of phosphates such as NaH_2PO_4 and Na_2HPO_4.

The hydroxide ion (OH^-) is classified as a Brønsted base because it can accept a proton as follows:

$$H^+(aq) + OH^-(aq) \longrightarrow H_2O(l)$$

Thus, any Arrhenius base (that is, any substance that yields hydroxide ions in solution) can be classified as a Brønsted base.

Ammonia (NH_3), which is not an Arrhenius base, is nevertheless classified as a Brønsted base because it can accept an H^+ ion from water (Figure 11.3):

$$NH_3(aq) + H_2O(l) \rightleftharpoons NH_4^+(aq) + OH^-(aq)$$

Ammonia is a ***weak base*** because only a small fraction of dissolved NH_3 molecules react with water to form NH_4^+ and OH^- ions. In contrast, sodium hydroxide and barium hydroxide are ***strong bases*** because they completely ionize in solution:

$$NaOH(s) \xrightarrow{H_2O} Na^+(aq) + OH^-(aq)$$
$$Ba(OH)_2(s) \xrightarrow{H_2O} Ba^{2+}(aq) + 2OH^-(aq)$$

The most commonly used strong base in the laboratory is sodium hydroxide. It is inexpensive and water soluble. (In fact, all the alkali metal hydroxides are water soluble.) The most commonly used weak base is aqueous ammonia solution, which is sometimes erroneously called ammonium hydroxide; there is no evidence that the species NH_4OH actually exists. All the Group 2A elements form hydroxides of the

Zn reacts more vigorously with a strong acid like HCl (left) than with a weak acid like CH_3COOH (right) of the same concentration because there are more H^+ ions in the former solution.

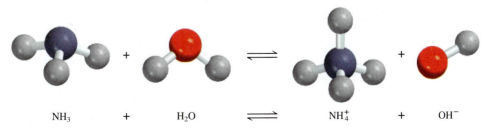

NH₃   +   H₂O   ⇌   NH₄⁺   +   OH⁻

Figure 11.3 The ionization of ammonia in water to form the ammonium ion and the hydroxide ion.

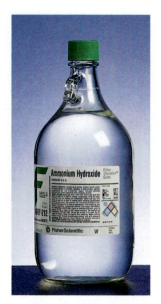

Note that this bottle of aqueous ammonia is erroneously labeled.

type $M(OH)_2$, where M denotes an alkaline earth metal. Of these hydroxides, only $Ba(OH)_2$ is water soluble. Magnesium and calcium hydroxides are used in medicine and industry. Hydroxides of other metals, such as $Al(OH)_3$ and $Zn(OH)_2$, are insoluble in water and are not used as bases.

Many substances can act as both a Brønsted acid and a Brønsted base and are called **amphoteric**. For example, the bicarbonate ion is a Brønsted acid because it ionizes in aqueous solution to yield a proton and the carbonate ion:

$$HCO_3^-(aq) \rightleftharpoons H^+(aq) + CO_3^{2-}$$

It is also a Brønsted base because it can accept a proton to form carbonic acid:

$$HCO_3^-(aq) + H^+(aq) \rightleftharpoons H_2CO_3(aq)$$

Water is amphoteric, too, because it can donate a proton to another water molecule:

$$H_2O(l) + H_2O(l) \rightleftharpoons OH^-(aq) + H_3O^+(aq)$$

Thus, the water molecule that donates the proton, acts as a Brønsted acid, whereas the water molecule that accepts the proton acts as a Brønsted base.

Acid-Base Neutralization

A **neutralization reaction** is *a reaction between an acid and a base.* Aqueous acid-base reactions generally produce water and a **salt,** which is *an ionic compound made up of a cation other than H^+ and an anion other than OH^- or O_2^-:*

$$acid + base \longrightarrow water + salt$$

All salts (such as table salt, NaCl) are **strong electrolytes,** because *they, for all practical purposes, completely ionize in aqueous solution.* NaCl is produced in the neutralization reaction between HCl and NaOH:

$$HCl(aq) + NaOH(aq) \xrightarrow{H_2O} NaCl(aq) + H_2O(l)$$

Acid-base neutralization reactions generally go to completion.

However, because both the acid and the base are strong electrolytes, they are completely dissociated into ions in solution. The ionic equation is

$$H^+(aq) + Cl^-(aq) + Na^+(aq) + OH^-(aq) \longrightarrow Na^+(aq) + Cl^-(aq) + H_2O(l)$$

Therefore, the reaction can be represented by the net ionic equation:

$$H^+(aq) + OH^-(aq) \longrightarrow H_2O(l)$$

because both Na^+ and Cl^- are **spectator ions,** that is, *they do not appear in the net equation.*

If we had started the preceding reaction with equal molar amounts of the acid and the base, at the end of the reaction we would have only the salt and water, with no leftover acid or base. This is a characteristic of acid-base neutralization reactions.

The following are other examples of acid-base neutralization reactions, represented by molecular equations:

$$HF(aq) + KOH(aq) \longrightarrow KF(aq) + H_2O(l)$$
$$H_2SO_4(aq) + 2NaOH(aq) \longrightarrow Na_2SO_4(aq) + 2H_2O(l)$$
$$HNO_3(aq) + NH_3(aq) \longrightarrow NH_4NO_3(aq)$$

The last equation looks different because it does not show water as a product. However, if we express $NH_3(aq)$ as $NH_4^+(aq)$ and $OH^-(aq)$, as discussed previously, then the equation becomes

$$HNO_3(aq) + NH_4^+(aq) + OH^-(aq) \longrightarrow NH_4NO_3(aq) + H_2O(l)$$

Conjugate Acid-Base Pairs

The concept of the ***conjugate acid-base pair,*** which can be defined as *an acid and its conjugate base or a base and its conjugate acid,* arises from the Brønsted definition of acids and bases. The conjugate base of a Brønsted acid is the species that remains after the acid donates a proton. Conversely, a conjugate acid results from the addition of a proton to a Brønsted base. Every Brønsted acid has a conjugate base, and every Brønsted base has a conjugate acid. For example, the chloride ion (Cl^-) is the conjugate base formed from the acid HCl, and H_2O is the conjugate base of the acid H_3O^+ (the hydronium ion). Similarly, the ionization of acetic acid can be represented as

The word *conjugate* means "joined together."

$$\underset{acid_1}{CH_3COOH(aq)} + \underset{base_2}{H_2O(l)} \rightleftharpoons \underset{base_1}{CH_3COO^-(aq)} + \underset{acid_2}{H_3O^+(aq)}$$

The subscripts 1 and 2 designate the two conjugate acid-base pairs. In other words, the acetate ion (CH_3COO^-) is the conjugate base of CH_3COOH and H_3O^+ is the conjugate acid of H_2O. Both the ionization of HCl and the ionization of CH_3COOH are examples of Brønsted acid-base reactions.

In Example 11.1 we practice identifying the conjugate pairs in an acid-base reaction.

Example 11.1

Identify the conjugate acid-base pairs in the reaction between ammonia and hydrofluoric acid in aqueous solution

$$NH_3(aq) + HF(aq) \rightleftharpoons NH_4^+(aq) + F^-(aq)$$

Strategy A conjugate base always has one fewer H atom and one more negative charge (or one fewer positive charge) than the formula of the corresponding acid. Similarly, a conjugate acid has one more H atom and one more positive charge (or one fewer negative charge) than the formula of the corresponding base.

Solution NH_3 has one fewer H atom and one fewer positive charge than NH_4^+, whereas F^- has one fewer H atom and one more negative charge than HF. Therefore, the conjugate acid-base pairs are (1) NH_4^+ and NH_3 and (2) HF and F^-.

Practice Exercise Identify the conjugate acid-base pairs for the reaction

$$CN^- + H_2O \rightleftharpoons HCN + OH^-$$

It is acceptable to represent the proton in aqueous solution either as H^+ or as H_3O^+. The formula H^+ is less cumbersome in calculations involving hydrogen ion concentrations and in calculations involving equilibrium constants. On the other hand, H_3O^+ is more useful in a discussion of Brønsted acid-base properties.

Lewis Definition of Acids and Bases

According to the Brønsted theory, a base must be able to accept protons. By this definition, both the hydroxide ion and ammonia are Brønsted bases:

$$H^+ + ^- : \overset{..}{\underset{..}{O}} - H \longrightarrow H - \overset{..}{\underset{..}{O}} - H$$

$$H^+ + : \overset{\displaystyle H}{\underset{\displaystyle H}{N}} - H \longrightarrow \left[H - \overset{\displaystyle H}{\underset{\displaystyle H}{N}} - H \right]^+$$

In each case, the atom to which the proton becomes attached possesses at least one unshared pair of electrons. This characteristic property of OH^-, NH_3, and other Brønsted bases suggests a more general definition of acids and bases.

In 1932 the American chemist G. N. Lewis formulated such a definition (see Section 3.4). He defined what we now call a ***Lewis base*** as *a substance that can donate a pair of electrons*. A ***Lewis acid*** is *a substance that can accept a pair of electrons*. For example, in the protonation of ammonia, NH_3 acts as a Lewis base because it donates a pair of electrons to the proton H^+, which acts as a Lewis acid by accepting the pair of electrons. A Lewis acid-base reaction, therefore, is one that involves the donation of a pair of electrons from one species to another. Such a reaction does not necessarily produce a salt and water.

The significance of the Lewis concept is that it is more general than the Arrhenius or Brønsted definitions. Lewis acid-base reactions include many reactions that do not involve Brønsted acids. Consider, for example, the reaction between boron trifluoride (BF_3) and ammonia to form an adduct compound (Figure 11.4):

$$\underset{\text{acid}}{F - \overset{\displaystyle F}{\underset{\displaystyle F}{B}}} + \underset{\text{base}}{: \overset{\displaystyle H}{\underset{\displaystyle H}{N}} - H} \longrightarrow F - \overset{\displaystyle F}{\underset{\displaystyle F}{B}} - \overset{\displaystyle H}{\underset{\displaystyle H}{N}} - H$$

In Section 3.4, we saw that the B atom in BF_3 is sp^2-hybridized. The vacant, unhybridized $2p_z$ orbital accepts a pair of electrons from NH_3. So BF_3 functions as an acid according to the Lewis definition, even though it does not contain an ionizable proton. A coordinate covalent bond (see Section 3.4) is formed between the B and N atoms, as is the case in all Lewis acid-base reactions.

Another Lewis acid containing boron is boric acid. Boric acid (a weak acid used in eyewash) is an oxoacid with the following structure:

$$\begin{array}{c} H \\ | \\ : \overset{..}{O} : \\ \overset{..}{O} \\ H - \overset{..}{\underset{..}{O}} - B - \overset{..}{\underset{..}{O}} - H \end{array}$$

Boric acid does not ionize in water to produce an H^+ ion. Its reaction with water is

$$B(OH)_3(aq) + H_2O(l) \rightleftharpoons B(OH)_4^-(aq) + H^+(aq)$$

In this Lewis acid-base reaction, boric acid accepts a pair of electrons from the hydroxide ion that is derived from the H_2O molecule.

Lewis acids are either deficient in electrons (cations) or the central atom has a vacant valence orbital.

Figure 11.4 A Lewis acid-base reaction involving BF_3 and NH_3.

A coordinate covalent bond (see page 197) is always formed in a Lewis acid-base reaction.

The hydration of carbon dioxide to produce carbonic acid

$$CO_2(g) + H_2O(l) \rightleftharpoons H_2CO_3(aq)$$

can be understood in the Lewis framework as follows: The first step involves dona-tion of a lone pair on the oxygen atom in H_2O to the carbon atom in CO_2. An orbital is vacated on the C atom to accommodate the lone pair by removal of the electron pair in the C—O π bond. These shifts of electrons are indicated by the curved arrows.

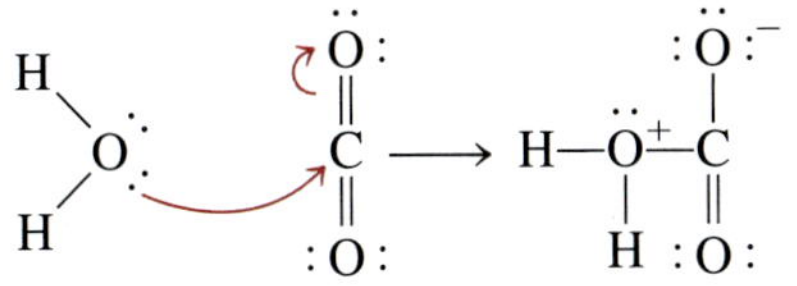

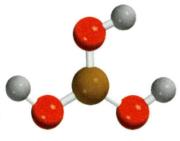

H_3BO_3

Therefore, H_2O is a Lewis base and CO_2 is a Lewis acid. Next, a proton is transferred onto the O atom bearing a negative charge to form H_2CO_3.

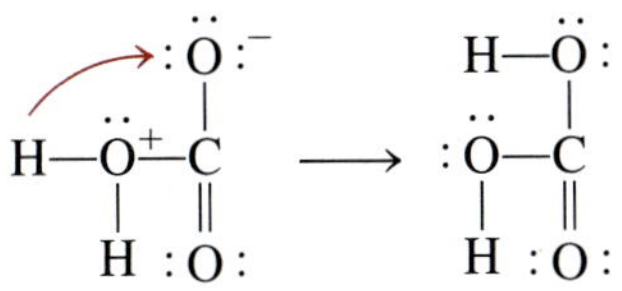

Other examples of Lewis acid-base reactions are

$$\underset{\text{acid}}{Ag^+(aq)} + \underset{\text{base}}{2NH_3(aq)} \rightleftharpoons Ag(NH_3)_2^+(aq)$$

$$\underset{\text{acid}}{Cd^{2+}(aq)} + \underset{\text{base}}{4I^-(aq)} \rightleftharpoons CdI_4^{2-}(aq)$$

$$\underset{\text{acid}}{Ni(s)} + \underset{\text{base}}{4CO(g)} \rightleftharpoons Ni(CO)_4(g)$$

The hydration of metal ions in aqueous solution is in itself a Lewis acid-base reaction. Thus, when copper(II) sulfate ($CuSO_4$) dissolves in water, each Cu^{2+} ion is associated with six water molecules as $Cu(H_2O)_6^{2+}$. In this case, the Cu^{2+} ion acts as the acid and the H_2O molecules act as the base.

Although the Lewis definition of acids and bases has greater significance because of its generality, we normally speak of "an acid" and "a base" in terms of the Brønsted definition. The term "Lewis acid" usually is reserved for substances that can accept a pair of electrons, but do not contain ionizable hydrogen atoms. Example 11.2 shows how to classify compounds as Lewis acids or Lewis bases.

Example 11.2

Identify the Lewis acid and Lewis base in each of the following reactions:

(a) $C_2H_5OC_2H_5 + AlCl_3 \rightleftharpoons (C_2H_5)_2OAlCl_3$

(b) $Hg^{2+}(aq) + 4CN^-(aq) \rightleftharpoons Hg(CN)_4^-(aq)$

—Continued

Continued—

Strategy In Lewis acid-base reactions, the acid (which accepts a lone pair of electrons) is usually a cation or an electron-deficient molecule, whereas the base (which donates a lone pair of electrons) is an anion or a molecule containing an atom with lone pairs. (a) Draw the Lewis structure for $C_2H_5OC_2H_5$ to see if it can donate or accept a lone pair of electrons. What is the hybridization state of Al in $AlCl_3$? Can it donate or accept a lone pair of electrons? (b) Which ion is likely to be an electron acceptor? An electron donor?

Solution (a) The Al is sp^2 hybridized in $AlCl_3$ with an empty $2p_z$ orbital. It is electron deficient, sharing only six electrons. Therefore, the Al atom has a tendency to gain two electrons to complete its octet. This property makes $AlCl_3$ a Lewis acid. On the other hand, the lone pairs on the oxygen atom in $C_2H_5OC_2H_5$ make the compound a Lewis base:

(b) Here the Hg^{2+} ion accepts four pairs of electrons from the CN^- ions. Therefore, Hg^{2+} is the Lewis acid and CN^- is the Lewis base.

Practice Exercise Identify the Lewis acid and Lewis base in the reactions

(a) $Co^{3+}(aq) + 6NH_3(aq) \rightleftharpoons Co(NH_3)_6^{3+}$

(b) $CO + BH_3 \rightleftharpoons BH_3CO$

11.2 | The Acid-Base Properties of Aqueous Solutions Are Governed by the Autoionization Equilibrium of Water

Water is amphoteric and can act either as an acid or as a base. Water functions as a base in reactions with acids such as HCl and CH_3COOH, and it functions as an acid in reactions with bases such as NH_3. Water is a very weak electrolyte and therefore a poor conductor of electricity, but it does undergo ionization to a small extent:

Tap water and water from underground sources do conduct electricity because they contain many dissolved ions.

$$H_2O(l) + H_2O(l) \rightleftharpoons H_3O^+(aq) + OH^-(aq) \qquad (11.1)$$

This reaction is sometimes called the *autoionization of water* (Figure 11.5). The acid-base conjugate pairs are (1) H_2O (acid) and OH^- (base) and (2) H_3O^+ (acid) and H_2O (base).

Figure 11.5 The reaction between two water molecules to form hydronium and hydroxide ions.

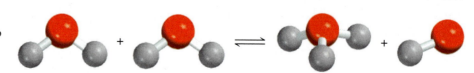

The Ion Product of Water

In the study of acid-base reactions in aqueous solution, the hydrogen ion concentration is key; its value indicates the acidity or basicity of the solution. Therefore, the equilibrium constant for the autoionization of water, according to Equation 9.2, is

$$K = \frac{a_{H_3O^+(aq)}\, a_{OH^-(aq)}}{a^2_{H_2O(l)}} \qquad \textbf{(11.2)}$$

As discussed in Section 10.1, if the H_3O^+ and OH^- concentrations are sufficiently small ($< 0.05\ M$) then the activities of these solute species are to a good approximation equal to their concentrations, relative to the standard concentration, $c_0 = 1\ M$. The activity of the nearly pure solvent (H_2O) can be taken to be 1 (Equation 10.5). As a result, the equilibrium constant for the autoionization of water (Equation 11.2) can be expressed[3] as

$$\boxed{K_w = [H^+][OH^-]} \qquad \textbf{(11.3)}$$

where, for simplicity, we use H^+ to represent H_3O^+ and where we have added the subscript "w" to represent "water." The constant K_w is called the ***ion-product constant,*** which is *the product of the molar concentrations of H^+ and OH^- ions at a particular temperature.*

In pure water at 25°C, the concentrations of H^+ and OH^- ions are equal and found to be $[H^+] = [OH^-] = 1.0 \times 10^{-7}\ M$. Thus, from Equation 11.3, at 25°C

$$K_w = (1.0 \times 10^{-7})(1.0 \times 10^{-7}) = 1.0 \times 10^{-14}$$

If you could randomly remove and examine 10 particles (H_2O, H^+, or OH^-) per second from a liter of water, it would take you two years, working nonstop to find one H^+ ion.

Whether we have pure water or an aqueous solution of dissolved species, the following relation *always* holds at 25°C:

$$K_w = [H^+][OH^-] = 1.00 \times 10^{-14}$$

Whenever $[H^+] = [OH^-]$, the aqueous solution is said to be neutral. In an acidic solution, there is an excess of H^+ ions and $[H^+] > [OH^-]$. In a basic solution, there is an excess of hydroxide ions, so $[H^+] < [OH^-]$. In practice, we can change the concentration of either H^+ or OH^- ions in solution, but we cannot vary both of them independently. If we adjust the solution so that $[H^+] = 1.0 \times 10^{-6}\ M$, the OH^- concentration *must* change to

$$[OH^-] = \frac{K_w}{[H^+]} = \frac{1.00 \times 10^{-14}}{1.0 \times 10^{-6}} = 1.0 \times 10^{-8}\ M$$

Equation 11.3 assumes that the solution is sufficiently dilute that it can be considered to be ideal. For more concentrated solutions ($> 0.05\ M$ for singly charged species), ion-pair formation and other types of intermolecular interactions can lead to nonideal solution behavior, and the ion activities ("effective concentrations") that go into Equation 11.2 will differ somewhat from the molarities. In that case, concentrations calculated from Equation 11.3 will deviate from the actual concentrations in the solution.

---

3. It is important to note that K_w is a dimensionless constant and that the actual expression (Equation 11.3) should be written

$$K_w = ([H^+]/c_0)\,([OH^-]/c_0)$$

where $c_0 = 1\ M$. For simplicity, we have omitted c_0 from the expression. Therefore, we must always be careful to convert the concentration units to M before using Equation 11.3.

However, for most purposes, it is acceptable to ignore these deviations from ideality. In most cases, therefore, this approach will closely approximate the chemical processes that actually take place in the solution phase.

An application of Equation 11.3 is presented in Example 11.3.

Example 11.3

The concentration of OH^- ions in a certain household ammonia cleaning solution is 0.0025 M. Calculate the concentration of H^+ ions.

Solution Rearranging Equation 11.3, we write

$$[H^+] = \frac{K_w}{[OH^-]} = \frac{1.00 \times 10^{-14}}{0.0025} = 4.0 \times 10^{-12} \, M$$

Check Because $[H^+] < [OH^-]$, the solution is basic, as we would expect from the earlier discussion of the reaction of ammonia with water.

Practice Exercise Calculate the concentration of OH^- ions in an HCl solution whose hydrogen ion concentration is $1.3 \times 10^{-4} \, M$.

---

pH—A Measure of Acidity

Because the concentrations of H^+ and OH^- ions in aqueous solutions are frequently very small numbers and therefore inconvenient to work with, Soren Sorensen[4] in 1909 proposed a more practical measure called pH. The *pH* of a solution is defined as *the negative logarithm (base 10) of the hydrogen ion activity*:

$$pH \doteq -\log_{10} a_{H_3O^+} \quad \text{or} \quad pH \doteq -\log_{10} a_{H^+} \qquad \textbf{(11.4)}$$

For dilute solutions in which the activity of the hydrogen ion can be well approximated by its concentration relative to the standard concentration, we have

$$pH = -\log_{10} a_{H^+} \approx -\log_{10}[H^+] \qquad \textbf{(11.5)}$$

The pH of concentrated acid solutions can be negative. For example, the pH of a 2.0 M HCl solution is approximately −0.3.

where $[H^+]$ is understood to be the *numerical* value of the hydrogen ion concentration in mol L^{-1} (without units). Equation 11.4 is simply a definition designed to give us convenient numbers to work with. The negative logarithm gives us a positive number for pH, which otherwise would be negative for most solutions due to the small value of $[H^+]$. Like the equilibrium constant, the pH of a solution is a dimensionless quantity.

Because pH is simply a way to express hydrogen ion concentration, acidic and basic solutions at 25°C can be distinguished by their pH values as follows:

Acidic solutions: $[H^+] > 1.00 \times 10^{-7} \, M$, pH < 7.00

Basic solutions: $[H^+] < 1.00 \times 10^{-7} \, M$, pH > 7.00

Neutral solutions: $[H^+] = 1.00 \times 10^{-7} \, M$, pH $= 7.00$

Thus, pH increases as $[H^+]$ decreases.

---

4. Soren Peer Lauritz Sorensen (1868–1939). Danish biochemist. Sorensen originally wrote the symbol as p_H and called p the "hydrogen ion exponent" (*Wasserstoffionexponent*); it is the initial letter of *Potenz* (German), *puissance* (French), and *power* (English). It is now customary to write the symbol as pH.

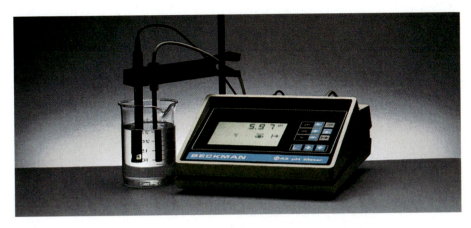

Figure 11.6 A pH meter is commonly used in the laboratory to determine the pH of a solution. Although many pH meters have scales marked with values from 1 to 14, pH values can, in fact, be less than 1 and greater than 14. The science behind the pH meter will be discussed in Chapter 13.

In the laboratory, the pH of a solution is measured with a pH meter (Figure 11.6), which measures the activity of the H^+ ion electrochemically. (The construction of a pH meter is discussed in more detail in Chapter 13). Table 11.1 lists the pH values of a number of common fluids at 25°C. As you can see, the pH of body fluids varies greatly, depending on location and function. The low pH (high acidity) of gastric juices facilitates digestion whereas the higher pH of blood is necessary for the transport of oxygen.

A pOH scale analogous to the pH scale can be devised using the negative logarithm (base 10) of the hydroxide ion activity of a solution. Thus, we define pOH as

$$\text{pOH} = -\log_{10} a_{OH^-} \tag{11.6}$$

As in the case of pH, the pOH for sufficiently dilute aqueous solutions can be written as

$$\text{pOH} \approx -\log_{10} [OH^-] \tag{11.7}$$

Now consider again the ion-product constant for water at 25°C:

$$[H^+][OH^-] = K_w = 1.01 \times 10^{-14}$$

Taking the negative logarithm of both sides, we obtain

$$-\log_{10}[H^+] - \log_{10}[OH^-] = -\log_{10}(1.01 \times 10^{-14})$$
$$= 14.00$$

From the definitions of pH and pOH we obtain

$$\text{pH} + \text{pOH} = 14.00 \tag{11.8}$$

at 25°C. Equation 11.8 provides us with another way to express the relationship between the H^+ ion concentration and the OH^- ion concentration.

Examples 11.4 to 11.6 illustrate calculations involving pH at 25°C.

Table 11.1	
The pH Values of Some Common Fluids	
Sample	**pH Value**
Gastric juice in the stomach	1.0–2.0
Lemon juice	2.4
Vinegar	3.0
Grapefruit juice	3.2
Orange juice	3.5
Urine	4.8–7.5
Water exposed to air*	5.5
Saliva	6.4–6.9
Milk	6.5
Pure water	7.0
Blood	7.35–7.45
Tears	7.4
Milk of magnesia	10.6
Household ammonia	11.5

*Water exposed to air for a long period of time absorbs atmospheric CO_2 to form carbonic acid, H_2CO_3.

Example 11.4

The concentration of H^+ ions in a bottle of table wine was 3.2×10^{-4} *M* right after the cork was removed. Only half of the wine was consumed. The other half, after it

—*Continued*

Continued—

had been standing open to the air for a month, was found to have a hydrogen ion concentration equal to 1.1×10^{-3} M. Calculate the pH of the wine on these two occasions.

Strategy We are given the H^+ ion concentration and asked to calculate the pH of the solution. We can use the definition of pH given in Equation 11.5—namely, pH = $-\log_{10} [H^+]$.

Solution When the bottle was first opened, $[H^+] = 3.2 \times 10^{-4}$ M, so substituting this value into Equation 11.5 gives the pH:

$$pH = -\log_{10} [H^+]$$
$$= -\log_{10} (3.2 \times 10^{-4}) = 3.49$$

On the second occasion, $[H^+] = 1.1 \times 10^{-3}$ M, so

$$pH = -\log_{10} (1.1 \times 10^{-3}) = 2.96$$

When taking the logarithm of a quantity x, the number of significant figures in x is the same as the number of significant figures to the right of the decimal point in $\log_{10} x$. For example, $\log_{10} 1.25 \times 10^{-3} = -2.903$. Because the quantity (1.25×10^{-3}) has three significant figures, the logarithm will have three significant figures to the right of the decimal point. See Appendix 1 for further discussion of significant figures.

Comment The increase in hydrogen ion concentration (or decrease in pH) is largely the result of the conversion of some of the alcohol (ethanol) to acetic acid, a reaction that takes place in the presence of molecular oxygen.

Practice Exercise Nitric acid (HNO_3) is used in the production of fertilizer, dyes, drugs, and explosives. Calculate the pH of an HNO_3 solution having a hydrogen ion concentration of 0.76 M.

Example 11.5

The pH of rainwater collected in a certain region of the northeastern United States on a particular day was 4.82. Calculate the H^+ ion concentration of the rainwater.

Strategy We are given the pH of a solution and asked to calculate $[H^+]$. Because pH is defined as pH = $-\log_{10} [H^+]$ (Equation 11.5), we can solve for $[H^+]$ by taking the antilog$_{10}$ of the pH; that is, $[H^+] = 10^{-pH}$.

Solution From Equation 11.5

$$pH = -\log_{10} [H^+] = 4.82$$

Therefore,

$$\log_{10} [H^+] = -4.82$$

To calculate $[H^+]$, we need to take the antilog of -4.82:

$$[H^+] = 10^{-4.82} = 1.5 \times 10^{-5} M$$

Check Because the pH is between 4 and 5, we can expect $[H^+]$ to be between 1×10^{-4} M and 1×10^{-5} M. Therefore, the answer is reasonable.

Practice Exercise The pH of a certain orange juice is 3.33. Calculate the H^+ ion concentration.

Example 11.6

In an NaOH solution at 25°C, $[OH^-]$ is 2.9×10^{-4} M. Calculate the pH of the solution.

Strategy Because NaOH is a strong base, it completely dissociates into ions, and so the concentration of NaOH will be equal to the OH^- concentration. Solving this problem takes two steps. First, we need to calculate pOH using Equation 11.7 ($pOH = \log_{10}[OH^-]$). Next, we use Equation 11.8 ($pH + pOH = 14.00$) to calculate the pH of the solution.

Solution According to Equation 11.7,

$$
\begin{aligned}
pOH &= -\log_{10}[OH^-] \\
&= -\log_{10}(2.9 \times 10^{-4}) \\
&= 3.54
\end{aligned}
$$

Now, to calculate the pH, we use Equation 11.8, which is valid at 25°C:

$$
\begin{aligned}
pH + pOH &= 14.00 \\
pH &= 14.00 - pOH \\
&= 14.00 - 3.54 = 10.46
\end{aligned}
$$

Alternatively, we can use the ion-product constant of water at 25°C, $K_w = [H^+][OH^-]$, to calculate $[H^+]$, and then we can calculate the pH from the $[H^+]$ using Equation 11.5. Try it.

Check The answer shows that the solution is basic (pH > 7), which is to be expected for an NaOH solution.

Practice Exercise The OH^- ion concentration of a blood sample at 25°C is 2.5×10^{-7} M. What is the pH of the blood?

Temperature Dependence of K_w and pH

The ion constant of water, K_w, like most equilibrium constants, depends on temperature. Recall from Section 10.4 that the temperature dependence of an equilibrium constant can be determined from the enthalpy of reaction through the van't Hoff equation (Equation 10.16). The enthalpy of reaction for the autoionization of water is $+55.84$ kJ mol^{-1}:

$$
H_2O(l) + H_2O(l) \rightleftharpoons H_3O^+(aq) + OH^-(aq) \qquad \Delta H° = +55.84 \text{ kJ mol}^{-1}
$$

That is, the autoionization of water is an endothermic process. According to Le Châtelier's principle (Section 10.4), the equilibrium of an endothermic process will shift in the direction of products when the temperature is increased, leading to a corresponding increase in the equilibrium constant. Therefore, the value of K_w will be larger than its value at 25°C (1.01×10^{-14}) for temperatures above 25°C and smaller than its value at 25°C for temperatures below 25°C. Table 11.2 shows the value of K_w at a variety of temperatures.

At 60°C (333 K), for example, K_w is 9.61×10^{-14}, nearly 10 times larger than at 25°C. Given this value of K_w, we can calculate the H^+ ion concentration of neutral water (where $[H^+] = [OH^-]$) at 60°C as follows:

$$
\begin{aligned}
K_w = [H^+][OH^-] = [H^+]^2 &= 9.61 \times 10^{-14} \\
[H^+] &= 3.10 \times 10^{-7}
\end{aligned}
$$

Table 11.2	Temperature Dependence of the Autoionization Constant of Water (K_w) and the pH of Neutral Water	
Temperature (K)	K_w	**pH of Neutral Water**
273	1.14×10^{-15}	7.47
298	1.00×10^{-14}	7.00
313	2.92×10^{-14}	6.77
333	9.61×10^{-14}	6.51
373	5.4×10^{-13}	6.13

This value is about three times larger than the corresponding value at 25°C. The pH of neutral water at this temperature is then

$$pH = -\log_{10}[H^+] = -\log_{10}(3.10 \times 10^{-7}) = 6.51$$

This number is lower than 7, but that does not mean that this is an acidic solution. The solution is still neutral because $[OH^-] = [H^+]$. At this temperature, the dividing pH between acidic and basic solutions is 6.51, not 7.00, as it is at 298 K. The pH of neutral water at a variety of temperatures is given in Table 11.2.

The values in Table 10.2 can be calculated using the van't Hoff equation (Equation 10.16) given the value of K_w at 298K and thermodynamic data from Appendix 2.

Example 11.7

A sample of water from a hot spring at 40°C has an $[H^+]$ concentration of 2.34×10^{-6} M. What is the $[OH^-]$ concentration in this sample?

Strategy To find the $[OH^-]$ using Equation 11.3, we need to know K_w at this temperature (40°C = 313 K). According to Table 11.2, $K_w = 2.92 \times 10^{-14}$ at 313 K.

Solution From Equation 11.3, we have

$$[H^+][OH^-] = K_w$$

$$[OH^-] = \frac{K_w}{[H^+]}$$

$$= \frac{2.92 \times 10^{-14}}{2.34 \times 10^{-6}}$$

$$= 1.25 \times 10^{-8}\ M$$

Practice Exercise The pOH of an aqueous solution is 4.32 at a temperature of 60°C. Determine the pH and H^+ ion concentration of this sample.

11.3 | The Strengths of Acids and Bases Are Measured by Their Ionization Constants

Most acids are *weak acids,* which ionize only to a limited extent in water (Figure 11.7). If we represent a nonionized weak acid by HA, with conjugate base A^-, the ionization of a weak acid in aqueous solution is represented by the following equilibrium:

$$HA(aq) + H_2O \rightleftharpoons A^-(aq) + H_3O^+(aq) \tag{11.9}$$

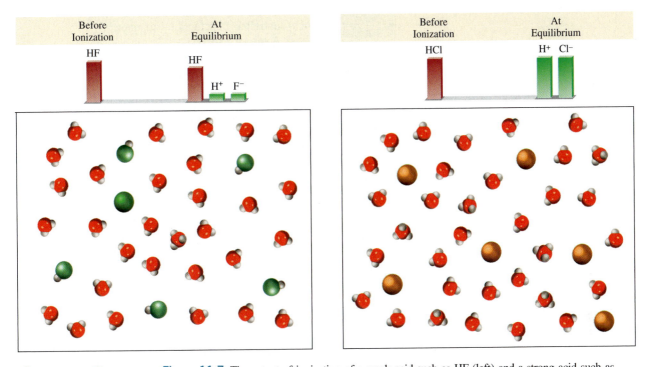

Figure 11.7 The extent of ionization of a weak acid such as HF (left) and a strong acid such as HCl (right). Initially, there were six HCl and six HF molecules present. The strong acid is assumed to be completely ionized in solution. The proton exists in solution as the hydronium ion (H_3O^+).

At equilibrium, aqueous solutions of weak acids contain a mixture of nonionized acid molecules (HA), H_3O^+ ions, and the conjugate base (A^-). Some examples of weak acids are hydrofluoric acid (HF), acetic acid (CH_3COOH), and the ammonium ion (NH_4^+).

The strength of a weak acid—that is, the extent to which it ionizes in aqueous solution—is determined by the magnitude of the equilibrium constant, called the ***acid ionization constant*** (K_a), of the acid ionization reaction (Equation 11.9):

$$K_a = \frac{a_{H_3O^+(aq)} \, a_{A^-(aq)}}{a_{HA(aq)} a_{H_2O(l)}} \qquad \textbf{(11.10)}$$

where a represents the activity of each species. As with our discussion of the ion constant of water, for sufficiently dilute solutions (<0.05 *M*), the activities of the solutes (HA, A^-, and H_3O^+) are well approximated by their concentrations (relative to the standard concentration of 1 *M*), and the activity of the solvent (water) can be assumed to be equal to 1, giving

$$K_a \approx \frac{[H_3O^+][A^-]}{[HA]} = \frac{[H^+][A^-]}{[HA]} \qquad \textbf{(11.11)}$$

For example, the K_a for the weak acid HF (HA = HF and $A^- = F^-$) is given by

$$K_a = \frac{[H^+][F^-]}{[HF]}$$

Table 11.3	Acid Ionization Constants of Common Weak Acids at 298 K	
Acid		K_a
HF (hydrofluoric acid)		7.1×10^{-4}
HCN (hydrocyanic acid)		4.9×10^{-10}
HNO_2 (nitrous acid)		4.5×10^{-4}
CH_3COOH (acetic acid)		1.75×10^{-5}
HCOOH (formic acid)		1.77×10^{-4}
$ClCH_2COOH$ (chloroacetic acid)		1.36×10^{-3}
$CH_3CH(OH)COOH$ (lactic acid)		1.39×10^{-4}
C_6H_5COOH (benzoic acid)		6.5×10^{-5}
C_6H_5OH (phenol)		1.3×10^{-10}
$C_9H_8O_4$ (acetylsalicylic acid; aspirin)		3.0×10^{-4}

The larger the K_a, the stronger the acid. Table 11.3 lists the K_a values for a number of common weak acids.

As defined in Section 11.1, *strong acids* can be assumed, for practical purposes, to ionize completely in water (see Figure 11.7). There are relatively few strong acids and most are inorganic acids, such as hydrochloric acid (HCl), nitric acid (HNO_3), perchloric acid ($HClO_4$), and sulfuric acid (H_2SO_4):

$$HCl(aq) + H_2O(l) \longrightarrow H_3O^+(aq) + Cl^-(aq)$$
$$HNO_3(aq) + H_2O(l) \longrightarrow H_3O^+(aq) + NO_3^-(aq)$$
$$HClO_4(aq) + H_2O(l) \longrightarrow H_3O^+(aq) + ClO_4^-(aq)$$
$$H_2SO_4(aq) + H_2O(l) \longrightarrow H_3O^+(aq) + HSO_4^-(aq)$$

H_2SO_4 is a diprotic acid, but we show only the first stage of ionization here because HSO_4^- is a weak acid. Other strong acids are hydrobromic acid (HBr) and hydroiodic acid (HI). At equilibrium, solutions of strong acids will not contain any unionized acid molecules.

The strongest acid that can exist in aqueous solution is the hydronium ion (H_3O^+). The K_a for the hydronium ion is the equilibrium constant for the following proton exchange reaction:

$$H_3O^+(aq) + H_2O(l) \rightleftharpoons H_2O(l) + H_3O^+(aq)$$

Because the species on either side of this equilibrium are the same, the equilibrium constant for this reaction is identically equal to 1, that is, the K_a for the hydronium ion is exactly 1. Acids that are stronger than the hydronium ion ($K_a > 1$) will react completely with water to yield hydronium ion and the conjugate base; therefore, they will be strong acids. This gives us a convenient definition of a strong acid as one with an acid dissociation constant greater than unity (in aqueous solution). All such acids will yield identical H_3O^+ concentrations in water, independent of their individual K_a values—a phenomenon known as ***solvent leveling.***

Like strong acids, *strong bases* are completely ionized in water (see Section 11.1). Hydroxides of alkali metals and certain alkaline earth metals are strong bases. [All alkali metal hydroxides are water soluble. Of the alkaline earth hydroxides, $Be(OH)_2$

and $Mg(OH)_2$ are insoluble; $Ca(OH)_2$ and $Sr(OH)_2$ are slightly soluble; and $Ba(OH)_2$ is soluble.] Some examples of strong bases are

$$NaOH(s) \xrightarrow{H_2O} Na^+(aq) + OH^-(aq)$$
$$KOH \xrightarrow{H_2O} K^+(aq) + OH^-(aq)$$
$$Ba(OH)_2(s) \xrightarrow{H_2O} Ba^{2+}(aq) + 2OH^-(aq)$$

Strictly speaking, these metal hydroxides are not Brønsted bases because they cannot accept a proton. The hydroxide ion (OH^-) formed when they ionize, however, *is* a Brønsted base because it can accept a proton:

$$H_3O^+(aq) + OH^-(aq) \longrightarrow 2H_2O(l)$$

Thus, when we call NaOH or any other metal hydroxide a base, we are actually referring to the OH^- species derived from the hydroxide.

Weak bases, like weak acids, *are weak electrolytes.* Ammonia is a weak base. It ionizes to a very limited extent in water:

$$NH_3(aq) + H_2O(l) \rightleftharpoons NH_4^+(aq) + OH^-(aq)$$

NH_3 does not ionize like an acid because it does not split up to form ions the way, say, HCl does. Instead, NH_3 accepts a proton from water to become the NH_4^+ ion.

The ionization of weak bases is treated in the same way as the ionization of weak bases. Consider the following reaction of a base A^- with water:

$$A^-(aq) + H_2O(aq) \rightleftharpoons HA(aq) + OH^-(aq) \qquad \textbf{(11.12)}$$

The equilibrium constant for this reaction, called the ***base ionization constant*** (K_b), is given by

$$K_b = \frac{a_{HA(aq)}\, a_{OH^-(aq)}}{a_{A^-(aq)}\, a_{H_2O(l)}} \approx \frac{[HA]\,[OH^-]}{[A^-]} \qquad \textbf{(11.13)}$$

where we have assumed that the solutions are sufficiently dilute so that the activities of the solutes can be approximated by their concentrations and that the activity of water is unity. For example, the K_b for NH_3 ($A^- = NH_3$ and $HA = NH_4^+$) is given by

$$K_b = \frac{[NH_4^+]\,[OH^-]}{[NH_3]}$$

The larger the K_b, the stronger the base. The K_b values for a number of common bases are listed in Table 11.4.

The OH^- ion is the strongest base that can exist in aqueous solution. Like the K_a for the hydronium ion, the K_b for the hydroxide ion is exactly 1 because it is the equilibrium constant for the reaction

$$OH^-(aq) + H_2O(l) \rightleftharpoons H_2O(l) + OH^-(l)$$

Bases stronger than OH^- ($K_b > 1$) react with water to produce OH^- and their conjugate acids. For example, the oxide ion (O^{2-}) is a stronger base than OH^-, so it reacts with water completely as follows:

$$O^{2-}(aq) + H_2O(l) \longrightarrow 2OH^-(aq)$$

For this reason the oxide ion does not exist in aqueous solution.

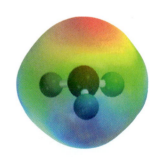

The lone pair (red color) on the N atom accounts for ammonia's basicity.

Table 11.4	Base Ionization Constants (K_b) for Select Weak Bases at 298 K
Base	K_b
$C_2H_5NH_2$ (ethylamine)	5.6×10^{-4}
CH_3NH_2 (methylamine)	4.4×10^{-4}
NH_3 (ammonia)	1.8×10^{-5}
$C_5H_5NH_2$ (pyridine)	1.7×10^{-9}
$C_6H_5NH_2$ (aniline)	3.8×10^{-10}
$C_8H_{10}N_2O_2$ (caffeine)	5.3×10^{-14}
$(NH_2)_2CO$ (urea)	1.5×10^{-14}

Like K_w, both the K_a for weak acids and the K_b for weak bases depend upon temperature. The temperature dependence of these quantities can be determined from the standard enthalpy change of the corresponding ionization reaction (see Equation 10.16). For example, the $\Delta H°$ for the ionization of the weak acid HF,

$$HF(aq) \xrightarrow{H_2O} H^+(aq) + F^-(aq)$$

-13 kJ mol^{-1}. Thus, this reaction is exothermic. According to Le Châtelier's principle, increasing the temperature of an exothermic reaction will shift the equilibrium toward the reactants—that is, the equilibrium constant will decrease. Therefore, we expect the K_a for HF to decrease with temperature (see Problem 11.35 at the end of the chapter). As another example, consider the weak base ammonia (NH_3). The $\Delta H°$ for the ionization of NH_3 in aqueous solution is $+2.4$ kJ mol^{-1}, indicating that this ionization is endothermic. According to Le Châtelier's principle, the K_b for NH_3 will increase with increasing temperature. This increase will be relatively weak, however, given the relatively small value of $\Delta H°$. To determine whether a given K_a or K_b will increase or decrease with temperature, you must first determine the $\Delta H°$ for the ionization reaction. If the reaction is endothermic, then the acid or base ionization constant will increase with increasing temperature. If it is exothermic, then the ionization constant will decrease with increasing temperature.

The Relationship Between the Ionization Constants of Acids and Their Conjugate Bases

An important relationship between the acid ionization constant and the ionization constant of its conjugate base can be derived as follows, using acetic acid as an example:

$$CH_3COOH(aq) \rightleftharpoons H^+(aq) + CH_3COO^-(aq)$$

$$K_a = \frac{[H^+][CH_3COO^-]}{[CH_3COOH]}$$

The conjugate base (CH_3COO^-), supplied, for example, by a sodium acetate (CH_3COONa) solution, reacts with water according to the equation

$$CH_3COO^-(aq) + H_2O(l) \rightleftharpoons CH_3COOH(aq) + OH^-(aq)$$

We can write the base ionization constant (Equation 11.13) as

$$K_b = \frac{[CH_3COOH][OH^-]}{[CH_3COO^-]}$$

The product of these two ionization constants is given by

$$K_a K_b = \frac{[H^+][CH_3COO^-]}{[CH_3COOH]} \times \frac{[CH_3COOH][OH^-]}{[CH_3COO^-]}$$

$$= [H^+][OH^-]$$

$$= K_w$$

This result may seem strange at first, but if we add the two chemical equations we see that the sum is simply the autoionization of water.

(1) $\quad\quad\quad\quad CH_3COOH(aq) \rightleftharpoons H^+(aq) + CH_3COO^-(aq) \quad\quad K_a$

(2) $\quad CH_3COO^-(aq) + H_2O(l) \rightleftharpoons CH_3COOH(aq) + OH^-(aq) \quad\quad K_b$

(3) $\quad\quad\quad\quad\quad\quad H_2O(l) \rightleftharpoons H^+(aq) + OH^-(aq) \quad\quad K_w$

This example illustrates one of the rules for chemical equilibria: When two reactions are added together to give a third reaction, the equilibrium constant for the third reaction is the product of the equilibrium constants for the two added reactions (see Section 10.1). Thus, for any conjugate acid-base pair, it is always true that

$$K_a K_b = K_w \quad\quad\quad\quad \textbf{(11.14)}$$

Expressing Equation 11.14 as

$$K_a = \frac{K_w}{K_b} \quad \text{or} \quad K_b = \frac{K_w}{K_a}$$

shows that the stronger the acid (the larger the K_a), the weaker its conjugage base (the smaller the K_b), and vice versa. This conclusion is illustrated in Example 11.8.

Example 11.8

The K_a values for hydrofluoric acid (HF) and acetic acid (CH_3COOH) at 25°C are 7.1×10^{-4} and 1.8×10^{-5}, respectively. Calculate the K_b values for their conjugate bases at this temperature.

Strategy The conjugate bases for HF and CH_3COOH are F^- and CH_3COO^- (acetate ion), respectively. The K_a's are given, so use Equation 11.14 ($K_w = K_a K_b$) to find the K_b's, using the value of K_w at 25°C (1.01×10^{-14}).

Solution The K_a for HF is 7.7×10^{-4}. Rearranging Equation 11.14 gives the K_b for F^- as

$$K_b = \frac{K_w}{K_a} = \frac{1.01 \times 10^{-14}}{7.1 \times 10^{-4}}$$

$$= 1.4 \times 10^{-11}$$

Similarly, the K_b for the acetate ion is given by

$$K_b = \frac{K_w}{K_a} = \frac{1.01 \times 10^{-14}}{1.8 \times 10^{-5}}$$

$$= 5.6 \times 10^{-10}$$

—Continued

Continued—

Check The K_b for the fluoride ion (F^-) is smaller than that for the acetate ion (CH_3COO^-), that is, F^- is a *weaker* base than CH_3COO^-. This is consistent with the fact that HF is a *stronger* acid than CH_3COOH.

Practice Exercise The values of K_b for ammonia (NH_3) and methylamine (CH_3NH_2) are 1.8×10^{-5} and 4.38×10^{-4}, respectively. Calculate the K_a values for the corresponding conjugate acids.

The values of K_a and K_b for a number of common conjugate acid-base pairs are shown in Table 11.5. Because K_a and K_b values can range over many orders of magnitude, they are often expressed on a logarithmic scale similar to pH. We define pK_a and pK_b as follows:

$$pK_a = -\log_{10} K_a \tag{11.15}$$

$$pK_b = -\log_{10} K_b \tag{11.16}$$

For example the pK_a of HF ($K_a = 7.1 \times 10^{-4}$) is $-\log_{10}(7.1 \times 10^{-4}) = 3.15$ and the pK_b of NH_3 ($K_b = 1.8 \times 10^{-5}$) is $-\log_{10}(1.8 \times 10^{-5}) = 4.74$. Taking the negative logarithm (base 10) of Equation 11.14 ($K_a K_b = K_w$) gives

$$-\log_{10} K_a K_b = -\log_{10} K_w$$
$$-\log_{10} K_a - \log_{10} K_b = -\log_{10} K_w$$
$$pK_a + pK_b = -\log_{10} K_w \tag{11.17}$$

Table 11.5	Relative Strengths of Select Conjugate Acid-Base Pairs					
Acid	K_a	pK_a	**Conjugate Base**	K_b	pK_b
HI (hydroiodic acid)*	$\sim 10^{10}$	-10	I^- (iodide ion)	$\sim 1 \times 10^{-24}$	24
HBr (hydrobromic acid)	$\sim 10^{9}$	-9	Br^- (bromide ion)	$\sim 1 \times 10^{-23}$	23
HCl (hydrochloric acid)*	$\sim 1 \times 10^{6}$	-6	Cl^- (chloride ion)	$\sim 1 \times 10^{-20}$	20
H_2SO_4 (sulfuric acid)*	$\sim 1 \times 10^{3}$	-3	HSO_4^- (hydrogen sulfate ion)	$\sim 1 \times 10^{-17}$	17
HNO_3 (nitric acid)*	$\sim 1 \times 10^{2}$	-2	NO_3^- (nitrate ion)	$\sim 1 \times 10^{-15}$	15
H_3O^+ (hydronium ion)	1	0.00	H_2O (water)	1.0×10^{-14}	14.00
HSO_4^- (hydrogen sulfate ion)	1.3×10^{-2}	1.89	SO_4^{2-} (sulfate ion)	7.7×10^{-13}	12.11
HF (hydrofluoric acid)	7.1×10^{-4}	3.15	F^- (fluoride ion)	1.4×10^{-11}	10.85
HNO_2 (nitrous acid)	4.5×10^{-4}	3.35	NO_2^- (nitrite ion)	2.2×10^{-11}	10.65
HCOOH (formic acid)	1.7×10^{-4}	3.77	$HCOO^-$ (formate ion)	5.9×10^{-11}	10.23
CH_3COOH (acetic acid)	1.8×10^{-5}	4.74	CH_3COO^- (acetate ion)	5.6×10^{-10}	9.26
NH_4^+ (ammonium ion)	5.6×10^{-10}	9.25	NH_3 (ammonia)	1.8×10^{-5}	4.75
HCN (hydrocyanic acid)	4.9×10^{-10}	9.31	CN^- (cyanide ion)	2.0×10^{-5}	4.69
H_2O (water)	1.0×10^{-14}	14.00	OH^- (hydroxyl ion)	1	0.00
NH_3 (ammonia)	$\sim 1.0 \times 10^{-34}$	34	NH_2^- (amide ion)	$\sim 1.0 \times 10^{20}$	-20

*The K_a for strong acids (and amide) given here are highly approximate values estimated from thermodynamic data. They are included here only to show trends in acid strength and should be used with caution.

So, at 25°C, $K_w = 1.01 \times 10^{-14}$ and

$$pK_a + pK_b = 14.00 \qquad \textbf{(11.18)}$$

The pK_a and pK_b values for a number of common conjugate acid-base pairs are also given in Table 11.5.

If we know the relative strengths of two acids, we can predict the extent of reaction between one of the acids and the conjugate base of the other, as illustrated in Example 11.9.

Example 11.9

Predict whether the following reaction will favor reactants or products at equilibrium:

$$HNO_2(aq) + CN^-(aq) \rightleftharpoons HCN(aq) + NO_2^-(aq)$$

Strategy The problem is to determine whether, at equilibrium, the reaction will be shifted to the right, favoring HCN and NO_2^-, or to the left, favoring HNO_2 and CN^-. Which of the two (HNO_2 or HCN) is a stronger acid and hence a stronger proton donor? Which of the two (CN^- or NO_2^-) is a stronger base and hence a stronger proton acceptor? Remember that the stronger the acid, the weaker its conjugate base. The equilibrium will favor the weaker acid and the weaker conjugate base, so use Table 11.5 to compare K_a values.

Solution According to Table 11.5, HNO_2 is a stronger acid than HCN because $K_a(HNO_2) > K_a(HCN)$. Thus, CN^- is a stronger base than NO_2^-. The net reaction will proceed from left to right as written because HNO_2 is a better proton donor than HCN (and CN^- is a better proton acceptor than NO_2^-).

Practice Exercise Predict whether the equilibrium constant for the following reaction is greater than or less than 1:

$$CH_3COOH(aq) + HCOO^-(aq) \rightleftharpoons CH_3COO^-(aq) + HCOOH(aq)$$

Ionization Constants for Diprotic and Triprotic Acids

The treatment of diprotic and triprotic acids is more involved than that of monoprotic acids because diprotic and triprotic acids may yield more than one hydrogen ion per molecule. Diprotic and triprotic acids ionize in a stepwise manner, that is, they lose one proton at a time. An ionization constant expression can be written for each ionization stage. Consequently, two or more equilibrium constant expressions must often be used to calculate the concentrations of species in the acid solution. For carbonic acid (H_2CO_3), for example, we write

$$H_3CO_3(aq) \rightleftharpoons H^+(aq) + HCO_3^-(aq) \quad K_{a_1} = \frac{[H^+][HCO_3^-]}{[H_2CO_3]}$$

$$HCO_3^-(aq) \rightleftharpoons H^+(aq) + CO_3^{2-}(aq) \quad K_{a_2} = \frac{[H^+][CO_3^{2-}]}{[HCO_3^-]}$$

The conjugate base in the first ionization stage becomes the acid in the second ionization stage.

Table 11.6 lists the ionization constants of several diprotic acids and one triprotic acid. For a given acid, the first ionization constant is much larger than the second ionization constant, and so on. This trend occurs because it is easier to remove an H^+ ion from a neutral molecule than to remove an additional H^+ ion from the resulting negatively charged ion.

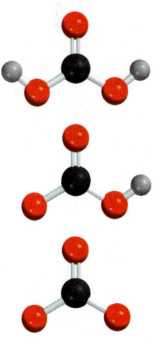

Top to bottom; H_2CO_3, HCO_3^-, and CO_3^{2-}.

Table 11.6 Ionization Constants of Some Diprotic Acids and a Triprotic Acid and Their Conjugate Bases at 25°C

Name of Acid	Formula	Structure	K_a	Conjugate Base	K_b
Sulfuric acid	H_2SO_4	H—O—S(=O)(=O)—O—H	very large	HSO_4^-	very small
Hydrogen sulfate ion	HSO_4^-	H—O—S(=O)(=O)—O$^-$	1.3×10^{-2}	SO_4^{2-}	7.7×10^{-13}
Oxalic acid	$C_2H_2O_4$	H—O—C(=O)—C(=O)—O—H	6.5×10^{-2}	$C_2HO_4^-$	1.5×10^{-13}
Hydrogen oxalate ion	$C_2HO_4^-$	H—O—C(=O)—C(=O)—O$^-$	6.1×10^{-5}	$C_2O_4^{2-}$	1.6×10^{-10}
Sulfurous acid*	H_2SO_3	H—O—S(=O)—O—H	1.3×10^{-2}	HSO_3^-	7.7×10^{-13}
Hydrogen sulfite ion	HSO_3^-	H—O—S(=O)—O$^-$	6.3×10^{-8}	SO_3^{2-}	1.6×10^{-7}
Carbonic acid	H_2CO_3	H—O—C(=O)—O—H	4.2×10^{-7}	HCO_3^-	2.4×10^{-8}
Hydrogen carbonate ion	HCO_3^-	H—O—C(=O)—O$^-$	4.8×10^{-11}	CO_3^{2-}	2.1×10^{-4}
Hydrosulfuric acid	H_2S	H—S—H	9.5×10^{-8}	HS^-	1.1×10^{-7}
Hydrogen sulfide ion[†]	HS^-	H—S$^-$	1×10^{-19}	S^{2-}	1×10^{5}
Phosphoric acid	H_3PO_4	H—O—P(=O)(—O—H)—O—H	7.5×10^{-3}	$H_2PO_4^-$	1.3×10^{-12}
Dihydrogen phosphate ion	$H_2PO_4^-$	H—O—P(=O)(—O—H)—O$^-$	6.2×10^{-8}	HPO_4^{2-}	1.6×10^{-7}
Hydrogen phosphate ion	HPO_4^{2-}	H—O—P(=O)(—O$^-$)—O$^-$	4.8×10^{-13}	PO_4^{3-}	2.1×10^{-2}

*H_2SO_3 has never been isolated and exists in only minute concentration in aqueous solution of SO_2. The K_a value here refers to the process $SO_2(g) + H_2O(l) \rightleftharpoons H^+(aq) + HSO_3^-(aq)$.

[†]The ionization constant of HS^- is very low and difficult to measure. The value listed here is only an estimate.

11.4 | The pH of an Acid or Base Can Be Calculated If Its Ionization Constant Is Known

Strong Acids and Bases

Calculating the pH of solutions of strong monoprotic acids or strong bases is straight-forward because they can be assumed to ionize completely in water. Therefore, the concentration of H_3O^+ (in the case of strong monoprotic acids) or OH^- (in the case of strong bases) will be equal to the initial concentration of acid or base, respectively, as long as the solution is sufficiently concentrated that the contribution from the autoionization of water is negligible.

Example 11.10 shows how to calculate the pH of a solution containing a strong monoprotic acid and a solution containing a strong base.

Example 11.10

Calculate the pH of (a) a 1.7×10^{-4} M HCl solution and (b) a 2.3×10^{-3} M $Ba(OH)_2$ solution.

Strategy HCl is a strong acid and $Ba(OH)_2$ is a strong base, so these species are completely ionized [no HCl or $Ba(OH)_2$ will remain] in aqueous solution. Once the $[H^+]$ is known in part (a), we can use Equation 11.5 ($pH = -\log_{10}[H^+]$) to calculate pH. Once $[OH^-]$ is known in part (b), we can use Equation 11.7 ($pOH = -\log_{10}[OH^-]$) to calculate pOH, and then Equation 11.8 ($pH + pOH = 14.00$) to calculate pH.

Solution (a) The ionization of HCl is

$$HCl(aq) \longrightarrow H^+(aq) + Cl^-(aq)$$

HCl is a strong monoprotic acid so dissociation is complete; therefore, the concentration of H^+ and Cl^- ions is the same as the initial HCl concentration:

$$[H^+] = [Cl^-] = [HCl]_0 = 1.7 \times 10^{-4} M$$

Thus,

$$pH = -\log_{10}[H^+] = -\log_{10}(1.7 \times 10^{-4})$$
$$= 3.77$$

(b) $Ba(OH)_2$ is a strong base, and each $Ba(OH)_2$ unit produces two OH^- ions:

$$Ba(OH)_2(aq) \longrightarrow Ba^{2+}(aq) + 2OH^-(aq)$$

Therefore, the concentration of OH^- ions in the solution will be twice the initial concentration of $Ba(OH)_2$:

$$[OH^-] = 2 \times [Ba(OH)_2] = 2 \times (2.3 \times 10^{-3} M)$$
$$= 4.6 \times 10^{-3} M$$

Thus,

$$pOH = -\log_{10}[OH^-] = -\log_{10}[4.6 \times 10^{-3}]$$
$$= 2.34$$

—Continued

Continued—
and from Equation 11.8,

$$pH = 14.00 - pOH$$
$$= 14.00 - 2.34$$
$$= 11.66$$

Comment In both parts (a) and (b), we have neglected the contribution of the autoionization of water to $[H^+]$ and $[OH^-]$ because 1.0×10^{-7} M is so small compared with 1.7×10^{-4} M and 4.6×10^{-3} M. Also, the calculated pH values may differ slightly from the actual values due to the nonideality of the solution. This will be most pronounced in the $Ba(OH)_2$ solution because it is less dilute and because a Ba^{2+} ion is highly charged and will interact strongly with other ions in the solution.

Practice Exercise Calculate the pH of a 1.8×10^{-5} M $Ba(OH)_2$ solution.

If the strong acid or strong base is very dilute ($< 10^{-6}$ M), then it is necessary to take into account the autoionization of water. This case is discussed later in this section.

Weak Monoprotic Acids

To calculate the pH of a weak monoprotic acid, it is necessary to take into account the ionization equilibrium

$$HA(aq) \rightleftharpoons A^-(aq) + H^+(aq)$$

where the acid and conjugate base are represented by HA and A^-, respectively. Ignoring the autoionization of water, the concentrations of HA, A^-, and H_3O^+ are related by the equilibrium constant expression (Equation 11.11):

$$K_a = \frac{[H^+][A^-]}{[HA]} \tag{11.11}$$

If we let $[HA]_0$ be the initial concentration of the acid and x be the concentrations of $[A^-]$ and $[H_3O^+]$ at equilibrium, then we have from the stoichiometry,

$$HA(aq) \rightleftharpoons A^-(aq) + H^+(aq)$$
$$([HA]_0 - x) \qquad x \qquad x$$

The equilibrium constant expression becomes

$$K_a = \frac{[H^+][A^-]}{[HA]} = \frac{(x)(x)}{[HA]_0 - x}$$

$$K_a = \frac{x^2}{[HA]_0 - x} \tag{11.19}$$

which gives the following equation for $x = [H_3O^+]$:

$$x^2 + K_a x - K_a[HA]_0 = 0 \tag{11.20}$$

Equation 11.20 is a quadratic equation of the form $ax^2 + bx + c = 0$, for which the general solution is given by the quadratic formula:

$$x = \frac{-b \pm \sqrt{b^2 - 4ac}}{2a}$$

If K_a is sufficiently small or $[HA]_0$ is sufficiently large, then the amount of acid dissociated, x, will be small compared to $[HA]_0$ and the x in the denominator of Equation 11.19 can be neglected to give

$$x \approx \sqrt{K_a[HA]_0} \qquad\qquad\qquad (11.21)$$

This approximation is generally considered valid as long as the value of x is less than 5 percent of $[HA]_0$ because 5 percent is typically the level of accuracy to which K_a is known. If this approximation is used, though, it is necessary to confirm that the resulting value of x is indeed less than 5 percent of $[HA]_0$.

Once $x = [H^+]$ is calculated, the pH can be found from Equation 11.5. Example 11.11 shows how to calculate the pH for a weak acid solution.

Example 11.11

Calculate the pH of (a) a 0.50 M HCN solution and (b) a 0.050 M HF solution.

Strategy To determine the pH of these weak acids, it is necessary to know their K_a values. According to Table 11.3, $K_a = 4.9 \times 10^{-10}$ for HCN and 7.1×10^{-4} for HF. Use the procedure outlined previously (which utilizes Equations 11.19 and 11.21) to determine pH.

Solution (a) The ionization of HCN is given by

$$HCN(aq) \rightleftharpoons CN^-(aq) + H^+(aq) \qquad K_a = 4.9 \times 10^{-10}$$

If we let $x = [H^+] = [CN^-]$, then Equation 11.19 gives (with $[HCN]_0 = 0.50$ M)

$$K_a = 4.9 \times 10^{-10} = \frac{x^2}{0.05 - x}$$

The magnitude of K_a is very small relative to $[HCN]_0$, so we can ignore the x in the denominator and use Equation 11.21 to give

$$x \approx \sqrt{K_a[HA]_0} = \sqrt{(4.9 \times 10^{-10})(0.50)}$$
$$= 1.6 \times 10^{-5} \, M$$

The value of x is well below 5 percent of $[HCN]_0 = 0.50$ M, so the approximation is justified. The pH of the solution can then be calculated from Equation 11.5:

$$pH = -\log_{10}[H_3O^+] = -\log_{10} x$$
$$= 4.80$$

(b) The ionization of HF is given by

$$HF(aq) \rightleftharpoons F^-(aq) + H^+(aq) \qquad K_a = 7.1 \times 10^{-4}$$

—Continued

Continued—

If we let $x = [H_3O^+] = [F^-]$, then Equation 11.19 gives (with $[HF]_0 = 0.050\ M$)

$$K_a = 7.1 \times 10^{-4} = \frac{x^2}{0.050 - x}$$

If we assume that x is small relative to $[HF]_0$, we obtain

$$x \approx \sqrt{K_a[HA]_0} = \sqrt{(7.1 \times 10^{-4})(0.050)}$$
$$= 5.9 \times 10^{-3}\ M$$

This value of x is greater than 5 percent of 0.050 M, so the approximation is not valid and we cannot ignore x in the denominator of the equilibrium constant expression. Starting from the ionization equilibrium expression for HF,

$$7.1 \times 10^{-4} = \frac{x^2}{0.050 - x}$$

we get

$$x^2 + (7.1 \times 10^{-4})x - 3.6 \times 10^{-5} = 0$$

The quadratic formula gives

$$x = \frac{-b \pm \sqrt{b^2 - 4ac}}{2a}$$
$$= \frac{-7.1 \times 10^{-4} \pm \sqrt{(7.1 \times 10^{-4})^2 - 4(1)(-3.6 \times 10^{-5})}}{2(1)}$$
$$= \frac{-7.1 \times 10^{-4} \pm 0.012}{2}$$
$$= 5.6 \times 10^{-3}\ M \quad \text{or} \quad -6.4 \times 10^{-3}\ M$$

The second solution ($-6.4 \times 10^{-3}\ M$) is physically impossible because the concentration of ions produced as a result of ionization cannot be negative. Choosing $x = 5.6 \times 10^{-3}\ M$, gives a pH of

$$\text{pH} = -\log_{10}(5.6 \times 10^{-3}) = 2.25$$

Practice Exercise Calculate the pH of (*a*) a 0.00100 M solution of nitrous acid (HNO$_2$), and (*b*) a 0.0153 M solution of formic acid (HCOOH).

Percent Ionization

The magnitude of K_a indicates the strength of an acid (as K_a increases, so does the strength of the acid). Another measure of the strength of an acid is its ***percent ionization,*** which is defined as

$$\text{percent ionization} = \frac{\text{ionized acid concentration at equilibrium}}{\text{initial concentration of acid}} \times 100\% \qquad \textbf{(11.22)}$$

We can compare the strengths of acids in terms of percent ionization only if the concentrations of the acids are the same.

The stronger the acid, the greater the percent ionization. For a monoprotic acid HA, the concentration of the acid that undergoes ionization is equal to the concentration of the H^+ ions or the concentration of the A^- ions at equilibrium. Therefore, we can write the percent ionization as

$$\text{percent ionization} = \frac{[H^+]}{[HA]_0} \times 100\%$$

where $[H^+]$ is the H^+ concentration at equilibrium and $[HA]_0$ is the initial concentration of weak acid HA.

Using the solution to Example 11.11(b), where $[H^+] = 5.6 \times 10^{-3}\ M$, the percent ionization of a 0.050 M HF solution is

$$\text{percent ionization} = \frac{5.6 \times 10^{-3}\ M}{0.050\ M} \times 100\% = 11\%$$

Thus, only about one out of every 9 HF molecules has ionized. This should be expected, though, because HF is a weak acid.

The extent to which a weak acid ionizes depends on the initial concentration of the acid. The more dilute the solution, the greater the percent ionization (Figure 11.8). In qualitative terms, when an acid is diluted, the number of particles (nonionized acid molecules plus ions) per unit volume is reduced. According to Le Châtelier's principle (see Section 10.4), to counteract this "stress" (that is, the dilution), the equilibrium shifts from nonionized acid to H^+ and its conjugate base to produce more particles (ions).

The dependence of percent ionization on initial concentration can be illustrated by the 0.050 M HF solution from Example 11.11(b), where the equilibrium concentration was found to be $5.6 \times 10^{-3}\ M$. We have just shown that the percent ionization of this solution is 11 percent. If we repeat the calculation in Example 11.11(b) for a concentration of 0.50 M, we obtain an equilibrium concentration of 0.019 M, which gives a percent ionization of

$$\frac{0.019\ M}{0.050\ M} \times 100\% = 3.8\%$$

Thus, a more dilute HF solution has a greater percent ionization of the acid.

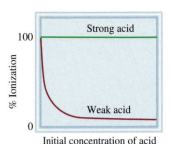

Figure 11.8 Dependence of percent ionization on the initial concentration of acid. Note that at very low concentrations, all acids (weak and strong) are almost completely ionized.

Weak Base Solutions

The calculation of the pH of weak base solutions is similar to the method outlined previously (see Example 11.11) for calculating the pH of weak acids. The determination of pH for a weak base is outlined in Example 11.12.

Example 11.12

Determine the pH of an ammonia solution that is (a) 0.40 M and (b) 0.004 M.

Strategy The procedure here is similar to the one used for a weak acid (see Example 11.11). For the ionization of ammonia, the major species in solution at equilibrium are NH_3, NH_4^+, and OH^+. The hydrogen ion concentration is very small as expected for a basic solution, so it will only be a minor species in solution. As in the case of weak acids, we can usually ignore contributions from the autoionization of water, unless the solution is very dilute.

—Continued

Continued—

Solution The ionization reaction is

$$NH_3(aq) + H_2O(l) \rightleftharpoons NH_4^+(aq) + OH^-(aq) \qquad K_b = 1.8 \times 10^{-5}$$

At equilibrium the concentrations of the various solutes are related by

$$K_b = \frac{[NH_4^+][OH^-]}{[NH_3]} = 1.8 \times 10^{-5}$$

If we ignore the $[OH^-]$ contribution from the autoionization of water, the $[NH_4^+]$ and $[OH^-]$ will be equal, given the stoichiometry of the NH_3 ionization reaction. If we let x be the equilibrium concentration of NH_4^+ and OH^- in mol L^{-1}, we can summarize as follows:

	$[NH_3]$	$[NH_4^+]$	$[OH^-]$
Initial (*M*)	$[NH_3]_0$	0.00	0.00
Change (*M*)	$-x$	$+x$	$+x$
Equilibrium (*M*)	$[NH_3]_0 - x$	x	x

Using these concentrations, the expression for K_b becomes

$$K_b = \frac{x^2}{[NH_3]_0 - x}$$

If we apply the approximation $[NH_3]_0 - x \approx [NH_3]_0$, assuming that x is small, we get

$$x^2 \approx K_b[NH_3]_0$$
$$x = \sqrt{K_a[NH_3]_0}$$

Otherwise, we have the following quadratic equation:

$$x^2 + K_b x - [NH_3]_0 K_b = 0$$

(a) Here $[NH_3]_0 = 0.40\ M$. If we assume that x is small, then

$$x = \sqrt{K_a[NH_3]_0} = \sqrt{(1.8 \times 10^{-5})(0.40)}$$
$$= 2.7 \times 10^{-3}\ M$$

To test the approximation, we write

$$\frac{2.7 \times 10^{-3}\ M}{0.40\ M} \times 100\% = 0.68\%$$

The approximation is valid, therefore, because this value is less than 5 percent.
    The pH of the solution can be calculated by first calculating its pOH. For a sufficiently dilute solution

$$pOH = -\log_{10}[OH^-] = -\log_{10} x$$
$$= -\log_{10}(2.7 \times 10^{-3})$$
$$= 2.57$$

The pH at 25°C is then given by

$$pH = 14.00 - pOH = 14.00 - 2.57 = 11.43$$

—Continued

Continued—

(b) Here $[NH_3]_0 = 0.0040\ M$. If we assume that x is small, then

$$x = \sqrt{K_a[NH_3]_0} = \sqrt{(1.8 \times 10^{-5})(0.0040)}$$
$$= 2.7 \times 10^{-4}\ M$$

To test the approximation, we write

$$\frac{2.7 \times 10^{-4}\ M}{0.0040\ M} \times 100\% = 6.8\%$$

This is larger than 5 percent, so the approximation is invalid and we must use the quadratic equation:

$$x^2 + K_b\,x - [NH_3]_0\,K_b = 0$$
$$x^2 + (1.8 \times 10^{-5})\,x - (7.2 \times 10^{-8}) = 0$$

The quadratic formula then gives

$$x = \frac{-1.8 \times 10^{-5} \pm \sqrt{(1.8 \times 10^{-5})^2 - 4(1)(-7.2 \times 10^{-8})}}{2}$$
$$= -2.8 \times 10^{-4}\ M\ \text{or}\ 2.6 \times 10^{-4}\ M$$

The first solution is physically impossible because concentrations must be positive, so

$$x = 2.6 \times 10^{-4}\ M$$

Proceeding as in part (a),

$$pOH = -\log_{10}[OH^-] = -\log_{10} x$$
$$= -\log_{10}(2.6 \times 10^{-4})$$
$$= 3.60$$

The pH at 25°C is then given by

$$pH = 14.00 - pOH = 14.00 - 3.60 = 10.40$$

Comment As with weak acids, the percent ionization of NH_3 (and weak bases, in general) in aqueous solution increases with decreasing concentration.

Practice Exercise Calculate the pH of a methylamine solution with an initial concentration of (a) 0.26 M and (b) 0.0026 M.

The pH of Dilute Acid and Base Solutions

In the previous calculations, we have assumed that the autoionization of water made a negligible contribution to the OH^- and H_3O^+ concentrations, relative to the contribution from the added acid or base. This is true as long as the solution of acid or base is not very dilute. To see this, consider a $1.0 \times 10^{-8}\ M$ in HCl solution at 25°C. Ignoring the autoionization of water, we would calculate the pH assuming that all the $[H_3O^+]$ in solution comes from the complete ionization of the strong acid HCl:

$$pH = -\log_{10}[H_3O^+] = -\log_{10}(1.0 \times 10^{-8}) = 8.00$$

Thus, we would calculate that this solution would be basic (pH > 7) at 25°C, which makes no sense physically—you cannot make a basic solution by adding a small amount of acid to neutral water.

Dilute Solutions of Weak Acids

Consider, for example, the ionization of a weak acid in aqueous solution:

$$HA(aq) + H_2O(l) \rightleftharpoons A^-(aq) + H_3O^+(aq)$$

For nondilute solutions, we assumed that the contribution to the H_3O^+ concentration from the autoionization of water was negligible, allowing us to equate $[H_3O^+]$ and $[A^-]$. However, because the autoionization of water also contributes H_3O^+ ions, this equality is no longer true if the autoionization of water cannot be ignored. Therefore, in addition to the expression for K_a (Equation 11.11),

$$K_a = \frac{[H^+][A^-]}{[HA]}$$

the equilibrium expression for the autoionization of water (Equation 11.3) must also be satisfied:

$$K_w = [H^+][OH^-]$$

These two equations contain four concentration variables ($[H^+]$, $[OH^-]$, $[HA]$, and $[A^-]$), so two additional equations are necessary to completely specify this system. One additional equation arises from the fact that the total amount of "A" is conserved (mass balance), so

$$[HA] + [A^-] = [HA]_0 \tag{11.23}$$

where $[HA]_0$ is the initial concentration of HA before dissociation. Also, the solution must be electrically neutral, that is, the ion concentrations must be such that the total number of positive charges equals the total number of negative charges. This condition (charge balance) requires that

$$[OH^-] + [A^-] = [H^+] \tag{11.24}$$

If we let $x = [H^+]$, these four simultaneous equations (Equation 11.3, 11.11, 11.23, and 11.24) can be solved (see Problem 11.102) to give a cubic equation for $x = [H^+]$:

$$x^3 + K_a x^2 - (K_w + K_a[HA]_0)\, x - K_a K_w = 0 \tag{11.25}$$

Equation 11.25 must be used if the acid is very dilute, very weak, or both, such that the H^+ ion concentration is expected to be less than 10^{-6} M. (If K_w can be assumed to be negligible, Equation 11.25 reduces to Equation 11.20, as expected.) Although there is no simple formula for the roots of a cubic equation, Equation 11.25 can be solved using a graphing calculator or a computer program such as Mathematica or Mathcad, as illustrated in Example 11.13.

Example 11.13

Calculate the pH of a 2.0×10^{-7} M acetic acid (CH_3COOH) solution at 25°C.

Strategy The solution is very dilute with an initial concentration of acetic acid that is very close to the H^+ concentration of neutral water, so the autoionization of water

—Continued

Continued—
cannot be neglected when determining the pH. Therefore, we must use Equation 11.25 to determine the H^+ ion concentration. K_a for acetic acid is 1.8×10^{-5}.

Solution Using $[HA]_0 = 2.0 \times 10^{-7} M$, $K_a = 1.8 \times 10^{-5}$, and $K_w = 1.0 \times 10^{-14}$, Equation 11.25 gives

$$x^3 + K_a x^2 - (K_w + K_a [HA]_0)\, x - K_a K_w = 0$$
$$x^3 + (1.8 \times 10^{-5})\, x^2 - (3.6 \times 10^{-12})\, x - (1.8 \times 10^{-19}) = 0$$

The coefficients of this equation are too small to be accurately solved using standard graphing calculator or computer software, so it is useful to recast the equation by transforming to a new variable $z = 10^7\, x$. (Because we expect the value of x to be between 10^{-6} and 10^{-7}, z should be a number between 1 and 10.) Applying this transformation and dividing both sides of the equation by 10^{-21} gives

$$z^3 + 180z^2 - 360z - 180 = 0$$

Using a graphing calculator (or appropriate computer program), we find the solutions (roots) of this cubic polynomial equation are -182, -0.414, and 2.310. The first two solutions are negative, so they are physically impossible. Therefore, the H^+ ion concentration is $2.3 \times 10^{-7} M$, after converting back to x from z. This value gives a pH of $-\log_{10}(2.3 \times 10^{-7}) = 6.64$.

Comment The acetic acid could only contribute, at a maximum, an H^+ concentration of $2.0 \times 10^{-7} M$ (the initial concentration of CH_3COOH) to this solution. The fact that the calculated $[H^+]$ is larger than $2.0 \times 10^{-7} M$ indicates that the autoionization of water is nonnegligible in this case.

Practice Exercise Calculate the pH of a $1.0 \times 10^{-4} M$ solution of HCN ($K_a = 4.9 \times 10^{-10}$).

Very Dilute Solutions of Strong Acids

Equation 11.25 can be simplified somewhat in the case of dilute strong acids. For a strong acid, the value of K_a is large and only the terms containing K_a in Equation 11.25 are significant. In this strong acid limit, Equation 11.25 becomes

$$x^2 - [HA]_0\, x - K_w = 0 \tag{11.26}$$

(If K_w is neglected, we get $x = [H^+] = [HA]_0$, which is the usual equation for the H^+ ion concentration of a strong acid.)

Example 11.14 shows how to use Equation 11.26 to determine the pH of a very dilute strong acid.

Example 11.14

Calculate the pH of a $1.0 \times 10^{-7} M$ solution of HCl.

Strategy The initial concentration of HCl is less than $10^{-6} M$; therefore, the autoionization of water cannot be ignored. Because this is a strong acid, we can use Equation 11.26 to calculate the pH.

Solution Using the initial concentration of HCl given, Equation 11.26 gives

$$x^2 - (1.0 \times 10^{-7})\, x - (1.0 \times 10^{-14}) = 0$$

—Continued

Continued—
This is a quadratic equation, so the solution can be determined using the quadratic formula:

$$x = \frac{1.0 \times 10^{-7} \pm \sqrt{(1.0 \times 10^{-7})^2 - 4(1)(-1.0 \times 10^{-14})}}{2}$$

$$= 1.6 \times 10^{-7}\,M \quad \text{or} \quad -6.2 \times 10^{-8}\,M$$

The second solution is physically impossible, so $x = [H^+] = 1.6 \times 10^{-7}\,M$, which gives a pH of 6.80.

Practice Exercise Calculate the pH of a $5.0 \times 10^{-8}\,M$ solution of nitric acid (HNO_3).

Dilute Solutions of Bases

The pH of dilute solutions of weak and strong bases for which the autoionization of water cannot be ignored can be determined by the same procedures used to derive Equations 11.25 and 11.26 for acids. The corresponding equations for dilute weak and strong bases are identical to Equations 11.25 and 11.26, except that K_a is replaced with K_b and x represents $[OH^-]$.

The pH of Diprotic and Triprotic Acids

Example 11.15 shows how to calculate the pH of diprotic acid by calculating the equilibrium concentrations of all the species present in aqueous solution.

$C_2H_2O_4$

Example 11.15

Oxalic acid ($C_2H_2O_4$) is a poisonous substance used chiefly as a bleaching and cleansing agent (for example, to remove bathtub rings). Calculate the concentrations of all the species present at equilibrium in a 0.10 M solution of oxalic acid.

Strategy Determining the equilibrium concentrations of the species of a diprotic acid in aqueous solution is more involved than for a monoprotic acid. Follow the same procedure as that used for a monoprotic acid for each stage, as in Example 11.11. Remember that the conjugate base from the first stage of ionization becomes the acid for the second stage of ionization.

Solution We proceed as follows:
First ionization stage: The first ionization of oxalic acid is given by

$$C_2H_2O_4(aq) \rightleftharpoons H^+(aq) + C_2HO_4^-(aq) \qquad K_{a_1} = 6.5 \times 10^{-2}$$

Letting x be the equilibrium concentration of H^+ and $C_2HO_4^-$, Equation 11.19 gives

$$K_{a_1} = \frac{x^2}{[C_2HO_4]_0 - x}$$

$$6.5 \times 10^{-2} = \frac{x^2}{0.10 - x}$$

If we ignore the x in the denominator (assuming it is small relative to 0.10), we obtain

$$x = 0.081\,M$$

—Continued

Continued—

This concentration is larger than 5 percent of the initial concentration, so we must recalculate x using the quadratic equation:

$$x^2 + (6.5 \times 10^{-2})\,x - 6.5 \times 10^{-3} = 0$$

The result is $x = 0.054\ M$.

The second ionization constant is significantly smaller than the first (6.1×10^{-5} versus 6.5×10^{-2}), so the second ionization will make very little additional contribution to the H^+ ion concentration. Therefore, we can calculate the pH of the solution at this point:

$$pH = -\log_{10}(0.054) = 1.27$$

Second ionization stage: The second ionization of oxalic acid is given by

$$C_2HO_4^-(aq) \rightleftharpoons H^+(aq) + C_2O_4^{2-}(aq) \quad K_{a_2} = 6.1 \times 10^{-5}$$

Letting y be the equilibrium concentration of H^+ and $C_2O_4^{2-}$ ions in mol L^{-1}, we construct the following table:

	$C_2HO_4^-$	H^+	$C_2O_4^{2-}$
Initial (*M*)	0.054	0.054	0
Change (*M*)	$-y$	$+y$	$+y$
Equilibrium (*M*)	$0.054 - y$	$0.054 + y$	y

Thus, the equilibrium expression for the second ionization stage becomes

$$K_{a_2} = \frac{[H^+][C_2O_4^{2-}]}{[C_2HO_4^-]}$$

$$6.1 \times 10^{-5} = \frac{(0.054 + y)(y)}{0.054 - y}$$

Applying the approximation $0.054 + y \approx 0.054$ and $0.054 - y \approx 0.054$, we obtain

$$6.1 \times 10^{-5} = \frac{(0.054)(y)}{0.054}$$

$$y = 6.1 \times 10^{-5}\ M$$

This value is less than 5 percent of 0.054, so the approximation is justified.

Summary:

$$[H^+] = [C_2HO_4^-] = 0.054\ M$$
$$[C_2H_2O_4] = 0.10\ M - 0.054\ M = 0.046\ M$$
$$[C_2O_4^{2-}] = 6.1 \times 10^{-5}$$
$$[OH^-] = K_w/[H^+] = 1.0 \times 10^{-14}/0.054 = 1.9 \times 10^{-13}\ M$$

Practice Exercise Calculate the concentrations of all species present in a 0.01 *M* solution of oxalic acid.

Example 11.15 shows that for diprotic acids, if $K_{a_1} \gg K_{a_2}$, then we can assume that the concentration of H^+ ions is the product of only the first stage of ionization. Furthermore, the concentration of the conjugate base for the second stage of ionization is *numerically* equal to K_{a_2}. One important exception to this is sulfuric acid (H_2SO_4). The first ionization,

$$H_2SO_4(aq) \rightleftharpoons H^+(aq) + HSO_4^-(aq)$$

is complete (sulfuric acid is a strong acid). In addition, the second ionization,

$$HSO_4^-(aq) \rightleftharpoons H^+(aq) + SO_4^{2-}(aq) \qquad\qquad K_{a_2} = 1.3 \times 10^{-2}$$

has a large enough K_a that it makes a nonnegligible contribution to the H^+ ion concentration.

The procedure for calculating the pH of a triprotic acid is similar to that for a diprotic acid. In this case of a triprotic acid, both the second and third ionizations do not generally affect the pH. For example, phosphoric acid (H_3PO_4) is a triprotic acid (i.e., it has three ionizable hydrogen atoms):

$$H_3PO_4(aq) \rightleftharpoons H^+(aq) + H_2PO_4^-(aq) \qquad K_{a_1} = 7.5 \times 10^{-3}$$
$$H_2PO_4^-(aq) \rightleftharpoons H^+(aq) + HPO_4^{2-}(aq) \qquad K_{a_2} = 6.2 \times 10^{-8}$$
$$HPO_4^{2-}(aq) \rightleftharpoons H^+(aq) + PO_4^{3-}(aq) \qquad K_{a_3} = 4.8 \times 10^{-13}$$

Phosphoric acid is a weak acid, though, and its ionization constants decrease markedly for the second and third stages. Thus, in a solution containing phosphoric acid, the concentration of the nonionized acid is the highest, and the only other species present in significant concentrations are H^+ and $H_2PO_4^-$ ions.

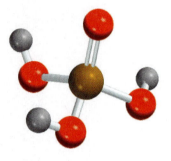

H_3PO_4

11.5 | The Strength of an Acid Is Determined in Part by Molecular Structure

The strength of an acid depends on a number of factors, such as the properties of the solvent, the temperature, and the molecular structure of the acid. When we compare the strengths of two acids, we can eliminate some variables by considering their properties in the same solvent and at the same temperature and concentration. Then we can focus on the structure of the acids.

Consider a certain acid HX. The strength of HX is measured by its tendency to ionize in solution:

$$HX \longrightarrow H^+ + X^-$$

Two factors influence the extent to which the acid ionizes. One is the strength of the H—X bond. The stronger the bond, the more difficult it is for the HX molecule to break up and hence the *weaker* the acid. The other factor is the polarity of the H—X bond. Any difference in the electronegativities between H and X results in a polar bond:

$$\overset{+\delta \quad -\delta}{H\!-\!X}$$

If the bond is highly polarized (that is, if there is a large accumulation of positive and negative charges on H and X^-), HX will tend to break up into H^+ and X^- ions. Thus, a high degree of polarity characterizes a *stronger* acid. In the following sections, we consider some examples in which either bond strength or bond polarity plays a prominent role in determining acid strength.

Table 11.7	Bond Enthapy and Acid Strengths for Hydrohalic Acids		
Acid	Bond Enthalpy (kJ mol^{-1})	K_a	Acid Strength
HF	568.2	7.1×10^{-4}	weak
HCl	431.9	$\sim 10^7$	strong
HBr	366.1	$\sim 10^9$	strong
HI	298.3	$\sim 10^{10}$	strong

Hydrohalic Acids

The halogens form a series of binary acids called the hydrohalic acids (HF, HCl, HBr, and HI). Of this series, which factor (bond strength or bond polarity) is predominant in determining the strength of the binary acids? Consider first the strength of the H—X bond in each of these acids. According to Table 11.7, HF has the highest bond enthalpy of the four hydrogen halides and HI has the lowest. It takes 568.2 kJ mol^{-1} to break the H—F bond and only 298.3 kJ mol^{-1} to break the H—I bond. Based on bond energy, HI should be the strongest acid because it has the easiest bond to break to form the H$^+$ and I$^-$ ions. Next, consider the polarity of the H—X bond. In this series of four acids, the polarity of the bond decreases from HF to HI because F is the most electronegative of the halogens (see Figure 3.9). Based on bond polarity, then, HF should be the strongest acid because it has the largest accumulation of positive and negative charges on the H and F atoms, respectively. Thus, we have two competing factors to consider when determining the strength of binary acids. The fact that HI is a strong acid and HF is a weak acid indicates that bond energy is the predominant factor in determining the acid strength of binary acids. In this series of binary acids, the weaker the bond, the stronger the acid, so the strength of the acids (Table 11.7) increases as follows:

$$HF \ll HCl < HBr < HI$$

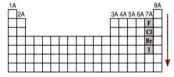

Strength of hydrohalic acids increases from HF to HI.

Oxoacids

Oxoacids (see Section 0.3) contain hydrogen, oxygen, and one other element Z, which occupies a central position. Figure 11.9 shows the Lewis structures of several common oxoacids. As you can see, these acids are characterized by the presence

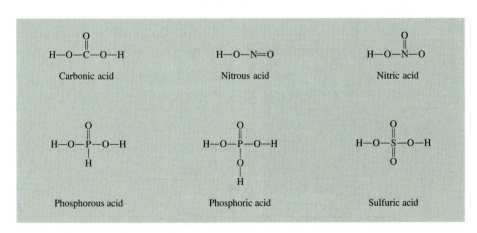

Figure 11.9 Lewis structures of some common oxoacids. For simplicity, the formal charges have been omitted.

of one or more O—H bonds. The central atom Z might also have other groups attached to it:

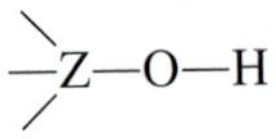

$$\diagup \mkern-10mu Z\mkern-2mu-\mkern-2mu O\mkern-2mu-\mkern-2mu H$$

If Z is an electronegative element (such as N or S) or is bonded to several other electronegative groups, it will attract electrons, thus making the Z—O bond more covalent and the O—H bond more polar. Consequently, the tendency for the hydrogen to be donated as an H^+ ion increases.

To compare their strengths, it helps to divide the oxoacids into two groups, those having different central atoms that are from the same group of the periodic table (and that have the same number of oxygen atoms around the central atom) and those having the same central atom but different numbers of attached groups.

1. *Oxoacids Having Different Central Atoms That Are from the Same Group of the Periodic Table and That Have the Same Number of Oxygen Atoms around the Central Atom.* The first group includes acids such as $HClO_3$ and $HBrO_3$ because both Cl and Br are from Group VIIA of the periodic table and both have the same number of oxygen atoms around the central atom:

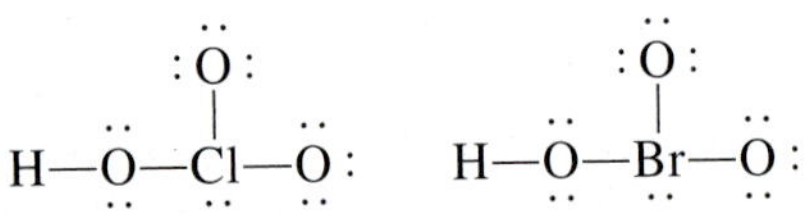

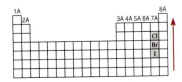

Strength of halogen-containing oxoacids having the same number of O atoms increases from bottom to top.

Within this group, acid strength increases with increasing electronegativity of the central atom. Cl is more electronegative than Br, so it attracts the electron pair it shares with oxygen (in the Cl—O—H group) to a greater extent than Br does. Consequently, the O—H bond is more polar in chloric acid than in bromic acid and ionizes more readily. Therefore, $HClO_3$ is a stronger acid than $HBrO_3$:

$$HClO_3 > HBrO_3$$

2. *Oxoacids Having the Same Central Atom but Different Numbers of Attached Groups.* The second group of acids includes the oxoacids of chlorine shown in Figure 11.10. In this series, the ability of chlorine (the central atom) to draw electrons away from the OH group (thus making the O—H bond more polar) increases with the number of electronegative O atoms attached to Cl. Thus, acid strength increases as the number of oxygen atoms around the central atom increases, making $HClO_4$ the strongest of these four acids (because it has the

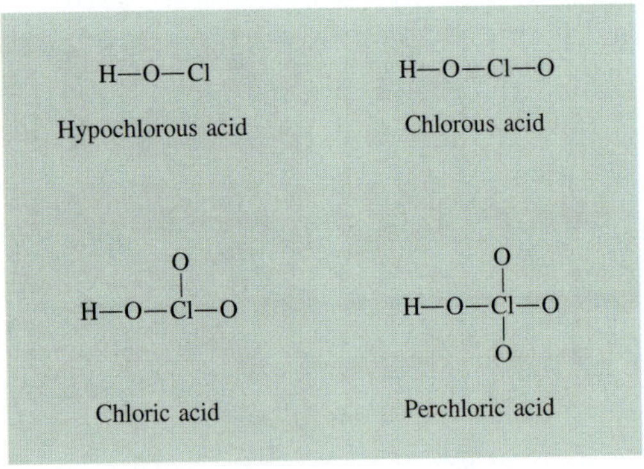

Figure 11.10 Lewis structures of the oxoacids of chlorine.

H—O—Cl	H—O—Cl—O
Hypochlorous acid	Chlorous acid

Chloric acid: H—O—Cl—O with O above Cl

Perchloric acid: H—O—Cl—O with O above and O below Cl

largest number of O atoms attached to Cl) and HClO the weakest (because it has the fewest number of O atoms attached to Cl):

$$HClO_4 > HClO_3 > HClO_2 > HClO$$

Example 11.16 reinforces these concepts by comparing the strengths of acids based on their molecular structures.

Example 11.16

Predict the relative strengths of the oxoacids in each of the following groups: (a) HClO, HBrO, and HIO; and (b) HNO_3 and HNO_2.

Strategy Examine the molecular structure. In part (a) the three acids have similar structures but differ only in the central atom (the halogens Cl, Br, and I). When oxoacids have different central atoms from the same group of the periodic table and their central atoms are surrounded by the same number of oxygen atoms, the strongest acid has the most electronegative central atom. In part (b) the acids have the same central atom (N) but differ in the number of O atoms. When oxoacids have the same central atom but different numbers of attached oxygen atoms, the strongest acid has the central atom with the most oxygen atoms.

Solution (a) HClO, HBrO, and HIO all have the same structure, and the halogens (the central atoms) all have the same number of oxygen atoms around the central atom. Because the electronegativity decreases from Cl to I, the polarity of the X—O bond (where X denotes a halogen atom) increases from HClO to HIO, and the polarity of the O—H bond decreases from HClO to HIO. Thus, the acid strength decreases as follows:

$$HClO > HBrO > HIO$$

(b) The structures of HNO_3 (nitric acid) and HNO_2 (nitrous acid) are shown in Figure 11.9. Because HNO_3 has more oxygen atoms around the nitrogen atom than HNO_2, HNO_3 is a stronger acid than HNO_2.

Practice Exercise Which of the following acids is weaker: H_3PO_3 or H_3PO_4?

Carboxylic Acids

The discussion so far has focused on inorganic acids. A group of organic acids that also deserves attention is the carboxylic acids, whose Lewis structures can be generalized as follows:

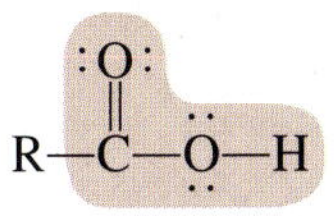

$$R-\overset{\overset{\displaystyle :O:}{\|}}{C}-\overset{..}{\underset{..}{O}}-H$$

where R is part of the acid molecule and the shaded portion represents the *carboxyl group* (—COOH). The strengths of carboxylic acids depend on the nature of the R group. Consider, for example, acetic acid and chloroacetic acid:

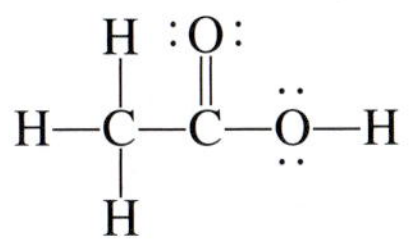

acetic acid ($K_a = 1.8 \times 10^{-5}$) 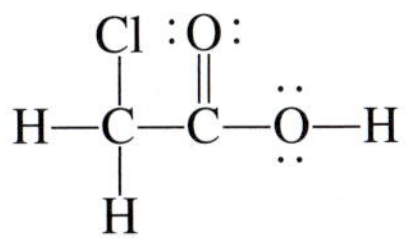   chloroacetic acid ($K_a = 1.4 \times 10^{-3}$)

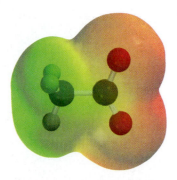

Electrostatic potential map of the acetate ion. The electron density is evenly distributed between the two O atoms.

The presence of the Cl atom in chloroacetic acid shifts electron density toward the R group because Cl is more electronegative than H, thereby making the O—H bond more polar. Consequently, there is a greater tendency for the acid to ionize:

$$CH_2ClCOOH(aq) \rightleftharpoons CH_2ClCOO^-(aq) + H^+(aq)$$

The conjugate base of the carboxylic acid, called the carboxylate anion ($RCOO^-$), can exhibit resonance:

$$R—\overset{\overset{\displaystyle :O:}{\|}}{C}—\overset{..}{\underset{..}{O}}:^- \longleftrightarrow R—\overset{\overset{\displaystyle :\overset{..}{O}:^-}{|}}{C}=\overset{..}{O}$$

In the language of molecular orbital theory, we attribute the stability of the anion to its ability to spread or delocalize the electron density over several atoms. The greater the extent of electron delocalization, the more stable the anion and the greater the tendency for the acid to ionize. Thus, benzoic acid (C_6H_5COOH, $K_a = 6.5 \times 10^{-5}$) is a stronger acid than acetic acid because the benzene ring (see pages 260–261) facilitates electron delocalization, so that the benzoate anion ($C_6H_5COO^-$) is more stable than the acetate anion (CH_3COO^-).

11.6 | Many Salts Have Acid-Base Properties in Aqueous Solution

As defined in Section 11.1, a salt is an ionic compound formed by the reaction between an acid and a base. Salts are strong electrolytes that completely dissociate into ions in water. The term **salt hydrolysis** describes *the reaction of an anion or a cation of a salt, or both, with water.* Salt hydrolysis may produce neutral solutions, basic solutions, or acidic solutions, depending on the salt.

The word *hydrolysis* is derived from the Greek words *hydro,* meaning "water," and *lysis,* meaning "to split apart."

Salts That Produce Neutral Solutions

It is generally true that salts containing an alkali metal ion or alkaline earth metal ion (except Be^{2+}) and the conjugate base of a strong acid (for example, Cl^-, Br^-, and NO_3^-) do not undergo hydrolysis to an appreciable extent, and their solutions are assumed to be neutral. For instance, when $NaNO_3$, a salt formed by the reaction of $NaOH$ with HNO_3, dissolves in water, it dissociates completely as follows:

In reality, all positive ions give acid solutions in water.

$$NaNO_3(s) \xrightarrow{\text{H}_2\text{O}} Na^+(aq) + NO_3^-(aq)$$

The hydrated Na^+ ion neither donates nor accepts H^+ ions. The NO_3^- ion is the conjugate base of the strong acid HNO_3, and it has no affinity for H^+ ions. Consequently, a solution containing Na^+ and NO_3^- ions is neutral, with a pH of about 7.

Salts That Produce Basic Solutions

The solution of a salt derived from a strong base and a weak acid is basic. For example, the dissociation of sodium acetate (CH_3COONa) in water is given by

$$CH_3COONa(s) \xrightarrow{\text{H}_2\text{O}} Na^+(aq) + CH_3COO^-(aq)$$

The hydrated Na^+ ion has no acidic or basic properties. The acetate ion (CH_3COO^-), however, is the conjugate base of the weak acid CH_3COOH and therefore has an affinity for H^+ ions. The hydrolysis reaction is given by

$$CH_3COO^-(aq) + H_2O(l) \rightleftharpoons CH_3COOH(aq) + OH^-(aq)$$

Because this reaction produces OH^- ions, the sodium acetate solution will be basic.

The equilibrium constant for this hydrolysis reaction is the base ionization constant expression for CH_3COO^-, so we use Equation 11.13 to write

$$K_b = \frac{[CH_3COOH][OH^-]}{[CH_3COO^-]}$$

Because each CH_3COO^- ion that hydrolyzes produces one OH^- ion, the concentration of OH^- at equilibrium is the same as the concentration of CH_3COO^- that hydrolyzed. We can define the *percent hydrolysis* as

$$\% \text{ hydrolysis} = \frac{[CH_3COO^-]_{\text{hydrolyzed}}}{[CH_3COO^-]_{\text{initial}}} \times 100\%$$

$$= \frac{[OH^-]_{\text{equilibrium}}}{[CH_3COO^-]_{\text{initial}}} \times 100\%$$

A calculation based on the hydrolysis of CH_3COONa is illustrated in Example 11.17. In solving salt hydrolysis problems, we follow the same procedure we used for weak acids and weak bases.

Example 11.17

Calculate the pH of a 0.15 M solution of sodium acetate (CH_3COONa). What is the percent hydrolysis?

Strategy In solution, the salt CH_3COONa dissociates completely into Na^+ and CH_3COO^- ions. The Na^+ ion, as we saw earlier, does not react with water and has no effect on the pH of the solution. The CH_3COO^- ion is the conjugate base of the weak acid CH_3COOH. Therefore, we expect that it will react to a certain extent with water to produce CH_3COOH and OH^-, and the solution will be basic.

Solution

Step 1: *Determine initial ion concentrations:* Because we started with a 0.15 M sodium acetate solution, the concentrations of the ions (Na^+ and CH_3COO^-) are also equal to 0.15 M after dissociation. Of these ions, only the acetate ion will react with water:

$$CH_3COO^-(aq) + H_2O(l) \rightleftharpoons CH_3COOH(aq) + OH^-(aq)$$

At equilibrium, the major species in solution are CH_3COOH, CH_3COO^-, and OH^-. The concentration of the H^+ ion is very small (as we would expect for a basic solution), so it is treated as a minor species. We ignore the ionization of water.

Step 2: *Determine equilibrium concentrations:* Because CH_3COO^- is a weak base ($K_b = 5.6 \times 10^{-10}$), we can find the concentrations of CH_3COOH, CH_3COO^-, and OH^- using the methods in Section 11.4. From Equation 11.13 we have

$$K_b = \frac{[CH_3COOH][OH^-]}{[CH_3COO^-]} = \frac{x^2}{0.15 - x} = 5.6 \times 10^{-10}$$

—Continued

Continued—

where x represents the equilibrium concentrations of CH_3COOH and OH^-. Because K_b is very small and the initial concentration of the base is large, we can apply the approximation $0.15 - x \approx 0.15$:

$$5.6 \times 10^{-10} = \frac{x^2}{0.15 - x} \approx \frac{x^2}{0.15}$$

$$x = 9.2 \times 10^{-6} \, M$$

The value of x is much less than 5 percent of the initial concentration (0.15 M), so our approximation is justified.

Step 3: *Determine the pH:* At equilibrium,

$$[OH^-] = 9.2 \times 10^{-6} \, M$$

so

$$pOH = -\log_{10}(9.2 \times 10^{-6})$$
$$= 5.04$$

and

$$pH = 14.00 - 5.04$$
$$= 8.96$$

Thus, the solution is basic, as we would expect. The percent hydrolysis is given by

$$\% \text{ hydrolysis} = \frac{9.2 \times 10^{-6} \, M}{0.15 \, M} \times 100\%$$
$$= 0.0061\%$$

Check The result shows that only a very small amount of the anion undergoes hydrolysis, as expected for a weak base.

Practice Exercise Calculate the pH of a 0.24 M solution of sodium formate (HCOONa).

Salts That Produce Acidic Solutions

When a salt derived from a strong acid such as HCl and a weak base such as NH_3 dissolves in water, the solution becomes acidic. For example, consider the process

$$NH_4Cl(s) \xrightarrow{\text{H}_2\text{O}} NH_4^+(aq) + Cl^-(aq)$$

The Cl^- ion, being the conjugate base of a strong acid, has no affinity for H^+ and no tendency to hydrolyze. The ammonium ion (NH_4^+) is the conjugate acid of the weak base NH_3 and ionizes as follows:

$$NH_4^+(aq) + H_2O(l) \rightleftharpoons NH_3(aq) + H_3O^+(aq)$$

Because hydronium ions are produced, the pH of the solution decreases. The equilibrium constant (or ionization constant) for this process is given by

$$K_a = \frac{[NH_3][H^+]}{[NH_4^+]} = \frac{K_w}{K_b} = \frac{1.0 \times 10^{-14}}{1.8 \times 10^{-5}} = 5.6 \times 10^{-10}$$

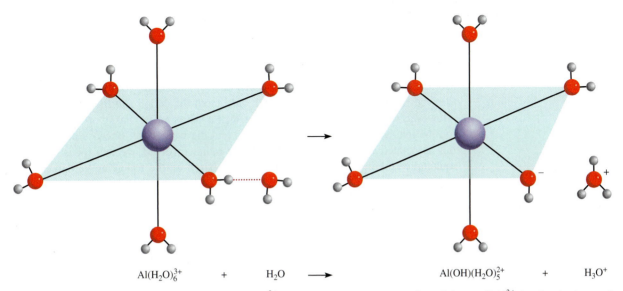

$$Al(H_2O)_6^{3+} \quad + \quad H_2O \quad \longrightarrow \quad Al(OH)(H_2O)_5^{2+} \quad + \quad H_3O^+$$

Figure 11.11 The six H_2O molecules surround the Al^{3+} ion octahedrally. The attraction of the small Al^{3+} ion for the lone pairs on the oxygen atoms is so great that the O—H bonds in an H_2O molecule attached to the metal cation are weakened, facilitating the loss of a proton (H^+) to an incoming H_2O molecule. This hydrolysis of the metal cation makes the solution acidic.

and we can calculate the pH of an ammonium chloride solution following the same procedure used in Example 11.17.

In principle, *all* metal ions react with water to produce an acidic solution. However, because the extent of hydrolysis is most pronounced for the small and highly charged metal cations such as Al^{3+}, Cr^{3+}, Fe^{3+}, Bi^{3+}, and Be^{3+}, we generally neglect the relatively small interaction of alkali metal ions and most alkaline earth metal ions with water. When aluminum chloride ($AlCl_3$) dissolves in water, the Al^{3+} ions act as Lewis acids to accept lone electron pairs from the water oxygen to generate the hydrated form, $Al(H_2O)_6^{3+}$, shown in Figure 11.11.

Consider one bond between the metal ion and an oxygen atom from one of the six water molecules in $Al(H_2O)_6^{3+}$. The positively charged Al^{3+} ion draws electron density toward itself, increasing the polarity of the O—H bonds. Consequently, the H atoms have a greater tendency to ionize than those in water molecules not involved in hydration. The resulting ionization process can be written as

$$Al(H_2O)_6^{3+} + H_2O(l) \rightleftharpoons Al(OH)(H_2O)_5^{2+} + H_3O^+(aq)$$

The equilibrium constant for the metal cation hydrolysis is given by

$$K_a = \frac{[Al(OH)(H_2O)_5^{2+}][H_3O^+]}{[Al(H_2O)_6^{3+}]} = 1.3 \times 10^{-5}$$

$Al(OH)(H_2O)_5^{2+}$ can undergo further ionization,

$$Al(OH)(H_2O)_5^{2+} + H_2O(l) \rightleftharpoons Al(OH)_2(H_2O)_4^+ + H_3O^+(aq)$$

until all six water molecules are ionized. However, it is generally sufficient to take into account only the first stage of hydrolysis.

The extent of hydrolysis is greatest for the smallest and most highly charged ions because a "compact" highly charged ion more effectively polarizes the O—H bond, thus facilitating ionization. This is why relatively large ions of low charge such as Na^+ and K^+ do not undergo appreciable hydrolysis.

Salts in Which Both the Cation and the Anion Hydrolyze

So far, we have considered salts in which only one ion undergoes hydrolysis. For salts derived from a weak acid and a weak base, both the cation and the anion hydrolyze. However, whether a solution containing such a salt is acidic, basic, or neutral depends on the relative strengths of the weak acid and the weak base. Because the mathematics associated with this type of system are rather involved, we will focus on making qualitative predictions about these solutions based on the following guidelines:

▸ $K_b > K_a$. If K_b for the anion is greater than K_a for the cation, then the solution must be basic because the anion will hydrolyze to a greater extent than the cation. At equilibrium, there will be more OH^- ions than H^+ ions.

▸ $K_b < K_a$. Conversely, if K_b for the anion is smaller than K_a for the cation, the solution will be acidic because cation hydrolysis will be more extensive than anion hydrolysis.

▸ $K_b \approx K_a$. If K_a is approximately equal to K_b, the solution will be nearly neutral.

The calculation of the pH of a salt in which both the anion and cation hydrolyze is complicated. There are six ion concentrations that must be determined—that of the anion and its conjugate acid, the cation and its conjugate base, $[H_3O^+]$, and $[OH^-]$. The procedure for this is similar to the calculation of the pH for dilute weak acids (Section 11.4). To solve for these six concentrations, six equations are necessary: Three of these equations come from the equilibrium expressions for K_a of the cation, K_b for the anion, and K_w for water. The other three equations come from ensuring charge neutrality and material balance for both the cation- and anion-derived species. Table 11.8 summarizes the behavior in aqueous solution of the salts discussed in this section.

Example 11.18 shows how to predict the acid-base properties of salt solutions.

Example 11.18

Predict whether the following solutions will be acidic, basic, or nearly neutral: (a) NH_4I, (b) $NaNO_2$, (c) $FeCl_3$, and (d) NH_4F.

Strategy In deciding whether a salt will undergo hydrolysis, ask yourself the following questions: Is the cation a highly charged metal ion or the conjugate acid of a weak base

—Continued

Table 11.8 Acid-Base Properties of Salts

Type of Salt	Examples	Ions That Undergo Hydrolysis	pH of Solution
Cation from strong base; anion from strong acid	$NaCl$, KI, KNO_3, $RbBr$, $BaCl_2$	None	≈ 7
Cation from strong base; anion from weak acid	CH_3COONa, KNO_2	Anion	>7
Cation from weak base, anion from strong acid	NH_4Cl, NH_4NO_3	Cation	<7
Cation from weak base; anion from weak acid	NH_4NO_2, CH_3COONH_4, NH_4CN	Anion and cation	<7 if $K_b < K_a$ ≈ 7 if $K_b \approx K_a$ >7 if $K_b > K_a$
Small, highly charged cation; anion from strong acid	$AlCl_3$, $Fe(NO_3)_3$	Hydrated cation	<7

Continued—
(such as the ammonium ion)? Is the anion the conjugate base of a weak acid? If the answer is yes to either question, then hydrolysis will occur. In cases where both the cation and the anion react with water, the pH of the solution will depend on the relative magnitudes of K_a for the cation and K_b for the anion (see Tables 11.3 through 11.5).

Solution We first break up the salt into its cation and anion components and then examine the possible reaction of each ion with water.

(a) The cation is NH_4^+, which will hydrolyze to produce NH_3 and H_3O^+. The I^- anion is the conjugate base of the strong acid HI. Therefore, I^- will not hydrolyze and the solution is acidic.

(b) The Na^+ cation does not hydrolyze to any appreciable extent. The NO_2^- anion is the conjugate base of the weak acid HNO_2 and will hydrolyze to give HNO_2 and OH^-. The solution will be basic.

(c) Fe^+ is a small metal ion with a high charge and hydrolyzes to produce H_3O^+ ions. The Cl^- ion does not hydrolyze because it is the conjugate base of a strong acid (HCl). The solution, therefore, will be acidic.

(d) Both the NH_4^+ and F^- ions will hydrolyze. According to Table 11.5, the K_a of NH_4^+ (5.6×10^{-10}) is greater than the K_b for F^- (1.4×10^{-11}). Therefore, the solution will be acidic.

Practice Exercise Predict whether the following solutions will be acidic, basic, or nearly neutral: (a) $LiClO_4$, (b) Na_3PO_4, (c) $Bi(NO_3)_2$, and (d) NH_4CN.

Salts of Amphoteric Ions

Some anions are amphoteric, that is, they can act as either an acid or a base. For example, the bicarbonate ion (HCO_3^-) can ionize or undergo hydrolysis as follows (see Table 11.6):

$$HCO_3^-(aq) \rightleftharpoons H^+(aq) + CO_3^{2-}(aq) \qquad K_{a_2} = 4.8 \times 10^{-11}$$
$$HCO_3^-(aq) + H_2O(l) \rightleftharpoons H_2CO_3(aq) + OH^-(aq) \quad K_b = K_w/K_{a_1} = 2.4 \times 10^{-8}$$

where K_{a_1} and K_{a_2} are the first and second ionization constants, respectively, for the diprotic acid H_2CO_3. Because $K_b > K_a$, the hydrolysis reaction will outweigh the ionization process. Thus, a solution of sodium bicarbonate ($NaHCO_3$) will be basic.

A full calculation of the pH of amphoteric ion salts is complicated. To determine the H^+ concentration, we must take into account the acid and base equilibria, the autoionization of water, and the charge and material balance—five equations in all for the five unknown concentrations ($[H^+]$, $[OH^-]$, $[CO_3^{2-}]$, $[HCO_3^-]$, and $[H_2CO_3]$). With a few approximations, however, it can be shown that the pH of an amphoteric salt can be approximated by

$$pH \approx \frac{pK_{a_1} + pK_{a_2}}{2} \qquad (11.27)$$

For the bicarbonate solution mentioned previously, the pK_a values calculated from the K_a values in Table 11.6 are $pK_{a_1} = 6.38$ and $pK_{a_2} = 10.32$, so Equation 11.27 gives

$$pH = \frac{6.38 + 10.32}{2} \approx 8.35$$

The solution is basic, consistent with the conclusion drawn previously from the relative sizes of K_{a_2} and $K_b = K_w/K_{a_1}$.

11.7 | Oxide and Hydroxide Compounds Can Be Acidic or Basic in Aqueous Solution Depending on Their Composition

Oxides can be generally classified as acidic, basic, or amphoteric. Our discussion of acid-base reactions would be incomplete if we did not examine the properties of these compounds. Figure 11.12 shows the formulas of a number of oxides of the representative elements. All alkali metal oxides and all alkaline earth metal oxides except BeO are basic. Beryllium oxide and several metallic oxides in Groups 3A and 4A are amphoteric. Nonmetallic oxides in which the number of oxygen atoms around a central representative element is high are acidic (e.g., N_2O_5, SO_3, and Cl_2O_7), but those in which the number of oxygen atoms is low (e.g., CO and NO) show no measurable acidic properties. No nonmetallic oxides are known to have basic properties.

The basic metallic oxides react with water to form metal hydroxides:

$$Na_2O(s) \xrightarrow{H_2O} 2NaOH(aq)$$

$$BaO(s) \xrightarrow{H_2O} Ba(OH)_2(aq)$$

The reactions between acidic oxides and water are as follows:

$$CO_2(g) + H_2O(l) \rightleftharpoons H_2CO_3(aq)$$
$$SO_3(g) + H_2O(l) \rightleftharpoons H_2SO_4(aq)$$
$$N_2O_2(g) + H_2O(l) \rightleftharpoons 2HNO_3(aq)$$
$$P_4O_{10}(s) + 6H_2O(l) \rightleftharpoons 4H_3PO_4(aq)$$
$$Cl_2O_7(l) + 6H_2O(l) \rightleftharpoons 2HClO_4(aq)$$

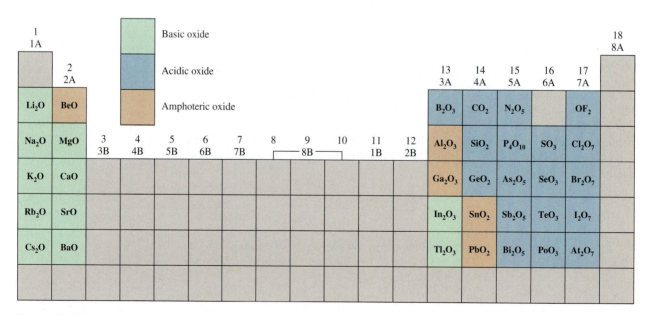

Figure 11.12 Oxides of the representative elements. For many elements, more than one oxide exists. For such elements, only the oxide with the largest number of attached oxygen atoms is shown.

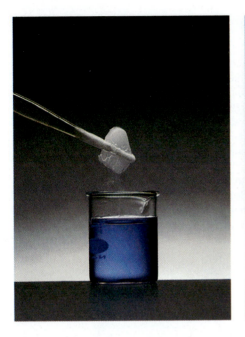 

Figure 11.13 (Left) A beaker of water to which a few drops of bromothymol blue indicator have been added. (Right) As dry ice is added to the water, the CO_2 reacts to form carbonic acid, which turns the solution acidic and changes the color from blue to yellow.

The reaction between CO_2 and H_2O explains why when pure water is exposed to air (which contains CO_2), it gradually reaches a pH of about 5.5 (Figure 11.13). The reaction between SO_3 and H_2O is largely responsible for acid rain (Figure 11.14).

Reactions between acidic oxides and bases and those between basic oxides and acids resemble normal acid-base reactions in that the products are a salt and water:

$$CO_2(g) + 2NaOH(aq) \longrightarrow Na_2CO_3(aq) + H_2O(l)$$

acidic oxide   base   salt   water

$$BaO(s) + 2HNO_3(aq) \longrightarrow Ba(NO_3)_2(aq) + H_2O(l)$$

basic oxide   acid   salt   water

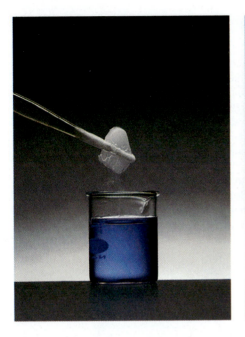

Figure 11.14 A forest damaged by acid rain.

Aluminum oxide (Al_2O_3) is amphoteric (see Figure 11.12). Depending on the reaction conditions, it can behave as either an acidic oxide or a basic oxide. For example, Al_2O_3 acts as a base with hydrochloric acid to produce a salt ($AlCl_3$) and water:

$$Al_2O_3(s) + 6HCl(aq) \longrightarrow 2AlCl_3(aq) + 3H_2O(l)$$

and acts as an acid with sodium hydroxide:

$$Al_2O_3(s) + 2NaOH(aq) + 3H_2O(l) \longrightarrow 2NaAl(OH)_4(aq)$$

Only a salt, $NaAl(OH)_4$ [containing the Na^+ and $Al(OH)_4^-$ ions], is formed in the latter reaction; no water is produced. Nevertheless, this can still be classified as an acid-base reaction because Al_2O_3 neutralizes NaOH.

Some transition metal oxides in which the oxygen to metal atom ratio is large are acidic. Two examples are manganese(VII) oxide (Mn_2O_7) and chromium(VI) oxide (CrO_3), both of which react with water to produce acids:

$$Mn_2O_7(l) + H_2O(l) \longrightarrow 2HMnO_4(aq)$$

permanganic acid

$$CrO_3(s) + H_2O(l) \longrightarrow H_2CrO_4(aq)$$

chromic acid

Antacids and the pH Balance in Your Stomach

An average adult produces between 2 and 3 L of gastric juice daily. Gastric juice is a thin, acidic digestive fluid secreted by glands in the mucous membrane that lines the stomach. It contains hydrochloric acid, among other substances. The pH of gastric juice is about 1.5, which corresponds to a hydrochloric acid concentration of 0.03 *M*—a concentration strong enough to dissolve zinc metal! What is the purpose of this highly acidic medium? Where do the H$^+$ ions come from? What happens when there is an excess of H$^+$ ions present in the stomach?

A simplified diagram of the stomach is shown on the right. The inside lining is made up of parietal cells, which are fused together to form tight junctions. The interiors of the cells are protected from the surroundings by cell membranes. These membranes allow water and neutral molecules to pass in and out of the stomach, but they usually block the movement of ions such as H$^+$, Na$^+$, K$^+$, and Cl$^-$. The H$^+$ ions come from carbonic acid (H$_2$CO$_3$) formed as a result of the hydration of CO$_2$, an end product of metabolism:

$$CO_2(g) + H_2O(l) \rightleftharpoons H_2CO_3(aq)$$
$$H_2CO_3(aq) \rightleftharpoons H(aq) + HCO_3^-(aq)$$

These reactions take place in the blood plasma bathing the cells in the mucosa. By a process known as *active transport,* H$^+$ ions move across the membrane into the stomach interior. (Active transport processes are aided by enzymes.) To maintain electrical balance, an equal number of Cl$^-$ ions also move from the blood plasma into the stomach. Once in the stomach, most of these ions are prevented from diffusing back into the blood plasma by cell membranes.

The purpose of the highly acidic medium within the stomach is to digest food and to activate certain digestive enzymes. Eating stimulates H$^+$ ion secretion. A small fraction

of these ions normally are reabsorbed by the mucosa, causing many tiny hemorrhages. About half a million cells are shed by the lining every minute, and a healthy stomach is completely relined every three days or so. If the acid content is excessively high, however, the constant influx of H$^+$ ions through the membrane back to the blood plasma can cause muscle contraction, pain, swelling, inflammation, and bleeding.

One way to temporarily reduce the H$^+$ ion concentration in the stomach is to take an antacid. The major function of antacids is to neutralize excess HCl in gastric juice. The table on page 603 lists the active ingredients of some popular antacids.

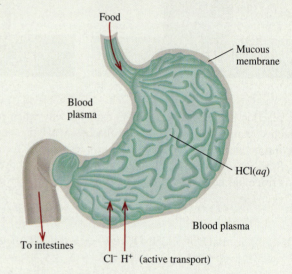

A simplified diagram of the human stomach.

Basic and Amphoteric Hydroxides

The alkali and alkaline earth metal hydroxides [except for Be(OH)$_2$] have basic properties. The following hydroxides are amphoteric: Be(OH)$_2$, Al(OH)$_3$, Sn(OH)$_2$, Pb(OH)$_2$, Cr(OH)$_3$, Cu(OH)$_2$, Zn(OH)$_2$, and Cd(OH)$_2$. For example, aluminum hydroxide reacts with both acids and bases:

$$Al(OH)_3(s) + 3H^+(aq) \longrightarrow Al^{3+}(aq) + 2H_2O(l)$$
$$Al(OH)_3(s) + OH^-(aq) \rightleftharpoons Al(OH)_4^-(aq)$$

All amphoteric hydroxides are insoluble in water.

Beryllium hydroxide, like aluminum hydroxide, exhibits amphoterism:

$$Be(OH)_2(s) + 3H^+(aq) \longrightarrow Be^{2+}(aq) + 2H_2O(l)$$
$$Be(OH)_2(s) + OH^-(aq) \rightleftharpoons Be(OH)_4^{2-}(aq)$$

This is another example of the diagonal relationship between beryllium and aluminum (see Section 2.5).

The reactions by which these antacids neutralize stomach acid are as follows:

$$NaHCO_3(aq) + HCl(aq) \longrightarrow$$
$$NaCl(aq) + H_2O(l) + CO_2(g)$$

$$CaCO_3(s) + 2HCl(aq) \longrightarrow$$
$$CaCl_2(aq) + H_2O(l) + CO_2(g)$$

$$MgCO_3(s) + 2HCl(aq) \longrightarrow$$
$$MgCl_2(aq) + H_2O(l) + CO_2(g)$$

$$Mg(OH)_2(s) + 2HCl(aq) \longrightarrow MgCl_2(aq) + 2H_2O(l)$$

$$Al(OH)_2NaCO_3(s) + 4HCl(aq) \longrightarrow$$
$$AlCl_3(aq) + NaCl(aq) + H_2O(l) + CO_2(g)$$

Some Common Commercial Antacid Preparations

Commercial Name	Active Ingredients
Alka-2	Calcium carbonate
Alka-Seltzer	Aspirin, sodium bicarbonate, citric acid
Bufferin	Aspirin, magnesium carbonate, aluminum glycinate
Buffered aspirin	Aspirin, magnesium carbonate, aluminum hydroxide-glycine
Milk of magnesia	Magnesium hydroxide
Rolaids	Dihydroxy aluminum sodium carbonate
Tums	Calcium carbonate

The CO_2 released by most of these reactions increases gas pressure in the stomach, causing the person to belch. The fizzing that takes place when an Alka-Seltzer tablet dissolves in water is caused by carbon dioxide, which is released by the reaction between citric acid and sodium bicarbonate:

$$C_4H_7O_5(COOH)(aq) + NaHCO_3(aq) \longrightarrow$$
citric acid
$$C_4H_7O_5COONa(aq) + H_2O(l) + CO_2(g)$$
sodium citrate

This action helps to disperse the ingredients and even enhances the taste of the solution.

The mucosa of the stomach can be damaged by the action of aspirin, the chemical name of which is acetylsalicylic acid. Aspirin is itself a moderately weak acid:

acetylsalicylic acid   acetylsalicylate ion

In the presence of the high concentration of H^+ ions in the stomach, this acid remains largely nonionized. A relatively nonpolar molecule, acetylsalicylic acid has the ability to penetrate membrane barriers that are also made up of nonpolar molecules. However, inside the membrane are many small water pockets, and when an acetylsalicylic acid molecule enters such a pocket, it ionizes into H^+ and acetylsalicylate ions. These ionic species become trapped in the interior regions of the membrane. The continued buildup of ions in this fashion weakens the structure of the membrane and eventually causes bleeding. Approximately 2 mL of blood are usually lost for every aspirin tablet taken, an amount not generally considered harmful. However, the action of aspirin can result in severe bleeding in some individuals. The presence of alcohol makes acetylsalicylic acid even more soluble in the membrane, and so further promotes the bleeding.

Summary of Facts and Concepts

Section 11.1

▶ Arrhenius acids ionize in water to give H^+ ions, and Arrhenius bases ionize in water to give OH^- ions.

▶ Brønsted acids donate protons, and Brønsted bases accept protons. These are the definitions that normally underlie the use of the terms *acid* and *base*.

▶ The reaction of an acid and a base is called neutralization.

▶ Lewis acids accept pairs of electrons and Lewis bases donate pairs of electrons. The term *Lewis acid* is generally reserved for substances that can accept electron pairs but do not contain ionizable hydrogen atoms.

Section 11.2

▶ The equilibrium constant for the autoionization of water is 1.0×10^{-14} at 25°C.

▶ The acidity of an aqueous solution is expressed as its pH, which is defined as the negative logarithm of the hydrogen ion activity. In a dilute solution the hydrogen ion activity can be well approximated by the numerical value of the hydrogen ion concentration in mol L^{-1}.

▶ At 25°C, an acidic solution has pH < 7, a basic solution has pH > 7, and a neutral solution has pH = 7.

Section 11.3

▶ In aqueous solution, the following are classified as strong acids: $HClO_4$, HI, HBr, HCl, H_2SO_4 (first stage of ionization), and HNO_3. Strong bases in aqueous solution include hydroxides of alkali metals and of alkaline earth metals (except beryllium).

▶ The acid ionization constant K_a increases with acid strength. K_b similarly expresses the strengths of bases.

Section 11.4

▶ Percent ionization is another measure of the strength of acids. The more dilute a solution of a weak acid, the greater the percent ionization of the acid.

▶ The product of the ionization constant of an acid and the ionization constant of its conjugate base is equal to the ion-product constant of water.

Section 11.5

▶ The relative strengths of acids can be explained qualitatively in terms of their molecular structures.

Section 11.6

▶ Most salts are strong electrolytes that dissociate completely into ions in solution. The reaction of these ions with water, called salt hydrolysis, can produce acidic or basic solutions. In salt hydrolysis, the conjugate bases of weak acids yield basic solutions, and the conjugate acids of weak bases yield acidic solutions.

▶ Small, highly charged metal ions, such as Al^{3+} and Fe^{3+}, hydrolyze to yield acidic solutions.

Section 11.7

▶ Most oxides can be classified as acidic, basic, or amphoteric. Metal hydroxides are either basic or amphoteric.

Key Words

acid, p. 557
acid ionization constant, p. 571
amphoteric, p. 560
autoionization of water, p. 564
base, p. 557
base ionization constants, p. 573
Brønsted acid, p. 558

Brønsted base, p. 558
conjugate acid-base pair, p. 561
diprotic acid, p. 559
hydronium ion, p. 558
ion-product constant, p. 565
Lewis acid, p. 562
Lewis base, p. 562

monoprotic acid, p. 558
neutralization reaction, p. 560
percent ionization, p. 582
pH, p. 566
salt, p. 560
salt hydrolysis, p. 594
solvent leveling, p. 572

spectator ions, p. 560
strong acid, p. 558
strong base, p. 559
strong electrolytes, p. 560
triprotic acid, p. 559
weak acid, p. 558
weak base, p. 559

Problems

Many Processes in Chemistry Are Acid-Base Reactions

11.1 Classify each of the following species as a Brønsted acid, a Brønsted base, or both: (a) H_2O, (b) OH^-, (c) H_3O, (d) NH_3, (e) NH_4^+, (f) NH_2^-, (g) NO_3^-, (h) CO_2 (i) HBr, and (j) HCN.

11.2 Write the formulas of the conjugate bases of the following acids: (a) HNO_2, (b) H_2SO_4, (c) H_2S, (d) HCN, and (e) $HCOOH$ (formic acid).

11.3 Identify the acid-base conjugate pairs in each of the following reactions:

(a) $CH_3COO^- + HCN \rightleftharpoons CH_3COOH + CN^-$
(b) $HCO_3^- + HCO_3^- \rightleftharpoons H_2CO_3 + CO_3^{2-}$
(c) $H_2PO_4^- + NH_3 \rightleftharpoons HPO_4^{2-} + NH_4^+$
(d) $HClO + CH_3NH_2 \rightleftharpoons CH_3NH_3^+ + ClO^-$
(e) $CO_3^{2-} + H_2O \rightleftharpoons HCO_3^- + OH^-$

11.4 Write the formula for the conjugate acid of each of the following bases: (a) HS^-, (b) HCO_3^-, (c) CO_3^{2-}, (d) $H_2PO_4^-$, (e) HPO_4^{2-}, (f) PO_4^{3-}, (g) HSO_4^-, (h) SO_4^{2-}, and (i) SO_3^{2-}.

11.5 Oxalic acid ($C_2H_2O_4$) has the following structure:

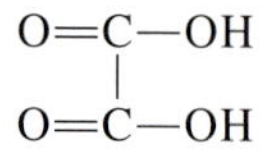

$$O{=}C{-}OH$$
$$|$$
$$O{=}C{-}OH$$

An oxalic acid solution contains the following species in varying concentrations: $C_2H_2O_4$, $C_2HO_4^-$, $C_2O_4^{2-}$, and H^+. (a) Draw Lewis structures of $C_2HO_4^-$, and $C_2O_4^{2-}$. (b) Which of the four species in an oxalic acid solution can act only as acids, which can act only as bases, and which can act as both acids and bases?

11.6 Write the formula for the conjugate base of each of the following acids: (a) $CH_2ClCOOH$, (b) HIO_4, (c) H_3PO_4, (d) $H_2PO_4^-$ (e) HPO_4^{2-} (f) H_2SO_4, (g) HSO_4^-, (h) HIO_3, (i) HSO_3^- (j) NH_4^+, (k) H_2S, (l) HS^-, and (m) $HClO_3$.

11.7 Classify each of the following species as a Lewis acid or a Lewis base: (a) CO_2, (b) H_2O, (c) I^-, (d) SO_2, (e) NH_3, (f) OH^-, (g) H^+, and (h) BCl_3.

11.8 Describe the following reaction in terms of the Lewis theory of acids and bases:

$$AlCl_3(s) + Cl^-(aq) \longrightarrow AlCl_4^-(aq)$$

11.9 Which would be considered a stronger Lewis acid: (a) BF_3 or BCl_3, and (b) Fe^{2+} or Fe^{3+}? Explain.

11.10 All Brønsted acids are Lewis acids, but not all Lewis acids are Brønsted acids. Give two examples of Lewis acids that are not Brønsted acids.

The Acid-Base Properties of Aqueous Solutions Are Governed by the Autoionization Equilibrium of Water

11.11 The pH of a solution is 6.7. From this statement alone, can you conclude that the solution is acidic? If not, what additional information would you need?

11.12 Can the pH of a solution be zero or negative? If so, give examples to illustrate these values.

11.13 Calculate the concentration of OH^- ions in a $1.4 \times 10^{-3}\ M$ HCl solution at 25°C.

11.14 Calculate the concentration of H^+ ions in a 0.62 M NaOH solution at 25°C.

11.15 Calculate the pH of each of the following solutions at 25°C: (a) 0.0010 M HCl, and (b) 0.76 M KOH.

11.16 Calculate the pH of each of the following solutions at 25°C: (a) $2.8 \times 10^{-4}\ M$ $Ba(OH)_2$, and (b) $5.2 \times 10^{-4}\ M$ HNO_3.

11.17 Calculate the hydrogen ion concentration in mol L^{-1} for solutions with the following pH values at 25°C: (a) 2.42, (b) 11.21, (c) 6.96, and (d) 15.00.

11.18 Calculate the hydrogen ion concentration in mol L^{-1} for each of the following solutions at 25°C: (a) a solution whose pH is 5.20, (b) a solution whose pH is 16.00, and (c) a solution whose hydroxide ion concentration is $3.7 \times 10^{-9}\ M$.

11.19 The pOH of a solution is 9.40 at 25°C. Calculate the hydrogen ion concentration of the solution.

11.20 Calculate the number of moles of KOH in 5.50 mL of a 0.360 M KOH solution at 25°C. What is the pOH of the solution?

11.21 How much NaOH (in grams) is needed to prepare 546 mL of solution with a pH of 10.00 at 25°C?

11.22 A solution is made by dissolving 18.4 g of HCl in 662 mL of water. Calculate the pH of the solution at 25°C. (Assume that the volume remains constant.)

11.23 For which of the following solutions would you expect the measured value of the pH to differ most from the pH calculated from the H^+ ion concentration: (a) a $2.0 \times 10^{-5}\ M$ solution of HCl, (b) a 0.1 M solution of HNO_3, and (c) a 0.00050 M solution of NaOH?

11.24 For which of the following solutions would you expect the measured value of the pH to differ most from the pH calculated from the H^+ ion concentration: (a) a $5.0 \times 10^{-4}\ M$ solution of KOH, (b) a $4.7 \times 10^{-6}\ M$ solution of HCl, and (c) a 0.0560 M solution of NaOH?

11.25 An aqueous solution at 60°C has an H^+ concentration of $1.5 \times 10^{-5}\ M$. What is the OH^- concentration of this solution?

11.26 What is the pH of an aqueous solution at 40°C that has a pOH of 4.5?

The Strengths of Acids and Bases Are Measured by Their Ionization Constants

11.27 Which of the following diagrams best represents a strong acid, such as HCl, dissolved in water? Which represents a weak acid? Which represents a very weak acid? (The hydrated proton is shown as a hydronium ion. Water molecules are omitted for clarity.)

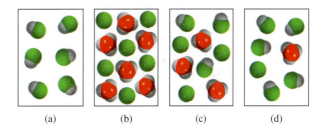

(a)   (b)   (c)   (d)

11.28 (1) Which of the following diagrams represents a solution of a weak diprotic acid? (2) Which diagrams represent chemically implausible situations? (The hydrated proton is shown as a hydronium ion. Water molecules are omitted for clarity.)

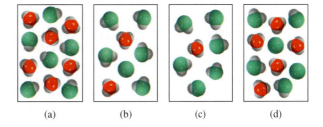

(a)   (b)   (c)   (d)

11.29 Classify each of the following species as a weak or strong acid: (a) HNO_3, (b) HF, (c) H_2SO_4, (d) HSO_4^-, (e) H_2CO_3, (f) HCO_3^-, (g) HCl, (h) HCN, and (i) HNO_2.

11.30 Classify each of the following species as a weak or strong base: (a) LiOH, (b) CN^-, (c) H_2O, (d) ClO_4^-, and (e) NH_2^-.

11.31 Which of the following statements is true for a 0.10 M solution of a weak acid HA?

(a) The pH is 1.00.

(b) $[H^+] \gg [A]$.

(c) $[H] = [A]$.

(d) The pH is less than 1.

11.32 Which of the following statements is true regarding a 1.0 M solution of a strong acid HA?

(a) $[A^-] > [H^+]$.

(b) The pH is approximately zero.

(c) $[H^+] = 1.0\ M$.

(d) $[HA] = 1.0\ M$.

11.33 Predict the direction that predominates in the following reaction:

$$F^-(aq) + H_2O(l) \rightleftharpoons HF(aq) + OH^-(aq)$$

11.34 Predict whether the following reaction will proceed from left to right to any measurable extent:

$$CH_3COOH(aq) + Cl^-(aq) \longrightarrow$$

11.35 The K_a for the weak acid HF is 7.1×10^{-4} at 25°C. Using the techniques from Chapter 10, determine the K_a for HF at 60°C. The $\Delta H°$ for the ionization of HF in aqueous solution is -61.53 kJ mol^{-1}.

The pH of an Acid or Base Can Be Calculated If Its Ionization Constant Is Known

11.36 Why do we normally not quote K_a values for strong acids such as HCl and HNO_3?

11.37 Why is it necessary to specify temperature when giving K_a values?

11.38 Which of the following solutions has the highest pH: (a) 0.40 M HCOOH, (b) 0.40 M $HClO_4$, or (c) 0.40 M CH_3COOH?

11.39 The K_a for benzoic acid is 6.5×10^{-5}. Calculate the pH of a 0.10 M benzoic acid solution.

11.40 A 0.0560-g quantity of acetic acid is dissolved in enough water to make 50.0 mL of solution. Calculate the concentrations of H, CH_3COO^-, and CH_3COOH at equilibrium. (K_a for acetic acid is 1.8×10^{-5}.)

11.41 The pH of an acid solution is 6.20. Calculate the K_a for the acid. The initial acid concentration is 0.010 M.

11.42 What is the original molarity of a solution of formic acid (HCOOH) whose pH is 3.26 at equilibrium?

11.43 Calculate the percent ionization of benzoic acid at the following concentrations: (a) 0.20 M, and (b) 0.00020 M.

11.44 Calculate the percent ionization of hydrofluoric acid at the following concentrations: (a) 0.60 M, (b) 0.0046 M, and (c) 0.00028 M. Comment on the trends.

11.45 A 0.040 M solution of a monoprotic acid is 14 percent ionized. Calculate the ionization constant of the acid.

11.46 (a) Calculate the percent ionization of a 0.20 M solution of the monoprotic acetylsalicylic acid (aspirin) for which $K_a = 3.0 \times 10^{-4}$. (b) The pH of gastric juice in the stomach of a certain individual is 1.00. After a few aspirin tablets have been swallowed, the concentration of acetylsalicylic acid in the person's stomach is 0.20 M. Calculate the percent ionization of the acid under these conditions. What effect does the nonionized acid have on the membranes lining the stomach? (*Hint:* See the essay starting on page 602)

11.47 Calculate the pH for each of the following solutions at 25°C: (a) 0.10 M NH_3, and (b) 0.050 M C_5H_5N (pyridine).

11.48 The pH of a 0.30 M solution of a weak base is 10.66. What is the K_b of the base?

11.49 What is the original molarity of a solution of ammonia whose pH is 11.22 at 25°C?

11.50 In a 0.080 M NH_3 solution at 25°C, what percent of the NH_3 is present as NH_4^+?

11.51 Write all the species (except water) that are present in a phosphoric acid solution. Indicate which species can act as a Brønsted acid, which as a Brønsted base, and which as both a Brønsted acid and a Brønsted base.

11.52 (a) Compare the pH of a 0.040 M HCl solution with that of a 0.040 M H_2SO_4 solution at 25°C. (b) What are the concentrations of HSO_4^-, SO_4^{2-}, and H^+ in a 0.20 M $KHSO_4$ solution? (*Hint:* H_2SO_4 is a strong acid; K_a for HSO_4^- is 1.3×10^{-2} at 25°C.)

11.53 Calculate the concentrations of H^+, HCO_3^-, and CO_3^{2-} in a 0.025 M H_2CO_3 solution at 25°C.

11.54 Calculate the pH of a 2.0×10^{-7} M solution of hydrochloric acid at 25°C.

11.55 Calculate the pH of a 2.0×10^{-7} M solution of formic acid at 25°C.

11.56 The first and second ionization constants of a diprotic acid H_2A are K_{a_1} and K_{a_2} at a certain temperature. Under what conditions will $[A^{2-}] = K_{a_2}$?

11.57 Calculate the pH of a 0.0015 M phosphoric acid solution at 25°C.

The Strength of an Acid Is Determined in Part by Molecular Structure

11.58 Predict the acid strengths of the following compounds: H_2O, H_2S, and H_2Se.

11.59 Compare the strengths of the following pairs of acids: (a) H_2SO_4 and H_2SeO_4, and (b) H_3PO_4 and H_3AsO_4.

11.60 Which of the following is the stronger acid: $CH_2ClCOOH$ or $CHCl_2COOH$? Explain your choice.

11.61 Consider phenol and methanol (both of which are alcohols):

phenol   methanol

Experimentally, phenol is found to be a stronger acid than methanol. Explain this difference in terms of the structures of the conjugate bases. (*Hint:* A more stable conjugate base favors ionization. Only one of these conjugate bases can be stabilized by resonance.)

Many Salts Have Acid-Base Properties in Aqueous Solution

11.62 Predict the pH (>7, $=7$, or <7) of aqueous solutions containing the following salts: (a) KBr, (b) $Al(NO_3)_3$, (c) $BaCl_2$, and (d) $Bi(NO_3)_3$.

11.63 Predict whether the following aqueous solutions are acidic, basic, or nearly neutral: (a) NaBr, (b) K_2SO_3, (c) NH_4NO_2, and (d) $Cr(NO_3)_3$.

11.64 A certain salt, MX (containing the M^+ and X^- ions), is dissolved in water, and the pH of the resulting solution is 7.0 at 25°C. Can you say anything about the strengths of the acid and the base from which the salt is derived?

11.65 In a certain experiment, a student finds that the pH values of 0.10 M solutions of three potassium salts KX, KY, and KZ are 7.0, 9.0, and 11.0, respectively. Arrange the acids HX, HY, and HZ in the order of increasing acid strength.

11.66 Calculate the pH of a 0.36 M CH_3COONa solution at 25°C.

11.67 Calculate the pH of a 0.42 M NH_4Cl solution at 25°C.

11.68 Predict the pH (>7, $=7$, <7) of an $NaHCO_3$ solution.

11.69 Predict whether a solution containing the salt K_2HPO_4 will be acidic, basic, or neutral.

11.70 Calculate the pH of a 0.015 M oxalic acid solution at 25°C?

Oxide and Hydroxide Compounds Can Be Acidic or Basic in Aqueous Solution Depending on Their Composition

11.71 Arrange the oxides in each of the following groups in order of increasing basicity: (a) K_2O, Al_2O_3, BaO, and (b) CrO_3, CrO, Cr_2O_3.

11.72 $Zn(OH)_2$ is an amphoteric hydroxide. Write balanced ionic equations to show its reaction with (a) HCl, and (b) NaOH [the product is $Zn(OH)_4^{2-}$].

11.73 $Al(OH)_3$ is insoluble in pure water, but it dissolves in excess NaOH in solution. Write a balanced ionic equation for this reaction. What type of reaction is this?

Additional Problems

11.74 The K_a of formic acid is 1.7×10^{-4} at 25°C. Will the acid become stronger or weaker at 40°C? Explain.

11.75 A typical reaction between an antacid and the hydrochloric acid in gastric juice is

$$NaHCO_3(s) + HCl(aq) \rightleftharpoons$$
$$NaCl(aq) + H_2O(l) + CO_2(g).$$

Calculate the volume (in L) of CO_2 generated from 0.350 g of $NaHCO_3$ and excess gastric juice at 1.00 bar and 37.0°C.

11.76 To which of the following would the addition of an equal volume of 0.60 M NaOH lead to a solution having a lower pH: (a) water, (b) 0.30 M HCl, (c) 0.70 M KOH, and (d) 0.40 M $NaNO_3$?

11.77 The pH of a 0.0642 M solution of a monoprotic acid is 3.86 at 25°C. Is this a strong acid?

11.78 Like water, liquid ammonia undergoes autoionization:

$$NH_3 + NH_3 \rightleftharpoons NH_4^+ + NH_2^-$$

(a) Identify the Brønsted acids and Brønsted bases in this reaction. (b) What species correspond to H^+ and OH^-, and what is the condition for a neutral solution?

11.79 HA and HB are both weak acids, although HB is the stronger of the two. Will it take a larger volume of a 0.10 M NaOH solution to neutralize 50.0 mL of 0.10 M HB or 50.0 mL of 0.10 M HA?

11.80 A solution contains a weak monoprotic acid HA and its sodium salt NaA, both at 0.1 M concentration. Show that $[OH^-] = K_w/K_a$.

11.81 The three common chromium oxides are CrO, Cr_2O_3, and CrO_3. If Cr_2O_3 is amphoteric, what can you say about the acid-base properties of CrO and CrO_3?

11.82 Use the data in Appendix 2 to determine $\Delta H°$, $\Delta S°$, and $\Delta G°$ for the autoionization of water at 25°C. Using the techniques discussed in Chapter 10, determine the value of K_w at 25°C and at 60°C. What assumptions must you make to do these calculations? Compare your results with the data in Table 11.2.

11.83 Use the data in Table 11.3 to calculate the equilibrium constant for the following reaction at 25°C:

$$HCOOH(aq) + OH^-(aq) \rightleftharpoons$$
$$HCOO^-(aq) + H_2O(l)$$

11.84 Use the data in Table 11.3 to calculate the equilibrium constant for the following reaction at 25°C:

$$CH_3COOH(aq) + NO_2^-(aq) \rightleftharpoons$$
$$CH_3COO^-(aq) + HNO_2(aq)$$

11.85 Most of the hydrides of Group 1A and Group 2A metals are ionic (the exceptions are BeH_2 and MgH_2, which are covalent compounds). Describe the reaction between the hydride ion (H^-) and water in terms of a Brønsted acid-base reaction.

11.86 The K_b for the weak base NH_3 is 1.8×10^{-5} at 25°C. Using the techniques from Chapter 10 and the data in Appendix 2, determine the K_b for NH_3 at 50°C.

11.87 Calculate the pH of a 0.20 M ammonium acetate (CH_3COONH_4) solution at 25°C.

11.88 Novocaine, used as a local anesthetic by dentists, is a weak base ($K_b = 8.91 \times 10^{-6}$ at 25°C). What is the ratio of the concentration of the base to that of its acid in the blood plasma (pH = 7.40) of a patient?

11.89 Which of the following is the stronger base: NF_3 or NH_3? (*Hint:* F is more electronegative than H.)

11.90 Which of the following is a stronger base: NH_3 or PH_3? (*Hint:* The N—H bond is stronger than the P—H bond.)

11.91 The ion product of D_2O is 1.35×10^{-15} at 25°C. (a) Calculate pD of neutral D_2O, where pD $= -\log_{10}[D]$. (b) For what values of pD will a solution be acidic in D_2O? (c) Derive a relation between pD and pOD.

11.92 Give a physical explanation as to why D_2O should have a smaller ion product constant than H_2O at 25°C (1.35×10^{-15} for D_2O versus 1.01×10^{-14} for H_2O).

11.93 Give an example of (a) a weak acid that contains oxygen atoms, (b) a weak acid that does not contain oxygen atoms, (c) a neutral molecule that acts as a Lewis acid, (d) a neutral molecule that acts as a Lewis base, (e) a weak acid that contains two ionizable H atoms, and (f) a conjugate acid-base pair, both of which react with HCl to give carbon dioxide gas.

11.94 What is the pH of 250.0 mL of an aqueous solution containing 0.616 g of the strong acid trifluoromethane sulfonic acid (CF_3SO_3H)?

11.95 (a) Use VSEPR to predict the geometry of the hydronium ion (H_3O^+). (b) The O atom in H_2O has two lone pairs and in principle can accept two H^+ ions. Explain why the species H_4O^{2+} does not exist. What would be its geometry if it did exist?

11.96 HF is a weak acid, but its strength increases with concentration. Explain. (*Hint:* F^- reacts with HF to form HF_2^-. The equilibrium constant for this reaction is 5.2 at 25°C.)

11.97 When chlorine reacts with water, the resulting solution is weakly acidic and reacts with $AgNO_3$ to give a white precipitate. Write balanced equations to represent these reactions. Explain why manufacturers of household bleaches add bases such as NaOH to their products to increase their effectiveness.

11.98 Hemoglobin (Hb) is a blood protein that is responsible for transporting oxygen. It can exist in the protonated form as HbH. The binding of oxygen can be represented by the simplified equation

$$HbH^+ + O_2 \rightleftharpoons HbO_2 + H^+$$

(a) What form of hemoglobin is favored in the lungs, where the oxygen concentration is highest? (b) In body tissues, where the cells release carbon dioxide produced by metabolism, the blood is more acidic due to the formation of carbonic acid. What form of hemoglobin is favored under this condition? (c) When a person hyperventilates, the concentration of CO_2 in his or her blood decreases. How does this action affect the given equilibrium? Frequently a person who is hyperventilating is advised to breathe into a paper bag. Why does this action help the individual?

11.99 Calculate the concentrations of all species in a 0.100 M H_3PO_4 solution at 25°C.

11.100 In the vapor phase, acetic acid molecules associate to a certain extent to form dimers:

$$2CH_3COOH(g) \rightleftharpoons (CH_3COOH)_2(g)$$

At 51°C, the pressure of a certain acetic acid vapor system is 0.0346 bar in a 360-mL flask. The vapor is condensed and neutralized with 13.8 mL of 0.0568 M NaOH. (a) Calculate the degree of dissociation (α) of the dimer under these conditions:

$$(CH_3COOH)_2 \rightleftharpoons 2CH_3COOH$$

(*Hint:* See Problem 10.66 for the general procedure.) (b) Calculate the equilibrium constant K_P for the reaction in part (a).

11.101 Calculate the concentrations of all the species in a 0.100 M Na_2CO_3 solution at 25°C.

11.102 Using Equations 11.3, 11.11, 11.23, and 11.24 as a starting point, derive the cubic equation for the H^+ ion concentration of a dilute weak acid given by Equation 11.25.

11.103 The Henry's law constant for CO_2 at 38°C is 2.30×10^{-3} mol L^{-1} bar^{-1}. Calculate the pH of a solution of CO_2 at 38°C in equilibrium with the gas at a partial pressure of 3.20 bar.

11.104 Hydrocyanic acid (HCN) is a weak acid and a deadly poisonous compound—in the gaseous form (hydrogen cyanide), it was used in gas chambers. Why is it dangerous to treat sodium cyanide with acids (such as HCl) without proper ventilation?

11.105 How many grams of NaCN would you need to dissolve in enough water to make exactly 250 mL of solution with a pH of 11.00?

11.106 A solution of formic acid (HCOOH) has a pH of 2.53 at 25°C. How many grams of formic acid are there in 100.0 mL of the solution?

11.107 Calculate the pH of a 1.0-L solution containing 0.150 mole of CH_3COOH and 0.100 mole of HCl at 25°C.

11.108 A 1.87-g sample of Mg reacts with 80.0 mL of an HCl solution whose pH is −0.544. What is the pH of the solution after all the Mg has reacted? Assume constant volume.

11.109 You are given two beakers, one containing an aqueous solution of strong acid (HA) and the other an aqueous solution of weak acid (HB) of the same concentration. Describe how you would compare the strengths of these two acids by (a) measuring the pH, (b) measuring electrical conductance, and (c) studying the rate of hydrogen gas evolution when these solutions are reacted with an active metal such as Mg or Zn.

11.110 Use Le Châtelier's principle to predict the effect of the following changes on the extent of hydrolysis of sodium nitrite ($NaNO_2$) solution: (a) HCl is added, (b) NaOH is added, (c) NaCl is added, and (d) the solution is diluted.

11.111 Describe the hydration of SO_2 as a Lewis acid-base reaction. (*Hint:* Refer to the discussion of the hydration of CO_2 on page 563.)

11.112 The disagreeable odor of fish is mainly due to organic compounds (RNH_2) containing an amino group, (—NH_2), where R is the rest of the molecule. These compounds are called amines, and they are bases just like ammonia. Explain why putting some lemon juice on fish can greatly reduce the odor.

11.113 A solution of methylamine (CH_3NH_2) has a pH of 10.64 at 25°C. How many grams of methylamine are there in 100.0 mL of the solution?

11.114 A 0.400 M formic acid (HCOOH) solution freezes at −0.758°C. Calculate the K_a of the acid at that temperature. (*Hint:* Assume that molarity is equal to molality. Carry your calculations to three significant figures and round off to two for K_a.)

11.115 Both the amide ion (NH_2^-) and the nitride ion (N^{3-}) are stronger bases than the hydroxide ion and hence do not exist in aqueous solutions. (a) Write equations showing the reactions of these ions with water, and identify the Brønsted acid and Brønsted base in each case. (b) Which of the two is the stronger base?

11.116 The atmospheric sulfur dioxide (SO_2) concentration over a certain region is 0.12 ppm by volume. Calculate the pH of the rainwater at 20°C due to this pollutant. Assume that the dissolution of SO_2 does not affect its pressure.

11.117 Calcium hypochlorite [$Ca(OCl)_2$] is used as a disinfectant for swimming pools. When dissolved in water it produces hypochlorous acid,

$$Ca(OCl)_2(s) + 2H_2O(l) \rightleftharpoons$$
$$2HClO(aq) + Ca(OH)_2(s)$$

which ionizes as follows:

$$HClO(aq) + H_2O(l) \rightleftharpoons H_3O^+(aq) + ClO^-(aq)$$
$$K_a = 3.0 \times 10^{-8}$$

As strong oxidizing agents, both HClO and ClO^- can kill bacteria by destroying their cellular components. However, too high an HClO concentration is irritating to the eyes of swimmers and too high a concentration of ClO^- will cause the ions to decompose in sunlight. The recommended pH for pool water is 7.8. Calculate the percent of these species present at this pH.

11.118 Explain the action of smelling salt, which is ammonium carbonate [$(NH_4)_2CO_3$]. (*Hint:* The thin film of aqueous solution that lines the nasal passage is slightly basic.)

11.119 About half of the hydrochloric acid produced annually in the United States (3.0 billion pounds) is used in metal pickling. This process involves the removal of metal oxide layers from metal surfaces to prepare them for coating. (a) Write the overall and net ionic equations for the reaction between iron(III) oxide, which represents the rust layer over iron, and HCl. Identify the Brønsted acid and Brønsted base. (b) Hydrochloric acid is also used to remove scale (which is mostly $CaCO_3$) from water pipes. Hydrochloric acid reacts with calcium carbonate in two stages: The first stage forms the bicarbonate ion, which then reacts further to form carbon dioxide. Write equations for these two stages and for the overall reaction. (c) Hydrochloric acid is used to recover oil from the ground. It dissolves rocks (often $CaCO_3$) so that the oil can flow more easily. In one process, a 15 percent (by mass) HCl solution is injected into an oil well to dissolve the rocks. If the density of the acid solution is 1.073 g mL^{-1}, what is the pH of the solution?

11.120 Which of the following does not represent a Lewis acid-base reaction?
(a) $H_2O + H^+ \longrightarrow H_3O^+$
(b) $NH_3 + BF_3 \longrightarrow H_3NBF_3$
(c) $PF_3 + F_2 \longrightarrow PF_5$
(d) $Al(OH)_3 + OH^- \longrightarrow Al(OH)_4^-$

11.121 True or false? If false, explain why the statement is wrong. (a) All Lewis acids are Brønsted acids. (b) The conjugate base of an acid always carries a negative charge. (c) The percent ionization of a base increases with its concentration in solution. (d) A solution of barium fluoride is acidic.

11.122 How many milliliters of a strong monoprotic acid solution at pH = 4.12 must be added to 528 mL of the same acid solution at pH 5.76 to change its pH to 5.34? Assume that the volumes are additive.

11.123   A 1.294-g sample of a metal carbonate (MCO_3) is reacted with 500 mL of a 0.100 M HCl solution. The excess HCl acid is then neutralized by 32.80 mL of 0.588 M NaOH. Identify M.

11.124   Prove the statement that when the concentration of a weak acid HA decreases by a factor of 10, its percent ionization increases by a factor of $\sqrt{10}$. State any assumptions.

11.125   Calculate the pH of a solution that is 1.00 M HCN and 1.00 M HF at 25°C. Compare the concentration (in molarity) of the CN^- ion in this solution with that in a 1.00 M HCN solution. Comment on the difference.

11.126   Tooth enamel is largely hydroxyapatite [$Ca_3(PO_4)_3OH$]. When it dissolves in water (a process called *demineralization*), it dissociates as follows:

$$Ca_5(PO_4)_3OH \longrightarrow 5Ca^{2+} + 3PO_4^{3-} + OH^-$$

The reverse process, called *remineralization,* is the body's natural defense against tooth decay. Acids produced from food remove the OH^- ions and thereby weaken the enamel layer. Most toothpastes contain a flouride compound such as NaF or SnF_2. What is the function of these compounds in preventing tooth decay?

Answers to Practice Exercises

11.1 (1) H_2O and OH^-. (2) HCN and CN^-. **11.2** (a) Lewis acid: C_O^{3+}, Lewis base: NH_3; (b) Lewis acid: BF_3; Lewis base: CO **11.3** 7.7×10^{-11} M **11.4** 0.12 **11.5** 4.7×10^{-4} M **11.6** 7.40 **11.7** 7.94, 1.1×10^{-8} M **11.8** 5.6×10^{-10}, 2.30×10^{-11} **11.9** Smaller than 1 **11.10** 9.56 **11.11** (a) 3.32, (b) 2.82 **11.12** (a) 11.53, (b) 10.50 **11.13** 6.91 **11.14** 6.89 **11.15** [$H_2C_2O_2$] = 0.0012 M, [$HC_2O_4^-$] = 0.0088 M, [$C_2O_4^{2-}$] = 1.5×10^{-13} M, [H^+] = 0.0088 M, [OH^-] = 1.1×10^{-12} M **11.16** H_3PO_3 **11.17** 8.58 **11.18** (a) pH ≈ 7, (b) pH > 7, (c) pH < 7, (d) pH > 7

12 Chapter

Acid-Base Equilibria and Solubility

In this chapter, we will continue the study of acid-base reactions with a discussion of buffer action and titrations. We will also look at another type of aqueous equilibrium—that between slightly soluble compounds and their ions in solution.

611

12.1 | The Ionization of Weak Acids and Bases Is Suppressed by the Addition of a Common Ion

Our discussion of acid-base ionization and salt hydrolysis in Chapter 11 was limited to solutions containing a single solute. In this section, we will consider the acid-base properties of a solution with two dissolved solutes that contain the same ion (cation or anion), called the *common ion.*

The presence of a common ion suppresses the ionization of a weak acid or a weak base. If sodium acetate and acetic acid are dissolved in the same solution, for example, they both dissociate and ionize to produce CH_3COO^- ions:

$$CH_3COOH(aq) \rightleftharpoons CH_3COO^-(aq) + H^+(aq)$$

$$CH_3COONa(s) \longrightarrow CH_3COO^-(aq) + Na^+(aq)$$

CH_3COONa is a strong electrolyte, so it dissociates completely in solution, but CH_3COOH, a weak acid, ionizes only slightly. According to Le Châtelier's principle, the addition of CH_3COO^- ions from CH_3COONa to a solution of CH_3COOH will suppress the ionization of CH_3COOH (that is, shift the equilibrium from right to left), thereby decreasing the hydrogen ion concentration. Thus, a solution containing both CH_3COOH and CH_3COONa will be *less* acidic than a solution containing only CH_3COOH at the same concentration. The shift in equilibrium of the acetic acid ionization is caused by the acetate ions from the salt. CH_3COO^- is the common ion because it is supplied by both CH_3COOH and CH_3COONa.

The ***common ion effect*** is *the shift in equilibrium caused by the addition of a compound having an ion in common with the dissolved substance.* The common ion effect plays an important role in determining the pH of a solution and the solubility of a slightly soluble salt (to be discussed later in Section 12.7). Here we will study the common ion effect as it relates to the pH of a solution. Keep in mind that despite its distinctive name, the common ion effect is simply a special case of Le Châtelier's principle.

Let us consider the pH of a solution containing a weak acid, HA, and a soluble salt of the weak acid, such as NaA. We start by writing

$$HA(aq) + H_2O(l) \rightleftharpoons H_3O^+(aq) + A^-(aq)$$

or simply

$$HA(aq) \rightleftharpoons H^+(aq) + A^-(aq)$$

The ionization constant K_a was defined in Equation 11.10:

$$K_a = \frac{a_{H_3O^+(aq)}\, a_{A^-(aq)}}{a_{HA(aq)}\, a_{H_2O(l)}}$$

If the concentrations are sufficiently low, then the activities of solutes in Equation 11.10 can be replaced with molarities (without units) and the activity of pure water can be assumed to be unity, giving (Equation 11.11)

$$K_a \approx \frac{[H^+][A^-]}{[HA]}$$

Rearranging this equation gives

$$[H^+] = \frac{K_a[HA]}{[A^-]}$$

Taking the negative logarithm (base 10) of both sides, we obtain

$$-\log_{10}[\text{H}^+] = -\log_{10}K_a - \log_{10}\frac{[\text{HA}]}{[\text{A}^-]}$$

or

$$-\log_{10}[\text{H}^+] = -\log_{10}K_a + \log_{10}\frac{[\text{A}^-]}{[\text{HA}]}$$

Using the definitions of pH (Equation 11.5) and pK_a (Equation 11.15) gives (assuming dilute solution behavior)

$$\text{pH} = pK_a + \log_{10}\frac{[\text{A}^-]}{[\text{HA}]} \tag{12.1}$$

Equation 12.1 is called the **Henderson-Hasselbalch equation.** A more general form of this expression is

$$\text{pH} = pK_a + \log_{10}\frac{[\text{conjugate base}]}{[\text{acid}]} \tag{12.2}$$

Keep in mind that pK_a is a constant, but the ratio of the two concentration terms in Equation 12.2 depends upon the particular solution.

In our example, HA is the acid and A^- is the conjugate base. Thus, if we know K_a and the concentrations of the acid and the salt of the acid, we can calculate the pH of the solution.

It is important to remember that the Henderson-Hasselbalch equation is derived from the equilibrium constant expression. It is valid regardless of the source of the conjugate base (that is, whether it comes from the acid alone or is supplied by both the acid and its salt).

In problems that involve the common ion effect, we are usually given the starting concentrations of a weak acid HA and its salt, such as NaA. If the concentrations of these species are reasonably high ($> 0.1\ M$), we can neglect the ionization of the acid and the hydrolysis of the salt. This is a valid approximation because HA is a weak acid and the extent of the hydrolysis of the A^- ion is generally very small. Moreover, the presence of A^- (from NaA) further suppresses the ionization of HA and the presence of HA further suppresses the hydrolysis of A^-. Thus, in such cases, we can use the *starting* concentrations as the equilibrium concentrations in Equation 12.1.

In Example 12.1 we calculate the pH of a solution containing a common ion.

Example 12.1

(a) Calculate the pH of a solution containing 0.20 M CH_3COOH and 0.30 M CH_3COONa.
(b) What would the pH of a 0.20 M CH_3COOH solution be if no salt were present?

Strategy (a) CH_3COOH is a weak acid ($CH_3COOH \rightleftharpoons CH_3COO^- + H^+$), and CH_3COONa is a soluble salt that is completely dissociated in solution ($CH_3COONa \longrightarrow Na^+ + CH_3COO^-$). The common ion here is the acetate ion (CH_3COO^-). At equilibrium, the major species in solution are CH_3COOH, CH_3COO^-, Na^+, H^+, and H_2O. The Na^+ ion has no acid or base properties, and we ignore the ionization of water. Because K_a is an

—Continued

Continued—

equilibrium constant, its value is the same whether we have just the acid or a mixture of the acid and its salt in solution. Therefore, we can calculate $[H^+]$ at equilibrium and hence pH if we know the equilibrium concentrations of both $[CH_3COOH]$ and $[CH_3COO]$. For CH_3COOH, K_a is 1.8×10^{-5} (see Table 11.3) giving a pK_a of 4.74. (b) For a solution of CH_3COOH, a weak acid, we calculate the $[H^+]$ at equilibrium, and hence the pH, using the initial concentration of CH_3COOH. This problem can be solved by following the same procedure as in Example 11.11.

Solution (a) Sodium acetate is a strong electrolyte, so it dissociates completely in solution:

$$CH_3COONa(aq) \longrightarrow Na^+(aq) + CH_3COO^-(aq)$$
$$ 0.30\ M 0.30\ M$$

The initial concentrations, changes, and final concentrations of the species involved in the equilibrium are

	CH_3COOH	**H^+**	**CH_3COO^-**
Initial (M)	0.20	0	0.30
Change (M)	$-x$	$+x$	$+x$
Equilibrium (M)	$0.20 - x$	x	$0.30 + x$

Assuming that x is small, then $0.30 + x \approx 0.30$ and $0.20 - x \approx 0.20$. The Henderson-Hasselbalch equation (Equation 12.1) then gives

$$pH = pK_a + \log_{10} \frac{[A^-]}{[HA]}$$

$$pH = 4.74 + \log_{10} \frac{0.30}{0.20}$$

$$= 4.92$$

The value of $x = [H^+]$ corresponding to this pH is between $10^{-4}\ M$ and $10^{-5}\ M$, which is significantly less than 5 percent of the initial concentrations of the acid or salt, so the approximation is valid.

(b) As in Example 11.11, $x = [H^+]$ is determined from

$$K_a = \frac{x^2}{[CH_3COOH]_0 - x}$$

Assuming that $x \ll [CH_3COOH]_0$ gives

$$x = \sqrt{K_a[CH_3COOH]_0} = \sqrt{(1.8 \times 10^{-5})(0.20)}$$
$$= 1.9 \times 10^{-3}$$

Thus,

$$pH = -\log_{10}(1.9 \times 10^{-3}) = 2.72$$

In this case, x is less than 5 percent of the initial acid concentration, so the approximation is valid.

Check Comparing the results in parts (a) and (b), we see that when the common ion (CH_3COO^-) is present, according to Le Châtelier's principle, the equilibrium shifts from right to left. This action decreases the extent of ionization of the weak acid. Consequently, fewer H^+ ions are produced in (a) and the pH of the solution is

—Continued

Continued—

higher than that in part (b). As always, you should check the validity of the assumptions.

Comment Note that, if the initial concentrations were small enough that the approximation used in part (a) were not valid, then the problem must be solved by substituting the concentrations $[CH_3COOH] = 0.20 - x$, $[CH_3COO^-] = 0.30 - x$, and $[H^+] = x$ into the expression for K_a and solving the resulting quadratic equation for x.

Practice Exercise What is the pH of a solution containing 0.30 M HCOOH and 0.52 M HCOOK?

The common ion effect also operates in a solution containing a weak base, such as NH_3, and a salt of the base, say NH_4Cl. At equilibrium

$$NH_4^+(aq) \rightleftharpoons NH_3(aq) + H^+(aq)$$

$$K_a = \frac{[NH_3][H^+]}{[NH_4^+]}$$

We can derive the Henderson-Hasselbalch equation for this system as follows: Rearranging the preceding equation we obtain

$$[H^+] = \frac{K_a[NH_4^+]}{[NH_3]}$$

Taking the negative logarithm of both sides gives

$$-\log_{10}[H^+] = -\log_{10} K_a - \log_{10}\frac{[NH_4^+]}{[NH_3]}$$

$$-\log_{10}[H^+] = -\log_{10} K_a + \log_{10}\frac{[NH_3]}{[NH_4^+]}$$

or

$$pH = pK_a + \log_{10}\frac{[NH_3]}{[NH_4^+]}$$

A solution containing both NH_3 and its salt NH_4Cl is *less* basic than a solution containing only NH_3 at the same concentration. The common ion NH_4^+ suppresses the ionization of NH_3 in the solution containing both the base and the salt.

12.2 | The pH of a Buffer Solution Is Resistant to Large Changes in pH

A *buffer solution* is *a solution of (1) a weak acid or a weak base and (2) its salt; both components must be present. The solution has the ability to resist changes in pH upon the addition of small amounts of either acid or base.* Buffers are very important to chemical and biological systems. The pH in the human body varies greatly from one fluid to another; for example, the pH of blood is about 7.4, whereas the gastric juice in our stomachs has a pH of about 1.5. These pH values, which are crucial for

Fluids for intravenous injection must include buffer systems to maintain the proper blood pH.

proper enzyme function and the balance of osmotic pressure, are maintained by buffers in most cases.

A buffer solution must contain a relatively large concentration of acid to react with any OH^- ions that are added to it, and it must contain a similar concentration of base to react with any added H^+ ions. Furthermore, the acid and the base components of the buffer must not consume each other in a neutralization reaction. These requirements are satisfied by an acid-base conjugate pair, for example, a weak acid and its conjugate base (supplied by a salt) or a weak base and its conjugate acid (supplied by a salt).

A simple buffer solution can be prepared by adding comparable molar amounts of acetic acid (CH_3COOH) and its salt sodium acetate (CH_3COONa) to water. The equilibrium concentrations of both the acid and the conjugate base (from CH_3COONa) can generally be assumed to be the same as the starting concentrations because both the acid (CH_3COOH) and its conjugate base (CH_3COO^-) are weak electrolytes. A solution containing these two substances has the ability to neutralize either added acid or added base. Sodium acetate, a strong electrolyte, dissociates completely in water:

$$CH_3COONa(aq) \longrightarrow CH_3COO^-(aq) + Na^+(aq)$$

If an acid is added, the H^+ ions will be neutralized by the conjugate base in the buffer (CH_3COO^-) according to the equation

$$CH_3COO^-(aq) + H^+(aq) \longrightarrow CH_3COOH(aq)$$

If a base is added to the buffer system, the OH^- ions will be neutralized by the acid in the buffer:

$$CH_3COOH(aq) + OH^-(aq) \longrightarrow CH_3COO^-(aq) + H_2O(l)$$

As you can see, the two reactions that characterize this buffer system are identical to those for the common ion effect described in Example 12.1. The *buffering capacity,* that is, the effectiveness of the buffer solution, depends on the amount of acid and conjugate base from which the buffer is made. The larger the amount, the greater the buffering capacity.

In general, a buffer system can be represented as salt-acid or conjugate base–acid. Thus, the sodium acetate–acetic acid buffer system discussed previously can be written as CH_3COONa/CH_3COOH or simply CH_3COO^-/CH_3COOH. Figure 12.1 shows this buffer system in action.

Example 12.2 distinguishes buffer systems from acid-salt combinations that do not function as buffers.

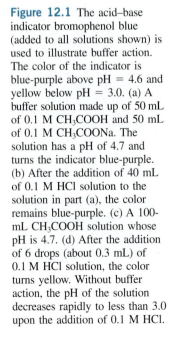

Figure 12.1 The acid–base indicator bromophenol blue (added to all solutions shown) is used to illustrate buffer action. The color of the indicator is blue-purple above pH = 4.6 and yellow below pH = 3.0. (a) A buffer solution made up of 50 mL of 0.1 M CH_3COOH and 50 mL of 0.1 M CH_3COONa. The solution has a pH of 4.7 and turns the indicator blue-purple. (b) After the addition of 40 mL of 0.1 M HCl solution to the solution in part (a), the color remains blue-purple. (c) A 100-mL CH_3COOH solution whose pH is 4.7. (d) After the addition of 6 drops (about 0.3 mL) of 0.1 M HCl solution, the color turns yellow. Without buffer action, the pH of the solution decreases rapidly to less than 3.0 upon the addition of 0.1 M HCl.

(a)   (b)   (c)   (d)

Example 12.2

Which of the following solutions can be classified as buffer systems: (a) KH_2PO_4/H_3PO_4, (b) $NaClO_4/HClO_4$, and (c) C_5H_5N/C_5H_5NHCl (C_5H_5N is pyridine; its K_b is given in Table 11.4)? Explain your answer.

Strategy What constitutes a buffer system? Which of the preceding solutions contains a weak acid and its salt (containing the weak conjugate base)? Which of the preceding solutions contains a weak base and its salt (containing the weak conjugate acid)? Why is the conjugate base of a strong acid not able to neutralize an added acid?

Solution The criteria for a buffer system are that we must have a weak acid and its salt (containing the weak conjugate base) or a weak base and its salt (containing the weak conjugate acid).

(a) H_3PO_4 is a weak acid, and its conjugate base, $H_2PO_4^-$, is a weak base (see Table 11.6). Therefore, this is a buffer system.

(b) Because $HClO_4$ is a strong acid, its conjugate base, ClO_4^-, is an extremely weak base. This means that the ClO_4^- ion will not combine with an H^+ ion in solution to form $HClO_4$. Thus, the system cannot act as a buffer system.

(c) As Table 11.4 shows, C_5H_5N is a weak base and its conjugate acid, $C_5H_5NH^+$ (the cation of the salt C_5H_5NHCl), is a weak acid. Therefore, this is a buffer system.

Practice Exercise Which of the following are buffer systems: (a) KF/HF, (b) KBr/HBr, and (c) $Na_2CO_3/NaHCO_3$?

The pH of a buffer solution is governed by the Henderson-Hasselbalch equation (Equation 12.2)

$$pH = pK_a + \log_{10} \frac{[\text{conjugate base}]}{[\text{acid}]}$$

The form of this equation helps us understand the stability of a buffer solution to changes in pH from a mathematical perspective. In Equation 12.2, the logarithmic dependence of the pH on the ratio of conjugate base to acid means that in order for the pH to change by 1, the ratio of [conjugate base]/[acid] has to change by a factor of 10.

The effect of a buffer solution on pH is illustrated by Example 12.3.

Example 12.3

(a) Calculate the pH of a buffer system containing 1.0 M CH_3COOH and 1.0 M CH_3COONa. (b) What is the pH of the buffer system after the addition of 0.10 mole of gaseous HCl to 1.0 L of the solution? Assume that the volume of the solution does not change when the HCl is added.

Strategy The pH of the buffer system before and after the addition of HCl can be calculated from the Henderson-Hasselbalch equation. The K_a of CH_3COOH is 1.8×10^{-5} (see Table 11.3) which gives a pK_a of 4.74.

Solution (a) Both the weak acid (CH_3COOH) and its conjugate base (CH_3COO^- from CH_3COONa) are weak electrolytes and will not dissociate appreciably $[CH_3COOH] \approx [CH_3COOH]_0 = 1.0$ M and $[CH_3COO^-] \approx [CH_3COO^-]_0 = 1.0$ M.

—Continued

Continued—

(We must check this approximation at the end of the calculation.) The Henderson-Hasselbalch equation (Equation 12.2) gives

$$pH = pK_a + \log_{10} \frac{[CH_3COO^-]}{[CH_3COOH]}$$

$$= 4.74 + \log_{10} \frac{1.0}{1.0}$$

$$= 4.74$$

When the concentrations of the acid and its conjugate base are the same, the pH of the buffer is equal to the pK_a.

This corresponds to an H^+ ion concentration of between 10^{-4} M and 10^{-5} M, so the approximation is valid.[1]

(b) When HCl (a strong acid) is added to the solution, it completely dissociates into H^+ ions and Cl^- ions. The initial changes are

$$HCl(aq) \longrightarrow H^+(aq) + Cl^-(aq)$$

	HCl	H^+	Cl^-
Initial (*M*)	0.10	0.0	0.0
Change (*M*)	−0.10	0.10	0.10
Final (*M*)	0.00	0.10	0.10

The H^+ ions are neutralized by the conjugate base (CH_3COO^-), and Cl^- is a spectator ion in solution because it is the conjugate base of a strong acid and will have no tendency to neutralize the weak acid present. At this point, it is more convenient to work with moles rather than molarity. The reason is that in some cases the volume of the solution may change when a substance is added. A change in volume will change the molarity, but not the number of moles. The neutralization reaction is summarized next:

$$CH_3COO^-(aq) + H^+(aq) \longrightarrow CH_3COOH(aq)$$

	CH_3COO^-	H^+	CH_3COOH
Initial (mol)	1.0	0.10	1.0
Change (mol)	−0.10	−0.10	+0.10
Final (mol)	0.90	0.00	1.10

Finally, to calculate the pH of the buffer after neutralization of the acid, we convert back to molarity by dividing moles by 1.0 L of solution. Thus, the weak acid concentration, [CH_3COOH] after addition of HCl is 1.10 M and the conjugate base concentration [CH_3COO^-] = 0.90 M. Reapplying the Henderson-Hasselbalch equation gives

$$pH = pK_a + \log_{10} \frac{[CH_3COO^-]}{[CH_3COOH]}$$

$$= 4.74 + \log_{10} \frac{0.90}{1.10}$$

$$= 4.65$$

Practice Exercise Calculate the pH of the 0.30 M NH_3/0.36 M NH_4Cl buffer system. What is the pH after the addition of 20.0 mL of 0.050 M NaOH to 80.0 mL of the buffer solution?

---

1. Although the approximation that the dissociation is small is valid, the results here will only be approximate because of the high concentrations of solute (~1 M). At such concentrations, the interactions among solute ions and molecules will be strong, and the solute activities will deviate somewhat from their concentrations. See the discussion in Section 10.1.

In the buffer solution examined in Example 12.3, there is a decrease in pH (the solution becomes more acidic) as a result of the added HCl. We can also compare the changes in H^+ ion concentration as follows

$$\text{before addition of HCl: } [H^+] = 10^{-4.74} = 1.8 \times 10^{-5}\,M$$
$$\text{after addition of HCl: } [H^+] = 10^{-4.65} = 2.2 \times 10^{-5}\,M$$

Thus, the H^+ ion concentration increases by a factor of

$$\frac{2.2 \times 10^{-5}}{1.8 \times 10^{-5}} = 1.2$$

To appreciate the effectiveness of the CH_3COONa/CH_3COOH buffer, let us find out what would happen if 0.10 mol HCl were added to 1.0 L of pure water (pH = 7), and compare the increase in H^+ ion concentration.

$$\text{before addition of HCl: } [H^+] = 1.0 \times 10^{-7}\,M$$
$$\text{after addition of HCl: } [H^+] = 0.10\,M$$

As a result of the addition of HCl, the H^+ ion concentration increases by a factor of 10^6, amounting to a million-fold increase! This comparison shows that a properly chosen buffer solution can maintain a fairly constant H^+ ion concentration, or pH (Figure 12.2).

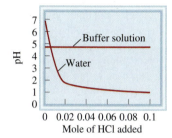

Figure 12.2 A comparison of the change in pH when 0.10 mol HCl is added to 1 L of pure water and to 1 L of an acetate buffer solution, as described in Example 12.3.

Preparing a Buffer Solution with a Specific pH

Now suppose we want to prepare a buffer solution with a specific pH. How do we go about it? Equation 12.2 indicates that if the molar concentrations of the acid and its conjugate base are approximately equal, that is, if [acid] ≈ [conjugate base], then

$$\log_{10}\frac{[\text{conjugate base}]}{[\text{acid}]} \approx 0$$

$$\text{pH} \approx \text{p}K_a$$

Thus, to prepare a buffer solution, we work backward. First we choose a weak acid whose $\text{p}K_a$ is close to the desired pH. Next, we substitute the pH and $\text{p}K_a$ values in Equation 12.2 to obtain the ratio [conjugate base]/[acid]. This ratio can then be converted to molar quantities for the preparation of the buffer solution. Example 12.4 shows this approach.

Example 12.4

Describe how you would prepare a "phosphate buffer" with a pH of about 7.40.

Strategy For a buffer to function effectively, the concentrations of the acid component must be roughly equal to the conjugate base component. According to Equation 12.2, when the desired pH is close to the $\text{p}K_a$ of the acid, pH ≈ $\text{p}K_a$,

$$\log_{10}\frac{[\text{conjugate base}]}{[\text{acid}]} \approx 0$$

$$\frac{[\text{conjugate base}]}{[\text{acid}]} \approx 1$$

—Continued

Maintaining the pH of Blood

All higher animals need a circulatory system to carry fuel and oxygen for their life processes and to remove wastes. In the human body, this vital exchange takes place in the versatile fluid known as blood, of which there are about 5 L in an average adult. Blood circulating deep in the tissues carries oxygen and nutrients to keep cells alive, and removes carbon dioxide and other waste materials. Using several buffer systems, nature has provided an extremely efficient method for the delivery of oxygen and the removal of carbon dioxide.

Blood is an enormously complex system, but for our purposes we need look at only two essential components: blood plasma and red blood cells, or *erythrocytes*. Blood plasma contains many compounds, including proteins, metal ions, and inorganic phosphates. The erythrocytes contain hemoglobin molecules, as well as the enzyme *carbonic anhydrase*, which catalyzes both the formation of carbonic acid (H_2CO_3) and its decomposition:

$$CO_2(aq) + H_2O(l) \rightleftharpoons H_2CO_3(aq)$$

The substances inside the erythrocyte are protected from extracellular fluid (blood plasma) by a cell membrane that allows only certain molecules to diffuse through it. The pH of blood plasma is maintained at about 7.40 by several buffer systems, the most important of which is the HCO_3^-/H_2CO_3 system. In the erythrocyte, where the pH is 7.25, the principal buffer systems are HCO_3^-/H_2CO_3 and hemoglobin. The hemoglobin molecule is a complex protein molecule (molar mass 65,000 g) that contains a number of ionizable protons. As a very rough approximation, we can treat it as a monoprotic acid of the form HHb:

$$HHb(aq) \rightleftharpoons H^+(aq) + Hb^-(aq)$$

where HHb represents the hemoglobin molecule and Hb^- the conjugate base of HHb. Oxyhemoglobin ($HHbO_2$), formed by the combination of oxygen with hemoglobin, is a stronger acid than HHb:

$$HHbO_2(aq) \rightleftharpoons H^+(aq) + HbO_2^-(aq)$$

As the figure on page 621 shows, carbon dioxide produced by metabolic processes diffuses into the erythrocyte, where it is rapidly converted to H_2CO_3 by carbonic anhydrase:

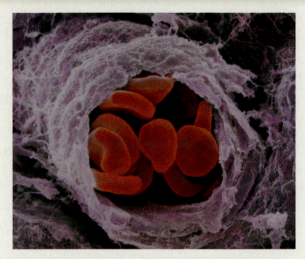

Electron micrograph of red blood cells in a small branch of an artery (color added).

$$CO_2(aq) + H_2O(l) \rightleftharpoons H_2CO_3(aq)$$

The ionization of the carbonic acid

$$H_2CO_3(aq) \rightleftharpoons H^+(aq) + HCO_3^-(aq)$$

has two important consequences. First, the bicarbonate ion diffuses out of the erythrocyte and is carried by the blood plasma to the lungs. This is the major mechanism for removing carbon dioxide. Second, the H^+ ions shift the equilibrium in favor of the nonionized oxyhemoglobin molecule:

$$H^+(aq) + HbO_2^-(aq) \rightleftharpoons HHbO_2(aq)$$

Because $HHbO_2$ releases oxygen more readily than does its conjugate base (HbO_2^-), the formation of the acid promotes the following reaction from left to right:

$$HHbO_2(aq) \rightleftharpoons HHb(aq) + O_2(aq)$$

The O_2 molecules diffuse out of the erythrocyte and are taken up by other cells to carry out metabolism. When the venous blood returns to the lungs, the preceding processes are reversed. The bicarbonate ions now diffuse into the

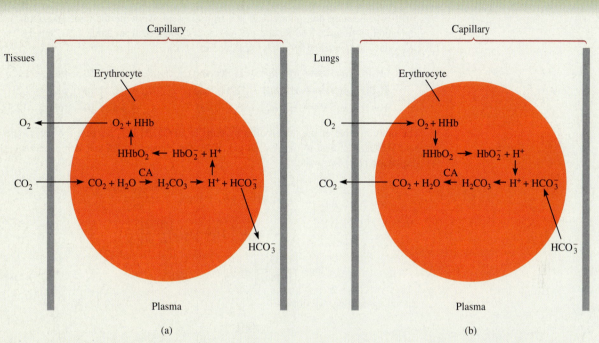

Oxygen–carbon dioxide transport and release by blood. (a) The partial pressure of CO_2 is higher in the metabolizing tissues than in the plasma. Thus, it diffuses into the blood capillaries and then into erythrocytes. There it is converted to carbonic acid by the enzyme carbonic anhydrase (CA). The protons provided by the carbonic acid then combine with the HbO_2 anions to form $HHbO_2$, which eventually dissociates into HHb and O_2. Because the partial pressure of O_2 is higher in the erythrocytes than in the tissues, oxygen molecules diffuse out of the erythrocytes and then into the tissues. The bicarbonate ions also diffuse out of the erythrocytes and are carried by the plasma to the lungs. (b) In the lungs, the processes are exactly reversed. Oxygen molecules diffuse from the lungs, where they have a higher partial pressure, into the erythrocytes. There they combine with HHb to form $HHbO_2$. The protons provided by $HHbO_2$ combine with the bicarbonate ions diffused into the erythrocytes from the plasma to form carbonic acid. In the presence of carbonic anhydrase, carbonic acid is converted to H_2O and CO_2. The CO_2 then diffuses out of the erythrocytes and into the lungs, where it is exhaled.

erythrocyte, where they react with hemoglobin to form carbonic acid:

$$HHb(aq) + HCO_3^-(aq) \rightleftharpoons Hb^-(aq) + H_2CO_3(aq)$$

Most of the acid is then converted to CO_2 by carbonic anhydrase:

$$H_2CO_3(aq) \rightleftharpoons H_2O(l) + CO_2(aq)$$

The carbon dioxide diffuses to the lungs and is eventually exhaled. The formation of the Hb^- ions (due to the reaction between HHb and HCO_3^-) also favors the uptake of oxygen at the lungs

$$Hb^-(aq) + O_2(aq) \rightleftharpoons HbO_2^-(aq)$$

because Hb^- has a greater affinity for oxygen than does HHb.

When the arterial blood flows back to the body tissues, the entire cycle is repeated.

Continued—

Solution Because phosphoric acid is a triprotic acid, we write the three stages of ionization as follows. The K_a and pK_a values are obtained from Table 11.6, and the pK_a values are obtained by applying Equation 11.15

$$H_3PO_4(aq) \rightleftharpoons H^+(aq) + H_2PO_4^-(aq) \qquad K_{a_1} = 7.5 \times 10^{-3}; pK_{a_1} = 2.12$$
$$H_2PO_4^-(aq) \rightleftharpoons H^+(aq) + HPO_4^{2-}(aq) \qquad K_{a_2} = 6.2 \times 10^{-8}; pK_{a_2} = 7.21$$
$$HPO_4^{2-}(aq) \rightleftharpoons H^+(aq) + PO_4^{3-}(aq) \qquad K_{a_3} = 4.8 \times 10^{-13}; pK_{a_3} = 12.32$$

The most suitable of the three buffer systems is $HPO_4^{2-}/H_2PO_4^-$ because the pK_a of the acid $H_2PO_4^-$ is closest to the desired pH. From the Henderson-Hasselbalch equation (Equation 12.2), we write

$$pH = pK_a + \log_{10} \frac{[\text{conjugate base}]}{[\text{acid}]}$$

$$7.40 = 7.21 + \log_{10} \frac{[HPO_4^{2-}]}{[H_2PO_4^-]}$$

$$\log_{10} \frac{[HPO_4^{2-}]}{[H_2PO_4^-]} = 0.19$$

Taking the antilog, we obtain

$$\frac{[HPO_4^{2-}]}{[H_2PO_4^-]} = 10^{0.19} = 1.5$$

Thus, one way to prepare a phosphate buffer with a pH of 7.40 is to dissolve disodium hydrogen phosphate (Na_2HPO_4) and sodium dihydrogen phosphate (NaH_2PO_4) in a mole ratio of 1.5:1.0 in water. For example, we could dissolve 1.5 moles of Na_2HPO_4 and 1.0 mole of NaH_2PO_4 in enough water to make up a 1-L solution.

Practice Exercise How would you prepare a liter of "carbonate buffer" at a pH of 10.10? You are provided with carbonic acid (H_2CO_3), sodium hydrogen carbonate ($NaHCO_3$), and sodium carbonate (Na_2CO_3). See Table 11.6 for K_a values.

12.3 | The Concentration of an Unknown Acid or Base Can Be Determined by Titration

Quantitative studies of acid-base neutralization reactions are most conveniently carried out using a technique known as titration. In a ***titration,*** *a solution of accurately known concentration,* called a ***standard solution,*** *is added gradually to another solution of unknown concentration, until the chemical reaction between the two solutions is complete.* If we know the volumes of the standard and unknown solutions used in the titration, along with the concentration of the standard solution, we can calculate the concentration of the unknown solution.

Sodium hydroxide is one of the bases commonly used in the laboratory. However, it is difficult to obtain solid sodium hydroxide in a pure form because it has a tendency to absorb water from air, and its solution reacts with carbon dioxide. For these reasons, a solution of sodium hydroxide must be *standardized* before it can be used in accurate analytical work. We can standardize the sodium hydroxide solution by titrating it against an acid solution of accurately known concentration. The acid often chosen for this task is a weak acid called potassium hydrogen phthalate (KHP), for which the

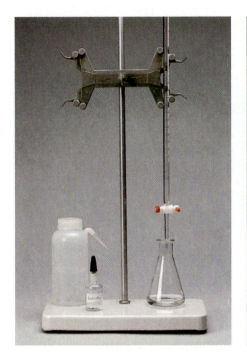 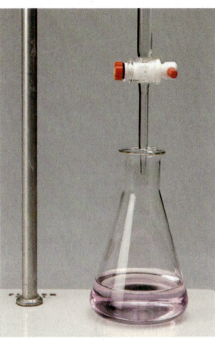

Figure 12.3 (a) Apparatus for an acid-base titration. An NaOH solution is added from the buret to a KHP solution in an Erlenmeyer flask. (b) A reddish-pink color appears when the equivalence point is reached. The color here has been intensified for visual display.

molecular formula is $KHC_8H_4O_4$. KHP is a white, soluble solid that is commercially available in highly pure form. The reaction between KHP and sodium hydroxide is

$$KHC_8H_4O_4(aq) + NaOH(aq) \longrightarrow KNaC_8H_4O_4(aq) + H_2O(l)$$

and the net ionic equation is

$$HC_8H_4O_4^-(aq) + OH^-(aq) \longrightarrow C_8H_4O_4^{2-}(aq) + H_2O(l)$$

The apparatus for the titration is shown in Figure 12.3. First, a known amount of KHP is transferred to an Erlenmeyer flask, and some distilled water is added to make up a solution. Next, an NaOH solution is carefully added to the KHP solution from a buret until we reach the **equivalence point,** that is, *the point at which the acid has completely reacted with or been neutralized by the base.* The equivalence point is usually signaled by a sharp change in the color of an indicator in the acid solution. In acid-base titrations, *indicators* are substances that have distinctly different colors in acidic and basic media. One commonly used indicator is phenolphthalein, which is colorless in acidic and neutral solutions but reddish pink in basic solutions. Indicators will be discussed in more detail in Section 12.4. At the equivalence point, all the KHP present has been neutralized by the added NaOH and the solution is still colorless. However, if we add just one more drop of NaOH solution from the buret, the solution will immediately turn pink because the solution is now basic.

Having discussed buffer solutions, we can now look in more detail at the quantitative aspects of acid-base titrations. We will consider three types of reactions: (1) titrations involving a strong acid and a strong base, (2) titrations involving a weak acid and a strong base, and (3) titrations involving a strong acid and a weak base. Titrations involving a weak acid and a weak base are complicated by the hydrolysis of both the cation and the anion of the salt formed. These titrations will not be dealt with in this text. Figure 12.4 shows the arrangement for monitoring the pH during the course of a titration.

Figure 12.4 A pH meter is used to monitor an acid-base titration.

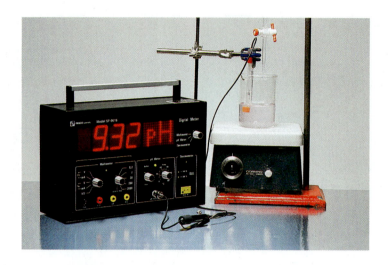

Strong Acid–Strong Base Titrations

The reaction between a strong acid (say, HCl) and a strong base (say, NaOH) can be represented by

$$NaOH(aq) + HCl(aq) \longrightarrow NaCl(aq) + H_2O(l)$$

or in terms of the net ionic equation

$$H^+(aq) + OH^-(aq) \longrightarrow H_2O(l)$$

Consider the addition of a 0.100 *M* NaOH solution (from a buret) to an Erlenmeyer flask containing 25.0 mL of 0.100 *M* HCl. Figure 12.5 shows the pH profile of the titration (also known as the *titration curve*). Before the addition of NaOH, the pH of the acid is given (approximately) by $-\log (0.100)$, or 1.00. When NaOH is added, the pH of the

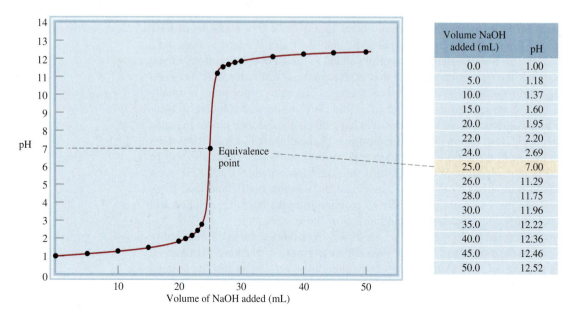

Volume NaOH added (mL)	pH
0.0	1.00
5.0	1.18
10.0	1.37
15.0	1.60
20.0	1.95
22.0	2.20
24.0	2.69
25.0	7.00
26.0	11.29
28.0	11.75
30.0	11.96
35.0	12.22
40.0	12.36
45.0	12.46
50.0	12.52

Figure 12.5 The pH profile for the titration of a strong acid with a strong base. A 0.100 *M* NaOH solution is added from a buret to 25.0 mL of a 0.100 *M* HCl solution in an Erlenmeyer flask (see Figure 12.3). This curve is sometimes referred to as a titration curve.

solution increases slowly at first. Near the equivalence point, the pH begins to rise steeply, and at the equivalence point (that is, the point at which equimolar amounts of acid and base have reacted) the curve rises almost vertically. In a strong acid–strong base titration, both the hydrogen ion and hydroxide ion concentrations are very small at the equivalence point (approximately 1×10^{-7} M); consequently, the addition of a single drop of the base can cause a large increase in $[OH^-]$ and in the pH of the solution. Beyond the equivalence point, the pH again increases slowly with the addition of NaOH.

It is possible to calculate the pH of the solution at every stage of titration. Here are three sample calculations.

1. *After the addition of 10.0 mL of 0.100 M NaOH to 25.0 mL of 0.100 M HCl.* The total volume of the solution is 35.0 mL. The number of moles of NaOH in 10.0 mL is

$$10.0 \text{ mL} \times \frac{0.100 \text{ mol NaOH}}{1 \text{ L NaOH}} \times \frac{1 \text{ L}}{1000 \text{ mL}} = 1.00 \times 10^{-3} \text{ mol}$$

The number of moles of HCl originally present in 25.0 mL of solution is

$$25.0 \text{ mL} \times \frac{0.100 \text{ mol HCl}}{1 \text{ L HCl}} \times \frac{1 \text{ L}}{1000 \text{ mL}} = 2.50 \times 10^{-3} \text{ mol}$$

Thus, the amount of HCl left after partial neutralization is (2.50×10^{-3}) − (1.00×10^{-3}), or 1.50×10^{-3} mol. Next, the concentration of H^+ ions in 35.0 mL of solution is found as follows:

$$\frac{1.50 \times 10^{-3} \text{ mol HCl}}{35.0 \text{ mL}} \times \frac{1000 \text{ mL}}{1 \text{ L}} = 0.0429 \text{ mol L}^{-1} = 0.0429 \text{ } M$$

Thus, $[H^+] = 0.0429$ M, and the pH of the solution is

$$\text{pH} \approx -\log_{10} 0.0429 = 1.37$$

2. *After the addition of 25.0 mL of 0.100 M NaOH to 25.0 mL of 0.100 M HCl.*
This is a simple calculation, because it involves a complete neutralization reaction and the salt (NaCl) does not undergo hydrolysis. At the equivalence point, Neither Na^+ nor Cl^- undergoes hydrolysis.

$$[H^+] = [OH^-] = 1.00 \times 10^{-7} \text{ } M$$

and the pH of the solution is 7.00.

3. *After the addition of 35.0 mL of 0.100 M NaOH to 25.0 mL of 0.100 M HCl.* The total volume of the solution is now 60.0 mL. The number of moles of NaOH added is

$$35.0 \text{ mL} \times \frac{0.100 \text{ mol NaOH}}{1 \text{ L NaOH}} \times \frac{1 \text{ L}}{1000 \text{ mL}} = 3.50 \times 10^{-3} \text{ mol}$$

The number of moles of HCl in 25.0 mL solution is 2.50×10^{-3} mol. After complete neutralization of HCl, the number of moles of NaOH left is (3.50×10^{-3}) − (2.50×10^{-3}), or 1.00×10^{-3} mol. The concentration of OH^- in 60.0 mL of solution is

$$\frac{1.00 \times 10^{-3} \text{ mol OH}^-}{60.0 \text{ mL}} \times \frac{1000 \text{ mL}}{1 \text{ L}} = 0.0167 \text{ mol L}^{-1} = 0.0167 \text{ } M$$

Thus, $[OH^-] = 0.0167\ M$ and $pOH = -\log_{10} 0.0167 = 1.78$. Hence, the pH of the solution is

$$pH = 14.00 - pOH = 14.00 - 1.78 = 12.22$$

Weak Acid–Strong Base Titrations

Consider the neutralization reaction between acetic acid (a weak acid) and sodium hydroxide (a strong base):

$$CH_3COOH(aq) + NaOH(aq) \longrightarrow CH_3COONa(aq) + H_2O(l)$$

This equation can be simplified to

$$CH_3COOH(aq) + OH^-(aq) \longrightarrow CH_3COO^-(aq) + H_2O(l)$$

The acetate ion undergoes hydrolysis as follows:

$$CH_3COO^-(aq) + H_2O(l) \rightleftharpoons CH_3COOH(aq) + OH^-(aq)$$

Therefore, at the equivalence point, when we only have sodium acetate present, the pH will be *greater than* 7 as a result of the excess OH^- ions formed (Figure 12.6). Note that this situation is analogous to the hydrolysis of sodium acetate (CH_3COONa) (see page 594).

Example 12.5 deals with the titration of a weak acid with a strong base.

Example 12.5

Calculate the pH in the titration of 25.0 mL of 0.100 M acetic acid by sodium hydroxide after the addition to the acid solution of (a) 10.0 mL of 0.100 M NaOH, (b) 25.0 mL of 0.100 M NaOH, or (c) 35.0 mL of 0.100 M NaOH.

—Continued

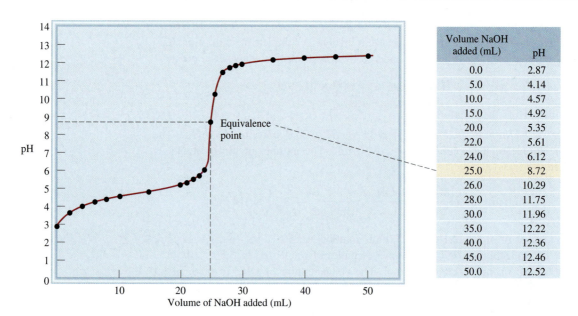

Volume NaOH added (mL)	pH
0.0	2.87
5.0	4.14
10.0	4.57
15.0	4.92
20.0	5.35
22.0	5.61
24.0	6.12
25.0	8.72
26.0	10.29
28.0	11.75
30.0	11.96
35.0	12.22
40.0	12.36
45.0	12.46
50.0	12.52

Figure 12.6 The pH profile of a weak acid–strong base titration. A 0.100 M NaOH solution is added from a buret to 25.0 mL of a 0.100 M CH$_3$COOH solution in an Erlenmeyer flask. Because of the hydrolysis of the salt formed, the pH at the equivalence point is greater than 7.

Continued—

Strategy The reaction between CH_3COOH and NaOH is

$$CH_3COOH(aq) + NaOH(aq) \longrightarrow CH_3COONa(aq) + H_2O(l)$$

We see that 1 mol $CH_3COOH \simeq 1$ mol NaOH. Therefore, at every stage of the titration, we can calculate the number of moles of base reacting with the acid, and the pH of the solution is determined by the excess acid or base left over. At the equivalence point, however, the neutralization is complete, and the pH of the solution will depend on the extent of the hydrolysis of the salt formed, which is CH_3COONa.

Solution (a) The number of moles of NaOH in 10.0 mL is

$$10.0 \text{ mL} \times \frac{0.100 \text{ mol NaOH}}{1 \text{ L NaOH}} \times \frac{1 \text{ L}}{1000 \text{ mL}} = 1.00 \times 10^{-3} \text{ mol}$$

The number of moles of CH_3COOH originally present in 25.0 mL of solution is

$$25.0 \text{ mL} \times \frac{0.100 \text{ mol } CH_3COOH}{1 \text{ L } CH_3COOH} \times \frac{1 \text{ L}}{1000 \text{ mL}} = 2.50 \times 10^{-3} \text{ mol}$$

We work with moles at this point because when two solutions are mixed, the solution volume increases. As the volume increases, molarity will change, but the number of moles will remain the same. The changes in the number of moles are summarized next:

	$CH_3COOH(aq)$ +	NaOH(aq) $\longrightarrow$	$CH_3COONa(aq)$ +	$H_2O(l)$
Initial (mol)	2.50×10^{-3}	1.00×10^{-3}	0	
Change (mol)	-1.00×10^{-3}	-1.00×10^{-3}	$+1.00 \times 10^{-3}$	
Final (mol)	1.50×10^{-3}	0	1.00×10^{-3}	

At this stage, we have a buffer system made up of CH_3COOH and CH_3COO^- (from the salt, CH_3COONa). To calculate the pH of the solution, we can use the Henderson-Hasselbalch equation (Equation 12.2):

$$pH = pK_a + \log_{10} \frac{[CH_3COO^-]}{[CH_3COOH]}$$
$$= 4.74 + \log_{10} \frac{1.00 \times 10^{-3}}{1.50 \times 10^{-3}}$$
$$= 4.74 - 0.18$$
$$= 4.56$$

Because the volume of the solution is the same for CH_3COOH and CH_3COO^- (35 mL), the ratio of the number of moles present is equal to the ratio of their molar concentrations.

(b) These quantities (that is, 25.0 mL of 0.100 M NaOH reacting with 25.0 mL of 0.100 M CH_3COOH) correspond to the equivalence point. The number of moles of NaOH in 25.0 mL of the solution is

$$25.0 \text{ mL} \times \frac{0.100 \text{ mol NaOH}}{1 \text{ L NaOH}} \times \frac{1 \text{ L}}{1000 \text{ mL}} = 2.50 \times 10^{-3} \text{ mol}$$

The changes in number of moles are summarized next:

	$CH_3COOH(aq)$ +	NaOH(aq) $\longrightarrow$	$CH_3COONa(aq)$ +	$H_2O(l)$
Initial (mol)	2.50×10^{-3}	2.50×10^{-3}	0	
Change (mol)	-2.50×10^{-3}	-2.50×10^{-3}	$+2.50 \times 10^{-3}$	
Final (mol)	0	0	2.50×10^{-3}	

—Continued

Continued—

At the equivalence point, the concentrations of both the acid and the base are zero. The total volume is (25.0 + 25.0) mL or 50.0 mL, so the concentration of the salt is

$$[CH_3COONa] = \frac{2.50 \times 10^{-3}\ mol}{50.0\ mL} \times \frac{1000\ ml}{1\ L}$$

$$= 0.0500\ mol\ L^{-1} = 0.0500\ M$$

The next step is to calculate the pH of the solution that results from the hydrolysis of the CH_3COO^- ions:

$$CH_3COO^-(aq) + H_2O(l) \rightleftharpoons CH_3COOH(aq) + OH^-(aq)$$

Following the procedure described in Example 11.17 and looking up the base ionization constant (K_b) for CH_3COO^- in Table 11.5, we write

$$K_b = 5.6 \times 10^{-10} = \frac{[CH_3COOH][OH^-]}{[CH_3COO^-]} = \frac{x^2}{0.0500 - x}$$

$$x = [OH^-] = 5.3 \times 10^{-6}\ M,\ pH = 8.72$$

(c) After the addition of 35.0 mL of NaOH, the solution is well past the equivalence point. The number of moles of NaOH originally present is

$$35.0\ mL \times \frac{0.100\ mol\ NaOH}{1\ L\ NaOH} \times \frac{1\ L}{1000\ mL} = 3.50 \times 10^{-3}\ mol$$

The changes in number of moles are summarized next:

	$CH_3COOH(aq)$ +	$NaOH(aq)$ ⟶	$CH_3COONa(aq)$ + $H_2O(l)$
Initial (mol)	2.50×10^{-3}	3.50×10^{-3}	0
Change (mol)	-2.50×10^{-3}	-2.50×10^{-3}	$+2.50 \times 10^{-3}$
Final (mol)	0	1.00×10^{-3}	2.50×10^{-3}

At this stage, we have two species in solution that are responsible for making the solution basic: OH^- and CH_3COO^- (from CH_3COONa). However, because OH^- is a much stronger base than CH_3COO^-, we can safely neglect the hydrolysis of the CH_3COO^- ions and calculate the pH of the solution using only the concentration of the OH^- ions. The total volume of the combined solutions is (25.0 + 35.0) mL or 60.0 mL, so we calculate OH^- concentration as follows:

$$[OH^-] = \frac{1.00 \times 10^{-3}\ mol}{60.0\ mL} \times \frac{1000\ mL}{1\ L} = 0.0167\ mol\ L^{-1} = 0.0167\ M$$

$$pOH = -\log_{10}[OH^-] = -\log_{10} 0.0167 = 1.78$$

$$pH = 14.00 - 1.78 = 12.22$$

Practice Exercise Exactly 100 mL of 0.10 M nitrous acid (HNO_2) is titrated with a 0.10 M NaOH solution. Calculate the pH for (a) the initial solution, (b) the point at which 80 mL of the base has been added, (c) the equivalence point, and (d) the point at which 105 mL of the base has been added.

Strong Acid–Weak Base Titrations

Consider the titration of HCl, a strong acid, with NH_3, a weak base:

$$HCl(aq) + NH_3(aq) \longrightarrow NH_4Cl(aq)$$

or simply

$$H^+(aq) + NH_3(aq) \longrightarrow NH_4^+(aq)$$

The pH at the equivalence point is *less than* 7 due to the hydrolysis of the NH_4^+ ion:

$$NH_4^+(aq) + H_2O(l) \rightleftharpoons NH_3(aq) + H_3O^+(aq)$$

or simply

$$NH_4^+(aq) \rightleftharpoons NH_3(aq) + H^+(aq)$$

Because of the volatility of an aqueous ammonia solution, it is more convenient to add hydrochloric acid from a buret to the ammonia solution. Figure 12.7 shows the titration curve for this experiment.

Example 12.6

Calculate the pH at the equivalence point when 25.0 mL of 0.100 *M* NH_3 is titrated by a 0.100 *M* HCl solution.

Strategy The reaction between NH_3 and HCl is

$$NH_3(aq) + HCl(aq) \longrightarrow NH_4Cl(aq)$$

—Continued

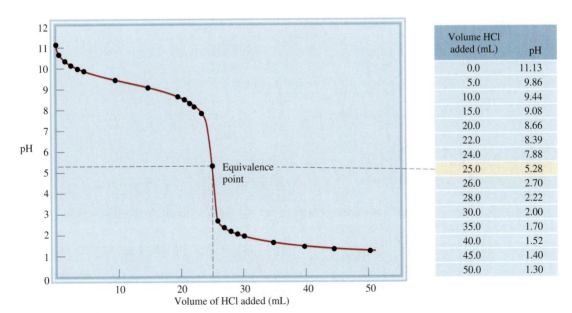

Volume HCl added (mL)	pH
0.0	11.13
5.0	9.86
10.0	9.44
15.0	9.08
20.0	8.66
22.0	8.39
24.0	7.88
25.0	5.28
26.0	2.70
28.0	2.22
30.0	2.00
35.0	1.70
40.0	1.52
45.0	1.40
50.0	1.30

Figure 12.7 The pH profile for the titration of a weak base with a strong acid. A 0.100 *M* HCl solution is added from a buret to 25.0 mL of a 0.100 *M* NH_3 solution in an Erlenmeyer flask. Because of the hydrolysis of the salt formed, the pH at the equivalence point is less than 7.

Continued—

We see that 1 mol $NH_3 \simeq$ 1 mol HCl. At the equivalence point, the major species in solution are the salt NH_4Cl (dissociated into NH_4^+ and Cl^- ions) and H_2O. First we determine the concentration of NH_4Cl formed. Then we calculate the pH as a result of the NH_4^+ ion hydrolysis. The Cl^- ion, being the conjugate base of a strong acid HCl, does not react with water. As usual, we ignore the ionization of water.

Solution The number of moles of NH_3 in 25.0 mL of 0.100 M solution is

$$25.0\text{mL} \times \frac{0.100 \text{ mol } NH_3}{1 \text{ L } NH_3} \times \frac{1 \text{ L}}{1000 \text{ mL}} = 2.50 \times 10^{-3} \text{ mol}$$

At the equivalence point, the number of moles of HCl added equals the number of moles of NH_3. The changes in the number of moles are summarized here:

	$NH_3(aq)$	+	$HCl(aq)$	$\longrightarrow$	$NH_4Cl(aq)$
Initial (mol)	2.50×10^{-3}		2.50×10^{-3}		0
Change (mol)	-2.50×10^{-3}		-2.50×10^{-3}		$+2.50 \times 10^{-3}$
Final (mol)	0		0		2.50×10^{-3}

At the equivalence point, the concentrations of both the acid and the base are zero. The total volume is (25.0 + 25.0) mL, or 50.0 mL, so the concentration of the salt is

$$[NH_4Cl] = \frac{2.50 \times 10^{-3} \text{ mol}}{50.0 \text{ mL}} \times \frac{1000 \text{ mL}}{1 \text{ L}} = 0.0500 \text{ mol L}^{-1} = 0.0500 \ M$$

The pH of the solution at the equivalence point is determined by the hydrolysis of NH_4^+ ions:

$$NH_4^+(aq) \rightleftharpoons NH_3(aq) + H^+(aq), \quad K_a = 5.6 \times 10^{-10}$$

The expression for K_a is

$$K_a = \frac{[NH_3][H^+]}{[NH_4^+]} = 5.6 \times 10^{-10}$$

If we let $x = [H^+] = [NH_3]$, then

$$K_a = \frac{x^2}{0.0500 - x} = 5.6 \times 10^{-10}$$

K_a is small, so we make the approximation $0.0500 - x \approx 0.0500$, giving

$$x = \sqrt{(5.6 \times 10^{-10})(0.0500)} = 5.3 \times 10^{-6} \ M$$

This is much less than 5 percent of 0.0500, so the approximation is valid.

Thus, the pH is

$$\text{pH} = -\log_{10}(5.3 \times 10^{-6}) = 5.28$$

Always check the validity of the approximation.

Check Note that the pH of the solution is acidic. This is what we would expect from the hydrolysis of the ammonium ion.

Practice Exercise Calculate the pH at the equivalence point in the titration of 50 mL of 0.10 M methylamine (see Table 11.4) with a 0.20 M HCl solution.

12.4 | An Acid-Base Indicator Is a Substance That Changes Color at a Specific pH

The equivalence point, as we have seen, is the point at which the number of moles of OH^- ions added to a solution is equal to the number of moles of H^+ ions originally present. To determine the equivalence point in a titration, then, we must know exactly how much volume of a base has been added from a buret to an acid in a flask. One way to achieve this goal is to add a few drops of an acid-base indicator to the acid solution at the start of the titration. You will recall from Section 12.3 that an indicator is used to determine the equivalence point of an acid-base titration. An ***indicator*** *is a substance (usually a weak organic acid or base) that has distinctly different colors in its nonionized and ionized forms.* Which of these two forms (ionized or nonionized) is dominant depends upon the pH of the solution in which the indicator is dissolved. The ***end point*** of a titration *occurs when the indicator changes color.* However, not all indicators change color at the same pH, so the choice of indicator for a particular titration depends on the nature of the acid and base used in the titration (that is, whether they are strong or weak). By choosing the proper indicator for a titration, we can use the end point to determine the equivalence point, as we will see in the following.

Let us consider a weak monoprotic acid that we will call HIn. To be an effective indicator, HIn and its conjugate base, In^-, must have distinctly different colors. In solution, the acid ionizes to a small extent:

$$HIn(aq) \rightleftharpoons H^+(aq) + In^-(aq)$$

If the indicator is in a sufficiently acidic medium, the equilibrium, according to Le Châtelier's principle, shifts to the left and the predominant color of the indicator is that of the nonionized form (HIn). On the other hand, in a basic medium, the equilibrium shifts to the right, and the color of the solution will be due mainly to that of the conjugate base (In^-). Roughly speaking, we can use the following concentration ratios to predict the perceived color of the indicator:

$$\frac{[HIn]}{[In^-]} \geq 10 \qquad \text{color of acid (HIn) predominates}$$

$$\frac{[HIn]}{[In^-]} \leq 0.1 \qquad \text{color of conjugate base (In}^-\text{) predominates}$$

If $[HIn] \approx [In^-]$, then the indicator color is a combination of the colors of HIn and In^-.

The end point of an indicator does not occur at a specific pH; rather, there is a range of pH within which the end point will occur. In practice, we choose an indicator whose end point lies on the steep part of the titration curve, the position of which depends upon the strength of the acid or base being titrated. Because the equivalence point also lies on the steep part of the curve, this choice ensures that the pH at the equivalence point will fall within the range over which the indicator changes color. In Section 12.3, we mentioned that phenolphthalein is a suitable indicator for the titration of NaOH and HCl. Phenolphthalein is colorless in acidic and neutral solutions, but reddish pink in basic solutions. Measurements show that at pH = 8.3 the indicator is colorless but that it begins to turn reddish pink when the pH exceeds 8.3. As shown in Figure 12.5, the steepness of the pH curve near the equivalence point means that the addition of a very small quantity of NaOH (say, 0.05 mL, which is

Figure 12.8 pH curve for the titration of a strong acid with a strong base. Because the indicators methyl red and phenolphthalein have color changes that lie along the steep portion of the curve, they can be used to monitor the equivalence point of the titration. Thymol blue cannot be used for this purpose (see Table 12.1).

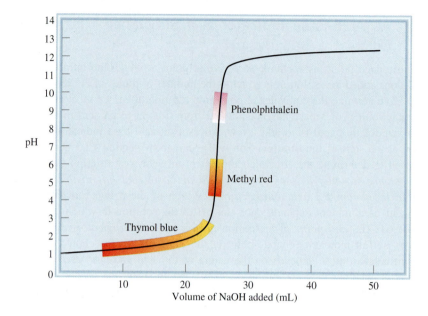

about the volume of a drop from the buret) brings about a large rise in the pH of the solution. What is important, however, is the fact that the steep portion of the pH profile includes the range over which phenolphthalein changes from colorless to reddish pink. Whenever such a correspondence occurs, the indicator can be used to locate the equivalence point of the titration (Figure 12.8).

Many acid-base indicators are plant pigments. For example, by boiling chopped red cabbage in water, we can extract pigments that exhibit many different colors at various pHs (Figure 12.9).

Table 12.1 lists a number of indicators commonly used in acid-base titrations. The choice of a particular indicator depends on the strength of the acid and base to be titrated. Example 12.7 illustrates this point.

Figure 12.9 Solutions containing extracts of red cabbage (obtained by boiling the cabbage in water) produce different colors when treated with an acid and a base. The pH of the solutions increases from left to right.

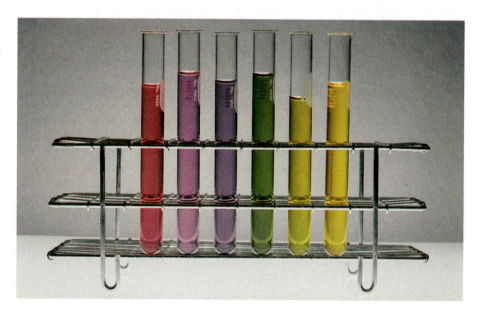

Table 12.1	Some Common Acid-Base Indicators		
	Color		
Indicator	**In Acid**	**In Base**	**pH Range***
Thymol blue	Red	Yellow	1.2–2.8
Bromophenol blue	Yellow	Bluish purple	3.0–4.6
Methyl orange	Orange	Yellow	3.1–4.4
Methyl red	Red	Yellow	4.2–6.3
Chlorophenol blue	Yellow	Red	4.8–6.4
Bromothymol blue	Yellow	Blue	6.0–7.6
Cresol red	Yellow	Red	7.2–8.8
Phenolphthalein	Colorless	Reddish pink	8.3–10.0

*The pH range is defined as the range over which the indicator changes from the acid color to the base color.

Example 12.7

Which indicator or indicators listed in Table 12.1 would you use for the acid-base titrations shown in (a) Figure 12.5, (b) Figure 12.6, and (c) Figure 12.7?

Strategy The choice of an indicator for a particular titration is based on the fact that its pH range for color change must overlap the steep portion of the titration curve. Otherwise, we cannot use the color change to locate the equivalence point.

Solution (a) Near the equivalence point, the pH of the solution changes abruptly from 4 to 10. Therefore all the indicators except thymol blue, bromophenol blue, and methyl orange are suitable for use in the titration.

(b) Here the steep portion covers the pH range between 7 and 10; therefore, the suitable indicators are cresol red and phenolphthalein.

(c) Here the steep portion of the pH curve covers the pH range between 3 and 7; therefore, the suitable indicators are bromophenol blue, methyl orange, methyl red, and chlorophenol blue.

Practice Exercise Referring to Table 12.1, specify which indicator or indicators you would use for the following titrations: (a) HBr versus CH_3NH_2, (b) HNO_3 versus NaOH, and (c) HNO_2 versus KOH.

12.5 | A Precipitation Reaction Occurs when a Reaction in Solution Leads to an Insoluble Product

One common type of reaction that occurs in aqueous solution is the **precipitation reaction,** which results in the formation of an insoluble product, or precipitate. A **precipitate** is an insoluble solid that separates from the solution. Precipitation reactions usually involve ionic compounds. For example, when an aqueous solution of lead nitrate [$Pb(NO_3)_2$] is added to an aqueous solution of sodium iodide (NaI), a yellow precipitate of lead iodide (PbI_2) is formed:

$$Pb(NO_3)_2(aq) + 2NaI(aq) \longrightarrow PbI_2(s) + 2NaNO_3(aq)$$

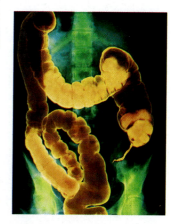

$BaSO_4$ imaging of human large intestine.

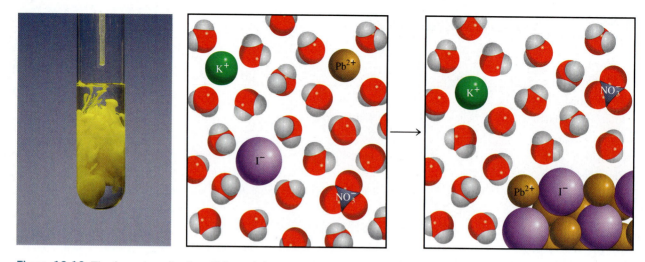

Figure 12.10 The formation of yellow PbI_2 precipitate as a solution of $Pb(NO_3)_2$ is added to a solution of NaI.

Sodium nitrate remains in solution. Figure 12.10 shows this reaction in progress.

Precipitation reactions are important in industry, medicine, and everyday life. For example, the preparation of many essential industrial chemicals such as sodium carbonate (Na_2CO_3) is based on precipitation reactions. The dissolving of tooth enamel, which is mainly made of hydroxyapatite [$Ca_5(PO_4)_3OH$], in an acidic medium leads to tooth decay. Barium sulfate ($BaSO_4$), an insoluble compound that is opaque to X rays, is used to diagnose ailments of the digestive tract. Stalactites and stalagmites, which consist of calcium carbonate ($CaCO_3$), are produced by a precipitation reaction, and so are many foods, such as fudge.

Qualitative Prediction of Solubility

How can we predict whether a precipitate will form when a compound is added to a solution or when two solutions are mixed? It depends on the *solubility* of the solute, which was defined in Section 9.2 as the maximum amount of solute that will dissolve in a given quantity of solvent at a specific temperature and pressure. All ionic compounds are strong electrolytes, but they are not equally soluble. Table 12.2

Table 12.2 Solubility Rules for Common Ionic Compounds in Water at 25°C	
Soluble Compounds	**Exceptions**
Compounds containing alkali metal ions (Li^+, Na^+, K^+, Rb^+, Cs^+) and the ammonium ion (NH_4^+) Nitrates (NO_3^-), bicarbonates (HCO_3^-), and chlorates (ClO_3^-)	
Halides (Cl^-, Br^-, I^-)	Halides of Ag^+, Hg_2^{2+}, and Pb^{2+}
Sulfates (SO_4^{2-})	Sulfates of Ag^+, Ca^{2+}, Sr^{2+}, Ba^{2+}, Hg_2^{2+}, and Pb^{2+}
Insoluble Compounds	**Exceptions**
Carbonates (CO_3^{2-}), phosphates (PO_4^{3-}), chromates (CrO_4^{2-}), sulfides (S^{2-})	Compounds containing alkali metal ions and the ammonium ion
Hydroxides (OH^-)	Compounds containing alkali metal ions and the Ba^{2+} ion

Figure 12.11 The appearance of several precipitates. From left to right: CdS, PbS, Ni(OH)$_2$, and Al(OH)$_3$.

classifies a number of common ionic compounds as soluble or insoluble. Keep in mind, however, that even insoluble compounds dissolve to a certain extent. Figure 12.11 shows several precipitates.

Example 12.8 applies the solubility rules in Table 12.2.

Example 12.8

Classify the following ionic compounds as soluble or insoluble: (a) silver sulfate (Ag$_2$SO$_4$), (b) calcium carbonate (CaCO$_3$), and (c) sodium phosphate (Na$_3$PO$_4$).

Strategy Although it is not necessary to memorize the solubilities of compounds, you should keep in mind the following useful rules: all ionic compounds containing alkali metal cations; the ammonium ion; and the nitrate, bicarbonate, and chlorate ions are soluble. For other compounds, we need to refer to Table 12.2.

Solution (a) According to Table 12.2, Ag$_2$SO$_4$ is insoluble.

(b) This is a carbonate and Ca is a Group 2A metal. Therefore, CaCO$_3$ is insoluble.

(c) Sodium is an alkali metal (Group 1A) so Na$_3$PO$_4$ is soluble.

Practice Exercise Classify the following ionic compounds as soluble or insoluble: (a) CuS, (b) Ca(OH)$_2$, and (c) Zn(NO$_3$)$_2$.

12.6 | The Solubility Product Is the Equilibrium Constant for the Dissolution Process

Although useful, the solubility rules discussed in Section 12.5 do not allow us to make quantitative predictions about how much of a given ionic compound will dissolve in water. To develop a quantitative approach, we start with what we already know about chemical equilibrium. Unless otherwise stated, in the following discussion the solvent is water and the temperature is 25°C.

The Solubility Product

Consider a saturated solution of silver chloride that is in contact with solid silver chloride. The solubility equilibrium can be represented as

$$AgCl(s) \rightleftharpoons Ag^+(aq) + Cl^-(aq)$$

Because salts such as AgCl are considered as strong electrolytes, all the AgCl that dissolves in water is assumed to dissociate completely into Ag^+ and Cl^- ions. Thus, we can write the equilibrium constant for the dissolution of AgCl in terms of the activities of the reactants and products (see Section 10.1):

$$K_{sp} = \frac{a_{Ag^+(aq)}\, a_{Cl^-(aq)}}{a_{AgCl(s)}} \tag{12.3}$$

where K_{sp} is called the solubility product constant or simply the *solubility product.* As we saw in Chapter 10, the activity of a pure solid in a reaction can be assumed to a very good approximation to be equal to 1. Also, if the concentrations of solutes are small enough that solute-solute interactions are negligible, then the solution can be considered ideal and we can approximate the activities of those solutes by their concentration relative to the standard concentration (1 mol L^{-1}). With these approximations, Equation 12.3 becomes

$$K_{sp} = [Ag^+][Cl^-] \tag{12.4}$$

(The concentrations in Equation 12.4 are dimensionless concentrations obtained by dividing the concentrations by 1 mol L^{-1}.) In general, the **solubility product** of a compound can be approximated as *the product of the molar concentrations of the constituent ions, each raised to the power of its stoichiometric coefficient in the equilibrium equation.* Keep in mind, however, that for solubility equilibria in which the concentration of ions is high ($> 0.001\ M$ for singly charged ions, much less for multiply charged ions), the solubilities calculated using Equation 12.4 will differ somewhat from experimentally measured solubilities. It is possible to correct Equation 12.4 by taking into account the deviation of the activity from the molar concentration, but such corrections are beyond the scope of this book.

Because each AgCl unit contains only one Ag^+ ion and one Cl^- ion, its solubility product expression is particularly simple to write. The following cases are slightly more complex:

▶ MgF_2

$$MgF_2(s) \rightleftharpoons Mg^{2+}(aq) + 2F^-(aq) \qquad K_{sp} = \frac{a_{Mg^{2+}(aq)}\, a^2_{F^-(aq)}}{a_{MgF_2(s)}} \approx [Mg^{2+}][F^-]^2$$

▶ Ag_2CO_3

$$Ag_2CO_3(s) \rightleftharpoons 2Ag^+(aq) + CO_3^{2-}(aq)$$

$$K_{sp} = \frac{a^2_{Ag^+(aq)}\, a_{CO_3^{2-}(aq)}}{a_{Ag_2CO_3(s)}} \approx [Ag^+]^2[CO_3^{2-}]$$

▶ $Ca_3(PO_4)_2$

$$Ca_3(PO_4)_2(s) \rightleftharpoons 3Ca^{2+}(aq) + 2PO_4^{3-}(aq)$$

$$K_{sp} = \frac{a^3_{Ca^{2+}(aq)}\, a^2_{PO_4^{3-}(aq)}}{a_{Ca_3(PO_4)_2(s)}} \approx [Ca^{2+}]^3[PO_4^{3-}]^2$$

Table 12.3	Solubility Products of Some Slightly Soluble Ionic Compounds at 25°C		
Compound	K_{sp}	**Compound**	K_{sp}
Aluminum hydroxide [Al(OH)$_3$]	1.8×10^{-33}	Lead(II) chromate (PbCrO$_4$)	2.0×10^{-14}
Barium carbonate (BaCO$_3$)	8.1×10^{-9}	Lead(II) fluoride (PbF$_2$)	4.1×10^{-8}
Barium fluoride (BaF$_2$)	1.7×10^{-6}	Lead(II) iodide (PbI$_2$)	1.4×10^{-8}
Barium sulfate (BaSO$_4$)	1.1×10^{-10}	Lead(II) sulfide (PbS)	3.4×10^{-28}
Bismuth sulfide (Bi$_2$S$_3$)	1.6×10^{-72}	Magnesium carbonate (MgCO$_3$)	4.0×10^{-5}
Cadmium sulfide (CdS)	8.0×10^{-28}	Magnesium hydroxide [Mg(OH)$_2$]	1.2×10^{-11}
Calcium carbonate (CaCO$_3$)	8.7×10^{-9}	Manganese(II) sulfide (MnS)	3.0×10^{-14}
Calcium fluoride (CaF$_2$)	4.0×10^{-11}	Mercury(I) chloride (Hg$_2$Cl$_2$)	3.5×10^{-18}
Calcium hydroxide [Ca(OH)$_2$]	8.0×10^{-6}	Mercury(II) sulfide (HgS)	4.0×10^{-54}
Calcium phosphate [Ca$_3$(PO$_4$)$_2$]	1.2×10^{-26}	Nickel(II) sulfide (NiS)	1.4×10^{-24}
Chromium(III) hydroxide [Cr(OH)$_3$]	3.0×10^{-29}	Silver bromide (AgBr)	7.7×10^{-13}
Cobalt(II) sulfide (CoS)	4.0×10^{-21}	Silver carbonate (Ag$_2$CO$_3$)	8.1×10^{-12}
Copper(I) bromide (CuBr)	4.2×10^{-8}	Silver chloride (AgCl)	1.6×10^{-10}
Copper(I) iodine (CuI)	5.1×10^{-12}	Silver iodide (AgI)	8.3×10^{-17}
Copper(II) hydroxide [Cu(OH)$_2$]	2.2×10^{-20}	Silver sulfate (Ag$_2$SO$_4$)	1.4×10^{-5}
Copper(II) sulfide (CuS)	6.0×10^{-37}	Silver sulfide (Ag$_2$S)	6.0×10^{-51}
Iron(II) hydroxide [Fe(OH)$_2$]	1.6×10^{-14}	Strontium carbonate (SrCO$_3$)	1.6×10^{-9}
Iron(III) hydroxide [Fe(OH)$_3$]	1.1×10^{-36}	Strontium sulfate (SrSO$_4$)	3.8×10^{-7}
Iron(II) sulfide (FeS)	6.0×10^{-19}	Tin(II) sulfide (SnS)	1.0×10^{-26}
Lead(II) carbonate (PbCO$_3$)	3.3×10^{-14}	Zinc hydroxide [Zn(OH)$_2$]	1.8×10^{-14}
Lead(II) chloride (PbCl$_2$)	2.4×10^{-4}	Zinc sulfide (ZnS)	3.0×10^{-23}

Table 12.3 lists the solubility products for a number of salts of low solubility.

Soluble salts such as NaCl and KNO$_3$, which have very large K_{sp} values, are not listed in the table for essentially the same reason that K_a values for strong acids are not often reported. The value of K_{sp} indicates the solubility of an ionic compound—the smaller the value, the less soluble the compound in water. However, in using K_{sp} values to compare solubilities, you should choose compounds that have similar formulas, such as AgCl and ZnS, or CaF$_2$ and Fe(OH)$_2$.

Another factor, in addition to deviations from ideal solution behavior, that can affect the use of K_{sp} in determining solubility is the fact that many anions in the ionic compounds listed in Table 12.3 are conjugate bases of weak acids. Consider copper sulfide (CuS). The S^{2-} ion can hydrolyze as follows

$$S^{2-}(aq) + H_2O(l) \rightleftharpoons HS^-(aq) + OH^-(aq)$$
$$HS^-(aq) + H_2O(l) \rightleftharpoons H_2S(aq) + OH^-(aq)$$

And highly charged small metal ions such as Al^{3+} and Bi^{3+} will undergo hydrolysis as discussed in Section 11.6. In such cases, it is necessary in accurate work to include the hydrolysis equilibrium expressions in addition to the solubility product to determine the equilibrium concentrations of all species in aqueous solution.

For concentrations of ions that do not correspond to equilibrium conditions, we use the reaction quotient (see Section 10.2), which in this case is called the *ion product (Q)*, to predict whether a precipitate will form. Note that Q has the same form as K_{sp} except that the concentrations of ions are *not* equilibrium concentrations. For example, if we mix a solution containing Ag^+ ions with one containing Cl^- ions, then the ion product, assuming ideal solution behavior, is given by

$$Q = [Ag^+]_0[Cl^-]_0$$

The subscript 0 reminds us that these are initial concentrations and do not necessarily correspond to those at equilibrium. The possible relationships between Q and K_{sp} are

$$Q < K_{sp}$$
$$[Ag^+]_0[Cl^-]_0 < 1.6 \times 10^{-10}$$
unsaturated solution

$$Q = K_{sp}$$
$$[Ag^+]_0[Cl^-]_0 = 1.6 \times 10^{-10}$$
saturated solution (equilibrium)

$$Q > K_{sp}$$
$$[Ag^+]_0[Cl^-]_0 > 1.6 \times 10^{-10}$$
supersaturated solution; AgCl will precipitate out until the product of the ionic concentrations is equal to 1.6×10^{-10}

Molar Solubility and Solubility

There are two other ways to express a substance's solubility: ***molar solubility,*** which is *the number of moles of solute in 1 L of a saturated solution (mol L^{-1}),* and ***solubility,*** which is *the number of grams of solute in 1 L of a saturated solution (g L^{-1}).* Note that both these expressions refer to the concentration of saturated solutions at some given temperature (usually 25°C).

Both molar solubility and solubility are convenient to use in the laboratory. We can use them to determine K_{sp} by following the steps outlined in Figure 12.12(a). Example 12.9 illustrates this procedure.

Example 12.9

The solubility of calcium sulfate ($CaSO_4$) is found to be 0.67 g L^{-1}. Calculate the value of K_{sp} for calcium sulfate.

Strategy We are given the solubility of $CaSO_4$ and asked to calculate its K_{sp}. The sequence of conversion steps, according to Figure 12.12(a), is

solubility of $CaSO_4$ in g L^{-1} $\longrightarrow$ molar solubility of $CaSO_4$ $\longrightarrow$
$[Ca^{2+}]$ and $[SO_4^{2-}]$ $\longrightarrow$ K_{sp} of $CaSO_4$

—Continued

Figure 12.12 The sequence of steps (a) for calculating K_{sp} from solubility data and (b) for calculating solubility from K_{sp} data.

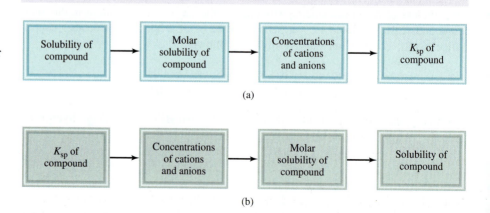

Continued—

Solution Consider the dissociation of $CaSO_4$ in water. Let s be the molar solubility (in mol L^{-1}) of $CaSO_4$.

	$CaSO_4(s) \rightleftharpoons$	$Ca^{2+}(aq) +$	$SO_4^{2-}(aq)$
Initial (M)		0	0
Change (M)	$-s$	$+s$	$+s$
Equilibrium (M)		s	s

The solubility product for $CaSO_4$ is

$$K_{sp} \approx [Ca^{2+}][SO_4^{2-}] = s^2$$

First, we calculate the number of moles of $CaSO_4$ dissolved in 1 L of solution

$$\frac{0.67 \text{ g } CaSO_4}{1 \text{ L soln}} \times \frac{1 \text{ mol } CaSO_4}{136.2 \text{ g } CaSO_4} = 4.9 \times 10^{-3} \text{ mol } L^{-1} = s$$

From the solubility equilibrium, we see that for every mole of $CaSO_4$ that dissolves, 1 mole of Ca^{2+} and 1 mole of SO_4^{2-} are produced. Thus, at equilibrium

$$[Ca^{2+}] = [SO_4^{2-}] = s = 4.9 \times 10^{-3} M$$

Now we can calculate K_{sp}:

$$K_{sp} = [Ca^{2+}][SO_4^{2-}] = (4.9 \times 10^{-3})(4.9 \times 10^{-3})$$
$$= 2.4 \times 10^{-5}$$

Practice Exercise The solubility of lead chromate ($PbCrO_4$) is 4.5×10^{-5} g L^{-1}. Calculate the solubility product of this compound.

Calcium sulfate is used as a drying agent and in the manufacture of paints, ceramics, and paper. A hydrated form of calcium sulfate, called plaster of Paris, is used to make casts for broken bones.

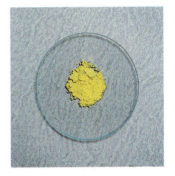

Silver bromide is used in photographic emulsions.

Sometimes we are given the value of K_{sp} for a compound and asked to calculate the compound's molar solubility. For example, the K_{sp} for the dissolution of silver bromide

$$AgBr(s) \rightleftharpoons Ag^+(aq) + Br^-(aq)$$

is 7.7×10^{-13}. We can calculate its molar solubility by the procedure outlined in Figure 12.12(b). First we identify the species present at equilibrium. Here we have Ag^+ and Br^- ions. Let s be the molar solubility (in mol L^{-1}) of $AgBr$. Because one unit of $AgBr$ yields one Ag^+ and one Br^- ion, at equilibrium, both $[Ag^+]$ and $[Br^-]$ are equal to s. Assuming ideal solution behavior, the solubility product expression for this reaction is given by

$$K_{sp} = [Ag^+][Br^-]$$
$$7.7 \times 10^{-13} = s \times s = s^2$$
$$s = \sqrt{7.7 \times 10^{-13}} = 8.8 \times 10^{-7} M$$

Therefore, at equilibrium

$$[Ag^+] = [Br^-] = 8.8 \times 10^{-7} M$$

Copper(II) hydroxide is used as a pesticide and to treat seeds.

Thus, the molar solubility of AgBr also is 8.8×10^{-7} M. This concentration is quite small, so the effect of nonideality will be minimal and the use of concentrations in the expression for K_{sp} is accurate.

Example 12.10 makes use of this approach.

Example 12.10

Using the data in Table 12.3, calculate the solubility of copper(II) hydroxide [$Cu(OH)_2$] in g L^{-1}.

Strategy We are given the K_{sp} of $Cu(OH)_2$ and asked to calculate its solubility in g L^{-1}. The sequence of conversion steps, according to Figure 12.12(b), is

K_{sp} of $Cu(OH)_2 \longrightarrow [Cu^{2+}]$ and $[OH^-] \longrightarrow$
molar solubility of $Cu(OH)_2 \longrightarrow$ solubility of $Cu(OH)_2$ in g L^{-1}

Solution Consider the dissociation of $Cu(OH)_2$ in water:

$$Cu(OH)_2(s) \rightleftharpoons Cu^{2+}(aq) + 2OH^-(aq)$$

Note that the molar concentration of OH^- is twice that of Cu^{2+}. The solubility product of $Cu(OH)_2$, assuming ideal solution behavior, is

$$K_{sp} = [Cu^{2+}][OH^-]^2$$
$$= (s)(2s)^2 = 4s^3$$

From the K_{sp} value in Table 12.3, we solve for the molar solubility of $Cu(OH)_2$ as follows:

$$2.2 \times 10^{-20} = 4s^3$$

and

$$s = \sqrt[3]{\frac{2.2 \times 10^{-20}}{4}} = 1.8 \times 10^{-7} M$$

Finally, from the molar mass of $Cu(OH)_2$ and its molar solubility, we calculate the solubility in g L^{-1}:

$$\text{solubility of } Cu(OH)_2 = \frac{1.8 \times 10^{-7} \text{ mol } Cu(OH)_2}{1 \text{ L sol}} \times \frac{97.57 \text{ g } Cu(OH)_2}{1 \text{ mol } Cu(OH)_2}$$
$$= 1.8 \times 10^{-5} \text{ g } L^{-1}$$

Practice Exercise Calculate the solubility of silver chloride (AgCl) in g L^{-1}.

As Examples 12.9 and 12.10 show, solubility and solubility product are related. If we know one, we can calculate the other, but each quantity provides different information. Table 12.4 shows the relationship between molar solubility and solubility product for a number of ionic compounds, assuming ideal solution behavior.

Predicting Precipitation Reactions

From knowledge of the solubility rules (see Section 12.5) and the solubility products listed in Table 12.3, we can predict whether a precipitate will form when we mix two solutions or add a soluble compound to a solution. This ability often has practical value.

Table 12.4	Relationship Between K_{sp} and Molar Solubility (s)			
Compound	**K_{sp} Expression**	**Cation**	**Anion**	**Relation between K_{sp} and s**
AgCl	$[Ag^+][Cl^-]$	s	s	$K_{sp} = s^2;\ s = (K_{sp})^{1/2}$
BaSO$_4$	$[Ba^{2+}][SO_4^{2-}]$	s	s	$K_{sp} = s^2;\ s = (K_{sp})^{1/2}$
Ag$_2$CO$_3$	$[Ag^+]^2[CO_3^{2-}]$	$2s$	s	$K_{sp} = 4s^3;\ s = \left(\dfrac{K_{sp}}{4}\right)^{1/3}$
PbF$_2$	$[Pb^{2+}][F^-]^2$	s	$2s$	$K_{sp} = 4s^3;\ s = \left(\dfrac{K_{sp}}{4}\right)^{1/3}$
Al(OH)$_3$	$[Al^{3+}][OH^-]^3$	s	$3s$	$K_{sp} = 27s^4;\ s = \left(\dfrac{K_{sp}}{27}\right)^{1/4}$
Ca$_3$(PO$_4$)$_2$	$[Ca^{2+}]^3[PO_4^{3-}]^2$	$3s$	$2s$	$K_{sp} = 108s^5;\ s = \left(\dfrac{K_{sp}}{108}\right)^{1/5}$

In industrial and laboratory preparations, we can adjust the concentrations of ions until the ion product exceeds K_{sp} in order to obtain a given compound (in the form of a precipitate). The ability to predict precipitation reactions is also useful in medicine. For example, kidney stones, which can be extremely painful, consist largely of calcium oxalate (CaC$_2$O$_4$, $K_{sp} = 2.3 \times 10^{-9}$). The normal physiological concentration of calcium ions in blood plasma is about 5 mM (1 m$M = 1 \times 10^{-3}$ M). Oxalate ions (C$_2$O$_4^{2-}$), derived from oxalic acid present in many vegetables such as rhubarb and spinach, react with the calcium ions to form insoluble calcium oxalate, which can gradually build up in the kidneys. Proper adjustment of a patient's diet can help to reduce precipitate formation. Example 12.11 illustrates the steps involved in predicting precipitation reactions.

Example 12.11

Exactly 200 mL of 0.0040 M BaCl$_2$ is added to exactly 600 mL of 0.0080 M K$_2$SO$_4$. Will a precipitate form?

Strategy Under what condition will an ionic compound precipitate from solution? The ions in solution are Ba^{2+}, Cl$^-$, K$^+$, and SO$_4^{2-}$. According to the solubility rules listed in Table 12.2, the only precipitate that can form is BaSO$_4$. From the information given, we can calculate [Ba^{2+}] and [SO$_4^{2-}$] because we know the number of moles of the ions in the original solutions and the volume of the combined solution.

Next we calculate the reaction quotient $Q = [Ba^{2+}]_0[SO_4^{2-}]_0$ and compare the value of Q with the K_{sp} of BaSO$_4$ to see if a precipitate will form, that is, if the solution is supersaturated.

Solution The number of moles of Ba^{2+} present in the original 200 mL of solution is

$$200\text{ mL} \times \frac{0.0040\text{ mol Ba}^{2+}}{1\text{ L soln}} \times \frac{1\text{ L}}{1000\text{ mL}} = 8.0 \times 10^{-4}\text{ mol Ba}^{2+}$$

The total volume after combining the two solutions is 800 mL. The concentration of Ba^{2+} in the 800-mL volume is

$$[Ba^{2+}] = \frac{8.0 \times 10^{-4}\text{ mol}}{800\text{ mL}} \times \frac{1000\text{ mL}}{1\text{ L}} = 1.0 \times 10^{-3}\ M$$

—Continued

Continued—

The number of moles of SO_4^{2-} in the original 600 mL of solution is

$$600 \text{ mL} \times \frac{0.0080 \text{ mol } SO_4^{2-}}{1 \text{ L soln}} \times \frac{1 \text{ L}}{1000 \text{ mL}} = 4.8 \times 10^{-3} \text{ mol } SO_4^{2-}$$

The concentration of SO_4^{2-} in the 800 mL of the combined solution is

$$[SO_4^{2-}] = \frac{4.8 \times 10^{-4} \text{ mol}}{800 \text{ mL}} \times \frac{1000 \text{ mL}}{1 \text{ L}} = 6.0 \times 10^{-3} \text{ } M$$

Now we must compare Q and K_{sp}. From Table 12.3,

$$BaSO_4(s) \rightleftharpoons Ba^{2+}(aq) + SO_4^{2-}(aq) \quad K_{sp} = 1.1 \times 10^{-10}$$

As for Q,

$$Q = [Ba^{2+}]_0[SO_4^{2-}]_0 = (1.0 \times 10^{-3})(6.0 \times 10^{-3}) = 6.0 \times 10^{-6}$$

Therefore,

$$Q > K_{sp}$$

The solution is supersaturated because the value of Q indicates that the concentrations of the ions are too large. Thus, some of the $BaSO_4$ will precipitate out of solution until

$$[Ba^{2}][SO_4^{2-}] = 1.1 \times 10^{-10}$$

Practice Exercise If 2.00 mL of 0.200 M NaOH is added to 1.00 L of 0.100 M $CaCl_2$, will precipitation occur?

Separation of Ions by Fractional Precipitation

In chemical analysis, it is sometimes desirable to remove one type of ion from solution by precipitation while leaving other ions in solution. For instance, the addition of sulfate ions to a solution containing both potassium and barium ions causes $BaSO_4$ to precipitate out, thereby removing most of the Ba^2 ions from the solution. The other "product," K_2SO_4, is soluble and will remain in solution. The $BaSO_4$ precipitate can be separated from the solution by filtration.

Even when *both* products are insoluble, we can still achieve some degree of separation by choosing the proper reagent to bring about precipitation. Consider a solution that contains Cl^-, Br^-, and I^- ions. One way to separate these ions is to convert them to insoluble silver halides. As the K_{sp} values in Table 12.3 show, the solubility of the halides decreases from AgCl to AgI. Thus, when a soluble compound such as silver nitrate is slowly added to this solution, AgI begins to precipitate first, followed by AgBr and then AgCl.

Example 12.12 describes the separation of only two ions (Cl^- and Br^-), but the procedure can be applied to a solution containing more than two different types of ions if precipitates of differing solubility can be formed.

Example 12.12

Silver nitrate is slowly added to a solution that is 0.020 M in Cl^- ions and 0.020 M in Br^- ions. Calculate the concentration of Ag^+ ions (in mol L^{-1}) required to initiate the precipitation of AgBr without precipitating AgCl.

—Continued

Continued—

Strategy In solution, $AgNO_3$ dissociates into Ag^+ and NO_3^- ions. The Ag^+ ions then combine with the Cl^- and Br^- ions to form AgCl and AgBr precipitates. Because AgBr is less soluble (it has a smaller K_{sp} than that of AgCl), it will precipitate first. Therefore, this is a fractional precipitation problem. Knowing the concentrations of Cl^- and Br^- ions, we can calculate $[Ag^+]$ from the K_{sp} values. Keep in mind that K_{sp} refers to a saturated solution. To initiate precipitation, $[Ag^+]$ must exceed the concentration in the saturated solution in each case.

Solution The solubility equilibrium for AgBr is

$$AgBr(s) \rightleftharpoons Ag^+(aq) + Br^-(aq) \qquad K_{sp} = [Ag^+][Br^-] = 7.7 \times 10^{-13}$$

Because $[Br^+] = 0.020 \ M$, the concentration of Ag^+ that must be exceeded to initiate the precipitation of AgBr is

$$[Ag^+] = \frac{K_{sp}}{[Br^-]} = \frac{7.7 \times 10^{-13}}{0.020} = 3.9 \times 10^{-11} \ M$$

Thus, $[Ag^+] > 3.9 \times 10^{-11} \ M$ is required to start the precipitation of AgBr.

The solubility equilibrium for AgCl is

$$AgCl(s) \rightleftharpoons Ag^+(aq) + Cl^-(aq) \qquad K_{sp} = [Ag^+][Cl^-] = 1.6 \times 10^{-10}$$

so

$$[Ag^+] = \frac{K_{sp}}{[Cl^-]} = \frac{1.6 \times 10^{-10}}{0.020} = 8.0 \times 10^{-9} \ M$$

Therefore, $[Ag^+] > 8.0 \times 10^{-9} \ M$ is required to start the precipitation of AgCl.

To precipitate AgBr without precipitating Cl^- ions then, $[Ag^+]$ must be greater than $3.9 \times 10^{-11} \ M$ and lower than $8.0 \times 10^{-9} \ M$.

Practice Exercise The solubility products of AgCl and Ag_3PO_4 are 1.6×10^{-10} and 1.8×10^{-18}, respectively. If Ag^+ is added (without changing the volume) to 1.00 L of a solution containing 0.10 mol Cl^- and 0.10 mol PO_4^{3-}, calculate the concentration of Ag^+ ions (in mol L^{-1}) required to initiate (a) the precipitation of AgCl and (b) the precipitation of Ag_3PO_4.

Example 12.12 raises the question: What is the concentration of Br^- ions remaining in solution just before AgCl begins to precipitate? To answer this question, we let $[Ag^+] = 8.0 \times 10^{-9} \ M$. Then

$$[Br^-] = \frac{K_{sp}}{[Ag^+]} = \frac{7.7 \times 10^{-13}}{8.0 \times 10^{-9}} = 9.6 \times 10^{-5} \ M$$

The percent of Br^- remaining in solution (the *unprecipitated* Br^-) at the critical concentration of Ag^+ is

$$\% \ Br^- = \frac{[Br^-]_{unppt'd}}{[Br^-]_{original}} \times 100\%$$

$$= \frac{9.6 \times 10^{-5} \ M}{0.020 \ M} \times 100\%$$

$$= 0.48\% \text{ unprecipitated}$$

Thus, $(100 - 0.48)$ percent, or 99.52 percent, of Br^- will have precipitated as AgBr just before AgCl begins to precipitate. By this procedure, the Br^- ions can be quantitatively separated from the Cl^- ions.

12.7 | The Solubility of a Substance Is Affected by a Number of Factors

The Common Ion Effect and Solubility

In Section 12.1, we discussed the effect of a common ion on acid and base ionizations. Here we will examine the relationship between the common ion effect and solubility. As we have noted, the solubility product is an equilibrium constant; precipitation of an ionic compound from solution occurs whenever the ion product exceeds K_{sp} for that substance. In a saturated solution of AgCl, for example, the ion product $[Ag^+][Cl^-]$ is, of course, equal to K_{sp}. Furthermore, simple stoichiometry tells us that $[Ag^+] = [Cl^-]$. But this equality does not hold in all situations.

Suppose we study a solution containing two dissolved substances that share a common ion, say, AgCl and $AgNO_3$. In addition to the dissociation of AgCl, the total concentration of the common silver ions in solution is also affected by the dissolution of silver nitrate:

$$AgNO_3(s) \xrightarrow{\text{H}_2\text{O}} Ag^+(aq) + NO_3^-(aq)$$

The solubility equilibrium of AgCl is

$$AgCl(s) \rightleftharpoons Ag^+(aq) + Cl^-(aq)$$

If $AgNO_3$ is added to a saturated AgCl solution, the increase in $[Ag^+]$ will make the ion product greater than the solubility product:

$$Q = [Ag^+]_0[Cl^-]_0 > K_{sp}$$

To reestablish equilibrium, some AgCl will precipitate out of the solution, as Le Châtelier's principle would predict, until the ion product is once again equal to K_{sp}. The effect of adding a common ion, then, is a *decrease* in the solubility of the salt (AgCl) in solution. Note that in this case $[Ag^+]$ is no longer equal to $[Cl^-]$ at equilibrium; rather, $[Ag^+] > [Cl^-]$.

Example 12.13 shows the common ion effect on solubility.

At a given temperature, only the solubility of a compound is altered (decreased) by the common ion effect. Its solubility product, which is an equilibrium constant, remains the same whether or not other substances are present in the solution.

Example 12.13

Calculate the solubility of silver chloride (in g L^{-1}) in a 6.5×10^{-3} M silver nitrate solution.

Strategy This is a common-ion problem. The common ion here is Ag^+, which is supplied by both AgCl and $AgNO_3$. Remember that the presence of the common ion will affect only the solubility of AgCl (in g L^{-1}), but not the K_{sp} value because it is an equilibrium constant.

Solution *Step 1:* The relevant species in solution are Ag^+ ions (from both AgCl and $AgNO_3$) and Cl^- ions. The NO_3^- ions are spectator ions.

—Continued

Continued—

Step 2: Because $AgNO_3$ is a soluble strong electrolyte, it dissociates completely:

$$AgNO_3(s) \xrightarrow{H_2O} Ag^+(aq) \quad + \quad NO_3^-(aq)$$
$$6.5 \times 10^{-3}\ M \qquad 6.5 \times 10^{-3}\ M$$

Let s be the molar solubility of AgCl in $AgNO_3$ solution. We summarize the changes in concentrations as follows:

	$AgCl(s)$	$\rightleftharpoons$	$Ag^+(aq)$	$+$	$Cl^-(aq)$
Initial (M)			6.5×10^{-3}		0.0
Change (M)	$-s$		$+s$		$+s$
Equilibrium (M)			$6.5 \times 10^{-3} + s$		s

Step 3:
$$K_{sp} = [Ag^+][Cl^-]$$
$$1.6 \times 10^{-10} = (6.5 \times 10^{-3} + s)\,s$$

Because AgCl is quite insoluble and the presence of Ag^+ ions from $AgNO_3$ further lowers the solubility of AgCl, s must be very small compared with 6.5×10^{-3}. Therefore, applying the approximation $6.5 \times 10^{-3} + s \approx 6.5 \times 10^{-3}$, we obtain

$$1.6 \times 10^{-10} = (6.5 \times 10^{-3})s$$
$$s = 2.5 \times 10^{-8}\ M$$

Step 4: At equilibrium

$$[Ag^+] = (6.5 \times 10^{-3} + 2.5 \times 10^{-8})\ M \approx 6.5 \times 10^{-3}\ M$$
$$[Cl^-] = 2.5 \times 10^{-8}\ M$$

and so our approximation was justified in step 3. Because all the Cl^- ions must come from AgCl, the amount of AgCl dissolved in $AgNO_3$ solution also is $2.5 \times 10^{-8}\ M$. Then, knowing the molar mass of AgCl (143.4 g), we can calculate the solubility of AgCl as follows:

$$\text{solubility of AgCl in } AgNO_3 \text{ solution} = \frac{2.5 \times 10^{-8}\ \text{mol AgCl}}{1\ \text{L soln}} \times \frac{143.4\ \text{g AgCl}}{1\ \text{mol AgCl}}$$
$$= 3.6 \times 10^{-6}\ \text{g L}^{-1}$$

Check The solubility of AgCl in pure water is 1.9×10^{-3} g L^{-1} (see the Practice Exercise in Example 12.10). Therefore, the lower solubility (3.6×10^{-6} g L^{-1}) in the presence of $AgNO_3$ is reasonable. You should also be able to predict the lower solubility using Le Châtelier's principle. Adding Ag^+ ions shifts the equilibrium to the left, thus decreasing the solubility of AgCl.

Practice Exercise Calculate the solubility in g L^{-1} of AgBr in (a) pure water and (b) 0.0010 M NaBr.

pH and Solubility

The solubilities of many substances also depend on the pH of the solution. Consider the solubility equilibrium of magnesium hydroxide:

$$Mg(OH)_2(s) \rightleftharpoons Mg^{2+}(aq) + 2OH^-(aq)$$

Adding OH^- ions (increasing the pH) shifts the equilibrium from right to left, thereby decreasing the solubility of $Mg(OH)_2$. (This is another example of the common ion effect.) On the other hand, adding H^+ ions (decreasing the pH) shifts the equilibrium from left to right, and the solubility of $Mg(OH)_2$ increases. Thus, insoluble bases tend to dissolve in acidic solutions. Similarly, insoluble acids tend to dissolve in basic solutions.

To explore the quantitative effect of pH on the solubility of $Mg(OH)_2$, let us first calculate the pH of a saturated $Mg(OH)_2$ solution. We write

$$K_{sp} = [Mg^{2+}][OH^-]^2 = 1.2 \times 10^{-11}$$

Let s be the molar solubility of $Mg(OH)_2$. Proceeding as in Example 12.10,

$$K_{sp} = (s)(2s)^2 = 4s^3$$
$$4s^3 = 1.2 \times 10^{-11}$$
$$s = \sqrt[3]{\frac{1.2 \times 10^{-11}}{4}} = 1.4 \times 10^{-4}$$

At equilibrium, therefore,

$$[OH^-] = 2 \times 1.4 \times 10^{-4}\, M = 2.8 \times 10^{-4}\, M$$
$$pOH = -\log_{10}(2.8 \times 10^{-4}) = 3.55$$
$$pH = 14.00 - 3.55 = 10.45$$

In a medium with a pH of less than 10.45, the solubility of $Mg(OH)_2$ would increase. This follows from the fact that a lower pH indicates a higher $[H^+]$ and thus a lower $[OH^-]$, as we would expect from $K_w = [H^+][OH^-]$. Consequently, $[Mg^{2+}]$ rises to maintain the a saturated solution, and more $Mg(OH)_2$ dissolves. The dissolution process and the effect of extra H^+ ions can be summarized as follows:

$$Mg(OH)_2(s) \rightleftharpoons Mg^{2+}(aq) + 2OH^-(aq)$$
$$\underline{2H^+(aq) + 2OH^-(aq) \rightleftharpoons 2H_2O(l)}$$
$$\text{overall} \quad Mg(OH)_2(s) + 2H^+(aq) \rightleftharpoons Mg^{2+}(aq) + 2H_2O(l)$$

If the pH of the medium were higher than 10.45, $[OH^-]$ would be higher and the solubility of $Mg(OH)_2$ would decrease because of the common ion (OH^-) effect.

The pH also influences the solubility of salts that contain a basic anion. For example, the solubility equilibrium for BaF_2 is

$$BaF_2(s) \rightleftharpoons Ba^{2+}(aq) + 2F^-(aq)$$

and

$$K_{sp} = [Ba^{2+}][F^-]^2$$

In an acidic medium, the high $[H^+]$ will shift the following equilibrium from left to right:

$$H^+(aq) + F^-(aq) \rightleftharpoons HF(aq)$$

Milk of magnesia, which contains $Mg(OH)_2$, is used to treat acid indigestion.

As $[F^-]$ decreases, $[Ba^{2+}]$ must increase to maintain the equilibrium condition. Thus, more BaF_2 dissolves. The dissolution process and the effect of pH on the solubility of BaF_2 can be summarized as follows:

$$BaF_2(s) \rightleftharpoons Ba^{2+}(aq) + 2F^-(aq)$$
$$\underline{2H^+(aq) + 2F^-(aq) \rightleftharpoons 2HF(aq)}$$
$$\text{overall} \quad BaF_2(s) + 2H^+(aq) \rightleftharpoons Ba^{2+}(aq) + 2HF(aq)$$

The solubilities of salts containing anions that do not hydrolyze are unaffected by pH. Examples of such anions are Cl^-, Br^-, and I^-.

Examples 12.14 and 12.15 deal with the effect of pH on solubility.

Example 12.14

Which of the following compounds will be more soluble in acidic solution than in water: (a) CuS, (b) AgCl, and (c) $PbSO_4$?

Strategy In each case write the dissociation reaction of the salt into its cation and anion. The cation will not interact with the H^+ ion because they both bear positive charges. The anion will act as a proton acceptor only if it is the conjugate base of a weak acid. How would the removal of the anion affect the solubility of the salt?

Solution (a) The solubility equilibrium for CuS is

$$CuS(s) \rightleftharpoons Cu^{2+}(aq) + S^{2-}(aq)$$

The sulfide ion is a weak base because it is the conjugate base of the weak acid HS^-. Therefore, the S^{2-} ion reacts with the H^+ ion as follows:

$$S^{2-}(aq) + H^-(aq) \rightleftharpoons HS^-(aq)$$

This reaction removes the S^{2-} ions from solution. According to Le Châtelier's principle, the equilibrium will shift to the right to replace some of the S^{2-} ions that were removed, thereby increasing the solubility of CuS.

(b) The solubility equilibrium is

$$AgCl(s) \rightleftharpoons Ag^+(aq) + Cl^-(aq)$$

Because Cl^- is the conjugate base of a strong acid (HCl), the solubility of AgCl is not affected by an acid solution.

(c) The solubility equilibrium for $PbSO_4$ is

$$PbSO_4(s) \rightleftharpoons Pb^{2+}(aq) + SO_4^{2-}(aq)$$

The sulfate ion is a weak base because it is the conjugate base of the weak acid HSO_4^-. Therefore, the SO_4^{2-} ion reacts with the H^+ ion as follows:

$$SO_4^{2-}(aq) + H^+(aq) \rightleftharpoons HSO_4^-(aq)$$

This reaction removes the SO_4^{2-} ions from solution. According to Le Châtelier's principle, the equilibrium will shift to the right to replace some of the SO_4^{2-} ions that were removed, thereby increasing the solubility of $PbSO_4$.

Practice Exercise Are the following compounds more soluble in water or in an acidic solution: (a) $Ca(OH)_2$, (b) $Mg_3(PO_4)_2$, and (c) $PbBr_2$?

Example 12.15

Calculate the concentration of aqueous ammonia necessary to initiate the precipitation of iron(II) hydroxide from a 0.0030 M solution of $FeCl_2$.

Strategy For iron(II) hydroxide to precipitate from solution, the product $[Fe^{2+}][OH^-]^2$ must be greater than its K_{sp}. First, we calculate $[OH^-]$ from the known

—Continued

Continued—

[Fe^{2+}] and the K_{sp} value listed in Table 12.3. This is the concentration of OH^- in a saturated solution of $Fe(OH)_2$. Next, we calculate the concentration of NH_3 that will supply this concentration of OH^- ions. Finally, any NH_3 concentration greater than the calculated value will initiate the precipitation of $Fe(OH)_2$ because the solution will become supersaturated.

Solution   Ammonia reacts with water to produce OH^- ions, which then react with Fe^{2+} to form $Fe(OH)_2$. The equilibria of interest are

$$NH_3(aq) + H_2O(l) \rightleftharpoons NH_4^+(aq) + OH^-(aq)$$
$$Fe^{2+}(aq) + 2OH^-(aq) \rightleftharpoons Fe(OH)_2(s)$$

First, we find the OH^- concentration above which $Fe(OH)_2$ begins to precipitate. We write

$$K_{sp} = [Fe^{2+}][OH^-]^2 = 1.6 \times 10^{-14}$$

Because $FeCl_2$ is a strong electrolyte, $[Fe^{2+}] = 0.0030\ M$ and

$$[OH^-]^2 = \frac{1.6 \times 10^{-14}}{0.0030} = 5.3 \times 10^{-12}$$
$$[OH^-] = 2.3 \times 10^{-6}\ M$$

Next, we calculate the concentration of the weak base NH_3 that will supply $2.3 \times 10^{-6}\ M\ OH^-$ ions.

Let x be the initial concentration of NH_3 in mol L^{-1}. We summarize the changes in concentrations resulting from the ionization of NH_3 as follows:

	$NH_3(aq)\ +\ H_2O(l)$	$\rightleftharpoons$	$NH_4^+(aq)$	$+$	$OH^-(aq)$
Initial (M)	x		0.00		0.00
Change (M)	-2.3×10^{-6}		$+2.3 \times 10^{-6}$		$+2.3 \times 10^{-6}$
Equilibrium (M)	$x - 2.3 \times 10^{-6}$		2.3×10^{-6}		2.3×10^{-6}

The equilibrium concentrations of NH_3, OH^-, and NH_4^+ are then related by the base ionization constant for NH_3 (which from Table 11.4 is equal to 1.8×10^{-5}):

$$K_b = \frac{[NH_4^+][OH^-]}{[NH_3]}$$
$$1.8 \times 10^{-5} = \frac{(2.3 \times 10^{-6})(2.3 \times 10^{-6})}{x - 2.3 \times 10^{-6}}$$

Solving for x, we obtain

$$x = 2.6 \times 10^{-6}\ M$$

Therefore, the concentration of NH_3 must be slightly greater than $2.6 \times 10^{-6}\ M$ to initiate the precipitation of $Fe(OH)_2$.

Practice Exercise   Calculate whether or not a precipitate will form if 2.0 mL of 0.60 M NH_3 is added to 1.0 L of $1.0 \times 10^{-3}\ M$ $FeSO_4$.

Complex Ion Equilibria and Solubility

Lewis acids and bases are discussed in Section 11.1.

Lewis acid-base reactions in which a metal cation combines with a Lewis base result in the formation of complex ions. Thus, we can define a ***complex ion*** as *an ion containing a central metal cation bonded to one or more molecules or ions.* Complex

Figure 12.13 Left: An aqueous cobalt(II) chloride solution. The pink color is due to the presence of $Co(H_2O)_6^{2+}$ ions. Right: After the addition of HCl solution, the solution turns blue because of the formation of the complex ion $CoCl_4^{2-}$.

ions are crucial to many chemical and biological processes. Here we will consider the effect of complex ion formation on solubility. In Chapter 15, we will discuss the chemistry of complex ions in more detail.

Transition metals have a particular tendency to form complex ions because they have more than one possible valence state. This property allows them to act effectively as Lewis acids in reactions with many molecules or ions that serve as electron donors (Lewis bases). For example, a solution of cobalt(II) chloride is pink because of the presence of the $Co(H_2O)_6^{2+}$ ions (Figure 12.13).

According to our definition, $Co(H_2O)_6^{2+}$ is itself a complex ion. When we write $Co(H_2O)_6^{2+}$, we mean the hydrated Co^{2+} ion.

When HCl is added, the solution turns blue as a result of the formation of the complex ion $CoCl_4^{2-}$:

$$Co^{2+}(aq) + 4Cl^-(aq) \rightleftharpoons CoCl_4^{2-}(aq)$$

Copper(II) sulfate ($CuSO_4$) dissolves in water to produce a blue solution. The hydrated copper(II) ions are responsible for this color; many other sulfates (Na_2SO_4, for example) are colorless. Adding a *few drops* of concentrated ammonia solution to a $CuSO_4$ solution causes the formation of a light-blue precipitate, copper(II) hydroxide:

$$Cu^{2+}(aq) + 2OH^-(aq) \longrightarrow Cu(OH)_2(s)$$

The OH^- ions are supplied by the ammonia solution. If more NH_3 is added, the blue precipitate redissolves to produce a beautiful dark-blue solution, this time due to the formation of the complex ion $Cu(NH_3)_4^{2+}$ (Figure 12.14):

$$Cu(OH)_2(s) + 4NH_3(aq) \rightleftharpoons Cu(NH_3)_4^{2+}(aq) + 2OH^-(aq)$$

Figure 12.14 Left: An aqueous solution of copper(II) sulfate. Center: After the addition of a few drops of a concentrated aqueous ammonia solution, a light-blue precipitate of $Cu(OH)_2$ is formed. Right: When more concentrated aqueous ammonia solution is added, the $Cu(OH)_2$ precipitate dissolves to form the dark-blue complex ion $Cu(NH_3)_4^{2+}$.

Table 12.5	Formation Constants of Select Complex Ions in Water at 25°C	
Complex Ion	Equilibrium Expression	Formation Constant (K_f)
$Ag(NH_3)_2^+$	$Ag^+ + 2NH_3 \rightleftharpoons Ag(NH_3)_2^+$	1.5×10^7
$Ag(CN)_2^-$	$Ag^+ + 2CN^- \rightleftharpoons Ag(CN)_2^-$	1.0×10^{21}
$Cu(CN)_4^{2-}$	$Cu^{2+} + 4CN^- \rightleftharpoons Cu(CN)_4^{2-}$	1.0×10^{25}
$Cu(NH_3)_4^{2+}$	$Cu^{2+} + 4NH_3 \rightleftharpoons Cu(NH_3)_4^{2+}$	5.0×10^{13}
$Cd(CN)_4^{2-}$	$Cd^{2+} + 4CN^- \rightleftharpoons Cd(CN)_4^{2-}$	7.1×10^{16}
CdI_4^{2-}	$Cd^{2+} + 4I^- \rightleftharpoons CdI_4^{2-}$	2.0×10^6
$HgCl_4^{2-}$	$Hg^{2+} + 4Cl^- \rightleftharpoons HgCl_4^{2-}$	1.7×10^{16}
HgI_4^{2-}	$Hg^{2+} + 4I^- \rightleftharpoons HgI_4^{2-}$	2.0×10^{30}
$Hg(CN)_4^{2-}$	$Hg^{2+} + 4CN^- \rightleftharpoons Hg(CN)_4^{2-}$	2.5×10^{41}
$Co(NH_3)_6^{3+}$	$Co^{3+} + 6NH_3 \rightleftharpoons Co(NH_3)_6^{3+}$	5.0×10^{31}
$Zn(NH_3)_4^{2+}$	$Zn^{2+} + 4NH_3 \rightleftharpoons Zn(NH_3)_4^{2+}$	2.9×10^9

Thus, the formation of the complex ion $Cu(NH_3)_4^{2+}$ increases the solubility of $Cu(OH)_2$.

A measure of the tendency of a metal ion to form a particular complex ion is given by the *formation constant K_f* (also called the *stability constant*), which is *the equilibrium constant for the complex ion formation*. The larger the value of K_f, the more stable the complex ion. Table 12.5 lists the formation constants of a number of complex ions.

The formation of the $Cu(NH_3)_4^{2+}$ ion can be expressed as

$$Cu^{2+}(aq) + 4NH_3(aq) \rightleftharpoons Cu(NH_3)_4^{2+}(aq)$$

for which the formation constant, assuming ideal solution behavior, is

$$K_f = \frac{[Cu(NH_3)_4^{2+}]}{[Cu^{2+}][NH_3]^4} = 5.0 \times 10^{13}$$

The very large value of K_f in this case indicates that the complex ion is quite stable in solution and accounts for the very low concentration of copper(II) ions at equilibrium.

The use of formation constants to determine aqueous equilibrium in solutions that exhibit complex ion formation is illustrated in Example 12.16.

Example 12.16

A 0.20-mole quantity of $CuSO_4$ is added to a liter of 1.20 *M* NH_3 solution. What is the concentration of Cu^{2+} ions at equilibrium?

Strategy The addition of $CuSO_4$ to the NH_3 solution results in complex ion formation:

$$Cu^{2+}(aq) + 4NH_3(aq) \rightleftharpoons Cu(NH_3)_4^{2+}(aq)$$

—Continued

Continued—

From Table 12.5, we see that the formation constant (K_f) for this reaction is very large; therefore, the reaction lies mostly to the right. At equilibrium, the concentration of Cu^{2+} will be very small. As a good approximation, we can assume that essentially all the dissolved Cu^{2+} ions end up as $Cu(NH_3)_4^{2+}$ ions. How many moles of NH_3 will react with 0.20 mole of Cu^{2+}? How many moles of $Cu(NH_3)_4^{2+}$ will be produced? A very small amount of Cu^{2+} will be present at equilibrium. Set up the K_f expression for the preceding equilibrium to solve for $[Cu^{2+}]$.

Solution The amount of NH_3 consumed in forming the complex ion is 4×0.20 mol, or 0.80 mol. (Note that 0.20 mol Cu^{2+} is initially present in solution and four NH_3 molecules are needed to form a complex ion with one Cu^{2+} ion.) The concentration of NH_3 at equilibrium is therefore $(1.20 - 0.80)$ mol L^{-1} soln or 0.40 M, and that of $Cu(NH_3)_4^{2+}$ is 0.20 mol L^{-1} soln or 0.20 M, the same as the initial concentration of Cu^{2+}. [There is a 1:1 mole ratio between the initial value of Cu^{2+} and the final value of $Cu(NH_3)_4^{2+}$.] Because $Cu(NH_3)_4^{2+}$ does dissociate to a slight extent, we call the concentration of Cu^{2+} at equilibrium x and write

$$K_f = \frac{[Cu(NH_3)_4^{2+}]}{[Cu^{2+}][NH_3]^4}$$

$$5.0 \times 10^{13} = \frac{0.20}{x(0.40)^4}$$

Solving for x and keeping in mind that the volume of the solution is 1 L, we obtain

$$x = 1.6 \times 10^{-13}\, M = [Cu^{2+}]$$

Check The small value of $[Cu^{2+}]$ at equilibrium, compared with 0.20 M, certainly justifies our approximation.

Practice Exercise If 2.50 g of $CuSO_4$ is dissolved in 9.0×10^2 mL of 0.30 M NH_3, what are the concentrations of Cu^{2+}, $Cu(NH_3)_4^{2+}$, and NH_3 at equilibrium?

The effect of complex ion formation generally is to *increase* the solubility of a substance, as Example 12.17 shows.

Example 12.17

Calculate the molar solubility of AgCl in a 1.0 M NH_3 solution.

Strategy AgCl is only slightly soluble in water:

$$AgCl(s) \rightleftharpoons Ag^+(aq) + Cl^-(aq)$$

The Ag^+ ions form a complex ion with NH_3 (see Table 12.5):

$$Ag^+(aq) + 2NH_3(aq) \rightleftharpoons Ag(NH_3)_2^+(aq)$$

Combining these two equilibria will give the overall equilibrium for the process.

Solution *Step 1:* Initially, the species in solution are Ag^+ and Cl^- ions and NH_3. The reaction between Ag^+ and NH_3 produces the complex ion $Ag(NH_3)_2^+$.

—Continued

Continued—

Step 2: The equilibrium reactions are

$$AgCl(s) \rightleftharpoons Ag^+(aq) + Cl^-(aq) \quad K_{sp} = [Ag^+][Cl^-] = 1.6 \times 10^{-10}$$

$$Ag^+(aq) + 2NH_3(aq) \rightleftharpoons Ag(NH_3)_2^+(aq) \quad K_f = \frac{[Ag(NH_3)_2^+]}{[Ag^+][NH_3]^2} = 1.5 \times 10^7$$

overall: $\quad AgCl(s) + 2NH_3(aq) \rightleftharpoons Ag(NH_3)_2^+(aq) + Cl^-(aq)$

The equilibrium constant K for the overall reaction is the product of the equilibrium constants of the individual reactions (see Section 10.2):

$$K = K_{sp}K_f = \frac{[Ag(NH_3)_2^+][Cl^-]}{[NH_3]^2}$$
$$= (1.6 \times 10^{-10})(1.5 \times 10^7)$$
$$= 2.4 \times 10^{-3}$$

Let s be the molar solubility of AgCl (mol L^{-1}). We summarize the changes in concentrations that result from the formation of the complex ion as follows:

	AgCl(s)	+ 2NH₃(aq)	⇌ Ag(NH₃)₂⁺(aq)	+ Cl⁻(aq)
Initial (*M*)		1.0	0.0	0.0
Change (*M*)	$-s$	$-2s$	$+s$	$+s$
Equilibrium (*M*)		$1.0 - 2s$	s	s

The formation constant for $Ag(NH_3)_2^+$ is quite large, so most of the silver ions exist in the complexed form. In the absence of ammonia we have, at equilibrium, $[Ag^+] = [Cl^-]$. As a result of complex ion formation, however, we can write $[Ag(NH_3)_2^+] = [Cl^-]$.

Step 3:
$$K = \frac{(s)(s)}{(1.0 - 2s)^2}$$
$$2.4 \times 10^{-3} = \frac{s^2}{(1.0 - 2s)^2}$$

Taking the square root of both sides, we obtain

$$0.049 = \frac{s}{1.0 - 2s}$$
$$s = 0.045 \ M$$

Therefore, at equilibrium, 0.045 mole of AgCl dissolves in 1 L of 1.0 *M* NH₃ solution.

Check The molar solubility of AgCl in pure water is $1.3 \times 10^{-5} \ M$. Thus, the formation of the complex ion $Ag(NH_3)_2^+$ enhances the solubility of AgCl (Figure 12.15).

Practice Exercise Calculate the molar solubility of AgBr in a 1.0 *M* NH₃ solution.

All amphoteric hydroxides are insoluble compounds.

Finally, we note that there is a class of hydroxides, called *amphoteric hydroxides,* that can react with both acids and bases. Examples are $Al(OH)_3$, $Pb(OH)_2$, $Cr(OH)_3$, $Zn(OH)_2$, and $Cd(OH)_2$. Thus, $Al(OH)_3$ reacts with acids and bases as follows:

$$Al(OH)_3(s) + 3H^+(aq) \longrightarrow Al^{3+}(aq) + 3H_2O(l)$$
$$Al(OH)_3(s) + OH^-(aq) \rightleftharpoons Al(OH)_4^-(aq)$$

Figure 12.15 From left to right: The formation of AgCl precipitate when AgNO$_3$ solution is added to NaCl solution. With the addition of NH$_3$ solution, the AgCl precipitate dissolves as the soluble Ag(NH$_3$)$_2^+$ forms.

The increase in solubility of Al(OH)$_3$ in a basic medium is the result of the formation of the complex ion Al(OH)$_4^-$ in which Al(OH)$_3$ acts as the Lewis acid and OH$^-$ acts as the Lewis base. Other amphoteric hydroxides behave in a similar manner.

12.8 | The Solubility Product Principle Can Be Applied to Qualitative Analysis

Here we will briefly discuss **qualitative analysis,** *the determination of the types of ions present in a solution.* We will focus on the cations. There are some 20 common cations that can be analyzed readily in aqueous solution. These cations can be divided into five groups according to the solubility products of their insoluble salts (Table 12.6). Because an unknown solution may contain from 1 to all 20 ions, any analysis must be carried out systematically from group 1 through group 5. Let us consider the general procedure for separating these 20 ions by adding precipitating reagents to an unknown solution.

Do not confuse the groups in Table 12.6, which are based on solubility products, with those in the periodic table, which are based on the electron configurations of the elements.

▶ **Group 1 Cations.** When dilute HCl is added to the unknown solution, only the Ag, Hg$_2^{2+}$, and Pb^{2+} ions precipitate as insoluble chlorides. The other ions, whose chlorides are soluble, remain in solution.

▶ **Group 2 Cations.** After the chloride precipitates have been removed by filtration, hydrogen sulfide is reacted with the unknown acidic solution. Under this condition, the concentration of the S^{2-} ion in solution is negligible. Therefore, the precipitation of metal sulfides is best represented as

$$M^{2+}(aq) + H_2S(aq) \rightleftharpoons MS(s) + 2H^+(aq)$$

Adding acid to the solution shifts this equilibrium to the left so that only the least soluble metal sulfides, that is, those with the smallest K_{sp} values, will precipitate out of solution. These are Bi$_2$S$_3$, CdS, CuS, HgS, and SnS (see Table 12.6).

▶ **Group 3 Cations.** At this stage, sodium hydroxide is added to the solution to make it basic. In a basic solution, the equilibrium between M^{2+} and MS shifts to the right. Therefore, the more soluble sulfides (CoS, FeS, MnS, NiS, and ZnS) now precipitate out of solution. Note that the Al^{3+} and Cr^{3+} ions actually precipitate as the hydroxides Al(OH)$_3$ and Cr(OH)$_3$, rather than as the sulfides, because the hydroxides are less soluble. The solution is then filtered to remove the insoluble sulfides and hydroxides.

▶ **Group 4 Cations.** After all the group 1, 2, and 3 cations have been removed from solution, sodium carbonate is added to the basic solution to precipitate Ba^{2+}, Ca^{2+}, and Sr^{2+} ions as BaCO$_3$, CaCO$_3$, and SrCO$_3$. These precipitates too are removed from solution by filtration.

Table 12.6	Separation of Cations into Groups According to Their Precipitation Reactions with Various Reagents			
Group	Cation	Precipitating Reagents	Insoluble Compound	K_{sp}
1	Ag^+	HCl	AgCl	1.6×10^{-10}
	Hg_2^{2+}		Hg_2Cl_2	3.5×10^{-18}
	Pb^{2+}	↓	$PbCl_2$	2.4×10^{-4}
2	Bi^{3+}	H_2S	Bi_2S_3	1.6×10^{-72}
	Cd^{2+}	in acidic	CdS	8.0×10^{-28}
	Cu^{2+}	solutions	CuS	6.0×10^{-37}
	Hg^{2+}		HgS	4.0×10^{-54}
	Sn^{2+}	↓	SnS	1.0×10^{-26}
3	Al^{3+}	H_2S	$Al(OH)_3$	1.8×10^{-33}
	Co^{2+}	in basic	CoS	4.0×10^{-21}
	Cr^{3+}	solutions	$Cr(OH)_3$	3.0×10^{-29}
	Fe^{2+}		FeS	6.0×10^{-19}
	Mn^{2+}		MnS	3.0×10^{-14}
	Ni^{2+}		NiS	1.4×10^{-24}
	Zn^{2+}	↓	ZnS	3.0×10^{-23}
4	Ba^{2+}	Na_2CO_3	$BaCO_3$	8.1×10^{-9}
	Ca^{2+}		$CaCO_3$	8.7×10^{-9}
	Sr^{2+}	↓	$SrCO_3$	1.6×10^{-9}
5	K^+	No precipitating	None	
	Na^+	reagent	None	
	NH_4^+		None	

▶ **Group 5 Cations.** At this stage, the only cations possibly remaining in solution are Na^+, K^+, and NH_4^+. The presence of NH_4^+ can be determined by adding sodium hydroxide:

$$NaOH(aq) + NH_4^+(aq) \longrightarrow Na^+(aq) + H_2O(l) + NH_3(g)$$

The ammonia gas is detected either by noting its characteristic odor or by observing a piece of wet red litmus paper turning blue when placed above (not in contact with) the solution. To confirm the presence of Na^+ and K^+ ions, we usually use a flame test, as follows: A piece of platinum wire (chosen because platinum is inert) is moistened with the solution and is then held over a Bunsen burner flame. Each type of metal ion gives a characteristic color when heated in this manner. For example, the color emitted by Na^+ ions is yellow, that of K^+ ions is violet, and that of Cu^{2+} ions is green (Figure 12.16).

Figure 12.17 summarizes this scheme for separating metal ions.

Two points regarding qualitative analysis must be mentioned. First, the separation of the cations into groups is made as selective as possible, that is, the anions that are added as reagents must be such that they will precipitate the fewest types of cations. For example, all the cations in group 1 also form insoluble sulfides. Thus, if H_2S were

Because NaOH is added in group 3 and Na_2CO_3 is added in group 4, the flame test for Na^+ ions must be carried out on the original solution.

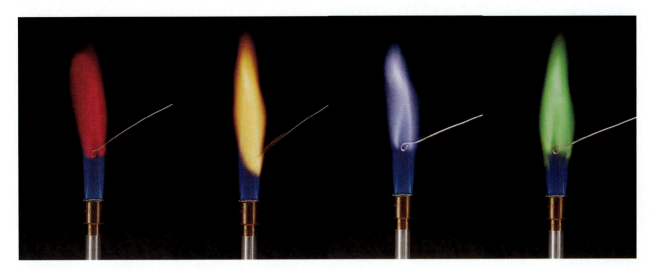

Figure 12.16 Left to right: Flame colors of lithium, sodium, potassium, and copper.

reacted with the solution at the start, as many as seven different sulfides might precipitate out of solution (group 1 *and* group 2 sulfides), an undesirable outcome. Second, the removal of cations at each step must be carried out as completely as possible. For example, if we do not add enough HCl to the unknown solution to remove all the group 1 cations, they will precipitate with the group 2 cations as insoluble sulfides, interfering with further chemical analysis and leading to erroneous conclusions.

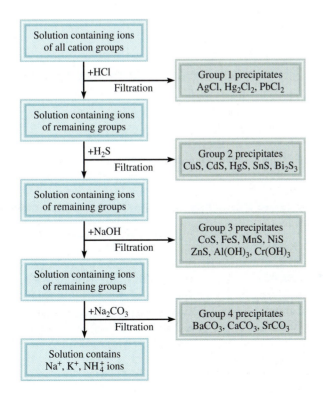

Figure 12.17 A flowchart for the separation of cations in qualitative analysis.

Summary of Facts and Concepts

Section 12.1

▶ The common ion effect tends to suppress the ionization of a weak acid or a weak base. This action can be explained by Le Châtelier's principle.

Section 12.2

▶ A buffer solution is a combination of either a weak acid and its weak conjugate base (supplied by a salt) or a weak base and its weak conjugate acid (supplied by a salt); the solution reacts with small amounts of added acid or base in such a way that the pH of the solution remains nearly constant. Buffer systems play a vital role in maintaining the pH of body fluids.

Section 12.3

▶ The pH at the equivalence point of an acid-base titration depends on hydrolysis of the salt formed in the neutralization reaction. For strong acid–strong base titrations, the pH at the equivalence point is 7; for weak acid–strong base titrations, the pH at the equivalence point is greater than 7; for strong acid–weak base titrations, the pH at the equivalence point is less than 7.

Section 12.4

▶ Acid-base indicators are weak organic acids or bases that change color near the equivalence point in an acid-base neutralization reaction.

▶ The specific indicator suitable for a particular titration will depend upon the pH of the end point.

Section 12.5

▶ From general rules about the solubilities of ionic compounds, we can predict qualitatively whether a precipitate will form in a reaction.

Section 12.6

▶ The solubility product K_{sp} is the equilibrium constant for the dissolution of a solid into a solution. Solubility can be estimated from K_{sp} and vice versa.

Section 12.7

▶ The presence of a common ion decreases the solubility of a slightly soluble salt.

▶ The solubility of slightly soluble salts containing basic anions increases as the hydrogen ion activity (or concentration) increases. The solubility of salts with anions derived from strong acids is unaffected by pH.

▶ Complex ions are formed in solution by the combination of a metal cation with a Lewis base. The formation constant K_f measures the tendency toward the formation of a specific complex ion. Complex ion formation can increase the solubility of an insoluble substance.

Section 12.8

▶ Qualitative analysis is used to identify cations and anions in solution.

Key Words

buffer solution, p. 615
common ion effect, p. 612
complex ion, p. 648
end point, p. 631
equivalence point, p. 623

formation constant, p. 650
Henderson-Hasselbalch
   equation, p. 613
indicator, p. 631
molar solubility, p. 638

precipitate, p. 633
precipitation reaction, p. 633
qualitative analysis, p. 653
solubility, p. 638
solubility product, p. 636

standard solution, p. 622
titration, p. 622

Problems[2]

The Ionization of Weak Acids and Bases Is Suppressed by the Addition of a Common Ion

12.1 Determine the pH of (a) a 0.40 M CH_3COOH solution, and (b) a solution that is 0.40 M CH_3COOH and 0.20 M CH_3COONa.

12.2 Determine the pH of (a) a 0.20 M NH_3 solution, and (b) a solution that is 0.20 M in NH_3 and 0.30 M NH_4Cl.

The pH of a Buffer Solution Is Resistant to Large Changes in pH

12.3 Which of the following solutions can act as a buffer: (a) KCl/HCl, (b) $KHSO_4/H_2SO_4$, (c) Na_2HPO_4/NaH_2PO_4, (d) KNO_2/HNO_2, and (e) $KHSO_3/K_2SO_4$?

12.4 Which of the following solutions can act as a buffer: (a) KCN/HCN, (b) $Na_2SO_4/NaHSO_4$, (c) NH_3/NH_4NO_3, and (d) NaI/HI?

---

2. Unless otherwise stated, assume that the temperature in these problems is 25°C.

12.5 Calculate the pH of the buffer system made up of 0.15 M NH_3/0.35 M NH_4Cl.

12.6 Calculate the pH of the following two buffer solutions: (a) 2.0 M CH_3COONa/2.0 M CH_3COOH, and (b) 0.20 M CH_3COONa/0.20 M CH_3COOH. State any approximations made in this calculation. Which is the more effective buffer? Why?

12.7 The pH of a bicarbonate–carbonic acid buffer is 8.00. Calculate the ratio of the concentration of carbonic acid (H_2CO_3) to that of the bicarbonate ion (HCO_3^-).

12.8 What is the pH of the buffer 0.10 M Na_2HPO_4/0.15 M KH_2PO_4?

12.9 The pH of a sodium acetate–acetic acid buffer is 4.50. Calculate the ratio $[CH_3COO^-]/[CH_3COOH]$.

12.10 The pH of blood plasma is 7.40. Assuming the principal buffer system is HCO_3^-/H_2CO_3, calculate the ratio $[HCO_3^-]/[H_2CO_3]$. Is this buffer more effective against an added acid or an added base?

12.11 Calculate the pH of the 0.20 M NH_3/0.20 M NH_4 Cl buffer. What is the pH of the buffer after the addition of 10.0 mL of 0.10 M HCl to 65.0 mL of the buffer?

12.12 Calculate the pH of 1.00 L of the buffer 1.00 M CH_3COONa/1.00 M CH_3COOH before and after the addition of (a) 0.080 mol NaOH, and (b) 0.12 mol HCl. (Assume that there is no change in volume.)

12.13 A diprotic acid, H_2A, has the following ionization constants: $K_{a_1} = 1.1 \times 10^{-3}$ and $K_{a_2} = 2.5 \times 10^{-6}$. In order to make up a buffer solution of pH 5.80, which combination would you choose: $NaHA$/H_2A or Na_2A/$NaHA$?

12.14 A student is asked to prepare a buffer solution at pH = 8.60, using one of the following weak acids: HA ($K_a = 2.7 \times 10^{-3}$), HB ($K_a = 4.4 \times 10^{-6}$), HC ($K_a = 2.6 \times 10^{-9}$). Which acid should she choose? Why?

The Concentration of an Unknown Acid or Base Can Be Determined by Titration

12.15 A 0.2688-g sample of a monoprotic acid neutralizes 16.4 mL of 0.08133 M KOH solution. Calculate the molar mass of the acid.

12.16 A 5.00-g quantity of a diprotic acid was dissolved in water and made up to exactly 250 mL. Calculate the molar mass of the acid if 25.0 mL of this solution required 11.1 mL of 1.00 M KOH for neutralization. Assume that both protons of the acid were titrated.

12.17 In a titration experiment, 12.5 mL of 0.500 M H_2SO_4 neutralizes 50.0 mL of NaOH. What is the concentration of the NaOH solution?

12.18 In a titration experiment, 20.4 mL of 0.883 M HCOOH neutralizes 19.3 mL of $Ba(OH)_2$. What is the concentration of the $Ba(OH)_2$ solution?

12.19 A 0.1276-g sample of an unknown monoprotic acid was dissolved in 25.0 mL of water and titrated with 0.0633 M NaOH solution. The volume of base required to bring the solution to the equivalence point was 18.4 mL. (a) Calculate the molar mass of the acid. (b) After 10.0 mL of base had been added during the titration, the pH was determined to be 5.87. What is the K_a of the unknown acid?

12.20 A solution is made by mixing exactly 500 mL of 0.167 M NaOH with exactly 500 mL of 0.100 M CH_3COOH. Calculate the equilibrium concentrations of H^+, CH_3COOH, CH_3COO^-, OH^-, and Na^+.

12.21 Calculate the pH at the equivalence point for the following titration: 0.20 M HCl versus 0.20 M methylamine (CH_3NH_2).

12.22 Calculate the pH at the equivalence point for the following titration: 0.10 M HCOOH versus 0.10 M NaOH.

12.23 A 25.0-mL solution of 0.100 M CH_3COOH is titrated with a 0.200 M KOH solution. Calculate the pH after the following additions of the KOH solution: (a) 0.0 mL, (b) 5.0 mL, (c) 10.0 mL, (d) 12.5 mL, and (e) 15.0 mL.

12.24 A 10.0-mL solution of 0.300 M NH_3 is titrated with a 0.100 M HCl solution. Calculate the pH after the following additions of the HCl solution: (a) 0.0 mL, (b) 10.0 mL, (c) 20.0 mL, (d) 30.0 mL, and (e) 40.0 mL.

An Acid-Base Indicator Is a Substance That Changes Color at a Specific pH

12.25 Referring to Table 12.1, specify which indicator or indicators you would use for the following titrations: (a) HCOOH versus NaOH, (b) HCl versus KOH, and (c) HNO_3 versus CH_3NH_2.

12.26 A student carried out an acid-base titration by adding NaOH solution from a buret to an Erlenmeyer flask containing HCl solution and using phenolphthalein as indicator. At the equivalence point, she observed a faint reddish-pink color. However, after a few minutes, the solution gradually turned colorless. What do you suppose happened?

12.27 The ionization constant K_a of an indicator HIn is 1.0×10^{-6}. The color of the nonionized form is red and that of the ionized form is yellow. What is the color of this indicator in a solution whose pH is 4.00?

12.28 The K_a of a certain indicator is 2.0×10^{-6}. The color of HIn is green and that of In^- is red. A few drops of the indicator are added to an HCl solution, which is then titrated against an NaOH solution. At what pH will the indicator change color?

A Precipitation Reaction Occurs when a Reaction in Solution Leads to an Insoluble Product

12.29 Characterize the following compounds as soluble or insoluble in water: (a) $Ca_3(PO_4)_2$, (b) $Mn(OH)_2$, (c) $AgClO_3$, and (d) K_2S.

12.30 Characterize the following compounds as soluble or insoluble in water: (a) $CaCO_3$, (b) $ZnSO_4$, (c) $Hg(NO_3)_2$, (d) $HgSO_4$, and (e) NH_4ClO_4.

12.31 Which of the following processes will likely result in a precipitation reaction? (a) Mixing an $NaNO_3$ solution with a $CuSO_4$ solution. (b) Mixing a $BaCl_2$ solution with a K_2SO_4 solution. Write a net ionic equation for the precipitation reaction.

12.32 An ionic compound X is only slightly soluble in water. What test would you employ to show that the compound does indeed dissolve in water to a certain extent?

12.33 A 0.9157-g mixture of $CaBr_2$ and $NaBr$ is dissolved in water, and $AgNO_3$ is added to the solution to form $AgBr$ precipitate. If the mass of the precipitate is 1.6930 g, what is the percent by mass of $NaBr$ in the original mixture?

The Solubility Product Is the Equilibrium Constant for the Dissolution Process

12.34 Calculate the concentration of ions in the following saturated solutions: (a) $[I^-]$ in AgI solution with $[Ag^+] = 9.1 \times 10^{-9} M$, and (b) $[Al^{3+}]$ in $Al(OH)_3$ solution with $[OH^-] = 2.9 \times 10^{-9} M$.

12.35 From the solubility data given, calculate the solubility products for the following compounds: (a) SrF_2, 7.3×10^{-2} g L^{-1}, and (b) Ag_3PO_4, 6.7×10^{-3} g L^{-1}.

12.36 The molar solubility of $MnCO_3$ is $4.2 \times 10^{-6} M$. What is the K_{sp} for this compound? The solubility of an ionic compound MX (molar mass = 346 g) is 4.63×10^{-3} g L^{-1}. What is the K_{sp} for the compound?

12.37 The solubility of an ionic compound M_2X_3 (molar mass = 288 g) is 3.6×10^{-17} g L^{-1}. What is the K_{sp} for the compound?

12.38 Using data from Table 12.3, calculate the molar solubility of CaF_2.

12.39 What is the pH of a saturated zinc hydroxide solution?

12.40 The pH of a saturated solution of a metal hydroxide MOH is 9.68. Calculate the K_{sp} for the compound.

12.41 If 20.0 mL of 0.10 M $Ba(NO_3)_2$ is added to 50.0 mL of 0.10 M Na_2CO_3, will $BaCO_3$ precipitate?

12.42 A volume of 75 mL of 0.060 M NaF is mixed with 25 mL of 0.15 M $Sr(NO_3)_2$. Calculate the concentrations in the final solution of NO_3^-, Na^+, Sr^{2+}, and F^-. (K_{sp} for SrF_2 is 2.0×10^{-10}.)

12.43 Solid NaI is slowly added to a solution that is 0.010 M in Cu^+ and 0.010 M in Ag^+. (a) Which compound will begin to precipitate first? (b) Calculate $[Ag^+]$ when CuI just begins to precipitate. (c) What percent of Ag^+ remains in solution at this point?

12.44 Find the approximate pH range suitable for the separation of Fe^{3+} and Zn^{2+} ions by precipitation of $Fe(OH)_3$ from a solution that is initially 0.010 M in both Fe^{3+} and Zn^{2+}.

The Solubility of a Substance Is Affected by a Number of Factors

12.45 How does the common ion effect influence solubility equilibria? Use Le Châtelier's principle to explain the decrease in solubility of $CaCO_3$ in an Na_2CO_3 solution.

12.46 The molar solubility of AgCl in $6.5 \times 10^{-3} M$ $AgNO_3$ is $2.5 \times 10^{-8} M$. In deriving K_{sp} from these data, which of the following assumptions are reasonable?
  (a) K_{sp} is the same as solubility.
  (b) K_{sp} of AgCl is the same in $6.5 \times 10^{-3} M$ $AgNO_3$ as in pure water.
  (c) Solubility of AgCl is independent of the concentration of $AgNO_3$.
  (d) $[Ag^+]$ in solution does not change significantly upon the addition of AgCl to $6.5 \times 10^{-3} M$ $AgNO_3$.
  (e) $[Ag^+]$ in solution after the addition of AgCl to $6.5 \times 10^{-3} M$ $AgNO_3$ is the same as it would be in pure water.

12.47 How many grams of $CaCO_3$ will dissolve in 3.0×10^{-2} mL of 0.050 M $Ca(NO_3)_2$?

12.48 The solubility product of $PbBr_2$ is 8.9×10^{-6}. Determine the molar solubility (a) in pure water, (b) in 0.20 M KBr solution, and (c) in 0.20 M $Pb(NO_3)_2$ solution.

12.49 Calculate the molar solubility of AgCl in a 1.00-L solution containing 10.0 g of dissolved $CaCl_2$.

12.50 Calculate the molar solubility of $BaSO_4$ (a) in water, and (b) in a solution containing 1.0 M SO_4^{2-} ions.

12.51 Which of the following ionic compounds will be more soluble in acid solution than in water: (a) $BaSO_4$, (b) $PbCl_2$, (c) $Fe(OH)_3$, or (d) $CaCO_3$?

12.52 Which of the following will be more soluble in acid solution than in pure water: (a) CuI, (b) Ag_2SO_4, (c) $Zn(OH)_2$, (d) BaC_2O_4, or (e) $Ca_3(PO_4)_2$?

12.53 Compare the molar solubility of $Mg(OH)_2$ in water and in a solution buffered at a pH of 9.0.

12.54 Calculate the molar solubility of $Fe(OH)_2$ in a solution buffered at (a) pH 8.00, and (b) pH 10.00.

12.55 The solubility product of $Mg(OH)_2$ is 1.2×10^{-11}. What minimum OH^- concentration must be attained (e.g., by adding NaOH) to decrease the Mg^{2+} concentration in a solution of $Mg(NO_3)_2$ to less than 1.0×10^{-10} M?

12.56 Calculate whether or not a precipitate will form if 2.00 mL of 0.60 M NH_3 are added to 1.0 L of 1.0×10^{-3} M $FeSO_4$.

12.57 If 2.50 g of $CuSO_4$ are dissolved in 9.0×10^{-2} mL of 0.30 M NH_3, what are the concentrations of Cu^{2+}, $Cu(NH_3)_4^{2-}$, and NH_3 at equilibrium? Calculate the concentrations of Cd^{2+}, $Cd(CN)_4^{2-}$, and CN^- at equilibrium when 0.50 g of $Cd(NO_3)_2$ dissolves in 5.0×10^{-2} mL of 0.50 M NaCN.

12.58 If NaOH is added to 0.010 M Al^{3+}, which will be the predominant species at equilibrium: $Al(OH)_3$ or $Al(OH)_4^-$? The pH of the solution is 14.00. [K_f for $Al(OH)_4^-$ is 2.0×10^{-33}.]

12.59 Calculate the molar solubility of AgI in a 1.0 M NH_3 solution.

12.60 Both Ag^+ and Zn^{2+} form complex ions with NH_3. Write balanced equations for the reactions. However, $Zn(OH)_2$ is soluble in 6 M NaOH, and AgOH is not. Explain.

12.61 Explain, with balanced ionic equations, why (a) CuI_2 dissolves in ammonia solution, (b) AgBr dissolves in NaCN solution, and (c) $HgCl_2$ dissolves in KCl solution.

The Solubility Product Principle Can Be Applied to Qualitative Analysis

12.62 In a group 1 analysis, a student obtained a precipitate containing both AgCl and $PbCl_2$. Suggest one reagent that would allow her to separate AgCl(s) from $PbCl_2(s)$.

12.63 In a group 1 analysis, a student adds HCl acid to the unknown solution to make $[Cl^-] = 0.15$ M. Some $PbCl_2$ precipitates. Calculate the concentration of Pb^{2+} remaining in solution.

12.64 Both KCl and NH_4Cl are white solids. Suggest one reagent that would enable you to distinguish between these two compounds.

12.65 Describe a simple test that would allow you to distinguish between $AgNO_3(s)$ and $Cu(NO_3)_2(s)$.

Additional Problems

12.66 The buffer range is defined by the equation

$$pH - pK_a = 1$$

Calculate the range of the ratio [conjugate base]/[acid] that corresponds to this equation.

12.67 The pK_a of the indicator methyl orange is 3.46. Over what pH range does this indicator change from 90 percent HIn to 90 percent In^-?

12.68 Sketch the titration curve of a weak acid versus a strong base like the one shown in Figure 12.6. On your graph indicate the volume of base used at the equivalence point and also at the half-equivalence point, that is, the point at which half of the acid has been neutralized. Show how you can measure the pH of the solution at the half-equivalence point. Using Equation 12.2, explain how you can determine the pK_a of the acid by this procedure.

12.69 A 200-mL volume of NaOH solution was added to 400 mL of a 2.00 M HNO_2 solution. The pH of the mixed solution was 1.50 units greater than that of the NaOH solution. Calculate the molarity of the NaOH solution.

12.70 The pK_a of butyric acid (HBut) is 4.7. Calculate K_b for the butyrate ion (But^-).

12.71 A solution is made by mixing exactly 500 mL of 0.167 M NaOH with exactly 500 mL 0.100 M HCOOH. Calculate the equilibrium concentrations of H^+, HCOOH, $HCOO^-$, OH^-, and Na^+.

12.72 $Cd(OH)_2$ is an insoluble compound. It dissolves in excess NaOH in solution. Write a balanced ionic equation for this reaction. What type of reaction is this?

12.73 A student mixes 50.0 mL of 1.00 M $Ba(OH)_2$ with 86.4 mL of 0.494 M H_2SO_4. Calculate the mass of $BaSO_4$ formed and the pH of the mixed solution.

12.74 For which of the following reactions is the equilibrium constant called a solubility product?

(a) $Zn(OH)_2(s) + 2OH^-(aq) \rightleftharpoons$
$$Zn(OH)_4^{2-}(aq)$$
(b) $3Ca^{2+}(aq) + 2PO_4^{3-}(aq) \rightleftharpoons$
$$Ca_3(PO_4)_2(s)$$
(c) $CaCO_3(s) + 2H^+(aq) \rightleftharpoons$
$$Ca^{2+}(aq) + H_2O(l) + CO_2(g)$$
(d) $PbI_2(s) \rightleftharpoons Pb^{2+}(aq) + 2I^-(aq)$

12.75 A 2.0-L kettle contains 116 g of boiler scale ($CaCO_3$). How many times would the kettle have to be completely filled with distilled water to remove all the deposit at 25°C?

12.76 Equal volumes of 0.12 M $AgNO_3$ and 0.14 M $ZnCl_2$ solution are mixed. Calculate the equilibrium concentrations of Ag^+, Cl^-, Zn^{2+}, and NO_3^-.

12.77 Calculate the solubility (in g L^{-1}) of Ag_2CO_3.

12.78 Find the approximate pH range suitable for separating Mg^{2+} and Zn^{2+} by the precipitation of $Zn(OH)_2$ from a solution that is initially 0.010 M in Mg^{2+} and Zn^{2+}.

12.79 A volume of 25.0 mL of 0.100 M HCl is titrated against a 0.100 M CH_3NH_2 solution added to it from a buret. Calculate the pH values of the solution (a) after 10.0 mL of CH_3NH_2 solution have been added, (b) after 25.0 mL of CH_3NH_2 solution have been added, and (c) after 35.0 mL of CH_3NH_2 solution have been added.

12.80 The molar solubility of $Pb(IO_3)_2$ in a 0.10 M $NaIO_3$ solution is 2.4×10^{-11} mol L^{-1}. What is the K_{sp} for $Pb(IO_3)_2$?

12.81 When a KI solution was added to a solution of mercury(II) chloride, a precipitate [mercury(II) iodide] formed. A student plotted the mass of the precipitate versus the volume of the KI solution added and obtained the following graph. Explain the appearance of the graph.

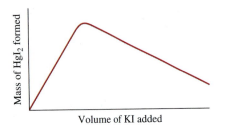

12.82 Barium is a toxic substance that can seriously impair heart function. For an X-ray of the gastrointestinal tract, a patient drinks an aqueous suspension of 20 g $BaSO_4$. If this substance were to equilibrate with the 5.0 L of the blood in the patient's body, what would be $[Ba^{2+}]$? For a good estimate, we may assume that the temperature is at 25°C. Why is $Ba(NO_3)_2$ not chosen for this procedure?

12.83 The pK_a of phenolphthalein is 9.10. Over what pH range does this indicator change from 95 percent HIn to 95 percent In⁻?

12.84 Solid NaBr is slowly added to a solution that is 0.010 M in Cu^{2+} and 0.010 M in Ag^+. (a) Which compound will begin to precipitate first? (b) Calculate $[Ag^+]$ when CuBr just begins to precipitate. (c) What percent of Ag^+ remains in solution at this point?

12.85 Cacodylic acid is $(CH_3)_2AsO_2H$. Its ionization constant is 6.4×10^{-7}. (a) Calculate the pH of 50.0 mL of a 0.10 M solution of the acid. (b) Calculate the pH of 25.0 mL of 0.15 M $(CH_3)_2AsO_2Na$. (c) Mix the solutions in parts (a) and (b). Calculate the pH of the resulting solution.

12.86 Radiochemical techniques are useful in estimating the solubility product of many compounds. In one experiment, 50.0 mL of a 0.010 M $AgNO_3$ solution containing a silver isotope with a radioactivity of 74,025 counts per min per mL were mixed with 100 mL of a 0.030 M $NaIO_3$ solution. The mixed solution was diluted to 500 mL and filtered to remove all the $AgIO_3$ precipitate. The remaining solution was found to have a radioactivity of 44.4 counts per min per mL. What is the K_{sp} of $AgIO_3$?

12.87 The molar mass of a certain metal carbonate, MCO_3, can be determined by adding an excess of HCl acid to react with all the carbonate and then "back-titrating" the remaining acid with NaOH. (a) Write an equation for these reactions. (b) In a certain MCO_3 experiment, 20.00 mL of 0.0800 M HCl were added to a 0.1022-g sample of MCO_3. The excess HCl required 5.64 mL of 0.1000 M NaOH for neutralization. Calculate the molar mass of the carbonate, and identify M.

12.88 Acid-base reactions usually go to completion. Confirm this statement by calculating the equilibrium constant for each of the following cases: (a) A strong acid reacting with a strong base. (b) A strong acid reacting with a weak base (NH_3). (c) A weak acid (CH_3COOH) reacting with a strong base. (d) A weak acid (CH_3COOH) reacting with a weak base (NH_3). (*Hint:* Strong acids exist as H^+ ions and strong bases exist as OH^- ions in solution. You need to look up K_a, K_b, and K_w.)

12.89 Calculate x, the number of molecules of water in oxalic acid hydrate ($H_2C_2O_4 \cdot xH_2O$) from the following data: 5.00 g of the compound is made up to exactly 250 mL of solution, and 25.0 mL of this solution requires 15.9 mL of 0.500 M NaOH solution for neutralization.

12.90 Describe how you would prepare a 1-L 0.20 M CH_3COONa/0.20 M CH_3COOH buffer system by (a) mixing a solution of CH_3COOH with a solution of CH_3COONa, (b) reacting a solution of CH_3COOH with a solution of NaOH, and (c) reacting a solution of CH_3COONa with a solution of HCl.

12.91 Phenolphthalein is the common indicator for the titration of a strong acid with a strong base. (a) If the pK_a of phenolphthalein is 9.10, what is the ratio of the nonionized form of the indicator (colorless) to the ionized form (reddish pink) at pH 8.00? (b) If 2 drops of 0.060 M phenolphthalein are used in a titration involving a 50.0-mL volume, what is the concentration of the ionized form at pH 8.00? (Assume that 1 drop = 0.050 mL.)

12.92 Oil paintings containing lead(II) compounds as constituents of their pigments darken over the years. Suggest a chemical reason for the color change.

12.93 What reagents would you employ to separate the following pairs of ions in solution: (a) Na^+ and Ba^{2+}, (b) K^+ and Pb^{2+}, and (c) Zn^{2+} and Hg^{2+}?

12.94 Look up the K_{sp} values for $BaSO_4$ and $SrSO_4$ in Table 12.3. Calculate the concentrations of Ba^{2+}, Sr^{2+}, and SO_4^{2-} in a solution that is saturated with both compounds.

12.95 In principle, amphoteric oxides, such as Al_2O_3 and BeO, can be used to prepare buffer solutions because they possess both acidic and basic properties (see Section 12.7). Explain why these compounds are of little practical use as buffer components.

12.96 $CaSO_4$ ($K_{sp} = 2.4 \times 10^{-5}$) has a larger K_{sp} value than that of Ag_2SO_4 ($K_{sp} = 1.4 \times 10^{-5}$). Does it follow that $CaSO_4$ also has greater solubility (g L^{-1})?

12.97 When lemon juice is squirted into tea, the color becomes lighter. In part, the color change is due to dilution, but the main reason for the change is an acid-base reaction. What is the reaction? (*Hint:* Tea contains "polyphenols" that are weak acids and lemon juice contains citric acid.)

12.98 How many milliliters of 1.0 *M* NaOH must be added to a 200 mL solution of 0.10 *M* NaH_2PO_4 to make a buffer solution with a pH of 7.50?

12.99 The maximum allowable concentration of Pb^{2+} ions in drinking water is 0.05 ppm (i.e., 0.05 g of Pb^{2+} in one million grams of water). Is this guideline exceeded if an underground water supply is at equilibrium with the mineral anglesite ($PbSO_4$, $K_{sp} = 1.6 \times 10^{-8}$)?

12.100 One of the most common antibiotics is penicillin G (benzylpenicillinic acid), which has the following structure:

It is a weak monoprotic acid:

$$HP \rightleftharpoons H^+ + P^- \qquad K_a = 1.64 \times 10^{-3}$$

where HP denotes the parent acid and P^- the conjugate base. Penicillin G is produced by growing molds in fermentation tanks at 25°C and a pH range of 4.5 to 5.0. The crude form of this antibiotic is obtained by extracting the fermentation broth with an organic solvent in which the acid is soluble. (a) Identify the acidic hydrogen atom. (b) In one stage of purification, the organic extract of the crude penicillin G is treated with a buffer solution at pH = 6.50. What is the ratio of the conjugate base of penicillin G to the acid at this pH? Would you expect the conjugate base to be more soluble in water than the acid? (c) Penicillin G is not suitable for oral administration, but the sodium salt (NaP) is because it is soluble. Calculate the pH of a 0.12 *M* NaP solution formed when a tablet containing the salt is dissolved in a glass of water.

12.101 Which of the following solutions has the highest [H^+]: (a) 0.10 *M* HF, (b) 0.10 *M* HF in 0.10 *M* NaF, or (c) 0.10 *M* HF in 0.10 *M* SbF_5? (*Hint:* SbF_5 reacts with F^- to form the complex ion SbF_6^-.)

12.102 Distribution curves show how the fractions of nonionized acid and its conjugate base vary as a function of pH of the medium. Plot distribution curves for CH_3COOH and its conjugate base CH_3COO^- in solution. Your graph should show fraction as the *y* axis and pH as the *x* axis. What are the fractions and pH at the point where these two curves intersect?

12.103 Water containing Ca^{2+} and Mg^{2+} ions is called *hard water* and is unsuitable for some household and industrial use because these ions react with soap to form insoluble salts, or curds. One way to remove the Ca^{2+} ions from hard water is by adding washing soda ($Na_2CO_3 \cdot 10H_2O$). (a) The molar solubility of $CaCO_3$ is 9.3×10^{-5} *M*. What is its molar solubility in a 0.050 *M* Na_2CO_3 solution? (b) Why are Mg^{2+} ions not removed by this procedure? (c) The Mg^{2+} ions are removed as $Mg(OH)_2$ by adding slaked lime [$Ca(OH)_2$] to the water to produce a saturated solution. Calculate the pH of a saturated $Ca(OH)_2$ solution. (d) What is the concentration of Mg^{2+} ions at this pH? (e) In general, which ion (Ca^{2+} or Mg^{2+}) would you remove first? Why?

12.104 Consider the ionization of the following acid-base indicator:

$$HIn(aq) \rightleftharpoons H^+(aq) + In^-(aq)$$

The indicator changes color according to the ratios of the concentrations of the acid to its conjugate base as described on p. 631. Show that the pH range over which the indicator changes from the acid color to the base color is pH = $pK_a \pm 1$, where K_a is the ionization constant of the acid.

12.105 Amino acids are the building blocks of proteins. These compounds contain at least one amino group (—NH_2) and one carboxyl group (—COOH). Consider glycine (NH_2CH_2COOH). Depending on the pH of the solution, glycine can exist in one of three possible forms:

Fully protonated: $\overset{+}{N}H_3$—CH_2—COOH
Dipolar ion: $\overset{+}{N}H_3$—CH_2—COO^-
Fully ionized: NH_2—CH_2—COO^-

Predict the predominant form of glycine at pH 1.0, 7.0, and 12.0. The pK_a of the carboxyl group is 2.3 and that of the ammonium group (—NH_3^+) is 9.6.

12.106 (a) Referring to Figure 12.7, describe how you would determine the pK_b of the base. (b) Derive an analogous Henderson-Hasselbalch equation relating pOH to pK_b of a weak base B and its conjugate acid HB^+. Sketch a titration curve showing the variation of the pOH of the base solution versus the volume of a strong acid added from a buret. Describe how you would determine the pK_b from this curve.

12.107 One way to distinguish a buffer solution from an acid solution is by dilution. (a) Consider a buffer solution made of 0.500 M CH_3COOH and 0.500 M CH_3COONa. Calculate its pH and the pH after it has been diluted 10-fold. (b) Compare the result in part (a) with the pHs of a 0.500 M CH_3COOH solution before and after it has been diluted in the same manner.

12.108 Draw distribution curves for an aqueous carbonic acid solution. Your graph should show fraction of species present as the y axis and pH as the x axis. Note that at any pH, only two of the three species (H_2CO_3, HCO_3^-, and CO_3^{2-}) are present in appreciable concentrations. Use the pK_a values in Table 11.3.

12.109 A sample of 0.96 L HCl at 372 mmHg and 22°C is bubbled into 0.034 L of 0.57 M NH_3. What is the pH of the resulting solution? Assume the volume of solution remains constant and that the HCl is totally dissolved in the solution.

12.110 Histidine is one of the 20 amino acids found in proteins. Shown here is a fully protonated histidine molecule where the numbers denote the pK_a values of the acidic groups.

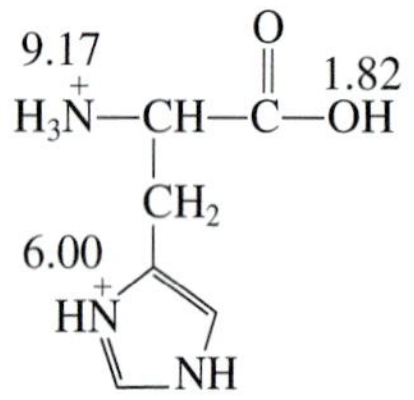

(a) Show stepwise the ionization of histadine in solution.

(b) A dipolar ion is one in which the species has an equal number of positive and negative charges. Identify the dipolar ion in part (a).

(c) The pH at which the dipolar ion predominates is called the isoelectric point, denoted by pI. The isoelectric point is the average of the pK_a values leading to and following the formation of the dipolar ion. Calculate the pI of histidine.

(d) The histidine group plays an important role in buffering blood. Which conjugate acid-base pair shown in part (a) is responsible for this action?

12.111 A 1.0-L saturated silver carbonate solution at 5°C is treated with enough hydrochloric acid to decompose the compound. The carbon dioxide generated is collected in a 19-mL vial and exerts a pressure of 114 mmHg at 25°C. What is the K_{sp} of Ag_2CO_3 at 5°C?

12.112 Using the value of K_{sp} for AgCl given in Table 12.3, estimate the value of K_{sp} for this compound at 50°C. What approximations did you make in this calculation? Does the solubility go up or down with increasing temperature? Does this trend follow from Le Châtelier's principle? (The enthalpy of solution for AgCl is 65.7 kJ mol^{-1} at 25°C.)

Answers to Practice Exercises

12.1 4.01 **12.2** (a) and (c) **12.3** 9.17, 9.20 **12.4** Weigh out Na_2CO_3 and $NaHCO_3$ in a mole ratio of 0.60 to 1.0. Dissolve in enough water to make up a 1-L solution. **12.5** (a) 2.19, (b) 3.95, (c) 8.02, (d) 11.39 **12.6** 5.92 **12.7** (a) Bromophenol blue, methyl orange, methyl red, and chlorophenol blue; (b) all except thymol blue, bromophenol blue, and methyl orange; (c) cresol red and phenolphthalein **12.8** (a) insoluble; (b) insoluble; (c) soluble **12.9** 2.0×10^{-14} **12.10** 1.9×10^{-3} g L^{-1} **12.11** No **12.12** (a) $> 1.6 \times 10^{-9} M$, (b) $> 2.6 \times 10^{-6} M$ **12.13** (a) 1.7×10^{-4} g L^{-1}, (b) 1.4×10^{-7} g L^{-1} **12.14** (a) More soluble in acid solution, (b) more soluble in acid solution, (c) about the same **12.15** $Fe(OH)_2$ precipitate will form. **12.16** $[Cu^{2+}] = 1.2 \times 10^{-13}$ M, $[Cu(NH_3)_4^{2+}] = 0.017$ M, $[NH_3] = 0.23$ M **12.17** 3.5×10^{-3} mol L^{-1}

13

Electrochemistry

13.1 | Oxidation-Reduction (Redox) Reactions Involve a Transfer of Electrons from One Species to Another

One form of energy that has tremendous practical significance is electric energy. A day without electricity from either the power company or batteries is unimaginable in our technological society. *The branch of chemistry that studies the interconversion of electric energy and chemical energy is called* ***electrochemistry.*** Electrochemical processes are ***oxidation-reduction reactions*** (or ***redox reactions***) in which the energy released by a spontaneous reaction is converted to electricity or in which electric energy is used to cause a nonspontaneous reaction to occur. In redox reactions, electrons are transferred from one substance to another. The reaction between magnesium metal and hydrochloric acid is an example of a redox reaction:

$$Mg(s) + 2HCl(aq) \longrightarrow MgCl_2 + H_2(g)$$

In this reaction, the Mg metal loses two electrons, and the two H^+ ions in the HCl solution gain one electron each. Species that lose electrons in a redox reaction are said to be ***oxidized,*** and those that gain electrons are said to be ***reduced.*** In the reaction of magnesium metal with hydrochloric acid, for example, Mg metal is oxidized and H^+ ions are reduced. The Cl^- ions are *spectator ions* because they do not participate directly in the redox reaction.

Oxidation-reduction reactions are very much a part of the world around us. They range from the burning of fossil fuels to the action of household bleach. Additionally, most metallic and nonmetallic elements are obtained from their ores by the process of oxidation or reduction. Many important redox reactions take place in water (rusting, for example), but not all redox reactions occur in aqueous solution. Some, such as the formation of calcium oxide (CaO) from calcium and oxygen, occur when a solid interacts with a gas:

$$2Ca(s) + O_2(g) \longrightarrow 2CaO(s)$$

CaO is an ionic compound made up of Ca^{2+} and O^{2-} ions. In this reaction, two Ca atoms give up or transfer four electrons to two O atoms (in O_2). For convenience, we can think of this process as two separate steps, one involving the loss of four electrons by the two Ca atoms and the other being the gain of four electrons by an O_2 molecule:

$$2Ca \longrightarrow 2Ca^{2+} + 4e^-$$
$$O_2 + 4e^- \longrightarrow 2O^{2-}$$

In an oxidation half-reaction, electrons appear as product; in a reduction half-reaction, electrons appear as a reactant.

Each of these steps is called a ***half-reaction,*** and each half-reaction *explicitly shows the electrons involved in the oxidation portion or the reduction portion of a redox reaction.* The sum of the oxidation half-reaction and the reduction half-reaction gives the overall redox reaction.

An ***oxidation reaction*** is the *half-reaction that involves a loss of electrons.* Chemists originally used *oxidation* to denote the combination of elements with oxygen. Now, however, "oxidation" has a broader meaning that includes reactions not involving oxygen. A ***reduction reaction*** is a *half-reaction that involves a gain of electrons.* In the formation of calcium oxide, calcium is oxidized. It acts as a ***reducing agent*** because it *donates electrons* to oxygen and causes oxygen to be reduced. Oxygen is reduced and acts as an ***oxidizing agent*** because it *accepts electrons* from calcium, causing calcium to be oxidized. An oxidation reaction cannot occur without a concomitant reduction reaction, and vice versa. The number of electrons lost by a reducing agent must equal the number of electrons gained by an oxidizing agent.

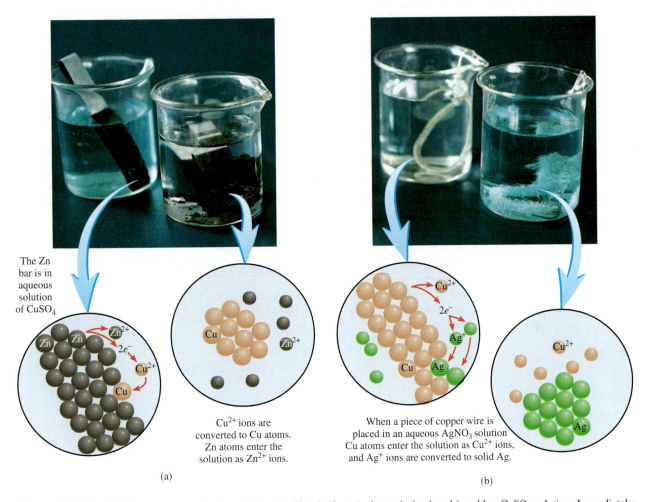

The Zn bar is in aqueous solution of CuSO$_4$

Cu^{2+} ions are converted to Cu atoms. Zn atoms enter the solution as Zn^{2+} ions.

(a)

When a piece of copper wire is placed in an aqueous AgNO$_3$ solution Cu atoms enter the solution as Cu^{2+} ions, and Ag^+ ions are converted to solid Ag.

(b)

Figure 13.1 Metal displacement reactions in solution. (a) First beaker: A zinc strip is placed in a blue CuSO$_4$ solution. Immediately Cu^{2+} ions are reduced to metallic Cu in the form of a dark layer. Second beaker: In time, most of the Cu^{2+} ions are reduced and the solution becomes colorless. (b) First beaker: A piece of Cu wire is placed in a colorless AgNO$_3$ solution. Ag^+ ions are reduced to metallic Ag. Second beaker: As time progresses, most of the Ag^+ ions are reduced, and the solution acquires the characteristic blue color of hydrated Cu^{2+} ions.

When metallic zinc is added to a solution containing copper(II) sulfate (CuSO$_4$), zinc donates two electrons, thus forming Zn^{2+} and reducing Cu^{2+} to copper metal:

$$Zn(s) + Cu^{2+}(aq) \longrightarrow Zn^{2+}(aq) + Cu(s)$$

In the process, the solution loses the blue color that is characteristic of hydrated Cu^{2+} ions and copper metal precipitates from the solution (Figure 13.1).

The oxidation and reduction half-reactions are

$$\text{oxidation:} \qquad Zn(s) \longrightarrow Zn^{2+}(aq) + 2e^-$$
$$\text{reduction:} \quad Cu^{2+}(aq) + 2e^- \longrightarrow Cu(s)$$

Similarly, metallic copper reduces silver ions in a solution of silver nitrate (AgNO$_3$):

$$Cu(s) + 2AgNO_3(aq) \longrightarrow Cu(NO_3)_2(aq) + 2Ag(s)$$

or

$$Cu(s) + 2Ag^+(aq) \longrightarrow Cu^{2+}(aq) + 2Ag(s)$$

The reactions shown in Figure 13.1 are examples of *metal displacement reactions*.

Oxidation Number

Defining oxidation and reduction in terms of the loss and gain of electrons, respectively, works well for the formation of ionic compounds such as CaO or the reduction of Cu^{2+} ions by Zn. These definitions, however, do not accurately characterize the formation of hydrogen chloride (HCl) and sulfur dioxide (SO_2):

$$H_2(g) + Cl_2(g) \longrightarrow 2HCl(g)$$
$$S(s) + O_2(g) \longrightarrow SO_2(g)$$

Because HCl and SO_2 are molecular compounds (not ionic), no electrons are actually transferred in the formation of these compounds. Nevertheless, chemists find it convenient to treat these reactions as redox reactions because experimental measurements show that there is a partial transfer of electrons (from H to Cl in HCl and from S to O in SO_2).

To keep track of electrons in redox reactions, it is useful to assign *oxidation numbers* to the reactants and products. The **oxidation number** of an atom, also called its **oxidation state,** signifies the *number of charges the atom would have in a molecule (or an ionic compound) if electrons were transferred completely.* For example, the oxidation numbers for the species involved in the formation of HCl and SO_2 (shown above the element symbols) are as follows:

$$\overset{0}{H_2}(g) + \overset{0}{Cl_2}(g) \longrightarrow 2\overset{+1\ -1}{HCl}(g)$$

$$\overset{0}{S}(s) + \overset{0}{O_2}(g) \longrightarrow \overset{+4\ -2}{SO_2}(g)$$

In both of these reactions, the oxidation numbers for the atoms in the reactants are zero because the reactant molecules are pure elements. For the product molecules, however, it is assumed that complete electron transfer has taken place and that atoms have gained or lost electrons.

The oxidation numbers reflect the number of electrons "transferred" and enable us to identify elements that are oxidized and reduced at a glance. The elements that show an increase in oxidation number—hydrogen and sulfur in the preceding examples—are oxidized. The elements that show a decrease in oxidation number—chlorine and oxygen—are reduced. Note that the sum of the oxidation numbers of H and Cl in HCl ($+1$ and -1) is zero. Likewise, if we add the charges on S ($+4$) and two atoms of O [$2 \times (-2)$], the total is zero. The totals are zero because the HCl and SO_2 molecules are neutral, so the charges must cancel.

Use the following rules to assign oxidation numbers:

1. In free elements (that is, elements in the uncombined state), each atom has an oxidation number of zero. Thus, each atom in H_2, Br_2, Na, Be, K, O_2, and P_4 has the same oxidation number: zero.

2. For ions composed of only one atom (that is, monatomic ions), the oxidation number is equal to the charge on the ion. Thus, the Li^+ ion has an oxidation number of $+1$; Ba^{2+} ion, $+2$; Fe^{3+} ion, $+3$; I^- ion, -1; O^{2-} ion, -2; and so on. All alkali metals have an oxidation number of $+1$, and all alkaline earth metals have an oxidation number of $+2$ in their compounds. Aluminum has an oxidation number of $+3$ in all its compounds.

3. The oxidation number of oxygen in most compounds (for example, MgO and H_2O) is -2. *Exceptions:* The oxidation number of oxygen in hydrogen peroxide

(H_2O_2) and peroxide ion (O_2^{2-}) is -1; in the superoxide anion (O_2^-), it is $-1/2$ (thus, oxidation numbers are generally, but not necessarily, integers).

4. The oxidation number of hydrogen is $+1$. *Exceptions:* When hydrogen is bonded to metals in binary compounds (for example, LiH, NaH, and CaH_2), its oxidation number is -1.

5. The oxidation number of halogens (F, Cl, Br, and I) is -1. *Exceptions:* When Cl, Br, and I are bonded to other halogens, only the most electronegative halogen in the bond has an oxidation number of -1. When combined with oxygen—for example, in oxoacids and oxoanions (see Section 0.3)—Cl, Br, and I have positive oxidation numbers because oxygen is more electronegative.

6. The sum of the oxidation numbers of all the elements in a molecule or polyatomic ion must equal the net charge on the molecule or ion. In a neutral molecule, for example, such as NH_3, the sum of the oxidation numbers of all the atoms must be zero. The oxidation number of N is -3 and of H is $+1$, so the sum of the oxidation numbers is $-3 + 3(+1) = 0$. For the polyatomic ammonium ion, NH_4^+, the oxidation numbers of N and H are still -3 and $+1$, respectively. The sum of the oxidation numbers, however, is $-3 + 4(+1) = +1$, which is equal to the net charge on the ion.

Example 13.1 shows how to apply these rules to assign oxidation numbers to neutral and charged species.

Example 13.1

Assign oxidation numbers to all the elements in the following compounds and ion: (a) Li_2O, (b) HNO_3, (c) $Cr_2O_7^{2-}$, and (d) ClF_3.

Strategy Use the rules for assigning oxidation numbers. All alkali metals have an oxidation number of $+1$, and in most cases hydrogen has an oxidation number of $+1$ and oxygen has an oxidation number of -2 in their compounds. Fluorine, the most electronegative element, has an oxidation number of -1 (except in F_2, where its oxidation number is 0).

Solution (a) According to rule 2, lithium has an oxidation number of $+1$ (Li^+) and oxygen has an oxidation number of -2 (O^{2-}).

(b) This is the formula for nitric acid, which yields an H^+ ion and an NO_3^- ion in solution. According to rule 4, H^+ has an oxidation number of $+1$, and according to rule 6, the nitrate ion must have a net oxidation number of -1. Oxygen has an oxidation number of -2 (rule 3), and if we use x to represent the oxidation number of nitrogen, then $x + 3(-2) = -1$ or $x = 5$.

(c) According to rule 6, the sum of the oxidation numbers in the dichromate ion $Cr_2O_7^{2-}$ must be -2. The oxidation number of O is -2 (rule 3), so we can determine the oxidation number of Cr by setting it equal to y: $2(y) + 7(-2) = -2$ or $y = 6$.

(d) In ClF_3 the three F atoms are bonded to the less electronegative Cl atom, which has an expanded octet. According to rule 5, each F has an oxidation number of -1. Because the overall molecule is neutral, the Cl atom must have an oxidation number of $+3$ so that the oxidation numbers sum to zero $[+3 + 3(-1) = 0]$.

Check In each case, does the sum of the oxidation numbers of all the atoms equal the net charge on the species?

Practice Exercise Assign oxidation numbers to all the elements in the following compound and ion: (a) PF_3, (b) MnO_4^-, (c) ClO_3^-, and (d) SF_6.

Periodic table of oxidation numbers (Figure 13.2)

Group 1 (1A)	2 (2A)	3 (3B)	4 (4B)	5 (5B)	6 (6B)	7 (7B)	8	9 (8B)	10	11 (1B)	12 (2B)	13 (3A)	14 (4A)	15 (5A)	16 (6A)	17 (7A)	18 (8A)
1 H (+1, −1)																	2 He
3 Li (+1)	4 Be (+2)											5 B (+3)	6 C (+4, +2, −4)	7 N (+5, +4, +3, +2, +1, −3)	8 O (+2, −$\frac{1}{2}$, −1, −2)	9 F (−1)	10 Ne
11 Na (+1)	12 Mg (+2)											13 Al (+3)	14 Si (+4, −4)	15 P (+5, +3, −3)	16 S (+6, +4, +2, −2)	17 Cl (+7, +6, +5, +4, +3, +1, −1)	18 Ar
19 K (+1)	20 Ca (+2)	21 Sc (+3)	22 Ti (+4, +3, +2)	23 V (+5, +4, +3, +2)	24 Cr (+6, +5, +4, +3, +2)	25 Mn (+7, +6, +4, +3, +2)	26 Fe (+3, +2)	27 Co (+3, +2)	28 Ni (+2)	29 Cu (+2, +1)	30 Zn (+2)	31 Ga (+3)	32 Ge (+4, −4)	33 As (+5, +3, −3)	34 Se (+6, +4, −2)	35 Br (+5, +3, +1, −1)	36 Kr (+4, +2)
37 Rb (+1)	38 Sr (+2)	39 Y (+3)	40 Zr (+4)	41 Nb (+5, +4)	42 Mo (+6, +4, +3)	43 Tc (+7, +6, +4)	44 Ru (+8, +6, +4, +3)	45 Rh (+4, +3, +2)	46 Pd (+4, +2)	47 Ag (+1)	48 Cd (+2)	49 In (+3, +1)	50 Sn (+4, +2)	51 Sb (+5, +3, −3)	52 Te (+6, +4, −2)	53 I (+7, +5, +1, −1)	54 Xe (+6, +4, +2)
55 Cs (+1)	56 Ba (+2)	57 La (+3)	72 Hf (+4)	73 Ta (+5)	74 W (+6, +4)	75 Re (+7, +6, +4)	76 Os (+8, +4)	77 Ir (+4, +3)	78 Pt (+4, +2)	79 Au (+3, +1)	80 Hg (+2, +1)	81 Tl (+3, +1)	82 Pb (+4, +2)	83 Bi (+5, +3)	84 Po (+2)	85 At (−1)	86 Rn

Figure 13.2 The oxidation numbers of elements in their compounds. The more common oxidation numbers are in red.

Figure 13.2 shows the known oxidation numbers of the familiar elements, arranged according to their positions in the periodic table. We can summarize these data as follows:

▸ Metallic elements have only positive oxidation numbers, whereas nonmetallic elements may have either positive or negative oxidation numbers.

▸ The highest oxidation number an element in groups 1A−7A can have is its group number. For example, the halogens are in group 7A, so their highest possible oxidation number is +7.

▸ The transition metals (groups 1B and 3B−8B) usually have several possible oxidation numbers.

Balancing Redox Equations

Equations for redox reactions, such as the oxidation of magnesium metal by chlorine gas discussed at the beginning of this section, are relatively straightforward to balance. In the laboratory, however, we often encounter more complex redox reactions involving oxoanions such as chromate (CrO_4^{2-}), dichromate ($Cr_2O_7^{2-}$), permanganate (MnO_4^{-}), nitrate (NO_3^-), and sulfate (SO_4^{2-}). In principle, we can balance any redox equation using the procedure outlined in Section 0.4, but there are some special techniques for handling redox reactions, techniques that also give us insight into electron transfer processes. We will discuss one such procedure here, called the *ion-electron method*. In this approach, the overall reaction is divided into two half-reactions, one for oxidation and one for reduction. The equations for the two half-reactions are balanced separately and then added together to give the overall balanced equation.

Suppose, for example, that we are asked to balance the equation for the oxidation of Fe^{2+} ions to Fe^{3+} ions by dichromate ions ($Cr_2O_7^{2-}$) in an acidic medium. In this reaction, the $Cr_2O_7^{2-}$ ions are reduced to Cr^{3+} ions. The following steps will help us balance the equation by the ion-electron method.

Step 1: *Write the unbalanced equation for the reaction in ionic form.*

$$Fe^{2+} + Cr_2O_7^{2-} \longrightarrow Fe^{3+} + Cr^{3+}$$

Step 2: *Separate the equation into two half-reactions.*

$$\text{oxidation:} \quad \overset{+2}{Fe^{2+}} \longrightarrow \overset{+3}{Fe^{3+}}$$

$$\text{reduction:} \quad \overset{+6}{Cr_2O_7^{2-}} \longrightarrow \overset{+3}{Cr^{3+}}$$

Step 3: *Balance each half-reaction for number and type of atoms and charges. For reactions in an acidic medium, add H_2O to balance the O atoms and H^+ to balance the H atoms.*

Oxidation half-reaction: The atoms are already balanced. To balance the charge, we add an electron to the right-hand side of the arrow:

$$Fe^{2+} \longrightarrow Fe^{3+} + e^-$$

Reduction half-reaction: Because the reaction takes place in an acidic medium, we add seven H_2O molecules to the right-hand side of the arrow to balance the O atoms:

$$Cr_2O_7^{2-} \longrightarrow 2Cr^{3+} + 7H_2O$$

To balance the H atoms, we add 14 H^+ ions to the left-hand side:

$$14H^+ + Cr_2O_7^{2-} \longrightarrow 2Cr^{3+} + 7H_2O$$

There are now 12 positive charges on the left-hand side and only 6 positive charges on the right-hand side. Therefore, we add six electrons on the left:

$$14H^+ + Cr_2O_7^{2-} + 6e^- \longrightarrow 2Cr^{3+} + 7H_2O$$

Step 4: *Add the two half-reactions together, and balance the final equation by inspection. The electrons on both sides must cancel. If the oxidation and reduction half-reactions contain different numbers of electrons, we need to multiply one or both half-reactions by a coefficient to equalize the number of electrons.*

Here we have only one electron for the oxidation half-reaction and six electrons for the reduction half-reaction, so we need to multiply the oxidation half-reaction by 6:

$$6(Fe^{2+} \longrightarrow Fe^{3+} + e^-)$$

$$\underline{14H^+ + Cr_2O_7^{2-} + 6e^- \longrightarrow 2Cr^{3+} + 7H_2O}$$

$$6Fe^{2+} + 14H^+ + Cr_2O_7^{2-} + 6\cancel{e^-} \longrightarrow 6Fe^{3+} + 2Cr^{3+} + 7H_2O + 6\cancel{e^-}$$

The electrons on both sides cancel, and we are left with the balanced net ionic equation:

$$6Fe^{2+} + 14H^+ + Cr_2O_7^{2-} \longrightarrow 6Fe^{3+} + 2Cr^{3+} + 7H_2O$$

Step 5: *Verify that the equation contains the same type and numbers of atoms and the same total charge on both sides of the equation.*

There are 6 Fe, 14 H, 2 Cr, and 7 O on both sides of the equation, and the charge on both sides is +24. The resulting equation, therefore, is "atomically" and "electrically" balanced.

For reactions in a basic medium, we proceed through step 4 as if the reaction were carried out in an acidic medium. Then, for every H^+ ion we add an equal number of OH^- ions to *both* sides of the equation. Where H^+ and OH^- ions appear on the same side of the equation, we combine the ions to give H_2O. Example 13.2 illustrates this procedure.

Example 13.2

Write a balanced equation for the oxidation of iodide ion (I^-) by permanganate ion (MnO_4^-) in basic solution to yield molecular iodine (I_2) and manganese(IV) oxide (MnO_2).

Strategy Follow the procedure for balancing redox equations by the ion-electron method. The reaction takes place in a basic medium, so any H^+ ions that appear in the two half-reactions must be neutralized by adding an equal number of OH^- ions to both sides of the equation.

Solution *Step 1:* The unbalanced equation is

$$MnO_4^- + I^- \longrightarrow MnO_2 + I_2$$

Step 2: The two half-reactions are

$$\text{oxidation:} \quad \overset{-1}{I^-} \longrightarrow \overset{0}{I_2}$$

$$\text{reduction:} \quad \overset{+7}{MnO_4^-} \longrightarrow \overset{+4}{MnO_2}$$

Step 3: Balance each half-reaction for the number and type of atoms and charges. For the oxidation half-reaction, we first balance the number of iodine atoms:

$$2I^- \longrightarrow I_2$$

To balance charges, add two electrons to the right-hand side of the equation:

$$2I^- \longrightarrow I_2 + 2e^-$$

For the reduction half-reaction, we balance the O atoms by adding two H_2O molecules on the right-hand side:

$$MnO_4^- \longrightarrow MnO_2 + 2H_2O$$

To balance the H atoms, add four H^+ ions on the left-hand side:

$$MnO_4^- + 4H^+ \longrightarrow MnO_2 + 2H_2O$$

—Continued

Continued—

There are three net positive charges on the left, so add three electrons to the same side to balance the charges:

$$MnO_4^- + 4H^+ + 3e^- \longrightarrow MnO_2 + 2H_2O$$

Step 4: Add the oxidation and reduction half-reactions to give the overall reaction. In order to equalize the number of electrons, multiply the oxidation half-reaction by 3 and the reduction half-reaction by 2:

$$3(2I^- \longrightarrow I_2 + 2e^-)$$
$$\underline{2(MnO_4^- + 4H^+ + 3e^- \longrightarrow MnO_2 + 2H_2O)}$$
$$6I^- + 2\,MnO_4^- + 8H^+ + 6e^- \longrightarrow 3I_2 + 2MnO_2 + 4H_2O + 6e^-$$

The electrons on both sides cancel, and we are left with the balanced net ionic equation:

$$6I^- + 2\,MnO_4^- + 8H^+ \longrightarrow 3I_2 + 2MnO_2 + 4H_2O$$

This is the balanced equation in an *acidic* medium, but the reaction is carried out in a *basic* medium. Thus, for every H^+ ion, we need to add an equal number of OH^- ions to *both* sides of the equation:

$$6I^- + 2\,MnO_4^- + 8H^+ + 8OH^- \longrightarrow 3I_2 + 2MnO_2 + 4H_2O + 8OH^-$$

Finally, combining the H^+ and OH^- ions to form water, we obtain

$$6I^- + 2MnO_4^- + 4H_2O \longrightarrow 3I_2 + 2MnO_2 + 8OH^-$$

Step 5: A final check shows that the equation is balanced in terms of both atoms and charges. There are 6 I, 2 Mn, 12 O, and 8 H atoms on both sides of the equation, and the charge on both sides is −8.

Practice Exercise Use the ion-electron method to balance the following equation for the reaction in an acidic medium:

$$Fe^{2+} + MnO_4^- \longrightarrow Fe^{3+} + Mn^{2+}$$

13.2 | Redox Reactions Can Be Used to Generate Electric Current in a Galvanic Cell

We saw in Section 13.1 that when a piece of zinc metal is placed in a $CuSO_4$ solution, Zn is oxidized to Zn^{2+} ions whereas Cu^{2+} ions are reduced to metallic copper (see Figure 13.1):

$$Zn(s) + Cu^{2+}(aq) \longrightarrow Zn^{2+}(aq) + Cu(s)$$

The electrons are transferred directly from the reducing agent (Zn) to the oxidizing agent (Cu^{2+}) in solution. However, if we physically separate the oxidizing agent from the reducing agent, the transfer of electrons can take place via an external conducting medium (a metal wire). As the reaction progresses, it sets up a constant flow of electrons and hence generates an electric current. The electric work produced by the current can be used to drive an electric motor or to operate a lightbulb.

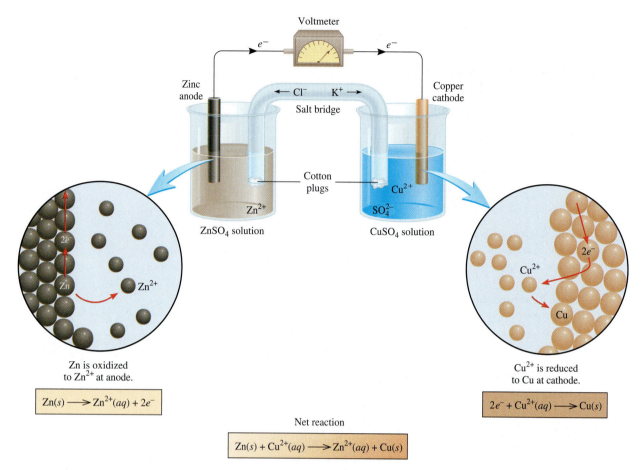

Figure 13.3 A galvanic cell. The salt bridge (an inverted U-shaped tube) containing a KCl solution provides an electrically conducting medium between the two solutions. The openings of the U-shaped tube are loosely plugged with cotton balls to prevent the KCl solution from flowing into the containers while allowing the anions and cations to move across. Electrons flow externally from the Zn electrode (anode) to the Cu electrode (cathode).

The experimental apparatus for generating electric current through the use of a spontaneous reaction is called a **galvanic cell** or *voltaic cell,* after the Italian scientists Luigi Galvani[1] and Alessandro Volta,[2] who constructed early versions of the device. Figure 13.3 shows the essential components of a galvanic cell. A zinc bar is immersed in a $ZnSO_4$ solution, and a copper bar is immersed in a $CuSO_4$ solution. The cell operates on the principle that the oxidation of Zn to Zn^{2+} and the reduction of Cu^{2+} to Cu can be made to take place simultaneously in separate locations with the transfer of electrons between them occurring through an external wire. The zinc and copper bars are called **electrodes.** This particular arrangement of electrodes (Zn and Cu) and solutions ($ZnSO_4$ and $CuSO_4$) is called the Daniell cell

---

1. Luigi Galvani (1737–1798). Italian biologist and physician. Galvani discovered that external electric impulses (sparks) applied to the leg muscles of a frog could make the muscles twitch (contract rapidly) even when the legs were detached from the body. His discoveries formed the basis for the science of neurology. Galvani's work provided much of the background and motivation for Volta's invention of the electric pile (battery).

2. Count Alessandro Giuseppe Antonio Anastasio Volta (1745–1827). Italian physicist. Volta showed that animal tissue was not necessary for the conduction of electricity, as had been hypothesized by Galvani. He invented the first electric (or voltaic) pile—a prototype battery consisting of alternating discs of dissimilar metals (for example, copper and zinc) separated by cardboard soaked in saltwater.

after the English chemist John Frederic Daniell.[3] By definition, the **anode** in a galvanic cell is *the electrode at which oxidation occurs* and the **cathode** is *the electrode at which reduction occurs.*

For the Daniell cell, the **half-cell reactions** (that is, *the oxidation and reduction reactions at the electrodes*) are

Zn electrode (anode): $\qquad\qquad$ $Zn(s) \longrightarrow Zn^{2+}(aq) + 2e^-$
Cu electrode (cathode): $\quad Cu^{2+}(aq) + 2e^- \longrightarrow Cu(s)$

Unless the two solutions are separated from each other, the Cu^{2+} ions will react directly with the zinc bar:

$$Cu^{2+}(aq) + Zn(s) \longrightarrow Cu(s) + Zn^{2+}(aq)$$

and no useful electric work will be obtained.

To complete the electric circuit, the solutions must be connected by a conducting medium through which cations and anions can move from one electrode compartment to the other. This requirement is satisfied by a *salt bridge,* which, in its simplest form, is an inverted U-shaped tube containing an inert electrolyte solution, such as KCl or NH_4NO_3, whose ions will not react with other ions in solution or with the electrodes (see Figure 13.3). During the course of the overall redox reaction, electrons flow externally from the anode (Zn electrode) through the wire and voltmeter to the cathode (Cu electrode). In the solution, the cations (Zn^{2+}, Cu^{2+}, and K^+) move toward the cathode, while the anions (SO_4^{2-} and Cl^-) move toward the anode. Without the salt bridge connecting the two solutions, the buildup of positive charge in the anode compartment (due to the formation of Zn^{2+} ions) and the buildup of negative charge in the cathode compartment (created when the Cu^{2+} ions are reduced to Cu) would quickly prevent the cell from operating.

An electric current flows from the anode to the cathode because there is a difference in electric potential energy between the electrodes. This flow of electric current is analogous to that of water over a waterfall, which occurs because there is a difference in gravitational potential energy between the top and bottom, or the flow of gas from a high-pressure region to a low-pressure region. Experimentally the *difference in electric potential between the anode and the cathode* is measured by a voltmeter (Figure 13.4) and the reading (in volts) is called **cell voltage** or *cell potential.* In the limit that the cell is operated reversibly,[4] the cell voltage is referred to as the **electromotive force** or **emf** (E). We will see that the voltage of a cell depends not only on the nature of the electrodes and the ions, but also on the concentrations of the ions and the temperature at which the cell is operated.

The conventional notation for representing galvanic cells is the **cell diagram.** The cell diagram for the Daniell cell shown in Figure 13.3, assuming that the concentrations of Zn^{2+} and Cu^{2+} ions are in their standard states (1 *M*, if the solutions can be approximated as ideal), is as follows:

$$Zn(s)|Zn^{2+}(1\ M)\|Cu^{2+}(1\ M)|Cu(s)$$

A useful pneumonic device to remember that oxidation occurs at the anode and reduction occurs at the cathode is to note that both *anode* and *oxidation* begin with vowels and both *cathode* and *reduction* begin with consonants.

Half-cell reactions are similar to the half-reactions discussed earlier.

---

3. John Frederic Daniell (1790–1845). English chemist and meteorologist. In addition to invention of the Daniell cell, an improvement over Volta's electric pile, Daniell made contributions to instrumental meteorology through his invention of a dew-point hygrometer that became a standard instrument for measuring relative humidity.

4. A cell is operated reversibly when a sufficiently large countervoltage is applied to the cell to make the net current go to zero. The electromotive force is then equal to the negative of this counter voltage. Because of the second law of thermodynamics, the cell voltage of a real cell will be greater than the electromotive force.

Figure 13.4 Practical setup of the galvanic cell described in Figure 13.3. Note the U-shaped tube (salt bridge) connecting the two beakers. When the aqueous solutions of $ZnSO_4$ and $CuSO_4$ are in their standard states at 25°C, the cell voltage is 1.10 V.

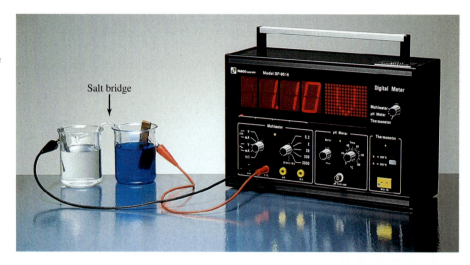

The single vertical line represents a phase boundary. The zinc electrode, for example, is a solid and the Zn^{2+} ions (from $ZnSO_4$) are in solution. Thus, we draw a line between Zn and Zn^{2+} to show the phase boundary. The double vertical lines denote the salt bridge. By convention, the anode is written first, to the left of the double lines, and the other components appear in the order in which we would encounter them in moving from the anode to the cathode.

13.3 | The Standard Emf of Any Electrochemical Cell Can Be Determined If the Standard Reduction Potentials for the Half-Reactions Are Known

Recall that the *emf* is the cell voltage when the cell is operated reversibly.

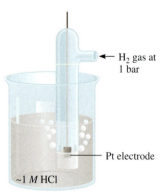

Figure 13.5 A hydrogen electrode operating under standard-state conditions. Hydrogen gas at 1 bar is bubbled through a HCl solution in which the H^+ activity is 1 (approximately 1 *M*). The platinum electrode is part of the hydrogen electrode.

When the Cu^{2+} and Zn^{2+} ions are in their standard states (that is, both have an activity of 1 in the solution), we find that the emf of a Daniell cell is 1.10 V at 25°C (see Figure 13.4). This emf must be related directly to the redox reactions, but how? Just as the overall cell reaction can be thought of as the sum of two half-cell reactions, the measured emf of the cell can be treated as the sum of the electric potentials at the Zn and Cu electrodes. Knowing one of these electrode potentials, we could obtain the other by subtraction (from 1.10 V). It is impossible to measure the potential of just a single electrode, but if we arbitrarily set the potential value of a particular electrode at zero, we can use it to determine the relative potentials of other electrodes. The hydrogen electrode, shown in Figure 13.5, serves as the reference for this purpose.

Hydrogen gas is bubbled into a hydrochloric acid solution at 25°C. The platinum electrode has two functions. First, it provides a surface on which the dissociation of hydrogen molecules can take place:

$$H_2(g) \longrightarrow 2H^+(aq) + 2e^-$$

Second, it serves as an electric conductor to the external circuit. Under standard-state conditions [when the pressure of $H_2(g)$ is 1 bar and the activity of H^+ is unity in the solution], the potential for the reduction of H^+ at 25°C is taken to be *exactly* zero:

$$2H^+(a_{H^+} = 1) + 2e^- \longrightarrow H_2(P = 1 \text{ bar}) \qquad \mathcal{E}° = 0 \text{ V}$$

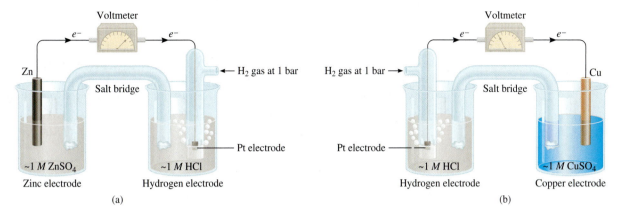

Figure 13.6 (a) A cell consisting of a zinc electrode and a hydrogen electrode. (b) A cell consisting of a copper electrode and a hydrogen electrode. Both cells are operating under standard-state conditions [unit activities (~1 M) for ions, $P \equiv 1$ bar for gases]. Note that in part (a) the standard hydrogen electrode acts as the cathode, but in part (b), it is the anode.

The superscript "°" denotes standard-state conditions, and $\mathcal{E}°$ is the **standard reduction potential,** or *the electromotive force associated with a reduction reaction at an electrode when reactant and product species are in their standard states.* Thus, the standard reduction potential of the hydrogen electrode is defined as zero. The hydrogen electrode is called the *standard hydrogen electrode (SHE).*

We can use the standard hydrogen electrode to measure the potentials of other kinds of electrodes. For example, Figure 13.6(a) shows a galvanic cell with a zinc electrode and a standard hydrogen electrode. In this case the zinc electrode is the anode and the standard hydrogen electrode is the cathode. We deduce this fact from the decrease in mass of the zinc electrode during the operation of the cell, which is consistent with the loss of zinc to the solution caused by the oxidation reaction:

$$Zn(s) \longrightarrow Zn^{2+}(aq) + 2e^-$$

The cell diagram is

$$Zn(s)|Zn^{2+}(a_{Zn^{2+}} = 1)\|H^+(a_{H^+} = 1)|H_2(1 \text{ bar})|Pt(s)$$

As mentioned earlier, the Pt electrode provides the surface on which the reduction takes place. When all the reactants are in their standard states (i.e., H_2 at 1 bar, H^+ and Zn^{2+} ions at unit activity), the emf of the cell is measured to be 0.76 V at 25°C. We can write the half-cell reactions as follows:

anode (oxidation): $\qquad\qquad\qquad Zn(s) \longrightarrow Zn^{2+}(a_{Zn^{2+}} = 1) + 2e^-$

cathode (reduction): $\quad \underline{2H^+(a_{H^+} = 1) + 2e^- \longrightarrow H_2(1 \text{ bar})}$

overall: $\quad Zn(s) + 2H(a_{H^+} = 1) \longrightarrow Zn^{2+}(a_{Zn^{2+}} = 1) + H_2(1 \text{ bar})$

By convention, the **standard emf** of the cell, $\mathcal{E}°_{cell}$, which is composed of a contribution from the anode and a contribution from the cathode, is given by

$$\mathcal{E}°_{cell} = \mathcal{E}°_{cathode} - \mathcal{E}°_{anode} \qquad\qquad (13.1)$$

where *both* $\mathcal{E}°_{cathode}$ and $\mathcal{E}°_{anode}$ are the standard reduction potentials of the electrodes. For the Zn-SHE cell, we write

$$\mathcal{E}°_{cell} = \mathcal{E}°_{H^+|H_2} - \mathcal{E}°_{Zn^{2+}|Zn} = 0 \text{ V} - 0.76 \text{ V} = -0.76 \text{ V}$$

where the subscript $H^+|H_2$ denote $2H^+ + 2e^- \longrightarrow H_2$ and the subscript $Zn^{2+}|Zn$ denotes $Zn^{2+} + 2e^- \longrightarrow Zn$. Thus, the standard reduction potential of zinc ($\mathcal{E}^{\circ}_{Zn^{2+}|Zn}$) is -0.76 V. The standard electrode potential of copper can be obtained in a similar fashion, by using a cell with a copper electrode and a standard hydrogen electrode [Figure 13.6(b)]. In this case, the copper electrode is the cathode because its mass increases during the operation of the cell, as is consistent with the reduction reaction:

$$Cu^{2+}(aq) + 2e^- \longrightarrow Cu(s)$$

The cell diagram is

$$Pt(s)|H_2(1\ bar)|H^+(a_{H^+} = 1)\|Cu^{2+}(a_{Cu^{2+}} = 1)|Cu(s)$$

and the half-cell reactions are

anode (oxidation): $\qquad\qquad\qquad H_2(1\ bar) \longrightarrow 2H^+(a_{H^+} = 1) + 2e^-$

cathode (reduction): $\qquad Cu^{2+}(a_{Cu^{2+}} = 1) + 2e^- \longrightarrow Cu(s)$

overall: $\quad H_2(1\ bar) + Cu^{2+}(a_{Cu^{2+}} = 1) \longrightarrow 2H^+(a_{H^+} = 1) + Cu(s)$

Under standard-state conditions and at 25°C, the emf of the cell is 0.34 V, so we write

$$\mathcal{E}^{\circ}_{cell} = \mathcal{E}^{\circ}_{cathode} - \mathcal{E}^{\circ}_{anode}$$
$$0.34\ V = \mathcal{E}^{\circ}_{Cu^{2+}|Cu} - \mathcal{E}^{\circ}_{H^+|H_2}$$
$$= \mathcal{E}^{\circ}_{Cu^{2+}|Cu} - 0\ V$$

In this case, the standard reduction potential of copper, $\mathcal{E}^{\circ}_{Cu^{2+}|Cu}$, is $+0.34$ V, where the subscript denotes $Cu^{2+} + 2e^- \longrightarrow Cu$.

For the Daniell cell shown in Figure 13.3, we can now write

anode (oxidation): $\qquad\qquad\qquad Zn(s) \longrightarrow Zn^{2+}(a_{Zn^{2+}} = 1) + 2e^-$

cathode (reduction): $\quad Cu^{2+}(a_{Cu^{2+}} = 1) + 2e^- \longrightarrow Cu(s)$

overall: $\quad Zn(s) + Cu^{2+}(a_{Cu^{2+}} = 1) \longrightarrow Zn^{2+}(a_{Zn^{2+}} = 1) + Cu(s)$

The emf of the cell is

$$\mathcal{E}^{\circ}_{cell} = \mathcal{E}^{\circ}_{cathode} - \mathcal{E}^{\circ}_{anode}$$
$$= \mathcal{E}^{\circ}_{Cu^{2+}|Cu} - \mathcal{E}^{\circ}_{Zn^{2+}|Zn}$$
$$= +0.34\ V - (-0.76\ V)$$
$$= +1.10\ V$$

As in the case of ΔG° (page 533), we can use the sign of $\mathcal{E}^{\circ}_{cell}$ to predict the extent of a redox reaction. A positive $\mathcal{E}^{\circ}_{cell}$ means the redox reaction will favor the formation of products at equilibrium. Conversely, a negative $\mathcal{E}^{\circ}_{cell}$ means that more reactants than products will be formed at equilibrium. We will examine the relationships among $\mathcal{E}^{\circ}_{cell}$, ΔG°, and K later in Section 13.4.

Table 13.1 lists standard reduction potentials for a number of half-cell reactions. By definition, the SHE has an $\mathcal{E}^{\circ}$ value of 0.00 V. Below the SHE, the negative standard reduction potentials increase, and above it, the positive standard reduction potentials increase.

Table 13.1	Standard Reduction Potentials at 25°C*	

Half-Reaction		$E°$ **(V)**
$F_2(g) + 2e^- \longrightarrow 2F^-(aq)$		+2.87
$O_3(g) + 2H^+(aq) + 2e^- \longrightarrow O_2(g) + H_2O$		+2.07
$Co^{3+}(aq) + e^- \longrightarrow Co^{2+}(aq)$		+1.82
$H_2O_2(aq) + 2H^+(aq) + 2e^- \longrightarrow 2H_2O$		+1.77
$PbO_2(s) + 4H^+(aq) + SO_4^{2-}(aq) + 2e^- \longrightarrow PbSO_4(s) + 2H_2O$		+1.70
$Ce^{4+}(aq) + e^- \longrightarrow Ce^{3+}(aq)$		+1.61
$MnO_4^-(aq) + 8H^+(aq) + 5e^- \longrightarrow Mn^{2+}(aq) + 4H_2O$		+1.51
$Au^{3+}(aq) + 3e^- \longrightarrow Au(s)$		+1.50
$Cl_2(g) + 2e^- \longrightarrow 2Cl^-(aq)$		+1.36
$Cr_2O_7^{2-}(aq) + 14H^+(aq) + 6e^- \longrightarrow 2Cr^{3+}(aq) + 7H_2O$		+1.33
$MnO_2(s) + 4H^+(aq) + 2e^- \longrightarrow Mn^{2+}(aq) + 2H_2O$		+1.23
$O_2(g) + 4H^+(aq) + 4e^- \longrightarrow 2H_2O$		+1.23
$Br_2(l) + 2e^- \longrightarrow 2Br^-(aq)$		+1.07
$NO_3^-(aq) + 4H^+(aq) + 3e^- \longrightarrow NO(g) + 2H_2O$		+0.96
$2Hg^{2+}(aq) + 2e^- \longrightarrow Hg_2^{2+}(aq)$		+0.92
$Hg_2^{2+}(aq) + 2e^- \longrightarrow 2Hg(l)$		+0.85
$Ag^+(aq) + e^- \longrightarrow Ag(s)$		+0.80
$Fe^{3+}(aq) + e^- \longrightarrow Fe^{2+}(aq)$		+0.77
$O_2(g) + 2H^+(aq) + 2e^- \longrightarrow H_2O_2(aq)$		+0.68
$MnO_4^-(aq) + 2H_2O + 3e^- \longrightarrow MnO_2(s) + 4OH^-(aq)$		+0.59
$I_2(s) + 2e^- \longrightarrow 2I^-(aq)$		+0.53
$O_2(g) + 2H_2O + 4e^- \longrightarrow 4OH^-(aq)$		+0.40
$Cu^{2+}(aq) + 2e^- \longrightarrow Cu(s)$		+0.34
$AgCl(s) + e^- \longrightarrow Ag(s) + Cl^-(aq)$		+0.22
$SO_4^{2-}(aq) + 4H^+(aq) + 2e^- \longrightarrow SO_2(g) + 2H_2O$		+0.20
$Cu^{2+}(aq) + e^- \longrightarrow Cu^+(aq)$		+0.15
$Sn^{4+}(aq) + 2e^- \longrightarrow Sn^{2+}(aq)$		+0.13
$2H^+(aq) + 2e^- \longrightarrow H_2(g)$		0.00
$Pb^{2+}(aq) + 2e^- \longrightarrow Pb(s)$		−0.13
$Sn^{2+}(aq) + 2e^- \longrightarrow Sn(s)$		−0.14
$Ni^{2+}(aq) + 2e^- \longrightarrow Ni(s)$		−0.25
$Co^{2+}(aq) + 2e^- \longrightarrow Co(s)$		−0.28
$PbSO_4(s) + 2e^- \longrightarrow Pb(s) + SO_4^{2-}(aq)$		−0.31
$Cd^{2+}(aq) + 2e^- \longrightarrow Cd(s)$		−0.40
$Fe^{2+}(aq) + 2e^- \longrightarrow Fe(s)$		−0.44
$Cr^{3+}(aq) + 3e^- \longrightarrow Cr(s)$		−0.74
$Zn^{2+}(aq) + 2e^- \longrightarrow Zn(s)$		−0.76
$2H_2O + 2e^- \longrightarrow H_2(g) + 2OH^-(aq)$		−0.83
$Mn^{2+}(aq) + 2e^- \longrightarrow Mn(s)$		−1.18
$Al^{3+}(aq) + 3e^- \longrightarrow Al(s)$		−1.66
$Be^{2+}(aq) + 2e^- \longrightarrow Be(s)$		−1.85
$Mg^{2+}(aq) + 2e^- \longrightarrow Mg(s)$		−2.37
$Na^+(aq) + e^- \longrightarrow Na(s)$		−2.71
$Ca^{2+}(aq) + 2e^- \longrightarrow Ca(s)$		−2.87
$Sr^{2+}(aq) + 2e^- \longrightarrow Sr(s)$		−2.89
$Ba^{2+}(aq) + 2e^- \longrightarrow Ba(s)$		−2.90
$K^+(aq) + e^- \longrightarrow K(s)$		−2.93
$Li^+(aq) + e^- \longrightarrow Li(s)$		−3.05

Increasing strength as oxidizing agent (left margin, upward arrow)

Increasing strength as reducing agent (right margin, downward arrow)

*For all half-reactions the activity is unity for dissolved species (~1 *M*) and the pressure is 1 bar for gases. These are the standard-state values.

It is important to know the following points about the table in calculations:

1. The $\mathcal{E}°$ values apply to the half-cell reactions as read in the forward (left to right) direction.

2. The more positive $\mathcal{E}°$ is, the greater the tendency for the substance to be reduced. For example, the half-cell reaction

$$F_2(1 \text{ bar}) + 2e^- \longrightarrow 2F^-(a_{F^-} = 1) \qquad \mathcal{E}° = +2.87 \text{ V}$$

has the highest positive $\mathcal{E}°$ value among all the half-cell reactions. Thus, F_2 is the *strongest* oxidizing agent because it has the greatest tendency to be reduced. At the other extreme is the reaction

$$Li^+(a_{Li^+} = 1) + e^- \longrightarrow Li(s) \qquad \mathcal{E}° = -3.05 \text{ V}$$

which has the most negative $\mathcal{E}°$ value. Thus, Li is the *weakest* oxidizing agent because it is the most difficult species to reduce. Conversely, we say that F^- is the weakest reducing agent and Li metal is the strongest reducing agent. Under standard-state conditions, oxidizing agents (the species on the left-hand side of the half-reactions in Table 13.1) increase in strength from bottom to top and the reducing agents (the species on the right-hand side of the half-reactions) increase in strength from top to bottom.

3. The half-cell reactions are reversible. Depending on the conditions, any electrode can act either as an anode or as a cathode. Earlier we saw that the standard hydrogen electrode is the cathode (H^+ is reduced to H_2) when coupled with zinc in a cell and that it becomes the anode (H_2 is oxidized to H^+) when used in a cell with copper.

4. Under standard-state conditions, any species on the left of a given half-cell reaction will react spontaneously with a species that appears on the right of any half-cell reaction located *below* it in Table 13.1. This principle is sometimes called the *diagonal rule*. In the case of the Daniell cell

$$Cu^{2+}(a_{Cu^{2+}} = 1) + 2e^- \longrightarrow Cu(s) \qquad \mathcal{E}° = +0.34 \text{ V}$$
$$Zn^{2+}(a_{Zn^{2+}} = 1) + 2e^- \longrightarrow Zn(s) \qquad \mathcal{E}° = -0.76 \text{ V}$$

We see that the substance on the left of the first half-cell reaction is Cu^{2+} and the substance on the right in the second half-cell reaction is Zn. Therefore, as we saw earlier, Zn spontaneously reduces Cu^{2+} to form Zn^{2+} and Cu.

5. Changing the stoichiometric coefficients of a half-cell reaction *does not* affect the value of $\mathcal{E}°$ because electrode potentials are intensive properties. This means that the value of $\mathcal{E}°$ is unaffected by the size of the electrodes or the amount of solutions present. For example,

$$I_2(s) + 2e^- \longrightarrow 2I^-(a_{I^-} = 1) \qquad \mathcal{E}° = +0.53 \text{ V}$$

but $\mathcal{E}°$ does not change if we multiply the half-reaction by 2:

$$2I_2(s) + 4e^- \longrightarrow 4I^-(a_{I^-} = 1) \qquad \mathcal{E}° = +0.53 \text{ V}$$

6. Like ΔH, ΔG, and ΔS, the sign of $\mathcal{E}°$ changes, but its magnitude remains the same when we reverse a reaction.

As Examples 13.3 and 13.4 show, Table 13.1 enables us to predict the outcome of redox reactions under standard-state conditions, whether they take place in a galvanic cell, where the reducing agent and oxidizing agent are physically separated from each other, or in a beaker, where the reactants are all mixed together.

Example 13.3

Predict what will happen if molecular bromine (Br_2) is added to a solution containing NaCl and NaI at 25°C. Assume all the species are in their standard states.

Strategy To predict what redox reaction(s) will take place, we need to compare the standard reduction potentials of Cl_2, Br_2, and I_2 and apply the diagonal rule.

Solution From Table 13.1, we write the standard reduction potentials as follows:

$$Cl_2(1 \text{ bar}) + 2e^- \longrightarrow 2Cl^-(a_{Cl^-} = 1) \qquad \mathscr{E}° = +1.36 \text{ V}$$
$$Br_2(l) + 2e^- \longrightarrow 2Br^-(a_{Br^-} = 1) \qquad \mathscr{E}° = +1.07 \text{ V}$$
$$I_2(s) + 2e^- \longrightarrow 2I^-(a_{I^-} = 1) \qquad \mathscr{E}° = +0.53 \text{ V}$$

Applying the diagonal rule, we see that Br_2 will oxidize I^- but will not oxidize Cl^-. Therefore, the only redox reaction that will occur appreciably under standard-state conditions is

oxidation: $2I^-(a_{I^-} = 1) \longrightarrow I_2(s) + 2e^-$

reduction: $Br_2(l) + 2e^- \longrightarrow 2Br^-(a_{Br^-} = 1)$

overall: $Br_2(l) + 2I^-(a_{I^-} = 1) \longrightarrow 2Br^-(a_{Br^-} = 1) + I_2(s)$

Check We can confirm our conclusion by calculating $\mathscr{E}°_{cell}$. Try it. Note that the Na^+ ions are inert and do not enter into the redox reaction.

Practice Exercise Can Sn reduce $Zn^{2+}(aq)$ under standard-state conditions?

Example 13.4

A galvanic cell consists of an Mg electrode in a 1.0 M $Mg(NO_3)_2$ solution and an Ag electrode in a 1.0 M $AgNO_3$ solution. Calculate the standard emf of this cell at 25°C. Assume ideal dilute solutions, that is, the activities of the ions in solution are approximated by $M/M°$, where $M° = 1 \text{ mol L}^{-1}$.

Strategy At first it may not be clear how to assign the electrodes in the galvanic cell. From Table 13.1, we write the standard reduction potentials of Ag^+ and Mg^{2+} and apply the diagonal rule to determine which is the anode and which is the cathode.

Solution The standard reduction potentials are

$$Ag^+(a_{Ag^+} = 1) + e^- \longrightarrow Ag(s) \qquad \mathscr{E}° = +0.80 \text{ V}$$
$$Mg^{2+}(a_{Mg^{2+}} = 1) + 2e^- \longrightarrow Mg(s) \qquad \mathscr{E}° = -2.37 \text{ V}$$

Applying the diagonal rule, we see that Ag will oxidize Mg:

anode (oxidation): $Mg(s) \longrightarrow Mg^{2+}(1 M) + 2e^-$

cathode (reduction): $2Ag^+(1 M) + 2e^- \longrightarrow 2Ag(s)$

overall: $Mg(s) + 2Ag^+(1 M) \longrightarrow Mg^{2+}(1 M) + 2Ag(s)$

—Continued

Dental Filling Discomfort

In modern dentistry, the material most commonly used to fill decaying teeth is known as *dental amalgam.* (An amalgam is a substance made by combining mercury with another metal or metals.) Dental amalgam actually consists of three solid phases having stoichiometries approximately corresponding to Ag_2Hg_3, Ag_3Sn, and Sn_8Hg. The standard reduction potentials for these solid phases are: $Hg_2^{2+}\|Ag_2Hg_3$, 0.85 V; $Sn^{2+}\|Ag_3Sn$, -0.05 V; and Sn^{2+}/Sn_8Hg, -0.13 V.

Anyone who bites a piece of aluminum foil (such as that used for wrapping candies) in such a way that the foil presses against a dental filling will probably experience a momentary sharp pain. In effect, an electrochemical cell has been created in the mouth, with aluminum ($\mathcal{E}° = -1.66$ V) as the anode, the filling as the cathode, and saliva as the electrolyte. Contact between the aluminum foil and the filling short-circuits the cell, causing a weak current to flow between the electrodes. This current stimulates the sensitive nerve of the tooth, causing an unpleasant sensation.

Another type of discomfort results when a less electropositive metal touches a dental filling. For example, if a filling makes contact with a gold inlay in a nearby tooth, corrosion of the filling will occur. In this case, the dental filling acts as the anode and the gold inlay as the cathode. Referring to the $\mathcal{E}°$ values for the three phases, we see that the Sn_8Hg phase is most likely to corrode. When that happens, release of Sn(II) ions in the mouth produces an unpleasant metallic taste. Prolonged corrosion will eventually result in another visit to the dentist for a replacement filling.

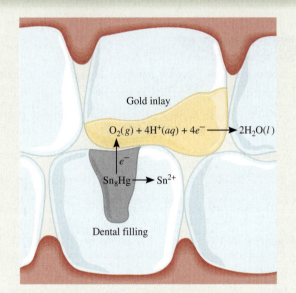

Corrosion of a dental filling brought about by contact with a gold inlay.

Continued—

where we have used the fact that, assuming ideal dilute solution behavior, the standard state for aqueous solutes is a 1 *M* solution. Note that to balance the overall equation we multiplied the reduction of Ag (but not $\mathcal{E}°$) by 2. We can do so because, as an intensive property, $\mathcal{E}°$ is not affected by this procedure. We find the emf of the cell by using Equation 13.1 and Table 13.1:

$$\begin{aligned}\mathcal{E}°_{cell} &= \mathcal{E}°_{cathode} - \mathcal{E}°_{anode} \\ &= \mathcal{E}°_{Ag^+|Ag} - \mathcal{E}°_{Mg^{2+}|Mg} \\ &= +0.80 \text{ V} - (-2.37 \text{ V}) \\ &= +3.17 \text{ V}\end{aligned}$$

Check The positive value of $\mathcal{E}°$ shows that the forward reaction is favored.

Practice Exercise What is the standard emf of a galvanic cell made of a Cd electrode in a 1.0 *M* $Cd(NO_3)_2$ solution and a Cr electrode in a 1.0 *M* $Cr(NO_3)_3$ solution at 25°C? Assume ideal dilute solution behavior.

13.4 | The Emf of an Electrochemical Cell Is Directly Related to Gibbs Free Energy Change of the Redox Reaction

In a galvanic cell, chemical energy is converted to electric energy. Electric energy in this case is the product of the emf of the cell and the total electric charge (in coulombs) that passes through the cell:

$$\text{electric energy} = \text{volts} \times \text{coulombs}$$
$$= \text{joules}$$

The total charge is determined by the number of moles of electrons (n) that pass through the circuit. By definition

$$\text{total charge} = nF$$

n is the number of moles of electrons exchanged between the reducing agent and the oxidizing agent in the overall redox reaction.

where F, the **Faraday**[5] **constant,** is the electric charge contained in 1 mole of electrons, which has been shown by experiment to be 96,485.3 coulombs. Thus,

$$F = 96{,}485.3 \text{ C mol}^{-1} \, e^{-}$$

(For most calculations it is sufficient to round F to three significant figures: $F = 96{,}500 \text{ C mol}^{-1}$.) Because

$$1 \text{ J} = 1 \text{ C} \times 1 \text{ V}$$

we can also express the units of the faraday constant as

$$F = 96{,}500 \text{ J V}^{-1} \text{ mol}^{-1}$$

The measured emf is the *maximum* voltage that the cell can achieve. This value is used to calculate the maximum amount of electric energy that can be obtained from the chemical reaction. This energy is used to do electric work (w_{ele}), so

$$w_{max} = -w_{ele}$$
$$= -nF\mathscr{E}_{cell}$$

where w_{max} is the maximum amount of work that can be done by the system. The negative sign on the right-hand side indicates that the electric work is done by the system on the surroundings.

The sign convention for electrical work is the same as that for P-V work, discussed in Section 7.1.

Relationship of $\mathscr{E}_{cell}^{\circ}$ to ΔG, ΔG°, and K

In Chapter 8, we showed that the change in Gibbs free energy for a reaction ΔG represents the maximum amount of nonexpansion work that can be obtained from that reaction:

$$\Delta G = w_{max}$$

---

5. Michael Faraday (1791–1867). English chemist and physicist. Faraday is regarded by many as the greatest experimental scientist of the nineteenth century. He started as an apprentice to a bookbinder at the age of 13 but became interested in science after reading a book on chemistry. Faraday invented the electric motor and was the first person to demonstrate the principle governing electrical generators. Besides making notable contributions to the fields of electricity and magnetism, Faraday also worked on optical activity and discovered and named benzene.

Therefore, we can write

$$\Delta G = -nF\mathscr{E}_{cell} \tag{13.2}$$

Both n and F are positive quantities, and ΔG is negative for a spontaneous process; so $\mathscr{E}_{cell}$ must be positive for a spontaneous process.

For reactions in which reactants and products are in their standard states, Equation 13.2 becomes

$$\Delta G° = -nF\mathscr{E}°_{cell} \tag{13.3}$$

Now we can relate $\mathscr{E}°_{cell}$ to the equilibrium constant (K) of a redox reaction. In Section 10.3, we saw that the standard free-energy change $\Delta G°$ for a reaction is related to its equilibrium constant as follows (Equation 10.12):

$$\Delta G° = -RT \ln K$$

Therefore, if we combine Equations 10.12 and 13.3, we obtain

$$-nF\mathscr{E}°_{cell} = -RT \ln K$$

Solving for $\mathscr{E}°_{cell}$,

$$\mathscr{E}°_{cell} = \frac{RT}{nF} \ln K \tag{13.4}$$

When $T = 298.15$ K, Equation 13.4 can be simplified by substituting in numerical values for R and F:

$$\mathscr{E}°_{cell} = \frac{(8.3145 \text{ J mol}^{-1} \text{ K}^{-1})(298.15 \text{ K})}{n(96485 \text{ J V}^{-1} \text{ mol}^{-1})} \ln K$$

$$\mathscr{E}°_{cell} = \frac{0.025693 \text{ V}}{n} \ln K \tag{13.5}$$

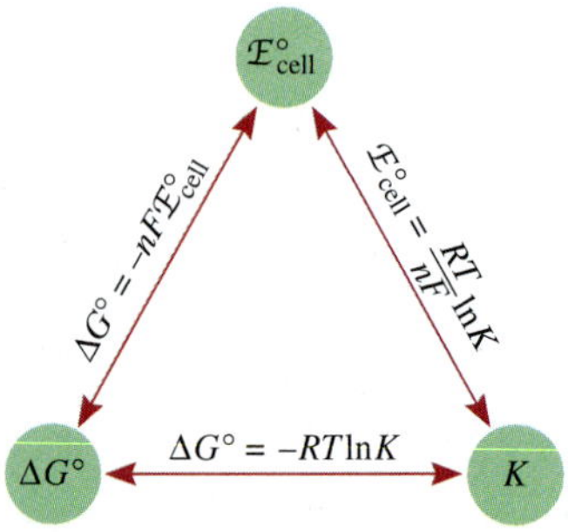

Figure 13.7 Relationships among $\mathscr{E}°_{cell}$, K, and $\Delta G°$.

Thus, if any one of the three quantities $\Delta G°$, K, or $\mathscr{E}°_{cell}$ is known, the other two can be calculated using Equations 10.13, 13.3, or 13.4 (Figure 13.7). We summarize the relationships among $\Delta G°$, K, and $\mathscr{E}°_{cell}$ and characterize the spontaneity of a redox reaction in Table 13.2.

Examples 13.5 and 13.6 apply Equations 13.3 through 13.5.

Table 13.2	Relationships Among $\Delta G°$, K, and $\mathscr{E}°_{cell}$		
$\Delta G°$	K	$\mathscr{E}°_{cell}$	**Reaction Under Standard-State Conditions**
Negative	>1	Positive	Favors formation of products.
0	=1	0	Reactants and products are equally favored.
Positive	<1	Negative	Favors formation of reactants.

Example 13.5

Calculate the equilibrium constant for the following reaction at 25°C:

$$Sn(s) + 2Cu^{2+}(aq) \rightleftharpoons Sn^{2+}(aq) + 2Cu^{+}(aq)$$

Strategy The general relationship between the equilibrium constant K and the standard emf is given by Equation 13.4; however, because the temperature is 25°C (298.15 K), we can use Equation 13.5:

$$\mathscr{E}^{\circ}_{\text{cell}} = \frac{0.025693 \text{ V}}{n} \ln K$$

Thus, if we can determine the standard emf, we can calculate the equilibrium constant. We can determine the $\mathscr{E}^{\circ}_{\text{cell}}$ of a hypothetical galvanic cell made up of two couples ($Sn^{2+}|Sn$ and $Cu^{2+}|Cu^{+}$) from the standard reduction potentials in Table 13.1.

Solution The half-cell reactions are

$$
\begin{aligned}
\text{anode (oxidation):} \quad & Sn(s) \longrightarrow Sn^{2+}(aq) + 2e^{-} \\
\text{cathode (reduction):} \quad & 2Cu^{2+}(aq) + 2e^{-} \longrightarrow 2Cu^{+}(aq)
\end{aligned}
$$

$$
\begin{aligned}
\mathscr{E}^{\circ}_{\text{cell}} &= \mathscr{E}^{\circ}_{\text{cathode}} - \mathscr{E}^{\circ}_{\text{anode}} \\
&= \mathscr{E}^{\circ}_{Cu^{2+}|Cu} - \mathscr{E}^{\circ}_{Sn^{2+}|Sn} \\
&= +0.15 \text{ V} - (-0.14 \text{ V}) \\
&= +0.29 \text{ V}
\end{aligned}
$$

Equation 13.5 can be written

$$\ln K = \frac{n\mathscr{E}^{\circ}_{\text{cell}}}{0.025693 \text{ V}}$$

In the overall reaction, two electrons are transferred, so $n = 2$. Therefore,

$$\ln K = \frac{2(0.29 \text{ V})}{0.025693 \text{ V}} = 22.6$$

$$K = e^{22.6} = 7 \times 10^{9}$$

Check The standard emf is positive, so the reaction is spontaneous toward the products under standard conditions; therefore, K should be larger than 1.

Practice Exercise Calculate the equilibrium constant for the following reaction at 25°C:

$$Fe^{2+}(aq) + 2Ag(s) \rightleftharpoons Fe(s) + 2Ag^{+}(aq)$$

Example 13.6

Calculate the standard free-energy change for the following reaction at 25°C:

$$2Au(s) + 3Ca^{2+}(a_{Ca^{2+}} = 1) \rightleftharpoons 2Au^{3+}(a_{Au^{3+}} = 1) + 3Ca(s)$$

Strategy The relationship between the standard free energy change and the standard emf of the cell is given by Equation 13.3: $\Delta G^{\circ} = -nF\mathscr{E}^{\circ}_{\text{cell}}$. Thus, if we can determine $\mathscr{E}^{\circ}_{\text{cell}}$, we can calculate ΔG°. We can determine the $\mathscr{E}^{\circ}_{\text{cell}}$ of a hypothetical galvanic cell

—Continued

Continued—
made up of two half reactions, $Au^{3+}|Au$ and $Ca^{2+}|Ca$, from the standard reduction potentials in Table 13.1.

Solution The half-cell reactions are

$$\text{anode (oxidation):}\quad 2Au(s) \longrightarrow 2Au^{3+}(aq) + 6e^-$$
$$\text{cathode (reduction):}\quad 3Ca^{2+}(aq) + 6e^- \longrightarrow 3Ca(s)$$

$$
\begin{aligned}
\mathscr{E}^\circ_{\text{cell}} &= \mathscr{E}^\circ_{\text{cathode}} - \mathscr{E}^\circ_{\text{anode}} \\
&= \mathscr{E}^\circ_{Ca^{2+}|Ca} - \mathscr{E}^\circ_{Au^{3+}|Au} \\
&= -2.87\text{ V} - 1.50\text{ V} \\
&= -4.37\text{ V}
\end{aligned}
$$

Now we use Equation 13.3:

$$\Delta G^\circ = -nF\mathscr{E}^\circ_{\text{cell}}$$

The overall reaction shows that $n = 6$, so

$$
\begin{aligned}
\Delta G^\circ &= -(6)(96{,}500\text{ J V}^{-1}\,\text{mol}^{-1})(-4.37\text{ V}) \\
&= 2.53 \times 10^6\text{ J mol}^{-1} \\
&= 2.53 \times 10^3\text{ kJ mol}^{-1}
\end{aligned}
$$

Check The large positive value of ΔG° tells us that the reaction favors the reactants at equilibrium. The result is consistent with the fact that $\mathscr{E}^\circ_{\text{cell}}$ for the galvanic cell is negative.

Practice Exercise Calculate ΔG° for the following reaction at 25°C:

$$2Al^{3+}(aq) + 3Mg(s) \rightleftharpoons 2Al(s) + 3Mg^{2+}(aq)$$

Temperature Dependence of Emf

As was the case for the equilibrium constant (see Section 10.4), the temperature dependence of the emf can also be determined from the thermodynamic properties of the cell. We start with the van't Hoff equation (Equation 10.16) for the temperature dependence of the equilibrium constant

$$\ln\frac{K_2}{K_1} = \frac{\Delta H^\circ}{R}\left(\frac{1}{T_1} - \frac{1}{T_2}\right)$$

Equation 10.16 is valid as long as the temperature dependence of the enthalpy of reaction ΔH° is small. Combining Equation 13.4 [$\mathscr{E}^\circ_{\text{cell}} = (RT/nF)\ln K$] with Equation 10.16 gives

$$\frac{nF\mathscr{E}^\circ_{\text{cell},\,T_2}}{RT_2} - \frac{nF\mathscr{E}^\circ_{\text{cell},\,T_1}}{RT_1} = \frac{\Delta H^\circ}{R}\left(\frac{1}{T_1} - \frac{1}{T_2}\right)$$

which, after rearranging and dividing by F, gives

$$\frac{\mathscr{E}^\circ_{\text{cell},\,T_2}}{T_2} = \frac{\mathscr{E}^\circ_{\text{cell},\,T_1}}{T_1} + \frac{\Delta H^\circ}{nF}\left(\frac{1}{T_1} - \frac{1}{T_2}\right) \tag{13.6}$$

Therefore, if the standard emf for a cell at a certain temperature is given, its value at another temperature can be determined using Equation 13.6 if the enthalpy change for the cell reaction is known. On the other hand, if the emf of a cell can be determined at two or more different temperatures, Equation 13.6 can be used to estimate the enthalpy change $\Delta H°$ for a cell reaction. Also, because $\Delta G°$ can be determined from one of the $\mathcal{E}°$ measurements, the entropy change $\Delta S°$ can be determined using $\Delta G° = \Delta H° - T\Delta S°$ (Equation 8.35). Thus, the standard enthalpy and entropy of a reaction can be determined directly from the emf as a function of temperature, which gives us a noncalorimetric method for the calculation of these important thermodynamic quantities.

Also, we can manipulate Equation 13.6 (Problem 13.116), assuming that both $\Delta H°$ and $\Delta S°$ are independent of temperature, to give

$$\mathcal{E}°_{\text{cell}, T_2} = \mathcal{E}°_{\text{cell}, T_1} + \frac{\Delta S°}{nF}(T_2 - T_1) \tag{13.7}$$

Thus, the change in the emf of a cell can be directly related to the standard entropy change for a cell reaction. Often the sign and magnitude of the entropy change for a reaction can be predicted qualitatively, so this gives us a way of estimating the direction and size of the temperature dependence for a reaction. For example, a cell reaction in which one of the products or reactants is a gas will generally have a larger standard entropy change than one in which all the reactants and products are in aqueous solution. Thus, we would expect the emf of the former reaction to be more affected by temperature change.

Example 13.7

Given that the $\Delta H°$ value for the cell reaction

$$2Ag^+(aq) + 2I^- \longrightarrow I_2 + 2Ag(s)$$

is -100.0 kJ mol^{-1} at 25°C, estimate the emf of this cell at 80°C. What assumptions must you make?

Strategy First find the emf of the cell at 25°C using the data in Table 13.1. For this cell, $n = 2$. To find the emf at 80°C, use Equation 13.6 and the given value of the standard enthalpy change at 25°C. Equation 13.6 assumes that $\Delta H°$ is independent of temperature over the relevant temperature range. Remember the conversion $1 \text{ V} = 1 \text{ J C}^{-1}$.

Solution From Table 13.1 the two half-reactions are

$$Ag^+(aq) + e^- \longrightarrow Ag(s) \qquad \mathcal{E}°_{\text{red}} = +0.80 \text{ V at } 25°C$$
$$I_2(s) + 2e^- \longrightarrow 2I^-(aq) \qquad \mathcal{E}°_{\text{red}} = +0.53 \text{ V at } 25°C$$

The cell emf at 25°C is then

$$\mathcal{E}°_{\text{cell}} = \mathcal{E}°_{\text{cathode}} - \mathcal{E}°_{\text{anode}} = +0.80 \text{ V} - 0.53 \text{ V} = +0.27 \text{ V}$$

Equation 13.6 is

$$\frac{\mathcal{E}°_{\text{cell}, T_2}}{T_2} = \frac{\mathcal{E}°_{\text{cell}, T_1}}{T_1} + \frac{\Delta H°}{nF}\left(\frac{1}{T_1} - \frac{1}{T_2}\right)$$

—Continued

Continued—

Substituting $T_1 = 25°C = 298$ K, $T_2 = 80°C = 353$ K, $n = 2$, and the given value of $\Delta H°$ into this equation gives

$$\frac{\mathcal{E}°_{cell,\,T_2}}{353\ \text{K}} = \frac{0.27\ \text{V}}{298\ \text{K}} + \frac{(-100.0\ \text{kJ mol}^{-1})(1000\ \text{J kJ}^{-1})}{2(96485\ \text{C mol}^{-1})}\left(\frac{1}{298\ \text{K}} - \frac{1}{353\ \text{K}}\right)$$

$$\mathcal{E}°_{cell,\,T_2} = 0.22\ \text{V K}^{-1}$$

Thus, the cell voltage decreases by 0.05 V when the temperature is raised from 25°C to 80°C.

Check A decrease in the cell voltage indicates a decrease in the equilibrium constant for this reaction. This reaction is exothermic. LeChâtelier's principle would predict that the equilibrium for an exothermic reaction would shift toward the reactants and decrease the equilibrium constant. Thus, our result is consistent with LeChâtelier's principle.

Practice Exercise Use the data in Table 13.1 and in Appendix 2 to determine the emf of the cell reaction

$$Cl_2(g) + 2Cu^+(aq) \longrightarrow 2Cl^-(aq) + 2Cu^{2+}(aq)$$

at 80°C.

13.5 | The Concentration Dependence of the Emf Can Be Determined Using the Nernst Equation

So far we have focused on redox reactions in which reactants and products are in their standard states, but there is no reason that a redox reaction of practical interest must be carried out under standard-state conditions, and, in general, the conditions will differ significantly from the standard state. Nevertheless, there is a mathematical relationship between the emf of a galvanic cell and the activities of reactants and products in a redox reaction under non–standard-state conditions.

Consider a redox reaction of the type

$$a\text{A} + b\text{B} \longrightarrow c\text{C} + d\text{D}$$

From Equation 10.11,

$$\Delta G = \Delta G° + RT \ln Q$$

where Q is the reaction quotient for this reaction

$$Q = \frac{a_C^c\, a_D^d}{a_A^a\, a_B^b}$$

Because $\Delta G = -nF\mathcal{E}$ (Equation 13.2) and $\Delta G° = -nF\mathcal{E}°$ (Equation 13.3), Equation 10.11 can be expressed as

$$-nF\mathcal{E} = -nF\mathcal{E}° + RT \ln Q \tag{13.8}$$

Dividing Equation 13.8 through by $-nF$, we get

$$\mathcal{E} = \mathcal{E}° - \frac{RT}{nF} \ln Q \tag{13.9}$$

Equation 13.9 is known as the **Nernst**[6] **equation.** At 25°C (298.15 K), Equation 13.9 can be rewritten as

$$E(25°C) = E° - \frac{0.025693 \text{ V}}{n} \ln Q \qquad \textbf{(13.10)}$$

During the operation of a galvanic cell, electrons flow from the anode to the cathode, resulting in product formation and a decrease in reactant concentration. Thus, Q increases, which means that E decreases. Eventually, the cell reaches equilibrium. At equilibrium, there is no net transfer of electrons, so $E = 0$, $Q = K$, where K is the equilibrium constant, and the Nernst equation (Equation 13.9) simplifies to Equation 13.4.

The Nernst equation can be used to calculate E as a function of reactant and product concentrations in a redox reaction if activities can be approximated by molar concentrations. For example, for the Daniell cell in Figure 13.3,

$$Zn(s) + Cu^{2+}(aq) \longrightarrow Zn^{2+}(aq) + Cu(s)$$

The Nernst equation for this cell at 25°C (Equation 13.10), assuming ideal solution behavior, can be written as

$$E = E° - \frac{0.025693 \text{ V}}{n} \ln \frac{[Zn^{2+}]}{[Cu^{2+}]}$$

where we have used the fact that the activity of pure solids can be assumed to be unity. If the ratio $[Zn^{2+}]/[Cu^{2+}]$ is less than 1, $\ln([Zn^{2+}]/[Cu^{2+}])$ is a negative number, so that the second term on the right-hand side of the preceding equation is positive. Under this condition, E is greater than the standard emf $E°$. If the ratio is greater than 1, E is smaller than $E°$.

Example 13.8 illustrates the use of the Nernst equation.

Example 13.8

Predict whether the following reaction would proceed spontaneously as written at 298 K:

$$Co(s) + Fe^{2+}(aq) \rightleftharpoons Co^{2+}(aq) + Fe(s)$$

given that $[Co^{2+}] = 0.15$ M and $[Fe^{2+}] = 0.68$ M. Assume ideal dilute solution behavior for the solutes.

Strategy Because the reaction is not run under standard-state conditions (concentrations are not 1 M under ideal dilute solution conditions), we need to use the Nernst equation (Equation 13.10) to calculate the emf (E) of a hypothetical galvanic cell and determine the spontaneity of the reaction. The standard emf ($E°$) can be calculated using the standard reduction potentials in Table 13.1. Remember that pure solids can be considered to have an activity of 1, so they can be neglected in the Q term of the Nernst equation. Note that 2 moles of electrons are transferred per mole of reaction, that is, $n = 2$.

—Continued

---

6. Walter Hermann Nernst (1864–1941). German chemist and physicist. Nernst's work was mainly in electrolyte solutions and thermodynamics. He also invented an electric piano. Nernst was awarded the Nobel Prize in Chemistry in 1920 for his contribution to thermodynamics.

Continued—

Solution The half-cell reactions are

$$\text{anode (oxidation):} \qquad\qquad\qquad Co(s) \longrightarrow Co^{2+}(aq) + 2e^-$$

$$\text{cathode (reduction):} \quad Fe^{2+}(aq) + 2e^- \longrightarrow Fe(s)$$

$$
\begin{aligned}
\mathscr{E}^\circ_{cell} &= \mathscr{E}^\circ_{cathode} - \mathscr{E}^\circ_{anode} \\
&= \mathscr{E}^\circ_{Fe^{2+}|Fe} - \mathscr{E}^\circ_{Co^{2+}|Co} \\
&= -0.44\ V - (-0.28\ V) \\
&= -0.16\ V
\end{aligned}
$$

From Equation 13.10, we write

$$
\begin{aligned}
\mathscr{E} &= \mathscr{E}^\circ - \frac{0.0257\ V}{n} \ln Q \\
&= \mathscr{E}^\circ - \frac{0.0257\ V}{n} \ln \frac{[Co^{2+}]}{[Fe^{2+}]} \\
&= -0.16\ V - \frac{0.025693\ V}{2} \ln \frac{0.15}{0.68} \\
&= -0.16\ V + 0.019\ V \\
&= -0.14\ V
\end{aligned}
$$

Because $\mathscr{E}$ is negative, the reaction is not spontaneous in the direction written.

Practice Exercise Will the following reaction occur spontaneously at 25°C, given that $[Fe^{2+}] = 0.60\ M$ and $[Cd^{2+}] = 0.010\ M$?

$$Cd(s) + Fe^{2+}(aq) \rightleftharpoons Cd^{2+}(aq) + Fe(s)$$

Assume ideal dilute solution conditions.

Now suppose we want to determine at what ratio of $[Co^{2+}]$ to $[Fe^{2+}]$ the reaction in Example 13.7 would become spontaneous at 25°C. We can use Equation 13.10 as follows:

$$\mathscr{E} = \mathscr{E}^\circ - \frac{0.025693\ V}{n} \ln Q$$

We set $\mathscr{E}$ equal to zero, which corresponds to the equilibrium situation.

$$0 = -0.16\ V - \frac{0.025693\ V}{2} \ln \frac{[Co^{2+}]}{[Zn^{2+}]}$$

$$\ln \frac{[Co^{2+}]}{[Zn^{2+}]} = -12.5$$

$$\frac{[Co^{2+}]}{[Zn^{2+}]} = e^{-12.5} = K$$

or
$$K = 4 \times 10^{-6}$$

Thus, for the reaction to be spontaneous ($\mathscr{E}$ is positive), the ratio $[Co^{2+}]/[Fe^{2+}]$ must be smaller than 4×10^{-6}.

As Example 13.9 shows, if gases are involved in the cell reaction, their concentrations should be expressed in bar.

Example 13.9

Consider the galvanic cell shown in Figure 13.6(a). In a certain experiment, the emf (E) of the cell is found to be 0.54 V at 25°C. Suppose that $[Zn^{2+}] = 1.00 \times 10^{-4}$ M and $P_{H_2} = 1.00$ bar. Calculate the molar concentration of H^+, assuming ideal solution behavior.

Strategy The equation that relates standard emf and nonstandard emf is the Nernst equation. Assuming ideal solution behavior, the standard states of Zn^{2+} and H^+ are 1 M solutions. The overall cell reaction is

$$Zn(s) + 2H^+(?\,M) \longrightarrow Zn^{2+}(1.00\,M) + H_2(1.00\,\text{bar})$$

Given the emf of the cell (E), we apply the Nernst equation to solve for $[H^+]$. Note that 2 moles of electrons are transferred per mole of reaction, that is, $n = 2$.

Solution As we saw earlier (page 675), the standard emf ($E°$) for the cell is 0.76 V. Because the system is at 25°C (298 K) we can use the form of the Nernst equation given in Equation 13.5:

The concentrations in Q are divided by their standard-state values of 1 M; and pressures are divided by 1 bar.

$$E = E° - \frac{0.0257\,\text{V}}{n}\ln Q$$

$$= E° - \frac{0.0257\,\text{V}}{n}\ln\frac{a_{Zn^{2+}}\,a_{H_2}}{a_{H^+}^2}$$

$$\approx E° - \frac{0.0257\,\text{V}}{n}\ln\frac{[Zn^{2+}]P_{H_2}}{[H^+]^2}$$

$$0.54\,\text{V} = 0.76\,\text{V} - \frac{0.0257\,\text{V}}{2}\ln\frac{(1.00 \times 10^{-4})(1.00)}{[H^+]^2}$$

$$-0.34\,\text{V} = \frac{0.0257\,\text{V}}{2}\ln\frac{1}{[H^+]^2}$$

$$26.4 = \ln\frac{1}{[H^+]^2}$$

$$e^{26.4} = \frac{1}{[H^+]^2}$$

$$[H^+] = \sqrt{\frac{1}{3.1 \times 10^{11}}} = 1.8 \times 10^{-6}\,M$$

Practice Exercise What is the emf of a galvanic cell consisting of a $Cd^{2+}|Cd$ half-cell and a $Pt|H^+|H_2$ half-cell if $[Cd^{2+}] = 0.20$ M, $[H^+] = 0.16$ M, and $P_{H_2} = 0.81$ bar? Assume ideal behavior.

Example 13.8 shows that a galvanic cell whose cell reaction involves H^+ ions can be used to measure the activity of H^+ or pH. In Section 10.2, we introduced the pH meter as a standard device for the measurement of pH. For the purpose of accurately and safely measuring pH in the laboratory, the hydrogen electrode itself is

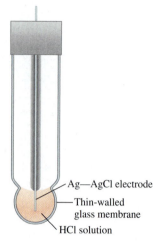

— Ag—AgCl electrode

— Thin-walled
glass membrane

— HCl solution

Figure 13.8 The widely used glass electrode is an example of an ion-specific electrode because it is specific to H^+. The electrode consists of a very thin bulb or membrane made of a special type of glass that is permeable to H^+ ions. An $Ag|AgCl$ electrode is immersed in a buffer solution (constant pH) containing Cl^- ions. When the electrode is placed in a solution whose pH is different from the buffer solution, the potential difference that develops between the two sides is a measure of the difference in the two pH values.

impractical because hydrogen is explosive and the platinum electrode is susceptible to contamination, so it is replaced by a *glass electrode* (Figure 13.8). This electrode consists of a very thin glass membrane that is permeable to H^+ ions. A silver wire coated with silver chloride is immersed in a dilute hydrochloric acid solution. When the electrode is placed in a solution whose pH is different from that of the inner solutions, the potential difference that develops between the two sides of the membrane can be monitored using a reference electrode. The emf of the cell made up of the glass electrode and the reference electrode is measured with a voltmeter that is calibrated in pH units.

Concentration Cells

Because electrode potential depends on ion concentrations, it is possible to construct a galvanic cell from two half-cells composed of the *same* material but differing in ion concentrations. Such a cell is called a ***concentration cell.***

Consider a situation in which zinc electrodes are put into two aqueous solutions of zinc sulfate at 0.10 *M* and 1.0 *M* concentrations. The two solutions are connected by a salt bridge, and the electrodes are joined by a piece of wire in an arrangement like that shown in Figure 13.1. According to Le Châtelier's principle, the tendency for the reduction

$$Zn^{2+}(aq) + 2e^- \longrightarrow Zn(s)$$

increases with increasing concentration of Zn^{2+} ions. Therefore, reduction should occur in the more concentrated compartment, and oxidation should take place on the more dilute side. The cell diagram is

$$Zn(s)|Zn^{2+}(0.10\ M)||Zn^{2+}(1.0\ M)|Zn(s)$$

and the half-reactions are

oxidation (anode): $Zn(s) \longrightarrow Zn^{2+}(0.10\ M) + 2e^-$
reduction (cathode): $Zn^{2+}(1.0\ M) + 2e^- \longrightarrow Zn(s)$
overall: $Zn^{2+}(1.0\ M) \longrightarrow Zn^{2+}(0.10\ M)$

The emf of the cell at 298 K (assuming ideal solution behavior), using Equation 13.5, is

$$\mathscr{E} = \mathscr{E}^\circ - \frac{0.0257\ V}{n} \ln \frac{[Zn^{2+}]_{dil}}{[Zn^{2+}]_{conc}}$$

where the subscripts "dil" and "conc" refer to the 0.10 *M* and 1.0 *M* concentrations, respectively. The $\mathscr{E}^\circ$ for this cell is zero (the same electrode and the same type of ions are involved), so

$$\mathscr{E} = 0\ V - \frac{0.0257\ V}{2} \ln \frac{0.10}{1.0}$$
$$= 0.0296\ V$$

The emf of concentration cells is usually small and decreases continually during the operation of the cell as the concentrations in the two compartments approach each other. When the concentrations of the ions in the two compartments are the same, $\mathscr{E}$ becomes zero, and no further change occurs.

A biological cell can be compared to a concentration cell for the purpose of calculating its *membrane potential*. Membrane potential is the electrical potential that exists across the membrane of various kinds of cells, including muscle cells and nerve cells. It is responsible for the propagation of nerve impulses and heartbeat. A membrane potential is established whenever there are unequal concentrations of the same type of ion in the interior and exterior of a cell. For example, the concentrations of K^+ ions in the interior and exterior of a nerve cell are 400 mM and 15 mM, respectively. Treating the situation as a concentration cell and applying the Nernst equation for just one kind of ions, we can write

$$\mathcal{E} = \mathcal{E}^\circ - \frac{0.0257 \text{ V}}{1} \ln \frac{[K^+]_{ex}}{[K^+]_{in}}$$

$$= -(0.0257 \text{ V}) \ln \frac{15}{400}$$

$$= 0.084 \text{ V} \quad \text{or} \quad 84 \text{ mV}$$

where "ex" and "in" denote exterior and interior, respectively. Note that we have set $\mathcal{E}^\circ = 0$ because the same type of ion is involved. Thus, an electrical potential of 84 mV exists across the membrane due to the unequal concentrations of K^+ ions.

Determining Activities from Emf Measurements

Emf measurements can be used to accurately determine the activity of ions in solution. For example, consider an electrochemical cell for the reduction of silver chloride by hydrogen gas:

$$Pt \,|\, H_2(1 \text{ bar}) \,|\, HCl(aq, m) \,|\, AgCl(s) \,|\, Ag$$

The overall reaction for this cell is

$$\frac{1}{2} H_2(g) + AgCl(s) \longrightarrow Ag(s) + H^+(aq) + Cl^-(aq)$$

Using the Nernst equation appropriate to 298 K (Equation 13.10), the emf of the cell at 298 K is given by

$$\mathcal{E} = \mathcal{E}^\circ - \frac{0.0257 \text{ V}}{n} \ln Q$$

$$= \mathcal{E}^\circ - \frac{0.0257 \text{ V}}{n} \ln \frac{a_{H^+} \, a_{Cl} \, a_{Ag}}{a_{H_2}^{1/2} \, a_{AgCl}}$$

Both silver (Ag) and silver chloride (AgCl) are solids so their activities are well-approximated by unity. In addition the activity of H_2 gas at 1 bar is also approximately unity. Thus, we have

$$\mathcal{E} = \mathcal{E}^\circ - \frac{0.0257 \text{ V}}{n} \ln a_{H^+} a_{Cl^-}$$

$$= \mathcal{E}^\circ - \frac{0.0257 \text{ V}}{n} \ln (\gamma_\pm \, m)^2$$

Here $n = 1$, and we define the *mean activity coefficient*[7] for a 1:1 electrolyte by

$$a_{H^+}\, a_{Cl^-} = \gamma_{\pm}^2\, m^2$$

Thus, the emf of the cell is

$$\mathcal{E} = \mathcal{E}^\circ - (0.0257\ \text{V}) \ln (\gamma_{\pm} m)^2$$
$$= \mathcal{E}^\circ - (0.0514\ \text{V}) \ln \gamma_{\pm} - (0.0514\ \text{V}) \ln m$$

The preceding equation can be rearranged to give

$$\ln \gamma_{\pm} = \frac{\mathcal{E}^\circ - \mathcal{E}}{0.0514\ \text{V}} - \ln m$$

This equation can be used to determine the mean activity for a given HCl concentration from a measurement of $\mathcal{E}$ for this cell, if the value of $\mathcal{E}^\circ$ is known. The value of $\mathcal{E}^\circ$ can be determined by measuring $\mathcal{E}$ for this cell at very low HCl concentrations. For a very dilute solution $\gamma_{\pm} \approx 1$ and $\mathcal{E}^\circ - \mathcal{E} + (0.0514\ \text{V}) \ln m$ (See Problem 13.35).

13.6 | Batteries Use Electrochemical Reactions to Produce a Ready Supply of Electric Current

A **battery** is *a galvanic cell, or a series of combined galvanic cells, that can be used as a source of direct electric current at a constant voltage.* Although the operation of a battery is similar in principle to that of the galvanic cells described in Section 13.2, a battery has the advantage of being completely self-contained and requiring no auxiliary components such as salt bridges. Here we will discuss several types of batteries that are in widespread use.

The Dry Cell Battery

The most common dry cell, that is, a cell without a fluid component, is the *Leclanché cell* used in flashlights and transistor radios. The anode of the cell consists of a zinc can or container that is in contact with manganese dioxide (MnO_2) and an electrolyte. The electrolyte consists of ammonium chloride and zinc chloride in water to which starch is added to thicken the solution to a pastelike consistency so that it is less likely to leak (Figure 13.9).

A carbon rod serves as the cathode, which is immersed in the electrolyte in the center of the cell. The cell reactions are

anode: $Zn(s) \longrightarrow Zn^{2+}(aq) + 2e^-$

cathode: $2NH_4^+(aq) + 2MnO_2(s) + 2e^- \longrightarrow Mn_2O_3(s) + 2NH_3(aq)$
$+ H_2O(l)$

overall: $Zn(s) + 2NH_4^+(aq) + 2MnO_2(s) \longrightarrow Zn^{2+}(aq) + Mn_2O_3(s)$
$+ 2NH_3(aq) + H_2O(l)$

— Paper spacer
— Moist paste of $ZnCl_2$ and NH_4Cl
— Layer of MnO_2
— Graphite cathode
— Zinc anode

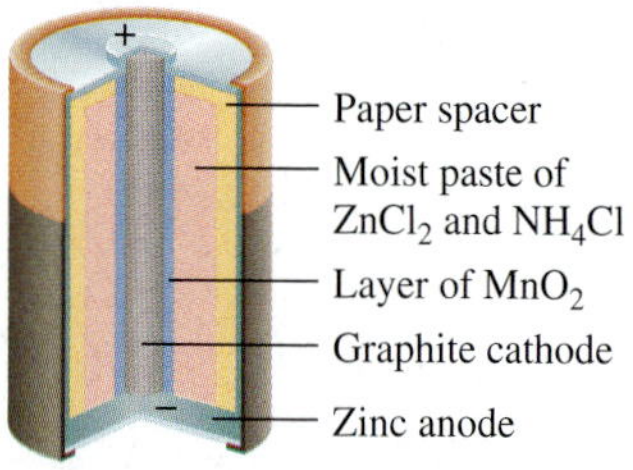

Figure 13.9 Interior section of a dry cell of the kind used in flashlights and transistor radios. Actually, the cell is not completely dry, because it contains a moist electrolyte paste.

7. Because the ions Ag^+ and Cl^- cannot exist separately in solution, the individual activity coefficients of these ions are not separately defined. Instead we define a mean activity $\gamma_{\pm}$, which in this case is the geometric mean of the individual activity coefficients $\gamma_{\pm} = \sqrt{\gamma_{Ag^+}\, \gamma_{Cl^-}}$ For a $m{:}n$ electrolyte (a compound composed of cations of charge $m+$ and anions of charge $n-$, for example, $MgCl_2$ is a 2:1 electrolyte), the mean activity coefficient is $\gamma_{\pm} = (\gamma_+^n \gamma_-^m)^{1/(m+n)}$. The mean activities for ionic compounds can be estimated using an approximation known as the Debye-Hückel theory.

Actually, this equation is an oversimplification of a complex process. The voltage produced by this dry cell is about 1.5 V, but this voltage will decrease somewhat over time because, as the reaction proceeds, the concentration $NH_4^+ (aq)$ will decrease and $NH_3(aq)$ will decrease, leading to an increase in the value of Q in the Nerst equation and a decrease in the emf of the cell. The other species in the overall reaction are present in pure liquid or solid form and do not affect Q.

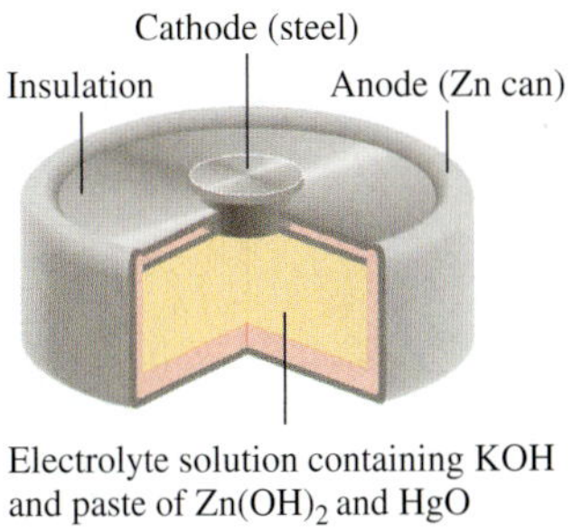

The Mercury Battery

The mercury battery is used extensively in medicine and the electronics industry and is more expensive than the common dry cell. Contained in a stainless-steel cylinder, the mercury battery consists of a zinc anode (amalgamated with mercury) in contact with a strongly alkaline electrolyte containing zinc oxide and mercury(II) oxide (Figure 13.10).

The cell reactions are

anode: $\quad Zn(Hg) + 2OH^-(aq) \longrightarrow ZnO(s) + H_2O(l) + 2e^-$

cathode: $\quad HgO(s) + H_2O(l) + 2e^- \longrightarrow Hg(l) + 2OH^-(aq)$

overall: $\qquad Zn(Hg) + HgO(s) \longrightarrow ZnO(s) + Hg(l)$

Because there is no change in electrolyte composition during operation—the overall cell reaction involves only solid substances—and because no aqueous species appear in the overall cell reaction, the mercury battery provides a more constant voltage (1.35 V) than the Leclanché cell. It also has a considerably higher capacity and longer life. These qualities make the mercury battery ideal for use in pacemakers, hearing aids, electric watches, and light meters.

The Lead Storage Battery

The lead storage battery commonly used in automobiles consists of six identical cells joined together in series. Each cell has a lead anode and a cathode made of lead dioxide (PbO_2) packed on a metal plate (Figure 13.11).

Cathode (steel)

Insulation

Anode (Zn can)

Electrolyte solution containing KOH and paste of $Zn(OH)_2$ and HgO

Figure 13.10 Interior section of a mercury battery.

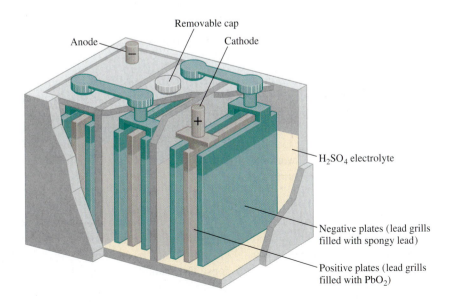

Removable cap

Anode

Cathode

H_2SO_4 electrolyte

Negative plates (lead grills filled with spongy lead)

Positive plates (lead grills filled with PbO_2)

Figure 13.11 Interior section of a lead storage battery. Under normal operating conditions, the concentration of sulfuric acid solution is about 38 percent by mass.

Both the cathode and the anode are immersed in an aqueous solution of sulfuric acid, which acts as the electrolyte. The cell reactions are

anode: $\qquad$ $Pb(s) + SO_4^{2-}(aq) \longrightarrow PbSO_4(s) + 2e^-$

cathode: $\qquad$ $PbO_2(s) + 4H^+(aq) + SO_4^{2-}(aq) + 2e^- \longrightarrow PbSO_4(s) + 2H_2O(l)$

overall: $\quad Pb(s) + PbO_2(s) + 4H^+(aq) + 2SO_4^{2-}(aq) \longrightarrow 2PbSO_4(s) + 2H_2O(l)$

Under normal operating conditions, each cell produces 2 V; a total of 12 V from the six cells is used to power the ignition circuit of the automobile and its other electric systems. The lead storage battery can deliver large amounts of current for a short time, such as the time it takes to start up the engine.

Unlike the Leclanché cell and the mercury battery, the lead storage battery is rechargeable. Recharging the battery means reversing the normal electrochemical reaction by applying an external voltage at the cathode and the anode. (This kind of process is called *electrolysis,* see Section 13.7.) The reactions that replenish the original materials are

anode: $\qquad$ $PbSO_4(s) + 2e^- \longrightarrow Pb(s) + SO_4^{2-}(aq)$

cathode: $\quad PbSO_4(s) + 2H_2O(l) \longrightarrow PbO_2(s) + 4H^+(aq) + SO_4^{2-}(aq) + 2e^-$

overall: $\quad 2PbSO_4(s) + 2H_2O(l) \longrightarrow Pb(s) + PbO_2(s) + 4H^+(aq) + 2SO_4^{2-}(aq)$

The overall reaction is exactly the reverse of the normal cell reaction.

Two aspects of the operation of a lead storage battery are worth noting. First, because the electrochemical reaction consumes sulfuric acid, the degree to which the battery has been discharged can be checked by measuring the density of the electrolyte with a hydrometer, as is usually done at gas stations. The density of the fluid in a "healthy," fully charged battery should be equal to or greater than 1.2 g mL^{-1}. Second, people living in cold climates sometimes have trouble starting their cars because the battery has "gone dead." In Section 13.4, we showed that the emf of a cell is temperature dependent. However, for a lead storage battery, there is a decrease in voltage of 1.5×10^{-4} V for every degree Celsius drop in temperature. Thus, even allowing for a 40°C change in temperature, the decrease in voltage amounts to only 6×10^{-3} V, which is about

$$\frac{6 \times 10^{-3} \text{ V}}{12 \text{ V}} \times 100\% = 0.05\%$$

of the operating voltage, an insignificant change. The real cause of an apparent breakdown of a battery is an increase in the viscosity of the electrolyte as the temperature decreases. For the battery to function properly, the electrolyte must be fully conducting. However, the ions move much more slowly in a viscous medium, so the resistance of the fluid increases, leading to a decrease in the power output of the battery. If an apparently "dead battery" is warmed to near room temperature on a frigid day, it recovers its ability to deliver normal power.

The Lithium-Ion Battery

Figure 13.12 shows a schematic solid-state lithium battery. The anode is made of a conducting carbonaceous material, usually graphite, which has tiny spaces in its structure that can hold both Li atoms and Li$^+$ ions. The cathode is made of a transition

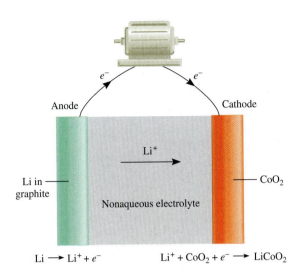

anode (oxidation):
cathode (reduction): $Li^+(aq) + CoO_2(s) + e^-$

overall:

$Li \longrightarrow Li^+ + e^-$ $Li^+ + CoO_2 + e^- \longrightarrow LiCoO_2$

Figure 13.12 A lithium-ion type battery. Lithium atoms are embedded in the graphite, which serves as the anode and CoO_2 is the cathode. During operation, Li^+ ions migrate through the nonaqueous electrolyte from the anode to the cathode while electrons flow externally from the anode to the cathode to complete the circuit.

metal oxide such as CoO_2, which can also hold Li^+ ions. Because of the high reactivity of the metal, a nonaqueous electrolyte (organic solvent plus dissolved salt) must be used. During the discharge of the battery, the half-cell reactions are

$$\text{anode (oxidation):} \qquad \qquad Li(s) \longrightarrow Li^+(aq) + e^-$$
$$\text{cathode (reduction):} \quad Li^+(aq) + CoO_2(s) + e^- \longrightarrow LiCoO_2(s)$$
$$\text{overall:} \qquad \qquad Li(s) + CoO_2(s) \longrightarrow LiCoO_2(s)$$

The advantage of the battery is that lithium has the most negative standard reduction potential value (see Table 13.1). Furthermore, lithium is the lightest metal, so that only 6.941 g of Li (its molar mass) is needed to produce 1 mole of electrons. A lithium-ion battery can be recharged literally hundreds of times without deterioration. These desirable characteristics make it suitable for use in cellular telephones, digital cameras, and laptop computers.

Fuel Cells

Fossil fuels are a major source of energy, but conversion of fossil fuel into electric energy is a highly inefficient process. Consider the combustion of methane:

$$CH_4(g) + 2O_2(g) \longrightarrow CO_2(g) + 2H_2O(l) + \text{energy}$$

To generate electricity, heat produced by the reaction is first used to convert water to steam, which then drives a turbine that drives a generator. An appreciable fraction of the energy released in the form of heat is lost to the surroundings at each step; even the most efficient power plant converts only about 40 percent of the original chemical energy into electricity. Because combustion reactions are redox reactions, it is more desirable to carry them out directly by electrochemical means, thereby greatly increasing the efficiency of power production. This objective can be accomplished by a device known as a *fuel cell, a galvanic cell that requires a continuous supply of reactants to keep functioning.*

In its simplest form, a hydrogen-oxygen fuel cell consists of an electrolyte solution, such as potassium hydroxide solution, and two inert electrodes. Hydrogen and

Figure 13.13 A hydrogen-oxygen fuel cell. The Ni and NiO embedded in the porous carbon electrodes are electrocatalysts.

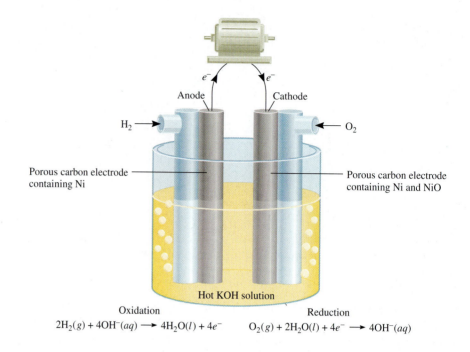

Oxidation
$$2H_2(g) + 4OH^-(aq) \longrightarrow 4H_2O(l) + 4e^-$$

Reduction
$$O_2(g) + 2H_2O(l) + 4e^- \longrightarrow 4OH^-(aq)$$

oxygen gases are bubbled through the anode and cathode compartments (Figure 13.13), where the following reactions take place:

anode: $2H_2(g) + 4OH^-(aq) \longrightarrow 4H_2O(l) + 4e^-$

cathode: $O_2(g) + 2H_2O(l) + 4e^- \longrightarrow 4OH^-(aq)$

overall: $2H_2(g) + O_2(g) \longrightarrow 2H_2O(l)$

The standard emf of the cell can be calculated as follows, with data from Table 13.1:

$$\mathscr{E}^\circ_{cell} = \mathscr{E}^\circ_{cathode} - \mathscr{E}^\circ_{anode}$$
$$= 0.40 \text{ V} - (-0.83 \text{ V})$$
$$= 1.23 \text{ V}$$

Thus, the cell reaction is spontaneous under standard-state conditions. Note that the reaction is the same as the hydrogen combustion reaction, but the oxidation and reduction are carried out separately at the anode and the cathode. Like platinum in the standard hydrogen electrode, the electrodes have a two-fold function. They serve as electric conductors, and they provide the necessary surfaces for the initial decomposition of the molecules into atomic species, prior to electron transfer. They are *electrocatalysts*. Metals such as platinum, nickel, and rhodium are good electrocatalysts.

In addition to the H_2-O_2 system, a number of other fuel cells have been developed. Among these is the propane-oxygen fuel cell. The half-cell reactions are

anode: $C_3H_8(g) + 6H_2O(l) \longrightarrow 3CO_2(g) + 2OH^-(aq) + 20e^-$

cathode: $5O_2(g) + 20H^+(aq) + 20e^- \longrightarrow H_2O(l)$

overall: $C_3H_8(g) + 5O_2(g) \longrightarrow 3CO_2(g) + 4H_2O(l)$

The overall reaction is identical to the burning of propane in oxygen.

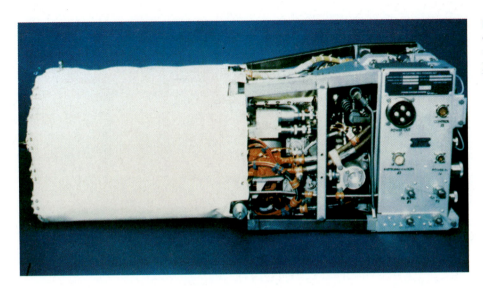

Figure 13.14 A hydrogen-oxygen fuel cell used in the space program. The pure water produced by the cell is consumed by the astronauts.

Unlike batteries, fuel cells do not store chemical energy. Reactants must be constantly resupplied, and products must be constantly removed from a fuel cell. In this respect, a fuel cell resembles an engine more than it does a battery. However, the fuel cell does not operate like a heat engine and therefore is not subject to the same kind of thermodynamic limitations in energy conversion.

Properly designed fuel cells may be as much as 70 percent efficient, about twice as efficient as an internal combustion engine. In addition, fuel-cell generators are free of the noise, vibration, heat transfer, thermal pollution, and other problems normally associated with conventional power plants. Nevertheless, fuel cells are not yet in widespread use. A major problem lies in the lack of cheap electrocatalysts able to function efficiently for long periods of time without contamination. The most successful application of fuel cells to date has been in space vehicles (Figure 13.14).

13.7 | In Electrolysis, an Electric Current Is Used to Drive a Nonspontaneous Reaction

In contrast to spontaneous redox reactions, which result in the conversion of chemical energy into electric energy, *electrolysis* is the process in which *electric energy is used to cause a nonspontaneous chemical reaction to occur.* An *electrolytic cell* is *an apparatus for carrying out electrolysis.* The same principles underlie electrolysis and the processes that take place in galvanic cells. Here we will discuss three examples of electrolysis based on those principles: electrolysis of molten NaCl, water, and aqueous NaCl. Then we will look at the quantitative aspects of electrolysis.

Electrolysis of Molten Sodium Chloride

In its molten state, sodium chloride, an ionic compound, can be electrolyzed to form sodium metal and chlorine. Figure 13.15(a) is a diagram of a *Downs cell,* which is used for large-scale electrolysis of NaCl. In molten NaCl, the cations and anions are the Na^+ and Cl^- ions, respectively.

Figure 13.15 (a) A Downs cell for the electrolysis of molten NaCl (m.p. 801°C). The sodium metal formed at the cathodes is in the liquid state. Because liquid sodium metal is less dense than molten NaCl, the sodium floats to the surface, as shown, and is collected. Chlorine gas forms at the anode and is collected at the top. (b) A simplified diagram showing the electrode reactions during the electrolysis of molten NaCl. The battery is needed to drive the nonspontaneous reactions.

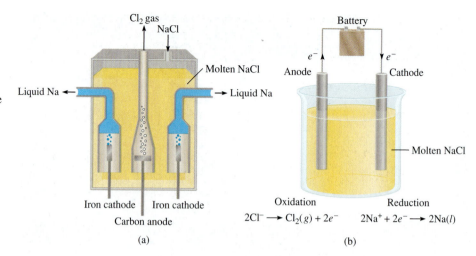

Oxidation
$2Cl^- \longrightarrow Cl_2(g) + 2e^-$

Reduction
$2Na^+ + 2e^- \longrightarrow 2Na(l)$

(a)   (b)

Figure 13.15(b) is a simplified diagram showing the reactions that occur at the electrodes. The electrolytic cell contains a pair of electrodes connected to the battery. The battery serves as an "electron pump," driving electrons to the cathode, where reduction occurs, and withdrawing electrons from the anode, where oxidation occurs. The reactions at the electrodes are

anode (oxidation):	$2Cl^-(l) \longrightarrow Cl_2(g) + 2e^-$
cathode (reduction):	$2Na^+(l) + 2e^- \longrightarrow 2Na(l)$
overall:	$2Na^+(l) + 2Cl^-(l) \longrightarrow 2Na(l) + Cl_2(g)$

This process is a major source of pure sodium metal and chlorine gas. Theoretical estimates show that the $\mathcal{E}°$ value for the overall process is about -4 V, which means that this is a nonspontaneous process. Therefore, a *minimum* of 4 V must be supplied by the battery to carry out the reaction. In practice, a higher voltage is necessary because of inefficiencies in the electrolytic process and because of overvoltage, to be discussed shortly.

Electrolysis of Water

Water in a beaker under standard ambient conditions (1 bar and 25°C) will not spontaneously decompose to form hydrogen and oxygen gas because the standard free-energy change for the reaction is a large positive quantity:

$$2H_2O(l) \longrightarrow 2H_2(g) + O_2(g) \quad \Delta G° = 474.2 \text{ kJ mol}^{-1}$$

However, this reaction can be induced in a cell like the one shown in Figure 13.16.

This electrolytic cell consists of a pair of electrodes made of a nonreactive metal, such as platinum, immersed in water. When the electrodes are connected to the battery, nothing happens because there are not enough ions in pure water to carry much of an electric current. (Remember that at 25°C, pure water has only 1×10^{-7} M H$^+$ ions and 1×10^{-7} M OH$^-$ ions.) On the other hand, the reaction occurs readily in a 0.1 M H$_2$SO$_4$ solution because there are a sufficient number of ions to conduct electricity. Immediately, gas bubbles begin to appear at both electrodes.

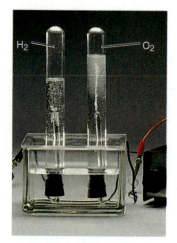

Figure 13.16 Apparatus for small-scale electrolysis of water. The volume of hydrogen gas generated is twice that of oxygen gas.

Figure 13.17 shows the electrode reactions. The process at the anode is

$$2H_2O(l) \longrightarrow O_2(g) + 4H^+(aq) + 4e^-$$

whereas, at the cathode we have

$$2H^+(aq) + 2e^- \longrightarrow H_2(g)$$

The overall reaction is given by

anode (oxidation): $\qquad 2H_2O(l) \longrightarrow O_2(g) + 4H^+(aq) + 4e^-$

cathode (reduction): $\quad 2[2H^+(aq) + 2e^- \longrightarrow H_2(g)]$

overall: $\qquad\qquad 2H_2O(l) \longrightarrow 2H_2(g) + O_2(g)$

Note that no H_2SO_4 is consumed in the overall reaction.

Electrolysis of an Aqueous Sodium Chloride Solution

This is the most complicated of the three examples of electrolysis considered here because aqueous sodium chloride solution contains several species that could be oxidized and reduced. The oxidation reactions that might occur at the anode are

(1) $\quad 2Cl^-(aq) \longrightarrow Cl_2(g) + 2e^-$
(2) $\quad 2H_2O(l) \longrightarrow O_2(g) + 4H^+(aq) + 4e^-$

Referring to Table 13.1, we find

$$Cl_2(g) + 2e^- \longrightarrow 2Cl^-(aq) \qquad \mathcal{E}° = 1.36 \text{ V}$$
$$O_2(g) + 4H^+(aq) + 4e^- \longrightarrow 2H_2O(l) \qquad \mathcal{E}° = 1.23 \text{ V}$$

The standard reduction potentials of (1) and (2) are not very different, but the values do suggest that H_2O should be preferentially oxidized at the anode. However, by experiment we find that the gas liberated at the anode is Cl_2, not O_2! In studying electrolytic processes, we sometimes find that the voltage required for a reaction is considerably higher than the electrode potential indicates. The **overvoltage** is *the difference between the electrode potential and the actual voltage required to cause electrolysis*. The overvoltage for O_2 formation is quite high. Therefore, under normal

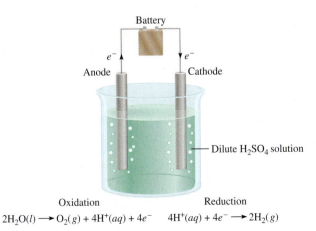

Oxidation
$$2H_2O(l) \longrightarrow O_2(g) + 4H^+(aq) + 4e^-$$

Reduction
$$4H^+(aq) + 4e^- \longrightarrow 2H_2(g)$$

Figure 13.17 A diagram showing the electrode reactions during the electrolysis of water.

operating conditions, Cl_2 gas is actually formed at the anode instead of O_2. The reductions that might occur at the cathode are

$$(3) \quad 2H^+(aq) + 2e^- \longrightarrow H_2(g) \qquad\qquad \mathcal{E}° = 0.00 \text{ V}$$

$$(4) \quad 2H_2O(l) + 2e^- \longrightarrow H_2(g) + 2OH^-(aq) \quad \mathcal{E}° = -0.83 \text{ V}$$

$$(5) \quad Na^+(aq) + e^- \longrightarrow Na(s) \qquad\qquad \mathcal{E}° = -2.71 \text{ V}$$

Reaction (5) is ruled out because it has a very negative standard reduction potential. Reaction (3) is preferred over (4) under standard-state conditions. At a pH of 7 (as is the case for an NaCl solution), however, they are equally probable. We generally use reaction (4) to describe the cathode reaction because the concentration of H^+ ions is too low (about 1×10^{-7} M) to make reaction (3) a reasonable choice. Thus, the half-cell reactions in the electrolysis of aqueous sodium chloride are

anode (oxidation): $2Cl^-(aq) \longrightarrow Cl_2(g) + 2e^-$

cathode (reduction): $2H_2O(l) + 2e^- \longrightarrow H_2(g) + 2OH^-(aq)$

overall: $2H_2O(l) + 2Cl^-(aq) \longrightarrow H_2(g) + Cl_2(g) + 2OH^-(aq)$

As the overall reaction shows, the concentration of the Cl^- ions decreases during electrolysis and that of the OH^- ions increases. Therefore, in addition to H_2 and Cl_2, the useful byproduct NaOH can be obtained by evaporating the aqueous solution at the end of the electrolysis.

Keep in mind the following from our analysis of electrolysis: cations are likely to be reduced at the cathode and anions are likely to be oxidized at the anode, and in aqueous solutions, water itself may be oxidized and/or reduced. The outcome depends on the nature of other species present.

Example 13.10 deals with the electrolysis of an aqueous solution of sodium sulfate (Na_2SO_4).

Example 13.10

An aqueous Na_2SO_4 solution is electrolyzed, using the apparatus shown in Figure 13.16. If the products formed at the anode and cathode are oxygen gas and hydrogen gas, respectively, describe the electrolysis in terms of the reactions at the electrodes. *Note:* Na_2SO_4 does not hydrolyze.

Strategy Before we look at the electrode reactions, we should consider the following facts: (1) Because Na_2SO_4 does not hydrolyze, the pH of the solution is close to 7. (2) The Na^+ ions are not reduced at the cathode, and the SO_4^{2-} ions are not oxidized at the anode. These conclusions are drawn from the electrolysis of water in the presence of sulfuric acid and in aqueous sodium chloride solution, as discussed earlier. Therefore, both the oxidation and reduction reactions involve only water molecules.

Solution The electrode reactions are

oxidation (anode): $2H_2O(l) \longrightarrow O_2(g) + 4H^+(aq) + 4e^-$

reduction (cathode): $2H_2O(l) + 2e^- \longrightarrow H_2(g) + 2OH^-(aq)$

The overall reaction, obtained by doubling the cathode reaction coefficients and adding the result to the anode reaction, is

$$6H_2O(l) \longrightarrow 2H_2(g) + O_2(g) + 4H^+(aq) + 4OH^-(aq)$$

—Continued

Continued—

If the H^+ and OH^- ions are allowed to mix, then

$$4H^+(aq) + 4OH^-(aq) \longrightarrow 4H_2O(l)$$

and the overall reaction becomes

$$2H_2O(l) \longrightarrow 2H_2(g) + O_2(g)$$

Practice Exercise An aqueous solution of $Mg(NO_3)_2$ is electrolyzed. What are the gaseous products at the anode and cathode?

Electrolysis has many important applications in industry, mainly in the extraction and purification of metals.

Quantitative Aspects of Electrolysis

The quantitative treatment of electrolysis was developed primarily by Faraday. He observed that the mass of product formed (or reactant consumed) at an electrode is proportional to both the amount of electricity transferred at the electrode and the molar mass of the substance in question. For example, in the electrolysis of molten NaCl, the cathode reaction tells us that one Na atom is produced when one Na^+ ion accepts an electron from the electrode. To reduce 1 mole of Na^+ ions, we must supply Avogadro's number (6.02×10^{23}) of electrons to the cathode. On the other hand, the stoichiometry of the anode reaction shows that oxidation of two Cl^- ions yields one chlorine molecule. Therefore, the formation of 1 mole of Cl_2 results in the transfer of 2 moles of electrons from the Cl ions to the anode. Similarly, it takes 2 moles of electrons to reduce 1 mole of Mg^{2+} ions and 3 moles of electrons to reduce 1 mole of Al^{3+} ions:

$$Mg^{2+} + 2e^- \longrightarrow Mg$$
$$Al^{3+} + 3e^- \longrightarrow Al$$

In an electrolysis experiment, we generally measure the current that passes through an electrolytic cell in a given period of time. The SI unit of current is the *ampere* (A) defined as the amount of current corresponding to 1 coulomb of charge per second.

Figure 13.18 shows the steps involved in calculating the quantities of substances produced in electrolysis. Let us illustrate the approach by considering molten $CaCl_2$ in an electrolytic cell. Suppose a current of 0.452 A is passed through the cell for 1.50 h. How much product will be formed at the anode and at the cathode? In solving electrolysis problems of this type, the first step is to determine which species will be oxidized at the anode and which species will be reduced at the cathode. Here the choice is straightforward because we only have Ca^{2+} and Cl^- ions in molten $CaCl_2$. Thus, we write the half- and overall cell reactions as

anode (oxidation): $\qquad\qquad 2Cl^- \longrightarrow Cl_2(g) + 2e^-$

cathode (reduction): $\quad Ca^{2+} + 2e^- \longrightarrow Ca(l)$

overall: $\quad Ca^{2+} + 2Cl^- \longrightarrow Ca(l) + Cl_2(g)$

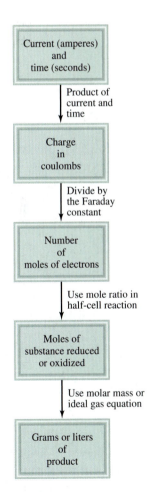

Figure 13.18 Steps involved in calculating amounts of substances reduced or oxidized in electrolysis.

The quantities of calcium metal and chlorine gas formed depend on the number of electrons that pass through the electrolytic cell, which in turn depends on current × time, or charge:

$$\text{charge (in C)} = 0.452 \text{ A} \times 1.50 \text{ h} \times \frac{3600 \text{ s}}{\text{h}} \times \frac{1 \text{ C}}{1 \text{ A} \cdot \text{s}} = 2.44 \times 10^3 \text{ C}$$

Because 1 mol e^- = 96,500 C and 2 mol e^- are required to reduce 1 mole of Ca^{2+} ions, the mass of Ca metal formed at the cathode is calculated as follows:

$$\text{mass of Ca} = 2.44 \times 10^3 \text{ C} \times \frac{1 \text{ mol } e^-}{96,500 \text{ C}} \times \frac{1 \text{ mol Ca}}{2 \text{ mol } e^-} \times \frac{40.08 \text{ g Ca}}{1 \text{ mol Ca}} = 0.507 \text{ g Ca}$$

The anode reaction indicates that 1 mole of chlorine gas is produced per 2 mol e^- of electricity. Hence, the mass of chlorine gas formed is

$$\text{mass of Cl}_2 = 2.44 \times 10^3 \text{C} \times \frac{1 \text{ mol } e^-}{96,500 \text{ C}} \times \frac{1 \text{ mol Ca}}{2 \text{ mol } e^-} \times \frac{70.90 \text{ g Cl}_2}{1 \text{ mol Cl}_2} = 0.896 \text{ g Cl}_2$$

Example 13.11 applies this approach to the electrolysis in an aqueous solution.

Example 13.11

A current of 1.26 A is passed through an electrolytic cell containing a dilute sulfuric acid solution for 7.44 h. Write the half-cell reactions and calculate the volume of gases generated at standard ambient temperature and pressure (SATP) (25°C and 1 bar pressure), assuming the gases behave ideally.

Strategy Earlier we saw that the half-cell reactions for the process are

anode (oxidation): $2H_2O(l) \longrightarrow O_2(g) + 4H^+(aq) + 4e^-$

cathode (reduction): $2[2H^+(aq) + 2e^- \longrightarrow H_2(g)]$

overall: $2H_2O(l) \longrightarrow 2H_2(g) + O_2(g)$

According to Figure 13.18, we carry out the following conversion steps to calculate the quantity of O_2 in moles:

current × time $\longrightarrow$ coulombs $\longrightarrow$ moles of e^- $\longrightarrow$ moles of O_2 and H_2

Then, using the ideal gas equation, we can calculate the volume of O_2 in liters at SATP. From the stoichiometry 2 moles of H_2 are produced for every mole of O_2, so by Avogadro's law the volume of hydrogen produced should be twice that of oxygen.

Solution First we calculate the number of coulombs of electricity that pass through the cell:

$$\text{charge (in C)} = 1.26 \text{ A} \times 7.44 \text{ h} \times \frac{3600 \text{ s}}{\text{h}} \times \frac{1 \text{ C}}{1 \text{ A} \cdot \text{s}} = 3.37 \times 10^4 \text{ C}$$

—Continued

Continued—

We see that for every mole of O_2 formed at the anode, 4 moles of electrons are generated so that

$$\text{mol } O_2 = 3.37 \times 10^4 C \times \frac{1 \text{ mol } e^-}{96,500 \text{ C}} \times \frac{1 \text{ mol Ca}}{4 \text{ mol } e^-} = 0.0873 \text{ mol } O_2$$

The volume of $0.0873 \text{ mol } O_2$ at SATP is given by

$$V = \frac{nRT}{P} = \frac{(0.0873 \text{ mol})(0.08314 \text{ L bar mol}^{-1} \text{ K}^{-1})(298 \text{ K})}{1 \text{ bar}} = 2.16 \text{ L}$$

The volume of H_2 produced is then $2 \times 2.16 \text{ L} = 4.32 \text{ L}$.

Check Recall from Chapter 5 that the volume of 1 mol of gas at SATP is 24.8 L. The amount of oxygen produced is slightly less than one tenth of a mole, so the volume should be slightly less than one-tenth of $24.8 = 2.48$, which is true.

Practice Exercise A constant current is passed through an electrolytic cell containing molten $MgCl_2$ for 18 h. If 4.8×10^5 g of Cl_2 are obtained, what is the current in amperes?

Summary of Facts and Concepts

Section 13.1

▶ Redox reactions involve the transfer of electrons. Equations representing redox processes can be balanced using the ion-electron method.

▶ All electrochemical reactions involve the transfer of electrons and therefore are redox reactions.

Section 13.2

▶ In a galvanic cell, electricity is produced by a spontaneous chemical reaction. Oxidation and reduction take place separately at the anode and cathode, respectively, and the electrons flow through an external circuit.

▶ The two parts of a galvanic cell are the half-cells, and the reactions at the electrodes are the half-cell reactions. A salt bridge allows ions to flow between the half-cells.

Section 13.3

▶ The voltage of a cell is the voltage difference between the two electrodes. In the external circuit, electrons flow from the anode to the cathode in a galvanic cell. In solution, the anions move toward the anode and the cations move toward the cathode. If the cell is operated reversibly, the cell voltage is called the electromotive force (emf).

▶ Standard reduction potentials show the relative likelihood of half-cell reduction reactions and can be used to predict the products, direction, and spontaneity of redox reactions between various substances.

Section 13.4

▶ The emf of an electrochemical cell is directly related to the Gibbs free energy change (ΔG) of the redox reaction.

▶ The quantity of electricity carried by 1 mole of electrons is called the Faraday constant (F), which is equal to $96,485 \text{ C mol}^{-1}$.

▶ The temperature dependence of the emf of a cell can be determined from the thermodynamic properties of the cell reaction.

Section 13.5

▶ The concentration dependence of the emf can be determined using the Nernst equation.

Section 13.6

▶ Batteries use a spontaneous electrochemical reaction to provide a ready supply of electric current.

Section 13.7

▶ In electrolysis, an electric current is used to drive an electrochemical reaction in a nonspontaneous direction.

Key Words

Problems

Oxidation-Reduction (Redox) Reactions Involve a Transfer of Electrons from One Species to Another

13.1 For the complete redox reactions given here, write the half-reactions and identify the oxidizing and reducing agents:

(a) $2Sr + O_2 \longrightarrow 2SrO$

(b) $2Li + H_2 \longrightarrow 2LiH$

(c) $2Cs + Br_2 \longrightarrow 2CsBr$

(d) $3Mg + N_2 \longrightarrow Mg_3N_2$

13.2 For the complete redox reactions given here, write the half-reactions and identify the oxidizing and reducing agents:

(a) $4Fe + 3O_2 \longrightarrow 2Fe_2O_3$

(b) $Cl_2 + 2NaBr \longrightarrow 2NaCl + Br_2$

(c) $Si + 2F_2 \longrightarrow SiF_4$

(d) $H_2 + Cl_2 \longrightarrow 2HCl$

13.3 Arrange the following species in order of increasing oxidation number of the sulfur atom: (a) H_2S, (b) S_8, (c) H_2SO_4, (d) S^{2-}, (e) HS^-, (f) SO_2, and (g) SO_3.

13.4 Phosphorus forms many oxoacids. Indicate the oxidation number of phosphorus in each of the following acids: (a) HPO_3, (b) H_3PO_2, (c) H_3PO_3, (d) H_3PO_4, (e) $H_4P_2O_7$, and (f) $H_5P_3O_{10}$.

13.5 Give oxidation numbers for all atoms in the following molecules and ions: (a) ClF, (b) IF_7, (c) CH_4, (d) C_2H_2, (e) C_2H_4, (f) K_2CrO_4, (g) $K_2Cr_2O_7$, (h) $KMnO_4$, (i) $NaHCO_3$, (j) Li_2, (k) $NaIO_3$, (l) KO_2, (m) PF_6^-, and (n) $KAuCl_4$.

13.6 Give oxidation numbers for all atoms in the following molecules and ions: (a) Cs_2O, (b) CaI_2, (c) Al_2O_3, (d) H_3AsO_3, (e) TiO_2, (f) MoO_4^{2-}, (g) $PtCl_4^{2-}$, (h) $PtCl_6^{2-}$, (i) SnF_2, (j) ClF_3, and (k) SbF_6^-.

13.7 Give oxidation numbers for all atoms in the following molecules and ions: (a) Mg_3N_2, (b) CsO_2, (c) CaC_2, (d) CO_3^{2-}, (e) $C_2O_4^{2-}$, (f) ZnO_2^{2-}, (g) $NaBH_4$, and (h) WO_4^{2-}.

13.8 Nitric acid is a strong oxidizing agent. State which of the following species is *least* likely to be produced when nitric acid reacts with a strong reducing agent such as zinc metal, and explain why: N_2O, NO, NO_2, N_2O_4, N_2O_5, and NH_4^+.

13.9 On the basis of oxidation number considerations, one of the following oxides would not react with molecular oxygen: NO, N_2O, SO_2, SO_3, and P_4O_6. Which one is it? Why?

13.10 Balance the following redox equations by the ion-electron method:

(a) $H_2O_2 + Fe^{2+} \longrightarrow Fe^{3+} + H_2O$ (in acidic solution)

(b) $Cu^+ + HNO_3 \longrightarrow Cu^{2+} + NO + H_2O$ (in acidic solution)

(c) $CN^- + MnO \longrightarrow CNO^- + MnO_2$ (in basic solution)

(d) $Br_2 \longrightarrow BrO_3^- + Br^-$ (in basic solution)

(e) $S_2O_3^{2-} + I_2 \longrightarrow I^- + S_4O_6^{2-}$ (in acidic solution)

13.11 Balance the following redox equations by the ion-electron method:

(a) $Mn^{2+} + H_2O_2 \longrightarrow MnO_2 + H_2O$ (in basic solution)

(b) $Bi(OH)_3 + SnO_2^{2-} \longrightarrow SnO_3^{2-} + Bi$ (in basic solution)

(c) $Cr_2O_7^{2-} + C_2O_4^{2-} \longrightarrow Cr^{3+} + CO_2$ (in acidic solution)

(d) $ClO_3^- + Cl^- \longrightarrow Cl_2 + ClO_2^-$ (in acidic solution)

The Standard Emf of Any Electrochemical Cell Can Be Determined If the Standard Reduction Potentials for the Half-Reactions Are Known

13.12 Calculate the standard emf for the reactions implied by the following cell diagrams at 298 K. For each, write the overall reaction and identify which electrode would be the cathode under standard-state conditions.

(a) $Fe(s)|Fe^{2+}(aq)||Cr^{3+}(aq)|Cr(s)$

(b) $Zn(s)|Zn^{2+}(aq)||Al^{3+}(aq)|Al(s)$

(c) $Mg(s)|Mg^{2+}(aq)||Pb^{2+}(aq)|Pb(s)$

13.13 Calculate the standard emf of a cell that uses $Ag\,|\,Ag^+$ and $Al\,|\,Al^{3+}$ half-cell reactions. Write the overall cell reaction that occurs under standard-state conditions.

13.14 Predict whether Fe^{3+} can oxidize I^- to I_2 under standard-state conditions.

13.15 Which of the following reagents can oxidize H_2O to $O_2(g)$ under standard-state conditions: (a) $H^+(aq)$, (b) $Cl^-(aq)$, (c) $Cl_2(g)$, (d) $Cu^{2+}(aq)$, (e) $Pb^{2+}(aq)$, or (f) $MnO_4^-(aq)$ (in acid)?

13.16 Consider the following half-reactions:

$$MnO_4^-(aq) + 8H^+(aq) + 5e^- \longrightarrow$$
$$Mn^{2+}(aq) + 4H_2O(l)$$

$$NO_3^-(aq) + 4H^+(aq) + 3e^{-1} \longrightarrow$$
$$NO(g) + 2H_2O(l)$$

Predict whether NO_3^- ions will oxidize Mn^{2+} to MnO_4^- under standard-state conditions.

13.17 Predict whether the following reactions would occur spontaneously in aqueous solution at 25°C. Assume all species are in their standard states.

(a) $Ca(s) + Cd^{2+}(aq) \longrightarrow Ca^{2+}(aq) + Cd(s)$
(b) $2Br(aq) + Sn^{2+}(aq) \longrightarrow Br_2(l) + Sn(s)$
(c) $2Ag(s) + Ni^{2+}(aq) \longrightarrow 2Ag^+(aq) + Ni(s)$
(d) $Cu(aq) + Fe^{3+}(aq) \longrightarrow Cu^{2+}(aq) + Fe^{2+}(aq)$

13.18 Which species in each pair is a better reducing agent under standard-state conditions: (a) Na or Li, (b) H_2 or I_2, (c) Fe^{2+} or Ag, or (d) Br^- or Co^{2+}?

The Emf of an Electrochemical Cell Is Directly Related to the Gibbs Free-Energy Change of the Redox Reaction

13.19 What is the equilibrium constant for the following reaction at 25°C?

$$Mg(s) + Zn^{2+}(aq) \rightleftharpoons Mg^{2+}(aq) + Zn(s)$$

13.20 The equilibrium constant for the reaction

$$Sr(s) + Mg^{2+}(aq) \rightleftharpoons Sr^{2+}(aq) + Mg(s)$$

is 2.69×10^{12} at 25°C. Calculate $\mathscr{E}°$ for a cell made up of $Sr\,|\,Sr^{2+}$ and $Mg\,|\,Mg^{2+}$ half-cells.

13.21 Use the standard reduction potentials to find the equilibrium constant for each of the following reactions at 25°C:

(a) $Br_2(l) + 2I^-(aq) \rightleftharpoons 2Br^-(aq) + I_2(s)$
(b) $2Ce^{4+}(aq) + 2Cl^-(aq) \rightleftharpoons$
$$Cl_2(g) + 2Ce^{3+}(aq)$$
(c) $5Fe^2(aq) + MnO_4^-(aq) + 8H^+(aq) \rightleftharpoons$
$$Mn^{2+}(aq) + 4H_2O + 5Fe^{3+}(aq)$$

13.22 Using data from Table 13.1, calculate $\Delta G°$ and K_c for the following reactions at 25°C:

(a) $Mg(s) + Pb^{2+}(aq) \rightleftharpoons Mg^{2+}(aq) + Pb(s)$
(b) $Br_2(l) + 2I^-(aq) \rightleftharpoons 2Br^-(aq) + I_2(s)$

(c) $O_2(g) + 4H^+(aq) + 4Fe^{2+}(aq) \rightleftharpoons$
$$2H_2O(l) + 4Fe^{3+}(aq)$$
(d) $2Al(s) + 3I_2(s) \rightleftharpoons 2Al^{3+}(aq) + 6I^-(aq)$

13.23 Under standard-state conditions, what spontaneous reaction will occur in aqueous solution among the ions Ce^{4+}, Ce^{3+}, Fe^{3+}, and Fe^{2+}? Calculate $\Delta G°$ and K_c for the reaction.

13.24 Given that $\mathscr{E}° = 0.52$ V for the reduction $Cu^+(aq) + e^- \longrightarrow Cu(s)$, calculate $\mathscr{E}°$, $\Delta G°$, and K for the following reaction at 25°C:

$$2Cu^+(aq) \longrightarrow Cu^{2+}(aq) + Cu(s)$$

13.25 Calculate the emf of the cell reaction

$$Br_2(l) + 2Cu^+(aq) \longrightarrow 2Br^-(aq) + 2Cu^{2+}(aq)$$

at 70°C. Explain any assumptions that you make.

13.26 For a certain redox reaction with $\Delta n = 1$, it is determined that the standard emf increases by 0.124 V if the temperature is changed from 25°C to 50°C. From this information, estimate the standard entropy change for this reaction. Explain any assumptions that you make.

The Concentration Dependence of the Emf Can Be Determined Using the Nernst Equation

13.27 What is the potential of a cell made up of $Zn\,|\,Zn^{2+}$ and $Cu\,|\,Cu^{2+}$ half-cells at 25°C if $[Zn^{2+}] = 0.25\ M$ and $[Cu^{2+}] = 0.15\ M$? Assume ideal behavior.

13.28 Calculate $\mathscr{E}°$, E, and ΔG for the following cell reactions. Assume ideal behavior.

(a) $Mg(s) + Sn^2(aq) \longrightarrow Mg^{2+}(aq) + Sn(s)$
$$[Mg^{2+}] = 0.045\ M, [Sn^{2+}] = 0.035\ M$$
(b) $3Zn(s) + 2Cr^{3+}(aq) \longrightarrow 3Zn^{2+}(aq) + 2Cr(s)$
$$[Cr^{3+}] = 0.010\ M, [Zn^{2+}] = 0.0085\ M$$

13.29 Calculate the standard potential of the cell consisting of the $Zn\,|\,Zn^{2+}$ half-cell and the standard hydrogen electrode. What will be the emf of the cell if $[Zn^2] = 0.45\ M$, $P_{H_2} = 2.0$ bar, and $[H^+] = 1.8\ M$? Discuss any assumptions that you make.

13.30 What is the emf of a cell consisting of a $Pb^2\,|\,Pb$ half-cell and a $Pt\,|\,H^+\,|\,H_2$ half-cell if $[Pb^{2+}] = 0.10\ M$, $[H^+] = 0.050\ M$, and $P_{H_2} = 1.0$ bar? Discuss any assumptions that you make.

13.31 Consider a Daniell cell operating under non–standard-state conditions. Suppose that the cell reaction is multiplied by 2. What effect does this have on each of the following quantities in the Nernst equation: (a) $\mathscr{E}°$, (b) E, (c) Q, (d) $\ln Q$, and (e) n?

13.32 Referring to the arrangement in Figure 13.1, calculate the $[Cu^{2+}]/[Zn^{2+}]$ ratio at which the following reaction is spontaneous at 25°C:

$$Cu(s) + Zn^{2+}(aq) \longrightarrow Cu^{2+}(aq) + Zn(s)$$

13.33 Calculate the emf of the following concentration cell (assuming ideal behavior):

$$Mg(s)\,|\,Mg^{2+}(0.24\ M)\,||\,Mg^{2+}(0.53\ M)\,|\,Mg(s)$$

13.34 Consider a concentration cell consisting of two hydrogen electrodes (Figure 13.5). At 25°C, the cell emf is found to be 0.0121 V. If the pressure of the gas at the anode is 2.5 bar, what is the pressure of hydrogen gas at the cathode?

13.35 Calculate the emf of the Daniell cell at 25°C when the concentrations of $CuSO_4$ and $ZnSO_4$ are 0.50 M and 0.10 M respectively. What would be the emf if activities were used instead of concentrations? (The $\gamma\pm$ for $CuSO_4$ and $ZnSO_4$ at their respective concentrations are 0.068 and 0.15, respectively.)

Batteries Use Electrochemical Reactions to Produce a Ready Supply of Electric Current

13.36 The hydrogen-oxygen fuel cell is described in Section 13.6. (a) What volume of $H_2(g)$, stored at 25°C at a pressure of 155 bar, would be needed to run an electric motor drawing a current of 8.5 A for 3.0 h? (b) What volume (liters) of air at 25°C and 1.00 bar will have to pass into the cell per minute to run the motor? Assume that air is 20 percent O_2 by volume and that all the O_2 is consumed in the cell. The other components of air do not affect the fuel-cell reactions. Assume ideal gas behavior.

13.37 Calculate the standard emf of the propane fuel cell discussed on page 696 at 25°C, given that ΔG_f° for propane is -23.5 kJ mol^{-1}.

In Electrolysis, an Electric Current Is Used to Drive a Nonspontaneous Reaction

13.38 The half-reaction at an electrode is

$$Mg^{2+}(molten) + 2e^- \longrightarrow Mg(s)$$

Calculate the number of grams of magnesium that can be produced by supplying 1.00 F to the electrode.

13.39 Consider the electrolysis of molten barium chloride ($BaCl_2$). (a) Write the half-reactions. (b) How many grams of barium metal can be produced by supplying 0.50 A for 30 min?

13.40 Considering only the cost of electricity, would it be cheaper to produce a ton of sodium or a ton of aluminum by electrolysis?

13.41 If the cost of electricity to produce magnesium by the electrolysis of molten magnesium chloride is $155 per ton of metal, what is the cost (in dollars) of the electricity necessary to produce (a) 10.0 tons of aluminum, (b) 30.0 tons of sodium, (c) 50.0 tons of calcium?

13.42 One of the half-reactions for the electrolysis of water is

$$2H_2O(l) \longrightarrow O_2(g) + 4H^+(aq) + 4e^-$$

If 0.076 L of O_2 is collected at 25°C and 755 mmHg, how many moles of electrons had to pass through the solution?

13.43 How many faradays of electricity are required to produce (a) 0.84 L of O_2 at exactly 1 bar and 25°C from aqueous H_2SO_4 solution; (b) 1.50 L of Cl_2 at 750 mmHg and 20°C from molten NaCl; and (c) 6.0 g of Sn from molten $SnCl_2$?

13.44 Calculate the amounts of Cu and Br_2 produced in 1.0 h at inert electrodes in a solution of $CuBr_2$ by a current of 4.50 A.

13.45 In the electrolysis of an aqueous $AgNO_3$ solution, 0.67 g of Ag is deposited after a certain period of time. (a) Write the half-reaction for the reduction of Ag^+. (b) What is the probable oxidation half-reaction? (c) Calculate the quantity of electricity used, in coulombs.

13.46 A steady current was passed through molten $CoSO_4$ until 2.35 g of metallic cobalt was produced. Calculate the number of coulombs of electricity used.

13.47 A constant electric current flows for 3.75 h through two electrolytic cells connected in series. One contains a solution of $AgNO_3$ and the second a solution of $CuCl_2$. During this time, 2.00 g of silver is deposited in the first cell. (a) How many grams of copper are deposited in the second cell? (b) What is the current flowing, in amperes?

13.48 What is the hourly production rate of chlorine gas (in kg) from an electrolytic cell using aqueous NaCl electrolyte and carrying a current of 1.500×10^3 A? The anode efficiency for the oxidation of Cl^- is 93.0 percent.

13.49 Chromium plating is applied by electrolysis to objects suspended in a dichromate solution, according to the following (unbalanced) half-reaction:

$$Cr_2O_7^{2-}(aq) + e^- + H^+(aq) \longrightarrow Cr(s) + H_2O(l)$$

How long (in hours) would it take to apply a chromium plating 1.0×10^{-2} mm thick to a car bumper with a surface area of 0.25 m^2 in an electrolytic cell carrying a current of 25.0 A? (The density of chromium is 7.19 g cm^{-3}.)

13.50 The passage of a current of 0.750 A for 25.0 min deposited 0.369 g of copper from a $CuSO_4$ solution. From this information, calculate the molar mass of copper.

13.51 A quantity of 0.300 g of copper was deposited from a $CuSO_4$ solution by passing a current of 3.00 A through the solution for 304 s. Calculate the value of the faraday constant.

13.52 In a certain electrolysis experiment, 1.44 g of Ag was deposited in one cell (containing an aqueous $AgNO_3$ solution), while 0.120 g of an unknown metal X was deposited in another cell (containing an aqueous XCl_3 solution) in series with the $AgNO_3$ cell. Calculate the molar mass of X.

13.53 One of the half-reactions for the electrolysis of water is

$$2H^+(aq) + 2e^- \longrightarrow H_2(g)$$

If 0.845 L of H_2 is collected at 25°C and 782 mmHg, how many moles of electrons had to pass through the solution?

Additional Problems

13.54 For each of the following redox reactions, (i) write the half-reactions, (ii) write a balanced equation for the whole reaction, and (iii) determine in which direction the reaction will proceed spontaneously under standard-state conditions:

(a) $H_2(g) + Ni^{2+}(aq) \longrightarrow H^+(aq) + Ni(s)$
(b) $MnO_4^-(aq) + Cl^-(aq) \longrightarrow Mn^{2+}(aq) +$
$\qquad\qquad Cl_2(g)$ (in acid solution)
(c) $Cr(s) + Zn^{2+}(aq) \longrightarrow Cr^{3+}(aq) + Zn(s)$

13.55 The oxidation of 25.0 mL of a solution containing Fe^{2+} requires 26.0 mL of 0.0250 M $K_2Cr_2O_7$ in acidic solution. Balance the following equation and calculate the molar concentration of Fe^{2+}:

$$Cr_2O_7^{2-} + Fe^{2+} + H^+ \longrightarrow Cr^{3+} + Fe^{3+}$$

13.56 The SO_2 present in air is mainly responsible for the phenomenon of acid rain. The concentration of SO_2 can be determined by titrating against a standard permanganate solution as follows:

$$5SO_2 + 2MnO_4^- + 2H_2O \longrightarrow 5SO_4^{2-} + 2Mn^{2+} + 4H^+$$

Calculate the number of grams of SO_2 in a sample of air if 7.37 mL of 0.00800 M $KMnO_4$ solution is required for the titration.

13.57 A sample of iron ore weighing 0.2792 g was dissolved in an excess of a dilute acid solution. All the iron was first converted to Fe(II) ions. The solution then required 23.30 mL of 0.0194 M $KMnO_4$ for oxidation to Fe(III) ions. Calculate the percent by mass of iron in the ore.

13.58 The concentration of a hydrogen peroxide solution can be conveniently determined by titration against a standardized potassium permanganate solution in an acidic medium according to the following unbalanced equation:

$$MnO_4^- + H_2O_2 \longrightarrow O_2 + Mn^{2+}$$

(a) Balance this equation. (b) If 36.44 mL of a 0.01652 M $KMnO_4$ solution is required to completely oxidize 25.00 mL of an H_2O_2 solution, calculate the molarity of the H_2O_2 solution.

13.59 Oxalic acid ($H_2C_2O_4$) is present in many plants and vegetables. (a) Balance the following equation in acid solution:

$$MnO_4^- + C_2O_4^{2-} \longrightarrow Mn^2 + CO_2$$

(b) If a 1.00-g sample of $H_2C_2O_4$ requires 24.0 mL of 0.0100 M $KMnO_4$ solution to reach the equivalence point, what is the percent by mass of $H_2C_2O_4$ in the sample?

13.60 Calcium oxalate (CaC_2O_4) is insoluble in water. This property has been used to determine the amount of Ca^{2+} ions in blood. The calcium oxalate isolated from blood is dissolved in acid and titrated against a standardized $KMnO_4$ solution as described in Problem 13.59. In one test, it is found that the calcium oxalate isolated from a 10.0-mL sample of blood requires 24.2 mL of 9.56×10^{-4} M $KMnO_4$ for titration. Calculate the number of milligrams of calcium per milliliter of blood.

13.61 From the following information, calculate the solubility product of AgBr:

$$Ag(aq) + e^- \longrightarrow Ag(s) \qquad \mathcal{E}° = 0.80 \text{ V}$$
$$AgBr(s) + e^- \longrightarrow Ag(s) + Br^-(aq) \qquad \mathcal{E}° = 0.07 \text{ V}$$

13.62 Consider a galvanic cell composed of the standard hydrogen electrode and a half-cell using the reaction $Ag(aq) + e^- \longrightarrow Ag(s)$. (a) Calculate the standard cell potential. (b) What is the spontaneous cell reaction under standard-state conditions? (c) Calculate the cell potential when $[H^+]$ in the hydrogen electrode is changed to (i) 1.0×10^{-2} M and (ii) 1.0×10^{-5} M, all other reagents being held at standard-state conditions. (d) Based on this cell arrangement, suggest a design for a pH meter.

13.63 A galvanic cell consists of a silver electrode in contact with 346 mL of 0.100 M $AgNO_3$ solution and a magnesium electrode in contact with 288 mL of 0.100 M $Mg(NO_3)_2$ solution. (a) Calculate $\mathcal{E}$ for the cell at 25°C. (b) A current is drawn from the cell until 1.20 g of silver have been deposited at the silver electrode. Calculate $\mathcal{E}$ for the cell at this stage of operation.

13.64 Explain why chlorine gas can be prepared by electrolyzing an aqueous solution of NaCl but fluorine gas cannot be prepared by electrolyzing an aqueous solution of NaF.

13.65 Assuming ideal behavior, calculate the emf of the following concentration cell at 25°C:

$$Cu(s)|Cu^{2+}(0.080 \text{ } M)\|Cu^{2+}(1.2 \text{ } M)|Cu(s)$$

13.66 The cathode reaction in the Leclanché cell is given by

$$2MnO_2(s) + Zn^{2+}(aq) + 2e^- \longrightarrow ZnMn_2O_4(s)$$

If a Leclanché cell produces a current of 0.0050 A, calculate how many hours this current supply will last if there are initially 4.0 g of MnO_2 present in the cell. Assume that there is an excess of Zn^{2+} ions.

13.67 For a number of years, it was not clear whether mercury(I) ions existed in solution as Hg^+ or as

Hg_2^{2+}. To distinguish between these two possibilities, we could set up the following system:

$$Hg(l) \mid \text{soln A} \parallel \text{soln B} \mid Hg(l)$$

where soln A contained 0.263 g mercury(I) nitrate per liter and soln B contained 2.63 g mercury(I) nitrate per liter. If the measured emf of such a cell is 0.0289 V at 18°C, what can you deduce about the nature of the mercury(I) ions?

13.68 An aqueous KI solution to which a few drops of phenolphthalein have been added is electrolyzed using an apparatus like the one shown here:

Describe what you would observe at the anode and the cathode. (*Hint:* Molecular iodine is only slightly soluble in water, but in the presence of I^- ions, it forms the brown color of I_3^- ions.)

13.69 A piece of magnesium metal with a mass of 1.56 g is placed in 100.0 mL of 0.100 M $AgNO_3$ at 25°C. Calculate $[Mg^{2+}]$ and $[Ag^+]$ in solution at equilibrium. What is the mass of the leftover magnesium? The volume remains constant.

13.70 Steel hardware, including nuts and bolts, is often coated with a thin plating of cadmium. Explain the function of the cadmium layer.

13.71 "Galvanized iron" is steel sheet that has been coated with zinc; "tin" cans are made of steel sheet coated with tin. Discuss the functions of these coatings and the electrochemistry of the corrosion reactions that occur if an electrolyte contacts the scratched surface of a galvanized iron sheet or a tin can.

13.72 Tarnished silver contains Ag_2S. The tarnish can be removed by placing silverware in an aluminum pan containing an inert electrolyte solution, such as NaCl. Explain the electrochemical principle for this procedure. [The standard reduction potential for the half-cell reaction

$$Ag_2S(s) + 2e^- \longrightarrow 2Ag(s) + S^{2-}(aq)$$

is 0.71 V.]

13.73 How does the tendency of iron to rust depend on the pH of the solution?

13.74 An acidified solution was electrolyzed using copper electrodes. A constant current of 1.18 A caused the anode to lose 0.584 g after 1.52×10^3 s. (a) What is the gas produced at the cathode, and what is its volume at SATP? (b) Given that the charge of an electron is 1.6022×10^{-19} C, calculate Avogadro's number. Assume that copper is oxidized to Cu^{2+} ions.

13.75 In a certain electrolysis experiment involving Al^{3+} ions, 60.2 g of Al is recovered when a current of 0.352 A is used. How many minutes did the electrolysis last?

13.76 Consider the oxidation of ammonia:

$$4NH_3(g) + 3O_2(g) \longrightarrow 2N_2(g) + 6H_2O(l)$$

(a) Calculate the $\Delta G°$ for the reaction. (b) If this reaction were used in a fuel cell, what would be the standard cell potential?

13.77 When an aqueous solution containing gold(III) salt is electrolyzed, metallic gold is deposited at the cathode and oxygen gas is generated at the anode. (a) If 9.26 g of Au is deposited at the cathode, calculate the volume (in liters) of O_2 generated at 23°C and 747 mmHg. (b) What is the current used if the electrolytic process took 2.00 h?

13.78 A galvanic cell is constructed by immersing a piece of copper wire in 25.0 mL of a 0.20 M $CuSO_4$ solution and a zinc strip in 25.0 mL of a 0.20 M $ZnSO_4$ solution. (a) Calculate the emf of the cell at 25°C and predict what would happen if a small amount of concentrated NH_3 solution were added to (i) the $CuSO_4$ solution and (ii) the $ZnSO_4$ solution. Assume that the volume in each compartment remains constant at 25.0 mL. (b) In a separate experiment, 25.0 mL of 3.00 M NH_3 is added to the $CuSO_4$ solution. If the emf of the cell is 0.68 V calculate the formation constant (K_f) of $Cu(NH_3)_4^{2+}$.

13.79 In an electrolysis experiment, a student passes the same quantity of electricity through two electrolytic cells, one containing a silver salt and the other a gold salt. Over a certain period of time, she finds that 2.64 g of Ag and 1.61 g of Au are deposited at the cathodes. What is the oxidation state of gold in the gold salt?

13.80 People living in cold-climate countries where there is plenty of snow are advised not to heat their garages in the winter. What is the electrochemical basis for this recommendation?

13.81 Given that

$$2Hg^{2+}(aq) + 2e^- \longrightarrow Hg_2^{2+}(aq) \quad \mathcal{E}° = 0.92V$$
$$Hg_2^{2+}(aq) + 2e^- \longrightarrow 2Hg(l) \quad \mathcal{E}° = 0.85V$$

calculate $\Delta G°$ and K for the following process at 25°C:

$$Hg_2^{2+}(aq) \longrightarrow Hg^{2+}(aq) + Hg(l)$$

(The preceding reaction is an example of a *disproportionation reaction* in which an element in one oxidation state is both oxidized and reduced.)

13.82 Fluorine (F_2) is obtained by the electrolysis of liquid hydrogen fluoride (HF) containing potassium fluoride (KF). (a) Write the half-cell reactions and the overall reaction for the process. (b) What is the purpose of KF? (c) Calculate the volume of F_2 (in liters) collected at 24.0°C and 1.2 bar after electrolyzing the solution for 15 h at a current of 502 A.

13.83 A 300-mL solution of NaCl was electrolyzed for 6.00 min. If the pH of the final solution was 12.24, calculate the average current used.

13.84 Industrially, copper is purified by electrolysis. The impure copper acts as the anode, and the cathode is made of pure copper. The electrodes are immersed in a $CuSO_4$ solution. During electrolysis, copper at the anode enters the solution as Cu^{2+} while Cu^{2+} ions are reduced at the cathode. (a) Write half-cell reactions and the overall reaction for the electrolytic process. (b) Suppose the anode was contaminated with Zn and Ag. Explain what happens to these impurities during electrolysis. (c) How many hours will it take to obtain 1.00 kg of Cu at a current of 18.9 A?

13.85 An aqueous solution of a platinum salt is electrolyzed at a current of 2.50 A for 2.00 h. As a result, 9.09 g of metallic Pt is formed at the cathode. Calculate the charge on the Pt ions in this solution.

13.86 Consider a galvanic cell consisting of a magnesium electrode in contact with 1.0 M $Mg(NO_3)_2$ and a cadmium electrode in contact with 1.0 M $Cd(NO_3)_2$. Calculate $E°$ for the cell, and draw a diagram showing the cathode, anode, and direction of electron flow.

13.87 A current of 6.00 A passes through an electrolytic cell containing dilute sulfuric acid for 3.40 h. If the volume of O_2 gas generated at the anode is 4.26 L (at 0°C and 1 bar pressure), calculate the charge (in coulombs) on an electron.

13.88 Gold will not dissolve in either concentrated nitric acid or concentrated hydrochloric acid. However, the metal does dissolve in a mixture of the acids (one part HNO_3 and three parts HCl by volume), called *aqua regia*. (a) Write a balanced equation for this reaction. (*Hint:* Among the products are $HAuCl_4$ and NO_2.) (b) What is the function of HCl?

13.89 Explain why most useful galvanic cells give voltages of no more than 1.5 to 2.5 V. What are the prospects for developing practical galvanic cells with voltages of 5 V or more?

13.90 A silver rod and a SHE are dipped into a saturated aqueous solution of silver oxalate ($Ag_2C_2O_4$) at 25°C. The measured potential difference between the rod and the SHE is 0.589 V, the rod being

positive. Calculate the solubility product constant for silver oxalate.

13.91 Zinc is an amphoteric metal, that is, it reacts with both acids and bases. The standard reduction potential is -1.36 V for the reaction

$$Zn(OH)_4^{2-}(aq) + 2e^- \longrightarrow Zn(s) + 4OH^-(aq)$$

Calculate the formation constant (K_f) for the reaction

$$Zn^{2+}(aq) + 4OH^-(aq) \rightleftharpoons Zn(OH)_4^{2-}(aq)$$

13.92 Use the data in Table 13.1 to determine whether or not hydrogen peroxide will undergo disproportionation in an acid medium:

$$2H_2O_2 \longrightarrow 2H_2O + O_2.$$

13.93 The magnitudes (but *not* the signs) of the standard reduction potentials of two metals X and Y are

$$Y^{2+} + 2e^- \longrightarrow Y \quad |E°| = 0.34 \text{ V}$$
$$X^{2+} + 2e^- \longrightarrow X \quad |E°| = 0.25 \text{ V}$$

where the notation $||$ denotes that only the magnitude (but not the sign) of the $E°$ value is shown. When the half-cells of X and Y are connected, electrons flow from X to Y. When X is connected to a SHE, electrons flow from X to the SHE. (a) Are the $E°$ values of the half-reactions positive or negative? (b) What is the standard emf of a cell made up of X and Y?

13.94 A galvanic cell is constructed as follows: One half-cell consists of a platinum wire immersed in a solution containing 1.0 M Sn^{2+} and 1.0 M Sn^{4+}; the other half-cell has a thallium rod immersed in a solution of 1.0 M Tl^+. (a) Write the half-cell reactions and the overall reaction. (b) What is the equilibrium constant at 25°C? (c) What is the cell voltage if the Tl^+ concentration is increased tenfold? ($E°_{Tl^+|Tl} = -0.34$ V.)

13.95 Given the standard reduction potential for Au^{3+} in Table 13.1 and

$$Au^+(aq) + e^- \longrightarrow Au(s) \quad E° = 1.69 \text{ V}$$

answer the following questions. (a) Why does gold not tarnish in air? (b) Will the following disproportionation occur spontaneously?

$$3Au^+(aq) \longrightarrow Au^{3+}(aq) + 2Au(s)$$

(c) Predict the reaction between gold and fluorine gas.

13.96 The ingestion of a very small quantity of mercury is not considered too harmful. Would this statement still hold if the gastric juice in your stomach were mostly nitric acid instead of hydrochloric acid?

13.97 When 25.0 mL of a solution containing both Fe^{2+} and Fe^{3+} ions is titrated with 23.0 mL of 0.0200 M $KMnO_4$ (in dilute sulfuric acid), all the Fe^{2+} ions are oxidized to Fe^{3+} ions. Next, the solution is treated with Zn metal to convert all the Fe^{3+} ions to Fe^{2+} ions. Finally, 40.0 mL of the same $KMnO_4$ solution is added to the solution in order to oxidize the Fe^{2+} ions to Fe^{3+}. Calculate the molar concentrations of Fe^{2+} and Fe^{3+} in the original solution.

13.98 Consider the Daniell cell in Figure 13.3. When viewed externally, the anode appears negative and the cathode positive (electrons are flowing from the anode to the cathode). Yet in solution anions are moving toward the anode, which means that it must appear positive to the anions. Because the anode cannot simultaneously be negative and positive, give an explanation for this apparently contradictory situation.

13.99 Use the data in Table 13.1 to show that the decomposition of H_2O_2 (a diaproportionation reaction) is spontaneous at 25°C:

$$2H_2O_2(aq) \longrightarrow 2H_2O(l) + O_2(g)$$

13.100 Lead storage batteries are rated by ampere hours, that is, the number of amperes they can deliver in an hour. (a) Show that 1 A h = 3600 C. (b) The lead anodes of a certain lead storage battery have a total mass of 406 g. Calculate the maximum theoretical capacity of the battery in ampere-hours. Explain why in practice we can never extract this much energy from the battery. (c) Calculate $\mathcal{E}°_{cell}$ and $\Delta G°$ for the battery.

13.101 The concentration of sulfuric acid in the lead storage battery of an automobile over a period of time has decreased from 38.0 percent by mass (density = 1.29 g mL^{-1}) to 26.0 percent by mass (1.19 g mL^{-1}). Assume the volume of the acid remains constant at 724 mL. (a) Calculate the total charge in coulombs supplied by the battery. (b) How long (in hours) will it take to recharge the battery back to the original sulfuric acid concentration using a current of 22.4 A.

13.102 A spoon was silver-plated electrolytically in an $AgNO_3$ solution. (a) Sketch a diagram for the process. (b) If 0.884 g of Ag was deposited on the spoon at a constant current of 18.5 mA, how long (in minutes) did the electrolysis take?

13.103 Comment on whether F_2 will become a stronger oxidizing agent with increasing H^+ concentration.

13.104 In recent years, there has been much interest in electric cars. List some advantages and disadvantages of electric cars compared to automobiles with internal combustion engines.

13.105 Calculate the pressure of H_2 (in bar) required to maintain equilibrium with respect to the following reaction at 25°C:

$$Pb(s) + 2H^+(aq) \rightleftharpoons Pb^{2+}(aq) + H_2(g)$$

given that $[Pb^{2+}] = 0.035$ M and the solution is buffered at pH 1.60.

13.106 A piece of magnesium ribbon and a copper wire are partially immersed in a 0.1 M HCl solution in a beaker. The metals are joined externally by another piece of metal wire. Bubbles are seen to evolve at both the Mg and Cu surfaces. (a) Write equations representing the reactions occurring at the metals. (b) What visual evidence would you seek to show that Cu is not oxidized to Cu^{2+}? (c) At some stage, NaOH solution is added to the beaker to neutralize the HCl. Upon further addition of NaOH, a white precipitate forms. What is it?

13.107 The zinc-air battery shows much promise for electric cars because it is lightweight and rechargeable:

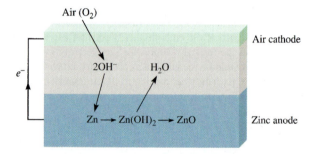

The net transformation is

$$Zn(s) + \frac{1}{2}O_2(g) \longrightarrow ZnO(s)$$

(a) Write the half-reactions at the zinc-air electrodes and calculate the standard emf of the battery at 25°C. (b) Calculate the emf under actual operating conditions when the partial pressure of oxygen is 0.21 bar. (c) What is the energy density (measured as the energy in kilojoules that can be obtained from 1 kg of the metal) of the zinc electrode? (d) If a current of 2.1×10^5 A is to be drawn from a zinc-air battery system, what volume of air (in liters) would need to be supplied to the battery every second? Assume that the temperature is 25°C and the partial pressure of oxygen is 0.21 bar.

13.108 Calculate $\mathcal{E}°$ for the reactions of mercury with (a) 1 M HCl and (b) 1 M HNO$_3$. Which acid will oxidize Hg to Hg_2^{2+} under standard-state conditions? Can you identify which test tube below contains HNO_3 and Hg and which contains HCl and Hg?

13.109 Because all alkali metals react with water, it is not possible to measure the standard reduction potentials of these metals directly as in the case of, say, zinc. An indirect method is to consider the following hypothetical reaction

$$Li^+(aq) + \frac{1}{2}H_2(g) \longrightarrow Li(s) + H^+(aq)$$

Use the appropriate equation presented in this chapter and the thermodynamic data in Appendix 2, to calculate $\mathcal{E}°$ for $Li^+(aq) + e^- \longrightarrow Li(s)$ at 298 K. Compare your result with that listed in Table 13.1.

13.110 Which of the following two redox reactions would you expect to have the largest change in standard emf when the temperature is increased?

$$Fe^{2+}(aq) + Cu^{2+}(aq) \longrightarrow Fe^{3+}(aq) + Cu^+(aq)$$
$$2I^-(aq) + Cl_2(g) \longrightarrow I_2(s) + 2Cl^-(aq)$$

Explain your answer.

13.111 Compare the pros and cons of a fuel cell, such as the hydrogen-oxygen fuel cell, and a coal-fired station for generating electricity.

13.112 Using the data from Table 13.1, calculate the equilibrium constant for the following reaction at 298 K:

$$Zn(s) + Cu^{2+}(aq) \longrightarrow Zn^{2+}(aq) + Cu(s)$$

13.113 Use Equation 13.6 to calculate the emf values of the Daniel cell at 80°C. Comment on your results. What assumptions are made in this calculation? (*Hint:* Use the thermodynamic data in Appendix 2.)

13.114 A construction company is installing an iron culvert (a long cylindrical tube) that is 40.0 m long with a radius of 0.900 m. To prevent corrosion, the culvert must be galvanized. This process is carried out by first passing an iron sheet of appropriate dimensions through an electrolytic cell containing Zn^{2+} ions, using graphite as the anode and the iron sheet as the cathode. If the voltage is 3.26 V, what is the cost of electricity for depositing a layer 0.200 mm thick if the efficiency of the process is 95 percent? The electricity rate is $0.12 per kilowatt hour (kWh), where $1 \text{ W} = 1 \text{ J s}^{-1}$, and the density of Zn is 7.14 g cm^{-3}.

13.115 A 9.00×10^{-2}-mL sample of 0.200 M MgI_2 was electrolyzed. As a result, hydrogen gas was generated at the cathode and iodine was formed at the anode. The volume of hydrogen collected at 26°C and 779 mmHg was 1.22×10^3 mL. (a) Calculate the charge in coulombs consumed in the process. (b) How long (in min) did the electrolysis last if a current of 7.55 A was used? (c) A white precipitate was formed in the process. What was it, and what was its mass in grams? Assume the volume of the solution was constant.

13.116 Assuming that both $\Delta H°$ and $\Delta S°$ are independent of temperature, derive Equation 13.7 starting from Equation 13.6.

13.117 Based on the following standard reduction potentials:

$$Fe^{2+}(aq) + 2e^- \longrightarrow Fe(s) \qquad \mathcal{E}_1° = -0.44 \text{ V}$$
$$Fe^{3+}(aq) + e^- \longrightarrow Fe^{2+}(aq) \quad \mathcal{E}_2° = 0.77 \text{ V}$$

calculate the standard reduction potential for the half-reaction

$$Fe^{3+}(aq) + 3e^- \longrightarrow Fe(s) \qquad \mathcal{E}_3° = ?$$

13.118 To remove the tarnish (Ag_2S) on a silver spoon, a student carried out the following steps. First, she placed the spoon in a large pan filled with water so the spoon was totally immersed. Next, she added a few tablespoonful of baking soda (sodium bicarbonate), which readily dissolved. Finally, she placed some aluminum foil at the bottom of the pan in contact with the spoon and then heated the solution to about 80°C. After a few minutes, the spoon was removed and rinsed with cold water. The tarnish was gone and the spoon had regained its original shiny appearance. (a) Describe with equations the electrochemical basis for the procedure. (b) Adding NaCl instead of $NaHCO_3$ would also work because both compounds are strong electrolytes. What is the added advantage of using $NaHCO_3$? (*Hint:* Consider the pH of the solution.) (c) What is the purpose of heating the solution? (d) Some commercial tarnish removers contain a fluid (or paste) that is a dilute HCl solution. Rubbing the spoon with the fluid will also remove the tarnish. Name two disadvantages of using this procedure compared to the one described.

Answers to Practice Exercises

13.1 (a) P +3, F −1; (b) Mn +7, O −2; (c) Cl +5, O −2; and (d) S + 6, F −1 **13.2** $5Fe^{2+} + MnO_4^- + 8H^+ \longrightarrow 5Fe^{3+} + Mn^{2+} + 4H_2O$ **13.3** No **13.4** 0.34 V **13.5** 1×10^{-42}

13.6 $\Delta G° = -4.1 \times 10^2$ kJ mol^{-1} **13.7** 1.10 V **13.8** Yes, $\mathcal{E} = +0.01$ V **13.9** 0.38 V **13.10** Anode, O_2; Cathode, H_2 **13.11** 2.0×10^4 A

Chemical Kinetics

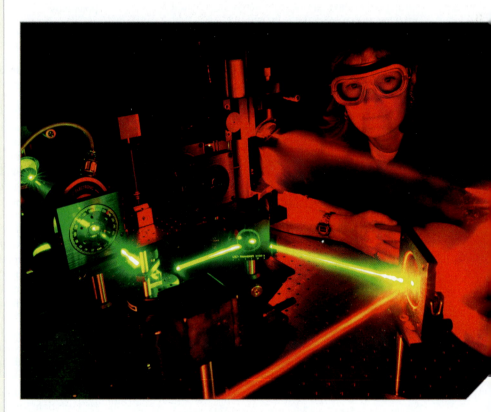

14.1 | Chemical Kinetics Is the Study of the Rates at Which Chemical Reactions Occur

The study of chemical equilibrium, explored in Chapters 10–13, can determine whether a particular reaction is possible under the given conditions and how far that reaction will proceed before equilibrium is reached. However, the laws of chemical thermodynamics and the principles of chemical equilibrium cannot determine how fast a reaction will proceed. The speed of a chemical reaction is obtained through the study of **chemical kinetics,** which is *the area of chemistry concerned with the rates at which chemical reactions occur.* The word *kinetic* suggests movement or change; recall from Chapter 5 that kinetic energy is the energy available because of the motion of an object. Here "kinetics" refers to the **reaction rate,** which is *the change in the concentration of a reactant or a product with time ($M\ s^{-1}$).*

Rates of Reaction

There are many reasons for studying the rate of a reaction. To begin with, there is an intrinsic curiosity about why reactions have such vastly different rates. Some processes, such as the initial steps in vision and photosynthesis and nuclear chain reactions, take place on a time scale as short as 10^{-12} to 10^{-6} s. Others, like the curing of cement and the conversion of graphite to diamond, take years or millions of years to complete. On a practical level, a knowledge of reaction rates is useful in drug design, pollution control, and food processing. Industrial chemists often place more emphasis on speeding up the rate of a reaction than they do on maximizing its yield.

Any chemical reaction can be represented by the general equation

$$\text{reactants} \longrightarrow \text{products}$$

This equation tells us that during the course of a reaction, reactants are consumed while products are formed. As a result, we can follow the progress of a reaction by monitoring either the decrease in concentration of the reactants or the increase in concentration of the products.

Figure 14.1 shows the progress of a simple reaction in which molecules of A (gray spheres) are converted to molecules of B (red spheres):

$$A \longrightarrow B$$

The decrease in the number of A molecules and the increase in the number of B molecules with time are graphed in Figure 14.2.

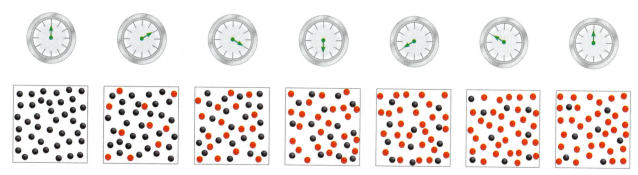

Figure 14.1 The progress of reaction A $\longrightarrow$ B at 10-s intervals over a period of 60 s. Initially, only molecules of A (gray spheres) are present. As time progresses, B molecules (red spheres) are formed.

Figure 14.2 The rate of reaction A $\longrightarrow$ B, represented as the decrease of A molecules with time and as the increase of B molecules with time.

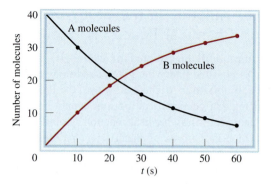

In general, it is more convenient to express the reaction rate in terms of the change in concentration with time. Thus, for the reaction A $\longrightarrow$ B, we can express the *average* rate over a time interval $\Delta t = t_2 - t_1$ as

$$\text{average rate} = -\frac{[A]_{t_2} - [A]_{t_1}}{t_2 - t_1} = -\frac{\Delta[A]}{\Delta t}$$

where $\Delta[A]$ represents the change in the concentration (molarity) of A over the time interval. Because the concentration of reactant A *decreases* during the time interval, $\Delta[A]$ is a negative quantity. The rate of a reaction is a positive quantity, so a minus sign is needed in the rate expression to make the rate positive. The rate can also be defined in terms of the rate of production of the product B:

$$\text{average rate} = \frac{[B]_{t_2} - [B]_{t_1}}{t_2 - t_1} = \frac{\Delta[B]}{\Delta t}$$

where no minus sign is required because $\Delta[B]$ is a positive quantity (the concentration of B *increases* with time).

In practice, we find that the quantity of interest is not the rate over a certain time interval (because this is only an average quantity whose value depends on the particular value of Δt); rather, we are interested in the instantaneous rate. In the language of calculus, as Δt becomes smaller and eventually approaches zero, the *instantaneous rate* of the reaction at a specific time t can be defined as

$$\text{instantaneous rate} = -\frac{d[A]}{dt} = +\frac{d[B]}{dt}$$

Using this definition, we do not need to worry about what time interval is used. Unless otherwise stated, we will refer to the instantaneous rate merely as "the rate."

To determine the rate of a reaction experimentally, we have to monitor the concentration of the reactant (or product) as a function of time. For reactions in solution, the concentration of a species can often be measured by spectroscopic means. If ions are involved, the change in concentration can also be detected by an electrical conductance measurement. Reactions involving gases are most conveniently followed by pressure measurements. We will consider two specific reactions for which different methods are used to measure the reaction rate.

Reaction of Bromine and Formic Acid

Consider, for example, the reaction of molecular bromine (Br_2) with formic acid (HCOOH) in aqueous solution:

$$Br_2(aq) + HCOOH(aq) \longrightarrow 2Br^-(aq) + 2H^+(aq) + CO_2(g)$$

Figure 14.3 The decrease in bromine concentration as time elapses shows up as a loss of color (from left to right).

Molecular bromine is reddish-brown in color. All the other species in the reaction are colorless. As the reaction progresses, the concentration of Br_2 steadily decreases and its color fades (Figure 14.3). This loss of color and hence concentration can be monitored easily with a spectrometer, which registers the amount of visible light absorbed by bromine (Figure 14.4).

Measuring the change (decrease) in bromine concentration at some initial time and then at some final time enables us to determine the average rate of the reaction during that interval:

$$\text{average rate} = -\frac{\Delta[Br_2]}{\Delta t} = -\frac{[Br_2]_{t_2} - [Br_2]_{t_1}}{t_2 - t_1}$$

The data in Table 14.1 show the experimental bromine concentration as a function of time for a reaction of Br_2 with a large excess of formic acid in aqueous solution. From these data, we can calculate the average rate over the first 50-s time interval as follows:

$$\text{average rate} = -\frac{(0.0101 - 0.0120)\ M}{50.0\ \text{s}} = 3.8 \times 10^{-5}\ M\ \text{s}^{-1}$$

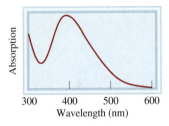

Figure 14.4 Plot of absorption of bromine versus wavelength. The maximum absorption of visible light by bromine occurs at 393 nm. As the reaction progresses, the absorption, which is proportional to $[Br_2]$, decreases with time, indicating a depletion in bromine.

Table 14.1	Measured Rates for the Reaction Between Molecular Bromine and Formic Acid at 25°C		
Time (s)	$[Br_2]$ (M)	Instantaneous Rate (M s^{-1})	$k = \dfrac{\text{rate}}{[Br_2]}$(s^{-1})
0.0	0.0120	4.20×10^{-5}	3.50×10^{-3}
50.0	0.0101	3.52×10^{-5}	3.49×10^{-3}
100.0	0.00846	2.96×10^{-5}	3.50×10^{-3}
150.0	0.00710	2.49×10^{-5}	3.51×10^{-3}
200.0	0.00596	2.09×10^{-5}	3.51×10^{-3}
250.0	0.00500	1.75×10^{-5}	3.50×10^{-3}
300.0	0.00420	1.48×10^{-5}	3.52×10^{-3}
350.0	0.00353	1.23×10^{-5}	3.48×10^{-3}
400.0	0.00296	1.04×10^{-5}	3.51×10^{-3}

Figure 14.5 The instantaneous rates of the reaction between molecular bromine and formic acid at $t = 100$ s, 200 s, and 300 s are given by the derivative (the slope of the tangent lines) at each time.

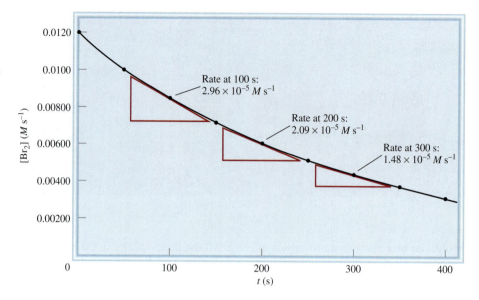

Rate at 100 s:
$2.96 \times 10^{-5}\ M\,\text{s}^{-1}$

Rate at 200 s:
$2.09 \times 10^{-5}\ M\,\text{s}^{-1}$

Rate at 300 s:
$1.48 \times 10^{-5}\ M\,\text{s}^{-1}$

If we had chosen the first 100 s as our time interval, the average rate would then be given by:

$$\text{average rate} = -\frac{(0.00846 - 0.0120)\ M}{100.0\ \text{s}} = 3.54 \times 10^{-5}\ M\,\text{s}^{-1}$$

These calculations demonstrate that the average rate of the reaction depends on the time interval we choose. By calculating the average reaction rate over shorter and shorter intervals, we can obtain the instantaneous rate of the reaction at each time (see Table 14.1). Figure 14.5 shows the plot of $[\text{Br}_2]$ versus time, based on the data listed in Table 14.1. Graphically, the instantaneous rate at a given time is given by the derivative of the curve (the slope of the tangent line) with respect to time (see Appendix 1). The instantaneous rate will always have the same value for the same concentrations of reactants, as long as the temperature is kept constant.

The rate of the bromine–formic acid reaction also depends on the concentration of formic acid. However, by adding a large excess of formic acid to the reaction mixture, we can ensure that the concentration of formic acid remains virtually constant throughout the course of the reaction. Under this condition the change in the amount of formic acid present in solution has no effect on the measured rate.

What effect does the bromine concentration have on the rate of reaction? According to Table 14.1, the bromine concentration is 0.0101 M at $t = 50$ s and the rate of the reaction is $3.52 \times 10^{-5}\ M\,\text{s}^{-1}$. At $t = 250$ s, the bromine concentration is 0.00500 M and the rate of reaction is $1.75 \times 10^{-5}\ M\,\text{s}^{-1}$. The concentration at $t = 50$ s is double the concentration at $t = 250$ s (0.0101 M versus 0.00500 M), and the rate of reaction at $t = 50$ s is double the rate at $t = 250$ s ($3.52 \times 10^{-5}\ M\,\text{s}^{-1}$ versus $1.75 \times 10^{-5}\ M\,\text{s}^{-1}$). In other words, the rate of reaction doubles as the concentration of bromine doubles. The rate is directly proportional to the Br_2 concentration:

The symbol "∝" means "is proportional to".

$$\text{rate} \propto [\text{Br}_2]$$
$$= k[\text{Br}_2]$$

where the term k is known as the ***rate constant***—*a constant of proportionality between the reaction rate and the concentration of reactant.*

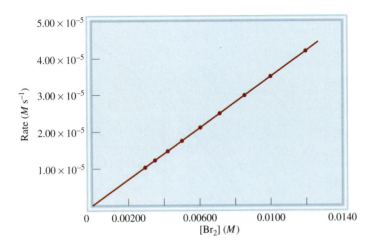

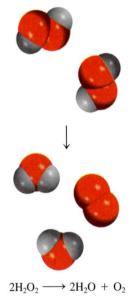

$$2H_2O_2 \longrightarrow 2H_2O + O_2$$

Figure 14.6 is a plot of the rate versus Br_2 concentration. The graph is a straight line because the rate is directly proportional to the concentration; the higher the concentration, the higher the rate. Rearranging the last equation gives

$$k = \frac{\text{rate}}{[Br_2]}$$

Because reaction rate has the units $M\ s^{-1}$, and $[Br_2]$ has units of M, the unit of k is s^{-1} in this case. The rate constant (k) is *unaffected* by the concentration of Br_2. The rate is greater at a higher concentration and smaller at a lower concentration of Br_2, but the *ratio* of rate/$[Br_2]$ remains the same provided the temperature does not change.

From Table 14.1, we can calculate the rate constant for the reaction. Taking the data for $t = 50$ s, we write

$$k = \frac{\text{rate}}{[Br_2]} = \frac{3.52 \times 10^{-5}\ M\ s^{-1}}{0.0101\ M} = 3.49 \times 10^{-3}\ s^{-1}$$

We can use the data for any t to calculate k. The slight variations in the values of k listed in Table 14.1 are due to experimental deviations in the rate measurements.

Decomposition of Hydrogen Peroxide

If one of the products or reactants is a gas, we can use a manometer to find the reaction rate. Consider, for example, the decomposition of hydrogen peroxide at 20°C:

$$2H_2O_2(aq) \longrightarrow 2H_2O(l) + O_2(g)$$

In this case, the rate of decomposition can be determined by monitoring the rate of oxygen evolution with a manometer (Figure 14.7).

The oxygen partial pressure can be readily converted to concentration by using the ideal gas equation:

$$P_{O_2}V = n_{O_2}RT$$

Figure 14.7 The rate of hydrogen peroxide decomposition can be measured with a manometer, which shows the increase in the oxygen gas pressure with time. The arrows show the mercury levels in the U tube.

Figure 14.8 The instantaneous
rate for the decomposition of
hydrogen peroxide at 400 min
is given by the slope of the
tangent multiplied by 1/RT.

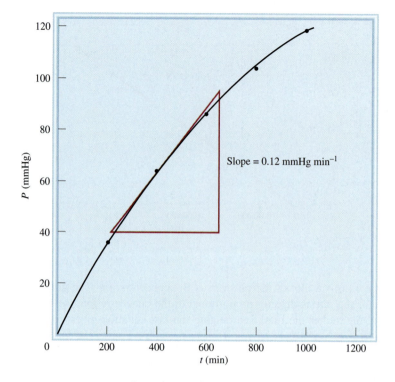

Figure 14.8 The instantaneous rate for the decomposition of hydrogen peroxide at 400 min is given by the slope of the tangent multiplied by 1/RT.

or

$$P_{O_2} = \frac{n_{O_2} RT}{V} = [O_2] RT$$

where n_{O_2}/V gives the molarity of oxygen gas. Rearranging the equation, we get

$$[O_2] = \frac{1}{RT} P_{O_2}$$

The reaction rate, which is given by the rate of oxygen production, can now be written as

$$\text{rate} = -\frac{d[O_2]}{dt} = \frac{-1}{RT} \frac{dP_{O_2}}{dt}$$

Figure 14.8 shows the increase in oxygen pressure with time and the determination of an instantaneous rate at 400 min. To express the rate in the usual units of $M \ s^{-1}$, we convert mmHg min^{-1} to bar s^{-1} and then multiply the slope of the tangent (dP_{O_2}/dt) by $1/RT$, as shown in the preceding equation.

Reaction Rates and Stoichiometry

We have just seen that for stoichiometrically simple reactions of the type A ⟶ B, the rate can be expressed in terms of the decrease in reactant concentration with time, $-d[A]/dt$, or the increase in product concentration with time, $d[B]/dt$. Now consider the slightly more complex reaction:

$$2A \longrightarrow B$$

Two moles of A disappear for each mole of B that forms, that is, the rate of disappearance of A is twice as fast as the rate of appearance of B. The rate can be written as either

$$\text{rate} = -\frac{1}{2}\frac{d[A]}{dt} \quad \text{or} \quad \text{rate} = +\frac{d[B]}{dt}$$

In general, for the reaction

$$aA + bB \longrightarrow cC + dD$$

the rate is given by

$$\text{rate} = -\frac{1}{a}\frac{d[A]}{dt} = -\frac{1}{b}\frac{d[B]}{dt} = +\frac{1}{c}\frac{d[C]}{dt} = +\frac{1}{d}\frac{d[D]}{dt} \qquad \textbf{(14.1)}$$

Example 14.1 shows how to write reaction rate expressions and calculate rates of product formation and reactant disappearance.

Example 14.1

Consider the reaction $4NO_2(g) + O_2(g) \longrightarrow 2N_2O_5(g)$. Suppose that, at a particular moment during the reaction, N_2O_5 is being formed at a rate of $0.048\ M\ s^{-1}$. (a) Write the rate expression in terms of the disappearance of the reactants, NO_2 and O_2, and the appearance of the product, N_2O_5. (b) What is the overall rate? (c) At what rate is O_2 reacting? (d) At what rate is NO_2 reacting?

Solution (a) Using Equation 14.1, we have

$$\text{rate} = -\frac{1}{4}\frac{d[NO_2]}{dt} = -\frac{d[O_2]}{dt} = +\frac{1}{2}\frac{d[N_2O_5]}{dt}$$

(b) We are given

$$\frac{d[N_2O_5]}{dt} = 0.048\ M\ s^{-1}$$

From the rate expression,

$$\text{rate} = \frac{1}{2}\frac{d[N_2O_5]}{dt} = \frac{1}{2}(0.048\ M\ s^{-1}) = 0.024\ M\ s^{-1}$$

(c) From part (a), $\text{rate} = -\dfrac{d[O_2]}{dt}$ and from part (b), rate $= 0.024\ M\ s^{-1}$, so

$$\frac{d[O_2]}{dt} = -\text{rate} = -0.024\ M\ s^{-1}$$

(d) From part (a) $\dfrac{d[NO_2]}{dt} = -4 \times \text{rate} = -4 \times (0.024\ M\ s^{-1}) = -0.096\ M\ s^{-1}$

Practice Exercise Consider the reaction $4PH_3(g) \longrightarrow P_4(g) + 6H_2(g)$. Suppose that, at a particular moment during the reaction, molecular hydrogen is being formed at the rate of $0.078\ M\ s^{-1}$. (a) What is the overall rate of the reaction? (b) At what rate is P_4 being formed? (c) At what rate is PH_3 reacting?

14.2 | The Rate Law Gives the Dependence of the Reaction Rate on the Reactant Concentrations

So far we have learned that the rate of a reaction is proportional to the concentration of reactants and that the proportionality constant k is called the rate constant. The *rate law expresses the relationship of the rate of a reaction to the rate constant and the concentrations of the reactants raised to some powers.* For the general reaction

$$aA + bB \longrightarrow cC + dD$$

the rate law takes the form

$$\text{rate} = k[A]^x [B]^y \qquad (14.2)$$

where x and y are numbers that must be determined experimentally. The exponents x and y are *not necessarily* equal to the stoichiometric coefficients a and b and must be obtained from experiment. When we know the values of x, y, and k, we can use Equation 14.2 to calculate the rate of the reaction, given the concentrations of A and B.

The exponents x and y specify the relationships between the concentrations of reactants A and B and the reaction rate. Added together, they give us the overall **reaction order,** defined as *the sum of the powers to which all reactant concentrations appearing in the rate law are raised.* For Equation 14.2, the overall reaction order is $x + y$. Alternatively, we can say that the reaction is x^{th} order in A, y^{th} order in B, and $(x + y)^{th}$ order overall.

To see how to determine the rate law of a reaction, consider the following reaction between fluorine and chlorine dioxide:

$$F_2(g) + 2ClO_2(g) \longrightarrow 2FClO_2(g)$$

One way to study the effect of reactant concentration on reaction rate is to determine how the *initial rate* depends on the starting concentrations. It is preferable to measure the initial rates because, as the reaction proceeds, the concentrations of the reactants decrease and it may become difficult to measure the changes accurately. Also, there may be a reverse reaction of the type

$$\text{products} \longrightarrow \text{reactants}$$

which would introduce error into the rate measurement. Both of these complications are virtually absent during the early stages of the reaction.

Table 14.2 lists three rate measurements for the formation of $FClO_2$. According to entries 1 and 3, the reaction rate doubles if we double $[F_2]$ while holding $[ClO_2]$ constant. Thus, the rate is directly proportional to $[F_2]$. Similarly, the data in entries

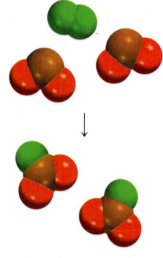

$$F_2 + 2ClO_2 \longrightarrow 2FClO_2$$

Table 14.2	Rate Data for the Reaction between F_2 and ClO_2	
$[F_2]$ (M)	$[ClO_2]$ (M)	Initial Rate ($M\ s^{-1}$)
1. 0.10	0.010	1.2×10^{-3}
2. 0.10	0.040	4.8×10^{-3}
3. 0.20	0.010	2.4×10^{-3}

1 and 2 show that the rate quadruples if we quadruple $[ClO_2]$ at constant $[F_2]$. Thus, the rate is also directly proportional to $[ClO_2]$. We can summarize our observations by writing the rate law as

$$\text{rate} = k[F_2][ClO_2]$$

Because both $[F_2]$ and $[ClO_2]$ are raised to the first power, the reaction is first order in F_2, first order in ClO_2, and second order $(1 + 1 = 2)$ overall. Note that $[ClO_2]$ is raised to the power of 1 in the rate equation, whereas its stoichiometric coefficient in the overall chemical equation is 2. For F_2, the reaction order (first) and stoichiometric coefficient (1) for F_2 are coincidentally the same.

From the reactant concentrations and the initial rate, we can also calculate the rate constant. Using the first entry of data in Table 14.2, we can write

$$
\begin{aligned}
k &= \frac{\text{rate}}{[F_2][ClO_2]} \\
&= \frac{1.2 \times 10^{-3} \, M \, s^{-1}}{(0.10 \, M)(0.010 \, M)} \\
&= 1.2 \, M^{-1} \, s^{-1}
\end{aligned}
$$

The units of k depend upon the overall order of the reaction. For this second-order reaction they are $M^{-1} \, s^{-1}$. For a first-order reaction, the units of k are s^{-1}. In general the units for k for an nth-order reaction are $M^{1-n} \, s^{-1}$.

Knowing the reaction order enables us to understand how the reaction depends on reactant concentration. Suppose, for example, that the general reaction

$$a\text{A} + b\text{B} \longrightarrow c\text{C} + d\text{D}$$

is first order in A and second order in B (that is, $x = 1$ and $y = 2$ in Equation 14.2). The rate law for the reaction is

$$\text{rate} = k[\text{A}][\text{B}]^2$$

This reaction is third order overall $(1 + 2 = 3)$. Let us assume that initially $[\text{A}] = 1.0 \, M$ and $[\text{B}] = 1.0 \, M$. The rate law tells us that if we double the concentration of A from $1.0 \, M$ to $2.0 \, M$ at constant $[\text{B}]$, we also double the reaction rate. On the other hand, if we double the concentration of B from $1.0 \, M$ to $2.0 \, M$ at constant $[\text{A}] = 1 \, M$, the rate will increase by a factor of four because of the power of two in the exponent.

If for another reaction $x = 0$ and $y = 1$, then the rate law is

$$\text{rate} = k[\text{A}]^0[\text{B}]^1 = k[\text{B}]$$

This reaction is zero order in A, first order in B, and first order overall. The exponent zero tells us that the rate of this reaction is *independent* of the concentration of A. Although we have only discussed reactions so far for which the reaction orders are whole numbers, reaction orders can also be fractional. In summary,

1. Rate laws are always determined experimentally. From the concentration of reactants and the initial reaction rates, we can determine the reaction order and then the rate constant of the reaction.

2. Reaction order is always defined in terms of reactant (not product) concentrations.

3. The order of a reactant is unrelated to the stoichiometric coefficient of the reactant in the overall balanced equation and must be determined from experiment.

Example 14.2 illustrates the procedure for determining the rate law of a reaction.

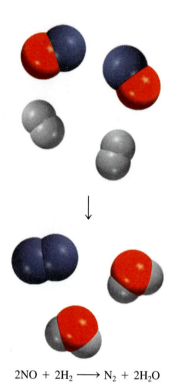

$$2NO + 2H_2 \longrightarrow N_2 + 2H_2O$$

Example 14.2

The reaction of nitric oxide with hydrogen at 1280°C is

$$2NO(g) + 2H_2(g) \longrightarrow N_2(g) + 2H_2O(g)$$

From the following data collected at this temperature, determine (a) the rate law, (b) the rate constant, and (c) the rate of the reaction when $[NO] = 13.0 \times 10^{-3}\ M$ and $[H_2] = 6.0 \times 10^{-3}\ M$.

Experiment	[NO] (M)	[H₂] (M)	Initial Rate ($M\,s^{-1}$)
1	10.0×10^{-3}	2.0×10^{-3}	5.0×10^{-5}
2	10.0×10^{-3}	4.0×10^{-3}	10.0×10^{-5}
3	14.0×10^{-3}	2.0×10^{-3}	9.8×10^{-5}

Strategy We are given a set of concentration and reaction rate data and asked to determine the rate law and the rate constant. We assume that the rate law takes the form

$$\text{rate} = k[NO]^x[H_2]^y$$

We need to compare the data from the different experiments to determine x and y. Once the orders of the reactants are known, we can calculate k from any set of rates and concentrations. Finally, the rate law enables us to calculate the rate at any concentrations of NO and H_2.

Solution (a) Experiments 1 and 2 show that the rate doubles when we double the concentration of H_2 at constant concentration of NO. Therefore, the rate law is first order in H_2 and $y = 1$.

Experiments 1 and 3 indicate that increasing [NO] from $10.0 \times 10^{-3}\ M$ to $14.0 \times 10^{-3}\ M$ at constant $[H_2]$ increases the rate from $5.0 \times 10^{-5}\ M\,s^{-1}$ to $9.8 \times 10^{-5}\ M\,s^{-1}$. Because the concentration is not increased by a simple integer factor, determining the power x is more complex and must be determined using logarithms. Using the rate law (with $y = 1$), the ratio of the rate in experiment 3 to that of experiment 1 is given by

$$\frac{\text{rate 3}}{\text{rate 1}} = \frac{k[NO]^x_{\exp 3}[H_2]^1_{\exp 3}}{k[NO]^x_{\exp 1}[H_2]^1_{\exp 1}}$$

$$\frac{9.8 \times 10^{-5}\ M\,s^{-1}}{5.0 \times 10^{-5}\ M\,s^{-1}} = \frac{(14.0 \times 10^{-3}\ M)^x(2.0 \times 10^{-3}\ M)}{(10.0 \times 10^{-3}\ M)^x(2.0 \times 10^{-3}\ M)}$$

$$1.96 = 1.4^x$$

Taking the natural log of both sides gives

$$\ln(1.96) = \ln(1.4)^x = x \times \ln(1.4)$$

$$0.673 = x \times 0.336$$

$$x = \frac{0.673}{0.336} = 2.00$$

so the reaction is second order in [NO]. Hence the rate law is given by

$$\text{rate} = k[NO]^2[H_2]$$

and the reaction is third order ($2 + 1 = 3$) overall.

—*Continued*

Continued—

(b) The rate constant k can be calculated using the data from any one of the experiments. Rearranging the rate law, we get

$$k = \frac{\text{rate}}{[\text{NO}]^2[\text{H}_2]}$$

The data from experiment 2 give us

$$k = \frac{5.0 \times 10^{-5}\,M\,\text{s}^{-1}}{(10.0 \times 10^{-3}\,M)^2(2.0 \times 10^{-3}\,M)}$$
$$= 2.5 \times 10^2\,M^{-2}\,\text{s}^{-1}$$

(c) Using the known rate constant and the given concentrations of NO and H_2, we write

$$\text{rate} = k[\text{NO}]^2[\text{H}_2]$$
$$= (2.5 \times 10^2\,M^{-2}\,\text{s}^{-1})(13.0 \times 10^{-3}\,M)^2(6.0 \times 10^{-3}\,M)$$
$$= 2.5 \times 10^{-4}\,M\,\text{s}^{-1}$$

Comment The reaction is first order in H_2, whereas the stoichiometric coefficient for H_2 in the balanced equation is 2. The order of a reactant is unrelated to its stoichiometric coefficient in the overall balanced equation.

Practice Exercise The reaction of peroxydisulfate ion ($\text{S}_2\text{O}_8^{2-}$) with iodide ion ($\text{I}^-$) is

$$\text{S}_2\text{O}_8^{2-}(aq) + 3\text{I}^-(aq) \longrightarrow 2\text{SO}_4^{2-}(aq) + \text{I}_3^-(aq)$$

From the following data collected at a certain temperature, determine the rate law and calculate the rate constant at that temperature.

Experiment	$[\text{S}_2\text{O}_8^{2-}]$ (M)	$[\text{I}^-]$ (M)	Initial Rate (M s^{-1})
1	0.080	0.034	2.2×10^{-4}
2	0.080	0.017	1.1×10^{-4}
3	0.190	0.025	3.8×10^{-4}

14.3 | Integrated Rate Laws Specify the Relationship Between Reactant Concentration and Time

Rate law expressions enable us to calculate the rate of a reaction from the rate constant and reactant concentrations. Using calculus, the rate laws can be transformed by the process of integration to give *integrated rate laws,* which tell us the concentrations of reactants at any time during the course of a reaction. We will illustrate these important concepts by considering two of the simplest kinds of rate laws—those applying to reactions that are first order overall and those applying to reactions that are second order overall.

First-Order Reactions

A *first-order reaction* is *a reaction whose rate depends on the reactant concentration raised to the first power*. In a first-order reaction of the type

$$\text{A} \longrightarrow \text{product}$$

the rate law is

$$\text{rate} = -\frac{d[A]}{dt} = k[A] \tag{14.3}$$

where k is the first-order rate constant. The units of k are s^{-1}. Rearranging Equation 14.3 gives

$$\frac{1}{[A]} d[A] = -k \, dt \tag{14.4}$$

The time dependence of [A] can be determined using calculus (see Appendix 1). Integrating Equation 14.4 from the initial concentration $[A]_0$ at $t = 0$, to the final concentration $[A]_t$ at $t = t$, we obtain

$$\int_{[A]_0}^{[A]_t} \frac{1}{[A]} d[A] = -\int_0^t k \, dt$$

$$\ln \frac{[A]_t}{[A]_0} = -kt \tag{14.5}$$

or

$$[A]_t = [A]_0 \, e^{-kt} \tag{14.6}$$

The time $t = 0$ need not correspond to the beginning of the experiment; it can be any time when we choose to start monitoring the change in the concentration of A. The exponential decay with time indicated by Equation 14.6 is a signature of first-order reactions [Figure 14.9(a)].

Equation 14.5 can be rearranged as follows

$$\ln \left(\frac{[A]_t}{c_0} \right) = -kt + \ln \left(\frac{[A]_0}{c_0} \right) \tag{14.7}$$

where c_0 is some reference concentration,[1] usually 1 M. Equation 14.7 has the form of a linear equation $y = mx + b$, where m is the slope of the line and b is the y-axis

---

1. The reference concentration, c_0, is necessary in Equation 14.7 because a quantity must be dimensionless in order to take its natural logarithm. As long as the same reference concentration is used for both $[A]_t$ and $[A]_0$, the choice of c_0 is arbitrary.

Figure 14.9 The time dependence of reactant concentration for a first-order reaction. (a) A plot of reactant concentration versus time showing the exponential decay of Equation 14.6. (b) A log-linear plot of reactant concentration versus time gives a straight line (Equation 14.7) with a slope of $-k$ and an intercept of $\ln[A]_0$.

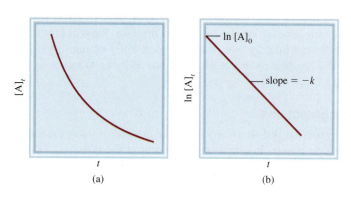

intercept. Thus, for a first-order reaction, a plot of $\ln([A]_t/c_0)$ versus time (t) will be linear with a slope of $-k$ and have a y-axis intercept of $\ln([A]_0/c_0)$. From a graph of this kind, illustrated in Figure 14.9(b), we can determine the rate constant from its slope.

First-order reactions are common. An example is the decomposition of ethane (C_2H_6) into highly reactive fragments called methyl radicals ($CH_3 \cdot$):

$$C_2H_6 \longrightarrow 2CH_3 \cdot$$

The decomposition of N_2O_5 is also a first-order reaction

$$2N_2O_5(g) \longrightarrow 4NO_2(g) + O_2(g)$$

Besides chemical reactions, a number of nonchemical processes in nature also obey first-order kinetics. A particularly important example is radioactive decay, which is discussed in Chapter 17.

In Example 14.3, we apply Equations 14.5 and 14.6 to an organic reaction.

Example 14.3

The conversion of cyclopropane to propene in the gas phase is a first-order reaction with a rate constant of 6.7×10^{-4} s^{-1} at 500°C.

$$\begin{array}{c} CH_2 \\ / \ \backslash \\ CH_2 - CH_2 \end{array} \longrightarrow CH_3 - CH = CH_2$$

cyclopropane   propene

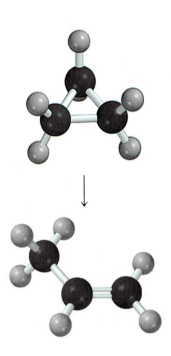

(a) If the initial concentration of cyclopropane was 0.25 M, what is the concentration after 8.8 min? (b) How long (in minutes) will it take for the concentration of cyclopropane to decrease from 0.25 M to 0.15 M? (c) How long (in minutes) will it take to convert 74 percent of the starting material?

Strategy The relationship between the concentrations of a reactant at different times in a first-order reaction is given by Equation 14.5 or 14.6. In part (a), we are given $[A]_0$ and t and asked to find $[A]_t$, which can be accomplished by a straightforward substitution into Equation 14.6. In part (b), we are given $[A]_0$ and $[A]_t$ and we are asked to find t. Equation 14.5 contains all these variables, but it must first be rearranged to isolate t on one side. In part (c), we are not given $[A]_t$ and $[A]_0$. However, if initially we have 100 percent of the compound and 74 percent has reacted, then what is left must be (100% − 74%), or 26 percent. Thus, the ratio $[A]_t/[A]_0$, which is all we need to determine the time, is equal to 0.26.

Solution (a) In applying Equation 14.6, note that k is given in units of s^{-1}, so we must convert 8.8 min to seconds in the calculation:

$$[A]_t = [A]_0 \, e^{-kt}$$
$$= 0.25 \, M \times \exp\left[-(6.7 \times 10^{-4} \, s^{-1})(8.8 \, min)\left(\frac{60 \, s}{1 \, min}\right)\right]$$
$$= 0.25 \, M \times 0.702$$
$$= 0.18 \, M$$

—Continued

Continued—

(b) Rearranging Equation 14.5 to isolate t gives

$$t = -k^{-1} \ln \frac{[A]_t}{[A]_0}$$

$$= -(6.7 \times 10^{-4} \, s^{-1})^{-1} \ln \frac{0.15 \, M}{0.25 \, M}$$

$$= 760 \, s \times \frac{1 \, min}{60 \, s}$$

$$= 13 \, min$$

(c) Using the rearranged Equation 14.5 given in part (b) and the fact that $[A]_t / [A]_0 = 0.26$, we have

$$t = -k^{-1} \ln \frac{[A]_t}{[A]_0}$$

$$= -(6.7 \times 10^{-4} \, s^{-1})^{-1} \ln (0.26)$$

$$= 2.0 \times 10^3 \, s \times \frac{1 \, min}{60 \, s}$$

$$= 33 \, min$$

Practice Exercise The reaction $2A \longrightarrow B$ is first order in A with a rate constant of $2.8 \times 10^{-2} \, s^{-1}$ at 80°C. (a) How long (in seconds) will it take for A to decrease from 0.88 M to 0.14 M? (b) Assuming that the initial concentration of [A] is 0.90 M and that the initial concentration of [B] is zero, how much time is necessary for the concentration of [B] to reach 0.25 M?

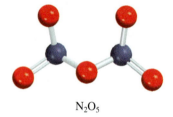

N_2O_5

N_2O_5 decomposes to give NO_2 (brown color).

Now let us determine graphically the order and rate constant of the decomposition of nitrogen pentoxide (N_2O_5) in carbon tetrachloride (CCl_4) solvent at 45°C:

$$2N_2O_5(\text{in } CCl_4) \longrightarrow 4NO_2(g) + O_2(g)$$

The table shows the variation of N_2O_5 concentration with time, and the corresponding ln [N_2O_5] values:

t (s)	[N_2O_5] (M)	ln ([N_2O_5])/1 M)
0	0.91	−0.094
300	0.75	−0.29
600	0.64	−0.45
1200	0.44	−0.82
3000	0.16	−1.83

Applying Equation 14.7, we plot ln ([N_2O_5]/1 M) versus t, as shown in Figure 14.10. Because this plot is linear, the rate law must be first order in [N_2O_5]. Next, there are several ways to estimate the rate constant from the slope, depending upon the accuracy

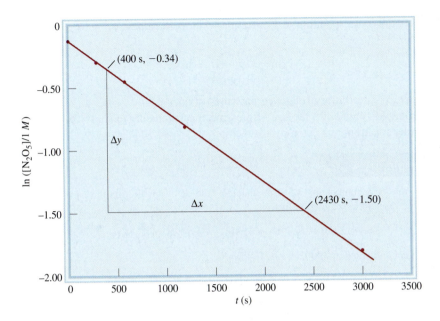

required. The simplest, but least accurate, way is to select two data points far apart on the line and subtract their *y* and *x* values as follows:

$$
\begin{aligned}
\text{slope} &= \frac{\Delta y}{\Delta x} = \frac{\Delta \ln\left([N_2O_5]/1\ M\right)}{\Delta t} \\
&= \frac{-1.83 - (-0.094)}{(3000 - 0)\ \text{s}} \\
&= -5.79 \times 10^{-4}\ \text{s}^{-1}
\end{aligned}
$$

Because slope $= -k$, we get $k = 5.79 \times 10^{-4}\ \text{s}^{-1}$. A more accurate method is to use least-squares linear regression (see Appendix 1) to find the line that best fits all the data. This method is tedious by hand, but can be performed readily either using a spreadsheet program or a simple scientific calculator. Using the data provided, the best-fit line from least squares linear regression gives $5.76 \times 10^{-4}\ \text{s}^{-1}$, which is slightly different than the preceding simple estimate.

For gas-phase reactions, we can replace the concentration terms in Equations 14.5 and 14.6 with the partial pressures of the gaseous reactant. Consider the first-order reaction

$$ A(g) \longrightarrow \text{product} $$

Using the ideal gas equation we write

$$ P_A V = n_A RT $$

or

$$ \frac{n_A}{V} = [A] = \frac{P_A}{RT} $$

Substituting $[A] = P/RT$ into Equation 14.5, we get

$$ \ln \frac{[A]_t}{[A]_0} = \ln \frac{P_{A,t}/RT}{P_{A,0}/RT} = \ln \frac{P_{A,t}}{P_{A,0}} = -kt $$

The equation corresponding to Equation 14.7 now becomes

$$\ln \frac{P_{A,t}}{P_{ref}} = -kt + \ln \frac{P_{A,0}}{P_{ref}} \tag{14.8}$$

where P_{ref} is an arbitrary reference pressure (often 1 bar).

Example 14.4 shows the use of pressure measurements to study the kinetics of a first-order reaction.

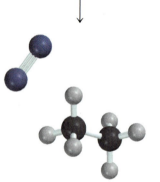

Example 14.4

The rate of decomposition of azomethane is studied by monitoring the partial pressure of the reactant as a function of time:

$$CH_3\!-\!N\!=\!N\!-\!CH_3(g) \longrightarrow N_2(g) + C_2H_6(g)$$

The data obtained at 300°C are as follows:

t (s)	P_{azo} (mmHg)	ln (P_{azo}/1 mmHg)
0	284	5.649
100	220	5.394
150	193	5.263
200	170	5.136
250	150	5.011
300	132	4.883

Are these values consistent with first-order kinetics? If so, determine the rate constant.

Strategy To test for first-order kinetics, determine whether the integrated first-order rate law (Equation 14.8 for pressure data) gives a straight line with the data provided. In this problem, pressure is measured in units of mmHg, so it is appropriate to use 1 mmHg as the reference pressure, P_{ref}. If the reaction is first order, then a plot of $\ln[P_{azo}(t)/(1\ \text{mmHg})]$ versus t will produce a straight line with a slope equal to $-k$.

Solution Figure 14.11, which is based on the data given in the table, shows that a plot of $\ln[P_{azo}(t)/(1\ \text{mmHg})]$ versus t yields a straight line, so the reaction is indeed first order. The slope of the line can be found by least-squares linear regression (see Appendix 1) and is equal to $2.55 \times 10^{-3}\ \text{s}^{-1}$. According to Equation 14.8, the slope is equal to k, so $k = 2.55 \times 10^{-3}\ \text{s}^{-1}$.

Practice Exercise Ethyl iodide (C_2H_5I) decomposes at a certain temperature in the gas phase as follows:

$$C_2H_5I(g) \longrightarrow C_2H_4(g) + HI(g)$$

From the following data, determine the order of the reaction and the rate constant.

t (min)	P_{azo}(bar)
0	0.36
15	0.30
30	0.25
48	0.19
75	0.13

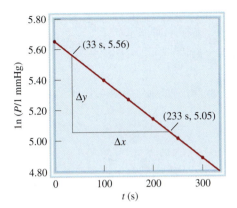

Figure 14.11 Plot of $\ln[P_{azo}(t)/(1 \text{ mmHg})]$ versus time for the decomposition of azomethane.

The Half-Life of a Reaction

A measure of considerable practical importance in kinetic studies is the **half-life** $(t_{1/2})$ of a reaction, which is *the time required for the concentration of a reactant to decrease to one-half of its initial concentration.* We can obtain an expression for $t_{1/2}$ for a first-order reaction as follows. At $t_{1/2}$, the concentration of reactant $[A]_t$ is equal to $[A]_0/2$, so Equation 14.5 becomes

$$\ln \frac{[A]_0/2}{[A]_0} = -kt_{1/2}$$
$$\ln (1/2) = -kt_{1/2}$$
$$-\ln(2) = -kt_{1/2}$$

so

$$t_{1/2} = \frac{\ln(2)}{k} \approx \frac{0.693}{k} \qquad (14.9)$$

The significance of Equation 14.9 is that the half-life of a first-order reaction is *independent* of the initial concentration of the reactant. Thus, it takes the same time for the concentration of the reactant to decrease from 1.0 M to 0.50 M, say, as it does for the concentration to decrease from 0.10 M to 0.050 M (Figure 14.12). Measuring the half-life of a reaction is one way to determine the rate constant of a first-order reaction. The shorter the half-life, the larger the k.

Example 14.5 shows how to calculate the half-life of a first-order reaction.

Example 14.5

The decomposition of ethane (C_2H_6) to methyl radicals is a first-order reaction with a rate constant of 5.36×10^{-4} s^{-1} at 700°C:

$$C_2H_6(g) \longrightarrow 2CH_3 \cdot (g)$$

Calculate the half-life of the reaction in minutes.

Strategy Using Equation 14.9, calculate the half-life of a first-order reaction. The rate constant is given in units of s^{-1}, so a conversion of units is needed to express the half-life in minutes.

—Continued

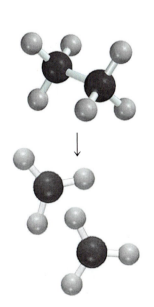

Figure 14.12 Plot of $[A]_t$ versus time for the first-order reaction A $\longrightarrow$ products. The half-life of the reaction is 1 min. After the elapse of each half-life, the concentration of A is halved.

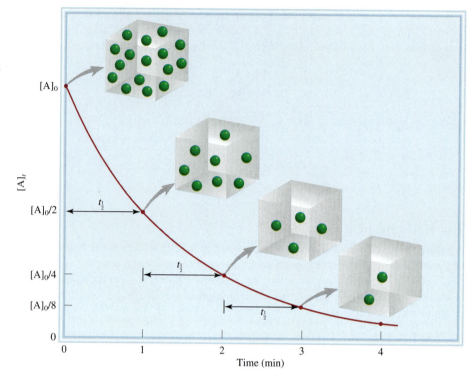

Continued—

Solution For a first-order reaction, we only need the rate constant to calculate the half-life of the reaction. From Equation 14.9,

$$t_{1/2} = \frac{\ln(2)}{k} \approx \frac{0.693}{k} = \frac{0.693}{5.36 \times 10^{-4}\,s^{-1}}$$

$$= 1.29 \times 10^3\,s \times \frac{1\,min}{60\,s}$$

$$= 21.5\,min$$

Practice Exercise Calculate the half-life of the decomposition of N_2O_5, discussed on page 726.

Second-Order Reactions

A *second-order reaction* is one whose rate depends on the concentration of one reactant raised to the second power or on the concentrations of two different reactants, each raised to the first power. The simplest type involves only one kind of reactant molecule:

$$A \longrightarrow product$$

From the rate law,

$$rate = -\frac{d[A]}{dt} = k[A]^2 \tag{14.10}$$

The units on k are M^{-1} s^{-1}. We can determine the dependence of [A] on time by rearranging Equation 14.10 and integrating from the initial time ($t = 0$) and initial concentration ([A]$_0$) to the final time t and final concentration ([A]$_t$):

$$\int_{[A]_0}^{[A]_t} \frac{d[A]}{[A]^2} = -\int_0^t k\,dt$$

$$\frac{1}{[A]_t} - \frac{1}{[A]_0} = kt$$

Rearranging gives the following integrated rate law for this second-order reaction:

$$\frac{1}{[A]_t} = kt + \frac{1}{[A]_0} \qquad (14.11)$$

We can obtain an equation for the half-life of a second-order reaction by setting [A]$_t$ = [A]$_0$/2 in Equation 14.11.

$$\frac{1}{[A]_0/2} = kt + \frac{1}{[A]_0}$$

Solving for $t_{1/2}$ we obtain

$$t_{1/2} = \frac{1}{k[A]_0} \qquad (14.12)$$

The half-life of a second-order reaction is inversely proportional to the initial reactant concentration. This result makes sense because the half-life should be shorter in the early stage of the reaction when more reactant molecules are present to collide (react) with each other. Because of its dependence on concentration, the half-life is much less useful in describing second-order reactions than first-order reactions. Measuring the half-lives at different initial concentrations is one way to distinguish between a first-order and a second-order reaction.

The kinetic analysis of a second-order reaction is shown in Example 14.6.

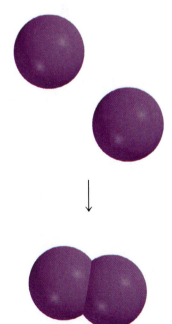

Example 14.6

Iodine atoms combine to form molecular iodine in the gas phase:

$$I(g) + I(g) \longrightarrow I_2(g)$$

This reaction follows second-order kinetics with $k = 7.0 \times 10^9$ M^{-1} s^{-1} at 23°C. (a) If the initial concentration of I was 0.086 M, calculate the concentration after 2.0 min. (b) Calculate the half-life of the reaction if the initial concentration of I is 0.60 M and if it is 0.42 M.

Strategy (a) The relationship between the concentrations of a reactant at different times is given by the integrated rate law. Because this is a second-order reaction with a single reactant, we must use Equation 14.11. (b) The half-life for a second-order reaction is given by Equation 14.12.

—Continued

Continued—

Solution (a) Given the initial concentration of I and the rate constant for the reaction, we can use Equation 14.11 to calculate the concentration of I at a later time:

$$\frac{1}{[I]_t} = kt + \frac{1}{[I]_0}$$

$$\frac{1}{[I]_t} = (7.0 \times 10^9 \, M^{-1} \, s^{-1})\left(2.0 \, \text{min} \times \frac{60 \, s}{\text{min}}\right) + \frac{1}{0.086 \, M}$$

where $[I]_t$ is the concentration at $t = 2.0$ min. Solving the equation, we get

$$[I]_t = 1.2 \times 10^{-12} \, M$$

This is such a low concentration that it is virtually undetectable. The very large rate constant for the reaction means that practically all the I atoms have combined only 2.0 min into the reaction.

(b) We need Equation 14.12 to calculate the half-life of the reaction. For $[I]_0 = 0.60 \, M$,

$$t_{1/2} = \frac{1}{k[I]_0} = \frac{1}{(7.0 \times 10^9 \, M^{-1} \, s^{-1})(0.60 \, M)} = 2.4 \times 10^{-10} \, s$$

For $[I]_0 = 0.42 \, M$,

$$t_{1/2} = \frac{1}{k[I]_0} = \frac{1}{(7.0 \times 10^9 \, M^{-1} \, s^{-1})(0.42 \, M)} = 3.4 \times 10^{-10} \, s$$

Comment These results confirm that the half-life of a second-order reaction, unlike that of a first-order reaction, is not a constant. Instead, the half-life of a second-order reaction depends on the initial concentration of the reactant(s).

Practice Exercise The reaction $2A \longrightarrow B$ is second order with a rate constant of $51 \, M^{-1} \, \text{min}^{-1}$ at 24°C. (a) Starting with $[A]_0 = 0.0092 \, M$, how long will it take to reach $[A]_t = 3.7 \times 10^{-3} \, M$? (b) Calculate the half-life of the reaction.

The second-order integrated rate law derived above (Equation 14.11) applies to a reaction that is second-order in a single species. One can also have reactions such as

$$A + B \longrightarrow \text{product}$$

that is second order overall, but first order in each of the reacting species (A and B). The second-order rate law for this reaction is

$$\text{rate} = -\frac{d[A]}{dt} = -\frac{d[B]}{dt} = k[A][B] \qquad \textbf{(14.13)}$$

This rate law can be integrated assuming $[A]_0 < [B]_0$ (Problem 14.92) to give an equation that relates the concentration of [A] and [B] to the initial concentrations and the elapsed time:

$$\frac{1}{[B]_0 - [A]_0} \ln \frac{[A]_0[B]}{[B]_0[A]} = kt \qquad \textbf{(14.14)}$$

Equation 14.14 assumes that the stoichiometric coefficients for A and B are both unity. For other reactions, such as A + 2B $\longrightarrow$ products or 2A + B $\longrightarrow$ products, the integrated second-order rate laws will be different, but can be derived in a similar manner.

Equation 14.14 becomes greatly simplified if one of the reactants is present in large excess. For example, if $[A]_0 \gg [B]_0$, then the concentration of A will change very little as the reaction proceeds, that is, $[A]_t \approx [A]_0$. Substituting this into Equation 14.14 and using the approximation that $[B]_0 - [A]_0 \approx -[A]_0$, gives

$$\ln \frac{[B]}{[B]_0} = -k[A]_0 t$$

which is identical to a first-order integrated rate law in B (Equation 14.5) with a rate constant k^* equal to $k[A]_0$. Thus, due to one of the reactants being present in large excess, this mixed second-order reaction behaves approximates as a first-order reaction and is referred to as a ***pseudo–first-order*** reaction.

Zero-Order Reactions

First- and second-order reactions are the most common reaction types. Reactions whose order is zero are rare. For a zero-order reaction of the form

$$A \longrightarrow \text{product}$$

the rate law is given by

$$\text{rate} = k[A]^0 = k$$

Thus, the rate of a zero-order reaction is a *constant,* independent of reactant concentration. The integrated rate law for zero-order reactions is

$$[A]_t = [A]_0 - kt \tag{14.15}$$

with a half−life of $[A]_0/2k$. Third-order and higher order reactions are quite complex; they are not discussed in this book. Table 14.3 summarizes the kinetics for zero-order, first-order, and second-order reactions.

Table 14.3	Summary of the Kinetics of Zero-Order, First-Order, and Second-Order Reactions		
Order	**Rate Law**	**Concentration-Time Equation**	**Half-Life**
0	Rate = k	$[A]_t = -kt + [A]_0$	$\dfrac{[A]_0}{2k}$
1	Rate = $k[A]$	$\ln \dfrac{[A]_t}{[A]_0} = -kt$	$\dfrac{\ln(2)}{k} \approx \dfrac{0.693}{k}$
2 (A $\longrightarrow$ products)	Rate = $k[A]^2$	$\dfrac{1}{[A]_t} = kt + \dfrac{1}{[A]_0}$	$\dfrac{1}{k[A]_0}$
2 (A + B $\longrightarrow$ products)	Rate = $k[A][B]$	$\dfrac{1}{[B]_0 - [A]_0} \ln \dfrac{[A]_0[B]}{[B]_0[A]} = kt$	—

Other More Complex Reactions

Parallel Reactions

Consider a reactant that is involved in two first-order reactions that are occurring simultaneously (*parallel reactions*).

$$A \xrightarrow{k_1} B$$
$$A \xrightarrow{k_2} C$$

The total rate of consumption of A, is the sum of the rates for each of these reactions:

$$\frac{d[A]}{dt} = -k_1[A] - k_2[A] = -(k_1 + k_2)[A]$$

Thus, the rate of consumption of reactant A participating in simultaneous parallel reactions is governed by a rate constant that is the sum of the rate constants of the individual reactions.

Consecutive Reactions

A *consecutive reaction* is one in which the product from the first step becomes the reactant for the second step, and so on. To examine such reactions, consider the following two-step first-order consecutive reaction

$$A \xrightarrow{k_1} B \xrightarrow{k_2} C$$

Each step is first order, so the rate law equations for the concentrations for the disappearance of A and B are

$$\frac{d[A]}{dt} = -k_1[A]$$

$$\frac{d[B]}{dt} = k_1[A] - k_2[B]$$

Assume that only A is present at time $t = 0$. Because the decrease in A is first order, Equation 14.6 holds. That is,

$$[A]_t = [A]_0\, e^{-kt}$$

The integrated rate laws for B and C can be obtained using calculus:

$$[B] = \frac{k_1[A]_0}{k_2 - k_1}(e^{-k_1 t} - e^{-k_2 t}); \quad [C] = [A]_0\left[1 + \frac{1}{k_2 - k_1}(k_2 e^{-k_1 t} - k_1 e^{-k_2 t})\right] \quad \textbf{(14.16)}$$

Plots of the concentrations of A, B, and C for a reaction in which $k_1 = k_2$ are shown in Figure 14.13.

Consider the concentration of C in the limit that one reaction is much slower than the other. If the first reaction is the slowest ($k_1 \ll k_2$) then Equation 14.16 reduces to

$$[C] \approx [A]_0(1 - e^{-k_1 t}) \quad k_1 \ll k_2$$

If the second reaction is the slowest ($k_2 \ll k_1$), then

$$[C] \approx [A]_0(1 - e^{-k_2 t}) \quad k_2 \ll k_1$$

In both of these limits, it is the rate of the *slowest* reaction that determines the rate of formation of the product. This is a general feature of consecutive reactions. If one reaction step in a consecutive reaction is considerably slower than the other steps, then the rate constant of the slowest reaction will govern the rate of product formation. This slowest reaction is referred to as the ***rate-determining step*** in a reaction sequence. If $k_1 \ll k_2$, for the preceding reaction sequence, then the reaction $A \longrightarrow B$ is the rate-determining step. If, $k_2 \ll k_1$, on the other hand, then the reaction $B \longrightarrow C$ is rate determining.

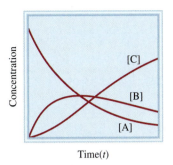

Figure 14.13 Variation in the concentrations of A, B, and C with time for the consecutive reaction $A \xrightarrow{k_1} B \xrightarrow{k_2} C$ for the case where $k_1 = k_2$.

Reversible Reactions

As we saw in Chapter 10, most reactions are, to some degree, reversible. To study the kinetics of a reversible reaction, it is necessary to take into account both the forward and the reverse rates. Consider, for example, a reversible first-order reaction of molecule A to form molecule B:

$$ A \underset{k_{-1}}{\overset{k_1}{\rightleftharpoons}} B $$

In this reaction, A is consumed in the forward reaction with rate constant k_1 and produced in the reverse reaction with a rate constant k_{-1}, so the net rate law is

$$ \frac{d[A]}{dt} = -k_1[A] + k_{-1}[B] $$

When the system reaches equilibrium, the concentration of A becomes constant in time, and the rate of production of A, given by d[A]/dt, will be equal to zero, giving

$$ 0 = -k_1[A] - k_{-1}[B] $$

Rearranging this equation gives

$$ \frac{[B]}{[A]} = \frac{k_1}{k_{-1}} \tag{14.17} $$

The quantity on the left-hand side of Equation 14.17 ([B]/[A]) is the equilibrium constant K for the reaction $A \longrightarrow B$, so we have the result that the equilibrium constant is equal to the ratio of the forward and reverse reactions:

$$ \frac{k_1}{k_{-1}} = K \tag{14.18} $$

The relationship between the rate constants of this reversible reaction and the equilibrium constant (Equation 14.18) is an example of the *principle of **detailed balance,** which states that, at equilibrium, the rates of forward and reverse processes are equal.*[2]

---

2. The principle of detailed balance is a consequence of *microscopic reversibility*—the fact that the fundamental equations governing molecular motion (i.e., Newton's laws or the Schrödinger equation) have the same form when time t is replaced with $-t$ and the sign of all velocities (or momenta) are also reversed.

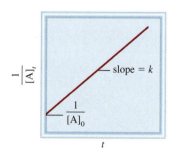

Figure 14.14 Typical dependence of the rate constant on temperature. The rate constants of most reactions increase with increasing temperature.

14.4 | The Arrhenius Equation Gives the Temperature Dependence of Rate Constants

With very few exceptions, reaction rates increase with increasing temperature. For example, the time required to hard-boil an egg in water is much shorter if the "reaction" is carried out at 100°C (about 10 min) than at 80°C (about 30 min). Conversely, an effective way to preserve foods is to store them at subzero temperatures, thereby slowing the rate of bacterial decay. Figure 14.14 shows a typical example of the relationship between the rate constant of a reaction and temperature. In order to explain this behavior, we must ask how reactions get started in the first place.

The Arrhenius Equation

In the late 1880s, the Swedish chemist Svante Arrhenius discovered that the temperature dependence of many reactions could be described empirically by the following equation, known as the **Arrhenius equation:**

$$k = Ae^{-E_a/RT} \tag{14.19}$$

where E_a (usually given in kJ mol^{-1}) is called the **activation energy** of the reaction, R is the gas constant (8.314 J K^{-1} mol^{-1}), T is the absolute temperature, and e is the base of the natural logarithm scale. The quantity A is called the **frequency factor** (or **preexponential factor**) and is related to the frequency of collisions between reactant molecules. It can often be treated as a constant for a given reacting system over a fairly wide temperature range. Because the exponential of a quantity is a dimensionless number, the units of the frequency factor A are the same as for k.

Equation 14.19 can be expressed in a more useful form by taking the natural logarithm of both sides

$$\ln k = \ln Ae^{-E_a/RT}$$

or

$$\ln k = \ln A - \frac{E_a}{RT} \tag{14.20}$$

In Equation 14.20, k and A are treated as dimensionless. This means that the values of k and A used in the equation represent the actual values of k or A divided by some reference value k_t. For example, if k is given as 1.2 s^{-1}, then we use 1 s^{-1} as k_t and the value of k that we use in the equation is the dimensionless number 1.2. This will work as long as we use the same reference value for both k and A.

Assuming A to be temperature independent, Equation 14.20 has the form of an equation of a straight line (i.e., $y = mx + b$). Thus, a plot of $\ln k$ versus $1/T$, called an *Arrhenius plot*, yields a straight line with slope equal to $-E_a/R$ with a y intercept of $\ln A$. Example 14.7 shows how the activation energy (E_a) can be determined graphically from an Arrhenius plot of the temperature dependence of k.

Example 14.7

The rate constants for the decomposition of acetaldehyde

$$CH_3CHO(g) \longrightarrow CH_4(g) + CO(g)$$

were measured at five different temperatures, as shown in the accompanying table. Use these data to plot $\ln k$ versus $1/T$, and determine the activation energy

—Continued

Continued—

(in kJ mol^{-1}) for the reaction. (Note: The reaction is 3/2 order in CH_3CHO, so k has units of $M^{-1/2}$ s^{-1}.)

k ($M^{-1/2}$ s^{-1})	T (K)
0.011	700
0.035	730
0.105	760
0.343	790
0.789	810

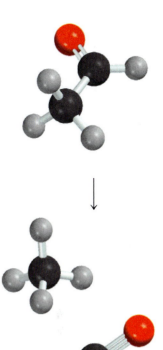

Strategy A plot of ln k versus $1/T$ will produce a straight line with a slope equal to $-E_a/R$. Thus, the activation energy can be determined from the slope of the plot.

Solution First, convert the data from k and T to ln k and $1/T$, respectively:

ln k	$1/T$ (K^{-1})
-4.51	1.43×10^{-3}
-3.35	1.37×10^{-3}
-2.254	1.32×10^{-3}
-1.070	1.27×10^{-3}
-0.237	1.23×10^{-3}

A plot of these data yields the graph in Figure 14.14, which is linear. Using least-squares linear regression (see Appendix 1), the slope of the line that best fits these data is

$$\text{slope} = -2.09 \times 10^4 \text{ K}$$

According to Equation 14.20, the slope is $-E_a/R$, so

$$-E_a/R = \text{slope} = -2.09 \times 10^4 \text{ K}$$
$$E_a = -(-2.09 \times 10^4 \text{ K})(8.314 \text{ J mol}^{-1} \text{ K}^{-1})$$
$$= 1.74 \times 10^5 \text{ J mol}^{-1}$$
$$= 1.74 \times 10^2 \text{ kJ mol}^{-1}$$

Practice Exercise The second-order rate constant for the decomposition of nitrous oxide (N_2O) into molecular nitrogen and an oxygen atom has been measured at different temperatures:

k (M^{-1} s^{-1})	T (°C)
1.87×10^{-3}	600
0.0113	650
0.0569	700
0.244	750

Determine graphically the activation energy for the reaction.

If the frequency factor A is treated as if it were independent of temperature, an equation relating the rate constants k_1 and k_2 at temperatures T_1 and T_2, respectively, can be used to calculate the activation energy or to find the rate constant at another temperature if the activation energy is known. To derive such an equation, we start with Equation 14.20:

$$\ln k_1 = \ln A - \frac{E_a}{RT_1}$$

$$\ln k_2 = \ln A - \frac{E_a}{RT_2}$$

Subtracting $\ln k_2$ from $\ln k_1$ gives

$$\ln k_2 - \ln k_1 = \frac{E_a}{R}\left(\frac{1}{T_1} - \frac{1}{T_2}\right)$$

$$\ln \frac{k_2}{k_1} = \frac{E_a}{R}\left(\frac{1}{T_1} - \frac{1}{T_2}\right)$$

$$\ln \frac{k_2}{k_1} = \frac{E_a}{R}\left(\frac{T_2 - T_1}{T_1 T_2}\right) \tag{14.21}$$

Example 14.8 shows how to use Equation 14.21.

Example 14.8

The rate constant of a first-order reaction is 3.46×10^{-2} s^{-1} at 298 K. What is the rate constant at 350 K if the activation energy for the reaction is 50.2 kJ mol^{-1}?

Strategy Use Equation 14.21, a modified form of the Arrhenius equation, to relate two rate constants at two different temperatures. Make sure the units of R and E_a are consistent.

Solution The data are

$$k_1 = 3.46 \times 10^{-2}\text{ s}^{-1} \qquad k_2 = ?$$
$$T_1 = 298\text{ K} \qquad\qquad T_2 = 350\text{ K}$$

Substituting the given temperatures and E_a into Equation 14.21 gives

$$\ln \frac{k_2}{k_1} = \frac{E_a}{R}\left(\frac{T_2 - T_1}{T_1 T_2}\right)$$
$$= \frac{50.2 \times 10^3\text{ J mol}^{-1}}{8.314\text{ J mol}^{-1}\text{ K}^{-1}}\left[\frac{350\text{ K} - 298\text{ K}}{(298\text{ K})(350\text{ K})}\right]$$
$$= 3.01$$

(We had to convert E_a to units of J mol^{-1} to match the units of R.) Exponentiating gives

$$\frac{k_2}{k_1} = e^{3.01} = 20.3$$
$$k_2 = 20.3 \times k_1 = 20.3 \times 3.46 \times 10^{-2}\text{ s}^{-1}$$
$$= 0.702\text{ s}^{-1}$$

—Continued

Continued—

Check The answer is reasonable because the rate constant is greater at a higher temperature.

Practice Exercise The first-order rate constant for the reaction of methyl chloride (CH$_3$Cl) with water to produce methanol (CH$_3$OH) and hydrochloric acid (HCl) is 3.32×10^{-10} s^{-1} at 25°C. Calculate the rate constant at 40°C if the activation energy is 116 kJ mol^{-1}.

Physical Meaning of the Activation Energy

In the course of a chemical reaction, chemical bonds in the reactants are broken and chemical bonds in the product are formed. The specific sequence in which these events occur is called the *reaction pathway*. Consider, for example, the simple gas-phase reaction of a hydrogen atom with molecular chlorine to form HCl and a Cl atom:

$$H(g) + Cl_2(g) \longrightarrow HCl(g) + Cl(g)$$

This reaction involves the breaking of a Cl—Cl bond and the formation of an H—Cl bond. To simplify the analysis, let us assume that the reaction occurs through the end-on collision of an H atom with the Cl$_2$ molecule as illustrated in Figure 14.15(a). When the hydrogen atom is well separated from the Cl$_2$ molecule, the system is in the initial reactant state [see Figure 14.15(a)] with a potential energy that is the sum of the potential energies for H and for Cl$_2$ [Figure 14.15(b)].

As the hydrogen atom nears the Cl—Cl molecule from the end [State 2 in Figure 14.15(b)], the electron clouds on both the H atom and the Cl$_2$ molecule will begin to distort and overlap, leading to the formation of a partial H—Cl bond and to a corresponding weakening of the Cl—Cl bond. Because we are replacing a full Cl—Cl bond with partial bonds, the potential energy of this intermediate state is higher than that of the initial reactant state as is shown in Figure 14.15(b). As the reaction proceeds, the potential energy continues to increase until it reaches a maximum at State 3, in which the central Cl atom is partially bonded to both the H and the other Cl. This *intermediate state, along the reaction pathway, at which the potential energy is a maximum*, is called the **transition state** (or **activated complex**). Continuing the reaction beyond the transition state, the potential energy decreases as the H—Cl bond strengthens with further weakening of the Cl—Cl bond (State 4), until the final product state (State 5) consisting of a fully formed H—Cl bond and a separated Cl atom is reached, and the potential energy is the sum of that for the separated products.

Because the bond enthalpy of the H—Cl bond (432 kJ mol^{-1}) is higher than that of the Cl—Cl bond (243 kJ mol^{-1}), energy is released in this reaction, so the potential energy of the product state is lower than that of the reactant state [$\Delta H = (243 - 432)$ kJ mol^{-1} = -189 kJ mol^{-1}]. However, even though the net reaction goes downhill in energy, the reaction can only proceed if the kinetic energy of the collision is sufficient to enable the system to reach the transition state—otherwise the H atom will simply rebound from the collision with no reaction occurring. The difference in potential energy between the transition state and the initial reactant state is precisely the *activation energy* (E_a) that appears in the Arrhenius equation (Equation 14.19), and represents *the minimum amount of energy required for a chemical reaction to proceed.*

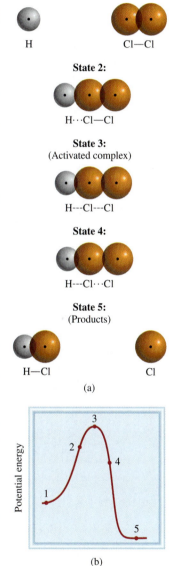

State 1:
(Reactants)

H   Cl—Cl

State 2:

H···Cl—Cl

State 3:
(Activated complex)

H---Cl---Cl

State 4:

H---Cl···Cl

State 5:
(Products)

H—Cl   Cl

(a)

(b)

Figure 14.15 (a) Stages of the gas-phase reaction of the end-on reaction of atomic hydrogen with molecular chlorine: H(g) + Cl$_2$(g) $\longrightarrow$ HCl(g) + Cl(g). (b) The potential energy profile of the reaction showing each of the stages in part (a).

Figure 14.16 Potential energy profiles for (a) exothermic and (b) endothermic reactions. These plots show the change in potential energy as reactants A and B are converted to products C and D. The transition state (AB‡) is a highly unstable species with a high potential energy. The activation energy is defined for the forward reaction in both (a) and (b). The products C and D are more stable than the reactants in (a) and less stable than the reactants in (b).

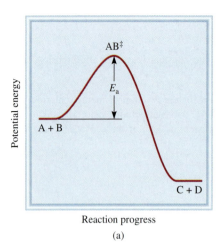

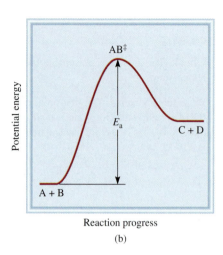

Reaction progress
(a)

Reaction progress
(b)

Figure 14.16 shows two different potential energy profiles for the generic reaction

$$A + B \longrightarrow C + D$$

If the products are more stable than the reactants, then the reaction will release heat, that is, the reaction will be exothermic [Figure 14.16(a)]. If the products are less stable than the reactants, on the other hand, then the reaction will absorb heat from the surroundings, and we have an endothermic reaction [Figure 14.16(b)]. In both cases, we plot the potential energy of the reacting system versus the progress of the reaction, with the transition state denoted by AB‡. Qualitatively, these graphs show how potential energy changes as reactants are converted to products.

We can think of activation energy as a barrier that prevents less energetic molecules from reacting. Because the number of reactant molecules in an ordinary reaction is very large, the speeds, and hence the kinetic energies of the molecules, vary greatly. Normally, only a small fraction of the colliding molecules—the fastest-moving ones—have enough kinetic energy to exceed the activation energy. These molecules can therefore take part in the reaction. The increase in the rate (or the rate constant) with temperature according to Equation 14.19 can now be explained: The speeds of the molecules obey the Maxwell-Boltzmann distributions shown in Figure 5.19(a) (reproduced in margin). If we compare the speed distributions at two different temperatures, we see that more high-energy molecules are present at the higher temperature. The rate of product formation is, therefore, also greater at the higher temperature.

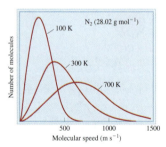

The Maxwell-Boltzmann distribution of molecular speeds for N$_2$ at three different temperatures.

The Collision Theory of Reaction Rates

For reactions taking place in the gas phase, both the Arrhenius equation and the dependence of the rate law on concentration can be accounted for using the kinetic theory of gases, giving what we call the *collision theory* of gas-phase chemical kinetics. In this theory, the rate of a reaction is directly proportional to the number of molecular collisions per second (i.e., to the frequency of molecular collisions):

$$\text{rate} \propto \frac{\text{number of collisions}}{\text{s}}$$

Consider the following gas-phase reaction involving the collision of two B molecules to form products:

$$B + B \longrightarrow \text{products}$$

Using the kinetic theory of gases (Chapter 5), the number of two-particle collisions (Z_{BB}) that take place per volume (V) per unit time can be determined to be

$$Z_{BB} = \frac{\sqrt{2}}{2}\pi d^2 \langle u \rangle \left(\frac{N_B}{V}\right)^2 \qquad (14.22)$$

where d is the diameter of molecule B (assumed to be a hard sphere), N_B is the number of B molecules, and $\langle u \rangle$ is the mean molecular speed. The mean molecular speed was given in Equation 5.42 as

$$\langle u \rangle = \sqrt{\frac{8k_B T}{\pi m}}$$

where m is the mass of molecule B, k_B is Boltzmann's constant, and T is the absolute temperature. Combining Equation 5.42 with the relation that $N_B/V = N_A[B]$, where N_A is Avogadro's number, gives

$$Z_{BB} = \left[2d^2 N_A^2 \sqrt{\frac{\pi k_B T}{m}}\right][B]^2 \qquad (14.23)$$

Now, if every collision produced a reaction, the rate of this reaction would equal Z_{BB}. However, generally only a small fraction of the collisions will be sufficiently energetic to overcome the activation barrier. As a result, the collision frequency in Equation 14.23 must be multiplied by the fraction of the collisions that have kinetic energies larger than E_A. Using the Maxwell-Boltzmann distribution of molecular speeds (Equation 5.38), it can be shown that the collision frequency must be multiplied by a factor $e^{-E_a/RT}$ to account for the fact that not all collisions will be of sufficient energy to react. The rate then becomes

$$\text{rate} = Z_{BB}e^{-E_a/RT} = \left[2d^2 N_A^2 \sqrt{\frac{\pi k_B T}{m}}\right] e^{-E_a/RT}[B]^2 \qquad (14.24)$$

This rate expression correctly identifies both the concentration dependence of the rate law (second order in B) and the temperature dependence of the rate constant, which from Equation 14.24 is

$$k = \left[2d^2 N_A^2 \sqrt{\frac{\pi k_B T}{m}}\right] e^{-E_a/RT} \qquad (14.25)$$

Comparing Equation 14.25 with the Arrhenius equation (Equation 14.19), we get

$$A = \left[2d^2 N_A^2 \sqrt{\frac{\pi k_B T}{m}}\right] \qquad (14.26)$$

Thus, the frequency factor, A, predicted by the collision theory of chemical kinetics is temperature dependent. In practice, we can usually treat it as a temperature-independent quantity in the calculation of E_a values. Doing so does not introduce any serious error because the exponential term ($e^{-E_a/RT}$) depends so much more strongly on temperature than the square-root term in A (Equation 14.26).

For a bimolecular reaction of the type

$$B + C \longrightarrow products$$

the collision frequency is similar, except that the mass m is replaced by the reduced mass, μ, of the B and C pair

$$\mu = \frac{m_B m_C}{m_B + m_C}$$

and the diameter d is replaced by the mean diameter $d_{BC} = (d_B + d_C)/2$. With these changes, the collision theory expression for the rate of this bimolecular reaction is

$$\text{rate} = Z_{BC} e^{-E_a/RT} = \left[d_{BC}^2 N_A^2 \sqrt{\frac{\pi k_B T}{\mu}} \right] e^{-E_a/RT} [B][C] \tag{14.27}$$

and

$$k = \left[d_{BC}^2 N_A^2 \sqrt{\frac{\pi k_B T}{\mu}} \right] e^{-E_a/RT} \tag{14.28}$$

which correctly predicts that the rate is first order in B, first order in C, and second order overall.

The collision theory expressions for the rate constants (Equations 14.25 and 14.28) do a good job of predicting rate constants for reactions that involve atomic species or simple radicals, given the activation energy. However, for more complex reactants, collision theory tends to overestimate the rate constants, often by a large factor. This discrepancy arises because our simple collision theory assumes that a reaction will occur if the collision has a kinetic energy that is sufficient to overcome the activation barrier. However, even if the collision is sufficiently energetic, a reaction may not occur because the reacting molecules are not oriented optimally relative to one another. This effect can be accounted for by modifying Equation 14.28 (or 14.25) to include an *orientation factor* (or *steric factor*) *P*:

$$k = PZ e^{-E_a/RT} \tag{14.29}$$

The factor P, which is often quite difficult to calculate, represents the fraction of colliding molecules that are properly oriented for reaction to occur.

As an example, consider the gas-phase reaction between potassium (K) atoms and methyl iodide (CH_3I) to form potassium iodide (KI) and a methyl radical ($CH_3\cdot$):

$$K(g) + CH_3I(g) \longrightarrow KI(g) + CH_3\cdot(g)$$

This reaction is most favorable when the K atom collides with the I atom in CH_3I head-on (Figure 14.17). Otherwise, few or no products are formed.

Transition-State Theory

Collision state theory is useful for gas-phase reactions of simple atoms and molecules, but it cannot adequately predict reaction rates for more complex molecules or molecules in solution. Another approach, called *transition-state theory* (or *activated-complex theory*), was developed by Henry Eyring and others in the 1930s. Because it is applicable to a wide range of reactions, transition-state theory has become the major theoretical tool in the prediction of chemical kinetics.

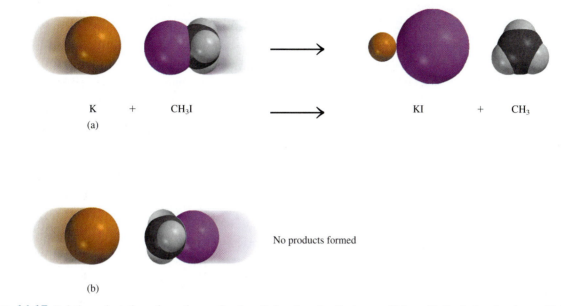

K   +   CH₃I   ⟶   KI   +   CH₃

(a)

No products formed

(b)

Figure 14.17 Relative orientation of reacting molecules. Only when the K atoms collides with the I atom head-on will the reaction most likely occur.

Consider a reaction between A and B that proceeds to products through a transition state labeled $AB^{\ddagger}$, as shown in Figure 13.6. The central assumption of transition-state theory is that, as the reaction proceeds, the reactants A and B are in equilibrium with the transition state $AB^{\ddagger}$, that is, we assume that the forward and backwards rate constants between A + B and $AB^{\ddagger}$ are much faster than the rate constant for the decomposition of the $AB^{\ddagger}$:

$$A + B \rightleftharpoons AB^{\ddagger} \longrightarrow \text{products}$$

The concentration equilibrium constant, $K_c^{\ddagger}$, between the reactants and the transition state can be written as

$$K_c^{\ddagger} = \frac{[AB^{\ddagger}]}{[A][B]} \qquad (14.30)$$

In terms of this equilibrium constant, it can be shown using statistical mechanics that the transition-state theory estimate for the rate constant is given by

$$k = \frac{k_B T}{h} K_c^{\ddagger} c_0^{1-m} \qquad (14.31)$$

where h is Planck's constant, k_B is Boltzmann's constant, T is the absolute temperature, m is the reaction order, and c_0 is the standard concentration (1.0 M). The factor of c_0^{1-m} is necessary to ensure that k has the appropriate units. The equilibrium constant $K_c^{\ddagger}$ can be calculated using standard computational chemistry software, or from physical properties of the reactants (bond lengths, bond angles and vibrational frequencies) of the reactants.

14.5 | The Reaction Mechanism Is the Sequence of Elementary Steps That Lead to Product Formation

An overall balanced chemical equation does not tell us much about how a reaction actually takes place. In many cases, it merely represents the sum of several *elementary steps* (or *elementary reactions*), *a series of simple reactions that represent the progress of the overall reaction at the molecular level.* The *sequence of elementary steps that leads to product formation* is called the **reaction mechanism.** The reaction mechanism is comparable to the route traveled during a trip; the overall chemical equation specifies only the origin and final destination. The details of the reaction mechanism (or pathway) connecting given initial and final states have profound effects on the rate of a reaction. This is in contrast to the situation in chemical thermodynamics, where we saw that the changes in thermodynamic state functions were independent of the path taken between initial and final states. The reaction mechanism cannot be deduced from the stoichiometry of the overall reaction but must be postulated based on experimental evidence.

To better understand the reaction mechanism, consider the reaction between nitric oxide and oxygen:

$$2NO(g) + O_2(g) \longrightarrow 2NO_2(g)$$

We know from experiment that the products are not formed directly from the collision of two NO molecules with an O_2 molecule because N_2O_2 is detected during the course of the reaction. Let us assume that the reaction actually takes place via two elementary steps as follows:

$$2NO(g) \longrightarrow N_2O_2(g)$$

$$N_2O_2(g) + O_2(g) \longrightarrow 2NO_2(g)$$

In the first elementary step, two NO molecules collide to form an N_2O_2 molecule. This event is followed by the reaction between N_2O_2 and O_2 to give two molecules of NO_2. The net chemical equation, which represents the overall change, is given by the sum of the elementary steps:

elementary step:	$NO + NO \longrightarrow N_2O_2$
elementary step:	$N_2O_2 + O_2 \longrightarrow 2NO_2$

overall reaction: $2NO + \cancel{N_2O_2} + O_2 \longrightarrow \cancel{N_2O_2} + 2NO_2$

Species such as N_2O_2 are called **reaction intermediates** because they *appear in the mechanism of the reaction (i.e., in the elementary steps) but not in the overall balanced equation.* An intermediate is always formed in an early elementary step and consumed in a later elementary step.

The **molecularity of a reaction** is *the number of molecules reacting in an elementary step.* These molecules may be of the same or different types. Each of the elementary steps just discussed is **bimolecular** *because each involves two molecules.* **Unimolecular** *reactions, elementary steps in which only one reacting molecule participates,* include the decomposition of N_2O_4 to form NO_2

$$N_2O_4(g) \longrightarrow 2NO_2(g)$$

and the conversion of cyclopropane to propene,

$$\underset{\text{cyclopropane}}{\overset{\displaystyle CH_2}{CH_2 \!-\! CH_2}} \longrightarrow \underset{\text{propene}}{CH_3 \!-\! CH \!=\! CH_2}$$

Very few *termolecular reactions, which are reactions that involve three molecules in one elementary step,* are known, because the simultaneous encounter of three molecules in the proper orientation to react is a far less likely event than a bimolecular collision or a unimolecular reaction.

Rate Laws for Elementary Steps

Knowing the elementary steps of a reaction enables us to deduce the rate law. Suppose, for example, that we have the following elementary reaction:

$$A \longrightarrow \text{products}$$

Because there is only one molecule present, this is a unimolecular elementary reaction. It follows that the larger the number of A molecules present, the faster the rate of product formation. Thus, the rate of a unimolecular elementary reaction is directly proportional to the concentration of A (that is, it is first order in A):

$$\text{rate} = k[A]$$

For a bimolecular elementary reaction involving A and B molecules,

$$A + B \longrightarrow \text{products}$$

the rate of product formation depends on how frequently A and B collide, which in turn depends on the concentrations of A and B. Thus, we can express the rate as

$$\text{rate} = k[A][B]$$

Similarly, for a bimolecular elementary reaction of the type

$$A + A \longrightarrow \text{products}$$

the rate becomes

$$\text{rate} = k[A]^2$$

The preceding examples show that the reaction order for each elementary reactant is equal to its stoichiometric coefficient in the chemical equation for that step. In general, we cannot tell by merely looking at the overall balanced equation whether the reaction occurs as shown or in a series of steps. This determination is made in the laboratory.

Rate Laws and Reaction Mechanisms

Experimental studies of reaction mechanisms begin with the collection of data (rate measurements). Next, we analyze the data to determine the rate constant and order of the reaction, and we write the rate law. Finally, we suggest a plausible mechanism for

Figure 14.18 Sequence of steps in the study of a reaction mechanism.

the reaction consisting of elementary steps (Figure 14.18). The elementary steps must satisfy two requirements:

▶ The sum of the elementary steps must give the overall balanced equation for the reaction.

▶ The rate-determining step should predict a rate law that is identical to the one determined experimentally.

For a proposed reaction mechanism, we must be able to detect the presence of any intermediate(s) formed in one or more elementary steps.

To evaluate whether the second criterion is satisfied, we must be able to construct the rate law based on a candidate reaction mechanism. This is often a complex procedure that requires approximations. If one step is significantly slower than the others, however, then the determination of the rate law is considerably simplified. In Section 14.3, we saw that when a reaction occurs in a series of consecutive individual steps, the rate of the overall process is governed by the rate of the slowest, or rate-determining, step. Consider, for example, the reaction

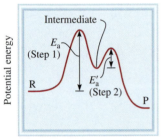

Figure 14.19 Potential energy profile for a two-step reaction in which the first step is rate determining. R and P represent reactants and products, respectively.

$$A + 2B \longrightarrow products$$

The reaction is observed to be first order in both A and B and second order overall. Thus, the reaction cannot take place in a single elementary step corresponding to the overall reaction because then it would be second order in B, contrary to observation. Therefore, multiple elementary steps must be involved in the reaction mechanism. A proposed mechanism is

$$A + B \xrightarrow{\ k_1\ } C \qquad \text{(slow)}$$
$$B + C \xrightarrow{\ k_2\ } products \qquad \text{(fast)}$$

Because $k_1 \ll k_2$, the first bimolecular elementary step is rate determining, and the rate law can be written as

$$\text{rate} = -\frac{d[A]}{dt} = k_1[A][B]$$

Figure 14.20 The decomposition of hydrogen peroxide is catalyzed by iodine ions. A few drops of liquid soap have been added to the solution to dramatize the evolution of oxygen gas. (Some of the iodide ions are oxidized to molecular iodine, which then reacts with iodide ions to form the brown triiodide ion I_3^-.)

which is consistent with the observed rate law. The potential energy profile for such a reaction is shown in Figure 14.19. The first step, which is rate determining, has a larger activation energy than the second step. The intermediate, although stable enough to be observed, reacts quickly to form the products.

Decomposition of Hydrogen Peroxide

An example of this type of reaction is the decomposition of hydrogen peroxide, which is facilitated by iodide ions (Figure 14.20). The overall reaction is

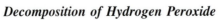

$$2H_2O_2(aq) \longrightarrow 2H_2O(l) + O_2(g)$$

Experimentally, the rate law is found to be

$$\text{rate} = k\,[H_2O_2][I^-]$$

Thus, the reaction is first order with respect to both H_2O_2 and I^-.

The H_2O_2 decomposition does not occur in a single elementary step corresponding to the overall balanced equation. If it did, the reaction would be second order in H_2O_2 (due to the collision of two H_2O_2 molecules). What's more, the I^- ion, which is not even part of the overall equation, appears in the rate law expression. How can we reconcile these facts? First, we can account for the observed rate law by assuming that the reaction takes place in two separate elementary steps, each of which is bimolecular:

step 1: $H_2O_2(aq) + I^-(aq) \xrightarrow{k_1} H_2O(l) + IO^-(aq)$

step 2: $H_2O_2(aq) + IO^-(aq) \xrightarrow{k_2} H_2O(l) + O_2(g) + I^-(aq)$

If we further assume that step 1 is rate determining, then the rate of the reaction can be determined from the first step alone:

$$\text{rate} = k_1[H_2O_2][I^-]$$

where $k_1 \ll k_2$. The IO^- ion is an intermediate because it does not appear in the overall balanced equation. Although the I^- ion also does not appear in the overall equation, I^- differs from IO^- because I^- is present at the start of the reaction and at its completion. The function of I^- is to speed up the reaction, that is, it is a *catalyst*. We discuss catalysis in Section 14.6.

Example 14.9 concerns the mechanistic study of a relatively simple reaction.

Example 14.9

The gas-phase decomposition of nitrous oxide (N_2O) is believed to occur via two elementary steps:

step 1: $N_2O \xrightarrow{k_1} N_2 + O$

step 2: $N_2O + O \xrightarrow{k_2} N_2 + O_2$

Experimentally, the rate law is found to be rate $= k\,[N_2O]$. (a) Write the equation for the overall reaction. (b) Identify the intermediates. (c) What can you say about the relative rates of steps 1 and 2?

Strategy (a) Because the overall reaction can be broken down into elementary steps, knowing the elementary steps would enable us to write the overall reaction. (b) Reaction intermediates appear as the product of one elementary reaction and as a reactant of a subsequent elementary reaction, so intermediates do not appear in the overall chemical reaction. (c) Based on the experimental rate law, determine which of the steps is rate determining.

Solution (a) Adding the equations for steps 1 and 2 gives the following overall reaction:

$$2N_2O \longrightarrow 2N_2 + O_2$$

—Continued

Continued—

(b) Because the O atom is produced in the first elementary step, is consumed in the second elementary step, and does not appear in the overall balanced equation, it is an intermediate.

(c) If we assume that step 1 is rate determining (that is, if $k_2 \gg k_1$), then the rate of the overall reaction is given by

$$\text{rate} = k_1[N_2O]$$

and $k = k_1$, which is consistent with the experimental rate law. If step 2 were rate determining, then the rate law would be second order, which is inconsistent with the experimental rate law, so we can conclude that step 1 is slow relative to step 2.

Practice Exercise The reaction between NO_2 and CO to produce NO and CO_2 is believed to occur via the following two steps:

$$\text{step 1:} \quad NO_2 + NO_2 \longrightarrow NO + NO_3$$
$$\text{step 2:} \quad NO_3 + CO \longrightarrow NO_2 + CO_2$$

The experimental rate law is rate $= k\,[NO_2]^2$. (a) Write the equation for the overall reaction. (b) Identify any reaction intermediate. (c) What can you say about the relative rates of steps 1 and 2?

The Formation of Hydrogen Iodide

Another common reaction mechanism is one that has at least two elementary steps, the first of which is very rapid in both the forward and reverse directions compared with the second step. This kind of reaction mechanism has a ***rapid preequilibrium.*** An example is the reaction between molecular hydrogen and molecular iodine to produce hydrogen iodide:

$$H_2(g) + I_2(g) \longrightarrow 2HI(g)$$

Experimentally, the rate law is found to be

$$\text{rate} = k\,[H_2][I_2]$$

For many years, it was thought that the reaction occurred just as written, that is, the reaction consists of a bimolecular elementary step involving a hydrogen molecule and an iodine molecule. In the 1960s, however, chemists proposed a more complicated two-step mechanism:

$$\text{step 1:} \qquad I_2 \underset{k_{-1}}{\overset{k_1}{\rightleftharpoons}} 2I \qquad \text{(rapid equilibrium)}$$

$$\text{step 2:} \quad H_2 + 2I \overset{k_2}{\longrightarrow} 2HI \quad \text{(slow)}$$

where k_1, k_{-1}, and k_2 are the rate constants for the reactions. The I atoms are intermediates in this reaction.

When the reaction begins, there are very few I atoms present. As I_2 dissociates, though, the concentration of I_2 decreases while that of I increases. Therefore, the forward rate of step 1 decreases and the reverse rate increases. Soon the two rates become equal, and a chemical equilibrium is established. Because the elementary reactions in step 1 are much faster than the one in step 2, equilibrium is reached before any significant reaction with hydrogen occurs, and it persists throughout the reaction.

The rate of the reaction is given by the slow, rate-determining step, which is step 2—a termolecular reaction between H_2 and two I atoms:

$$\text{rate} = k_2[H_2][I]^2$$

This rate law is of limited usefulness because it contains the concentration of I, which is difficult to monitor because I is an intermediate. To eliminate the concentration of the intermediate from the rate law, we can use the principle of detailed balance (Section 14.3). According to this principle, the forward rate at equilibrium must equal the reverse rate, that is,

$$k_1[I_2] = k_{-1}[I]^2$$

or

$$[I]^2 = \frac{k_1}{k_{-1}}[I_2] = K[I_2]$$

where $K = k_1/k_{-1}$ is the equilibrium constant for step 1 (see Equation 14.18). Substituting this expression for $[I]^2$ into the rate law, we obtain

$$\text{rate} = \frac{k_2 k_1}{k_{-1}}[H_2][I_2] = k_2 K[H_2][I_2]$$
$$= k[H_2][I_2]$$

where $k = k_1 k_2/k_{-1} = k_2 K$. As you can see, this two-step mechanism also gives the correct rate law for the reaction. This agreement along with the experimental observation of intermediate I atoms provides evidence that the mechanism is correct. However, keep in mind that demonstrating that a particular reaction mechanism is consistent with the experimental rate law does not prove conclusively that the mechanism is correct, as there may be other, yet unknown, mechanisms that also give the correct rate law.

Example 14.10 examines another example of a reaction mechanism with a pre-equilibrium condition.

Example 14.10

Under certain conditions, the experimental rate law for the gas-phase reaction of molecular hydrogen with molecular bromine

$$H_2(g) + Br_2(g) \longrightarrow 2HBr(g)$$

is given by

$$\text{rate} = k[H_2][Br_2]^{1/2}$$

Show that the unusual half-reaction order for Br_2 can be explained by the following mechanism:

step 1: $\quad Br_2 \underset{k_{-1}}{\overset{k_1}{\rightleftharpoons}} 2Br \quad$ (rapid equilibrium)

step 2: $\quad Br + H_2 \overset{k_2}{\longrightarrow} HBr + H \quad$ (slow)

step 3: $\quad H + Br_2 \overset{k_2}{\longrightarrow} HBr + Br \quad$ (fast)

—Continued

Continued—

Strategy The second step is rate determining, but its rate law contains the intermediate Br atom. To eliminate [Br] from the rate law, use the fact that the forward and back reactions in step 1 are in rapid equilibrium.

Solution Because the second step is slow and rate determining, the rate law is given by

$$\text{rate} = k_2[\text{Br}][\text{H}_2]$$

Because Br is an intermediate, it is necessary to eliminate it from the rate law. To do this we note that the the first step is in rapid equilibrium, so we can assume that

$$K = \frac{k_1}{k_{-1}} = \frac{[\text{Br}]^2}{[\text{Br}_2]}$$

Solving for [Br] gives

$$[\text{Br}]^2 = \frac{k_1}{k_{-1}}[\text{Br}_2]$$

or

$$[\text{Br}] = \sqrt{\frac{k_1}{k_{-1}}}[\text{Br}_2]^{1/2}$$

Substituting this expression for the intermediate concentration [Br] into the rate law for the rate-determining step (step 2) gives

$$\text{rate} = k_2[\text{Br}][\text{H}_2] = k_2\sqrt{\frac{k_1}{k_{-1}}}[\text{Br}_2]^{1/2}[\text{H}_2]$$

$$= k[\text{Br}_2]^{1/2}[\text{H}_2]$$

where $k = k_2(k_1/k_{-1})^{1/2}$, which is consistent with the observed rate law.

Practice Exercise A reaction A + 2B $\longrightarrow$ products has the rate law

$$\text{rate} = k[\text{A}][\text{B}]^2$$

Show that this rate law is consistent with the following mechanism:

$$\text{step 1:} \quad \text{A} + \text{B} \underset{k_{-1}}{\overset{k_1}{\rightleftharpoons}} \text{C} \qquad \text{(rapid equilibrium)}$$

$$\text{step 2:} \quad \text{C} + \text{B} \xrightarrow{k_2} \text{products} \quad \text{(slow)}$$

What is the rate constant k in terms of k_1, k_{-1}, and k_2? Can you suggest an experiment that could distinguish between this proposed mechanism and a direct termolecular combination of one A molecule and two B molecules, which would have a rate that was also consistent with the experimental rate law?

The Steady-State Approximation

Not all reactions have a single rate-determining step. A reaction may have two or more comparably slow steps. The kinetic analysis of such reactions is generally more involved. One approximation that is useful in such cases is to assume that the concentration of intermediates is constant over much of the reaction progress. Consider, for example, the following mechanism for the decomposition of A to products that was proposed by Frederick Lindemann[3] in 1922 to explain the observed pressure dependence of the rate law for many unimolecular reactions:

$$\text{step 1:} \quad A + M \underset{k_{-1}}{\overset{k_1}{\rightleftharpoons}} A^* + M$$

$$\text{step 2:} \quad A^* \xrightarrow{k_2} \text{products}$$

In the first step, the molecule A collides with another molecule in the forward direction and is excited to a high energy state denoted as A^*. The excited molecule (A^*) can then collide with another molecule to become deexcited back to the low energy form (A) in the reverse of step 1, or, in the second step, it can fall apart to form the products.

From this mechanism the rate of formation of the products is

$$\text{rate} = k_2[A^*] \tag{14.32}$$

If we do not know the relative magnitudes of k_1, k_{-1}, and k_2, we cannot assume rapid preequilibrium to eliminate the concentration of the intermediate A^* from the rate law. Instead, we assume that after an initial rise, the concentration of the intermediate A^* becomes constant as the rates of activation, deactivation, and decomposition become balanced. Thus,

$$\text{rate of } A^* \text{ production} = \frac{d[A^*]}{dt} \approx 0$$

That is, the concentration of A^* achieves a *steady state* during the reaction. From the reaction mechanism, the rate of A^* production is given by the rate of activation in step 1 ($k_1[A][M]$) minus the rate of deactivation in step 1 ($k_{-1}[A^*][M]$) and minus the rate of decomposition ($k_2[A^*]$):

$$\frac{d[A^*]}{dt} = k_1[A][M] - k_{-1}[A^*][M] - k_2[A^*]$$

Setting the rate of production of A^* to zero yields

$$k_1[A][M] - k_{-1}[A^*][M] - k_2[A^*] \approx 0$$

which, after arrangement becomes

$$[A^*] = \frac{k_1[A][M]}{k_2 + k_{-1}[M]} \tag{14.33}$$

Equation 14.33 gives the **steady-state approximation** for the concentration of A^*. Substituting this into the rate law (Equation 14.32) gives the **steady-state rate law**

$$\text{rate} = \frac{k_1 k_2[A][M]}{k_2 + k_{-1}[M]} \tag{14.34}$$

---

3. Frederick Alexander Lindemann (1886–1957). German/British physicist. Lindemann was born and educated in Germany, but spent most of his scientific and political career in England. He made numerous contributions to the thermodynamics of materials and to chemical reaction dynamics. While in the Royal Air Corps during World War I, he developed a mathematical theory of airplane rotation that enabled pilots to recover from uncontrolled spin. He was very active in British politics, serving as scientific advisor to Winston Churchill during World War II. Lindemann was instrumental in the creation of the United Kingdom Atomic Energy Agency.

As Equation 14.34 shows, rate laws derived from the steady-state approximation are generally more complicated than those in which one step can be assumed to be rate limiting. Let's examine the dependence of this rate law in two limits. For an ideal gas system, the concentration of molecules with which A can collide is directly proportional to the total pressure. When the system is at low pressure, there are relatively few molecules with which to collide and [M] is small. In this limit, $k_2 \gg k_{-1}[M]$, and Equation 14.34 becomes

$$\text{low pressure:} \quad \text{rate} = k_1 k_2[A][M]$$

Thus, the rate is second order overall and shows a linear increase with total pressure. At high pressures, [M] is large and $k_2 \ll k_{-1}[M]$, so Equation 14.34 becomes

$$\text{high pressure:} \quad \text{rate} = \frac{k_1}{k_{-1}}[A]$$

In this case, the rate law is first order in A and independent of the total pressure. A plot of the pressure dependence of a unimolecular decay that obeys the Lindemann mechanism is shown in Figure 14.21. This dependence is consistent with that which is experimentally observed in many unimolecular decompositions in the gas phase.

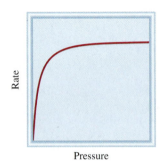

Figure 14.21 Plot of the pressure dependence of the rate for a unimolecular decomposition that follows the Lindemann mechanism.

Experimental Support for Reaction Mechanisms

How can we find out whether the proposed mechanism for a particular reaction is correct? In the case of hydrogen peroxide decomposition, we might try to detect the presence of the IO^- ions by spectroscopic means. Evidence of their presence would support the reaction mechanism previously discussed. Similarly, for the hydrogen iodide reaction, the detection of iodine atoms would lend support to the two-step mechanism. For example, I_2 dissociates into I atoms when it is irradiated with visible light. Thus, we might predict that the formation of HI from H_2 and I_2 would speed up as the intensity of light is increased because that should increase the concentration of I atoms. Indeed, this is just what is observed.

In one case, chemists wanted to know which C—O bond was broken in the reaction between methyl acetate and water in order to better understand the reaction mechanism

$$\underset{\text{methyl acetate}}{CH_3-\overset{\overset{\textstyle O}{\|}}{C}-O-CH_3} + H_2O \longrightarrow \underset{\text{acetic acid}}{CH_3-\overset{\overset{\textstyle O}{\|}}{C}-OH} + \underset{\text{methanol}}{CH_3OH}$$

The two possibilities are

$$\underset{(a)}{CH_3-\overset{\overset{\textstyle O}{\|}}{C}\!\cancel{\;}\!O-CH_3} \qquad \underset{(b)}{CH_3-\overset{\overset{\textstyle O}{\|}}{C}-O\!\cancel{\;}\!CH_3}$$

To distinguish between mechanisms (a) and (b), chemists used water containing the oxygen-18 isotope instead of ordinary water (which contains the oxygen-16 isotope). When the oxygen-18 water was used, only the acetic acid formed contained oxygen-18:

$$CH_3-\overset{\overset{\textstyle O}{\|}}{C}-{}^{18}O-H$$

Femtochemistry

The ability to follow chemical reactions at the molecular level has been one of the most relentlessly pursued goals in chemistry. Accomplishing this goal will allow chemists to understand when a certain reaction occurs and the dependence of its rate of reaction on temperature and other parameters. On the practical side, this information will help chemists control reaction rates and increase reaction yields. A complete understanding of reaction mechanisms requires a detailed knowledge of the activated complex (also called the transition state). The transition state, however, is a highly energetic species that could not be isolated because of its extremely short lifetime.

The situation changed in the 1980s when researchers in the group of Ahmed Zewail at the California Institute of Technology began to use very short laser pulses to probe chemical reactions. Because transition states last only 10 to 1000 femtoseconds, the laser pulses needed to probe them must be extraordinarily short. [One femtosecond (1 fs) is 1×10^{-15} s. To appreciate how short this time duration is, note that there are as many femtoseconds in a second as there are seconds in about 32 million years!] One of the reactions studied was the decomposition of cyclobutane (C_4H_8) to ethylene (C_2H_4). There are two possible mechanisms. The first is a single-step process in which two carbon-carbon bonds break simultaneously to form the product:

$$\begin{array}{c} CH_2-CH_2 \\ | \quad\quad | \\ CH_2-CH_2 \end{array} \longrightarrow 2CH_2{=}CH_2$$

The second mechanism has two steps, with an intermediate where the dot represents an unpaired electron:

$$\begin{array}{c} CH_2-CH_2 \\ | \quad\quad | \\ CH_2-CH_2 \end{array} \longrightarrow \begin{array}{c} \overset{\bullet}{C}H_2 \ \ \overset{\bullet}{C}H_2 \\ | \quad\quad | \\ CH_2-CH_2 \end{array} \longrightarrow 2CH_2{=}CH_2$$

The Caltech researchers initiated the reaction with a pump laser pulse, which energized the reactant. The first probe pulse hit the molecules a few femtoseconds later and was followed by many thousands more, every 10 fs or so, for the duration of the reaction. Each probe pulse resulted in an absorption spectrum, and changes in the spectrum revealed the motion of the molecule and the state of the chemical bonds. In this way, the researchers were effectively equipped with a camera having different shutter speeds to capture the progress of the reaction. The results showed that cyclobutane decomposed to ethylene via the second (two-step) mechanism. The lifetime of the intermediate was about 700 fs.

The femtosecond laser technique has been used to unravel the mechanisms of many chemical reactions and biological processes such as photosynthesis and vision. It has created a new area in chemical kinetics that has become known as *femtochemistry*. For the development of the field of femtochemistry, Professor Zewail was awarded the Nobel Prize in Chemistry in 1999.

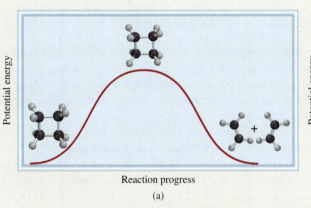

Reaction progress

(a)

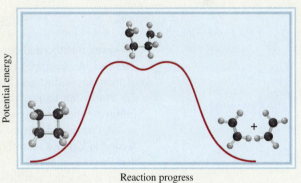

Reaction progress

(b)

The decomposition of cyclobutane to form two ethylene molecules can take place in one of two ways. (a) The reaction proceeds via a single step, which involves the breaking of two C—C bonds simultaneously. (b) The reaction proceeds in two steps, with the formation of a shortlived intermediate in which just one bond is broken. There is only a small energy barrier for the intermediate to proceed to the final products. The correct mechanism is (b).

Thus, the reaction must have occurred via bond-breaking mechanism (a) because the product formed via mechanism (b) would retain both of its original oxygen atoms.

Now consider photosynthesis, the process by which green plants produce glucose from carbon dioxide and water:

$$6CO_2 + 6H_2O \longrightarrow C_6H_{12}O_6 + 6O_2$$

A question that arose early in studies of photosynthesis was whether the molecular oxygen was derived from water, carbon dioxide, or both. By using water containing only the oxygen-18 isotope, it was demonstrated that the evolved oxygen came from water, and none came from carbon dioxide, because the O_2 contained only the ^{18}O isotope. This result supported the mechanism in which water molecules are "split" by light:

$$2H_2O + h\nu \longrightarrow O_2 + 4H^+ + 4e^-$$

where $h\nu$ represents the energy of a photon. The protons and electrons are used to drive energetically unfavorable reactions that are necessary for plant growth and function.

These examples give some idea of how inventive chemists must be in studying reaction mechanisms. For complex reactions, however, it is virtually impossible to prove the uniqueness of any particular mechanism.

14.6 | Reaction Rates Can Often Be Increased by the Addition of a Catalyst

We saw in Section 14.5 that the reaction rate for the decomposition of hydrogen peroxide depends on the concentration of iodide ions even though I^- does not appear in the overall equation. Instead I^- acts as a catalyst for the reaction. A *catalyst* is *a substance that increases the rate of a chemical reaction without itself being consumed.*

The catalyst may react to form an intermediate, but it is regenerated in a subsequent step of the reaction. In the laboratory preparation of molecular oxygen, for example, a sample of potassium chlorate is decomposed by heating, as follows

$$2KClO_3(s) \longrightarrow 2KCl(s) + 3O_2(g)$$

However, this thermal decomposition process is very slow in the absence of a catalyst. The rate of decomposition can be increased dramatically by adding a small amount of manganese dioxide (MnO_2), a black powdery substance that acts as a catalyst. All the MnO_2 can be recovered at the end of the reaction, just as all the I^- ions remain following the decomposition of H_2O_2.

A catalyst speeds up a reaction by providing a set of elementary steps with more favorable kinetics than those that exist in its absence. From Equation 14.19, we know that the rate constant k (and hence the rate) of a reaction depends on the frequency factor A and the activation energy E_a—the larger the A or the smaller the E_a, the greater the rate. In many cases, a catalyst increases the rate by lowering the activation energy for the reaction.

Let us assume that the following reaction has a rate constant k and an activation energy E_a:

$$A + B \xrightarrow{k} C + D$$

In the presence of a catalyst, however, the rate constant is k_c (called the *catalytic rate constant*):

$$A + B \xrightarrow{k_c} C + D$$

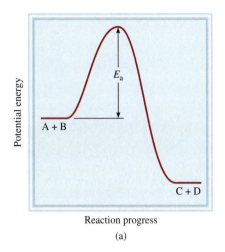

(a)

Reaction progress

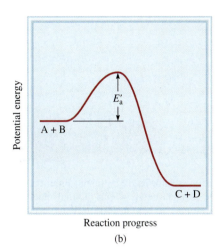

(b)

Reaction progress

Figure 14.22 Comparison of the activation energy barriers of an uncatalyzed reaction and the same reaction with a catalyst. The catalyst lowers the energy barrier but does not affect the actual energies of the reactants or products. Although the reactants and products are the same in both cases, the reaction mechanisms and rate laws are different in (a) and (b).

By the definition of a catalyst,

$$\text{rate}_{\text{catalyzed}} > \text{rate}_{\text{uncatalyzed}}$$

Figure 14.22 shows the potential energy profiles for both reactions. The total energies of the reactants (A and B) and those of the products (C and D) are unaffected by the catalyst; the only difference between the two reactions is a lowering of the activation energy from E_a to E_a' Because the activation energy for the reverse reaction is also lowered, a catalyst enhances the rates of the forward and reverse reactions equally.

There are three general types of catalysis, depending on the nature of the rate−increasing substance: heterogeneous catalysis, homogeneous catalysis, and enzyme catalysis.

Heterogeneous Catalysis

In *heterogeneous catalysis,* the reactants and the catalyst are in different phases. Usually the catalyst is a solid and the reactants are either gases or liquids. Heterogeneous catalysis is by far the most important type of catalysis in industrial chemistry, especially in the synthesis of many key chemicals. Here we describe three specific examples of heterogeneous catalysis that account for millions of tons of chemicals produced annually on an industrial scale.

The Haber Synthesis of Ammonia

Ammonia is an extremely valuable inorganic substance used in the fertilizer industry, the manufacture of explosives, and many other applications. Around the turn of the century, many chemists strove to synthesize ammonia from nitrogen and hydrogen. The supply of atmospheric nitrogen is virtually inexhaustible, and hydrogen gas can be produced readily by passing steam over heated coal:

$$H_2O(g) + C(s) \longrightarrow CO(g) + H_2(g)$$

Hydrogen is also a by-product of petroleum refining.

The formation of NH_3 from N_2 and H_2 is exothermic:

$$N_2 + 3H_2(g) \longrightarrow 2NH_3(g) \qquad\qquad \Delta H° = -91.8 \text{ kJ/mol}$$

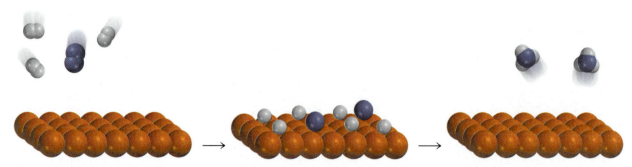

Figure 14.23 The catalytic action in the synthesis of ammonia. First, the H_2 and N_2 molecules bind to the surface of the catalyst. This interaction weakens the covalent bonds within the molecules and eventually causes the molecules to dissociate. Second, the highly reactive H and N atoms combine to form NH_3 molecules, which then leave the surface.

The reaction rate, however, is extremely slow at room temperature. To be practical on a large scale, a reaction must occur at an appreciable rate *and* it must have a high yield of the desired product. Raising the temperature does accelerate the reaction of N_2 with H_2, but at the same time, it promotes the decomposition of NH_3 molecules back into N_2 and H_2, thus lowering the yield of NH_3.

In 1905, after testing literally hundreds of compounds at various temperatures and pressures, Fritz Haber discovered that iron plus a few percent of oxides of potassium and aluminum catalyze the reaction of hydrogen with nitrogen to yield ammonia at about 500°C. This procedure is known as the *Haber process*.

In heterogeneous catalysis, the surface of the solid catalyst is usually the site of the reaction. The initial step in the Haber process is the dissociation of N_2 and H_2 on the metal surface (Figure 14.23). Although the dissociated species are not truly free atoms because they are bonded to the metal surface, they are highly reactive. The two reactant molecules behave very differently on the catalyst surface. Studies show that H_2 dissociates into atomic hydrogen at temperatures as low as −196°C (the boiling point of liquid nitrogen). Nitrogen molecules, on the other hand, dissociate at about 500°C. The highly reactive N and H atoms combine rapidly at high temperatures to produce the desired NH_3 molecules:

$$N + 3H \longrightarrow NH_3$$

The Manufacture of Nitric Acid

Nitric acid is one of the most important inorganic acids. It is used in the production of fertilizers, dyes, drugs, and explosives. The major industrial method of producing nitric acid is the *Ostwald*[4] *process*. The starting materials, ammonia and molecular oxygen, are heated in the presence of a platinum-rhodium catalyst (Figure 14.24) to about 800°C to produce nitric acid:

$$4NH_3(g) + 5O_2(g) \longrightarrow 4NO(g) + 6H_2O(g)$$

The nitric oxide readily oxidizes (without catalysis) to nitrogen dioxide:

$$2NO(g) + O_2(g) \longrightarrow 2NO_2(g)$$

4. Wilhelm Ostwald (1853–1932). German chemist. Ostwald made important contributions to chemical kinetics, thermodynamics, and electrochemistry. He developed the industrial process for preparing nitric acid that now bears his name. He received the Nobel Prize in Chemistry in 1909.

Figure 14.24 The platinum-rhodium catalyst used in the Ostwald process.

When dissolved in water, NO_2 forms both nitrous acid and nitric acid:

$$2NO_2(g) + H_2O(l) \longrightarrow HNO_2(aq) + HNO_3(aq)$$

On heating, nitrous acid is converted to nitric acid as follows:

$$3HNO_2(aq) \longrightarrow HNO_3(aq) + H_2O(l) + 2NO(g)$$

The NO thus generated can be recycled to produce NO_2 in the second step.

Catalytic Converters

At the high temperatures inside a running car engine, nitrogen and oxygen gases react to form nitric oxide:

$$N_2(g) + O_2(g) \longrightarrow 2NO(g)$$

When released into the atmosphere, NO rapidly combines with O_2 to form NO_2. Nitrogen dioxide and other gases emitted by automobiles, such as carbon monoxide (CO) and various unburned hydrocarbons, make automobile exhaust a major source of air pollution.

All new cars are equipped with catalytic converters (Figure 14.25). An efficient catalytic converter serves two purposes: It oxidizes CO and unburned hydrocarbons to CO_2 and H_2O, and it reduces NO and NO_2 to N_2 and O_2. Hot exhaust gases into which air has been injected are passed through the first chamber of one converter to

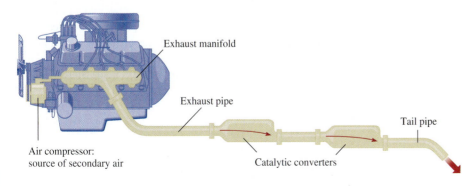

Exhaust manifold

Exhaust pipe

Tail pipe

Air compressor:
source of secondary air

Catalytic converters

Figure 14.25 A two-stage catalytic converter for an automobile.

Figure 14.26 A catalytic converter (cross-sectional view). The beads contain platinum, palladium, and rhodium, which catalyze the combustion of CO and hydrocarbons.

accelerate the complete burning of hydrocarbons and to decrease CO emission. (A cross section of a catalytic converter is shown in Figure 14.26.) However, because high temperatures increase NO production, a second chamber containing a different catalyst (a transition metal or a transition metal oxide such as CuO or Cr_2O_3) and operating at a lower temperature is required to dissociate NO into N_2 and O_2 before the exhaust is discharged through the tailpipe.

Homogeneous Catalysis

In **homogeneous catalysis,** *the reactants and catalyst are dispersed in a single phase, usually liquid.* Acid catalysis and base catalysis are the most important types of homogeneous catalysis in liquid solution. For example, the reaction of ethyl acetate with water to form acetic acid and ethanol normally occurs too slowly to be measured

$$CH_3 - \overset{\overset{O}{\|}}{C} - O - C_2H_5 + H_2O \longrightarrow CH_3 - \overset{\overset{O}{\|}}{C} - OH + C_2H_5OH$$

ethyl acetate   acetic acid   ethanol

In the absence of the catalyst, the rate law is given by

$$rate = k[CH_3COOC_2H_5]$$

The reaction, however, can be catalyzed by an acid. In the presence of hydrochloric acid, the rate is faster and the rate law is given by

$$rate = k_c[CH_3COOC_2H_5][H^+]$$

Because $k_c > k$, the rate is determined solely by the catalyzed portion of the reaction.

Homogeneous catalysis can also take place in the gas phase. A well-known example of catalyzed gas-phase reactions is the lead chamber process, which for many years was the primary method of manufacturing sulfuric acid. Starting with sulfur, one might expect the production of sulfuric acid to occur in the following steps:

$$S(s) + O_2(g) \longrightarrow SO_2(g)$$
$$2SO_2(g) + O_2(g) \longrightarrow 2SO_3(g)$$
$$H_2O(l) + SO_3(g) \longrightarrow H_2SO_4(aq)$$

In reality, however, sulfur dioxide is not converted directly to sulfur trioxide; rather, the oxidation is more efficiently carried out in the presence of a nitrogen dioxide catalyst:

$$2SO_2(g) + 2NO_2(g) \longrightarrow 2SO_3(g) + 2NO(g)$$
$$\underline{2NO(g) + O_2(g) \longrightarrow 2NO_2(g)}$$

overall reaction: $\qquad 2SO_2(g) + O_2(g) \longrightarrow 2SO_3(g)$

There is no net loss of NO_2 in the overall reaction, so NO_2 meets the criteria for a catalyst.

In recent years, chemists have devoted much effort to developing a class of metallic compounds to serve as homogeneous catalysts. These compounds are soluble in various organic solvents, so they can catalyze reactions in the same phase as the dissolved reactants. Many of the processes they catalyze are organic. For example, a red-violet compound of rhodium ($[(C_6H_5)_3P]_3RhCl$) catalyzes the conversion of a carbon-carbon double bond to a carbon-carbon single bond as follows:

$$\overset{|}{\underset{|}{C}}=\overset{|}{\underset{|}{C}} + H_2 \longrightarrow -\overset{|}{\underset{|}{\underset{H}{C}}}-\overset{|}{\underset{|}{\underset{H}{C}}}-$$

Homogeneous catalysis has several advantages over heterogeneous catalysis. For one thing, the reactions can often be carried out under atmospheric conditions, thus reducing production costs and minimizing the decomposition of products at high temperatures. In addition, homogeneous catalysts can be designed to function selectively for a particular type of reaction, and homogeneous catalysts cost less than the precious metals (for example, platinum and gold) used in heterogeneous catalysis.

Enzyme Catalysis

Of all the intricate processes that have evolved in living systems, none is more striking or more essential than enzyme catalysis. *Enzymes* are *biological catalysts*. The amazing fact about enzymes is that not only can they increase the rate of biochemical reactions by factors ranging from 10^6 to 10^{18}, but they are also highly specific. An enzyme acts only on certain molecules, called *substrates* (that is, reactants), while leaving the rest of the system unaffected. It has been estimated that an average living *cell* may contain some 3000 different enzymes, each of them catalyzing a specific reaction in which a substrate is converted into the appropriate products. Enzyme catalysis is usually homogeneous because the substrate and enzyme are present in the aqueous solution of the cell.

An enzyme is typically a large protein molecule that contains one or more *active sites* where interactions with substrates take place. These sites are structurally compatible with specific substrate molecules, in much the same way as a key fits a particular lock. In fact, the notion of a rigid enzyme structure that binds only to molecules whose shape exactly matches that of the active site was the basis of an early theory of enzyme catalysis, the so-called lock-and-key theory developed by the German chemist Emil Fischer[5] in 1894 (Figure 14.27).

Fischer's hypothesis accounts for the specificity of enzymes, but it contradicts more recent research evidence that a single enzyme binds to substrates of different

---

5. Emil Fischer (1852–1919). German chemist. Regarded by many as the greatest organic chemist of the nineteenth century, Fischer made many significant contributions in the synthesis of sugars and other important molecules. He was awarded the Nobel Prize in Chemistry in 1902.

Figure 14.27 The lock-and-key model of the specificity of an enzyme for substrate molecules.

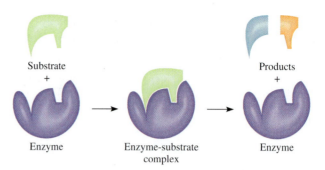

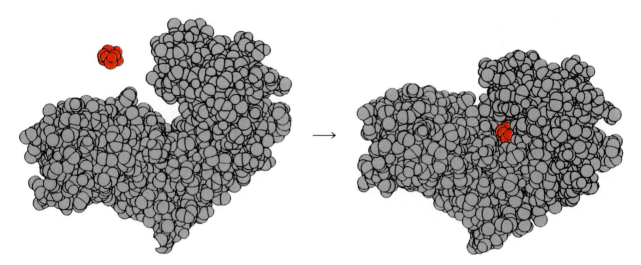

Figure 14.28 Left to right: The binding of a glucose molecule (red) to hexokinase (an enzyme in a metabolic pathway). Note how the region at the active site closes around glucose after binding. Frequently, the geometries of both the substrate and the active site are altered to fit each other. All enzymes possess a fair amount of flexibility.

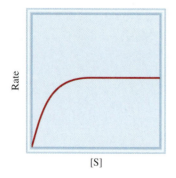

Figure 14.29 Plot of the initial rate (v_0) of an enzyme-catalyzed reaction versus substrate concentration.

sizes and shapes. Chemists now know that an enzyme molecule (or at least its active site) has a fair amount of structural flexibility and can modify its shape to accommodate more than one type of substrate. Figure 14.28 shows a molecular model of an enzyme in action.

In enzyme kinetics, it is customary to measure the *initial rate* (v_0) of a reaction to minimize the effect of reversible reactions and the inhibition of enzymes by products on the reaction rate. Furthermore, the initial rate corresponds to a known fixed substrate concentration. As time proceeds, the substrate concentration decreases.

Figure 14.29 shows the variation of the initial rate (v_0) of an enzyme-catalyzed reaction with substrate (S) concentration. The rate increases rapidly and linearly with [S] at low substrate concentrations, but it gradually levels off toward a limiting value at high substrate concentrations. In this region, all of the enzyme molecules are bound to the substrate molecules, and the rate becomes zero order in substrate concentration.

The mathematical treatment of enzyme kinetics is quite complex, even when we know the basic steps involved in the reaction. The following simplified scheme, which

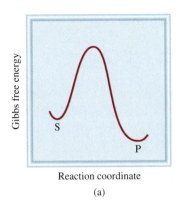

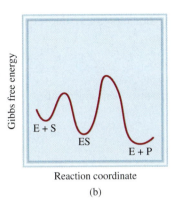

Reaction coordinate
(a)

Reaction coordinate
(b)

Figure 14.30 Comparison of the free energy profile for (a) an uncatalyzed reaction and (b) the same reaction catalyzed by an enzyme. The plot in (b) assumes that the catalyzed reaction has a two-step mechanism, in which the second step (ES $\longrightarrow$ E + P) is rate determining.

can explain the dependence of the rate on the substrate concentration (as shown in Figure 14.29) was proposed in 1913 by Michaelis[6] and Menten[7]

$$ \mathrm{E} + \mathrm{S} \underset{k_{-1}}{\overset{k_1}{\rightleftharpoons}} \mathrm{ES} $$

$$ \mathrm{ES} \overset{k_2}{\longrightarrow} \mathrm{E} + \mathrm{P} $$

where E, S, and P represent enzyme, substrate, and product, respectively, and ES is the enzyme-substrate intermediate. Figure 14.30 shows the free energy profile for the reaction compared to the reaction without a enzyme catalyst.

From the mechanism, the rate of formation of the product P is given by

$$ \text{rate} = \frac{d[\mathrm{P}]}{dt} = k_2[\mathrm{ES}] \tag{14.35} $$

To give a useable rate law, we need to eliminate the concentration of the intermediate from Equation 14.35 using the steady-state approximation (Section 14.5). We start by setting the rate of formation of the intermediate ES to zero:

$$ \frac{d[\mathrm{ES}]}{dt} = k_1[\mathrm{E}][\mathrm{S}] - k_{-1}[\mathrm{ES}] - k_2[\mathrm{ES}] = 0 $$

Solving this equation for [ES] gives

$$ [\mathrm{ES}] = \frac{k_1[\mathrm{E}][\mathrm{S}]}{k_{-1} + k_2} \tag{14.36} $$

Equation 14.36 can be written in terms of the easily measured initial concentration of enzyme $[\mathrm{E}]_0$ by noting that the total amount of enzyme (bound and unbound) is conserved:

$$ [\mathrm{E}]_0 = [\mathrm{E}] + [\mathrm{ES}] \tag{14.37} $$

---

6. Leonor Michaelis (1875–1949). German biochemist and physician. In addition to his work with Menten on enzyme kinetics he made advances in microbiology and medicine. He made possible the permanent wave technique in hair care through his discovery that thioglycolic acid could dissolve keratin, the structural component of hair.

7. Maud L. Menten (1879–1960) Canadian biochemist. In addition to her work with Michaelis on enzyme kinetics, she characterized a number of bacterial toxins and was the first to separate proteins using electrophoresis.

Substituting Equation 14.37 into Equation 14.36 gives

$$[ES] = \frac{k_1}{k_{-1} + k_2}([E]_0 - [ES])[S]$$

which gives, after rearrangement,

$$[ES] = \frac{[E]_0[S]}{K_M + [S]} \tag{14.38}$$

where

$$K_M = \frac{k_{-1} + k_2}{k_1}$$

is called the *Michaelis constant*. Substituting Equation 14.38 into the rate law (Equation 14.35) gives the *Michaelis-Menten equation* for enzyme kinetics:

$$\text{rate} = \frac{k_2[E]_0[S]}{K_M + [S]} \tag{14.39}$$

Thus, the rate is always proportional to the initial concentration of enzyme. If the substrate concentration is low (that is, $[S] \ll K_M$), then the Michaelis-Menten equation (Equation 14.39) reduces to

$$\text{rate} = (k_2/K_M)[E]_0[S]$$

which says that at low substrate concentration, the rate is directly proportional to the substrate concentration. This corresponds to the initial linear rise seen in Figure 14.29. On the other hand, if the substrate concentration is high (that is, $[S] \gg K_M$), Equation 14.39 becomes

$$\text{rate} = k_2[E]_0$$

that is, the rate becomes independent of [S]. This is seen in Figure 14.29 as a leveling off of the rate at high [S]. Thus, we see that the Michaelis-Menten reaction mechanism correctly describes the experimentally observed rate data for enzyme kinetics.

Summary of Facts and Concepts

Section 14.1

▶ The rate of a chemical reaction is the change in the concentration of reactants or products over time. The rate is not constant, but varies continuously as concentrations change.

Section 14.2

▶ The rate law expresses the relationship of the rate of a reaction to the rate constant and the concentrations of the reactants raised to appropriate powers. The rate constant k for a given reaction changes only with temperature.

▶ Reaction order is the power to which the concentration of a given reactant is raised in the rate law. The overall reaction order is the sum of the powers to which reactant concentrations are raised in the rate law. The rate law and the reaction order cannot be determined from the stoichiometry of the overall equation for a reaction; they must be determined by experiment. For a zero-order reaction, the reaction rate is equal to the rate constant.

Section 14.3

▶ The rate law for a reaction can be integrated to determine the concentrations of reactants and products as functions of time.

▶ The half-life of a reaction (the time it takes for the concentration of a reactant to decrease by one-half) can be used to determine the rate constant of a first-order reaction.

Section 14.4

▸ In terms of collision theory, a reaction occurs when molecules collide with sufficient energy, called the activation energy, to break the bonds and initiate the reaction. The rate constant and the activation energy are related by the Arrhenius equation.

▸ The activation energy can be determined from the temperature dependence of the rate constant.

▸ In a reversible reaction, the equilibrium constant can be determined from the ratio of the forward and reverse reaction rates.

Section 14.5

▸ The overall balanced equation for a reaction may be the sum of a series of simple reactions, called elementary steps. The complete series of elementary steps for a reaction is the reaction mechanism.

▸ The rate law for a given reaction can be determined if the mechanism is known; however, because a number of possible mechanisms can give the same rate law, it is not generally possible to determine the mechanism directly from the rate law alone—additional experiments are necessary.

▸ If one step in a reaction mechanism is much slower than all other steps, it is the rate-determining step.

Section 14.6

▸ A catalyst speeds up a reaction usually by lowering the value of E_a without being consumed in the net reaction.

▸ In heterogeneous catalysis, the catalyst is a solid and the reactants are gases or liquids. In homogeneous catalysis, the catalyst and the reactants are in the same phase. Enzymes are catalysts in living systems.

Key Words

activated complex, p. 739
activated-complex theory, p. 742
activation energy, p. 736
bimolecular, p. 744
catalyst, p. 754
chemical kinetics, p. 713
detailed balance, p. 735
elementary step, p. 744
enzymes, p. 759

first-order reaction, p. 723
frequency factor, p. 736
half-life, p. 729
heterogeneous catalysis, p. 755
homogeneous catalysis, p. 758
integrate rate laws, p. 723
molecularity of a
    reaction, p. 744
orientation factor, p. 742

pseudo-first-order
    reaction, p. 733
rapid preequilibrium, p. 748
rate constant, p. 716
rate-determining step, p. 735
reaction intermediate, p. 744
reaction mechanism, p. 744
reaction order, p. 720
reaction rate, p. 713

second-order reaction, p. 730
steady-state
    approximation, p. 751
steady-state rate law, p. 751
steric factor, p. 742
termolecular, p. 745
transition state, p. 739
transition-state theory, p. 742
unimolecular, p. 744

Problems

Chemical Kinetics Is the Study of the Rates at Which Chemical Reactions Occur

14.1 Write the reaction rate expressions for the following reactions in terms of the disappearance of the reactants and the appearance of products:

(a) $H_2(g) + I_2(g) \longrightarrow 2HI(g)$

(b) $5Br^-(aq) + BrO_3^-(aq) + 6H^+(aq) \longrightarrow$
$$Br_2(aq) + 3H_2O(l)$$

14.2 Write the reaction rate expressions for the following reactions in terms of the disappearance of the reactants and the appearance of products:

(a) $2H_2(g) + O_2(g) \longrightarrow 2H_2O(g)$

(b) $4NH_3(g) + 5O_2(g) \longrightarrow$
$$4NO(g) + 6H_2O(g)$$

14.3 Consider the reaction

$$2NO(g) + O_2(g) \longrightarrow 2NO_2(g)$$

Suppose that at a particular moment during the reaction nitric oxide (NO) is reacting at the rate of $0.066\ M\ s^{-1}$. (a) At what rate is NO_2 being formed? (b) At what rate is molecular oxygen reacting?

14.4 Consider the reaction

$$N_2(g) + 3H_2(g) \longrightarrow 2NH_3(g)$$

Suppose that at a particular moment during the reaction molecular hydrogen is reacting at the rate of $0.074\ M\ s^{-1}$. (a) At what rate is ammonia being formed? (b) At what rate is molecular nitrogen reacting?

The Rate Law Gives the Dependence of the Reaction Rate on the Reactant Concentrations

14.5 The rate law for the reaction

$$NH_4^+(aq) + NO_2^-(aq) \longrightarrow N_2(g) + 2H_2O(l)$$

is given by rate = k [NH_4^+] [NO_2^-]. At 25°C, the rate constant is $3.0 \times 10^{-4}\,M^{-1}\,s^{-1}$. Calculate the rate of the reaction at this temperature if [NH_4^+] = 0.26 M and [NO_2^-] = 0.080 M.

14.6 Use the data in Table 14.2 to calculate the rate of the reaction at the time when [F_2] = 0.010 M and [ClO_2] = 0.020 M.

14.7 Consider the reaction

$$A + B \longrightarrow products$$

From the following data obtained at a certain temperature, determine the order of the reaction and calculate the rate constant:

[A] (M)	[B] (M)	Rate ($M\,s^{-1}$)
1.50	1.50	3.20×10^{-1}
1.50	2.50	3.20×10^{-1}
3.00	1.50	6.40×10^{-1}

14.8 Consider the reaction

$$X + Y \longrightarrow Z$$

From the following data, obtained at 360 K, (a) determine the order of the reaction, and (b) determine the initial rate of disappearance of X when the concentration of X is 0.30 M and the concentration of Y is 0.40 M.

[X] (M)	[Y] (M)	Initial Rate of Disappearance of X ($M\,s^{-1}$)
0.10	0.50	0.053
0.20	0.30	0.127
0.40	0.60	1.02
0.20	0.60	0.254
0.40	0.30	0.509

14.9 Determine the overall orders of the reactions to which the following rate laws apply: (a) rate = $k[NO_2]^2$, (b) rate = k, (c) rate = $k[H_2][Br_2]^{1/2}$, and (d) rate = $k[NO]^2[O_2]$.

14.10 What are the units of the rate constant for a third-order reaction?

14.11 Consider the reaction A $\longrightarrow$ B. The rate of the reaction is $1.6 \times 10^{-2}\,M\,s^{-1}$ when the concentration of A is 0.35 M. Calculate the rate constant if the reaction is (a) first order in A and (b) second order in A.

14.12 Cyclobutane decomposes to ethylene according to the following equation:

$$C_4H_8(g) \longrightarrow 2C_2H_4(g)$$

Determine the order of the reaction and the rate constant based on the following pressures, which were recorded when the reaction was carried out at 430°C in a constant-volume vessel.

Time (s)	$P_{C_4H_8}$ (mmHg)
0	400
2,000	316
4,000	248
6,000	196
8,000	155
10,000	122

14.13 The following gas-phase reaction was studied at 290°C by observing the change in pressure as a function of time in a constant-volume vessel:

$$ClCO_2CCl_3(g) \longrightarrow 2COCl_2(g)$$

Determine the order of the reaction and the rate constant based on the following data:

Time (s)	P (mmHg)
0	15.76
181	18.88
513	22.79
1164	27.08

where P is the total pressure.

Integrated Rate Laws Specify the Relationship Between Reactant Concentration and Time

14.14 What is the half-life of a compound if 75 percent of a given sample of the compound decomposes in 60 min? Assume first-order kinetics.

14.15 A certain first-order reaction is 35.5 percent complete in 49 min at 298 K. What is the rate constant at this temperature?

14.16 The thermal decomposition of phosphine (PH_3) into phosphorus and molecular hydrogen is a first-order reaction:

$$4PH_3(g) \longrightarrow P_4(g) + 6H_2(g)$$

The half-life of the reaction is 35.0 s at 680°C. Calculate (a) the first-order rate constant for the reaction and (b) the time required for 95 percent of the phosphine to decompose.

14.17 The rate constant for the second-order reaction

$$2NOBr(g) \longrightarrow 2NO(g) + Br_2(g)$$

is 0.80 $M^{-1}\,s^{-1}$ at 10°C. (a) Starting with a concentration of 0.086 M, calculate the concentration of NOBr after 22 s. (b) Calculate the half-lives when [NOBr]$_0$ = 0.072 M and when [NOBr]$_0$ = 0.054 M.

14.18   The rate constant for the second-order reaction

$$2NO_2(g) \longrightarrow 2NO(g) + O_2(g)$$

is 0.54 M^{-1} s^{-1} at 300°C. How long (in seconds) would it take for the concentration of NO_2 to decrease from 0.62 M to 0.28 M?

14.19   If the first half-life of a zero-order reaction is 200 s, what will be the duration of the next half-life?

14.20   A reaction A + B $\longrightarrow$ C is first order in both A and B and second order overall with a rate constant of 0.0500 M^{-1} s^{-1}. If the initial concentration of A is 1.00 M, the initial concentration of B is 0.500 M, and no product is initially present, what is the concentration of the product C after 2 minutes? After 10 minutes?

14.21   Water is a very weak electrolyte that undergoes the following ionization (called autoionization):

$$H_2O(l) \underset{k_{-1}}{\overset{k_1}{\rightleftharpoons}} H^+(aq) + OH^-(aq)$$

(a) If k_1 = 2.3 × 10^{-5} s^{-1} and k_{-1} = 1.3 × 10^{11} M^{-1} s^{-1}, calculate the value of the equilibrium constant K = $[H^+][OH^-]/[H_2O]$. (b) Calculate the product $[H^+][OH^-]$ and $[H^+]$ and $[OH^-]$.

14.22   Consider the following reaction, which takes place in a single elementary step:

$$2A + B \underset{k_{-1}}{\overset{k_1}{\rightleftharpoons}} A_2B$$

If the equilibrium constant K_c is 12.6 at a certain temperature and k_{-1} is 5.1 × 10^{-2} s^{-1}, calculate the value of k_1.

The Arrhenius Equation Gives the Temperature Dependence of Rate Constants

14.23   The variation of the rate constant with temperature for the first-order reaction

$$2N_2O_5(g) \longrightarrow 2N_2O_4(g) + O_2(g)$$

is given in the table. Determine graphically the activation energy for the reaction.

T (K)	k (s^{-1})
298	1.74 × 10^{-5}
308	6.61 × 10^{-5}
318	2.51 × 10^{-4}
328	7.59 × 10^{-4}
338	2.40 × 10^{-3}

14.24   Given the same reactant concentrations, the reaction

$$CO(g) + Cl_2(g) \longrightarrow COCl_2(g)$$

at 250°C is 1.50 × 10^3 times as fast as the same reaction at 150°C. Calculate the activation energy for this reaction. Assume that the frequency factor is constant.

14.25   For the reaction

$$NO(g) + O_3(g) \longrightarrow NO_2(g) + O_2(g)$$

the frequency factor A is 8.7 × 10^{12} s^{-1} and the activation energy is 63 kJ mol^{-1}. What is the rate constant for the reaction at 75°C?

14.26   The rate constant of a first-order reaction is 4.60 × 10^{-4} s^{-1} at 350°C. If the activation energy is 104 kJ mol^{-1}, calculate the temperature at which its rate constant is 8.80 × 10^{-4} s^{-1}.

14.27   The rate constants of some reactions double with every 10°C rise in temperature. Assume that a reaction takes place at 295 K and 305 K. What must the activation energy be for the rate constant to double as described?

14.28   The rate at which tree crickets chirp is 2.0 × 10^2 per minute at 27°C but only 39.6 per minute at 5°C. From these data, calculate the "activation energy" for the chirping process. (*Hint:* The ratio of rates is equal to the ratio of rate constants.)

14.29   The rate of bacterial hydrolysis of fish muscle is twice as great at 2.2°C as at −1.1°C. Estimate the value of the activation energy for this reaction. Is there any relation to the problem of storing fish for food? [Source: J.A. Campbell, Eco-Chem. *J. Chem. Ed.* 52, 390 (1975).]

14.30   The gas-phase reaction X $\longrightarrow$ Y has a reaction enthalpy of −64 kJ mol^{-1} and an activation energy of 22 kJ mol^{-1}. What is the activation energy of the reaction Y $\longrightarrow$ X?

The Reaction Mechanism Is the Sequence of Elementary Steps That Lead to Product Formation

14.31   The rate law for the reaction

$$2NO(g) + Cl_2(g) \longrightarrow 2NOCl(g)$$

is given by rate $k[NO][Cl_2]$. (a) What is the order of the reaction? (b) A mechanism involving the following steps has been proposed for the reaction:

$$NO(g) + Cl_2(g) \longrightarrow NOCl_2(g)$$
$$NOCl_2(g) + NO(g) \longrightarrow 2NOCl(g)$$

If this mechanism is correct, what does it imply about the relative rates of these two steps?

14.32   For the reaction

$$X_2 + Y + Z \longrightarrow XY + XZ$$

it is found that doubling the concentration of X_2 doubles the reaction rate, tripling the concentration of Y triples the rate, and doubling the concentration of Z has no effect. (a) What is the rate law for this reaction? (b) Why is it that the change in the concentration of Z has no effect on the rate? (c) Suggest a mechanism for the reaction that is consistent with the rate law.

14.33 The rate law for the decomposition of ozone to molecular oxygen

$$2O_3(g) \longrightarrow 3O_2(g)$$

is

$$\text{rate} = k\frac{[O_3]^2}{[O_2]}$$

The mechanism proposed for this process is

$$O_3 \underset{k_{-1}}{\overset{k_1}{\rightleftharpoons}} O + O_2$$

$$O + O_3 \overset{k_2}{\longrightarrow} 2O_2$$

(a) Derive the rate law from these elementary steps assuming that rapid preequilibrium is established in the first step. Explain why the rate decreases with increasing O_2 concentration. (b) Repeat the derivation using the steady-state approximation. What does the experimental rate law tell you about the relative magnitudes of the rate constants in this case?

14.34 The rate law for the reaction

$$2H_2(g) + 2NO(g) \longrightarrow N_2(g) = 2H_2O(g)$$

is

$$\text{rate} = k[H_2][NO]^2$$

Which of the following mechanisms can be ruled out on the basis of the observed rate expression?

mechanism I:

$H_2 + NO \longrightarrow H_2O + N$	(slow)
$N + NO \longrightarrow N_2 + O$	(fast)
$O + H_2 \longrightarrow H_2O$	(fast)

mechanism II:

$H_2 + 2NO \longrightarrow N_2O + H_2O$	(slow)
$N_2O + H_2 \longrightarrow N_2 + H_2O$	(fast)

mechanism III:

$2NO \rightleftharpoons N_2O_2$	(fast equilibrium)
$N_2O_2 + H_2 \longrightarrow N_2O + H_2O$	(slow)
$N_2O + H_2 \longrightarrow N_2 + H_2O$	(fast)

Reaction Rates Can Often Be Increased by the Addition of a Catalyst

14.35 Most reactions, including enzyme-catalyzed reactions, proceed faster at higher temperatures. For a given enzyme, however, the rate drops off abruptly at a certain temperature. Account for this behavior.

14.36 Use the rapid preequilibrium approximation to derive the Michaelis-Menten mechanism for enzyme catalysis (Equation 14.39).

Additional Problems

14.37 Suggest experimental means by which the rates of the following reactions could be followed:

(a) $CaCO_3(s) \longrightarrow CaO(s) + CO_2(g)$

(b) $Cl_2(g) + 2Br^-(aq) \longrightarrow$
$$Br_2(aq) + 2Cl^-(aq)$$

(c) $C_2H_6(g) \longrightarrow C_2H_4(g) + H_2(g)$

(d) $C_2H_5I(g) + H_2O(l) \longrightarrow$
$$C_2H_5OH(aq) + H^+(aq) + I^-(aq)$$

14.38 List four factors that influence the rate of a reaction.

14.39 "The rate constant for the reaction

$$NO_2(g) + CO(g) \longrightarrow NO(g) + CO_2(g)$$

is 1.64×10^{-6} M^{-1} s." What is incomplete about this statement?

14.40 In a certain industrial process involving a heterogeneous catalyst, the volume of the catalyst (in the shape of a sphere) is 10.0 cm^3. Calculate the surface area of the catalyst. If the sphere is broken down into eight spheres, each having a volume of 1.25 cm^3, what is the total surface area of the spheres? Which of the two geometric configurations of the catalyst is more effective? (The surface area of a sphere is $4\pi r^2$, where r is the radius of the sphere.) Based on your analysis here, explain why it is sometimes dangerous to work in grain elevators.

14.41 Use the data in Example 14.4 to determine graphically the half-life of the reaction.

14.42 The following data were collected for the reaction between hydrogen and nitric oxide at 700°C:

$$2H_2(g) + 2NO(g) \longrightarrow 2H_2O(g) + N_2(g)$$

Experiment	[H$_2$]	[NO]	Initial Rate (M s^{-1})
1	0.010	0.025	2.4×10^{-6}
2	0.0050	0.025	1.2×10^{-6}
3	0.010	0.0125	0.60×10^{-6}

(a) Determine the order of the reaction. (b) Calculate the rate constant. (c) Suggest a plausible mechanism that is consistent with the rate law. (*Hint:* Assume that the oxygen atom is the intermediate.)

14.43 When methyl phosphate is heated in acid solution, it reacts with water:

$$CH_3OPO_3H_2 + H_2O \longrightarrow CH_3OH + H_3PO_4$$

If the reaction is carried out in water enriched with ^{18}O, the oxygen-18 isotope is found in the phosphoric acid product but not in the methanol. What does this tell us about the mechanism of the reaction?

14.44 The rate of the reaction

$$CH_3COOC_2H_5(aq) + H_2O(l) \longrightarrow$$
$$CH_3COOH(aq) + C_2H_5OH(aq)$$

shows first-order characteristics—that is, rate = $k[CH_3COOC_2H_5]$—even though it is a second-order reaction (first order in $CH_3COOC_2H_5$ and first order in H_2O). Explain.

14.45 Explain why most metals used in catalysis are transition metals.

14.46 The reaction $2A + 3B \longrightarrow C$ is first order with respect to A and B. When the initial concentrations are $[A] = 1.6 \times 10^{-2}\ M$ and $[B] = 2.4 \times 10^{-3}\ M$, the rate is $4.1 \times 10^{-4}\ M\ s^{-1}$. Calculate the rate constant of the reaction.

14.47 The bromination of acetone is acid-catalyzed:

$$CH_3COCH_3 + Br_2 \xrightarrow{\ H^+\ catalyst'\ } CH_3COCH_2Br + HBr$$

The rate of disappearance of bromine was measured for several different concentrations of acetone, bromine, and H^+ ions at a certain temperature:

Experiment	$[CH_3COCH_3]$	$[Br_2]$	$[H^+]$	Rate of Disappearance of Br_2 $(M\ s^{-1})$
1	0.30	0.050	0.050	5.7×10^{-5}
2	0.30	0.10	0.050	5.7×10^{-5}
3	0.30	0.050	0.10	1.2×10^{-4}
4	0.40	0.050	0.20	3.1×10^{-4}
5	0.40	0.050	0.050	7.6×10^{-5}

(a) What is the rate law for the reaction? (b) Determine the rate constant. (c) The following mechanism has been proposed for the reaction:

$$CH_3{-}\overset{\overset{\textstyle O}{\|}}{C}{-}CH_3 + H_3O^+ \rightleftharpoons CH_3{-}\overset{\overset{\textstyle +OH}{\|}}{C}{-}CH_3 + H_2O$$
(fast equilibrium)

$$CH_3{-}\overset{\overset{\textstyle +OH}{\|}}{C}{-}CH_3 + H_2O \longrightarrow CH_3{-}\overset{\overset{\textstyle OH}{|}}{C}{=}CH_2 + H_3O^+ \ \text{(slow)}$$

$$CH_3{-}\overset{\overset{\textstyle OH}{|}}{C}{=}CH_2 + Br_2 \longrightarrow CH_3{-}\overset{\overset{\textstyle O}{\|}}{C}{-}CH_2Br + HBr \ \text{(fast)}$$

Show that the rate law deduced from the mechanism is consistent with that shown in (a).

14.48 The decomposition of N_2O to N_2 and O_2 is a first-order reaction. At 730°C, the half-life of the reaction is 3.58×10^3 min. If the initial pressure of N_2O is 2.10 bar at 730°C, calculate the total gas pressure after one half-life. Assume ideal gas behavior and that the volume remains constant.

14.49 The reaction

$$S_2O_8^{2-} + 2I^- \longrightarrow 2SO_4^{2-} + I_2$$

proceeds slowly in aqueous solution, but it can be catalyzed by the Fe^{3+} ion. Given that Fe^+ can oxidize I^- and Fe^{3+} can reduce $S_2O_8^{2-}$, write a plausible two-step mechanism for this reaction. Explain why the uncatalyzed reaction is slow.

14.50 The rate constants for the first-order decomposition of a organic compound in solution has been measured at several temperatures

$k(s^{-1})$	4.92×10^{-3}	0.0216	0.0950	0.326	1.15
$T(°C)$	5.0	15	25	35	45

Determine graphically the preexponential factor and the energy of activation for the reaction.

14.51 A flask contains a mixture of compounds A and B. Both compounds decompose by first-order kinetics. The half-lives are 50.0 min for A and 18.0 min for B. If the concentrations of A and B are equal initially, how long will it take for the concentration of A to be four times that of B?

14.52 Referring to Example 14.4, explain how you would measure the partial pressure of azomethane experimentally as a function of time.

14.53 The rate law for the reaction

$$2NO_2(g) \longrightarrow N_2O_4(g)$$

is rate $= k[NO_2]^2$. Which of the following changes will change the value of k? (a) The pressure of NO_2 is doubled. (b) The reaction is run in an organic solvent. (c) The volume of the container is doubled. (d) The temperature is decreased. (e) A catalyst is added to the container.

14.54 Consider the following parallel reactions

$$A \xrightarrow{\ k_1\ } B$$
$$A \xrightarrow{\ k_2\ } C$$

The activation energies are 27.2 kJ mol^{-1} for k_1 and 49.6 kJ mol^{-1} for k_2. At what temperature would $k_1/k_2 = 1.50$ given that the rate constants are equal at 400 K.

14.55 Consider the parallel reactions in Problem 14.54. If $[A]_0$ denotes the initial concentration of A, determine the ratio $[B]/[C]$ upon completion of the reaction in terms of the rate constants k_1 and k_2, assuming that B and C are not present initially.

14.56 The reaction of G_2 with E_2 to form 2EG is exothermic, and the reaction of G_2 with X_2 to form 2XG is endothermic. The activation energy of the exothermic reaction is greater than that of the endothermic reaction. Sketch the potential energy profile diagrams for these two reactions on the same graph.

14.57 Briefly comment on the effect of a catalyst on each of the following: (a) activation energy, (b) reaction mechanism, (c) enthalpy of reaction, (d) rate of forward step, and (e) rate of reverse step.

14.58 When 6 g of granulated Zn is added to a solution of $2\ M$ HCl in a beaker at room temperature, hydrogen gas is generated. For each of the following changes (at constant volume of the acid), state whether the rate of hydrogen gas evolution will be increased, decreased, or unchanged: (a) 6 g of powdered Zn is used; (b) 4 g of granulated Zn is used; (c) 2 M acetic acid is used instead of 2 M HCl; and (d) temperature is raised to 40°C.

14.59 Strictly speaking, the rate law derived for the reaction in Problem 14.42 applies only to certain concentrations of H_2. The general rate law for the reaction takes the form where k_1 and k_2 are constants. Derive rate law expressions under the conditions of very high and very low hydrogen concentrations. Does the result from Problem 14.42 agree with one of the rate expressions here?

14.60 The decomposition of dinitrogen pentoxide

$$2N_2O_5 \longrightarrow 4NO_2 + O_2$$

has been studied in carbon tetrachloride (CCl_4) solvent at a certain temperature and the following results were obtained:

[N$_2$O$_5$]	Initial Rate ($M\,s^{-1}$)
0.92	0.95×10^{-5}
1.23	1.20×10^{-5}
1.79	1.93×10^{-5}
2.00	2.10×10^{-5}
2.21	2.26×10^{-5}

Determine graphically the rate law for the reaction and calculate the rate constant.

14.61 The thermal decomposition of N_2O_5 obeys first-order kinetics. At 45°C, a plot of $\ln[N_2O_5]$ versus t gives a slope of $6.18 \times 10^{-4}\,min^{-1}$. What is the half-life of the reaction?

14.62 When a mixture of methane and bromine is exposed to light, the following reaction occurs slowly:

$$CH_4(g) + Br_2(g) \longrightarrow CH_3Br(g) + HBr(g)$$

Suggest a reasonable mechanism for this reaction. (*Hint:* Bromine vapor is deep red; methane is colorless.)

14.63 The rate of the reaction between H_2 and I_2 to form HI (discussed on p. 748) increases with the intensity of visible light. (a) Explain why this fact supports the two-step mechanism given. (The color of I_2 vapor is shown on p. 748.) (b) Explain why visible light has no effect on the formation of H atoms.

14.64 Consider the following elementary step:

$$X + 2Y \longrightarrow XY_2$$

(a) Write a rate law for this reaction. (b) If the initial rate of formation of XY_2 is $3.8 \times 10^{-3}\,M\,s^{-1}$ and the initial concentrations of X and Y are 0.26 M and 0.88 M, respectively, what is the rate constant of the reaction?

14.65 In recent years, ozone in the stratosphere has been depleted at an alarmingly fast rate by chlorofluorocarbons (CFCs). A CFC molecule such as $CFCl_3$ is first decomposed by UV radiation:

$$CFCl_3 \longrightarrow CFCl_2 + Cl$$

The chlorine radical then reacts with ozone as follows:

$$Cl + O_3 \longrightarrow ClO + O_2$$
$$ClO + O \longrightarrow Cl + O_2$$

(a) Write the overall reaction for the last two steps. (b) What are the roles of Cl and ClO? (c) Why is the fluorine radical not important in this mechanism? (d) One suggestion to reduce the concentration of chlorine radicals is to add hydrocarbons such as ethane (C_2H_6) to the stratosphere. How will this work? (e) Draw potential energy versus reaction progress diagrams for the uncatalyzed and catalyzed (by Cl) destruction of ozone: $O_3 + O \longrightarrow 2O_2$. Use the thermodynamic data in Appendix 2 to determine whether the reaction is exothermic or endothermic.

14.66 Chlorine oxide (ClO), which plays an important role in the depletion of stratospheric ozone (see Problem 14.65), decays rapidly at room temperature according to the following equation

$$2ClO(g) \longrightarrow Cl_2(g) + O_2(g)$$

From the following data, determine the reaction order and calculate the rate constant of the reaction

Time (s)	[ClO] (M)
0.12×10^{-3}	8.49×10^{-6}
0.96×10^{-3}	7.10×10^{-6}
2.24×10^{-3}	5.79×10^{-6}
3.20×10^{-3}	5.20×10^{-6}
4.00×10^{-3}	4.77×10^{-6}

14.67 A compound X undergoes two *parallel* first-order reactions as follows: X $\longrightarrow$ Y with rate constant k_1 and X $\longrightarrow$ Z with rate constant k_2. The ratio of k_1/k_2 at 40°C is 8.0. What is the ratio at 300°C? Assume that the frequency factors of the two reactions are the same.

14.68 Consider a car fitted with a catalytic converter. The first 5 min or so after it is started are the most polluting. Why?

14.69 (a) What can you deduce about the activation energy of a reaction if its rate constant changes significantly with a small change in temperature? (b) If a bimolecular reaction occurs every time an A and a B molecule collide, what can you say about the orientation factor and activation energy of the reaction?

14.70 The rate law for the reaction

$$CO(g) + NO_2(g) \longrightarrow CO_2(g) + NO(g)$$

is

$$rate = k[NO_2]^2$$

Suggest a plausible mechanism for the reaction, given that the unstable species NO_3 is an intermediate.

14.71 Many reactions involving heterogeneous catalysts are zero order, that is, rate = k. An example is the decomposition of phosphine (PH_3) over tungsten (W):

$$PH_3(g) \longrightarrow P_4(g) + 6H_2(g)$$

It is found that the reaction is independent of $[PH_3]$ as long as phosphine's pressure is sufficiently high (≥ 1 bar). Explain.

14.72 Thallium(I) is oxidized by cerium(IV) as follows:

$$Tl^+ + 2Ce^{4+} \longrightarrow Tl^{3+} + 2Ce^{3+}$$

The elementary steps, in the presence of Mn(II), are as follows:

$$Ce^{4+} + Mn^{2+} \longrightarrow Ce^{3+} + Mn^{3+}$$
$$Ce^{4+} + Mn^{3+} \longrightarrow Ce^{3+} + Mn^{4+}$$
$$Tl^+ + Mn^{4+} \longrightarrow Tl^{3+} + Mn^{2+}$$

(a) Identify the catalyst, intermediates, and the rate-determining step if the rate law is rate = $k[Ce^{4+}][Mn^{2+}]$. (b) Explain why the reaction is slow without the catalyst. (c) Is the catalysis homogeneous or heterogeneous?

14.73 Sucrose ($C_{12}H_{22}O_{11}$), commonly called table sugar, undergoes hydrolysis (reaction with water) to produce fructose ($C_6H_{12}O_6$) and glucose ($C_6H_{12}O_6$):

$$\underset{sucrose}{C_{12}H_{22}O_{11}} + H_2O \longrightarrow \underset{fructose}{C_6H_{12}O_6} + \underset{glucose}{C_6H_{12}O_6}$$

This reaction is of considerable importance in the candy industry for at least two reasons. First, fructose is sweeter than sucrose, so the same level of sweetness can be achieved with fewer raw materials if fructose is substituted for sucrose. Second, a mixture of fructose and glucose, called *invert sugar,* does not crystallize, so the candy containing this sugar would be chewy rather than brittle as candy containing sucrose crystals would be. (a) From the following data, determine the order of the reaction. (b) How long does it take to hydrolyze 95 percent of the sucrose? (c) Explain why the rate law does not include $[H_2O]$ even though water is a reactant.

Time (min)	$[C_{12}H_{22}O_{11}]$ (M)
0	0.500
60.0	0.400
96.4	0.350
157.5	0.280

14.74 The first-order rate constant for the decomposition of dimethyl ether

$$(CH_3)_2O(g) \longrightarrow CH_4(g) + H_2(g) + CO(g)$$

is 3.2×10^{-4} s^{-1} at 450°C. The reaction is carried out in a constant-volume flask. Initially only dimethyl ether is present and the pressure is 0.350 bar. What is the pressure of the system after 8.0 min? Assume ideal gas behavior.

14.75 Polyethylene is used in many items, including water pipes, bottles, electrical insulation, toys, and mailer envelopes. It is a *polymer,* a molecule with a very high molar mass made by joining many ethylene molecules together. (Ethylene is the basic unit, or monomer, for polyethylene.) The initiation step is

$$R_2 \xrightarrow{k_i} 2R \cdot \qquad \text{initiation}$$

The R· species (called a radical) reacts with an ethylene molecule (M) to generate another radical

$$R \cdot + M \longrightarrow M_1 \cdot$$

The reaction of M_1 with another monomer leads to the growth or propagation of the polymer chain:

$$M_1 \cdot + M \xrightarrow{k_p} M_2 \cdot \quad \text{(propagation)}$$

This step can be repeated with hundreds of monomer units. The propagation terminates when two radicals combine

$$M' \cdot + M'' \cdot \xrightarrow{k_t} M'{-}M'' \quad \text{(termination)}$$

(a) The initiator frequently used in the polymerization of ethylene is benzoyl peroxide $[(C_6H_5COO)_2]$:

$$(C_6H_5COO)_2 \longrightarrow 2C_6H_5COO \cdot$$

This is a first-order reaction. The half-life of benzoyl peroxide at 100°C is 19.8 min. Calculate the rate constant (in min^{-1}) of the reaction. (b) If the half-life of benzoyl peroxide is 7.30 h (438 min) at 70°C, what is the activation energy (in kJ mol^{-1}) for the decomposition of benzoyl peroxide? (c) Write the rate laws for the elementary steps in the polymerization of ethylene, and identify the reactant, product, and intermediates. (d) What condition would favor the growth of long, high–molar-mass polyethylenes?

14.76 Consider the following elementary steps for a consecutive reaction:

$$A \xrightarrow{k_1} B \xrightarrow{k_2} C$$

(a) Write an expression for the rate of change of B. (b) Derive an expression for the concentration of B under steady-state conditions that is, when B is decomposing to C at the same rate as it is formed from A. (c) Compare the expression derived in (b) to the Equation 14.16 derived in Section 14.3.

14.77 Ethanol is a toxic substance that, when consumed in excess, can impair respiratory and cardiac functions by interferring with the neurotransmitters of the nervous system. In the human body, ethanol is metabolized by the enzyme alcohol dehydrogenase to acetaldehyde, which causes "hangovers." (a) Based on your knowledge of enzyme kinetics, explain why binge drinking (i.e., consuming too much alcohol too fast) can prove fatal. (b) Methanol is even more toxic than ethanol. It is also metabolized by alcohol dehydrogenase, and the product, form-aldehyde, can cause blindness or death. An antidote to methanol poisoning is ethanol. Explain how this procedure works.

14.78 Consider the potential energy profiles for the following three reactions (from left to right). (a) Rank the rates of the reactions from slowest to fastest. (b) Calculate ΔH for each reaction and determine which reaction(s) are exothermic and which reaction(s) are endothermic. Assume the reactions have roughly the same frequency factors.

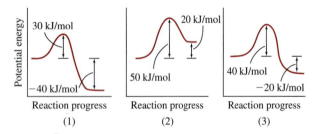

14.79 Consider the following potential energy profile for the A $\longrightarrow$ D reaction. (a) How many elementary steps are there? (b) How many intermediates are formed? (c) Which step is rate determining? (d) Is the overall reaction exothermic or endothermic?

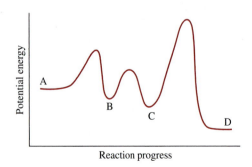

14.80 A factory that specializes in the refinement of transition metals such as titanium was on fire. The firefighters were advised not to douse the fire with water. Why?

14.81 The activation energy for the decomposition of hydrogen peroxide

$$2H_2O_2(aq) \longrightarrow 2H_2O_2(l) + O_2(g)$$

is 42 kJ mol^{-1}, whereas it is 7.0 kJ mol^{-1} when the reaction is catalyzed by the enzyme catalase. Calculate the temperature that would cause the nonenzymatic catalysis to proceed as rapidly as the enzyme-catalyzed decomposition at 20°C. Assume the frequency factor A is the same in both cases.

14.82 To carry out metabolism, oxygen is taken up by hemoglobin (Hb) to form oxyhemoglobin (HbO$_2$) according to the following simplified equation

$$Hb(aq) + O_2(aq) \xrightarrow{k} HbO_2(aq)$$

where the second-order rate constant is 2.1×10^{-6} M^{-1} s^{-1} at 37°C. For an average adult, the con-centrations of Hb and O$_2$ in the blood at the lungs are 8.0×10^{-6} M and 1.5×10^{-6} M, respectively. (a) Calculate the rate of formation of HbO$_2$. (b) Calculate the rate of consumption of O$_2$. (c) The rate of formation of HbO$_2$ increases to 1.4×10^{-4} M s^{-1} during exercise to meet the demand of an increased metabolism rate. Assuming the Hb concentration remains the same, what must the oxygen concentration be to sustain this rate of HbO$_2$ formation?

14.83 At a certain elevated temperature, ammonia decomposes on the surface of tungsten metal as follows: 2NH$_3$ $\longrightarrow$ N$_2$ + 3H$_2$. From the following plot of the rate of the reaction versus the pressure of NH$_3$, propose a mechanism for the reaction.

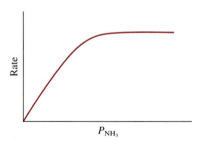

14.84 The rate of a reaction was followed by measuring the absorption of light by the reactants and products as a function of wavelength (λ_1, λ_2, λ_3) as time progressed (see Figure on p. 771). Which of the following mechanisms is consistent with the experimental data?

(a) A $\longrightarrow$ B, A $\longrightarrow$ C

(b) A $\longrightarrow$ B + C

(c) A $\longrightarrow$ B, B $\longrightarrow$ C + D

(d) A $\longrightarrow$ B, B $\longrightarrow$ C

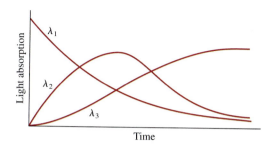

Light absorption vs Time graph with curves labeled λ_1, λ_2, λ_3.

14.85 The following expression shows the dependence of the half-life ($t_{\frac{1}{2}}$) of a reaction

$$A \longrightarrow B$$

on the initial reactant concentration, $[A]_0$:

$$t_{\frac{1}{2}} \propto \frac{1}{[A]_0^{n-1}}$$

where n is the order of the reaction. (a) Derive this result. (b) Verify this result for zeroth-, first-, and second-order reactions.

14.86 The rate constant for the gaseous reaction

$$H_2(g) + I_2(g) \longrightarrow 2HI(g)$$

is $2.42 \times 10^{-2} \, M^{-1} \, s^{-1}$ at 400°C. Initially an equimolar sample of H_2 and I_2 is placed in a vessel at 400°C and the total pressure is 1658 mmHg. (a) What is the initial rate ($M \, min^{-1}$) of formation of HI? (b) What are the rate of formation of HI and the concentration of HI (in molarity) after 10.0 min.

14.87 A protein molecule, P, of molar mass $\mathcal{M}$ dimerizes when it is allowed to stand in solution at room temperature. A plausible mechanism is that the protein molecule is first denatured (i.e., loses its activity due to a change in overall structure) before it dimerizes:

$$P \xrightarrow{k} P^* \text{ (denatured)} \qquad \text{slow}$$
$$2P^* \longrightarrow P_2 \qquad \text{fast}$$

where the asterisk denotes a denatured protein molecule. Derive an expression for the average molar mass (of P and P_2), $\bar{\mathcal{M}}$, in terms of the initial protein concentration, $[P]_0$; the concentration at time t, $[P]_t$; and $\mathcal{M}$. Describe how you would determine k from molar mass experiments.

14.88 When the concentration of A in the reaction $A \longrightarrow B$ was changed from 1.20 M to 0.60 M, the half-life increased from 2.0 min to 4.0 min at 25°C. Calculate the order of the reaction and the rate constant. (*Hint:* Use the equation in Problem 14.85.)

14.89 At a certain elevated temperature, ammonia decomposes on the surface of tungsten metal as follows:

$$NH_3 \longrightarrow \tfrac{1}{2}N_2 + \tfrac{3}{2}H_2$$

The kinetics data are expressed as the variation of the half-life with the initial pressure of NH_3:

P (mmHg)	264	130	59	16
$t_{\frac{1}{2}}(s)$	456	228	102	60

(a) Determine the order of the reaction. (b) How does the order depend on the initial pressure? (c) How does the mechanism of the reaction vary with pressure. (*Hint:* Use the equation in Problem 14.85.)

14.90 The "boat" form and "chair" form of cyclohexane (C_6H_{12}) interconvert as shown here:

Boat   Chair

In this representation, the H atoms are omitted and a C atom is assumed to be at each intersection of two lines (bonds). The conversion is first order in each direction. The activation energy for the chair to boat conversion is 41 kJ mol^{-1}. If the frequency factor is $1.0 \times 10^{12} \, s^{-1}$, what is k_1 at 298 K? The equilibrium constant K_c for the reaction is 9.83×10^3 at 298 K.

14.91 The activation energy for the reaction

$$N_2O(g) \longrightarrow N_2(g) + O(g)$$

is 2.4×10^2 kg mol^{-1} at 600 K. Calculate the percentage of the increase in rate from 600 K to 606 K. Comment on your results.

14.92 Derive Equation 14.14 for a mixed second order reaction

$$A + B \longrightarrow \text{product}$$

that is first order in both A and B.

Answers to Practice Exercises

14.1 (a) 0.013 $M \, s^{-1}$, (b) 0.013 $M \, s^{-1}$ (c) $-0.052 \, M \, s^{-1}$
14.2 rate $= k[S_2O_8^{-2}][I^-]$; $k = 8.1 \times 10^{-2} \, M^{-1} \, s^{-1}$.
14.3 (a) 66 s. (b) 29 s. **14.4** First order; $k = 1.4 \times 10^{-2}$ min^{-1}
14.5 1.2×10^3 s **14.6** (a) 3.2 min, (b) 2.1 min
14.7 240 kJ mol^{-1} **14.8** $k = 3.13 \times 10^{-9} \, s^{-1}$

14.9 (a) $NO_2 + CO \longrightarrow NO + CO_2$, (b) NO_3, (c) The first step is rate determining.

14.10 $k = \dfrac{k_1 k_2}{k_{-1}}$; an experiment to detect the presence of C could distinguish between the two proposed mechanisms.

Chapter 15

The Chemistry of Transition Metals

In this chapter, we examine the properties of transition metal elements, that is, elements with unfilled *d* or *f* orbitals. There are about 50 transition elements, and they have widely varying and fascinating properties. To present even one interesting feature of each transition element is beyond the scope of this book. We will therefore limit our discussion to the transition elements that have incompletely filled *d* subshells and to their most commonly encountered property—the tendency to form complex ions.

15.1 | Transition Metals Have Electronic Configurations with Incomplete *d* or *f* Shells

Transition metals have incompletely filled *d* subshells or readily give rise to ions with incompletely filled *d* subshells (Figure 15.1) (The Group 2B metals—Zn, Cd, and Hg—are sometimes treated as transition metals, but they do not have this characteristic electron configuration, so they really do not belong in this category.) This attribute is responsible for several notable properties, including distinctive coloring, formation of paramagnetic compounds, catalytic activity, and especially a strong tendency to form complex ions. In this chapter, we focus on the first-row elements from scandium to copper, the most common transition metals. Table 15.1 lists some of their properties.

As we read across any period from left to right, atomic numbers increase, electrons are added to the outer shell, and the nuclear charge increases by the addition of protons. In the third-period elements—sodium to argon—the outer electrons weakly shield one another from the extra nuclear charge. Consequently, atomic radii decrease rapidly from sodium to argon, and ionization energies and electronegativities increase steadily (see Figures 2.17, 2.22, and 3.39).

For the transition metals, the trends are different. According to Table 15.1, the nuclear charge increases from scandium to copper, but electrons are being added to the inner 3*d* subshell. These 3*d* electrons shield the 4*s* electrons from the increasing nuclear charge somewhat more effectively than outer-shell electrons can shield one another, so the atomic radii decrease less rapidly. For the same reason, electronegativities and ionization energies increase only slightly from scandium across to copper compared with the increases from sodium to argon.

Although the transition metals are less electropositive (more electronegative) than the alkali and alkaline earth metals, their standard reduction potentials suggest that all of them except copper should react with strong acids such as hydrochloric acid to produce hydrogen gas. However, most transition metals are inert toward acids or react slowly with them because of a protective layer of oxide. A case in point is chromium: Despite a rather negative standard reduction potential, it is quite inert chemically because of the formation

1 1A																	18 8A
1 **H**	2 2A											13 3A	14 4A	15 5A	16 6A	17 7A	2 **He**
3 **Li**	4 **Be**											5 **B**	6 **C**	7 **N**	8 **O**	9 **F**	10 **Ne**
11 **Na**	12 **Mg**	3 3B	4 4B	5 5B	6 6B	7 7B	8	9 ⎯8B⎯	10	11 1B	12 2B	13 **Al**	14 **Si**	15 **P**	16 **S**	17 **Cl**	18 **Ar**
19 **K**	20 **Ca**	21 **Sc**	22 **Ti**	23 **V**	24 **Cr**	25 **Mn**	26 **Fe**	27 **Co**	28 **Ni**	29 **Cu**	30 **Zn**	31 **Ga**	32 **Ge**	33 **As**	34 **Se**	35 **Br**	36 **Kr**
37 **Rb**	38 **Sr**	39 **Y**	40 **Zr**	41 **Nb**	42 **Mo**	43 **Tc**	44 **Ru**	45 **Rh**	46 **Pd**	47 **Ag**	48 **Cd**	49 **In**	50 **Sn**	51 **Sb**	52 **Te**	53 **I**	54 **Xe**
55 **Cs**	56 **Ba**	57 **La**	72 **Hf**	73 **Ta**	74 **W**	75 **Re**	76 **Os**	77 **Ir**	78 **Pt**	79 **Au**	80 **Hg**	81 **Tl**	82 **Pb**	83 **Bi**	84 **Po**	85 **At**	86 **Rn**
87 **Fr**	88 **Ra**	89 **Ac**	104 **Rf**	105 **Db**	106 **Sg**	107 **Bh**	108 **Hs**	109 **Mt**	110 **Ds**	111 **Rg**	112	113	114	115	116	(117)	118

Figure 15.1 The transition metals (blue squares). Note that although the Group 2B elements (Zn, Cd, Hg) are described as transition metals by some chemists, neither the metals nor their ions possess incompletely filled *d* subshells.

Table 15.1	Electron Configurations and Other Properties of the First-Row Transition Metals								
	Sc	**Ti**	**V**	**Cr**	**Mn**	**Fe**	**Co**	**Ni**	**Cu**
Electron configuration									
M	$4s^2 3d^1$	$4s^2 3d^2$	$4s^2 3d^3$	$4s^1 3d^5$	$4s^2 3d^5$	$4s^2 3d^6$	$4s^2 3d^7$	$4s^2 3d^8$	$4s^1 3d^{10}$
M^{2+}	—	$3d^2$	$3d^3$	$3d^4$	$3d^5$	$3d^6$	$3d^7$	$3d^8$	$3d^9$
M^{3+}	[Ar]	$3d^1$	$3d^2$	$3d^3$	$3d^4$	$3d^5$	$3d^6$	$3d^7$	$3d^8$
Electronegativity	1.3	1.5	1.6	1.6	1.5	1.8	1.9	1.9	1.9
Ionization energy (kJ mol^{-1})									
First	631	658	650	652	717	759	760	736	745
Second	1235	1309	1413	1591	1509	1561	1645	1751	1958
Third	2389	2650	2828	2986	3250	2956	3231	3393	3578
Radius (pm)									
M	162	147	134	130	135	126	125	124	128
M^{2+}	—	90	88	85	80	77	75	69	72
M^{3+}	81	77	74	64	66	60	64	—	—
Standard reduction potential (V)*	−2.08	−1.63	−1.2	−0.74	−1.18	−0.44	−0.28	−0.25	0.34

*The half-reaction is M^{2+} (aq) + 2e$^-$ ⟶ M(s) (except for Sc and Cr, where the ions are Sc^{3+} and Cr^{3+}, respectively).

on its surface of chromium(III) oxide (Cr_2O_3). Consequently, chromium is commonly used as a protective and noncorrosive plating on other metals. On automobile bumpers and trim, chromium plating serves a decorative as well as a functional purpose.

General Physical Properties

Most of the transition metals have relatively small atomic radii and a close-packed structure (see Figure 6.28) in which each atom has a coordination number of 12. The combined effect of small atomic size and close packing result in strong metallic bonds. Therefore, transition metals have higher densities, higher melting and boiling points, and higher heats of fusion and vaporization than the Group 1A, 2A, and 2B metals (Table 15.2).

Table 15.2	Physical Properties of Elements K to Zn											
	1A	**2A**	**Transition Metals**									**2B**
	K	**Ca**	**Sc**	**Ti**	**V**	**Cr**	**Mn**	**Fe**	**Co**	**Ni**	**Cu**	**Zn**
Atomic radius (pm)	235	197	162	147	134	130	135	126	125	124	128	138
Melting Point (°C)	63.7	838	1539	1668	1900	1875	1245	1536	1495	1453	1083	419.5
Boiling Point (°C)	760	1440	2730	3260	3450	2665	2150	3000	2900	2730	2595	906
Density (g cm^{-3})	0.86	1.54	3.0	4.51	6.1	7.19	7.43	7.86	8.9	8.9	8.96	7.14

Electron Configurations

The electron configurations of the first-row transition metals were discussed in Section 2.2. Calcium has the electron configuration $[Ar]4s^2$. From scandium across to copper, electrons are added to the $3d$ orbitals. Thus, the outer electron configuration of scandium is $4s^23d^1$, that of titanium is $4s^23d^2$, and so on. The two exceptions are chromium and copper, whose outer electron configurations are $4s^13d^5$ and $4s^13d^{10}$, respectively. These irregularities are the result of the extra stability associated with half-filled and completely filled $3d$ subshells.

When the first-row transition metals form cations, electrons are removed first from the $4s$ orbitals and then from the $3d$ orbitals. (This is the opposite of the order in which orbitals are filled in atoms.) For example, the outer electron configuration of Fe^{2+} is $3d^6$, not $4s^23d^4$.

Oxidation States

Transition metals exhibit variable oxidation states in their compounds. Figure 15.2 shows that the common oxidation states for each element from scandium to copper include +2, +3, or both. The +3 oxidation states are more stable at the beginning of the series, whereas the +2 oxidation states are more stable toward the end. To understand this trend, you must examine the ionization energy plots in Figure 15.3. In general, the ionization energies increase gradually from left to right. However, the third ionization energy (when an electron is removed from the $3d$ orbital) increases more rapidly than the first and second ionization energies. Because it takes more energy to remove the third electron from the metals near the end of the row than from those near the beginning, the metals near the end tend to form M^{2+} ions rather than M^{3+} ions.

The highest oxidation state for a transition metal, that of manganese ($4s^23d^5$), is +7. For elements to the right of Mn (Fe to Cu), the oxidation numbers are lower. Transition metals usually exhibit their highest oxidation states in compounds with very electronegative elements such as oxygen and fluorine. Examples include V_2O_5, CrO_3, and Mn_2O_7.

Sc	Ti	V	Cr	Mn	Fe	Co	Ni	Cu
				+7				
			+6	+6	+6			
		+5	+5	+5	+5			
		+4	+4	+4	+4	+4	+4	
+3	+3	+3	+3	+3	+3	+3	+3	+3
		+2	+2	+2	+2	+2	+2	+2
								+1

Figure 15.2 Oxidation states of the first-row transition metals. The most stable oxidation numbers are shown in color. The zero oxidation state is encountered in some compounds, such as $Ni(CO)_4$ and $Fe(CO)_5$.

Figure 15.3 Variation of the first, second, and third ionization energies for the first-row transition metals.

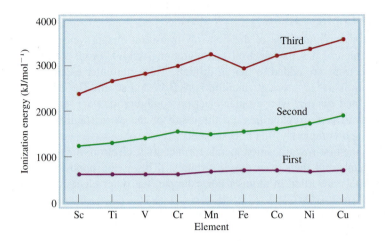

Two Examples: The Chemistry of Iron and Copper

Figure 15.4 shows samples of the first-row transition metals. Here we will briefly survey the chemistry of two of these elements—iron and copper—paying particular attention to their occurrence, preparation, uses, and important compounds.

Iron

After aluminum, iron is the most abundant metal in the Earth's crust (6.2 percent by mass). It is found in many ores; some of the economically important ones are *hematite* (Fe_2O_3), *siderite* ($FeCO_3$), and *magnetite* (Fe_3O_4) (Figure 15.5). Pure iron is a gray metal and is not particularly hard. Its ion is essential in living systems because it reversibly binds oxygen to hemoglobin, the protein in blood that carries oxygen from the lungs to the rest of the tissues of the body.

Iron reacts with hydrochloric acid to give hydrogen gas:

$$Fe(s) + 2H^+(aq) \longrightarrow Fe^{2+}(aq) + H_2(g)$$

Concentrated sulfuric acid oxidizes the metal to Fe^{3+}, but concentrated nitric acid renders the metal "passive" by forming a thin layer of Fe_3O_4 over the surface. One of the best-known reactions of iron is rust formation. The two oxidation states of iron are +2 and +3. Iron(II) compounds include FeO (black), $FeSO_4 \cdot 7H_2O$ (green), $FeCl_2$ (yellow), and FeS (black). In the presence of oxygen, Fe^{2+} ions in solution are readily oxidized to Fe^{3+} ions. Iron(III) oxide is reddish brown, and iron(III) chloride is brownish black.

Copper

Copper, a rarer element than iron (6.8×10^{-3} percent of Earth's crust by mass), is found in nature in the uncombined state as well as in ores such as chalcopyrite ($CuFeS_2$) (Figure 15.6). The reddish-brown metal is obtained by roasting the ore to give Cu_2S and then metallic copper:

$$2CuFeS_2(s) + 4O_2(g) \longrightarrow Cu_2S(s) + 2FeO(s) + 3SO_2(g)$$
$$Cu_2S(s) + O_2(g) \longrightarrow 2Cu(l) + SO_2(g)$$

Impure copper can be purified by electrolysis. After silver, which is too expensive for large-scale use, copper has the highest electrical conductivity. It is also a good thermal conductor. Copper is used in alloys, electrical cables, plumbing (pipes), and coins.

Scandium (Sc)   Titanium (Ti)   Vanadium (V)

Chromium (Cr)   Manganese (Mn)   Iron (Fe)

Cobalt (Co)   Nickel (Ni)   Copper (Cu)

Figure 15.4 The first-row transition metals.

Copper reacts only with hot concentrated sulfuric acid and nitric acid. Its two important oxidation states are $+1$ and $+2$. The $+1$ state is less stable and disproportionates in solution:

$$2Cu^+(aq) \longrightarrow Cu(s) + Cu^{2+}(aq)$$

All compounds of Cu(I) are diamagnetic and colorless except for Cu_2O, which is red. The Cu(II) compounds are all paramagnetic and colored. The hydrated Cu^{2+} ion is blue. Some important Cu(II) compounds are CuO (black), $CuSO_4 \cdot 5H_2O$ (blue), and CuS (black).

Figure 15.5 The iron ore magnetite (Fe_3O_4).

15.2 | Transition Metals Can Form a Variety of Coordination Compounds

Transition metals have a stong tendency to form complex ions (see p. 648). A ***coordination compound*** *typically consists of a complex ion and counter ion.* (Some coordination compounds such as $Fe(CO)_5$ do not contain complex ions.) Much of what we now know about coordination compounds stems from the classic work of

Figure 15.6 The copper ore chalcopyrite ($CuFeS_2$).

Recall that a complex ion contains a central metal ion bonded to one or more ions or molecules (see Section 11.7).

Alfred Werner,[1] who prepared and characterized many coordination compounds. In 1893, at the age of 26, Werner proposed what is now commonly referred to as *Werner's coordination theory.*

Nineteenth-century chemists were puzzled by a certain class of reactions that seemed to violate valence theory. For example, the valences of the elements in cobalt(III) chloride and in ammonia seem to be completely satisfied, and yet these two substances react to form a stable compound having the formula $CoCl_3 \cdot 6NH_3$. To explain this behavior, Werner postulated that most elements exhibit two types of valence: *primary valence* and *secondary valence.* In modern terminology, primary valence corresponds to the oxidation number and secondary valence to the coordination number of the element. In $CoCl_3 \cdot 6NH_3$, according to Werner, cobalt has a primary valence of $+3$ and a secondary valence of $+6$.

Today we use the formula $[Co(NH_3)_6]Cl_3$ to indicate that the ammonia molecules and the cobalt atom form a complex ion; the chloride ions are not part of the complex but are held to it by ionic forces. Most, but not all, of the metals in coordination compounds are transition metals.

The molecules or ions that surround the metal in a complex ion are called **ligands** (Table 15.3). Every ligand has at least one unshared pair of valence electrons, as these following examples demonstrate:

$$ \overset{\cdot\cdot}{\underset{H \qquad H}{O}} \qquad \overset{\cdot\cdot}{\underset{H \;\; \underset{H}{N} \;\; H}{N}} \qquad :\!\overset{\cdot\cdot}{\underset{\cdot\cdot}{Cl}}\!:^{-} \qquad :\!C\!\equiv\!O\!: $$

Ligands, therefore, act as Lewis bases (see Section 3.4), donating one or more electron pairs to the metal. On the other hand, the transition metal atom (in either its neutral or positively charged state) acts as a Lewis acid, accepting (and sharing) pairs of electrons from the ligands. As a result, the metal-ligand bonds are usually coordinate covalent bonds (see Section 3.4).

The atom in a ligand that is bound directly to the metal atom is known as the **donor atom.** For example, nitrogen is the donor atom of the NH_3 ligand in the $[Cu(NH_3)_4]_2$ complex ion. The **coordination number** in coordination compounds is *the number of donor atoms surrounding the central metal atom in a complex ion.* For example, the coordination number of Ag in $[Ag(NH_3)_2]^+$ is 2, of Cu^{2+} in $[Cu(NH_3)_4]^{2+}$ is 4, and of Fe^{3+} in $[Fe(CN)_6]^{3-}$ is 6. The most common coordination numbers are 4 and 6, but other coordination numbers, such as 2 or 5, are also known.

In a crystal lattice, the coordination number of an atom (or ion) is defined as the number of neighboring atoms (or ions) surrounding the atom (or ion).

Depending on the number of donor atoms present, ligands are classified as *monodentate, bidentate,* or *polydentate* (see Table 15.3). H_2O and NH_3 are monodentate ligands because they have only one donor atom each. Ethylenediamine (sometimes abbreviated "en") is a bidentate ligand:

$$ H_2\overset{\cdot\cdot}{N}\!-\!CH_2\!-\!CH_2\!-\!\overset{\cdot\cdot}{N}H_2 $$

The two nitrogen atoms can coordinate with a metal atom as shown in Figure 15.7.

Bidentate and polydentate ligands are also called **chelating agents** *because they can hold the metal atom like a claw* (from the Greek *chele*, meaning "claw"). Ethylenediaminetetraacetate ion (EDTA) is a polydentate ligand used to treat metal poisoning (Figure 15.8). Six donor atoms enable EDTA to form a very stable, water-soluble complex ion with lead. In this form, lead is removed from the blood and tissues and excreted from the body. EDTA is also used to clean up spills of radioactive metals.

---

1. Alfred Werner (1866–1919). Swiss chemist. Werner started as an organic chemist but became interested in coordination chemistry. For his theory of coordination compounds, Werner was awarded the Nobel Prize in Chemistry in 1913.

Table 15.3	Some Common Ligands
Name	**Structure**

Monodentate ligands

Ammonia

$$H-\overset{\cdot\cdot}{N}-H$$
$$\mid$$
$$H$$

Carbon monoxide

$$:C\equiv O:$$

Chloride ion

$$:\overset{\cdot\cdot}{\underset{\cdot\cdot}{Cl}}:^{-}$$

Cyanide ion

$$[:C\equiv N:]^{-}$$

Thiocyanate ion

$$[:\overset{\cdot\cdot}{\underset{\cdot\cdot}{S}}-C\equiv N:]^{-}$$

Water

$$H-\overset{\cdot\cdot}{\underset{\cdot\cdot}{O}}-H$$

Bidentate ligands

Ethylenediamine

$$H_2\overset{\cdot\cdot}{N}-CH_2-CH_2-\overset{\cdot\cdot}{N}H_2$$

Oxalate ion

$$\left[\begin{array}{cc} \overset{\cdot\cdot}{\underset{}{O}} & \overset{\cdot\cdot}{\underset{}{O}} \\ \parallel & \parallel \\ \cdot\cdot O & C-C & O\cdot\cdot \\ \end{array}\right]^{2-}$$

Polydentate ligand

Ethylenediaminetetraacetate ion (EDTA)

$$\left[\quad\right]^{4-}$$

Oxidation Numbers of Metals in Coordination Compounds

Another important property of coordination compounds is the oxidation number of the central metal atom. The net charge of a complex ion is the sum of the charges on the central metal atom and its surrounding ligands. In the $[PtCl_6]^{2-}$ ion, for example, each chloride ion has an oxidation number of -1, so the oxidation number of Pt must be $+4$. If the ligands do not bear net charges, the oxidation number of the metal is

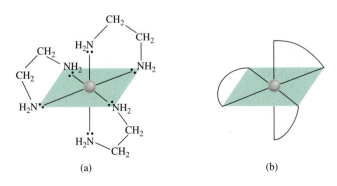

(a)   (b)

Figure 15.7 (a) Structure of metal-ethylenediamine complex. Each ethylenediamine molecule provides two N donor atoms and is therefore a bidentate ligand. (b) Simplified structure of the same complex.

Figure 15.8 (a) EDTA complex of lead. The complex bears a net charge of -2 because each O donor atom has one negative charge and the lead ion carries two positive charges. Note the octahedral geometry around the Pb^{2+} ion. (b) Molecular model of the Pb^{2+}-EDTA complex. The yellow sphere is the Pb^{2+} ion.

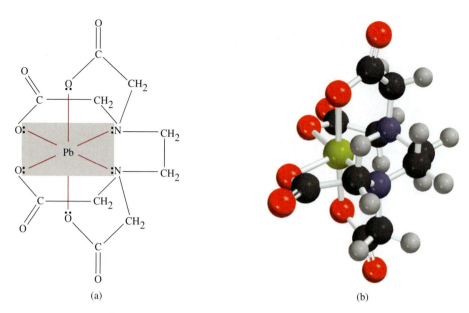

(a)   (b)

equal to the charge of the complex ion. Thus, in $[Cu(NH_3)_4]^{2+}$ each NH_3 is neutral, so the oxidation number of Cu is $+2$.

Example 15.1 shows how to determine the oxidation numbers of metals in coordination compounds.

Example 15.1

Specify the oxidation number of the central metal atom in each of the following compounds: (a) $[Ru(NH_3)_5(H_2O)]Cl_2$, (b) $[Cr(NH_3)_6](NO_3)_3$, (c) $[Fe(CO)_5]$, and (d) $K_4[Fe(CN)_6]$.

Strategy The oxidation number of the metal atom is equal to its charge. First, we total the charges on the anions or the cations that electrically balance the complex ion. This gives us the net charge of the complex ion. Next, based on whether the ligands are charged or neutral, we can deduce the net charge of the metal and hence its oxidation number.

Solution (a) Each of the two chloride ions carries a -1 charge, so the charge on the complex ion is $+2$. The oxidation number of Ru must also be $+2$ because both NH_3 and H_2O are neutral species.

(b) Each of the three nitrate ions has a charge of -1, so the complex cation must be $[Cr(NH_3)_6]^{3+}$. NH_3 is a neutral ligand, so the oxidation number of Cr is $+3$.

(c) The complex is neutral (there are no anions or cations to electrically balance its charge) and the CO ligands are neutral, too, so the oxidation number of Fe is zero.

(d) Each of the four potassium ions has a charge of $+1$, so the complex anion is $[Fe(CN)_6]^{4-}$. Next, each cyanide ligand bears a charge of -1, so Fe must have an oxidation number of $+2$.

Practice Exercise Write the oxidation numbers of the transition metals in the following compounds: (a) $[CuBr_4]Cl_2$, (b) $[Cr(NH_3)SO_4]Br$, and (c) $[Pt(en)_2Cl_2](NO_3)_2$.

Table 15.4	Some Common Ligands in Coordination Compounds
Ligand	**Name of Ligand in Coordination Compound**
Bromide, Br^-	Bromo
Chloride, Cl^-	Chloro
Cyanide, CN^-	Cyano
Hydroxide, OH^-	Hydroxo
Oxide, O^{2-}	Oxo
Carbonate, CO_3^{2-}	Carbonato
Nitrite, NO_2^-	Nitro
Oxalate, $C_2O_4^{2-}$	Oxalato
Ammonia, NH_3	Ammine
Carbon monoxide, CO	Carbonyl
Water, H_2O	Aquo
Ethylenediamine	Ethylenediamine
Ethylenediaminetetraacetate	Ethylenediaminetetraacetato

Naming Coordination Compounds

Having discussed the various types of ligands and the oxidation numbers of metals, it is now time to learn how to name coordination compounds. The rules for naming coordination compounds are as follows:

1. The cation is named before the anion, as in other ionic compounds. The rule holds regardless of whether the complex ion bears a net positive or a net negative charge. In $K_3[Fe(CN)_6]$ and $[Co(NH_3)_4Cl_2]Cl$, for example, we name the K^+ and $[Co(NH_3)_4Cl_2]^+$ cations first, respectively.

2. Within a complex ion, the ligands are named first, in alphabetical order, and the metal ion is named last.

3. The names of anionic ligands end with the letter *o*, whereas a neutral ligand is usually called by the name of the molecule. The exceptions are H_2O (aquo), CO (carbonyl), and NH_3 (ammine). Table 15.4 lists the names of some common ligands.

4. When several ligands of a particular kind are present, we use the Greek prefixes *di-*, *tri-*, *tetra-*, *penta-*, and *hexa-* to name them. Thus, the ligands in the cation $[Co(NH_3)_4Cl_2]^+$ are "tetraamminedichloro." (Note that prefixes are ignored when alphabetizing the ligands.) If the ligand itself contains a Greek prefix, we use the prefixes *bis* (2), *tris* (3), and *tetrakis* (4) to indicate the number of ligands present. For example, the ethylenediamine ligand already contains the prefix *di*, so if a complex contains two such ligands then the name is *bis(ethylenediamine)*.

5. The oxidation number of the metal is written in Roman numerals following the name of the metal. For example, the Roman numeral III is used to indicate the +3 oxidation state of chromium in $[Cr(NH_3)_4Cl_2]^+$, which is called tetraamminedichlorochromium(III) ion.

6. If the complex is an anion, its name ends in *-ate*. For example, in $K_4[Fe(CN)_6]$ the anion $[Fe(CN)_6]^{4-}$ is called hexacyanoferrate(II) ion. The Roman numeral II indicates the oxidation state of iron. Table 15.5 gives the names of anions containing metal atoms.

Table 15.5	
Names of Anions Containing Metal Atoms	
Metal	**Name of Metal in Anionic Complex**
Aluminum	Aluminate
Chromium	Chromate
Cobalt	Cobaltate
Copper	Cuprate
Gold	Aurate
Iron	Ferrate
Lead	Plumbate
Manganese	Managanate
Molybdenum	Molybdate
Nickel	Nickelate
Silver	Argentate
Tin	Stannate
Tungsten	Tungstate
Zinc	Zincate

Examples 15.2 and 15.3 show how to apply these rules to the naming of coordination compounds.

Example 15.2

Write the systematic names of the following coordination compounds: (a) $Ni(CO)_4$, (b) $[Co(NH_3)_4Cl_2]Cl$, (c) $K_3[Fe(CN)_6]$, and (d) $[Cr(en)_3]Cl_3$.

Strategy Follow the preceding procedure for naming coordination compounds and refer to Tables 15.4 and 15.5 for the names of ligands and anions containing metal atoms, respectively.

Solution (a) The CO ligands are neutral, so the Ni atom bears no net charge. The compound is called tetracarbonylnickel(0), or more commonly, nickel tetracarbonyl.

(b) The chloride ion has a -1 charge, so the complex ion has a $+1$ charge, namely, $[Co(NH_3)_4Cl_2]^+$. The ammonia molecule is neutral and each of the two chloride ions bears a -1 charge. As a result, the oxidation number of Co must be $+3$ (to give a net positive charge to the complex ion). The compound is called tetraamminedichlorocobalt(III) chloride.

(c) The complex ion is the anion, and it bears a -3 charge because each of the potassium ions bear a $+1$ charge. The oxidation number of Fe must be $+3$ because each of the six cyanide ions in $[Fe(CN)_6]^{3-}$ bears a -1 charge. The compound is potassium hexacyanoferrate(III)—commonly called potassium ferricyanide.

(d) Each of the three chloride ions bears a -1 charge, so the cation is $[Cr(en)_3]^{3+}$, where *en* is the abbreviation for the ethylenediamine ligand. The *en* ligands are neutral, so the oxidation number of Cr must be $+3$. Because there are three *en* groups present and the name of the ligand already contains the prefix *di* (rule 4), the compound is called *tris*(ethylenediamine)chromium(III) chloride.

Practice Exercise What are the systematic names for (a) $[Cr(H_2O)_4Cl_2]Cl$, (b) $[Pt(NH_3)_3Br]Cl$, (c) $K_3[CoF_6]$, and (d) $[Co(NH_3)_4(NO_2)Cl]NO_3$?

Example 15.3

Write the formulas for the following compounds: (a) pentaamminechlorocobalt(III) chloride, (b) dichlorobis(ethylenediamine)platinum(IV) nitrate, and (c) sodium hexanitrocobaltate(III).

Strategy Follow the preceding procedure, and refer to Tables 15.4 and 15.5 for the names of ligands and anions containing metal atoms, respectively.

Solution (a) The complex cation contains five NH_3 groups, a Cl^- ion, and a Co ion having a $+3$ oxidation number. The net charge of the cation must be $+2$, $[Co(NH_3)_5Cl]^{2+}$. Two chloride anions are needed to balance the positive charges. Therefore, the formula of the compound is $[Co(NH_3)_5Cl]Cl_2$.

(b) There are two chloride ions (-1 each), two *en* groups (neutral), and a Pt ion with an oxidation number of $+4$. The net charge on the cation must be $+2$, $[Pt(en)_2Cl_2]^{2+}$. Two nitrate ions are needed to balance the $+2$ charge of the complex cation. Therefore, the formula of the compound is $[Pt(en)_2Cl_2](NO_3)_2$.

(c) The complex anion contains six nitro groups (-1 each) and a cobalt ion with an oxidation number of $+3$. The net charge on the complex anion must be -3, $[Co(NO_2)_6]^{3-}$. Three sodium cations are needed to balance the -3 charge of the complex anion. Therefore, the formula of the compound is $Na_3[Co(NO_2)_6]$.

—Continued

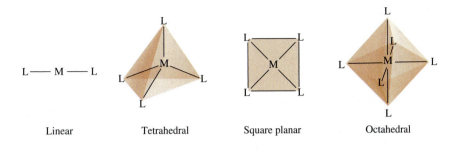

Figure 15.9 Common geometries of complex ions. In each case, M is a metal and L is a monodentate ligand.

Linear Tetrahedral Square planar Octahedral

Continued—

Practice Exercise Write the formulas for the following compounds:
(a) *tris*(ethylenediamine)cobalt(III) sulfate, (b) potassium hexacyanoferrate(III),
(c) potassium tetracyanonickelate(II), and (d) calcium tetrabromodichloroplatinate(II).

Structure of Coordination Compounds

In studying the geometry of coordination compounds, there is often more than one way to arrange the ligands around the central atom. Compounds rearranged in this fashion have distinctly different physical and chemical properties. Figure 15.9 shows four different geometric arrangements for metal atoms with monodentate ligands.

In these diagrams the structure and coordination number of the metal atom relate to each other as follows:

Coordination Number	Structure
2	Linear
4	Tetrahedral or square planar
6	Octahedral

The three-dimensional spatial arrangements in coordination compounds can lead to stereoisomerism. Recall from Section 4.4 that *stereoisomers* are compounds that are made up of the same types and numbers of atoms bonded together in the same sequence, but with different three-dimensional structures. The two types of stereoisomers are *geometric isomers* (isomers that cannot be interconverted without breaking a chemical bond) and *optical isomers* (isomers that are nonsuperimposable mirror images). Coordination compounds may exhibit one or both types of stereoisomerism. Many coordination compounds, however, do not have stereoisomers.

We use the terms *cis* and *trans* to distinguish one geometric isomer of a compound from the other (Section 4.4). *Cis* means that two particular atoms (or groups of atoms) are adjacent to each other, and *trans* means that the atoms (or groups of atoms) are on opposite sides in the structural formula. The *cis* and *trans* isomers of coordination compounds generally have quite different colors, melting points, dipole moments, and chemical reactivities. Figure 15.10 shows the *cis* and *trans* isomers of

cis-tetraamminedichlorocobalt(III) chloride (left) and *trans*-tetraamminedichlorocobalt(III) chloride (right).

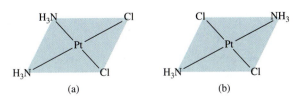

(a) (b)

Figure 15.10 The (a) *cis* and (b) *trans* isomers of diamminedichloroplatinum(II). Note that the two Cl atoms are adjacent to each other in the *cis* isomer and diagonal from each other in the *trans* isomer.

Coordination Compounds in Living Systems

Coordination compounds play many important roles in animals and plants. They are essential in the storage and transport of oxygen, as electron transfer agents, as catalysts, and in photosynthesis. Here we focus on coordination compounds containing iron and magnesium.

Because of its central function as an oxygen carrier for metabolic processes, hemoglobin is probably the most studied of all the proteins. The molecule contains four long, folded protein chains called *subunits*. Hemoglobin carries oxygen in the blood from the lungs to the tissues, where it delivers the oxygen molecules to myoglobin. Myoglobin, which is made up of only one subunit, stores oxygen for metabolic processes in muscle.

The porphine molecule forms an important part of the hemoglobin structure. Upon coordination to a metal, the H^+ ions that are bonded to two of the four nitrogen atoms in porphine are displaced. Complexes derived from porphine are

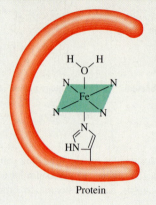

The heme group in hemoglobin. The Fe^{2+} ion is coordinated with the nitrogen atoms of the heme group. The ligand below the porphyrin is the histidine group, which is attached to the protein. The sixth ligand is a water molecule.

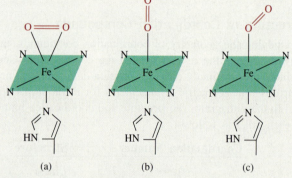

(a) (b) (c)

Three possible ways for molecular oxygen to bind to the heme group in hemoglobin. The structure shown in (a) would have a coordination number of 7, which is considered unlikely for Fe(II) complexes. Although the end-on arrangement in (b) seems the most reasonable, evidence points to the structure in (c) as the correct one.

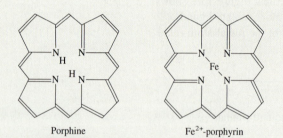

Porphine Fe^{2+}-porphyrin

Simplified structures of the porphine molecule and the Fe^{2+}-porphyrin complex. The dashed lines represent coordinate covalent bonds.

---

diamminedichloroplatinum(II). Although the types of bonds are the same in both isomers (two Pt—N and two Pt—Cl bonds), the spatial arrangements are different. Another example is tetraamminedichlorocobalt(III) ion, shown in Figure 15.11.

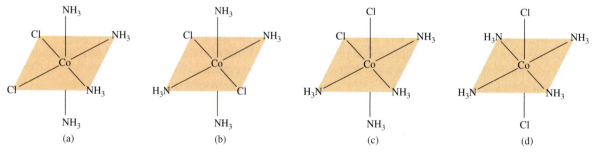

(a)   (b)   (c)   (d)

Figure 15.11 The (a) *cis* and (b) *trans* isomers of the tetraamminedichlorocobalt(III) ion. The structure shown in (c) can be generated by rotating that in (a), and the structure shown in (d) can be generated by rotating that in (b). The ion has only two geometric isomers, (a) [or (c)] and (b) [or (d)].

called *porphyrins,* and the iron-porphyrin combination is called the *heme* group. The iron in the heme group has an oxidation number of +2; it is coordinated to the four nitrogen atoms in the porphine group and also to a nitrogen donor atom in a ligand that is attached to the protein. The sixth ligand is a water molecule, which binds to the Fe^{2+} ion on the other side of the ring to complete the octahedral complex. This hemoglobin molecule is called *deoxyhemoglobin* and imparts a bluish tinge to venous blood. The water ligand can be replaced readily by molecular oxygen to form the red oxyhemoglobin found in arterial blood. Each of the four subunits contains a heme group, so each hemoglobin molecule can bind up to four O_2 molecules. Most modern experimental evidence suggests that the bond between O and Fe in oxyhemoglobin is bent relative to the heme group.

The porphyrin group is a very effective chelating agent that occurs in a number of biological systems. The iron-heme complex is present in another class of proteins, called the *cytochromes.* The iron forms an octahedral complex in these proteins, but because both a histidine and a methionine group are firmly bound to the metal ion, these two amino acids cannot be displaced by oxygen or other ligands. Instead, the cytochromes act as electron carriers, which are essential to metabolic processes. In cytochromes, iron undergoes rapid reversible redox reactions:

$$Fe^{3+} + e^- \rightleftharpoons Fe^{2+}$$

which are coupled to the oxidation of organic molecules such as the carbohydrates.

The chlorophyll molecule, which is necessary for plant photosynthesis, also contains a porphyrin ring, but in this case the metal ion is Mg^{2+} rather than Fe^{2+}.

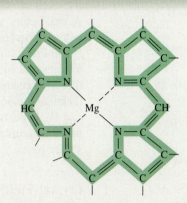

The porphyrin structure in chlorophyll. The dotted lines indicate the coordinate covalent bonds. The electron-delocalized portion of the molecule is shown in color.

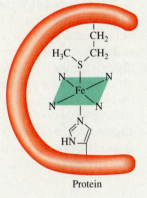

The heme group in cytochrome c. The ligands above and below the porphyrin are a methionine group and a histidine group of the protein, respectively.

As an example of optical isomerism (see Section 4.4) in coordination compounds, Figure 15.12 shows the *cis* and *trans* isomers of dichlorobis(ethylenediamine)cobalt (III) ion and their mirror images. The *trans* isomer and its mirror image are superimposable, but the *cis* isomer and its mirror image are not. The *cis* isomer and its mirror image are, therefore, optical isomers. Unlike geometric isomers, optical isomers have identical physical and chemical properties, except for the way in which they interact with polarized light and in the way they react with other chiral molecules.

Figure 15.12 The (a) *cis* and (b) *trans* isomers of the dichloro bis(ethylenediamine)cobalt(III) ion and their mirror images. If you could rotate the mirror image in (b) 90° clockwise about the vertical position and place the ion over the *trans* isomer, you would find that the two are superimposable. No matter how you rotated the *cis* isomer and its mirror image, however, you could not superimpose one on the other.

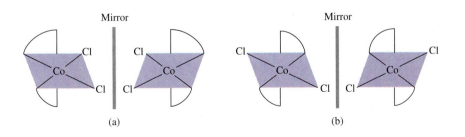

15.3 | Bonding in Coordination Compounds Can Be Described by Crystal Field Theory

A satisfactory theory of bonding in coordination compounds must account for their color and magnetism as well as their stereochemistry and bond strength. So far, no single theory can do all that. Instead, several different approaches have been applied to transition metal complexes. We will discuss only one of them here—crystal field theory—because it accounts for both the color and magnetic properties of many coordination compounds.

We begin the discussion of crystal field theory with the most straightforward case, namely, complex ions with octahedral geometry. We then apply it to tetrahedral and square-planar complexes.

Crystal Field Splitting in Octahedral Complexes

The name "crystal field" is associated with the theory used to explain the properties of solid, crystalline substances. The same theory is used to study coordination, compounds.

Crystal field theory explains the bonding in complex ions purely in terms of electrostatic forces. In a complex ion, two types of electrostatic interaction come into play. One is the attraction between the positive metal ion and the negatively charged ligand or the partially negatively charged end of a polar ligand. This is the force that binds the ligands to the metal. The second type of interaction is electrostatic repulsion between the lone pairs on the ligands and the electrons in the *d* orbitals of the metals.

As we saw in Chapter 3, *d* orbitals have different orientations, but in the absence of external disturbance, they all have the same energy. In an octahedral complex, a central metal atom is surrounded by six lone pairs of electrons (on the six ligands), so all five *d* orbitals experience electrostatic repulsion. The magnitude of this repulsion depends on the orientation of the particular *d* orbital. Figure 15.13 shows, for example, that the lobes of the $d_{x^2-y^2}$ orbital point toward the corners of the octahedron along the *x* and *y* axes, where the lone-pair electrons are positioned. Thus, an electron residing in this orbital would experience a greater repulsion from the ligands than an electron would in, say, the d_{xy} orbital. For this reason, the energy of the $d_{x^2-y^2}$ orbital is increased relative to the d_{xy}, d_{yz}, and d_{xz} orbitals. The energy of the d_{z^2} orbital is also greater, because its lobes are pointed at the ligands along the *z* axis.

Figure 15.13 The five *d* orbitals in an octahedral environment. The metal atom (or ion) is at the center of the octahedron, and the six lone pairs on the donor atoms of the ligands are at the corners.

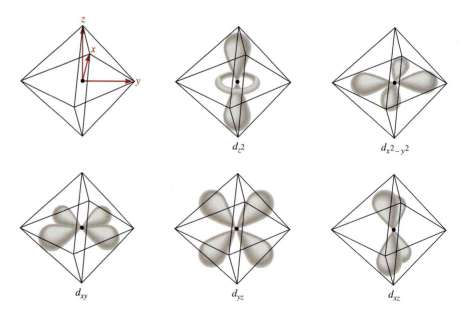

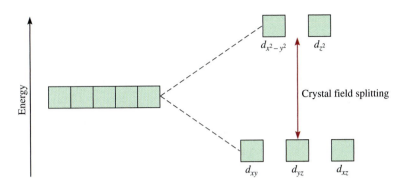

Figure 15.14 Crystal field splitting between *d* orbitals in an octahedral complex.

As a result of these metal-ligand interactions, the five *d* orbitals in an octahedral complex are split between two sets of energy levels: a higher level with two orbitals ($d_{x^2-y^2}$ and d_{z^2}) having the same energy and a lower level with three equal energy orbitals (d_{xy}, d_{yz}, and d_{xz}), as shown in Figure 15.14. The ***crystal field splitting*** (Δ) is *the energy difference between the two sets of d orbitals in a metal atom when ligands are present.* The magnitude of Δ, which depends on the metal and the nature of the ligands, has a direct effect on the color and magnetic properties of complex ions.

Color

In Chapter 1, we learned that white light, such as sunlight, is a combination of all colors. A substance appears black if it absorbs all the visible light that strikes it. It appears white or colorless if it absorbs no visible light. An object appears green if it absorbs all light but reflects the green component. An object also looks green if it reflects all colors except red, the *complementary* color of green (Figure 15.15).

What is true for reflected light also applies to transmitted light (that is, the light that passes through a medium, such as a solution). Consider the hydrated cupric ion ($[Cu(H_2O)_6]^{2+}$), for example. It absorbs light in the orange region of the spectrum, so an aqueous solution of $CuSO_4$ appears blue to us. Recall from Chapter 1 that when the energy of a photon is equal to the difference between the ground state and an excited state, absorption occurs as the photon strikes the atom (or ion or compound) and an electron is promoted to a higher level. This concept makes it possible to calculate the energy change involved in the electron transition. The energy of a photon, given by Equation 1.3, is:

$$\Delta E = h\nu$$

where *h* represents Planck's constant (6.626×10^{-34} J s) and *ν* is the frequency of the radiation, which is 5.00×10^{14} s^{-1} for the wavelength of orange light (600 nm). Here $E = \Delta$, the crystal field splitting, so we have

$$\begin{aligned}
\Delta &= h\nu \\
&= (6.626 \times 10^{-34} \text{ J s})(5.00 \times 10^{14} \text{ s}^{-1}) \\
&= 3.32 \times 10^{-19} \text{ J}
\end{aligned}$$

(This is the energy absorbed by *one* ion.) If the wavelength of the photon absorbed by an ion lies outside the visible region, then the transmitted light looks the same (to us) as the incident light—white—and the ion appears colorless.

The best way to measure crystal field splitting is to use spectroscopy to determine the wavelength at which light is absorbed. The $[Ti(H_2O)_6]^{3+}$ ion provides a

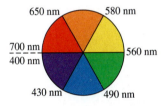

Figure 15.15 A color wheel showing appropriate wavelengths. Complementary colors, such as red and green, are on opposite sides of the wheel.

Figure 15.16 (a) The process of photon absorption, and (b) a graph of the absorption spectrum of $[Ti(H_2O)_6]^{3+}$. The energy of the incoming photon is equal to the crystal field splitting. The maximum absorption peak in the visible region occurs at 498 nm.

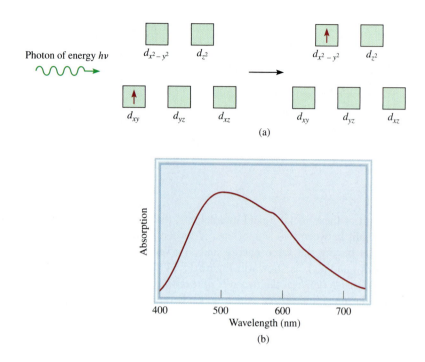

(a)

(b)

straightforward example because Ti^{3+} has only one $3d$ electron (Figure 15.16). The $[Ti(H_2O)_6]^{3+}$ ion absorbs light in the visible region of the spectrum (Figure 15.17).

The wavelength corresponding to maximum absorption in Figure 15.16(b) is 498 nm. Thus, the crystal field splitting can be calculated as follows. We start by writing

$$\Delta = h\nu \tag{15.1}$$

Also

$$\nu = \frac{c}{\lambda}$$

where c is the speed of light and λ is the wavelength. Therefore,

$$\Delta = \frac{hc}{\nu} = \frac{(6.626 \times 10^{-34} \text{ J s})(3.00 \times 10^8 \text{ m s}^{-1})}{(498 \text{ nm})(1 \times 10^{-9}\text{m nm}^{-1})}$$

$$= 3.99 \times 10^{-19} \text{ J}$$

Figure 15.17 The colors of some of the first-row transition metal ions in solution. From left to right: Ti^{3+}, Cr^{3+}, Mn^{2+}, Fe^{3+}, Co^{2+}, Ni^{2+}, and Cu^{2+}. The Sc^{3+} and V^{5+} ions are colorless.

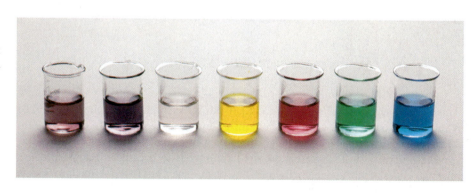

This is the energy required to excite *one* $[Ti(H_2O)_6]^{3+}$ ion. To express this energy difference in the more convenient units of kilojoules per mole, we write

$$\Delta = (3.99 \times 10^{-19} \text{ J ion}^{-1})(6.022 \times 10^{23} \text{ ions mol}^{-1})$$
$$= 2.40 \times 10^5 \text{ J mol}^{-1}$$
$$= 240 \text{ kJ mol}^{-1}$$

Aided by spectroscopic data for a number of complexes, all having the same metal ion but different ligands, chemists have calculated the crystal field splitting for each ligand and established a ***spectrochemical series,*** which is *a list of ligands arranged in increasing order of their abilities to split the d-orbital energy levels:*

The order of the spectrochemical series is the same no matter which metal atom (or ion) is present.

$$I^- < Br^- < Cl^- < OH^- < F^- < H_2O < NH_3 < en < CN^- < CO$$

These ligands are arranged in the order of increasing value of Δ. CO and CN^-, which cause a large splitting of the *d*-orbital energy levels, are called *strong-field ligands.* The halide ions and hydroxide ion, which split the *d* orbitals to a lesser extent, are called *weak-field ligands.*

Magnetic Properties

The magnitude of the crystal field splitting also determines the magnetic properties of a complex ion. The $[Ti(H_2O)_6]^{3+}$ ion, having only one *d* electron, is always paramagnetic. For an ion with several *d* electrons, however, the situation is not as clear cut. Consider, for example, the octahedral complexes $[FeF_6]^{3-}$ and $[Fe(CN)_6]^{3-}$ (Figure 15.18).

Both contain Fe^{+3}, the electron configuration of which is $[Ar]3d^5$, but there are two possible ways to distribute the five *d* electrons among the *d* orbitals. According to Hund's rule (see Section 2.2), maximum stability is reached when the electrons are placed in five separate orbitals with parallel spins. This arrangement can be achieved only if two of the five electrons are promoted to the higher-energy $d_{x^2-y^2}$ and d_{z^2} orbitals. No such energy input is needed if all five electrons enter the d_{xy}, d_{yz}, and d_{xz} orbitals. According to the Pauli exclusion principle (Section 2.1), four of these electrons must pair up, leaving only one unpaired electron in this arrangement.

Figure 15.19 shows the distribution of electrons among the *d* orbitals that results in low-spin and high-spin complexes. The actual arrangement of the electrons is determined by the amount of stability gained by having maximum parallel spins versus the investment in energy required to promote electrons to higher *d* orbitals. Because F^- is a weak-field ligand, the five *d* electrons enter five separate *d* orbitals with parallel spins to create a high-spin complex (see Figure 15.19). The cyanide ion is a strong-field

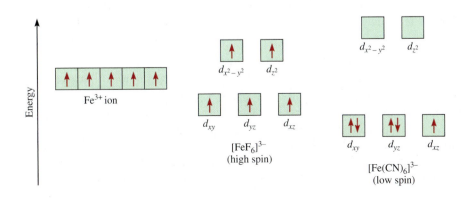

Figure 15.18 Energy-level diagrams for the Fe^{3+} ion and for the $[FeF_6]^{3-}$ and $[Fe(CN)_6]^{3-}$ complex ions.

Figure 15.19 Orbital diagrams for the high-spin and low-spin octahedral complexes corresponding to the electron configurations d^4, d^5, d^6, and d^7. No such distinctions can be made for d^1, d^2, d^3, d^8, d^9, and d^{10}.

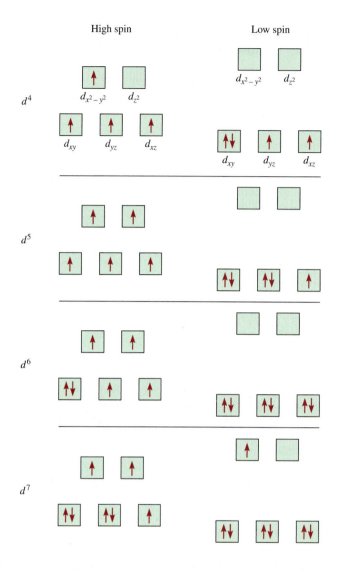

ligand, on the other hand, so the energy needed to promote two of the five d electrons to higher d orbitals is too much, and therefore a low-spin complex is formed.

The actual number of unpaired electrons (or spins) in a complex ion can be found by using a technique called *electron spin resonance spectroscopy* (ESR), and in general, experimental findings support predictions based on crystal field splitting. However, a distinction between low- and high-spin complexes can be made only if the metal ion contains more than three and fewer than eight d electrons, as shown in Figure 15.19.

The magnetic properties of a complex ion depend upon the number of unpaired electrons present.

Tetrahedral and Square-Planar Complexes

So far we have concentrated on octahedral complexes. The splitting of the d orbital energy levels in tetrahedral and square-planar complexes can also be accounted for satisfactorily by the crystal field theory. In fact, the splitting pattern for a tetrahedral ion is just the reverse of that for octahedral complexes. In this case, the d_{xy}, d_{yz}, and d_{xz} orbitals are more closely directed at the ligands and therefore have more energy than the $d_{x^2-y^2}$ and d_{z^2} orbitals (Figure 15.20). Most tetrahedral complexes are high spin. Presumably, the tetrahedral arrangement reduces the magnitude of the metal-ligand

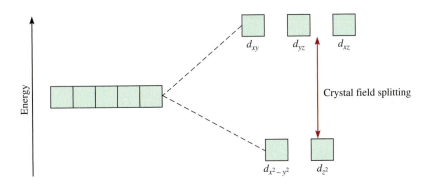

Figure 15.20 Crystal field splitting between d orbitals in a tetrahedral complex.

interactions, resulting in a smaller Δ value. This is a reasonable assumption because the number of ligands is smaller in a tetrahedral complex.

As Figure 15.21 shows, the splitting pattern for square-planar complexes is the most complicated. The $d_{x^2-y^2}$ orbital possesses the highest energy (as in the octahedral case), and the d_{xy} orbital the next highest. However, the relative placement of the d_{z^2} and the d_{xz} and d_{yz} orbitals cannot be determined simply by inspection and must be calculated.

Ligand Field Theory

Crystal field theory, although quite successful at explaining the spectral and magnetic properties of many coordination compounds, ignores the covalent character of metal-ligand bonds because it is based solely on electrostatic interactions. Thus, it would be unable to predict the fact that a neutral ligand, such as CO or ethylenediamine, can have a larger crystal field splitting than an ionic ligand, such as Cl^- or F^-. In order to take into account the covalent character of metal-ligand bonds, a molecular orbital approach (see Sections 3.5 and 4.5) to the electronic structure of coordination compounds, known as *ligand field theory,* has been developed.

Ligand field theory is based on the idea that atomic orbitals that are close in energy will mix more effectively in molecular orbitals than those that are far apart. Consider, for example, the octahedral complex ion FeF_6^{3-}. From Figure 15.13, we saw that the $d_{x^2-y^2}$ and d_{z^2} atomic orbitals are oriented toward the ligands. As such, the $3d_{x^2-y^2}$ and $3d_{z^2}$ orbitals on the metal center will mix with the ligand lone-pair orbitals to form two bonding and two antibonding σ molecular orbitals. The remaining $3d$ orbitals—$3d_{xy}$, $3d_{yz}$, and $3d_{xz}$—are oriented in between the ligands and, because of

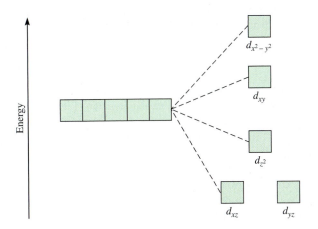

Figure 15.21 Energy-level diagram for a square-planar complex. Because there are more than two energy levels, we cannot define crystal field splitting as we can for the octahedral and tetrahedral complexes.

Figure 15.22 Molecular-orbital energy-level diagram for the octahedral complex FeF_6^{3-}. The σ_d^* molecular orbitals are essentially pure $d_{x^2-y^2}$ and d_{z^2} orbitals of the central metal ion. In this complex, the crystal field splitting (Δ) between the three non-bonding orbitals (d_{xy}, d_{yz}, and d_{xz}) and the two σ_d^* orbitals is small (F^- *is a weak-field ligand*) so Hund's rule applies, and these five orbitals each contain one electron with all the spins parallel.

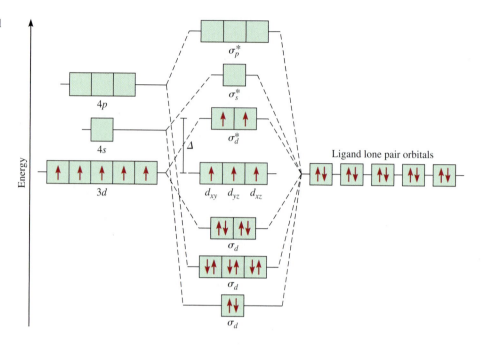

minimal overlap, they will not mix with the ligand lone-pair orbitals, that is, they will be nonbonding orbitals. The $4s$ orbital on the metal center is spherical in shape and will overlap with all of the lone-pair ligand orbitals to form a pair of bonding and antibonding σ molecular orbitals. The remaining three $4p$ orbitals on the metal center will overlap individually with the lone-pair ligand orbitals along the x, y, and z axes. So we have the six $4s$, $4p$, $3d_{x^2-y^2}$, and $3d_{z^2}$ orbitals of the metal center mixing with the ligand lone-pair orbitals to form six bonding and six antibonding σ molecular orbitals oriented along the six vertices of an octahedron, with three nonbonding orbitals corresponding to the $3d_{xy}$, $3d_{yz}$, and $3d_{xz}$ orbitals.

The MO energy-level diagram for FeF_6^{3-} is shown in Figure 15.22. The arrangement of the highest occupied molecular orbitals ($3d_{xy}$, $3d_{yz}$, $3d_{xz}$, and the two σ_d^* orbitals) is identical to that predicted by crystal field theory with the $3d_{x^2-y^2}$ and $3d_{z^2}$ orbitals replaced with the two σ_d^* molecular orbitals. However, the MO-based ligand field theory is more complete as it provides an understanding of the dependence of the crystal field splitting on the ligand type. In this octahedral complex, the $3d_{xy}$, $3d_{yz}$, and $3d_{xz}$ orbitals are not oriented along the metal-ligand bond axes, but instead are oriented in between the ligands. As such, they will interact most strongly with the orbitals of the ligand (p or π) that are perpendicular to the metal-ligand bond. In the case of monatomic ionic ligands, such as OH^-, F^-, Cl^-, Br^-, and I^-, the electrostatic repulsion from the full p orbitals on these ligands raises the energy of the $3d_{xy}$, $3d_{yz}$, and $3d_{xz}$ orbitals resulting in a reduction of the crystal field splitting (Δ)—thus, these species are weak-field ligands. This repulsion is stronger if the p electrons are less tightly bound to the ligand, that is, if the ligand is less electronegative, thus explaining why I^- has a lower crystal field splitting than F^- in the spectrochemical series. On the other hand, ligands with unoccupied antibonding π^* orbitals, such as CO or CN^-, tend to lower the energy of the $3d_{xy}$, $3d_{yz}$, and $3d_{xz}$ orbitals, leading to an increase in the crystal field splitting. Such species are strong-field ligands. The lowering of the $3d_{xy}$, $3d_{yz}$, and $3d_{xz}$ orbital energies in this fashion is called **back bonding** and is due to mixing with the unoccupied π^* orbitals on the ligand, which are oriented away from the ligand double (or triple bond) and toward the $3d_{xy}$, $3d_{yz}$, and $3d_{xz}$ orbitals of the metal center (see Figure 3.18). Neutral ligands, such as NH_3 and H_2O, neither have a strong repulsion from p orbitals nor back bonding of the π orbitals and so are generally intermediate between weak- and strong-field ligands in the spectrochemical series.

15.4 | The Reactions of Coordination Compounds Have a Wide Number of Useful Applications

Complex ions undergo ligand exchange (or substitution) reactions in solution. The rates of these reactions vary widely, depending on the nature of the metal ion and the ligands.

In studying ligand exchange reactions, it is often useful to distinguish between the stability of a complex ion and its tendency to react, which we call *kinetic lability*. Stability in this context is a thermodynamic property, which is measured in terms of the species' formation constant K_f. For example, the complex ion tetracyanonickelate(II) is stable because it has a large formation constant ($K_f \approx 1 \times 10^{30}$):

$$Ni^{2+} + 4CN^- \rightleftharpoons [Ni(CN_4)]^{2-}$$

At equilibrium, there is a distribution of *CN$^-$ ions in the complex ion.

By using cyanide ions labeled with the radioactive isotope carbon-14, chemists have shown that $[Ni(CN)_4]^{2-}$ undergoes ligand exchange very rapidly in solution. The following equilibrium is established almost as soon as the species are mixed:

$$[Ni(CN)_4]^{2-} + 4*CN^- \rightleftharpoons [Ni(*CN)_4]^{2-} + 4CN^-$$

where the asterisk denotes a C-14 atom. Complexes like the tetracyanonickelate(II) ion are **labile complexes** because they *undergo rapid ligand exchange reactions*. Thus, a thermodynamically stable species (that is, one that has a large formation constant) is not necessarily unreactive.

A complex that is thermodynamically *unstable* in acidic solution is $[Co(NH_3)_6]^{3+}$. The equilibrium constant for the following reaction is about 1×10^{20}:

$$[Co(NH_3)_6]^{3+} + 6H^+ + 6H_2O \rightleftharpoons [Co(H_2O)_6]^{3+} + 6NH_4^+$$

When equilibrium is reached, the concentration of the $[Co(NH_3)_6]^{3+}$ ion is very low. It takes several days to achieve equilibrium, however, because the $[Co(NH_3)_6]^{3+}$ ion is an **inert complex**—*a complex ion that undergoes very slow exchange reactions* (on the order of hours or even days). Thus, a thermodynamically unstable species is not necessarily chemically reactive. The rate of reaction is determined by the energy of activation, which is high in this case.

Most complex ions containing Co^{3+}, Cr^{3+}, and Pt^{2+} are kinetically inert. Because they exchange ligands very slowly, they are easy to study in solution. As a result, our knowledge of the bonding, structure, and isomerism of coordination compounds has come largely from studies of these compounds with these ligands.

Applications of Coordination Compounds

Coordination compounds are found in living systems and have many uses in the home, in industry, and in medicine. We describe a few examples here and in the insert on page 784.

Metallurgy

Coordination compounds can be used to extract and purify metals, such as gold and silver. Although gold and silver are usually found in the uncombined state in nature, in other metal ores they may be present in relatively small concentrations and are more difficult to extract. In a typical process, the crushed ore is treated with an aqueous

Figure 15.23 A cyanide pond for extracting gold from metal ore.

cyanide solution in the presence of air to dissolve the gold by forming the soluble complex ion $[Au(CN)_2]^-$:

$$4Au(s) + 8CN^-(aq) + O_2(g) + 2H_2O(l) \longrightarrow 4[Au(CN)_2]^-(aq) + 4OH^-(aq)$$

The complex ion $[Au(CN)_2]^-$ (along with some cation, such as Na^+) is separated from other insoluble materials by filtration and treated with an electropositive metal such as zinc to recover the gold:

$$Zn(s) + 2[Au(CN)_2]^-(aq) \longrightarrow [Zn(CN)_4]^{2-}(aq) + 2Au(s)$$

Figure 15.23 shows an aerial view of a "cyanide pond" used for the extraction of gold.

Therapeutic Chelating Agents

Recall from Section 15.2 that the chelating agent EDTA is used in the treatment of lead poisoning. Certain platinum-containing compounds can effectively inhibit the growth of cancerous cells. A specific case is discussed in the insert on p. 795.

Chemical Analysis

Although EDTA has a great affinity for a large number of metal ions (especially +2 and +3 ions), other chelates are more selective in their binding. For example, dimethylglyoxime,

$$
\begin{array}{c}
H_3C \\
\diagdown \\
C{=}N{-}OH \\
| \\
C{=}N{-}OH \\
\diagup \\
H_3C
\end{array}
$$

forms an insoluble brick-red solid with Ni^{2+} and an insoluble bright-yellow solid with Pd^{2+}. These characteristic colors are used in qualitative analysis to identify nickel and palladium. Furthermore, the quantities of ions present can be determined by gravimetric

Cisplatin—An Anticancer Drug

Luck often plays a role in major scientific breakthroughs, but it takes an alert and well-trained person to recognize the significance of an accidental discovery and to take full advantage of it. Such was the case when, in 1964, the biophysicist Barnett Rosenberg and his research group at Michigan State University were studying the effect of an electric field on the growth of bacteria. They suspended a bacterial culture between two platinum electrodes and passed an electric current through it. To their surprise, they found that after an hour or so the bacteria cells ceased dividing. It did not take long for the group to determine that a platinum-containing substance extracted from the bacterial culture inhibited cell division.

Rosenberg reasoned that the platinum compound might be useful as an anticancer agent because cancer involves uncontrolled division of the affected cells, so he set out to identify the substance. Given the presence of ammonia and chloride ions in solution during electrolysis, Rosenberg synthesized a number of platinum compounds containing ammonia and chlorine. The one that proved most effective at inhibiting cell division was *cis*-diamminedichloroplatinum(II) [$Pt(NH_3)_2Cl_2$], also called cisplatin.

Cisplatin works by chelating DNA (deoxyribonucleic acid), the molecule that contains the genetic code. During cell division, the double-stranded DNA splits into two single strands, which must be accurately copied in order for the new cells to be identical to their parent cell. X-ray studies show that cisplatin binds to DNA by forming cross-links in which the two chlorides on cisplatin are replaced by nitrogen atoms in the adjacent guanine bases on the *same* strand of the DNA. (Guanine is one of the four bases in DNA. See Figure 16.27.) Consequently, the double-stranded structure assumes a bent configuration at the binding site. Scientists believe that this structural distortion is a key factor in inhibiting replication. The damaged cell is then destroyed by the immune system of the body. Because the binding of cisplatin to DNA requires both Cl atoms to be on the same side of the complex, the *trans* isomer of the compound is totally ineffective as an anticancer drug. Unfortunately, cisplatin can cause serious side effects, including severe kidney damage. Therefore, ongoing research efforts are directed toward finding related complexes that destroy cancer cells with less harm to healthy tissue.

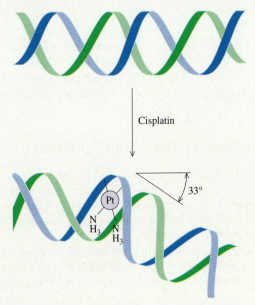

Cisplatin destroys the cancer cells' ability to reproduce by changing the configuration of their DNA. It binds to two sites on a strand of DNA, causing it to bend about 33° away from the rest of the strand.

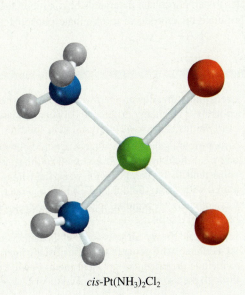

cis-$Pt(NH_3)_2Cl_2$

Figure 15.24 Structure of nickel dimethylglyoxime. Note that the overall structure is stabilized by hydrogen bonds.

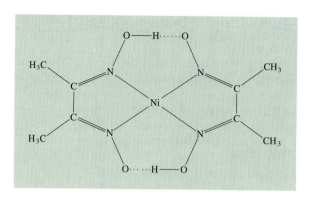

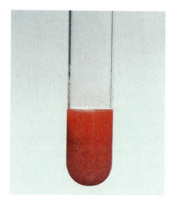

An aqueous suspension of *bis*(dimethylglyoximato)nickel(II)

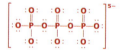

Tripolyphosphate ion

analysis as follows: To a solution containing Ni^{2+} ions, we add an excess of dimethylglyoxime reagent, and a brick-red precipitate forms. The precipitate is then filtered, dried, and weighed. Knowing the formula of the complex (Figure 15.24), we can readily calculate the amount of nickel present in the original solution.

Detergents

The cleansing action of soap in hard water is hampered by the reaction of the Ca^{2+} ions in the water with the soap molecules to form insoluble salts or curds. In the late 1940s, the detergent industry introduced a "builder," usually sodium tripolyphosphate, to circumvent this problem. The tripolyphosphate ion is an effective chelating agent that forms stable, soluble complexes with Ca^{2+} ions. Sodium tripolyphosphate revolutionized the detergent industry. However, because phosphates are plant nutrients, phosphate-containing wastewater discharged into rivers and lakes caused algae to grow, depleting the waters of oxygen. As a result, most or all aquatic life eventually died. This process, called *eutrophication,* led many states to ban phosphate detergents beginning in the 1970s, thus forcing manufacturers to reformulate their products to eliminate phosphates.

Summary of Facts and Concepts

Section 15.1

▶ Transition metals usually have incompletely filled *d* subshells and a pronounced tendency to form complexes. Compounds that contain complex ions are called coordination compounds.

▶ The first-row transition metals (scandium to copper) are the most common of all the transition metals; their chemistry is characteristic, in many ways, of the entire group.

Section 15.2

▶ Complex ions consist of a metal ion surrounded by ligands. The donor atoms in the ligands each contribute an electron pair to the central metal ion in a complex.

▶ Coordination compounds may display geometric and/or optical isomerism.

Section 15.3

▶ Crystal field theory explains bonding in complexes in terms of electrostatic interactions. According to crystal field theory, the *d* orbitals are split into two higher-

energy and three lower-energy orbitals in an octahedral complex. The energy difference between these two sets of *d* orbitals is the basis of crystal field splitting.

▶ Strong-field ligands cause a large crystal field splitting, and weak-field ligands cause a small splitting. Electron spins tend to be parallel with weak-field ligands and paired with strong-field ligands, where a greater investment of energy is required to promote electrons into the high-lying *d* orbitals.

▶ Ligand field theory corrects some of the defects of crystal field theory by going beyond electostatic interactions to include the effects of covalent metal-ligand bonding through the use of molecular orbital theory.

Section 15.4

▶ Complex ions undergo ligand exchange reactions in solution.

▶ Coordination compounds find application in many different areas, for example, as antidotes for metal poisoning and in chemical analysis.

Key Words

back bonding, p. 792
chelating agent, p. 778
coordination compound, p. 777

coordination number, p. 778
crystal field splitting, p. 787
crystal field theory, p. 786

donor atom, p. 778
inert complex, p. 793
labile complex, p. 793

ligand, p. 778
ligand field theory, p. 791
spectrochemical series, p. 789

Problems

Transition Metals Have Electron Configurations with Incomplete *d* or *f* Shells

15.1 Why is zinc not considered a transition metal?

15.2 Explain why atomic radii decrease very gradually from scandium to copper.

15.3 Without referring to the text, write the ground-state electron configurations of the first-row transition metals. Explain any irregularities.

15.4 Write the electron configurations of the following ions: V^{5+}, Cr^{3+}, Mn^{2+}, Fe^{2+}, Cu^{2+}, Sc^{3+}, and Ti^{4+}.

15.5 Why do transitional metals have more oxidation states than other elements?

15.6 Why does chromium seem to be less reactive than its standard reduction potential suggests?

Transition Metals Can Form a Variety of Coordination Compounds

15.7 Complete the following statements for the complex ion $[Co(en)_2(H_2O)CN]^{2+}$. (a) Among the ligands, *en* is the abbreviation for _____. (b) The oxidation number of Co is _____. (c) The coordination number of Co is _____. (d) _____ is a bidentate ligand.

15.8 Complete the following statements for the complex ion $[Cr(C_2O_4)_2(H_2O)_2]^-$. (a) The oxidation number of Cr is _____. (b) The coordination number of Cr is _____. (c) _____ is a bidentate ligand.

15.9 Calculate the oxidation numbers of the metals in the following species: (a) $K_3[Fe(CN)_6]$, (b) $K_3[Cr(C_2O_4)_3]$, and (c) $[Ni(CN)_4]_2$.

15.10 Give the oxidation numbers of the metals in the following species: (a) Na_2MoO_4, (b) $MgWO_4$, and (c) $Fe(CO)_5$.

15.11 What are the systematic names for the following ions and compounds: (a) $[Co(NH_3)_4Cl_2]$, (b) $[Co(en)_2Br_2]$, (c) $Cr(NH_3)_3Cl_3$, and (d) $[Co(NH_3)_6]Cl_3$?

15.12 What are the systematic names for the following ions and compounds: (a) $[cis\text{-}Co(en)_2Cl_2]$, (b) $[Pt(NH_3)_5Cl]Cl_3$, and (c) $[Co(NH_3)_5Cl]Cl_2$?

15.13 Write the formulas for each of the following ions and compounds: (a) tetrahydroxozincate(II), (b) pentaaquochlorochromium(III) chloride, (c) tetrabromocuprate(II), and (d) ethylenediaminetetraacetatoferrate(II).

15.14 Write the formulas for each of the following ions and compounds: (a) bis(ethylenediamine)dichlorochromium(III), (b) pentacarbonyliron(0), (c) potassium tetracyanocuprate(II), and (d) tetraammineaquochlorocobalt(III) chloride.

15.15 The complex ion $[Ni(CN)_2Br_2]^{2-}$ has a square-planar geometry. Draw the structures of the geometric isomers of this complex.

15.16 How many geometric isomers are possible for each of the following species: (a) $[Co(NH_3)_2Cl_4]^-$ and (b) $[Co(NH_3)_3Cl_3]$?

15.17 Draw structures of all the geometric and optical isomers of each of the following cobalt complexes: (a) $[Co(NH_3)_6]^{3+}$, (b) $[Co(NH_3)_5Cl]^{2+}$, and (c) $[Co(C_2O_4)_3]^{3-}$

15.18 Draw structures of all the geometric and optical isomers of each of the following cobalt complexes: (a) $[Co(NH_3)_4Cl_2]^+$ and (b) $[Co(en)_3]^{3+}$.

Bonding in Coordination Compounds Can Be Described by Crystal Field Theory

15.19 The $[Ni(CN)_4]^{2-}$ ion, which has square-planar geometry, is diamagnetic, whereas the $[NiCl_4]^{2-}$ ion, which has tetrahedral geometry, is paramagnetic. Show the crystal field splitting diagrams for those two complexes.

15.20 Transition metal complexes containing CN^- ligands are often yellow, whereas those containing H_2O ligands are often green or blue. Explain.

15.21 Predict the number of unpaired electrons in the following complex ions: (a) $[Cr(CN)_6]^{4-}$ and (b) $[Cr(H_2O)_6]^{2+}$.

15.22 The absorption maximum for the complex ion $[Co(NH_3)_6]^{3+}$ occurs at 470 nm. (a) Predict the color of the complex, and (b) calculate the crystal field splitting in kJ mol^{-1}.

15.23 From each of the following pairs, choose the complex that absorbs light at a longer wavelength: (a) $[Co(NH_3)_6]^{2+}$, $[Co(H_2O)_6]^{2+}$; (b) $[FeF_6]^{3-}$, $[Fe(CN)_6]^{3-}$; and (c) $[Cu(NH_3)_4]^{2+}$, $[CuCl_4]^{2-}$.

15.24 A solution made by dissolving 0.875 g of $Co(NH_3)_4Cl_3$ in 25.0 g of water freezes at –0.56°C. Calculate the number of moles of ions produced when 1 mole of $Co(NH_3)_4Cl_3$ is dissolved in water and suggest a structure for the complex ion present in this compound.

The Reactions of Coordination Compounds Have a Wide Number of Useful Applications

15.25 Oxalic acid ($H_2C_2O_4$) is sometimes used to clean rust stains from sinks and bathtubs. Explain the chemistry behind this cleaning action.

15.26 The $[Fe(CN)_6]^{3-}$ complex is more labile than the $[Fe(CN)_6]^{4-}$ complex. Suggest an experiment that would prove that $[Fe(CN)_6]^{3-}$ is a labile complex.

15.27 Aqueous copper(II) sulfate solution is blue in color, but a green precipitate forms when aqueous potassium fluoride is added. When aqueous potassium chloride is added instead, a bright-green solution forms. Explain what is happening in these two cases.

15.28 When aqueous potassium cyanide is added to a solution of copper(II) sulfate, a white precipitate, soluble in an excess of potassium cyanide, forms. No precipitate forms when hydrogen sulfide is bubbled through the solution at this point. Explain.

15.29 A concentrated aqueous copper(II) chloride solution is bright green, but the solution turns light blue, when diluted with water. Explain.

15.30 In a dilute nitric acid solution, Fe^{3+} reacts with thiocyanate ion (SCN^-) to form a dark-red complex:

$$[Fe(H_2O)_6]^{3+} + SCN^- \rightleftharpoons H_2O + [Fe(H_2O)_5NCS]^{2+}$$

The equilibrium concentration of $[Fe(H_2O)_5NCS]^{2+}$ may be determined by the strength of the dark color of the solution (measured by a spectrometer). In one such experiment, 1.0 mL of 0.20 M $Fe(NO_3)_3$ was mixed with 1.0 mL of 1.0×10^{-3} M KSCN and 8.0 mL of dilute HNO_3. The color of the solution quantitatively indicated that the $[Fe(H_2O)_5NCS]^{2+}$ concentration was 7.3×10^5 M. Calculate the formation constant for $[Fe(H_2O)_5NCS]^{2+}$.

Additional Problems

15.31 As we read across the first-row transition metals from left to right, the +2 oxidation state becomes more stable in comparison with the +3 state. Why?

15.32 Which is a stronger oxidizing agent in aqueous solution, Mn^{3+} or Cr^{3+}? Explain your choice.

15.33 Carbon monoxide binds to Fe in hemoglobin some 200 times more strongly than oxygen. This is why CO is so toxic. The metal-to-ligand sigma bond is formed by donating a lone pair from the donor atom to an empty sp^3d^2 orbital on Fe. (a) On the basis of electronegativities, would you expect the C or O atom to form the bond to Fe? (b) Draw a diagram illustrating the overlap of the orbitals involved in the bonding.

15.34 What are the oxidation states of Fe and Ti in the ore ilmenite ($FeTiO_3$)? (*Hint:* Look up the ionization energies of Fe and Ti in Table 15.1; the fourth ionization energy of Ti is 4180 kJ mol^{-1}.)

15.35 A student has prepared a cobalt complex that has one of the following three structures: $[Co(NH_3)_6]Cl_3$, $[Co(NH_3)_5Cl]Cl_2$, or $[Co(NH_3)_4Cl_2]Cl$. Explain how the student could distinguish between these possibilities by performing an electrical conductance experiment. At the student's disposal are three

strong electrolytes—NaCl, $MgCl_2$, and $FeCl_3$—which may be used for comparison purposes.

15.36 Chemical analysis shows that hemoglobin contains 0.34 percent of Fe by mass. What is the minimum possible molar mass of hemoglobin? The actual molar mass of hemoglobin is about 65,000 g. How do you account for the discrepancy between your minimum value and the actual value?

15.37 Explain the following facts: (a) Copper and iron have several oxidation states, whereas zinc has only one. (b) Copper and iron form colored ions, whereas zinc does not.

15.38 A student in 1895 prepared three coordination compounds containing chromium that had the following properties:

Formula	Color	Cl⁻ Ions in Solution per Formula Unit
$CrCl_3 \cdot 6H_2O$	Violet	3
$CrCl_3 \cdot 6H_2O$	Light green	2
$CrCl_3 \cdot 6H_2O$	Dark green	1

Write modern formulas for these compounds and suggest a method for confirming the number of Cl^- ions present in solution in each case. (*Hint:* Some of the compounds may exist as hydrates.)

15.39 The formation constant for the reaction

$$Ag^+ + 2NH_3 \rightleftharpoons [Ag(NH_3)_2]$$

is 1.5×10^7 and that for the reaction

$$Ag + 2CN^- \rightleftharpoons [Ag(CN)_2]$$

is 1.0×10^{21} at 25°C. Calculate the equilibrium constant and $\Delta G°$ at 25°C for the reaction

$$[Ag(NH_3)_2] + 2CN^- \rightleftharpoons [Ag(CN)_2] + 2NH_3$$

15.40 From the standard reduction potentials listed in Table 13.1 for Zn/Zn^{2+} and Cu/Cu^{2+}, calculate $\Delta G°$ and the equilibrium constant for the following reaction:

$$Zn(s) + 2Cu^{2+}(aq) \longrightarrow Zn^{2+}(aq) + 2Cu^+(aq)$$

15.41 Using the standard reduction potentials listed in Table 13.1 and the *Handbook of Chemistry and Physics,* show that the following reaction is favorable under standard-state conditions:

$$2Ag(s) + Pt^{2+}(aq) \longrightarrow 2Ag^+(aq) + Pt(s)$$

What is the equilibrium constant of this reaction at 25°C?

15.42 The Co^{2+}-porphyrin complex is more stable than the Fe^{2+}-porphyrin complex. Why, then, is iron the metal ion in hemoglobin (and other heme-containing proteins)?

15.43 Oxyhemoglobin is bright red, whereas deoxy-hemoglobin is purple. Show that the difference in color can be accounted for qualitatively on the basis of high-spin and low-spin complexes. (*Hint:* O_2 is a strong-field ligand; see the inset on p. 784.)

15.44 Hydrated Mn^{2+} ions are practically colorless (see Figure 15.17) even though they possess five $3d$ electrons. Explain. (*Hint:* Electronic transitions in which there is a change in the number of unpaired electrons do not occur readily.)

15.45 Which of the following hydrated cations are colorless: $Fe^{2+}(aq)$, $Zn^{2+}(aq)$, $Cu^+(aq)$, $Cu^{2+}(aq)$, $V^{5+}(aq)$, $Ca^{2+}(aq)$, $Co^{2+}(aq)$, $Sc^{2+}(aq)$, or $Pb^{2+}(aq)$? Explain your choice.

15.46 Aqueous solutions of $CoCl_2$ are generally either light pink or blue. Low concentrations and low temperatures favor the pink form, whereas high concentrations and high temperatures favor the blue form. Adding hydrochloric acid to a pink solution of $CoCl_2$ causes the solution to turn blue; the pink color is restored by the addition of $HgCl_2$. Account for these observations.

15.47 Suggest a method that would allow you to distinguish between *cis*-$Pt(NH_3)_2Cl_2$ and *trans*-$Pt(NH_3)_2Cl_2$.

15.48 You are given two solutions containing $FeCl_2$ and $FeCl_3$ at the same concentration. One solution is light yellow, and the other one is brown. Identify these solutions based only on color.

15.49 The label of a certain brand of mayonnaise lists EDTA as a food preservative. How does EDTA prevent the spoilage of mayonnaise?

15.50 The compound 1,1,1-trifluoroacetylacetone (tfa) is a bidentate ligand:

$$CF_3CCH_2CCH_3$$
(with two O double bonds)

It forms a tetrahedral complex with Be^{2+} and a square-planar complex with Cu^{2+}. Draw structures of these complex ions, and identify the type of isomerism exhibited by them.

15.51 How many geometric isomers can the following square planar complex have?

15.52 $[Pt(NH_3)_2Cl_2]$ is found to exist in two geometric isomers designated I and II, which react with oxalic acid as follows:

$$I + H_2C_2O_4 \longrightarrow [Pt(NH_3)_2C_2O_4]$$
$$II + H_2C_2O_4 \longrightarrow [Pt(NH_3)_2(HC_2O_4)_2]$$

Comment on the structures of I and II.

15.53 Commercial silver-plating operations frequently use a solution containing the complex $Ag(CN)_2^-$ ion. Because the formation constant (K_f) is quite large, this procedure ensures that the free Ag^+ concentration in solution is low for uniform electrodeposition. In one process, a chemist added 9.0 L of 5.0 M NaCN to 90.0 L of 0.20 M $AgNO_3$. Calculate the concentration of free Ag^+ ions at equilibrium. See Table 11.5 for K_f values.

15.54 Draw qualitative diagrams for the crystal field splittings in (a) a linear complex ion ML_2, (b) a trigonal-planar complex ion ML_3, and (c) a trigonal-bipyramidal complex ion ML_5.

15.55 (a) The free Cu(I) ion is unstable in solution and has a tendency to disproportionate:

$$2Cu^+(aq) \rightleftharpoons Cu^{2+}(aq) + Cu(s)$$

Use the information in Table 13.1 to calculate the equilibrium constant for the reaction. (b) Based on your result in (a), explain why most Cu(I) compounds are insoluble.

15.56 Consider the following two ligand exchange reactions:

$$[Co(H_2O)_6]^{3+} + 6NH_3 \rightleftharpoons [Co(NH_3)_6]^{3+} + 6H_2O$$
$$[Co(H_2O)_6]^{3+} + 3\,en \rightleftharpoons [Co(en)_6]^{3+} + 6H_2O$$

(a) Which of the reactions should have the larger $\Delta S°$?

(b) Given that the Co—N bond strength is approximately the same in both complexes, which reaction will have a larger equilibrium constant? Explain your choices.

15.57 Copper is known to exist in a +3 oxidation state, which is believed to be involved in some biological electron transfer reactions. (a) Would you expect this oxidation state of copper to be stable? Explain. (b) Name the compound K_3CuF_6, and predict the geometry of the complex ion and its magnetic properties. (c) Most of the known Cu(III) compounds have square-planar geometry. Are these compounds diamagnetic or paramagnetic?

Answers to Practice Exercises

15.1 (a) +2, (b) +3, (c) +4 **15.2** (a) tetraaquodichlorochromium(III) chloride; (b) triamminebromoplatinum(II) chloride; (c) potassium hexafluorocobaltate(III); and (d) tetraamminechloronitrocobalt(III) nitrate **15.3** (a) $[Co(en)_3]_2(SO_4)_3$; (b) $K_3[Fe(CN)_6]$; (c) $K_2[Ni(CN)_4]$; and (d) $Ca_2[PtBr_4Cl_2]$

Organic and Polymer Chemistry

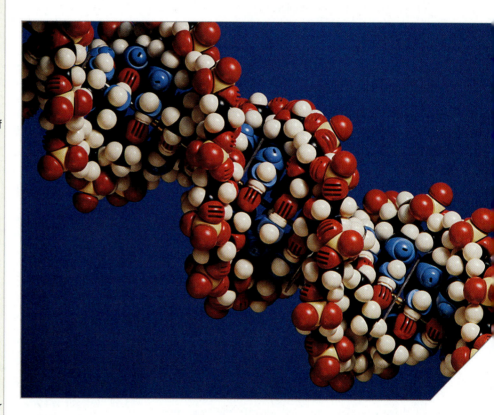

Organic chemistry is the study of carbon compounds. The word *organic* was originally used by eighteenth-century chemists to describe substances obtained from living sources—plants and animals. These chemists believed that nature possessed a certain vital force and that only living things could produce organic compounds. This romantic notion was disproved in 1828 by Friedrich Wöhler, a German chemist who prepared urea, an organic compound, from the reaction between the inorganic compounds lead cyanate and aqueous ammonia:

$$Pb(OCN)_2 + 2NH_3 + 2H_2O \longrightarrow 2(NH_2)_2CO + Pb(OH)_2$$

urea

Today, well over 14 million synthetic and natural organic compounds are known. This number is significantly greater than the 100,000 or so known inorganic compounds.

16.1 | Hydrocarbons Are Organic Compounds Containing Only Hydrogen and Carbon

Carbon can form more compounds with itself and/or other atoms than any other element because carbon atoms can form single, double, and triple bonds to other carbon atoms and because carbon atoms can link up with each other in chains and ring structures. Classes of organic compounds can be distinguished based on the functional groups they contain. A ***functional group*** is *an atom or group of atoms that gives characteristic chemical properties to the parent molecule.* Different molecules containing the same kind of functional group or groups undergo similar reactions. Thus, by learning the characteristic properties of a few functional groups, we can understand the properties of many organic compounds. In Section 16.3, we will discuss the functional groups known as alcohols, ethers, aldehydes, ketones, carboxylic acids, and amines.

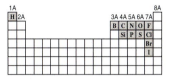

Common elements in organic compounds

Aliphatic Hydrocarbons

All organic compounds are derived from a group of compounds known as ***hydrocarbons*** because they are *made up only of hydrogen and carbon.* On the basis of structure, hydrocarbons can be classified as aliphatic or aromatic. ***Aliphatic hydrocarbons*** *do not contain the benzene group, or the benzene ring,* whereas ***aromatic hydrocarbons*** *contain one or more benzene rings.* In addition, aliphatic hydrocarbons are further divided into alkanes, alkenes, and alkynes (Figure 16.1).

Alkanes

Alkanes *have the general formula* C_nH_{2n+2}, *where* n *is a positive integer* ($n = 1, 2, 3, \ldots$). They contain *only C—H and C—C single covalent bonds.* Alkanes are also known as ***saturated hydrocarbons*** because they *contain the maximum number of hydrogen atoms that can bond with the number of carbon atoms present.*

The simplest alkane (that is, with $n = 1$) is methane (CH_4), which, in addition to several natural sources, is a natural product of the anaerobic bacterial decomposition of vegetable matter under water. Because it was first collected in marshes, methane became known as "marsh gas." A rather improbable but proven source of methane is termites. When these voracious insects consume wood, the microorganisms that inhabit their digestive system break down cellulose (the major component of wood) into methane, carbon dioxide, and other compounds. It is estimated that termites produce 170 million tons of methane annually! Methane is also produced in some sewage treatment processes. Commercially, methane is obtained from natural gas.

Termites are a natural source of methane.

Figure 16.1 Classification of hydrocarbons.

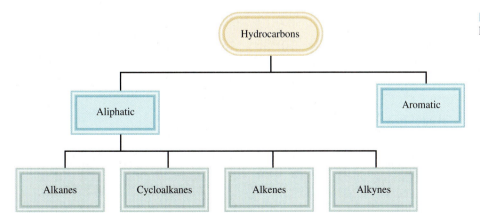

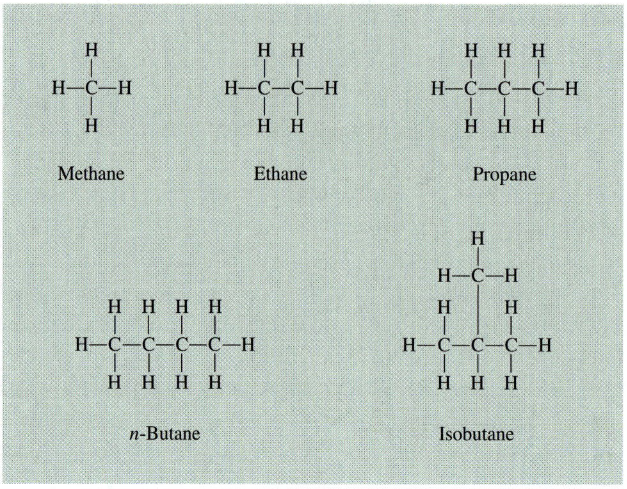

Figure 16.2 Structures of the first four alkanes. Butane has two different structural isomers, as discussed in Section 4.4.

Methane   Ethane   Propane

n-Butane   Isobutane

Figure 16.2 shows the structures of methane, ethane, propane, and butane, the first four alkanes ($n = 1$ to $n = 4$). Natural gas is a mixture of methane, ethane, and a small amount of propane. Recall from Section 4.3 that the carbon atom of methane is sp^3-hybridized. In fact, the carbon atoms in all the alkanes are sp^3-hybridized. There is only one way to join the carbon atoms in ethane and propane, but there are two possible ways in butane, resulting in the *structural isomers* n-butane and isobutane. (See Section 4.4 to review structural isomers.) All alkanes with $n \geq 4$ can exist in a variety of different structural isomers with the number of possiblities increasing rapidly with increasing n.

The nomenclature of alkanes and all other organic compounds is based on the recommendations of the International Union of Pure and Applied Chemistry (IUPAC). The first four alkanes (methane, ethane, propane, and butane) have nonsystematic names. As Table 16.1 shows, the number of carbon atoms is reflected in the Greek prefixes for the alkanes containing 5 to 10 carbons.

Table 16.1	The First Ten Straight-Chain Alkanes			
Name of Hydrocarbon	Molecular Formula	Number of Carbon Atoms	Melting Point (°C)	Boiling Point (°C)
Methane	CH_4	1	−182.5	−161.6
Ethane	CH_3—CH_3	2	−183.3	−88.6
Propane	CH_3—CH_2—CH_3	3	−189.7	−42.1
Butane	CH_3—$(CH_2)_2$—CH_3	4	−138.3	−0.5
Pentane	CH_3—$(CH_2)_3$—CH_3	5	−129.8	36.1
Hexane	CH_3—$(CH_2)_4$—CH_3	6	−95.3	68.7
Heptane	CH_3—$(CH_2)_5$—CH_3	7	−90.6	98.4
Octane	CH_3—$(CH_2)_6$—CH_3	8	−56.8	125.7
Nonane	CH_3—$(CH_2)_7$—CH_3	9	−53.5	150.8
Decane	CH_3—$(CH_2)_8$—CH_3	10	−29.7	174.0

Use the following IUPAC rules to name alkanes:

1. The parent name of the hydrocarbon is that given to the longest continuous chain of carbon atoms in the molecule. Thus, the parent name of the following compound is heptane because there are seven carbon atoms in the longest chain.

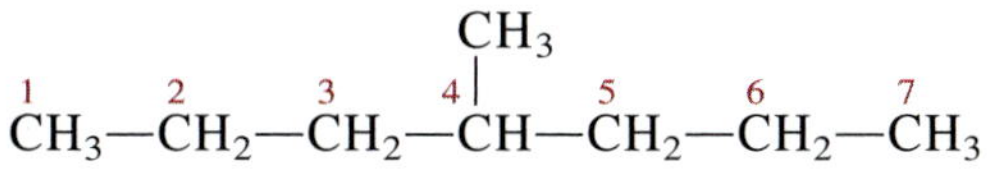

$$\overset{\displaystyle CH_3}{\underset{\displaystyle CH_3-CH_2-CH_2-CH-CH_2-CH_2-CH_3}{|}}$$

2. An alkane less one hydrogen atom is an **alkyl group.** When a hydrogen atom is removed from methane, for example, the CH_3 fragment remains, which is called a *methyl* group. Similarly, removing a hydrogen atom from ethane leaves an *ethyl* group (C_2H_5). Table 16.2 lists the names of several common alkyl groups. Any chain branching off the longest chain is named as an alkyl group.

3. When one or more hydrogen atoms are replaced by other groups, the name of the compound must indicate the locations of the carbon atoms where the replacements are made. The procedure is to number each carbon atom on the longest chain in the direction that gives the smaller numbers for the locations of all branches. Consider the two different systems for the *same* compound shown here:

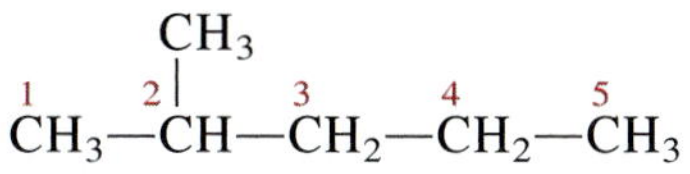

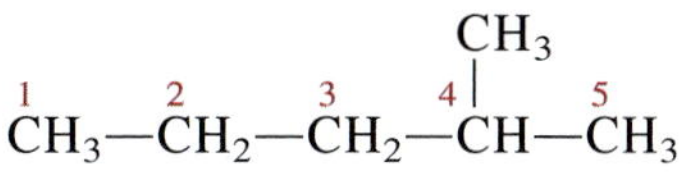

The compound on the left is numbered correctly because the methyl group is located at carbon 2 of the pentane chain; in the compound on the right, the methyl group is located at carbon 4. Thus, the name of the compound is 2-methylpentane, not 4-methylpentane. Note that the branch name and the parent name are written as a single word, and a hyphen follows the number.

Table 16.2	Common Alkyl Groups
Name	**Formula**
Methyl	$-CH_3$
Ethyl	$-CH_2-CH_3$
n-Propyl	$-CH_2-CH_2-CH_3$
n-Butyl	$-CH_2-CH_2-CH_2-CH_3$
Isopropyl	$-\overset{\displaystyle CH_3}{\underset{\displaystyle CH_3}{\overset{|}{\underset{|}{C}}}}-H$
t-Butyl*	$-\overset{\displaystyle CH_3}{\underset{\displaystyle CH_3}{\overset{|}{\underset{|}{C}}}}-CH_3$

*The letter *t* stands for tertiary.

Table 16.3

Names of Some Common Functional Groups

Functional Group	Name
—NH_2	Amino
—F	Fluoro
—Cl	Chloro
—Br	Bromo
—I	Iodo
—NO_2	Nitro
—$CH{=}CH_2$	Vinyl

4. When there is more than one alkyl branch of the same kind present, we use a prefix such as *di-*, *tri-*, or *tetra-* with the name of the alkyl group. Consider the following examples:

$$\overset{1}{CH_3}-\overset{2}{\underset{\displaystyle CH_3}{CH}}-\overset{3}{\underset{\displaystyle CH_3}{CH}}-\overset{4}{CH_2}-\overset{5}{CH_2}-\overset{6}{CH_3}$$

2,3-dimethylhexane

$$\overset{1}{CH_3}-\overset{2}{CH_2}-\overset{3}{\underset{\displaystyle\underset{\displaystyle CH_3}{|}}{\overset{\displaystyle CH_3}{\overset{\displaystyle |}{C}}}}-\overset{4}{CH_2}-\overset{5}{CH_2}-\overset{6}{CH_3}$$

3,3-dimethylhexane

When there are two or more different alkyl groups, the names of the groups are listed alphabetically. For example,

$$\overset{1}{CH_3}-\overset{2}{CH_2}-\overset{3}{\underset{\displaystyle CH_3}{CH}}-\overset{4}{\underset{\displaystyle C_2H_5}{CH}}-\overset{5}{CH_2}-\overset{6}{CH_2}-\overset{7}{CH_3}$$

4-ethyl-3-methylheptane

5. Alkanes can have many different types of substituents. Table 16.3 lists the names of some substituents, including nitro- and bromo-. Thus, the compound

$$\overset{1}{CH_3}-\overset{2}{\underset{\displaystyle NO_2}{CH}}-\overset{3}{\underset{\displaystyle Br}{CH}}-\overset{4}{CH_2}-\overset{5}{CH_2}-\overset{6}{CH_3}$$

is called 3-bromo-2-nitrohexane. Note that the substituent groups are listed alphabetically in the name and the chain is numbered in the direction that gives the lowest number to the first substituted carbon. Substituted alkanes, like 3-bromo-2-nitrobutane, that contain asymmetric carbon atoms (that is, carbon atoms that are bonded to four different substituents) can exhibit optical isomerism (Section 4.3).

Example 16.1

Give the IUPAC name of the following compound:

$$CH_3-\underset{\displaystyle\underset{\displaystyle CH_3}{|}}{\overset{\displaystyle CH_3}{\overset{\displaystyle |}{C}}}-CH_2-\underset{\displaystyle CH_3}{\overset{\displaystyle CH_3}{CH}}-CH_2-CH_3$$

Strategy Follow the IUPAC rules and use the information in Table 16.2 to name the compound. How many C atoms are there in the longest continuous chain?

Solution The longest continuous chain has six C atoms so the parent compound is called hexane. The chain is numbered from left to right, putting the two methyl groups on carbon 2 and the other methyl group on carbon 4, because numbering the chain from right to left would give these substituents higher numbers.

$$\overset{1}{CH_3}-\overset{2}{\underset{\displaystyle\underset{\displaystyle CH_3}{|}}{\overset{\displaystyle CH_3}{\overset{\displaystyle |}{C}}}}-\overset{3}{CH_2}-\overset{4}{\underset{\displaystyle CH_3}{\overset{\displaystyle CH_3}{CH}}}-\overset{5}{CH_2}-\overset{6}{CH_3}$$

—Continued

Continued—
Therefore, the IUPAC name of the compound is 2,2,4-trimethylhexane.

Practice Exercise Give the IUPAC name of the following compound:

$$\overset{\overset{\displaystyle CH_3}{|}}{CH_3}-\overset{\overset{\displaystyle }{|}}{CH}-CH_2-\overset{\overset{\displaystyle C_2H_5}{|}}{CH}-CH_2-\overset{\overset{\displaystyle C_2H_5}{|}}{CH}-CH_2-CH_3$$

Example 16.2 shows that prefixes such as di-, tri-, and tetra- are used as needed, but are ignored when alphabetizing.

Example 16.2

Write the structural formula of 3-ethyl-2,2-dimethylpentane.

Strategy Follow the preceding procedure and the information in Table 16.2 to write the structural formula of the compound. How many C atoms are there in the longest continuous chain?

Solution The parent compound is pentane, so the longest chain has five C atoms. There are two methyl groups attached to carbon 2 and one ethyl group attached to carbon 3. Therefore, the structure of the compound is

$$\overset{\overset{\displaystyle CH_3\ \ C_2H_5}{|\ \ \ \ |}}{\underset{\underset{\displaystyle CH_3}{|}}{\overset{1}{CH_3}-\overset{2}{C}-\overset{3}{CH}-\overset{4}{CH_2}-\overset{5}{CH_3}}}$$

Practice Exercise Write the structural formula of 5-ethyl-2,4,6-trimethyloctane.

Before proceeding further, it is necessary to learn different ways of drawing the structure of organic compounds. To see how atoms are connected in 2-methylbutane (C_5H_{12}), for example, we must first write the "expanded" molecular formula, $CH_3CH(CH_3)CH_2CH_3$ and draw its structural formula, shown in Figure 16.3(a).

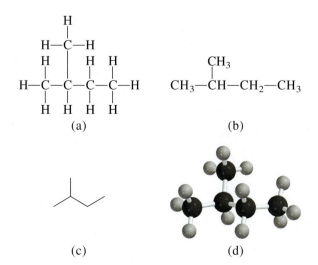

(a)　　　　　　(b)

(c)　　　　　　(d)

Figure 16.3 Four representations of the structure of 2-methylbutane: (a) structural formula, (b) abbreviated structural formula, (c) skeletal structure, (d) ball-and-stick model.

Although informative, the structural formula is time-consuming to draw. Therefore, chemists have devised ways to simplify the representation. Figure 16.3(b) is an abbreviated version, and the structure shown in Figure 16.3(c) is called the *skeletal structure* in which all the C and H letters are omitted. A carbon atom is assumed to be at each intersection of two lines (bond) and at the end of each line. Because every C atom forms four bonds, we can always deduce the number of H atoms bonded to any C atom. The element symbols for substituent atoms in the skeletal structure are always shown explicitly, such as in 1-bromo-3-chlorobutane [C_4H_8BrCl]:

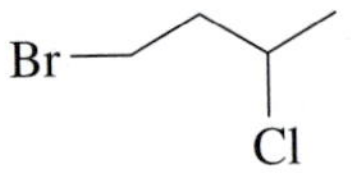

What is lacking in these structures, however, is the three dimensionality of the molecule, which for 2-methylbutane is shown by the ball-and-stick molecular model in Figure 16.3(d). Depending on the purpose of the discussion, any of these representations can be used to describe the properties of the molecule.

Cycloalkanes

Alkanes whose carbon atoms are joined in rings are known as **cycloalkanes.** They have the general formula C_nH_{2n}, where $n = 3, 4, \ldots$, giving them two fewer hydrogen atoms than the acyclic alkane with the same number of carbon atoms. The simplest cycloalkane is cyclopropane (C_3H_6). The structures for the first four cycloalkanes are shown in Figure 16.4. Many biologically significant substances such as cholesterol, testosterone, and progesterone contain a ring system.

Alkenes

Alkenes (also called *olefins*) *contain at least one carbon-carbon double bond. Alkenes with one C=C bond have the general formula C_nH_{2n}, where $n = 2, 3, \ldots$*, giving them the same general formula as cycloalkanes (that is, two fewer hydrogen atoms than the acyclic alkane with the same number of carbon atoms). The simplest alkene is ethene (C_2H_4), in which both carbon atoms are sp^2-hybridized and the double bond is made up of a sigma (σ) bond and a pi (π) bond (see Section 4.4). Ethene is also commonly known as ethylene. In fact, many organic compounds have common names in addition to their IUPAC names and, when appropriate, we will list both (with the common name in parenthesis).

As with alkanes, the name of the parent alkene is determined by the number of carbon atoms in the longest chain containing the carbon-carbon double bond (see Table 16.1). Consider two examples:

$$CH_2\!=\!CH\!-\!CH_2\!-\!CH_3 \qquad\qquad H_3C\!-\!CH\!=\!CH\!-\!CH_3$$

1-butene   2-butene

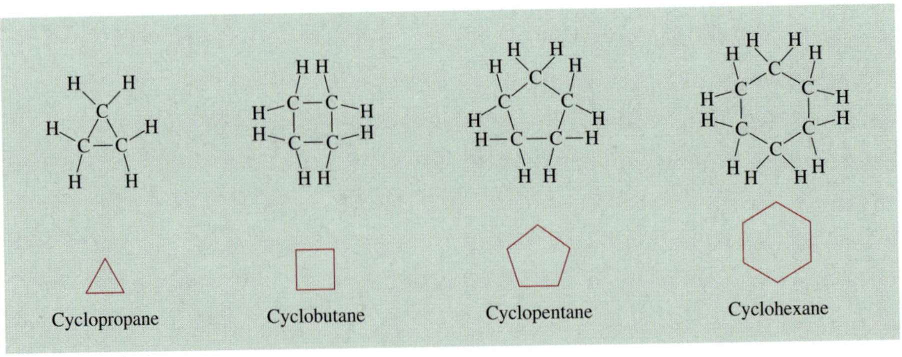

Figure 16.4 Structures of the first four cycloalkanes and their simplified skeletal forms.

As these two examples demonstrate, the names of alkanes end with *-ene* and numbers are used to indicate the position of the double bond. The parent chain is numbered so that the lowest number possible is given to one of the carbon atoms in the double bond, regardless of any other substituents present in the compound (for example, alkyl groups or halides). The numbers in the names of alkenes refer to the lowest numbered carbon atom in the chain that is part of the C=C bond of the alkene. The name *butene* means that there are four carbon atoms in the longest chain. Because of restricted rotation about the carbon-carbon double bond, alkenes can form geometric isomers (Section 4.4). In this case, the name of an alkene must also specify whether the isomer is *cis* or *trans*:

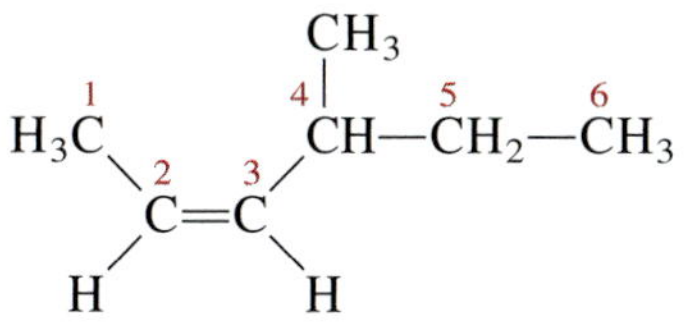

4-methyl-*cis*-2-hexene

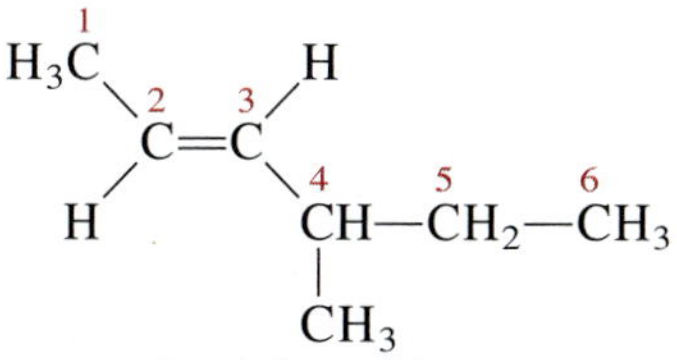

4-methyl-*trans*-2-hexene

In the *cis* isomer, the two H atoms are on the same side of the C=C bond. In the *trans* isomer, the two H atoms are across from each other.

If more than one double bond is present in an alkene, the prefixes *di-*, *tri-*, etc. precede the *-ene* suffix, as shown for the following isomers of pentadiene:

$$CH_2{=}CH{-}CH{=}CH{-}CH_3 \qquad\qquad CH_2{=}CH{-}CH_2{-}CH{=}CH_2$$

1,3-pentadiene   1,4-pentadiene

(Note: 1,3-Pentadiene will have *cis* and *trans* geometric isomers.)

Example 16.3

Draw and name the geometric isomers of 2-butene.

Strategy This is an alkene, so geometric *cis* and *trans* isomers are possible. First draw the structural formula and examine the groups attached to the double bonded carbon atoms to see if you can construct *cis* and *trans* isomers.

Solution The structural formula is

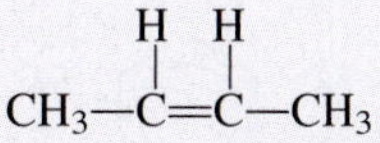

$$\begin{array}{cc} & \text{H} \quad \text{H} \\ & | \quad\;\; | \\ CH_3{-}&C{=}C{-}CH_3 \end{array}$$

In this formula we could put the methyl (—CH_3) groups on either the same (*cis*) or opposite (*trans*) sides of the double bond:

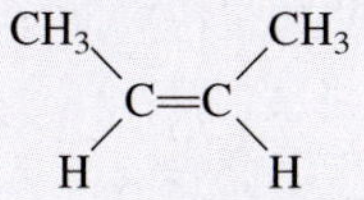

cis-2-butene 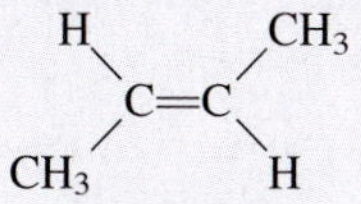   *trans*-2-butene

Because there is only one double bond, these are the only geometric isomers possible.

Practice Exercise Draw and name the geometric isomers of 2,6-dimethyl-3-heptene.

Alkynes

Alkynes *contain at least one carbon-carbon triple bond.* Those that have one carbon-carbon triple bond have the *general formula* C_nH_{2n-2}*, where n = 2, 3, ...* As a result, alkynes have two fewer hydrogen atoms than the alkene or cycloalkane with the same number of carbon atoms, and four fewer hydrogen atoms than the corresponding alkane.

Names of compounds containing $C \equiv C$ bonds end with -*yne*. The name of the parent compound is determined by the number of carbon atoms in the longest chain containing the triple bond (see Table 16.1 for names of alkane counterparts). The simplest alkyne is ethyne (C_2H_2), commonly known as acetylene. As with alkenes, the parent chain of alkynes is numbered so that the position of the carbon-carbon triple bond is given the lowest number possible:

$$HC \equiv C - CH_2 - CH_3 \qquad H_3C - C \equiv C - CH_3$$

$$\text{1-butyne} \qquad\qquad\qquad \text{2-butyne}$$

Aromatic Hydrocarbons

Benzene (C_6H_6), the parent compound of the aromatic hydrocarbons, was discovered by Michael Faraday in 1826. Over the next 40 years, chemists were preoccupied with determining its molecular structure. Despite the small number of atoms in the molecule, there are quite a few ways to represent the structure of benzene without violating the tetravalency of carbon. Most of the proposed structures were rejected, however, because they did not explain the known properties of benzene. Finally, in 1865, August Kekulé[1] deduced that the benzene molecule could be best represented by a ring structure—a cyclic compound consisting of six carbon atoms:

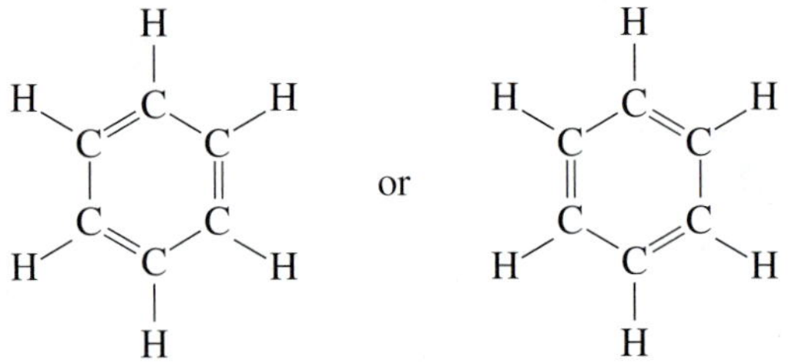

As discussed in Section 4.3, the properties of benzene are best represented by both of the preceeding resonance structures. Alternatively, the properties of benzene can be explained in terms of delocalized molecular orbitals (see Section 4.5):

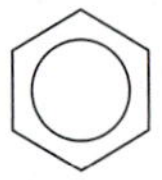

The naming of monosubstituted benzenes (that is, benzenes in which one H atom has been replaced by another atom or a group of atoms) consists of the name of the substituent followed by *benzene*:

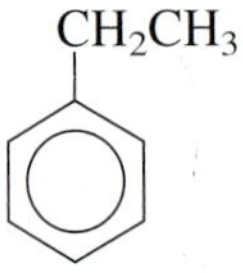

ethylbenzene 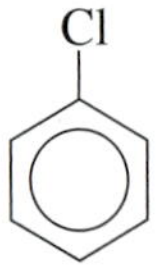   chlorobenzene 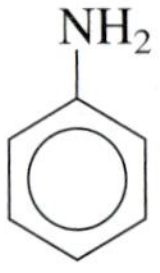   aminobenzene   nitrobenzene
                                 (aniline)

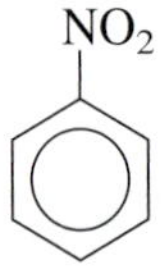

---

1. August Kekulé (1829–1896). German chemist. Kekulé was a student of architecture before he became interested in chemistry. He supposedly solved the riddle of the structure of the benzene molecule after having a dream in which dancing snakes bit their own tails. Kekulé's work is regarded by many as the crowning achievement of theoretical organic chemistry of the nineteenth century.

If more than one substituent is present, the location of the second group must be indicated relative to the first. The systematic way to accomplish this is to number the carbon atoms as follows:

Three different dibromobenzenes are possible:

| 1,2-dibromobenzene | 1,3-dibromobenzene | 1,4-dibromobenzene |
| (*o*-dibromobenzene) | (*m*-dibromobenzene) | (*p*-dibromobenzene) |

The prefixes *o*- (*ortho*-), *m*- (*meta*-), and *p*- (*para*-) are also used to denote the relative positions of the two substituted groups, as shown previously for the dibromobenzenes. Compounds in which the two substituted groups are different are named accordingly. Thus,

is named 3-bromonitrobenzene, or *m*-bromonitrobenzene.

Finally, the group consisting of benzene minus a hydrogen atom (C_6H_5) is called the *phenyl* group. Thus, the following molecule is called 2-phenylpropane:

$$CH_3 - CH - CH_3$$

This compound is also called isopropyl benzene (see Table 16.2).

An enormously large number of compounds can be generated from substances in which benzene rings are fused together. Some of these *polycyclic* aromatic hydrocarbons are shown in Figure 16.5. The best known of these compounds is naphthalene, which is used in mothballs. These and many other similar compounds are present in coal tar. Some of the compounds with several rings are powerful carcinogens.

Configurations of Organic Molecules

Molecular geometry gives the spatial arrangement of atoms in a molecule. However, because of internal molecular motions, atoms are not held rigidly in position. For this reason, even a simple molecule like ethane may be structurally more complicated than we might think.

The two C atoms in ethane are sp^3-hybridized and they are joined by a sigma bond. As discussed in Chapter 4, sigma bonds have cylindrical symmetry, that is, the overlap of the sp^3 orbitals is the same regardless of the rotation about the C—C bond.

Figure 16.5 Some polycyclic aromatic hydrocarbons. Compounds denoted by * are potent carcinogens. An enormous number of such compounds exist in nature.

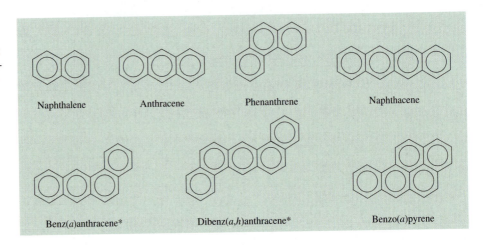

Naphthalene   Anthracene   Phenanthrene   Naphthacene

Benz(*a*)anthracene*   Dibenz(*a,h*)anthracene*   Benzo(*a*)pyrene

This bond rotation is not totally free, however, because there are interactions between the H atoms on the two different C atoms. Figure 16.6 shows the two extreme *conformations* of ethane. **Conformations** are *different spatial arrangements of a molecule that are generated by rotation about single bonds.* In the staggered configuration, the three H atoms on one C atom are pointing away from the three H atoms on the other C atom, whereas in the eclipsed conformation the two groups of H atoms are aligned parallel to one another.

A simpler and more effective way of viewing these two conformations is by using a **Newman projection,** also shown in Figure 16.6. When looking at the C—C bond end-on, the rear carbon atom is represented by a circle and the front carbon atom is represented by a dot in the center of the circle. The C—H bonds attached to the front carbon are the lines emanating from the dot at the center of the circle, and the C—H bonds attached to the rear carbon appear as lines emanating from the edge of the circle. The eclipsed form of ethane is less stable than the staggered form. Figure 16.7 shows the variation of the potential energy of ethane as a function of rotation. The rotation of one CH_3 group relative to the other is described in terms of the angle between the C—H bonds on the front and back carbon atoms, called the *dihedral angle*. The dihedral angle for the first eclipsed configuration is zero. A clockwise rotation of 60° about the C—C bond generates a staggered configuration, which is converted to another eclipsed confirmation by a similar rotation, and so on.

Figure 16.6 Staggered and eclipsed conformations of ethane shown in both ball-and-stick representations (top) and Newman projections (bottom).

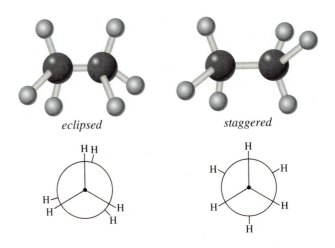

eclipsed   staggered

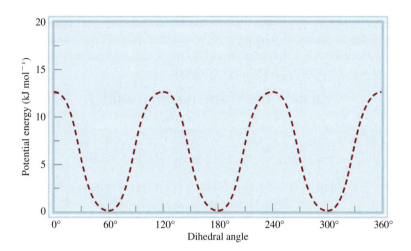

Figure 16.7 Potential energy for ethane conformations as a function of dihedral angle.

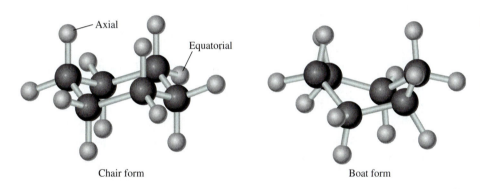

Chair form

Boat form

Figure 16.8 Cyclohexane can exist in various shapes (or conformations). The most stable conformation is the chair form and a less stable one is the boat form. The two types of H atoms are labeled axial and equatorial, respectively.

Another molecule with more than one stable conformation is cyclohexane. Theoretical analysis shows that cyclohexane can assume two different conformations that are relatively free of strain (Figure 16.8). By *strain,* we mean that bonds are compressed, stretched, or twisted out of normal geometric shapes as predicted by sp^3 hybridization. The most stable configuration of cyclohexane is the *chair* form.

Configurational analysis of molecules is of great importance in understanding the details of reactions ranging from simple hydrocarbons to proteins and DNA.

16.2 | Hydrocarbons Undergo a Number of Important Chemical Reactions

Reactions of Alkanes

Alkanes are not very reactive substances. Under suitable conditions, however, they do react. Methane and ethane, for example, are alkanes that undergo highly exothermic combustion reactions:

$$CH_4(g) + 2O_2(g) \longrightarrow CO_2(g) + 2H_2O(l) \quad \Delta H^\circ = -890.4 \text{ kJ mol}^{-1}$$
$$2C_2H_6(g) + 7O_2(g) \longrightarrow 4CO_2(g) + 6H_2O(l) \quad \Delta H^\circ = -3119 \text{ kJ mol}^{-1}$$

These, and similar combustion reactions, such as those for gasoline (a mixture of alkanes) and fuel oil, have long been utilized in industrial processes and in domestic heating and cooking.

The *halogenation* of alkanes, that is, the replacement of one or more hydrogen atoms by halogen atoms, is another type of reaction that alkanes undergo. When a mixture of methane and chlorine is heated above 100°C or irradiated with light of a suitable wavelength, methyl chloride is produced:

$$CH_4(g) + Cl_2(g) \longrightarrow CH_3Cl(g) + HCl(g)$$
chloromethane
(methyl chloride)

If excess chlorine gas is present, the reaction can proceed further:

$$CH_3Cl(g) + Cl_2(g) \longrightarrow CH_2Cl_2(l) + HCl(g)$$
dichloromethane
(methylene chloride)

$$CH_2Cl_2(g) + Cl_2(g) \longrightarrow CHCl_3(l) + HCl(g)$$
trichloromethane
(chloroform)

$$CHCl_3(l) + Cl_2(g) \longrightarrow CCl_4(l) + HCl(g)$$
carbon tetrachloride

A great deal of experimental evidence suggests that the initial step of the first halogenation reaction occurs as follows:

$$Cl_2 + energy \longrightarrow Cl\cdot + Cl\cdot$$

that is, the covalent bond in Cl_2 breaks and two chlorine atoms form. We know it is the Cl—Cl bond that breaks when the mixture is heated or irradiated because the bond enthalpy of Cl_2 is 242.7 kJ mol^{-1} (see Table 6.6), whereas about 414 kJ mol^{-1} is needed to break each of the C—H bonds in CH_4.

A chlorine atom is a *free radical,* which contains an unpaired electron (shown by a single dot). Chlorine atoms are highly reactive and attack methane molecules according to the following reaction

$$CH_4 + Cl\cdot \longrightarrow \cdot CH_3 + HCl$$

This reaction produces hydrogen chloride and the methyl radical ($\cdot CH_3$). The methyl radical is another reactive species; it combines with molecular chlorine to give chloromethane (methyl chloride) and a chlorine atom:

$$\cdot CH_3 + Cl_2 \longrightarrow CH_3Cl + Cl\cdot$$

The production of dichloromethane (methylene chloride) from chloromethane (methyl chloride) and any further reactions can be explained in the same way. The actual mechanism is more complex than the scheme we have shown because "side reactions" that do not lead to the desired products often take place, such as

$$Cl\cdot + Cl\cdot \longrightarrow Cl_2$$
$$\cdot CH_3 + \cdot CH_3 \longrightarrow C_2H_6$$

Alkanes in which one or more hydrogen atoms have been replaced by a halogen atom are called *alkyl halides.* Among the large number of alkyl halides, the best known are trichloromethane (chloroform) ($CHCl_3$), carbon tetrachloride (CCl_4), dichloromethane (methylene chloride) (CH_2Cl_2), and the chlorofluorocarbons (CFCs).

Trichloromethane (chloroform) is a volatile, sweet-tasting liquid that was used for many years as an anesthetic. However, because of its toxicity (it can severely damage the liver, kidneys, and heart), it has been replaced by other compounds. Carbon tetrachloride, also a toxic substance, was once used as a solvent in the dry cleaning industry for its ability to remove grease stains from clothing. Dichloromethane (methylene chloride) is used as a paint remover and was once used to decaffeinate coffee. Chlorofluorocarbons are used as refrigerant gases but are being phased out because of their role in the depletion of stratospheric ozone.

Reactions of Alkenes and Alkynes

Ethene (or ethylene), the simplest alkene, is used in large quantities for the manufacture to organic polymers (to be discussed in Section 16.4) and to prepare many other organic chemicals. Ethylene is prepared industrially by the *cracking* process, that is, by the thermal decomposition of a large hydrocarbon into smaller molecules. When ethane is heated to about 800°C in the presence of a catalyst, it undergoes the following reaction:

$$C_2H_6(g) \xrightarrow{\text{Pt catalyst}} CH_2{=}CH_2(g) + H_2(g)$$

Other alkenes can be prepared by cracking the higher members of the alkane family.

Alkenes and alkynes are called ***unsaturated hydrocarbons*** *because they can add hydrogen atoms to their carbon-carbon double and triple bonds.* Unsaturated hydrocarbons commonly undergo ***addition reactions*** in which *one molecule adds to another to form a single product.* An industrially useful example of an addition reaction is ***hydrogenation,*** *the addition of hydrogen to compounds containing multiple bonds.* A simple hydrogenation reaction is the conversion of ethylene to ethane:

$$CH_2{=}CH_2(g) + H_2(g) \longrightarrow C_2H_6(g)$$

The addition reaction between HCl and ethylene. The initial interaction is between the positive end of HCl (blue) and the electron-rich region of ethylene (red), which is associated with the pi electrons of the C=C bond.

Other addition reactions to the C=C bond include

$$C_2H_4(g) + HX(g) \longrightarrow CH_3{-}CH_2X(g)$$
$$C_2H_4(g) + X_2(g) \longrightarrow CH_2X{-}CH_2X(g)$$

where X represents a halogen (Cl, Br, or I).

The addition of a hydrogen halide to an unsymmetrical alkene such as propene (propylene) is more complicated because two products are possible:

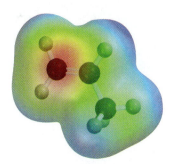

The electron density is higher on the carbon atom in the CH_2 group in propene.

In reality, however, only 2-bromopropane is formed. This phenomenon was observed in all reactions between unsymmetrical reagents and alkenes. In 1871, Vladimir Markovnikov[2] postulated a generalization that enables us to predict the outcome of such an addition reaction: in the addition of unsymmetrical (i.e., polar) reagents to alkenes, the positive portion of the reagent (usually hydrogen) adds to the carbon atom that already has the most hydrogen atoms. This generalization, now known as *Markovnikov's rule,* makes it possible to predict the outcome of addition reactions to alkenes.

2. Vladimir W. Markovnikov (1838–1904). Russian chemist. Markovnikov's observations of the addition reactions to alkenes were published a year after his death.

The reaction of calcium carbide with water produces acetylene, a flammable gas.

Ethyne (C_2H_2), better known as acetylene, is a colorless gas (b.p. $-84°C$) prepared by the reaction between calcium carbide and water:

$$CaC_2(s) + H_2O(l) \longrightarrow C_2H_2(g) + Ca(OH)_2(aq)$$

Acetylene has many important uses in industry. Because of its high heat of combustion

$$2C_2H_2(g) + 5O_2(g) \longrightarrow 4CO_2(g) + 2H_2O(l) \quad \Delta H° = -2599.2 \text{ kJ mol}^{-1}$$

acetylene burned in an "oxyacetylene torch" gives an extremely hot flame (about 3000°C). Thus, oxyacetylene torches are used to weld metals. The standard free energy of formation of acetylene is positive ($\Delta G_f° = +209.2$ kJ mol^{-1}), unlike that of the alkanes. This means that acetylene is unstable (relative to its elements) and has a tendency to decompose:

$$C_2H_2(g) \longrightarrow 2C(s) + H_2(g) \quad \Delta G_f° = -209.2 \text{ kJ mol}^{-1}$$

In the presence of a suitable catalyst or when the gas is kept under pressure, this reaction can occur with explosive violence. To be transported safely, the gas must be dissolved in an inert organic solvent such as acetone at moderate pressure. In the liquid state, acetylene is very sensitive to shock and is highly explosive.

Acetylene, an unsaturated hydrocarbon, can be hydrogenated to yield ethylene:

$$C_2H_2(g) + H_2(g) \longrightarrow C_2H_4(g)$$

It also undergoes the following addition reactions with hydrogen halides and halogens:

$$C_2H_2(g) + HX(g) \longrightarrow CH_2{=}CHX(g)$$
$$C_2H_2(g) + X_2(g) \longrightarrow CHX{=}CHX(g)$$
$$C_2H_2(g) + 2X_2(g) \longrightarrow CHX_2{-}CHX_2(g)$$

Propyne (methylacetylene) ($CH_3{-}C{\equiv}C{-}H$) is the next member in the alkyne family. It undergoes reactions similar to those of acetylene. The addition reactions of propyne also obey Markovnikov's rule:

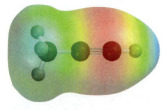

Propyne. Can you account for Markovnikov's rule in this molecule?

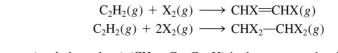

propyne   2-bromopropene

Reactions of Aromatic Hydrocarbons

Benzene is a colorless, flammable liquid (b.p. 80.1°C) obtained chiefly from petroleum and coal tar. Although it has the same empirical formula as acetylene (CH) and a high degree of unsaturation, it is much less reactive than either ethylene or acetylene. The stability of benzene is due to electron delocalization. In fact, benzene can be hydrogenated, but only with difficulty. The following reaction is carried out at significantly higher temperatures and pressures than are needed for similar reactions of alkenes:

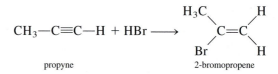

cyclohexane

Recall that alkenes react readily with halogens to form addition products because the pi bond portion of the carbon-carbon double bond can be broken easily. The most common reaction of halogens with benzene is the **substitution reaction,** in which *an atom or group of atoms replaces an atom or groups of atoms in another molecule.* For example,

bromobenzene

If this were an addition reaction, the electron delocalization of the benzene ring would be destroyed in the product and the molecule would lack the stability characteristic of aromatic compounds.

Alkyl groups can be introduced onto the ring system by reacting benzene with an alkyl halide in the presence of a Lewis acid catalyst, such as $AlCl_3$:

ethyl chloride   ethylbenzene

16.3 | The Structure and Properties of Organic Compounds Are Greatly Influenced by the Presence of Functional Groups

Organic functional groups are responsible for most of the reactions of the parent compounds. In this section, we focus on oxygen-containing functional groups (alcohols, ethers, aldehydes, ketones, carboxylic acids, and esters) and nitrogen-containing functional groups (amines).

Alcohols

All **alcohols** contain *the hydroxyl* functional *group,* —*OH.* Some common alcohols are shown in Figure 16.9. Like hydrocarbons, alcohols are named according to the number of carbon atoms in the longest continuous chain containing the hydroxy group.

The prefixes from Table 16.1 are used, combined with the suffix *-ol.* Thus, the simplest alcohol is methanol (methyl alcohol, CH_3OH) If the —OH group is not on the terminal carbon, then a numerical prefix is used to indicate the specific carbon to which the hydroxyl group is attached. The parent chain is numbered so as to give the carbon with the hydroxyl group the lowest possible number. For example, there are

Figure 16.9 Common alcohols. All of them contain the —OH group. The properties of phenol are quite different from those of aliphatic alcohols.

Methanol
(methyl alcohol)

Ethanol
(ethyl alcohol)

2-Propanol
(isopropyl alcohol)

Phenol

Ethylene glycol

two isomers of the alcohol C_3H_7OH: 1-propanol (or just propanol) and 2-propanol (also called isopropyl alcohol):

propanol

2-propanol
(isopropyl alcohol)

Alcohols containing substituents are named in a similar manner as for hydrocarbons. For example, the molecule

is called 2-methylpropanol.

Ethanol (ethyl alcohol, C_2H_5OH) is by far the best-known alcohol. It is produced biologically by the fermentation of sugar or starch. In the absence of oxygen, the enzymes present in bacterial cultures or yeast catalyze the reaction

$$C_6H_{12}O_6(aq) \xrightarrow{\text{enzymes}} 2CH_3CH_2OH(aq) + 2CO_2(g)$$

sugar   ethanol

This process gives off energy, which microorganisms, in turn, use for growth and other functions.

Commercially, ethanol is prepared by an addition reaction in which water is combined with ethene at about 280°C and 300 bar

$$CH_2{=}CH_2(g) + H_2O(g) \xrightarrow{H_2SO_4} CH_3CH_2OH(g)$$

Ethanol has countless applications as a solvent for organic chemicals and as a starting compound for the manufacture of dyes, synthetic drugs, cosmetics, and explosives. It is also a constituent of alcoholic beverages. Ethanol is the only nontoxic (more properly, the least toxic) of the straight-chain alcohols; our bodies produce an enzyme, called *alcohol dehydrogenase,* that helps metabolize ethanol by oxidizing it to ethanal (acetaldehyde):

$$CH_3CH_2OH \xrightarrow{\text{alcohol dehydrogenase}} CH_3CHO + H_2$$

ethanal

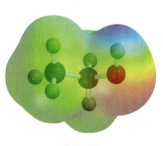

C_2H_5OH

This equation is a simplified version of what actually takes place; in your body the H atoms are taken up by other molecules, so that no H_2 gas is evolved.

Ethanol can also be oxidized by inorganic oxidizing agents, such as acidified dichromate, to ethanal (acetaldehyde) and then to ethanoic acid (acetic acid):

$$CH_3CH_2OH \xrightarrow{Cr_2O_7^{2-},\, H^+} CH_3CHO \xrightarrow{Cr_2O_7^{2-},\, H^+} CH_3COOH$$

ethanal   ethanoic acid
(acetaldehyde)   (acetic acid)

Ethanol is called an aliphatic alcohol because it is derived from an alkane (ethane). The simplest aliphatic alcohol is methanol (CH_3OH). Methanol is sometimes called *wood alcohol* because it was prepared at one time by the dry distillation of wood. It is now synthesized industrially by the reaction of carbon monoxide and molecular hydrogen at high temperatures and pressures:

$$CO(g) + 2H_2(g) \xrightarrow{Fe_2O_3\ catalyst} CH_3OH(l)$$

methanol

Methanol is highly toxic. Ingestion of only a few milliliters can cause nausea and blindness. Ethanol intended for industrial use is often mixed with methanol to prevent people from drinking it. Ethanol containing methanol or other toxic substances is called *denatured alcohol.*

Alcohols are very weakly acidic, but they do not react with strong bases, such as NaOH. The alkali metals react with alcohols to produce hydrogen:

$$2CH_3OH + 2Na \longrightarrow 2CH_3ONa + H_2$$

sodium methoxide

Although the reaction is vigorous, it is much less violent than the analogous reaction between Na and water:

$$2H_2O + 2Na \longrightarrow 2NaOH + H_2$$

Two other aliphatic alcohols used in everyday products are 2-propanol (isopropyl alcohol), commonly known as rubbing alcohol, and, ethylene glycol (see Figure 16.9), which is used in commercial automobile antifreeze,[3] Ethylene glycol has two —OH groups and so can form hydrogen bonds with water more effectively than compounds that have only a single —OH group. Most alcohols—especially those with low molar masses—are highly flammable.

Ethers

Ethers have the general structure R—O—R', where R and R' are a hydrocarbon (aliphatic or aromatic) group. To name an ether, we chose the hydrocarbon group (R or R') with the longest carbon chain as the parent group and then treat the remaining R—O— group as a substituent on the parent hydrocarbon using the suffix *-oxy.* Thus, the ether CH_3—O—CH_2CH_3 is called methoxyethane. Ethers can be formed by the reaction between two alcohols in the presence of a strong acid catalyst (such as H_2SO_4):

$$CH_3OH + HOCH_3 \xrightarrow{H_2SO_4\ catalyst} CH_3OCH_3 + H_2O$$

methoxymethane
(dimethyl ether)

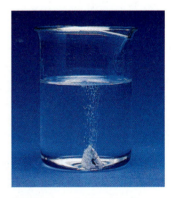

Alcohols react more slowly with sodium metal than does water.

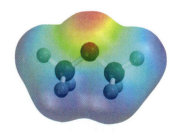

CH_3OCH_3

---

3. Some commercial automobile antifreeze products use propylene glycol ($C_3H_8O_2$) instead of ethylene glycol, because of its lower toxicity (especially in pets).

Ethoxyethane (diethyl ether) is prepared on an industrial scale by heating ethanol with sulfuric acid at 140°C

$$C_2H_5OH + C_2H_5OH \xrightarrow{H_2SO_4} C_2H_5OC_2H_5 + H_2O$$

The formation of ethers is an example of a **condensation reaction,** in which *two molecules are joined together and a small molecule, usually water, is eliminated.*

Like alcohols, ethers are extremely flammable. When left standing in air, they have a tendency to slowly form explosive peroxides:

$$C_2H_5OC_2H_5 + O_2 \longrightarrow \begin{array}{c} CH_3 \\ | \\ C_2H_5O-C-O-O-H \\ | \\ H \end{array}$$

ethoxyethane
(diethyl ether)

1-ethyoxyethyl hydroperoxide

Peroxides contain the —O—O— linkage. The simplest peroxide is hydrogen peroxide (H_2O_2). Ethoxyethane ($C_2H_5OC_2H_5$), commonly known as diethyl ether (or just "ether"), was used as an anesthetic for many years. It produces unconsciousness by depressing the activity of the central nervous system. The major disadvantages of diethyl ether are its irritating effects on the respiratory system and the occurrence of postanesthetic nausea and vomiting. "Neothyl," or methoxypropane (methyl propyl ether, $CH_3OCH_2CH_2CH_3$), is currently favored as an anesthetic because it is relatively free of side effects.

Aldehydes and Ketones

Under mild oxidation conditions, it is possible to convert alcohols to aldehydes and ketones:

$$CH_3OH + \tfrac{1}{2}O_2 \longrightarrow H_2C{=}O + H_2O$$

methanal
(formaldehyde)

$$C_2H_5OH + \tfrac{1}{2}O_2 \longrightarrow \begin{array}{c} H_3C \\ \diagdown \\ \quad C{=}O + H_2O \\ \diagup \\ H \end{array}$$

ethanal
(acetaldehyde)

$$\begin{array}{c} H \\ | \\ CH_3-C-CH_3 \\ | \\ OH \end{array} + \tfrac{1}{2}O_2 \longrightarrow \begin{array}{c} H_3C \\ \diagdown \\ \quad C{=}O + H_2O \\ \diagup \\ H_3C \end{array}$$

2-propanone
(acetone)

CH$_3$CHO

The functional group in these compounds is the *carbonyl group,* $\diagdown C{=}O$. In an *aldehyde,* at least one hydrogen atom is bonded to the carbon in the carbonyl group. In a **ketone,** the carbon atom in the carbonyl group is bonded to two hydrocarbon groups. The official naming conventions for aldehydes and ketones are similar to those for alcohols, except that aldehydes carry the suffix *-al* and ketones have the suffix *-one*. For example, the compound

$$CH_3-CH_2-C \begin{smallmatrix} O \\ \\ H \end{smallmatrix}$$

is an aldehyde with three carbons and is called *propanal*. Similarly, the molecule

$$CH_3 \begin{smallmatrix} O \\ \| \\ C \end{smallmatrix} CH_3$$

is a three-carbon ketone with the name 2-propanone (acetone). In an aldehyde, the carbon atom of the carbonyl group is always number 1. In a ketone, though, the carbon chain is numbered so as to give the carbonyl carbon the lowest possible number.

The simplest aldehyde, methanal (formaldehyde, $H_2C=O$) has a tendency to *polymerize*, that is, the individual molecules join together to form a compound of high molar mass. This reaction gives off tremendous amounts of heat and is often explosive, so methanal is usually prepared and stored in aqueous solution (to reduce the concentration). This rather disagreeable-smelling liquid is used as a starting material in the polymer industry (see Section 16.4) and in the laboratory as a preservative for animal specimens. Interestingly, the higher molar mass aldehydes, such as cinnamic aldehyde,

Cinnamic aldehyde gives cinnamon its characteristic aroma.

$$\langle\bigcirc\rangle-CH=CH-C\begin{smallmatrix} H \\ \\ O \end{smallmatrix}$$

have a pleasant odor and are used in the manufacture of perfumes.

Ketones generally are less reactive than aldehydes. The simplest ketone is 2-propanone (acetone), a volatile liquid that is used mainly as a solvent for organic compounds and nail polish remover.

Carboxylic Acids

Under appropriate conditions, both alcohols and aldehydes can be oxidized to ***carboxylic acids****, acids that contain the carboxyl group,* —*COOH*. For example, ethanol and ethanal can be oxidized to form ethanoic acid (acetic acid, CH_3COOH):

$$CH_3CH_2OH + O_2 \longrightarrow CH_3COOH + H_2O$$
$$CH_3CHO + \tfrac{1}{2}O_2 \longrightarrow CH_3COOH$$

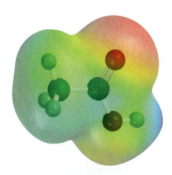

CH_3COOH

These reactions occur so readily, in fact, that wine must be protected from atmospheric oxygen while in storage. Otherwise, it would soon turn to vinegar as a result of the formation of acetic acid. Figure 16.10 shows the structure of some of the common carboxylic acids. The IUPAC names for carboxylic acids are formed by adding the suffix *-oic acid* to the parent hydrocarbon. The parent hydrocarbon includes the carboxyl carbon, so the compounds CH_3COOH and $CH_3CH_2CH_2COOH$ are ethanoic acid (acetic acid) and butanoic acid (butyric acid), respectively.

Carboxylic acids are widely distributed in nature; they are found in both the plant and animal kingdoms. All protein molecules are made of amino acids, a special kind of carboxylic acid containing an amino group (—NH_2) and a carboxyl group (—COOH).

The oxidation of ethanol to acetic acid in wine is catalyzed to enzymes.

Figure 16.10 Some common carboxylic acids. They all contain the —COOH group. (Glycine is one of the amino acids found in proteins.)

methanoic acid (formic acid)

ethanoic acid (acetic acid)

butanoic acid (butyric acid)

benzoic acid

glycine

oxalic acid

citric acid

This is a condensation reaction.

Unlike the inorganic acids HCl, HNO$_3$, and H$_2$SO$_4$, carboxylic acids are usually weak acids. They react with alcohols to form pleasant-smelling esters in a process called *esterification*:

$$CH_3COOH + HOCH_2CH_3 \longrightarrow CH_3-\overset{O}{\underset{\|}{C}}-O-CH_2CH_3 + H_2O$$

acetic acid   ethanol   ethyl acetate

Esterification reactions are condensation reactions because two reactants (in this case, an alcohol and a carboxylic acid) are joined together and a small molecule (water) is eliminated.

Other common reactions of carboxylic acids include neutralization

$$CH_3COOH + NaOH \longrightarrow CH_3COONa + H_2O$$

and the formation of acid halides, such as acetyl chloride,

$$CH_3COOH + PCl_5 \longrightarrow CH_3COCl + HCl + POCl_3$$

acetyl chloride

Acid halides are reactive compounds used as intermediates in the preparation of many other organic compounds. They hydrolyze in much the same way as many nonmetallic halides, such as SiCl$_4$:

$$CH_3COCl(l) + H_2O(l) \longrightarrow CH_3COOH(aq) + HCl(g)$$
$$SiCl_4(l) + 3H_2O(l) \longrightarrow H_2SiO_3(s) + 4HCl(g)$$

silicic acid

Esters

Esters have the general formula R′COOR, where R′ can be H or a hydrocarbon group and R is a hydrocarbon group. Esters are used in the manufacture of perfumes and as flavoring agents in the confectionery and soft-drink industries. Many fruits owe their characteristic smell and flavor to the presence of small quantities of esters. For example, bananas contain 3-methylbutyl acetate [CH$_3$COOCH$_2$CH$_2$CH(CH$_3$)$_2$], oranges contain octyl acetate (CH$_3$COOC$_8$H$_{17}$), and apples contain methyl butyrate (CH$_3$CH$_2$CH$_2$COOCH$_3$). As mentioned previously, esters are formed through the esterification (condensation reaction) of a carboxylic acid with an alcohol.

The odor of fruits is mainly due to the ester compounds in them.

The functional group in esters is the —COOR group. In the presence of an acid catalyst, such as HCl, esters hydrolyze to yield a carboxylic acid and an alcohol (the opposite of esterification). For example, in an acid solution, ethyl acetate hydrolyzes to ethanoic acid (acetic acid) and ethanol:

$$CH_3COOC_2H_5 + H_2O \rightleftharpoons CH_3COOH + C_2H_5OH$$

ethyl acetate   ethanoic acid   ethanol
(acetic acid)

The hydrolysis of an ester does not go to completion, however, because the reverse reaction, that is, the formation of an ester from an alcohol and an acid, also occurs to an appreciable extent. When an ester is hydrolyzed in basic solution (such as aqueous NaOH), on the other hand, the reaction does go to completion from left to right because the sodium acetate that is formed does not react with ethanol:

$$CH_3COOC_2H_5 + NaOH \longrightarrow CH_3COO^-Na^+ + C_2H_5OH$$

ethyl acetate   sodium acetate   ethanol

As a result, ester hydrolysis is usually carried out in basic solutions. Note that NaOH is a reactant in this reaction and not a catalyst because NaOH is consumed (not regenerated) in the reaction. The term **saponification** (meaning *soap making*) was originally used to describe the alkaline hydrolysis of fatty acid esters to yield soap molecules (sodium stearate):

$$C_{17}H_{35}COOC_2H_5 + NaOH \longrightarrow C_{17}H_{35}COO^-Na^+ + C_2H_5OH$$

ethyl stearate   sodium stearate

Saponification has now become *a general term for alkaline hydrolysis of any type of ester.*

Amines

Amines are *organic bases with the general formula R_3N, where R may be H or a hydrocarbon group.* As with ammonia, the reaction of amines with water is

$$R-NH_2 + H_2O \longrightarrow R-NH_3^+ + OH^-$$

where R represents a hydrocarbon group. Like all bases, amines form salts when they react with acids:

$$CH_3CH_2NH_2 + HCl \longrightarrow CH_3CH_2NH_3^+Cl^-$$

ethanamine   ethylammonium
(ethylamine)   chloride

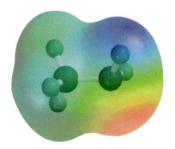

CH_3NH_2

These salts are usually colorless, odorless solids.

Aromatic amines are used mainly in the manufacture of dyes. Benzenamine (aniline), the simplest aromatic amine, is a toxic compound; a number of other aromatic amines such as 2-naphthylamine and benzidine are potent carcinogens:

benzenamine   2-naphthylamine   benzidine
(aniline)

Example 16.4

Draw structures for the following organic compounds: (*a*) 3-pentanol (*b*) 2-methylbutanal, (*c*) 1,2-dimethyoxyethane, and (*d*) 2,2-dimethylpropanoic acid.

Strategy Find the longest carbon chain containing the functional group and use the naming rules discussed previously.

Solution (a) The suffix -*ol* indicates that the compound is an alcohol and the prefix *pentan-* indicates that the parent hydrocarbon is pentane. The numerical prefix (3-) tells us that the —OH group is attached to the third carbon in the pentane chain, so the structure is

$$\underset{1}{CH_3}-\underset{2}{CH_2}-\underset{3}{\overset{\overset{\displaystyle OH}{|}}{CH}}-\underset{4}{CH_2}-\underset{5}{CH_3}$$

(b) The suffix -*al* means that the compound is an aldehyde, so it has a carbonyl (—C=O) group attached to a terminal carbon of the parent hydrocarbon chain. The parent chain is four carbons long because the prefix is *butan-*. Additionally, a methyl group is attached to the second carbon of the parent (the carbon adjacent to the carbonyl carbon because the carbonyl carbon is always the first carbon in an aldehyde):

$$\underset{H}{\overset{O}{\diagdown}}\underset{1}{C}-\underset{2}{\overset{\overset{\displaystyle CH_3}{|}}{CH}}-\underset{3}{CH_2}-\underset{4}{CH_3}$$

(c) This compound is an ether and the parent hydrocarbon chain is ethane. The *dimethoxy-* prefix indicates that there are two methoxy groups (CH_3—O—) groups attached to the parent ethane at the first and second carbon positions:

$$CH_3-O-\underset{1}{CH_2}-\underset{2}{CH_2}-OCH_3$$

(d) This compound is a carboxylic acid and the longest chain (including the carboxyl carbon) consists of three carbons (*propan-*). Attached to this chain at the second carbon position (the carbon adjacent to the carboxyl carbon) are two methyl groups, giving a structure of

$$\underset{3}{CH_3}-\underset{2}{\overset{\overset{\displaystyle CH_3}{|}}{\underset{\underset{\displaystyle CH_3}{|}}{C}}}-\underset{1}{C}\overset{\diagup O}{\diagdown_{OH}}$$

The common name for this compound is *pivalic acid*.

Practice Exercise Write structural formulas for the following compounds: (*a*) 1,3-dibromo-2-pentanol, (*b*) 2-methoxy-3-methylpentane, (*c*) 2-methyl-3-pentanone, and (*d*) 3-phenyl-pentanoic acid.

Summary of Functional Groups

Table 16.4 summarizes the common functional groups, including the C=C and C≡C groups. Organic compounds commonly contain more than one functional group. Generally, the reactivity of a compound is determined by the number and types of functional groups it contains.

Example 16.5 shows how we can use the functional groups to predict reactions.

16.3 *The Structure and Properties of Organic Compounds Are Greatly Influenced by the Presence of Functional Groups*

823

Table 16.4	Important Functional Groups and Their Reactions	
Functional Group	**Name**	**Typical Reactions**
$\overset{\diagdown}{\diagup}C{=}C\overset{\diagup}{\diagdown}$	Carbon-carbon double bond	Addition reactions with halogens, hydrogen halides, and water, hydrogenation to yield alkanes
$-C{\equiv}C-$	Carbon-carbon triple bond	Addition reactions with halogens, hydrogen halides; hydrogenation to yield alkenes and alkanes
$-\ddot{X}\!:$ (X = F, Cl, Br, I)	Halogen	Exchange reactions: $CH_3CH_2Br + KI \longrightarrow CH_3CH_2I + KBr$
$-\ddot{O}-H$	Hydroxyl	Esterification (formation of an ester) with carboxylic acids; oxidation to aldehydes, ketones, and carboxylic acids
$\overset{\diagdown}{\diagup}C{=}\ddot{O}$	Carbonyl	Reduction to yield alcohols; oxidation of aldehydes to yield carboxylic acids
$-\overset{\overset{\textstyle :O:}{\|}}{C}-\ddot{O}-H$	Carboxyl	Esterification with alcohols; reaction with phosphorus pentachloride to yield acid chlorides
$-\overset{\overset{\textstyle :O:}{\|}}{C}-\ddot{O}-R$ (R = hydrocarbon)	Ester	Hydrolysis to yield acids and alcohols
$-\ddot{N}\overset{\diagup R}{\diagdown_{R}}$ (R = H or hydrocarbon)	Amine	Formation of ammonium salts with acids

Example 16.5

Cholesterol is a major component of gallstones, and it is believed that the cholesterol level in the blood is a contributing factor in certain types of heart disease. From the following structure of the compound, predict its reaction with (*a*) Br_2, (*b*) H_2 (in the presence of a Pt catalyst), and (*c*) CH_3COOH.

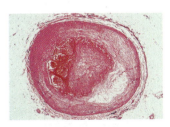

An artery becoming blocked by cholesterol.

Strategy To predict the type of reactions a molecule may undergo, we must first identify the functional groups present (see Table 16.4).

—Continued

Important Experimental Technique: Nuclear Magnetic Resonance Spectroscopy

There are a number of experimental techniques that are useful in the identification and characterization of organic compounds. Two techniques, mass spectrometry and IR spectroscopy, were discussed in Chapters 0 and 4, respectively. Here, we will discuss another important technique, *nuclear magnetic resonance (NMR) spectroscopy*.

Like electrons (Section 1.4), the nucleus of an atom has an intrinsic spin and a corresponding magnetic moment. Unlike electrons, which have a total spin of 1/2 (with possible orientations of $+1/2$ and $-1/2$), the total spin (I) of a nucleus can take on both integer and half-integer values:

$$I = 0, \frac{1}{2}, 1, \frac{3}{2}, 2, \ldots$$

The value of I depends upon the number of protons (atomic number) and neutrons present in the nucleus:

- If an even number of both protons and neutrons are present (for example, carbon-12) the nucleus will have zero spin ($I = 0$).

- If the number of protons is odd and the number of neutrons is even (or vice versa), the nucleus will have half-integer spin (for example, carbon-13 with $I = 1/2$).

- If both protons and neutrons are present in odd numbers, the nuclear spin will be a positive integer (for example, nitrogen-14 has $I = 1$).

A nucleus with a total spin I possesses $(2I + 1)$ possible spin orientations, each denoted by a different value of the nuclear spin quantum number, m_I, which can take on any value in the set $\{-I, -I + 1, \ldots, I - 1, I\}$. In the absence of a magnetic field, these $2I + 1$ spin orientation states corresponding to a nucleus of spin I are degenerate, that is, they are equal in energy. In the presence of an external magnetic field of strength B_0, this degeneracy is lifted and the nuclear spin energy (E_I) is proportional to the value of m_I:

$$E_I = -m_I B_0 \gamma \hbar \tag{1}$$

where γ is called the *gyromagnetic ratio* and $\hbar = h/(2\pi)$.

The proton (^{1}H), for example, has a nuclear spin $I = 1/2$, with two possible orientations: $m_I = +1/2$ or $m_I = -1/2$. In the presence of a magnetic field of strength B_0, the difference in energy (ΔE) between the two spin states ($+1/2$ and $-1/2$) is

$$\Delta E = E_{-1/2} - E_{1/2} = -B_0\left[\left(-\frac{1}{2}\right) - \left(+\frac{1}{2}\right)\right]\gamma \hbar = \gamma B_0 \hbar$$

A transition from the $m_I = +\frac{1}{2}$ level to the $m_I = -\frac{1}{2}$ level (called a NMR) can be induced by electromagnetic radiation with a frequency corresponding to the energy gap:

$$\nu = \Delta E/h = \frac{\gamma B_0}{2\pi} \tag{2}$$

This resonance frequency is called the *Larmor frequency.* In nuclear magnetic resonance spectroscopy, the resonance is determined either by varying the radiation frequency (typically in the radio wave region) or by varying the applied magnetic field strength.

The strength of the magnetic field is measured in *tesla* (T), after the Greek engineer and inventor Nikola Tesla.[4] Another common unit is the *gauss* (10^4 gauss = 1 T). The gyromagnetic ratio has units of $T^{-1} s^{-1}$ and will vary depending on the nucleus type. The most common isotopes used in NMR spectroscopy are ^{1}H, ^{19}F, ^{31}P, and ^{13}C. The most abundant isotope of carbon, ^{12}C, cannot be studied with NMR because it has a nuclear spin of zero.)

NMR Spectra of Organic Molecules

The usefulness of NMR in organic chemistry arises from the fact that the resonance frequency of a given nucleus is affected by its local environment within the molecule; that is, nuclei with different environments will have slightly different resonance frequencies. Thus, the pattern of nuclear magnetic resonance frequencies observed (the NMR spectrum) will differ from compound to compound, which allows NMR to be used as a method for identifying and characterizing chemical substances. The shape of an NMR spectrum for a given compound is determined primarily by three effects: *chemical shift, spin-spin coupling,* and *kinetic exchange processes.*

---

4. Nikola Tesla (1856–1943) Serbian-American physicist, engineer and inventor. Tesla immigrated to the United States in 1884. He worked for a short time in the laboratory of Thomas Edison but soon left to devote himself full time to independent experimentation and invention. In 1888, he devised a method to produce and transmit alternating electric current. He is credited with a number of inventions including the Tesla coil, a type of transformer.

An NMR spectrometer. The instrument is surrounded by a insulated container of liquid nitrogen, which is used to cool the large electromagnets required to generate the magnetic field.

Chemical Shift

The resonance frequency of a nucleus depends on the *local* magnetic field, which is a combination of the applied external field and the field due to neighboring nuclei and electrons. The difference between the actual resonance frequency and that predicted by Equation 2 is called the *chemical shift*. The figure below shows a low-resolution proton (^{1}H) NMR spectrum of ethanol (CH_3CH_2OH). Three separate peaks are observed because there are three different types of H atoms in ethanol with respect to local environment: the hydroxyl hydrogen (—OH), two methylene hydrogen atoms (—CH_2) and three methyl (—CH_3) hydrogen atoms. The relative areas under the hydroxyl, methylene, and methyl hydrogen peaks have relative areas in the ratio 1:2:3, respectively, reflecting the relative number of hydrogen atoms of each type.

In practice the chemical shift of a nucleus is generally measured relative to the chemical shift for a standard reference nucleus. We define the chemical shift parameter (δ) as the relative difference in resonance frequencies between a nucleus of interest (ν) and that of a reference nucleus (ν_{ref})

$$\delta = \frac{\nu - \nu_{ref}}{\nu_{ref}} \times 10^6 \qquad (3)$$

Because of the factor of 10^6 present in Equation 3, chemical shifts are reported as parts per million (ppm). For proton (^{1}H) NMR, the reference compound is usually tetramethylsilane (TMS), $(CH_3)_4Si$.

Spin-Spin Coupling

By increasing the magnetic field strength, the resolution of an NMR spectrum can be improved. The high-resolution ^{1}H spectrum of ethanol is shown.

In high-resolution, we see that the —CH_2 and —CH_3 peaks in the low-resolution spectrum are made up of four and three peaks, respectively. Because the hydrogen atoms in each group are equivalent the splitting of the lines at high resolution is not due to chemical shift. Instead, the splitting is caused by an interaction between nuclear spins and nearby nonequivalent spins that is transmitted through the bonding electrons. This phenomenon is called *spin-spin coupling*. In

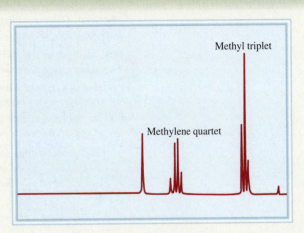

High-resolution NMR spectrum of ethanol.

the methylene group in ethanol, each nuclear spin has two possible orientations. The methyl peak is split into two peaks by interaction with the magnetic field generated by the first methylene proton. Interaction with the second methylene proton further splits these two peaks to give a total of four lines. Two of these peaks are coincident so only three peaks are observed—a pattern called a *triplet*. In a similar manner, spin-spin coupling with the methyl group splits the methylene peak into four separate peaks, called a *quartet*. When examining an unknown compound, the observed peak patterns (singlet, triplet, etc.) can be used to deduce the local bonding environment.

Kinetic Exchange Processes

In ethanol, the hydroxyl peak should be split by spin-spin interaction with the methylene protons. However, the high-resolution NMR spectrum shows only a single hydroxyl peak. The lack of apparent spin-spin coupling in this peak is due to the fact that this spectrum was taken on a sample that contained a small amount of water. In the NMR spectrum of pure ethanol, the hydroxyl peak will be a triplet as we expect. However, if a small amount of water is present, a rapid proton-exchange reaction between the —OH proton in ethanol and H_3O^+ in the water will occur that has the effect of broadening the three peaks in the —OH triplet so that they coalesce into a single observed peak.

If the rate of exchange in a proton-exchange reaction is slow, the split peaks will not completely coalesce, but would be broader and less distinct than in the absence of exchange. In such cases, the rate constant for the exchange can be determined. Therefore, NMR spectroscopy can be used as a tool to study the kinetics of proton-exchange reactions and of many other processes such as the rotation about a chemical bond and diffusion.

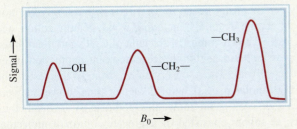

The low-resolution proton NMR spectrum of ethanol

Continued—

Solution Despite a relatively complex structure, there are only two functional groups in cholesterol: the hydroxyl group and the carbon-carbon double bond.

(a) The reaction with bromine results in the addition of bromine to the double-bonded carbons, which become single-bonded.

(b) This is a hydrogenation reaction. Again, the carbon-carbon double bond is converted to a carbon-carbon single bond.

(c) The acid reacts with the hydroxyl group to form an ester and water (esterification, a type of condensation reaction).

The products formed in (*a*), (*b*), and (*c*) are

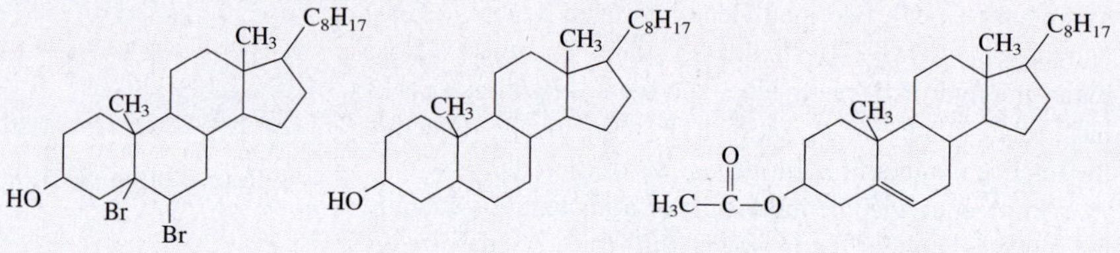

Practice Exercise Predict the products of the following reaction:

$$CH_3OH + CH_3CH_2COOH \longrightarrow ?$$

16.4 | Polymers Are Very Large Molecular Weight Compounds Formed from the Joining Together of Many Subunits Called Monomers

A **polymer** is *a molecular compound with a high molar mass, ranging into thousands and millions of grams, and made up of many repeating units called* **monomers.** The word *polymer* is a combination of the Greek prefix *poly-*, meaning "many," and the Greek suffix *-mer*, meaning "parts." The physical properties of these so-called **macromolecules** differ greatly from those of small molecules, and special techniques are required to study them.

Naturally occurring polymers include proteins, nucleic acids, cellulose (polysaccharides), and rubber (polyisoprene). Most synthetic polymers are organic compounds. Familiar examples are nylon, poly(hexamethylene adipamide); Dacron, poly(ethylene terephthalate); and Lucite or Plexiglas, poly(methyl methacrylate).

The development of polymer chemistry began in the 1920s with the investigation into some unusual properties of certain materials, including wood, gelatin, cotton, and rubber. Rubber, for example, had a known empirical formula of C_5H_8, but it exhibited high viscosity, low osmotic pressure, and negligible freezing-point depression when it was dissolved in an organic solvent. These observations strongly suggested the presence of solutes of very high molar mass. Chemists at that time, however, were not ready to accept the idea that such giant molecules could exist. Instead, they postulated that materials such as rubber consisted of aggregates of small molecular units, like C_5H_8 or $C_{10}H_{16}$, held together by intermolecular forces. This misconception persisted until Hermann Staudinger[5] clearly showed that these so-called aggregates were, in fact, enormously large molecules, each of which contained many thousands of atoms held together by covalent bonds.

Once the structures of these macromolecules were understood, the way was open for manufacturing polymers, which now pervade almost every aspect of our daily lives. About 90 percent of today's chemists, including biochemists, work with polymers.

Because of their size, we might expect molecules containing thousands of carbon and hydrogen atoms to form an enormous number of structural and geometric isomers (if C=C bonds are present). Polymers, however, are made up of *monomers,* simple repeating units, and this composition severely restricts the number of possible isomers. Synthetic polymers are created by joining monomers together, one at a time, by means of addition reactions and condensation reactions.

Addition Reactions

Addition reactions involve the addition of atoms to compounds containing double or triple bonds, particularly C=C and C≡C. Examples include hydrogenation and the reactions of hydrogen halides and halogens with alkenes and alkynes.

Polyethylene, a very stable polymer used in packaging wraps, is made by joining ethylene monomers via an addition-reaction mechanism. First an *initiator* molecule (R_2) is heated to produce two radicals:

$$R_2 \longrightarrow 2R \cdot$$

The reactive radical attacks an ethylene molecule to generate a new radical:

$$R \cdot + CH_2{=}CH_2 \longrightarrow R{-}CH_2{-}CH_2 \cdot$$

which further reacts with another ethylene molecule, and so on:

$$R{-}CH_2{-}CH_2 \cdot + CH_2{=}CH_2 \longrightarrow R{-}CH_2{-}CH_2{-}CH_2{-}CH_2 \cdot$$

Very quickly a long chain of CH_2 groups is built. Eventually, this process is terminated by the combination of two long-chain radicals to give the polymer called polyethylene:

$$R{-}(CH_2{-}CH_2)_n\,CH_2CH_2 \cdot + R{-}(CH_2{-}CH_2)_n\,CH_2CH_2 \cdot \longrightarrow$$
$$R{-}(CH_2{-}CH_2)_n\,CH_2CH_2{-}CH_2CH_2{-}(CH_2{-}CH_2)_n\,R$$

where $-(CH_2{-}CH_2)_n-$ is a convenient shorthand convention for representing the repeating unit in the polymer. The value of n is understood to be very large (on the order of hundreds).

The individual chains of polyethylene pack together well, giving it distinctive crystalline properties (Figure 16.11). Polyethylene is mainly used in films in frozen food packaging and other product wrappings. A specially treated type of polyethylene called Tyvek is used for home insulation.

---

5. Hermann Staudinger (1881–1963). German chemist. One of the pioneers in polymer chemistry. Staudinger was awarded the Nobel Prize in Chemistry in 1953.

Figure 16.11 The structure of polyethylene. Each carbon atom is sp^3-hybridized.

Figure 16.12 A cooking
utensil coated with Silverstone,
which contains polytetrafluoroethylene (Teflon).

Polyethylene is a **homopolymer** because it is *a polymer made up of only one type of monomer.* Other homopolymers that are synthesized by a radical mechanism are polytetrafluoroethylene (Teflon; Figure 16.12) and poly(vinyl chloride) (PVC):

$$\text{\footnotesize(}CF_2{-}CF_2\text{\footnotesize)}_n \qquad \text{\footnotesize(}CH_2{-}CH\text{\footnotesize)}_n$$
$$\qquad\qquad\qquad\qquad\quad\ \ |$$
$$\qquad\qquad\qquad\qquad\quad\ Cl$$

Teflon   PVC

The chemistry of polymers is more complex if the monomers are asymmetric:

$$\overset{H_3C}{}\diagdown\underset{H}{\overset{}{C}}{=}\underset{H}{\overset{H}{C}}\diagup \qquad \left(\overset{CH_3}{\underset{H}{\overset{}{C}}}\underset{H}{\overset{H}{C}}\right)_n$$

propene   polypropene

Several stereoisomers can result from an addition reaction of propylenes (Figure 16.13). If the additions occur randomly, *atactic* polypropylenes form, which do not pack together well. These polymers are rubbery, amorphous, and relatively weak. Two other possibilities are an *isotactic* structure, in which the R groups are all on the same side of the asymmetric carbon atoms, and a *syndiotactic* form, in which the R groups alternate to the left and right of the asymmetric carbons. Of these, the isotactic isomer has the highest melting point and greatest crystallinity and is endowed with superior mechanical properties.

A major problem that the polymer industry faced in the beginning was how to selectively synthesize either the isotactic or syndiotactic polymer without having it contaminated by other products. Giulio Natta[6] and Karl Ziegler[7] discovered, however, that certain catalysts, including triethylaluminum $[Al(C_2H_5)_3]$ and titanium trichloride $(TiCl_3)$, could promote the formation of specific isomers only. Using Natta-Ziegler catalysts, chemists have been able to design polymers to suit almost any purpose.

---

6. Giulio Natta (1903–1979). Italian chemist. Natta received the Nobel Prize in Chemistry in 1963 for discovering stereospecific catalysts for polymer synthesis.

7. Karl Ziegler (1898–1976). German chemist. Ziegler shared the Nobel Prize in Chemistry in 1963 with Natta for his work in polymer synthesis.

Figure 16.13 Stereoisomers of polymers. When the R group (green sphere) is CH_3, the polymer is polypropylene. (a) When the R groups are all on one side of the chain, the polymer is isotactic. (b) When the R groups alternate from side to side, the polymer is syndiotactic. (c) When the R groups are disposed at random, the polymer is atactic.

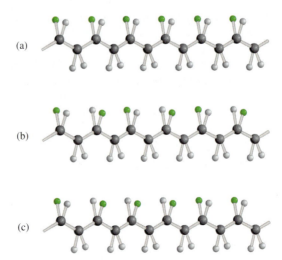

Figure 16.14 Latex (an aqueous suspension of rubber particles) being collected from a rubber tree.

Rubber

Rubber is probably the best-known organic polymer and the only true hydrocarbon polymer found in nature. It is formed by the radical addition of the monomer isoprene. Actually, polymerization can result in poly-*cis*-isoprene, poly-*trans*-isoprene, or a mixture of both, depending on reaction conditions:

$$n\text{CH}_2=\overset{\overset{\text{CH}_3}{|}}{\text{C}}-\text{CH}=\text{CH}_2 \longrightarrow \left(\underset{\overset{\displaystyle -\text{CH}_2}{}}{\overset{\displaystyle \text{CH}_3}{}}\text{C}=\text{C}\underset{\overset{\displaystyle \text{CH}_2}{}}{\overset{\displaystyle \text{H}}{}} \right)_n \quad \text{or} \quad \left(\underset{\overset{\displaystyle \text{CH}_3}{}}{\overset{\displaystyle -\text{CH}_2}{}}\text{C}=\text{C}\underset{\overset{\displaystyle \text{CH}_2}{}}{\overset{\displaystyle \text{H}}{}} \right)_n$$

isoprene   poly-*cis*-isoprene   poly-*trans*-isoprene

Note that in the *cis* isomer, the two CH_2 groups are on the same side of the C=C bond, whereas the same groups are across from each other in the *trans* isomer. Natural rubber is poly-*cis*-isoprene, which is extracted from the tree *Hevea brasiliensis* (Figure 16.14).

An unusual and very useful property of rubber is its elasticity. Rubber will stretch up to 10 times its length and, if released, will return to its original size. In contrast, a piece of copper wire can be stretched only a small percentage of its length and still return to its original size. Unstretched rubber has no regular X-ray diffraction pattern and is therefore amorphous. Stretched rubber, however, possesses a fair amount of crystallinity and order. The elasticity of rubber is due to the flexibility of each individual long-chain molecule. In the bulk state, however, where rubber is a tangle of polymeric chains, a strong external force can make individual chains slip past one another instead of stretch, thereby causing the rubber to lose most of its elasticity. In 1839, Charles Goodyear[8] discovered that natural rubber could be cross-linked with sulfur (using zinc oxide as the catalyst) to prevent chain slippage (Figure 16.15). His process, known as *vulcanization*, paved the way for many practical and commercial uses of rubber, such as in automobile tires and dentures.

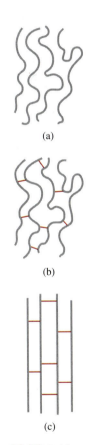

(a)

(b)

(c)

Figure 16.15 Rubber molecules ordinarily are bent and convoluted. Parts (a) and (b) represent the long chains before and after vulcanization, respectively; (c) shows the alignment of molecules when stretched. Without vulcanization these molecules would slip past one another, and rubber would lose its elastic properties.

8. Charles Goodyear (1800–1860). American chemist. Goodyear was the first person to realize the potential of natural rubber. His vulcanization process made rubber usable in countless ways and opened the way for the development of the automobile industry.

Table 16.5	Some Monomers and Their Common Synthetic Polymers		
Monomer		**Polymer**	
Formula	**Name**	**Name and Formula**	**Uses**
$H_2C{=}CH_2$	Ethylene	Polyethylene $+CH_2{-}CH_2+_n$	Plastic piping, bottles, electrical insulation, toys
$H_2C{=}\overset{\displaystyle H}{\underset{\displaystyle CH_3}{C}}$	Propene	Polyethylene $\left(\!CH{-}CH_2{-}CH{-}CH_2\!\right)_n$ with CH_3 CH_3	Packaging film, carpets, crates for soft-drink bottles, lab wares, toys
$H_2C{=}\overset{\displaystyle H}{\underset{\displaystyle Cl}{C}}$	Vinyl chloride	Poly(vinyl chloride) (PVC) $+CH_2{-}\underset{\displaystyle Cl}{CH}+_n$	Piping, siding, gutters, floor tile, clothing, toys
$H_2C{=}\overset{\displaystyle H}{\underset{\displaystyle CN}{C}}$	Acrylonitrile	Polyacrylonitrile (PAN) $\left(\!CH_2{-}\underset{\displaystyle CN}{CH}\!\right)_n$	Carpets, knitwear
$F_2C{=}CF_2$	Tetrafluoro- ethylene	Polytetrafluoroethylene (Teflon) $+CF_2{-}CF_2+_n$	Coating on cooking utensils, electrical insulation, bearings
$H_2C{=}\overset{\displaystyle COOCH_3}{\underset{\displaystyle CH_3}{C}}$	Methyl methacrylabe	Poly(methyl methacrylate) (Plexiglas) $+CH_2{-}\overset{\displaystyle COOCH_3}{\underset{\displaystyle CH_3}{C}}+_n$	Optical equipment, home furnishings
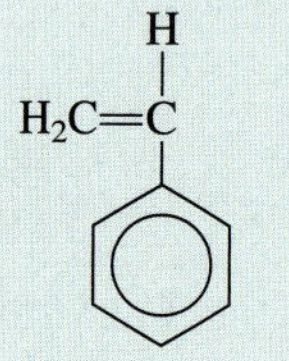 $H_2C{=}\overset{\displaystyle H}{\underset{\displaystyle \bigcirc}{C}}$	Styrene	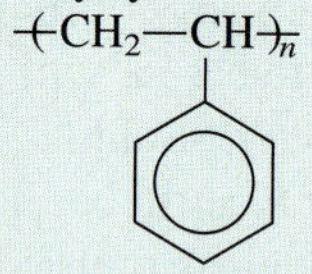 Polystyrene $+CH_2{-}\underset{\displaystyle \bigcirc}{CH}+_n$	Containers, thermal insulation (ice buckets, water coolers), toys
$H_2C{=}\overset{\displaystyle H}{C}{-}\overset{\displaystyle H}{C}{=}CH_2$	Butadiene	Polybutadiene $+CH_2CH{=}CHCH_2+_n$	Tire tread, coating resin
See above structures	Butadiene and styrene	Styrene-butadiene rubber (SBR) 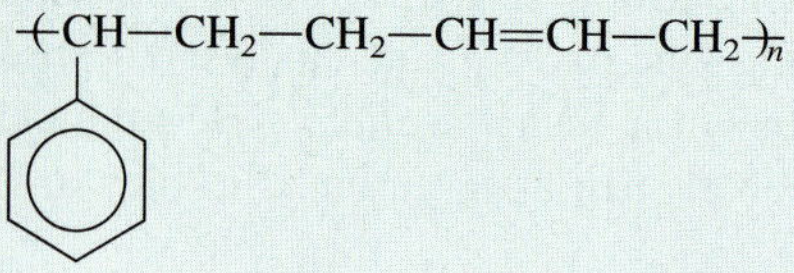 $+CH{-}CH_2{-}CH_2{-}CH{=}CH{-}CH_2+_n$ with $\bigcirc$	Synthetic rubber

During World War II, a shortage of natural rubber in the United States prompted an intensive program to produce synthetic rubber. Most synthetic rubbers (called *elastomers*) are made from petroleum products such as ethylene, propylene, and butadiene. For example, chloroprene molecules polymerize readily to form polychloroprene, commonly known as *neoprene,* which has properties that are comparable or even superior to those of natural rubber:

$$H_2C=CH-CCl=CCl_2$$

chloroprene

$$\left(\begin{array}{c} -CH_2 \qquad H \\ \diagdown C=C \diagup \\ Cl \qquad CH_2- \end{array}\right)_n$$

polychloroprene
(neoprene)

Another important synthetic rubber is formed by the addition of butadiene to styrene in a 3:1 ratio to give styrene-butadiene rubber (SBR). SBR is a ***copolymer*** because it is *a polymer containing two or more different monomers.* Table 16.5 lists some common homopolymers and one copolymer produced by addition reactions.

Condensation Reactions

One of the best-known polymer condensation processes is the reaction between hexamethylenediamine and adipic acid, shown in Figure 16.16. The final product, called nylon-66 (because there are six carbon atoms each in hexamethylenediamine and adipic acid), was first made by Wallace Carothers[9] at Du Pont in 1931. The versatility of nylons is so great that the annual production of nylons and related substances now amounts to several billion pounds. Figure 16.17 shows how nylon–66 is prepared in the laboratory.

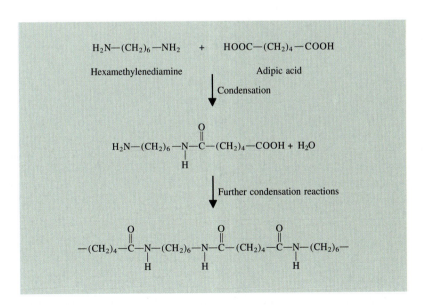

Figure 16.16 The formation of nylon by the condensation reaction between hexamethylenediamine and adipic acid.

Figure 16.17 The nylon rope trick. Adding a solution of adipoyl chloride (an adipic acid derivative in which the OH groups have been replaced by Cl groups) in cyclohexane to an aqueous solution of hexamethylenediamine causes nylon to form at the interface of the two solutions, which do not mix. It can then be drawn off.

---

9. Wallace H. Carothers (1896–1937). American chemist. Besides its enormous commercial success, Carothers' work on nylon is ranked with that of Staudinger in clearly elucidating macromolecular structure and properties. Depressed by the death of his sister and convinced that his life's work was a failure, Carothers committed suicide at the age of 41.

Condensation reactions are also used to manufacture Dacron (polyester).

$$n\text{HO}-\overset{\overset{\text{O}}{\|}}{\text{C}}-\underset{\text{terephthalic acid}}{\bigcirc}-\overset{\overset{\text{O}}{\|}}{\text{C}}-\text{OH} + n\text{HO}-(\text{CH}_2)_2-\text{OH} \longrightarrow \underset{\text{Dacron}}{\left(\overset{\overset{\text{O}}{\|}}{\text{C}}-\bigcirc-\overset{\overset{\text{O}}{\|}}{\text{C}}-\text{O}-\text{CH}_2\text{CH}_2-\text{O}\right)_n} + n\text{H}_2\text{O}$$

Polyesters are used in fibers, films, and plastic bottles.

Conducting Polymers

Recall from Chapter 6 that the conductivity in metals and semiconductors is due to electrons moving within the delocalized electron states of the conduction band. In contrast to metals and semiconductors, most organic compounds do not conduct electricity to any appreciable extent, that is, they are electrical insulators. **Conducting polymers,** however, are organic materials that exhibit significant electrical conduction. Although their conductivity is generally low relative to metals, such as aluminum and copper, conducting polymers have a number of properties (including low cost, low density, and flexibility) that give them significant potential for future technological use.

Electrical conduction in conducting polymers, like that in metals and semiconductors, occurs because some of the electrons in the material occupy a partially filled band of electron states called a *conduction band* (see Section 6.4). In most conducting polymers, the conduction band arises from delocalized molecular orbitals in a *conjugated π system*—an alternating carbon-carbon single and carbon-carbon double bond pattern that gives rise to a geometry in which there is significant overlap between the π orbitals forming the double bonds. The simplest conducting polymer is *trans*-polyacetylene

Just as π-orbital overlap between the alternating single and double bonds in benzene (C_6H_6) leads to delocalized molecular orbitals in the benzene ring (see Section 4.5), the π-orbital overlap in *trans*-polyacetylene gives rise to delocalized molecular orbitals along the conjugated π chain. Polyacetylene is unstable in the presence of air and moisture, so its commercial potential is limited, but other more stable (and thus more commercially promising) polymers with conjugated π systems include polypyrole, polythiophene, and poly(*p*–phenylene):

polypyrole   polythiophene   poly(*p*-phenylene)

The conductivity of materials, σ, is measured in units of siemens (S) per centimeter, where a siemen is the inverse of the ohm, the SI unit of electrical resistance, that is, $1\text{ S} = 1\text{ }\Omega^{-1}$. Materials can be classified as conducting ($\sigma > 10^2\text{ S cm}^{-1}$), semiconducting ($10^{-7}\text{ S cm}^{-1} < \sigma < 10^2\text{ S cm}^{-1}$), or as insulating ($\sigma < 10^{-8}\text{ S cm}^{-1}$). Pure conducting polymers are generally semiconductors. For example, pure *trans*-polyacetylene has a conductivity of 4.4×10^{-5} S cm^{-1}, which makes it weakly semiconducting. Metal-like conductivity in conducting polymers can be achieved by doping the polymer with impurities, such as I_2 or AsI_5. As discussed in Section 6.4, the presence of these dopants acts to increase the number of charge carriers on the conducting polymer backbone by either adding electrons (*n*-type) or removing electrons (*p*-type). Doped polyacetylene can achieve conductivities approaching 10^5 S cm^{-1}, which is on the order of that for pure copper ($\sigma = 4.1 \times 10^5$ S cm^{-1}). For this reason, doped conducting polymers are often referred to as "synthetic metals."

16.5 | Proteins Are Polymer Chains Composed of Amino Acid Monomers

Proteins, which play a key role in nearly all biological processes, are *polymers of amino acids.* Enzymes, the catalysts of biochemical reactions, are mostly proteins. Proteins also facilitate a wide range of other functions, such as the transport and storage of vital substances, coordinated motion, mechanical support, and protection against diseases.

The human body contains an estimated 100,000 different kinds of proteins, each of which has a specific physiological function. As we will see in this section, the chemical composition and structure of these complex natural polymers are the basis of their specificity.

Amino Acids

Proteins have high molar masses, ranging from about 5000 g to 1×10^7 g, and yet the percent composition by mass of the elements in proteins is remarkably constant: carbon, approximately 53 percent; hydrogen, 7 percent; oxygen, 23 percent; nitrogen, 16 percent; and sulfur, 1 percent.

The basic structural units of proteins are *amino acids.* An ***amino acid*** is *a compound that contains at least one amino group* (—NH_2) *and at least one carboxyl group* (—*COOH*):

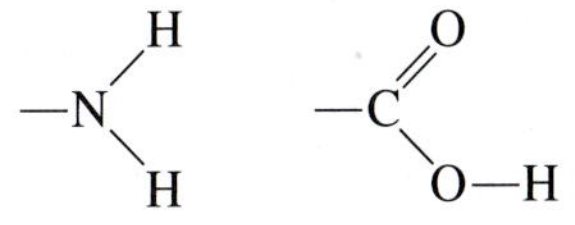

$$-N\begin{matrix} H \\ \\ H \end{matrix} \qquad -C\begin{matrix} O \\ \\ O-H \end{matrix}$$

amino group   carboxyl group

Twenty different amino acids are the building blocks of all the proteins in the human body. Table 16.6 shows the structures of these vital compounds along with their three-letter abbreviations.

Amino acids in solution at neutral pH exist as *dipolar ions* because the proton on the carboxyl group has migrated to the amino group. Compare, for example, the unionized form and the dipolar ion form of glycine, the simplest amino acid, shown here:

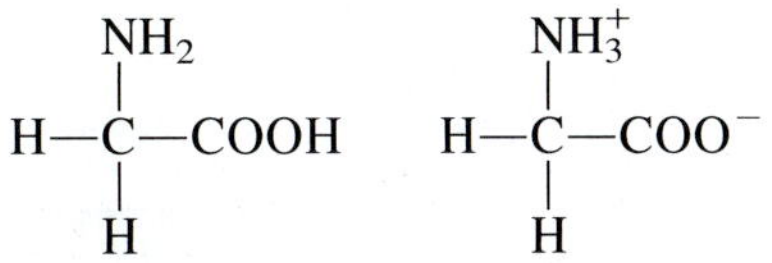

$$\underset{\text{unionized form}}{H-\underset{\underset{H}{|}}{\overset{\overset{NH_2}{|}}{C}}-COOH} \qquad \underset{\text{dipolar ion}}{H-\underset{\underset{H}{|}}{\overset{\overset{NH_3^+}{|}}{C}}-COO^-}$$

The first step in the synthesis of a protein molecule is a condensation reaction between an amino group on one amino acid and a carboxyl group on another. The

Table 16.6	The 20 Amino Acids Essential to Living Organisms*	
Name	**Abbreviation**	**Structure**
Alanine	Ala	$H_3C-\overset{\overset{\displaystyle H}{\displaystyle |}}{\underset{\underset{\displaystyle NH_3^+}{\displaystyle |}}{C}}-COO^-$
Arginine	Arg	$H_2N-\overset{\overset{\displaystyle H}{\displaystyle |}}{\underset{\underset{\displaystyle NH}{\displaystyle ||}}{C}}-N-CH_2-CH_2-CH_2-\overset{\overset{\displaystyle H}{\displaystyle |}}{\underset{\underset{\displaystyle NH_3^+}{\displaystyle |}}{C}}-COO^-$
Asparagine	Asn	$H_2N-\overset{\overset{\displaystyle O}{\displaystyle ||}}{C}-CH_2-\overset{\overset{\displaystyle H}{\displaystyle |}}{\underset{\underset{\displaystyle NH_3^+}{\displaystyle |}}{C}}-COO^-$
Aspartic acid	Asp	$HOOC-CH_2-\overset{\overset{\displaystyle H}{\displaystyle |}}{\underset{\underset{\displaystyle NH_3^+}{\displaystyle |}}{C}}-COO^-$
Cysteine	Cys	$HS-CH_2-\overset{\overset{\displaystyle H}{\displaystyle |}}{\underset{\underset{\displaystyle NH_3^+}{\displaystyle |}}{C}}-COO^-$
Glutamic acid	Glu	$HOOC-CH_2-CH_2-\overset{\overset{\displaystyle H}{\displaystyle |}}{\underset{\underset{\displaystyle NH_3^+}{\displaystyle |}}{C}}-COO^-$
Glutamine	Gln	$H_2N-\overset{\overset{\displaystyle O}{\displaystyle ||}}{C}-CH_2-CH_2-\overset{\overset{\displaystyle H}{\displaystyle |}}{\underset{\underset{\displaystyle NH_3^+}{\displaystyle |}}{C}}-COO^-$
Glycine	Gly	$H-\overset{\overset{\displaystyle H}{\displaystyle |}}{\underset{\underset{\displaystyle NH_3^+}{\displaystyle |}}{C}}-COO^-$
Histidine	His	$HC=C-CH_2-\overset{\overset{\displaystyle H}{\displaystyle |}}{\underset{\underset{\displaystyle NH_3^+}{\displaystyle |}}{C}}-COO^-$ with ring $\underset{\underset{\displaystyle C}{}}{N}\diagdown\diagup NH$, $\underset{\displaystyle H}{}$
Isoleucine	Ile	$H_3C-CH_2-\overset{\overset{\displaystyle CH_3}{\displaystyle |}}{\underset{\underset{\displaystyle H}{\displaystyle |}}{C}}-\overset{\overset{\displaystyle H}{\displaystyle |}}{\underset{\underset{\displaystyle NH_3^+}{\displaystyle |}}{C}}-COO^-$

(Continued)

*The shaded portion is the R group of the amino acid.

Table 16.6	The 20 Amino Acids Essential to Living Organisms—Cont.	
Name	**Abbreviation**	**Structure**
Leucine	Leu	$\begin{array}{l} H_3C \\ \quad\ \ CH-CH_2-\overset{\displaystyle H}{\underset{\displaystyle NH_3^+}{C}}-COO^- \\ H_3C \end{array}$
Lysine	Lys	$H_2N-CH_2-CH_2-CH_2-CH_2-\overset{\displaystyle H}{\underset{\displaystyle NH_3^+}{C}}-COO^-$
Methionine	Met	$H_3C-S-CH_2-CH_2-\overset{\displaystyle H}{\underset{\displaystyle NH_3^+}{C}}-COO^-$
Phenylalanine	Phe	$\bigcirc-CH_2-\overset{\displaystyle H}{\underset{\displaystyle NH_3^+}{C}}-COO^-$
Proline	Pro	$\begin{array}{c} \overset{+}{H_2N}-\overset{\displaystyle H}{\underset{\displaystyle |}{C}}-COO^- \\ \ \ \ \ |\qquad\quad | \\ H_2C\qquad CH_2 \\ \ \ \ \ \ \ CH_2 \end{array}$
Serine	Ser	$HO-CH_2-\overset{\displaystyle H}{\underset{\displaystyle NH_3^+}{C}}-COO^-$
Threonine	Thr	$\begin{array}{c} \ \ \ \ OH\ \ \ H \\ H_3C-\overset{\displaystyle |}{\underset{\displaystyle |}{C}}-\overset{\displaystyle |}{\underset{\displaystyle |}{C}}-COO^- \\ \ \ \ \ \ H\quad NH_3^+ \end{array}$
Tryptophan	Trp	$\begin{array}{c} \bigcirc\!\!=\!\!\overset{\displaystyle C}{\parallel}-CH_2-\overset{\displaystyle H}{\underset{\displaystyle NH_3^+}{C}}-COO^- \\ \quad\ \ CH \\ \quad\ N \\ \quad\ H \end{array}$
Tyrosine	Tyr	$HO-\bigcirc-CH_2-\overset{\displaystyle H}{\underset{\displaystyle NH_3^+}{C}}-COO^-$
Valine	Val	$\begin{array}{l} H_3C \\ \quad\ \ CH-\overset{\displaystyle H}{\underset{\displaystyle NH_3^+}{C}}-COO^- \\ H_3C \end{array}$

molecule formed from the condensation reaction between two amino acids is called a *dipeptide,* and the bond joining them together is a *peptide bond:*

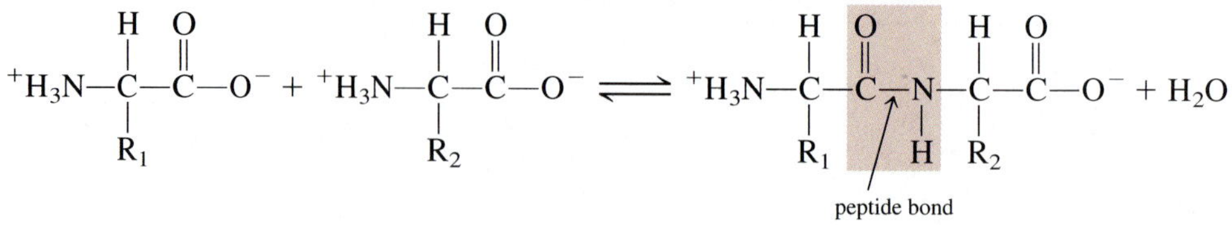

$$^+H_3N-\underset{\underset{R_1}{|}}{\overset{\overset{H}{|}}{C}}-\overset{\overset{O}{\|}}{C}-O^- \; + \; {}^+H_3N-\underset{\underset{R_2}{|}}{\overset{\overset{H}{|}}{C}}-\overset{\overset{O}{\|}}{C}-O^- \; \rightleftharpoons \; {}^+H_3N-\underset{\underset{R_1}{|}}{\overset{\overset{H}{|}}{C}}-\overset{\overset{O}{\|}}{C}-\underset{\underset{H}{|}}{N}-\underset{\underset{R_2}{|}}{\overset{\overset{H}{|}}{C}}-\overset{\overset{O}{\|}}{C}-O^- \; + \; H_2O$$

peptide bond

where R_1 and R_2 represent an H atom or some other group, and —CO—NH— is called the *amide group.* Because the equilibrium of the reaction joining two amino acids lies to the left, the process is coupled to the hydrolysis of ATP (see p. 459).

Either end of a dipeptide can engage in a condensation reaction with another amino acid to form a *tripeptide,* a *tetrapeptide,* and so on. The final product, the protein molecule, is a *polypeptide;* it can also be thought of as a polymer of amino acids. An amino acid unit in a polypeptide chain is called a *residue.* Typically, a polypeptide chain contains 100 or more amino acid residues. The sequence of amino acids in a polypeptide chain is written conventionally from left to right, starting with the amino-terminal residue and ending with the carboxyl-terminal residue. Figure 16.18 shows that alanylglycine and glycylalanine, two dipeptides that can form from alanine and glycine, are different molecules.

With 20 different amino acids to choose from, $20^2 = 400$ different dipeptides can be generated. Even for a very small protein such as insulin, which contains only 50 amino acid residues, the number of chemically different structures that are possible is on the order of 20^{50} or 10^{65}! This is an incredibly large number when you consider that the total number of atoms in our galaxy is about 10^{68}. With so many possibilities for protein synthesis, it is remarkable that generation after generation of cells can produce identical proteins for specific physiological functions.

Protein Structure

The type and number of amino acids in a given protein along with the sequence or order in which these amino acids are joined together determine the structure of the protein. In the 1930s, Linus Pauling and coworkers conducted a systematic investigation of protein structure. First they studied the geometry of the basic repeating

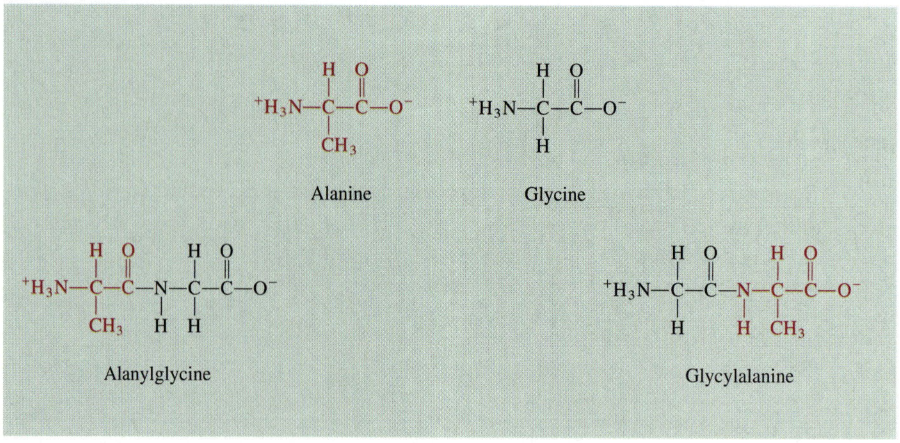

Figure 16.18 The formation of two dipeptides from two different amino acids.

group (that is, the amide group), which is represented by the following resonance structures:

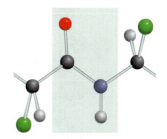

Figure 16.19 The planar amide group in proteins. Rotation about the peptide bond in the amide group is hindered by its double-bond character. The black atoms represent carbon; blue, nitrogen; red, oxygen; green, R group; and gray, hydrogen.

Because it is more difficult (i.e., it would take more energy) to twist a double bond than a single bond, the four atoms in the amide group become locked in the same plane (Figure 16.19). Figure 16.20 depicts the repeating amide group in a polypeptide chain.

On the basis of models and X-ray diffraction data, Pauling deduced that there are two common structures for protein molecules, called the *α helix* and the *β-pleated* sheet. The α-helical structure of a polypeptide chain is shown in Figure 16.21. The helix is stabilized by *intramolecular* hydrogen bonds between the NH and CO groups of the main chain, giving rise to an overall rodlike shape. The CO group of each amino acid is hydrogen-bonded to the NH group of the amino acid that is four residues away in the sequence. In this manner all the main-chain CO and NH groups take part in hydrogen bonding. X-ray studies have shown that the structure of a number of proteins, including myoglobin and hemoglobin, consist of α helices to a great extent.

The β-pleated structure gives the protein a sheet-like structure rather than the rod-like structure of the α helix. The polypeptide chain is almost fully extended, and each chain forms many *intermolecular* hydrogen bonds with adjacent chains. Figure 16.22 shows the two different types of β-pleated structures, called *parallel* and *antiparallel*. Silk molecules possess the antiparallel structure. Because its polypeptide chains are already in an extended form, silk lacks elasticity and extensibility, but it is quite strong due to the many intermolecular hydrogen bonds.

It is customary to divide protein structure into four levels of organization. The *primary structure* refers to the unique amino acid sequence of the polypeptide chain. The *secondary structure* includes those parts of the polypeptide chain that are stabilized by a regular pattern of hydrogen bonds between the CO and NH groups of the backbone (for example, the α helix or β-pleated sheet). The term *tertiary structure* applies to the three-dimensional structure stabilized by dispersion forces, hydrogen bonding, and other intermolecular forces. It differs from secondary structure because the amino acids taking part in these interactions may be far apart in the polypeptide chain.

A protein molecule may be made up of more than one polypeptide chain. Thus, in addition to the various interactions *within* a chain that give rise to the secondary and tertiary structures, we must also consider the interaction *between* chains. The

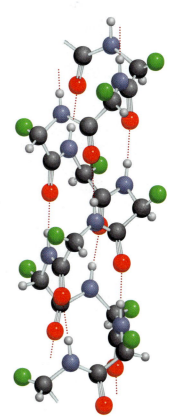

Figure 16.21 The α-helical structure of a polypeptide chain. The structure is held in position by intramolecular hydrogen bonds, shown as dotted lines. For color key, see Figure 16.19.

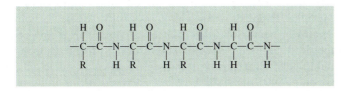

Figure 16.20 A polypeptide chain. Note the repeating units of the amide group. The symbol R represents part of the structure characteristic of the individual amino acids. For glycine, R is simply an H atom.

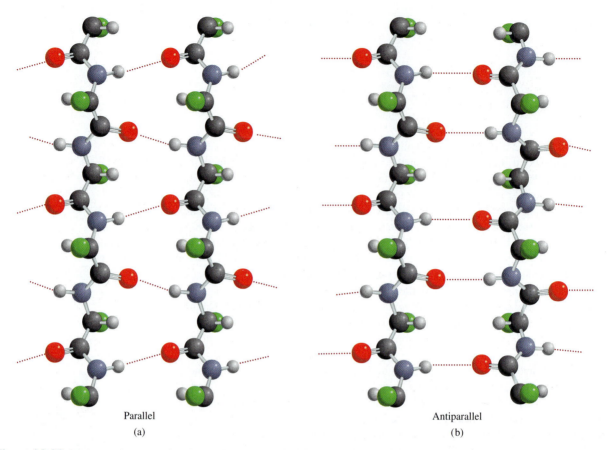

Parallel
(a)

Antiparallel
(b)

Figure 16.22 Hydrogen bonds (a) in a parallel β-pleated sheet structure, in which all the polypeptide chains are oriented in the same direction, and (b) in an antiparallel β-pleated sheet, in which adjacent polypeptide chains run in opposite directions. For color key, see Figure 16.19.

overall arrangement of the polypeptide chains is called the *quaternary structure*. Hemoglobin, for example, consists of four separate polypeptide chains, or *subunits*. These subunits are held together by van der Waals forces and ionic (electrostatic) forces (Figure 16.23).

Pauling's work was a great triumph in protein chemistry. It showed for the first time how to predict a protein structure purely from knowledge of the geometry of its fundamental building blocks—its amino acids. There are many proteins, however, whose structures do not consist of α helices or β-pleated sheets. Chemists now know that the three-dimensional structures of these biopolymers are maintained by several types of intermolecular forces in addition to hydrogen bonding (Figure 16.24). The delicate balance of the various interactions can be appreciated by considering what happens when a glutamic acid residue in two of the four polypeptide chains in hemoglobin is replaced by valine, an amino acid with very different properties than glutamic acid. The resulting protein molecules aggregate to form insoluble polymers, causing the disease known as sickle cell anemia (see the boxed text on p. 840).

In spite of all the forces that give proteins their structural stability, most proteins have a certain amount of flexibility. Enzymes, for example, are flexible enough to change their geometries to fit substrates of various sizes and shapes. Another interesting

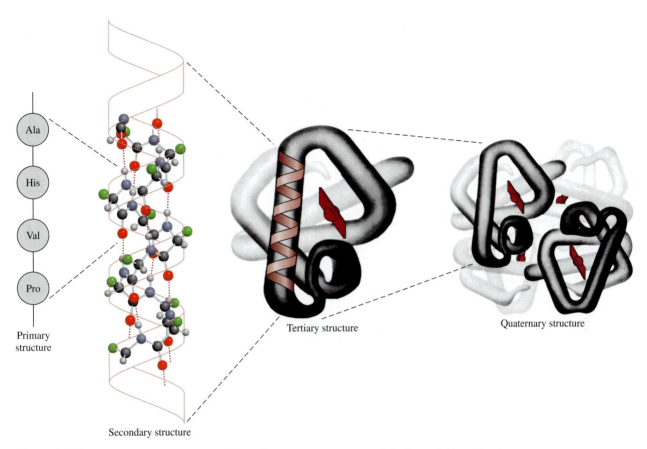

Primary structure

Secondary structure

Tertiary structure

Quaternary structure

Figure 16.23 The primary, secondary, tertiary, and quaternary structures of the hemoglobin molecule.

of protein flexibility is found in the binding of hemoglobin to oxygen. Each of the four polypeptide chains in hemoglobin contains a heme group that can bind to an oxygen molecule. In deoxyhemoglobin, the affinity of each of the heme groups for oxygen is about the same. As soon as one of the heme groups becomes oxygenated, however, the affinity of the other three hemes for oxygen is greatly enhanced. This

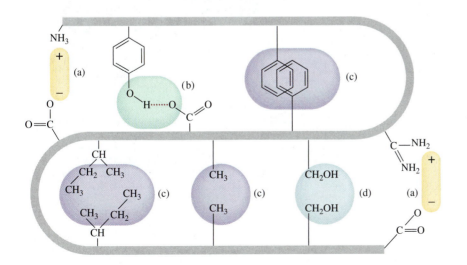

Figure 16.24 Intermolecular forces in a protein molecule: (a) ionic forces, (b) hydrogen bonding, (c) dispersion forces, and (d) dipole-dipole forces.

Sickle Cell Anemia: A Molecular Disease

Sickle cell anemia is a hereditary disease in which abnormally shaped red blood cells restrict the flow of blood to vital organs in the human body, causing swelling, severe pain, and in many cases a shortened life span. There is currently no cure for this condition, but its painful symptoms are known to be caused by a defect in hemoglobin, the oxygen-carrying protein in red blood cells.

The hemoglobin molecule is a large protein with a molar mass of about 65,000 g. Normal human hemoglobin (HbA) consists of two α chains, each containing 141 amino acids, and two β chains made up of 146 amino acids each. These four polypeptide chains, or subunits, are held together by ionic and van der Waals forces.

There are many mutant hemoglobin molecules—molecules with an amino acid sequence that differs somewhat from the sequence in HbA. Most mutant hemoglobins are harmless, but sickle cell hemoglobin (HbS) and others are known to cause serious diseases. HbS differs from HbA in only one very small detail. A valine molecule replaces a glutamic acid molecule on each of the two β chains:

$$\underset{\text{glutamic acid}}{HOOC-CH_2-CH_2-\overset{\overset{\displaystyle H}{|}}{\underset{\underset{\displaystyle NH_3^+}{|}}{C}}-COO^-} \qquad \underset{\text{valine}}{\overset{\displaystyle H_3C}{\underset{\displaystyle H_3C}{>}}CH-\overset{\overset{\displaystyle H}{|}}{\underset{\underset{\displaystyle NH_3^+}{|}}{C}}-COO^-}$$

This change may seem small (two amino acids out of 292), but it has a profound effect on the stability of HbS in solution. The valine groups are located at the bottom outside of the molecule to form a protruding "key" on each of the β chains. The nonpolar portion of valine

$$\overset{\displaystyle H_3C}{\underset{\displaystyle H_3C}{>}}CH-$$

can attract another nonpolar group in the α chain of an adjacent HbS molecule through dispersion forces. Biochemists often refer to this kind of attraction between nonpolar groups as a *hydrophobic* (see Chapter 9) interaction. Gradually, enough HbS molecules aggregate to form a "superpolymer."

Generally speaking, the solubility of a substance decreases as its size increases because the solvation process becomes unfavorable with increasing molecular surface area. As a result, proteins are not usually very soluble in water. In fact, the aggregated HbS molecules eventually precipitate out of solution. The precipitate causes normal disk-shaped red blood cells to assume a warped crescent or sickle shape

(see the figure on page 109). These deformed cells clog the narrow capillaries, thereby restricting blood flow to the organs of the body. It is the reduced blood flow that gives rise to the symptoms of sickle cell anemia. Sickle cell anemia has been termed a molecular disease by Linus Pauling, who did some of the early important chemical research on the nature of the affliction, because the destructive action occurs at the molecular level and the disease is, in effect, due to a molecular defect.

Some substances, such as urea and the cyanate ion,

$$\underset{\text{urea}}{H_2N-\overset{\overset{\displaystyle}{\underset{\underset{\displaystyle O}{\|}}{C}}}{}-NH_2} \qquad \underset{\text{cyanate ion}}{O=C=N^-}$$

can break up the hydrophobic interaction between HbS molecules and have been applied with some success to reverse the "sickling" of red blood cells. This approach may alleviate the pain and suffering of sickle cell patients, but it does not prevent the body from making more HbS. To cure sickle cell anemia, researchers must find a way to alter the genetic machinery that directs the production of HbS.

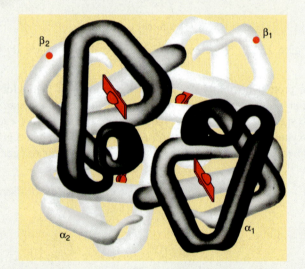

The overall structure of hemoglobin. Each hemoglobin molecule contains two α chains and two β chains. Each of the four chains is similar to a myoglobin molecule in structure, and each also contains a heme group for binding oxygen. In sickle cell hemoglobin, the defective regions (the valine groups) are located near the ends of the β chains, as indicated by the dots.

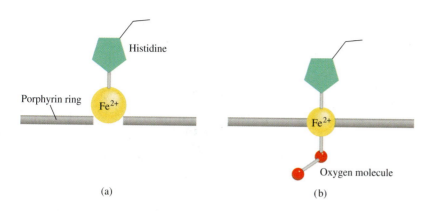

phenomenon, called *cooperativity,* makes hemoglobin a particularly suitable substance for the uptake of oxygen in the lungs. By the same token, once a fully oxygenated hemoglobin molecule releases an oxygen molecule (to myoglobin in the tissues), the other three oxygen molecules will depart with increasing ease. The cooperative nature of the binding is such that information about the presence (or absence) of oxygen molecules is transmitted from one subunit to another along the polypeptide chains, a process made possible by the flexibility of the three-dimensional structure (Figure 16.25).

The Fe^{2+} ion at the center of the heme group may have too large a radius to fit into the porphyrin ring of deoxyhemoglobin. When O_2 binds to Fe^{2+}, however, the ion shrinks somewhat so that it can fit into the plane of the ring. As the ion slips into the ring, it pulls the histidine residue opposite O_2 toward the ring and thereby sets off a sequence of structural changes from one subunit to another. Although the details of the changes are unclear, biochemists believe that this is how the binding of an oxygen molecule to one heme group affects binding to the remaining three heme groups.

When proteins are heated above body temperature or when they are subjected to unusual acid or base conditions or treated with special reagents called *denaturants,* they lose some or all of their tertiary and secondary structure. Called **denatured proteins,** proteins in this state *no longer exhibit normal biological activity.* Figure 16.26 shows the variation of rate with temperature for a typical enzyme-catalyzed reaction. Initially, the rate increases with increasing temperature, as we would expect. Beyond the optimum temperature, however, the enzyme begins to denature and the rate decreases rapidly. If a protein is denatured under mild conditions, its original structure can often be regenerated by removing the denaturant or by restoring the temperature to normal conditions. This process is called *reversible denaturation.*

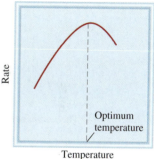

Figure 16.26 Dependence of the rate of an enzyme-catalyzed reaction on temperature. Above the optimum temperature at which an enzyme is most effective, its activity drops off as a consequence of denaturation.

16.6 | DNA and RNA Are Polymers Composed of Nucleic Acids

Nucleic acids are *high-molar-mass polymers that play an essential role in protein synthesis.* **Deoxyribonucleic acid (DNA)** and **ribonucleic acid (RNA)** are *the two types of nucleic acid.* DNA molecules are among the largest molecules known; they have molar masses of up to tens of billions of grams. RNA molecules, on the other hand, vary greatly in size, some having a molar mass of about 25,000 g. Compared with proteins, which are made of up to 20 different amino acids, nucleic acids are fairly simple in composition. A DNA or RNA molecule contains only four types of building

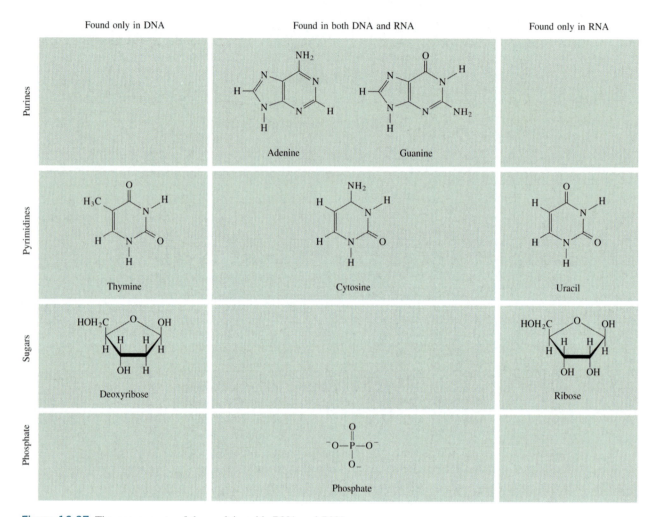

Figure 16.27 The components of the nucleic acids DNA and RNA.

blocks: purines, pyrimidines, furanose sugars, and phosphate groups (Figure 16.27). Each purine or pyrimidine is called a *base.*

In the 1940s, Erwin Chargaff[10] studied DNA molecules obtained from various sources and observed certain regularities. *Chargaff's rules,* as his findings are now known, describe these patterns:

1. The amount of adenine (a purine) is equal to that of thymine (a pyrimidine), that is, A = T, or A/T = 1.

2. The amount of cytosine (a pyrimidine) is equal to that of guanine (a purine), that is, C = G, or C/G = 1.

3. The total number of purine bases is equal to the total number of pyrimidine bases, that is, A + G = C + T.

---

10. Erwin Chargaff (1905–2002). American biochemist of Austrian origin. Chargaff was the first to show that different biological species contain different DNA molecules.

DNA Fingerprinting

The human genetic makeup, or *genome,* consists of about 3 billion nucleotides. These 3 billion units compose the 23 pairs of chromosomes, which are continuous strands of DNA ranging in length from 50 million to 500 million nucleotides. Encoded in this DNA and stored in units called *genes* are the instructions for protein synthesis. Each of about 100,000 genes is responsible for the synthesis of a particular protein. In addition to instructions for protein synthesis, each gene contains a sequence of bases, repeated several times, that has no known function. What is interesting about these sequences, called *minisatellites,* is that they appear many times in different locations, not just in a particular gene. Furthermore, each person has a unique number of repeats. Only identical twins have the same number of minisatellite sequences.

In 1985, a British chemist named Alec Jeffreys suggested that minisatellite sequences provide a means of identification, much like fingerprints. *DNA fingerprinting* has since gained prominence with law enforcement officials as a way to identify crime suspects.

To make a DNA fingerprint, a chemist needs a sample of any tissue, such as blood or semen; even hair and saliva contain DNA. The DNA is extracted from cell nuclei and cut into fragments by the addition of so-called restriction enzymes. These fragments, which are negatively charged, are separated by an electric field in a gel. The smaller fragments move faster than the larger ones, so they eventually separate into bands. The bands of DNA fragments are transferred from the gel to a plastic membrane, and their position is thereby fixed. Then a DNA probe—a DNA fragment that has been tagged with a radioactive label—is added. The probe binds to the fragments that have a complementary DNA sequence. An X-ray film is laid directly over the plastic sheet, and bands appear on the exposed film in the positions corresponding to the fragments recognized by the probe. About four different probes are needed to obtain a profile that is unique to just one individual. The probability of finding identical patterns in the DNA of two randomly selected individuals is estimated to be on the order of 1 in 10 billion.

The first U.S. case in which a person was convicted of a crime with the help of DNA fingerprints was tried in 1987. Today, DNA fingerprinting has become an indispensable tool of law enforcement. It is also used by those who believe they were falsely accused or convicted to prove their innocence.

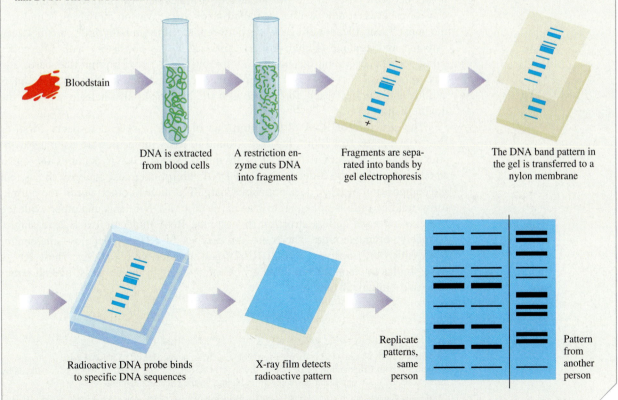

Procedure for obtaining a DNA fingerprint. The developed film shows the DNA fingerprint, which is compared with patterns from known subjects.

Figure 16.28 Structure of a nucleotide, one of the repeating units in DNA.

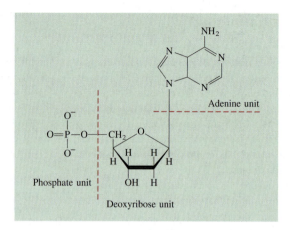

Based on chemical analyses and information obtained from X-ray diffraction measurements, James Watson[11] and Francis Crick[12] formulated the double-helical structure for the DNA molecule in 1953. Watson and Crick determined that the DNA molecule has two helical strands. Each strand is made up of **nucleotides,** which *consist of a base, a deoxyribose, and a phosphate group linked together* (Figure 16.28).

The key to the double-helical structure of DNA is the formation of hydrogen bonds between bases in the two strands of a molecule. Although hydrogen bonds can form between any two bases, called *base pairs,* Watson and Crick found that the most favorable couplings were between adenine and thymine and between cytosine and guanine (Figure 16.29). This scheme is consistent with Chargaff's rules because every purine base is hydrogen-bonded to a pyrimidine base, and vice versa (A + G = C + T). Other attractive forces such as dipole-dipole interactions and van der Waals forces between the base pairs also help to stabilize the double helix.

The structure of RNA differs from that of DNA in several respects. First, as shown in Figure 16.27, the four bases found in RNA molecules are adenine, cytosine, guanine, and uracil (instead of thymine). Second, RNA contains the sugar ribose rather than the 2-deoxyribose of DNA. Third, chemical analysis shows that the composition of RNA does not obey Chargaff's rules. In other words, the purine-to-pyrimidine ratio is not equal to 1 as in the case of DNA. This and other evidence rule out a double-helical structure. In fact, the RNA molecule exists as a single-stranded polynucleotide. There are three types of RNA molecules—messenger RNA (mRNA), ribosomal RNA (rRNA), and transfer RNA (tRNA). These RNAs have similar nucleotides but differ from one another in molar mass, overall structure, and biological functions.

---

11. James Dewey Watson (1928–). American biologist. Watson shared the 1962 Nobel Prize in Physiology or Medicine with Crick and Maurice Wilkins for their work on the DNA structure, which is considered by many to be the most significant development in biology in the twentieth century.

12. Francis Harry Compton Crick (1916–2004). British biologist. Crick started as a physicist but became interested in biology after reading the book *What Is Life?* by Erwin Schrödinger (see Chapter 1). In addition to elucidating the structure of DNA, for which he was a corecipient of the Nobel Prize in Physiology or Medicine in 1962, Crick has made many significant contributions to molecular biology.

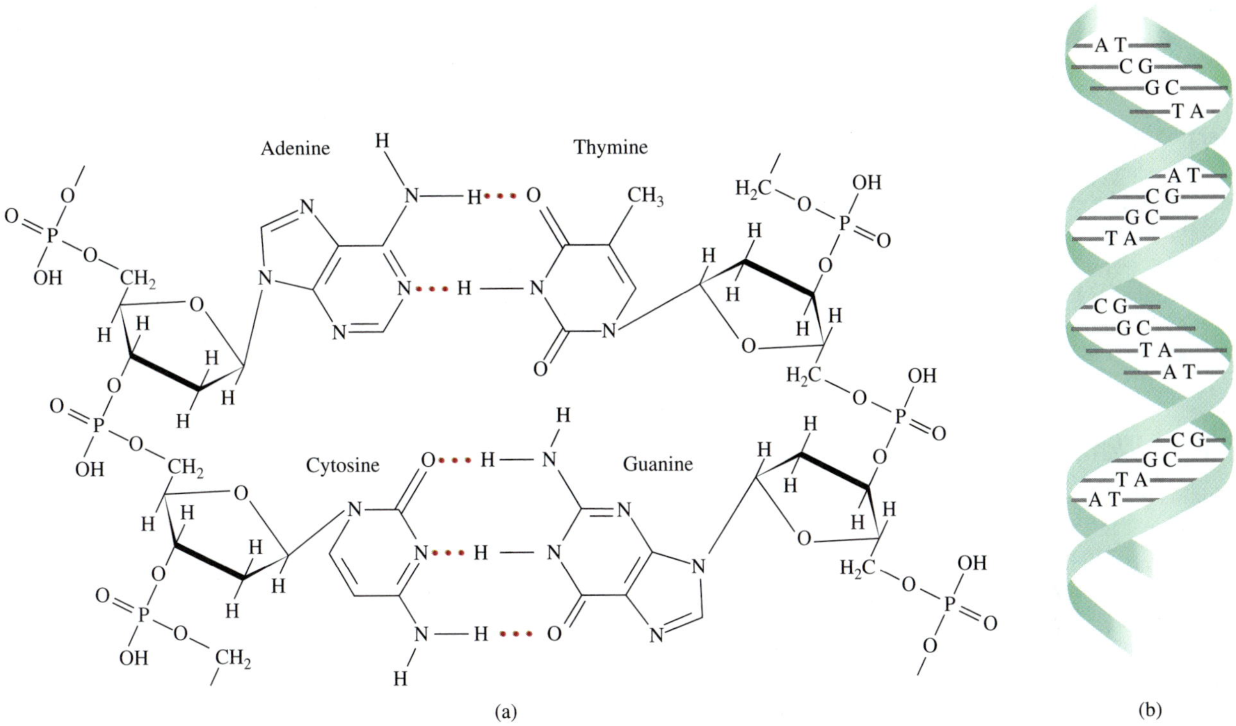

Figure 16.29 (a) Base-pair formation by adenine and thymine and by cytosine and guanine. (b) The double-helical strand of a DNA molecule held together by hydrogen bonds (and other intermolecular forces) between base pairs A-T and C-G.

Summary of Facts and Concepts

Sections 16.1 and 16.2

▶ Because carbon atoms can link up with other carbon atoms in straight and branched chains, carbon can form more compounds than any other element. The naming of organic compounds follows systematic guidelines set down by the IUPAC, although many compounds have common names that are different than their systematic names.

▶ Organic compounds are derived from two types of hydrocarbons: aliphatic hydrocarbons and aromatic hydrocarbons. Methane (CH_4) is the simplest of the alkanes, a family of hydrocarbons with the general formula C_nH_{2n+2}.

▶ Cyclopropane (C_3H_6) is the simplest of the cycloalkanes, a family of alkanes whose carbon atoms are joined in a ring. Alkanes and cycloalkanes are saturated hydrocarbons.

▶ Ethylene ($CH_2 = CH_2$) is the simplest of the olefins, or alkenes, a class of hydrocarbons containing carbon-carbon double bonds and having the general formula C_nH_{2n}.

▶ Acetylene ($CH \equiv CH$) is the simplest of the alkynes, which are compounds that have the general formula C_nH_{2n-2} and contain carbon-carbon triple bonds.

▶ Compounds that contain one or more benzene rings are called aromatic hydrocarbons. These compounds undergo substitution by halogens and alkyl groups.

Section 16.3

▶ Functional groups impart specific types of chemical reactivity to molecules.

▶ Classes of compounds characterized by their functional groups include alcohols, ethers, aldehydes and ketones, carboxylic acids and esters, and amines.

Section 16.4

▶ Polymers are large molecules made up of small, repeating units called monomers.

▶ Proteins, nucleic acids, cellulose, and rubber are natural polymers. Nylon, Dacron, and Lucite are examples of synthetic polymers.

▶ Organic polymers can be synthesized via addition reactions or condensation reactions.

▶ Stereoisomers of a polymer made up of asymmetric monomers have different properties, depending on how the starting units are joined together.

▶ Synthetic rubbers include polychloroprene and styrene-butadiene rubber, which is a copolymer of styrene and butadiene.

are the three-dimensional folded arrangements of proteins that are stabilized by hydrogen bonds and other intermolecular forces.

Section 16.5

▶ Structure determines the function and properties of proteins. To a great extent, hydrogen bonding and other intermolecular forces determine the structure of proteins.

▶ The primary structure of a protein is its amino acid sequence. Secondary structure is the shape defined by hydrogen bonds joining the CO and NH groups of the amino acid backbone. Tertiary and quaternary structures

Section 16.6

▶ Nucleic acids—DNA and RNA—are high-molar-mass polymers that carry genetic instructions for protein synthesis in cells. Nucleotides are the building blocks of DNA and RNA. DNA nucleotides contain a purine or pyrimidine base, a deoxyribose molecule, and a phosphate group. RNA nucleotides are similar but contain different bases and ribose instead of deoxyribose.

Key Words

addition reactions, p. 813	carboxylic acid, p. 819	functional group, p. 801	nucleotides, p. 844
alcohol, p. 815	condensation reaction, p. 818	homopolymer, p. 828	organic chemistry, p. 800
aldehyde, p. 818	conducting polymers, p. 832	hydrocarbon, p. 801	polymer, p. 826
aliphatic hydrocarbon, p. 801	conformations, p. 810	hydrogenation, p. 813	proteins, p. 833
alkane, p. 801	copolymer, p. 831	ketone, p. 818	ribonucleic acid (RNA), p. 841
alkene, p. 806	cycloalkane, p. 806	macromolecules, p. 826	saponification, p. 821
alkyl group, p. 803	denatured proteins, p. 841	monomer, p. 826	saturated hydrocarbon, p. 801
alkyne, p. 807	deoxyribonucleic acids	Newman projection, p. 810	substitution reaction, p. 815
amine, p. 821	(DNA), p. 841	nuclear magnetic resonance	unsaturated
amino acid, p. 833	ester, p. 820	spectroscopy, p. 824	hydrocarbon, p. 813
aromatic hydrocarbon, p. 801	ether, p. 817	nucleic acids, p. 841	

Problems

Hydrocarbons Are Organic Compounds Containing Only Hydrogen and Carbon

16.1 What do *saturated* and *unsaturated* mean when applied to hydrocarbons? Give examples of a saturated hydrocarbon and an unsaturated hydrocarbon.

16.2 Alkenes have geometric isomers, so why is it that alkanes and alkynes do not?

16.3 Give examples of a chiral substituted alkane and an achiral substituted alkane.

16.4 Draw and name all possible structural isomers for alkanes with the formula C_7H_{16}.

16.5 How many distinct chloropentanes ($C_5H_{11}Cl$) could be produced in the direct chlorination of *n*-pentane $[CH_3(CH_2)_3CH_3]$? Draw the structure of each molecule.

16.6 Draw all possible isomers for the molecule C_4H_8.

16.7 Draw all possible isomers for the molecule C_3H_5Br.

16.8 The structural isomers of pentane (C_5H_{12}) have quite different boiling points. Explain the observed variation in boiling point, in terms of their different structures.

16.9 Discuss how you can determine which of the following compounds might be alkanes, cycloal-

kanes, alkenes, or alkynes, without drawing their structures: (a) C_6H_{12}, (b) C_4H_6, (c) C_5H_{12}, (d) C_7H_{14}, and (e) C_3H_4.

16.10 Draw the structures of *cis*-2-butene and *trans*-2-butene. Which of the two compounds has the higher heat of hydrogenation? Explain.

16.11 Would you expect cyclobutadiene to be a stable molecule? Explain.

16.12 How many different isomers can be derived from ethylene if two hydrogen atoms are replaced by a fluorine atom and a chlorine atom? Draw their structures and name them. Indicate which are structural isomers and which are geometric isomers.

16.13 Geometric isomers are not restricted to compounds containing the C=C bond. For example, certain disubstituted cycloalkanes can exist in the *cis* and the *trans* forms. Label the following molecules as the *cis* and *trans* isomers of the same compound:

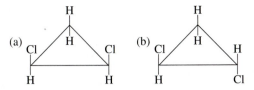

(a)   (b)

16.14 Which of the following amino acids are chiral: (a) $CH_3CH(NH_2)COOH$, (b) $CH_2(NH_2)COOH$, (c) $CH_2(OH)CH(NH_2)COOH$?

16.15 Name the following compounds:

(a) $CH_3-\overset{\overset{\displaystyle CH_3}{|}}{CH}-CH_2-CH_2-CH_3$

(b) $CH_3-\overset{\overset{\displaystyle C_2H_5}{|}}{CH}-\overset{\overset{\displaystyle CH_3}{|}}{CH}-\overset{\overset{\displaystyle CH_3}{|}}{CH}-CH_3$

(c) $CH_3-CH_2-\underset{\underset{\displaystyle CH_2-CH_2-CH_3}{|}}{CH}-CH_2-CH_3$

(d) $CH_2{=}CH-\overset{\overset{\displaystyle CH_3}{|}}{CH}-CH{=}CH_2$

(e) $CH_3-C{\equiv}C-CH_2-CH_3$

(f) $CH_3-CH_2-CH-CH{=}CH_2$ (phenyl substituent)

16.16 Write structural formulas for the following organic compounds: (a) 3-methylhexane, (b) 1,3,5-trichlorocyclohexane, (c) 2,3-dimethylpentane, (d) 2-bromo-4-phenylpentane, and (e) 3,4,5-trimethyloctane.

16.17 Write structural formulas for the following compounds: (a) *trans*-2-pentene, (b) 2-ethyl-1-butene, (c) 4-ethyl-*trans*-2-heptene, and (d) 3-phenylbutyne.

16.18 Draw and name the two geometric isomers of 1,3-pentadiene.

16.19 Benzene and cyclohexane both contain six-membered rings. Benzene is planar, whereas cyclohexane is nonplanar. Explain.

16.20 Write structures for the following compounds: (a) 1-bromo-3-methylbenzene, (b) 1-chloro-2-propyl-benzene, and (c) 1,2,4,5-tetramethylbenzene.

16.21 Name the following compounds:

(a)

(b)

(c)

Hydrocarbons Undergo a Number of Important Chemical Reactions

16.22 Suggest two chemical tests that would help you distinguish between $CH_3CH_2CH_2CH_2CH_3$ and $CH_3CH_2CH_2CH{=}CH_2$.

16.23 Sulfuric acid (H_2SO_4) adds to the double bond of alkenes as H^+ and $^-OSO_3H$. Predict the products when sulfuric acid reacts with (a) ethylene and (b) propylene.

16.24 Acetylene is an unstable compound. It has a tendency to form benzene as follows:

$$3C_2H_2(g) \longrightarrow C_6H_6(l)$$

Calculate the standard enthalpy change in kJ mol^{-1} for this reaction at 25°C.

16.25 Predict the products when HBr is added to (a) 1-butene and (b) 2-butene.

The Structure and Properties of Organic Compounds Are Greatly Influenced by the Presence of Functional Groups

16.26 Draw the Lewis structure for each of the following functional groups: alcohol, ether, aldehyde, ketone, carboxylic acid, ester, and amine.

16.27 Draw structures for molecules with the following formulas: (a) CH_4O, (b) C_2H_6O, (c) $C_3H_6O_2$, and (d) C_3H_8O.

16.28 Classify each of the following molecules as an alcohol, aldehyde, ketone, carboxylic acid, amine, or ether:

(a) $CH_3-O-CH_2-CH_3$

(b) $CH_3-CH_2-NH_2$

(c) $CH_3-CH_2-\overset{\overset{\displaystyle O}{\|}}{C}{\diagdown}_H$

(d) $CH_3-\underset{\underset{\displaystyle O}{\|}}{C}-CH_2-CH_3$

(e) $H-\overset{\overset{\displaystyle O}{\|}}{C}-OH$

(f) $CH_3-CH_2CH_2-OH$

(g) phenyl$-CH_2-\overset{\overset{\displaystyle NH_2}{|}}{\underset{\underset{\displaystyle H}{|}}{C}}-\overset{\overset{\displaystyle O}{\|}}{C}-OH$

16.29 Aldehydes are generally more susceptible to oxidation in air than are ketones. Use ethanal (or acetaldehyde) and 2-propanone (or acetone) as examples and show why ketones are more stable than aldehydes in this respect.

16.30 Draw structures for the following compounds: (a) 3-hexanol, (b) 2,2-dibromo-3-pentanol, (c) 3-pentanone, (d) 2-phenyl–butanal, and (e) ethoxypentane.

16.31 Draw structures for the following compounds: (a) 1,2-diethoxybutane, (b) pentanoic acid, (c) 1,2-dimethoxybenzene, (d) 3-bromo-2-chloro-4-heptanol, and (e) dichloroethanoic acid.

16.32 Give IUPAC names for the following compounds:

(a) $CH_3-O-CH_2-CH_3$

(b) $CH_3-CH_2-NH_2$

(c) $CH_3-CH_2-\overset{\displaystyle O}{\underset{\displaystyle |}{C}}\diagdown H$

(d) $CH_3-\overset{\displaystyle ||}{\underset{\displaystyle O}{C}}-CH_2-CH_3$

(e) $H-\overset{\displaystyle O}{\overset{\displaystyle ||}{C}}-OH$

(f) $CH_3-CH_2CH_2-OH$

16.33 Give IUPAC names for the following compounds:

(a) $CH_3-\overset{\displaystyle CH_3}{\underset{\displaystyle |}{CH}}-\overset{\displaystyle O}{\underset{\displaystyle OH}{C}}$

(b) $CH_3-CH-\overset{\displaystyle OH}{\underset{\displaystyle |}{CH}}-CH_2-CH_3$ with CH_3

(c) $CH_3-CH_2-\overset{\displaystyle CH_3}{\underset{\displaystyle |}{CH}}-\overset{\displaystyle O}{C}\diagdown H$

(d) $CH_3-CH_2-C\overset{\diagup O}{\diagdown O-CH_2-CH_2-CH_3}$

(e) $CH_3-\overset{\displaystyle C_2H_5}{\underset{\displaystyle |}{CH}}-CH_2-CH_2-NH_2$
(Be careful with this one!)

16.34 Complete the following equation and identify the products:

$$HCOOH + CH_3OH \longrightarrow ?$$

16.35 A compound has the empirical formula $C_5H_{12}O$. Upon controlled oxidation, it is converted into a compound of empirical formula $C_5H_{10}O$, which behaves as a ketone. Draw possible structures for the original compound and the final compound.

16.36 A compound having the molecular formula $C_4H_{10}O$ does not react with sodium metal. In the presence of light, the compound reacts with Cl_2 to form three compounds having the formula C_4H_9OCl. Draw a structure for the original compound that is consistent with this information.

16.37 Predict the product or products of each of the following reactions:

(a) $CH_3CH_2OH + HCOOH \longrightarrow$

(b) $H-C\equiv C-CH_3 + H_2 \longrightarrow$

(c) $\overset{C_2H_5}{\underset{H}{\diagdown}}C=C\overset{H}{\underset{H}{\diagup}} + HBr \longrightarrow$

16.38 Identify the functional groups in each of the following molecules:

(a) $CH_3CH_2COCH_2CH_2CH_3$

(b) $CH_3COOC_2H_5$

(c) $CH_3CH_2OCH_2CH_2CH_2CH_3$

Polymers Are Very Large Molecular Weight Compounds Formed from the Joining Together of Many Subunits Called Monomers

16.39 Calculate the molar mass of a particular polyethylene sample, $(-CH_2-CH_2-)_n$, where $n = 4600$.

16.40 In Chapter 8, you learned about the colligative properties of solutions. Which of the colligative properties is best suited for determining the molar mass of a polymer? Why?

16.41 Teflon is formed by a radical addition reaction involving the monomer tetrafluoroethylene. Show the mechanism for this reaction.

16.42 Vinyl chloride ($H_2C=CHCl$) undergoes copolymerization with 1,1-dichloroethylene ($H_2C=CCl_2$) to form a polymer commercially known as Saran. Draw the structure of the polymer, showing the repeating monomer units.

16.43 Kevlar is a copolymer used in bulletproof vests. It is formed in a condensation reaction between 1,4-diaminobenzene and 1,4-dicarboxybenzene.

Sketch a portion of the polymer chain showing several monomer units. Write the overall equation for the condensation reaction.

16.44 Describe the formation of polystyrene.

16.45 Deduce plausible monomers for polymers with the following repeating units:

(a) $+CH_2-CF_2\rightarrow_n$

(b) $\left(CO-\bigcirc-CONH-\bigcirc-NH\right)_n$

16.46 Deduce plausible monomers for polymers with the following repeating units:

(a) $+CH_2-CH=CH-CH_2\rightarrow_n$

(b) $+CO+CH_2\rightarrow_6NH\rightarrow_n$

16.47 The conductivity of *cis*-polyacetylene (1.7×10^{-9} S cm^{-1}) is much smaller than that of *tran*-polyacetylene (see page 832). Can you explain this difference by examining the structures of these two polymers? The structure of *cis*-polyacetylene is

Proteins Are Polymer Chains Composed of Amino Acid Monomers

16.48 Discuss the characteristics of an amide group and its importance in protein structure.

16.49 Briefly explain the phenomenon of cooperativity exhibited by the hemoglobin molecule in binding oxygen.

16.50 Why is sickle cell anemia called a molecular disease?

16.51 Draw the structures of the dipeptides that can be formed from the reaction between the amino acids glycine and valine.

16.52 Draw the structures of the dipeptides that can be formed from the reaction between the amino acids glycine and lysine.

16.53 The amino acid glycine can be condensed to form a polymer called polyglycine. Draw the repeating monomer unit.

16.54 The folding of a polypeptide chain depends not only on its amino acid sequence but also on the nature of the solvent. Discuss the types of interactions that might occur between water molecules and the amino acid residues of the polypeptide chain. Which groups would be exposed on the exterior of the protein in contact with water and which groups would be buried in the interior of the protein?

16.55 The following are data obtained on the rate of product formation of an enzyme-catalyzed reaction:

Temperature (°C)	Rate of Product Formation (M s^{-1})
10	0.0025
20	0.0048
30	0.0090
35	0.0086
45	0.0012

Comment on the dependence of the rate on temperature. (No calculations are required.)

16.56 The pitch of a helix is the distance required for the helix to make a full turn. The average pitch of an α helix (Figure 16.21) is 5.4 Å. Assuming this pitch is the same for human hair and that hair grows at the rate of 0.6 inches per month, how many turns of the α helix are generated each second (assume 1 month = 30 days).

DNA and RNA Are Polymers Composed of Nucleic Acids

16.57 What is the difference between ribose and deoxyribose?

16.58 A standard DVD can store approximately 4.7 gigabytes of information, where each byte consists of 8 bits. Each bit can be either "0" or "1". Could the entire human genome be stored on a single DVD?

16.59 Describe the role of hydrogen bonding in maintaining the double-helical structure of DNA.

Additional Problems

16.60 Draw all the possible structural isomers for the molecule having the formula C_7H_7Cl. The molecule contains one benzene ring.

16.61 Given

$$C_2H_4(g) + 3O_2(g) \longrightarrow 2CO_2(g) + 2H_2O(l)$$
$$\Delta H° = -1411 \text{ kJ mol}^{-1}$$

$$2C_2H_2(g) + 5O_2(g) \longrightarrow 4CO_2(g) + 2H_2O(l)$$
$$\Delta H° = -2599 \text{ kJ mol}^{-1}$$

$$H_2(g) + O_2(g) \longrightarrow H_2O(l)$$
$$\Delta H° = -285.8 \text{ kJ mol}^{-1}$$

calculate the enthalpy change of hydrogenation for acetylene:

$$C_2H_2(g) + H_2(g) \longrightarrow C_2H_4(g)$$

16.62 State which member of each of the following pairs of compounds is the more reactive and explain why: (a) propane and cyclopropane, (b) ethylene and methane, and (c) acetaldehyde and acetone.

16.63 State which of the following types of compounds can form hydrogen bonds with water molecules: (a) carboxylic acids, (b) alkenes, (c) ethers, (d) aldehydes, (e) alkanes, (f) amines.

16.64 An organic compound is found to contain 37.5 percent carbon, 3.2 percent hydrogen, and 59.3 percent fluorine by mass. The following pressure and volume data were obtained for 1.00 g of this substance at 90°C:

P (bar)	V (L)
2.0	0.332
1.5	0.409
1.0	0.564
0.50	1.028

The molecule is known to have a dipole moment of zero. (a) What is the empirical formula of this substance? (b) Does this substance behave as an ideal gas? (c) What is its molecular formula? (d) Draw the Lewis structure of this molecule and describe its geometry. (e) What is the systematic name of this compound?

16.65 State at least one commercial use for each of the following compounds: (a) 2-propanol (isopropanol), (b) acetic acid, (c) naphthalene, (d) methanol, (e) ethanol, (f) ethylene glycol, (g) methane, and (h) ethylene.

16.66 How many liters of air (78 percent N_2, 22 percent O_2 by volume) at 20°C and 1.00 bar are needed for the complete combustion of 1.0 L of octane, C_8H_{18}, a typical gasoline component that has a density of 0.70 g mL^{-1}?

16.67 How many carbon-carbon sigma bonds are present in each of the following molecules? (a) 2-butyne, (b) anthracene (see Figure 16.5), and (c) 2,3-dimethylpentane.

16.68 How many carbon-carbon sigma bonds are present in each of the following molecules: (a) benzene, (b) cyclobutane, and (c) 2-methyl-3-ethylpentane?

16.69 The combustion of 20.63 mg of compound Y, which contains only C, H, and O, with excess oxygen gave 57.94 mg of CO_2 and 11.85 mg of H_2O. (a) Calculate how many milligrams of C, H, and O were present in the original sample of Y. (b) Derive the empirical formula of Y. (c) Suggest a plausible structure for Y if the empirical formula is the same as the molecular formula.

16.70 Draw all the structural isomers of compounds with the formula $C_4H_8Cl_2$. Indicate which isomers are chiral, and give them systematic names.

16.71 Discuss the importance of hydrogen bonding in biological systems. Use proteins and nucleic acids as examples.

16.72 Proteins vary widely in structure, whereas nucleic acids have rather uniform structures. How do you account for this major difference?

16.73 If untreated, fevers of 40°C or higher may lead to brain damage. Why?

16.74 The "melting point" of a DNA molecule is the temperature at which the double-helical strand breaks apart. Suppose you are given two DNA samples. One sample contains 45 percent C-G base pairs, and the other contains 64 percent C-G base pairs. The total number of bases is the same in each sample. Which of the two samples has a higher melting point? Why?

16.75 When fruits such as apples and pears are cut, the exposed parts begin to turn brown. This is the result of an oxidation reaction catalyzed by enzymes present in the fruit. Often the browning action can be prevented or slowed by adding a few drops of lemon juice to the exposed areas. What is the chemical basis for this treatment?

16.76 "Dark meat" and "white meat" are one's choices when eating a turkey. Explain what causes the meat to assume different colors. (*Hint:* The more active muscles in a turkey have a higher rate of metabolism and need more oxygen.)

16.77 Nylon can be destroyed easily by strong acids. Explain the chemical basis for the destruction. (*Hint:* The products are the starting materials of the polymerization reaction.)

16.78 Despite what you may have read in science fiction novels or seen in horror movies, it is extremely unlikely that insects can ever grow to human size. Why? (*Hint:* Insects do not have hemoglobin molecules in their blood.)

16.79 How many different tripeptides can be formed by lysine and alanine?

16.80 Chemical analysis shows that hemoglobin contains 0.34 percent Fe by mass. What is the minimum possible molar mass of hemoglobin? The actual molar mass of hemoglobin is four times this minimum value. What conclusion can you draw from these data?

16.81 What kind of intermolecular forces are responsible for the aggregation of hemoglobin molecules that leads to sickle cell anemia? Draw structures of the nucleotides containing the following components: (a) deoxyribose and cytosine, and (b) ribose and uracil.

16.82 When a nonapeptide (containing nine amino acid residues) isolated from rat brains was hydrolyzed, it gave the following smaller peptides as identifiable products: Gly-Ala-Phe, Ala-Leu-Val, Gly-Ala-Leu, Phe-Glu-His, and His-Gly-Ala. Reconstruct the amino acid sequence in the nonapeptide, giving your reasons. (Remember the convention for writing peptides.)

16.83 At neutral pH, amino acids exist as dipolar ions. Using glycine as an example, and given that the pK_a of the carboxyl group is 2.3 and that of the ammonium group is 9.6, predict the predominant form of the molecule at pH 1, 7, and 12. Justify your answers using the Henderson-Hasselbalch equation (Equation 12.2).

16.84 The combustion of 3.795 mg of liquid B, which contains only C, H, and O, with excess oxygen gave 9.708 mg of CO_2 and 3.969 mg of H_2O. In a molar mass determination, 0.205 g of B vaporized at 1.00 bar and 200.0°C and occupied a volume of 89.8 mL. Derive the empirical formula, molar mass, and molecular formula of B, and draw three plausible structures.

16.85 Beginning with 3-methyl-1-butyne, show how you would prepare the following compounds:

(a) $CH_2\!=\!\underset{\underset{Br}{|}}{C}\!-\!\underset{\underset{CH_3}{|}}{CH}\!-\!CH_3$

(b) $CH_2Br\!-\!CBr_2\!-\!\underset{\underset{CH_3}{|}}{CH}\!-\!CH_3$

(c) $CH_3\!-\!\underset{\underset{Br}{|}}{CH}\!-\!\underset{\underset{CH_3}{|}}{CH}\!-\!CH_3$

16.86 Indicate the asymmetric carbon atoms in the following compounds:

(a) $CH_3\!-\!CH_2\!-\!\underset{\underset{NH_2}{|}}{\overset{\overset{CH_3}{|}}{CH}}\!-\!\underset{}{\overset{\overset{O}{\|}}{C}}\!-\!NH_2$

(b)

16.87 Suppose benzene contained three distinct single bonds and three distinct double bonds. How many different isomers would there be for dichlorobenzene ($C_6H_4Cl_2$)? Draw all your proposed structures.

16.88 Write the structural formula of an aldehyde that is a structural isomer of acetone.

16.89 Draw structures for the following compounds: (a) cyclopentane, (b) *cis*-2-butene, (c) 2-hexanol, (d) 1,4-dibromobenzene, and (e) 2-butyne.

16.90 Name the classes to which the following compounds belong: (a) C_4H_9OH, (b) $CH_3OC_2H_5$, (c) C_2H_5CHO, (d) C_6H_5COOH, and (e) CH_3NH_2.

16.91 Ethanol (C_2H_5OH) and dimethyl ether, (CH_3OCH_3) are structural isomers. Compare their melting points, boiling points, and solubilities in water. Give a molecular explanation of any differences.

16.92 Amines are Brønsted bases. The unpleasant smell of fish is due to the presence of certain amines. Explain why cooks often add lemon juice to suppress the odor of fish (in addition to enhancing the flavor).

16.93 You are given two bottles, each containing a colorless liquid. You are told that one liquid is cyclohexane and the other is benzene. Suggest one chemical test that would allow you to distinguish between these two liquids.

16.94 Give the chemical names of the following organic compounds, and write their formulas: marsh gas, grain alcohol, wood alcohol, rubbing alcohol, antifreeze, mothballs, and the chief ingredient of vinegar.

16.95 The compound $CH_3\!-\!C\!\equiv\!C\!-\!CH_3$ is hydrogenated to an alkene using platinum as the catalyst. Predict whether the product is the pure *trans* isomer, the pure *cis* isomer, or a mixture of *cis* and *trans* isomers. Based on your prediction, comment on the mechanism of the heterogeneous catalysis.

16.96 How many asymmetric carbon atoms are present in each of the following compounds?

(a) $H\!-\!\underset{\underset{H}{|}}{\overset{\overset{H}{|}}{C}}\!-\!\underset{\underset{Cl}{|}}{\overset{\overset{H}{|}}{C}}\!-\!\underset{\underset{H}{|}}{\overset{\overset{H}{|}}{C}}\!-\!Cl$

(b) $H_3C\!-\!\underset{\underset{H}{|}}{\overset{\overset{OH}{|}}{C}}\!-\!\underset{\underset{H}{|}}{\overset{\overset{CH_3}{|}}{C}}\!-\!CH_2OH$

(c)

16.97 2-Propanol (or isopropanol) is prepared by reacting propylene (CH_3CHCH_2) with sulfuric acid, followed by treatment with water. (a) Show the sequence of steps leading to the product. What is the role of the sulfuric acid? (b) Draw the structure of an alcohol that is an isomer of isopropanol. (c) Is isopropanol a chiral molecule? (d) What property of isopropanol makes it useful as a rubbing alcohol?

16.98 When a mixture of methane and bromine vapor is exposed to light, the following reaction occurs slowly: $CH_4(g) + Br_2(g) \longrightarrow CH_3Br(g) + HBr(g)$. Suggest a mechanism for this reaction. (*Hint:* Bromine vapor is deep red; methane is colorless.)

16.99 What is the field strength (in tesla) needed to generate a ^1H frequency of 600 MHz. The gyromagnetic ratio for ^1H is $26.75 \times 10^7\ T^{-1}\ s^{-1}$.

16.100 For a 400 MHz NMR spectrometer, what is the difference in resonance frequencies for two protons whose δ values differ by 2.5 ppm.

16.101 For each of the following molecules, state how many ^1H NMR peaks occur and whether each peak is a singlet, doublet, triplet, etc. (a) CH_3OCH_3, (b) $C_2H_5OC_2H_5$, (c) C_2H_6, (d) CH_3F, and (e) $CH_3COOC_2H_5$.

16.102 Sketch the ^1H NMR spectrum (including splittings) of 2-methylpropanol [isobutyl alcohol—$(CH_3)_2CHCH_2OH$], given the following chemical shift data: $-CH_3$: 0.89 ppm, $-CH$: 1.67 ppm, $-CH_2$: 3.27 ppm, and $-OH$: 4.50 ppm.

16.103 The ^{1}H NMR spectrum of methylbenzene (or toluene) has been recorded at 60 MHz and 1.41 T. It consists of two peaks, one due to the methyl protons and one due to the aromatic protons. (a) What would be the magnetic field at 300 MHz? (b) At 60 MHz, the resonance frequencies are 140 MHz for the methyl protons and 430 MHz for the aromatic protons. What would the frequencies be if the spectrum were recorded on a 300-MHz spectrometer?

16.104 2-Butanone can be reduced to 2-butanol by reagents such as lithium aluminum hydroxide (LiAlH$_4$). (a) Write the formula of the product. Is it chiral? (b) In reality, the product does not exhibit optical activity. Explain.

16.105 In Lewis Carroll's tale *Through the Looking Glass,* Alice wonders whether "looking-glass milk" on the other side of the mirror would be fit to drink. Based on your knowledge of chirality and enzyme action, comment on the validity of Alice's concern.

16.106 Nylon was designed to be a synthetic silk. (a) The average molar mass of a batch of nylon-66 is 12,000 g mol^{-1}. How many monomer units are there in this sample? (b) Which part of nylon's structure is similar to a polypeptide's structure? (c) How many different tripeptides (made up of three amino acids) can be formed from the amino acids alanine (Ala), glycine (Gly), and serine (Ser), which account for most of the amino acids in silk?

16.107 The enthalpy change in the denaturation of a certain protein is 125 kJ mol^{-1}. If the entropy change is 397 J K^{-1} mol^{-1}, calculate the minimum temperature at which the protein would denature spontaneously.

16.108 When deoxyhemoglobin crystals are exposed to oxygen, they shatter. On the other hand, deoxymyoglobin crystals are unaffected by oxygen. Explain. (Myoglobin is made up of only one of the four subunits, or polypeptide chains, of hemoglobin.)

16.109 Depending on the experimental conditions, the measurement of the molar mass of hemoglobin in an aqueous solution may show that the solution is monodisperse or polydisperse. Explain.

16.110 In protein synthesis, the selection of a particular amino acid is determined by the so-called genetic code, or a sequence of three bases in DNA. Could a sequence of only two bases unambiguously determine the selection of 20 amino acids found in proteins? Explain.

16.111 Consider the fully protonated amino acid valine:

$$\underset{\substack{| \\ CH_3}}{\overset{\substack{+ \ 9.62 \\ CH_3\ NH_3 \\ |}}{H-C-C-COOH}}\ ^{2.32}$$

where the numbers denote the pK_a values. (a) Which of the two groups ($-\overset{+}{N}H_3$ or $-COOH$) is more acidic? (b) Calculate the predominant form of valine at pH 1.0, 7.0, and 12.0. (c) Calculate the isoelectric point of valine. (Hint: See Problem 12.110).

16.112 Consider the formation of a dimeric protein

$$2P \longrightarrow P_2$$

At 25°C, we have $\Delta H° = 17$ kJ mol^{-1} and $\Delta S° = 65$ J mol^{-1} K^{-1}. Is the dimerization favored at this temperature? Comment on the effect of lowering the temperature. Does your result explain why some enzymes lose their activities under cold conditions?

16.113 The left side of the accompanying diagram shows the structure of the enzyme ribonuclease in its native form. The three-dimensional protein structure is maintained in part by the disulfide bond ($-S-S-$) between the amino acid residues (each color sphere represents an S atom). Using certain denaturants, the compact structure is destroyed and the disulfide bonds are converted to sulfhydryl group ($-SH$) shown on the right side of the arrow. (a) Describe the bonding scheme in the disulfide bond in terms of hybridization. (b) Which amino acid in Table 16.6 contains the $-SH$ group? (c) Predict the signs of ΔH and ΔS for the denaturation process. If denaturation is induced by a change in temperature, show why a rise in temperature would favor denaturation. (d) The sulfhydryl groups can be oxidized (i.e., removing the H atoms) to form the disulfide bonds. If the formation of the disulfide bonds is totally random between any two $-SH$ groups, what is the fraction of the regenerated protein structures that corresponds to the native form. (e) An effective remedy to deodorize a dog that has been sprayed by a skunk is to rub the affected areas with a solution of an oxidizing agent such as hydrogen peroxide. What is the chemical basis for this action? (*Hint:* An odiferous component of a skunk's secretion is 2-butene-1-thiol, $CH_3CH=CHCH_2SH$.)

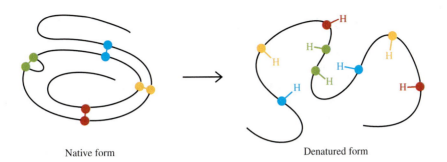

Native form → Denatured form

16.114 Write the structures of three alkenes that yield 2-methylbutane on hydrogenation.

16.115 An alcohol was converted to a carboxylic acid with acidic potassium dichromate. A 4.46-g sample of the acid was added to 50.0 mL of 2.27 M NaOH, and the excess NaOH required 28.7 mL of 1.86 M HCl for neutralization. What is the molecular formula of the alcohol?

16.116 Write the skeletal structural formulas of the alcohols with the formula $C_6H_{13}O$ and indicate those that are chiral.

16.117 Fat and oil are names for the same class of compounds, called triglycerides, which contain three ester groups:

$$
\begin{array}{c}
\quad\quad\quad\ O \\
\quad\quad\quad\ \| \\
CH_2-O-C-R \\
\quad\quad\quad\ O \\
\quad\quad\quad\ \| \\
CH-O-C-R' \\
\quad\quad\quad\ O \\
\quad\quad\quad\ \| \\
CH_2-O-C-R''
\end{array}
$$

A fat or oil

where R, R', and R'' represent long hydrocarbon chains. (a) Suggest a reaction that leads to the formation of a triglyceride molecule, starting with glycerol (see page 338 for the structure of glycerol) and carboxylic acids. (b) In the old days, soaps were made by hydrolyzing animal fat with lye (a sodium hydroxide solution). Write an equation for this reaction. (c) The difference between fats and oils is that fats are solids at room temperature, whereas the oils are liquids. Fats are usually produced by animals, whereas oils are commonly found in plants. The melting points of these substances are determined by the number of C=C bonds (or the extent of unsaturation) present—the larger the number of C=C bonds, the lower the melting point and the more likely that the substance is a liquid. Explain. (d) One way to convert liquid oil to solid fat is to hydrogenate the oil, a process by which some or all of the C=C bonds are converted to C—C bonds. This procedure prolongs the shelf life of the oil by removing the more reactive C=C group and facilitates packaging. How would you carry out such a process (that is, what reagents and catalyst would you employ)? (e) The degree of unsaturation of oil can be determined by reacting the oil with iodine, which reacts with the C=C bond as follows:

$$
\begin{array}{c}
\quad\ |\ \ \ \ \ \ \ \ |\quad\quad\quad\quad\quad\quad\ I\quad I \\
\quad\ |\ \ \ \ \ \ \ \ |\quad\quad\quad\quad\quad\quad\ |\quad | \\
-C-C=C-C-\ +\ I_2\ \longrightarrow\ -C-C-C-C- \\
\quad\ |\ \ \ \ \ \ \ \ |\quad\quad\quad\quad\quad\quad\ |\quad |
\end{array}
$$

The procedure is to add a known amount of iodine to the oil and allow the reaction to go to completion. The amount of excess (unreacted) iodine is determined by titrating the remaining iodine with a standard sodium thiosulfate ($Na_2S_2O_3$) solution:

$$I_2 + 2Na_2S_2O_3 \longrightarrow Na_2S_4O_6 + 2NaI$$

The number of grams of iodine that react with 100 g of oil is called the *iodine number*. In one case, 43.8 g of I_2 was treated with 35.3 g of corn oil. The excess iodine required 20.6 mL of a 0.142 M $Na_2S_2O_3$ for neutralization. Calculate the iodine number of the corn oil.

Answers to Practice Exercises

16.1 4,6-diethyl-2-methyloctane

16.2

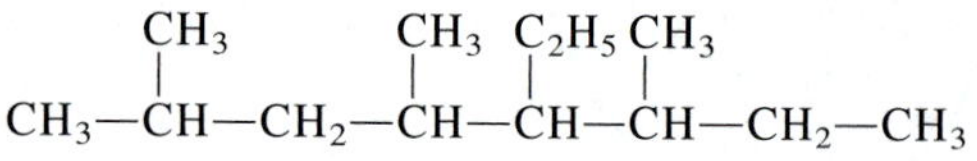

$$CH_3-\underset{\underset{\displaystyle CH_3}{|}}{CH}-CH_2-\underset{\underset{\displaystyle CH_3}{|}}{CH}-\underset{\underset{\displaystyle C_2H_5}{|}}{CH}-\underset{\underset{\displaystyle CH_3}{|}}{CH}-CH_2-CH_3$$

16.3

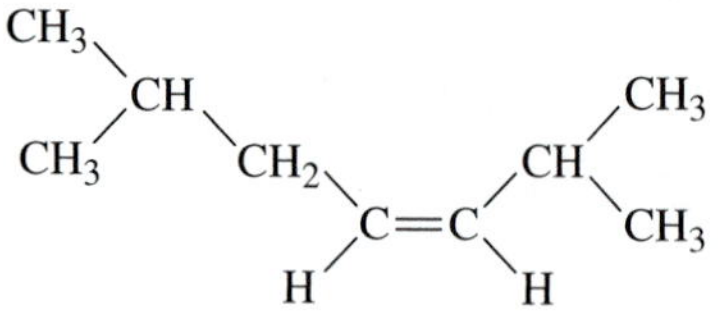

2,6-dimethyl-*cis*-3-heptene

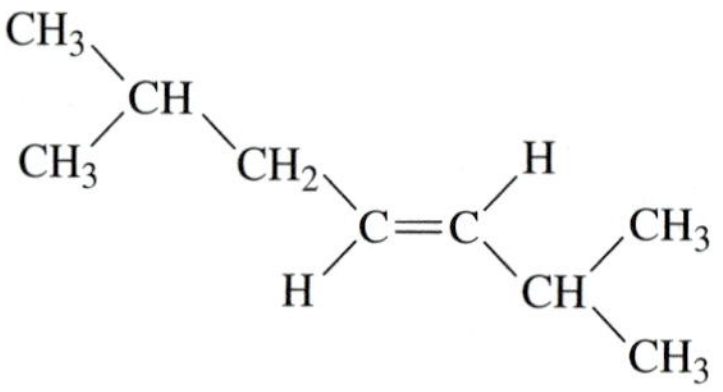

2,6-dimethyl-*trans*-3-heptene

16.4

(a) 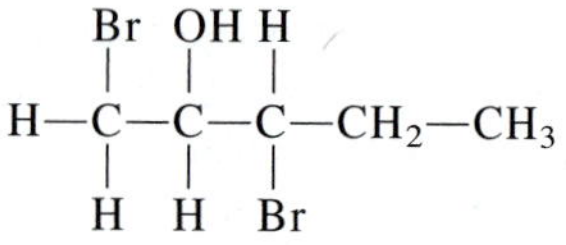
$$H-\underset{\underset{\displaystyle H}{|}}{\overset{\overset{\displaystyle Br}{|}}{C}}-\underset{\underset{\displaystyle H}{|}}{\overset{\overset{\displaystyle OH}{|}}{C}}-\underset{\underset{\displaystyle Br}{|}}{\overset{\overset{\displaystyle H}{|}}{C}}-CH_2-CH_3$$

(b)
$$CH_3-\underset{\underset{\displaystyle H}{|}}{\overset{\overset{\displaystyle CH_3}{\overset{|}{\overset{\displaystyle O}{|}}}}{C}}-\underset{\underset{\displaystyle CH_3}{|}}{\overset{\overset{\displaystyle H}{|}}{C}}-CH_2-CH_3$$

(c)
$$CH_3-\underset{\underset{\displaystyle H}{|}}{\overset{\overset{\displaystyle CH_3}{|}}{C}}-\overset{\overset{\displaystyle O}{\|}}{C}-CH_2-CH_3$$

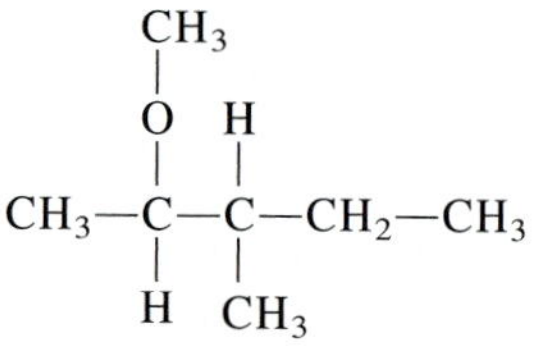

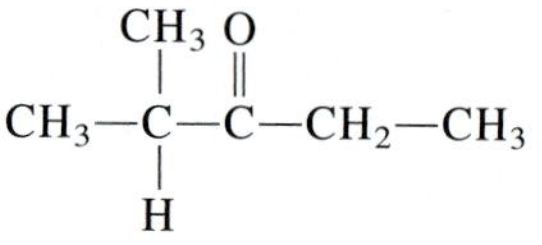

(d) 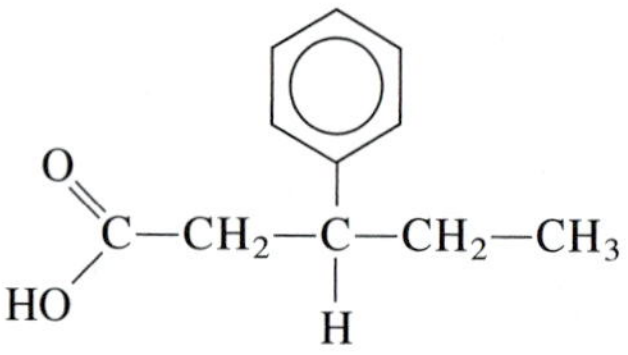
$$\underset{\displaystyle HO}{\overset{\displaystyle O}{\|}}{C}-CH_2-\underset{\underset{\displaystyle H}{|}}{C}-CH_2-CH_3$$

16.5 $CH_3CH_2COOCH_3$ and H_2O

17

Nuclear Chemistry

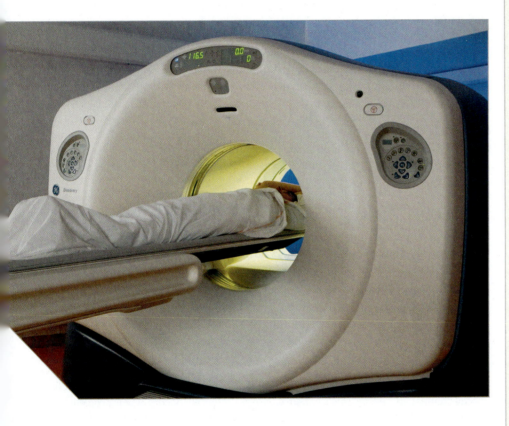

17.1 | Nuclear Chemistry Is the Study of Changes Involving Atomic Nuclei

Radioactivity and Nuclear Decay

In 1895, the German physicist Wilhelm Röntgen[1] noticed that cathode rays (Section 0.2) caused glass and metals to emit highly energetic radiation that penetrated matter, darkened covered photographic plates, and caused a variety of substances to fluoresce. Because these rays could not be deflected by a magnet, they could not contain charged particles as cathode rays do. Röntgen called them X-rays because their nature was unknown. Not long after Röntgen's discovery, Antoine Becquerel,[2] a professor of physics in Paris, began to study the fluorescent properties of substances. Purely by accident, he found that exposing thickly wrapped photographic plates to a certain uranium compound caused them to darken, even without the stimulation of cathode rays. Like X-rays, the rays from the uranium compound were highly energetic and could not be deflected by a magnet, but they differed from X-rays because they arose spontaneously. One of Becquerel's students, Marie Curie,[3] suggested the name **radioactivity** to describe this *spontaneous emission of particles and/or radiation.* Since then, any element that spontaneously emits radiation is said to be *radioactive.*

Certain elements are radioactive because their nuclei are unstable and spontaneously decay through the ejection (or capture) of subatomic particles, often accompanied by the emission of high-energy electromagnetic radiation. The principal radioactive decay processes are alpha decay, beta decay, gamma emission, and spontaneous fission.

▶ **Alpha (α) decay** is the emission by an unstable nucleus of *alpha (α) particles,* which are helium-4 nuclei. Alpha decay occurs only in some very heavy nuclei in which the mass number (the total number of neutrons and protons) exceeds about 140. Emission of an α particle decreases the atomic number (Z) of the nucleus by 2 and its atomic mass number (A) by 4. Polonium-210 ($^{210}_{84}Po$), for example, decays spontaneously to lead-206 ($^{206}_{82}Pb$), through the emission of an α particle [Figure 17.1(a)].

▶ **Beta (β) decay** can occur in two different ways. In β^- *decay,* a neutron in the nucleus spontaneously decays into a proton with the simultaneous emission from the nucleus of an electron and a massless, neutral subatomic particle called an *antineutrino* ($\overline{\nu}$). Carbon-14 ($^{14}_6C$), for example, spontaneously decays into nitrogen-14 ($^{14}_7N$) through the emission of an electron (β^- particle). In contrast, β^+ decay (also called *positron emission*) occurs when a proton in the nucleus spontaneously decays into a neutron with the simultaneous

---

1. Wilhelm Conrad Röntgen (1845–1923). German physicist. He received the Nobel Prize in Physics in 1901 for the discovery of X rays.

2. Antoine Henri Becquerel (1852–1908). French physicist. He was awarded the Nobel Prize in Physics in 1903 for discovering radioactivity in uranium.

3. Marie (Marya Sklodowska) Curie (1867–1934). Polish-born chemist and physicist. In 1903, she and her French husband, Pierre Curie, were awarded the Nobel Prize in Physics for their work on radioactivity. In 1911, she again received the Nobel Prize, this time in chemistry, for her work on the radioactive elements radium and polonium. She is one of only three people to have received two Nobel prizes in science. Despite her great contribution to science, her nomination to the French Academy of Sciences in 1911 was rejected by one vote because she was a woman! Her daughter Irene and son-in-law Frédéric Joliot-Curie shared the Nobel Prize in Chemistry in 1935.

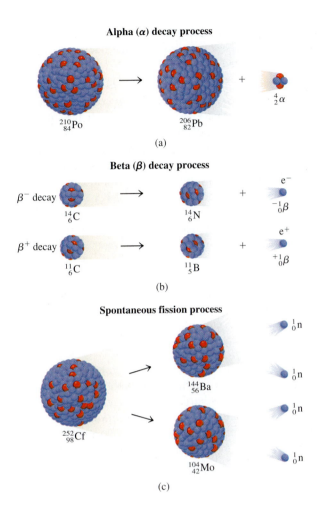

Figure 17.1 Examples of nuclear decay processes: (a) alpha decay, (b) beta decay, and (c) spontaneous fission.

ejection from the nucleus of a ***positron*** (or *antielectron*)[4] and a *neutrino* (ν). Carbon-11 ($^{11}_{6}C$), for example, decays to boron-11 ($^{11}_{5}B$) through the emission of a positron. Figure 17.1(b) illustrates these two beta decay processes. The process of β^- decay increases the atomic number (Z) of the nucleus by 1 (a neutron is converted into a proton) but leaves the mass number (N) unchanged. In writing nuclear equations, the β^- particle is denoted by the symbol $^{0}_{-1}\beta$. Likewise, the emission of a positron (denoted by $^{0}_{+1}\beta$) in β^+ decay also leaves the mass number unchanged but *decreases* the atomic number by 1. A decay process closely related to positron emission is

---

4. In the Standard Model of particle physics, every particle of ordinary matter has an *antimatter* counterpart, called an *antiparticle.* The antiparticle has some properties in common with its matter counterpart and some properties that are exactly opposite. For example, the antiparticles of the proton and electron, the *antiproton* and *positron,* are opposite in charge to their matter counterparts but identical in mass and intrinsic spin. Neutral particles, such as the neutron and the massless neutrino have antimatter counterparts (the antineutron and antineutrino) that are identical in mass and spin but differ in the precise way they interact with other particles. Some neutral particles, such as the *neutral pion,* are their own antiparticle. When a particle and its corresponding antiparticle collide, they annihilate one another and their combined mass-energy is converted to another form—either into a photon of high-energy electromagnetic radiation or into lighter subatomic particles with greater kinetic energy.

orbital electron capture in which the nucleus captures an orbital electron. The orbital electron combines with a proton in the nucleus to form a neutron and a neutrino, which is ejected.

▸ *Gamma (γ) emission* is *the emission of very high energy electromagnetic radiation by an unstable nucleus during the process of nuclear decay.* Like electrons in atoms, nuclei have an energy-shell structure, that is, a nucleus can only exist in certain well-defined quantum states. When a nucleus decays (for example, through an α- or β-decay process), the transformed nucleus is generally left in an excited quantum state, which then decays to the ground state through the emission of a high-energy photon called *gamma radiation.* Although γ radiation is generally higher in energy and frequency than X-rays, the term refers primarily to its origin from nuclear emission rather than its specific location on the electromagnetic spectrum. Deexcitation of the nucleus can also occur through the related process of **internal conversion** in which the excess energy is transferred to a core orbital electron, which is subsequently ejected from the atom.

▸ *Spontaneous fission* is *the spontaneous decay of heavy nuclei into two or more smaller nuclei and some neutrons* [Figure 17.1(c)]. The smaller nuclei are generally radioactive and decay through a series of β-decay processes until a stable nuclear state is obtained.

These nuclear decay processes describe the *spontaneous decay* of a nucleus, that is, they occur without external input of energy or matter. Radioactive decay can also be induced by *the bombardment of nuclei by neutrons, protons, or other nuclei,* in a process called **nuclear transmutation.** An important example of nuclear transmutation is the conversion of atmospheric $^{14}_{7}\text{N}$ into $^{14}_{6}\text{C}$ and $^{1}_{1}\text{H}$, which results when the nitrogen isotope captures a neutron (from the sun). In some cases, heavier elements are synthesized from lighter elements. This type of transmutation occurs naturally in outer space but can also be achieved artificially, as discussed in Section 17.4.

Spontaneous radioactive decay and nuclear transmutation are *nuclear reactions,* which differ significantly from ordinary chemical reactions. Table 17.1 summarizes the differences.

Table 17.1	Comparison of Chemical Reactions and Nuclear Reactions
Chemical Reactions	**Nuclear Reactions**
1. Atoms are rearranged by the breaking and forming of chemical bonds.	1. Elements (or isotopes of the same elements) are converted from one to another.
2. Only electrons in atomic or molecular orbitals are involved in the breaking and forming of bonds.	2. Protons, neutrons, electrons, and other elementary particles may be involved.
3. Reactions are accompanied by absorption or release of relatively small amounts of energy.	3. Reactions are accompanied by absorption or release of tremendous amounts of energy.
4. Rates of reaction are influenced by temperature, pressure, concentration, and catalysts.	4. Rates of reaction normally are not affected by temperature, pressure, and catalysts.

Balancing Nuclear Reactions

To discuss nuclear reactions in any depth, we need to understand how to write and balance the equations. Writing a nuclear equation differs somewhat from writing equations for chemical reactions. In addition to writing the symbols for the various chemical elements, we must also explicitly indicate protons, neutrons, and electrons. In fact, we must show the numbers of protons and neutrons present in *every* species in such an equation.

In writing nuclear reactions, the symbols for the elementary particles are as follows:

$$\underset{proton}{^1_1\text{p or }^1_1\text{H}} \qquad \underset{neutron}{^1_0\text{n}} \qquad \underset{electron}{^0_{-1}e \text{ or } ^0_{-1}\beta} \qquad \underset{positron}{^0_{+1}e \text{ or } ^0_{+1}\beta} \qquad \underset{\alpha \text{ particle}}{^4_2\text{He or } ^4_2\alpha}$$

In accordance with the notation used in Section 0.2, the superscript in each case denotes the mass number A (the total number of neutrons and protons present) and the subscript is the atomic number Z (the number of protons). Thus, the atomic number of a proton is 1, because there is one proton present, and the mass number is also 1, because there is one proton but no neutrons present. On the other hand, the mass number of a neutron is 1, but its atomic number is 0 because there are no protons present. For the electron, the mass number is 0 (there are neither protons nor neutrons present), but the "atomic number" is assigned a value -1, because the electron possesses a unit negative charge.

The symbol $^0_{-1}e$ represents an electron in or from an atomic orbital. The symbol $^0_{-1}\beta$ is reserved for an electron that, although physically identical to any other electron, is emitted from a nucleus in β^- decay and does not originate from an atomic orbital. Similarly, the symbol $^0_{+1}\beta$ specifically denotes a positron emitted by a nucleus in β^+ decay.

In balancing any nuclear equation, we must observe the following rules:

▶ The total number of protons plus neutrons in the products and in the reactants must be the same (conservation of mass number).

▶ The total number of nuclear charges in the products and in the reactants must be the same (conservation of atomic number).

If we know the atomic numbers and mass numbers of all the species but one in a nuclear equation, we can identify the unknown species by applying these rules, as shown in Example 17.1, which illustrates how to balance nuclear decay equations.

Example 17.1

Identify the product X by balancing the following nuclear equations:

$$\text{(a) } ^{212}_{84}\text{Po} \longrightarrow ^{208}_{82}\text{Pb} + \text{X} \quad \text{and (b) } ^{137}_{55}\text{Po} \longrightarrow ^{137}_{56}\text{Pb} + \text{X}$$

Strategy In a balanced nuclear equation, the sum of the atomic numbers and the sum of the mass numbers must match on both sides of the equation. To identify X, therefore, subtract the atomic numbers and mass numbers of the known products from those of the reactants.

—Continued

Continued—

Solution (a) The mass number is 212 and the atomic number is 84 on the left-hand side of the equation, and 208 and 82, respectively, on the right-hand side. Thus, X must have a mass number of $212 - 208 = 4$ and an atomic number of $84 - 82 = 2$, which means that it is an α particle. The balanced equation is

$$^{212}_{84}\text{Po} \longrightarrow\ ^{208}_{82}\text{Pb} + ^{4}_{2}\alpha$$

(b) In this case, the mass number is the same on both sides of the equation, but the atomic number of the product is 1 more than that of the reactant. Thus, X must have a mass number of $137 - 137 = 0$ and an atomic number of $55 - 56 = -1$, which means that it is a β^- particle. Because β^- decay increases the atomic number of the nucleus by 1, the balanced equation is

$$^{137}_{55}\text{Cs} \longrightarrow\ ^{137}_{56}\text{Ba} + ^{0}_{-1}\beta$$

Check The equations in (a) and (b) are balanced for nuclear particles but not for electrical charges. To balance the charges, we would need to add two electrons on the right-hand side of (a) and express barium as a cation (Ba^+) in part (b).

Practice Exercise Identify X in the following nuclear equation:

$$^{22}_{11}\text{Na} \longrightarrow\ ^{0}_{+1}\beta + \text{X}$$

17.2 | The Stability of a Nucleus Is Determined Primarily by Its Neutron-to-Proton Ratio

The nucleus occupies a very small portion of the total volume of an atom, but it contains most of the mass of the atom because both the protons and the neutrons reside there. In studying the stability of the atomic nucleus, it is helpful to know something about its density, because it reveals how tightly the particles are packed together. As a sample calculation, assume that a nucleus has a radius of approximately 5 fm (1 femtometer = 1 fm = 10^{-15} m) and a mass of 1×10^{-22} g. These values correspond roughly to a nucleus containing 30 protons and 30 neutrons. Density is mass/volume, and we can calculate the volume from the known radius (the volume of a sphere is $4\pi r^3/3$, where r is the radius of the sphere). First, convert the femtometer units to centimeter (1 fm = 10^{-13} cm), calculate the density in g cm^{-3}:

$$\text{density} = \frac{\text{mass}}{\text{volume}} = \frac{1 \times 10^{-22}\,\text{g}}{4\pi(5 \times 10^{-13}\,\text{cm})^3/3} = 2 \times 10^{13}\,\text{g cm}^{-3}$$

This is exceedingly dense. The highest density known for an element is 22.6 g cm^{-3}, for osmium (Os). Thus, the average atomic nucleus is roughly 9×10^{12} (or 9 trillion) times more dense than the densest element known!

The enormously high density of the nucleus prompts us to wonder what holds the protons and neutrons (collectively called **nucleons**) together so tightly. According to *Coulomb's law,* we know that like charges repel and unlike charges attract one another. We would thus expect the protons to repel one another strongly, particularly when we consider how close they must be to each other. The explanation for this seeming paradox is that, in addition to the repulsive coulomb force, there are also very short-range attractions between nucleons. These short-range attractions are due to the

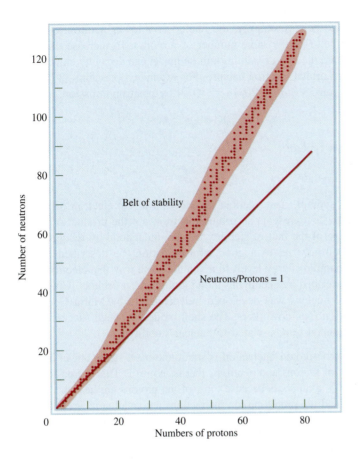

Figure 17.2 Plot of the number of neutrons versus the number of protons in the nucleus for various stable isotopes, represented by dots. The straight line represents the points at which the neutron-to-proton ratio equals 1. The shaded area shows the belt of stability.

strong force—the strongest of the four fundamental forces of nature (see Section 0.1). The stability of any nucleus depends on the relative magnitudes of the coulombic repulsion and the strong force attraction. If repulsion exceeds attraction, the nucleus disintegrates, emitting particles and/or radiation. If strong forces prevail, the nucleus is stable. As in chemical reactions, there is generally a significant activation barrier to nuclear decay, so unstable nuclei do not decay instantaneously, even if the net decay process is energetically favorable.

The graph in Figure 17.2 plots the number of neutrons versus the number of protons in various isotopes. The stable nuclei are located in an area of the graph known as the *belt of stability*. Most radioactive nuclei lie outside this belt. The principal factor in predicting whether a given nucleus is stable is the *neutron-to-proton ratio* (n/p), which is the number of neutrons divided by the number of protons in a given nucleus. For stable atoms of elements having low atomic number, the n/p values are close to 1. As the atomic number increases, the neutron-to-proton ratios of the stable nuclei become greater than 1. Above the belt of stability, the nuclei have higher neutron-to-proton ratios than those within the belt (for the same number of protons). To lower this ratio (and hence move down toward the belt of stability), these nuclei can undergo beta decay, which leads to an increase in the number of protons in the nucleus and a simultaneous decrease in the number of neutrons. Some examples are

$$^{14}_{6}\text{C} \longrightarrow \,^{14}_{7}\text{N} + \,^{0}_{-1}\beta$$

$$^{40}_{19}\text{K} \longrightarrow \,^{40}_{20}\text{Ca} + \,^{0}_{-1}\beta$$

In the beta decay of carbon-14, the n/p ratio decreases from $8/6 = 4/3$ to $7/7 = 1$.

Below the belt of stability, the nuclei have lower neutron-to-proton ratios than those in the belt (for the same number of protons). To increase this ratio (and hence move up toward the belt of stability), these nuclei can decay by either positron emission (β^+ decay) or orbital electron capture. For example, potassium-38 (n/p = 19/19 = 1) decays to argon-38 (n/p = 20/18 = 10/9) by positron emission,

$$^{38}_{19}\text{K} \longrightarrow {}^{38}_{18}\text{Ar} + {}^{0}_{+1}\beta$$

or by electron capture,

$$^{38}_{19}\text{K} + {}^{0}_{-1}e \longrightarrow {}^{38}_{18}\text{Ar}$$

(We use the symbol ${}^{0}_{-1}e$ to denote the electron in the electron capture equation, not ${}^{0}_{-1}\beta$, because ${}^{0}_{-1}e$ represents an orbital electron, not the product of beta decay.)

The shape of the belt of stability can be understood by recognizing that the protons and neutrons can, to a good approximation, be described as if they occupy individual quantum energy levels in much the same way that electrons in multielectron atoms are described as occupying individual electron orbitals (Chapter 2). This *shell model* of the nucleus was developed independently by Maria Goeppert-Mayer[5] and by Hans Jensen[6] and coworkers. The shell model has been successful at explaining a number of general trends related to nuclear stability:

▸ If the proton-proton interaction is small, the most stable nuclei have nearly equal numbers of protons and neutrons, that is, n/p ≈ 1. This is true for nuclei with small atomic numbers because most of the protons are near the surface of the nucleus and the proton-proton interactions are small compared to the strong-force attractions.[7] The trend is toward larger n/p ratios at higher atomic numbers because proton-proton interaction is large and a larger number of neutrons are needed to counteract the electrostatic repulsion among the protons and stabilize the nucleus.

▸ Nuclei that contain 2, 8, 20, 50, 82, or 126 protons or neutrons are generally more stable than nuclei that do not possess these numbers. For example, there are 10 stable isotopes of tin (Sn) with the atomic number 50 and only 2 stable isotopes of antimony (Sb) with the atomic number 51. The numbers 2, 8, 20, 50, 82, and 126 are called *magic numbers*. The magic numbers for nuclear stability correspond to filled nuclear shells and are analogous to the numbers of electrons associated with the very stable noble gases (that is, 2, 10, 18, 36, 54, and 86 electrons).

▸ Nuclei with even numbers of both protons and neutrons are generally more stable than those with odd numbers of these particles (Table 17.2).

▸ All isotopes of the elements with Z > 83 are radioactive. Also, all isotopes of technetium (Tc, Z = 43) and promethium (Pm, Z = 61) are radioactive.

---

5. Maria Goeppert-Mayer (1906–1972). German-American theoretical physicist. Born in the then-German province of Silesia (now Poland), she received her Ph.D. at the University of Göttingen. She emigrated to the United States in 1930. For her development of the shell model of the nucleus in the late 1940s, she was awarded the Nobel Prize in Physics in 1963 together with Johannes Jensen and Eugene Wigner.

6. Johannes Hans Daniel Jensen (1907–1973). German physicist. While a professor at the University of Heidelberg, he developed the shell model of the nucleus in 1949. For this work, he shared the 1963 Nobel Prize in Physics with Maria Goeppert-Mayer and Eugene Wigner.

7. Nuclei are, to a good approximation, spherical. For small nuclei, a large fraction of the protons will be near the surface. As the size (radius) of the nucleus increases, the fraction of protons in the interior will increase. Because protons in the interior are completely surrounded by other nucleons, they will experience a greater Coulomb repulsion per proton than those at the surface, which are only partially surrounded by other nucleons.

Table 17.2	Number of Stable Isotopes with Even and Odd Numbers of Protons and Neutrons	
Protons	**Neutrons**	**Number of Stable Isotopes**
Odd	Odd	4
Odd	Even	50
Even	Odd	53
Even	Even	164

Nuclear Binding Energy

A quantitative measure of nuclear stability is the ***nuclear binding energy,*** which is *the energy required to break up a nucleus into its component protons and neutrons.* This quantity represents the conversion of mass to energy that occurs during an exothermic nuclear reaction. The concept of nuclear binding energy evolved from studies of nuclear properties in which the masses of nuclei were always found to be less than the sum of the masses of the nucleons. The $^{19}_{9}F$ isotope, for example, which consists of 9 protons, 10 neutrons, and 9 electrons has an atomic mass of 18.9984 u. Using the known masses of the $^{1}_{1}H$ atom (1.007825 u) and the neutron (1.008665 u), we can carry out the following analysis. The mass of 9 $^{1}_{1}H$ atoms (that is, the mass of 9 protons and 9 electrons) is 9×1.007825 u = 9.070425 u, and the mass of 10 neutrons is 10×1.008665 u = 10.08665 u. Therefore, the atomic mass of an $^{19}_{9}F$ atom calculated from the known numbers of electrons, protons, and neutrons is

$$9.070425 \text{ u} + 10.08665 \text{ u} = 19.15708 \text{ u}$$

This value is larger than 18.9984 u (the measured mass of $^{19}_{9}F$) by 0.1587 u. *The difference between the mass of an atom and the sum of the masses of its protons, neutrons, and electrons is called the **mass defect.*** According to the theory of special relativity, the loss in mass shows up as energy given off to the surroundings. Thus, the formation of $^{19}_{9}F$ is exothermic. Using *Einstein's mass-energy equivalence relationship*

$$E = mc^2$$

where E is energy, m is mass, and c is the velocity of light, we can calculate the amount of energy released. Begin by writing

$$\Delta E = \Delta mc^2 \qquad\qquad \textbf{(17.1)}$$

where

$$\Delta E = \text{energy of products} - \text{energy of reactants}$$
$$\Delta m = \text{mass of products} - \text{mass of reactants}$$

Thus, the change in mass is

$$\Delta m = 18.9984 \text{ u} - 19.15708 \text{ u} = -0.1587 \text{ u}$$

Because the mass of fluorine-19 is less than the mass calculated from the number of electrons and nucleons present, Δm is a negative quantity. Consequently, ΔE is also negative; that is, energy is released to the surroundings as a result of the formation of the fluorine-19 nucleus. We calculate ΔE as follows:

$$\Delta E = (-0.1587 \text{ u}) \left(\frac{1 \text{ kg}}{6.022 \times 10^{26} \text{ u}} \right)(2.998 \times 10^8 \text{ m s}^{-1})^2$$
$$= -2.369 \times 10^{-11} \text{ kg m}^2 \text{ s}^{-2}$$
$$= -2.369 \times 10^{-11} \text{ J}$$

This is the amount of energy released when one fluorine-19 nucleus is formed from 9 protons and 10 neutrons. The nuclear binding energy of this nucleus is 2.369×10^{-11} J, which is the amount of energy needed to decompose the nucleus into individual protons and neutrons. For convenience, energies for nuclear processes are generally reported in units of *electron volts* (1 eV $= 1.60218 \times 10^{-19}$ J) or, more commonly, in *million electron volts* (MeV). In MeV, the binding energy of the fluorine nucleus is

$$\text{binding energy} = (2.369 \times 10^{-11} \text{ J}) \left(\frac{1 \text{ eV}}{1.602 \times 10^{-19} \text{ J}} \right) \left(\frac{1 \text{ MeV}}{10^6 \text{ eV}} \right)$$
$$= 147.9 \text{ MeV}$$

The enthalpy changes in ordinary chemical reactions are on the order of 200 kJ mol^{-1}. In contrast, the energy released in the formation of 1 mole of fluorine nuclei is

$$\Delta E = (-2.369 \times 10^{-11} \text{ J})(6.022 \times 10^{22} \text{ mol}^{-1})$$
$$= -1.427 \times 10^{13} \text{ J mol}^{-1}$$
$$= -1.427 \times 10^{10} \text{ kJ mol}^{-1}$$

The nuclear binding energy, therefore, is 1.43×10^{10} kJ mol^{-1} for fluorine-19 nuclei, which is nearly 100 million (10^8) times greater than the typical energy changes in chemical reactions.

Although the nuclear binding energy is an indication of the stability of a nucleus, we must account for the different numbers of nucleons in two nuclei when comparing their stabilities. For this reason, it is more meaningful to use the *nuclear binding energy per nucleon,* defined as

$$\text{nuclear binding energy per nucleon} = \frac{\text{nuclear binding energy}}{\text{number of nucleons}}$$

For the fluorine-19 nucleus,

$$\text{nuclear binding energy per nucleon} = \frac{147.9 \text{ MeV}}{19 \text{ nucleons}}$$
$$= 7.784 \text{ MeV nucleon}^{-1}$$

The nuclear binding energy per nucleon makes it possible to compare the stability of all nuclei on a common basis. The graph in Figure 17.3 plots the variation of the nuclear binding energy per nucleon versus mass number. The curve rises rather steeply, such that the highest binding energies per nucleon belong to elements with intermediate mass number (between 40 and 100) and are greatest for elements in the iron, cobalt, and nickel region (the Group 8B elements) of the periodic table. Thus, the *net* attractive forces among the particles (protons and neutrons) are greatest for the nuclei of these elements.

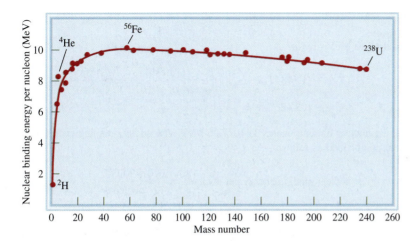

Figure 17.3 Plot of nuclear binding energy per nucleon versus mass number.

The nuclear binding energy and the nuclear binding energy per nucleon are calculated for an iodine nucleus in Example 17.2.

Example 17.2

The atomic mass of $^{127}_{53}\text{I}$ is 126.90447 u. Calculate the nuclear binding energy in MeV of this nucleus and the corresponding nuclear binding energy per nucleon.

Strategy To calculate the nuclear binding energy, we must first determine the difference between the mass of the nucleus and the mass of the individual protons and neutrons, which gives us the mass defect. Next, we apply Einstein's mass-energy relationship ($\Delta E = \Delta mc^2$) to calculate ΔE in joules, and then convert that value to MeV. Finally, we divide ΔE by the number of nucleons to get the corresponding nuclear binding energy per nucleon.

Solution There are 53 protons and 74 neutrons in the iodine nucleus. The mass of 53 $^{1}_{1}\text{H}$ atoms is

$$53 \times 1.007825 \text{ u} = 53.41473 \text{ u}$$

and the mass of 74 neutrons is

$$74 \times 1.008665 \text{ u} = 74.64121 \text{ u}$$

Therefore, the predicted mass for $^{127}_{53}\text{I}$ is 53.41473 u + 74.64121 u = 128.05594 u, and the mass deficit is

$$\Delta m = 126.90447 \text{ u} - 128.05594 \text{ u} = -1.15147 \text{ u}$$

The energy released is

$$\Delta E = \Delta mc^2$$

$$= \left(-1.15147 \text{ u} \times \frac{1 \text{ kg}}{6.022142 \times 10^{26} \text{ u}} \right) (2.99792 \times 10^8 \text{ m s}^{-1})^2$$

$$= -1.71847 \times 10^{-10} \text{ kg m}^2 \text{ s}^{-2}$$

$$= -1.71847 \times 10^{-10} \text{ J}$$

—*Continued*

Continued—
Converting to MeV gives

$$\Delta E = -(1.71847 \times 10^{-10}\,\text{J})\left(\frac{1\,\text{eV}}{1.60218 \times 10^{-19}\,\text{J}}\right)\left(\frac{1\,\text{MeV}}{10^6\,\text{eV}}\right)$$

$$= -1077.26\,\text{MeV}$$

Thus, the nuclear binding energy is 1077.26 MeV. The nuclear binding energy per nucleon is obtained as follows

$$\text{nuclear binding energy per nucleon} = \frac{1077\,\text{MeV}}{127\,\text{nucleons}}$$

$$= 8.48\,\text{MeV nucleon}^{-1}$$

Practice Exercise Calculate the nuclear binding energy (in MeV) and the nuclear binding energy per nucleon of $^{209}_{83}\text{Bi}$ (208.9804 u).

The mass-energy relationship can be used to predict the energy release in nuclear decay reactions, as illustrated in Example 17.3.

Example 17.3

Calculate the energy released (in joules and kJ mol^{-1}) when a nitrogen-12 nucleus undergoes β^+ decay (positron emission):

$$^{12}_{7}\text{N} \longrightarrow {}^{12}_{6}\text{C} + {}^{0}_{+1}\beta + \nu$$

The masses of $^{12}_{7}\text{N}$ and $^{12}_{6}\text{C}$ atoms are 12.018709 u and 12.00000 u, respectively. The mass of the positron is the same as of an electron ($m_e = 9.1093826 \times 10^{-31}$ kg = 5.485799×10^{-4} u) and the neutrino is massless.

Strategy Calculate the change in mass, Δm, for the reaction, and convert to energy using Equation 17.1. Note that the masses given are for the neutral atoms not the nuclei, so we have to subtract the mass of the orbital electrons.

Solution The change in mass is

$$\begin{aligned}
\Delta m &= \text{mass of products} - \text{mass of reactants} \\
&= \left[m(^{12}_{6}\text{C}) - 6\,m_e\right] + m_e - \left[m(^{12}_{7}\text{N}) - 7m_e\right] \\
&= m(^{12}_{6}\text{C}) - m(^{12}_{7}\text{N}) + 2m_e \\
&= 12.000000\,\text{u} - 12.018709\,\text{u} + 2(5.485799 \times 10^{-4}\,\text{u}) \\
&= -0.017611\,\text{u}
\end{aligned}$$

Converting to kilograms yields

$$\Delta m = (-0.017611\,\text{u})(1.66053886 \times 10^{-27}\,\text{kg u}^{-1}) = -2.9245 \times 10^{-29}\,\text{kg}$$

Using Equation 17.1 to convert from mass to energy,

$$\begin{aligned}
\Delta E &= \Delta mc^2 \\
&= (-3.01554 \times 10^{-29}\,\text{kg})(2.99792458 \times 10^8\,\text{m s}^{-1})^2 \\
&= -2.6284 \times 10^{-12}\,\text{J}
\end{aligned}$$

—Continued

Continued—

Thus, the energy released per ^{12}N atom in this decay is 2.6284×10^{-12} J. Multiplying by Avagadro's number and converting to kilojoules gives

$$\Delta E = (2.6284 \times 10^{-15} \text{ kJ})(6.0221367 \times 10^{23} \text{ mol}^{-1}) = 1.5829 \times 10^9 \text{ kJ mol}^{-1}$$

Comment The energy released in this reaction goes into the kinetic energies of the positron and the carbon nucleus, as well as into the energy of the neutrino. The exact distribution of this energy among the product particles will vary.

Practice Exercise Calculate the energy released (in joules and kJ mol^{-1}) for the beta decay of a ^{12}B nucleus:

$$^{12}_{5}\text{B} \longrightarrow ^{12}_{6}\text{C} + {}^{0}_{-1}\beta + \bar{\nu}$$

The masses of ^{12}B and ^{12}C are 12.014353 u and 12.000000 u, respectively.

Radioactive Decay Series

Nuclei outside the belt of stability, as well as all nuclei with $Z > 83$, tend to be unstable and will undergo radioactive decay through the emission of particles or radiation. The disintegration of a radioactive nucleus is often the beginning of a ***radioactive decay series,*** which is *a sequence of nuclear reactions that ultimately form a stable isotope.* Table 17.3 shows the 14-step decay series of naturally occurring uranium-238. This decay scheme is known as the *uranium decay series.* Table 17.3 also shows the *half-lives* (see Chapter 14) of all the products of the decay series.

It is important to be able to balance the nuclear equations for each of the steps in a radioactive decay series. For example, the first step in the uranium decay series is the decay of uranium-238 to thorium-234, with the emission of an α particle. Hence, the equation for this reaction is

$$^{238}_{92}\text{U} \longrightarrow ^{234}_{90}\text{Th} + {}^{4}_{2}\alpha$$

The next step is the β decay of thorium-234 to protactinium-234:

$$^{234}_{90}\text{Th} \longrightarrow ^{234}_{91}\text{Pa} + {}^{0}_{-1}\beta$$

and so on. When discussing radioactive decay steps, the beginning radioactive isotope is called the *parent* and the product is called the *daughter.*

17.3 | Radioactive Decay Is a First-Order Kinetic Process

The radioactive decay of an unstable nucleus is a random process. In any given interval of time, there is a well-defined probability that a given nucleus will decay. This probability is independent of time and is the same for all nuclei of a given type, but is different for different isotopes. The number of nuclei decaying per unit time is the rate of nuclear decay (or ***activity***), which can be measured using devices, such as the Geiger-Mueller counter (Figure 17.4).

In a collection of N identical nuclei, the number of nuclei decaying in a short time interval is proportional to the number of nuclei, which defines a first-order decay process (Section 14.2). We have

$$\text{activity} = -\frac{dN}{dt} = \lambda N \tag{17.2}$$

Table 17.3   The Uranium Decay Series*

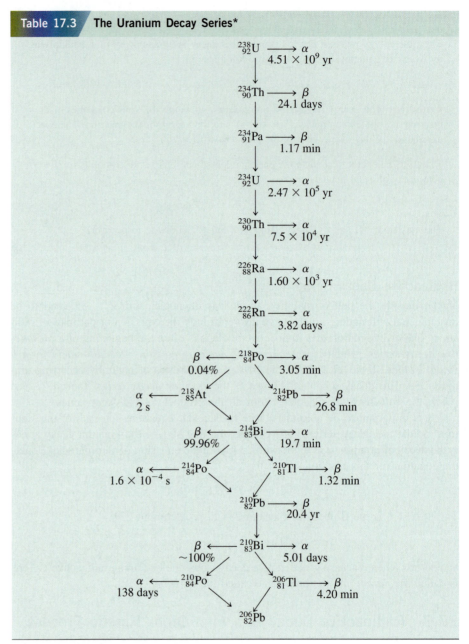

*The times denote the half-lives.

where λ is the nuclear decay rate constant and the negative sign is there to guarantee that activity is a positive quantity. We use λ instead of k because λ is the notation used by nuclear scientists. The standard SI unit of radioactive activity is the *becquerel (Bq)* (named for Henri Becquerel, the discoverer of radioactivity), which is defined as a decay rate of 1 disintegration per second:

$$1 \text{ Bq} \equiv 1 \text{ disintegration s}^{-1}$$

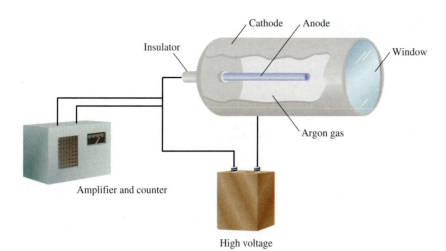

Cathode   Anode

Insulator

Window

Amplifier and counter

Argon gas

High voltage

Figure 17.4 Schematic diagram of a Geiger-Mueller counter. Radiation (α-, β-, or γ-rays) entering through the window ionizes the argon gas to generate a small current between the electrodes. This current is amplified and is used to flash a light or operate a counter with a clicking sound.

An older unit of activity that is commonly used, especially in the United States, is the *curie (Ci)* (named for Marie Curie), which is defined as 3.7×10^{10} Bq. The curie is the decay rate produced by 1 g of radium-226 (an isotope of one of the elements discovered by Marie Curie).

Equation 17.2 is identical in form to the rate equation describing first-order chemical reaction kinetics (Equation 14.3), and the corresponding solution for $N(t)$ is given by (see Equation 14.6)

$$N(t) = N_0 e^{-\lambda t} \tag{17.3}$$

where N_0 is the number of nuclei initially present. As in first-order chemical reactions, the half-life of a radioactive material is independent of the amount of material present and is given by (see Equation 14.9)

$$t_{\frac{1}{2}} = \frac{\ln 2}{\lambda} \approx \frac{0.693}{\lambda} \tag{17.4}$$

The half-lives (hence the rate constants) of radioactive isotopes vary greatly from isotope to isotope. Two extreme cases listed in Table 17.3, for example, are uranium-238 and polonium-214:

$$^{238}_{92}\text{U} \longrightarrow {}^{234}_{90}\text{Th} + {}^{4}_{2}\alpha \qquad\qquad t_{\frac{1}{2}} = 4.51 \times 10^9 \text{ yr}$$

$$^{214}_{84}\text{Po} \longrightarrow {}^{210}_{82}\text{Pb} + {}^{4}_{2}\alpha \qquad\qquad t_{\frac{1}{2}} = 1.6 \times 10^{-4} \text{ s}$$

The half-life of the uranium isotope is about 1×10^{21} times larger than the half-life of the polonium isotope. Unlike the rate constants for chemical reactions, moreover, the rate constants for nuclear decay are unaffected by changes in environmental conditions, such as temperature and pressure (see Table 17.1).

Calculations using Equations 17.3 and 17.4 are similar to those that we encountered in Section 14.3 for first-order chemical reactions. One major difference is that although it is the rate constant that is generally provided for chemical reactions, the half-life is more commonly given for nuclear reactions. In addition, in chemical reactions, we generally measure the concentration as a function of time, whereas in radioactive decay, it is the rate (or activity) that is measured.

Example 17.4 shows a typical nuclear decay calculation.

Example 17.4

The isotope $^{234}_{90}\text{Th}$ undergoes beta decay with a half-life of 24.1 days. Calculate the amount of $^{234}_{90}\text{Th}$ left after 30.0 days from an initial sample of 0.539 g.

Strategy Use Equation 17.4 to find the rate constant, and substitute that into Equation 17.2 to find the amount remaining.

Solution Rearranging Equation 17.4 gives an equation for the rate constant in terms of the half-life:

$$\lambda = \frac{\ln 2}{t_{\frac{1}{2}}}$$

so

$$\lambda = \frac{\ln 2}{24.1 \text{ days}} = 0.0288 \text{ days}^{-1}$$

Substituting the calculated value of λ into Equation 17.3 together with the given initial amount of thorium-234 and the elapsed time gives

$$N = N_0 e^{-\lambda t}$$
$$= 0.539 \text{ g } e^{-(0.0288 \text{ days}^{-1})(30.0 \text{ days})}$$
$$= 0.227 \text{ g}$$

Thus, 0.227 g of the original 0.539 g of $^{234}_{90}\text{Th}$ will be left after 30.0 days.

Check The result (30.0 days) is longer than one half-life, but shorter than two half-lives, so the calculated amount should be less than one-half of the original amount but greater than one-quarter. This is indeed so.

Practice Exercise Iodine-131 undergoes beta decay with a half-life of 8.04 days and is commonly used to test the activity of the thyroid gland. If a patient is given a dose of 0.146 mg of iodine-131, how long would it take for the iodine-131 to decay to 0.001 mg, assuming none is excreted?

Dating Based on Radioactive Decay

The predictable behavior of radioactive decay as a function of time can in many cases be used as an "atomic clock" to estimate the age of objects. If the initial amount of a radioactive isotope in a sample can be accurately estimated, we can determine the time that has elapsed from that initial time by measuring the current amount of material present and applying Equation 17.3. Some examples of dating by radioactive decay measurements are described here.

Radiocarbon Dating

The atmosphere of Earth is constantly being bombarded by cosmic rays of extremely high penetrating power. These rays, which originate in outer space, consist of electrons, neutrons, and atomic nuclei. One of the important reactions between the atmosphere and cosmic rays is the capture of neutrons by atmospheric nitrogen (nitrogen-14 isotope) to produce the radioactive carbon-14 isotope and hydrogen:

$$^{14}_{7}\text{N} + ^{1}_{0}\text{n} \longrightarrow ^{14}_{6}\text{C} + ^{1}_{1}\text{H}$$

The carbon-14 atoms so produced are unstable and decay according to the following equation:

$$^{14}_{7}\text{C} \longrightarrow \, ^{14}_{7}\text{N} + \, ^{0}_{-1}\beta$$

The half-life of carbon-14 is 5730 years, which, according to Equation 17.4, gives a decay rate constant of

$$\lambda = \frac{\ln 2}{t_{\frac{1}{2}}} = \frac{0.693}{5739 \text{ yr}} = 1.21 \times 10^{-4} \text{ yr}^{-1}$$

Using the conversion factor 1 year $= 3.19 \times 10^{7}$ s gives

$$\lambda = (1.21 \times 10^{-4} \text{ yr}^{-1}) \left(\frac{1 \text{ yr}}{3.19 \times 10^{7} \text{ s}} \right) = 3.79 \times 10^{-12} \text{ s}^{-1}$$

The unstable carbon atoms eventually form $^{14}\text{CO}_2$, which mixes with ordinary $^{12}\text{CO}_2$ in the air. Because carbon-14 in the atmosphere is continually produced (by neutron bombardment) and destroyed (by radioactive decay), a dynamic equilibrium is reached in which the ratio of carbon-14 to carbon-12 in the atmosphere remains constant over time at a value of 1.3×10^{-12}. Thus, in a 1-g sample of carbon from the atmosphere, we should find 1.3×10^{-12} g of carbon-14. Using Equation 17.2 we can calculate the expected activity of one gram of atmospheric carbon:

$$\text{activity} = \lambda N$$
$$= (3.79 \times 10^{-12} \text{ s}^{-1}) \left(\frac{1.3 \times 10^{-12} \text{ g}}{12.00 \text{ g mol}^{-1}} \right) (6.022 \times 10^{22} \text{ mol}^{-1})$$
$$= 0.25 \text{ s}^{-1}$$
$$= 0.25 \text{ Bq}$$

The carbon-14 isotopes enter the biosphere when carbon dioxide is taken up in plant photosynthesis. Plants are eaten by animals, which exhale carbon-14 in CO_2. Eventually, carbon-14 participates in many aspects of the carbon cycle. The continual exchange between carbon in living tissue and carbon in the atmosphere means that the ratio of carbon-14 to carbon-12 in living matter will be the same as that in the atmosphere. When an individual plant or animal dies, however, the carbon-14 isotope in its cells continues to decay, but is no longer replenished, so the ratio of ^{14}C to ^{12}C decreases. These same processes occur when carbon atoms are trapped in coal, petroleum, or wood preserved underground, and in Egyptian mummies. As the years pass, there are proportionately fewer ^{14}C nuclei in a mummy than in a living person.

In 1955, Willard F. Libby[8] suggested that the ^{14}C to ^{12}C ratio could be used to estimate the length of time the carbon-14 isotope in a particular specimen has been decaying without replenishment. This ingenious technique is based on a remarkably simple idea. By measuring the activity of carbon extracted from the sample, we can calculate the time elapsed since the sample was living tissue by comparing this activity with the equilibrium activity of atmospheric carbon, calculated earlier to be 0.25 Bq g^{-1}. Rearranging Equation 17.3 gives

$$t = \frac{1}{\lambda} \ln \frac{N_0}{N} \qquad\qquad \textbf{(17.5)}$$

---

8. Willard Frank Libby (1908–1980). American chemist. Libby received the Nobel Prize in Chemistry in 1960 for his work on radiocarbon dating.

Equation 17.4 implies that the ratio N_0/N is the same as the ratio of the equilibrium atmospheric activity (the activity at $t = 0$) to the current activity of the sample giving

$$t = (8.3 \times 10^3 \text{ yr}) \ln\left(\frac{0.25 \text{ Bq g}^{-1}}{\text{activity per 1-g sample}}\right) \tag{17.6}$$

The success of radiocarbon dating depends on the accuracy with which we can measure the activity. Precision is more difficult with older samples because they contain fewer ^{14}C nuclei. Nevertheless, radiocarbon dating has become an extremely valuable tool for estimating the age of archaeological artifacts, paintings, and other objects dating back 1000 to 50,000 years. Its predictions have been found to agree very well with other independent measures of age, such as counting the annual rings on trees.

Dating Using Uranium-238 Isotopes

Because some of the intermediate products in the uranium decay series have very long half-lives (see Table 17.3), this series is particularly suitable for estimating the age of rocks in the earth and of extraterrestrial objects. The half-life for the first step ($^{238}_{92}U$ to $^{234}_{90}Th$) is 4.51×10^9 years. This is about 20,000 times the second largest value (that is, 2.47×10^5 yr), which is the half-life for $^{234}_{92}U$ to $^{230}_{90}Th$. As a good approximation, therefore, we can assume that the half-life for the overall process (that is, from $^{238}_{92}U$ to $^{206}_{82}Pb$) is governed solely by the first step:

$$^{238}_{92}U \longrightarrow {}^{206}_{82}Pb + 8\,{}^4_2\alpha + 6\,{}^0_{-1}\beta \qquad\qquad t_{\frac{1}{2}} = 4.51 \times 10^9 \text{ yr}$$

(In the language of chemical kinetics, the $^{234}_{92}U$ to $^{230}_{90}Th$ step is *rate limiting*.)

Naturally occurring uranium minerals contain some lead-206 isotopes formed by radioactive decay. Assuming that no lead was present when the mineral was formed and that the mineral has not undergone chemical changes that would allow the lead-206 isotope to be separated from the parent uranium-238, it is possible to estimate the age of the rocks from the mass ratio of $^{206}_{82}Pb$ to $^{238}_{92}U$. According to the previous equation, for every mole (238 g) of uranium that undergoes complete decay, 1 mole (206 g) of lead is formed. If only half a mole of uranium-238 has decayed, the mass ratio of ^{206}Pb to ^{238}U becomes

$$\frac{206 \text{ g}/2}{238 \text{ g}/2} = 0.866$$

and the process would have taken a half-life of 4.51×10^9 years to complete (Figure 17.5). Ratios lower than 0.866 mean that the rocks are less than 4.51×10^9 years old, and higher ratios suggest a greater age. Studies based on the uranium series (as well as other decay series) put the age of the oldest rocks and, therefore, probably the age of Earth itself, at 4.5×10^9, or 4.5 billion, years.

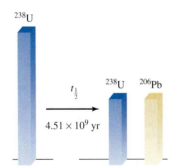

Figure 17.5 After one half-life, half of the original uranium-238 is converted to lead-206.

Dating Using Potassium-40 Isotopes

Radioactive dating using potassium-40 is one of the most important techniques in geochemistry. The radioactive potassium-40 isotope decays by several different modes, but the one most relevant for dating is electron capture:

$$^{40}_{19}K + {}^0_{-1}e \longrightarrow {}^{40}_{18}Ar \qquad\qquad t_{\frac{1}{2}} = 1.2 \times 10^9 \text{ yr}$$

The accumulation of gaseous argon-40 is used to gauge the age of a specimen. When a potassium-40 atom in a mineral decays, argon-40 is trapped in the lattice of the mineral and can escape only if the material is melted. Melting, therefore, is the procedure for analyzing a mineral sample in the laboratory. The amount of argon-40 present can be conveniently measured with a mass spectrometer (see page 46). Knowing the ratio of argon-40 to potassium-40 in the mineral and the half-life of decay makes it possible to establish the ages of rocks ranging from millions to billions of years old.

Example 17.5 gives an illustration of a radioisotope calculation of age.

Example 17.5

The activity of a 0.126-g sample of carbon taken from a piece of linen cloth from an archeological dig was measured to be 1.87×10^{-2} Bq. Estimate the age of the sample.

Strategy Find the activity of a 1-g sample of carbon and use Equation 17.6 to estimate the age.

Solution The activity per gram of the carbon sample is $(1.87 \times 10^{-2}$ Bq$)/(0.126$ g$)$ or 0.148 Bq g^{-1}. Using Equation 17.6 gives

$$t = (8.3 \times 10^3 \text{ yr}) \ln\left(\frac{0.25 \text{ Bq } g^{-1}}{\text{activity per 1-g sample}}\right)$$

$$= (8.3 \times 10^3 \text{ yr}) \ln\left(\frac{0.25 \text{ Bq } g^{-1}}{0.148 \text{ Bq } g^{-1}}\right)$$

$$= 4400 \text{ yr}$$

Practice Exercise Estimate the age of a rock for which the $^{206}_{82}$Pb to $^{238}_{92}$U mass ratio is 0.390. What assumptions do you have to make to perform this calculation?

17.4 | New Isotopes Can Be Produced Through the Process of Nuclear Transmutation

The scope of nuclear chemistry would be rather narrow if it were limited to natural radioactive elements. An experiment performed by Rutherford in 1919, however, suggested the possibility of producing radioactivity artificially. He bombarded a sample of nitrogen-14 with α particles yielding oxygen-17 and protons:

$$^{14}_{7}\text{N} + {}^{4}_{2}\alpha \longrightarrow {}^{17}_{8}\text{O} + {}^{1}_{1}\text{p}$$

This reaction demonstrated for the first time that it was feasible to convert one element into another, by the process of *nuclear transmutation*. Nuclear transmutation differs from radioactive decay in that the transmutation is brought about by the collision of two particles, not by the inherent instability of an isotope.

The reaction of nitrogen-14 with an α particle to yield oxygen-17 and a proton can be abbreviated as $^{14}_{7}\text{N}(\alpha,\text{p})^{17}_{8}\text{O}$. Note that in the parentheses the bombarding particle is written first, followed by the ejected particle. Example 17.6 shows how to use this notation to represent nuclear transmutations.

Example 17.6

Write the balanced equation for the nuclear reaction $^{56}_{26}\text{Fe}(d,\alpha)^{54}_{25}\text{Mn}$, where d represents the deuterium nucleus (i.e., ^2_1H).

Strategy To write the balanced nuclear equation, remember that the first isotope ($^{56}_{26}\text{Fe}$) is the reactant and the second isotope ($^{54}_{25}\text{Mn}$) is the product. Furthermore, the first symbol in the parentheses (d) is the bombarding particle, and the second (α) is the particle emitted as a result of nuclear transmutation.

Solution According to the abbreviated nuclear equation, iron-56 is bombarded with a deuterium nucleus, producing manganese-54 nucleus plus an α particle (^4_2He):

$$^{56}_{26}\text{Fe} + ^2_1\text{H} \longrightarrow ^{54}_{25}\text{Mn} + ^4_2\alpha$$

Check Make sure that the sum of the mass numbers and the sum of the atomic numbers are the same on both sides of the equation.

Practice Exercise Write the balanced equation for $^{106}_{46}\text{Pd}(\alpha,p)^{109}_{47}\text{Ag}$.

Particle accelerators have made it possible to synthesize *transuranium elements—elements with atomic numbers greater than 92*. Neptunium ($Z = 93$) was first prepared in 1940. Since then, 23 other transuranium elements have been synthesized. All isotopes of these elements are radioactive. Table 17.4 lists the transuranium elements up to $Z = 111$ and the reactions by which they are formed.

Table 17.4	The Transuranium Elements		
Atomic Number	**Name**	**Symbol**	**Preparation**
93	Neptunium	Np	$^{238}_{92}\text{U} + ^1_0\text{n} \longrightarrow ^{239}_{93}\text{Np} + ^{0}_{-1}\beta$
94	Plutonium	Pu	$^{239}_{93}\text{Np} \longrightarrow ^{239}_{94}\text{Pu} + ^{0}_{-1}\beta$
95	Americium	Am	$^{239}_{94}\text{Pu} + ^1_0\text{n} \longrightarrow ^{240}_{95}\text{Am} + ^{0}_{-1}\beta$
96	Curium	Cm	$^{239}_{94}\text{Pu} + ^4_2\alpha \longrightarrow ^{242}_{96}\text{Cm} + ^1_0\text{n}$
97	Berkelium	Bk	$^{241}_{95}\text{Am} + ^4_2\alpha \longrightarrow ^{243}_{97}\text{Bk} + 2^1_0\text{n}$
98	Californium	Cf	$^{242}_{96}\text{Cm} + ^4_2\alpha \longrightarrow ^{245}_{98}\text{Cf} + ^1_0\text{n}$
99	Einsteinium	Es	$^{238}_{92}\text{U} + 15^1_0\text{n} \longrightarrow ^{253}_{99}\text{Es} + 7^{0}_{-1}\beta$
100	Fermium	Fm	$^{238}_{92}\text{U} + 17^1_0\text{n} \longrightarrow ^{255}_{100}\text{Fm} + 8^{0}_{-1}\beta$
101	Mendelevium	Md	$^{253}_{99}\text{Es} + ^4_2\alpha \longrightarrow ^{256}_{101}\text{Md} + ^1_0\text{n}$
102	Nobelium	No	$^{246}_{96}\text{Cm} + ^{12}_6\text{C} \longrightarrow ^{254}_{102}\text{No} + 4^1_0\text{n}$
103	Lawrencium	Lr	$^{252}_{98}\text{Cf} + ^{10}_5\text{B} \longrightarrow ^{257}_{103}\text{Lr} + 5^1_0\text{n}$
104	Rutherfordium	Rf	$^{249}_{98}\text{Cf} + ^{12}_6\text{C} \longrightarrow ^{257}_{104}\text{Rf} + 4^1_0\text{n}$
105	Dubnium	Db	$^{249}_{98}\text{Cf} + ^{15}_7\text{N} \longrightarrow ^{260}_{105}\text{Db} + 4^1_0\text{n}$
106	Seaborgium	Sg	$^{249}_{98}\text{Cf} + ^{18}_8\text{O} \longrightarrow ^{263}_{106}\text{Sg} + 4^1_0\text{n}$
107	Bohrium	Bh	$^{209}_{83}\text{Bi} + ^{54}_{24}\text{Cr} \longrightarrow ^{262}_{107}\text{Bh} + ^1_0\text{n}$
108	Hassium	Hs	$^{208}_{82}\text{Pb} + ^{58}_{26}\text{Fe} \longrightarrow ^{265}_{108}\text{Hs} + ^1_0\text{n}$
109	Meitnerium	Mt	$^{209}_{83}\text{Bi} + ^{58}_{26}\text{Fe} \longrightarrow ^{266}_{109}\text{Mt} + ^1_0\text{n}$
110	Darmstadtium	Ds	$^{208}_{82}\text{Pb} + ^{62}_{28}\text{Ni} \longrightarrow ^{269}_{110}\text{Ds} + ^1_0\text{n}$
111	Roentgenium	Rg	$^{209}_{82}\text{Bi} + ^{64}_{28}\text{Ni} \longrightarrow ^{272}_{111}\text{Rg} + ^1_0\text{n}$

Although light elements are generally not radioactive, they can be made so by bombarding their nuclei with appropriate particles. As we saw earlier, the radioactive carbon-14 isotope can be prepared by bombarding nitrogen-14 with neutrons. Tritium, $_1^3\text{H}$, is prepared by bombarding lithium-6 with neutrons:

$$_3^6\text{Li} + {_0^1}\text{n} \longrightarrow {_1^3}\text{H} + {_2^4}\alpha$$

Tritium then decays to helium-3 with the emission of $_{-1}^{0}\beta$ particles:

$$_1^3\text{H} \longrightarrow {_2^3}\text{He} + {_{-1}^{0}}\beta$$

Many synthetic isotopes are prepared by using neutrons as projectiles. This approach is particularly convenient because neutrons carry no charges and therefore are not repelled by the targets—the nuclei. In contrast, when the projectiles are positively charged particles (such as, protons or α particles), they must have considerable kinetic energy to overcome the electrostatic repulsion between themselves and the target nuclei. The synthesis of phosphorus from aluminum is one example:

$$_{13}^{27}\text{Al} + {_2^4}\alpha \longrightarrow {_{15}^{30}}\text{P} + {_0^1}\text{n}$$

A *particle accelerator* uses electric and magnetic fields to increase the kinetic energy of charged species so that a reaction will occur (Figure 17.6). Alternating the polarity (that is, + and −) on specially constructed plates causes the particles to accelerate along a spiral path. When they have the energy necessary to initiate the desired nuclear reaction, they are guided out of the accelerator into a collision with a target substance.

Various designs have been developed for particle accelerators, one of which accelerates particles along a linear path of about 3 km (Figure 17.7). It is now possible to accelerate particles to a speed well above 90 percent of the speed of light. (According to Einstein's theory of relativity, it is impossible for a particle with nonzero rest mass to move *at* the speed of light.) Physicists use the extremely energetic particles produced in accelerators to smash atomic nuclei to fragments. Studying the debris from such disintegrations provides valuable information about nuclear structure and binding forces.

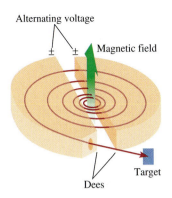

Figure 17.6 Schematic diagram of a cyclotron particle accelerator. The particle (an ion) to be accelerated starts out at the center and is forced to move in a spiral path through the influence of electric and magnetic fields until it emerges at high velocity. The magnetic fields are perpendicular to the plane of the "dees" (so called because of their shape), which are hollow and serve as electrodes.

Figure 17.7 A section of a particle accelerator.

17.5 | In Nuclear Fission, a Large Nucleus Is Split into Smaller Nuclei

Nuclear fission occurs when *a heavy nucleus (A > 200) divides to form smaller nuclei of intermediate mass and one or more neutrons.* Because the heavy nucleus is less stable than its products (see Figure 17.2), fission releases a large amount of energy.

The first nuclear fission reaction to be studied was that of uranium-235 bombarded with slow neutrons, whose speed was comparable to that of air molecules at room temperature. Under these conditions, uranium-235 undergoes fission (Figure 17.8), yielding more than 30 different elements among the possible fission products (Figure 17.9). A representative reaction is

$$^{235}_{92}\text{U} + ^{1}_{0}\text{n} \longrightarrow ^{90}_{38}\text{Sr} + ^{143}_{54}\text{Xe} + 3^{1}_{0}\text{n}$$

Although many heavy nuclei can be made to undergo fission, only the fission of naturally occurring uranium-235 and of the artificial isotope plutonium-239 have any practical importance.

Table 17.5 lists the nuclear binding energies of uranium-235 and its fission products in the preceding reaction. As the table shows, the binding energy per nucleon for uranium-235 is less than the sum of the binding energies for strontium-90 and xenon-143. As a result, energy is released when a uranium-235 nucleus is split into two smaller nuclei. The difference between the binding energies of the reactants and products is (768 + 1200) MeV − 1760 MeV = 208 MeV per uranium-235 nucleus. For 1 mole of uranium-235, the energy released would be

$$(208 \text{ MeV})(1.602 \times 10^{-13} \text{ J MeV}^{-1})(6.02 \times 10^{23} \text{ mol}^{-1}) = 2.0 \times 10^{13} \text{ J mol}^{-1}$$

This is an extremely exothermic reaction, considering that the heat of combustion of 1 ton of coal is only about 5×10^7 J.

Uranium-235 fission not only releases an enormous amount of energy, but it also produces more neutrons than were originally captured in the process. This feature makes possible a **nuclear chain reaction,** which is *a self-sustaining sequence of nuclear fission reactions.* The neutrons generated during the initial stages of fission can induce fission in other uranium-235 nuclei, which in turn produce more neutrons, and so on. In less than a second, the reaction can become uncontrollable, liberating a tremendous amount of heat to the surroundings.

Figure 17.10 shows two types of fission reactions. For a chain reaction to occur, enough uranium-235 must be present in the sample to capture the neutrons. Otherwise,

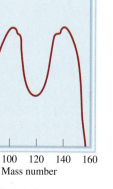

Figure 17.9 Relative yields of the products resulting from the fission of U-235, as a function of mass number.

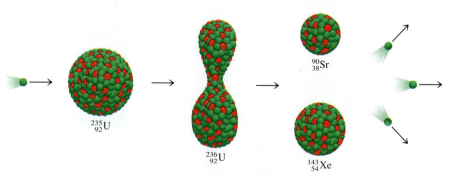

Figure 17.8 Nuclear fission of U-235. When a U-235 nucleus captures a neutron (red dot), it undergoes fission to yield two smaller nuclei. On average, 2.4 neutrons are emitted for every U-235 nucleus that divides.

Table 17.5	Nuclear Binding Energies of ^{235}U and Two Typical Fission Products	
	Nuclear Binding Energy	
	J	*MeV*
^{235}U	2.82×10^{-10}	1.76×10^{3}
^{90}Sr	1.23×10^{-10}	7.68×10^{2}
^{143}Xe	1.92×10^{-10}	1.20×10^{3}

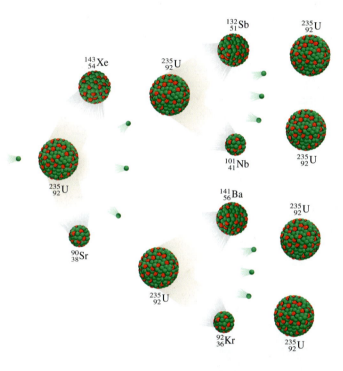

Figure 17.10 If a critical mass is present, many of the neutrons emitted during the fission process will be captured by other ^{235}U nuclei and a chain reaction will occur.

many of the neutrons will escape from the sample and the chain reaction will not occur. In this situation, the mass of the sample is said to be *subcritical*. Figure 17.10 shows what happens when the amount of the fissionable material is equal to or greater than the **critical mass,** *the minimum mass of fissionable material required to generate a self-sustaining nuclear chain reaction.* In this case, most of the neutrons will be captured by uranium-235 nuclei, and a chain reaction will occur.

The Atomic Bomb

The first application of nuclear fission was in the development of the atomic bomb in 1945. How is such a bomb made and detonated? The crucial factor in the design of the bomb is determining the critical mass necessary. A small atomic bomb is equivalent to 20,000 tons of TNT (trinitrotoluene). Because 1 ton of TNT releases about 4×10^{9} J of energy, 20,000 tons (or 20 kilotons) produces 8×10^{13} J. Recall that 1 mole, (235 g) of uranium-235 liberates 2.0×10^{13} J of energy when it undergoes fission. Thus, the mass of the isotope present in a small bomb must be at least

$$(235 \text{ g})\left(\frac{8 \times 10^{13} \text{ J}}{2.0 \times 10^{13} \text{ J}}\right) \approx 1 \text{ kg}$$

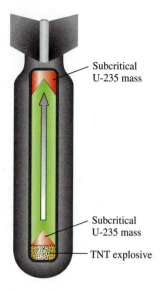

Figure 17.11 Schematic cross section of an atomic bomb. The TNT explosives are set off first. The explosion forces the subcritical sections of fissionable material together to form an amount considerably larger than the critical mass.

An atomic bomb is never assembled with the critical mass already in place. Instead, the critical mass is formed by using a conventional explosive, such as TNT, to force the fissionable sections together, as shown in Figure 17.11. Neutrons from a source at the center of the device trigger the nuclear chain reaction. Uranium-235 was the fissionable material in the bomb dropped on Hiroshima, Japan, on August 6, 1945. Plutonium-239 was used in the bomb exploded over Nagasaki, Japan, 3 days later. The fission reactions generated were similar in these two cases, as was the extent of the destruction.

Nuclear Reactors

A peaceful, but controversial, application of nuclear fission is the generation of electricity using heat from a controlled chain reaction in a nuclear reactor. Currently, nuclear reactors provide about 20 percent of the electric energy in the United States. This is a small but by no means negligible contribution to the nation's energy production. Several different types of nuclear reactors are in operation; we will briefly discuss the main features of three of them—namely, light water reactors, heavy water reactors, and breeder reactors—along with their advantages and disadvantages.

Most of the nuclear reactors in the United States are *light water reactors,* an example of which is shown schematically in Figure 17.12. Figure 17.13 shows the refueling process in the core of such a nuclear reactor.

The speed of the neutrons is a crucial aspect of the fission reaction. Slow neutrons split uranium-235 nuclei more efficiently than do fast ones. Because fission reactions are highly exothermic, the neutrons produced usually move at high velocities. To improve the efficiency of the reaction, therefore, the neutrons must be slowed down with *moderators* (*substances that can reduce the kinetic energy* of neutrons) before they can be used to induce nuclear disintegration. A good moderator must be nontoxic

Figure 17.12 Schematic diagram of a nuclear fission reactor. The fission reaction is controlled by cadmium or boron rods. The heat generated by the reaction is used to produce steam for the generation of electricity.

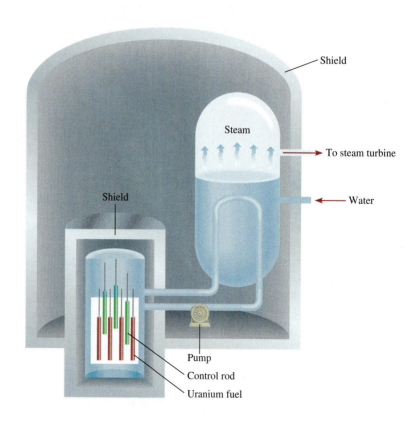

and inexpensive (because very large quantities of it are needed), and it should resist conversion into a radioactive substance by neutron bombardment. Furthermore, it is helpful if the moderator is a liquid so that it can also be used as a coolant. No substance fulfills all these requirements, although water comes closer than many others that have been considered. Nuclear reactors that use ^1H_2O as a moderator are called light water reactors because ^1_1H is the lightest isotope of the element hydrogen.

The nuclear fuel consists of uranium, usually in the form of its oxide (U_3O_8) (Figure 17.14). Naturally occurring uranium contains about 0.7 percent of the uranium-235 isotope, which is too low a concentration to sustain a small-scale chain reaction. For effective operation of a light water reactor, uranium-235 must be enriched to a concentration of 3 or 4 percent. In principle, the main difference between an atomic bomb and a nuclear reactor is that the chain reaction that takes place in a nuclear reactor is kept under control at all times. The factor limiting the rate of the reaction is the number of neutrons present. This can be controlled by lowering cadmium or boron control rods between the fuel elements. These rods capture neutrons according to the following equations:

$$^{113}_{48}Cd + ^1_0n \longrightarrow ^{114}_{48}Cd + \gamma$$
$$^{10}_5B + ^1_0n \longrightarrow ^7_3Li + ^4_2\alpha$$

Figure 17.13 Refueling the core of a nuclear reactor.

where γ denotes gamma emission. Without the control rods, the reaction could not be controlled and the reactor core would melt from the heat generated and release radioactive materials into the environment.

Nuclear reactors have rather elaborate cooling systems that absorb the heat given off by the nuclear reaction and transfer it outside the reactor core, where it is used to produce enough steam to drive an electric generator. In this respect, a nuclear power plant is similar to a conventional power plant that burns fossil fuel. In both cases, large quantities of cooling water are needed to condense steam for reuse. Thus, most nuclear power plants are built near a river or a lake. Unfortunately, this method of cooling causes thermal pollution.

A *heavy water reactor* uses D_2O (^2H_2O or *heavy water*) as the moderator rather than H_2O. Deuterium absorbs neutrons much less efficiently than ordinary hydrogen. Because fewer neutrons are absorbed, the reactor is more efficient and does not require enriched uranium. On the other hand, more neutrons leak out of the reactor, too, though this is not a serious disadvantage.

Figure 17.14 Uranium oxide (U_3O_8).

The main advantage of a heavy water reactor is that it eliminates the need for building expensive uranium enrichment facilities. However, D_2O must be prepared by either fractional distillation or electrolysis of ordinary water, which can be very expensive considering the amount of water used in a nuclear reactor. In countries where hydroelectric power is abundant, the cost of producing D_2O by electrolysis can be reasonably low. At present, Canada is the only nation successfully using heavy water nuclear reactors. Because no enriched uranium is required in a heavy water reactor, Canada enjoys the benefits of nuclear power without undertaking work that is closely associated with weapons technology.

A *breeder reactor* uses uranium fuel, but unlike a conventional nuclear reactor, it *produces more fissionable materials than it uses.* Recall from Table 17.3 that uranium-238, when bombarded with fast neutrons, undergoes the following reactions:

$$^{238}_{92}U + ^1_0n \longrightarrow ^{239}_{92}U$$
$$^{239}_{92}U \longrightarrow ^{239}_{93}Np + ^0_{-1}\beta$$
$$^{239}_{93}Np \longrightarrow ^{239}_{94}Pu + ^0_{-1}\beta$$

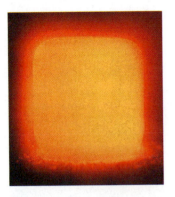

Figure 17.15 The red glow of the radioactive plutonium oxide, (PuO$_2$).

In this manner, the nonfissionable uranium-238 is transmuted into the fissionable isotope plutonium-239 (Figure 17.15).

In a typical breeder reactor, nuclear fuel containing uranium-235 or plutonium-239 is mixed with uranium-238 so that breeding takes place within the core. For every uranium-235 (or plutonium-239) nucleus undergoing fission, more than one neutron is captured by uranium-238 to generate plutonium-239. Thus, the stockpile of fissionable material can be steadily increased as the starting nuclear fuels are consumed. It takes about 7 to 10 years to regenerate the sizable amount of material needed to refuel the original reactor and to fuel another reactor of comparable size. This interval is called the *doubling time*.

Another fertile isotope is $^{232}_{90}$Th. Upon capturing slow neutrons, thorium is transmuted to uranium-233, which, like uranium-235, is a fissionable isotope:

$$^{232}_{90}\text{Th} + ^{1}_{0}\text{n} \longrightarrow ^{233}_{90}\text{Th}$$

$$^{233}_{90}\text{Th} \longrightarrow ^{233}_{91}\text{Pa} + ^{0}_{-1}\beta$$

$$^{233}_{91}\text{Pa} \longrightarrow ^{233}_{92}\text{U} + ^{0}_{-1}\beta$$

Uranium-233 ($t_{\frac{1}{2}} = 1.6 \times 10^5$ yr) is stable enough for long-term storage.

Although the amounts of uranium-238 and thorium-232 in the Earth's crust are relatively plentiful (4 ppm and 12 ppm by mass, respectively), the development of breeder reactors has been very slow. To date, the United States does not have a single operating breeder reactor, and only a few have been built in other countries, such as France and Russia. One problem is economics; breeder reactors are more expensive to build than conventional reactors. There are also more technical difficulties associated with the construction of such reactors. As a result, the future of breeder reactors, in the United States at least, is rather uncertain.

The Hazards of Nuclear Energy

Many people, not just environmentalists, regard nuclear fission as a highly undesirable method of energy production. Many fission products such as strontium-90 are dangerous radioactive isotopes with long half-lives. Plutonium-239, used as a nuclear fuel and produced in breeder reactors, is one of the most toxic substances known. It is an α-emitter with a half-life of 24,400 years.

Accidents, too, can present many dangers. An accident at the Three Mile Island reactor near Harrisburg, Pennsylvania, on March 28, 1979, first brought the potential hazards of nuclear plants to public attention. In this instance, very little radiation escaped the reactor, but the plant remained closed for more than a decade while repairs were made and safety issues addressed. Several years later, on April 26, 1986, a reactor at the Chernobyl nuclear plant in Ukraine surged out of control. The fire and explosion that followed released large amounts of radioactive material into the environment. People working near the plant died within weeks as a result of the exposure to the intense radiation. The long-term effects of the radioactive fallout from this incident have not yet been clearly assessed, although agriculture and dairy farming were affected by it. The number of potential cancer cases attributable to the radiation contamination is estimated to be between a few thousand and more than 100,000.

In addition to the risk of accidents, the problem of radioactive waste disposal has not been satisfactorily resolved even for safely operated nuclear plants. Many suggestions have been made as to where to store or dispose of nuclear waste, including burial underground, burial beneath the ocean floor, and storage in deep geologic formations. None

Nature's Own Fission Reactor

It all started with a routine analysis in May 1972 at the nuclear fuel processing plant in Pierrelatte, France. A staff member was checking the isotope ratio of U-235 to U-238 in a uranium ore and obtained a puzzling result. It had long been known that the relative natural occurrence of U-235 and U-238 is 0.7202 percent and 99.2798 percent, respectively. In this case, however, the amount of U-235 present was only 0.7171 percent. This may seem like a very small deviation, but the measurements were so precise that this difference was considered highly significant. The ore had come from the Oklo mine in the Gabon Republic, a small country on the west coast of Africa. Subsequent analyses of other samples showed that some contained even less U-235, in some cases as little as 0.44 percent.

The logical explanation for the low percentages of U-235 was that a nuclear fission reaction at the mine must have consumed some of the U-235 isotopes. But how did this happen? There are several conditions under which such a nuclear fission reaction could take place. In the presence of heavy water, for example, a chain reaction is possible with unenriched uranium. Without heavy water, such a fission reaction could still occur if the uranium ore and the moderator were arranged according to some specific geometric constraints at the site of the reaction. Both of the possibilities seem rather farfetched. The most plausible explanation was that the uranium ore originally present in the mine was enriched with U-235 and that a nuclear fission reaction took place with light water, as in a conventional nuclear reactor.

As mentioned earlier, the natural abundance of U-235 is 0.7202 percent, but it has not always been that low. The half-lives of U-235 and U-238 are 700 million and 4.51 billion years, respectively. This means that U-235 must have been *more* abundant in the past, because it has a shorter half-life. In fact, at the time Earth was formed, the natural abundance of U-235 was as high as 17 percent! Because the lowest concentration of U-235 required for the operation of a fission reactor is 1 percent, a nuclear chain reaction could have taken place as recently as 400 million years ago. By analyzing the amounts of radioactive fission products left in the ore, scientists concluded that the Gabon "reactor" operated about 2 billion years ago.

Having an enriched uranium sample is only one of the requirements for starting a controlled chain reaction. There must also have been a sufficient amount of the ore and an appropriate moderator present. It appears that as a result of a geologic transformation, uranium ore was continually being washed into the Oklo region to yield concentrated deposits. The moderator needed for the fission process was largely water, present as water of crystallization in the sedimentary ore.

Thus, in a series of extraordinary events, a natural nuclear fission reactor operated at the time when the first life forms appeared on Earth. As is often the case in scientific endeavors, humans are not necessarily the innovators but merely the imitators of nature.

Photo showing the natural nuclear reactor site (lower right-hand corner) at Oklo, Gabon Republic.

of these sites, however, has proved absolutely safe in the long run. Radioactive wastes could leak into underground water, for example, thus endangering nearby communities.

The ideal disposal site would seem to be the sun, where a bit more radiation would make little difference, but this kind of operation would be enormously expensive and complicated and would require space technology that was 100 percent reliable (so that the waste did not accidentally return to Earth).

17.6 | In Nuclear Fusion, Energy Is Produced When Light Nuclei Combine to Form Heavier Ones

Recall from Figure 17.2 that for the lightest elements, nuclear stability increases with increasing mass number. Thus, if two light nuclei combine or fuse together to form a larger, more stable nucleus, an appreciable amount of energy should be released in the process. *The combining of light nuclei to form a heavier one is called **nuclear fusion.***

Although nuclear fusion is not yet feasible for producing energy on Earth, it occurs constantly in the sun. The sun is made up mostly of hydrogen and helium. In its interior, where temperatures reach about 15 million degrees Celsius, the following fusion reactions are believed to take place:

$$\begin{aligned}
{}^{1}_{1}\text{H} + {}^{2}_{1}\text{H} &\longrightarrow {}^{3}_{2}\text{He} \\
{}^{3}_{2}\text{He} + {}^{3}_{2}\text{He} &\longrightarrow {}^{4}_{2}\text{He} + 2\,{}^{1}_{1}\text{H} \\
{}^{1}_{1}\text{H} + {}^{1}_{1}\text{H} &\longrightarrow {}^{2}_{1}\text{H} + {}^{0}_{+1}\beta
\end{aligned}$$

Because *fusion reactions take place only at very high temperatures,* they are often called ***thermonuclear reactions.***

Fusion Reactors

A major concern in choosing the proper nuclear fusion process for energy production is the temperature necessary to carry out the process. Some promising reactions are

Reaction	Energy Released
${}^{2}_{1}\text{H} + {}^{2}_{1}\text{H} \longrightarrow {}^{3}_{1}\text{H} + {}^{1}_{1}\text{H}$	3.9 MeV (6.3×10^{-13} J)
${}^{2}_{1}\text{H} + {}^{3}_{1}\text{H} \longrightarrow {}^{4}_{2}\text{He} + {}^{1}_{0}\text{n}$	17 MeV (2.8×10^{-12} J)
${}^{6}_{3}\text{Li} + {}^{2}_{1}\text{H} \longrightarrow 2\,{}^{4}_{2}\text{He}$	22 MeV (3.6×10^{-12} J)

These reactions take place at extremely high temperatures—on the order of 100 million degrees Celsius—to overcome the repulsive forces between the nuclei. The first reaction is particularly attractive because the world's supply of deuterium is virtually inexhaustible. The total volume of water on Earth is about 1.5×10^{21} L. Because the natural abundance of deuterium is 1.5×10^{-2} percent, the total amount of deuterium present is roughly 4.5×10^{21} g, or 5.0×10^{15} tons. The cost of preparing deuterium is minimal compared with the value of the energy released by the reaction.

In contrast to the nuclear fission, nuclear fusion seems very promising as an energy source, at least "on paper." Although thermal pollution would be a problem, fusion fuels are cheap and almost inexhaustible and the fusion process produces little radioactive waste. If a fusion generator were turned off, it would shut down completely and instantly, without any danger of a meltdown.

If nuclear fusion is so advantageous, why isn't there even one fusion reactor producing energy? Although we command the scientific knowledge to design such a reactor, the technical difficulties have not yet been solved. The basic problem is finding a way to hold the nuclei together long enough, and at the appropriate temperature, for fusion to occur. At temperatures of about 100 million degrees Celsius, molecules cannot exist, and most or all of the atoms are stripped of their electrons. This *state of matter, a gaseous mixture of positive ions and electrons,* is called ***plasma.*** The problem of containing this plasma is formidable. What solid container can exist at such temperatures? None, unless the amount of plasma is small, but then the solid surface would immediately cool the sample and quench the fusion reaction. One approach to solving this problem is to use *magnetic confinement.* Because a plasma consists of charged particles moving at high

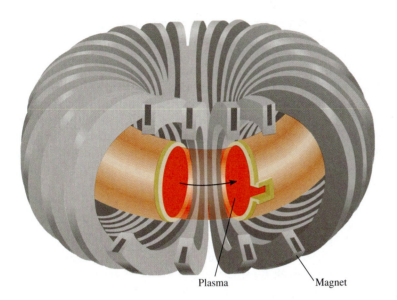

Figure 17.16 A magnetic plasma containment design called a tokamak.

Plasma   Magnet

speeds, a magnetic field will exert a force on it. As shown in Figure 17.16, the plasma moves through a doughnut-shaped tunnel, confined by a complex magnetic field, so the plasma never comes in contact with the walls of the container.

Another promising design employs high-powered lasers to initiate the fusion reaction. In test runs, a number of laser beams transfer energy to a small fuel pellet, heating it and causing it to *implode,* that is, to collapse inward from all sides and compress into a small volume (Figure 17.17), thus causing fusion to occur. Like the magnetic confinement approach, though, laser fusion presents a number of technical difficulties that still need to be overcome before it can be put to practical use on a large scale.

The Hydrogen Bomb

Nuclear fusion is the main source of power in a hydrogen bomb (also called a *thermonuclear bomb*), but the technical problems inherent in the design of a nuclear fusion reactor do not hinder the production of hydrogen bombs. This is because, unlike a

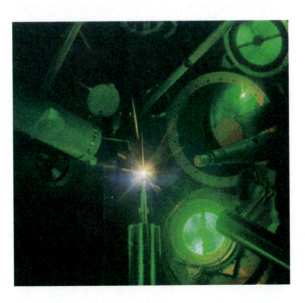

Figure 17.17 A small-scale fusion reaction was created at Lawrence Livermore National Laboratory using Nova, the world's most powerful laser.

Figure 17.18 The explosion of a thermonuclear bomb.

reactor, which must release power in a controlled fashion, the objective of a bomb is to unleash as much power as possible with very little control, so containing the reaction is unnecessary (or even counterproductive).

Hydrogen bombs do not contain gaseous hydrogen or gaseous deuterium; instead, they contain solid lithium deuteride (LiD), which can be packed very tightly. Thus, the detonation of a hydrogen bomb occurs in two stages—first a fission reaction and then a fusion reaction. The required temperature for fusion is achieved with a fission reaction generated by an atomic bomb. Immediately after the atomic bomb explodes, the following fusion reactions occur, releasing vast amounts of energy (Figure 17.18):

$$^6_3\text{Li} + {}^2_1\text{H} \longrightarrow 2{}^4_2\alpha$$
$$^2_1\text{H} + {}^2_1\text{H} \longrightarrow {}^3_1\text{H} + {}^1_1\text{H}$$

There is no critical mass in a fusion bomb, and the force of the explosion is limited only by the quantity of reactants present. Thermonuclear bombs are described as being "cleaner" than atomic bombs because the only radioactive isotopes they produce are tritium, which is a weak β-particle emitter ($t_{\frac{1}{2}} = 12.5$ years), and the products of the fission starter. Their damaging effects on the environment can be aggravated, however, by incorporating in the construction some nonfissionable material such as cobalt. Upon bombardment by neutrons, cobalt-59 is converted to cobalt-60, which is a very strong γ-ray emitter with a half-life of 5.2 years. The presence of radioactive cobalt isotopes in the debris or fallout from a thermonuclear explosion would be fatal to those who survived the initial blast.

17.7 | Radioactive and Stable Isotopes Alike Have Many Applications in Science and Medicine

We have previously described the use of isotopes in the study of reaction mechanisms (see Section 14.5) and in dating artifacts (see Section 17.3). In this section, we will describe the use of isotopes in determining chemical structures, in unraveling the reactions in photosynthesis, and in diagnosing diseases.

Structural Determination

The formula of the thiosulfate ion is $S_2O_3^{2-}$, but for many years chemists were uncertain whether the two sulfur atoms occupied equivalent positions in the ion. The thiosulfate ion is prepared by treatment of the sulfite ion with elemental sulfur:

$$\text{SO}_3^{2-}(aq) + \text{S}(s) \longrightarrow \text{S}_2\text{O}_3^{2-}(aq)$$

When thiosulfate is treated with dilute acid, the reaction is reversed. The sulfite ion is reformed, and elemental sulfur precipitates:

$$\text{S}_2\text{O}_3^{2-}(aq) \xrightarrow{\text{H}^+} \text{SO}_3^{2-}(aq) + \text{S}(s) \tag{17.7}$$

If $\text{SO}_3^{2-}(aq)$ is treated with elemental sulfur enriched with the radioactive sulfur-35 isotope, the isotope acts as a "label" for the S atoms in $\text{S}_2\text{O}_3^{2-}(aq)$. When $\text{S}_2\text{O}_3^{2-}(aq)$ is subsequently treated with dilute acid (Equation 17.7), all the labels are found in the sulfur precipitate; none of them appears in the $\text{SO}_3^{2-}(aq)$. As a result, the two atoms of sulfur in $\text{S}_2\text{O}_3^{2-}$ cannot be structurally equivalent, as would be the case if the structure were

$$[\text{O—S—O—S—O}]^{2-}$$

Otherwise, the radioactive isotope would be present in both the elemental sulfur precipitate and the sulfite ion. Based on spectroscopic studies, we now know that the structure of the thiosulfate ion is

$$\left[\begin{array}{c} :\!\ddot{S}\!: \\ \| \\ :\!\ddot{O}\!-\!\overset{}{\underset{}{S}}\!-\!\ddot{O}\!: \\ \| \\ :\!\ddot{O}\!: \end{array}\right]^{2-}$$

Study of Photosynthesis

The study of photosynthesis is also rich with isotope applications. The overall photosynthesis reaction can be represented as

$$6CO_2 + 6H_2O \longrightarrow C_6H_{12}O_6 + 6O_2$$

Recall from Section 14.5 that the ^{18}O isotope was used to determine the source of O_2 in this reaction. The radioactive ^{14}C isotope helped to determine the path of carbon in photosynthesis. Starting with $^{14}CO_2$, it was possible to isolate the intermediate products during photosynthesis and measure the amount of radioactivity of each carbon-containing compound. In this manner, the path from CO_2 through various intermediate compounds to carbohydrate could be clearly charted. *Isotopes, especially radioactive isotopes that are used to trace the path of the atoms of an element in a chemical or biological process* are called ***tracers.***

Isotopes in Medicine

Tracers are used also for diagnosis in medicine. Sodium-24 (a β-emitter with a half-life of 14.8 h) can be injected into the bloodstream as a salt solution and can be then monitored to trace the flow of blood and detect possible constrictions or obstructions in the circulatory system. Iodine-131 (a β-emitter with a half-life of 8 days) is commonly used to test the activity of the thyroid gland. A malfunctioning thyroid can be detected by giving the patient a drink of a solution containing a known amount of $Na^{131}I$ and measuring the radioactivity just above the thyroid to see if the iodine is absorbed at the normal rate. The amounts of radioisotope used to diagnose diseases in the human body must always be kept small; otherwise, the patient might suffer permanent damage from the high-energy radiation. Another radioactive isotope of iodine, iodine-123 (a γ-ray emitter), is used to image the thyroid gland (Figure 17.19).

Technetium, the first artificially prepared element, is one of the most useful elements in nuclear medicine. Although technetium is a transition metal, all its isotopes are radioactive. In the laboratory, it is prepared by the following nuclear reactions:

$$^{98}_{42}Mo + ^{1}_{0}n \longrightarrow ^{99}_{42}Mo$$
$$^{99}_{42}Mo \longrightarrow ^{99m}_{43}Tc + ^{0}_{-1}\beta$$

where the superscript "m" denotes that the technetium-99 isotope is produced in its excited nuclear state. This isotope has a half-life of about 6 h, decaying by γ radiation to technetium-99 in its nuclear ground state. To use technetium as a diagnostic tool, the patient either drinks or is injected with a solution containing ^{99m}Tc. By detecting the γ rays emitted by ^{99m}Tc, doctors can obtain images of organs such as the heart, liver, and lungs.

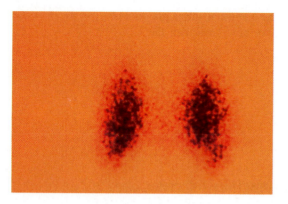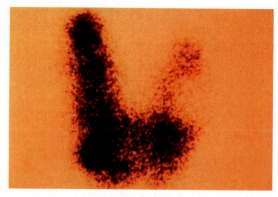

Figure 17.19 After ingesting $Na^{131}I$, the uptake of the radioactive iodine by the thyroid gland in a patient is monitored with a scanner. The photos show a normal thyroid gland (left) and an enlarged thyroid gland (right).

A major advantage of using radioactive isotopes as tracers is that they are easy to detect. Their presence even in very small amounts can be detected by photographic techniques or by devices known as counters, such as the Geiger-Mueller counter pictured in Figure 17.4.

17.8 | The Biological Effects of Radiation Can Be Quite Dramatic

In this section, we will examine briefly the effects of radiation on biological systems. But first we need to define quantitative measures of radiation relevant to biological systems. As discussed in Section 17.3, the fundamental unit of radioactivity is the becquerel (Bq), which corresponds to 1 disintegration per second. This unit only measures the quantity of radiation, whereas the intensity of radiation depends on the energy and type of radiation being emitted as well. One common unit for the absorbed dose of radiation is the *rad* (radiation *a*bsorbed *d*ose), which is the amount of radiation that results in the absorption of 1×10^{-5} J per gram of irradiated material. The biological effect of radiation depends on the part of the body irradiated and the type of radiation. For this reason, the rad is often multiplied by a factor called the *RBE* (*r*elative *b*iological *e*ffectiveness). The product is called a *rem* (*r*oentgen *e*quivalent for *m*an):

$$1 \text{ rem } = 1 \text{ rad } \times 1 \text{ RBE}$$

Of the three types of nuclear radiation, α particles usually have the least penetrating power. Beta particles are more penetrating than α particles but less so than γ rays. Gamma rays have very short wavelengths and high energies. Furthermore, because they carry no charge, they cannot be stopped by shielding materials as easily as α and β particles. If α- or β-emitters are ingested, however, their damaging effects are greatly aggravated because the organs will be constantly subjected to damaging radiation at close range. For example, strontium-90 (a β-emitter) can replace calcium in bones, where it does the greatest damage.

Table 17.6 lists the average amounts of radiation an American receives every year. For short-term exposures to radiation, a dosage of 50–200 rem will cause a decrease in white blood cell counts and other complications, while a dosage of 500 rem or

Table 17.6	Average Yearly Radiation Doses for Americans
Source	**Dose (mrem/yr)***
Cosmic rays	20–50
Ground and surroundings	25
Human body†	26
Medical and dental X rays	50–75
Air travel	5
Fallout from weapons tests	5
Nuclear waste	2
Total	133–188

*1 mrem = 1 millirem = 1×10^{-3} rem.
†The radioactivity in the body comes from food and air.

greater may result in death within weeks. Current safety standards permit nuclear workers to be exposed to no more than 5 rem per year and specify a maximum of 0.5 rem of human-made radiation per year for the general public.

The chemical basis of radiation damage is ionizing radiation. Radiation by either α or β particles or γ rays can remove electrons from atoms and molecules in its path, leading to the formation of ions and radicals. **Radicals** (also called *free radicals*) are *molecular fragments having one or more unpaired electrons; they are usually short-lived and highly reactive.* When water is irradiated with γ rays, for example, the following reactions take place:

$$H_2O \xrightarrow{\text{radiation}} H_2O^+ + e^-$$
$$H_2O^+ + H_2O \longrightarrow H_3O^+ + \cdot OH$$

The electron (in the hydrated form) can subsequently react with water or with a hydrogen ion to form atomic hydrogen, and with oxygen to produce the superoxide ion, O_2^- (a radical):

$$e^- + O_2 \longrightarrow \cdot O_2^-$$

In tissues, the superoxide ions and other free radicals attack cell membranes and a host of organic compounds, such as enzymes and DNA molecules. Organic compounds can themselves be directly ionized and destroyed by high-energy radiation.

Exposure to high-energy radiation can induce cancer in humans and other animals. Cancer is characterized by uncontrolled cellular growth. On the other hand, cancer cells also can be destroyed by proper radiation treatment. In radiation therapy, a compromise is sought. The radiation to which the patient is exposed must be sufficient to destroy cancer cells without killing too many normal cells and, it is hoped, without inducing another form of cancer.

Radiation damage to living systems is generally classified as *somatic* or *genetic*. Somatic injuries are those that affect the organism during its own lifetime. Sunburn, skin rash, cancer, and cataracts are examples of somatic damage. Genetic damage is inheritable changes or gene mutations. For example, a person whose chromosomes have been damaged or altered by radiation may have deformed offspring.

Food Irradiation

If you eat processed food, you have probably eaten ingredients exposed to radiation. In the United States, up to 10 percent of herbs and spices are irradiated with X-rays to control mold. A typical dose is equivalent to 60 million chest X-rays. Although food irradiation has been used in one way or another for more than 40 years, it faces an uncertain future in this country.

Back in 1953, the U.S. Army started an experimental program of food irradiation so that deployed troops could have fresh food without refrigeration. The procedure is a simple one. Food is exposed to high levels of radiation to kill insects and harmful bacteria. It is then packaged in airtight containers, in which it can be stored for months without deterioration. The radiation sources for most food preservation are cobalt-60 and cesium-137, both of which are γ-emitters, although X-rays and electron beams can also be used to irradiate food.

Food irradiation reduces energy demand by eliminating the need for refrigeration, and it prolongs the shelf life of various foods, which is of vital importance for poor countries. Nevertheless, there is considerable opposition to this procedure. There is a concern, for example, that irradiated food may itself become radioactive. So far, no such evidence has been found. A more serious objection is that irradiation can destroy the nutrients in food, such as vitamins and amino

Strawberries irradiated at 200 kilorads (right) are still fresh after 15 days storage at 4°C; those not irradiated are moldy.

acids. Additionally, the ionizing radiation used to irradiate food produces reactive species, such as the hydroxyl radical, which then react with the organic molecules in food to produce potentially harmful substances. The same effects, however, are produced when food is cooked by heat.

Food Irradiation Dosages and Their Effects*

Dosage	Effect
Low dose (up to 100 kilorad)	Inhibits sprouting of potatoes, onions, garlic.
	Inactivates trichinae in pork.
	Kills or prevents insects from reproducing in grains, fruits, and vegetables after harvest.
Medium dose (100–1000 kilorads)	Delays spoilage of meat, poultry, and fish by killing spoilage microorganisms.
	Reduces salmonella and other food-borne pathogens in meat, fish, and poultry.
	Extends shelf life by delaying mold growth on strawberries and some other fruits.
High dose (1000–10,000 kilorads)	Sterilizes meat, poultry, fish, and some other foods.
	Kills microorganisms and insects in spices and seasonings.

*Source: *Chemical & Engineering News,* May 5, 1986.

Summary of Facts and Concepts

Section 17.1

▸ Certain elements are radioactive because their nuclei are unstable and spontaneously decay through the ejection (or capture) of subatomic particles, often accompanied by the emission of high-energy electromagnetic radiation.

▸ The principal radioactive decay processes are alpha decay, beta decay, gamma emission, and spontaneous fission. Beta decay can occur through the emission of an electron (β^- decay) or a positron (β^+ decay) from the nucleus. Closely related to positron emission is orbital electron capture.

- The equation for a nuclear reaction includes the particles emitted, and both the mass numbers and the atomic numbers must balance.

Section 17.2

- For stable nuclei of low atomic number, the neutron-to-proton ratio is close to 1. For heavier stable nuclei, the ratio becomes greater than 1. All nuclei with 84 or more protons are unstable and radioactive. Nuclei with even atomic numbers tend to have a greater number of stable isotopes than those with odd atomic numbers.
- Nuclear binding energy is a quantitative measure of nuclear stability. It can be calculated from a knowledge of the mass defect of the nucleus.
- Uranium-238 is the parent of a natural radioactive decay series.

Section 17.3

- Radioactive decay is a first-order kinetic process that can be used to determine the age of objects.

Section 17.4

- Artificial radioactive elements can be created by bombarding other elements with accelerated neutrons, protons, or α particles.

Section 17.5

- Nuclear fission is the splitting of a large nucleus into two smaller nuclei and one or more neutrons. When the free neutrons are captured efficiently by other nuclei, a chain reaction can occur.
- Nuclear reactors use the heat from a controlled nuclear fission reaction to produce power. The three important types of reactors are light water reactors, heavy water reactors, and breeder reactors.

Section 17.6

- Nuclear fusion, the type of reaction that occurs in the sun, is the combination of two light nuclei to form one heavy nucleus. Fusion takes place only at very high temperatures, so high that controlled large-scale nuclear fusion has so far not been achieved.

Section 17.7

- Radioactive isotopes are easy to detect and thus make excellent tracers in chemical reactions and in medical practice.

Section 17.8

- High-energy radiation damages living systems by causing ionization and the formation of free radicals.

Key Words

activity, p. 867
alpha (α) decay, p. 856
beta (β) decay, p. 856
breeder reactor, p. 879
critical mass, p. 877
gamma (γ) emission, p. 858
heavy water reactor, p. 879

internal conversion, p. 858
light water reactor, p. 878
mass defect, p. 863
moderators, p. 878
nuclear binding energy, p. 863
nuclear chain reaction, p. 876
nuclear fission, p. 876

nuclear fusion, p. 882
nuclear transmutation, p. 858
nucleons, p. 860
orbital electron capture, p. 858
plasma, p. 882
positron, p.857
radicals, p. 887

radioactive decay series, p. 867
radioactivity, p. 856
shell model, p. 862
spontaneous fission, p. 858
thermonuclear reactions, p. 882
tracers, p. 885
transuranium elements, p. 874

Problems

Nuclear Chemistry Is the Study of Changes Involving Atomic Nuclei

17.1 Balance the following nuclear equations by identifying X in each case:

(a) $^{26}_{12}Mg + ^{1}_{1}p \longrightarrow ^{4}_{2}\alpha + X$

(b) $^{59}_{27}Co + ^{2}_{1}H \longrightarrow ^{60}_{27}Co + X$

(c) $^{235}_{92}U + ^{1}_{0}n \longrightarrow ^{94}_{36}Kr + ^{139}_{56}Ba + 3X$

(d) $^{53}_{24}Cr + ^{4}_{2}\alpha \longrightarrow ^{1}_{0}n + X$

(e) $^{20}_{8}O \longrightarrow ^{20}_{9}F + X$

17.2 Balance the following nuclear equations by identifying X in each case:

(a) $^{135}_{53}I \longrightarrow ^{135}_{54}Xe + X$

(b) $^{40}_{19}K \longrightarrow ^{0}_{-1}\beta + X$

(c) $^{59}_{27}Co + ^{1}_{0}n \longrightarrow ^{56}_{25}Mn + X$

(d) $^{235}_{92}U + ^{1}_{0}n \longrightarrow ^{99}_{40}Zr + ^{135}_{52}Sr + 2X$

17.3 Write balanced nuclear equations for the following processes:

(a) Positron emission of $^{13}_{7}N$

(b) β^{-} decay of $^{20}_{9}F$

(c) Electron capture by $^{37}_{18}Ar$

(d) α decay of $^{148}_{64}Gd$

17.4 Write balanced nuclear equations for the following processes:

(a) α decay of $^{235}_{92}U$

(b) β^+ decay of $^{15}_{8}O$

(c) β^- decay of $^{11}_{4}Be$

(d) β^- decay of $^{238}_{92}U$

The Stability of a Nucleus Is Determined Primarily by Its Neutron-to-Proton Ratio

17.5 The radius of a uranium-235 nucleus is about 7.0×10^{-3} pm. Calculate the density of the nucleus in g cm^{-3}. (Assume the atomic mass is 235 u.)

17.6 For each of the following pairs of isotopes, predict which one is least stable: (a) ^6_3Li or ^9_3Li, (b) $^{23}_{11}Na$ or $^{25}_{11}Na$, and (c) $^{48}_{20}Ca$ or $^{48}_{21}Sc$.

17.7 For each of the following pairs of elements, predict which one has more stable isotopes: (a) Co or Ni, (b) F or Se, and (c) Ag or Cd.

17.8 For each of the following pairs of isotopes, indicate which one you think would be radioactive: (a) $^{20}_{10}Ne$ or $^{17}_{10}Ne$, (b) $^{40}_{20}Ca$ or $^{45}_{20}Ca$, (c) $^{95}_{42}Mo$ or $^{92}_{43}Tc$, (d) $^{195}_{80}Hg$ or $^{196}_{80}Hg$, and (e) $^{209}_{83}Bi$ or $^{242}_{96}Cm$.

17.9 Given that $\Delta H° = -436.4$ kJ mol^{-1} for

$$H(g) + H(g) \longrightarrow H_2(g)$$

calculate the change in mass (in kg) per mole of H_2 formed.

17.10 The total energy output of the sun is estimated to be 5×10^{26} J s^{-1}. What is the corresponding mass loss in kg s^{-1} of the sun?

17.11 Calculate the nuclear binding energy (in joules and MeV) and the binding energy per nucleon of the following isotopes: (a) ^7_3Li (7.01600 u) and (b) $^{35}_{17}Cl$ (34.95952 u).

17.12 Calculate the nuclear binding energy (in joules and MeV) and the binding energy per nucleon of the following isotopes: (a) ^4_2He (4.0026 u) and (b) $^{184}_{74}W$ (183.9510 u).

17.13 Carbon-11 and carbon-14 are both unstable isotopes of carbon. Which would be more likely to undergo positron emission? Explain.

17.14 Neon-23 and neon-19 are both unstable isotopes of neon. Which would be more likely to undergo β^- decay? Explain.

17.15 Calculate the energy released (in J and J mol^{-1}) when a fluorine-17 nucleus undergoes positron emission (β^+ decay) to form oxygen-17. The masses of ^{17}F and ^{17}O are 17.002098 and 16.999133 u, respectively.

17.16 Polonium-210 (mass = 209.98286 u) undergoes alpha decay to form lead-206 (mass = 205.97444 u):

$$^{210}_{84}Po \longrightarrow ^{206}_{82}Pb + ^4_2\alpha + \text{energy}$$

This nuclear reaction is sometimes used as an atomic heat source to power thermoelectric generators in satellites and unmanned space probes.

(a) Calculate the total energy (in kJ) available from the decay of 1.0 g of Po-210. The mass of the alpha particle is 4.001507 u. (b) What mass of coal would have to be burned to generate the same amount of energy? Assume that the reaction for the burning of coal is

$$C(\text{graphite}) + O_2(g) \longrightarrow CO_2(g)$$

Radioactive Decay Is a First-Order Kinetic Process

17.17 A radioactive substance undergoes decay as follows:

Time (Days)	Mass (g)
0	500
1	389
2	303
3	236
4	184
5	143
6	112

Calculate the first-order decay constant and the half-life of the reaction.

17.18 The radioactive decay of Tl-206 to Pb-206 has a half-life of 4.20 min. Starting with 5.00×10^{22} atoms of Tl-206, calculate the number of such atoms left after 42.0 min.

17.19 A freshly isolated sample of ^{90}Y was found to have an activity of 9.8×10^5 disintegrations per minute (dpm) at 1:00 P.M. on December 3, 2003. At 2:15 P.M. on December 17, 2003, its activity was redetermined and found to be 2.6×10^4 dpm. Calculate the half-life of ^{90}Y.

17.20 Why do radioactive decay series obey first-order kinetics?

17.21 In the thorium decay series, thorium-232 loses a total of six α particles and four β^- particles in a 10-stage process. What is the final isotope produced?

17.22 Strontium-90 is one of the products of the fission of uranium-235. This strontium isotope is radioactive, with a half-life of 28.1 years. Calculate how long (in years) it will take for 1.00 g of the isotope to be reduced to 0.200 g by decay.

17.23 Consider the decay series

$$A \longrightarrow B \longrightarrow C \longrightarrow D$$

where A, B, and C are radioactive isotopes with half-lives of 4.50 s, 15.0 days, and 1.00 s, respectively, and D is nonradioactive. Starting with 1.00 mole of A, and none of B, C, or D, calculate the number of moles of A, B, C, and D present after 30 days.

New Isotopes Can Be Produced Through the Process of Nuclear Transmutation

17.24 Write balanced nuclear equations for the following reactions and identify X:

(a) $X(p,\alpha)_6^{12}C$, (b) $_{13}^{27}Al(d,\alpha)X$, and (c) $_{25}^{55}Mn(n,\gamma)X$.

17.25 Write balanced nuclear equations for the following reactions and identify X:

(a) $_{34}^{80}Se(d,p)X$, (b) $X(d,2p)_3^9Li$, and (c) $_5^{10}B(n,\alpha)X$.

17.26 Describe how you would prepare astatine-211, starting with bismuth-209.

17.27 A long-cherished dream of alchemists was to produce gold from cheaper and more abundant elements. This dream was finally realized when $_{80}^{198}Hg$ was converted into gold by neutron bombardment. Write a balanced equation for this reaction.

Radioactive and Stable Isotopes Alike Have Many Applications in Science and Medicine

17.28 Describe how you would use a radioactive iodine isotope to demonstrate that the following process is in dynamic equilibrium:

$$PbI_2(s) \rightleftharpoons Pb^{2+}(aq) + 2I^-(aq)$$

17.29 Consider the following oxidation-reduction reaction:

$$IO_4^-(aq) + 2I^-(aq) + H_2O(l) \longrightarrow$$
$$I_2(s) + IO_3^-(aq) + 2OH^-(aq)$$

When KIO_4 is added to a solution containing iodide ions labeled with radioactive iodine-128, all the radioactivity appears in I_2 and none in the IO_3^- ion. What can you deduce about the mechanism for the reaction?

17.30 Explain how you might use a radioactive tracer to show that ions diffuse even in crystals.

17.31 Each molecule of hemoglobin, the oxygen carrier in blood, contains four Fe atoms. Explain how you would use the radioactive isotope $_{26}^{59}Fe$ ($t_{\frac{1}{2}} = 46$ days) to show that the iron in a certain food is converted into hemoglobin.

Additional Problems

17.32 How does a Geiger-Mueller counter work?

17.33 Nuclei with an even number of protons and an even number of neutrons are more stable than those with an odd number of protons and/or an odd number of neutrons. What is the significance of the even numbers of protons and neutrons in this case?

17.34 Tritium ($_1^3H$) is radioactive and decays by β^- emission with a half-life of 12.5 years In ordinary water, the ratio of $_1^1H$ to $_1^3H$ atoms is 1.0×10^{17} to 1. (a) Write a balanced nuclear equation for tritium decay. (b) How many disintegrations will be observed per minute in a 1.00-kg sample of water?

17.35 (a) What is the activity, in millicuries, of a 0.500-g sample of $_{93}^{237}Np$? (This isotope decays by α-particle emission and has a half-life of 2.20×10^6 yr.) (b) Write a balanced nuclear equation for the decay of $_{93}^{237}Np$.

17.36 The following equations are for nuclear reactions that are known to occur in the explosion of an atomic bomb. Identify X in each case.

(a) $_{92}^{235}U + _0^1n \longrightarrow _{56}^{140}Ba + 3_0^1n + X$
(b) $_{92}^{235}U + _0^1n \longrightarrow _{55}^{144}Cs + _{37}^{90}Rb + 2X$
(c) $_{92}^{235}U + _0^1n \longrightarrow _{35}^{87}Br + 3_0^1n + X$
(d) $_{92}^{235}U + _0^1n \longrightarrow _{62}^{160}Cs + _{30}^{72}Zn + 4X$

17.37 Calculate the nuclear binding energies (in joules per nucleon) for the following species: (a) ^{10}B (10.0129 u), (b) ^{11}B (11.00931 u), (c) ^{14}N (14.00307 u), and (d) ^{56}Fe (55.9349 u).

17.38 Write complete nuclear equations for the following processes: (a) tritium, ^3H, undergoes β^- decay, (b) ^{242}Pu undergoes α-particle emission, (c) ^{131}I undergoes β^- decay, (d) ^{251}Cf emits an α particle, and (e) ^8B undergoes positron decay.

17.39 The nucleus of nitrogen-18 lies above the stability belt. Write an equation for a nuclear reaction by which nitrogen-18 can achieve stability.

17.40 Why is strontium-90 a particularly dangerous isotope for humans?

17.41 How are scientists able to tell the age of a fossil?

17.42 After the Chernobyl accident, people living close to the nuclear reactor site were urged to take large amounts of potassium iodide as a safety precaution. What is the chemical basis for this action?

17.43 Astatine, the last known member of Group 7A, can be prepared by bombarding bismuth-209 with α particles. (a) Write an equation for the reaction. (b) Represent the equation in the abbreviated form introduced in Section 17.4.

17.44 To detect bombs that may be smuggled onto airplanes, the Federal Aviation Administration (FAA) requires all major airports in the United States to install thermal neutron analyzers. The thermal neutron analyzer bombards baggage with low-energy neutrons, converting some of the nitrogen-14 nuclei to nitrogen-15, with simultaneous emission of γ rays. Because nitrogen content is usually high in explosives, detection of a high dosage of γ rays will suggest that a bomb may be present. (a) Write an equation for the nuclear process. (b) Compare this technique with the conventional X-ray detection method.

17.45 Explain why achieving nuclear fusion in the laboratory requires a temperature of about 100 million degrees Celsius, which is much higher than that in the interior of the sun (15 million degrees Celsius).

17.46 Tritium contains one proton and two neutrons. There is no proton-proton repulsion present in the nucleus. Why, then, is tritium radioactive?

17.47 The carbon-14 decay rate of a sample obtained from a young tree is 0.260 disintegrations per second per gram of the sample. Another wood sample prepared from an object recovered at an archaeological excavation gives a decay rate of 0.186 disintegrations per second per gram of the sample. What is the age of the object?

17.48 The usefulness of radiocarbon dating is limited to objects no older than 50,000 years. What percent of the carbon-14, originally present in the sample, remains after this period of time?

17.49 The radioactive potassium-40 isotope decays to argon-40 with a half-life of 1.2×10^9 years. (a) Write a balanced equation for the reaction. (b) A sample of moon rock is found to contain 18 percent potassium-40 and 82 percent argon by mass. Calculate the age of the rock in years.

17.50 Both barium (Ba) and radium (Ra) are members of Group 2A and are expected to exhibit similar chemical properties. Ra, however, is not found in barium ores. Instead, it is found in uranium ores. Explain.

17.51 Nuclear waste disposal is one of the major concerns of the nuclear industry. In choosing a safe and stable environment to store nuclear wastes, consideration must be given to the heat released during nuclear decay. As an example, consider the β decay of ^{90}Sr (89.907738 u):

$$^{90}_{38}\text{Sr} \longrightarrow \; ^{90}_{39}\text{Y} + \; ^{0}_{-1}\beta \qquad t_{\frac{1}{2}} = 28.1 \text{ yr}$$

The ^{90}Y (89.907152 u) further decays as follows:

$$^{90}_{39}\text{Y} \longrightarrow \; ^{90}_{40}\text{Zr} + \; ^{0}_{-1}\beta \qquad t_{\frac{1}{2}} = 64 \text{ h}$$

Zirconium-90 (89.904703 u) is a stable isotope. (a) Use the mass defect to calculate the energy released (in joules) in each of the decays of ^{90}Sr and ^{90}Y. (The mass of the electron is 5.4857×10^{-4} u.) (b) Starting with 1 mole of ^{90}Sr, calculate the number of moles of ^{90}Sr that will decay in a year. (c) Calculate the amount of heat released (in kJ) corresponding to the number of moles of ^{90}Sr that decayed to ^{90}Zr in part (b).

17.52 Which of the following poses a greater health hazard: a radioactive isotope with a short half-life or a radioactive isotope with a long half-life? Explain. [Assume the isotopes have the same type of radiation (α or β^-) and comparable energetics per particle emitted.]

17.53 As a result of being exposed to the radiation released during the Chernobyl nuclear accident, the dose of

iodine-131 in a person's body is 7.4 mCi (1 mCi = 1×10^3 Ci). Use the relationship rate = λN to calculate the number of atoms of iodine-131 to which this radioactivity corresponds. (The half-life of ^{131}I is 8.1 d.)

17.54 Referring to the Food Irradiation inset on page 888, why is it highly unlikely that irradiated food would become radioactive?

17.55 From the definition of a curie, calculate Avogadro's number, given that the molar mass of ^{226}Ra is 226.03 g mol^{-1} and that it decays with a half-life of 1.6×10^3 years.

17.56 Since 1994, elements 110 (darmstadtium, Ds), 111 (roentgenium, Rg), 112 (ununbium, Uub), and 114 (ununquadium, Uuq) have been synthesized. Element 110 was created by bombarding ^{208}Pb with ^{62}Ni; element 111 was created by bombarding ^{209}Bi with ^{64}Ni; element 112 was created by bombarding ^{208}Pb with ^{66}Zn, and element 114 was created by bombarding ^{244}Pu with ^{48}Ca. Write an equation for each synthesis. Predict the chemical properties of these elements.

17.57 Sources of energy on Earth include fossil fuels, geothermal, gravitational, hydroelectric, nuclear fission, nuclear fusion, solar, and wind. Which of these have a "nuclear origin," either directly or indirectly?

17.58 A person received an anonymous gift of a decorative cube, which he placed on his desk. A few months later he became ill and died shortly afterward. After investigation, the cause of his death was linked to the box. The box was airtight and had no toxic chemicals on it. What might have killed the man?

17.59 Identify two of the most abundant radioactive elements that exist on Earth. Explain why they are still present? (You may need to consult a handbook of chemistry.)

17.60 (a) Calculate the energy released (in MeV and joules) when a U-238 isotope decays to Th-234. The atomic masses are 238.0508 u for U-238, 234.0436 u for Th-234, and 4.0026 u for He-4. (b) The energy released in (a) is transformed into the kinetic energy of the recoiling Th-234 nucleus and the α particle. Which of the two will move away faster? Explain.

17.61 Cobalt-60 is an isotope used in diagnostic medicine and cancer treatment. It decays with γ-ray emission. Calculate the wavelength of the radiation in nanometers if the energy of the γ ray is 2.4×10^{-13} J per photon.

17.62 Americium-241 is used in smoke detectors because it has a long half-life (458 years) and because the α particles it emits are energetic enough to ionize air molecules. Given the following schematic diagram of a smoke detector, explain how it works.

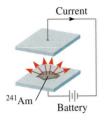

Current

^{241}Am

Battery

17.63 The constituents of wine contain carbon, hydrogen, and oxygen atoms, among others. A bottle of wine was sealed about 6 years ago. To confirm its age, which of the isotopes would you choose in a radioactive dating study? The half-lives of the isotopes are 5730 years for ^{13}C, 124 s for ^{5}O, and 12.5 years for ^{3}H. Assume that the activities of the isotopes were known at the time the bottle was sealed.

17.64 Name two advantages of a nuclear-powered submarine over a conventional submarine.

17.65 In 1997, a scientist at a nuclear research center in Russia placed a thin shell of copper on a sphere of highly enriched uranium-235. Suddenly, there was a huge burst of radiation, which turned the air blue. Three days later, the scientist died of radiation poisoning. Explain what caused the accident. (*Hint:* Copper is an effective metal for reflecting neutrons.)

17.66 A radioactive isotope of copper decays as follows:

$$^{64}_{29}\text{Cr} \longrightarrow {}^{64}_{30}\text{Zn} + {}^{0}_{-1}\beta \qquad t_{\frac{1}{2}} = 12.8 \text{ h}$$

Starting with 84.0 g of ^{64}Cu, calculate the quantity of ^{64}Zn produced after 18.4 h.

17.67 A 0.0100-g sample of a radioactive isotope with a half-life of 1.3×10^{9} years decays at the rate of 2.9×10^{4} dpm. Calculate the molar mass of the isotope.

17.68 An isolated neutron will undergo spontaneous β^{-} decay ($^{1}_{0}n \longrightarrow {}^{1}_{+1}p + {}^{0}_{-1}\beta + \bar{\nu}$) with a half-life

of about 10.5 min; however, spontaneous β^{+} decay of an isolated proton ($^{1}_{+1}p \longrightarrow {}^{1}_{0}n + {}^{0}_{-1}\beta + \nu$) is not observed. (a) Give an explanation for this observation. (*Hint:* What are the masses of protons and neutrons?) (b) Why is β^{+} (positron emission) from a nucleus possible even though it does not occur for an isolated proton.

17.69 The half-life of ^{27}Mg is 9.50 min. (a) Initially there were 4.20×10^{12} ^{27}Mg nuclei present. How many ^{27}Mg are left 30.0 min later? (b) Calculate the ^{27}Mg activities (in Ci) at $t = 0$ and $t = 30.0$ min. (c) What is the probability that any one ^{27}Mg nucleus decays during a 1-s interval? What assumption is made in this calculation?

17.70 The radioactive isotope ^{238}Pu, used in pacemakers, decays by emitting an α particle with a half-life of 86 years. (a) Write an equation for the decay process. (b) The energy of the emitted α particle is 9.0×10^{-13} J, which is the energy per decay. Assume that all the α particle energy is used to run the pacemaker, calculate the power output at $t = 0$ and $t = 10$ yr. Initially 1.0 mg of ^{238}Pu was present in the pacemaker. (*Hint:* after 10 years, the activity of the isotope decreases by 8.0 percent. Power is measured in watts or J s^{-1}.)

17.71 (a) Assume nuclei are spherical in shape, show that the radius (r) of a nucleus is proportional to the cube root of mass number (A). (b) In general, the radius of a nucleus is given by $r = r_0 A^{1/3}$, where r_0, the proprotionality constant, is given by 1.2×10^{-15} m. Calculate the volume of the ^{238}U nucleus.

17.72 The quantity of a radioactive material is often measured by its activity (measured in curies or millicuries) rather than by its mass. In a brain scan procedure, a 70-kg patient is injected with 15.0 mCi of ^{99m}Tc, which decays by emitting γ-ray photons with a half-life of 6.0 h. Given that the RBE of these photons is 0.98 and only two-third of the photons are absorbed by the body, calculate the rem dose received by the patient. Assume all the ^{99m}Tc nuclei decay while in the body.

Answers to Practice Exercises

17.1 $^{22}_{10}$Ne **17.2** 1640.4 MeV, 7.8488 MeV nucleon^{-1}
17.3 2.1421×10^{-12} J, 1.2900×10^{9} kJ mol^{-1} **17.4** 57.8 days
17.5 2.41×10^{9} yr; assumptions: (1) no lead initially present,

and (2) no chemical reaction occurs.
17.6 $^{106}_{46}$Pd + $^{4}_{2}\alpha \longrightarrow {}^{109}_{47}$Ag + $^{1}_{1}$p

Appendix 1

Measurement and Mathematical Background

A1.1 | Measurement

The measurements chemists make are often used in calculations to obtain other related quantities. Different instruments enable us to measure a substance's properties: The meter stick measures length or scale; the buret, the pipet, the graduated cylinder, and the volumetric flask measure volume; the balance measures mass; and the thermometer measures temperature. Any measured quantity should always be written as a number with an appropriate unit. To say that the distance between New York and San Francisco by car along a certain route is 5166 is meaningless. We must specify that the distance is 5166 kilometers. The same is true in chemistry; units are essential to stating measurements correctly.

Units

For many years, scientists recorded measurements in *metric units,* which are related decimally, that is, by powers of 10. In 1960, however, the General Conference of Weights and Measures, the international authority on units, proposed a revised metric system called the **International System of Units** (abbreviated **SI,** from the French *Système Internationale d'Unites*). The SI system of units consists of seven basic units:

- ▶ **Time:** The SI unit of time is the *second* (symbol: s), defined as the length of time required for 9,192,631,770 periods of the radiation corresponding to the transition between two specific electronic states of a cesium-133 atom at 0 K.

- ▶ **Length:** The SI unit of length is the *meter* (symbol: m). A meter is defined such that the speed of light in a vacuum is 299,792,458 m s^{-1} exactly.

- ▶ **Mass:** The unit of mass in the SI system is the *kilogram* (symbol: kg). The kilogram is defined as the mass of a platinum-iridium alloy cylinder called the *International Prototype Kilogram* (IPK) that is stored in a vault near Paris.

- ▶ **Electric Current:** The SI unit of current is the *ampere* (symbol: A). The ampere is defined as the current carried by two parallel, straight wires of infinite length and negligible cross section that yields a force of 2×10^{-7} newtons (or 2×10^{-7} kg m s^{-1}) when the wires are held exactly 1 m apart.

- ▶ **Thermodynamic temperature:** The SI unit of temperature is the kelvin (symbol: K), defined such that the triple point of water is exactly 273.16 K.

- ▶ **Amount of substance:** The *mole* (symbol: mol) is the SI unit for the amount of a substance and is equal to the number of atoms in 0.012 kg of carbon-12 atoms.

- ▶ **Luminous intensity:** The SI unit for the intensity of light is the *candela* (symbol: cd). The candela is the luminous intensity, in a given direction, of a light source that emits monochromatic light of frequency 540×10^{12} s^{-1} with a radiant intensity of 1/683 watts per steradian.

All other units of measurement can be derived from these seven base units. SI units are modified in decimal fashion by a series of prefixes, as shown in Table A1.1. Table A1.2 lists the definitions of a number of common units derived from the SI system.

In addition to the SI units listed in Tables A1.1 and A1.2, a number of non-SI units are utilized commonly in scientific work. Table A1.3 lists those non-SI units that are used in this text.

Table A1.1	Prefixes Used with SI Units			
Prefix	**Symbol**	**Meaning**		**Example**
tera-	T	1,000,000,000,000, or 10^{12}		1 terameter (Tm) = 1×10^{12} m
giga-	G	1,000,000,000, or 10^9		1 gigameter (Gm) = 1×10^9 m
mega-	M	1,000,000, or 10^6		1 megameter (Mm) = 1×10^6 m
kilo-	k	1,000, or 10^3		1 kilometer (km) = 1×10^3 m
hecto-	h	100 or 10^2		1 hectometer (hm) = 100 m
deca-	da	10 or 10^1		1 decameter (dam) = 10 m
deci-	d	1/10, or 10^{-1}		1 decimeter (dm) = 0.1 m
centi-	c	1/100, or 10^{-2}		1 centimeter (cm) = 0.01 m
milli-	m	1/1,000, or 10^{-3}		1 millimeter (mm) = 0.001 m
micro-	μ	1/1,000,000, or 10^{-6}		1 micrometer (μm) = 1×10^{-6} m
nano-	n	1/1,000,000,000, or 10^{-9}		1 nanometer (nm) = 1×10^{-9} m
pico-	p	1/1,000,000,000,000 or 10^{-12}		1 picometer (pm) = 1×10^{-12} m
femto-	f	10^{-15}		1 femtometer (fm) = 1×10^{-15} m
atto-	a	10^{-18}		1 attometer (am) = 1×10^{-18} m
zepto-	z	10^{-21}		1 zeptometer (zm) = 10^{-21} m

Table A1.2	Common SI Derived Units		
Quantity	**Unit**	**Symbol**	**Definition**
Energy	joule	J	1 kg m^2 s^{-2}
Force	newton	N	1 kg m s^{-1}
Volume	cubic meter	m^3	1 m^3
Temperature (T)	degree celsius	°C	T(°C) = T(K) − 273.15
Power	watt	W	1 J s^{-1}
Electric charge	coulomb	C	1 A s
Pressure	pascal	Pa	1 N m^{-2}
Frequency	hertz	Hz	1 s^{-1}
Electromotive force	volt	V	1 W A^{-1}
Electric conductance	siemens	S	1 A V^{-1}
Magnetic flux density	tesla	T	1 kg s^{-2} A^{-1}

Scientific Notation

Chemists often deal with numbers that are either extremely large or extremely small. For example, in 1 g of the element hydrogen there are roughly

$$602,200,000,000,000,000,000,000$$

hydrogen atoms. Each hydrogen atom has a mass of only

$$0.00000000000000000000000166 \text{ g}$$

These numbers are cumbersome to handle, and it is easy to make mistakes when using them in arithmetic computations. Consider the following multiplication:

$$0.0000000056 \times 0.00000000048 = 0.000000000000000002688$$

Table A1.3	Common Non-SI Derived Units Used in Scientific Work		
Quantity	**Unit**	**Symbol**	**Definition**
Volume	liter	L	$1\ L = 1\ dm^3 = 10^{-3}\ m^3$
Mass	atomic mass unit	u	$1\ u = 1.66053886 \times 10^{-27}\ kg$
Concentration	molar	M	$1\ M = 1\ mol\ (L\ of\ solution)^{-1}$
	molal	m	$1\ m = 1\ mol\ (kg\ of\ solute)^{-1}$
Length	Ångstrom	Å	$1\ \text{Å} = 10^{-10}\ m$
Energy	electron volt	eV	$1\ eV = 1.60217646 \times 10^{-19}\ J$
	erg	erg	$1\ erg = 1\ g\ cm^2\ s^{-2}$
	calorie	cal	$1\ cal = 4.184\ J$
Pressure	bar	bar	$1\ bar = 10^5\ Pa$
	atmosphere	atm	$1\ atm = 1.01325\ bar$
	torr	torr	$1\ torr = 101325/760\ Pa$ $\approx 133.3223684\ Pa$
	mm of mercury	mmHg	$1\ mmHg = 133.3223874\ Pa$ $\approx 1\ torr$

It would be easy for us to miss one zero or add one more zero after the decimal point. Consequently, when working with very large and very small numbers, we use a system called *scientific notation*. Regardless of their magnitude, all numbers can be expressed in the form

$$N \times 10^n$$

where N is a number between 1 and 10 and n, the exponent, is a positive or negative integer (whole number). Any number expressed in this way is said to be written in scientific notation.

Suppose that we are given a certain number and asked to express it in scientific notation. Basically, this assignment calls for us to find n. We count the number of places that the decimal point must be moved to give the number N (which is between 1 and 10). If the decimal point has to be moved to the left, then n is a positive integer; if it has to be moved to the right, n is a negative integer. The following examples illustrate the use of scientific notation:

▸ Express 568.762 in scientific notation:

$$568.762 = 5.68762 \times 10^2$$

Note that the decimal point is moved to the left by two places and $n = 2$.

▸ Express 0.00000772 in scientific notation:

$$0.00000772 = 7.72 \times 10^{-6}$$

▸ Here the decimal point is moved to the right by six places and $n = 6$.

Next, we consider how scientific notation is handled in arithmetic operations.

Addition and Subtraction: To add or subtract using scientific notation, we first write each quantity—say N_1 and N_2—with the same exponent n. Then we combine N_1 and N_2; the exponents remain the same. Consider the following examples:

$$(7.4 \times 10^3) + (2.1 \times 10^3) = 9.5 \times 10^3$$
$$(4.31 \times 10^4) + (3.9 \times 10^3) = (4.31 \times 10^4) + (0.39 \times 10^4) = 4.70 \times 10^4$$
$$(2.22 \times 10^{-2}) - (4.10 \times 10^{-3}) = (2.22 + 10^{-2}) - (0.410 + 10^{-2}) = 1.81 \times 10^{-2}$$

Multiplication and Division: To multiply numbers expressed in scientific notation, we multiply N_1 and N_2 in the usual way, but *add* the exponents together. To divide using scientific notation, we divide N_1 and N_2 as usual and subtract the exponents.

The following examples show how these operations are performed:

$$(8.0 \times 10^4) \times (5.0 \times 10^2) = (8.0 \times 5.0)(10^{4+2}) = 40 \times 10^6$$
$$= 4.0 \times 10^7$$

$$(4.0 \times 10^{-5}) \times (7.0 \times 10^2) = (4.0 \times 7.0)(10^{-5+2}) = 28 \times 10^{-2}$$
$$= 2.8 \times 10^{-1}$$

$$\frac{6.9 \times 10^7}{3.0 \times 10^{-5}} = \frac{6.9}{3.0} \times 10^{7-(-5)} = 2.3 \times 10^{12}$$

$$\frac{2.8 \times 10^4}{5.6 \times 10^9} = \frac{2.8}{5.6} \times 10^{4-9} = 0.50 \times 10^{-5}$$
$$= 5.0 \times 10^{-6}$$

Significant Figures

Except when all the numbers involved are integers (for example, in counting the number of students in a class), it is often impossible to obtain the exact value of the quantity under investigation. For this reason, it is important to indicate the margin of error in a measurement by clearly indicating the number of *significant figures,* which are *the meaningful digits in a measured or calculated quantity.* When significant figures are used, the last digit is understood to be uncertain. For example, we might measure the volume of a given amount of liquid using a graduated cylinder with a scale that gives an uncertainty of 1 mL in the measurement. If the volume is found to be 6 mL, then the actual volume is in the range of 5 mL to 7 mL. We may represent the volume of the liquid as (6 ± 1) mL. In this case, there is only one significant figure (the digit 6) that is uncertain by either plus or minus 1 mL. For greater accuracy, we might use a graduated cylinder that has finer divisions, so that the volume we measure is now uncertain by only 0.1 mL. If the volume of the liquid is now found to be 6.0 mL, we may express the quantity as (6.0 ± 0.1) mL, and the actual value is somewhere between 5.9 mL and 6.1 mL. We can further improve the measuring device and obtain more significant figures, but in every case, the last digit is always uncertain; the amount of this uncertainty depends on the particular measuring device we use.

Guidelines for Using Significant Figures

We must always be careful in scientific work to write the proper number of significant figures. In general, it is fairly easy to determine how many significant figures a number has by following these rules:

1. Any digit that is not zero is significant. Thus, 845 cm has three significant figures, 1.234 kg has four significant figures, and so on.

2. Zeros between nonzero digits are significant. Thus, 606 m contains three significant figures, 40,501 kg contains five significant figures, and so on.

3. Zeros to the left of the first nonzero digit are not significant. Their purpose is to indicate the placement of the decimal point. For example, 0.08 L contains one significant figure, 0.0000349 g contains three significant figures, and so on.

4. If a number is greater than 1, then all the zeros written to the right of the decimal point count as significant figures. Thus, 2.0 mg has two significant figures, 40.062 mL has five significant figures, and 3.040 dm has four significant figures. If a number is less than 1, then only the zeros that are at the end of the number and the zeros that are between nonzero digits are significant. This means that 0.090 kg has two significant figures, 0.3005 L has four significant figures, 0.00420 min has three significant figures, and so on.

5. For numbers that do not contain decimal points, the trailing zeros (that is, zeros after the last nonzero digit) may or may not be significant. Thus, 400 cm may have one significant figure (the digit 4), two significant figures (40), or three significant figures (400). We cannot know which is correct without more information. By using scientific notation, however, we avoid this ambiguity. In this particular case, we can express the number 400

as 4×10^2 for one significant figure, 4.0×10^2 for two significant figures, or 4.00×10^2 for three significant figures.

Example A1.1 shows the determination of significant figures.

Example A1.1

Determine the number of significant figures in the following measurements: (a) 478 cm, (b) 6.01 g, (c) 0.825 m, (d) 0.043 kg, (e) 1.310×10^{22} atoms, and (f) 7000 mL.

Solution (a) Three, because each digit is a nonzero digit. (b) Three, because zeros between nonzero digits are significant. (c) Three, because zeros to the left of the first nonzero digit do not count as significant figures. (d) Two. Same reason as in (c). (e) Four, because the number is greater than one so all the zeros written to the right of the decimal point count as significant figures. (f) This is an ambiguous case. The number of significant figures may be four (7.000×10^3), three (7.00×10^3), two (7.0×10^3), or one (7×10^3). This example illustrates why scientific notation must be used to show the proper number of significant figures.

Practice Exercise Determine the number of significant figures in each of the following measurements: (a) 24 mL, (b) 3001 g, (c) 0.0320 m^3, (d) 6.4×10^4 molecules, and (e) 560 kg.

A second set of rules specifies how to handle significant figures in calculations:

1. In addition and subtraction, the answer cannot have more digits to the right of the decimal point than either of the original numbers. Consider these examples:

$$\begin{array}{r} 89.332 \\ + 1.1 \\ \hline 90.432 \end{array}$$ ⟵ one digit after the decimal point
⟵ round off to 90.4

$$\begin{array}{r} 2.097 \\ - 0.12 \\ \hline 1.977 \end{array}$$ ⟵ two digits after the decimal point
⟵ round off to 1.98

The rounding-off procedure is as follows. To round off a number at a certain point, we simply drop the digits that follow if the first of them is less than 5. Thus, 8.724 rounds off to 8.72 if we want only two digits after the decimal point. If the first digit following the point of rounding off is equal to or greater than 5, we add 1 to the preceding digit. Thus, 8.727 rounds off to 8.73, and 0.425 rounds off to 0.43.

2. In multiplication and division, the number of significant figures in the final product or quotient is determined by the original number that has the *smallest* number of significant figures. The following examples illustrate this rule:

$$2.8 \times 4.5039 = 12.62092$$ ⟵ round off to 1.3

$$\frac{6.85}{112.04} = 0.0611388789$$ ⟵ round off to 0.0611

3. When taking the natural or base-10 logarithm of a measured value, the number of significant figures to the right of the decimal point in the result is the same as the number of significant figures in the original measurement. For example,

$$\log_{10}(29.23) = 1.465828815$$ ⟵ round to 1.4658
$$\ln(7.9 \times 10^{17}) = 41.21080934$$ ⟵ round to 41.21

4. When taking the natural or base-10 antilogarithm (exponentiation), the number of significant figures in the result is the same as the number of significant figures to the right of the decimal point in the original measurement. For example,

$$e^{5.875} = 356.02466 \qquad \longleftarrow \text{ round to } 3.56 \times 10^2$$
$$10^{3.3} = 1995.26231 \qquad \longleftarrow \text{ round to } 2 \times 10^3$$

5. Keep in mind that exact numbers obtained from definitions or by counting numbers of objects can be considered to have an infinite number of significant figures. If an object has a mass of 0.2786 g, then the mass of eight such objects is

$$0.2786 \text{ g} \times 8 = 2.229 \text{ g}$$

We do *not* round off this product to one significant figure because the number 8 is 8.00000 . . . by definition. Similarly, to take the average of two measured lengths 6.64 cm and 6.68 cm, we write

$$\text{average} = \frac{6.68 \text{ cm} + 6.64 \text{ cm}}{2} = 6.66 \text{ cm}$$

because the number 2 is 2.00000 . . ., by definition.

Example A1.2 shows how significant figures are handled in arithmetic operations.

Example A1.2

Carry out the following arithmetic operations to the correct number of significant figures: (a) 11,254.1 g + 0.1983 g, (b) 66.59 L − 3.113 L, (c) 8.16 m × 5.1355, (d) (0.0154 kg)/(88.3 mL), (e) $\log_{10}(3.485 \times 10^7)$, and (f) $e^{2.34}$.

Solution

(a) $\quad$ 11,245.1 g
$\quad$ + $\quad$ 0.1983 g
$\quad$ $\overline{11,254.2983 \text{ g}}$ $\quad \longleftarrow$ round off to 11,254.3

(b) $\quad$ 66.59 L
$\quad$ − 3.113 L
$\quad$ $\overline{63.477 \text{ L}}$ $\quad \longleftarrow$ round off to 63.58 L

(c) 8.16 m × 5.1355 = 41.90568 m $\quad \longleftarrow$ round off to 41.9 m

(d) (0.0154 kg)/(88.3 mL) = 1.7440544×10^{-4} kg mL^{-1}
$\qquad\qquad\qquad\qquad \longleftarrow$ round off to 1.74×10^{-4} kg mL^{-1}

(e) $\log_{10}(3.485 \times 10^7) = 7.542202782$ $\quad \longleftarrow$ round off to 7.5422

(f) $e^{2.34} = 10.38123656$ $\quad \longleftarrow$ round to 1.0×10^1

Practice Exercise Carry out the following arithmetic operations to the correct number of significant figures: (a) 26.5862 L + 0.17 L, (b) 9.1 g − 4.682 g, (c) 7.1 × 10^4 dm × 2.2654 10^2 dm, (d) (6.54 g)/(86.5542 mL), (e) ln(109.3), and (f) $10^{17.234}$.

The preceding rounding-off procedure applies to one-step calculations. In *chain calculations,* that is, calculations involving more than one step, we use a modified procedure. Consider the following two-step calculation:

$$\text{first step:} \quad A \times B = C$$
$$\text{second step:} \quad C \times D = E$$

Let us suppose that A = 3.66, B = 8.45, and D = 2.11. Depending on whether we round off C to three or four significant figures, we obtain a different number for E:

Method 1	Method 2
$3.66 \times 8.45 = 30.9$	$3.66 \times 8.45 = 30.93$
$30.9 \times 2.11 = 65.2$	$30.93 \times 2.11 = 65.3$

However, if we had carried out the calculation as $3.66 \times 8.45 \times 2.11$ on a calculator without rounding off the intermediate result, we would have obtained 65.3 as the answer for E. In general, in this text we will show the correct number of significant figures in each step of a multistep calculation.

Accuracy and Precision

In discussing measurements and significant figures it is useful to distinguish between *accuracy* and *precision*. **Accuracy** tells us *how close a measurement is to the true value of the quantity that was measured.* To a scientist, there is a distinction between accuracy and precision. **Precision** *refers to how closely two or more measurements of the same quantity agree with one another.*

The difference between accuracy and precision is a subtle but important one. Suppose, for example, that three students are asked to determine the mass of a piece of copper wire. The results of two successive weighings by each student are

	Student A	Student B	Student C
	1.964 g	1.972 g	2.000 g
	1.978 g	1.968 g	2.002 g
average	1.971 g	1.970 g	2.001 g

The true mass of the wire is 2.000 g. Therefore, Student B's results are more *precise* than those of Student A (1.972 g and 1.968 g deviate less from 1.970 g than 1.964 g and 1.978 g from 1.971 g), but neither set of results is very *accurate*. Student C's results are not only the most *precise*, but also the most *accurate*, because the average value is closest to the true value. Highly accurate measurements are usually precise too. On the other hand, highly precise measurements do not necessarily guarantee accurate results. For example, an improperly calibrated meter stick or a faulty balance may give precise readings that are in error.

A1.2 | Mathematical Background

Differential Calculus

In the physical sciences, we are often interested in understanding how a physical property changes when another property of the system is varied. For example, in Section 10.4, we discuss how the equilibrium constant for a reaction changes when the temperature is changed. This analysis is especially simple if the relationship between the property of interest (y) and the property to be varied (x) is a straight line (Figure A1.1):

$$y = mx + b$$

where m is the *slope* and b is the *intercept* of the line. Because the slope of this line is independent of x, we can obtain both m and b (and thus the value of y at any x) from a measurement of y at two points x_1 and x_2:

$$m = \frac{y_2 - y_1}{x_2 - x_1} \quad \text{and} \quad b = y_2 - mx_2$$

Figure A1.1 The slope of a
line $y = mx + b$ can be found
if two points (x_1, y_1) and (x_2, y_2) are known.

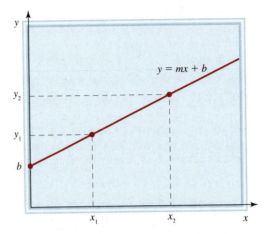

Figure A1.1 The slope of a line $y = mx + b$ can be found if two points (x_1, y_1) and (x_2, y_2) are known.

However, if the relationship between y and x is not a straight line, the slope of the curve will not be constant (i.e., independent of x). For example, consider the curve

$$y = f(x)$$

The rate of change of this function is not necessarily constant, but is in general a function of x. We define the *first derivative* (df/dx) of a function f as the slope of the line tangent to the curve $y = f(x)$ at point x. This slope can be found by the following procedure (Figure A1.2)

1. Choose a nearby point x'.
2. Calculate the slope of the line connecting the points $f(x)$ and $f(x')$

$$\text{slope} = \frac{f(x') - f(x)}{x' - x}$$

3. The first derivative is found as the value of this slope in the limit that x' approaches x:

$$\frac{df}{dx} = \lim_{x' \to x} \frac{f(x') - f(x)}{x' - x} \tag{A1.1}$$

The first derivative of a function $f(x)$ at point x gives the instantaneous rate of change of the function with respect to variations in x and represents a generalization of the slope for

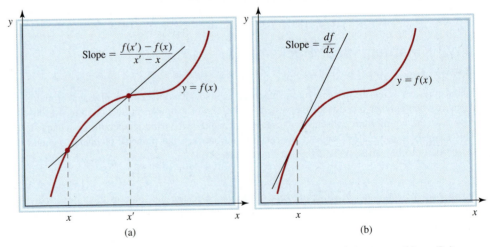

Figure A1.2 (a) The slope between two points x and x' on a curve $f(x)$ separated by a finite distance ($x' - x$) is determined using Equation A1.1. (b) In the limit that x' approaches x, the slope defined by Equation A1.1 approaches the first derivative of $f(x)$ at the point x.

Table A1.4	Derivatives of Some Common Functions		
$f(x)$	$\dfrac{df}{dx}$	$f(x)$	$\dfrac{df}{dx}$
c (a constant)	0	$\ln x$	$1/x$
x	1	$\sin x$	$\cos x$
x^n	$n\,x^{n-1}$	$\cos x$	$-\sin x$
e^x	e^x		

curves that are not straight lines. Use of the derivative (Equation A1.1) in describing the rate of change of functions is called *differential calculus*. The values of the derivative for a number of functions used in this text are listed in Table A1.4.

For more complicated expressions involving the functions above, the following rules can be used to evaluate the first derivative.

1. **Additivity**: $\dfrac{d(f+g)}{dx} = \dfrac{df}{dx} + \dfrac{dg}{dx}$ (A1.2)

   Example: $\dfrac{d(x^3 + x^2 + 5)}{dx} = \dfrac{d(x^3)}{dx} + \dfrac{d(x^2)}{dx} + \dfrac{d(5)}{dx} = 3x^2 + 2x$

2. **Product Rule**: $\dfrac{d(f \times g)}{dx} = \dfrac{df}{dx} \times g + f \times \dfrac{dg}{dx}$ (A1.3)

   Example: $\dfrac{d(x^2 e^x)}{dx} = \dfrac{d(x^2)}{dx}e^x + x^2\dfrac{d(e^x)}{dx} = 2xe^x + x^2 e^x$

3. **Chain Rule**: $\dfrac{d[f(g)]}{dx} = \dfrac{df}{dg} \times \dfrac{dg}{dx}$ (A1.4)

   Example: If $f(x) = \ln(x)$ and $g(x) = \sin x$, then
   $$\frac{d[\ln(\sin x)]}{dx} = \frac{d(\ln g)}{dg} \times \frac{d(\sin x)}{dx} = \frac{1}{g} \times \cos x = \frac{\cos x}{\sin x}$$

Partial Derivatives

Often a function of interest will depend upon more than one variable. For example, the function

$$f(x,y) = 3x^2 y + 3y^3$$

is a function of both x and y. We can differentiate this function two ways:

1. We can differentiate with respect to x and hold y constant.

   $$\left(\frac{\partial f}{\partial x}\right)_y = \left(\frac{\partial[3x^2 y + 3y^2]}{\partial x}\right)_y = 6xy$$

2. We can differentiate with respect to y and hold x constant.

   $$\left(\frac{\partial f}{\partial y}\right)_x = \left(\frac{\partial[3x^2 y + 3y^2]}{\partial y}\right)_y = 3x^2 + 6y$$

Such derivatives are called *partial derivatives* because we are differentiating with respect to only one of the dependent variables. To distinguish partial derivatives from ordinary derivatives, we use the "∂" symbol instead of "d." The variables that are held constant are indicated as subscripts. For example, suppose we have the function

$$g(x,y,z) = 3x^2yz^3$$

The partial derivative of g with respect to x (with y and z held constant) is

$$\left(\frac{\partial g}{\partial x}\right)_{y,z} = \left(\frac{\partial [3x^2yz^3]}{\partial x}\right)_{y,z} = 3xyz^3$$

Integral Calculus

In scientific work, it is often more convenient to measure the rate of change of a quantity rather than the quantity itself. For example, a Geiger counter directly measures the rate of decay of a radioactive material, not the amount of material present. In another example, the speedometer on a car measures the rate of change of the cars position (that is, its velocity) and not its absolute position. The central problem of *integral calculus* is to determine the value of a particular property given that its derivative (rate of change) is known.

Indefinite Integration

We define the *indefinite integral* (or *antiderivative*) of a function $f(x)$ as the function $F(x)$ that, when differentiated, yields the original function $f(x)$. That is,

$$f(x) = \frac{dF(x)}{dx}$$

The process of integration is denoted symbolically by

$$\int f(x)\,dx = F(x) + C \tag{A1.5}$$

where the constant C is necessary because adding a constant to any function does not change the value of its derivative, which is why Equation A1.5 is called an indefinite integral. The symbol "dx" indicates that the integration is carried out over the variable x. Table A1.5 lists the indefinite integrals of some common functions used in this text.

Definite Integration

Because the indefinite integral $F(x)$ is only defined with an arbitrary constant C (Equation A1.5), it is not possible to fix the exact value of a $F(x)$ given only its rate of change $f(x)$. However, if, in addition to $f(x)$, the value of $F(x)$ at some initial value $x = a$ is known, it is then possible to determine the definite value of $F(x)$ at any other point $x = b$. This is because the constant C

Table A1.5   Indefinite Integrals of Common Functions

$f(x)$	$F(x) = \int f(x)\,dx$	$f(x)$	$F(x) = \int f(x)\,dx$
c (a constant)	cx	$\sin x$	$-\cos x$
x	$\dfrac{x^2}{2}$	$\cos x$	$\sin x$
x^n	$\dfrac{x^{n+1}}{n+1}$	e^x	e^x

cancels when we calculate the difference between $F(b)$ and $F(a)$. To do this we define the *definite integral* of $f(x)$ between $x = a$ and $x = b$ as

$$\int_a^b f(x)\, dx = F(x)\big|_a^b = F(b) - F(a) \tag{A1.6}$$

where we used the notation $F(x)\big|_a^b$ to denote $F(b) - F(a)$.

As an example, suppose that we are given that the rate of change of a function F with respect to x is equal to the square of x:

$$\frac{dF(x)}{dx} = f(x) = x^2$$

and that the value of F at $x = 1$ is equal to 5. We can then calculate the value of F at $x = 2$ using Equation A1.6:

$$F(2) = F(1) + \int_1^2 x^2\, dx$$

$$= 5 + \frac{x^3}{3}\bigg|_1^2 = 5 + \left(\frac{2^3}{3} - \frac{1^3}{3}\right) = \frac{22}{3}$$

where we have used Table A1.5 to determine the indefinite integral of x^2 to be $x^3/3$.

The definite integral also has a geometrical interpretation. If we divide the interval between $x = a$ and $x = b$ into N equal segments of length $\Delta x = (b - a)/N$, we can approximate the area under a curve $f(x)$ by summing up the areas of N rectangles of height $f(x_i)$ and width Δx, where $i = 1,\ldots,N$ [Figure A1.3(a)]:

$$\text{area under curve} \approx \sum_{i=1}^N f(x_i)\, \Delta x \tag{A1.7}$$

This approximation will become exact in the limit that the segment spacing Δx approaches zero. In this limit, the sum in Equation A1.7 approaches the definite integral of $f(x)$ between $x = a$ and $x = b$ [Figure A1.3(b)]:

$$\text{area under curve} = \lim_{\Delta x \to 0} \sum_{i=1}^N f(x_i)\, \Delta x = \int_a^b f(x)\, dx = F(b) - F(a) \tag{A1.8}$$

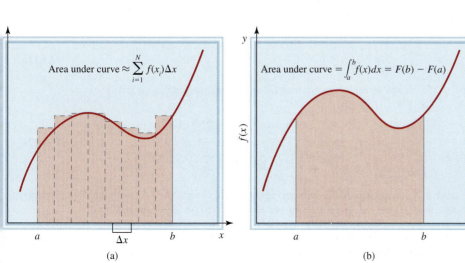

(a)   (b)

Figure A1.3 (a) The area under the curve $f(x)$ can be approximated by the sum of rectangular areas of width Δx and height $f(x_i)$. (b) In the limit that $\Delta x \to 0$, the area under the rectangles approaches the exact area under the curve given by the definite integral

$$\int_a^b f(x)\, dx.$$

Linear Regression

Relationships between physical properties can, in many cases, be described by a straight line

$$y = mx + b \tag{A1.9}$$

One example is the van't Hoff equation (Equation 10.16), which expresses the linear relationship between the logarithm of an equilibrium constant and inverse temperature ($1/T$), with a slope proportional to the enthalpy of reaction $\Delta H°$:

$$\ln \frac{K}{K_1} = \frac{\Delta H°}{R}\left(\frac{1}{T_1} - \frac{1}{T}\right)$$

By plotting experimental data for the equilibrium constant versus $1/T$, the enthalpy of reaction can be found from the slope, which according to the van't Hoff equation is equal to $-\Delta H°/R$, where R is the gas constant. However, expressions, such as the van't Hoff equation, are often approximate, so experimental data will not conform exactly to a straight line. Also, experimental measurements always have a certain level of uncertainty that depends upon the precision of the experimental technique. As a result, we often desire to determine the best straight line through a set of data points. The standard method for accomplishing this is known as *least-squares linear regression*.

To see how least-squares linear regression works, consider that we are given a set of N experimental data values $\{(x_i, y_i), i = 1,...,N\}$. We wish to find the best straight line (Equation A1.9) that fits this data. The deviation of each data point from a line of slope m and intercept b is given by

$$\Delta y_i = [y_i - (mx_i + b)]$$

In least-squares linear regression, the best line fit is determined by finding the values of m and b for which the sum of the squared deviations is as small as possible, that is, we wish to minimize the quantity

$$F(m,b) = \sum_{i=1}^{N} [y_i - (mx_i + b)]^2$$

Using differential calculus, the values of m and b for the best line through the data points are found to be

$$m = \frac{N\sum_{i=1}^{N} x_i y_i - \left(\sum_{i=1}^{N} x_i\right)\left(\sum_{i=1}^{N} y_i\right)}{N\sum_{i=1}^{N} x_i^2 - \left(\sum_{i=1}^{N} x_i\right)^2} \tag{A1.10}$$

$$b = \frac{\sum_{i=1}^{N} y_i - m\sum_{i=1}^{N} x_i}{N} \tag{A1.11}$$

The quality of the fit can be determined by calculating the correlation coefficient

$$r = \frac{N\sum_{i=1}^{N} x_i y_i - \left(\sum_{i=1}^{N} x_i\right)\left(\sum_{i=1}^{N} y_i\right)}{\sqrt{\left[N\sum_{i=1}^{N} x_i^2 - \left(\sum_{i=1}^{N} x_i\right)^2\right]\left[N\sum_{i=1}^{N} y_i^2 - \left(\sum_{i=1}^{N} y_i\right)^2\right]}} \tag{A1.12}$$

A value of r that is close to 1.0 indicates that the data is well fitted by a straight line, whereas a value near 0.0 tells us that a straight line does not describe the data very well.

Evaluating Equations A1.10 to A1.12 by hand can be tedious, especially for large data sets. Fortunately, most scientific calculators and spreadsheet programs have built-in least-squares linear regression functions that can do most of the work for you.

Solving Polynomial Equations

A polynomial equation is any equation of the form

$$a_n x^n + a_{n-1} x^{n-1} + \cdots + a_2 x^2 + a_1 x + a_0 = 0$$

where the coefficients $\{a_i; i = 0, 1, \ldots, n\}$ are real numbers.

Quadratic Equations ($n = 2$)

A significant fraction of polynomial equations that we encounter in chemistry have $n = 2$. Such equations are called *quadratic equations* and are of the form

$$ax^2 + mx + b = 0$$

The two roots of this equation are computed using the *quadratic formula*:

$$x = \frac{-b \pm \sqrt{b^2 - 4ac}}{2a} \qquad \textbf{(A1.13)}$$

In most physical problems, a single answer is required. In such cases, one can generally exclude one of the two roots in Equation A1.13 as unphysical. For example, if we are calculating a molarity and Equation A1.13 gives values of 0.534 and -0.287, then we can eliminate the latter value (-0.287) because molarities must be positive.

Higher-Order Polynomial Equations

For polynomial equations with $n = 3$ or higher, analytical solutions analogous to the quadratic equation for $n = 2$ either do not exist ($n \geq 5$) or are quite complex ($n = 3$ or 4). For such equations, the roots are found numerically using a computer program or spreadsheet or a graphing calculator.

Appendix 2

Thermodynamic Data at 1 Bar and 25°C*

Inorganic Substances

Substance	ΔH_f° (kJ mol^{-1})	ΔG_f° (kJ mol^{-1})	S° (J mol^{-1} K^{-1})	$\bar{C}_P^\circ$ (J mol^{-1} K^{-1})
Ag(s)	0	0	42.6	27.2
Ag$^+$(aq)	105.6	77.1	72.7	21.8
AgCl(s)	−127.0	−109.7	96.1	50.8
AgBr(s)	−100.4	−96.9	107.1	52.4
AgI(s)	−61.8	−66.3	115.5	56.8
AgNO$_3$(s)	−124.4	−33.4	140.9	93.1
Al(s)	0	0	28.3	24.2
Al^{3+}(aq)	−531.0	−485.0	−321.7	—
Al$_2$O$_3$(s)	−1675.7	−1582.3	50.9	79
As(s)	0	0	35.15	24.6
AsO$_4^{3-}$(aq)	−888.1	−648.4	−162.8	—
AsH$_3$(g)	66.4	68.9	222.8	38.1
H$_2$AsO$_4$(s)	−906.3	—	—	—
Au(s)	0	0	47.4	25.4
Au$_2$O$_3$(s)	80.8	163.2	125.5	—
AuCl(s)	−34.7	—	—	—
AuCl$_3$(s)	−117.6	—	—	—
B(s)	0	0	5.9	11.1
B$_2$O$_3$(s)	−1273.5	−1194.3	54.0	62.8
H$_3$BO$_3$(s)	−1094.3	−968.9	90	86.1
H$_3$BO$_3$(aq)	−1067.8	−963.3	159.8	—
Ba(s)	0	0	62.5	28.1
Ba^{2+}(aq)	−537.6	−560.8	9.6	—
BaO(s)	−558.2	−528.4	70.3	—
BaCl$_2$(s)	−855	−806.7	123.7	75.1
BaSO$_4$(s)	−1464.4	−1353.1	132.2	—
BaCO$_3$(s)	−1213	−1134.4	112.1	86
Be(s)	0	0	9.5	16.4
BeO(s)	−609.4	−580.1	13.8	25.6
Br$_2$(l)	0	0	152.3	75.7
Br$^-$(aq)	−121.6	−104.0	82.4	−141.8
HBr(g)	−36.2	−53.2	198.70	29.1
C(graphite)	0	0	5.69	8.5
C(diamond)	1.90	2.87	2.4	6.1
CO(g)	−110.5	−137.3	197.9	29.1

*The thermodynamic quantities of ions are based on the reference states that $\Delta H_f^\circ[\text{H}^+(aq) = 0]$, $\Delta G_f^\circ[\text{H}^+(aq) = 0]$, and $S^\circ[\text{H}^+(aq)] = 0$.

(Continued)

Substance	ΔH_f° (kJ mol^{-1})	ΔG_f° (kJ mol^{-1})	S° (J mol^{-1} K^{-1})	$\bar{C}_P^\circ$ (J mol^{-1} K^{-1})
$CO_2(g)$	−393.5	−394.4	213.6	37.1
$CO_2(aq)$	−412.9	−386.2	121.3	—
$CO_3^{2-}(aq)$	−677.1	−527.8	−56.9	—
$HCO_3^-(aq)$	−691.1	−587.1	94.98	—
$H_2CO_3(aq)$	−699.7	−623.2	187.4	—
$CS_2(g)$	116.7	67.1	237.8	45.1
$CS_2(l)$	89	64.6	151.3	76.4
$HCN(aq)$	105.4	112.1	128.9	—
$CN^-(aq)$	150.6	172.4	94.1	—
$(NH_2)_2CO(s)$	−333.19	−197.15	104.6	—
$(NH_2)_2CO(aq)$	−319.2	−203.84	173.85	—
$Ca(s)$	0	0	41.6	25.9
$Ca^{2+}(aq)$	−542.8	−553.6	−55.3	—
$CaO(s)$	−634.9	−603.3	38.1	42
$Ca(OH)_2(s)$	−985.2	−897.5	83.4	87.5
$CaF_2(s)$	−1228.1	−1175.1	68.5	67
$CaCl_2(s)$	−795.4	−748.8	108.4	72.9
$CaSO_4(s)$	−1434.5	−1322	106.5	99.7
$CaCO_3(s)$	−1207.6	−1129.1	91.7	83.5
$Cd(s)$	0	0	51.8	26
$Cd^{2+}(aq)$	−75.9	−77.6	−73.2	—
$CaO(s)$	−258.4	−228.6	43.4	—
$CdCl_2(s)$	−391.5	−343.9	115.3	74.7
$CdSO_4(s)$	−933.3	−822.7	123	99.6
$Cl_2(g)$	0	0	223.0	33.9
$Cl^-(aq)$	−167.2	−131.2	56.5	−136.4
$HCl(g)$	−92.3	−95.27	187.0	29.1
$Co(s)$	0	0	30	24.8
$Co^{2+}(aq)$	−58.2	−54.4	−113.0	—
$CoO(s)$	−237.9	−214.2	53	55.2
$Cr(s)$	0	0	23.8	23.4
$Cr^{2+}(aq)$	−143.5	—	—	—
$Cr_2O_3(s)$	−1139.7	−1058.1	81.2	118.7
$CrO_4^{2-}(aq)$	−881.2	−727.8	50.2	—
$Cr_2O_7^{2-}(aq)$	−1460.6	−1257.29	213.8	—
$Cs(s)$	0	0	85.2	32.2
$Cs^+(aq)$	−258.3	−292.2	133.1	−10.5
$Cu(s)$	0	0	33.2	24.4
$Cu^+(aq)$	71.7	50.0	40.6	—
$Cu^{2+}(aq)$	64.8	65.5	−99.6	—
$CuO(s)$	−157.3	−129.7	42.6	42.3
$Cu_2O(s)$	−168.6	−146	93.1	63.6

(Continued)

Substance	ΔH_f° (kJ mol^{-1})	ΔG_f° (kJ mol^{-1})	S° (J mol^{-1} K^{-1})	$\bar{C}_P^\circ$ (J mol^{-1} K^{-1})
CuCl(s)	−137.2	−119.9	86.2	48.5
CuCl$_2$(s)	−220.1	−175.7	108.1	71.9
CuS(s)	−53.1	−53.6	66.5	47.8
CuSO$_4$(s)	−771.4	−662.2	109.2	—
F$_2$(g)	0	0	202.8	31.3
F$^-$(aq)	−332.6	−278.8	−13.8	—
HF(g)	−273.3	−275.4	173.8	—
Fe(s)	0	0	27.3	25.1
Fe^{2+}(aq)	−87.86	−84.9	−113.39	—
Fe^{3+}(aq)	−47.7	−10.5	−293.3	—
FeO(s)	−272.0	−255.2	60.8	—
Fe$_2$O$_3$(s)	−824.2	−742.2	87.4	103.9
Fe(OH)$_2$(s)	−568.19	−483.55	79.5	—
Fe(OH)$_3$(s)	−824.25	—	—	—
H(g)	218.2	203.2	114.6	20.8
H$_2$(g)	0	0	130.7	28.8
H$^+$(aq)	0	0	0	—
OH$^-$(aq)	−229.94	−157.30	−10.5	—
H$_2$O(g)	−241.83	−228.6	188.8	33.6
H$_2$O(l)	−285.8	−237.1	69.95	75.3
H$_2$O$_2$(l)	−187.8	−120.4	109.6	89.1
Hg(l)	0	0	75.9	28
Hg^{2+}(aq)	—	−164.38	—	—
HgO(s)	−90.7	−58.5	72.0	44.1
HgCl$_2$(s)	−224.3	−178.6	144.5	—
Hg$_2$Cl$_2$(s)	−265.4	−210.7	191.6	—
HgS(s)	−58.16	−48.8	77.8	48.8
HgSO$_4$(s)	−704.17	—	—	—
Hg$_2$SO$_4$(s)	−741.99	−623.92	200.75	132
I$_2$(s)	0	0	116.7	54.4
I$^-$(aq)	−55.9	−51.67	109.37	—
HI(g)	25.9	1.30	206.3	29.2
K(s)	0	0	64.7	29.6
K$^+$(aq)	−251.2	−282.28	102.5	—
KOH(s)	−424.6	−379.4	81.2	68.9
KCl(s)	−436.5	−408.5	82.6	51.3
KClO$_3$(s)	−391.20	−289.9	142.97	100.3
KClO$_4$(s)	−432.8	−304.18	151.0	112.4
KBr(s)	−392.17	−379.2	96.4	52.3
KI(s)	−327.65	−322.29	104.35	52.9
KNO$_3$(s)	−492.7	−393.1	132.9	96.4
Li(s)	0	0	29.1	24.8
Li$^+$(aq)	−278.46	−293.8	14.2	68.6

(Continued)

Substance	ΔH_f° (kJ mol^{-1})	ΔG_f° (kJ mol^{-1})	S° (J mol^{-1} K^{-1})	$\bar{C}_P^\circ$ (J mol^{-1} K^{-1})
Li$_2$O(s)	-595.8	-561.2	37.6	54.1
LiOH(s)	-487.5	-441.5	42.8	49.6
Mg(s)	0	0	32.7	24.9
Mg^{2+}(aq)	-461.96	-456.0	-117.99	—
MgO(s)	-601.8	-569.6	26.78	—
Mg(OH)$_2$(s)	-924.5	-833.5	63.2	77
MgCl$_2$(s)	-641.8	-592.3	89.5	71.4
MgSO$_4$(s)	-1278.2	-1173.6	91.6	96.5
MgCO$_3$(s)	-1112.9	-1029.3	65.69	75.5
Mn(s)	0	0	31.76	26.3
Mn^{2+}(aq)	-218.8	-223.4	-83.68	50.0
MnO$_2$(s)	-520.9	-466.1	53.1	54.1
N$_2$(g)	0	0	191.6	29.1
N$_3^-$(aq)	245.18	—	—	—
NH$_3$(g)	-45.9	-16.4	192.8	35.1
NH$_4^+$(aq)	-132.80	-79.5	112.8	—
NH$_4$Cl(s)	-314.4	-202.9	94.6	84.1
NH$_3$(aq)	-80.3	-26.5	111.3	—
N$_2$H$_4$(l)	50.6	149.3	121.2	98.9
NO(g)	91.3	87.6	210.8	29.9
NO$_2$(g)	33.85	51.8	240.46	37.2
N$_2$O$_4$(g)	9.66	98.29	304.3	79.2
N$_2$O(g)	81.56	103.6	219.99	38.6
HNO$_2$(aq)	-118.8	-53.6	—	—
HNO$_3$(l)	-174.1	-80.7	155.6	109.9
NO$_3^-$(aq)	-207.4	-111.3	146.4	—
Na(s)	0	0	51.3	28.2
Na$^+$(aq)	-240.1	-261.9	59.0	46.4
Na$_2$O(s)	-414.2	-375.5	75.1	69.1
NaCl(s)	-411.2	-384.1	72.1	50.5
NaI(s)	-287.8	-286.1	98.5	52.1
Na$_2$SO$_4$(s)	-1387.1	-1270.2	149.6	128.2
NaNO$_3$(s)	-467.9	-367	116.5	92.9
Na$_2$CO$_3$(s)	-1130.7	-1044.4	135	112.3
NaHCO$_3$(s)	-950.8	-851	101.7	87.6
Ni(s)	0	0	29.9	26.1
Ni^{2+}(aq)	-54.0	-45.6	-128.9	—
NiO(s)	-244.35	-216.3	38.58	—
Ni(OH)$_2$(s)	-529.7	-447.2	88	—
O(g)	249.2	231.7	161.1	21.9
O$_2$(g)	0	0	205.0	29.4
O$_3$(aq)	-12.09	16.3	110.88	—
O$_3$(g)	142.7	163.2	238.9	39.2

(Continued)

Substance	ΔH_f° (kJ mol^{-1})	ΔG_f° (kJ mol^{-1})	S° (J mol^{-1} K^{-1})	$\bar{C}_P^\circ$ (J mol^{-1} K^{-1})
P(white)	0	0	41.1	23.8
P(red)	-18.4	13.8	29.3	21.2
$PO_4^{3-}(aq)$	-1277.4	-1018.7	-220.5	—
$P_4O_{10}(s)$	-3012.48	—	—	—
$PH_3(g)$	5.4	13.5	210.2	37.1
$HPO_4^{2-}(aq)$	-1292.1	-1089.2	-33.5	—
$H_2PO_4^-(aq)$	-1296.3	-1130.2	90.4	—
Pb(s)	0	0	64.89	26.4
$Pb^{2+}(aq)$	-1.7	-24.4	10.5	—
PbO(s)	-217.3	-187.9	68.5	45.8
$PbO_2(s)$	-277.4	-217.3	68.6	64.6
$PbCl_2(s)$	-359.4	-314.1	136	—
PbS(s)	-100.4	-98.7	91.2	49.5
$PbSO_4(s)$	-920	-813	148.5	103.2
Pt(s)	0	0	41.6	25.9
$PtCl_4^{2-}(aq)$	-516.3	-384.5	175.7	—
Rb(s)	0	0	76.8	31.1
$Rb^+(aq)$	-251.2	-284.0	121.5	—
S(rhombic)	0	0	32.1	22.6
S(monoclinic)	0.30	0.10	32.55	—
$SO_2(g)$	-296.8	-300.1	248.2	39.9
$SO_3(g)$	-395.7	-371.1	256.8	50.7
$SO_3^{2-}(aq)$	-635.5	-486.5	-29.0	—
$SO_4^{2-}(aq)$	-909.3	-744.5	20.1	-293.0
$H_2S(g)$	-20.6	-33.4	205.8	34.2
$HSO_3^-(aq)$	-626.2	-527.7	139.7	—
$HSO_4^-(aq)$	-887.3	-755.9	131.8	-84.0
$H_2SO_4(l)$	-814	-690	156.9	138.9
$SF_6(g)$	-1220.5	-1116.5	291.5	97
Si(s)	0	0	18.8	20
$SiO_2(s)$	-910.7	-856.3	41.5	44.4
Sr(s)	0	0	55	26.8
$Sr^{2+}(aq)$	-545.8	-559.5	-32.6	—
$SrCl_2(s)$	-828.9	-781.1	114.9	75.6
$SrSO_4(s)$	-1453.1	-1340.9	117	—
$SrCO_3(s)$	-1220.1	-1140.1	97.1	81.4
Zn(s)	0	0	41.6	25.4
$Zn^{2+}(aq)$	-153.9	-147.1	-112.1	46.0
ZnO(s)	-350.5	-320.5	43.7	40.3
$ZnCl_2(s)$	-415.1	-369.4	111.5	71.3
ZnS(s)	-206	-201.3	57.7	46
$ZnSO_4(s)$	-982.8	-871.5	110.5	99.2

Organic Substances

Substance	Formula	ΔH_f° (kJ mol^{-1})	ΔG_f° (kJ mol^{-1})	S° (J mol^{-1} K^{-1})	$\bar{C}_P$ (J mol^{-1} K^{-1})
Ethanoic acid (acetic acid)(l)	CH_3COOH	−484.3	−389.9	159.8	123.3
Ethanal (acetaldehyde)(g)	CH_3CHO	−166.2	−133	263.8	55.3
2-Propanone (acetone)(l)	CH_3COCH_3	−246.8	−153.55	198.7	126.3
Ethyne (acetylene)(g)	C_2H_2	227.4	209.9	200.9	44
Benzene(l)	C_6H_6	49.1	124.5	173.4	136
Butane(g)	C_4H_{10}	−124.7	−15.7	310.0	140.9
Ethanol(l)	C_2H_5OH	−277.6	−174.8	160.7	112.3
Ethane(g)	C_2H_6	−84	−32	229.2	52.5
Ethene (ethylene)(g)	C_2H_4	52.4	68.4	219.3	42.9
Methanoic acid (formic acid)(l)	$HCOOH$	−425	−361.4	129.0	99
Glucose(s)	$C_6H_{12}O_6$	−1274.5	−910.56	212.1	—
Methane(g)	CH_4	−74.6	−50.8	186.3	35.7
Methanol(l)	CH_3OH	−239.2	−166.6	126.8	81.1
Propane(g)	C_3H_8	−103.8	−23.4	270.3	73.6
Sucrose(s)	$C_{12}H_{22}O_{11}$	−2221.7	−1544.3	360.2	—

Appendix 3

Derivation of the Names of Elements*

Element	Symbol	Atomic No.	Atomic Mass[†]	Date of Discovery	Discoverer and Nationality[‡]	Derivation
Actinium	Ac	89	(227)	1899	A. Debierne (Fr.)	Gr. *aktis,* beam or ray
Aluminum	Al	13	26.98	1827	F. Woehler (Ge.)	Alum, the aluminum compound in which it was discovered; derived from L. *alumen,* astringent taste
Americium	Am	95	(243)	1944	A. Ghiorso (USA) R. A. James (USA) G. T. Seaborg (USA) S. G. Thompson (USA)	The Americas
Antimony	Sb	51	121.8	Ancient		L. *antimonium* (*anti,* opposite of; *monium,* isolated condition), so named because it is a tangible (metallic) substance which combines readily; symbol L. *stibium,* mark
Argon	Ar	18	39.95	1894	Lord Raleigh (GB) Sir William Ramsay (GB)	Gr. *argos,* inactive
Arsenic	As	33	74.92	1250	Albertus Magnus (Ge.)	Gr. *aksenikon,* yellow pigment; L. *arsenicum,* orpiment; the Greeks once used arsenic trisulfide as a pigment
Astatine	At	85	(210)	1940	D. R. Corson (USA) K. R. MacKenzie (USA) E. Segre (USA)	Gr. *astatos,* unstable
Barium	Ba	56	137.3	1808	Sir Humphry Davy (GB)	Barite, a heavy spar, derived from Gr. *barys,* heavy
Berkelium	Bk	97	(247)	1950	G. T. Seaborg (USA) S. G. Thompson (USA) A. Ghiorso (USA)	Berkeley, Calif.
Beryllium	Be	4	9.012	1828	F. Woehler (Ge.) A. A. B. Bussy (Fr.)	Fr. L. *beryl,* sweet

Source: Reprinted with permission from "The Elements and Derivation of Their Names and Symbols." G. P. Dinga, *Chemistry* **41** (2), 20–22 (1968). Copyright by the American Chemical Society.

*At the time this table was drawn up, only 103 elements were known to exist.

[†]The atomic masses given here correspond to the 1961 values of the Commission on Atomic Weights. Masses in parentheses are those of the most stable or most common isotopes.

[‡]The abbreviations are (Ar.) Arabic; (Au.) Austrian; (Du.) Dutch; (Fr.) French; (Ge.) German; (GB) British; (Gr.) Greek; (H.) Hungarian; (I.) Italian; (L.) Latin: (P.) Polish; (R.) Russian, (Sp.) Spanish; (Swe.) Swedish; (USA) American.

(*Continued*)

Element	Symbol	Atomic No.	Atomic Mass[†]	Date of Discovery	Discoverer and Nationality[‡]	Derivation
Bismuth	Bi	83	209.0	1753	Claude Geoffroy (Fr.)	Ge. *bismuth,* probably a distortion of *weisse masse* (white mass) in which it was found
Boron	B	5	10.81	1808	Sir Humphry Davy (GB) J. L. Gay-Lussac (Fr.) L. J. Thenard (Fr.)	The compound borax, derived from Ar. *buraq,* white
Bromine	Br	35	79.90	1826	A. J. Balard (Fr.)	Gr. *bromos,* stench
Cadmium	Cd	48	112.4	1817	Fr. Stromeyer (Ge.)	Gr. *kadmia,* earth; L. *cadmia,* calamine (because it is found along with calamine)
Calcium	Ca	20	40.08	1808	Sir Humphry Davy (GB)	L. *calx,* lime
Californium	Cf	98	(249)	1950	G. T. Seaborg (USA) S. G. Thompson (USA) A. Ghiorso (USA) K. Street, Jr. (USA)	California
Carbon	C	6	12.01	Ancient		L. *carbo,* charcoal
Cerium	Ce	58	140.1	1803	J. J. Berzelius (Swe.) William Hisinger (Swe.) M. H. Klaproth (Ge.)	Asteroid Ceres
Cesium	Cs	55	132.9	1860	R. Bunsen (Ge.) G. R. Kirchhoff (Ge.)	L. *caesium,* blue (cesium was discovered by its spectral lines, which are blue)
Chlorine	Cl	17	35.45	1774	K. W. Scheele (Swe.)	Gr. *chloros,* light green
Chromium	Cr	24	52.00	1797	L. N. Vauquelin (Fr.)	Gr. *chroma,* color (because it is used in pigments)
Cobalt	Co	27	58.93	1735	G. Brandt (Ge.)	Ge. *Kobold,* goblin (because the ore yielded cobalt instead of the expected metal, copper, it was attributed to goblins)
Copper	Cu	29	63.55	Ancient		L. *cuprum, copper,* derived from *cyprium,* Island of Cyprus, the main source of ancient copper
Curium	Cm	96	(247)	1944	G. T. Seaborg (USA) R. A. James (USA) A. Ghiorso (USA)	Pierre and Marie Curie
Dysprosium	Dy	66	162.5	1886	Lecoq de Boisbaudran (Fr.)	Gr. *dysprositos,* hard to get at
Einsteinium	Es	99	(254)	1952	A. Ghiorso (USA)	Albert Einstein

(Continued)

Element	Symbol	Atomic No.	Atomic Mass[†]	Date of Discovery	Discoverer and Nationality[‡]	Derivation
Erbium	Er	68	167.3	1843	C. G. Mosander (Swe.)	Ytterby, Sweden, where many rare earths were discovered
Europium	Eu	63	152.0	1896	E. Demarcay (Fr.)	Europe
Fermium	Fm	100	(253)	1953	A. Ghiorso (USA)	Enrico Fermi
Fluorine	F	9	19.00	1886	H. Moissan (Fr.)	Mineral fluorspar, from L. *fluere*, flow (because fluorspar was used as a flux)
Francium	Fr	87	(223)	1939	Marguerite Perey (Fr.)	France
Gadolinium	Gd	64	157.3	1880	J. C. Marignac (Fr.)	Johan Gadolin, Finnish rare earth chemist
Gallium	Ga	31	69.72	1875	Lecoq de Boisbaudran (Fr.)	L. *Gallia*, France
Germanium	Ge	32	72.59	1886	Clemens Winkler (Ge.)	L. *Germania*, Germany
Gold	Au	79	197.0	Ancient		L. *aurum*, shining dawn
Hafnium	Hf	72	178.5	1923	D. Coster (Du.) G. von Hevesey (H.)	L. *Hafnia*, Copenhagen
Helium	He	2	4.003	1868	P. Janssen (spectr) (Fr.) Sir William Ramsay (isolated) (GB)	Gr. *helios*, sun (because it was first discovered in the sun's spectrum)
Holmium	Ho	67	164.9	1879	P. T. Cleve (Swe.)	L. *Holmia*, Stockholm
Hydrogen	H	1	1.008	1766	Sir Henry Cavendish (GB)	Gr. *hydro*, water; *genes*, forming (because it produces water when burned with oxygen)
Indium	In	49	114.8	1863	F. Reich (Ge.) T. Richter (Ge.)	Indigo, because of its indigo blue lines in the spectrum
Iodine	I	53	126.9	1811	B. Courtois (Fr.)	Gr. *iodes*, violet
Iridium	Ir	77	192.2	1803	S. Tennant (GB)	L. *iris*, rainbow
Iron	Fe	26	55.85	Ancient		L. *ferrum*, iron
Krypton	Kr	36	83.80	1898	Sir William Ramsay (GB) M. W. Travers (GB)	Gr. *kryptos*, hidden
Lanthanum	La	57	138.9	1839	C. G. Mosander (Swe.)	Gr. *lanthanein*, concealed
Lawrencium	Lr	103	(257)	1961	A. Ghiorso (USA) T. Sikkeland (USA) A. E. Larsh (USA) R. M. Latimer (USA)	E. O. Lawrence (USA), inventor of the cyclotron
Lead	Pb	82	207.2	Ancient		Symbol, L. *plumbum*, lead, meaning heavy
Lithium	Li	3	6.941	1817	A. Arfvedson (Swe.)	Gr. *lithos*, rock (because it occurs in rocks)
Lutetium	Lu	71	175.0	1907	G. Urbain (Fr.) C. A. von Welsbach (Au.)	*Luteria*, ancient name for Paris

(Continued)

Element	Symbol	Atomic No.	Atomic Mass[†]	Date of Discovery	Discoverer and Nationality[‡]	Derivation
Magnesium	Mg	12	24.31	1808	Sir Humphry Davy (GB)	*Magnesia*, a district in Thessaly; possibly derived from L. *magnesia*
Manganese	Mn	25	54.94	1774	J. G. Gahn (Swe.)	L. *magnes*, magnet
Mendelevium	Md	101	(256)	1955	A. Ghiorso (USA) G. R. Choppin (USA) G. T. Seaborg (USA) B. G. Harvey (USA) S. G. Thompson (USA)	Mendeleev, Russian chemist who prepared the periodic chart and predicted properties of undiscovered elements
Mercury	Hg	80	200.6	Ancient		Symbol, L. *hydrargyrum*, liquid silver
Molybdenum	Mo	42	95.94	1778	G. W. Scheele (Swe.)	Gr. *molybdos*, lead
Neodymium	Nd	60	144.2	1885	C. A. von Welsbach (Au.)	Gr. *neos*, new; *didymos*, twin
Neon	Ne	10	20.18	1898	Sir William Ramsay (GB) M. W. Travers (GB)	Gr. *neos*, new
Neptunium	Np	93	(237)	1940	E. M. McMillan (USA) P. H. Abelson (USA)	Planet Neptune
Nickel	Ni	28	58.69	1751	A. F. Cronstedt (Swe.)	Swe. *kopparnickel*, false copper; also Ge. *nickel*, referring to the devil that prevented copper from being extracted from nickel ores
Niobium	Nb	41	92.91	1801	Charles Hatchett (GB)	Gr. *Niobe*, daughter of Tantalus (niobium was considered identical to tantalum, named after *Tantalus*, until 1884)
Nitrogen	N	7	14.01	1772	Daniel Rutherford (GB)	Fr. *nitrogene*, derived from L. *nitrum*, native soda, or Gr. *nitron*, native soda, and Gr. *genes*, forming
Nobelium	No	102	(253)	1958	A. Ghiorso (USA) T. Sikkeland (USA) J. R. Walton (USA) G. T. Seaborg (USA)	Alfred Nobel
Osmium	Os	76	190.2	1803	S. Tennant (GB)	Gr. *osme*, odor
Oxygen	O	8	16.00	1774	Joseph Priestley (GB) C. W. Scheele (Swe.)	Fr. *oxygene*, generator of acid, derived from Gr. *oxys*, acid, and L. *genes*, forming (because it was once thought to be a part of all acids)

(Continued)

Element	Symbol	Atomic No.	Atomic Mass[†]	Date of Discovery	Discoverer and Nationality[‡]	Derivation
Palladium	Pd	46	106.4	1803	W. H. Wollaston (GB)	Asteroid Pallas
Phosphorus	P	15	30.97	1669	H. Brandt (Ge.)	Gr. *phosphoros*, light bearing
Platinum	Pt	78	195.1	1735 1741	A. de Ulloa (Sp.) Charles Wood (GB)	Sp. *platina*, silver
Plutonium	Pu	94	(242)	1940	G. T. Seaborg (USA) E. M. McMillan (USA) J. W. Kennedy (USA) A. C. Wahl (USA)	Planet Pluto
Polonium	Po	84	(210)	1898	Marie Curie (P.)	Poland
Potassium	K	19	39.10	1807	Sir Humphry Davy (GB)	Symbol, L. *kalium*, potash
Praseodymium	Pr	59	140.9	1885	C. A. von Welsbach (Au.)	Gr. *prasios*, green; *didymos*, twin
Promethium	Pm	61	(147)	1945	J. A. Marinsky (USA) L. E. Glendenin (USA) C. D. Coryell (USA)	Gr. mythology, *Prometheus*, the Greek Titan who stole fire from heaven
Protactinium	Pa	91	(231)	1917	O. Hahn (Ge.) L. Meitner (Au.)	Gr. *protos*, first; *actinium* (because it distintegrates into actinium)
Radium	Ra	88	(226)	1898	Pierre and Marie Curie (Fr., P.)	L. *radius*, ray
Radon	Rn	86	(222)	1900	F. E. Dorn (Ge.)	Derived from radium
Rhenium	Re	75	186.2	1925	W. Noddack (Ge.) I. Tacke (Ge.) Otto Berg (Ge.)	L. *Rhenus*, Rhine
Rhodium	Rh	45	102.9	1804	W. H. Wollaston (GB)	Gr. *rhodon*, rose (because some of its salts are rose-colored)
Rubidium	Rb	37	85.47	1861	R. W. Bunsen (Ge.) G. Kirchhoff (Ge.)	L. *rubidus*, dark red (discovered with the spectroscope, its spectrum shows red lines)
Ruthenium	Ru	44	101.1	1844	K. K. Klaus (R.)	L. *Ruthenia*, Russia
Samarium	Sm	62	150.4	1879	Lecoq de Boisbaurdran (Fr.)	Samarskite, after Samarski, a Russian engineer
Scandium	Sc	21	44.96	1879	L. F. Nilson (Swe.)	Scandinavia
Selenium	Se	34	78.96	1817	J. J. Berzelius (Swe.)	Gr. *selene*, moon (because it resembles tellurium, named for the earth)
Silicon	Si	14	28.09	1824	J. J. Berzelius (Swe.)	L. *silex, silicis*, flint
Silver	Ag	47	107.9	Ancient		Symbol, L. *argentum*, silver

(*Continued*)

Element	Symbol	Atomic No.	Atomic Mass[†]	Date of Discovery	Discoverer and Nationality[‡]	Derivation
Sodium	Na	11	22.99	1807	Sir Humphry Davy (GB)	L. *sodanum,* headache remedy; symbol, L. *natrium,* soda
Strontium	Sr	38	87.62	1808	Sir Humphry Davy (GB)	Strontian, Scotland, derived from mineral strontionite
Sulfur	S	16	32.07	Ancient		L. *sulphurium* (Sanskrit, *sulvere*)
Tantalum	Ta	73	180.9	1802	A. G. Ekeberg (Swe.)	Gr. mythology, *Tantalus,* because of difficulty in isolating it
Technetium	Tc	43	(99)	1937	C. Perrier (I.)	Gr. *technetos,* artificial (because it was the first artificial element)
Tellurium	Te	52	127.6	1782	F. J. Müller (Au.)	L. *tellus,* earth
Terbium	Tb	65	158.9	1843	C. G. Mosander (Swe.)	Ytterby, Sweden
Thallium	Tl	81	204.4	1861	Sir William Crookes (GB)	Gr. *thallos,* a budding twig (because its spectrum shows a bright green line)
Thorium	Th	90	232.0	1828	J. J. Berzelius (Swe.)	Mineral thorite, derived from *Thor,* Norse god of war
Thulium	Tm	69	168.9	1879	P. T. Cleve (Swe.)	*Thule,* early name for Scandinavia
Tin	Sn	50	118.7	Ancient		Symbol, L. *stannum,* tin
Titanium	Ti	22	47.88	1791	W. Gregor (GB)	Gr. giants, the Titans, and L. titans, giant deities
Tungsten	W	74	183.9	1783	J. J. and F. de Elhuyar (Sp.)	Swe. *tung sten,* heavy stone; symbol, wolframite, a mineral
Uranium	U	92	238.0	1789 1841	M. H. Klaproth (Ge.) E. M. Peligot (Fr.)	Planet Uranus
Vanadium	V	23	50.94	1801 1830	A. M. del Rio (Sp.) N. G. Sefstrom (Swe.)	*Vanadis,* Norse goddess of love and beauty
Xenon	Xe	54	131.3	1898	Sir William Ramsay (GB) M. W. Travers (GB)	Gr. *xenos,* stranger
Ytterbium	Yb	70	173.0	1907	G. Urbain (Fr.)	Ytterby, Sweden
Yttrium	Y	39	88.91	1843	C. G. Mosander (Swe.)	Ytterby, Sweden
Zinc	Zn	30	65.39	1746	A. S. Marggraf (Ge.)	Ge. *zink,* of obscure origin
Zirconium	Zr	40	91.22	1789	M. H. Klaproth (Ge.)	Zircon, in which it was found, derived from *Ar. zargum,* gold color

Appendix 4

Isotopes of the First Ten Elements

Element	Isotope*	Mass (in u)	Natural Abundance (%)	Primary Decay Process	Half-Life
Hydrogen	^{1}H	1.0078250	99.9885		
	^{2}H	2.0141018	0.0115		
	^{3}H	3.0160493		β^- to ^{3}He	12.33 y
Helium	^{3}He	3.0160293	0.000137		
	^{4}He	4.0026032	99.999863		
	^{6}He	6.018886		β^- to ^{6}Li	807 ms
	^{8}He	8.033922		β^- to ^{8}Li	119 ms
Lithium	^{6}Li	6.015122	7.6		
	^{7}Li	7.016004	92.4		
	^{8}Li	8.022486		β^- to ^{8}Be	838 ms
	^{9}Li	9.026789		β^- to ^{9}Be	178.3 ms
	^{11}Li	11.043796		β^- to ^{11}Be	8.5 ms
Beryllium	^{7}Be	7.016880		EC† to ^{7}Li	53.28 d
	^{9}Be	9.012182	100		
	^{10}Be	10.013534		β^- to ^{10}B	1.52×10^6 y
	^{11}Be	11.021658		β^- to ^{11}B	13.81 s
	^{12}Be	12.026921		β^- to ^{12}B	23.6 ms
Boron	^{8}B	8.024607		β^+ (or EC) + α to ^{4}He	770 ms
	^{9}B	9.013339		$2 \times \alpha$ to ^{1}H	8×10^{-19} s
	^{10}B	10.012937	19.9		
	^{11}B	11.009305	80.1		
	^{12}B	12.014352		β^- to ^{12}C	20.20 ms
	^{13}B	13.017780		β^- to ^{13}C	17.36 ms
	^{14}B	14.025404		β^- to ^{14}C	13.8 ms
	^{15}B	15.031097		β^- to ^{15}C	10.5 ms
Carbon	^{10}C	10.016853		β^+ (or EC) to ^{10}B	19.255 s
	^{11}C	11.011434		β^+ (or EC) to ^{11}B	20.39 min
	^{12}C	12.000000	98.89		
	^{13}C	13.003355	1.11		
	^{14}C	14.003242		β^- to ^{14}N	5730 y
	^{15}C	15.010599		β^- to ^{15}N	2.449 s
	^{16}C	16.014701		β^- to ^{16}N	747 ms
	^{17}C	17.033584		β^- to ^{17}N	193 ms
	^{18}C	18.026757		β^- to ^{18}N	95 ms

*Naturally occurring isotopes in black, radioisotopes in red.
†EC = electron capture

(*Continued*)

Element	Isotope*	Mass (in u)	Natural Abundance (%)	Primary Decay Process	Half-Life
Nitrogen	^{12}N	12.018613		β^+ (or EC) to ^{12}C	11 ms
	^{13}N	13.005739		β^+ (or EC) to ^{13}C	9.965 min
	^{14}N	14.003074	99.634		
	^{15}N	15.000109	0.366		
	^{16}N	16.006101		β^- to ^{16}O	7.13 s
	^{17}N	17.008450		β^- to ^{17}O	4.173 s
	^{18}N	18.014082		β^- to ^{18}O	624 ms
	^{19}N	19.017927		β^- to ^{19}O	290 ms
Oxygen	^{13}O	13.024810		β^+ (or EC) to ^{13}N	8.58 ms
	^{14}O	14.008595		β^+ (or EC) to ^{14}N	70.61 s
	^{15}O	15.003065		β^+ (or EC) to ^{15}N	122.24 s
	^{16}O	15.994915	99.762		
	^{17}O	16.999132	0.038		
	^{18}O	17.999160	0.200		
	^{19}O	19.003579		β^- to ^{19}F	26.91 s
	^{20}O	20.004076		β^- to ^{20}F	13.51 s
	^{21}O	21.008655		β^- to ^{21}F	3.42 s
	^{22}O	22.009967		β^- to ^{22}F	2.25 s
	^{23}O	23.015691		β^- to ^{23}F	82 ms
Fluorine	^{17}F	17.002095		β^+ (or EC) to ^{17}O	64.49 s
	^{18}F	18.000938		β^+ (or EC) to ^{18}O	109.77 min
	^{19}F	18.998403	100		
	^{20}F	19.999981		β^- to ^{20}Ne	11.16 s
	^{21}F	20.999949		β^- to ^{21}Ne	4.158 s
	^{22}F	22.002999		β^- to ^{22}Ne	4.23 s
	^{23}F	23.003574		β^- to ^{23}Ne	2.23 s
	^{24}F	24.008099		β^- to ^{24}Ne	0.34 s
	^{25}F	25.012095		β^- to ^{25}Ne	87 ms
Neon	^{18}Ne	18.005697		β^+ (or EC) to ^{18}F	1672 ms
	^{19}Ne	19.001880		β^+ (or EC) to ^{19}F	17.22 s
	^{20}Ne	19.992440	90.48		
	^{21}Ne	20.993847	0.27		
	^{22}Ne	21.991386	9.25		
	^{23}Ne	22.994467		β^- to ^{23}Na	37.24 s
	^{24}Ne	23.993615		β^- to ^{24}Na	3.38 min
	^{25}Ne	24.997790		β^- to ^{25}Na	602 ms
	^{26}Ne	26.000462		β^- to ^{26}Na	0.23 s
	^{27}Ne	27.007615		β^- to ^{27}Na	32 ms

Glossary

The number in parentheses is the number of the section in which the term first appears.

A

absolute zero of temperature. Theoretically the lowest attainable temperature. (5.1)

acceptor impurities. Impurities that can accept electrons from semiconductors. (6.4)

accuracy. The closeness of a measurement to the true value of the quantity that is measured. (A.1)

acid. A substance that yields hydrogen ions (H) when dissolved in water. (0.3, 11.1)

acid ionization constant (K_a). The equilibrium constant for the acid ionization. (11.3)

actinide series. Elements that have incompletely filled $5f$ subshells or readily give rise to cations that have incompletely filled $5f$ subshells. (2.2)

activated complex. The species temporarily formed by the reactant molecules as a result of the collision before they form the product. (14.4)

activated-complex theory. See transition-state theory. (14.4)

activation energy (E_a). The minimum amount of energy required to initiate a chemical reaction. (14.4)

activity. (1) A measure of the effective concentration of a species in an equilibrium constant expression (9.2); (2) In nuclear chemistry, the activity is defined as the number of nuclei decaying per unit time. (10.1, 17.3)

activity series. A summary of the results of many possible displacement reactions. (4.4)

actual yield. The amount of product actually obtained in a reaction. (0.6)

addition reaction. Reactions involving the addition of atoms to compounds at the site of double or triple bonds. (16.2, 16.4)

adhesion. Attraction between unlike molecules. (6.1)

adiabatic process. A process for which the heat inflow and outflow is zero (that is, $q = 0$). (7.1)

alcohol. An organic compound containing the hydroxyl group—OH. (16.3)

aldehydes. Compounds with a carbonyl functional group and the general formula RCHO, where R is an H atom, an alkyl, or an aromatic group. (16.3)

aliphatic hydrocarbons. Hydrocarbons that do not contain the benzene group or the benzene ring. (16.1)

alkali metals. The Group 1A elements (Li, Na, K, Rb, Cs, and Fr). (0.2)

alkaline earth metals. The Group 2A elements (Be, Mg, Ca, Sr, Ba, and Ra). (0.2)

alkanes. Hydrocarbons having the general formula C_nH_{2n2}, where n 1, 2, ... (16.1)

alkenes. Hydrocarbons that contain one or more carbon-carbon double bonds. They have the general formula C_nH_{2n}, where $n = 2, 3, . . .$ (16.1)

alkyl group. An organic functional group consisting of an alkane less one hydrogen. (16.1)

alkynes. Hydrocarbons that contain one or more carbon-carbon triple bonds. They have the general formula C_nH_{2n2}, where $n = 2, 3, . . .$ (16.1)

allotropes. Two or more forms of the same element that differ significantly in chemical and physical properties. (0.3)

alpha decay. The emission of an alpha particle by an unstable nucleus. (17.1)

alpha (α) particles. Helium nuclei with a positive charge of 2. (17.1)

amines. Organic bases that have the functional group—NR_2, where R may be H, an alkyl group, or an aromatic group. (16.3)

amino acids. A compound that contains at least one amino group and at least one carboxyl group. (16.5)

amorphous solid. A solid that lacks a regular three-dimensional arrangement of atoms or molecules. (6.3)

amphoteric oxide. An oxide that exhibits both acidic and basic properties. (11.1)

amplitude. The vertical distance between the midline of a wave and its peak or trough. (1.1)

angular momentum. The vector cross product of momentum and force: $\mathbf{L} = \mathbf{p} \times \mathbf{F}$. (1.2)

angular momentum quantum number. The quantum number l that determines the shape of atomic orbitals. (1.4)

anion. An ion with a net negative charge. (0.2)

anode. The electrode at which oxidation occurs. (13.2)

antibonding molecular orbital. A molecular orbital that is of higher energy and lower stability than the atomic orbitals from which it was formed. (3.5)

aqueous solution. A solution in which the solvent is water. (0.4)

aqueous species. Substances dissolved in water solution. (0.4)

aromatic hydrocarbon. A hydrocarbon that contains one or more benzene rings. (16.1)

Arrhenius equation. An equation describing the temperature dependence of reaction rates. (14.4)

atmospheric pressure. The pressure exerted by Earth's atmosphere. (5.1)

atom. The basic unit of an element that can enter into chemical combination. (0.2)

atomic mass. The mass of an atom in atomic mass units. (0.5)

atomic mass unit (u). A mass exactly equal to $1/12^{th}$ the mass of one carbon-12 atom. (0.5)

atomic number (Z). The number of protons in the nucleus of an atom. (0.2)

atomic orbital. The wave function ϕ of a single electron in an atom. (1.4)

atomic radius. One-half the distance between the two nuclei in two adjacent atoms of the same element in a metal. For elements that exist as diatomic units, the atomic radius is one-half the distance between the nuclei of the two atoms in a particular molecule. (2.5)

Aufbau principle. As protons are added one by one to the nucleus to build up the elements, electrons similarly are added to the atomic orbitals. (2.2)

autoionization of water. The reaction of two water molecules to form an OH⁻ ion and an H_3O^+ ion. (11.2)

average atomic mass. The average of the atomic masses of the different isotopes of an element, weighted by their natural abundances. (0.5)

Avogadro's law. At constant pressure and temperature, the volume of a gas is directly proportional to the number of moles of the gas present. (5.3)

Avogadro's number (N_A). 6.022×10^{23}; the number of particles in a mole. (0.5)

azeotrope. A point of phase coexistence in a binary mixture in which each phase has identical composition. (9.3)

B

back bonding. The lowering of $3d_{xy}$, $3d_{xz}$, and $3d_{yz}$ orbital energies of a metal ion center in a coordination compound due to mixing with unoccupied pi-antibonding orbitals on the ligand.

band theory of conductivity. Delocalized electrons move freely through "bands" formed by overlapping molecular orbitals. (6.4)

bar. 10^5 Pa. (5.1, A.1)

barometer. An instrument that measures atmospheric pressure. (5.1)

base. A substance that yields hydroxide ions (OH) when dissolved in water. (0.3, 11.1)

base ionization constant (K_b). The equilibrium constant for the base ionization. (11.3)

battery. A galvanic cell, or a series of combined galvanic cells, that can be used as a source of direct electric current at a constant voltage. (13.6)

beta decay. Emission of a beta particle (electron or positron) by an unstable nucleus. (17.1)

beta (β) particles. Electrons (β^-) or positrons (β^+) emitted by an unstable nucleus. (17.1)

bimolecular reaction. An elementary step involving two reactant molecules. (14.5)

binary compounds. Compounds formed from just two elements. (0.3)

blackbody radiation. Radiation emitted by an object solely as a consequence of its temperature. (1.1)

body-centered cubic (bcc). Having a crystal unit cell with lattice points at the vertices and at the center of a cube. (6.2)

Bohr model. An early quantum model of the hydrogen atom. (1.2)

Bohr radius. A constant equal to about 0.59 Å. (1.2)

boiling point. The temperature at which the vapor pressure of a liquid is equal to the external atmospheric pressure. (5.2)

boiling-point elevation. The raising of the boiling point of a liquid through the addition of a solute. (9.4)

bond dissociation energy. The energy required to break a bond. (3.1)

bond enthalpy. The enthalpy change required to break a bond in a mole of gaseous molecules. (7.5)

bond length. The distance between the nuclei of two bonded atoms in a molecule. (3.1)

bond order. The difference between the numbers of electrons in bonding molecular orbitals and antibonding molecular orbitals, divided by two. (3.2, 3.6)

bonding molecular orbital. A molecular orbital that is of lower energy and greater stability than the atomic orbitals from which it was formed. (3.5)

boundary surface diagram. A surface containing 90 percent of the electron density in an orbital. (1.4)

Boyle's law. The volume of a fixed amount of gas maintained at constant temperature is inversely proportional to the gas pressure. (5.3)

Boyle temperature (T_B). The temperature at which the second virial coefficient of a real gas is equal to zero. (5.5)

Bravais lattices. The 14 basic types of unit cells for a crystal lattice. (6.2)

breeder reactor. A nuclear reactor that produces more fissionable materials than it uses. (17.5)

Brønsted acid. A substance capable of donating a proton. (11.1)

Brønsted base. A substance capable of accepting a proton. (11.1)

buffer solution. A solution of (a) a weak acid or base and (b) its salt; both components must be present. The solution has the ability to resist changes in pH upon the addition of small amounts of either acid or base. (12.2)

C

calorimetry. The measurement of heat changes. (7.3)

carboxylic acids. Acids that contain the carboxyl group —COOH. (16.3)

catalyst. A substance that increases the rate of a chemical reaction without itself being consumed. (14.6)

cathode. The electrode at which reduction occurs. (13.2)

cation. An ion with a net positive charge. (0.2)

cell diagram. A notation for representing electrochemical cells. (13.2)

cell voltage. Difference in electrical potential between the anode and the cathode of a galvanic cell. (13.2)

Charles's and Gay-Lussac's law. See Charles's law.

Charles's law. The volume of a fixed amount of gas maintained at constant pressure is directly proportional to the absolute temperature of the gas. (5.3)

chelating agent. A substance that forms complex ions with metal ions in solution. (15.1)

chemical bond. An attractive force between atoms in a molecule that is strong enough for the atoms to not be pulled apart in the course of normal interactions with the environment. (3.1)

chemical energy. Energy stored within the structural units of chemical substances. (0.1)

chemical equation. An equation that uses chemical symbols to show what happens during a chemical reaction. (0.4)

chemical equilibrium. A state in which the rates of the forward and reverse reactions are equal. (10.1)

chemical formula. An expression showing the chemical composition of a compound in terms of the symbols for the atoms of the elements involved. (0.3)

chemical kinetics. The area of chemistry concerned with the speeds, or rates, at which chemical reactions occur. (14.1)

chemical property. Any property of a substance that cannot be studied without converting the substance into some other substance. (0.1)

chemical reaction. A process in which a substance (or substances) is changed into one or more new substances. (0.4)

chemistry. The study of matter and the changes it undergoes. (0.1)

chiral. Compounds or ions that are not superimposable with their mirror images. (22.4)

Clapeyron equation. An equation relating the slope (dP/dT) of a phase diagram boundary to the enthalpy and volume changes of the phase transition. (9.1)

Clausius-Clapeyron equation. An equation describing the temperature dependence of vapor pressure. (9.1)

Clausius inequality. The heat absorbed by a system in a process is always less than or equal to the change in entropy of the system during the process. (8.2)

closed system. A system that enables the exchange of energy (usually in the form of heat), but not of mass, with its surroundings. (6.2)

closest packing. The most efficient arrangements for packing atoms, molecules, or ions in a crystal. (6.2)

cohesion. The intermolecular attraction between like molecules. (6.1)

colligative properties. Properties of solutions that depend on the number of solute particles in solution and not on the nature of the solute particles. (9.4)

colloid. A dispersion of particles of one substance (the dispersed phase) throughout a dispersing medium made of another substance. (12.8)

common ion effect. The shift in equilibrium caused by the addition of a compound having an ion in common with the dissolved substances. (12.1)

complex ion. An ion containing a central metal cation bonded to one or more molecules or ions. (12.7)

compound. A substance composed of atoms of two or more elements chemically united in fixed proportions. (0.1)

compression factor. The quantity $Z = PV/nRT$. (5.3)

concentration equilibrium constant (K_c). The equilibrium constant for a reaction expressed as a ratio of reactant and product concentrations. (10.1)

concentration of a solution. The amount of solute present in a given quantity of solvent or solution. (0.5)

concentration cell. An electrochemical cell that generates a potential difference from a gradient in concentration. (13.5)

condensation. The phenomenon of going from the gaseous state to the liquid state. (5.2)

condensation reaction. A reaction in which two smaller molecules combine to form a larger molecule. Water is invariably one of the products of such a reaction. (16.3)

conducting polymer. A polymer that conducts electricity. (16.4)

conduction band. A band of closely spaced unfilled molecular orbitals just above the valence band. (6.4)

conductor. Substance capable of conducting electric current. (6.4)

conformations. Different spatial arrangements of a molecule that are generated by rotation about single bonds. (16.1)

conjugate acid-base pair. An acid and its conjugate base or a base and its conjugate acid. (11.1)

constant-pressure heat capacity (C_P). The heat capacity of a substance in a constant-pressure process. (7.3)

constant-pressure process. A process that takes place at constant pressure. (7.1)

constant-volume heat capacity (C_V). The heat capacity of a substance in a constant-volume process. (7.3)

constant-volume process. A process that takes place at constant volume. (7.1)

coordinate covalent bond. A bond in which the pair of electrons is supplied by one of the two bonded atoms; also called a dative bond. (3.4)

coordination compound. A neutral species containing one or more complex ions. (15.2)

coordination number. In a crystal lattice it is defined as the number of atoms (or ions) surrounding an atom (or ion) (6.2). In coordination compounds it is defined as the number of donor atoms surrounding the central metal atom in a complex. (15.2)

copolymer. A polymer containing two or more different monomers. (16.4)

core electrons. The inner shell electrons of an atom. (2.4)

Coulomb's law. The potential energy between two ions is directly proportional to the product of their charges and inversely proportional to the distance between them. (0.1)

coupled reactions. A pair of reactions in which a nonspontaneous reaction is driven by a spontaneous one. (8.6)

covalent bond. A bond in which two electrons are shared by two atoms. (3.2)

covalent compounds. Compounds containing only covalent bonds. (3.2)

covalent crystal. A crystal in which atoms are held together with an extensive three-dimensional network of covalent bonds. (6.3)

covalent radius. A measure of the size of an atom based on the lengths of the covalent bonds that it forms with other atoms. (3.2)

critical mass. The minimum mass of fissionable material required to generate a self-sustaining nuclear chain reaction. (17.5)

critical point. A point on a phase diagram where a coexistence line ends. (5.2)

critical pressure. The minimum pressure necessary to bring about liquefaction at the critical temperature. (5.2)

critical temperature. The temperature above which a gas will not liquefy. (5.2)

crystal field splitting (Δ). The energy difference between two sets of d orbitals in a metal atom when ligands are present. (15.3)

crystal field theory. A theory explaining the bonding in complex ions purely in terms of electrostatic forces. (15.3)

crystalline solid. A solid that possesses rigid and long-range order; its atoms, molecules, or ions occupy specific positions. (6.2)

crystallization. The process in which dissolved solute comes out of solution and forms crystals. (9.2)

cycloalkanes. Alkanes whose carbon atoms are joined in rings. (16.1)

D

Dalton's law of partial pressures. The total pressure of a mixture of gases is just the sum of the pressures that each gas would exert if it were present alone. (5.3)

deBroglie wavelength. The wavelength associated with a particle due to its momentum. (1.3)

degenerate orbitals. Atomic orbitals with equal energy. (1.4)

delocalized molecular orbitals. Molecular orbitals that are not confined between two adjacent bonding atoms but actually extend over three or more atoms. (4.5)

denatured protein. Protein that does not exhibit normal biological activities. (16.5)

density. The mass of a substance divided by its volume. (0.5)

deoxyribonucleic acids (DNA). A type of nucleic acid. (16.6)

deposition. The process in which the molecules go directly from the vapor into the solid phase. (5.2)

detailed balance. The principle that states that, at equilibrium, the rates of forward and reverse processes are equal. (14.3)

diagonal relationship. Similarities between pairs of elements in different groups and periods of the periodic table. (2.5)

diamagnetic. Repelled by a magnet; a diamagnetic substance contains only paired electrons. (2.1)

diatomic molecule. A molecule that consists of two atoms. (0.2)

diffusion. The gradual mixing of molecules of one gas with molecules of another by virtue of their kinetic motion. (5.5)

dipole-dipole interactions. Forces that act between polar molecules. (4.6)

dipole-induced dipole interaction. The interaction between a polar molecule with the dipole moment it induces in a nearby atom or molecule. (4.6)

dipole moment (μ). A vector measuring the degree of charge separation in a molecule. The magnitude (μ) is the product of charge separated times the charge separation distance in a molecule. The dipole moment vector points from the negative to the positive end of the molecule. (4.2)

diprotic acid. Each unit of the acid yields two hydrogen ions upon ionization. (11.1)

dispersion interactions. The attractive forces that arise as a result of temporary dipoles induced in the atoms or molecules; also called London forces. (4.6)

donor atom. The atom in a ligand that is bonded directly to the metal atom. (15.2)

donor impurities. Impurities that provide conduction electrons to semiconductors. (6.4)

double bond. Two atoms are held together by two pairs of electrons. (3.2)

duet. The two electrons in the electronic configuration in helium. (3.2)

dynamic equilibrium. The condition in which the rate of a forward process is exactly balanced by the rate of a reverse process. (5.2)

E

effective nuclear charge. The effective charge felt by an electron in an atom due to interactions with the nucleus and all other electrons. (2.5)

effusion. Escape of a gas through a small hole in a container. (5.5)

electrochemistry. The branch of chemistry that deals with the interconversion of electrical energy and chemical energy. (13.1)

electrode The parts of an electrochemical cell where electron transfer (oxidation or reduction) takes place. (13.2)

electrolysis. A process in which electrical energy is used to cause a nonspontaneous chemical reaction to occur. (13.7)

electrolyte. A substance that dissociates into ions in water to form a conducting solution. (9.4)

electrolytic cell. An apparatus for carrying out electrolysis. (13.7)

electromagnetic force. Force between electrically charged objects or magnetic materials. (0.1)

electromagnetic radiation. The emission and transmission of energy in the form of electromagnetic waves. (1.1)

electromagnetic wave. A wave that has an electric field component and a mutually perpendicular magnetic field component. (1.1)

electromotive force (emf). The voltage difference between electrodes. (13.2)

electron. A subatomic particle that has a very low mass and carries a single negative electric charge. (0.2)

electron affinity. The negative of the energy change when an electron is accepted by an atom in the gaseous state to form an anion. (2.5)

electron configuration. The distribution of electrons among the various orbitals in an atom or molecule. (2.1)

electron spin. An intrinsic angular momentum associated with an electron. (1.4)

electron spin quantum number. The quantum number (m_s) associated with electron spin. The value of m_s can be $+1/2$ (up) or $-1/2$ (down). (1.4)

electronegativity. The ability of an atom to attract electrons toward itself in a chemical bond. (3.3)

element. A substance that cannot be separated into simpler substances by chemical means. (0.1)

elementary steps. A series of simple reactions that represent the progress of the overall reaction at the molecular level. (14.5)

emission spectra. Continuous or line spectra emitted by substances. (1.2)

empirical formula. An expression showing the types of elements present and the simplest ratios of the different kinds of atoms. (0.3)

enantiomers. Optical isomers, that is, compounds and their nonsuperimposable mirror images. (4.4)

endothermic processes. Processes that absorb heat from the surroundings. (7.1)

end point. The pH at which the indicator changes color. (12.4)

energy. The capacity to do work or to produce change. (0.1)

energy level. One of the quantized energy states of a molecule or atom. (1.4)

energy shell. The set of degenerate orbitals making up a particular energy level in an atom. (1.4)

enthalpy (H). A thermodynamic quantity used to describe heat changes taking place at constant pressure. (7.2)

enthalpy of dilution (ΔH_{dil}). The enthalpy associated with the dilution process. (7.6)

enthalpy of hydration (ΔH_{hydr}). The enthalpy associated with the hydration process. (7.6)

enthalpy of reaction (ΔH). The difference between the enthalpies of the products and the enthalpies of the reactants. (7.2)

enthalpy of solution (ΔH_{soln}). The heat generated or absorbed when a certain amount of solute is dissolved in a certain amount of solvent. (7.6)

entropy (S). A direct measure of the number of molecular states consistent with the macroscopic state of a system. (8.1)

entropy of mixing. The entropy change associated with the mixing of two substances to form a mixture. (8.5)

enzyme. A biological catalyst. (14.6)

equation of state. The pressure of a material as a function of temperature, volume, and composition. (5.3)

equilibrium constant (K). A number equal to the ratio of the equilibrium concentrations

of products to the equilibrium concentrations of reactants, each raised to the power of its stoichiometric coefficient. (10.1)

equilibrium state. A thermodynamic state in which all macroscopic properties of the system are unchanging in time and remain so even if the system is disconnected from its surroundings. (7.1)

equilibrium vapor pressure. The vapor pressure measured under dynamic equilibrium of condensation and evaporation at some temperature. (5.2)

equipartition of energy theorem. The energy in a molecule is, on average, distributed evenly among all types of molecular motion (degrees of freedom). (7.3)

equivalence point. The point at which the acid has completely reacted with or been neutralized by the base. (12.3)

esters. Compounds that have the general formula R'COOR, where R' can be H or an alkyl group or an aromatic group and R is an alkyl group or an aromatic group. (16.3)

ether. An organic compound containing the R—O—R' linkage, where R and R' are alkyl and/or aromatic groups. (16.3)

evaporation. The process in which a liquid is transformed into a gas; also called vaporization. (5.2)

excess reagents. One or more reactants present in quantities greater than necessary to react with the quantity of the limiting reagent. (0.6)

excited state (or level). A state that has higher energy than the ground state. (1.2)

exothermic processes. Processes that give off heat to the surroundings. (7.1)

extensive property. A property that depends on how much matter is being considered. (0.1)

F

face-centered cubic (fcc). Having a crystal unit cell with lattice points both at the vertices of a cube and in the center of each face. (6.2)

Faraday constant. Charge contained in 1 mole of electrons, equivalent to 96,485.3 coulombs. (13.4)

first law of thermodynamics. Energy can be converted from one form to another, but cannot be created or destroyed. (7.1)

first-order reaction. A reaction whose rate depends on reactant concentration raised to the first power. (14.3)

force. The quantity that causes an object to change its course of motion (either in direction or speed). (0.1)

force constant. The proportionality constant between the force on a Hooke's-law spring and its displacement. (4.2)

formal charge. The difference between the valence electrons in an isolated atom and the number of electrons assigned to that atom in a Lewis structure. (3.4)

formation constant (K_f). The equilibrium constant for the complex ion formation. (12.7)

fractional crystallization. The separation of a mixture of substances into pure components on the basis of their different solubilities. (9.2)

fractional distillation. A procedure for separating liquid components of a solution that is based on their different boiling points. (9.3)

freezing. The transformation of a liquid into a solid. (5.2)

freezing point. The temperature at which the solid and liquid phases of a substance coexist at equilibrium. (5.2)

freezing-point depression (ΔT_f). The freezing point of the pure solvent ($T°f$) minus the freezing point of the solution (Tf). (9.4)

frequency (γ). The number of waves that pass through a particular point per unit time. (1.1)

frequency factor. The preexponential factor in the Arrhenius equation. (14.4)

fuel cell. A galvanic cell that requires a continuous supply of reactants to keep functioning. (13.6)

functional group. That part of a molecule characterized by a special arrangement of atoms that is largely responsible for the chemical behavior of the parent molecule. (16.1)

G

galvanic cell. The experimental apparatus for generating electricity through the use of a spontaneous redox reaction. (13.2)

gamma (γ) emission. Emission of very high-energy electromagnetic radiation by an unstable nucleus. (17.1)

gas. A substance that resists neither changes in volume nor changes in shape. (0.1)

Gay-Lussac's law. At a constant amount of gas and volume, the pressure of an ideal gas is proportional to temperature. (5.3)

geometric isomers. Compounds with the same type and number of atoms and the same chemical bonds but different spatial arrangements; such isomers cannot be interconverted without breaking a chemical bond. (4.4)

glass. An amorphous solid formed by supercooling a liquid well below its melting temperature. (11.7)

Gibbs free energy (G). A quantity equal to the enthalpy less the entropy times absolute temperature ($H - TS$). The maximum amount of nonexpansion that can be extracted from a process is equal to the change in Gibbs free energy for that process. (8.4)

Gibbs free energy of mixing. The Gibbs free energy change associated with the mixing of two substances to form a mixture. (8.5)

Gibbs phase rule. For a system with c components and r phases in equilibrium, the dimensionality of the phase coexistence region is given by $f = c - r + 2$. (5.2)

gravitational force. The force between objects due to their mass. (0.1)

ground state (or level). The lowest energy state of a system. (1.2)

group. The elements in a vertical column of the periodic table. (0.2)

H

half-cell reactions. Oxidation and reduction reactions at the electrodes. (13.2)

half-life ($t_{1/2}$). The time required for the concentration of a reactant to decrease to half of its initial concentration. (14.3)

half-reaction. A reaction that explicitly shows electrons involved in either oxidation or reduction. (13.1)

halogens. The nonmetallic elements in Group 7A (F, Cl, Br, I, and At). (0.2)

heat. Transfer of energy between two bodies that are at different temperatures. (7.1)

heat capacity (C). The amount of heat required to raise the temperature of a given quantity of the substance by one degree Celsius. (7.3)

heat of dilution. See *enthalpy of dilution.* (7.6)

heat of hydration. See *enthalpy of hydration.* (7.6)

heat of solution. See *enthalpy of solution.* (7.6)

heavy water reactor. A nuclear reactor that uses D_2O (heavy water) as a moderator. (17.5)

Heisenberg uncertainty principle. It is impossible to know simultaneously both the momentum and the position of a particle with certainty. (1.3)

Henderson-Hasselbalch equation. An equation relating the pH to the pK_a in terms of the concentrations (or activities) of an acid and its conjugate base. (12.1)

Henry's law. The solubility of a gas in a liquid is proportional to the pressure of the gas over the solution. (9.2)

Hess's law. When reactants are converted to products, the change in enthalpy is the same whether the reaction takes place in one step or in a series of steps. (7.4)

heterogeneous catalyst. A catalyst that is in a different physical phase than the reactants. (14.6)

heterogeneous equilibrium. An equilibrium state in which the reacting species are not all in the same phase. (10.1)

heterogeneous mixture. The individual components of a mixture remain physically separated and can be seen as separate components. (0.1)

high-boiling azeotrope. An azeotrope occurring at a minimum of the temperature-composition diagram. (9.3)

highest occupied molecular orbital (HOMO). The highest energy molecular orbital in a molecule that contains electrons. (4.5)

homogeneous equilibrium. An equilibrium state in which all reacting species are in the same phase. (10.1)

homogeneous catalysis. A catalyst that is in the same physical phase as the reactants. (14.6)

homogeneous mixture. The composition of a mixture, after sufficient stirring, is the same throughout the solution. (0.1)

homopolymer. A polymer that is made from only one type of monomer. (16.4)

Hund's rule. The most stable arrangement of electrons in subshells is the one with the greatest number of parallel spins. (2.2)

hybrid orbitals. Atomic orbitals obtained when two or more nonequivalent orbitals of the same atom combine. (4.3)

hybridization. The process of mixing the atomic orbitals in an atom (usually the central atom) to generate a set of new atomic orbitals. (4.3)

hydrates. Compounds that have a specific number of water molecules attached to them. (0.3)

hydration. Solvation by water. (9.2)

hydrocarbons. Compounds made up only of carbon and hydrogen. (16.1)

hydrogenation. The addition of hydrogen to compounds containing multiple bonds. (16.2)

hydrogen bond. A special type of dipole-dipole interaction between the hydrogen atom bonded to an atom of a very electronegative element (F, N, O) and another atom of one of the three electronegative elements. (4.6)

hydrogenlike ion. A monatomic ion with one electron. (1.2)

hydronium ion. The hydrated proton, H_3O^- (11.1)

hydrophilic. Water-liking. (9.2)

hydrophobic. Water-fearing. (9.2)

I

ideal gas. A hypothetical gas whose pressure-volume-temperature behavior can be completely accounted for by the ideal gas equation. (5.3)

ideal gas equation of state. An equation expressing the relationships among pressure, volume, temperature, and amount of gas ($PV = nRT$, where R is the gas constant). (5.3)

ideal solution. A solution for which the enthalpy of mixing is zero and the Gibbs free energy of mixing is the same as that for an ideal gas. An ideal liquid solution will obey Raoult's law. (8.5)

indicators. Substances that have distinctly different colors in acidic and basic media. (12.4)

induced dipole. The separation of positive and negative charges in a neutral atom (or a nonpolar molecule) caused by the proximity of an ion or a polar molecule. (4.6)

inert complex. A complex ion that undergoes very slow ligand exchange reactions. (15.4)

inert pair effect. The tendency of heavier elements in group 3A to form both 1+ and 2+ ions; similarly for group 4A elements to form 2+ and 4+ ions. (2.3)

infrared spectroscopy. The study of the infrared frequencies that are absorbed by a particular material. (4.2)

inorganic compounds. Compounds other than organic compounds. (0.3)

insulator. A substance incapable of conducting electricity. (6.4)

integrated rate law. An equation relating the concentration of a species directly to time. (14.3)

intensive property. A property that does not depend on how much matter is being considered. (0.1)

intermolecular interactions Attractive forces that exist between molecules. (4.6)

internal conversion. The process by which the excess energy of a nucleus is transferred to a core orbital electron, which is subsequently ejected from the atom. (17.1)

internal energy. Energy of a system that is left when the kinetic energy of overall motion is subtracted out. (7.1)

International System of Units (SI). A system of units based on metric units. (0.1)

intramolecular interaction Forces that hold atoms together in a molecule. (4.6)

ion. An atom or a group of atoms that has a net positive or negative charge. (0.2)

ion pair. One or more cations and one or more anions held together by electrostatic forces. (9.4)

ionic bond. The electrostatic force that holds ions together in an ionic compound. (3.3)

ionic compound. Any neutral compound containing cations and anions. (0.3)

ion-induced dipole interaction. The interaction between an ion and the dipole that it induces in a nearby atom or nonpolar molecule. (4.6)

ion-ion interactions. The electrostatic interactions between ions. (4.6)

ionic crystal. A crystal composed of ionic species. (6.3)

ionic radius. The radius of a cation or an anion as measured in an ionic compound. (2.5)

ionization. Removal of an electron (or electrons) from an atom, molecule, or ion. (1.2)

ionization energy. The minimum energy required to remove an electron from an isolated atom (or an ion) in its ground state. (1.2)

ion-dipole interactions. Forces that operate between an ion and a dipole. (4.6)

ion-product constant. Product of hydrogen ion concentration and hydroxide ion concentration (both in molarity) at a particular temperature. (11.2)

irreversible process. A process that is not reversible. An irreversible process run in the forward direction will not be the exact opposite of the process run in the reverse direction. (7.1)

isobaric process. See *constant-pressure process*. (7.1)

isochoric process. See *constant-volume process*. (7.2)

isoelectronic. Ions, ions and atoms, or ions and molecules that possess the same number of electrons, and hence the same ground-state electron configuration, are said to be isoelectronic. (2.3)

isolated system. A system that does not allow the transfer of either mass or energy to or from its surroundings. (7.1)

isomers. Two or more different compounds that share the same molecular formula. (4.4)

isothermal process. A process that takes place at constant temperature (7.1)

isotopes. Atoms having the same atomic number but different mass numbers. (0.2)

J

joule (J). Unit of energy given by kg m^2 s^{-2}. (0.1, A.1)

K

kelvin (K). The SI base unit of temperature. (A.1)

ketones. Compounds with a carbonyl functional group and the general formula $R'RCO$, where R and R' are alkyl and/or aromatic groups. (16.3)

kinetic energy (KE). Energy available because of the motion of an object. (0.1)

kinetic molecular theory of gases. Treatment of gas behavior in terms of the random motion of molecules. (5.4)

Kirchhoff's law. An equation (Eq. 7.49) governing the temperature dependence of enthalpy changes. (7.7)

L

labile complex. Complexes that undergo rapid ligand exchange reactions. (15.4)

lanthanide (rare earth) series. Elements that have incompletely filled 4f subshells or readily give rise to cations that have incompletely filled 4f subshells. (2.2)

latent heat. The enthalpy change for a phase transition. (7.6)

lattice constant. One of the edge lengths of a unit cell in a crystal lattice. (6.2)

lattice energy. The energy required to completely separate one mole of a solid ionic compound into gaseous ions. (7.6)

law of conservation of energy. The total quantity of energy in the universe is constant. (0.1)

law of conservation of mass. Matter can be neither created nor destroyed. (0.2)

law of definite proportions. Different samples of the same compound always contain its constituent elements in the same proportions by mass. (0.2)

law of mass action. For a reversible reaction at equilibrium and a constant temperature, a certain ratio of reactant and product concentrations has a constant value, K (the equilibrium constant). (10.1)

law of multiple proportions. If two elements can combine to form more than one type of compound, the masses of one element that combine with a fixed mass of the other element are in ratios of small whole numbers. (0.2)

Le Châtelier's principle. If an external stress is applied to a system at equilibrium, the system will adjust itself in such a way as to partially offset the stress as the system reaches a new equilibrium position. (10.4)

Lennard-Jones potential. A model for the interaction between nonpolar molecules that includes both a strong repulsion at short distances and a van der Waals interaction at long distances. (4.6)

lever rule. An equation for determining the fraction of different phases in a region of phase coexistence. (9.3)

Lewis acid. A substance that can accept a pair of electrons. (11.1)

Lewis base. A substance that can donate a pair of electrons. (11.1)

Lewis dot symbol. The symbol of an element with one or more dots that represent the number of valence electrons in an atom of the element. (3.2)

Lewis structure. A representation of covalent bonding using Lewis symbols. Shared electron pairs are shown either as lines or as pairs of dots between two atoms, and lone pairs are shown as pairs of dots on individual atoms. (3.2)

ligand. A molecule or an ion that is bonded to the metal ion in a complex ion. (15.2)

ligand field theory. A theory for bonding in coordination compounds based on molecular orbital theory. (15.3)

light water reactor. A nuclear reactor that uses H_2O as a moderator. (17.5)

limiting law. A scientific law that becomes exact only in some well-defined limit. (5.2)

limiting reagent. The reactant used up first in a reaction. (0.6)

line spectra. Spectra produced when radiation is absorbed or emitted by substances only at some wavelengths. (1.2)

linear combination of atomic orbitals (LCAO). A molecular orbital constructed as a linear combination of atomic orbitals. (3.5)

liquid. A substance that resists changes in volume but not of shape. (0.1)

liter. The volume occupied by one cubic decimeter. (A.1)

London interactions. See dispersion interactions.

lone pairs. Valence electrons that are not involved in covalent bond formation. (3.2)

low-boiling azeotrope. An azeotrope occurring at a minimum of the temperature-composition diagram. (9.3)

lowest unoccupied molecular orbital (LUMO). The lowest-energy molecular orbital in a molecule that does not contain electrons. (4.5)

M

macromolecules. Molecules with very large molecular masses, for example, polymers, proteins, and DNA. (16.4)

macroscopic Pertaining to objects that are large compared to the molecular scale. (5.1)

macrostate. See *thermodynamic state*. (7.1)

magnetic quantum number. The quantum number m_l that determines the orientation of an atomic orbital. (1.4)

manometer. A device used to measure the pressure of gases. (5.1)

many-electron atoms. Atoms that contain two or more electrons. (2.1)

mass. A measure of the quantity of matter contained in an object. (0.1)

mass defect. The difference between the mass of an atom and the sum of the masses of its protons, neutrons, and electrons. (17.2)

mass number (A). The total number of neutrons and protons present in the nucleus of an atom. (0.2)

mass spectrometer. Determination of the masses of atoms, molecules, and ions using a mass spectrometer, a device that uses magnetic fields to separate atomic and molecular species by mass. (0.5).

matter. Anything that occupies space and possesses mass. (0.1)

Maxwell-Boltzmann distribution of molecular speeds. The distribution of molecular speeds in a gas, given by Equation 5.36. (5.4)

melting. The transformation of a solid into a liquid. (5.2)

melting point. The temperature at which solid and liquid phases coexist in equilibrium. (5.2)

metallic crystal. A crystal lattice held together by metallic bonds. (6.3)

metalloid. An element with properties intermediate between those of metals and nonmetals. (0.2)

metals. Elements that are good conductors of heat and electricity and have the tendency to form positive ions in ionic compounds. (0.2)

microstate. See *molecular state.* (7.1)

microwave spectroscopy. The study of the microwave frequencies that are absorbed by a material. (3.3)

miscible. Two liquids that are completely soluble in each other in all proportions are said to be miscible. (9.2)

mixture. A combination of two or more substances in which the substances retain their identity. (0.1)

moderator. A substance that can reduce the kinetic energy of neutrons. (17.5)

molal boiling-point elevation constant (K_b). The constant of proportionality between the molality of a dilute solution and its boiling point. (9.4)

molal freezing-point depression constant (K_f). The constant of proportionality between the molality of a dilute solution and its freezing point. (9.4)

molality (m). The number of moles of solute dissolved in one kilogram of solvent. (0.5)

molar concentration. See *molarity.*

molar enthalpy of fusion(ΔH_{fus}). The enthalpy change associated with the melting (fusion) of 1 mole of a liquid. (7.6)

molar enthalpy of sublimation (ΔH_{sub}). The enthalpy change associated with the sublimation of 1 mole of a liquid. (7.6)

molar enthalpy of vaporization (ΔH_{vap}). The enthalpy change associated with the vaporization of 1 mole of a liquid. (7.6)

molar heat capacity ($\bar{C}$). The heat capacity of 1 mole of a substance. (7.3)

molar mass (M). The mass (in grams or kilograms) of one mole of atoms, molecules, or other particles. (0.5)

molar solubility. The number of moles of solute in one liter of a saturated solution (mol L^{-1}). (12.6)

molarity (M). The number of moles of solute in one liter of solution. (0.5)

mole (mol). The amount of substance that contains as many elementary entities (atoms, molecules, or other particles) as there are atoms in exactly 12 grams (or 0.012 kilograms) of the carbon-12 isotope. (0.5)

molecular crystal. A crystal in which the molecules are held together by attractive intermolecular (van der Waals) interactions. (6.3)

molecular formula. An expression showing the exact numbers of atoms of each element in a molecule. (0.3)

molecular geometry. The three-dimensional arrangement of atoms in a molecule. (4.1)

molecular mass. The sum of the atomic masses (in amu) present in the molecule. (0.5)

molecular model. A three-dimensional representation of a molecule. (0.3)

molecular orbital. An orbital that is not associated with a particular atom but extends over the entire molecule or ion. (3.5)

molecular orbital energy-level diagram (MO diagram). A diagram illustrating the molecular orbitals of a molecule in terms of their relative energy. (3.5)

molecular orbital theory. The use of molecular orbitals to described bonding in ions and molecules. (3.5)

molecularity of a reaction. The number of molecules reacting in an elementary step. (14.5)

molecular speed. The magnitude of a velocity of a molecule. (5.4)

molecular state. A description of a system in terms of the properties (for example, positions and velocities) of its constituent molecules. (7.1)

molecule. An aggregate of at least two atoms in a definite arrangement held together by special forces. (0.2)

monatomic ion. An ion that contains only one atom. (0.2)

monomer. The single repeating unit of a polymer. (16.4)

monoprotic acid. Each unit of the acid yields one hydrogen ion upon ionization. (11.1)

multiple bonds. Bonds formed when two atoms share two or more pairs of electrons. (3.2)

N

Nernst equation. The relation between the emf of a galvanic cell and the standard emf and the concentrations of the oxidizing and reducing agents. (13.5)

net ionic equation. An equation that indicates only the ionic species that actually take part in the reaction. (4.2)

neutralization reaction. A reaction between an acid and a base. (11.1)

neutron. A subatomic particle that bears no net electric charge. Its mass is slightly greater than a proton's. (0.2)

Newman projection. A scheme for illustrating conformations in organic molecules. (16.1)

newton (N). The SI unit for force. (0.1, A.1)

noble gas core. The electron configuration of the noble gas element that most nearly precedes the element being considered. (2.2)

noble gases. Nonmetallic elements in Group 8A (He, Ne, Ar, Kr, Xe, and Rn). (2.4)

node. The point at which the amplitude of the wave is zero. (1.3)

nonelectrolytes. Molecules that do not dissociate in aqueous solution. (9.4)

nonmetals. Elements that are usually poor conductors of heat and electricity. (0.2)

nonpolar molecule. A molecule with a zero dipole moment. (4.2)

nonvolatile. Liquids not having a measurable vapor pressure. (9.4)

normal boiling point. The temperature at which the vapor pressure of a liquid is equal to 1 atm. (5.2)

normal melting point. The melting point of a substance at 1 atm pressure. (5.2)

***n*-type semiconductors.** Semiconductors that contain donor impurities. (6.4)

nuclear binding energy. The energy required to break up a nucleus into its protons and neutrons. (17.2)

nuclear chain reaction. A self-sustaining sequence of nuclear fission reactions. (17.5)

nuclear fission. A heavy nucleus (mass number 200) divides to form smaller nuclei of intermediate mass and one or more neutrons. (17.5)

nuclear fusion. The combining of small nuclei into larger ones. (17.6)

nuclear magnetic resonance (NMR) spectroscopy. A type of spectroscopy that studies the excitation of nuclear spin with an electromagnetic field (radio waves). (16.3)

nuclear transmutation. The change undergone by a nucleus as a result of bombardment by neutrons or other particles. (17.1)

nucleic acids. High molar mass polymers that play an essential role in protein synthesis. (16.6)

nucleon. A general term for the protons and neutrons in a nucleus. (17.2)

nucleotide. The repeating unit in each strand of a DNA molecule which consists of a base-deoxyribose-phosphate linkage. (16.6)

nucleus. The central core of an atom. (0.2)

O

octet rule. An atom other than hydrogen tends to form bonds until it is surrounded by eight valence electrons. (3.2)

open system. A system that can exchange mass and energy (usually in the form of heat) with its surroundings. (7.1)

optical isomers. Compounds that are nonsuperimposable mirror images. (4.4)

orbital. See atomic orbital and molecular orbital.

orbital electron capture. Radioactive decay process closely related to β^+ decay in which an orbital electron is captured by the nucleus and combined with a proton to produce a neutron. (17.1)

orbital energy. The negative of the amount of energy required to remove an electron from a particular atomic or molecular orbital. (2.2)

orbital diagram. A pictorial representation of the electron configuration that shows the spin of the electron. (2.1)

organic chemistry. The branch of chemistry that deals with carbon compounds. (16.1)

organic compounds. Compounds that contain carbon, usually in combination with elements such as hydrogen, oxygen, nitrogen, and sulfur. (0.3)

orientation factor. A factor used in the collision theory of reaction rates that reflects the probability of two molecules colliding in an orientation that is favorable to reaction. (14.4)

osmosis. The net movement of solvent molecules through a semipermeable membrane from a pure solvent or from a dilute solution to a more π concentrated solution. (9.4)

osmotic pressure (Π). The pressure required to stop osmosis. (9.4)

overvoltage. The difference between the electrode potential and the actual voltage required to cause electrolysis. (13.7)

oxidation. The loss of electrons by a species in a reaction. (13.1)

oxidation number. The number of charges an atom would have in a molecule if electrons were transferred completely in the direction indicated by the difference in electronegativity. (13.1)

oxidation reaction. The half-reaction that involves the loss of electrons. (13.1)

oxidation-reduction reaction. A reaction that involves the transfer of electron(s) or the change in the oxidation state of reactants. (13.1)

oxidation state. See oxidation number.

oxidizing agent. A substance that can accept electrons from another substance or increase the oxidation numbers in another substance. (13.1)

oxoacid. An acid containing hydrogen, oxygen, and another element (the central element). (0.3)

oxoanion. An anion derived from an oxoacid. (0.3)

P

paramagnetic. Attracted by a magnet. A paramagnetic substance contains one or more unpaired electrons. (2.1)

partial pressure. Pressure of one component in a mixture of gases. (5.3)

particle-in-a-box model. A simple model of a quantum particle that is confined completely within a region of space. (1.3)

pascal (Pa). A pressure of one newton per square meter (1 N m^{-2}). (5.1, A.1)

Pauli exclusion principle. No two electrons in an atom can have the same four quantum numbers. (2.1)

percent by mass. The ratio of the mass of a solute to the mass of the solution, multiplied by 100 percent. (0.5)

percent composition by mass. The percent by mass of each element in a compound. (0.5)

percent ionization. Ratio of ionized acid concentration at equilibrium to the initial concentration of acid. (11.4)

percent yield. The ratio of actual yield to theoretical yield, multiplied by 100 percent. (0.6)

period. A horizontal row of the periodic table. (0.2)

periodic table. A tabular arrangement of the elements. (0.2)

pH. The negative logarithm of the hydrogen ion concentration. (11.2)

phase. A homogeneous part of a system in contact with other parts of the system but separated from them by a well-defined boundary. (5.2)

phase transition. Transformation from one phase to another. (5.2)

phase boundary. The boundaries separating different phases in a phase diagram. (5.2)

phase diagram. A diagram showing the conditions at which a substance exists as a solid, liquid, or vapor. (5.2)

photoelectric effect. A phenomenon in which electrons are ejected from the surface of certain metals exposed to light of at least a certain minimum frequency. (1.1)

photoelectron spectroscopy. An experimental technique to determine the orbital energies of an atom or molecule. (2.2)

photons. Quanta of electromagnetic radiation. (1.1)

physical equilibrium. An equilibrium in which only physical properties change. (5.2)

physical property. Any property of a substance that can be observed without transforming the substance into some other substance. (0.1)

pi bond (π). A covalent bond between two atoms in which the electron wavefunction between the bonded atoms changes sign when rotated by 180° about the bond axis. (3.2)

pi molecular orbital. A molecular orbital in which the electron density is concentrated above and below the line joining the two nuclei of the bonding atoms. (3.5)

plasma. A gaseous mixture of positive ions and electrons. (17.6)

polar covalent bond. In such a bond, the electrons spend more time in the vicinity of one atom than the other. (3.3)

polar molecule. A molecule with a nonzero dipole moment. (4.2)

polarimeter. The instrument for measuring the rotation of polarized light by optical isomers. (4.4)

polarizability. A measure of the degree to which the electron cloud around an atom, molecule, or ion can be deformed in the presence of an external electric field. (4.6)

polyatomic ion. An ion that contains more than one atom. (0.2)

polyatomic molecule. A molecule that consists of more than two atoms. (0.2)

polymer. A compound distinguished by a high molar mass, ranging into thousands and millions of grams, and made up of many repeating units. (16.4)

polymorphism. The existence of a solid in more than one crystal form. (5.2)

positron. A particle that has the same mass as the electron, but bears a charge of +e. (17.1)

potential energy. Energy available by virtue of an object's position. (0.1)

precipitate. An insoluble solid that separates from the solution. (12.5)

precipitation reaction. A reaction that results in the formation of a precipitate. (12.5)

precision. The closeness of agreement of two or more measurements of the same quantity. (A.1)

pressure. Force applied per unit area. (5.1)

pressure-composition diagram. A fixed temperature phase diagram for a binary mixture. (9.3)

principal quantum number. The quantum number n that determines the size (and to some extent) the energy of an atomic orbital. (1.4)

product. The substance formed as a result of a chemical reaction. (0.4)

protein. Polymers of amino acids. (16.5)

proton. A subatomic particle having a single positive electric charge. The mass of a proton is about 1840 times that of an electron. (0.2)

pseudo-first-order reaction. A second-order reaction in which one reactant is present in such a large concentration that the rate law appears to be first order. (14.3)

_p_-type semiconductors. Semiconductors that contain acceptor impurities. (6.4)

Q

qualitative analysis. The determination of the types of ions present in a solution. (12.8)

quantitative analysis. The determination of the amount of substances present in a sample. (12.8)

quantum. The smallest quantity of energy that can be emitted (or absorbed) in the form of electromagnetic radiation. (1.1)

quantum-mechanical tunneling. The ability of a quantum particle to exist in regions of space that are energetically forbidden in classical mechanics. (1.3)

quantum mechanics. The study of the wave-like properties of matter at the molecular scale. (1.3)

quantum numbers. Numbers that describe the distribution of electrons in hydrogen and other atoms. (1.2)

R

racemic mixture. An equimolar mixture of the two enantiomers. (4.4)

radial distribution function. (1) A plot of the distribution of electron probability about the nucleus; (2) In liquids, it is a

measure of the local density about a chosen central atom. (6.1)

radial probability function. A function $P(r)$ that shows the probability of finding an electron at a distance r away from the nucleus. (1.4)

radiant energy. Energy transmitted in the form of waves. (0.1)

radical. Any neutral fragment of a molecule containing an unpaired electron. (17.8)

radioactive decay series. A sequence of nuclear reactions that ultimately result in the formation of a stable isotope. (17.2)

radioactivity. The spontaneous breakdown of an atom by emission of particles and/or radiation. (17.1)

Raoult's law. The partial pressure of the solvent over a solution is given by the product of the vapor pressure of the pure solvent and the mole fraction of the solvent in the solution. (9.3)

rapid preequilibrium. A reaction in which the mechanism contains a rapid and reversible step prior to the rate determining step. (14.5)

rare earth series. See *lanthanide series.*

rate constant (k). Constant of proportionality between the reaction rate and the concentrations of reactants. (14.1)

rate law. An expression relating the rate of a reaction to the rate constant and the concentrations of the reactants. (14.2)

rate-determining step. The slowest step in the sequence of steps leading to the formation of products. (14.3)

reactants. The starting substances in a chemical reaction. (0.4)

reaction intermediate. A species that appears in the mechanism of the reaction (that is, the elementary steps) but not in the overall balanced equation. (14.5)

reaction mechanism. The sequence of elementary steps that leads to product formation. (14.5)

reaction order. The sum of the powers to which all reactant concentrations appearing in the rate law are raised. (14.2)

reaction quotient (Q). A number equal to the ratio of product concentrations to reactant concentrations, each raised to the power of its stoichiometric coefficient at some point other than at equilibrium. (10.2)

reaction rate. The change in the concentration of reactant or product with time. (14.1)

redox reaction. A reaction in which there is either a transfer of electrons or a change in the oxidation numbers of the substances taking part in the reaction. (13.1)

reducing agent. A substance that can donate electrons to another substance or decrease the oxidation numbers in another substance. (13.1)

reduction. The gain of electrons by a species in a reaction. (13.1)

reduction reaction. The half-reaction that involves the gain of electrons. (13.1)

representative elements. Elements in Groups 1A through 7A, all of which have incompletely filled *s* or *p* subshells of highest principal quantum number. (2.4)

residual entropy. The value of the entropy at 0 K for systems for which the third law of thermodynamics is not applicable. (8.3)

resonance. The use of two or more Lewis structures to represent a particular molecule. (3.4)

resonance structure. One of two or more alternative Lewis structures for a molecule that cannot be described fully with a single Lewis structure. (3.4)

reversible process. A process that proceeds so slowly that the system always remains infinitesimally close to equilibrium with itself and its surroundings. A reversible process run in the forward direction will be the exact opposite of the process run in the reverse direction. (7.1)

ribonucleic acid (RNA). A form of nucleic acid. (16.6)

root-mean-square (rms) speed (u_{rms}). A measure of the average molecular speed at a given temperature. (5.7)

S

salt. An ionic compound made up of a cation other than H^+ and an anion other than OH^- or O_2^-. (11.1)

salt hydrolysis. The reaction of the anion or cation, or both, of a salt with water. (11.16)

saponification. Soapmaking. (16.3)

saturated hydrocarbons. Hydrocarbons that contain the maximum number of hydrogen atoms that can bond with the number of carbon atoms present. (16.1)

saturated solution. At a given temperature, the solution that results when the maximum amount of a substance has dissolved in a solvent. (9.2)

scanning tunneling microscope (STM). A molecular-level imaging device that uses quantum-mechanical tunneling to produce an image of the atoms at the surface of a sample. (1.4)

second law of thermodynamics. The entropy of any isolated system increases in a spontaneous process and remains unchanged in an equilibrium (reversible) process. (8.1)

second-order reaction. A reaction whose rate depends on reactant concentration raised to the second power or on the concentrations of two different reactants, each raised to the first power. (14.3)

self-consistent field (SCF) method. An iterative method for optimizing the atomic (or molecular) orbitals in a multielectron atom (or molecule). (2.1)

semiconductors. Elements that normally cannot conduct electricity, but can have their conductivity greatly enhanced either by raising the temperature or by adding certain impurities. (6.4)

semipermeable membrane. A membrane that enables solvent molecules to pass through, but blocks the movement of solute molecules. (9.4)

shell model. A model of the nucleus in which the nucleons in a nucleus populate independent quantum orbitals in much the same way as electrons occupy atomic orbitals in an atom. (17.2)

sigma bond (σ). A covalent bond between two atoms for which the electron wavefunction in the bonding region is cylindrically symmetric about the bond axis. (3.2)

sigma molecular orbital. A molecular orbital in which the electron density is concentrated around a line between the two nuclei of the bonding atoms. (3.5)

significant figures. The number of meaningful digits in a measured or calculated quantity. (A.1)

simple cubic (SC). Having a crystal unit cell with lattice points at the vertices of a cube. (6.2)

single bond. Two atoms are held together by one electron pair. (3.2)

solid. A substance that resists changes in both volume and shape. (0.1)

solubility. The maximum amount of solute that can be dissolved in a given quantity of solvent at a specific temperature. (9.2,12.6)

solubility product (K_{sp}). The product of the molar concentrations of the constituent ions, each raised to the power of its stoichiometric coefficient in the equilibrium equation. (12.6)

solute. The substance present in smaller amount in a solution. (0.1)

solution. A homogeneous mixture of two or more substances. (0.1)

solvation. The process in which an ion or a molecule is surrounded by solvent molecules arranged in a specific manner. (9.2)

solvent. The substance present in larger amount in a solution. (0.1)

solvent leveling. The phenomenon that strong acids will yield identical H_3O^+ concentrations in water, independent of their individual K_a values. (11.3)

specific heat capacity (C_s). The heat capacity of 1 gram of a substance. (7.3)

spectator ions. Ions that are not involved in the overall reaction. (11.1)

spectrochemical series. A list of ligands arranged in increasing order of their abilities to split the *d*-orbital energy levels. (15.3)

spontaneous fission. The spontaneous decay of a heavy unstable nucleus into two or more smaller nuclei and some neutrons. (17.1)

standard atmospheric pressure (atm). A unit of pressure equal to 1.01325 bar, exactly. (5.1)

spontaneous process. A process that occurs without external intervention under a given set of conditions. (8.1)

standard ambient temperature and pressure (SATP). 25°C and 1 bar. (5.3)

standard boiling point. The temperature at which the vapor pressure of a liquid is equal to 1 bar. (5.2)

standard emf. The difference of the standard reduction potential of the substance that undergoes reduction and the standard reduction potential of the substance that undergoes oxidation. (13.3)

standard enthalpy of reaction. The enthalpy change when the reaction is carried out under standard-state conditions. (7.4)

standard entropy change of reaction. The entropy change when the reaction is carried out under standard-state conditions. (8.3)

standard freezing point. The temperature at which a solid and its melt coexist at 1 bar pressure. (5.2)

standard Gibbs free-energy of reaction. The Gibbs free-energy change when the reaction is carried out under standard-state conditions. (8.4)

standard Gibbs free-energy of formation. The Gibbs free-energy change when 1 mole of a compound is synthesized from its elements in their standard states. (8.4)

standard melting point. The melting point of a substance at 1 bar pressure. (5.2)

standard molar enthalpy of formation (ΔH_f°). The enthalpy change that results when 1 mole of a compound in its standard state is formed from its elements in their standard states.

standard molar entropy. Absolute entropies measured at the standard pressure of 1 bar. (8.3)

standard reduction potential. The voltage measured as a reduction reaction occurs at the electrode when all solutes are 1 M and all gases are at 1 atm. (13.3)

standard solution. A solution of accurately known concentration. (12.3)

standard state. The condition of 1 atm of pressure. (7.4)

standard temperature and pressure (STP). 0°C and 1 atm. (5.3)

standing wave. A wave that does not propagate through space. (1.3)

state function. A property that is determined by the state of the system. (7.1)

statistical mechanics. The study of the molecular origin of thermodynamics. (7.3)

steady state. A thermodynamic state in which all macroscopic properties of the system are unchanging in time, but requires a continual inflow and outflow of energy and/or matter to remain unchanging. (7.1)

steady-state approximation. The approximation that the concentration of reaction intermediates is constant (in steady state) during a reaction. (14.5)

steady-state rate law. The rate law resulting from the steady-state approximation. (14.5)

stereoisomers. Compounds that are made up of the same types and numbers of atoms bonded together in the same sequence but with different spatial arrangements. (4.4)

steric factor. See *orientation factor.* (14.4)

steric number. The total number of electron pair sets surrounding a central atom in a molecule. (4.1)

Stock system. A systematic method for naming inorganic compounds (0.3)

stoichiometric amounts. The exact molar amounts of reactants and products that appear in the balanced chemical equation. (0.6)

stoichiometry. The quantitative study of reactants and products in a chemical reaction. (0.6)

stoichiometric coefficient. The constant multiplying a given element or compound in a chemical equation. (0.4)

strong acids. Strong electrolytes that are assumed to ionize completely in water. (11.1)

strong bases. Strong electrolytes that are assumed to ionize completely in water. (11.1)

strong electrolyte. A species that, for all practical purposes, dissociates completely in aqueous solution. (11.1)

strong force. The force that binds protons and neutrons together in the nucleus. (0.1)

structural formula. A chemical formula that shows how atoms are bonded to one another in a molecule. (0.3)

structural isomers. Molecules that have the same molecular formula but different structures. (4.4)

sublimation. The process in which molecules go directly from the solid into the vapor phase. (5.2)

subshell. The set of distinct orbitals within an energy shell that possess the same value of the quantum number l. (1.4)

substance. A form of matter that has a definite or constant composition (the number and type of basic units present) and distinct properties. (0.1)

substitution reaction. A reaction in which an atom or group of atoms replaces an atom or groups of atoms in another molecule. (16.3)

supercooling. Cooling of a liquid below its freezing point without forming the solid. (6.3)

supercritical fluid (SCF). A substance above its critical pressure and temperature. (5.2)

supersaturated solution. A solution that contains more of the solute than is present in a saturated solution. (9.2)

surface tension. The amount of energy required to stretch or increase the surface of a liquid by a unit area. (6.1)

surroundings. The rest of the universe outside a system. (7.1)

system. Any specific part of the universe that is of interest to us. (7.1)

T

temperature. The quantity that controls the direction of energy flow between two systems. (5.1)

temperature-composition phase diagram. A fixed pressure phase diagram for a binary mixture. (9.3)

termolecular reaction. Elementary reactions involving three reactant molecules. (14.5)

ternary compounds. Compounds consisting of three elements. (0.3)

theoretical yield. The amount of product predicted by the balanced equation when all of the limiting reagent has reacted. (0.6)

thermal energy. Energy associated with the random motion of atoms and molecules. (0.1)

thermal equilibrium. Equilibrium with respect to heat flow. (5.1)

thermochemical equation. An equation that shows both the mass and enthalpy relations. (7.2)

thermochemistry. The study of heat changes in chemical reactions. (7.1)

thermodynamics. The scientific study of the interconversion of heat and other forms of energy. (7.1)

thermodynamic state. A description of a system in terms of macroscopic variables (temperature, volume, pressure, etc.). (7.1)

thermometer. A device for measuring temperature. (5.1)

thermonuclear reactions. Nuclear fusion reactions that occur at very high temperatures. (17.6)

third law of thermodynamics. The entropy of a pure substance in its thermodynamically most stable form is zero at the absolute zero of temperature. (8.3)

titration. The gradual addition of a solution of accurately known concentration to another solution of unknown concentration until the chemical reaction between the two solutions is complete. (12.3)

tracers. Isotopes, especially radioactive isotopes, that are used to trace the path of the atoms of an element in a chemical or biological process. (17.7)

transition elements. Elements that have incompletely filled d subshells or readily give rise to cations that have incompletely filled d subshells. (2.2)

transition metals. See *transition elements.* (2.2)

transition state. See activated complex. (14.4)

transition-state theory. A theory for reaction rates that assumes that the reactants are in equilibrium with the transition state during the reaction. (14.4)

transuranium elements. Elements with atomic numbers greater than 92. (17.4)

triple bond. Two atoms are held together by three pairs of electrons. (3.2)

triple point. The point at which the vapor, liquid, and solid states of a substance are in equilibrium. (5.2)

triprotic acid. Each unit of the acid yields three protons upon ionization. (11.1)

U

unimolecular reaction. An elementary step in which only one reacting molecule participates. (14.5)

unit cell. The basic repeating unit of the arrangement of atoms, molecules, or ions in a crystalline solid. (6.2)

universal gas constant. The constant R that appears in the ideal gas equation. It is usually expressed as $0.08314 \text{ L mol}^{-1} \text{K}^{-1}$ or $8.314 \text{ J mol}^{-1} \text{K}^{-1}$. R is equal to Boltzmann's constant times Avagadro's number. (5.2)

unsaturated hydrocarbons. Hydrocarbons that contain carbon-carbon double bonds or carbon-carbon triple bonds. (16.2)

unsaturated solution. A solution that contains less solute than it has the capacity to dissolve. (9.2)

V

valence band. A band of closely spaced filled orbitals in a crystalline material. (6.4)

valence bond theory. A theory of bonding in which bonds are formed by the overlap of unfilled atomic orbitals on the bonded atoms. (3.2)

valence electrons. The outer electrons of an atom, which are those involved in chemical bonding. (2.4)

valence shell. The outermost electron-occupied shell of an atom, which holds the electrons that are usually involved in bonding. (4.1)

valence-shell electron-pair repulsion (VSEPR) model. A model that accounts for the geometrical arrangements of shared and unshared electron pairs around a central atom in terms of the repulsions between electron pairs. (4.1)

van der Waals equation. An equation that describes the P, V, and T of a nonideal gas. (5.5)

van der Waals interactions. Intermolecular interactions that fall off as $1/r^6$. Includes dispersion forces as well as dipole-dipole and dipole-induced dipole interactions in liquids and gases. (4.6)

van't Hoff equation. An equation (Eq. 10.16) relating the temperature dependence of an equilibrium constant with its enthalpy of reaction. (10.4)

van't Hoff factor. The ratio of actual number of particles in solution after dissociation to the number of formula units initially dissolved in solution. (9.4)

vapor-pressure lowering. The lowering of the vapor pressure of a liquid through the addition of a solute. (9.4)

vaporization. The escape of molecules from the surface of a liquid; also called evaporation. (5.2)

virial equation of state. An equation of state for real gases given by Equation 5.44. (5.5)

viscosity. A measure of a fluid's resistance to flow. (6.1)

volatile. Has a measurable vapor pressure. (9.4)

volume. It is the length cubed. (0.1)

W

wave. A vibrating disturbance by which energy is transmitted. (1.1)

wavefunction. A function that describes the quantum state of a particle. (1.3)

wavelength (λ). The distance between identical points on successive waves. (1.1)

weak acids. Weak electrolytes that ionize only to a limited extent in water. (11.1)

weak bases. Weak electrolytes that ionize only to a limited extent in water. (11.3)

weak force. The force responsible for the beta decay of radioactive nuclei. (0.1, 17.2).

work. Directed energy change resulting from a process. (7.1)

X

X-ray diffraction. The scattering of X rays by the units of a regular crystalline solid. (6.2)

Z

zero-point energy. The energy of the lowest energy level of a quantum system. (1.3)

zeroth law of thermodynamics. If two systems are in thermal equilibrium with a third system, then they are also in thermal equilibrium with each other. (5.1)

Answers
to Even-Numbered Problems

Chapter 0

0.2 (a) Physical. (b) Chemical. (c) Physical. (d) Chemical. (e) Chemical. **0.4** (a) homogeneous mixture. (b) element. (c) compound. (d) homogeneous mixture. (e) heterogeneous mixture. (f) homogeneous mixture. (g) heterogeneous mixture. **0.6** 8.1 kJ. **0.8** 24.5 m s^{-1}. **0.10** (a) Isotope $^4_2$He, $^{20}_{10}$Ne, $^{40}_{18}$Ar, $^{84}_{36}$Kr, $^{132}_{54}$Xe, $^{222}_{86}$Rn; No. protons 2, 10, 18, 36, 54, 86; No. neutrons 2, 10, 22, 48, 78, 136. (b) n/p ratio 1.00, 1.00, 1.22, 1.33, 1.44, 1.58. The n/p ratio increases with increasing atomic number. (c) This trend does hold for other elements. **0.12** 0.20 km. **0.14** 145. **0.16** Isotope $^{15}_7$N, $^{33}_{16}$S, $^{63}_{29}$Cu, $^{84}_{38}$Sr, $^{130}_{56}$Ba, $^{186}_{74}$W, $^{202}_{80}$Hg; No. protons 7, 16, 29, 38, 56, 74, 80; No. neutrons 8, 17, 34, 46, 74, 112, 122; No. electrons 7, 16, 29, 38, 56, 74, 80. **0.18** (a) $^{186}_{74}$W. (b) $^{201}_{80}$Hg. **0.20** (a) H$_2$ and N$_2$. (b) CO and HF. (c) S$_8$ and C$_{60}$. (d) HClO and HIO. **0.22** Ion K$^+$, Mg^{2+}, Fe^{3+}, Br$^-$, Mn^{2+}, C^{4-}, Cu^{2+}, No. protons 19, 12, 26, 35, 25, 6, 29; No. electrons 18, 10, 23, 36, 23, 10, 27. **0.24** The molecular formula of ethanol is C$_2$H$_6$O. **0.26** (a) potassium dihydrogen phosphate. (b) potassium hydrogen phosphate. (c) hydrogen bromide (molecular compound). (d) lithium carbonate. (e) sodium dichromate. (f) ammonium sulfate. (g) iodic acid. (h) phosphorus pentafluoride. (i) tetraphosphorus hexoxide. (j) cadmium iodide. (k) strontium sulfate. (l) aluminum hydroxide. **0.28** (a) cobalt(II) nitrate hexahydrate. (b) nickel(II) bromide trihydrate. (c) zinc(II) sulfate heptahydrate. (d) sodium sulfate monohydrate. **0.30** (a) Na$_2$SO$_4$·10H$_2$O. (b) CaSO$_4$·H$_2$O. (c) MgCl$_2$·6H$_2$O. (d) Zn(NO$_3$)$_2$·6H$_2$O. **0.32** (a) 2C + O$_2$→2CO. (b) 2CO + O$_2$→ 2CO$_2$. (c) H$_2$ + Br$_2$→2HBr. (d) 2H$_2$O$_2$→2H$_2$O + O$_2$. (e) 2C$_4$H$_{10}$ + 13O$_2$→8CO$_2$ + 10H$_2$O. (f) 3KOH + H$_3$PO$_4$→K$_3$PO$_4$ + 3H$_2$O. (g) N$_2$ + 3H$_2$→2NH$_3$. (h) NH$_4$NO$_3$→N$_2$O + 2H$_2$O. (i) 2C$_6$H$_6$ + 15O$_2$→12CO$_2$ + 6H$_2$O. (j) 3Na$_2$O + 2Al→Al$_2$O$_3$ + 6Na. **0.34** 35.45 u. **0.36** (a) 16.04 u. (b) 46.01 u. (c) 80.07 u. (d) 78.11 u. (e) 149.9 u. (f) 174.27 u. (g) 310.18 u. **0.38** 9.96 × 10^{-15} mol. **0.40** 3.44 × 10^{-10} g Pb. **0.42** (a) 73.89 g. (b) 76.15 g. (c) 119.37 g. (d) 176.12 g. (e) 101.11 g. (f) 352.0 g. (g) 84.01 g. **0.44** 3.01 × 10^{22} C atoms; 3.01 × 10^{22} O atoms; 6.02 × 10^{22} H atoms. **0.46** 8.56 × 10^{22} molecules. **0.48** (a) CH$_2$O. (b) KCN. **0.50** The molecular formula and the empirical formula are C$_6$H$_{10}$S$_2$O. **0.52** There will be seven peaks of the following mass numbers: 34, 35, 36, 37, 38, 39, and 40. **0.54** 288 g I$_2$. **0.56** (a) NH$_4$NO$_3$(s) ⟶ N$_2$O(g) + 2H$_2$O(g). (b) 20 g. **0.58** 0.709 g NO$_2$. **0.60** 93.0%. **0.62** 53.85%. **0.64** (a) 4.41 × 10^{-40}. (b) 4.6 × 10^{-18} J. **0.66** 2C$_2$H$_6$ + 7O$_2$→4CO$_2$ + 6H$_2$O. **0.68** 9.58 × 10^5 g. **0.70** (a) The formula should be (NH$_4$)$_2$CO$_3$. (b) The formula should be Ca(OH)$_2$. (c) The correct formula is CdS. (d) The correct formula is ZnCr$_2$O$_7$. **0.72** 103.3 g mol^{-1}. **0.74** C:30.20%, H:5.069%, Cl:44.57%, S:20.16%. **0.76** 700. g. **0.78** (a) Zn(s) + H$_2$SO$_4$(aq) ⟶ ZnSO$_4$(aq) + H$_2$(g). (b) 64.2%. (c) We assumed that the impurities are inert and do not react with the sulfuric acid to produce hydrogen. **0.80** 89.5%. **0.82** (a) 6.532 × 10^4 g mol^{-1}. (b) 7.6 × 10^2 g.

0.84 NaCl: 32.17%; Na$_2$SO$_4$: 20.09%; NaNO$_3$: 47.75%. **0.86** (a) 16 u, CH$_4$ 17 u, NH$_3$ 18 u, H$_2$O 64 u, SO$_2$. (b) A CH$_3$ fragment could break off from this C$_3$H$_8$ giving a peak at 15 u. (c) ±0.030 u. **0.88** 2.01 × 10^{21}. **0.90** 16.00 u. **0.92** (e) 0.50 mol Cl$_2$. **0.94** X$_2$O$_3$(s) + 3CO(g) ⟶ 2X(s) + 3CO$_2$(g). **0.96** 54.57% of the molecules. **0.98** (a) Compound X: MnO$_2$, compound Y: Mn$_3$O$_4$. (b) 3MnO$_2$ ⟶ Mn$_3$O$_4$ + O$_2$. **0.100** 6.1 × 10^5. **0.102** PbC$_8$H$_{20}$. **0.104** 1.85 × 10^5 kg. **0.106** (a) 2.0 × 10^{24} cm^3. (b) 3.9 × 10^3 m. **0.108** (a) 4.3 × 10^{23}. (b) 160 pm. **0.110** (a) C$_3$H$_8$(g) + 3H$_2$O(g) ⟶ 3CO(g) + 7H$_2$(g). (b) 909 kg. **0.112** Gently heat the liquid to see if any solid remains after the liquid evaporates. Also, collect the vapor and then compare the densities of the condensed liquid with the original liquid. The composition of a mixed liquid would change with evaporation along with its density. **0.114** The predicted change (loss) in mass is only 1.91 × 10^{-8} g, which is too small a quantity to measure. Therefore, for all practical purposes, the law of conservation of mass is assumed to hold for ordinary chemical processes. **0.116** The acids, from left to right, are chloric acid, nitrous acid, hydrocyanic acid, and sulfuric acid. **0.118** (a) 0.307 g. (b) 0.410 g. (c) 90.1%. **0.120** 6.022 g of Mg$_3$N$_2$, 25.66 g MgO. **0.122** 9.6 g Fe$_2$O$_3$, 7.4 g KClO$_3$.

Chapter 1

1.2 (a) 6.58 × 10^{14} Hz. (b) 1.22 × 10^8 nm. **1.4** 4.95 × 10^{14} Hz. **1.6** (a) 4.0 × 10^2 nm. (b) 5.0 × 10^{-19} J. **1.8** 1.2 × 10^2 nm, which is in the ultraviolet region. **1.10** (a) 370 nm, 3.70 × 10^{-7} m. (b) ultraviolet region. (c) 5.38 × 10^{-19} J. **1.12** 3.3 × 10^{28}. **1.14** The radiation emitted by the star is measured as a function of wavelength (λ). The wavelength corresponding to the maximum intensity is then plugged in to Wien's law ($T\lambda_{max}$=1.44 × 10^{-2} K m) to determine the surface temperature. **1.16** Each laser will eject the same number of electrons. The electrons ejected by the blue laser will have higher kinetic energy because the photons from the blue laser have higher energy. **1.18** (a) 3.28 × 10^{-19} J. (b) 5.9 × 10^{-20} J. **1.20** The emitted light could be analyzed by passing it through a prism. **1.22** Excited atoms of the chemical elements emit the same characteristic frequencies or lines in a terrestrial laboratory, in the sun, or in a star many light-years distant from Earth. **1.24** 3.027 × 10^{-19} J. **1.26** 6.17 × 10^{14} Hz and 486 nm. **1.28** 21.17 nm and 47.64 nm. **1.30** 3.289832496 × 10^{15} Hz. **1.32** 0.565 nm. **1.34** 1.99 × 10^{-31} cm. **1.36** 1 × 10^6 m s^{-1}. **1.38** 1.20 × 10^{-18} J. **1.40**

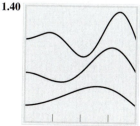

The average value for x will be greater than $L/2$.
1.42 (a) $n = 2$, $l = 1$, $m_l = 1$, 0, or -1. (b) $n = 6$, $l = 0$, $m_l = 0$ (only allowed value). (c) $n = 5$, $l = 2$, $m_l = 2$, 1, 0, -1, or -2.
1.44 The allowed values of l are 0, 1, 2, 3, 4, and 5 [$l = 0$ to $(n - 1)$, integer values]. These l values correspond to the $6s$, $6p$, $6d$, $6f$, $6g$, and $6h$ subshells. These subshells each have 1, 3, 5, 7, 9, and 11 orbitals, respectively (number of orbitals = $2l + 1$).
1.46 A $2s$ orbital is larger than a $1s$ orbital. Both have the same spherical shape. The $1s$ orbital is lower in energy than the $2s$. The $1s$ orbital does not have a node while the $2s$ orbital does.
1.48 He^+ has an ionization energy of 5.248×10^3 kJ mol^{-1}. Li^{2+} has an ionization energy of 1.181×10^4 kJ mol^{-1}.
1.50 (a) False. (b) False. (c) True. (d) False. (e) True.
1.52 The color of the flame would be blue. **1.54** 4.10×10^{23}.
1.56 The longest wavelength for the Lyman series is 121 nm. The shortest wavelength for the Balmer series is 365 nm. **1.58** 483 nm.
1.60 For hydrogen, the percent change of the Rydberg constant is 0.0547%. For the He-4, the percent change of the Rydberg constant is 0.014%. **1.62** 2.8×10^6 K. **1.64** 2.0×10^{-5} m s^{-1}.
1.66 1.8×10^5 J. **1.68** 2.76×10^{-11} m. **1.70** 17.4 pm.
1.72 (a) 6.171×10^{-21} J. (b) 1.635×10^{-18} J. (c) 7.90×10^4 K.
1.74 An electron confined to the nucleus would have an uncertainty in its velocity of 5.8×10^{10} m s^{-1}, which is impossible because it far exceeds the speed of light. The uncertainty of the velocity of a proton confined to the nucleus is 3.2×10^7 m s^{-1}, which is less than the speed of light.
1.76 For the $1s$ orbital, the average value for r is $\frac{2}{3} a_o$ and the most likely value of r is a_o. For the $2s$ orbital, the average value for r is $6a_o$ and the most likely value of r is $(3 + \sqrt{5})a_o$.
1.78 The five lowest energy levels correspond to the following sets of quantum numbers: (n_x, n_y): (1,1); (1,2) and (2,1); (2,2); (1,3) and (3,1); (2,3) and (3,2). There are more than five states here because the second, fourth, and fifth energy levels are degenerate.

Chapter 2

2.2 There are three coordinates for each electron. Sodium has 11 electrons, hence its wavefunction would be a function of 33 (11 × 3 = 33) variables. The sodium ion has one less electron, hence its wavefunction would be a function of 30 (10 × 3 = 30) variables.
2.4 0.66. **2.6** In the many-electron atom, the $3p$ orbital electrons are more effectively shielded by the inner electrons of the atom (that is, the $1s$, $2s$, and $2p$ electrons) than the $3s$ electrons. The $3s$ orbital is said to be more "penetrating" than the $3p$ and $3d$ orbitals. In the hydrogen atom, there is only one electron and hence there cannot be any shielding, so the $3s$, $3p$, and $3d$ orbitals have the same energy. **2.8** For aluminum, there are not enough electrons in the $2p$ subshell. (The $2p$ subshell holds six electrons.) The number of electrons (13) is correct. The electron configuration should be $1s^2 2s^2 2p^6 3s^2 3p^1$. The configuration shown might be an excited state of an aluminum atom. For boron, there are too many electrons. (Boron only has five electrons.) The electron configuration should be $1s^2 2s^2 2p^1$. For fluorine, there are also too many electrons. (Fluorine only has nine electrons.) The configuration shown is that of the F^- ion. The correct electron configuration is $1s^2 2s^2 2p^5$. **2.10** Tc: $[Kr]5s^2 4d^5$. **2.12** Ge is: $[Ar]4s^2 3d^{10} 4p^2$; Fe: $[Ar]4s^2 3d^6$; Zn: $[Ar]4s^2 3d^{10}$; Ni: $[Ar]4s^2 3d^8$; W: $[Xe]6s^2 4f^{14} 5d^4$; Tl: $[Xe]6s^2 4f^{14} 5d^{10} 6p^1$. **2.14** S^+ has the most unpaired electrons. It has 3 unpaired electrons in the $3p$ orbitals.
2.16 (a) $[Ne]3s^2 3p^5$. (b) paramagnetic. **2.18** Cr. **2.20** Na^+: [Ne]; Mg^{2+}: [Ne]; Cl^-: [Ar]; K^+: [Ar]; Ca^{2+}: [Ar]; Cu^+: $[Ar]3d^9$; Zn^{2+}:

$[Ar]3d^{10}$. **2.22** C and B^-; Mn^{2+} and Fe^{3+}; Ar and Cl^-; Zn and Ge^{2+}. **2.24** B: 1; Ne: 0; P: 3; Sc: 1; Mn: 5; Se: 2; Kr: 0; Fe: 4; Cd: 0; I: 1; Pb: 2. **2.26** Fe^{2+}: 4; Fe^{3+}: 5; Cr^{3+}: 3; Cr^{4+}: 2; Cu^+: 0; Cu^{2+}: 1. **2.28** When palladium forms a cation, it loses two or four d-electrons, obtaining an incompletely filled d subshell.
2.30 P(s) extensive three-dimensional structure; $I_2(s)$ molecular; Mg(s) extensive three-dimensional structure; Ne(g) atomic; C(s) extensive three-dimensional structure; S(s) molecular; Cs(s) extensive three-dimensional structure; $O_2(g)$ molecular.
2.32 Zn does not have incompletely filled d subshells and does not readily give rise to cations that have incompletely filled subshells. Zn forms a cation by losing its outer s electrons.
2.34 The electron configuration of nitrogen is $[He] 2s^2 2p^3$. The half-filled $2p$ subshell is extra stable, which is why the electron affinity for nitrogen is nearly zero. **2.36** (a) Cs. (b) Ba. (c) Sb. (d) Br. (e) Xe. **2.38** $Mg^{2+} < Na^+ < F^- < O^{2-} < N^{3-}$. **2.40** Te^{2-}
2.42 Cs < Na < Al < S < Cl. **2.44** Apart from the small irregularities, the ionization energies of elements in a period increase with increasing atomic number. We can explain this trend by referring to the increase in effective nuclear charge from left to right. A larger effective nuclear charge means a more tightly held outer electron, and hence a higher first ionization energy. Thus, in the third period, sodium has the lowest and neon has the highest first ionization energy. **2.46** To form the $+2$ ion of calcium, it is only necessary to remove two valence electrons. For potassium, however, the second electron must come from the atom's noble gas core, which accounts for the much higher second ionization energy. **2.48** (a) 5.25×10^3 kJ mol^{-1}. (b) 1.18×10^4 kJ mol^{-1}.
2.50 (a) K < Na < Li. (b) I < Br < F < Cl. **2.52** Based on electron affinity values, we would not expect the alkali metals to form anions. However, in the early seventies a chemist named J.L. Dye at Michigan State University discovered that under very special circumstances alkali metals could be coaxed into accepting an electron to form negative ions! These ions are called alkalide ions. **2.54** All of the noble gases have completely filled shells, which leads to greater stability. Consequently, these elements are very stable chemically. **2.56** (a) He: $1s^2$. (b) N:$1s^2 2s^2 2p^3$. (c) Na:$1s^2 2s^2 2p^6 3s^1$. (d) As: $[Ar]4s^2 3d^{10} 4p^3$. (e) Cl: $[Ne]3s^2 3p^5$. **2.58** 242 nm. **2.60** (a) $Mg^{2+} < Na^+ < F^- < O^{2-}$. (b) $O^{2-} < F^- < Na^+ < Mg^{2+}$. **2.62** Bromine, iodine, chlorine, fluorine.
2.64 Fluorine. **2.66** Noble gases have filled shells or subshells. Therefore, they have little tendency to accept electrons (endothermic). **2.68** 574%.
2.70 O^+ and N, Ar and S^{2-}, Ne and N^{3-}, Zn and As^{3+}, Cs^+ and Xe. **2.72** LiH (lithium hydride), $LiH + H_2O \rightarrow LiOH + H_2$; CH_4 (methane), $CH_4 + H_2O \rightarrow$ no reaction at room temperature; NH_3 (ammonia), $NH_3 + H_2O \rightarrow NH_4^+ + OH^-$; H_2O (water), $H_2O + H_2O \rightarrow H_3O^+ + OH^-$; HF (hydrogen fluoride), $HF + H_2O \rightarrow H_3O^+ + F^-$. **2.74** The heat generated from the radioactive decay can break bonds; therefore, few radon compounds exist.
2.76 (a) It was determined that the periodic table was based on atomic number, not atomic mass. (b) Argon: 39.95 u, Potassium: 39.10 u. **2.78** Both ionization energy and electron affinity are affected by atomic size—the smaller the atom, the greater the attraction between the electrons and the nucleus. If it is difficult to remove an electron from an atom (that is, high ionization energy), then it follows that it would also be favorable to add an electron to the atom (large electron affinity). Noble gases are an exception to this generalization because they do not tend to gain electrons or lose electrons. **2.80** Helium should be named helon to match the other noble gases: neon, argon, xenon, krypton,

and radon. The ending, -ium, suggests that helium has properties similar to some metals (i.e., sodium, magnesium, barium, etc.). **2.82** In general, as the effective nuclear charge increases, the outer-shell electrons are held more strongly, and hence the atomic radius decreases. **2.84** 7.28×10^3 kJ mol^{-1}. **2.86** The plot is: (a) I_1 corresponds to the electron in $3s^1$; I_2 corresponds to the first electron in $2p^6$; I_3 corresponds to the first electron in $2p^5$; I_4 corresponds to the first electron in $2p^4$; I_5 corresponds to the first electron in $2p^3$; I_6 corresponds to the first electron in $2p^2$; I_7 corresponds to the electron in $2p^1$; I_8 corresponds to the first electron in $2s^2$; I_9 corresponds to the electron in $2s^1$; I_{10} corresponds to the first electron in $1s^2$; I_{11} corresponds to the electron in $1s^1$. (b) It requires more energy to remove an electron from a closed shell. The breaks indicate electrons in different shells and subshells.

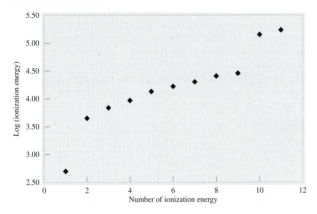

2.88 419 nm. **2.90** In He, r is greater than that in H. Also, the shielding in He makes Z_{eff} less than 2. Therefore, $I_1(\text{He}) < 2I(\text{H})$. In He$^+$, there is only one electron, so there is no shielding. The greater attraction between the nucleus and the lone electron reduces r to less than the r of hydrogen. Therefore, $I_2(\text{He}) > 2I(\text{H})$.

Chapter 3

3.2 Sr: Si: :Ö: :Br:

Be: K·   :Se: Al:

3.4 Ö=C=S̈

3.6 The triple bond is composed of a sigma (σ) and two pi (π) bonds. The σ bond is a result of the head-on overlap of $2p_x$ orbitals. The two pi bonds formed by the overlap of the $2p_y$ and $2p_z$ orbitals on the carbon atom with their counterparts on the nitrogen atom. The two lone pairs of electrons remain in the $2s$ orbitals of each atom. **3.8** Because of the electron spin, electrons tend to be paired. Hence, a single bond is composed of a pair of electrons. If electrons did not have a magnetic moment, it is possible that a bond would consist of a single electron and atoms could form twice as many bonds and still obtain the noble gas configuration.
3.10 C—H < F—H < B—H, Li—Cl < Na—Cl < K—F.
3.12 Cl—Cl < Br—Cl < Si—C < Cs—F. **3.14** (a) covalent.
(b) polar covalent. (c) ionic. (d) polar covalent. **3.16** 358.0 kJ mol^{-1}, which is close to the value of 363.6 kJ mol^{-1} given in Table 3.3.
3.18 (a) energy decrease. (b) energy is tripled. (c) energy is

quadrupled. (d) energy is doubled. **3.20** (a) 225 kJ mol^{-1}.
(b) 163 kJ mol^{-1}. (c) 71 kJ mol^{-1}.

3.22 (a) :F̈—Ö—F̈: (b) :F̈—N̈=N̈—F̈: (c) H—Si—Si—H (with H's)

(d) $^-$:Ö—H (e) H—C—C—Ö:$^-$ (f) H—C—N$^+$—H

3.24 (a) $^-$:Ö—Ö:$^-$ (b) $^-$:C≡C:$^-$

(c) :N≡O$^+$: (d) H—N$^+$—H (with H)

3.26 Ö=Cl$^+$—Ö:$^-$ ⟷ $^-$:Ö—Cl$^+$—Ö:$^-$ ⟷ $^-$:Ö—Cl=Ö

3.28 H—C=N$^+$=N:$^-$ ⟷ H—C—N≡N: (with H)

3.30 $^-$N̈=N̈=Ö ⟷ :N≡N̈—Ö:$^-$ ⟷ $^{2-}$:N̈—N≡O$^+$:

3.32 $^-$:Ö—P$^+$—F: ⟷ $^-$:Ö—P—F: ⟷ Ö=P—F: ⟷ $^-$:Ö—P—F:

3.34 The H$_2$ molecule has two electrons in the σ_{1s} orbital, giving it a bond order of one. The H$_2^+$ ion has a single electron in the σ_{1s} orbital, giving it a bond order of one-half. The H$_2^{2+}$ ion has no electrons. Hence, in terms of bond distances: $H_2 < H_2^+ < H_2^{2+}$.
3.36 $Li_2^- = Li_2^+ < Li_2$

	Li_2^+	Li_2	Li_2^-
σ_{2s}^*	☐	☐	↑
σ_{2s}	↑	↑↓	↑↓

3.38 The bond order of the carbide ion
$(\sigma_{1s})^2(\sigma_{1s}^*)^2(\sigma_{2s})^2(\sigma_{2s}^*)^2(\pi_{2p_y})^2(\pi_{2p_z})^2(\sigma_{2p_x})^2$ is 3 and that of C$_2$
$(\sigma_{1s})^2(\sigma_{1s}^*)^2(\sigma_{2s})^2(\sigma_{2s}^*)^2(\pi_{2p_y})^2(\pi_{2p_z})^2$ is only 2.
3.40 In forming the N$_2^+$ from N$_2$, an electron is removed from the sigma *bonding* molecular orbital. Consequently, the bond order decreases to 2.5 from 3.0. In forming the O$_2^+$ ion from O$_2$, an electron is removed from the pi *antibonding* molecular orbital. Consequently, the bond order increases to 2.5 from 2.0.
3.42 The bond order of
$F_2^+: (\sigma_{1s})^2(\sigma_{1s}^*)^2(\sigma_{2s})^2(\sigma_{2s}^*)^2(\sigma_{2p_x})^2(\pi_{2p_y})^2(\pi_{2p_z})^2(\pi_{2p_y}^*)^2(\pi_{2p_z}^*)^1$
is 1.5 compared to 1 for
$F_2: (\sigma_{1s})^2(\sigma_{1s}^*)^2(\sigma_{2s})^2(\sigma_{2s}^*)^2(\sigma_{2p_x})^2(\pi_{2p_y})^2(\pi_{2p_z})^2(\pi_{2p_y}^*)^2(\pi_{2p_z}^*)^2$.
Therefore, F$_2^+$ should have a shorter bond length.
3.44 In general, the more s-character of the bonding electrons, the shorter the bond. The carbon atoms in C$_2$H$_2$ are sp hybridized and hence the bonding electrons in C$_2$H$_2$ has the most s-character and the C—H bonds in C$_2$H$_2$ are the shortest. The carbon atoms in C$_2$H$_6$ are sp^3 hybridized and hence the bonding electrons in C$_2$H$_6$ have the least amount of s-character and the C—H bonds in C$_2$H$_6$ are the longest.

3.46 $^-\ddot{\text{N}}{=}\overset{+}{\text{N}}{=}\ddot{\text{N}}^- \longleftrightarrow :\text{N}{\equiv}\overset{+}{\text{N}}{-}\ddot{\text{N}}:^{2-} \longleftrightarrow {}^{2-}:\ddot{\text{N}}{-}\overset{+}{\text{N}}{\equiv}\text{N}:$

3.48 (a) $AlCl_4^-$. (b) AlF_6^{3}. (c) $AlCl_3$. **3.50** If the central atom were more electronegative, there would be a concentration of negative charges at the central atom. This would lead to instability.

3.52 (a) $-\overset{\displaystyle :\text{O}:}{\underset{\displaystyle \|}{\text{C}}}{-}\ddot{\text{O}}{-}\text{H}$ (b) $-\overset{\displaystyle :\text{O}:}{\underset{\displaystyle \|}{\text{C}}}{-}\ddot{\text{O}}:^- \longleftrightarrow -\overset{\displaystyle :\text{O}:}{\underset{\displaystyle |}{\text{C}}}{=}\ddot{\text{O}}:$

3.54 (a) $\cdot\dot{\text{C}}{-}\text{H}$ paramagnetic (b) $\cdot\ddot{\text{O}}{-}\text{H}$ paramagnetic

(c) $:\text{C}{=}\text{C}:$ diamagnetic (d) $\text{H}{-}\overset{+}{\text{N}}{\equiv}\ddot{\text{C}}:$ diamagnetic

(e) $\text{H}{-}\dot{\text{C}}{=}\ddot{\text{O}}$ paramagnetic

3.56 It is much easier to dissociate F_2 into two neutral F atoms (154.4 kJ mol^{-1}) than it is to dissociate it into a fluorine cation and anion (1506 kJ mol^{-1}).

3.58 $:\ddot{\text{Cl}}{-}\ddot{\text{O}}{-}\underset{+}{\text{N}}{-}\overset{\displaystyle :\text{O}:}{\underset{\displaystyle \|}{}}\ddot{\text{O}}:^-$

3.60 C_2H_6 C_4H_{10}

$$\text{H}{-}\underset{\text{H}}{\overset{\text{H}}{\text{C}}}{-}\underset{\text{H}}{\overset{\text{H}}{\text{C}}}{-}\text{H} \qquad \text{H}{-}\underset{\text{H}}{\overset{\text{H}}{\text{C}}}{-}\underset{\text{H}}{\overset{\text{H}}{\text{C}}}{-}\underset{\text{H}}{\overset{\text{H}}{\text{C}}}{-}\underset{\text{H}}{\overset{\text{H}}{\text{C}}}{-}\text{H}$$

C_5H_{12}

$$\text{H}{-}\underset{\text{H}}{\overset{\text{H}}{\text{C}}}{-}\underset{\text{H}}{\overset{\text{H}}{\text{C}}}{-}\underset{\text{H}}{\overset{\text{H}}{\text{C}}}{-}\underset{\text{H}}{\overset{\text{H}}{\text{C}}}{-}\underset{\text{H}}{\overset{\text{H}}{\text{C}}}{-}\text{H}$$

3.62 The structures are (the nonbonding electron pairs on fluorine have been omitted for simplicity):

$$\underset{\text{H}}{\overset{\text{H}}{\text{C}}}{=}\underset{\text{F}}{\overset{\text{H}}{\text{C}}} \qquad \underset{\text{H}}{\overset{\text{H}}{\text{C}}}{=}\underset{\text{H}}{\overset{\text{H}}{\text{C}}}{-}\underset{\text{H}}{\overset{\text{H}}{\text{C}}}{-}\text{H} \qquad \underset{\text{H}}{\overset{\text{H}}{\text{C}}}{=}\underset{\text{H}}{\overset{\text{H}}{\text{C}}}{-}\underset{\text{H}}{\overset{\text{H}}{\text{C}}}{-}\underset{\text{H}}{\overset{\text{H}}{\text{C}}}{-}\text{H}$$

3.64 (a) $^-:\text{C}{\equiv}\text{O}:^+$ (b) $:\text{N}{\equiv}\text{O}:^+$ (c) $^-:\text{C}{\equiv}\text{N}:$ (d) $:\text{N}{\equiv}\text{N}:$
3.66 True. Each noble gas atom already has completely filled ns and np subshells. If a noble gas shares electrons with other atoms, it will have more than an octet. **3.68** The description involving a griffin and a unicorn is more appropriate. Mules and donkeys are real animals, whereas resonance structures do not exist.
3.70 The unpaired electron on each N atom will be shared to form a covalent bond:

$O_2N\cdot + \cdot NO_2 \longrightarrow N_2O_4$

$$\underset{:\ddot{\text{O}}:}{\overset{:\ddot{\text{O}}:}{\underset{+}{\text{N}}}}{-}\underset{:\ddot{\text{O}}:^-}{\overset{\ddot{\text{O}}.}{\underset{+}{\text{N}}}} \longleftrightarrow \underset{:\ddot{\text{O}}:}{\overset{:\ddot{\text{O}}.}{\underset{+}{\text{N}}}}{-}\underset{:\ddot{\text{O}}:^-}{\overset{:\ddot{\text{O}}:^-}{\underset{+}{\text{N}}}} \longleftrightarrow$$

$$\underset{:\ddot{\text{O}}:}{\overset{:\ddot{\text{O}}:^-}{\underset{+}{\text{N}}}}{-}\underset{:\ddot{\text{O}}:}{\overset{:\ddot{\text{O}}.}{\underset{+}{\text{N}}}} \longleftrightarrow \underset{:\ddot{\text{O}}:^-}{\overset{:\ddot{\text{O}}.}{\underset{+}{\text{N}}}}{-}\underset{:\ddot{\text{O}}:^-}{\overset{:\ddot{\text{O}}.}{\underset{+}{\text{N}}}}$$

3.72 $\underset{:\ddot{\text{O}}:^-}{\overset{:\ddot{\text{O}}.}{\underset{+}{\text{N}}}}{-}\ddot{\text{O}}{-}\underset{:\ddot{\text{O}}:^-}{\overset{:\ddot{\text{O}}.}{\underset{+}{\text{N}}}}$

3.74 (a) $:\ddot{\text{O}}{-}\text{H}$ (b) The O—H bond is quite strong (460 kJ mol^{-1}). To complete its octet, the OH radical has a strong tendency to form a bond with a H atom. (c) 260 nm.

3.76 $\ddot{\text{O}}{=}\text{S}{=}\ddot{\text{O}} \longleftrightarrow {}^-:\ddot{\text{O}}{-}\overset{+}{\text{S}}{=}\ddot{\text{O}} \longleftrightarrow \ddot{\text{O}}{=}\overset{+}{\text{S}}{-}\ddot{\text{O}}:^-$

The resonance structure with formal charges of zero (the one on the left) is the most plausible.
3.78 2×10^2 kJ mol^{-1}. **3.80** (1) You could determine the magnetic properties of the solid. An Mg^+O^- solid would be paramagnetic while $Mg^{2+}O^{2-}$ solid is diamagnetic. (2) You could determine the lattice energy of the solid. Mg^+O^- would have a lattice energy similar to Na^+Cl^-. This lattice energy is much lower than the lattice energy of $Mg^{2+}O^{2-}$.
3.82 (a) $[Ne_2](\sigma_{3s})^2(\sigma_{3s}^*)^2(\pi_{3p_y})^2(\pi_{3p_z})^2(\sigma_{3p_x})^2$. (b) 3.
(c) diamagnetic. **3.84** CN^+. **3.86** The wavefunction determined from molecular orbital (MO) theory contains the same terms as the wavefunction determined from valence bond (VB) theory. The wavefunction determined from MO theory also contains terms corresponding to both electrons being on the same atom. If the likelihood of the two electrons being on the same atom is negligible (such as for a homonuclear diatomic molecule), then the wavefunctions from the two theories are identical.

3.88

	BN$^-$	CO$^-$	OF$^+$
$\sigma_{2p_x}^*$	☐	☐	☐
$\pi_{2p_y}^*, \pi_{2p_z}^*$	☐ ☐	↑ ☐	↑ ↑
σ_{2p_x}	↑	↑↓	↑↓ ↑↓
π_{2p_y}, π_{2p_z}	↑↓ ↑↓	↑↓ ↑↓	↑↓
σ_{2s}^*	↑↓	↑↓	↑↓
σ_{2s}	↑↓	↑↓	↑↓
Bond order:	2.5	2.5	2

3.90 163.8 pm.

3.92

$$\underset{\text{H}}{\overset{\text{H}}{\diagdown}}\overset{+}{\text{H}} \longleftrightarrow \overset{+}{\text{H}}\underset{\text{H}}{\overset{\text{H}}{\diagdown}} \longleftrightarrow \overset{\text{H}^+}{\text{H}{-}\text{H}}$$

3.94 $:\text{N}{\equiv}\overset{+}{\text{N}}{-}\overset{-}{\ddot{\text{N}}}{-}\overset{+}{\text{N}}{=}\ddot{\text{N}}^- \longleftrightarrow {}^-\ddot{\text{N}}{=}\overset{+}{\text{N}}{=}\overset{+}{\text{N}}{-}\ddot{\text{N}}{=}\text{N}: \longleftrightarrow$

$: \text{N}{\equiv}\overset{+}{\text{N}}{-}\overset{-}{\ddot{\text{N}}}{-}\overset{+}{\text{N}}{\equiv}\text{N}:$

3.96 (a) H_2. (b) N_2. (c) O. (d) F.

3.98 (a)

$\pi_{2p_y}^*$ or $\pi_{2p_z}^*$ First excited state: N_2^*

σ_{2p_x} Ground state: N_2

(b) The bond order for N_2 is 3. The bond order for N_2^* is 2. N_2^* should have a longer bond length than N_2. (c) diamagnetic. (d) 4.23×10^{-19} J.

Chapter 4

4.2 (a) trigonal planar. (b) linear. (c) tetrahedral. **4.4** (a) tetrahedral. (b) T-shaped. (c) bent. (d) trigonal planar. (e) trigonal pyramidal. **4.6** (a) tetrahedral. (b) bent. (c) trigonal planar. (d) linear.

(e) square planar. (f) tetrahedral. (g) trigonal bipyramidal. (h) trigonal pyramidal. (i) tetrahedral. **4.8** The carbon in the CH_3 group has a tetrahedral geometry. The carbon bound to the two oxygen has a trigonal planar geometry. The oxygen bound to the hydrogen has a bent geometry. **4.10** (a) 180°. (b) 120°. (c) 109.5°. (d) 180°. (e) 120°. (f) 109.5°. (g) 109.5°. **4.12** (c). **4.14** H_2Te < H_2Se < H_2S < H_2O. **4.16** OCS, OCl_2, PCl_3, SF_4. **4.18** While CS_2 is nonpolar, OCS is polar because oxygen is more electronegative than carbon.

4.20

4.22 6.98×10^{-20} C. **4.24** (b) = (d) < (c) < (a). **4.26** sp^3 for both. **4.28** Before the reaction, boron is sp^2 hybridized and nitrogen is sp^3 hybridized. After the reaction, boron and nitrogen are both sp^3. **4.30** (a) sp^3. (b) sp^3. (c) sp^2. **4.32** (a) sp. (b) sp. (c) sp. **4.34** The two end carbons are sp^2 hybridized. The central carbon is sp hybridized.

4.36 (a) 4 σ bonds and 0 π bonds. (b) 5 σ bonds and 1 π bond. (c) 10 σ bonds and 3 π bonds.

4.38 $CH_3CH_2CH_2CH_2CH_2CH_3$

$CH_3CHCH_2CH_2CH_3$ $CH_3CH_2CHCH_2CH_3$
(with CH_3 branches)

$CH_3CCH_2CH_3$ $CH_3CHCHCH_3$
(with CH_3 branches)

4.40

The first and second, and first and third are structural isomers. The second and third are geometric isomers.

4.42 The nitrate ion has a trigonal planar geometry. The $2p_z$ orbitals on the oxygen and nitrogen atoms overlap producing a molecular orbital that is above and below all three N—O bonds. The molecular orbital can be written as a linear combination of the $2p_z$ orbitals. From the symmetry of the molecule, we know that the coefficient for the oxygen atomic orbital are the same.

4.44 NO_2^- has a bent geometry and has two electrons in the lowest energy π orbital. This π orbital is formed from the $2p_z$ orbitals of the nitrogen and oxygen atoms and is above and below both nitrogen-oxygen bonds. NO_2^+ has a linear geometry and has four electrons in π orbitals. The lowest energy π orbital of NO_2^+ is similar to that of NO_2^-, in that it could be composed of the $2p_z$ orbitals of the nitrogen and oxygen atoms and is above and below both nitrogen-oxygen bonds. The lowest energy π orbital of NO_2^+ could also be composed of the $2p_y$ orbitals of the nitrogen and oxygen atoms. Hence, NO_2^+ has two degenerate π orbitals. Notice that because of the symmetry of both NO_2^-

and NO_2^+, the coefficients for the oxygen atomic orbitals are the same.

$$\pi_z = c_1\phi_{O_1,2p_z} + c_2\phi_{N,2p_z} + c_1\phi_{O_2,2p_z}$$
$$\pi_y = c_1\phi_{O_1,2p_y} + c_2\phi_{N,2p_y} + c_1\phi_{O_2,2p_y}$$

4.46 The ion contains 24 valence electrons. Of these, six are involved in three sigma bonds between the nitrogen and oxygen atoms. The hybridization of the nitrogen atom is sp^2. There are 16 nonbonding electrons on the oxygen atoms. The remaining two electrons are in a delocalized π molecular orbital which results from the overlap of the p_z orbital of nitrogen and the p_z orbitals of the three oxygen atoms. **4.48** ICl has a dipole moment and Br_2 does not. The dipole moment increases the intermolecular attractions between ICl molecules and causes that substance to have a higher melting point than bromine. **4.50** CO_2 < CH_3Br < CH_3OH < RbF. CO_2 is a nonpolar molecular compound. CH_3Br is a polar molecule. CH_3OH is polar and can form hydrogen bonds. RbF is an ionic compound. **4.52** (a) Ne. (b) CO_2. (c) CH_4. (d) F_2. (e) PH_3. **4.54** LiF, ionic bonding and dispersion forces; BeF_2, ionic bonding and dispersion forces; BF_3, dispersion forces; CF_4, dispersion forces; NF_3, dipole-dipole interaction and dispersion forces; OF_2, dipole-dipole interaction and dispersion forces; F_2, dispersion forces. **4.56** -3.3×10^{-21} J. **4.58** The minimum of the potential energy curve for Xe is lower (more negative) than for Ar, indicating that the Xe–Xe interatomic interactions are stronger than the Ar–Ar interactions. Because of this, Xe will have a higher boiling point than Ar.

4.60 (a) F—B—F (with F below) planar
(b) [O—Cl—O (with double bond O below)] nonplanar
(c) H—O—H (with lone pairs, dipole arrow down)
(d) F—O—F (with lone pairs) polar
(e) N—O, O (with lone pair on N) Greater than 120°

4.62 sp^3d. **4.64** yes.

4.66 (a) O=S=O (with lone pairs) nonpolar
(b) F—P—F (with F above and below) polar
(c) F—Si—F (with F above and below); SiF_4 structure
(d) [H—Si—H (with lone pair)]$^-$ trigonal planar
(e) Br—C—Br (with H above and below) polar

4.68

$:\overset{..}{\underset{..}{Cl}}—\overset{..}{\underset{..}{O}}—\overset{+}{N}\overset{O}{\underset{}{||}}\overset{..}{\underset{..}{O}}:^-$

The nitrogen is sp^2 and has a trigonal planar geometry resulting in O—N—O bond angles of 120°. The oxygen is sp^3 and has a bent geometry resulting in a Cl—O—N bond angle of less than 109.5°.

4.70 XeF_3^+: T-shaped; XeF_5^+: square pyramidal; SbF_6^-: octahedral.
4.72 Only ICl_2^- and $CdBr_2$ will be linear. The rest are bent.
4.74 The azide ion (N_3^-), has a linear geometry and has four electrons in σ orbitals and four electrons in π orbitals. The two σ orbitals connect adjacent nitrogen. The two π orbitals are degenerate and both lie above and below both nitrogen-nitrogen bonds. One of the π orbitals is composed of the $2p_z$ orbitals of the three nitrogen atoms. The other π orbital is composed of the $2p_y$ orbitals of the three nitrogen atoms.

4.76

polar   polar   nonpolar

4.78

The four P atoms occupy the corners of a tetrahedron. Each phosphorus atom is sp^3 hybridized. The P_2 molecule has a triple bond, which is composed of one sigma bond and two pi bonds. The pi bonds are formed by sideways overlap of $3p$ orbitals. Because the atomic radius of P is larger than that of N, the overlap is not as extensive as that in the N_2 molecule. Consequently, the P_2 molecule is much less stable than N_2 and also less stable than P_4, which contains only sigma bonds. **4.80** (a) -3.35×10^{-2} J; (b) -4.60×10^{-3} J. **4.82** The carbons are in sp^2 hybridization states. The nitrogens are in the sp^3 hybridization state, except for the ring nitrogen double-bonded to a carbon that is sp^2 hybridized. The oxygen atom is sp^2 hybridized. **4.84** Butane should have the higher boiling point because the butane molecule has a larger surface area resulting in more intermolecular interactions.
4.86 SiO_2 has an extensive three-dimensional structure. CO_2 exists as discrete molecules. It will take much more energy to break the strong network covalent bonds of SiO_2; therefore, SiO_2 has a much higher boiling point than CO_2. **4.88** (a) Each nitrogen is sp^2 hybridized. (b) The structure on the right has a dipole moment. **4.90** Each of the oxygen atoms in O_2 is sp^2 hybridized. Two of the hybrid orbitals on each oxygen contain lone pairs of electrons. The remaining hybrid orbital on each oxygen overlaps with each other forming a σ bond. The two unhybridized $2p$ orbitals on each oxygen overlap forming a π bond. If we think about the bonding of O_2 in terms of unhybridized orbitals, we can imagine that the σ and π bonds are formed by the overlap of $2p$ orbitals and that one lone pair on each atom remains in a $2s$ orbital and the other in a $2p$ orbital. **4.92** 180Å.
4.94

The four carbons marked with an asterisk are sp^2 hybridized. The remaining carbons are sp^3 hybridized.

Chapter 5

5.2 1.2 bar. **5.4** Two phase changes occur in this process. First, the liquid is turned to solid (freezing), then the solid ice is turned to gas (sublimation). **5.6** Initially, the ice melts because of the increase in pressure. As the wire sinks into the ice, the water above the wire refreezes. Eventually the wire actually moves completely through the ice block without cutting it in half. **5.8** Region labels: The region containing point A is the solid region. The region containing point B is the liquid region. The region containing point C is the gas region. (a) Raising the temperature at constant pressure beginning at A implies starting with solid ice and warming until melting occurs. If the warming continued, the liquid water would eventually boil and change to steam. Further warming would increase the temperature of the steam. (b) At point C water is in the gas phase. Cooling without changing the pressure would eventually result in the formation of solid ice. Liquid water would never form. (c) At B the water is in the liquid phase. Lowering the pressure without changing the temperature would eventually result in boiling and conversion to water in the gas phase. **5.10** 2, a plane.
5.12 45.1 torr. **5.14** 44.1 L. **5.16** 6.3 bar. **5.18** 472°C. **5.20** 1.30×10^3 mL. **5.22** 8.4×10^2 L. **5.24** 9.0 L. **5.26** 71 mL. **5.28** 6.2×10^{-3} bar. **5.30** 35.1 g mol^{-1}. **5.32** N_2: 1.9×10^{22}; O_2: 5.1×10^{21}; Ar: 2.4×10^{20}. **5.34** 2.98 g L^{-1}. **5.36** SF_4. **5.38** 370 L. **5.40** 87.7%.
5.42 M_2O_3 and $M_2(SO_4)_3$. **5.44** 94.7%, assuming that none of the impurities reacted with HCl to produce CO_2 gas.
5.46 $C_2H_5OH(l) + 3O_2(g) \longrightarrow 2CO_2(g) + 3H_2O(l)$; 1.71×10^3 L. **5.48** Temperature is a microscopic concept, in that it is a measure of motion at the atomic level. **5.50** N_2: 472 m s^{-1}; O_2: 441 m s^{-1}; O_3: 360 m s^{-1}. **5.52** 5.3×10^{-2} J; 5.3×10^{-5} bar. **5.54** -235.9°C. **5.56** 6.07×10^{-19} J; 3.65×10^5 J. **5.58** 1.0043.
5.60 van der Waals: 18.2 bar; ideal: 18.7 bar. **5.62** SO_2 should behave less ideally because it has stronger intermolecular forces (it is polar) and it is larger. **5.64** virial: 1.04 L; ideal: 1.11 L.
5.66 Neon has the smallest van der Waals constants of the three gases and hence should behave the most ideally. **5.68** C_6H_6.
5.70 (a) 0.89 bar. (b) 1.5 L. **5.72** 349 torr. **5.74** 4.8%. **5.76** 451 mL.
5.78 The partial pressure of carbon dioxide is higher in the winter because carbon dioxide is utilized less by photosynthesis in plants.
5.80 1.28×10^{22}; CO_2, O_2, N_2, and H_2O.
5.82 $X_{CH_4} = \dfrac{232}{294} = 0.789 \qquad X_{C_2H_6} = \dfrac{62}{294} = 0.211$.
5.84 50.1%. **5.86** 5.25×10^{18} kg. **5.88** 0.0701 M. **5.90** He: 0.16 bar; Ne: 2.0 bar. **5.92** When the water enters the flask from the dropper, some hydrogen chloride dissolves, creating a partial vacuum. Pressure from the atmosphere forces more water up the vertical tube. **5.94** 7. **5.96** (a) 61.2 m s^{-1}. (b) 4.58×10^{-4} s. (c) The Bi atoms traveled at 328 m s^{-1}, compared to the average root-mean-square speed of 366 m s^{-1}. These are comparable. It is not surprising that they are a little different. **5.98** The fruit ripens more rapidly because the quantity (partial pressure) of ethylene gas inside the bag increases. **5.100** As the pen is used, the amount of ink decreases, increasing the volume inside the pen. As the volume increases, the pressure inside the pen decreases. The hole is needed to equalize the pressure as the volume inside the pen increases. **5.102** 4.2×10^5 m^2 s^{-2}. **5.104** 86.0 g mol^{-1}.
5.106 (a) $CaO(s) + CO_2(g) \longrightarrow CaCO_3(s)$; $BaO(s) + CO_2(g) \longrightarrow BaCO_3(s)$. (b) CaO: 10.5%, BaO: 89.5%. **5.108** (a) 0.112 mol CO min^{-1}; 20 min. **5.110** (a) -44°C; (b) 72.4%. **5.112** 1.7×10^{12}. **5.114** The nitrogen oxide is most likely NO_2, although N_2O cannot be completely ruled out.

5.116 7.04×10^{-3} m s^{-1}, 3.52×10^{-30} J.

5.118 (a) $\dfrac{PV}{nRT} = \dfrac{1}{1 - \frac{b}{v}} - \dfrac{a}{VRT}$; $B = b - \dfrac{a}{RT}$; $C = b^2$, $D = b^3$.

(b) $T = \dfrac{a}{bR}$. (c) 1090°C.

5.120 The ice condenses the water vapor inside. Since the water is still hot, it will begin to boil at reduced pressure. **5.122** 1.4 g.
5.124 $P_{CO_2} = 8.7 \times 10^6$ Pa; $P_{N_2} = 3.2 \times 10^5$ Pa;
$P_{SO_2} = 1.4 \times 10^3$ Pa. **5.126** (a) (i) Since the two He samples are at the same temperature, their rms speeds and the average kinetic energies are the same; (ii) The He atoms in V_1 (smaller volume) collide with the walls more frequently. Since the average kinetic energies are the same, the force exerted in the collisions is the same in both flasks. (b) (i) The rms speed is greater at the higher temperature, T_2; (ii) The He atoms at the higher temperature, T_2, collide with the walls with greater frequency and with greater force. (c) (i) False; (ii) True; (iii) True. **5.128** (a) The plots dip due to intermolecular attractions between gas particles. Consider the approach of a particular molecule toward the wall of a container. The intermolecular attractions exerted by its neighbors tend to soften the impact made by this molecule against the wall. The overall effect is a lower gas pressure than we would expect for an ideal gas. Thus, PV/RT decreases. The plots rise because at higher pressures (smaller volumes), the molecules are close together and repulsive forces among them become dominant. Repulsive forces increase the force of impact of the gas molecules with the walls of the container. The overall effect is a greater gas pressure than we would expect for an ideal gas. Hence, $PV/RT > 1$ and the curves rise above the horizontal line. (b) For 1 mole of an ideal gas, $PV/RT = 1$, no matter what the pressure of the gas. At very low pressures, all gases behave ideally; therefore, PV/RT converges to 1 as the pressure approaches zero. As the pressure of a gas approaches zero, a gas behaves more and more like an ideal gas. (c) The intercept on the ideal gas line means that $PV/RT = 1$. However, this does *not* mean that the gas behaves ideally. It just means that at this particular pressure, molecular attraction is equal to molecular repulsion so the net interaction is zero. **5.130** 10.1 bar.
5.132 (a) Two triple points: Diamond/graphite/liquid and graphite/liquid/vapor. (b) Diamond. (c) Apply high pressure at high temperature. **5.134** CO. **5.136** 1200 K.

Chapter 6

6.2 Past the critical point (higher temperature and higher pressure) the liquid can no longer exist, so gases can no longer be liquified.
6.4 Due to the water's relatively strong intermolecular forces (hydrogen bonding) water molecules at the surface are attracted to other water molecules resulting in an unusually strong surface tension. **6.6** As temperature increases, viscosity decreases. The higher the temperature, the faster the motion at the atomic-level and hence molecules can more easily move relative to one another leading to a lower viscosity. **6.8** Because a Kr atom is larger than an Ar atom, the maximum in the radial distribution function for Kr will at a larger value of r than for that of Ar. Because krypton has stronger intermolecular forces, liquid krypton should have a more defined structure than liquid argon, resulting in fluctuations in krypton's radial distribution at larger r. **6.10** (a) Six. (b) Eight. (c) Twelve. **6.12** 2 atoms/unit cell. **6.14** Two. **6.16** 8 atoms/unit cell. **6.18** XY_3. **6.20** 6.020×10^{23}. **6.22** Molecular solid. **6.24** Molecular solids: Se_8, HBr, CO_2, P_4O_6, and SiH_4. Covalent

solids: Si, C. **6.26** Na, Mg, and Al are metals. Si is a covalent solid. P, S, Cl, and Ar are molecular solids. The metals should have the highest boiling and melting points. The molecular solids should have the lowest boiling and melting points. **6.28** Each C atom in diamond is covalently bonded to four other C atoms. Graphite has delocalized electrons in two dimensions. **6.30** The ability of a metal to conduct electricity *decreases* with increasing temperature, because the enhanced vibration of atoms at higher temperatures tends to disrupt the flow of electrons. **6.32** Doping silicon with Ga and Al would form p-type semiconductors, while doping with Sb and As would form n-type. **6.34** Metals are shiny because they reflect light. Conduction electrons are able to absorb and then re-emit light. **6.36** Covalent crystal. **6.38** Oil is made up of nonpolar molecules and therefore does not mix with water. To minimize contact, the oil drop assumes a spherical shape, which has a small surface area. **6.40** (a) Si and SiO_2 for covalent solids. (b) 5.00×10^9, yes. **6.42** That metals can conduct electricity indicates that some of their electrons are delocalized. **6.44** 172 pm. **6.46** 458 pm. **6.48** 1.69 g cm^{-3}. **6.50** gas: 34 Å, liquid: 0.15 Å. Molecules in a liquid are much closer than molecules in a gas. **6.52** The gas phase LiCl molecules share electrons enabling them to be closer. **6.54** Na (186 pm and 0.965 g cm^{-3}). **6.56** 6.020×10^{23}. **6.58** The critical point occurs because at high enough temperature and pressure, gases and liquids cannot be distinguished. A solid and a liquid never become so similar that they cannot be differentiated.

Chapter 7

7.2 (a) 0 J. (b) 9.4 J. (c) 18 J. **7.4** -1.04×10^3 J.

7.6 van der Waals: $P = \dfrac{nRT}{V - nb} - a\left(\dfrac{n}{V}\right)^2$

$w = \displaystyle\int_{V_1}^{V_2} Pdv = -nRT\int_{V_1}^{V_2} \dfrac{dV}{V - nb} + an^2\int_{V_1}^{V_2}\dfrac{dV}{V^2}$

$w = nRT\ln\left(\dfrac{V_1 - nb}{V_2 - nb}\right) + an^2\left(\dfrac{1}{V_1} - \dfrac{1}{V_2}\right)$

7.8 As energy consumers, we are interested in the availability of *usable* energy. **7.10** -1.57×10^4 kJ. **7.12** No work was done. The pressure is the same before and after the reaction. $\Delta U = -553.8$ kJ. **7.14** -3.31×10^3 J $= -3.31$ kJ. **7.16** 31°C.
7.18 26.3°C. **7.20** Because the humidity is very low in deserts, there is little water vapor in the air to trap and hold the heat radiated back from the ground during the day. Once the sun goes down, the temperature drops dramatically. 40°F temperature drops between day and night are common in desert climates. Coastal regions have much higher humidity levels compared to deserts. The water vapor in the air retains heat, which keeps the temperature at a more constant level during the night. In addition, sand and rocks in the desert have small specific heats compared with water in the ocean. The water absorbs much more heat during the day compared to sand and rocks, which keep the temperature warmer at night.
7.22 (a) $\Delta H_f^\circ[Br_2(l)] = 0$; $\Delta H_f^\circ[Br_2(g)] > 0$.
(b) $\Delta H_f^\circ[I_2(s)] = 0.$; $\Delta H_f^\circ[I_2(g)] > 0$. **7.24** Measure ΔH° for the formation of Ag_2O from Ag and O_2 and of $CaCl_2$ from Ca and Cl_2.
7.26 (a) -167.2 kJ mol^{-1}. (b) -56.2 kJ mol^{-1}.
7.28 (a) -1411 kJ mol^{-1}. (b) -1124 kJ mol^{-1}.
(c) -2044.0 kJ mol^{-1}. **7.30** 218.2 kJ mol^{-1}. **7.32** 0.30 kJ mol^{-1}.
7.34 -238.7 kJ mol^{-1}. **7.36** The bond enthalpy is a measure of the strength of the chemical bond. It has to be defined in the gas phase where intermolecular forces are negligible. Otherwise, the

difference in intermolecular forces of the molecule versus the fragments would affect the size of the bond enthalpy.
7.38 (a) -234 kJ mol^{-1} **7.40** $\Delta H^{\circ}_c(C_nH_{2n+2}) = [(n-1)$ 347 kJ mol^{-1} + $(2n+2)$ 414 kJ mol^{-1} + $\frac{1}{2}(n+1)$ 498.7 kJ mol^{-1}] $- [2n (799$ kJ mol$^{-1}) + (2n+2)$ 460 kJ mol^{-1}] **7.42** 2.68×10^3 kJ.
7.44 50.02 kJ mol^{-1}. **7.46** -1.12 kJ mol^{-1}. **7.48** 5.62 kJ mol^{-1}.
7.50 7.60%. **7.52** $\Delta H^{\circ}_f(AgNO_2) = -44.35$ kJ mol^{-1}.
7.54 -173.7 kJ mol^{-1}. **7.56** (a) Although we cannot measure ΔH°_{rxn} for this reaction, the reverse process is the combustion of glucose. We could easily measure ΔH°_{rxn} for this combustion in a bomb calorimeter. (b) 1.1×10^{19} kJ. **7.58** -17 J. **7.60** 0.492 J g^{-1}°C^{-1}.
7.62 The first reaction, which is exothermic, can be used to promote the second reaction, which is endothermic.
7.64 2.92×10^3 L. **7.66** 5.60 kJ mol^{-1}. **7.68** (a) The heat capacity of the food is greater than the heat capacity of air; hence, the cold in the freezer will be retained longer. (b) Tea or coffee has a greater amount of water, which has a higher specific heat than noodles.
7.70 1.64×10^3 g H_2O. **7.72** 1.07 kJ. **7.74** (a) 758°C.
(b) -986.6 kJ mol^{-1}. **7.76** 25 kg. **7.78** -360 kJ mol^{-1} Zn.
7.80 Carbon dioxide is linear and hence has only two modes of rotation and a lower heat capacity than ammonia. So, the container with the carbon dioxide will have the higher temperature.
7.82 (a) -65.2 kJ mol^{-1}. (b) -9.0 kJ mol^{-1}. **7.84** 5.8×10^2 m.
7.86 6.9 g assuming that all of the kinetic energy of the probe was converted into heat, that the liquid methane was at its boiling point (-164°C), and that the enthalpy of vaporization is not affected by the different pressure on the surface of Titan.
7.88 -277.0 kJ mol^{-1}. **7.90** 104 g. **7.92** (a) 0°C. (b) 100 g of Al, 41.7 g of ice, 58.3 g of water. **7.94** Energy must be supplied to break a chemical bond. By the same token, energy is released when a bond is formed. **7.96** $\Delta E = -5153$ kJ mol^{-1}; $\Delta H = -5158$ kJ mol^{-1}. **7.98** The heat capacity of the reactants should be larger than that of the products because the reactants are composed of more gas particles. Hence, the enthalpy change for the reaction should decrease with increasing temperature.
7.100

	q	w	ΔU	ΔH
(a)	$-$	$-$	$-$	$-$
(b)	$-$	$+$	0	0
(c)	$-$	$-$	$-$	$-$
(d)	$+$	$-$	$+$	$+$
(e)	$+$	0	$+$	$+$
(f)	$+$	$+$	$+$	$+$

7.102 (a) 48°C. (b) 4.1×10^3 g H_2O. **7.104** -1 L·bar; -100 J.
7.106 43 kJ mol^{-1}.

Chapter 8

8.4 Least probable is all balls in one box. **8.6** (a) no. (b) yes.
8.8 112 J mol K^{-1}. **8.10** 5.8 J K^{-1}. **8.12** 9.1 J K^{-1}.
8.14 (a) $\Delta S_{sys} = 5.8$ J K^{-1}; $\Delta S_{surr} = -5.8$ J K^{-1}; $\Delta S_{univ} = 0$ J K^{-1}.
(b) $\Delta S_{sys} = 5.8$ J K^{-1}; $\Delta S_{surr} = -4.1$ J K^{-1}; $\Delta S_{univ} = 1.7$ J K^{-1}.
8.16 92.8 J mol^{-1} K^{-1}. **8.18** (a) 9.13 J mol^{-1} K^{-1}. (b) 11.5 J mol^{-1} K^{-1}. (c) 13.4 J mol^{-1} K^{-1}. **8.22** (a) 48.7 J mol^{-1} K^{-1}. (b) -12.0 J mol^{-1} K^{-1}. (c) -242.8 J mol^{-1} K^{-1}. **8.24** (a) negative.
(b) negative. (c) positive. (d) ~zero (small). **8.26** $\Delta S_{sys} = 25$ J K^{-1} mol^{-1}; $\Delta S_{surr} = -613$ J K^{-1} mol^{-1}; $\Delta S_{univ} = -588$ J K^{-1} mol^{-1}. If this reaction had a positive ΔS_{univ}, it could consume almost all of the oxygen gas in the atmosphere. **8.28** 352°C.
8.30 (a) 175.2 kJ mol^{-1}. (b) 8.5 kJ mol^{-1}. (c) -2471.6 kJ mol^{-1}.
8.32 (b) is spontaneous, 350 K. **8.34** 979 K. **8.36** In all cases,

$\Delta H > 0$ and $\Delta S > 0$. $\Delta G < 0$ for (a), $= 0$ for (b), and > 0 for (c).
8.38 (a) 57.5 kJ mol^{-1}. (b) 57.5 kJ mol^{-1}. **8.40** (a) -11.5 J K^{-1} -3.44 kJ. (b) 58.4 J K^{-1}; -17.4 kJ. **8.42** (a) 0.4 kJ. (b) 3.4 kJ.
8.44 79.1 kJ. **8.46** This only tells us whether it is possible for the reaction to occur under these conditions. It tells us nothing about how fast the reaction will occur. **8.48** 0.055 J K. **8.52** 47°C.
8.54 (d). **8.56** 21.1 J K^{-1}. **8.58** See p. 456. **8.62** 7420 m.

8.64 (a) $\Delta S^{\circ} = \int_{T_1}^{T_2} \frac{C}{T} dT = \int_{T_1}^{T_2} \frac{\alpha T^3}{T} dT = \alpha \int_0^T T^2 dT = \alpha \frac{T^3}{3}$.

(b) 5.88×10^{-5} J mol^{-1} K^{-4}. (c) 2.0×10^{-5} J K^{-1}; 2.4×10^{-3} J K^{-1}; 3.39×10^{-2} J K^{-1}.

Chapter 9

9.2 (a) 59.8 kJ mol^{-1}. (b) 356°C, which agrees with the experimental value of 357°C. **9.4** X < Y. **9.6** no. **9.8** Cyclohexane cannot form hydrogen bonds. **9.10** The longer chains become more nonpolar. **9.12** 45.9 g. **9.14** Some helium gas will dissolve into the solution. This will not change the concentration of air initially dissolved in the solution. **9.16** 1.0×10^{-5}. **9.18** Ethanol: 30.0 mmHg; 1-propanol: 26.3 mmHg. **9.20** A = 1.9×10^2 mmHg. B = 4.0×10^2 mmHg. **9.22** Ethanol water mixture is a nonideal solution that exhibits positive deviations from Raoult's law. Therefore, this system has a minimum boiling point and will form a low boiling azeotrope that cannot be separated by fractional distillation. **9.24** 1.3×10^3 g. **9.26** 128 g. **9.28** 0.59 m.
9.30 120 g mol^{-1}. $C_4H_8O_4$. **9.32** -8.6°C. **9.34** 4.3×10^2 g mol^{-1}. $C_{24}H_{20}P_4$. **9.36** 1.75×10^4 g mol^{-1}. **9.38** 342 g mol^{-1}. **9.40** They also have the same boiling point, vapor pressure, and osmotic pressure. **9.42** 0.50 m glucose > 0.50 m acetic acid > 0.50 m HCl.
9.44 0.9420 m. **9.46** 7.2 bar. **9.48** 1.6 bar. **9.50** A: Steam; B: Water vapor. **9.52** 2.05×10^{-5} mmHg; 8.90×10^{-5} °C; 2.5×10^{-5} °C; 1.17×10^{-3} atm = 0.889 mmHg. **9.54** Water migrates through the semipermiable cell walls of the cucumber into the concentrated salt solution. **9.56** 3.5. **9.58** X is pure water; Y is NaCl solution; and Z is urea solution. **9.60** The increase in pressure pushes the elastic membrane to the right, causing the drug to exit through the small holes at a constant rate. **9.62** 33 bar.
9.64 $\mathcal{M}_z(B) = 248$ g mol^{-1}; $\mathcal{M}_z(A) = 124$ g mol^{-1} = $\mathcal{M}_z(B)/2$; A dimerization process. **9.66** (a) Boiling under reduced pressure.
(b) CO_2 boils off, expands and cools, condensing water vapor to form fog. **9.68** 33.3 mL; 66.7 mL. **9.70** To protect the red blood cells and other cells from shrinking (in a hypertonic solution) or expanding (in a hypotonic solution). **9.72** 14.3%.
9.74 (a) Decreases with lattice energy. (b) Increases with polarity of solvent. (c) Increase with enthalpy of hydration. **9.76** 0.815.
9.78 NH_3 can form hydrogen bonds with water. **9.80** 3%.
9.82 1.2×10^2 g mol^{-1}. It forms a dimer in benzene.
9.86 -0.737°C.

9.88 (a)

$$H-\overset{\underset{|}{H}}{\underset{\underset{|}{H}}{C}}-\overset{:O:}{\overset{\|}{C}}-\overset{\underset{|}{H}}{\underset{\underset{|}{H}}{C}}-H \qquad \ddot{S}=C=\ddot{S}$$

The vapor pressure of the solution is greater than the sum of the vapor pressures as predicted by Raoult's law for the same concentration. (b) 410 mmHg. (c) Positive. **9.90** 168 m.
9.92 0.853, 64.5 torr. Ideal solution and ideal gas were assumed.
9.94 3.1 mol. **9.96** 600 K. **9.98** The values increase with increasing atomic mass, which is consistent with the Henry's law constants

increasing with increasing intermolecular attraction between the gas particles and the water molecules. **9.100** (a) 43.3 kJ mol^{-1}.

Chapter 10

10.2 (a) A + C $\rightleftharpoons$ AC. (b) A + D $\rightleftharpoons$ AD. **10.4** $K_c = 1.08 \times 10^7$, $K_P = 1.33 \times 10^5$. **10.6** 3.5×10^{-7}. **10.8** (a) 0.082. (b) 0.29. **10.10** $K_P = 0.106$; $K_c = 2.05 \times 10^{-3}$. **10.12** 7.32×10^{-3}. **10.14** 3.3. **10.16** 3.53×10^{-2}. **10.18** 4.0×10^{-6}. **10.20** 5.6×10^{23}. **10.22** [NH$_3$] will increase and [N$_2$] and [H$_2$] will decrease. **10.24** $P_{NO} = 0.50$ bar; $P_{NO_2} = 0.020$ bar. **10.26** [I] $= 8.58 \times 10^{-4}$ M; [I$_2$] = 0.0194 M. **10.28** (a) 0.52. (b) [CO$_2$] = 0.48 M; [H$_2$] = 0.020 M; [CO] = 0.075 M; [H$_2$O] = 0.065 M. **10.30** [H$_2$] = [CO$_2$] = 0.05 M; [H$_2$O] = [CO] = 0.11 M. **10.32** 8.0×10^1 J mol^{-1}. **10.34** $\Delta G^{\circ}_{rxn} = 457.2$ kJ. $K = 7.27 \times 10^{-81}$. **10.36** -24.6 kJ mol^{-1} (b) -950 J mol^{-1}. **10.38** -341 kJ mol^{-1}. **10.40** 3.1×10^{-2} bar. **10.42** (a) Shift position of equilibrium to the right. (b) No effect. (c) No effect. **10.44** (a) No effect. (b) No effect. (c) Shift the position of equilibrium to the left. (d) No effect. (c) To the left. **10.46** (a) To the right. (b) To the left. (c) To the right. (d) To the left. (e) No effect. **10.48** (a) To the left. (b) No change. **10.50** (a) 1.67. (b) $P_A = 0.69$ bar; $P_B = 0.81$ bar. **10.52** (a) $K_P = 7.76 \times 10^{24}$. (b) $K_P = 7.58 \times 10^{19}$, assuming ΔH and ΔS are temperature independent in this temperature range. **10.54** -36.2 kJ, assuming ΔH and ΔS are temperature independent in this temperature range. **10.56** $P_{NO} = 0.24$ bar; $P_{Cl_2} = 0.12$ bar. (b) $K_P = 0.016$. **10.58** (a) No effect. (b) More CO$_2$ and H$_2$O will form. **10.60** (a) 8×10^{-44}. (b) The reaction has very large activation energy. **10.62** (a) 0.134. (b) 49.2%. (c) decrease. **10.64** $K_P = 4.95 \times 10^{-3}$.

10.66 $K_P = \dfrac{4\alpha^2}{1 - \alpha^2}P_{total}$.

10.68 2.34×10^{-3} M. **10.70** $P_{total} = 330$ bar. **10.72** Decrease in concentration of CO$_2$ shifts equilibrium to left. Either cool the environment or feed carbonated water. **10.74** 0.56 and 0.44. (b) $K_P = 8.0$. **10.76** (a) 1.16. (b) 53.7%. **10.78** (a) 0.50. (b) 0.23. (c) 0.037. (d) Greater than 0.037 mol. **10.80** [H$_2$] = 0.070 M, [I$_2$] = 0.182 M, [HI] = 0825 M. **10.82** 2.3×10^{-9}. **10.84** (a) 4.2×10^{-4}. (b) 0.83. (c) 1.1. (d) In (b): 2.3×10^3; in (c): 0.021. **10.86** $K_P = 2.34 \times 10^{-2}$; $K_c = 9.61 \times 10^{-4}$. **10.88** $P_{N_2O_4} = 0.14$ bar; $P_{NO_2} = 1.2$ bar; $K_P = 9.4$. **10.90** (a) The equilibrium will shift to the right. (b) To the right. (c) No change. (d) No change. (e) No change. (f) To the left. **10.92** $P_{N_2O_4} = 0.088$ bar; $P_{NO_2} = 0.10$ bar. **10.94** (a) 1.04 bar. (b) 0.40 bar. (c) 1.69 bar. (d) 0.62. **10.96** (a) $K_c = 1.1 \times 10^{-7}$; $K_P = 2.7 \times 10^{-6}$. (b) 22 mg m^{-3}. Yes. **10.98** Temporary dynamic equilibrium between the melting ice cubes and the freezing of the water between the ice cubes. **10.100** [NH$_3$] = 0.042 M; [N$_2$] = 0.086 M; [H$_2$] = 0.26 M. **10.102** The equilibrium constant for vaporization is equal to the pressure of the gas. **10.104** $\Delta G^{\circ} = 62.5$ kJ mol^{-1}; $\Delta H^{\circ} = 157.8$ kJ mol^{-1}; $\Delta S^{\circ} = 109$ kJ mol^{-1} K^{-1}.

Chapter 11

11.2 (a) NO$_2^-$. (b) HSO$_4^-$. (c) HS$^-$. (d) CN$^-$. (e) HCOO$^-$. **11.4** (a) H$_2$S. (b) H$_2$CO$_3$. (c) HCO$_3^-$. (d) H$_3$PO$_4$. (e) H$_2$PO$_4^-$. (f) HPO$_4^{2-}$. (g) H$_2$SO$_4$. (h) HSO$_4^-$. (i) HSO$_3^-$. **11.6** (a) CH$_2$ClCOO$^-$. (b) IO$_4^-$. (c) H$_2$PO$_4^-$. (d) HPO$_4^{2-}$. (e) PO$_4^{3-}$. (f) HSO$_4^-$. (g) SO$_4^{2-}$.

(h) IO$_3^-$. (i) SO$_3^{2-}$. (j) NH$_3$. (k) HS$^-$. (l) S^{2-}. (m) ClO$_3^-$. **11.8** AlCl$_3$ is the Lewis acid; Cl$^-$ is the Lewis base. **11.10** CO$_2$ and BF$_3$. **11.12** Yes. 1.0 M HCl(aq): pH = ~0; 2.0 M HCl(aq): pH = ~ -0.3. **11.14** 1.6×10^{-14} M. **11.16** (a) 10.75. (b) 3.28. **11.18** (a) 6.3×10^{-6} M. (b) 1.0×10^{-16} M. (c) 2.7×10^{-6} M. **11.20** 1.98×10^{-3} mol, 0.444. **11.22** 0.118. **11.24** (c). **11.26** 9.0. **11.28** (1) (c). (2) (b) and (d). **11.30** (a) Strong. (b) Weak. (c) Weak. (d) Weak. (e) Strong. **11.32** (b) and (c). **11.34** It will not proceed to any measurable extent. **11.36** They are assumed to completely ionize. **11.38** (c). **11.40** [H$^+$] = [CH$_3$COO$^-$] = 5.8×10^{-4} M; [CH$_3$COOH] = 0.0187 M. **11.42** 1.7×10^{-3} M. **11.44** 3.5%. (b) 33%. (c) 79%. **11.46** (a) 3.9%. (b) 0.030%. **11.48** 7.0×10^{-7}. **11.50** 1.5%. **11.52** (a) HCl: 1.40; H$_2$SO$_4$: 1.31. (b) [H$^+$] = [SO$_4^{2-}$] = 0.045 M; [HSO$_4^-$] = 0.16 M. **11.54** 6.6. **11.56** $K_{a_1} \gg K_{a_2}$. **11.58** H$_2$Se > H$_2$S > H$_2$O. **11.60** CHCl$_2$COOH. **11.62** (a) Neutral. (b) Acidic. (c) Neutral. (d) Acidic. **11.66** pH = 9.15. **11.68** pH > 7. **11.70** pH = 1.9. **11.72** (a) 2HCl(aq) + Zn(OH)$_2$(s) $\rightarrow$ ZnCl$_2$(aq) + 2H$_2$O(l). (b) 2OH$^-$(aq) + Zn(OH)$_2$(s) $\rightarrow$ Zn(OH)$_4^{2-}$(aq). **11.74** stronger. **11.76** (c). **11.78** (a) Forward reaction, NH$_3$ (Brønsted acid), NH$_3$ (Brønsted base); Reverse reaction, NH$_4^+$ (Brønsted acid) and NH$_2^-$ (Bronsted base). (b) NH$_4^+$ corresponds to H$^+$ and NH$_2^-$ corresponds to OH$^-$.

11.80 [OH$^-$] = $\dfrac{K_w}{K_a}$.

11.82 $\Delta H^{\circ} = 55.9$ kJ mol^{-1}; $\Delta S^{\circ} = -80.4$ J K^{-1} mol^{-1}; $\Delta G^{\circ} = 80$ kJ mol^{-1}; at 25°C $K_w = 1.01 \times 10^{-14}$; at 60°C $K_w = 1.08 \times 10^{-13}$. **11.84** 1.77×10^{10}. **11.86** 2.0×10^{-5}. **11.88** 0.028. **11.90** NH$_3$. **11.92** Because D is heavier than H, the quantum-zero point energy of the O—D vibration will be lower than that for H, and more energy would be required to break the O—D bond than the O—H bond. **11.94** 1.79. **11.96** F$^-$ reacts with HF to form HF$_2^-$, thereby shifting the ionization of HF to the right. **11.98** (a) HbO$_2$. (b) HbH$^+$. (c) To the right. **11.100** $\alpha = 0.180$; $K_P = 4.62 \times 10^{-3}$. **11.104** The H$^+$ ions covert CN$^-$ to HCN, which escapes as a gas. **11.106** 0.25 g. **11.108** -0.20. **11.110** (a) Equilibrium will shift to the right. (b) To the left. (c) No effect. (d) To the right. **11.112** The amines are converted to their salts RNH$_3^+$. **11.114** 1.4×10^{-4}. **11.116** 4.40. **11.118** In a basic medium, the ammonium salt is converted to the pungent-smelling ammonia. **11.120** (c). **11.122** 21 mL. **11.126** The fluoride ion replaces the hydroxide ion during remineralization, making the enamel more resistant to acids.

Chapter 12

12.2 (a) 11.28. (b) 9.08. **12.4** (a), (b), and (c). **12.6** 4.74 for both; (a) because it has a higher concentration. **12.8** 7.03. **12.10** 10; More effective against the acid. **12.12** (a) 4.82. (b) 4.64. **12.14** HC. **12.16** 90.1 g mol^{-1}. **12.18** 0.467 M. **12.20** [H$^+$] = 3.0×10^{-13} M; [OH$^-$] = 0.0335 M; [Na$^+$] = 0.0835 M; [CH$_3$COO$^-$] = 0.0500 M; [CH$_3$COOH] = 8.4×10^{-10} M. **12.22** 8.23. **12.24** (a) 11.36. (b) 9.55. (c) 8.95. (d) 5.19. (e) 1.70. **12.26** CO$_2$ dissolves in water to form H$_2$CO$_3$, which neutralizes NaOH. **12.28** 5.70. **12.30** (a) Insoluble. (b) Soluble. (c) Soluble. (c) Insoluble. (d) Soluble. **12.34** (a) 9.1×10^{-9} M. (b) 7.4×10^{-8} M. **12.36** 1.80×10^{-11}. **12.38** 2.2×10^{-4} mol L^{-1}. **12.40** 2.3×10^{-9}. **12.42** [Na$^+$] = 0.045 M; [NO$_3^-$] = 0.076 M; [Sr^{2+}] = 0.016 M; [F$^-$] = 1.1×10^{-4} M. **12.44** 2.68 < pH < 8.11. **12.46** (c) and (e). **12.48** (a) 0.013 M. (b) 2.2×10^{-4} M. (c) 3.3×10^{-3} M.

12.50 (a) $1.0 \times 10^{-5} M$. (b) $1.0 \times 10^{-10} M$. **12.52** (b), (c), (d), and (e). **12.54** (a) $0.016 M$. (b) $1.6 \times 10^{-6} M$. **12.56** Yes. **12.58** $Al(OH)_3$. **12.60** $Ag^+(aq) + 2NH_3(aq) \rightleftharpoons Ag(NH_3)_2^+(aq)$; $Zn^{2+}(aq) + 4NH_3(aq) \rightleftharpoons Zn(NH_3)_4^{2+}(aq)$. Zinc hydroxide forms a complex ion with excess OH^-, and silver hydroxide does not. **12.62** Silver chloride will dissolve in aqueous ammonia because of the formation of a complex ion. Lead chloride will not dissolve; it doesn't form an ammonia complex. **12.64** Sodium hydroxide. **12.66** $0.1 < 10$. **12.68** $pH = pK_a$. **12.70** $K_b = 5 \times 10^{-10}$. **12.72** $Cd(OH)_2(s) + 2OH^- \rightleftharpoons Cd(OH)_4^{2-}(aq)$; Lewis acid-base reaction. **12.74** (d). **12.76** $[Ag^+] = 2.0 \times 10^{-9} M$; $[Cl^-] = 0.080 M$; $[Zn^{2+}] = 0.070 M$; $[NO_3^-] = 0.060 M$. **12.78** > 8.11 but < 9.54. **12.80** $K_{sp} = 2.4 \times 10^{-13}$. **12.82** 1.0×10^{-5} M. **12.84** (a) AgBr. (b) $[Ag^+] = 1.8 \times 10^{-7} M$. (c) % $Ag^+(aq) = 1.8 \times 10^{-3}$ %. **12.86** 3.0×10^{-8}. **12.88** (a) 1.0×10^{14}. (b) 1.8×10^9. (c) 3.2×10^4. **12.92** The sulfur-containing air-pollutants react with Pb^{2+} to form PbS, which gives paintings a darkened look. **12.94** $[Ba^{2+}] = 1.8 \times 10^{-7} M$; $[Sr^{2+}] = [SO_4^{2-}] = 6.2 \times 10^{-4} M$. **12.96** No. **12.98** $V_{NaOH} = 13$ mL. **12.100** (a) The acidic hydrogen is from the carboxyl group. (b) 5.2×10^3, Less soluble. (c) $pH = 7.93$. **12.104** $pH = pK_a \pm 1$. **12.106** (b) $pK_a + pK_b = 14.00$.

Chapter 13

13.2 (a) $Fe \longrightarrow Fe^{3+} + 3e^-$; $O_2 + 4e^- \longrightarrow 2O^{2-}$. Oxidizing agent: O_2; reducing agent, Fe. (b) $2Br^- \longrightarrow Br_2 + 2e^-$; $Cl_2 + 2e^- \longrightarrow 2Cl^-$. Oxidizing agent; Cl_2; reducing agent, Br^-. (c) $Si \longrightarrow Si^{4+} + 4e^-$; $F_2 + 2e^- \longrightarrow 2F^-$. Oxidizing agent, F_2; reducing agent, Si. (d) $H_2 \longrightarrow 2H^+ + 2e^-$; $Cl_2 + 2e^- \longrightarrow 2Cl^-$. Oxidizing agent, Cl_2; reducing agent, H_2. **13.4** (a) $+5$. (b) $+1$. (c) $+3$. (d) $+5$. (e) $+5$. (f) $+5$. **13.6** (a) Cs: $+1$, O: -2. (b) Ca: $+2$, I: -1. (c) Al: $+3$, O: -2. (d) H: $+1$, As: $+3$, O: -2. (e) Ti: $+4$, O: -2. (f) Mo: $+6$, O: -2. (g) Pt: $+2$, Cl: -1. (h) Pt: $+4$, Cl: -1. (i) Sn: $+2$, F: -1. (j) Cl: $+3$, F: -1. (k) Sb: $+5$, F: -1. **13.8** N_2O_5. **13.10** (a) $2Fe^{2+} + H_2O_2 + 2H^+ \rightarrow 2Fe^{2+} + 2H_2O$. (b) $3Cu + 6H^+ + 2HNO_3 \rightarrow 3Cu^{2+} + 2NO + 4H_2O$. (c) $3CN^- + 2MnO_4^- + H_2O \rightarrow 3CNO^- + 2MnO_2 + 2OH^-$. (d) $3Br_2 + 6OH^- \rightarrow BrO_3^- + 5Br^- + 3H_2O$. (e) $2S_2O_3^{2-} + I_2 \rightarrow S_4O_6^{2-} + 2I^-$. **13.12** (a) $\mathcal{E} = -0.013V$, Fe (cathode), Cr (anode). (b) $\mathcal{E} = -0.943$, Zn (cathode), Al (anode). (c) $\mathcal{E} = +2.2467$, Mg (anode) and Pb (cathode). **13.14** Yes. **13.16** No. **13.18** (a) Li. (b) H_2. (c) Fe^{2+}. (d) Br^-. **13.20** 0.368 V. **13.22** (a) -432 kJ mol^{-1}, 5×10^{75}. (b) -104 kJ mol^{-1}, 2×10^{18}. (c) -178 kJ mol^{-1}, 1×10^{31}. (d) -1.27×10^3 kJ mol^{-1}, 8×10^{211}. **13.24** 0.37 V, -36 kJ mol^{-1}, 2×10^6. **13.26** 478 J mol^{-1} K^{-1}. **13.28** (a) 2.23 V, 2.23 V, -430 kJ mol^{-1}. (b) 0.02 V, 0.04 V, -23 kJ mol^{-1}. **13.30** 0.0789 V, ideal solution, can replace activities with concentrations. **13.32** 6.7×10^{-38}. **13.34** 0.97 bar. **13.36** (a) 0.076 L. (b) 0.16 L. **13.38** 0.64 g of Ba. **13.40** sodium. **13.42** 0.012 moles e^-. **13.44** 0.084 moles of each. **13.46** $7.70 \times 10^3 C$. **13.48** 1.84 kg h^{-1}. **13.50** 63.1 g mol^{-1}. **13.52** 27.1 g mol^{-1}. **13.56** 9.44×10^{-3} g SO_2. **13.58** (a) $2MnO_4^- + 5H_2O_2 + 6H^+ \rightarrow 5O_2 + 2Mn^{2+} + 8H_2O$. (b) 0.0602 M. **13.60** 0.231 mg Ca^{2+}/mL blood. **13.62** (a) 0.80 V. (b) $2Ag^+(aq) + H_2(g) \longrightarrow 2Ag(s) + 2H^+(aq)$. (c) (i) 0.92 V; (ii) 1.10 V. (d) The cell operates as a pH meter. **13.64** Fluorine gas reacts with water. **13.66** 2.5×10^2 h. **13.68** The solution surrounding the anode will become brown because of the formation of the triiodide ion. The solution around

the cathode will become basic, which will cause the phenolphthalein indicator to turn red. **13.74** (a) 0.222 L H_2. (b) 6.09×10^{23}/mol e^-. **13.76** (a) -1357 kJ mol^{-1}. (b) 1.17 V. **13.78** 1.10 V. (b) 5.3×10^3. **13.80** Heating the garage will melt the snow on the car, which is contaminated with salt. The aqueous salt will hasten corrosion. **13.82** (a) Anode $2F^- \rightarrow F_2(g) + 2e^-$; Cathode $2H^+ + 2e^- \rightarrow H_2(g)$; Overall: $2H^+ + 2F^- \rightarrow H_2(g) + F_2(g)$. (b) The K^+ is not reduced. (c) 2.8×10^3 L. **13.84** (c) 4.44 hr. **13.86** 1.97 V. **13.88** (a) $Au(s) + 3HNO_3(aq) + 4HCl(aq) \rightarrow HAuCl_4(aq) + 3H_2O(l) + 3NO_2(g)$. (b) To increase the acidity and to form the stable complex ion, $AuCl_4^-$. **13.90** 9.8×10^{-12}. **13.92** $\mathcal{E} = 1.09$ V; therefore, products are favored. **13.94** (a) $Sn^{4+}(aq) + 2e^- \rightarrow Sn^{2+}(aq)$; $2Tl(s) \rightarrow Tl^+(aq) + e^-$; $Sn^{4+}(aq) + 2Tl(s) \rightarrow Sn^{2+}(aq) + 2Tl^+(aq)$. (b) $K = 8 \times 10^{15}$. (c) 0.41 V. **13.96** Yes. **13.100** (b) 104 A h. The concentration of H_2SO_4 keeps decreasing. (c) 2.01 B; -3.88×10^2 kJ mol^{-1}. **13.102** (b) 11.9 hr. **13.108** $E° = -0.85V$. **13.110** The second one because it will have a larger change in entropy. **13.112** 2×10^{37}. **13.114** $217. **13.118** (a) $Al + 3Ag^+ \rightarrow Al^{3+} + 3Ag$. (b) A $NaHCO_3$ solution is basic.

Chapter 14

14.2 (a) rate $= -\dfrac{d[H_2]}{dt} = -\dfrac{d[I_2]}{dt} = \dfrac{1}{2}\dfrac{d[HI]}{dt}$.

(b) rate $= -\dfrac{1}{5}\dfrac{d[Br^-]}{dt} = -\dfrac{d[BrO_3^-]}{dt} = -\dfrac{1}{6}\dfrac{d[H^+]}{dt} = \dfrac{1}{3}\dfrac{d[Br_2]}{dt}$.

14.4 0.049 M s^{-1}. (b) -0.025 M s^{-1}. **14.6** 2.4×10^{-4} M s^{-1}. **14.8** (a) Third order. (b) 0.38 M s^{-1}. **14.10** M^{-2} s^{-1}. **14.12** $k = 1.19 \times 10^{-4}$ s^{-1}. **14.14** 30 min. **14.16** (a) 0.01998 s^{-1}. (b) 151 s. **14.18** 3.6 s. **14.20** (a) 0.487 M. (b) 0.500 M. **14.22** 0.64 M^{-2} s^{-1}. **14.24** 135 kJ mol^{-1}. **14.26** 644 K. **14.28** 51.0 kJ mol^{-1}. **14.30** 86 kJ mol^{-1}. **14.32** rate $= k[X_2][Y]$. (b) Reaction is zero order in Z. (c) $X_2 + Y \longrightarrow XY + X$ (slow) $X + Z \longrightarrow XZ$(fast). **14.34** Mechanism I. **14.38** Temperature, energy of activation, concentration of reactants, catalyst. **14.40** 22.6 cm^2; 44.9 cm^2. The large surface area of grain dust can result in a violent explosion. **14.42** (a) Third order. (b) 0.38 M^{-2} s^{-1}. (c) $H_2 + 2NO \rightarrow N_2 + H_2O + O$ (slow) $O + H_2 \rightarrow H_2O$ (fast). **14.44** Water is present in excess so its concentration does not change appreciably. **14.46** 10.7 M^{-1} s^{-1}. **14.48** 2.63 bar. **14.50** 100 kJ mol^{-1}; 3.4×10^{16}. **14.52** $P_{azomethane} = P_T x_{azomethane}$. **14.54** 377 K. **14.58** (a) Increase. (b) Decrease. (c) Decrease. (d) Increase. **14.60** $k = 1.0 \times 10^{-5}$ s^{-1}. **14.62** $Br_2 \rightarrow 2Br\cdot$; $Br\cdot + CH_4 \rightarrow HBr + \cdot CH_3$; $CH_3 + Br_2 \rightarrow CH_3Br + Br\cdot$; $Br\cdot + CH_4 \rightarrow HBr + \cdot CH_3$ and so on. **14.64** (a) rate $= k[X][Y]^2$. (b) 1.9×10^{-2} M^{-2} s^{-1}. **14.66** Second order. 2.4×10^7 M^{-1} s^{-1}. **14.68** Because the engine is relatively cold so the exhaust gases will not fully react with the catalytic converter. **14.70** $NO_2 + NO_2 \rightarrow NO_3 + NO$ (slow), $NO_3 + CO \rightarrow NO_2 + CO_2$ (fast). **14.72** (a) Mn^{2+}; Mn^{3+}; first step. (b) Without the catalyst, reaction would be termolecular. (c) Homogeneous. **14.74** 0.45 bar. **14.76** (a) $k_1[A] - k_2[B]$. (b) $[B] = (k_1/k_2)[A]$. **14.78** (1) (b) $<$ (c) $<$ (a); (2) (a) and (c) are exothermic, and (b) is endothermic. **14.80** Titanium acts as a catalyst to decompose steam. **14.82** (a) 2.5×10^{-5} M s^{-1}. (b) Same as (a). (c) 8.3×10^{-6} M. **14.84** d. **14.86** (a) 1.13×10^{-3} M min^{-1}. (b) 8.8×10^{-3} M. **14.88** Second order, 0.42 M^{-1} min^{-1}. **14.90** 6.4×10^8 s^{-1}.

Chapter 15

15.4 V^{5+}:[Ar]; Cr^{3+}:[Ar] $3d^1$; Mn^{2+}: [Ar]$3d^3$; Fe^{2+}: [Ar]$3d^5$; Cu^{2+}: [Ar]$3d^9$; Sc^{3+}: [Ar]; Ti^{4+}:[Ar]. **15.8** (a) +3. (b) 6. (c) oxalte.
15.10 (a) Na: +1, Mo: +6. (b) Mg: +2, W: +6. (c) Fe: 0.
15.12 (a) *cis*–dichlorobis(ethylenediammine)cobalt(III).
(b) pentaamminechloroplatinum(IV) chloride.
(c) pentaamminechlorocobalt(III) chloride.
15.14 (a) $[Cr(en)_2Cl_2]^+$. (b) $Fe(CO)_5$.
(c) $K_2[Cu(CN)_4]$. (d) $[Co(NH_3)_4(H_2O)Cl]Cl_2$.
15.16 (a) 2. (b) 2.

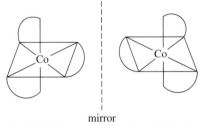

trans   *cis*

15.18 (a) Two geometric isomers. (b) Two optical isomers.

[structure: two Co complexes with mirror plane]

mirror

15.20 CN^- is a strong-field ligand. Absorbs near UV (blue) so appears yellow. **15.22** (a) Orange. (b) 255 kJ mol^{-1}.
15.24 $[Co(NH_3)_4Cl_2]Cl$. Two moles. **15.26** Use $^{14}CN^-$ label (in NaCN). **15.28** First $Cu(CN)_2$ (white) is formed. It redissolves as $Cu(CN)_4^{2-}$. **15.30** 1.4×10^2. **15.32** Mn^{3+}. $3d^5$ (Cr^{3+}) is a stable electron configuration. **15.34** Ti: +3; Fe: +3. **15.36** Four Fe atoms per hemoglobin molecule. 1.6×10^4 g mol^{-1}.
15.38 (a) $[Cr(H_2O)_6]Cl_3$. (b) $[Cr(H_2O)_5Cl]Cl_2 \cdot H_2O$.
(c) $[Cr(H_2O)_4Cl_2]Cl \cdot 2H_2O$. Compare electrical conductance with solutions of NaCl, MgCl$_2$, and FeCl$_3$. **15.40** -1.8×10^2 kJ mol^{-1}; 6×10^{30}. **15.42** Iron is more abundant. **15.44** $Mn(H_2O)_6^{2+}$.
15.46 Low temperature and low concentration of Cl$^-$ ions favor the formation of $Co(H_2O)_6^{2+}$ ions. Adding HCl (more Cl$^-$ ions) favors the formation of $CoCl_4^{2-}$. This reaction decreases [Cl$^-$], so the pink color is restored. **15.48** Fe^{3+} is nearly colorless, so it must be light yellow in color. **15.50** The Be complex exhibits optical isomerism. The Cu complex exhibits geometric isomerism.

[Be complex structures with mirror plane]

mirror

[Cu complex structures]

cis   *trans*

15.52 Isomer I must be the *cis* isomer. The chlorines must be *cis* to each other for one oxalate ion to complex with Pt; Isomer II must be the *trans* isomer. With the chlorines on opposite sides of the molecule, each Cl is replaced with a hydrogen oxalate ion.
15.54 (a) If the bonds are along the z-axis, the d_{z^2} orbital will have the highest energy. The $d_{x^2-y^2}$ and d_{xy} orbitals will have the lowest energy. (b) If the trigonal plane is in the xy-plane, then the $d_{x^2-y^2}$ and d_{xy} orbitals will have the highest energy, and d_{z^2} will have the lowest energy. (c) If the axial positions are along the z-axis, the d_{z^2} orbital will have the highest energy. The trigonal plane will be in the xy-plane, and thus the $d_{x^2-y^2}$ and d_{xy} orbitals will be next highest in energy. **15.56** (a) Second equation. (b) Second equation.

Chapter 16

16.4 $CH_3—CH_2—CH_2—CH_2—CH_2—CH_2—CH_3$

$CH_3—CH_2—CH_2—CH—CH_2—CH_3$
                                            |
                                          CH_3

$CH_3—CH_2—CH—CH—CH_3$
                    | |
                  CH_3 CH_3

$CH_3—CH_2—CH_2—CH_2—CH—CH_3$
                                      |
                                    CH_3

$CH_3—CH_2—C—CH_2—CH_3$ (with CH$_3$ above and CH$_3$ below central C)

$CH_3—CH_2—CH_2—C—CH_3$ (with CH$_3$ above and CH$_3$ below central C)

$CH_3—CH—CH_2—CH—CH_3$ (with CH$_3$ below each substituted carbon)

$CH_3—C—CH—CH_3$ (CH$_3$, CH$_3$ above; CH$_3$ below)

$CH_3—CH_2—CH—CH_2—CH_3$
                    |
                  CH_2
                    |
                  CH_3

16.6

[structures showing alkenes/alkynes with H and C]

16.10

[H, CH$_3$ / H$_3$C, H alkene] *trans*   [H$_3$C, CH$_3$ / H, H alkene] *cis*

16.12

cis-chlorofluoroethylene   *trans*-chlorofluoroethylene

[Cl, F / H, H structure]   [Cl, H / H, F structure]
(a)   (b)

1,1-chlorofluoroethylene

[Cl, H / F, H structure]
(c)

(a) and (b) are geometric isomers. (c) is a structural isomer of both (a) and (b). **16.14** (a) and (c).

16.18

trans-1,3-pentadiene   cis-1,3-pentadiene

16.20

16.22 $CH_3CH_2CH_2CH_2CH_3$ is an alkane and $CH_3CH_2CH_2CH=CH_2$ is an alkene. Only an alkene would react with a hydrogen halide and hydrogen. **16.24** -633.1 kJ mol^{-1}. **16.28** (a) Ether. (b) Amine. (c) Aldehyde. (d) Ketone. (e) Carboxylic acid. (f) Alcohol. (g) Amino acid.

16.30

(e) CH_3CH_2—O—$CH_2CH_2CH_2CH_2CH_3$

16.32 (a) methoxyethane. (b) ethanamine. (c) propanal. (d) 2-butanone. (e) methanoic acid. (f) 1-proponol.
16.34 $HCOOH + CH_3OH \longrightarrow HCOOCH_3 + H_2O$. Methyl formate. **16.36** $(CH_3)_2CH$—O—CH_3. **16.38** (a) Ketone. (b) Ester. (c) Ether. **16.40** osmotic pressure.
16.42 —CH_2—$CHCl$—CH_2—CCl_2—. **16.44** By an addition reaction involving styrene monomers.
16.46 (a) $H_2C=CH$—$CH=CH_2$. (b) $HO_2C(CH_2)_6NH_2$.

16.52

16.56 11 turns s^{-1}. **16.58** yes.
16.60

16.62 (a) Cyclopropane because of the strained bond angles. (The C—C—C angle is 60° instead of 109.5°). (b) Ethylene because of the C=C bond. (c) Acetaldehyde (susceptible to oxidation).
16.64 (a) HCF. (b) No. (c) $C_2H_2F_2$. (d) tetrahedral. (e) trans–difluoroethylene. **16.66** 8.4×10^3 L air. **16.68** (a) Six. (b) Four. (c) Seven. **16.72** Proteins are made of 20 amino acids. Nucleic acids are made of four building blocks (purines, pyrimidines, sugar, phosphate group) only. **16.74** C-G base pairs have three hydrogen bonds and higher boiling point; A-T base pairs have two hydrogen bonds. **16.76** Leg muscles are active, have a high metabolic rate and hence a high concentration of myoglobin. The iron content in Mb makes the meat look dark. **16.78** Insects have blood that contains no hemoglobin. It is unlikely that a human-sized insect could obtain sufficient oxygen for metabolism by diffusion. **16.80** There are four Fe atoms per hemoglobin molecule; 1.6×10^4 g mol^{-1}.
16.82 Gly-Ala-Phe-Glu-His-Gly-Ala-Leu-Val.
16.84 Empirical and molecular formula: $C_5H_{10}O$; 88.7 g/mol.

$CH_2=CH$—CH_2—O—CH_2—CH_3.
16.86 The C atoms bonded to the methyl group and the amino group and the H atom. (b) The C atoms bonded to Br.
16.88 CH_3CH_2CHO. **16.90** (a) Alcohol. (b) Ether. (c) Aldehyde. (d) Carboxylic acid. (e) Amine. **16.92** The acids in lemon juice convert the amines to the ammonium salts, which have very low vapor pressures. **16.94** Methane (CH_4), ethanol (C_2H_5OH), methanol (CH_3OH), isopropanol (C_3H_7OH), ethylene glycol (CH_2OHCH_2O), naphthalene ($C_{10}H_8$), acetic acid (CH_3COOH).

16.96 (a) 1. (b) 2. (c) 5. **16.98** Br_2 dissociates into Br atoms, which react with CH_4 to form CH_3Br and HBR.

16.100 1.0×10^3 Hz.

16.104 (a)

. The compound is chiral. (b) the product is a racemic mixture. **16.106** (a) 53. (c) 27.

16.110

16.112 (a) The —COOH group. (b) pH = 1.0: The valine is in the fully protonated form. pH = 7.0: Only the —COOH group is ionized. pH = 12.0: Both groups are ionized.

16.114

16.116

*Chiral

Chapter 17

17.2 (a) $_{-1}^{0}\beta$. (b) $_{20}^{40}Ca$. (c) $_{2}^{4}\alpha$. (d) $_{0}^{1}n$. **17.4** (a) $_{92}^{235}U \rightarrow {}_{90}^{231}Th + {}_{2}^{4}\alpha$.
(b) $_{8}^{15}O \rightarrow {}_{7}^{15}N + {}_{+1}^{0}\beta$. (c) $_{4}^{11}Be \rightarrow {}_{5}^{11}B + {}_{-1}^{0}\beta$.
(d) $_{92}^{235}U \rightarrow {}_{93}^{235}Np + {}_{-1}^{0}\beta$. **17.6** (a) Lithium-9. (b) Sodium-25.
(c) Scandium-48. **17.8** (a) Neon-17. (b) Calcium-45.
(c) Technetium. (d) Mercury-195. (e) Curium.
17.10 6×10^9 kg s^{-1}. **17.12** (a) 1.14×10^{-12} J/nucleon.
(b) 2.36×10^{-10} J. **17.14** β^- decay increases the n/p ratio, so would be more likely to occur in nuclei that are neutron deficient, such as Neon-19, than in nuclei with a larger n/p ratio, such as Neon-23. **17.16** (a) 2.489×10^6 kJ. (b) 278 kg.
17.18 4.89×10^{19} atoms. **17.20** Radiactive decay of a nucleus does not depend upon collisions with other atoms. Therefore, the mechanism consists of a single "unimolecular" elementary step, which implies pure first-order decay. **17.22** 65.2 yr.
17.24 (a) $_{7}^{15}N + {}_{1}^{1}p \rightarrow {}_{6}^{12}C + {}_{2}^{4}\alpha$; X is $_{7}^{15}N$. (b) $_{13}^{27}Al + {}_{1}^{2}d \rightarrow {}_{12}^{25}Mg + {}_{2}^{4}\alpha$; X is $_{12}^{25}Mg$. (c) $_{25}^{55}Mn + {}_{0}^{1}n \rightarrow {}_{25}^{56}Mn + \gamma$; X is $_{25}^{56}Mn$.
17.26 All you need is a high–intensity alpha particle emitter. Any heavy element like plutonium or curium will do. Place the bismuth–209 sample next to the alpha emitter and wait.
17.28 The easiest experiment would be to add a small amount of aqueous iodide containing some radioactive iodine to a saturated solution of lead(II) iodide. If the equilibrium is dynamic, radioactive iodine will eventually be detected in the solid lead(II) iodide. **17.30** If one were to dope part of a crystal with a radioactive tracer, one could demonstrate diffusion in the solid

state by detecting the tracer in a different part of the crystal at a later time. This actually happens with many substances.
17.32 The design and operation of a Geiger counter are discussed in Figure 17.4 of the text. **17.34** 70.5 disintegrations/min.
17.36 (a) $_{92}^{235}U + {}_{0}^{1}n \rightarrow {}_{56}^{140}Ba + 3{}_{0}^{1}n + {}_{36}^{93}Kr$.
(b) $_{92}^{235}U + {}_{0}^{1}n \rightarrow {}_{55}^{144}Cs + {}_{37}^{90}Rb + 2{}_{0}^{1}n$.
(c) $_{92}^{235}U + {}_{0}^{1}n \rightarrow {}_{35}^{87}Br + {}_{57}^{146}La + 3{}_{0}^{1}n$.
(d) $_{92}^{235}U + {}_{0}^{1}n \rightarrow {}_{62}^{160}Sm + {}_{30}^{72}Zn + 4{}_{0}^{1}n$.
17.38 (a) $_{1}^{3}H \rightarrow {}_{2}^{3}He + {}_{-1}^{0}\beta$. (b) $_{94}^{242}Pu \rightarrow {}_{2}^{4}\alpha + {}_{92}^{238}U$.
(c) $_{53}^{131}I \rightarrow {}_{54}^{131}Xe + {}_{-1}^{0}\beta$. (d) $_{98}^{251}Cf \rightarrow {}_{96}^{247}Cm + {}_{2}^{4}\alpha$.
17.40 Because both Ca and Sr belong to Group 2A, radioactive strontium that has been ingested into the human body becomes concentrated in bones (replacing Ca) and can damage blood cell production. **17.42** Normally the human body concentrates iodine in the thyroid gland. The purpose of the large doses of KI is to displace radioactive iodine from the thyroid and allow its excretion from the body. **17.44** (a) $_{7}^{14}N + {}_{0}^{1}n \rightarrow {}_{7}^{15}N + \gamma$. (b) X-ray analysis only detects shapes, particularly of metal objects. Bombs can be made in a variety of shapes and sizes and can be constructed of "plastic" explosives. Thermal neutron analysis is much more specific than X-ray analysis. **17.46** The neutron-to-proton ratio for tritium equals 2 and is thus outside the belt of stability. In a more elaborate analysis, it can be shown that the decay of tritium to ^{3}He is exothermic; thus, the total energy of the products is less than the reactant. **17.48** 0.24%. **17.50** All isotopes of radium are radioactive; therefore, radium is not naturally occurring and would not be found with barium. However, radium is a decay product of uranium–238, so it is found in uranium ores. **17.52** A radioactive isotope with a shorter half-life because more radiation would be emitted over a certain period of time. **17.54** The energy of irradiation is not sufficient to bring about nuclear transmutation.
17.56 Ds and Rg are transition metals: $_{82}^{208}Pb + {}_{28}^{62}Ni \rightarrow {}_{110}^{270}Ds$; $_{83}^{209}Bi + {}_{28}^{64}Ni \rightarrow {}_{111}^{273}Rg$. Uub resembles Zn, Cd, and Hg: $_{82}^{208}Pb + {}_{30}^{66}Zn \rightarrow {}_{112}^{274}Uub$. Uuq is in the carbon family: $_{94}^{244}Pu + {}_{20}^{48}Zn \rightarrow {}_{114}^{289}Uuq$. **17.58** There was radioactive material inside the box.
17.60 (a) 6.87×10^{-13} J. (b) The smaller particle (α) will move away at a greater speed due to its lighter mass. **17.62** The α particles emitted by ^{241}Am ionize the air molecules between the plates. The voltage from the battery makes one plate positive and the other negative, so each plate attracts ions of opposite charge. This creates a current in the circuit attached to the plates. The presence of smoke particles between the plates reduces the current, because the ions that collide with smoke particles (or steam) are usually absorbed (and neutralized) by the particles. This drop in current triggers the alarm. **17.64** (a) The nuclear submarine can be submerged for a long period without refueling. (b) Conventional diesel engines receive an input of oxygen. A nuclear reactor does not. **17.66** 53.0 g Zn. **17.68** (a) Δm for neutron decay is -1.39×10^{-30} g (that is, it is exothermic) whereas, for proton decay, Δm is $+3.22 \times 10^{-30}$ g (that is, it is endothermic). Thus, proton decay can only occur with input of energy, which is not readily available to an isolated proton. (b) In a nucleus, energy for proton decay can be supplied by the decrease in the repulsive electrostatic energy due to replacing one proton with a neutron or, if by changing into a neutron, the proton can occupy a lower energy shell.
17.70 (a) $_{94}^{238}Pu \rightarrow {}_{2}^{4}He + {}_{92}^{234}U$. (b) $t = 10$ yr: 0.53 mW.
17.72 0.49 rem.

Credits

Chapter 0

Opener: © Jeffrey Coolidge/Iconica/Getty Images; **p. 3:** © B.A.E. Inc./Alamy; **0.2a, 0.2b, p.6:** © McGraw-Hill Higher Education, Inc./Ken Karp, Photographer; **p. 9:** © Jacques Jangoux/Photo Researchers; **0.6 (all):** © Richard Megna/Fundamental Photographs; **p. 14:** © Jerry Driendl/The Image Bank/Getty Images; **0.12c:** © E.R. Degginger/Color-Pic Inc.; **p. 22:** © Andrew Lambert/Photo Researchers; **p. 23, p. 30, p. 32:** © McGraw-Hill Higher Education, Inc./Ken Karp, Photographer; **0.15:** © McGraw-Hill Higher Education, Inc./Stephen Frisch, Photographer; **p. 38:** Courtesy of the National Scientific Balloon Facility, Palestine, TX; **p. 39:** © E.R. Degginger/Color-Pic Inc.; **p. 40:** © L.V. Bergman/The Bergman Collection; **p. 42:** © Getty Images; **p. 49, p. 54:** © McGraw-Hill Higher Education, Inc./Stephen Frisch, Photographer; **p. 59:** Merlin Metalworks, Inc.

Chapter 1

Opener: IBM Almaden Research Center; **p. 76:** © Sciencephotos/Alamy; **p. 77 (bottom):** © Steve Callahan/Visuals Unlimited; **p. 79:** © Hulton Archive/Getty Images; **p. 81:** © Popperfoto/Alamy; **p. 84:** © McGraw-Hill Higher Education, Inc./Stephen Frisch, Photographer; **1.13:** Courtesy of Sargent-Welch; **p. 86:** © Interfoto Pressebildagentur/Alamy; **p. 93:** Professor Ahmed H. Zewail/California Institute of Technology, Department of Chemistry; **p. 94:** © Meggers Gallery/AIP/Photo Researchers, Inc.; **1.19a, 1.19b:** © Dr. Stanley Flegler/Visuals Unlimited; **p. 97:** © AFP/Getty Images; **p. 99:** © Francis Simon/AIP/Photo Researchers, Inc.; **p. 109 (left):** © Dr. Stanley Flegler/Visuals Unlimited; **p. 109 (right):** Reprint Courtesy of International Business Machines Corporation copyright 1993 © International Business Machines Corporation.

Chapter 2

Opener: © Dr. Mark J. Winter/Photo Researchers, Inc.; **p. 144 (top):** © Bettmann/Corbis; **p. 144 (bottom):** © McGraw-Hill Higher Education, Inc./Charles D. Winters, Photographer; **2.21 (sodium, chlorine, argon):** © McGraw-Hill Higher Education, Inc./Ken Karp, Photographer; **2.21 (magnesium, aluminum, sulfur):** © L.V. Bergman/The Bergman Collection; **2.21 (silicon):** © Frank Wing/Stock Boston; **2.21 (phosphorus):** © Al Fenn/Time & Life Pictures/Getty Images; **p. 163:** © Corbis.

Chapter 3

Opener: © Joel Gordon; **p. 195 (top):** © Brand X Pictures/PunchStock ; **p. 195 (top, middle):** © Bildarchiv Monheim GmbH/Alamy; **p. 195 (bottom, middle):** © Topham/The Image Works; **p. 198:** © McGraw-Hill Higher Education, Inc./Ken Karp, Photographer; **3.13:** © Donald Clegg.

Chapter 4

Opener: © agefotostock; **4.21:** © Editorial Image, LLC/Alamy; **p. 262:** © Royalty-Free/Corbis.

Chapter 5

Opener: © James Randklev/Stone/Getty Images; **p. 287 (top):** © Karl Weatherly/Getty Images ; **p. 287 (bottom), 5.6 (all), 5.9, 5.11:** © McGraw-Hill Higher Education, Inc./Ken Karp, Photographer; **p. 302:** Courtesy of the National Scientific Balloon Facility, Palestine, TX; **p. 303:** Courtesy of General Motors; **p. 304:** © Fred J. Maroon/Photo Researchers; **p. 315:** © NASA.

Chapter 6

Opener: © Connie Coleman/Getty Images; **p. 337:** © Hermann Eisenbeiss/Photo Researchers; **6.6, 6.8:** © McGraw-Hill Higher Education, Inc./Ken Karp, Photographer; **p. 340: :** © Alan Carey/The Image Works; **p. 345:** © Tony Mendoza/Stock Boston; **p. 351:** © Byron Quintard/Jacqueline McBride/Lawrence Livermore National Labs; **p. 353 (top):** © Siede Preis/Getty Images ; **p. 353 (bottom):** © L.V. Bergman/The Bergman Collection; **p. 358:** © Royalty-Free/Corbis; **p. 359:** Courtesy of The Railway Technical Research Institute.

Chapter 7

Opener: © Richard Megna/Fundamental Photographs; **p. 377:** © McGraw-Hill Higher Education, Inc./Stephen Frisch, Photographer; **p. 379:** © Richard Megna/Fundamental Photographs; **p. 380:** © McGraw-Hill Higher Education, Inc./Stephen Frisch, Photographer; **p. 396 (top):** © McGraw-Hill Higher Education, Inc./Ken Karp, Photographer; **p. 396 (bottom):** © Comstock/Jupiter Images; **p. 397, p. 399:** © E.R. Degginger/Color-Pic Inc.

Chapter 8

Opener: © Rick Gayle/Corbis; **p. 425:** © McGraw-Hill Higher Education, Inc./Charles D. Winters, Photographer; **p. 428:** © Matthias K. Gobbert/University of Maryland, Baltimore County/Dept. of Mathematics and Statistics; **p. 431:** © McGraw-Hill Higher Education, Inc./Ken Karp, Photographer; **p. 447:** U.S. Government.

Chapter 9

Opener: Royalty-Free/CORBIS; **9.4 (all):** © McGraw-Hill Higher Education, Inc./Ken Karp, Photographer; **p. 483:** Bill Evans/USGS; **p. 488:** © Steve Allen/Getty Images ; **p. 495 (top):** © Hank Morgan/Photo Researchers; **p. 495 (bottom):** © McGraw-Hill Higher Education, Inc./Ken Karp, Photographer; **9.20d (all):** © David Phillips/Photo Researchers, Inc.; **p. 499:** © John Meed/Photo Researchers, Inc.; **p. 508 (both):** © McGraw-Hill Higher Education, Inc./Ken Karp, Photographer.

Chapter 10

Opener: © Videodiscovery; **p. 512:** © McGraw-Hill Higher Education, Inc./Ken Karp, Photographer; **p. 519:** © Collection Varin-Visage/Photo Researchers; **10.8, 10.11 (both):** © McGraw-Hill Higher Education, Inc./Ken Karp, Photographer; **p. 545:** © Melissa McManus/Stone/Getty; **10.12:** © McGraw-Hill Higher Education, Inc./Ken Karp, Photographer.

Chapter 11

Opener: © E.R. Degginger/Color-Pic Inc.; **11.1:** © McGraw-Hill Higher Education, Inc./Ken Karp, Photographer; **p. 559:** © McGraw-Hill Higher Education, Inc./Stephen Frisch, Photographer; **p. 560, 11.4, 11.6, 11.13 (both):** © McGraw-Hill Higher Education, Inc./Ken Karp, Photographer; **11.14:** © Michael Melford.

Chapter 12

Opener, p. 615, 12.1: © McGraw-Hill Higher Education, Inc./Ken Karp, Photographer;

Index